# Lecture Notes in Computer Science    1593
Edited by G. Goos, J. Hartmanis and J. van Leeuwen

**Springer**
*Berlin
Heidelberg
New York
Barcelona
Hong Kong
London
Milan
Paris
Singapore
Tokyo*

Peter Sloot   Marian Bubak
Alfons Hoekstra   Bob Hertzberger (Eds.)

# High-Performance Computing and Networking

7th International Conference, HPCN Europe 1999
Amsterdam, The Netherlands, April 12-14, 1999
Proceedings

 Springer

Series Editors

Gerhard Goos, Karlsruhe University, Germany
Juris Hartmanis, Cornell University, NY, USA
Jan van Leeuwen, Utrecht University, The Netherlands

Volume Editors

Peter Sloot
Alfons Hoekstra
Bob Hertzberger
University of Amsterdam
Faculty of Mathematics, Computer Science, Physics, and Astronomy
Kruislaan 403, 1098 SJ Amsterdam, The Netherlands
E-mail: {sloot, alfons, bob}@wins.uva.nl

Marian Bubak
University of Mining and Metallurgy (AGH)
Institute of Computer Science and Academic Computer Center CYFRONET
al. Mickiewicza 30, 30-059 Cracow, Poland
E-mail: bubak@uci.agh.edu.pl

Cataloging-in-Publication data applied for

Die Deutsche Bibliothek - CIP-Einheitsaufnahme

**High performance computing and networking** : 7th international conference / HPCN Europe 1999, Amsterdam, The Netherlands, April 12 - 14, 1999. Peter Sloot ... (ed.). - Berlin ; Heidelberg ; New York ; Barcelona ; Hong Kong ; London ; Milan ; Paris ; Singapore ; Tokyo : Springer, 1999
 (Lecture notes in computer science ; Vol. 1593)
 ISBN 3-540-65821-1


CR Subject Classification (1998): C.2.4, D.1-2, E.4, F.2, G.1-2, J.1-2, J.3, J.6, K.6

ISSN 0302-9743
ISBN 3-540-65821-1 Springer-Verlag Berlin Heidelberg New York

This work is subject to copyright. All rights are reserved, whether the whole or part of the material is concerned, specifically the rights of translation, reprinting, re-use of illustrations, recitation, broadcasting, reproduction on microfilms or in any other way, and storage in data banks. Duplication of this publication or parts thereof is permitted only under the provisions of the German Copyright Law of September 9, 1965, in its current version, and permission for use must always be obtained from Springer-Verlag. Violations are liable for prosecution under the German Copyright Law.

© Springer-Verlag Berlin Heidelberg 1999
Printed in Germany

Typesetting: Camera-ready by author
SPIN: 10704672   06/3142 – 5 4 3 2 1 0    Printed on acid-free paper

# PREFACE

This volume contains the Proceedings of the international HPCN Europe 1999 event that was held in Amsterdam, the Netherlands, April 12 -- 14, 1999. HPCN (*High Performance Computing and Networking*) Europe was organized for the first time in 1993 in Amsterdam as the result of several initiatives in Europe, the United States, and Japan. HPCN Europe events were then held in Munich (1994), Milan (1995), Brussels (1996), Vienna (1997), and it returned Amsterdam in 1998.

The HPCN Europe keeps changing its image. HPCN 1999 is a conference and multi-workshop event. At the conference there will be three conference tracks presenting 84 selected papers in the following areas: Industrial and General Applications of HPCN, Computational Science, Computer Science, as well as two poster sessions with 40 presentations.

Three renowned speakers will present the HPCN 1999 keynote lectures: Ralf Gruber (EPFL, Switzerland), Carl Kesselman (Caltech, USA), and Pieter Adriaans (Syllogic BV, the Netherlands).

Since research in HPCN is progressing rapidly, newly emerging domains of this field and their present and possible future applications are covered by 13 thematic workshops. Over 40 well-known experts from Europe and the United States participated in the organization of the workshops and have agreed to present invited lectures demonstrating the current trends in their fields of interest.

Methods of solving computing-intensive problems are considered at the High Performance Numerical Computations with Applications Workshop and the High Performance Computing on Very Large Data Sets Workshop. The High Performance Data Mining for Large-Scale Data Sets Workshop focuses on algorithms for mining massive/distributed data and on integration of data mining with other systems and applications. Performance measurement, analysis and prediction are the concern of the EuroTools Workshop. The importance of new approaches to efficient and easy use of computer resources is illustrated at the Distributed and MetaComputing Workshop and the Java in HPC Workshop. The Virtual Reality Workshop discusses the state of the art in this rapidly developing domain of presentation of large datasets generated in simulations in industrial and scientific applications. The 1355 Interconnect and Routing Workshop is dedicated to a specific technology of networking. Recent applications of HPCN in medicine are discussed at the Hospital of the Future Workshop and at the second International Mini Conference on Telemedical Information Society. Other practical aspects of HPCN are given much attention at the Traffic Simulation Workshop, the Applications in Finance Workshop, and the Enabling Technology, Hard- and Software Workshop. Some of the papers presented at the workshops are also included in these Proceedings.

The papers included in the Proceedings reflect the multidisciplinary character and broad spectrum of the field. We thank all contributors for their cooperation. Due to the high quality of almost 200 submitted contributions the selection of papers for oral and poster presentation was not simple. We are very grateful to the reviewers for their efforts in evaluating so many papers in a very short time over Christmas. The best conference papers will be published in a special issue of the journal, *Future Generation Computer Systems*.

The drawing up of an interesting program for the Conference would not be possible without invaluable suggestions and contributions of the members of the HPCN 1999 Program Committee. We highly appreciate the personal effort of the members of the local organizing committee and the conference secretariat. Special thanks to Lodewijk Bos, Karin Zuur, and David Dubbeldam who prepared these proceedings. We would like to express our sincere thanks to the computer support group for setting up the paper submission engine, and Diederik Burer and Martin Bergman for creating the HPCN Europe web pages. We would like to thank Piotr Luszczek and ACC CYFRONET for help in collecting review reports.

The organizers acknowledge the help of the Dutch HPCN foundation, the Dutch Organization for Scientific Research, the University of Amsterdam, and the council of the city of Amsterdam for supporting the event. Finally we thank all the attendees and contributors to the conference who made this conference and multi-workshop a high quality event!

Peter Sloot
Marian Bubak
Alfons Hoekstra
Bob Hertzberger

**Committees:**

Prof. dr. Bob Hertzberger, Universiteit van Amsterdam
    HPCN Europe 99 event chair

**Scientific Organizing Committee:**

P.M.A. Sloot, University of Amsterdam (chair)
M. Bubak, University of Cracow, Poland
A.G. Hoekstra, University of Amsterdam
V. Sunderam, Emory University, Altanta, USA

**Programme Committee:**

A. Barak, Hebrew University of Jerusalem
A. Bode, Univeristy of Munich
A.V. Bogdanov, Institute for High Performance Computing and Databases, St. Petersburg
P. Brezany, University of Vienna
D. Caromel, Univ. of Nice, INRIA Sophia Antipolis
J. Dongarra, University of Tennessee
I. Duff, DRAL
W. Gentzsch, University of Regensburg
R. Gruber, EPFL, Switzerland
A. Hey, University of Southamptoon
P. Kacsuk, KFKI-MSZKI Research Institute, Hungary
H. Liddell, QMC London
B. Madahar, GEC Marconi, UK
J. Murphy, BAE, UK
J. Reeve, University of Southampton
A. Reinefeld, University of Paderborn
D. Roose, University of Leuven
G. Serazzi, Politecnico di Milano
H. Sips, Technical University Delft
O. Thomas, GMD, Germany
C. Upstill, PAC, UK
H. van der Vorst, Utrecht University
R. Williams, California Institute of Technology
K-f Wong, Hong Kong University

# Referees

Paul Allen
Nick Allsopp
Patrick Amestoy
Leon Aronson
Konstantin Baev
Amnon Barak
Francoise Baude
Robert Belleman
Arndt Bode
Alexander Bogdanov
K. M. Bossley
Alexander Boukhanovsky
A. Braverman
Peter Brezany
Matthias Brune
Marian Bubak
Denis Caromel
Przemyslaw Czerwinski
Michel Dayde
Jack Dongarra
Philipp Drum
O. Duer
Iain Duff
Alistair Dunlop
Detlef Fliegl
Josef Froemcke
Nathalie Furmento
D. Garti
Wolfgang Gentzsch
I. Gilderman
Luc Giraud
Paul Gordon
Monika Grobecker
Ralf Gruber
Ronan Guivarch
Vitaly Gursky
Pieter Hartel
A. Hey
Alfons Hoekstra
Peter Kacsuk
Drona Kandhai

Wolfgang Karl
M. Kemelmakher
A. Keren
Jacek Kitowski
Jacko Koster
O. Kremien
Dmitry Kretchman
Frits Kuijlman
Szu-wen Kuo
Ihor Kuz
O. Laadan
Erwin Laure
R. Lavi
Hong-va Leong
Heather Liddell
Hai-Xiang Lin
Markus Lindermeier
Peter Lockey
Thomas Ludwig
Peter Luksch
Bob Madahar
Ursula Maier
Dmitry Malashonok
J. Maresky
Michael May
Eduard Mehofer
K.E. Meirman
Helen Meng
J. Murphy
Ekaterina Myasnikova
Istvan Nadas
Zsolt Nemeth
Alexander Nikolaev
Steve O'Connell
A. Ofir
S Dominique Orban
Benno Overeinder
Stephane Perennes
A.D. Pimentel
Norbert Podhorszki
Guenther Rackl
Jeff Reeve
A. Reinefeld
Harald Richter

M. Roest
Dirk Roose
David Sagnol
Maria Samsonova
Annick Sartenaer
Erich Schikuta
Arjen Schoneveld
Giuseppe Serazzi
Viera Sipkova
Henk Sips
E. Sokolov
Natalie Sokolova
Krzysztof Sowa
Piero Spinnato
Elena Stankova
J. Stijnen
Kurt Stockinger
Mike Surridge
O. Thomas
Tran Trach-Minh
Joerg Trinitis
C. Upstill
G.D. van Albada
H. van der Vorst
Arjan van Gemund
Kees van Reeuwijk
K. Vuik
Helmut Wanek
Matthias Weidmann
Andrew Wendelborn
Bernd Wender
Roy Williams
Roland Wismueller
Ivan Wolton
Kam-Fai Wong
Pavel Yakutseni
EJ Zaluska

## Organising Committee

Walther Hesselink and Rutger Hamelynck, Conference Office, University of Amsterdam
Lodewijk Bos
Karin Zuur, University of Amsterdam
Diederik Burer, University of Amsterdam
Martin Bergman, University of Amsterdam
David Dubbeldam, University of Amsterdam

## HPCN Europe '99 Workshop Chairs

J. Kaandorp and M. Göbel (Virtual Reality)
D. Epema (Distributed Computing and Metacomputing)
A. Visser and M. Schreckenberg (Traffic Simulation)
Vladimir Getov (Java in HPC)
A. Marsh (IEEE EMBS ITIS-ITAB '99)
J.W. Tellegen (Healthcare of the future European Technology Transfer)
J. Hollenberg and J. Murphy (Enabling Technology, Hard- and Software)
J.-L. Pazat (EuroTools)
T. Yang H.-X. Lin (High Performance Numerical Computations with Applications)
P. Brezany (High Performance Computation on Very Large Data Sets)
J. Topper (Applications in Finance)
P.W. Thompson and B. Dobinson (1355 interconnect and routing workshop)
Y. Guo (High Performance Data Mining for Large Scale Data Sets)

# Table of Contents

## (Industrial) End-User Applications of HPCN

High Performance Integer Optimization for Crew Scheduling  3
*P. Sanders, T. Takkula, D. Wedelin*

Simulating Synthetic Polymer Chains in Parallel  13
*B. Jung, H.-P. Lenhof, P. Müller, C. Rüb*

Real-Time Signal Processing in a Collision Avoidance Radar System Using Parallel Computing  23
*G.L. Reijns, A.J.C. van Gemund, J. Schier, P.J.F. Swart*

The Impact of Workload on Simulation Results for Distributed Transaction Processing  33
*R. Riedl*

Computer Simulation of Ageing with an Extended Penna Model  43
*A. Z. Maksymowicz, M. Bubak, K. Zając, M. Magdoń*

The Scenario Management Tool SMARTFED for Real-Time Interactive High Performance Networked Simulations  50
*R.P. van Sterkenburg, A. A. ten Dam*

OPERA: An HPCN Architecture for Distributed Component -Based Real-Time Simulations  60
*F.-X.Lebas, T. Usländer*

Airport Simulation Using CORBA and DIS  70
*G. Rackl, F. de Stefani, F. Héran, A. Pasquarelli, T. Ludwig*

Intelligent Routing for Global Broadband Satellite Internet  80
*C.-H. Chang, H.-K. Wu, M.-H. Jin, Y.-O. Tseng*

Integrated CAD/CFD Visualisation of a Generic Formula 1 Car Front Wheel Flowfield  90
*W.P. Kellar, A.M. Savill, W.N. Dawes*

Adaptive Scheduling Strategy Optimizer for Parallel Rolling Bearing Simulation  99
*D. Fritzson, P. Nordling*

MPI-Based Parallel Implementation of a Lithography Pattern Simulation
Algorithm   109
   *H. Radhakrishna, S. Divakar, N. Magotra, S.R.J. Brueck, A. Waters*

Parallelizing an High Resolution Operational Ocean Model   120
   *J. Schüle, T. Wilhelmsson*

Weather and Climate Forecasts and Analyses at MHPCC   130
   *J. Roads, S. Chen, C. McCord, W. Smith, D. Stevens, H. Juang,
F. Fujioka*

GeoFEM: High Performance Parallel FEM for Solid Earth   133
   *K. Garatani, H. Nakamura, H. Okuda, G. Yagawa*

Elastic Matching of Very Large Digital Images on High Performance
Clusters   141
   *J. Modersitzki, W. Obelöer, O. Schmitt, G. Lustig*

Data Intensive Distributed Computing: A Medical Application Example   150
   *J. Lee, B. Tierney, W. Johnston*

Utilizing HPC Technology in 3D Cardiac Modeling   159
   *N. Papazis, D. Dimitrelos*

A Parallel Algorithm for 3D Reconstruction of Angiographic Images   168
   *R. Rivas, M.B. Ibáñez, Y. Cardinale, P. Windyga*

A Diffraction Tomography Method for Medical Imaging Implemented on
High Performance Computing Environment   178
   *T.A. Maniatis, K.S. Nikita, K. Voliotis*

**Computational Science**

Heterogeneous Distribution of Computations While Solving Linear Algebra
Problems on Networks of Heterogeneous Computers   191
   *A. Kalinov, A. Lastovetsky*

Modeling and Improving Locality for Irregular Problems: Sparse Matrix-
Vector Product on Cache Memories as a Case Study   201
   *D.B. Heras, V.B. Pérez, J.C.C. Domínguez, F.F. Rivera*

Parallelization of Sparse Cholesky Factorization on an SMP Cluster   211
   *S. Satoh, K. Kusano, Y. Tanaka, M. Matsuda, M. Sato*

Scalable Parallel Sparse Factorization with Left-Right Looking Strategy on
Shared Memory Multiprocessors  221
   *O. Schenk, K. Gärtner, W. Fichtner*

Algorithm of Two-Level Parallelization for Direct Simulation Monte Carlo
of Unsteady Flows in Molecular Gasdynamics  231
   *A.V. Bogdanov, I.A. Grishin, G.O. Khanlarov, G.A. Lukianov,
V.V. Zakharov*

Parallelization of Gridless Finite-Size-Particle Plasma Simulation Codes  241
   *S. Briguglio, G. Vlad, G. Fogaccia, B. Di Martino*

A Distributed Object-Oriented Method for Particle Simulations on Clusters  251
   *Y. Sun, Z. Liang, C.-L. Wang*

A Simple Dynamic Load-Balancing Scheme for Parallel Molecular
Dynamics Simulation on Distributed Memory Machines  260
   *N. Sato, J.-M. Jézéquel*

Resource Management for High-Performance PC Clusters  270
   *A. Keller, M. Brune, A. Reinefeld*

The Web as a Global Computing Platform  281
   *Q.H. Mahmoud*

WebFlow: A Framework for Web Based Metacomputing  291
   *T. Haupt, E. Akarsu, G. Fox*

Dynamite - Blasting Obstacles to Parallel Cluster Computing  300
   *G.D. van Albada, J. Clinckemaillie, A.H.L. Emmen, J. Gehring,
O. Heinz, F. van der Linden, B.J. Overeinder, A. Reinefeld, P.M.A. Sloot*

Iterative Momentum Relaxation for Fast Lattice-Boltzmann Simulations  311
   *D. Kandhai, A. Koponen, A. Hoekstra, P.M.A. Sloot*

Lattice Gas: An Efficient and Reusable Parallel Library Based on a Graph
Partitioning Technique  319
   *A. Dupuis, B. Chopard*

Algorithms of Parallel Realisation of the PIC Method with Assembly
Technology  329
   *M.A. Kraeva, V.E. Malyshkin*

Computational Aspects of Multi-species Lattice-Gas Automata  339
   *D. Dubbeldam, A.G. Hoekstra, P.M.A. Sloot*

Towards a Scalable Metacomputing Storage Service . . . . . . . . . . . . . . . . . . . 350
   *C.J. Patten, K.A. Hawick, J.F. Hercus*

Dynamic Visualization of Computations on the Internet . . . . . . . . . . . . . . . . . 360
   *J. Li, E. de Doncker*

A Flexible Security System for Metacomputing Environments . . . . . . . . . . . . . . 370
   *A. Ferrari, F. Knabe, M. Humphrey, S. Chapin, A. Grimshaw*

Computational Experiments Using Distributed Tools in a Web-Based
Electronic Notebook Environment . . . . . . . . . . . . . . . . . . . . . . . . . . . . . 381
   *A.D. Malony, J.L. Skidmore, M.J. Sottile*

A Parallel/Distributed Architecture for Hierarchically Heterogeneous
Web-Based Cooperative Applications . . . . . . . . . . . . . . . . . . . . . . . . . . . 391
   *T.L. Casavant, T.E. Scheetz, T.A. Braun, K.J. Munn, S. Kaliannan*

Effective Dynamic Load Balancing of the UKMO Tracer Advection
Routines . . . . . . . . . . . . . . . . . . . . . . . . . . . . . . . . . . . . . . . . . . 402
   *D.A. Smith*

Dynamic Load Balancing in Parallel Finite Element Simulations . . . . . . . . . . . . . 409
   *A. Schoneveld, M. Lees, E. Karyadi, P.M.A. Sloot*

A Load Balancing Routine for the NAG Parallel Library . . . . . . . . . . . . . . . . . 420
   *R.W. Ford, M. O'Brien*

Performance Assessment of Parallel Spectral Analysis: Towards a Practical
Performance Model for Parallel Medical Applications . . . . . . . . . . . . . . . . . . 430
   *F. Munz, T. Ludwig, S. Ziegler, P. Bartenstein, M. Schwaiger, A. Bode*

Parallel Algorithm and Processor Selection Based on Fuzzy Logic . . . . . . . . . . . . 440
   *S. Yu, M. Clement, Q. Snell, B. Morse*

Recurrent Neural Network Approach for Partitioning Irregular Graphs . . . . . . . . . . 450
   *M-T. Kechadi*

## Computer Science

JIAJIA: A Software DSM System Based on a New Cache Coherence
Protocol . . . . . . . . . . . . . . . . . . . . . . . . . . . . . . . . . . . . . . . . . . 463
   *W. Hu, W. Shi, Z. Tang*

Efficient Analytical Modelling of Multi-level Set-Associative Caches . . . . . . . . . . 473
   *J.S. Harper, D.J. Kerbyson, G.R. Nudd*

Buffer Management in Wormhole-Routed Torus Multicomputer Networks  483
  *K. Kotapati, S.P. Dandamudi*

Performance Analysis of Broadcast in Synchronized Multihop Wireless
Networks  493
  *K.-H. Pan, H.-K. Wu, R.-J. Shang, F. Lai*

EARL - A Programmable and Extensible Toolkit for Analyzing Event Traces
of Message Passing Programs  503
  *F. Wolf, B. Mohr*

XSIL: Extensible Scientific Interchange Language  513
  *K. Blackburn, A. Lazzarini, T. Prince, R. Williams*

ForkLight: A Control-Synchronous Parallel Programming Language  525
  *C. W. Keßler, H. Seidl*

HPF Parallelization of a Molecular Dynamics Code: Strategies and
Performances  535
  *B. Di Martino, M. Celino, V. Rosato*

Design of High-Performance C++ Package for Handling of Multidimensional
Histograms  543
  *M. Bubak, J. T. Mościcki, J. Shiers*

Dynamic Remote Memory Acquiring for Parallel Data Mining on PC
Cluster: Preliminary Performance Results  553
  *M. Oguchi, M. Kitsuregawa*

The Digital Puglia Project: An Active Digital Library of Remote Sensing
Data  563
  *G. Aloisio, M. Cafaro, R. Williams*

An Architecture for Distributed Enterprise Data Mining  573
  *J. Chattratichat, J. Darlington, Y. Guo, S. Hedvall, M. Köhler, J. Syed*

Representatives Selection in Multicast Group  583
  *S. Li, A.L. Ananda*

Deadlock Prevention in Incremental Replay of Message-Passing Programs  593
  *F. Zambonelli*

Remote and Concurrent Process Duplication for SPMD Based Parallel
Processing on COWs  603
  *M. Hobbs, A. Goscinski*

Using BSP to Optimize Data Distribution in Skeleton Programs     613
    *A. Zavenella, S. Pelagatti*

Swiss-Tx Communication Libraries     623
    *S. Brauss, M. Frey, A. Gunzinger, M. Lienhard, J. Nemecek*

Finding the Optimal Unroll-and-Jam     633
    *N. Zingirian, M. Maresca*

A Linker for Effective Whole-Program Optimizations     643
    *A G.M. Cilio, H. Corporaal*

The Nestor Library: A Tool for Implementing Fortran Source to Source Transformations     653
    *G.-A. Silber, A. Darte*

Performance Measurements on Sandglass-Type Parallelization of Doacross Loops     663
    *M. Takabatake, H. Honda, T. Yuba*

Transforming and Parallelizing ANSI C Programs Using Pattern Recognition     673
    *M. Boekhold, I. Karkowski, H. Corporaal*

Centralized Architecture for Parallel Query Processing on Networks of Workstations     683
    *S. Zeng, S.P. Dandamudi*

Object-Oriented Database System for Large-Scale Molecular Dynamics Simulations     693
    *J. Kitowski, D. Wajs, P. Trzeciak*

Virtual Engineering of Multi-disciplinary Applications and the Significance of Seamless Accessibility of Geometry Data     702
    *V. Deshpande, L. Fornasier, E.A. Gerteisen, N. Hilbrink, A. Mezentsev, S. Merazzi, T. Woehler*

Some Results from a New Technique for Response Time Estimation in Parallel DBMS     713
    *N. Tomov, E. Dempster, M.H. Williams, A. Burger, H. Taylor, P.J.B. King, P. Broughton*

PastSet- A Distributed Structured Shared Memory System     722
    *B. Vinter, O.J. Anshus, T. Larsen*

Optimal Scheduling of Iterative Data-Flow Programs onto Multiprocessors
with Non-negligible Interprocessor Communication  732
    D.A.L. Piriyakumar, P. Levi, C.S.R. Murthy

Overlapping Communication with Computation in Distributed Object
Systems  744
    F. Baude, D. Caromel, N. Furmento, D. Sagnol

Exploiting Speculative Thread-Level Parallelism on a SMT Processor  754
    P. Marcuello, A. González

Network Interface Active Messages for Low Overhead Communication on
SMP PC Clusters  764
    M. Matsuda, Y. Tanaka, K. Kubota, M. Sato

Experimental Results about MPI Collective Communication Operations  774
    M. Bernaschi, G. Iannello, M. Lauria

MaDCoWS: A Scalable Distributed Shared Memory Environment for
Massively Parallel Multiprocessors  784
    D. Dimitrelos, C. Halatsis

**Virtual Reality**

VisualExpresso: Generating a Virtual Reality Internet  797
    D. Cleary, D. O'Donoghue

VIVRE: User-Centred Visualization  807
    D.R.S. Boyd, J.R. Gallop, K.E.V. Palmen, R.T. Platon, C.D. Seelig

GEOPROVE: Geometric Probes for Virtual Environments  817
    R.G. Belleman, J.A. Kaandorp, D. Dijkman, P.M.A. Sloot

**Distributed Computing and Metacomputing**

A Gang-Scheduling System for ASCI Blue-Pacific  831
    J.E. Moreira, H. Franke, W. Chan, L.L. Fong, M.A. Jette, A. Yoo

Towards Quality of Service for Parallel Computing: An Overview of the
MILAN Project  841
    H. Karl

Resource Allocation and Scheduling in Metasystems  851
    U. Schwiegelshohn, R. Yahyapour

## Java in HPC

The Use of Java in High Performance Computing: A Data Mining Example  863
    D. Walker, O. Rana

Interfaces and Implementations of Random Number Generators for Java
Grande Applications  873
    P.D. Coddington, J.A. Mathew, K.A. Hawick

Java as a Basis for Parallel Data Mining in Workstation Clusters  884
    M. Gimbel, M. Philippsen, B. Haumacher, P.C. Lockemann, W.F. Tichy

Garbage Collection for Large Memory Java Applications  895
    A. Krall, P. Tomsich

## IEEE EMBS ITIS-ITAB '99

The Emergence of Virtual Medical Worlds  909
    A. Marsh, T. Kauranne, G. Zahlmann, A. Emmen, L. Versweyveld

Characteristics of Users of Medical Innovations  912
    M. Moore

Security Analysis and Design Based on a General Conceptual Security
Model and UML  919
    B. Blobel, P. Pharow, F. Roger-France

3DHeartView: Introducing 3-Dimensional Angiographical Modelling  931
    The 3DHeartView Team

WWW Based Service for Automated Interpretation of Diagnostic Images:
The AIDI-Heart Project  941
    M. Ohlsson, A. Järund, L. Edenbrandt

Decision Trees – A CIM Tool in Nursing Education  951
    P. Kokol, M. Zorman, V. Podgorelec, A. Habjanič, T. Medoš, M. Brumec

HealthLine : Integrated Information Provision to Telemedicine Networks  959
    Y. Samiotakis, S. Anagnostopoulou, A. Alexakis

Multi Modal Presentation in Virtual Telemedical Environments  964
    E. Jovanov, D. Starcevic, A. Marsh, Z. Obrenovic, V. Radivojevic,
    A. Samardzic

Using Web Technologies and Meta-Computing to Visualise a Simplified
Simulation Model of Tumor Growth in Vitro 973
    G.S. Stamatakos, E.I. Zacharaki, N.A. Mouravliansky, K.K. Delibasis,
    K.S. Nikita, N.K. Uzunoglu, A. Marsh

The Electronic Commerce Component in Telemedicine 983
    D. Polemi, A. Marsh

Efficient Implementation of the Marching Cubes Algorithm for Rendering
Medical Data 989
    K.K. Delibasis, G.K. Matsopoulos, N.A. Mouravliansky, K.S. Nikita

**High Performance Numerical Computation and Applications**

Multilevel Algebraic Elliptic Solvers 1001
    T.F. Chan, P. Vaněk

Parallel Performance of Chimera Overlapping Mesh Technique 1015
    J. Rokicki, D. Drikakis, J. Majewski, J. Żołtak

Electromagnetic Scattering with the Boundary Integral Method on MIMD
Systems 1025
    T. Jacques, L. Nicolas, C. Vollaire

Decomposition of Complex Numerical Software into Cooperating
Components 1032
    M.R.T. Roest, E.A.H. Vollebregt

Multi-block Parallel Simulation of Fluid Flow in a Fuel Cell 1042
    S. Baird, J.J. McGuirk

A Parallel Implementation of the Block Preconditioned GCR method 1052
    C. Vuik, J. Frank

Comparison of Two Parallel Analytic Simulation Models of Inhomogeneous
Distributed Parameter Systems 1061
    S.W. Brok, L. Dekker

Case Studies of Four Industrial Meta-Applications 1077
    T. Cooper

A Parallel Approach for Solving a Lubrication Problem in Industrial Devices 1087
    M. Arenaz, R. Doallo, J. Touriño, C. Vázquez

## High Performance Computing on Very Large Datasets

Restructuring I/O-Intensive Computations for Locality 1097
  M. Kandemir, A. Choudhary, J. Ramanujam

Virtual Memory Management in Data Parallel Applications 1107
  E. Caron, O. Cozette, D. Lazure, G. Utard

High Performance Parallel I/O Schemes for Irregular Applications on Clusters of Workstations 1117
  J. No, J. Carretero, A. Choudhary

Advanced Data Repository Support for Java Scientific Programming 1127
  P. Brezany, M. Winslett

## Posters

Advanced Communication Optimizations for Data-Parallel Programs 1139
  G. Agrawal

A Cellular Automata Simulation Environment for Modelling Soil Bioremediation 1143
  S. D. Telford

High Efficient Parallel Computation of Resonant Frequencies of Waveguide Loaded Cavities on JIAJIA Software DSMs 1147
  W. Shi, J. Ma, Z. Tang

A Study of Parallel Image Processing in a Distributed Processing Environment 1151
  M. Iikura, K. Kobayashi, T. Yoshioka, S. Ihara

Data Prefetching for Digital Alpha 1155
  S. Manoharan

Circuit-Switched Broadcast in Multi-port 2D Tori 1159
  S.-Y. Wang, Y.-C. Tseng, S.-Y. Ni, J.-P. Sheu

A Distributed Algorithm for the Estimation of Average Switching Activity in Combinational Circuits 1163
  S. Koranne

BVIEW: A Tool for Monitoring Distributed Systems 1167
  U.A. Ranawake, J.E. Dorband

The Queue System within PHASE  1171
  S. Girona, S. Bello, J. Labarta, P. Ribes, R. Martin, J. Soto, G. Laffitte

Support Tools for Supercomputing and Networking  1175
  V.N. Kasyanov, V.A. Evstigneev, J.V. Malinina, J.V. Birjukova,
  V.A. Markin, E.V. Haritonov, S.G. Tsikoza

Ultra High-Speed Superconductor System Design: Phase 2  1179
  M. Dorojevets, K. Likharev

Lilith Lights: A Network Traffic Visualization Tool for High Performance
Clusters  1183
  D.A. Evensky, A.C. Gentile, P. Wyckoff

DSMC of Inner Atmosphere of a Comet on Shared Memory Multiprocessors  1187
  G.O. Khanlarov, G.A. Lukianov

Data Mining and Simulation Applied to a Staff Scheduling Problem  1190
  K. Smyllie

Neural Network Software for Unfolding Positron Lifetime Spectra  1194
  P. Lindén, R. Chakarova, T. Faxén, I. Pázsit

MPVisualizer: A General Tool to Debug Message Passing Parallel
Applications  1199
  A.P. Cláudio, J.D Cunha, M.B. Carmo

Effect of Multicycle Instructions on the Integer Performance of the
Dynamically Trace Scheduled VLIW Architecture  1203
  A.F. de Souza, P. Rounce

MAD - A Top Down Approach to Parallel Program Debugging  1207
  D. Kranzlmüller, R. Hügl, J. Volkert

High-Performance Programming Support for Multimedia Document
DataBase Management  1211
  P. Adam, H. Essafi, M.-P. Gayrard, M. Pic

Behavioral Objects and Layered Services: The Application Programming
Style in the Harness Metacomputing System  1215
  M. Migliardi, V. Sunderam

Coordination Models and Facilities Could Be Parallel Software Accelerators  1219
  A.E. Doroshenko, L.-E. Thorelli, V.Vlassov

Delays in Asynchronous Communication Domain Decomposition . . . . . . . . . . . 1223
   *M.D. Gubitoso, C. Humes Jr.*

Experimenting Reflection for Programming Concurrent Objects Scheduling
Strategies . . . . . . . . . . . 1227
   *L. Bray, J.-P. Arcangeli, P. Sallé*

Virtual User Account System for Distributed Batch Processing . . . . . . . . . . . 1231
   *W. Dymaczewski, N. Meyer, M. Stroiński, P. Wolniewicz*

Content-Based Multimedia Data Retrieval on Cluster System Environment . . . . . . . . . . . 1235
   *S. Srakaew, N. Alexandridis, P.-N. Punpit, G. Blankenship*

Industrial Supercomputing Center in Hungary - Pre-feasibility Study . . . . . . . . . . . 1242
   *S. Forrai, P. Kacsuk*

Reducing Memory Traffic Via Redundant Store Instructions . . . . . . . . . . . 1246
   *C. Molina, A. González, J. Tubella*

3D-Visualization for Presenting Results of Numerical Simulation . . . . . . . . . . . 1250
   *Y. E. Gorbachev, E. V. Zudilova*

Optimal Distribution of Loops Containing No Dependence Cycles . . . . . . . . . . . 1254
   *Z. Szczerbinski*

Solving Maximum Clique and Independent Set of Graphs Based on Hopfield
Network . . . . . . . . . . . 1258
   *Y. Zhang, C.H. Chi*

Storing Large Volumes of Structured Scientific Data on Tertiary Storage . . . . . . . . . . . 1262
   *J. Mościński, D. Nikolow, M. Pogoda, R. Słota*

Modelling Http Traffic Generated by Community of Users . . . . . . . . . . . 1266
   *G. Bilchev, I. Marshall, S. Olafsson, C. Roadknight*

OCM-Based Tools for Performance Monitoring of Message Passing
Applications . . . . . . . . . . . 1270
   *M. Bubak, W. Funika, K. Iskra, R. Maruszewski*

Enhancing OCM to Support MPI Applications . . . . . . . . . . . 1274
   *M. Bubak, W. Funika, R. Gembarowski, P. Hodurek, R. Wismüller*

Symbol Table Management in a HPF Debugger . . . . . . . . . . . 1278
   *M. Bubak, W. Funika, G. Młynarczyk, K. Sowa, R. Wismüller*

Tuple Counting Data Flow Analysis and Its Use in Communication
Optimization 1282
    J.B. Fenwick Jr., L.L. Pollock

Investigation of High-Performance Algorithms for Numerical Calculations of
Evolution of Quantum Systems Based on Their Intrinsic Properties 1286
    A.V. Bogdanov, A.S. Gevorkyan, A.G. Grigoryan, S.A. Matveev

Processor Allocation and Task Scheduling Strategy to Minimize the
Distributed Sparse Cholesky Factorization Time 1292
    T.-T. Kan, C.-L. Chen

**Late Papers**

Implementation of MPI over HTTP 1299
    S. Lakshminarayanan, S. S. Ghosh, N. Balakrishnan

A Distributed Vision Network for Industrial Packaging Inspection 1303
    A. Meliones, D. Baltas, P. Kammenos, K. Spinnler, A. Kuleschow,
    G. Vardangalos, P. Lambadaris

Implementation of Montgomery Exponentiation on Fine Grained FPGAs 1308
    A. Tiountchik, E. Trichina

List of Authors 1313

# Track C1:

# (Industrial) End-User Applications of HPCN

# High Performance Integer Optimization for Crew Scheduling*

Peter Sanders[1], Tuomo Takkula[2], Dag Wedelin[2]

[1] Max-Planck-Institute for Computer Science, Saarbrücken, Germany
[2] Chalmers University of Technology, Computing Science, Göteborg, Sweden

**Abstract.** Performance aspects of a Lagrangian relaxation based heuristic for solving large 0-1 integer linear programs are discussed. In particular, we look at its application to airline and railway crew scheduling problems. We present a scalable parallelization of the original algorithm used in production at Carmen Systems AB, Göteborg, Sweden, based on distributing the variables and a new sequential *active set strategy* which requires less work and is better adapted to the memory hierachy properties of modern RISC processors. The active set strategy can even be parallelized on networks of workstations.

## 1 Introduction

In this paper we describe our work on improving the performance of the integer optimizer of the Carmen system [2,5,18], which is in use at most major airlines in Europe. The optimizer can solve a wide class of integer linear programming problems (ILP), but here we focus on *pairing optimization*, which is a crucial step in the scheduling process. The optimization problems are then of the set partitioning or set covering type [2]

$$\min c^T x$$
$$\text{s.t.} \quad Ax \sim 1 \qquad (1)$$
$$x \in \{0,1\}^n$$

where $A \in \{0,1\}^{m \times n}, c \in \mathbb{Q}_+^n$ and 1 is a vector of all ones. The relation $\sim$ is either '=' in the set partitioning case or '$\geq$' in the set covering case. The rows of $A$ correspond to non-stop flights (*legs*) which are operated by the carrier and which need to be staffed with crews. The columns of $A$ correspond to so-called *pairings*. A pairing corresponds to a crew schedule in terms of a sequence of legs, starting at a home base, and returning (sometimes after a few days) to the home base. Thus $A_{ij} = 1$ if and only if leg $i$ is operated by pairing $j$. The pairings and their associated cost coefficients are usually computed in a complicated process taking crew utilization, regulations, union agreements, overnight cost, work plan robustness, credit time and many other factors into account. The

---
* This work has been supported by the ESPRIT HPCN program, project PAROS.

process of generating the matrix $A$ is called *pairing generation*. It is very time-consuming and discussed in [11]. In addition, there are usually a few additional *base constraints* which have a more general form and model the availability of personnel at different home bases of the company.

Let $N$ denote the number of nonzero entries in the $m \times n$-matrix $A$. The large problems we are most interested in have up to several hundred thousand variables and typically between a few hundred to a few thousand constraints. They are very sparse, usually having only 5 to 10 nonzeros per column. These problems are among the largest 0-1 problems solved in commercial applications, and generally available commercial solvers such as CPLEX cannot handle problems of this size. Our algorithm does not use branch & bound, but can be viewed as a Lagrangian based heuristic, which can efficiently find solutions to most of these problems in a few hours to within a thousandth of the optimal solution. For other Lagrangian relaxation methods applied to pairing problems in railway industry see for instance [4, 6], which also address problems of similar size using specialized heuristics. Recent work in airline crew scheduling include [3, 8, 12–14, 17]. Usually, the ILPs considered there are smaller than those considered in this paper.

For our machine model we assume $P$ processing elements (PEs) interconnected by some network. Each PE is a high performance RISC processor with at least two levels of cache and its own local memory. It will turn out that the communication cost of our algorithms can be modeled quite abstractly by the cost of a global broadcast or reduction for operands of length $m$. Let $T_{\text{coll}}(m)$ be a common bound for this time. This communication cost is compared with the cost for internal computations. We use $T_{\text{nz}}$, the time spent per iteration divided by $N$, for this comparison.

In Sect. 2 we briefly describe the existing optimizer and then discuss performance improvements and a new variant of the algorithm which needs less work and is additionally better adapted to the memory hierarchy. Sect. 3 introduces two promising approaches to parallelization. The implementation and first measurements are presented in Sect. 4. The last two sections are dedicated to conclusions and future work.

## 2 Sequential Algorithms

### 2.1 The basic algorithm

Our approach has been described in detail in [18], to which we refer for full detail and for quality comparisons with other algorithms. A simple way to understand the algorithm is to view it as a heuristic based on Lagrangian relaxation (see [9]). Consider problem (1) with a general non-negative integer right hand side $b$. Then the corresponding Lagrangian relaxation subproblem with equality [1] is

$$\min_{x \in \{0,1\}^n} c^T x + y^T(b - Ax), \quad y \in \mathbb{R}^m. \qquad (2)$$

---
[1] The "$\geq$" variant differs from (2) only by requiring $y$ to be nonnegative.

A general property of the relaxation is that if we can find a vector $y$ so that the solution to (2) is feasible for (1), then we have an optimal solution to (1) (This is related to the dual problem of maximizing (2) with respect to $y$). The relaxed unconstrained problem (2) can easily be solved optimally by considering $\bar{c}^T := c^T - y^T A$ and setting $x_j := 1$ if and only if $\bar{c}_j$ is negative. In most cases however it is not possible to find a relaxation (defined by $y$) that has this desirable property. The algorithm therefore introduces a heuristic scheme that makes this possible.

The algorithm proceeds by modifying $\bar{c}$ by considering one constraint. We call this process *iterating* a constraint. For constraint $i$ we define a row index set $z^i := \{j \mid A_{ij} = 1\}$ and a sparse vector $s^i$ representing its contribution to $\bar{c}$, i.e., $\bar{c} = c + \sum_{i=1}^{m} s^i$. The algorithm differs from pure Lagrangian relaxation in the way in which $s^i$ is computed, and the fact that these vectors are kept between iterations.

First, the change in $\bar{c}$ from the last update is cancelled out by computing $r := \bar{c} - s^i$. We then determine *critical values* $r^-$ and $r^+$ as the $b$-smallest and the $(b+1)$-smallest elements of $r$ (considering only indices in $z^i$). The variables associated with these values are called *critical variables*. We compute the *Lagrangian dual* corresponding to constraint $i$ as $y_i := \frac{r^- + r^+}{2}$ and shifted values

$$y^- := y_i - \frac{\kappa}{1-\kappa}(r^+ - r^-), \quad y^+ := y_i + \frac{\kappa}{1-\kappa}(r^+ - r^-)$$

where $\kappa$ is a control parameter in $[0, 1)$. The new contribution $s^i$ to $\bar{c}$ is updated to

$$s^i_j := \begin{cases} -y^- & \text{if } \bar{r}_j \leq r^-, \\ -y^+ & \text{if } \bar{r}_j \geq r^+, \end{cases} \quad \text{for } j \in z^i.$$

and $\bar{c}$ is set to its new value $\bar{c} := r + s^i$.

On the top level, this constraint iteration is done for every constraint over and over again. During these iterations $\kappa$ is slowly increased from 0, until a feasible solution is obtained from $\bar{c}$. Here is a high-level description of the algorithm:

$\bar{c} := c$, $\kappa := 0$, $x := 0$
**while** there are infeasible constraints **do**
    increase $\kappa$
    **for** each constraint in some random order **do** iterate constraint
**for** $1 \leq j \leq n$ **do if** $\bar{c}_j < 0$ **then** $x_j := 1$ **else** $x_j := 0$

The algorithm above is repeated with refined schedules for the increase of $\kappa$. Usually, four to five trials are enough to yield excellent solutions for large scale problems. An extension to constraint matrices in $\{-1, 0, +1\}^{m \times n}$ is easy [18].

## 2.2 Basic Performance Issues

The original implementation used a structured programming approach with a uniform representation of all constraint types which included an explicit (sparse) representation of the $s$-vectors. This performed well on machines where floating point operations dominated the execution time. But on today's machines memory accesses have almost completely taken over this role. Therefore our new code is object oriented with specialized representations for important constraint types.

A specialized pseudo-code for a set partitioning constraint (set covering is very similar) is given below. Only the set of nonzero indices $z$ and the previous shifted duals $y^+$, $y^-$ together with their indices of occurrence, $j^+$, $j^-$ are stored for each constraint – but no $s$-vector. By patching $\bar{c}_{j^-}$ we can compute $\bar{c} - s$ in constant time up to a collective shift by the old $y^+$, and we can then avoid the explicit use of the intermediate $r$ vector. The new critical elements can be found by locating the two minimal elements of $\bar{c}$ restricted to indices in $z$ plus a constant number of scalar operations. Finding the two minimal elements and their indices (in the function "minIndex2Indirect") is almost as easy as finding one minimum alone and can be done using $|z| + O(\log|z|)$ comparisons and an equal number of double indirect memory accesses on the average (refer to [7, Problem 6.2] for some discussion). All other operations have negligible cost. Computing the new (shifted) duals needs only some scalar operations now. Finally, $\bar{c}$ can be updated by subtracting a constant offset for all indices in $z$ and patching $\bar{c}_{j^-_{new}}$. Effectively, we have fused the two vector additions from the abstract iteration scheme into a single offset computation. The cost for this is dominated by $|z|$ double indirect memory read and write operations.

**Specialized code for a set partitioning constraint:**
**method** SetPartitioningConstraint::iterate(**var** $\bar{c} : \mathbb{R}^n$, $\kappa : [0,1)$, $z$ : array of $\mathbb{Z}$ )
$\quad \bar{c}_{j^-} := \bar{c}_{j^-} - y^- + y^+$ \quad\quad -- make $\bar{c}_{j^-}$ comparable with other elements
$\quad (r^-, r^+, j^-_{new}, j^+_{new}) := \text{minIndex2Indirect}(\bar{c}, z, k)$
$\quad (r^-, r^+) := (r^- - y^+, r^+ - y^+)$ \quad\quad -- undo offset
$\quad y_i = \frac{1}{2}(r^+ + r^-)$
$\quad (y^-_{new}, y^+_{new}) := (y_i, y_i) + \frac{\kappa}{1-\kappa}(r^+ - r^-)(1, -1)$
$\quad$ **for** $1 \leq i \leq |z|$ **do** $\bar{c}_{z[i]} := \bar{c}_{z[i]} - (y^+_{new} - y^+)$
$\quad \bar{c}_{j^-_{new}} := \bar{c}_{j^-} - (y^+ - y^-_{new})$ \quad\quad -- patch
$\quad (y^-, y^+, j^-, j^+) := (y^-_{new}, y^+_{new}, j^-_{new}, j^+_{new})$

## 2.3 An Active Set Strategy

Since $n \gg m$, it seems to be wasteful to go through all the variables all the time. Observations show that the set of variables which has recently been used for critical variables – we will call this the *active set* – remains quite stable most of the time. Since only critical variables affect the numerical progress of the algorithm, one would be tempted to forget about the rest of the variables and only work in the active set. However, closer inspection shows that from time to time variables never considered before make their way into the active set. In particular, towards the end, just before the iteration converges, a number of new variables is pulled in to create a feasible solution.

Out of several logically useful ways to exploit this idea consider the following one: Most iterations of the algorithm from Sect. 2.1 work on a copy of the problem containing only the information relevant for the active set. Note that often this reduced problem or at least its $\bar{c}$-vector will fit into the (second level) cache. From time to time we make a *global scan* to inspect all variables: First, we update the global $\bar{c}$-vector. Then we allow a number of previously inactive variables to become active. More precisely, we consider all those variables which would be critical for some constraint if a true global iteration were performed – in a sense we make a dry run of a global iteration. In order to avoid an uncontrolled growth of the active set for long running ill-behaved problems, we periodically deactivate all variables which have not been critical for a long time.

So far we have already harvested two advantages: We perform less work per iteration and most iterations are more cache friendly. However, going through all the constraints to update $\bar{c}$ and then again checking all the constraints in order to find critical inactive variables would be as expensive as an iteration of the old algorithm. But we can do better now. For the global scan we store the nonzero entries of all variables in a column wise fashion and also traverse the data in this order. We now work one variable at a time, update its $\bar{c}$ entry and then immediately check whether it is a new (second) minimum for some constraint. Before, accessing $\bar{c}$ implied a cache miss – now the current entry is held in a register. We pay for this by having to hold the Lagrangian duals and the minima for all constraints in a frequently accessed array now. These arrays are so small however that they are likely to fit into the (first level) cache. The nonzeros of the current column will always fit into first level cache and often our code can even hold them in registers. Compared to a global iteration of the basic algorithm we are down from $2N+O(n)$ cache faults to $N/C+O(n/C)$ where $C$ is number of `ints` fitting into a cache line.[2] The remaining cache faults stem from sequentially reading indices of nonzero elements and $\bar{c}$. These faults can therefore be hidden by prefetching. Experiments show that column wise traversal is about three times faster than row wise traversal for large problems on a 140MHz Sun Ultra-1. On newer machines with faster clock even higher differences should be expected.

## 3 Parallelization

Parallelizing the basic algorithm is not trivial because its iterative nature only allows parallelization within an iteration. As discussed above, a single iteration is so fast that we have only rather fine grained parallelism available. Even worse, parallelizing the outer loop is very difficult since many constraints are coupled by common entries of $\bar{c}$. So, we mostly have to rely on the very fine grained parallelism in the innermost loops. In Sect. 3.1 we show how we can achieve useful speed-ups nevertheless and in Sect. 3.2 we extend this approach for the global scan.

---

[2] Exploiting that a column index fits into 2 bytes we could even cut that in half.

## 3.1 Parallelizing by Distributing Variables

We make PE $k$ responsible for some subset $V_k$ of variables, i.e., PE $k$ stores those entries of $\bar{c}$ and those nonzeros which refer to the variables in $V_k$. We parallelize the innermost loops, i.e., finding minima and adding a constant offset: The latter is easy – just broadcast the offset and perform the remaining operations locally. Finding the critical elements is only slightly more difficult. First determine the two locally minimal elements $r^-, r^+$ and their positions $j^-, j^+$ and then compute the global critical elements using a global reduction with the associative operator

$$(r_1^-, r_1^+, j_1^-, j_1^+) \otimes (r_2^-, r_2^+, j_2^-, j_2^+) := \begin{cases} (r_1^-, r_1^+, j_1^-, j_1^+) & \text{if } r_1^+ \leq r_2^- \\ (r_1^-, r_2^-, j_1^-, j_2^-) & \text{if } r_1^- \leq r_2^- < r_1^+ \\ (r_2^-, r_1^-, j_2^-, j_1^-) & \text{if } r_2^- < r_1^- \leq r_2^+ \\ (r_2^-, r_2^+, j_2^-, j_2^+) & \text{if } r_2^+ < r_1^- \end{cases}.$$

**Analysis:** We also have to specify *how* the variables should be partitioned. Let $l_{ik}$ denote the number of nonzeros on PE $k$ for constraint $i$. $W := \sum_i \max_k l_{ik}$ should be close to $N/P$ in order to achieve good load balancing. This looks like a nontrivial problem since a single partitioning should exhibit good load balancing for all constraints. Fortunately, randomization saves the situation. We simply distribute the variables randomly. Using Chernoff bounds (e.g., [15]) it can be shown that $W = N/P + O(\sqrt{N/P \log N})$ with high probability. As long as $N/m \gg P \log P$ this implies good load balance. This condition is fulfilled for all problem instances under consideration.

A more severe condition on the problem size is that the local computations should dominate the communication time, i.e., we need $T_{nz} \frac{N}{P} \gg m T_{coll}$ if we iterate one constraint at a time. Unfortunately, this will only be the case for very large problems even on tightly coupled parallel machines.

**Parallelizing over both constraints and variables:** We do not really need a separate collective operation for each constraint. Constraints which do not share nonzeros can be iterated independently. If we consider the constraint dependence graph where the nodes are constraint numbers and edges connect constraints which share variables, we can identify subsets of independent constraints using a graph coloring algorithm. All the constraints colored with the same color are independent and can therefore be iterated using only one vector valued reduction and broadcast operation.

We have made an experimental implementation of this algorithm for the SGI Origin 2000 using its native compiler `#pragmas` for parallelization. For coloring we used a simple $O(N + m^2)$ time implementation of the first fit heuristic. For the set covering constraints of the (large but not huge) problem instance `lh_d126_09` with $m = 176$, $n = 464222$ and $N = 4048428$ the graph coloring heuristic colored the constraints using 83 colors. We obtained a speedup of 7 on 8 PEs but 16 PEs were no faster. This could be further improved by tuning the reduction operation.

Several lessons can be learned from this simple experiment: Good performance can be achieved for large problem instances on machines which allow low latency interaction between PEs. Bandwidth is of secondary importance in this case. Graph coloring reduces the synchronization overhead by a dearly needed constant factor but more should not be expected since large problem instances have a quite dense dependence graph. In a future version we plan to investigate the possibility to achieve coarser granularity by *multi-coloring*: Color each constraint with $k$ colors. This suffices to iterate each constraint $k$ times in unspecified order. This should yield more constraint parallelism since we relax the requirement to iterate each constraint in each iteration.

## 3.2 Parallel Active Set

The active set heuristic opens the way to a more coarse grained parallelization – at least for the global scan. We perform the same operations – adding offsets and finding minima – as the basic algorithm. But we do it in a batched way for all constraints at once. Therefore we can broadcast all $m$ duals together and we only need to perform a reduction operation for a vector valued input of length $m$.

This implies message lengths of several kilobytes so that the startup overhead for communication is no longer the limiting factor, even on networks of workstations. The bandwidth of the network now becomes an issue but is unproblematic for our case where the work per PE, $T_{nz}N/P$, is large compared to the communication volume $O(m)$. Using a pipelined implementation of the collective communication operations we will have $T_{\text{coll}}(m) \in O(m)$ on most networks.

Even on a slow shared medium like a 1MByte/s Ethernet the situation is not too bad. For example, on 4 PEs the instance lh_dl26_09 used above needs about 30ms for communication and about 160ms for computations on a Sun Ultra-1.

In its simplest form, all variables are (randomly) distributed to the PEs and PE 0 is additionally responsible for the active set. So during the active set iterations the other PEs remain idle. This works well if the active set is so small that the global scan dominates the execution time.

Currently, we are working on variants with a dedicated fast PE for the active set (possibly even a multiprocessor). Concurrently, the other PEs scan their variables using slightly outdated duals or generate new pairings [10].

## 4 Implementation and Experiments

For the implementation, all parts of the parallel active set code depending on a particular parallel environment were isolated in a small module. By avoiding any global or `static` variables, care was taken to be compatible with thread libraries like POSIX threads. However, MPI [16] was chosen as the first parallelization platform. The functions `MPI_Barrier` and `MPI_Reduce` proved to be a perfect match for the operations required by the global scan of the parallel active set algorithm. These operations are not only simpler to use than shared memory primitives but a good MPI implementation can also come close to the peak performance of the hardware for the long inputs we use. Only setting up the problem is more cumbersome than in our previous experiments using a shared memory machine. It proved to be unproblematic to port the code to LAM, mpich and the native implementations from SGI and SUN. The code works on machines from Sun, SGI and HP.

As expected, the global scan scales well now even on slow networks but this can only be exploited if this task dominates the computation time. Fig. 1 shows how much larger $n$ is compared to the active set. We see that for constraint matrices with large aspect ratio $n/m$ the active set is very small.

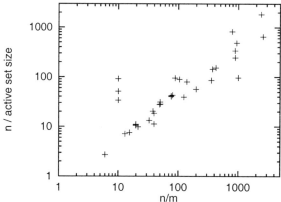

**Fig. 1.** Double-logarithmic plot of ratio between total number of variables and active set size (averaged over a run). The problem suite used mainly consists of problems from Lufthansa and Swedish Railways. The outliers for $n/m = 10$ are artificial problems which are much denser than typical crew scheduling problems [19].

How exactly the parallel active set can be used in practice depends on the character of the problem. For some problems all variables have to be inspected very rarely and the sequential active set algorithm will do much less work than the original algorithm. Otherwise, we have a significant parallelization potential for the global scan. For our crew scheduling problems, and for our preferred parameter settings, the truth lies somewhere in the middle. Sixteen active set iterations per global scan work well. This value is not very sensitive. This gives us a significant sequential improvement plus a moderate parallelism.

Table 4 shows some typical running time results on a Sun Ultra-1 network connected by a shared Fast Ethernet for a cross-section of our problem suite. We note that several problems are so large that a code like CPLEX is not able to find a solution even to the LP-relaxation. The code "prob1" is currently used in production at Carmen. Times in seconds are given for the optimized sequential code without active set, and then for the active set strategy running on 1, 2 and 4 workstations. Objective function values are also given for the new code with and without the active set. Since the active set code works in a different way it is here not possible just to compare time per iteration as before. This also means that the number of required iterations in individual runs can vary considerably, so that very large and very small overall speedups are possible. Even if all variables are forced to be active, our sequential tuning efforts roughly double the speed for instances where $\bar{c}$ fits in the cache. We also see that we achieve a performance improvement by a factor of 3–10 due to the improved sequential code and an additional speedup of three for large problems using four networked workstations. We finally note that there is no significant decrease in the quality due to the active set strategy.

**Table 1.** Solution quality and optimization time comparison for the code without active set and the parallel active set code.

| problem name | $m$ | $n$ | $prob1$ time | no active set | | active set, no. of PEs | | | | best |
|---|---|---|---|---|---|---|---|---|---|---|
| | | | | time | obj | 1 | 2 | 4 | obj | known obj |
| sj_daily04sc | 429 | 38148 | 296 | 94 | 261358 | 37 | 31 | 24 | 261358 | 261358 |
| sj_daily34sc | 419 | 156197 | 1510 | 885 | 259896 | 139 | 106 | 50 | 259896 | 259896 |
| lh_dl26_02 | 682 | 642613 | 9962 | 4071 | 733110 | 843 | 412 | 294 | 733110 | 733110 |
| lh_dl26_04 | 154 | 121714 | 1256 | 373 | 339220 | 216 | 105 | 60 | 339220 | 339220 |
| lh_dt1_11 | 5287 | 266966 | 1560 | 765 | 16758592 | 298 | 78 | 66 | 16758625 | 16758592 |
| lh_dt58_02 | 5339 | 409350 | 2655 | 2305 | 16538051 | 924 | 406 | 207 | 16537995 | 16537995 |

## 5 Conclusions

Based on the "industrial strength" Lagrangian heuristic for solving large sparse 0/1 integer programs in the Carmen system, we have achieved a number of significant performance improvements. On the sequential side we have not only reformulated the necessary mathematics to better fit modern CPUs with multi-level caches, but with the active set strategy we also have a new algorithm which can handle problems with many variables much more efficiently.

Both the original and the active set approach have been parallelized in different ways. The former scales well on tightly coupled machines and using the lazy update strategy it also achieves some speedup even on networks of workstations. The parallel active set code is even better suited for loosely coupled machines.

The new and much faster implementation is an important step towards significantly reducing one of the main time critical parts of the crew scheduling process, where shorter and more flexible planning cycles can be directly translated into economic benefits for the airlines. The fast and reliable solution of very large problems also opens up for new modelling possibilities, both in scheduling, as well as in other applications where large integer optimization problems have to be solved.

Within the PAROS project, the optimizer is not the bottleneck in the system any more, and the immediate task for Carmen and its partners will be an efficient integration of the parallel optimizer with the parallel pairing generator [11, 1] which also runs on a network of workstations. Other open questions have to do with parallelization strategies for the more general non set covering constraints, which is not a problem with the active set strategy, but more difficult for the variable based parallelizations.

## References

1. P. Alefragis, C. Goumopoulos, E. Housos, P. Sanders, T. Takkula, and D. Wedelin. Parallel crew scheduling in PAROS. In *EUROPAR'98*, Lecture Notes in Computer Science, 1998. to appear.
2. E. Andersson, E. Housos, N. Kohl, and D. Wedelin. *OR in the Airline Industry*, chapter Crew Pairing Optimization. Kluwer Academic Publishers, Boston, London, Dordrecht, 1997.
3. C. Barnhart and R. G. Shenoi. An alternate model and solution approach for the long-haul crew pairing problem. Jul 1996.
4. A. Caprara, M. Fischetti, and P. Toth. A heuristic algorithm for the set covering problem. In *Lecture Notes in Computer Science*, pages 72–84, 1996.
5. The Carmen System, version 5.1. Carmen Systems AB, Göteborg, Sweden.
6. S. Ceria, P. Nobili, and A. Sassano. A Lagrangian-based heuristic for large-scale set covering problems. Technical report, Dipartimento di Informatica e Sistemistica, Università di Roma, La Sapienza, Italy, 1995.
7. T. H. Cormen, C. E. Leiserson, and R. L. Rivest. *Introduction to Algorithms*. McGraw-Hill, 1990.
8. G. Desaulniers, J. Desrosiers, Y. Dumas, S. Marc, B. Rioux, M. Solomon, and F. Soumis. Crew pairing at Air France. *European Journal of Operational Research*, 97:245–259, 1997.
9. M. L. Fisher. The Lagrangian relaxation method for solving integer programming problems. *Management Science*, 27(1):1–18, 1981.
10. C. Goumopoulos, P. Alefragis, and E. Housos. Parallel algorithms for airline crew planning on networks of workstations. In *International Conference on Parallel Processing*, Minneapolis, 1998.
11. C. Goumopoulos, E. Housos, and O. Liljenzin. Parallel crew scheduling on workstation networks using PVM. In *EuroPVM-MPI*, number 1332 in LNCS, Cracow, Poland, 1997.
12. K. L. Hoffman and M. Padberg. Solving airline crew scheduling problems by branch-and-cut. *Management Science*, 39(6):657–682, 1993.
13. S. Lavoie, M. Minoux, and E. Odier. A new approach for crew pairing problems by column generation with an application to air transportation. *European Journal of Operations Research*, 35:45–58, 1988.
14. R. Marsten. RALPH: Crew Planning at Delta Air Lines. *Technical Report. Cutting Edge Optimization*, 1997.
15. J. Motwani and P. Raghavan. *Randomized Algorithms*. Cambridge University Press, 1995.
16. M. Snir, S. W. Otto, S. Huss-Lederman, D. W. Walker, and J. Dongarra. *MPI - the Complete Reference*. MIT Press, 1996.
17. P. H. Vance. *Crew Scheduling, Cutting Stock, and Column Generation: Solving Huge Integer Programs*. PhD thesis, Georgia Institute of Technology, August 1993.
18. D. Wedelin. An algorithm for large scale 0-1 integer programming with application to airline crew scheduling. *Annals of Operations Research*, 57:283–301, 1995.
19. A. Wool and T. Grossman. Computational experience with approxima- tion algorithms for the set covering problem. Technical Report CS94-25, Weizmann Institute of Science, Faculty of Mathematical Sciences, Jan. 1, 1994.

# Simulating Synthetic Polymer Chains in Parallel[*]

Bernd Jung
Editorial Office "Macromolecular Chemistry and Physics"
Hegelstraße 45, 55122 Mainz, Germany

Hans–Peter Lenhof, Peter Müller, Christine Rüb
Max-Planck-Institut für Informatik
Im Stadtwald, 66123 Saarbrücken, Germany

### Abstract

We have investigated algorithms that are particularly suited for the parallel MD simulations of synthetic polymers. These algorithms distribute the atoms of the polymer among the processors. Dynamic non–bonded interactions, which are the difficult part of an MD simulation, are realised with the help of a special coarse–grained representation of the chain structure. We have devised and compared a master version and a distributed version of the algorithm. Surprisingly, the master version is competitive for a relatively large number of processors. We also investigated methods to improve load balancing. The resulting simulation package will be made available in the near future.

## 1 Introduction

Simulations have become a fruitful field for the use of computers. A good example is the technique of molecular dynamics (MD) simulations where the motion of all atoms in a molecular system is calculated for some desired simulation time $T$. This is done by evaluating energetic interactions between atoms at time $t$ according to a physical model called force field. The resulting forces are applied to the atoms using Newton's equations of motion, thus obtaining slightly different atom positions at time $t + \Delta t$. This procedure is repeated for as many time steps $\Delta t$ as necessary. If such an MD simulation is done with a sufficient number of atoms and if the interaction model captures enough of the relevant physics, then the computed trajectories of the individual atoms give insight into the overall, i.e., macroscopic properties of the molecule(s) (for an introduction of MD simulations, see [1, 4]).

The most limiting factor when doing MD simulations is the stupendous amount of computing time that is needed. This stems from the fact that the time step

---

[*]Funded by the German DFG–Research–Cluster "Efficient Algorithms For Discrete Problems And Their Applications", grant LE 952/1–2.

$\Delta t$ is normally in the order of $10^{-15}s$ whereas the desired simulation time is up to microseconds, seconds or even more, i.e., a huge number of iterations have to be executed. Thus MD simulations are a natural candidate for parallelisation. In the following sections we will describe the basic ideas of our parallel MD simulation package[1] which is adapted to a special class of molecules called *synthetic polymers*. Target platforms are parallel computers with distributed memory connected via a communication network, for instance, Cray T3E or Intel Paragon XP. Message exchange is done by using the MPI message–passing–library ([3]).

## 2  Simulating Synthetic Polymers

Synthetic polymers are macromolecules which do not occur in nature. These industrially manufactured macromolecules turn up in everyday life as countless variants of plastic, like nylon, polyester, etc. During the polymerisation process, small basic units (the monomers) composed of only a handful of atoms are forced to link together to form huge macromolecules. The typical structure (called conformation) of such a molecule is a winding chain, or a loosely expanded coil as a chemist would say. For both theoretical and practical studies of the many interesting properties of such polymers, it is adequate to simulate single chains (the influence of surrounding molecules is taken into account by appropriate stochastic forces). Therefore, the typical problem size for an MD simulation of a polymer chain is relatively small compared to protein simulations where tens or even hundreds of thousands of atoms are modelled. Furthermore, for most investigations, it is sufficient to look at quite short chain lengths (a few hundred monomers, i.e., rarely more than 4000-5000 atoms) as this normally reveals the asymptotic behaviour. Nonetheless, even simulating such short chains becomes very time consuming as relaxation takes place very slowly, requiring microseconds, seconds or even hours (e.g., near glass transition temperatures). Even on high–performance computers, this can exceed all available resources, easily requiring thousands of CPU hours.

One way to speed up simulations is to use faster force fields. The force field which models the potential energy of the polymer chain usually splits up into several components that describe the energetic contributions. In our MD–package, we use the widespread AMBER force field ([9],[10]) which differentiates bonds, bond angles, torsion angles and non–bonded Van–der–Waals (VDW) interactions. The sine and cosine terms in this force field have been replaced by terms with excellent numerical approximation but with less computational effort (see [5]). This optimised force field allows much faster evaluation of bonded interactions (by a factor of 2–7 compared to AMBER) and yields a very fast sequential version which is the basis of our parallel algorithm.

Accelerating the evaluation of the chemical model by approximating or by decreasing the level of detail in these models can be done only to a certain extent, if the simulation shall remain useful. The remaining resort is parallelisation. Over the last 10–15 years, a vast number of parallel MD simulation algorithms

---
[1] A first release will be made available to the public in spring 1999.

have been presented and some of them have been implemented and are available as more or less ready–to–use packages ([7, 8, 11, 2]). Almost all of them concentrate on the parallel simulation of biopolymers and very few (e.g. [8]) also explicitly target at synthetic polymers. This is difficult to understand and regrettable. Synthetic polymers are increasingly penetrating all aspects of modern life and there is a huge demand for compute–intensive theoretical investigations of polymer models and real–world polymers.

As most parallel MD–packages have been developed with biopolymers in mind, using them on synthetic polymers is often not satisfactory, especially with respect to running time. Biopolymers and synthetic polymers differ dramatically (see Figure 1): the former show dense and compact collapsed structures whereas polymers form extended random coils.

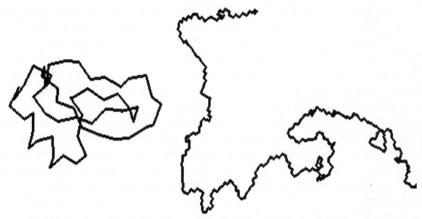

Figure 1: Backbone representation of a protein (left) and a polymer (right), both of roughly the same size

Furthermore, their dynamics are totally different. Biopolymers just tremble on the spot and are largely stationary (only a few side chains in active sites show a bit more agitation) but synthetic polymers undergo both considerable movement in space and conformational change.

The standard technique of parallel MD simulation is *spatial decomposition* where simulation space is subdivided into big cells which are allocated to processors (PE). This approach reflects data locality in a natural way. Atoms close together will be mapped to the same PE or to neighbouring PEs. Communication for the exchange of atomic data is therefore limited to direct neighbours. As biopolymers have dense conformations and very limited dynamics, this is an efficient approach. For synthetic polymers, it is less adequate as the molecule rapidly dashes off to completely new positions in simulation space, making extensive reallocations necessary.

For these reasons, we decided to use the idea of *atom decomposition* where the subdivision of the molecule is done with regard to the atoms and not their position. This approach creates the problem that communication is no longer limited to neighbouring processors. In addition, it is not even known which processors have to exchange messages at all.

In the following sections, we will describe how these problems can be tackled without sacrificing good parallel efficiency.

# 3 Parallelisation

A force field differentiates interactions into bonded and non–bonded. For our purposes, it is more convenient to use a slightly different distinction: *static* and *dynamic* interactions.

All interaction pairs that do not change during the simulation are static, i.e., bonds, bond angles and torsion angles and a number of atom pairs due to VDW interaction. These atom pairs form the vast majority of pairs that have to be considered. The set of dynamic interaction pairs is due to the VDW potential between non–bonded atoms. As this potential drops very rapidly with distance, only pairs within a given cutoff–radius $r_{cut}$ are considered. For a certain subset of atom pairs, the cutoff condition is always fulfilled because of the atoms' positions in the polymer chain (the chain never breaks up). These pairs therefore belong to the set of static pairs.

In the case of a synthetic polymer, all static pairs and most of the dynamic pairs occur along the chain, i.e., the atoms involved are not far away with respect to the molecule structure. Only a small part turns up between remote regions of the polymer chain. This is favourable for an atom decomposition along the chain. We therefore subdivide the polymer chain into segments of roughly equal size and allocate them to the available processors. Each PE is responsible for all atoms in "its" segment. In particular, each PE evaluates the subset of static pairs that occur within its piece of the chain, generates random numbers for stochastic forces and performs the numerical integration to compute new atom positions. As the static interaction pairs are defined by the molecule's structure, appropriate pair lists can be generated at the beginning of the simulation. Some static pairs may involve atoms that are allocated to different processors, so there is a need for communication between processors which are neighbours along the chain. This communication is fixed, i.e., it is known in advance which atoms in a limited "border region" a processor must exchange with its left and right neighbour. This kind of communication comes partly free if we use non-blocking communication: in every iteration, the sending of necessary data for static atom pairs is initiated, then internal computations are performed where no information from other processors is needed and finally, when the data have arrived, the remaining static pairs can be evaluated. This interleaving of local computation and communication with neighbours works very well ([6]) (provided that the hardware supports pure non-blocking communication primitives). As a sufficient amount of local work per PE must be at hand in order to be able to partially hide ongoing communication, this requires either the simulation of very long chains (which is normally not necessary from the chemical view) or the use of only a moderate number of processors (say up to 32).

Let us now face the problem of dynamic pairs. As mentioned above, most of them appear along the polymer chain, i.e., within the segment of a processor, so we can use a standard nearest–neighbour–search technique like a grid of cubes (with appropriate edge length) that each processor lays over its piece of the chain. Finding dynamic pairs for some atom is then straightforward and quickly achieved by looking at all atoms in the immediately surrounding cubes [2]. The main difficulty arises in dynamic pairs that occur between chain segments of different processors[3]. Each PE must find out which parts of the polymer chain are so close to its segment that relevent dynamic pairs occur. Of course, the trivial approach of broadcasting all atom coordinates among all processors (as [8] do) is prohibitively expensive for input sizes of more than a few hundred atoms. The idea of seldom doing such a costly operation and using pair lists during the next $x$ iterations is also normally not advisable for synthetic polymers. Their considerable conformational change during the simulation makes such a list quickly obsolete, leading to inaccuracies. We are therefore left with the task of doing a global search for such pairs in every iteration. We have devised the following two–step method to do this. First, the processors exchange coarse-grained information about their segments with other processors. Then, based on this coarse-grained data, candidate regions are identified where dynamic pairs may exist. Finally, coordinates of all necessary atoms are exchanged so that relevant interactions which occur due to dynamic pairs can be evaluated.

We have examined and implemented two variants of this idea. Both variants interleave local computation and non-blocking send/receive operations.

**Variant 1:** One processor acts as a master. In every iteration, all processors send coarse–grained information about their segments to the master which identifies candidate regions with dynamic pairs between different PE and informs them.

**Variant 2:** This version is truly distributed. All processors send the coarse–grained data about their segment to all other processors. In order to keep overall communication low, each PE commmunicates only with those processors whose segment may be close enough.

## 3.1 A coarse–grained chain representation

In a precursor to our current MD package ([6]), we used the local mesh of cubes as coarse–grained representation of the chain, i.e., these cubes were sent to a master which examined them. Though this approach yielded acceptable results, it became clear that further reduction of the data being sent was necessary. We therefore now use axis–parallel bounding boxes in order to give a rough approximation of the current overall structure. Each processor determines in every iteration the number and form of bounding boxes which are necessary

---

[2] Actually, because of symmetry only half of these cubes have to be looked at.

[3] Pairs that fall into border regions do not require additional communication as the necessary data becomes known due to the static pairs communication. The following discussion does not deal with these pairs.

to yield a good approximation of its chain segment. The quality of such an approximation is measured by comparing the total volume of these boxes to a lower bound. If we imagine a chain segment to be completely stretched and aligned with respect to a coordinate axis (say x–axis), then the volume of the corresponding bounding box may act as such a lower bound. Each PE starts out with one bounding box for its entire segment and increases the number of bounding boxes until the total volume is within a quality factor $q$ with respect to the lower bound. Our experiments have shown that rather crude approximations suffice, $q$ about 4–6. This leads to just a handful of bounding boxes per PE and cuts down the amount of data by as much as 90 % without giving up too much information about the overall chain structure. Because of using axis–parallel boxes, the computational overhead is also very modest as the box boundaries can be determined by simple comparisons of atom coordinates.

The bounding boxes are used to determine regions with possible dynamic pairs by simple straightforward tests (see Figure 2). For such a test between two boxes, the boxes are enlarged by the cutoff–radius of the VDW potential and then intersected. If the boxes are sufficiently close, two intersection boxes arise. They indicate which part of the corresponding bounding boxes might contain atoms that fall into the cutoff–radius of atoms in the other box (and hence processor). Only atoms that reside in these intersection boxes must be exchanged.

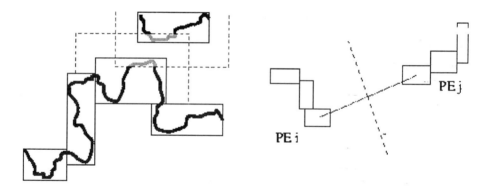

Figure 2: Left: bounding boxes allow a compact description of the chain structure. Right: a separating plane between two PEs

## 3.2 Master vs. distributed

In Variant 1, all bounding boxes are sent to a master processor which performs for each pair of processors and their bounding boxes the above overlap tests. It then sends the resulting intersection boxes back to the processors. Each PE now knows whether one or more parts of its segment are so close to other processors' segments that dynamic pairs may occur. If this is the case, the processor copies all atoms which currently reside in the intersection box(es) into communication buffers and sends them to the appropriate processor(s). Due to the symmetrical nature of this method, the processor also knows that corresponding messages will be sent by the other side. It receives this data and has now the necessary information to find and evaluate all remaining dynamic pairs.

Variant 2 is essentially the same, apart from the distribution of bounding boxes. Now, each processor broadcasts its boxes to all other processors. Then, it will do the relevant part of overlap tests as the master before and communicate with those processors whose bounding boxes have led to non–empty intersections, as in Variant 1. The broadcasting step is done only once at the beginning of the simulation. In subsequent iterations, most bounding box exchange can be omitted by a look–ahead approach. If a pair of processors determine that no intersections between their bounding boxes exist, they calculate a separating plane (see Figure 2) based on the current bounding boxes. In following iterations, the processors then check if their own current boxes are still on the safe side of this plane. If so, it is futile to send boxes as no intersections can occur. If the segment of a processor and hence the bounding boxes have moved so much that the distance between them and the separating plane is $\leq \frac{1}{2} r_{cut}$, the exchange of bounding boxes is resumed by sending a short notification message to the corresponding partner processor. This guarantees that the algorithm does not fail to detect possible box overlaps.

## 3.3 Dynamic load balancing

An equal distribution of load among processors is a prerequisite for achieving high speedups. Our measurements have shown that a considerable amount of time (up to 15 %) is spent waiting for messages despite using non–blocking message passing. The simulation has to deal with two different kinds of load: static load due to the evaluation of static pairs, and the calculation of random numbers and new atom positions. This work is well balanced as all PEs are allocated segments of equal size. Load fluctuations arise only from dynamic pairs as their number and hence the necessary work of finding and evaluating them depends on the current chain conformation. This imbalance is one reason for the high amount of waiting time. Another reason is the limited amount of local work because of the polymer chain lengths that we consider. Even (artificially) perfectly balanced simulation runs showed a significant loss to waiting time as there is simply not enough local work to hide communication completely. Whereas nothing can be done for the latter, some kind of dynamic load balancing may help to reduce imbalances due to dynamic pairs.

We therefore implemented such a technique based on the idea of nearest–neighbour–balancing. At regular intervals, a barrier synchronisation is done, i.e., a code point is introduced which all processors must reach before they can carry on. Each PE measures its waiting time at the barrier. The slowest PE will have a very small waiting time because it reaches the barrier last; the fastest PE will have to wait longest. These timing results are then exchanged between neighbouring PEs. If the differences in waiting times exceed a certain threshold, one repeat unit of a monomer is reallocated between neighbouring processors. This approach measures load not directly (via timing code blocks) but indirectly. Remember that all communication is done in non–blocking mode. Absolute values of local work are therefore not very informative. As long as local work helps in bridging waiting times, it is useful. Because of this effect, waiting times are a better choice in deciding whether imbalances between PEs have become too large. Unfortunately, reallocating little pieces of the chain between neighbouring PEs equalises load imbalances due to dynamic pairs but it also impedes the initial balance with respect to static work. Using this way of load balancing thus allows only limited accelerations. Because of that, we also considered the following approach which tries to balance load over time.

As our force field model needs stochastic forces, a fixed number of random numbers must be generated in every iteration. This work shows some nice properties: it is divisible and independent in the exact moment of generation. The idea is now to transform the fixed amount of work for random numbers into a dynamic form, by linking random number generation to the amount of waiting time. Whenever a PE reaches a point where it needs data from other processors, it checks whether the message has already arrived. If the message is there, it can go on to process the new data. Otherwise — instead of being idle — the PE bridges this time by generating a block of $B$ random numbers. If the processor is far ahead of other processors because it undergoes a phase of little dynamic work, this will lead to the generation of more random numbers than actually needed for the current iteration: a surplus which is stored away for "bad times". Later on, during phases of excessive dynamic work, the PE might not have many opportunities to generate numbers for the current iteration and it can take the deficit from the buffer[4], thus balancing its load over time.

In practice, this kind of load balancing has to struggle with a few difficulties. The ideal block size $B$ is hard to determine. It depends on the speed of the underlying random number generator and the length of waiting times that are to be bridged. Furthermore, the algorithm must have enough memory to build up a reserve of random numbers over periods with less dynamic work. Experiments indicate that at least 20 MBytes/PE are necessary to allow improvements. Finally, if a PE uses waiting times to generate numbers in advance, it must have the chance to eventually use them in order to realise a gain in running time, i.e., a sufficient alternation of phases with too much and too few dynamic work must occur.

# 4 Results and Conclusion

The following table gives some running times for the simulation of a polyethylene chain (3002 atoms) in good solvent during $4 \cdot 10^{-10}$ s on a Cray T3E–600. Short-range Van-der-Waals interactions are considered with a cutoff radius of 3.6 Å ( 1 Å = $10^{-10} m$). The time step used is $\Delta t = 0.5 \cdot 10^{-15} s$. All runs use the same random seed in order to prevent fluctuations due to different random numbers. The results show that our atom–decomposition algorithm achieves good parallel efficieny for both variants. It was a little surprising to us that the master paradigm (Variant 1) which is often somewhat disdained in the literature, is

---

[4] Unless the buffer is empty as well. Then the PE cannot help but generating the numbers on the spot.

|      | no load balancing |           |           |           |
|------|-----------|-----------|-----------|-----------|
|      | Variant 1 |           | Variant 2 |           |
| #PE  | time      | eff. [%]  | time      | eff. [%]  |
| 1    | 24495     | 100.0     | 24495     | 100.0     |
| 2    | 12459     | 98.3      | 12472     | 98.2      |
| 4    | 6287      | 97.4      | 6306      | 97.1      |
| 8    | 3223      | 95.0      | 3233      | 94.7      |
| 16   | 1657      | 92.4      | 1655      | 92.5      |
| 24   | 1330      | 76.7      | 1329      | 76.8      |
| 32   | 1063      | 72.0      | 1052      | 72.8      |

|      | nearest–neighbour balancing |          |           |          | balancing over time |          |           |          |
|------|-----------|----------|-----------|----------|-----------|----------|-----------|----------|
|      | Variant 1 |          | Variant 2 |          | Variant 1 |          | Variant 2 |          |
| #PE  | time      | eff. [%] | time      | eff. [%] | time      | eff. [%] | time      | eff. [%] |
| 1    | 24495     | 100.0    | 24495     | 100.0    | 24495     | 100.0    | 24495     | 100.0    |
| 2    | 12459     | 98.3     | 12485     | 98.1     | 12447     | 98.4     | 12472     | 98.2     |
| 4    | 6307      | 97.1     | 6300      | 97.2     | 6287      | 97.4     | 6290      | 97.4     |
| 8    | 3216      | 95.2     | 3213      | 95.3     | 3216      | 95.2     | 3213      | 95.3     |
| 16   | 1657      | 92.4     | 1655      | 92.5     | 1655      | 92.5     | 1653      | 92.6     |
| 24   | 1273      | 80.2     | 1268      | 80.5     | 1296      | 78.8     | 1271      | 80.3     |
| 32   | 1037      | 73.8     | 1025      | 74.7     | 1040      | 73.6     | 1019      | 75.1     |

Table 1: **Running times (in seconds) on Cray T3E–600**

able to compete with the distributed approach even at a relatively large number of PEs (16–32). Variant 1 has the natural adantage of requiring only $2(p-1)$ messages ($p = \#$PEs) in order to inform PEs whether their segment may be involved in dynamic pairs with a different processor's segment. Of course, the master constitutes a potential communication bottleneck as all these messages have to pour in and out of it. But this drawback is greatly diminished in our implementation by using non–blocking communication and does not have much effect for a moderate number of processors. In Variant 2, we do not have such a bottleneck as it is truly distributed. But much more effort is required in order to keep the number of messages acceptable (the number scales roughly as $p \log p$). The additional overhead of creating and testing separating planes reduces the benefits of a distributed approach. Besides, implementing this variant required much more coding and debugging.

The presented load balancing schemes presented both achieve only a moderate acceleration of the simulation. Because of the extensive dynamics of polymer chains, fast changing load imbalances are a natural concomitant for this kind of simulation. It is our belief that there is not much more room for improvement with respect to dynamic load balancing. But as our two approaches do not exclude one another, we will also try to combine them.

During this work it became clear to us that there is no single algorithmic paradigm equally suited for all types of molecules. The realm of MD simulation is far away from a single all-purpose MD-package. As long as computing resources are still dwarfed by CPU-time demands of production runs, the only way of coping is to employ chemical models and algorithms that are tailored to the special needs and peculiarities of the simulation target.

# References

[1] M. P. Allen and D. J. Tildesley. *Computer simulation of liquids*. Clarendon Press, Oxford, 1989.

[2] H. Berendsen, D. van der Spoel, and R. van Drunen. A message passing parallel MD implementation. *Computational Physics Communications*, 91:43–56, 1995. Package GROMACS: http://rugmd0.chem.rug.nl:80/~ gmx/.

[3] M. Forum. MPI: A message-passing interface standard. *International Journal of Supercomputer Applications*, 8:165–416, 1994.

[4] J. M. Haile. *Molecular Dynamics Simulation*. Wiley, 1992.

[5] B. Jung. Fast force field expressions for computer simulations. *Macromol. Chem. Theory and Simulation*, 2:673–684, 1993.

[6] B. Jung, H.-P. Lenhof, P. Müller, and C. Rüb. Parallel MD-simulations of synthetic polymers. In *Proc. of the 8th SIAM Conference on Parallel Processing for Scientific Computing*, 1997.

[7] M. Nelson, W. Humphrey, A. Gursoy, A. Dalke, L. Kale, R. Skeel, and K. Schulten. NAMD - A parallel object–oriented molecular dynamics program. *Journal of Supercomputing Applications and High Performance computing*, 10:251–268, 1996. Package NAMD, http://www.ks.uiuc.edu/Research/namd/namd.html.

[8] W. Smith and T. R. Forester. Parallel macromolecular simulations and the replicated data strategy. *Computer Physics Communications*, 79:63–77, 1994. Package DL-POLY http://gserv1.dl.ac.uk/TCSC/Software/DL_POLY/main.html.

[9] S. J. Weiner and K. A. Kollman. A new force field for molecular mechanical simulation of nucleic acids and proteins. *Journal of American Chemical Society*, 106:765–784, 1984.

[10] S. J. Weiner, K. A. Kollman, D. T. Nguyen, and D. A. Case. An all atom force field for simulations of proteins and nucleic acids. *Journal of Computational Chemistry*, 7:230–252, 1986.

[11] A. Windemuth. Advanced algorithms for molecular dynamics simulation: The program PMD. In T. G. Mattson, editor, *Parallel Computing in Computational Chemistry*, pages 151–168. ACS Books, 1995. Package PMD, http://tincan.bioc.columbia.edu/Lab/pmd/.

# Real-Time Signal Processing in a Collision Avoidance Radar System Using Parallel Computing

G.L. Reijns[1], A.J.C. van Gemund[1], J. Schier[2], and P.J.F. Swart[3]

[1] Dept. of Information Technology and Systems
Lab. of Computer Architecture and Digital Technique
Delft University of Technology
The Netherlands

[2] Inst. of Information Theory and Automation
Czech Academy of Sciences
Czechia

[3] Int. Research Inst. for Telecom-transmission and Radar
Delft University of Technology
The Netherlands

**Abstract.** COLARADO (an acronym for collision avoidance radar able to distinguish obstacles) is a system for use in the control of autonomously driven vehicles, having static radar antennas. By combining the echos, resulting from 2 continuously emitting transmitters and 3 receivers, it is possible to compute the locations of obstacles in 3 dimensions. The main problem is to separate the real obstacles from ghosts amongst noise and clutter, up to a distance of 12.8 m. This paper discusses the signal processing algorithms as well as the related real-time parallel computer hardware and system software.

## 1 Introduction

Colarado is a multistatic frequency-modulated continuous wave (FM-CW) radar system developed for guidance of an autonomously driven vehicle [6, 9]. The objectives of the project are to demonstrate that the principles of Colarado can indeed be used to detect obstacles (targets) in the neighbourhood of the vehicle. The next phase in the development will aim at a cost reduction of the radar and processing hardware. The advantage of using radar is that, unlike sensors based upon other physical principles (optical, ultrasound), its performance is insensitive to atmospheric disturbances (fog) or acoustic noise. The principle of the system is shown in figure 1. The radar part of the system provides distances of paths between the transmitting antenna's, via the obstacles, and the receiving antennas, but no directional data. Our FM-CW radar uses two static transmitting-antennas and three static receiving-antennas. The RF signal is sent alternately to each of the transmitting-antennas on a per sweep basis. During 256 $\mu$s of each sweep time, the transmit frequency increases linearly from 9 to 10.5 GHz. The radar antennas are mounted on the front side of a vehicle on a frame

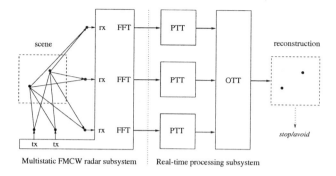

**Fig. 1.** Colarado System.

with an area of 1×2.5 m, such that all antennas are optimally spaced apart [5]. The antenna diagrams of the 2 transmit and 3 receive antennas all point in the forward direction with each antenna having an opening angle of 108 degrees in azimuth and 80 degrees in elevation. The echo signals from the obstacles are mixed with the transmitted signal and the results are processed on Fast Fourier Transform hardware (one chip per receiver). In two sweeps, 2 times 3 spectra are collected, related to 2 times 3 propagation paths. The obstacles will appear as peaks in the spectra. The output of each receiver FFT chain delivers peak amplitudes as function of time, which are representative of the propagated path lengths. This paper discusses the algorithms, the system software and the (parallel) computer hardware of Colarado, depicted in the blocks PTT and OTT of figure 1. Each of the PTT and OTT blocks contains a processor. The PTT (Peak Tracing and Tracking) processors identify and follow the peaks in the spectrum. The OTT (Object Tracing and Tracking) block combines the collections of the peaks to construct the locations of real objects and reject the ghosts. The block Reconstruction, which comprises a PC, displays an image of the object locations.

## 2 Signal processing

### 2.1 Introduction

The objectives of the processing are the real-time reconstruction of 3D scenes as sensed by the system. The frequency spectrum of a sweep comprises 128 bins of 10 cm propagation range, corresponding to a maximum distance of 12.8 m. Obstacles situated on a common 3D spatial ellipsoide with the transmit antenna and receive antenna as focus points, all have the same path length. In Colarado one has to cope with the problem of ghost targets, which results from combining spectral peaks belonging to different objects. Based upon a single transmitter the reconstruction process would yield in principle $N^3$ possible solutions for $N$ actual targets. The use of a single transmitter and 3 receiving antennas does not provide sufficient information to determine which peak combinations represent real objects and which represent ghosts. To eliminate the ghost peak combinations two different transmitting antennas are being used, which are activated alternately on a per sweep basis. Switching between both transmitting antennas, results in two target maps that are matched in order to eliminate the $N^3 - N$ ghost combinations. The processing of the peak combinations and the determination of the object coordinates, is executed by the OTT (object tracing and tracking) software. Ideally, the matching results in $N$ real targets. However, due to the restricted bandwidth of the radar and the limited resolution of the FFT, the coordinates of an object as produced by both transmitters might differ a little bit. Point objects, which are situated along a line that cuts perpendicularly through the middle of the frame with antennas, are detected with a resolution of 10 cm along that line and 4.5 degrees in azimuth. For point objects, which are 47 degrees line off side, the azimuth resolution decreases to 6 degrees in the horizontal plane [5]. The quality of the process that separates obstacles from ghosts will be highly dependent on the proper setting of the so-called tolerance window that is used in the matching process. A suitable compromise must be found between a too small window, which results in "blind areas" where objects cannot be detected, and a too large window, which results in numerous ghost peak combinations, that do not represent any real objects. It is a major challenge for the signal processing to optimize detection under real-time constraints.

## 2.2 Peak Tracing and Tracking (PTT)

The purpose of the PTT block is to detect peaks in the output of the FFT data [8]. Typically, the signal spectrum is contaminated by various types of noise and ground reflections. The number of erroneous peaks, caused by noise and clutter, must be reduced to an absolute minimum in order to enable the executing of OTT processing in real-time. The block diagram of the peak de-

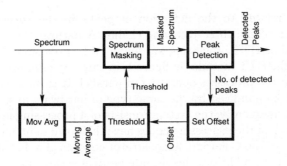

**Fig. 2.** Block diagram of peak detection.

tection is given in figure 2. The filtering in the peak detection is based on a threshold derived from the moving average of the spectrum and an additional offset. This offset is determined by the number of detected peaks in a sweep. In the current system, 128 range bins are used, hence it is not unreasonable to limit the detection to a maximum of 30 peaks per sweep. The moving average is computed in the MovAvg block. Together with the offset signal, it is used to compute the threshold in the Threshold block. Spectrum Masking filters the spectrum with the computed threshold. The Peak detection block counts the number of detected peaks in a sweep. This count is used in the Set Offset block to (re)adjust the offset of the threshold above the moving average.

## 2.3 Object Tracing and Tracking (OTT)

The adopted solution for ghost elimination is summarized below [1]. Based upon the signals received during the emission of the first of the two antennas, the OTT algorithm generates all possible combinations of peaks detected by the 3 PTT modules $c^{(1)}$ in a sweep. Using the coordinates of the receiving antennas and of the first transmitting antenna, these combinations are transformed ($xfm$) to potential object solutions in $o^{(1)}$, which in turn are transformed back ($inv$) to the "peak space", now based on the coordinates of the second transmit antenna and the same receiving antennas $c'^{(2)}$. The result is matched with the peaks actually received during the emission sweep of the second transmitting antenna ($c^{(2)}$). The process is represented by:

$$\left. \begin{array}{c} c^{(1)} \stackrel{xfm}{\rightarrow} o^{(1)} \stackrel{inv}{\rightarrow} c'^{(2)} \\ c^{(2)} \end{array} \right\} \text{match}$$

# 3 Hardware and software input routines

## 3.1 Four-node parallel processing system

The processing related to the detection of peaks in the spectrum (PTT) is identical for each of the three receiving chains. A natural solution to the high capacity demanding PTT processing was found in the provision of identical hardware to each of the PTT modules in figure 1, in this way minimizing the communication between the processes operating in parallel. It turned out that also the OTT task could be handeled by the same type of hardware as provided for one PTT process. Consequently, the system consists of 4 processing nodes, operating in parallel. As each of the processing nodes have their own main memory, the processing hardware of the Colarado demonstrator consists of a distributed memory parallel computer. In addition, this 4-node parallel system has a connection to a host (a PC), which acts also as a graphical user interface.

The execution time of the first version of the PTT program to detect 10 peaks in a sweep, was measured on a transputer T805 and also on a 486 dx2-66 processor. In addition a prediction was made of the execution time of the PTT program on four other types of processors, based on their instruction-sets and instruction execution times, I/O times, main memory and some published benchmarks [2]. The MPC601-80 processor turned out to give the highest performance at that time (1997) and would meet the predicted real-time execution requirements of both PTT and OTT. Moreover, complete boards with so called power-trams were commercially available, each power-tram consisting of an MPC601-80, a T425 transputer co-operating as communication processor, and 8 M bytes of memory. Each PC 601 is locally interfaced through shared memory to the T425 transputer, which provides communication between the Power PC nodes as well as between the nodes and the host. The transputers T425 are also used in the communication with FFT units of the radar links. As one power-tram should have the capability to handle the PTT processing of one receiving chain, three power-trams are required for the PTT processing. A fourth power-tram is then needed for the OTT processing. The total signal processing hardware comprises therefore four power-trams hosted by a Pentium PC, which is also used as a graphical user interface.

## 3.2 Input from the FFT units

An important system requirement is the capability to handle the input streams of the radar data after the FFT. A radar sweep of 256 $\mu$s is followed by a "dead" time of 32 $\mu$s. So, the repetition time is 288 $\mu$s. Per sweep, the data of 128 range bins are routed to a power-tram. A range bin contains 12 bits of amplitude data, which occupies 2 bytes. So, a minimum data rate of 0.89 Mbytes/s must be supported. The transputer link can be configured to operate at a maximum of 20 Mbits/s. However, a transfer occurs in bytes, which are preceded by a double length start bit and ended by a stop bit. After transmitting one data byte, which takes a minimum of 550 ns, the sender waits for an acknowledge, which consists of a start bit followed by a zero bit. For that reason, 256 data bytes could be transferred in a minimum time of 171 $\mu$s. In practice we have found 190 $\mu$s which is equivalent to a data rate of 1.35 Mbytes/s. In addition to radar amplitude data, radar phase data is sent on a separate input link to each of the three PTT power-trams. The stream of phase data has the same data rate as the amplitude stream. A transputer has the capability to simultaneously input/output data on its four serial links. Storage of input data in the transputer memory is based on cycle stealing. So, although four input/output processes can run in parallel, their maximum input/output rate reduces with an increase of the number of active processes that run in parallel. Moreover, as the input block transfer is based on DMA, one has also to take into account the time of the routines to initialize and terminate the DMA.

## 3.3 Input/output routines

From a software developers point of view, the power-tram modules are very similar to the original INMOS transputer modules, because communication to the outside is hidden in the operating system. This operating system, delivered with the hardware, is called Power Tools. In order to determine whether one could use the input/output routines of this operating system for the transfer of the radar data to each of the three PTT power-trams, a communication test was carried out, based on the well-known ping pong strategy. In this strategy the communication block rate of a link between two nodes is measured by means of timing the round trip of a data block. First a block of data is sent from node 0 to node 1, then the same block is returned. The average time for one block transfer is the measured time divided by two. The measurements show a relatively large startup time of 160 $\mu$s for the communication and packetizing effect and an asymptotical communication bandwidth of 1 Mbytes/s. As there are only 288 $\mu$s available to transfer 256 bytes, it can be concluded that the input routines of Power Tools can not be used for the transfer of radar data to the PTT Power Trams. For this reason, we needed to develop our own T425 low level block transfer routines to read and store the radar data into a PTT transputer memory [4]. Each block transfer is initiated by a strobe signal, emitted by a PTT power-tram on each input channel to the radar hardware. Upon reception of this strobe, the block transfer is started if the radar hardware is ready for it. Each stream uses the following code, executed endlessly: `send(stobe); receive(256); send(stobe); receive(256);` .... As the phase data channel contains data, which is related to the amplitude channel, the two data streams need to be synchronized. This synchronisation is implemented by the radar hardware, which waits before starting the data transmission until the strobes of both related streams have been received.

The output of a PTT module to the OTT power-tram has a relatively low data rate, as only the distances of a restricted number of detected peaks in a sweep must be transferred. For that reason, this transfer has been executed under the control of the I/O routines of the operating system Power Tools. Whereas the transputer of a power-tram handles the communication, the high processing demand for the detection of peaks is carried out by the PC601 module of the PTT power-tram, which operates under Power Tools. It was demonstrated that the PTT units can handle all I/O data delivered by two input channels and one output channel and execute the rather complex peak detection software. However, to fully adhere with future requirements, which will include besides amplitude data processing also phase data processing, the computing capacity of a node must be increased. This could be done by replacing the PC601 by a higher speed version. The operation of the OTT power-tram (including I/O) is likewise executed under the control of Power Tools. The output of the OTT provides the coordinates of the detected obstacles, which data are routed to the Pentium host for 2D display.

## 4 System software architecture

The computational algorithms for the Colarado project have been developed using two different hardware platforms, a Unix/Linux workstation and the parallel system, described in section 2, called the Demonstator. The first system, called Simulator, is used for off-line developing and testing of the signal processing algorithms where researchers can work concurrently, and is not connected to the real-time hardware. Using the Unix workstation, the radar data is generated either by simulation software or the data is read from a disk. This disk contains spectrum data, recorded from the FFT output. In principle, software develop-

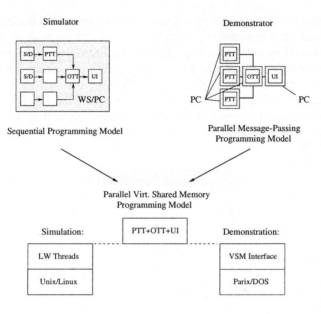

**Fig. 3.** Programming models.

ment for these two hardware platforms is faced with having to deal with two different programming models [4]. The workstation executes software based on a sequential programming model while the demonstrator parallel system executes software based on a parallel message-passing model. In order to enable a software implementation of an algorithm, developed on the Simulator, to run also on the Demonstrator, a third, unified programming model has been designed. This third programming model is a parallel virtual shared memory model and is implemented on top of the two other programming models (figure 3, UI = user interface). This unification allows the algorithm designer to use the same source code for the Simulator and real-time Demonstrator while compilation and linkage is differentiated as function of the target platform using a small number of compile time switches in the make file. With respect to the Simulator the parallel virtual shared memory model is implemented using light weight threads with negligible overhead. The virtual shared memory implementation for the Demonstator is based on a few functions on top of the native message-passing library. In this case the overhead is larger since the implementation of e.g. a one-sided data read call must be implemented in terms of a send plus a receive call. Although the high asymptotic bandwidth is maintained, startup costs are approximately doubled. Given the relatively large data sizes involved, however, overall performance remains practically unaffected.

## 4.1 Experiments

Many experiments have been carried out with the aim to optimize the different algorithms and their parameter-settings [8, 10]. In the beginning, the radar output was simulated by defining artificial moving point objects and noise. However, this was not satisfactory as the FFT processed radar spectrum showed to be quite different. One example of the many field experiments which were executed is shown in figure 4. The picture shows the output of the OTT when 5 persons are standing in a row in front of the radar at a distance of 6 m, after 100 sweeps are processed. The antennas are located at the upper edge of the plot. The per-

**Fig. 4.** OTT output, 5 persons standing in a row.

sons are represented by crosses. The solutions found by the system are marked by boxes. The measurements were carried out in a sports hall, which resulted also in some ground clutter and back-wall reflections. In the picture, we see that there are a few ghost solutions at the left and right hand side of the plot and in addition some solutions resulting from the background. The ghosts are probably due to the limited resolution of the radar and FFT, of which the effect can be shown to be more pronounced on the left and right hand side than around the middle line in the picture of figure 4 (see also section 2.1).

## 5 Future improvements

### 5.1 Algorithms

The computing time required by the described OTT algorithm in section 2.3 can still be reduced. Tracing has been defined to be related with the detection of peak combinations of objects that were not detected before. OTT tracking is then restricted to peak combinations that were already accounted for. The improved OTT concept is based upon the following: The two consecutive instances of the combination space $C(n)$ and $C(n+1)$ ($n$ is the number of a sweep-pair) can be divided into three sets. $C_{old} = C(n) \setminus C(n+1)$ contains an image of "old" objects that have left the scene between two instances. $C_{cur} = C(n) \cap C(n+1)$ contains objects, which are found in both instances ("current" region) $C_{new} = C(n+1) \setminus C(n)$ contains objects that have just entered the scene. With this division of the combination space, the OTT algorithm has been split into "Trace" and "Track". The trace part searches for legal combinations in $C_{new}$. The track part will only check the current set of combinations for validity (it will check whether or not an object has left the scene, i.e. moved from $C_{cur}$ to $C_{old}$). Let $O(n+1)$ be the next set of objects. The incremental approach to object selection is then characterised by:

$$O(n+1) \leftarrow \underbrace{(O(n) \cap C_{cur})}_{track} \cup \underbrace{(C^{(1)}_{new} \cap C^{(2)}_{new})}_{trace}$$

in which $C^{(1)}_{new}$ and $C^{(2)}_{new}$ are the sets of peak combinations not yet accounted for and occurring during the emissions of the sweep pairs (1) and (2) respectively. It has been shown that the track part of the algorithm has an complexity $0(N)$ [1, 3].

## 5.2 Use of phase data

Apart from amplitude data, the radar FFT output contains also phase data. The use of this phase data will provide an improvement in the resolution of the coordinates of obstacles, which in turn will enable a better distinction between obstacles and ghosts. Experiments with better resolution have shown to provide a substantial improvement [1]. In the present hardware, provisions have been made to input phase data, but the system hardware need to be upgraded and new software has to be developed in order to utilize a combination of amplitude data and phase data. It is estimated that the additional software will be at least a factor 3 more complex.

# 6 Conclusions

The signal processing part of a short distance collision avoidance radar system (Colarado) has been described. Colarado will be installed on vehicles which have a moderate speed. The radar part of the FM-CW (frequency modulated continuous wave) of the system provides distances of the paths between transmitting antennas-obstacle-receiving antennas, but does not provide directional data. The real-time processing hardware consists of a 4 node parallel processing system and an additional PC as host. A node consists of a Power processor, which executes the calculations and a transputer for the inter-node communication. The objectives are to locate the obstacles and distinguish them from the ghosts among a background of noise and clutter. The radar system provides a limited resolution of the distances of the obstacles. The lower this resolution, the more difficult it is to differentiate the obstacles from the ghosts. The algorithms have been developed with the aim to reduce the computing complexity to an acceptable level, retaining an optimum detection of obstacles with a high suppression probability of ghosts. The algorithms were gradually improved, using real field data. A demonstration model of the real-time system has been completed and operates in a laboratory environment. The software was developed on a Unix workstation. In order to enable the developed code to execute also on the 4 node real-time parallel computer, a virtual shared memory model was built. This model hides the properties of the sequential programming model of the workstation and of the message-passing model of the parallel computer.

## Acknowledgements

The authors wish to acknowledge the support from the Dutch STW agency, grant No. DEL 22.2733.

## References

1. H.J. Agterkamp. "An efficient system for real-time object recognition with bistatic radars," Tech. Report 1-68340-44(1995)07, Delft Univ., Fac. of Electrical engineering, CARDIT Lab., October 1995.
2. H.J. Agterkamp. "Multiple target trace and track software design," CARDIT, TU Delft, internal memo.
3. A.J.C. van Gemund, G.J. Agterkamp and G.L. Reijns, "An approach to real-time object recognition in radar-based collision avoidance systems," Technical report, Delft Univ., Fac. of Electrical engineering, CARDIT Lab., May 1995.
4. J.C. Joosse. "A unified programming model for the computational system of Colarado," MSc. thesis, Delft Univ., Fac. of Electrical engineering, CARDIT Lab., March 1997.
5. J.P. Karelse. "Multistatic antenna configuration optimization for the Colarado system," MSc thesis, IRCTR, TU Delft, Feb. 1998.
6. L.P. Ligthart, L.R. Nieuwkerk, J.S. van Sinttrueyen, "Basic characteristics of FM-CW radar systems," Avionics panel symposium on multifunction on radar for airborne applications NATO conference , Toulouse, 1985.
7. J. Schier, H.X. Lin and A.J.C. van Gemund, "Colarado system;Peak tracking and tracing algorithm and its parallel implementation," Technical Report 95-101, Delft Univ., Fac. of Applied Mathematics and Informatics, Delft, 1995.
8. J. Schier, A.J.C. van Gemund, "PTT and OTT enhancement," Technical Report, Delft Univ., Fac. of Electrical engineering, CARDIT Lab., April 1998.
9. P.J.F. Swart, L.R. Nieuwkerk, "Collision avoidance radar able to differentiate objects," *Proc. European Microwave Conference*, Israel, Sept. 1997, pp. 45-50.
10. P.J. Swart, J. Schier, A.J.C. van Gemund, W.F. van der Zwan, J.P. Karelse, G.L. Reijns, P. van Genderen, L.P. Ligthart, H.T. Steenstra. "The Colarado multistatic FMCW radar system," *Second European Microwave Conference*, Amsterdam 1998, pp. 449-454.

# The Impact of Workload on Simulation Results for Distributed Transaction Processing

Reinhard Riedl

Department of Computer Science, University of Zurich, Winterthurerstr. 190
CH-8057 Zurich, Switzerland
riedl@ifi.unizh.ch
http://www.ifi.unizh.ch/ riedl

**Abstract.** To use benchmarks or to use real database traces (for the simulation of high performance distributed transaction processing), that is the question. Whether it is preferable to deal with special real workload, or to ignore the rich diversity observable with real measurements by restricting oneself to standardized benchmarks. We demonstrate how the evaluation and comparison of different load balancing algorithms depends on the workload considered. For that purpose we compare the TPC-B benchmark with two real workload traces: We analyze their properties with respect to workload characteristics on a theoretical level and we indicate how the differences observed influence performance evaluations in simulation experiments. The results imply that performance of different load distribution algorithms essentially depends on the workload processed.

## 1 Introduction

In this paper we discuss the question, how the choice of workload for simulation experiments influences their results as well as the conclusions which are drawn from these simulation experiments for the design of load balancing algorithms for real world distributed systems. Hereby, data-affinity aspects are at the center of our interest, because they relate most directly to workload properties.

## 2 The Scenario

The question of the abstract [7] indicates the answer though. In this section we discuss the scenario which is of interest to us.

We consider transaction processing in a high performance distributed database/ data communication system. We assume that there are a considerably large number of processing units (addressed as nodes), each one with its own main memory, address space, and operating system, and that there is a hard disk environment, which is shared between all nodes. The data communication system supports with basic communication functions to remote processing nodes and with the terminals at the front-end of the system. The database is partitioned into database pages and it is accessed via a database handler. Data shipping is applied in order to access data residing at a remote node. Users interact with the database system by initiating transactions at the terminals. Furthermore, we assume that locking is applied to guarantee the ACID-properties for the transactions.

Users usually experience the quality of service of such a system as response time behavior. Response times essentially depend on the following cost factors:

- CPU-processing time
- page access time
- waiting times due to lock contention
- waiting times for CPU-processing
- context transfer time

Hereby, costs for context transfer arise only in case transaction instances sharing some context data - e.g. user identity - are processed at different nodes.

The data access costs have a major impact on the observed response times. Local accesses to data residing in the local page buffer are cheapest, while remote accesses to data residing at a remote page buffer are most expensive. Therefore, the optimal load distribution with respect to data access times is that load distribution which guarantees a maximal percentage of local data accesses.

Unfortunately, we do not know a priori, which data a transaction instance is going to access. Moreover the situation is further complicated by the fact that data-shipping moves the database pages through the system (and there may exist various copies of the same page in case replication is allowed in order to reduce READ access times). Furthermore, database pages are accessed concurrently by different transactions, which makes their location even less predictable. For work on workload characterization with respect to data access behavior see [2] and [5].

This paper continues former work on workload modeling and and data-affinity based load balancing carried out in collaboration with SNI Munich - in particular Eike Born and Thomas Delica - in the LYDIA Esprit project 8144. Check out [9] for more information and project deliverables.

## 3 Workload Representation Schemes

In this section we first consider various schemes for the representation of workload with respect to data access behavior. Then we present three workload characteristics which can be calculated from these schemes

We model workload as a stream of incoming transaction instances $T_i$, enumerated by the natural numbers. The transactions reference database pages $d_j$ from the set $D$ of all database pages. We assume that both the set of transactions and the set of pages are finite and denote the number of transactions by $N$.

Each transaction may then be represented by a representation vector with an a priori component and an a posteriori component. The a priori component describes what is known about the transaction before its processing is started, i.e.

that information which can be used for load distribution. We ignore particular values of input variables for that purpose and we consider only the following general a priori observables (which are assumed to take positive integer values)

1. TAC = transaction code
2. TID = terminal identity
3. UID = user identity
4. CID = conversation identity
5. AID = application identity

The a posteriori components describes the data access behavior of the instance. Hereby we do not distinguish between READ and WRITE accesses. These a posteriori observables may be modeled by the footprint vector

$$H(T_i) = (h_j(T_i))_{j \in D}$$

, whose dimension is equal to $|D|$, and whose entries are either 0 or 1 depending on whether the database page with the number of the entry is referenced by the transaction at least once or not at all: $h_j(T_i) = 0 \Leftrightarrow$ transaction $i$ does not reference page $j$.

As we are concerned with load distribution (and because the system component responsible for load distribution cannot distinguish between transaction instances with identical a priori observables), we further collapse this representation of workload to obtain the footprint matrix and the reference matrix.

Let $N_{quint}$ denote the number of different observed quintuples of a priori observables, $N_{quint} \leq N$. Then both matrices have $N_{quint}$ rows indexed by the different quintuples and $|D|$ columns corresponding to the database pages. In case of the footprint matrix

$$F = (f_{i,j})$$

different transactions are collapsed in a Boolean way, while in case of the reference matrix

$$R = (r_{i,j})$$

collapsing is done additively. That means an entry in $F$ is equal to 1 iff at least one transaction with a priori observables coinciding with the index of the row has referenced the page corresponding to the index of the column and 0 otherwise, while entries in $R$ denote the corresponding sum of such transaction instances, i.e. $f_{i,j} = 1$ iff $r_{i,j} \geq 1$.

From these representations we may derive various more simple load characteristics. Before we do that let us shortly indicate the size of matrices in case of real workloads. In a typical high performance transaction processing system we may expect about $10^{5.5}$ transactions to be processed in one hour with about $10^{7.5}$ page accesses to some $10^{6.5}$ different database pages. This means, that the reference matrices for a one-hour-workload will probably have about $10^{12}$ entries, less than one in ten thousand of which is nonzero though.

## 3.1 The Footprint Growth Function

Let us first assume that transactions are enumerated according to their finishing or abortion time. Then we may define the footprint growth function

$$FG : \{0, 1, ..., N\} \mapsto D$$

$$FG(n) := |\{j | \exists i \leq n, (h_j(T_i)) = 1\}|$$

We say that a workload has footprint growth of order $\alpha \geq 0$ in case that

$$FG(n) \approx Cn^\alpha$$

for some appropriate constant $C$. The footprint growth describes the dynamics of the workload. The higher $\alpha$ is, the less static page accesses may be expected and the less chances we have to predict 'somewhat' which pages a transaction with given a priori observables will access. Of course, a fitting of FG(n) with a function of type $x^\alpha$ will not always be possible. However, even then the footprint growth function yields some insight into the dynamics of the workload.

In general, one would expect a strictly sub-linear footprint growth function, which may even be nearly constant, i.e. have order 0. In systems with a strong growth rate, though, one might even expect a super-linear growth rate, i.e. an order strictly larger than 1. Nevertheless, this is unlikely and in case such an observation is made the reasons should be analyzed carefully and in depth. Super-linear growth may not only point to exceedingly large overall growth rates, but also to the fact that the workload trace is too short in order to obtain steady states in simulations based on it.

## 3.2 The Gelenbe-Zipf Function

Erol Gelenbe has posed the question, whether database traces obey a Zipf-like law. We define the Gelenbe-Zipf function as follows

$$GZ : \{1, 2, ..., N\} \mapsto \{0, 1, ..., |D|\}$$

$$GZ(n) := |\{j | \sum_i r_{i,j} = n\}|$$

For natural languages it is known that

$$nGZ(n)$$

is approximately constant. We say that a workload fulfills the Gelenbe-Zipf law with the Gelebe-Zipf exponent $\alpha$ in case that

$$n^\alpha GZ(n)$$

is approximately constant.

Obviously, a workload does not necessarily obey to some Gelenbe-Zipf law. In case it does, however, this does not only imply some rather special distribution of page references, but it also gives some implicit entropy-type meta-information about the workload.

## 3.3 The Block-Diagonal Matrix Function

As seen from a mathematical perspective, a natural question for a given workload is how many remote data accesses would result in case of an optimal static load allocation and function shipping with an optimal distribution of database pages. In the following we explain how this can be calculated. (Obviously, the result depends on the number of nodes.)

As a motivation, consider the following problem. Let $C$ be an arbitrary $n \times m$ matrix with non-negative entries and $M \leq \min(n, m)$. Search for that permutation of both rows and columns, such that you get an $M \times M$ block-diagonal matrix, corresponding to a linear mapping with $M$ non-trivial invariant subspaces. Such a permutation would correspond to an assignment of workload and data in our system, which yields no remote data accesses.

In general, there is no solution to that problem. The question therefore is, whether we can find a solution which is as good as possible, that is a solution which yields nearly a block-diagonal structure with as few entries outside the block-diagonal as possible. The latter is equivalent to a permutation of rows and columns plus the imposition of a block structure such that the sum of entries outside the block-diagonal is as low as possible.

That sum of off-block-diagonal entries corresponds to the sum of remote accesses for the optimal assignment of workload and database pages. Thus we can proceed to calculate the minimal percentage of remote accesses as follows. We first take an arbitrary permutation $\pi$ of rows and and arbitrary permutation $\sigma$ of columns and obtain a resulting matrix $C^M_{\pi,\sigma}$. For any such matrix $C^M_{\pi,\sigma}$ and any $M \times M$ block-partitioning we define its distance

$$d^M_{block}(C_{\pi,\sigma})$$

as the sum of all entries outside the block-diagonal. Then we define the $(\pi, \sigma)$-distance of $C$ to the set of all $M \times M$ block-diagonal matrices as

$$d^M(C_{\pi,\sigma}) := \min_{block} d^M_{block}(C_{\pi,\sigma}).$$

Finally we define the badness function $B^M$ of $C$ as

$$B^M(C) := \min_{\pi,\sigma} d^M_\pi(C_{\pi\sigma}).$$

$B^M$ then counts the minimal number of remote database accesses necessary in case that it is evaluated for the reference matrix $C$ defined above (and we have $M$ nodes in the system). Moreover,

$$g^M(R) := 1 - \frac{1}{\sum_{i,j} r_{i,j}} B^M(R)$$

denotes the maximal percentage of local accesses.

Unfortunately, for real workloads the effort for the calculation of $d^N$ is tremendous. However, for a given workload partitioning $P$, i.e. a clustering of

quintuples of a priori observables into $M$ workload classes, we may calculate the 'badness'
$$B_P^M(R) := B^M(R_P)$$
of the static workload allocation corresponding to P. The allocation corresponding to $P$ assigns all transaction instances belonging to the same workload class (via their quintuple of a priori observables) to the same node. Its 'badness' is simply the minimal number of remote accesses, where the minimum is taken over all possible distributions of database page over the set of nodes and it is achieved for the optimal distribution of database pages with respect to the given clustering of transactions. Hereby, $R_P^M = (r_{P\ l,j}^M)$ appearing in the definition of $B_P^M$ is the reference matrix with respect to the clustering:
$$r_{P\ l,j}^M := \sum_{i \in l} r_{i,j},$$
where summation is carried out over all quintuples $i$ which belong to the workload class $l$.

The fact that in general it is a rather hard task to calculate $g^M(R)$ corresponds to the well-known observation that clustering in high-dimensional spaces is a hard task either. There seems to be no way to calculate $g^M(R)$ without an explicit calculation of the optimal clustering of transactions and so far it is even unknown how good estimates for $g^M(R)$ can be obtained without that clustering at hands.

Still the approach depicted above may be exploited to compare two different workload partitionings. Furthermore, if workload partitioning according to single a priori observables is coarse enough we can try to calculate good clusterings starting from such a partitioning, that is clusterings, for which that partitioning is a refinement. Such clusterings may then be used to derive upper estimates for $g^M(R)$.

## 4 Comparison of TPC-B and Real Traces

In this section we compare the industrial benchmark TPC-B and real customer traces with respect to the workload characteristics defined above and with respect to their behavior in simulation experiments.

### 4.1 TPC-B

In this subsection we shortly indicate the definition of TPC-B.

The benchmark TPC-B assumes a large banking company with $b$ branches, each branch containing $t$ terminals and managing $a$ accounts. One database transaction type is specified, requiring that an account owner, drawn uniformly from all $b \cdot a$ account owners, visits a branch, with probability 0.85 the one managing his own account and with probability 0.15 a branch drawn uniformly from those branches not managing his account. Then he/she requires a modification (debit/credit) of his account. For this operation, an update of the account database page, the local teller database page, the local branch database page and a sequential write of a history is required. In usual implementations, additional locking of a free pool management database page and a trace-table database page are necessary

The specification of the TPC-B benchmark neither requires the existence of a transaction monitor nor does it assumes the existence of different applications, such that only the terminal (teller) identification TID or equivalently the user identification can be used for workload partitioning.

## 4.2 Real Traces

We have compared the TPC-B with 2 real customers traces that have been supplied to us by SNI Munich. They were drawn in a BS-2000 environment with a CODASYL database system. For that purpose, firstly information collected by monitoring tools available with the transaction monitor and the database handler was merged and secondly resulting primary traces were filtered in order to obtain workload representations with a priori and a posteriori observables as depicted above. The real traces were given to us in that form (though not as matrices but as listings of page references, of course). The following table gives the main figures of the traces.
m

|  | no. of transactions | no. of accessed pages | no. of page accesses |
|---|---|---|---|
| Customer Trace A | 10 360 | 41 003 | 998 355 |
| Customer Trace B | 1223 | 63 940 | 781 152 |

## 4.3 Comparison with respect to Workload Characteristics

We have evaluated both the real traces and TCP-B with respect to the workload characteristics depicted above. The table below lists the results. Hereby $M$ denotes the number of branches of TPC-B. Please note that we do not know what the optimal clusterings are and which badness they would yield. The reason why it is nearly impossible to decide whether clusterings obtained by standard clustering procedures are good approximations of the optimal clustering or not is the extremely high dimension of the clustering space. Readers interested in clustering methods may check [3] or [1].

We can only give rough estimates for the badness function of the real traces based on the results of our clustering experiments. We have experimented with the single linkage clustering algorithm, the complete linkage clustering algorithm, and the Ward clustering algorithm, starting from various different partitionings. These experiments have shown that partitionings according to single a priori observables are good first guesses and it is rather hard to improve them considerably, that is our 'improvements' were insignificant so far;-(. 'TPC-B*' in the following table refers to TPC-B without history pages.

|  | growth rate | GZ-exponent | $g^M$ |
|---|---|---|---|
| Customer Trace A | 0.6 | 2.5 | $\sim 0.6$ |
| Customer Trace B | 1.3 | 1.5 | $\sim 0.4$ |
| TPC-B | 1 | - | 0.025 |
| TPC-B* | 0 | - | 0.03 |

The table shows that except for the history pages the footprint size of TPC-B eventually becomes constant, while it appears to grow unrestrictedly in the real world measurements during the finite interval, when the measurement was carried out. Moreover, nothing like a Gelenbe-Zipf law may be observed with TPC-B, for which we always have

$$|\{n| \sum_i GZ(n) \neq 0\}| \leq 6.$$

Furthermore, the distance to the block-diagonal matrices is nearly zero for TPC-B. This means that there are static load distributions for TPC-B, which will yield nearly no remote data accesses, while such allocations do not exist for the real customer traces.

The reasons for the super-linear footprint growth in case of Trace B are unclear: it might be due to its small number of transactions. In fact there is little polynomial footprint growth observable for Trace A in the interval $[0, 1200]$, too.

### 4.4 Comparison with Respect to Simulations

Theoretical analysis clearly indicates that in simulation experiments the synthetic workload TPC-B will exhibit a behavior different from the two real world traces. Putting it differently, theoretical analysis implies that comparison of load balancing algorithms with simulation experiments using the TPC-B benchmark specification and comparison with simulation experiments using workload generated from the real world traces will yield different results.

Thus it comes as no surprise that we have made exactly these observations in simulation experiments. We have used a simulation model with 4 nodes, each of which has one processor and an unlimited page buffer, and all of which are directly interconnected and share the disks. Page replication is possible and a lock protocol guarantees consistency. The simulation considers page transfer and invalidation costs, costs for disk-I/O, and costs for context-transfer, but it neglects other cost factors such as costs for lock management. In this simulation environment we compared 'Join the Shortest Queue'-Routing (JSQ) with data-affinity based routing strategies. For more information on that subject see [4], [8], or [6].

The obtained results differ significantly for differ intensities of input streams (and for different choices of parameters in data affinity-based routing). Thus it difficult to obtain We just state the following results:

- While for simulations with real traces remote data accesses can be reduced by affinity-based by a factor around 3, they are reduced by factor more than 10 for TPC-B.

- While affinity based routing does not always outperform JSQ in simulations with real traces, it significantly outperforms JSQ in case of simulations with TPC-B.

This is not meant as a comparison of JSQ with data-affinity based routing, but it demonstrates that indeed such a comparison can be made only with respect to a given workload, since there is no statement like 'This load distribution algorithm is better than that one' unless one the two is really a 'bad' load distribution algorithm.

By folklore, there is no free lunch for any fair business. This statement is valid in our scenario, too. And it may be read the other way round: for any load distribution algorithm, which makes sense, and for any simulation experiment, there is a workload which 'optimizes' that load distribution algorithm. The latter means that the load distribution algorithm will outperform all other algorithms when the simulation is carried out with that workload.

## 5 Conclusions and Further Work

We conclude from our analysis that the real workload as measured in customer traces strongly differs in various workload characteristics from the synthetic industrial benchmark TPC-B.

While TPC-B has a fixed number of database traces, in real workload traces we observed a steady footprint growth even after various tens of thousands of database pages had been accessed. The main consequence of this fact for simulations is that an 'intelligent' distribution algorithm could easily 'learn' the workload characteristics of TCP-B. The latter would be at least much more complex a task for real workload. Benchmarks TPC-C and TPC-D are more complex, though, and a more detailed analysis of the workload characteristics of these benchmarks will be presented elsewhere.

It rests as an open engineering question how database traces of arbitrary length may be obtained, which in the large exhibit a polynomial footprint growth as it was observed in Trace A and which look like measured traces locally.

While the workload structure of TPC-B may be very effectively exploited by data-affinity based routing in order to reduce response times, the latter is much more difficult for real workloads and the success is usually much more moderate.

While TPC-B does not know any Gelenbe-Zipf law the latter is true for the analyzed real workloads. So far we have not yet fully understood the meaning of this observation. However, we conjecture that the lower the Gelenbe-Zipf exponent is, the more effective is data-affinity based routing. Further analysis on this question has to be done though.

Finally, the evaluations teach us once more the lesson, that the success of data-affinity based routing primarily depends on the workload patterns. There is no general statement like JSQ is better or worse than data-affinity based routing. If one has to decide on whether to implement it for a distributed system, one must first analyze the workload which the system is supposed to process. A more detailed comparison of these two load balancing strategies for a wider range of workloads will be presented elsewhere.

## References

1. N. Auerbach: Lastclusterung für verteilte Hochleistungsdatenbanksysteme, Diplomarbeit, University of Zurich, 1997;
2. E. Born, T. Delica, W. Ehrl, L. Richter, R. Riedl: Characterization of Workloads for Distributed DB/DC-Processing, Journal on Information Science, 1997;
3. R. Dubes, A. Jain: Algorithms for Clustering Data, Prentice-Hall, 1988;
4. S. Haldar and D. Subrammanian: An affinity-based dynamic load balancing protocol for distributed transaction processing systems, , Performance Evaluation 17, 1993;
5. Load Balancing in High-Performance Distributed DB/DC-Systems: Modeling, Analysis and Clustering of Workload for Data-Affinity Based Load Distribution, ALV Workshop Munich, 1998;
6. R. Riedl and R. Salomon; Proc. 2nd ICSC Symp. Fuzzy Control of Load-Balancing for Distributed DB/DC Transaction Processing, Fuzzy Logic and Applications, 1997;
7. W. Shakespeare; The Tragedy of Hamlet, Prince of Denmark, 3.1;
8. Yu P.S., Cornell C., Dias M., Balakrishna I.: Performance Analysis of Affinity Clustering on Transaction Processing Coupling Architectures, IEEE Trans. Knowledge and Data Engineering, Vol 6, No 5, 1994;
9. http://www.ics.forth.gr/pleiades/projects/LYDIA/

# Computer Simulation of Ageing with an Extended Penna Model

A. Z. Maksymowicz[1], M. Bubak[2,3], K. Zając[2], and M. Magdoń[4]

[1] Department of Physics and Nuclear Techniques, AGH Mickiewicza 30, 30-059 Kraków, Poland
[2] Institute of Computer Science, AGH, Mickiewicza 30, 30-059 Kraków, Poland
[3] Academic Computer Centre CYFRONET, Nawojki 11, 30-950 Kraków, Poland
[4] Department of Mathematical Statistics, Agriculture University, Mickiewicza 21, 31-120 Kraków, Poland
*email:* bubak@uci.agh.edu.pl
*phone:* (+48 12) 617 39 64, *fax:* (+48 12) 633 80 54

**Abstract.** This paper presents the results of computer simulation obtained with two modifications of the Penna bit-string model of biological ageing. Extinction of population may be caused by number of reasons, overhunting or too many harmful mutations inherited by offsprings among them. In this work we concentrate on population growth dynamics and their age distribution characteristics, such as number of mutations or survival rate, for different hunting and/or inherited harmful mutations rates, and discuss the role of bad mutations threshold fluctuations for possible improvement in health of the final population.

## 1 Introduction

Methods elaborated in the field of computational physics, which are based on large-scale simulation with integer operation (Ising model, cellular automata), are very promising as a tool for investigation of biological systems [1, 2]. Recently, a combination of statistical mechanics and elements of genetic population theory has succeeded in reproducing basic features of ageing processes [3].

The ageing is characterised by harmful mutation distribution in the population which also shows that older individuals have smaller survival probability than the younger ones. The age distribution may be introduced in the Penna bit-string model [4, 5] which offers more detailed information that simple theories based on a general concept of birth and death rates balancing the dynamics of the population growth. For each time step, the birth is an event with a given rate $B$ for members of population that survived elimination and has reached minimum reproduction age $R$. The death rate is controlled by the Verhulst factor [6],

$$V = \frac{pop}{maxpop}, \qquad (1)$$

where $pop(t)$ is the actual population at time $t$ and $maxpop$ is the environmental capacity. The Verhulst factor is just the fraction of population eliminated in one iteration step. However, an individual may die for other reasons such as hunting or bad mutations which cause illnesses. The latter is duly introduced in the Penna model.

The key assumption of the Penna model is that age of each individual is the factor which activates consecutive mutations prescribed in its inherited genome. This enables insight into the age distribution characteristics of the society during the evolution. On entry to next era we scan over all members and eliminate the ones which are too old, or having too many mutations, or hunted out, or perhaps for limited environmental capacity. If the individual survives, it may give birth if the reproduction age is reached. Thus dynamics of population, $pop(t)$, is governed by a genetic load, environmental capacity, the birth rate $B$ and other parameters that make the rules on how the system evolves. A strong advantage of the Penna model model is its flexibility which makes it easy to produce specific versions of the model devoted to some picked out scenario. This allows to study the role of different factors such as influence of sexual selection, parental care, overfishing or hunting etc., on the population [3, 7].

It is worth to mention that the logistic equation results, a perhaps expected follow-up of the model at the limiting case when only the Verhulst factor and birth at any age is declared, is not to be recovered in the model presented. This is so since the logistic equation is a deterministic rule while the Verhulst deaths and the birth rates are probabilistic rules and so they act similarly as noise. It is only when we force the deterministic version (the strict birth control and strict deaths rate due due limited environmental capacity) that we recover predictions of the logistic mapping such as periodic or chaotic solutions.

The computer simulations (similar to those with Ising model) is based mainly on bit and integer operations. Genome of an individual is implemented as a computer word with 32 bits. Bad mutation corresponds to setting up a bit. The position of the set bit in the word is related to *age* of the individual at which this mutation becomes active. The temporal order makes that we disclose the first *age* bits to get the number of bad mutations and compare it against the $T$ threshold. Each time step we increase *age* by one and add one more bad mutation on the individual account if it is found on the just opened bit position. Once a disease (bad mutation) arises, it affects the individual for its whole life.

In this paper we concentrate on catastrophic senescence when whole population vanishes due to one of the named causes. We systematically studied both factors and obtained a critical line on the *harmful mutation – hunting rate* plane above which population is extinct.

We also studied the role of fluctuations in threshold $T$ for maximum number of bad mutations on population growth dynamics. Such fluctuations are expected in the population as result of the factor that biological clock starts at embryos stage, before the birth. We show that these fluctuation produce individuals of higher resistance to mutations, and this fraction of population is more likely to survive, as well as less resistant ones. A distribution of $T$ around its average value with some width $\Delta$ may reflect either individual resistance to illnesses, or may result from some non-inherited somatic mutations in the pregnancy period.

When fluctuations in threshold $T$ are included, structural changes in age distribution in the final population (below the critical values) are observed. Higher number of mutations is allowed at the age given, and the survival rate is also larger. We conclude that fluctuations help the population to survive.

The simulation program was written in C and parallelised on the instruction level and with pragmas and compiler directives. The simulations were carried out on HP Exemplar S2000.

## 2  Impact of Hunting

As discussed in *Introduction*, in the Penna model a computer word represents a *genome* and each bit corresponds to presence (1) or lack (0) of harmful mutation. With each evolution step the age increases by one and next bit is disclosed bringing up a risk of another harmful mutation which may be terminal if the total number of these bad mutations reaches a threshold value $T$. Then the individual dies. It also may die for other reasons such as limited environmental capacity or hunting. If it survives, however, it gives birth to $B$ offsprings after reaching reproduction age $R$, which compensates the losses in population. Each baby takes on the bit-string after parents, yet it also gets $M$ more mutations randomly distributed over its lifespan which is introduced as a mutation rate in the model parameters set. If hunting is included, each individual dies for that reason at each iteration step with probability $H$.

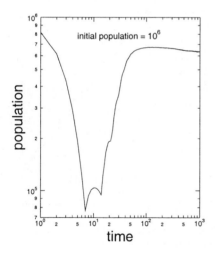 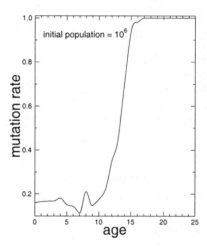

**Fig. 1.** Population *vs* time.        **Fig. 2.** Mutation rate *vs* age.

First, we used same set of parameters as in [3]

- initial population $pop(0) = 10^6$,

- maximum environmental capacity $maxpop = 5 \cdot 10^6$,
- threshold number of bad mutations $T = 3$,
- minimum reproduction rate $R = 8$,
- birth rate $B = 1$,
- mutation rate $M = 1$.

in order to recover their results for the limiting case $H = 0$.

For independent runs with the same initial parameters, we recover the same results as produced in Figs 1 – 6 so that the graphs practically coincide.

Fig. 1 illustrates the dependence of the population on time. The characteristic dip occuring there results from the randomly distributed mutations in the generated initial population. This population has randomly chosen genomes, that is as many as 50 per cent of bad mutations uniformly distributed over the whole lifespan. As result, the population rapidly decays since it is easy to find individuals which exceed the critical number $T = 3$ of bad mutations at their young ages. The decrease lasts until minimum reproduction age $R$ is reached when newly born individuals make up for the losses. Then the population grows until some equilibrium is established. This final state is independent on the initial conditions and only the model parameters control the final population.

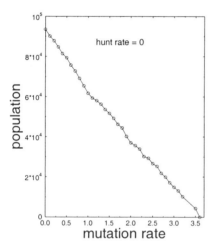 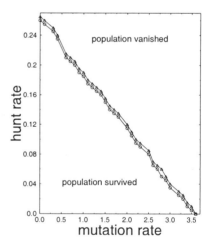

**Fig. 3.** Population as a function of mutation rate.

**Fig. 4.** Critical hunting rate *vs* mutation rate.

Fig. 2 presents the dependence of the mutation rate on age. The equilibrium population *pop* in the Penna model depends on the mutation rate $M$. Above a certain critical value $M_c$, $M > M_c$, no individuals survive. Dependence of *pop* on $M$ is shown on Fig. 3. If hunting with non-zero $Ht$ rate is included, the population decreases. In fact, the population vanishes for $H$ above a critical value which depend on the mutation rate $M$. The Fig. 4 show that critical line below which $pop > 0$ and above which $pop = 0$. The double line in Fig. 4 represents the accuracy with which this diagram line may be drawn, roughly two significant digits are guaranteed for critical $H$.

In our earlier work [8] (simulation without hunting) we concluded that too many mutations lead to a phase transition with suggestion of the critical exponent

$$\beta = 1/2, \quad pop \sim (M_c - M)^\beta. \tag{2}$$

In this work we observe that population ($pop$) *versus* hunting rate ($H$) reproduces a second order phase transitions, yet the exponent coefficient in

$$pop \sim (H_c - H)^\beta \tag{3}$$

seems to be closer to one which makes this transition smoother. $pop$ *versus* mutation rate $M$ for fixed hunting rate $H$ has smaller exponent coefficient than $pop$ $vs$ hunting rate $H$ for fixed mutation rate $M$.

## 3  Mutation Threshold Fluctuations

Penna model assumes one threshold value $T_0$ for all population at which genetic death happens, and this value is constant over time. However, death due to harmful mutation may take place at the younger age, $age < T_0$, before the bad mutation gets active. In fact, it is claimed that as many as around 0.6 of human embryos may die before birth [9].

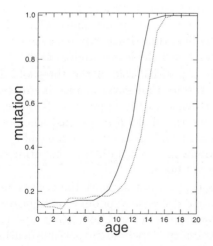
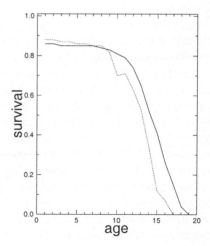

**Fig. 5.** Mutation $vs$ age, dotted line: $T = 3$, solid line: $T_0 = 0$, $\Delta = 6$.

**Fig. 6.** Survival $vs$ age, dotted line: $T = 3$, solid line: $T_0 = 0$, $\Delta = 6$.

We simulate this situation by introducing a uniform distribution of $T's$ around $T_0$, and $T$ fluctuates from $T_0 - \Delta > 0$ to $T_0 + \Delta > 0$. If we choose $T_0 = 0$ and $\Delta = 6$, then $T$ is in the range from $-6$ to $6$. Therefore the fraction $7/13$ of the offsprings gets negative or zero $T$ and dies before birth, the born individuals have distribution of $T's$ from 1 to 6. This way we generate $7/13 = 0.54$ fraction of dead offsprings, reasonably close to the claimed fraction 0.6 of the eliminated human embryos [9], and the average $T$ of those born ones is just 3 as in the standard model. Simultaneously we need to readjust the birth rate $B$ so as to compensate for those losses and get the same inflow of newly born individuals as for $T_0 = 3$ and $\Delta = 0$.

With the parameters $T_0 = 0$ and $\Delta = 6$ for fluctuating threshold $T$ we get the same final population, as expected. The mutation distribution in the final population is slightly altered and we observe greater accumulation of bad mutations for older individuals as illustrated in Fig. 5. Also distribution of population with given number of mutations is less pronounced, as the individuals with higher threshold for bad mutations are now allowed and so the distribution is smeared out over wider range of actually active mutations. The survival rate distribution is more favourable for older individuals, see Fig. 6, which may be due to presence in the population some members with exceptionally large $T$, possible in our model.

## 4 Summary and Conclusions

We conlude that population is rapidly vanishing if either the number $M$ of (somatic) mutations which are assimilated by a baby (above the ones inherited) approaches a critical value $M_c$, or if the hunting rate $H$ goes to $H_c$.

Actually, one can draw a critical line on the $(H, M)$ plane above which the population is extinct. While approaching the line, the critical exponent seems to be different depending on the direction from which the line is approached.

Another finding from our simulation is a positive role of the threshold $T$ fluctuations, a fact which comes from the results that survival rate is greater at given age, and also the maximum age reached in the population is higher, see Fig. 6. The idea of the fluctuations in the threshold $T$ is justified when we account for the fact that the organism starts its biological clock at the embryos stage rather then at birth, and then individuals may reach different counting of the already activated mutations when they are born.

Our findings related to influence of hunting do not support the results reported in [3] on overfishing when rather a rapid decrease in population indicates the opposite tendency, perhaps of the first order type of transition. The proposed modification of the Penna model offers a possibility to include the possible deaths caused by bad mutations of embryos before the birth, or born individuals at ages before the threshold $T$. The modification provides some different characteristics of the final population which may as well be used for verification of the model, if relevant statistical data are made available.

At present, we are developing a new extension of the Penna model which is designed to study the influence of nonuniform spatial distribution of environmental features on the ageing process.

**Acknowledgements.** We are grateful to Dietrich Stauffer for introducing to ageing and to Stanisław Cebrat for new ideas on modifications of Penna model. The simulations were carried out at ACC CYFRONET.
This research was supported in part by the KBN grant 8 T11C 006 15.

# References

1. Stauffer, D.: Computer simulation with integer operations: from Ising model and cellular automata to amphiphilic systems and biology, in: Borcherds, P., Bubak, M., and Maksymowicz, A., (Eds.), Proceedings of the 8 Joint EPS-APS International Conference on Physics Computing, Kraków, September 17-21, 1996, Kraków, Poland, pp. 494-499.
2. Bernaschi, M., Castiglionne, Succi, S.: A parallel simulator of the immune response, in: Sloot, P., Bubak, M., Hertzberger, B., (Eds.), Proc. Int. Conf. High Performance Computing and Networking, Amsterdam, April 21-23, 1998, Lecture Notes in Computer Science 1401, Springer, 1998, pp. 163-172.
3. Bernardes A. T.: Monte Carlo Simulations of Biological Ageing, *Ann. Rev. of Computational Physics* **4** (1996) 359.
4. Penna T. J. P.: A Bit-String Model for Biological Ageing, *J. Stat. Phys.* **78** (1995) 1629.
5. Penna T. J. P., Stauffer D.: Efficient Monte Carlo Simulations for Biological Ageing, *Int. J. Mod. Phys.* **C6** (1995) 233.
6. Brown D. and Rolhery P.: *Models in Biology: Mathematics, Statistics and Computing*, Wiley, New York, 1993.
7. Martins, S.G.F., Penna, T.J.P.: Computer simulation of sexual selection on age-structured population, Int. J. Modern Physics C, **9** (1998) 491-496.
8. Maksymowicz A. Z., Bubak M., Sitkowski T., Stauffer D., and Kopeć M.: Simulation of Biological Ageing for Penna Model, *Suppl. of Medical & Biological Engineering & Computing* **35** (1997) 599.
9. Copp A. J., *Trends Genet.* **11** (1995) 87.

# The Scenario Management Tool SMARTFED for Real-Time Interactive High Performance Networked Simulations [1]

R.P. van Sterkenburg and A.A. ten Dam

National Aerospace Laboratory NLR, Informatics Division,
P.O. Box 90502, 1006 BM Amsterdam, The Netherlands
Tel: +31 20 511 3166/3447, Fax: +31 20 5113210
{sterkenb, tendam}@nlr.nl

**Abstract.** The National Aerospace Laboratory NLR is the central aerospace research and development organization in the Netherlands, and is actively involved in international simulation projects. In the HPCN project SIMULTAAN, NLR developes a generic scenario management tool named SMARTFED for real-time monitoring and control of networked simulations. The scenario management tool also offers functionalities to define and execute scenarios. This paper describes SMARTFED and its use in networked real-time high performance simulations. The envisaged use of SMARTFED in aerospace applications is illustrated by NLR's DELTA federation.

## 1 Introduction

The world of builders of training or engineering simulators is a complex one. Firstly, a simulator for real world sceneries must provide an accurate approximation of part of the real world. Secondly, a simulator must be delivered on time, on budget, and satisfy stringent performance requirements, see [1]. The ICT community actively pursues and promotes international standards, and simulators need to comply with these standards.

In the real-time simulator industry emphasis has for a long time been put on stand alone real-time high performance simulators. The operational parts of these simulators are placed close to each other at the same location and are connected through high speed local area networks. In recent years, networked simulations have emerged. A networked simulation consists of a number of simulators (or parts of simulators) that can be situated at geographically different locations, possibly far apart. Nevertheless, the simulators operate in a shared simulation.

In order to realise a networked simulation, at least two things must be available: an architecture to communicate between simulators and a tool that controls and monitors the simulation, i.e. the scenario management tool. Both architecture and tool are developed within the SIMULTAAN [2] project. The SIMULTAAN project is a

---

[1] This work has been carried out in the framework of the SIMULTAAN project, which is partly funded by the Dutch Foundation for High Performance Computing and Networking (HPCN).

project in the Netherlands in which industry, institutes and a university collaborate to achieve a common view on the development of training simulators using state-of-the-art technology.

A novel approach is taken by NLR with respect to scenario management functions of networked interactive high performance simulations. This approach has led to the Scenario Management tool SMARTFED [2]. The main requirements that are satisfied by this tool are that it provides control over a networked simulation and that it can operate in a real-time environment. With SMARTFED, all simulators within a networked simulation are forced to co-operate in the execution of a particular scenario, i.e. a plan of actions and events to be performed during a real-time high performance simulation run. Here, the simulators within the simulation may be running on geographically different locations.

The concepts that are introduced in this paper will be illustrated by the so-called DELTA federation of NLR. The DELTA federation is a realisation of a networked simulation connecting four state-of-the-art real-time simulators. In the simulation a full-flight simulator, a real aircraft, a tower simulator and an air traffic control simulator are connected.

The remainder of this paper is organised as follows. Section 2 gives a brief description of networked real-time simulations. Section 3 gives a global description of SMARTFED, its modular setup and its communication with the simulation. Section 4 describes in more detail the federation management tasks. Section 5 focuses on the monitoring task and section 6 describes the scenario definition and execution tasks. In section 7 a more detailed description of the use of SMARTFED in the DELTA federation at NLR is given. Section 8 contains the concluding remarks. Acronyms can be found in section 9 and references in section 10.

## 2 Networked Training Simulations

Networked simulations consist of a number of simulators that collaborate to achieve a common goal. These simulators are usually located at geographically different sites and are connected through networks. A number of standardised communication protocols exist, but in general these protocols are based on the broadcasting principle, which causes a significant amount of overhead.

To reduce the amount of overhead, new architectures have been designed that use a publish and subscribe mechanism. With this mechanism, data is only sent from one simulator to another if the receiving simulator has declared its interest in (i.e. subscribed to) that type of data. A new upcoming international standard is known as the High Level Architecture (HLA), see [3]. Implementations of HLA are built to work on heterogeneous networks. They offer real-time support for distributed simulations.

In HLA terminology a simulator is called a federate and a collection of federates is called a federation. This terminology will be used throughout the remainder of this paper.

With HLA, communication is realised through the use of the Run-Time Infrastructure (RTI). An important collection of data made available through the publish and subscribe mechanism, is the Management Object Model (MOM) data. The MOM provides global information about the federation (e.g. the list of participating federates) and about the federates (e.g. the hostname of the federate) in the federation.

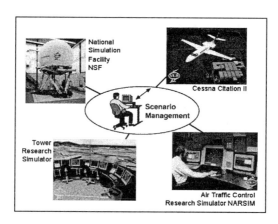

**Fig. 1.** NLR's DELTA federation: an example of a networked simulation

Figure 1 shows NLR's DELTA federation, a networked simulation that connects a number of simulators. The participating federates are a tower simulator, NLR's air traffic control simulator NARSIM, NLR's full-flight simulator NSF (see [4]) and a real aircraft. Scenario Management plays the central role in the federation as it controls and monitors the simulation.

The combination of SMARTFED and HLA allows the reuse of existing simulators on heterogeneous environments whilst making efficient use of available bandwidth. Moreover, also HLA itself is available on several platforms, which is an important asset. To operate in a federation, a federate only needs to comply with the HLA interface standard, which is a relatively simple addition to an existing simulator.

Use of HLA allows each federate in a federation to subscribe to and publish two types of data, namely (portions of) the federate state, and interactions between federates. For the DELTA federation, an example of federate state data is position information of a simulated airplane. Examples of interactions between federates are early collision warnings between two airplanes and a request for trajectory change, sent to air traffic control by an airplane.

## 3  Overview of SMARTFED capabilities

Within the HPCN project SIMULTAAN, NLR is tasked with the development of the generic Scenario Management tool SMARTFED. An important benefit of SMARTFED is that it allows engineers to concentrate on specific aspects in a project while ensuring reuse of available technology in the Netherlands. For instance, SMARTFED paves the way to incorporate existing federates in complex training scenarios.

Control over the execution of a simulation is desirable for a number of reasons. It is needed to ensure cooperation between federates that participate in a simulation. Another reason is that the same scenario may need to be performed more than once. For example, in training simulations the environmental conditions must be the same in multiple runs to ensure that trainees practise under similar circumstances.

In a networked simulation, SMARTFED controls and monitors all federates that participate in a particular scenario run, and executes the scenario. This is illustrated in Figure 2.

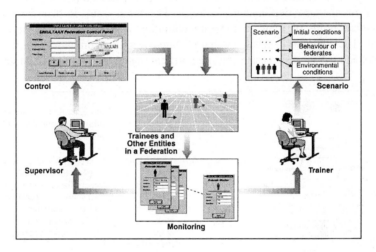

**Fig. 2.** Tasks of the Scenario Management tool

SMARTFED performs three tasks. Firstly, there is the control task. In SMARTFED this is realised by the Federation Manager. The Federation Manager controls the execution state of all federates in the entire simulation. Secondly, an operator (e.g. the supervisor or instructor) needs to be able to monitor the federation. This task is realised in SMARTFED by the Federation Monitor that enables him/her to watch the entire simulation on a screen. Different views on the federation and participating federates are provided. When the DELTA federation is operational, the operator may want to watch the position of all aircraft and the interactions between the aircraft. The third task concerns definition and execution of scenarios. In SMARTFED this is realised by Scenario Definition and Execution Manager. In the DELTA federation, the scenario may contain the initial positions of all aircraft and unprecedented events like the generation of an engine failure at a certain point in the simulation.

All these tasks are made available to the users through graphical user interfaces. In the following sections we discuss in more detail how SMARTFED realises the above mentioned scenario management tasks.

## 4 Federation Manager

The main functionality of the Federation Manager is to provide central control over the networked simulation. The Federation Manager is operated by a supervisor. The supervisor decides when certain commands are sent to the federation. As depicted in Figure 2 the supervisor can react on signals displayed by the Federation Monitor (see section 4). The Federation Manager and the Federation Monitor will usually be used by one and the same person.

A general state-transition diagram has been designed for SIMULTAAN federates [5], see Figure 3. In principle each federate must comply with this state-transition diagram. However, federates may well possess an internal state-transition diagram that differs from the one depicted in figure 3. The main issue is that from a scenario management point of view, a federate complies with the depicted state-transition diagram. The Federation Manager sends state-transition commands to the federates. Federates in turn send success or failure notifications to the Federation Manager.

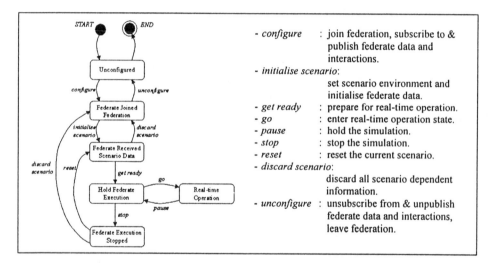

**Fig. 3.** The state transition diagram of a federation

The user of the Federation Manager (e.g. the supervisor) decides whether a scenario shall be executed or not and when a scenario execution will start and stop. The Federation Manager subscribes to the federate data and federation data provided by HLA's MOM in order to know the states of federates that are present in the federation. The Federation Manager may only send a state-transition to the federation when all federates are in the same state. State-transition commands are made available to the supervisor by means of a graphical user interface. Also a message window is available to notify the supervisor of warnings or errors that occur during the federation execution, for instance loss of a network connection of a federate to the rest of the federation.

Another functionality of the Federation Manager is the initiation of snapshots. A snapshot contains a dump of the entire internal state of a federate. Of course this is only possible if a federate itself is able to take a snapshot. Snapshots can be generated only when a federate is in the 'Hold Federate Execution' state. The same holds for restoring a federate by means of previously created snapshot.

It is also possible to register bookmarks during a scenario execution. During 'After Action Reviews', parts of the scenario can be replayed and analysed. The registered bookmarks serve as easy to find starting points for the review. A supervisor adds a bookmark at a point in time during the scenario run that might be of interest to the evaluation of the scenario afterwards.

## 5 Federation Monitor

The Federation Monitor provides the user with a view of the entire federation. This includes both a graphical overview showing the positions of the simulated entities with respect to each other and a textual view of the federation containing detailed information of the participating federates.

As mentioned earlier a federate can publish and subscribe to two types of data: federate data and interaction data. Published data can be viewed using the Federation Monitor. As soon as data from a federate is updated by its owner, the Federation Monitor will receive the new values. The data attributes are viewed in textual format, in addition numerical attributes can be viewed graphically.

The data collection that contains the information of all data and interactions available in a federation is called the FOM (Federation Object Model), conform HLA terminology. A FOM is composed of the collection of available federate data and interaction data. The data and interactions subscribed to and published by a single federate is called the SOM (Simulation Object Model). The FOM is used as a basis for the implementation of the Federation Monitor. For HLA a Backus-Naur Form notation of a FOM is defined, see [3]. FOM files that comply with this notation can be read by the Federation Monitor. The Federation Monitor will enable the user to browse graphically through the FOM. With this browser the user can subscribe and unsubscribe to federate data and federate interactions whenever he/she wants.

When new data is published by a federate, it will appear as an icon in the Federation Monitor using information contained in the FOM. Detailed information on federate data and its attributes will be displayed. The Federation Monitor will subscribe to federate interactions and display those incoming events that the user is interested in. A separate interaction view provides the user with an overview of all interactions that have occurred during the federation execution. For each interaction the detailed information on its parameters are available.

The application of the High Performance technology is closely related to the development of high-capacity networks. Since in principle, the Federation Monitor can subscribe to all data that are made available in the federation, network congestion may occur when a user actually does so. The level of accuracy of the monitoring depends on the network load and may decrease as the number of monitored objects increases, see also [6]. It is expected that the use of SMARTFED will make evident new requirements for capabilities of high-capacity networks in terms of network bandwidth and latency.

## 6 Scenario Definition and Execution Manager

The Scenario Definition and Execution Manager obviously has two main tasks: scenario definition and scenario execution.

Scenario Definition enables the user to specify a scenario. A scenario is defined for a particular federation. A scenario consists of the following parts:
- composition of the federation: what is the name of the federation and which federates participate in the federation for the scenario.
- definition of environmental conditions: in which geographical environment is the federation operating (e.g. European airspace) and what are the meteorological conditions.
- definition of initial conditions of federates: what are the initial values of the attributes of the data of a federate (e.g. position, speed).
- definition of stimuli during the scenario: which events shall occur at what time during the scenario (e.g. engine failure at $t=10:30:00$).

Scenario Execution reads a predefined scenario and sends the initial conditions to the federation when the 'initialise scenario' command is generated by the Federation Manager. During the 'Real-time Operation' state (see Fig. 3) the Scenario Execution component will send events to the federation at times specified in the scenario.

The implementation of the Scenario Definition and Execution Manager is based on the implementation of the Federation Monitor. While the Federation Monitor allows the user only to watch the federate data and interactions, Scenario Definition allows the user to set the values of federate data available in the FOM and generate interactions during the scenario definition phase.

## 7 The DELTA federation

The realisation of SMARTFED is currently driven by SIMULTAAN requirements, especially with respect to the tailoring to HLA as a communication standard between networked simulators. In the near future it will be studied how SMARTFED can be extended to include other communication protocols, among which communication protocols used in aerospace.

As the central aerospace research and development organization in the Netherlands, NLR operates a number of advanced facilities, among which are the Full Flight Simulator NSF, the Cessna Citation II aircraft, the Air Traffic Control Research Simulator NARSIM, and the Tower Research Simulator TRS.

Figure 1, see section 2, depicts the DELTA concept: the unification of part of NLR's facilities in a joint simulation. Use of SMARTFED will effectively result in a

DELTA federation proper. The use of SMARTFED for scenario management in the DELTA federation will require a number of extensions. Since not all communication is digital, for instance a voice link between the pilot and an Air Traffic Controller must be supported.

In order to appreciate the complexity of scenario management of these real-time high performance facilities, a brief description of each facility is given.

The National Simulation Facility NSF is NLR's versatile flight simulation facility. The simulator equipment consists of many modules, such as cockpits, visual, motion and computer systems, and a large set of simulation software modules and tools. With NSF virtually any vehicle can be simulated and its modular and versatile set-up enables efficient interchange of aircraft models, equipment, etc.

The Cessna Citation II PH-LAB is one of the research aircraft operated by the NLR. The Citation II is a twin jet certified in accordance with FAR Part 25 airworthiness standards and may be operated in known icing conditions. The PH-LAB can be used amongst others, for testing a variety of high-accuracy sensors and other equipment or complete systems in actual flight conditions

NARSIM simulates in real-time the Air Traffic Control (ATC) process, having both the air traffic controller and the pilot in the loop. NARSIM is concerned with approach/enroute ATC activities. In the past several years, NARSIM has been used in research programmes for a variety of customers. The NARSIM facility is also used in projects within European Commission's Fourth Framework Programme.

A Tower Research Simulator (TRS) is currently under development. The TRS will be capable of simulating the tower/ground ATC activities at an airfield, in real time and in a realistic operational and visual environment. This environment will contain a 360° field-of-view projection area where visual cues are simulated highly realistically. A key item in the research applications of the tower simulator is the role of the human controller, both the pilot and the air traffic controller, in the automated environment. At present, parts of the future Tower Research Simulator are already applied in various projects of the European Commission.

In the DELTA federation at least four persons are involved that may use the federation for training purposes. These persons are a pilot in the Cessna Citation, a pilot in the simulated aircraft in NSF, and one (or more) traffic controllers using NARSIM and/or the TRS. A possible scenario would include interaction between air traffic controllers and pilots, where the supervisor uses SMARTFED to sent stimuli to the trainees. The DELTA federation can be used to guide pilots all the way from the gate at one airport to the gate at another airport. Several interactions are possible apart from the ones mentioned in section 2, for example the transfer of an aircraft from one air traffic controller to another, since each air traffic controller is responsible for a particular area only.

NLR provides its services also to foreign aerospace industries, operators and other industries. Clearly, SMARTFED will stimulate the use of available facilities and simulators, both inside NLR and outside NLR, in a combined simulation, where each facility or simulator remains responsible for its own internal affairs.

## 8 Concluding remarks and future work

In this paper we have presented the generic Scenario Management tool SMARTFED as developed by NLR in the SIMULTAAN project. SMARTFED paves the way to incorporate existing federates in complex real-time high performance training scenarios.

SMARTFED is implemented for application in a networked simulation that communicates through HLA. HLA has proven to be well suited as communication architecture for controlled real-time networked simulations. For a simulator to be HLA compliant, it only needs to satisfy the interface requirements as specified by HLA. This means that equipping a simulator with a HLA interface is sufficient for the simulator to participate in a distributed real-time simulation that is controlled by SMARTFED. An asset of HLA is the availability on a number of platforms, allowing SMARTFED to interact with federates in a heterogeneous network.

In the near future, SMARTFED will be generalised to communicate through CORBA and through high-speed networks such as SCRAMNET. Further research will be done this year to tailor SMARTFED to these types of communication.

A faster than real-time option will be available to a user to execute a scenario as fast as possible. This option supports the use of dedicated HPCN facilities especially in the pre- and postprocessing phase, making it possible for a user to have access to computationally demanding visualisation facilities located at other sites. The fast-time option can also be used during replay to skip less important parts of a simulation.

Furthermore, the central role of SMARTFED in a networked simulation makes it possible to answer specific research questions with respect to network bandwidth and latency in real-time networked simulations. The DELTA federation is a good candidate to investigate these questions.

## 9 Acronyms

| | |
|---|---|
| ATC | Air Traffic Control |
| CORBA | Common Object Request Broker Architecture |
| FOM | Federation Object Mode! |
| HPCN | High Performance Computing and Networking |
| ICT | Information and Communication Technology |
| MOM | Management Object Model |
| NLR | National Aerospace Laboratory NLR |
| NSF | National Simulation Facility |
| OMT | Object Model Template |
| RTI | Run-time Infrastructure |
| SMARTFED | Scenario MAnager for Real-Time FEderation Directing |
| SOM | Simulation Object Model |
| TRS | Tower Research Simulator |

# 10 References

1. J. Kos, A.A. ten Dam, G.W. Pruis and W.J. Vankan, 1998, Efficient Harmonisation of Simulation Competence in a CACE Working Environment, *Proceedings of the 5$^{th}$ International Workshop on Simulation for European Space Programmes (SESP'98)*, ESTEC, The Netherlands.
2. SIMULTAAN web-site: http://www.nlr.nl/public/hosted-sites/simultaan
3. HLA web-site: http://www.dmso.mil/hla
4. A.A. ten Dam, P. Schrap and W. Brouwer, 1994, Programme and Real-time Operations Simulation Support Tool PROSIM: The Simulation Program of the Dutch National Simulation Facility NSF, *Proceedings of the 3$^{rd}$ International Workshop on Simulation for European Space Programmes (SESP'94)*, ESTEC, The Netherlands.
5. R.P. van Sterkenburg, 1998, SIMULTAAN State Transition Diagram for Federations, *NLR Technical Report*, Amsterdam, The Netherlands.
6. D. Prochnow, E.H. Page and B. Youmans, 1998, Development of a Federation Management Tool: Implications for HLA, *Simulator Interoperability Workshop (SIW) Spring 98*.

# OPERA: An HPCN Architecture for Distributed Component-Based Real-Time Simulations

François-Xavier Lebas[1], Thomas Usländer[2]

[1] Thomson Training & Simulation, System Architect,
F-95523 Cergy-Pontoise , France
lebas@tts.thomson-csf.com
[2] Fraunhofer IITB, Technical Manager
D-76131 Karlsruhe, Germany
usl@iitb.fhg.de

**Abstract.** The primary mission of the EU-funded OPERA project is the training of operators in the area of chemical industry by means of simulation applications that run in a cost-effective standard LAN environment. As a second goal, OPERA also aims at achieving great flexibility to ease changes in the released simulator. To achieve its primary goal, OPERA had to solve the problem of real-time combined with the requirement for high computing. To achieve its secondary goal, OPERA noticeably had 1) to provide a way to integrate hybrid simulation models, 2) to be highly scalable and configurable (being able to run from a single computer to multi-processors clusters). As a matter of fact, OPERA results appear to be valuable in many areas, far beyond the initial requirements in the domain of distributed simulation.

## 1. Introduction

The primary mission of the OPERA (Operators Training Distributed Real-Time Simulations) [4] project that is funded by the European Commission in its 4$^{th}$ Framework Programme is the support of the training of chemical plants operators by means of simulation applications. The design of a training simulation system in the area of the chemical industry must satisfy the following strong design requirements:

- Need for real-time operations
- High performance
- Flexibility to easily adapt to changes made in the actual chemical plant
- Ability to integrate heterogeneous legacy (commercial or in-house) software

Real-time is not only required to give an acceptable and realistic system view to operators, but also to allow the coupling of real-time stimulated equipment such as actual control room (DCS based or not) or process device(s). High performance is required to meet real-time deadlines in spite of complex computing process models. Flexibility requirement was expressed by many end-users as an important requirement able to differentiate the OPERA approach when being compared to currently available systems. Requirement to facilitate the integration of legacy software results from the industrial need to reuse existing process simulation models and code and strongly contributes to realise the final OPERA prototypes within a reasonable cost and time frame.

Beyond these functional requirements, the OPERA technical approach added several strong design commitments such as to release a *cost-effective* solution (at least one order of magnitude cheaper than actual), 2) heavily rely on *widespread* standards, 3) offer *open* interfaces enabling a close integration with Web-based solutions as well as Microsoft Windows™ based office systems.

## 2. Logical Architecture

### 2.1. Hybrid Modelling Integration Framework

After several years of research and development in the modelling of physical phenomena, two general models have emerged [1]: Discrete-Event systems and Continuous systems.

Despite the fact that the discrete-event paradigm has proven to be theoretically more general [2], when implemented, each model revealed strengths and weaknesses to cope with full reality while giving optimal results. So currently, both are still in use mixed with in-house approaches.

Furthermore, all models have been implemented by research projects and commercial products in different ways (ex: using distinct strategies, languages, frameworks ...), what makes their integration tricky. However, there is a general growing requirement for broad simulations using several models concurrently.

Rather than trying to design the ultimate (utopic ?) framework, recent developments in simulation [3], have shown the feasibility to achieve value-added simulations by coupling heterogeneous simulator units together. This approach is the first level of the OPERA strategy to integrate legacy codes. However, although some concepts of [3] were reused (e.g. time management) , the OPERA design differs from the HLA Run-time Infrastructure specification in order to achieve high performance, real-time and flexibility requirements.

For the coupling of autonomous simulators, a generic interface to so-called Proprietary Simulation Environments (PSE) has been defined that allows 1) at configuration time to select simulation components to load and run, 2) at run-time to control and monitor the simulation run.

The PSE run-time interface mainly controls simulation states (e.g., run, freeze, pause) and allows the synchronisation of time. The PSE configuration-time interface helps to define executable and possibly distributed simulation configurations. The PSE interface mainly relies on the components described below.

## 2.2. Component-based architecture

The second level of the OPERA integration strategy was to define a general component-based framework that is able to design native object-oriented models and/or to integrate existing code whether being object-oriented or not (e.g.: often written in FORTRAN).

Roughly spoken, the component-based approach promotes the idea that software components should be developed in the same way as hardware components. Some component-based systems are already available on the market but for very specific purposes and generally do not rely on standards.

Object-orientation does not necessarily imply a component-based architecture but a component-based architecture requires object-oriented development. That's why an OO approach based on UML was selected for OPERA to support the whole design.

The OPERA OO framework only relies on a little set of concepts that can be easily mapped upon a broad range of legacy code, through the code-wrapping technique.

**Fig. 1.** Basic elements of the OPERA integration framework

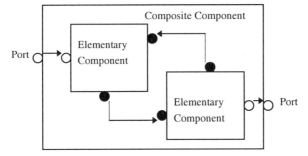

OPERA has conceived a so-called Distributed Simulation Object Model (DiSOM). The DiSOM object types and their relationships are (see figure 1):

- Elementary Components typically model an atomic component (i.e., a component that **cannot** or does **not need** to be broken in smaller parts). Such an elementary component is also the unit of distribution in a distributed OPERA run-time environment. It is called an **Atomic DiSOM Component (ADC)**.
- ADC Ports model an interface point with other ADCs.
- Composite Component comprise a configuration of ADCs and/or other Composite Components.

ADCs can be connected through their type-compatible ports to build bigger simulations systems, called DiSOM Configurations. Connected ADC ports share the same data.
Beyond their initial purpose as a unit of distribution, the ADCs:
- Configuration properties enable to statically or dynamically configure an ADC (e.g., load Initial Conditions, activate malfunctions, tune physical parameters ...),
- Monitoring properties used by the OPERA framework to monitor the performance of an ADC and whole DiSOM configurations.

Albeit simple, these concepts reveal to be powerful when being applied to process domains involved in a training simulator (see examples below):

Table 2. Mapping of OPERA generic concepts to actual pilot plants

| Domain | Port | Elementary Component | Composite Component |
|---|---|---|---|
| Instrumentation & Control | Plug or Pin | Electronic component | Device |
| Chemicals | Mixture | Distillation tray | Distillation column |
| Fluid networks | Junction | Control volume | Fluid Network |

*Visual programming* : Note that this component-based view eases the development of visual tools that enable to incrementally (top-down and bottom-up) build simulations by just dragging objects from libraries, connecting and initialising.
Note also that a similar concepts are used by several projects among which is the Brite-Euram CAPE-OPEN project that aim to promote simulation codes interoperability.

## 2.3. Time management

Within OPERA, time management is a key point for the following reasons: 1) from the external user point of view, the time perception of a simulation system is manyfold as the internal time may be warped in various ways: run faster, slower ..., 2) time synchronisation is a cornerstone to enable interoperability of hybrid models : models must checkpoint at some points in time, 3) Message must be delivered in order, 4) in a possibly distributed system, clock management is a difficult task due to the fact that in a network, hardware-based clock cannot be used, 5) the real-time requirement complicates everything by defeating usual distributed synchronisation algorithms.

4) and 5) has been solved by using a central time authority.

To solve 1) 2) and 3), OPERA simulation components just work on the basis of a logical time concept. The logical time is a non-linear function of the scaled wall clock time [3]. It is changed at discrete time steps controlled a central component in the

OPERA system called the Central Conductor. Logical time advances only when all participating simulation components have reached a new time-coherent state. For time-stepped simulation components, this is at the end of a simulation cycle. For event-driven simulation components, this is after the computation related to an event of the current logical time has been carried out and it is ready to receive an event for a future logical time point. The cycle duration is chosen according to the following compromise:
- supply MMI and stimulated equipment's with updated values often enough to give a continuous-like view of the state,
- give sufficient delay to compute models.

When models computation exceed the cycle duration, a time lag phenomenon appears. The OPERA system allows to detect the occurrence of such lag phenomena through the built-in monitoring tool. It then forwards a real-time violation warning message to the user who may take the appropriate action (e.g. restart the simulation with a changed DiSOM configuration).

## 2.4. Distributed Parallel Simulation

A simulation can be seen as the execution of interconnected simulation components within one logical time space. Normally this execution is done by one scheduler who is responsible for the co-ordinated time advancement.

Basically, a distributed simulation uses the same approach with the difference that the execution is performed by several computing hosts. This does not imply that two events will be executed at the same universal time. The challenge comes with the concept of parallisation. Here the execution of two events can be execution at the same point in universal time[1].

In distributed parallel simulations two simulation servers are working in synchronised, but different logical time spaces. This shall be illustrated by the following example:

Simulator 1 (S1) processes model components A and B and simulator 2 (S2) processes model component C. Within S1 the value exchange between A and B can be handled without considering time, because the execution of A and B is performed within the same logical time space. However, in order to transfer values to the model component S2/C, a time-stamped message is required. This message must contain the value for the variable and the logical time for which this value is valid.

The time advancement in both simulators must guarantee, that a message from simulator 1 to 2 will not cause a time violation in the sense that in simulator 2 arrives a message for time $t_n$ where the logical time in time space 2 is already advanced to time $t_{n+1}$.

---

[1] Because distribution without parallelisation rarely makes sense, the term "distributed simulation" is often used as a synonym for "distributed parallel simulation".

## 2.5. High performance

With respect to high performance, the OPERA approach is
- to ease the distribution of models over a LAN (e.g., Ethernet 10/100 Mbit/s) using best state-of-the-art but standardised design to reduce latency times,
- to enable the integration of models that internally use fine-grained as well as coarse-grained parallelisation techniques,
- to provide on-line monitoring tools for the measurement of the distribution efficiency,
- to enable on-site reconfigurations.

The following figures exhibit network Latency and Throughput for couples of Wintel-based stations exchanging through Orbix on a 10baseT Ethernet and running at different frequencies (from 166 to 300 MHz[2]) :

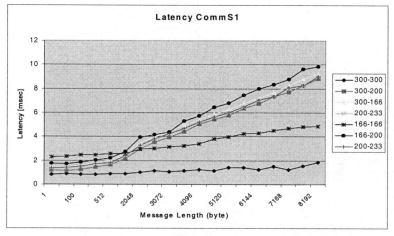

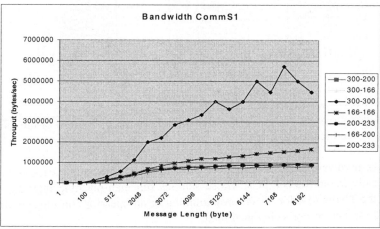

---

[2] Note that the 300MHz is a Pentium II equipped with 196Mb (vs Pentium and 64Mb)

## 3. Physical Architecture

The physical OPERA architecture is sketched in the figure below where OPERA developments are figured in grey, while COTS are in white.

**Fig. 2.** Physical Architecture

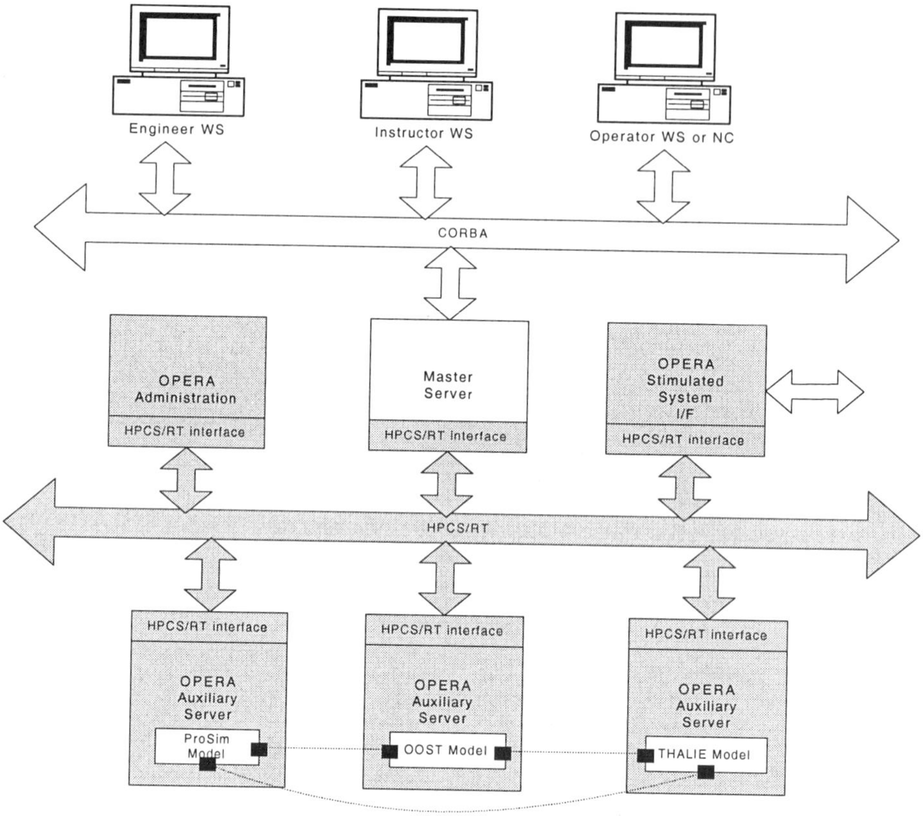

In the figure above the following components (from top to bottom) of the OPERA Physical Architecture are shown:

- End-users hardware stations (Engineer, Instructor and Operators stations) which are typically Windows/NT PCs. Those stations run man-machine interface tools such as the Thalie tool for visual programming or an operator console specifically designed for a customer's chemical plant.

- The first-level communication bus based on CORBA technology over a standard LAN. This layer has moderate real-time requirements ( ~200ms) due to the fact that 1) it has only to handle requests initiated by manual (so rare) actions, and 2) replies are time-stamped and do not require a strict timing policy. This layer uses the native CORBA/Orbix implementation with little design constraints. Thanks to Orbix, the OPERA API can be used both by Java and OLE/COM client applications.
- The second-level communication bus is called HPCS/RT (High Performance Computing Simulations in Real Time). It has more stringent real-time requirements (~10ms latency) and thus shall run on a standard but high-speed LAN (e.g., 100 Mbit/s Ethernet). In order to obey overall real-time constraints, the OPERA environment needs an exclusive and controlled access to this LAN at run-time. This layer exhibits a specialised API built on top of CORBA.
- The OPERA Master Server manages sessions and requests between clients and HPCS/RT.
- Auxiliary Servers are CORBA wrappers around PSE and/or legacy codes (e.g., from the OPERA partners ProSim, Fraunhofer-IITB (OOST) and TT&S (Thalie)).

HPCS/RT is built on a master/slave approach suitable for such a system:
- The Central Conductor (CC), located in Master, is the primary access point of OPERA user applications (e.g. instructor and operator station applications). It mediates these user operations to its counterpart, the Local Conductors, running on the OPERA Auxiliary server machines. The CC manages the synchronisation of operations and the advancement of the logical time.
- Local Conductors (LC), each located in Auxiliaries, relay forth and back requests to the Proprietary Simulation Environments (PSEs).

**Configuration description** : executable OPERA configurations are described in the so-called DiSOM model database which contains 1) a description of executable components along with their assigned resources, 2) rules for the execution of ADCs in the distributed environment, i.e., the mapping of ADCs to computing resources and an execution plan of ADCs on a particular computing resource. Execution plan allow to assign CPU to ADCs.

**Time Synchronisation**

The CC is the global synchronisation authority for both time-stepped and event-driven simulators. The LCs request the time advancement of their logical time at the CC and the CC grants the time advancement. For an LC representing a time-stepped simulator this is done with a *timeAdvanceRequest/timeAdvanceGrant* sequence, for an LC representing an event-driven simulator this is done with a *nextEventRequest/timeAdvanceGrant* sequence between the LCs and the CC. The time advance grant means that the LC (respectively the PSE Scheduler behind the LC) can do its simulation work independently until its logical time reaches the granted time point.

After reaching this point of time, the next global synchronisation must follow. If the simulation shall be stopped, paused, frozen or synchronised with the real time, this is controlled only in the central conductor.

## 4. Achievements

**Real-time** : As a training simulator, OPERA support statistical realtime (e.g: can afford to miss deadlines). Whether using not-realtime proven COTS [6], OPERA achieve a fair realtime behaviour 1) by wrapping and carefully selecting Orbix features  2) by pre-allocating as much resources as possible (memory and threads). Approach is similar to the RTSORAC project [5].
**Exception handling** : HPCS/RT trap all exceptions raised by legacy process models wrapped into components. Default behaviours are configured for each exception category (ex: out-of-domain trap suspend the simulation run).
**Performance** : At synchronisation points (see Time Synchonisation), remote ports updates are collected and grouped before being sent to other servers in order to reduce latency. Also CORBA unions have been preferred to (costly) CORBA anys.
**Tuning** : OPERA uses a pragmatic feedback approach to optimise the distribution of simulation components. Starting from a first configuration, user has tools:
- to monitor the network, system and application performance (e.g. consumed bandwidth, request load) at run-time.
- to (re)configure simulations components over the network.

**Pilot Applications at User Sites**

From early to mid 1999, the OPERA system will be applied to three pilot plants :
- An acrylic production plant operated by ICI in England,
- A polyester film production line operated by DuPont in Scotland,
- A training plant producing ammonium sulfate owned by TCL (Technology Centrum Limburg) in Netherlands and co-operated by TCL and VIA.

As a demonstration of OPERA ability to integrate heterogeneous modelling software and simulation frameworks, pilot applications use the following COTS:
- Thalie™ ( Thomson Training & Simulation ) : a simulation system mainly using the continuous paradigm and a time-stepped approach,
- OOST™ (Fraunhofer-IITB) : a discrete-event simulation system based on the function block paradigm,
- CheOps™ ( ProSim) : a modelling framework for chemical processing,
- And several other simulation softwares and engineering models from OPERA technology providers (IITB, TT&S and ProSim) and end-users: ICI and DuPont.

Note that most of these models were wrapped quite easily despite their heterogeneity.

# 5. Conclusion

As a result,the OPERA architecture supports the following features:
- A *scalable*, *on-site configurable* architecture that is able to run from a single computer up to a LAN-based cluster of multi-processor computers,
- An high performance architecture settled on widespread *standard* hardware(s) and software(s) able to take profit of newest mass-market improvement(s),
- An architecture *open* to today's standards ( WWW, Active-X/COM ™),
- An *integration* framework able to host (reuse) heterogeneous legacy codes and simulation frameworks,
- A statistical soft *real-time* capability,
- Tools for *measuring* and *optimising* possibly distributed execution of OPERA configurations.

By combining this unique set of features, OPERA opens the way to a new generation of distributed real-time systems running in low-cost LAN environments.

# References

[1]     Y. Monsef, , Complex Systems Simulation Modelling, Lavoisier Tec Doc.
[2]     B. Ziegler, Object Oriented Simulation with Hierarchical Modular Models : Intelligent Agents and Endomorphic Systems, Academic Press, San Diego.
[3]     US Department of Defense, Defense Modeling and Simulation Office: "Department of Defense High-Level Architecture Interface Specification", http://www.dmso.mil/projects/hla
[4]     OPERA Projekte http://tes.iitb.fhg.de/Projekte/opera.html
[5]     RTSORAC project - http://www.cs.uri.edu
[6]     TRDF - LeLann

# Glossary

| | |
|---|---|
| ADC | OPERA Atomic DiSOM Component |
| CC | OPERA Central Conductor |
| COTS | Commercial Off The Shelf |
| DCS | Distributed Control System |
| DiSOM | OPERA Distributed Simulation Object Model |
| HLA | High-Level Architecture |
| HPCS/RT | OPERA High Performance Computing Simulations in Real Time |
| LC | OPERA Local Conductor |
| OPERA | OPERAtors training distributed real-time simulations |
| PSE | Proprietary Simulation Environments |
| UML | Unified Modelling Language - OMG standard |

# Airport Simulation Using CORBA and DIS

Günther Rackl[1], Filippo de Stefani[2], Francois Héran[3], Antonello Pasquarelli[2], and Thomas Ludwig[1]

[1] LRR-TUM
Institut für Informatik, Technische Universität München, 80290 München, Germany
Email: {rackl,ludwig}@in.tum.de

[2] ALENIA MARCONI Systems
Via Tiburtina km 12.400, 00131 Roma, Italia
Email: {fdestefani,apasquarelli}@lti.alenia.it

[3] SOGITEC Division Electronique
4 rue Marcel-Monge, 92158 SURESNES CEDEX, France
Email: fheran@sogitec.fr

**Abstract.** This paper presents the SEEDS simulation environment for the evaluation of distributed traffic control systems. Starting with an overview of the general simulator architecture, performance measurements of the simulation environment carried out with a prototype for airport ground-traffic simulation are described.
The main aspects of the performance analysis are the attained application performance using CORBA and DIS as communication middleware, and the scalability of the overall approach.
The evaluation shows that CORBA and DIS are well suited for distributed interactive simulation purposes because of their adequate performance, high scalability, and the high-level programming model which allows to rapidly develop and maintain complex distributed applications.

## 1 SEEDS Overview

SEEDS[1] (Simulation Environment for the Evaluation of Distributed traffic control Systems) is a distributed HPCN simulation environment [1] composed of powerful workstations connected by a local area network and it is targeted to the evaluation of Advanced Surface Movement Guidance and Control Systems (A-SMGCS). The simulation environment allows the definition and evaluation of technologies and performance requirements needed to implement new functions and procedures of A-SMGCS, to mould new roles in the airport, and to introduce new automatic tools and interfaces to support A-SMGCS operator decisions.

The SEEDS consortium is composed of Alenia (I), as coordinator, Sogitec (F), Artec (B), as industrial partners, University of Siena (I) and LRR-TUM at Technische Universität München (D), as associated partners, and Sicta (I) as final user-partner. An European User Group (UG), composed of airport service and flight assistance administrations, participates to all the phases of the project.

---
[1] SEEDS (European Project Number 22691) is partially funded by EC DG III in the area of HPCN

The software architecture of the simulator, defined using the Unified Modeling Language (UML) notation, is based on a communication middleware using the OMG's CORBA (Common Object Request Broker Architecture) standard combined with the DIS (Distributed Interactive Simulation) protocol. For airport traffic and image generation, centralized and distributed software architectures have been analyzed, and the distributed solution was adopted in SEEDS for performance and scalability reasons.

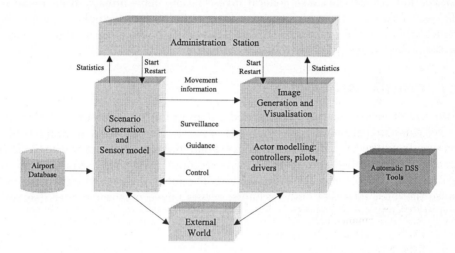

**Fig. 1.** SEEDS Simulator Architecture

The simulation environment is composed of commercial off-the-shelf components and of some proprietary software modules, and it is open to be connected to other ATM (Air Traffic Management) simulators. The SEEDS architecture is illustrated in Figure 1. The main software modules are: Scenario Generation and 2D/3D Visualization; Sensor Models; Airport Database; Surveillance, Control, Guidance and Planning Modules; Controller, Pilot and Driver models; Administration Station modules. The scenario generator is in charge of animating the scene according to the stimuli coming from the simulation actors and from the External World. The actors can be simulated (software processes) or real (human operators). The 3D visualization reproduces the scene as seen by the actor's eye (eye model); the 2D visualization reproduces the scene as seen by the sensors present in the airport (ASDE radar, GPS, DGPS, Magnetic/Dynamic sensor error models). The controllers have a set of DSS (Decision Support Tools) which help them to plan the aircraft surface movements. The other functions of A-SMGCS (Surveillance, Control and Guidance) are also implemented. The Administration Station is used to configure, start-up, stop, restart the simulation, and it collects application level and system level statistics.

## 2 Description of the Performance Measurements

This section describes the measurements carried out in order to evaluate the performances of CORBA and DIS in the SEEDS prototype. Within the simulator, CORBA [5] is generally applied for all kinds of communication except the distribution of the 3D scenarios. This includes simulator startup and control, actor interaction, and distribution of 2D images. DIS [4] in contrast is used for the 3D visualization of the airport. All the measurements carried out are very application-oriented because general investigations have already been made in the past (see e.g. [3]), and for scalability predictions very application-specific values are needed. Furthermore, the analysis carried out considers the two most performance critical paths within the simulator.

### 2.1 CORBA Measurements

Two sets of measurements were carried out in order to characterize the client-server communication over a high-speed network. The CORBA network traffic overhead and the CORBA latency were evaluated. With reference to the SEEDS architecture described above, the software modules involved in the measurement were Sensor Models and 2D Visualization. The focus was on these modules because the most intensive data exchange occurs between sensor models (server) and 2D Visualization (client).

### 2.2 DIS Measurements

The use of the DIS protocol (IEEE1278.1 1994) has been selected for the 3D visualization of the airport for scalability, i.e. the possibility of adding any number of visual channels without having to increase the bandwidth of the LAN, and its low bandwidth requirement allowing to process a large number of mobile objects. These two advantages are inherent to the DIS mechanisms due to the following features:

- *Autonomous simulation nodes:* All events are broadcast and are available to all interested objects; a receiving node is responsible for calculating the effects of an event on the entities it is simulating, such that the sending node is not concerned by a high number of receivers.
- *Transmission of "ground truth" information:* Each node transmits the absolute truth about the state of the object(s) it represents. Receiving nodes are solely responsible for determining whether their objects can perceive an event and whether they are affected by it. Degradation of information is performed by the receiving node in accordance with an appropriate model of sensor characteristics before it is presented to human crew members or automated crews.
- *Transmission of state change information:* Nodes transmit only changes in the behavior of the entities they represent: If an entity continues to do the same thing (e.g. straight and level flight at a constant velocity), the update

rate drops to a predetermined minimum level. Additionally, *dead reckoning* algorithms are used to extrapolate entity states.
- *Object/event simulation architecture:* Information about non-changing objects in the virtual world is assumed to be known to all simulations and need not be transmitted. Dynamic objects keep each other informed of their movements and the events that they cause through the transmission of *Protocol Data Units* (PDUs).

Protocol Data Units are exchanged between networked simulations to convey messages about entities and events. PDUs provide data concerning simulated entity states, the types of entity interactions that take place in a DIS exercise, and they provide data for the management and control of a DIS exercise.

In SEEDS, the kinds of PDUs transmitted are SIMANs (SIMulation Management PDUs), ESPDUs (Entity State PDUs) for mobile objects, Signal PDUs (Radio Communication family) for AIC (switched lights, bars etc, and commands by controllers), and customized PDUs for meteorological data. The content of an Entity State PDU is summarized in Table 1.

| Field Size (bytes) | Entity State PDU Fields | | | |
|---|---|---|---|---|
| | Record | Field | Value | Comment |
| 12 | PDU Header | Protocol, Type, Time etc. | | |
| 6 | Entity ID | | | |
| 1 | Force ID | | 0 | N/A |
| 1 | # of Articulation Parameters (n) | 0 | | |
| 8 | Entity Type | Kind, category etc. | | |
| 8 | Alt. Entity Type | All fields = 0 | | N/A |
| 12 | Ent. Lin. Velocity | X, Y, Z | | |
| 24 | Entity Location | X, Y, Z | | |
| 12 | Entity Orientation | Psi, Theta, Phi | | |
| 4 | Entity Appearance | | 0 | |
| 1 | Dead Reckoning Parameters | DR Algorithm | | V/H = 2 A/C = 3 |
| 15 | | Other Parameters | | |
| 12 | | Entity Lin. Acceleration | | |
| 12 | | Entity Angular Velocity | | |
| 12 | Entity Marking | character set, string | | |
| 4 | Capabilities | | 0 | N/A |

(*) Total Entity State PDU size = (144 + 16n) bytes where n = number of articulation parameters.

**Table 1.** Entity State PDU

*Dead reckoning* algorithms are used to decrease the network load needed to transmit state information. Each simulation node maintains a simplified representation of the state of remote entities, and extrapolates from their last reported states until the next state update information arrives. Nodes simulating each entity are responsible for transmitting new state information when the discrepancy between their "ground truth" and the extrapolated approximations being generated at remote nodes exceeds thresholds. Each node maintains a dead reckoning model of its own objects that corresponds to the model being used by the other nodes, and it continuously compares its "ground truth" information (or high-fidelity model) with the approximations being used by the other nodes. If the thresholds are not exceeded, a new state information is sent after a given amount of time (corresponding to the so-called *heartbeat rate*). When a state update is transmitted, it includes not only the correct position and orientation, but also the velocity vectors and other derivatives that can be used to initiate a new extrapolation.

## 3 CORBA Performance

This section describes the test bed used to evaluate CORBA performance and outlines the operative conditions and experimental methods. Here, the CORBA performance is characterized in terms of the CORBA traffic overhead and the CORBA latency.

The network traffic overhead is defined as the data overhead introduced by CORBA in terms of bytes exchanged over the network. To evaluate it, the size of the application message (theoretical bytes) are compared to the measured bytes exchanged over the network between client and server.

The CORBA latency is defined as the communication delay for the remote method call introduced by the CORBA middleware. The measured time does not take into account the elaboration time on the server side. Thus, the measured time is a two-way delay which takes into account the marshalling and unmarshalling time on both client and server side, and the network transmission time.

### 3.1 Measurement Conditions and Test Bed

The measurements were conducted using two Pentium PCs 233MHz with Windows NT 4.0 connected by a Fast Ethernet network. The implementation of CORBA used during the tests was Orbix 2.3 by Iona Technologies. The measurements were carried out considering the client and the server on different machines. The client requests were implemented using the two-way static invocation whose IDL is reported in Figure 2. After the two way request is invoked, the client blocks until the server replies by returning a sequence of structures. In our case, the request does not pass parameters to the server and waits for the answer.

The structure contains the information associated to each track, which represents a moving object present in the simulation, as seen by the sensors. Different measures were conducted considering a different number of moving objects; in

```
// Track association
enum ObjType { NotAssociated, Aircraft, Vehicle };

// Track information
typedef struct TrackInfo_tag
{
    ObjType         trackType;
    unsigned long   trackNum;
    octet           trackStatus;
    unsigned long   time;
    double          x, y, altitude, speed, heading;
    string          WVCat;
    boolean         departing;
    string          callsign, type, RWY_stand, CTD_ETA, routing;
} TrackInfo;

typedef sequence<TrackInfo> TracksList;

interface TRACKS { TracksList getTracksList(); };
```

**Fig. 2.** CORBA IDL for Receiving a List of Tracks

particular, the measurements were carried out for 1, 10, 50, 100 moving objects. In order to have a common base to compare the results of the different tests, the string fields in the structure were chosen with a fixed length; in this condition the structure size was 84 bytes.

### 3.2 CORBA Network Traffic Overhead

The measurements were carried out considering scenarios with a different number of moving objects. The transfered bytes using CORBA as a middleware and the theoretical bytes (application data) exchanged are shown in Figure 3. It can be seen that the data overhead stays at about 30% at a number of 50–100 moving objects. It must be taken into account that, in addition to the CORBA overhead, this overhead also encompasses the underlying TCP/IP connection. Theoretical data, confirmed by tests with packets of 1024 bytes, indicate that the TCP/IP communication overhead for the same number of moving objects is about 4%.

### 3.3 CORBA Latency Time

The latency time measurements were made with the GetSystemTime system call available on Windows NT. This system call returns the system time with a milli-second resolution. The IDL reported in Figure 2 was modified in order to take into account the elaboration time on the server side.

Again, different scenarios with a different number of moving objects were considered. The obtained results are illustrated in Figure 4. It can be seen that for 100 moving objects a latency of approx. 10 ms arises.

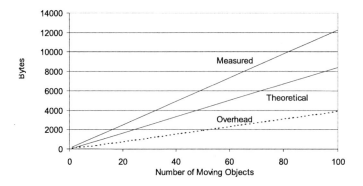

**Fig. 3.** CORBA Data Overhead

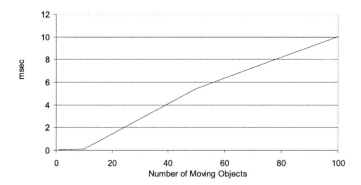

**Fig. 4.** CORBA Latency

## 4 DIS Performance

### 4.1 Measurement Conditions and Test Bed

The main output produced by the traffic generator (ADE) is the mobile object ground truth. According to the various inputs, the module computes the mechanical behavior of the objects on the basis of their known kinematics, and manages their position, attitude, velocity, acceleration. ADE maintains a CORBA object server that is used by the sensors and the 3D interface (3DI). Whenever the position of an entity is updated, the DIS interface (DISMOD) is called by 3DI to evaluate the necessity to broadcast a new PDU. On the visual generators, requests to DISMOD are made when the state of an entity is needed for displaying. The processes of animation and visualization are thus asynchronous. The test bed uses two PCs on which the processes depicted in Figure 5 are running.

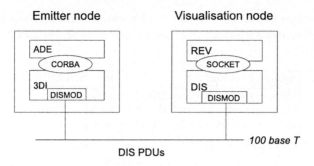

**Fig. 5.** DIS Implementation

The goal of the measurements is to check the factors which may influence the scalability, network load and protocol processing times.

### 4.2 DIS Network Load

Testing for scalability requires to simulate increasing numbers of mobile objects which belong to two classes: aircraft and ground vehicles. A limited number of aircraft is moving on the airport and airborne aircraft are less than 10, whereas a large number of ground vehicles may be moving, servicing the aircraft at the gates. Bandwidth estimations can be calculated by knowing the (fixed) size of the data packets and by measuring their update rate. In SEEDS, the major component of the PDU traffic is represented by Entity State PDUs (ESPDUs). A basic DIS ESPDU has 144 bytes of "simulation data" (see Table 1) and is 186 bytes on the wire. The highest load would be achieved if ESPDUs were sent over the network to update the entity tables on the image generation systems at a constant rate. Therefore, dead-reckoning mechanisms are used, for which statistical reduction rate have been measured depending on the entity type (See Table 2). The estimation of the PDU traffic is based on the following values.

|  | Midsize Airport (Naples Int'l) | Large Airport (Roma L. da Vinci) | # ESPDU / s per entity (with DR) |
|---|---|---|---|
| # moving A/C | 5 | 18 | 0.25 |
| # airborne A/C | 4 | 6 | 1 |
| # static A/C | 10 | 60 | 0.2 |
| # moving V/H | 8 | 30 | 0.5 |
| # static V/H | 20 | 120 | 0.2 |
| # entities (total) | 47 | 234 |  |

**Table 2.** Estimation of the PDU traffic

Table 3 presents the ESPDU traffic load without dead-reckoning, for an update rate of 5 Hz, and with dead reckoning thresholds of 1m and 3 degrees.

| Airport type and DR | Load | ESPDU/s |
|---|---|---|
| Midsize airport without DR | 70 kbit/s | 235 |
| Large airport without DR | 348 kbit/s | 1170 |
| Midsize airport with DR | 23 kbit/s | 15 |
| Large airport with DR | 91 kbit/s | 62 |

**Table 3.** Estimation of the ESPDU traffic load

### 4.3 DIS Protocol Processing Time

Processing times includes the send processing time at the emitter node (3DI) and the receive processing time at the visualization nodes (pilot or controller 3D view). These times have been measured at 0.53 ms and 0.42 ms per ESPDU, using the profiler available with our development suite. Table 4 presents the resulting processing time required for both nodes.

| Airport type and DR | ESPDU/s | Send processing time | Receive processing time |
|---|---|---|---|
| Midsize airport with DR | 15 | 8 ms | 6 ms |
| Large airport with DR | 62 | 33 ms | 26 ms |

**Table 4.** Protocol Processing Time

Due to the asynchronous issuance of the ESPDUs relative to the entities, the computational overhead is distributed in time and can be handled by a single PC in our implementation. Latency is not a critical issue for 3D visualization as it remains relatively short in view of the decision processes that are triggered by the visual outputs.

## 5 Evaluation and Alternative Approaches

The CORBA and DIS performance data described before show that the implemented approach is very well suited for the distributed simulation environment. A CORBA latency of approx. 10 ms for a track list of 100 moving objects is absolutely sufficient for 2D visualization. Even for large airports with a higher number of actors and moving objects no problems can be expected with these results. Concerning DIS, it can be seen that a network load of 100 kbit/s and a protocol processing time of 1 ms is very low due to the very simulation-specific DIS protocol design.

Alternative approaches could use other lower-level communication middleware in order to reach an even higher performance. Possible protocols could be PVM [2], MPI, or pure socket communication. For example, the latency of PVM or raw sockets over Fast Ethernet can reach values of approx. 1 ms [6]. But, to be able to compare these values with our measured performance data, detailed investigations would be necessary because of the complex data structures that are transfered within the SEEDS simulator.

Moreover, an essential aspect is the software development effort using different communication middleware. Here, the CORBA/DIS approach is absolutely preferable to a lower-level middleware layer. The high-level object-oriented modeling approach allows to build flexible, extensible, and maintainable simulators, whereas a lower-level approach introduces several difficulties, especially in large heterogeneous distributed environments.

## 6 Conclusion

This paper has presented and evaluated the SEEDS simulation environment for distributed traffic control systems. The chosen approach using CORBA and DIS to model a distributed, interactive simulator has shown to be appropriate. In particular, high-performance criteria have been fulfilled, and the scalability of the overall approach has been proved by the mentioned results. Therefore, higher-level object-oriented modeling of HPCN systems can be a solution for realizing future HPCN applications, especially in the field of distributed interactive simulations. Additionally, the interoperability of the CORBA and DIS platforms has successfully been carried out. This integration of general purpose middleware like CORBA and domain-specific protocols like DIS can be a major step towards building high-performing interoperable systems in the future.

## References

1. S. Bottalico, F. de Stefani, T. Ludwig, and G. Rackl. SEEDS - Simulation Environment for the Evaluation of Distributed Traffic Control Systems. In C. Lengauer, M. Griebel, and S. Gorlatch, editors, *Euro-Par'97 — Parallel Processing*, number 1300 in Lecture Notes in Computer Science, pages 1357–1362, Berlin, 1997. Springer.
2. A. Geist, A. Beguelin, J. Dongarra, et al. *PVM: Parallel Virtual Machine*. MIT Press, London, 1994.
3. A. Gokhale and D. C. Schmidt. Measuring the Performance of Communication Middleware on High-Speed Networks. In *Proceedings of the SIGCOMM Conference*. Stanford University, 1996.
4. B. Goldiez. DIS Technology. In *Proceedings of the 13th Workshop on the Standards for the Interoperability of Distributed Simulations*. Institute for Simulation and Training (IST), 1995.
5. OMG (Object Management Group). The Common Object Request Broker: Architecture and Specification — Revision 2.2. Technical report, February 1998.
6. I. Zoraja, H. Hellwagner, and V. Sunderam. SCIPVM: Parallel Distributed Computing on SCI Workstation Clusters. *Concurrency: Practice and Experience*, 1999. To appear.

# Intelligent Routing for Global Broadband Satellite Internet

Chao-Hsu Chang, Hsiao-kuang Wu, Ming-Hui Jin, Yueh-O Tseng

Department of Computer Science and Information Engineering
National Central University, Taiwan
cschang@nw58.csie.ncu.edu.tw

**Abstract.** With the fast development of internet, the multicasting applications (such as Distant Learning) have become more and more important. Since multicasting multimedia information over the wired links might generate heavy traffic streams to cause congestion, the utilization of internet will be reduced dramatically. The satellite with broadcasting capability and high bandwidth could provide an alternate transmission way. Most of applications using satellite are unidirectional. However, the bi-directional interactive applications will be needed to provide instant interaction. However the transmission latency through satellite is considerably long and might lead to the loss of interactive ability. Therefore, we propose an intelligent routing strategy to overcome the high latency and enhance the interactive capability over satellite links. Especially, this routing method is compatible with Mbone. We compare this routing method with one-to-one and one-to-many transmission and analyze the result. The results show transmission efficiency has improved.

## 1. Introduction

The interactive multimedia applications such as videoconference、team work software、distance learning ..,etc have become more and more important, due to the fast development of the internet. However, the multimedia information of high capacity could suffer the congestion when carried over the wired connection, especially multicasting. After that, the traffic over the wired session will get worse and worse and the network performance will degrade. However, the satellite link with high bandwidth [1] can be an internet extension to offer an alternate way for transmission while wired congestion occurs. Moreover, the satellite with broadcasting capability can achieve the requirements of multicasting applications.

With the high capacity of over 45M bits/s data rate and unlimited coverage area, the satellite can carry the continuous multimedia information to everywhere. Once the earthquake or typhoon damages the wired network infrastructure, the connection can be reconstructed via satellite links immediately. Recent development [5-7] in GEO (geo-synchronous satellite) network demonstrates the promise of ubiquitous access to the internet. With the fast development of satellite equipment, such as Direct PC[11] and VSAT system[12], many applications such as Direct TV、Broadcasting TV all utilize the satellites. However, most of them are used in the nature of unidirectional transmission and with the lack of bi-directional transmission. That is the interactive communication. The high latency through the satellite is the main reason for this condition. To overcome the high latency problem, we plan to divide the interactive multimedia information into two kinds of traffics, control traffic and data traffic. The data traffic consists of the high bandwidth multimedia information and appropriate for satellite transmission due to the high capacity of the satellite channels. The control traffic includes the short control messages with time bounded requirements. Since this kinds of packets contains little information, they wouldn't occupy a lot of bandwidth and are suitable for wired link transmission. To support the interactive transmission, transmitting the data traffic over the satellite links will make traffic load loose in the wired links. Relatively, delivering the control traffic over the wired links will be smooth. By this way, the interactive capability can be improved further and also the wired networks could avoid the unexpected congestion.

We construct an intelligent routing scheme based on the integrated platform between the satellite and wired network to achieve the interactive support, balance the traffic load, reduce the probability of the congestion in wired network and promote the internet utilization. Furthermore, this scheme is compatible with Mbone, the internet multicast scheme. The Mbone [8-9] uses the Mbone router to deliver the group messages and the satellite links with the broadcast capability can easily meet the multicasting requirement of the Mbone. The rest of this paper is organized as follows. In section 2, we describe the details about the intelligent routing scheme. The simulation and result analysis are presented in section 3. We propose our future plans and conclude with a summary in section 4.

## 2. Intelligent Routing Design

### 2.1 Intelligent Routing Framework

For constructing the intelligent routing scheme [10] we propose an intelligent router and use the VSAT system with the capability of transmitting information over satellite links bi-directionally to integrate the satellite into the terrestrial internet. The integrated platform is described as figure 1. The VSAT system [12] composed of an antenna receiver, an information sender and the intelligent router connects the satellite and wired network for providing the broadband internet service. The intelligent engine of the intelligent router within the VSAT system will decide the appropriate forwarding path, wired or satellite links, for each incoming packet according to the network environment.

To avoid receiving the redundant packets to the same destination by many VSAT systems from the satellite links due to the satellite broadcast, each VSAT system has its own control range, called "cluster". The VSAT system named as "home VSAT" takes over the transmission between the wired and satellite links within its own cluster. Each member station in the cluster could send packets to the destination either through the wired links or through the satellite links by the home VSAT.

The intelligent router could balance the traffic loads between satellite and internet links for network performance promotion. It decides the routing path of the delivered packets by identifying the routing information in the packet or dynamic network environment. Once a new packet arrives, the intelligent router will examine the routing information of IP packet format and adopt the intelligent routing scheme to make a best forwarding decision. The intelligent router will forward the packets appropriate for satellite transmission to the home VSAT by the encapsulation scheme. Otherwise the intelligent router will route them through wired links. If the encapsulated packet can't be fitted into a new IP packet due to the large data, the fragmentation scheme could be used. Once the intelligent router within the VSAT system receives the encapsulation packets, it will extract the original packet and forward it through the satellite links.

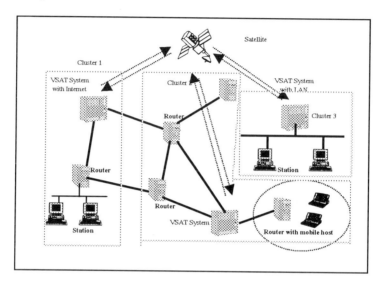

**Fig. 1.** The integrated platform with three clusters

To support the intelligent routing, the intelligent router must provide the capability of deciding the best forwarding path for each incoming packet. The best method is to obtain the routing information from the incoming packets. Therefore, the source uses the "IP option" field of the IP packet format to represent the routing information. However, most of the existing applications will not mark the packets with the routing information. To support these applications, the intelligent router will maintain a routing database, collected from the analysis of the dynamic network environment and routing table, for deciding a proper forwarding path. The database consists of the distances (hop numbers), average transmission delay time to each nodes. To obtain the distance and latency, we use the hello packet with timestamp and hop count fields to record the sending time and the distant information in link state routing. The intelligent router will monitor the hello packet, and adds the hop count value and packet starting time before forwarding it. This scheme is named as hop count technique.

When the intelligent router receives a incoming packet, it will follow the routing decision flow char shown in figure 2 to choose the routing paths. The intelligent router will extract the packet routing information from the "IP Option" field or use the routing database to compute the destination distance in this flow. If the destination is in the same cluster, the intelligent router will forward the packet via the wired links.

Otherwise, it will put this packet into the estimating queue for routing by intelligent engine.

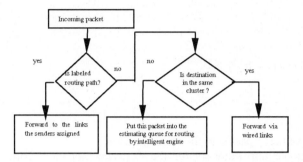

**Fig. 2.** Routing decision flow chart using by intelligent router

The function diagram of intelligent router is shown in figure 3. It includes the dispatcher, the intelligent engine and the encapsulated function. The dispatcher follows the routing decision flow shown in figure 2 to process the incoming packets in the receiving queue and dispatch them into the suitable queues. The wired queue stores the packets routed to the wired links. The satellite queue stores the packets for the satellite link transmission by encapsulation. The estimation queue stores the packets without routing information. The intelligent engine uses the routing information database and the traffic cost model to analyze the transmission cost, and obtains the best forwarding path for each packet in the estimating queue further.

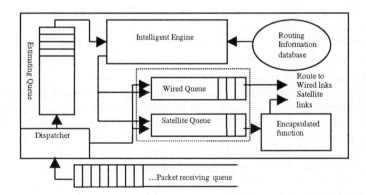

**Fig. 3.** The function diagram of intelligent router

## 2.2 Traffic Cost model

The routing strategy for the packets without routing information in the estimating queue is balancing the traffic between the satellite and wired links. When the traffic load is heavy in the wired links, routing packets to the satellite is better. Therefore, we propose a traffic cost model to predict the satellite and wired transmission cost. The optimal solution is choosing the best routing decision for each packets according to the minimal total cost. The traffic cost model is shown as follows.

$$\min \sum_{i=1}^{K} (P_i \times h_i \times d_i \times N_i \times X_i + \frac{P_i}{b} \times 2D \times (1-X_i)) \quad (1)$$

subject to

$$\sum_{i=1}^{K} (\frac{P_i}{b} \times 2D \times (1-X_i)) \leq B \quad (2)$$

where

$P_i$ : The size of the $i^{th}$ packet in the waiting queue.
$h_i$ : The estimate hop numbers for the $i^{th}$ packet obtained from the hop count technique.
$d_i$ : The average delay time (Considering the congestion loading, processing time, network loading, ..., etc) of each hop transmission for the $i^{th}$ packet.
$N_i$ : The number of multicast fan-out parameter.( $N_i$=1 represents the unicast )
$X_i$ : The domain of $X_i$ is {0, 1}. $X_i$ = 1 express the routing path decision for the $i^{th}$ packet is wired links. Otherwise, the routing path decision for the $i^{th}$ packet is satellite links.
B : The available bandwidth of satellite links in VSAT system.
b : The average number of packets transmitted to satellite in each transmission.
D : The latency time between earth-station and satellite.
K : The number of packets in the estimating queue in a fixed period

The traffic cost model includes two parts of traffic cost, the wired transmission cost and the satellite transmission cost. The $P_i$ *$h_i$ *$d_i$ *$N_i$ model expresses the wired transmission cost of the $i^{th}$ packet. Because the cost is in proportion to packet size, distance and average transmission delay. The $(P_i/b)$* 2D represents the satellite transmission cost. The satellite links with high bandwidth can pack the packets together for transmission once. Thus, the b parameter represents the packing ratio of satellite channel. This cost model can be used to sense the traffic condition in the network and offers the decision information to the intelligent engine. If the wired network meet congestion, it is obvious that the wired transmission cost computed by wired cost model is large relatively. In addition, the satellite uses the TDMA (Time Division Multiple Access) for channel allocation and bandwidth guarantee. This scheme can reserve a fixed bandwidth for each satellite channel and make the transmission over satellite smooth.

To obtain the suitable routing path, we find out the $X_i$ for each packet in this cost model. The K represents the packet numbers decided the forwarding path once. For meeting the real environment of the network traffic and preventing the database information from being out of date, the value of K can't be set too large. Then, a heuristic algorithm can solve this cost model.

Each host can assign routing information in the option field of IP packet format [3,4]. We use the Option field to represent the packet data type or routing information. The "Bit 0" of the option filed represents the processing method when a packet was fragmented into several segments. When setting "1" represents to copy the option fields into all of the segments. When setting "0" represents to copy the option fields into the first segments. To inform the intelligent router the routing choice, the "copy" bit, must be set as "1" for avoiding the data lost of option field when fragmentation is happen. "Bit 1,2" represents the option class and "Bit 3-7" represents the option number. IP option is used for testing and debugging the network and the option class "1" and "3" is reserved. We use the option class "3" and option number "1" to represent the packets wished to be routed via satellite. This packet is appropriate to carry the multimedia information. Using the option class "3" and option number "2" to represent the packets wished to be routed via wired links. The control data is suitable for this kind of packets.

To handle the redundant packets due to the satellite broadcast, the home VSAT must maintain a set of group member information called "cluster subnet_id". When receiving a packet from satellite links, the home VSAT with the intelligent router has to check whether the destination of the packet is in its cluster by comparing the network identification of the packet. If the destination is in its cluster, the intelligent router will route it through the wired links. Otherwise, intelligent router will take this packet as a redundant packet and drop it.

*The Algorithm of intelligent router in VSAT system*
Step 0: When a new packet has arrived from wired links, then goto step 1, else goto step 5 (receive packets from satellite links)
Step 1: The intelligent router will identify its destination. If the destination is in the same cluster then go to step 4 (destination is in the same cluster), else goto step 2(destination is in different cluster)
Step 2: If the packet is a encapsulated packet. Then extract the original packet and forward it to the satellite links. Else, identify the routing information in the option fields of the IP packet. If the packet is assigned for satellite transmission, then goto step 3 else goto step 4. Else, put the packet without any routing information in the estimating queue.
Step 3: send this packet via satellite links by this VSAT system. END
Step 4: pass this packet to the router and route it to the wired network. END
Step 5: If the destination is in the same cluster then goto step 4, else drop this packet (it is the redundant packet) and END.

**2.3 Compatible with mbone**

Internetwork forwarding of IP multicast datagram is handled by "multicast routers" which may be co-resident with, or separate from, internet gateways. A host transmits an IP multicast datagram as a local network multicast which reaches all immediately-neighboring members of the destination host group. The "multicast router" uses the tunneling technique to achieve the group information transmission.

To integrate our intelligent router into Mbone system for support of multicasting, it is necessary to combine our intelligent router with "multicast router". When receiving a new multicasting packet, the intelligent router will decide whether the group users are all within the same cluster or not. If yes, the intelligent router passes this packet to multicast router and let it to route the packets through wired links by using original Mbone technique. Otherwise, the intelligent router will encapsulate this packet and delivery it to the home VSAT for satellite multicasting transmission. When VSAT system receives the multicasting packet from satellite links, they will send this packet to the "multicast router" in its cluster for the multicast transmission. From above description, it is obvious that the satellite is compatible with the "multicast router".

## 3. Simulation and Result Analysis

To study the enhancement of the intelligent routing scheme, we use the BONes simulation, an integrated software package for simulating network data transfer systems, to simulate the transmission relationship between satellite and wired links.

To investigate the condition of the packet loss rate and transmission latency, we simulate the one-to-one and one-to-many transmission between wired and satellite links. We generate the traffic with average transmission rate of 512K bytes over the satellite and wired links. The hop numbers between source and destination is 10 links and the congestion probability in wired links is 25%.

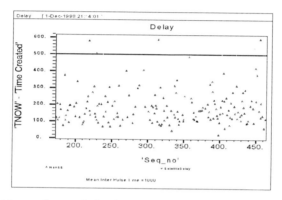

**Fig. 4.** The packet loss and transmission latency comparison between satellite and wired link

The figure 4 shows the traffic loss and latency result between the satellite and wired links and we conclude as follows
(1) The average transmission latency in wired links is 10ms and the latency will increase up to 50ms due to the congestion. However, the latency in satellite links with low congestion probability is 500ms.
(2) The variance of delay via wired links is 72.505ms
(3) The packet loss rate in wired is 34.8468% and 41.3237% for congestion probability 20% and 25% individually. Oppositely, the packet loss rate through satellite links is almost 0%.

The figure 5 shows the receiving time of one-to-many transmission between the satellite links and wired links. Figure 6 expresses the simulation diagram of intelligent routing scheme compared with satellite transmission. We use 10000 units as the satellite buffer and 5000 units as the wired buffer. The satellite links latency is 500ms and the wired latency is 10ms. The traffic rate is 1/Mean inter-arrive time.

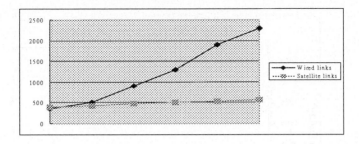

**Fig. 5.** The receiving time of one-to-many transmission between the satellite and wired links

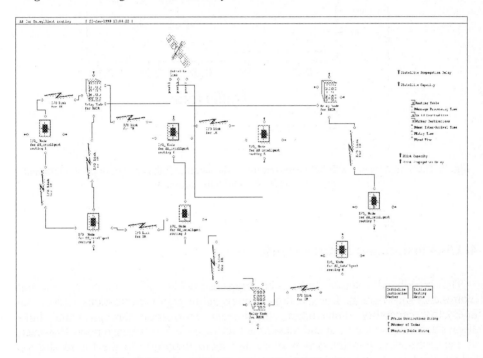

**Fig. 6.** The simulation diagram of the intelligent routing

The comparison result between routing on the satellite links and routing by intelligent engine is shown in figure 7. The receiving time based on the intelligent routing is :

$$\sum_{i=1}^{K}(P_i \times h_i \times d_i \times N_i \times X_i + \frac{P_i}{b} \times 2D \times (1-X_i)) \qquad (3)$$

According to this equation and the simulation result below, the intelligent routing transmission strategy will be better than transmission all through the satellite links. Moreover, this strategy can integrate the internet to provide an alternate transmission way. Now, the load of the existing internet is getting heavy. To balance the traffic between the internet and satellite links is getting important.

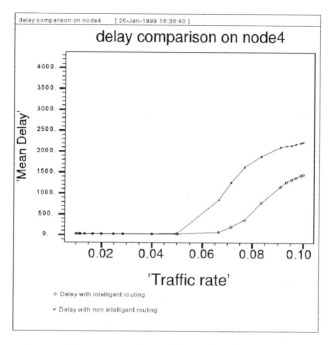

**Fig. 7.** The comparison result between routing on the satellite links and routing by intelligent engine at node of intelligent routing 4

## 4. Conclusion and Future Work

The broadband satellite links with high bandwidth are suitable to transmit multimedia information. But the high latency might cause applications to lose the interactive capability. Transmitting multimedia information through wired links always generates congestion and makes the transmission performance poor. However, as the multimedia transmission is more and more important in the future and we believe the satellite is a appropriate solution for offering the high bandwidth.

This paper proposes an intelligent routing method to support multimedia transmission and solve the problem of the lack of the interactive ability caused by satellite high latency. We propose a traffic cost model to offer the intelligent routing strategies. This method is transparent to applications and the transmission performance will promote significantly. Moreover, the intelligent routing method is compatible with the Mbone. By using this method the performance of the multicast transmission can be improved immediately. Furthermore, the method can enhance the integration between the wired and satellite links and own the fault tolerant when the wired links failed.

To evaluate the efficiency of this routing method, we do several simulations to investigate its performance according the different parameters. The efficiency can be improved furthermore by adjusting the cluster size. If we increase the cluster size, the traffic via satellite links will reduce accordingly. On the other hand, the traffic within the cluster via wired links will increase. The influence of the different cluster size will be investigated in the future work. Also, in order to support the QOS(Quality of Service) for multimedia transmissions, reservation schemes such as RSVP will be addressed on as well in the future work.

## References

1. Y. Zhang and S. K. Dao, "HBX: High bandwidth X for satellite internetworking." Proceeding of the 10th X Technical Conference, X Resource, Issues 17, February 1996.
2. Y. Zhang, D.D. Lucia and B. Ryu, S. K. Dao," Satellite Communication in the Globa; Internet: Issues, Pitfalls, and Potential ",Hughes Research Laboratories USA
3. W. R. Stevens. '"'TCP/IP Illustrated, Volume 3 ",Addison-Wesley Publishing Company
4. DOUGLAS E.COMER, "Internetworking with TCP/IP VoII: Principles, Protocols, and Architecture ",Prentice-Hall Inc.
5. H. Inoue, K. Kanchanasut and S. Yamaguchi, "An Adaptive WWW cache Mechanism in the AI3 Network", the Proceedings of INET97, Internet Society, June 1997.
6. Y. Zhang, and S. K. Dao, "Integrating Direct Broadcast Satellite with Local Wireless Access ", In Proceeding of the First International Workshop on Satellite-Based Information Services(WOSBIS), Rye New York, November 13,1996.
7. S. McCanne andV. Jacobson, "A flexible framework for packet video", ACM Multimedia, San Francisco, California,p511-522, November 1995.
8. E.Amir. A map of the Mbone: August $5^{th}$,1996. http://www.cs.berkeley.edu/~elan/mbone.html.
9. M. R. Macedonia and D. P. Brutzman, "Mbone Provides Audio and Video across the Internet", IEEE Computer,27(4):30-36,April 1994
10. D. Zappala, B. Braden, D. Estrin and S. Shenker, "Interdomain Multicast Routing Support for Integrated Service Networks", Internet-Draft, draft-zappala-multicast-routing-ar-00.ps, March 1997, Work in progress.
11. The Direct PC, http://www.hughes.com.
12. The VSAT System, http://www.vsat.com.
13. BONes DESIGNER handbook. Alta Group of Cadence Design Systems, Inc.,March 1996.

# Integrated CAD/CFD Visualisation of a Generic Formula 1 Car Front Wheel Flowfield

W P Kellar, A M Savill, and W N Dawes

Computational Fluid Dynamics Laboratory,
Cambridge University Engineering Dept.,
Trumpington Street, Cambridge CB2 1PZ, UK
wpk20@eng.cam.ac.uk
http://www2.eng.cam.ac.uk/~mea/fluid/cfdlab/cfdlab.html

**Abstract.** As part of a detailed project investigating racing car wheel aerodynamics, a CAD/CFD *(Computer Aided Design - Computational Fluid Dynamics)* interface has been used to enable CFD numerical flow visualisation for a generic racing car geometry. The interface was developed and improved through assessment and integration of industrial and baseline CAD formats with the requirements of an in-house state-of-the-art unstructured Navier-Stokes CFD package. The resulting CFD solution was proved to correlate well with experimental visualisation, and was useful in optimising a front spoiler. The time taken to move from an initial geometry definition to a CFD solution was reduced from typical times, for the types of geometry tested, by a factor of about five.

## 1 Introduction: CFD in Automotive Applications

Governing bodies of many categories of motor racing specify that the wheels of the racing cars must be exposed as to oncoming flow. Such regulations have a significant effect on the vehicle aerodynamics, causing a drag coefficient rise of $\approx 150\%$ [1] due to the addition of unshrouded wheels to a streamlined car body. The complex flowfield nature [2] suggests the value of 3D flow visualisation to gain insight into key characteristics. Such visualisation methods are surprisingly not often undertaken by motor racing teams, who concentrate on experimental parametric measurements of the overall aerodynamic performance complemented by local CFD analyses.

A CFD analysis improves the aerodynamic design process by allowing preliminary investigation of the performance of a large number of designs, without the associated cost and time required for prototype manufacture and experimental testing. Those designs which are indicated by the CFD to be the most promising can then be taken to the physical prototype stage. The turnover time of any CFD analysis is therefore an issue, to make it as useful a design tool as possible.

Current industrial CFD codes obtain 3D flow solutions by solving steady viscous Navier-Stokes equations within a discretised domain. Appropriate turbulence modelling is important in automotive applications due to the significant wakes and separated flows. Unstructured grids are almost universally used for the surface and domain discretisation [3] [4] due to the typically complex nature of the component surface geometries. Use of CFD in Formula 1 is particularly well advanced.

However, a major practical bottleneck is the CAD/CFD interface. Commercial CAD packages do not support the precise surface definitions necessary for the construction of a CFD mesh. The generation of any CFD mesh [4] requires that the component surface be uniquely and consistently defined topologically; although this is at least cosmetically possible with commercial CAD, the true underlying geometry data is deficient in a number of areas.

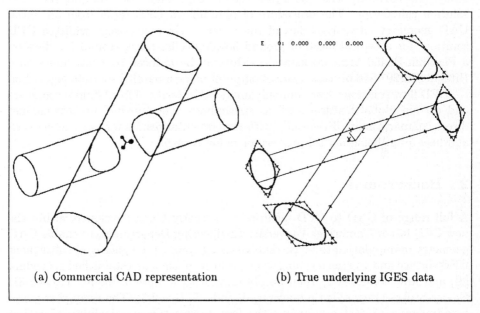

(a) Commercial CAD representation    (b) True underlying IGES data

**Fig. 1.** CAD data for a pair of intersecting cylinders. The cosmetic representation of the surface intersections is clear in (a); an inspection of the true IGES data (b) reveals this information to be absent in the actual data file.

An example of this problem is intersection data; a series of complex intersecting surfaces can be simulated as such in commercial CAD, but the underlying data (typically IGES format - *International Graphics Exchange Standard*) does not contain the intersection information (Fig. 1). Instead, the intersection geometry is re-calculated by the CAD package every time the component is accessed. The reason for this method of data handling is to enable the complex intersection information to be stored in an efficient manner - however, all packages wishing to use the geometry must then also have some intersection algorithm to interpret the data. The developing nature of CFD tools means that such complex data analysis algorithms are at an early stage. Thus the use of IGES data in CFD currently involves significant manual intervention, and constitutes the major time constraint in the CAD/CFD process.

Commercial CAD/CFD interfaces currently treat the CAD definition by the manual fitting of mesh patches to the underlying CAD entity; i.e. by computer operator intervention rather than a fully automatic procedure. This takes place at the pre-processing end of the CFD package. The entirety of the CAD entity must be mapped to give a surface which is uniquely defined topologically - this process is essentially the manual definition of the surface intersections. Individual CAD components are often treated as separate units, so that mesh sections representing these components can be investigated individually.

These manual processes are known as 'rubber-banding' and 'boxing' respectively, and currently take significantly more time than, for example, the CFD solution calculation. The procedure of obtaining a CFD mesh from an initial CAD geometry currently takes of the order of six man-weeks, whilst a CFD solution (for a typical mesh of around 500,000 cells) takes around 1-2 days on a Pentium II 450 MHz workstation - this can be reduced to a few hours on a Hitachi SR2201 256 processor supercomputer using a parallel solution algorithm.

CFD flow solutions are commonly analysed in terms of local fluid parameters and certain global features such as streamlines. The 'boxing' process enables direct comparison of CFD results with experimental results for parameters such as wheel drag, being a vital process for many reasons.

## 2 Background

A full range of CAD to CFD facilities is currently being developed within the new CFD lab at Cambridge University Engineering Department, to enable CAD geometry manipulation to be carried out using commercial packages, subsequent CFD surface and volume meshing to be performed using an established procedure [4], and flow solutions to then be obtained with a state-of-the-art viscous 3D Navier-Stokes (Reynolds time-averaged) code [5] (see 2.3). The racing car wheel aerodynamics project was among the first to test the practicability of such a CAD/CFD interface, and played an important role in defining and perfecting the best approach.

### 2.1 Geometrical (CAD) Modelling

The CAD model comprised a 40% scale representation of the front right-hand portion of a generic Formula 1 car. The modelling was carried out in a representative commercial package and in a more basic in-house format [4] to facilitate detailed investigation of the geometry characteristics, and to determine which format would provide the quickest route to a CFD mesh and solution.

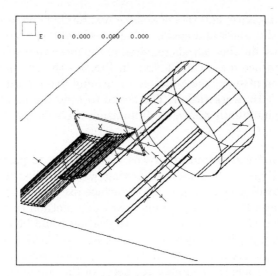

**Fig. 2.** In-house CAD format representation of the front right-hand corner of a generic Formula 1 car. The arrows indicate the unit normals to the CAD patches - necessary information for the surface meshing procedure. The parametric surface definitions are shown as rectangular nets.

The in-house format was the most likely to successfully interface with the surface mesher, whilst the commercial package was the easiest on which to create realistic geometries. The in-house format could be summarised as a format in which a component is defined as a series of patches with $(x,y,z)$ vertex coordinates, edge coordinates, surface parametric definitions and topological surface designations. The format was structured intrinsically to contain the rigorous topological information necessary for mesh construction [6]. This basic CAD definition represented the wheel as a cylinder, with only essential additional model features such as the front spoiler arrangement - see Fig. 2. The relatively user-friendly interface of the commercial CAD package enabled extra features (specifically the car body) to be modelled. However, the lack of patch intersection data and the subsequently necessary manual intervention prevented further development of the commercial CAD file at this stage. Development of a rigorous geometry intersection algorithm for the CFD package is currently underway [7], to fully automate the handling of commercial CAD geometry data.

## 2.2 Surface and Volume Meshing - Domain Discretisation

A surface mesh was constructed from the rigorous geometry definition of the baseline CAD file. The triangulation was automatic, based on a Delaunay triangulation algorithm [4] [8]. This represents a significant advance over typical commercial CFD meshing procedures, in which fundamental mesh parameters such as edge node spacing must be specified by hand for each edge - a very time-consuming process. However, a degree of manual mesh optimisation is still possible in this instance, and this enabled a fine mesh to be formed over regions of interest, e.g. the wheel and spoiler. The wheel only was modelled as viscous, to reduce the total number of cells required in the mesh to resolve viscous effects. The resultant surface mesh can be seen in Fig. 3 - the computer requirements for the generation of such triangulation are around five minutes CPU time on a Pentium II 450 MHz workstation. The desire to keep the number of mesh cells to an optimal minimum is aimed at minimising the solution computation time (see 2.3).

Further geometrical optimisation was required due to particular flow modelling considerations. For example, the wing was modelled with a continuous patch over the leading edge to give good surface discretisation in a fluid region of high significance. Development is underway of a method to parameterise such geometries so that similar geometrical optimisation could be automatically carried out. A surface mesh was also obtained of a hybrid CAD file - a combination of commercial and baseline CAD entities. This mesh required highly detailed triangulation to resolve the complex geometry of the commercial CAD patches, and optimisation of the number of cells in the mesh was not considered in the project.

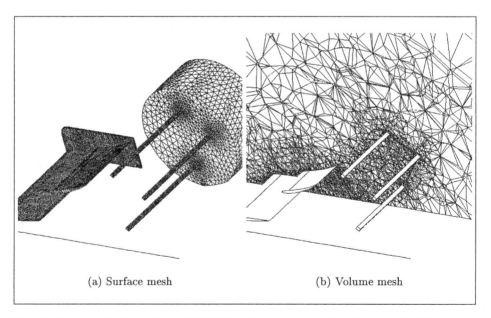

**Fig. 3.** Meshes of in-house format CAD file. The volume mesh is illustrated with a characteristic slice through the domain. The meshes show good resolution of the underlying CAD, good equilateral triangulation in the surface mesh, and good tetrahedralisation in the volume mesh. The ground plane mesh is omitted for clarity.

Volume meshing was based directly upon the surface mesh triangulation. The volume meshing algorithm was essentially a 'black box', giving a volume mesh output for a surface mesh input. The method used was Delaunay tetrahdralisation, which gives a unique and optimal mesh solution [4]; consideration was given in the algorithm to sliver cells which are not helpful for a CFD solution. The volume mesh used for the flow solution is shown in Fig. 3. The complexity of the mesh supports the use of an automatic unstructured algorithm, compared with the alternative of producing a structured mesh of the same quality. The normal computing requirements for the generation of such a mesh are 1-2 hours CPU time on a Pentium II 450 MHz workstation.

## 2.3 Obtaining A Flow Solution

The numerical method used in this study was the unstructured tetrahedral mesh method of Dawes [5] called NEWT. This method has been used successfully in a number of applications, ranging from unsteady turbomachinery flows to the simulation of oil rig explosion scenarios. The equations solved are the fully 3D unsteady Reynolds time-averaged Navier-Stokes equations expressed in a strong conservation form. Turbulence closure is provided by the k-$\epsilon$ model together with a modified low Reynolds number model handling near-wall regions and transition. Assuming a piecewise linear variation of variables over the faces of the tetrahedral cells, a second-order-accurate discretisation of the convective fluxes is achieved. The net flux imbalance into each cell is used to update the flow field variables at the mesh nodes, using four-stage Runge-Kutta time integration.

The solution code is being developed within the CFD laboratory at Cambridge University Engineering Department into a parallel version to run on multiprocessor workstations, which will enable upwards of 1,000,000 cells to be solved in a reasonable time. A mesh of 650,000 cells has been solved with this parallel version (using non-optimal mesh decomposition) in eight hours on a Hitachi SR2201 256 processor supercomputer.

In this study, the flow was calculated at a nominal Mach number of 0.3. This was to avoid convergence problems in a low-speed simulation calculated with a compressible code.

## 2.4 Associated Experimental Work

As an important corroboration of the numerical results, a complementary experimental study was carried out on the model geometry [9]. This work comprised wheel drag coefficient measurement for a number of typical geometry configurations, and corresponding smoke flow visualisation. The nature of the smoke flow visualisation was limited by practical considerations, e.g. blockage effects and the low tunnel speed needed to keep the smoke visible.

## 3 Results With Discussion

### 3.1 CFD Numerical Results

A flow solution for the in-house CAD file was obtained successfully. The final volume mesh (Fig. 3) had 340,524 cells. The CPU time was approximately fifteen hours on a Pentium II 450 MHz workstation. The net mass flow error between inlet and outlet was 0.08%, illustrating good solution convergence. The time taken to reach this solution from the initial CAD stage was reduced by a factor of around five from typical CAD/CFD interface times.

Graphic post-processing enabled fluid parameters, e.g. Mach number, to be shown as contours, and streamlines through the flow (Fig. 4) could be plotted to show features of interest.

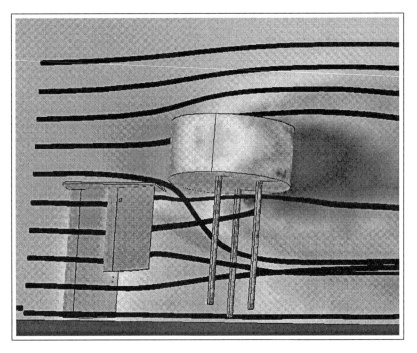

**Fig. 4.** CFD flow solution - sample plan view of wheel flowfield. Flow direction is left-right. The streamlines start in the left of the figure, in a plane parallel to the ground that passes through the middle of the front spoiler.

### 3.2 Corresponding Experimental Results

The smoke flow results were recorded photographically and on video. The results were summarised graphically (Fig. 5) to show the major features of the flowfield.

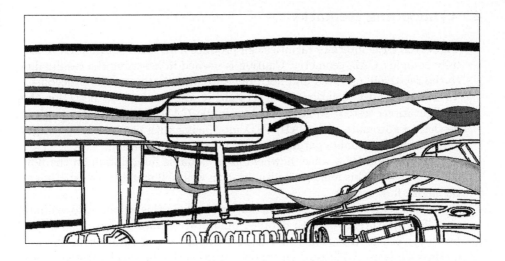

**Fig. 5.** Smoke flow visualisation results - plan view of wheel flowfield. The flow is left-right. The streamlines start in similar locations to those in the CFD solution, only slightly nearer the ground plane.

### 3.3 Results Comparisons

The flowfield was clearly similar in the two visualisation techniques, despite the different flow conditions considered. The results were largely qualitative due to the limitations specified previously.

The CFD solution illustrated the wheel flowfield to be described by an asymmetric wake, which was formed by crown separation and horseshoe vortices from the lower downstream part of the wheel. The wake symmetry was influenced by the strongly vortical flow leaving the front spoiler arrangement. In terms of local flow parameters, the CFD results gave indications of, for example, stagnation points on solid surfaces. Detailed information such as this was obtained most easily from CFD results.

Full quantitative CFD results were not achievable due to the relatively high velocities of the simulation, and the lack of an automatic 'boxing' procedure. However, a simple integration of pressure distributions acting on the front spoiler elements suggested that the simulation lift coefficient compared very well with experimental results [9].

The numerical visualisation of the flowfield characteristics enabled the development of experimental geometry features intended to optimise aerodynamic performance. A significant experimental wheel drag reduction was achieved based on a spoiler modification design deduced directly from the CFD results.

## 4 Concluding Remarks

- The CAD/CFD interface facilities can produce a 3D CFD solution for a usefully complex CAD geometry. Caution is needed in assessing the results due to the relatively untested and not fully automatic nature of the procedure.
- To obtain a CFD solution from scratch, it appeared to be best (for the configurations tested) to start from a simple CAD format. Future work is needed to automate such CAD/CFD interfacing with commercial packages - this could also enable parametric CAD design optimisation.
- A CFD solution for a wheel flowfield gave valuable information on the flow characteristics. This numerical visualisation was well matched to experimental smoke flow results.
- It was necessary, although only partially possible in this instance, to validate the CFD measurements with detailed quantitative information.
- Further work in this area could hope to apply wheel rotation in the numerical analysis. Additional geometry variations could also be tested, along with effects such as wheel yaw under steering which are difficult to simulate experimentally.

## References

1. Morelli, A.: 'Aerodynamic Actions on an Automobile Wheel'. Road Vehicle Aerodynamics. London (1969)
2. Sawley, M. L.: Numerical Simulation of the Flow Around a Formula 1 Car. EPFL Supercomputing Review. (11/1997) 11–17
3. Hanna, R. K.: The Role of Unstructured CFD in the Development Process for Formula 1 Racing Cars. Autotech (1995). C498/36/244
4. Dawes, W. N.: The Generation of 3D, Stretched, Viscous Unstructured Meshes for Arbitrary Domains. ASME (1996) 96-GT-55
5. Dawes, W. N.: The Practical Application of Solution-Adaption to the Numerical Simulation of Complex Turbomachinery Problems. Prog. Aerospace Sci. (1992). vol. 29 pp. 221–269
6. Connell, S. D.; Braaten M. E.: Semi-Structured Mesh Generation for 3D Navier-Stokes Calculations. AIAA Journal (1995). vol. 33 no 6 pp 1017–1024
7. Sinclair, F. M.: From CAD to CFD - Interior Volume Removal as a Data Conversion Tool. Cambridge University Engineering Dept., Cambridge, UK.
8. Dhanasekaran, P. C.; Demargne, A. A. J.: Surface Mesh Generator (version 1.8) manual. Whittle Laboratory, Cambridge University Engineering Dept., Cambridge, UK.
9. Pearse, S. R. G.: Effects of Formula 1 Style Forebodies on Exposed Racing Car Wheels. MEng submission (1998). Cambridge University Engineering Dept., Cambridge, UK.

# Adaptive Scheduling Strategy Optimizer for Parallel Rolling Bearing Simulation

Dag Fritzson[1] and Patrik Nordling[2]

[1] SKF Engineering & Research Centre B.V.
Postbus 2350, 3430 DT Nieuwegein, The Netherlands
[2] Department of Computer and Information Science
Linköping University, S-581 83 Linköping, Sweden

**Abstract.** Rolling bearing simulations are very computationally intensive and need to utilize the potential of parallel computing.
The load distribution over the processors in a rolling bearing simulation is very dynamic. In this paper we present the Adaptive Scheduling Strategy Optimizer (ASSO) for scheduling parallel simulations. The result of this is that the application can automatically select a near optimal scheduling strategy (with respect to the available scheduling strategies). The ASSO is used daily in real bearing simulations.

## 1 Introduction

Rolling bearings are one of the most common type of machine elements. Typically there are several in any type of rotating machinery. Bearings are built in a wide range of sizes, from a few millimetres to several metres in diameter. Figure 1 shows the essential components.

From the end user point of view, a turn-around time of less than 24 hours for a bearing simulation cycle is desirable. This cycle includes generation of input data, analysis of output data, and calculation outside office hours. This means that about 16 hours of calculation time is available for one or more bearing simulations. A 24 hour cycle enables efficient use of simulation techniques in bearing development. This means that we need to have efficient parallel computing.

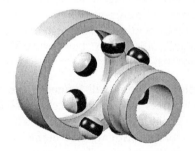

**Fig. 1.** *A Deep Grove Ball Bearing. The components are outer ring, inner ring, cage (the contact surfaces are shown), and rolling elements (balls).*

## 2 What is Bearing Simulation?

Bearing simulation is the simultaneous solution of Newton's equations of motion for every body in the bearing, modeled as a multi-body system. The Newton equilibrium equations results in a system of *second order ordinary differential equations* (ODEs) which is rewritten as a first order system. Typical characteristics of such ODEs are: stiffness, very high numerical precision needed for the solution, and computationally expensive evaluation of the derivatives.

Efficient and accurate contact calculations are important in bearing simulation. The contact force and moment calculations are mainly based on elastohydrodynamic lubrication theory (EHL). A detailed geometric description is also very important, since changes in the order of 0.1 micrometre may have a significant effect on bearing performance. Almost all time in the the evaluation of the right hand side (RHS) of the ODEs, and the calculation of the Jacobian is contact calculations.

Our code, written in C++ and C, runs on almost every platform available. Ports have been made to Cray T3E, IBM SP-2, cluster of workstations, i.e. any machine with a C++ compiler running a UNIX operating system. We have used PVM[1] as the underlying communication library. Simulation time on a modest parallel computer is in the order of days.

## 3 What influence simulation time?

In order to run the simulation efficiently on parallel computers, it is necessary to choose a good scheduling strategy, i.e. how to distribute the computation load on the parallel machine.

This has turned out to be a difficult task. The application knowledge is very important and governs what can be parallelized. It takes a lot of knowledge of the platform used. It is also necessary to be able to predict the behaviour of the bearing simulation. The person that will do/start the simulation must choose a good scheduling strategy with three things in mind:

**The platform** will influence the choice of scheduling strategy. Different computers have different characteristics.

If the machine is a multi user system (such as a work station cluster), the computational performance/availability of the processors might vary during the simulation. In our case, if the user of the workstation wants to utilize the work station fully, it can temporarily be removed/disabled from the cluster.

**The bearing types** such as Deep Grove Ball Bearing (DGBB), Compact Aligning Roller Bearing (CARB), or Cylindrical Roller Bearing (CRB) etc., have different "computational" characteristics. This is mainly due to different contact situations. A bearing model having many "large" rolling element/cage contacts will benefit from a large parallel machine; the opposite is true for a bearing model with a few "small" contact segments.

**The input data** which describes the bearing and its loading situation is very important. One of the most important parameters that can be specified (from a scheduling point of view) in the input data is how many rolling elements there are in the bearing.

---

[1] Parallel Virtual Machine

Even if a very good scheduling strategy could be found with respect to the whole simulation, it may be possible to do better. The explanation for this is that there are phases when there are of lot of contacts, for which a certain scheduling strategy is the best, and there are other phases where there are a few contacts where another scheduling strategy is best. This feature is highly dependent on input data and bearing type.

## 4 Basic Scheduling Strategies

In a rolling bearing simulation an RHS computation calculates the time derivative of the state variables. The main part of the work in the RHS and the Jacobian are the contact force calculations. We have here identified the following levels of granularity in the parallelization, where $n_W$ is the number of rolling elements.

- Parallelization over the rolling elements. The tasks for the slaves will be to evaluate the RHS for one specific rolling element, or to calculate the sub Jacobian belonging to a rolling element. The slave task will include all the contact force calculations between the rolling elements and the other bodies. The maximum number of processors that we can utilize is limited by the number of rolling elements and is $n_W + 1$, where the extra processor is used for the master.
- Parallelization over the contact force calculations at the *body level*. Every rolling element can be in contact with several bodies and we can do the contact calculations for bodies in parallel for every rolling element. If $n_B$ is the number of bodies that the rolling elements can interact with, then the maximum number of processors that can be used is $n_B \cdot n_W + 1$.
- Parallelization over the contact force calculations at the *segment level*. A segment is defined in the model as an area where physical contact is possible. Every possible contact that can occur in the bearing is considered for the parallelization. If $n_S$ is the number of segments each roller is in contact with then the maximum number of processors that can be used is $n_S \cdot n_W + 1$.

Here we exploit the first and third of these granularity levels in a two step scheduling algorithm. The first level of granularity is used at the first scheduling level, and the third level of granularity is used at the second scheduling level.

Using the two step scheduling algorithm, the master/slave paradigm which is employed gives rise to a two level communication tree (see figure 2). The master is located on one node, doing the actual integration of the ordinary differential equations. From this node we distribute work to the slaves on the parallel machine.

## 5 How, When, and What to Measure?

Measurements of calculation times (mainly from contact force calculations) from the previous evaluation of the RHS are used to schedule the next evaluation. It

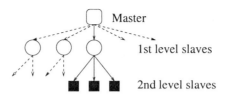

**Fig. 2.** *The two level communication structure. The arrows indicates communication. The communication is reversed when the contact computation is completed. The first level is typically rolling elements and the second level contact segments.*

**Fig. 3.** *The dashed line indicates the use of the best methods for a number of calls. Then a test suite occurs, and all the methods are tested. Based on the result of the test suit one of the methods is chosen.*

is only possible to use the time measurements if the simulation does not vary significantly between two consecutive calls to evaluate the RHSs. This is the case for the rolling bearing application.

The use of any algorithm to select an optimal scheduling scheme should interfere as little as possible with the simulation. Our approach is to do the measurements as the simulation progress. This only works well if the computational time for contacts varies slowly. Our experience, after studing a large amount of output data, is that this is true for rolling bearing simulation.

The basic idea is to test a number of scheduling strategies (in a *test suite*) and, once the testing is done, to use the best of those for a certain number of calls to the scheduler. This is illustrated in figure 3, where six scheduling strategies are tested. The best method is the used for a reasonably large number of RHS calls.

## 6 The Scheduling Strategies utilized here

In the presented implementation six methods are tested in every test suite. Two of those are based on scheduling at the rolling element level and four at the segment level scheduling.

We have access to a large number of scheduling strategies. We introduce the following concepts:

**Dynamic.** First send jobs to all the slaves, wait for the first to finish, send the next job to the one that finishes first. This goes on until all tasks have been completed. This can be done in any order.

**Static.** Statically assign computation to the slaves, send the jobs once to each slave. This is a static assignment for the particular RHS.

**LPT-Scheduling** Assigns computational request to the slaves according to the *Largest Processing Time* algorithm,[1], and guarantees a upper bound for the solution with respect to the optimal solution.

## 6.1 First Level Scheduling over Rolling Elements

The first level scheduling is done at the rolling element level. We schedule the elements on the slaves so that we obtain a load which is as equally distributed as possible. The two strategies are:

- Static LPT-Scheduling over the rolling elements. It is a static distribution of the tasks to the slaves, within one scheduling.
- Dynamic scheduling of the rolling elements. This method has a dynamical distribution over the rolling elements. The tasks are distributed in a descending order with respect to expected computational time for the rolling elements.

## 6.2 Two Level Scheduling Algorithms

Parallelization over contact forces on a segment level (second level scheduling), theoretically gives the best result if we do not consider communication time. The disadvantage of directly doing a second level scheduling is that we may get a lot of communication, which may slow down the simulation. The parallelization over the rolling elements gives, in some sense, minimal communication. We therefore developed two step scheduling algorithms. The idea is first to do a scheduling over the rolling elements and then reschedule some of the segment contact force calculations.

In order to try to minimize the communication the algorithm is designed to be controlled by parameters that should be selected in a optimal way.

The segment level scheduling strategies have the following parameters that can be set:

- The cutoff, $c$ [ms], identifies which segment computations we can reschedule in the second phase of the two level algorithm. Segment computations that are less then the value $c$ will not be rescheduled.
- The baseline modifier, $t$ [-]. The *baseline value* is obtained by: computing the total sum of all the timings from all computations divided by the number of processors available. The baseline is the theoretical optimal computation time. The baseline is used in the following way: for each processor we allocate a number of rolling elements (from the first phase in the strategy). The rolling elements allocated on one processor form a "pile" of work to be done. The

segments of the rolling elements that lie above the baseline in this pile of work will be placed in a pool of possible re-scheduable parts. The parameter $t$ is used for scaling the baseline up or down, see figure 4.

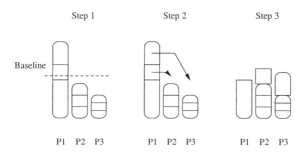

**Fig. 4.** *Illustration of the two step scheduling algorithm with three rolling elements and three processors. The height of the "piles" corresponds to the computation time in the previous RHS call. Each rolling element consists of three segment computations. Step 1 shows the result from an LPT scheduling over the rolling elements. Step 2 reschedules the segment computations above the baseline.*

The choice of $c$ and $t$ is as hard as choosing a scheduling strategy. We therefore regard the choice of these parameters as an optimization problem with two degrees of freedom.

### 6.3 The parameter optimization for two level scheduling

This optimization of the parameters $c$ and $t$ is complicated since:

- The function (i.e. the computation time) is rather noisy, and contains many local minima.
- The optimum is a "moving target". The computational demand will vary during the simulation, and for the global optimum the parameters $c$ and $t$ may also vary.

We choose a simple strategy to test the parameter space with two different values of $c$ and two values of $t$. With these four parameter pairs we choose the best $(c, t)$ pair. The following heuristics are used:

- the baseline modifier is set to two discrete constant values, $t_0 = 0.1$ and $t_1 = 1.0$. The lower value is a rather bold choice, which will "release" almost all segments for the part scheduling. The higher value could be regarded as a conservative choice of the $t$ parameter.
- the cutoff parameter is initially set to $c_0 = 0$ and to $c_1 = \Delta c$. In order to take different platforms into account the default value of $\Delta c$ is set equal to the latency for the underlying machine and communication library. From a simplistic theoretical point of view no computational tasks smaller than two latencies will benefit from being moved to another processor.

The default values of $c$ and $\Delta c$ are chosen with the philosophy: we want to be bold and aggressive, we therefore chose the initial value of $c_0 = 0$, this will essentially release all segments with $t_0 = 0.1$. In order to avoid communication congestion and moving tasks which are too small, we search in $c$ with an increment $\Delta c$ that will in a reasonable number of tries "reach" the theoretical best value of $c$.

We obtain four "two level strategies" because we have four combinations of the parameters for the segment level scheduling strategy.

In addition to the selection of one of the six scheduling methods, we need also to search for optimal values of the constants $c_0$ and $c_1$. This is done by comparing the "two level strategies" and increase or decrease $c_0$ and $c_1$ with $\Delta c$ given the result of the test suite.

## 7 The Machines and Test Cases

The computers used for the investigations in this report are:

- a Parsytec GC/128 with 64 nodes, and two 80MHz PowerPC processors in each node,
- a DEC Alpha Workstation Cluster, with four workstations of model 2000, each with two Ev-5/250 Alpha processors
- a SPARCCenter 2000 with shared memory, and 20 50Mhz super Sparc processors,

For the simulations we have used two rolling bearing models:

- a DGBB with 7 balls, referred to as case DGBB,
- a CRB with 19 rollers, referred to as case CRB.

We sample the real simulation at ten equally spaced point in time. At each of these points we do a short test simulation numbered 1 to 10. With these short simulations spread over the whole interval, we can still draw some conclusions that are valid for the whole simulation.

## 8 The Performance of the ASSO

Here we present the qualitative behaviour of the ASSO module compared with available scheduling strategies in our bearing application. These results were obtained for the Parsytec GC/128 using 16 processors, the SPARCCenter 2000 with 20 processors, and the DEC Alpha Workstation Cluster with 8 processors.

We will here refer to the use of the adaptive scheduling strategy optimizer as a method. But it is really using a lot of other scheduling strategies.

For the sake of evaluating the adaptive scheduling strategy optimizer it is sufficient to compare method "**ASSO**" (see description below) with the other scheduling strategies. This method is indicated with large plus signs, as compared to the other methods that have a more reasonable size of the plotting symbols.

On the x-axis in the figures the test run is indicated (i.e. 1–10). The y-axis is the ratio of the total computation time for the methods in relation to method "**m1**", i.e. **m1** serves as a norm for the comparison.

For the basic scheduling concepts see page 4. Other used definitions are:

**Via-communication** use the rolling element slave to compute common parts that are used by the contact segment (second level) slaves.

**Direct-communication** bypass the rolling element slave and compute the common parts on each contact segment (second level) slaves. The result is sent back to the rolling element slave.

**Default parameters** for the LPT-Scheduling the default values of $c$ and $t$ are 0 and 1.0 respectively.

In the figures below the methods are:

**m1** Dynamic, rolling element scheduling in increasing rolling element order.
**m2** Dynamic, rolling element scheduling in random order.
**m3** Static, LPT-Scheduling on rolling element level.
**m4** Dynamic, rolling element scheduling in rolling element order sorted with respect to computation time.
**m5** Static, LPT-Scheduling on contact segment level with Direct-communication, Default parameters.
**m6** Static, LPT-Scheduling on contact segment level with Via-communication, Default parameters.
**m8** Static, LPT-Scheduling on contact segment level with Via-communication, with both $c$ and $t$ parameters set to 0.
**ASSO** is the adaptive scheduling strategy optimizer. $\Delta c$ is the latency as measured by the scheduler.

The results are presented in figure 5. We note that on the Parsytec GC/128 where we have utilized 16 nodes, i.e. 15 slaves, that the two-level scheduling strategy is much better than the one-level strategy. This is hardly surprising since, for example, the `DGBB` has only 7 balls, the best speed up we can obtain with a one-level strategy is 7, but for a two-level scheduling strategy there is no such limit. The adaptive scheduling strategy optimizer makes choices that give us a performance close to the optimum for that test case. A similar effect appears in the `CRB` test cases.

The DEC Alpha Workstation Cluster has only 8 processors and the choice of scheduling strategy seems not to be that vital. For all test cases the performance of different scheduling strategies only fluctuates in the order of 10%. In the `CRB` case, the best scheduling strategy is method 4, which is a dynamic one-level scheduling strategy. The two-level scheduling strategy is inferior but not entirely worthless.

The choice of a decent scheduling strategy on the SPARCCenter 2000 or Parsytec GC/128 is more important. Since these platforms have more processors than rolling elements, the two-level scheduling strategies must be considered, and the ASSO module does exactly that.

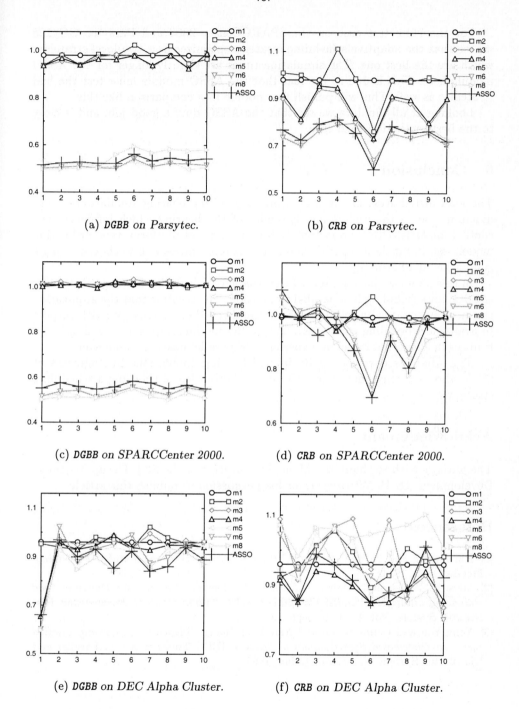

**Fig. 5.** The $y$ axis the timing ratio relative to **m1**, where lowest ratio is highest performance, for the different scheduling methods. The $x$ axis indicates the test case. See page 8 for definition of the labels.

For the DGBB test case on the SPARCCenter 2000 and Parsytec GC/128 we see that the adaptive scheduling strategy optimizer uses two-level strategies which are the best one. The simulation times are slightly longer for the ASSO module here. This is due to the fact that the ASSO module must test the bad strategies as well. This will punish the module in a comparison like this.

Looking at all the data we see that the ASSO does a good job, and is close to the best method in each case.

## 9 Conclusion

The load distribution of the processors in a rolling bearing simulation is very dynamic due to the underlying dynamics of the bearing which influences the contact analysis. Other important factors are type of platform, type of bearing model, and input data. All this makes it virtually impossible to choose a single good scheduling strategy.

In this paper we have presented the Adaptive Scheduling Strategy Optimizer (ASSO) for scheduling in a parallel simulation. The result is that the application can now automatically select a near-optimal scheduling strategy (with respect to available scheduling strategies). The ASSO module implements a simple optimization algorithm based on experience of parallel bearing simulation.

The ASSO is now used daily in real bearing simulations by engineers at SKF, who as a consequence are relieved of the burden of choosing a scheduling strategy.

## Acknowledgments

The authors wish to thank the Managing Director of the SKF Group Technical Development, Dr H. Wittmeyer, for his permission to publish this article.

## References

[1] Kai Hwang and Fayé A. Briggs. *Computer Architecture and Parallel Processing.* McGraw Hill, 1984.
[2] Marc H. Willebeek-LeMair and Anthony P. Reeves. *Strategies for Dynamic Load Balancing on Highly Parallel Computers.* IEEE Transactions on Parallel and Distributed System. Vol. 4, No. 9, Sept. 1993.
[3] Yong Yan and Canming Jin and Xiaodong Zhang. *Adaptively Scheduling Parallel Loops in Distributed Shared-Memory Systems.* IEEE Transactions on Parallel and Distributed System. Vol. 8, No. 1, Jan. 1997.

# MPI-Based Parallel Implementation of a Lithography Pattern Simulation Algorithm

H. Radhakrishna[1], S. Divakar[2], N. Magotra[2], S.R.J. Brueck[3], A. Waters[1]

[1]Albuquerque High Performance Computing Center (AHPCC), University of New Mexico (UNM), Albuquerque, NM 87131

[2]EECE Dept., UNM, Albuquerque, NM 87131

[3]Center for High Technology Materials, UNM, Albuquerque, NM 87131

**Abstract.** This paper presents the parallelization of a pattern simulation algorithm for Imaging Interferometric Lithography (IIL), a Very Large Scale Integration (VLSI) process technology for producing sub-micron features. The approach uses Message Passing Interface (MPI) libraries [1]. We also discuss some modifications to the basic parallel implementation that will result in efficient memory utilization and reduced communications among the processors. The scalability of runtime with degree of parallelism is also demonstrated. The algorithm was tested on three different platforms: IBM SP-2 running AIX, SGI Onyx 2 running IRIX 6.4, and a LINUX cluster of Pentium-233 workstations. The paper presents the results of these tests and also provides a comparison with those obtained with Mathcad (on Windows 95) and serial C (on Unix) implementations.

## Introduction

Simulation studies and scientific data analysis tend to be computation and memory intensive. Very often, many such simulations/data analyses have to be performed over short periods of time to study the effect of a particular parameter variation on the final outcome or to ensure that the results are reliable. For instance, analysis of meteorological data accumulated from satellite imagery, nuclear test simulations, heat transfer and fluid flow modeling and large scale optimizations all involve vast amounts of data and a large number of operations involving them. These requirements necessitate a fast, compact and reliable simulation package. It is almost impossible to accomplish these goals using serial (sequential) programs, which have obvious limitations in terms of computational speed and memory usage. The last decade or so has seen the evolution of many general high performance computing (HPC) platforms like Cray, IBM SP2, SGI and HPC languages like HPF and Split C, which are specifically designed to address such issues.

In this paper, we describe a novel application of HPC in the VLSI area. Specifically, a parallelization approach to a VLSI patterning simulation problem is described, which was originally implemented using Mathcad, an interpretive programming environment. The main idea behind the digital simulation of this problem is to study the effect of parameter variations in a systematic fashion and then translate these into feedback parameters to be used in the optical setup. The optical experiment involves printing sub-micron (<180 nm) features using a mix and match interferometric technique involving traditional optical lithography as well as interferometric lithography [2,3,4]. This technique however results in contrast degradation, which manifests itself as reduced dynamic range (*process latitude* in lithography parlance) of the image that etches the mask pattern on to the wafer. The digital signal and image processing (DSIP) lab in the EECE department, University of New Mexico is addressing this contrast degradation problem and research is ongoing to develop contrast enhancement techniques [5] to counter this problem. The simulation setup is shown in Figure 1.

As shown in Figure 1, the digital simulation reproduces the optical experiment, the outputs of which are then analyzed to arrive at feedback parameters for the actual optical experiment. Thus there is also a need for optimizing the parameters which control the lithographic process. For small sized test patterns, the simulation code can be executed many times over to arrive at an optimal parameter set. However, real-world patterns are much larger and hence the sequential simulation code (interpretive or compiled) cannot produce useful results in a reasonable amount of time. Also, the complexity of simulation increases as the square of the increase in linear dimensions. These factors motivate the MPI-based parallelization approach discussed in this paper.

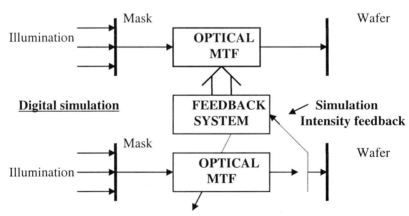

**Figure 1**. Digital simulation setup

Simulation speeds can be increased by either employing faster, more powerful processors or by dividing the task among many, relatively less powerful processors. The latter approach is more suitable if the problem being parallelized has independent code segments that can be assigned to different processors. For larger patterns, a second degree of parallelization can be introduced by breaking the bigger pattern down into smaller segments and then carrying out simulations on these segments separately.

The paper is organized as follows: the first section is a brief introduction to IIL, a novel VLSI processing technology being investigated at the Center for High Technology Materials (CHTM), University of New Mexico, Albuquerque. The next section describes the patterning algorithm in detail in relation to the parallelization approach. Subsequent sections describe the serial code, the task of parallelization, a discussion about efficient memory utilization, run-time results and conclusions, in that order.

## Imaging Interferometric Lithography

Over the past three decades, progress in the area of VLSI technology has been characterized by an ever-decreasing feature size with optical lithography being the dominant technique for volume manufacturing applications. At present optical lithography is the only proven volume-manufacturing technique. It now appears certain that optics will be used for the 130-nm critical dimension (CD) node (2003) [6]. It is highly likely that 193-nm ArF-laser-based optical tools will extend it at least to the initial stages of the 70-nm CD node as well (2009). This requirement is fundamentally beyond the reach of existing resolution enhancement techniques such as optical proximity correction and phase shift masks. Imaging interferometric lithography (IIL) has recently been introduced as an innovative and powerful approach employing optical lithography to push the fundamental limits of optics. IIL is based on a detailed understanding of the ultimate capabilities of optical lithography derived from a spatial frequency perspective [7].

The thrust of the IIL research being conducted at CHTM is to elucidate the fundamental scientific issues for deep sub-wavelength optical lithography, down to at least $\lambda/3$ or ~ 65 nm at a 193-nm exposure wavelength. Figure 2 shows a test pattern used to demonstrate the concept of IIL. This pattern can be reconstructed using a combination of optical and interferometric exposures. The optical exposures contribute to the low frequency information and the interferometric exposures add the high frequency details ("edges"). The interferometric exposures are referred to as "offset exposures" owing to the offset that is introduced in the final frequency domain representation.

The optical system employs a Fourier transforming lens pair as shown in Figure 3 which ensures imaging while preserving waveform flatness. Inherently, the optical system acts like a low-pass filter blocking mask pattern Fourier components above a certain critical frequency, which is a function of the wavelength of the source and numerical aperture of the lens. The mask Fourier components are therefore first downshifted in frequency using off-axis illumination so that they pass through the lens. The interferometric optics employed will then upshift the frequencies back to their original values. The communication systems analogy is one of wavelength-division multiplexing wherein the mask spectrum (or the transmitted mask spectrum) is constructed using frequency slices transmitted by optics. The output of the Fourier lens system is an intensity, which etches the pattern onto the wafer that has a photoresist coating. The resist performs a thresholding function, and converts the intensity into a 1/0 bitmap reproducing the original mask pattern, within the limitations imposed by optics.

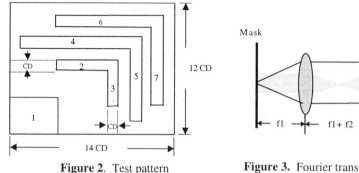

**Figure 2.** Test pattern  **Figure 3.** Fourier transforming lens pair

## Exposure Synthesis Algorithm

The exposure synthesis algorithm is a systematic procedure to break up an arbitrary patterning problem into a combination of optical and interferometric exposures. Specifically, it involves the realization of any given pattern in terms of a coherent optical exposure and offset (x and y) interferometric exposures. Most real world patterns possess the so-called manhattan geometry (only 0 or $90^0$ edges), allowing them to be decomposed into their constituent rectangles. As illustrated in Figure 2, the mask of a nested L structure can be broken up into a number of horizontal and vertical rectangles. The dimensions of the rectangles and the patterns are defined in the units of Critical Dimension (CD) which is the smallest feature that can be resolved. The mask pattern can be conveniently described in the Fourier domain in terms of the Fourier transform instead of a spatial zero/one bitmap.

We can recall here that the Fourier transform ($T(k_x,k_y;a,b)$) of a rectangle is given by a two-dimensional Sinc function,

$$T(k_x,k_y;a,b) = a \cdot b \cdot \frac{\sin(2\pi k_x a/2)}{2\pi k_x a/2} \cdot \frac{\sin(2\pi k_y b/2)}{2\pi k_y b/2}$$

where a and b are the length and width of the rectangle respectively and $k_x$, $k_y$ are the x and y frequency variables.

For a positive resist (wherein the resist is *etched* for light intensities greater than a particular threshold), the Fourier transform (FT) of the desired structure (Figure 2) is given by

$$FT(\text{Desired structure}) = FT(\text{Outer rectangle}) - \sum_i FT(\text{inner rectangles})$$

The maximum number of frequency components in the Fourier representation is dictated by optics. For a wavelength of $\lambda = 364$ nm and a numerical aperture of NA =

.5, the maximum radial frequency transmitted by optics is $f_{opt} = (NA/\lambda)$. The lens limit is $2*f_{opt}$. The periodicity of the mask pattern and the discrete nature of the spectrum imply that the Fourier series representation has frequency components dictated by the extent of the pattern given by $N_x$ and $N_y$ where

$$N_x = \lfloor Px \cdot fil \rfloor$$
$$N_y = \lfloor Py \cdot fil \rfloor$$

where $P_x$ and $P_y$ represent the x and y dimensions of the pattern in Figure 2 and *fil* is the fundamental optics limit ($2/\lambda$).

The optical system imposes a modulation transfer function (MTF) on the propagation of the mask Fourier components onto the image plane. The modulated Fourier transform of the mask pattern is given by the product

$$A_{n,m} = F\left(\frac{n-N_x}{P_x}, \frac{m-N_y}{P_y}\right) \cdot \text{MTFI}\left(f\left(\frac{n-N_x}{P_x}, \frac{m-N_y}{P_y}\right)\right)$$

where $f = \sqrt{f_x^2 + f_y^2}$, the radial frequency. n and m vary between 0 to $2*N_x$ and 0 to $2*N_y$ respectively, and MTFI, the interferometric modulation transfer function is given by

$$\text{MTFC} = 1 \text{ for } f \leq f_{il}$$
$$= 0 \text{ for } f \geq f_{il}$$

In other words, the MTF acts like a lowpass filter, filtering out frequencies beyond the lens limit. Since coherent illumination is used in interferometric lithography, a coherent imaging analysis is appropriate. The MTF of a coherent optical system is given by

$$\text{MTFC} = 1 \text{ for } f \leq f_{opt}$$
$$= 0 \text{ for } f \geq f_{opt}$$

The electric field at the wafer is given by

$$\text{Ecoh}_{n,m} = A_{n,m} \cdot \text{MTFC}\left(f\left(\frac{n-N_x}{P_x}, \frac{m-N_y}{P_y}\right)\right)$$

Intensity and field have a square-law relationship. In the frequency domain, this means that the intensity Fourier transform can be computed by convolving the conjugate field Fourier transform terms as

$$\text{Icoh}_{n,m} = \sum_{kx}\sum_{ky} \text{Ecoh}_{kx,ky} * \overline{\text{Ecoh}_{kx+Nx-n,ky+Ny-m}}$$

The intensity distribution at the wafer can then be obtained by computing an inverse Fourier transform,

$$I(x,y) = \sum_n \sum_m \text{Icoh}_{n,m}\, e^{i2\pi\left(\frac{n-Nx}{Px} \cdot x + \frac{m-Ny}{Py} \cdot y\right)}$$

The photoresist performs a thresholding operation on these intensity levels as given by

$$P_{ii,jj} = \sigma(I) \text{ where } \sigma(I) = 1 \text{ if } I>1 \text{ else } 0.$$

Interferometric optics is introduced at this stage to bring the frequencies back to their original position in frequency space. In other words, it accounts for the frequency shift introduced by off-axis illumination.

## Serial C Code

The exposure synthesis algorithm was first implemented using Mathcad on Windows platform. However, the turn around time was long for even small sized pattern simulations. Hence the development of a compiled code version was initiated. A serial program was developed in C on Unix. The program has four major sections as shown in the Figure 4,

```
/* Section 1 */
Preliminary calculations of values
{ User Inputs, simple arithmetic,
  Fourier transform of the mask.}

/* Section 2 :: Modulated mask Fourier
transform (multiplication) */
  for( n=0; n<=2*Nx; n++)
    for(m=0;m<=2*Ny; m++)
      calculate An,m and Ecohn,m

/* Section 3 :: Image intensity fourier
transform (convolution) */
  for ( n=0; n<=2*Nx; n++)
    for ( m=0; m<=2*Ny; m++)
    {
      if(n<nx)
        kxmin=0; kxmax=n+Nx;
      else
        kxmin=n-Nx; kxmax=2*Nx;
      if(m<ny)
        kymin=0; kymax=m+Ny;
      else
        kymin=m-Ny; kymax=2*Ny;
      for (kx=kxmin; kx<=kxmax; kx++)
        for (ky=kymin; ky<=kymax; ky++)
          calculate Icohn,m and IILn,
    }
```

```
        /* Section 4 :: Reconstruction of the
    image(Inverse Discrete Fourier Series and
    Thresholding) */
        for ( x=0; x<=pixel * Px; x++)
            {for ( y=0; y<=pixel * Py; y++)
                {for ( n=0; n<=2*Nx; n++)
                    {for ( m=0; m<=2*Ny; m++)
                        calculate I(x,y)
                    }
                    apply threshold on each pixel and
                    store the data
                }
            }
```

**Figure 4.** Serial C code

A time-complexity analysis of each section was done. The Inverse Discrete Fourier Series (IDFS) in section 4 was the most time consuming task. The per pixel reconstruction cost is $O(nm)$. Moreover thresholding has to be performed pixel-by-pixel. The runtime also increases with the reconstruction resolution expressed in terms of pixels/CD. Parallelization of the inverse DFS computation automatically results in a speed improvement for this thresholding operation, as the task is now divided among $N$ processors ($N$ is between 2 and 8 in our simulations). The only overhead involved is in combining the "slices" of the thresholded images by the main processor (rank zero). In other words, the non-rank zero processors write their respective slices to the disk, which are then combined in the right order by the rank zero processor.

## Task of Parallelization

As evident from the code listing in the Figure 4, the algorithm lends itself to a parallel implementation. Parallelization is accomplished in the reconstruction section, as this is the most time consuming part of the algorithm. Parallelization of this section is particularly easy due to mutual independence of data, which also means practically no communication among the tasks. Parallelization is achieved using the message passing paradigm via the MPI libraries. Use of MPI libraries also facilitates architecture independence and hence ensures portability.

Since the parallel code is implemented on homogenous systems, load sharing is not a problem. The load is distributed equally among the available processors. As mentioned earlier only section 4 was computationally intensive and hence the other sections did not warrant a parallel implementation. In section 4, the *for* loop for *x,* is divided into N independent segments, with each of the N processors performing *reconstruction* as well as *thresholding* over its specified region. This is pictorially depicted in Figure 5 below.

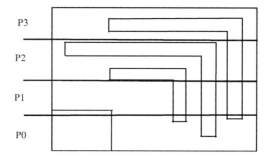

**Figure 5.** Division of pattern simulation among 4 processors

If the reconstruction resolution is 4 pixels per CD (***pixel = 4***), and N = 4, then each processor can operate on one quarter of the data, independently. This reduces the runtime by almost a factor of four in comparison to the serial code (Table 1). The runtime reduction is not exactly a factor of four due to the fact that there exists a small segment of serial code before the parallel implementations in section-4. However it was observed that the parallelized segment was scalable linearly with the degree of parallelization (N).

As mentioned earlier, the computation time increases with increase in resolution for sequential implementations. However, in the parallel implementation, if an equivalent increase in N accompanies increase in the resolution, runtime remains unchanged. In other words, if the variable ***pixel*** is increased to 8 and N is also set to 8, the runtime remains almost the same as before. In any case, multiple processors ensure load-sharing resulting in a much better runtime performance as compared to serial C code and interpreted Mathcad code.

## Discussion on Memory Utilization

The implementation of this algorithm requires large amounts of memory for real-world patterns (Pentium chip!). Two modes of operation are possible: duplication mode and non-duplication mode. In the duplication mode, the rank zero processor broadcasts the preliminary data to all the other processors. This, though suitable for a Parallel RAM (PRAM) configuration, is not advisable for a multiprocessor homogenous system with a shared memory (the exact configuration may vary). In the non-duplication mode, the rank-zero processor does all the computations up to section 4, after which it broadcasts the data to the remaining processors, which can then carry out independent parallel computations. Further refinements are possible: only the data segment relevant to a given processor (in an equal-load sharing configuration, 1/N times the total data) can be sent. This will also reduce the communication load on the shared bus, which can be substantial for large patterns with very high resolutions. Shared memory implementations can also eliminate communications altogether by smart memory management with the use of pointers and offsets. However, this may complicate memory accessing, necessitating scheduling schemes.

# Results

Results of the experiments conducted on the test pattern are summarized in this section. The experiments were conducted under no load conditions. A brief description of the computing environments used in each case is presented below.

- IBM SP2 with sixty-four RS6000 processor (running at 66MHz) with 64MB RAM using IBM AIX 4.1.5 operating system with PSSP 2.2. Nodes are interconnected by High Performance switch with a data transfer rate of 40MB/sec.
- SGI Onyx 2, a four node 8 processor homogenous system. Has R10000 processors (running at 195MHz) with 1GB RAM using IRIX 6.5 operating system.
- Linux cluster of 10 nodes. Each node has a Pentium MMX (running at 233MHz) with 128MB RAM, using LINUX 2.0 operating system. Nodes are interconnected via 10/100Base T Ethernet switches.
- IBM Thinkpad with a 300MHz Pentium, running on Windows 95 operating system. Mathcad code implementation was done on this system.

The test pattern with 7 inner rectangles is as shown in Figure 2. The reconstructed image resolution is set to 12 pixels/CD. SP2 has 66MHz nodes which is slower when compared to the other environments of experimentation. Nevertheless, it should be mentioned that the large number of nodes in SP2 makes it a better system to implement efficient parallelization. Of the 64 nodes, 4 nodes are configured for interactive jobs and others are configured for batch processing. The experiments were conducted on the batch processing nodes.

Table 1 shows the comparison of runtimes in various environments. All times are in seconds. As the pattern size increases the simulation time starts to increase. The turn around time is large for a real life pattern (which is typically greater than 240 x 240 ). These cases may take hours together to simulate. Parallelization of such process is thereby essential for reduced turn around time and faster results.

| 14CD x 12CD 12pixels/CD | SGI Onyx 2 | Linux Cluster | IBM SP2 | Mathcad |
|---|---|---|---|---|
| Serial C code | 154 | 341 | 1208 | 180 |
| Two Processor | 83 | 189 | 660 | - |
| Four Processor | 49 | 121 | 360 | - |
| Eight Processor | 35 | 82 | 205 | - |

Table 1. Comparision of runtimes on various environments

| 14CD x 12CD | SGI Onyx | Linux cluster |
|---|---|---|
| 4 pixels/CD and 4 processors | 30 | 76 |
| 8 pixels/CD and 8 processors | 32 | 79 |

Table 2. Variation of resolution with N.

Table 2 illustrates the almost constant simulation times (in seconds), if N is increased by the same factor as the required reconstruction resolution.

The graph of runtime against degree of parallelization (1/N) is as shown below. The graph illustrates the linear dependence of the simulation runtime with degree of parallelization.

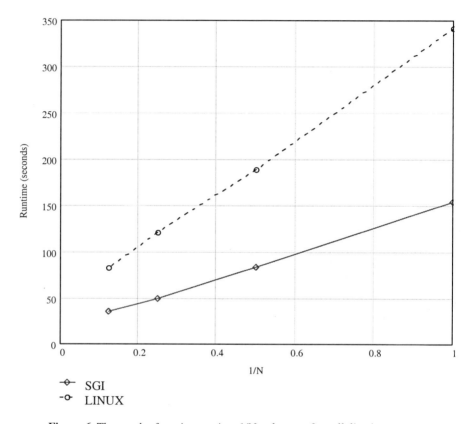

**Figure 6.** The graph of runtime against 1/N – degree of parallelization

Figure 7. below shows a comparison of a simulated pattern with a practical experiment pattern.

**Figure 7.** Comparison of an experimental result with its corresponding simulation result.

## Conclusion

An MPI parallel programming paradigm-based scheme was implemented for a VLSI simulation patterning problem. The implementation was successful in reducing the computation time by a factor determined by the degree of parallelization ($N$). The linear dependence between runtime and degree of parallelization was also illustrated. Strategies for efficient resource utilization were also discussed. Further improvements in speed can be achieved by employing parallel FFT routines. The techniques described here being generic can be applied to similar lithographic problems.

## References

[1] P. S. Pacheco, "Parallel Programming with MPI", Morgan-Kaufmann Publishers, 1997.

[2] S. H. Zaidi, S. R. J. Brueck, "Multiple Exposure Interferometric Lithography," J. Vac. Sc.,Tech, B11, 658-666, 1993.

[3] X. Chen, Z. Zhang, S. R. J. Brueck, R. A. Carpio, J. S. Peterson, "Process Development for 180 nm Structures using Interferometric Lithography and I-Line Photoresist," Proc. SPIE 3048, 309, 1997.

[4] X. Chen, S. R. J. Brueck, "Imaging Interferometric Lithography – A Wavelength Division Multiplex Approach to Extending Optical Lithography," J. Vac. Sc. Technol., to be published Nov./Dec. 1998.

[5] N. Magotra, S. Divakar, C. Tu, S.R.J. Brueck, X. Chen, "Digital Image Processing Applied to Imaging Interferometric Lithography", IEEE Asilomar Conference on Signals and Systems, Monterey Calif., 1998.

[6] Semiconductors Industries Association (http://www.sematech.org/), "National Technology Roadmap for Semiconductors", 1997.

[7] X. Chen, "A study of Interferometric lithography – Approaching the linear systems of optics", Ph.D. dissertation, EECE department, UNM, Aug 1998.

# Parallelizing a High Resolution Operational Ocean Model

Josef Schüle[1] and Tomas Wilhelmsson[2]

[1] Institute for Scientific Computing, Technical University Braunschweig, D-38092 Braunschweig, Germany, j.schuele@tu-bs.de
[2] Department of Numerical Analysis and Computing Science, Royal Institute of Technology, S-100 44 Stockholm, Sweden, towil@nada.kth.se

**Abstract.** The Swedish Meterological and Hydrological Institute (SMHI) makes daily forecasts of temperature, salinity, water level, and ice conditions in the Baltic Sea. These forecasts are based on data from a High Resolution Operational Model for the Baltic (HIROMB). This application has been parallelized and ported from a CRAY C90 to a CRAY T3E.

Our parallelization strategy is based on a subdivision of the computational grid into a set of smaller rectangular grid blocks which are distributed onto the parallel processors. The model will run with three grid resolutions, where the coarser grids produce boundary values for the finer. The linear equation systems for water level and ice dynamics are solved with a distributed multi-frontal solver.

We find that the production of HIROMB forecasts can successfully be moved from C90 to T3E while increasing resolution from 3 to 1 nautical mile. Though 5 processors of the T3E are 2.2 times faster than a C90 vector processor, speedup and load balance could be further improved.

## 1 Introduction

The Swedish Meterological and Hydrological Institute (SMHI) makes daily forecasts of currents, temperature, salinity, water level, and ice conditions in the Baltic Sea. These forecasts are based on data from a High Resolution Operational Model of the Baltic Sea (HIROMB) currently calculated on one processor CRAY C90 parallel shared memory vector computer. Within the HIROMB project [1], the German Federal Maritime and Hydrographic Agency (BSH) and SMHI have developed an operational ocean model, which covers the North Sea and the Baltic Sea region with a horizontal resolution from 3 to 12 nautical miles (nm). This application has to be ported to the distributed memory parallel CRAY T3E. This will save operation expenses and even more important allow a refinement of the grid resolution to 1 nm within acceptable execution times.

# 2 HIROMB: A High Resolution Operational Model for the Baltic Sea

The operational HIROMB code is loosely coupled via disk I/O with the atmospheric model HIRLAM [2]. Atmospheric pressure, velocity and direction of wind, humidity and temperature, all at sea level, together with cloud coverage are input, while sea level, currents, salinity, temperature, and coverage, thickness and direction of ice are output. HIROMB is run once daily and uses the latest forecast from HIRLAM as input. There are plans to couple the models more tightly together in the future.

Specifying tidal level and sea level at the open boundary between the North Sea and the North Atlantic account for the influence of the water level in the North Atlantic. The sea level is provided by a storm surge model covering the North Atlantic. Fresh water inflow is regarded at 80 major river outlets.

## 2.1 Grids

The 3 nm grid, see Figure 1, covers the waters east of 6° E and includes the Skagerrak, Kattegat, Belt Sea and Baltic Sea. Boundary values for the open western border is provided by a coarser 12 nm grid covering the whole North Sea and Baltic Sea region, see Figure 2. In the vertical, there is a variable resolution starting at 4 m for the mixed layer and gradually increasing to 60 m for the deeper layers.

All interaction between the two grids is taking place at the western edge of the finer grid where values for flux, temperature, salinity, and ice properties are interpolated and exchanged.

## 2.2 Model Description

We can essentially identify three parts in the model, the baroclinic part, the barotropic part, and the ice dynamics.

**Baroclinic Part** Water temperature and salinity are calculated for the whole sea including all depth levels. Explicit two-level time-stepping is used for horizontal diffusion and advection. Vertical exchange of momentum, salinity, and temperature are computed implicitly. Temperature and salinity both obey the same physical laws and are advected using the same subroutine tflow. Assuming an average depth $\bar{k}$, the baroclinic part has a complexity of

$$\mathcal{O}(surface\_points \times \bar{k}).$$

**Barotropic Part** A semi-implicit scheme is used for the vertically integrated flow, resulting in a system of linear equations (the Helmholtz equations) over the whole surface for water level changes. This system is sparse and non-symmetric, reflecting the 9-point stencil used to discretize the differential equations over

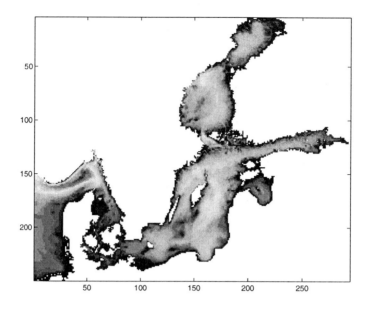

**Figure 1.** The grid with 3 nautical mile resolution covers the Baltic Sea together with Skagerrak, Kattegat, and Belt Sea. The maximum depth south of the Norwegian coast is over 700 meters. There are 1 785 168 active grid points of which 74 382 are on the surface.

the water surface. It is factorized with a direct solver once at the start of the simulation and then solved for a new right-hand side in each time step. The complexity of the barotropic part is estimated to

$$\mathcal{O}(surface\_points).$$

**Ice Dynamics** Ice dynamics include ice formation and melting, changes in ice thickness and compactness, and is taking place on a very slow time scale. The equations are highly nonlinear and are solved with Newton iterations using a sequence of linearizations. A new equation system is factorized and solved in each iteration using a direct sparse solver. Convergence is achieved after at most a dozen iterations. The linear equation systems are typically small, containing only surface points in the eastern and northern part of the Baltic Sea. In the mid winter season however, the time spent in ice dynamics calculations may dominate the whole computation time.

Neither the ice coverage nor the number of iterations necessary to solve the nonlinear equation system is known a priori. Thus, the effective complexity, that is linear dependent on the number of ice points,

$$\mathcal{O}(ice\_points)$$

may vary on a large scale.

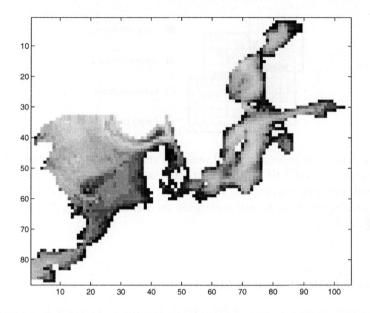

**Figure 2.** The grid with 12 nautical mile resolution covers the Baltic Sea and the whole North Sea. It is used to provide boundary values to the 3 nautical mile grid. There are 221 760 active grid points of which 9 240 are on the surface.

## 3 Reorganization of the Code

As an initial effort it was decided to keep all algorithms and numerics in the parallel version as close as possible to the sequential version of HIROMB. This was possible since we had access to a direct sparse solver of unsymmetric matrices for distributed memory machines, see Section 3.1. Our parallelization strategy is based on a subdivision of the computational grid into a set of smaller rectangular grid blocks which are distributed onto the parallel processors, see Section 3.3.

Serial HIROMB is written in Fortran 77. All new code added for parallel HIROMB make use of Fortran 90 features while much of the remaining original code is still Fortran 77. Fortran 90 pointers are used to switch context between blocks, see Section 3.4. The Message Passing Interface (MPI) library [3] is used for updating block boundaries.

Software tools for handling grids divided into blocks in this manner are available, notably the KeLP package from UCSD [4]. KeLP is a C++ class library providing run-time support for irregular decompositions of data organized into blocks. Unfortunately, we were not aware of this package when the HIROMB parallelization project started.

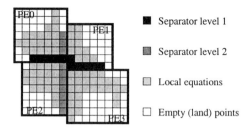

**Figure 3.** This four processor (PE) example shows how equations are classified. Equations on the first half of the PEs that involve equations on the second half are denoted level one separators. Recursively, equations on PE 0 involving PE 1 are level two separators as are equations on PE 2 involving PE 3. All other equations are local.

### 3.1 Parallel Direct Sparse Matrix Solution

In the original, serial version of HIROMB, the linear systems of equations for the water level and the ice dynamics were solved with a direct solver from the Yale Sparse Matrix Package (YSMP) [5]. For parallel HIROMB we employ a distributed multi-frontal solver written by Bruce Herndon at Stanford University [6]. This solver was originally used in a semiconductor device simulator.

The multi-frontal method seeks to take a poorly structured sparse factorization and transform it into a series of smaller dense factorizations. These dense eliminations can then efficiently be done by well known methods for dense systems. Many distributed sparse solvers have their own preferred decomposition. An advantage of Herndon's solver is that it accepts and uses the application's decomposition directly. Even with this non-optimal decomposition, the solver provides excellent solution capabilities. A disadvantage of the solver, however, is that the number of processors has to be a power of two.

The distributed factorization is based on a power-of-2 decomposition of the matrix. Equations that refer to values on remote processors are identified and ordered hierarchically into a global elimination tree as shown in Figures 3 and 4. These so called separator equations are shared among all processors located below in the hierarchy of separator levels.

The processors factorize their own local portions of the matrix independently and then cooperate with each other to factorize the shared portions of the matrix. One half of the processors will take part in factorizing shared equations at the lowest separator level, then a quarter of them will factorize the next separator level, and so on, until only one processor factorizes the remaining equations at the root of the elimination tree.

If the fraction of shared equations is large, performance will suffer due to lack of parallelism when factorizing shared equations at the higher separator levels. New software is becoming available which parallelizes this stage further. The PARASOL library [7] put special emphasis on the parallel factorization of equations remaining at the root of the elimination tree.

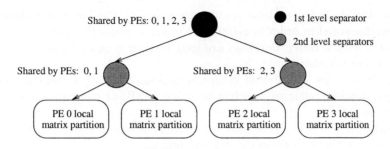

**Figure 4.** The separator equations are arranged hierarchically into an elimination tree.

## 3.2 Rectangular Arrays

The original serial vector code uses an indirect addressing storage scheme to store and treat only the grid points containing water. While efficient in terms of memory, this scheme gives low computational performance due to its irregular memory access pattern. As a starting point for parallelization, this scheme was replaced with a direct scheme storing the whole three-dimensional grid block. Land points are marked as inactive.

In a first version, the water depth was kept constant at the maximal value. In a refined version, the depth is corrected to the maximal value for each block. This refinement reduced memory requirements considerably and increased performance (see Table 1).

## 3.3 Decomposition into Blocks

In a rectangular block covering the whole 3 nm grid only 25% of the surface points and less than 10% of the volume points are active. By decomposing the grid into a set of smaller rectangular blocks and discarding all blocks without surface water points, it is possible to reduce the fraction of inactive points significantly. These remaining blocks are assigned to the processors in a parallel environment. We only cut the grid along horizontal dimensions and not between the surface and the sea bed.

The geometry of the Baltic Sea forces the blocks to have different sizes, shapes, and depths in order to obtain a good load balance and a minimal number of inter-block communication points. The more blocks the grid is decomposed into, the more inactive points can be discarded. On the other hand, overhead from switching block context and updating block boundaries will increase as the surface to volume ratio increases. Hence there is an optimal number of blocks which minimizes the total execution time.

We use a domain decomposition package written by Jarmo Rantakokko at Uppsala University [8]. The algorithm is summarized below:

1. The domain is halved into equally sized blocks. The blocks are shrinked or even discarded in order to reduce the number of inactive points.
2. Blocks with a heavy workload are split further in order to get a more even workload ($\rightarrow$ step 1).
3. The blocks are distributed onto the processors using a data distribution algorithm like recursive spectral bisection (RSB) or recursive coordinate bisection (RCB).
4. Blocks on the same processor are combined and merged together if it is possible without introducing inactive points.
5. As a final step, the number of inactive points are reduced further by shrinking the blocks where it is possible.

The estimated workload in each block is based on a performance model that takes into account both the number of surface as well as volume grid points. In the winter the fraction of ice coverage is included as well.

Currently we use the same decomposition for both water and ice calculations. Thus the decomposition during winters will have to be a compromise with unavoidable load inbalances. This is subject to change. Usage of a separate decomposition for the ice dynamics is in progress. Still, the decomposition will be a compromise between the baroclinic and the barotropic part that show different complexities (see 2.2).

An example of the 3 nm grid decomposed with this algorithm into 24 blocks on 16 processors is shown in Figure 5. One layer of ghost points is added around each block.

## 3.4 Multi-Block Memory Management

For parallel HIROMB we wanted to introduce the possibility to treat several blocks per task while still keeping as much as possible of the original code intact. This is done by replacing all fields in the original code by Fortran 90 pointers. By reassigning these pointers, the current block context can be switched.

The `ExchangeBoundaries` routine updates ghost points among blocks. If a neighbor block is on the same processor, boundary values are copied without communication calls. For remote neighbor blocks we use buffered sends followed by receives to avoid dead-locks.

When ghost points need to be updated within an original routine, the routine is split into several parts. As an example, the large routine `flux` had to be split into six pieces with calls to `ExchangeBoundaries` in between. Note that each of the `flux[1-6]` routines consists of unmodified Fortran 77 code, and may still be compiled with a Fortran 77 compiler.

All arrays with the same dimension and type are stored together. This allows us to minimize the number of MPI calls when exchanging boundaries between blocks without explicitly packing of messages. The MPI derived type mechanism could have been used instead if separate data types are created for each combination of fields present in the calls to `ExchangeBoundaries`.

**Figure 5.** This is a decomposition of the 3 nautical mile grid into 24 blocks distributed onto 16 color coded processors. No ice was present in this case.

In parallel HIROMB the block assignment is accompanied by a grid assignment. The routine `SetCurrentGrid` switches context between grids. Data exchanges between different grids are performed by the master processor. The decompositions for each grid are completely independent and do not regard geographical localizations. More grids can be added with just minor code changes.

### 3.5 Distributing input data and assembling output data

Parallel HIROMB uses the same input and output files as serial HIROMB. File I/O is performed by the master processor and is to a large extent overlapped with computations by the workers. This is very important to achieve parallel efficiency. Due the constraints of Herndon's solver (see 3.1) the workers need to be a power of two. Hence, including the master, $2^n + 1$ processors are used in the simulations.

The master process also updates and distributes external influences to worker data. This prevents an entire overlapping of file I/O with computations by the workers.

## 4 Timings and Results

Table 1 illustrates the performance obtained so far on a CRAY T3E-600. These timings were done for a full 12 hour simulation with data output every 3 hours.

| | One time step only | | Full simulation with I/O | | | |
|---|---|---|---|---|---|---|
| PEs | Time (sec) | S | T_1 (min) | T_2 (min) | S | Ideal S |
| 5 | 7.74 | 1.0 | 10.0 | 8.8 | 1.0 | 1.0 |
| 9 | 4.32 | 1.8 | 5.9 | 5.6 | 1.7 | 2.0 |
| 17 | 2.76 | 2.8 | 4.1 | 3.4 | 2.5 | 4.0 |
| 33 | 1.85 | 4.2 | 3.0 | | 3.4 | 8.0 |
| 65 | 1.31 | 5.9 | 2.3 | | 4.3 | 16.0 |

**Table 1.** These timings were done for one time step without I/O and for a full 12 hour simulation with data output every 3 hours using a CRAY T3E-600. At least 4 worker processors (5 with master) were necessary due to memory constraints. No ice was present in this data set from August 11, 1998. S refers to the speedup normalized to the 5 PE configuration. T_1 refers to calculations with the global maximal depth $k_{max}$, while for T_2 each block used its minimal depth.

No ice was present in this data set.

The original serial HIROMB code runs the same example in 20.2 minutes on one C90 processor. This rather poor performance is partly due to the indirect addressing scheme employed in the original code. During the course of porting to the T3E, we were able to optimize one of the dominating routines in the original HIROMB version, tflow, making it eight times faster. On a SGI Power Challenge XL, this reduced the total simulation time by 50%. This optimization has not been introduced in HIROMB on the C90.

Table 1 shows, that parallel HIROMB on 5 T3E-processors runs twice as fast as the original code on one C90 processor with global depth (T_1) and another 10% faster when the minimal depth for each block is used instead (T_2).

However, the speedup is far from ideal. This has several causes. The most important one is that the parameters in the decomposition algorithm has not been tuned optimally yet which creates a non-optimal load balance. This will remain challenging, as the baroclinic and barotropic parts do have different complexities. During winters the inhomogeneous workload from the ice dynamics makes the situation even worse, since ice formation is normally confined only to the eastern and northern part of the Baltic Sea. A separate decomposition for the ice dynamics is currently in progress. Furthermore, the coarse 12 nm grid is only distributed onto 8 processors and takes 0.49 seconds per time step. In the 65 processor case, this amounts to 37% of the total computation time. Finally, I/O operations on the master processor is currently not fully overlapped with computations on the worker processors. This is obvious when comparing the speedups with and without I/O.

## 5 Conclusions and Future Work

Moving the production of HIROMB forecasts from one C90 processor to 17 T3E-processors will reduce the elapsed time for making a forecast by almost a factor of six. The timings obtained so far and the estimated memory requirements indicate that a 1 nm grid computation can be successfully run in production on the T3E. There are still many possible enhancements that can be considered, for example:

- A better workload model for the decomposition algorithm.
- Introduce a separate distribution of the ice coverage to get a good load balance also for the ice dynamics calculation. This is currently in progress.
- Switch to the PARASOL sparse solver when it becomes available.
- Extend parallelism by running ice dynamics and water properties and the different grids concurrently instead of sequentially.

## Acknowledgments

The authors would like to thank Lennart Funkquist at SMHI and Eckhard Kleine at BSH for their help with explaining the HIROMB model. Computer time was provided by the National Supercomputer Centre in Linköping, Sweden and by the Institute for Scientific Computing in Braunschweig, Germany.

## References

1. Lennart Funkquist and Eckhard Kleine. HIROMB, an introduction to an operational baroclinic model for the North Sea and Baltic Sea. Technical report, SMHI, Norrköping, Sweden, 199X. In manuscript.
2. Nils Gustafsson, editor. *The HIRLAM 2 Final Report*, HIRLAM Tech Rept. 9, Available from SMHI. S-60176 Norrköping, Sweden, 1993.
3. Mark Snir, Steve Otto, Steven Huss-Lederman, David W. Walker, and Jack Dongarra. *MPI: The Complete Reference*. MIT Press, Cambridge, Massachusetts, 1996. ISBN 0-262-69184-1.
4. S. J. Fink, S. R. Kohn, and S. B. Baden. Efficient run-time support for irregular block-structured applications. *J. Parallel and Distributed Computing*, 1998. To appear.
5. S. C. Eisenstat, H. C. Elman, M. H. Schultz, and A. H. Sherman. The (new) Yale sparse matrix package. In G. Birkhoff and A. Schoenstadt, editors, *Elliptic Problem Solvers II*, pages 45–52. Academic Press, 1994.
6. Bruce P. Herndon. *A Methodology for the Parallelization of PDE Solvers: Application to Semiconductor Device Physics*. PhD thesis, Sanford University, January 1996.
7. Patrick Amestoy, Iain Duff, Jean Yves L'Excellent, and Petr Plecháč. PARASOL An integrated programming environment for parallel sparse matrix solvers. Technical report, Department of Computation and Information, 1998.
8. Jarmo Rantakokko. A framework for partitioining domains with inhomogenous workload. Technical Report Report No. 194, Department of Scientific Computing, Uppsala University, Uppsala, Sweden, 1997.

# Weather and Climate Forecasts and Analyses at MHPCC

J. Roads[1], S. Chen[1], C. McCord[2], W. Smith[2],
D. Stevens[3], H. Juang[4], F. Fujioka[5]

[1] Scripps Institution of Oceanography
Experimental Climate Prediction Center
UCSD, 0224
La Jolla, CA 92093
jroads@ucsd.edu, schen@ucsd.edu

[2] Maui High Performance Computing Center
Kihei, Maui, HI
carol@mhpcc.edu, williams@mhpcc.edu

[3] Univ. of Hawaii
Honolulu, HI
dstevens@soest.hawaii.edu

[4] National Centers for Environmental Prediction
Camp Springs, MD
Henry.Juang@noaa.gov

[5] US Forest Service
Riverside CA
ffujioka/psw_rfl@fs.fed.us

In Hawaii, where weather and climate variations are strongly affected by the steep island topography, there is a clear and acknowledged need for improved weather and climate forecasts at increased spatial resolution. The Hawaii Weather, Climate,Modeling Ohana (HWCMO) was therefore formed at the Maui High Performance Computing Center in 1997 to establish an experimental weather/climate mesoscale modeling effort for near real-time support to the local National Weather service (NWS), decision makers in federal and state agencies, and local researchers. This operational mesoscale forecasting effort is currently providing on an almost regular schedule, daily forecast products (http://www.mhpcc.edu/ wswx/), using 5 nodesof the MHPCC multi-node IBM SP2 cluster.

## 1   Introduction

The goal of the Hawaii Weather, Climate, Modeling Ohana (HWCMO) is to foster experimentation and development of applications for weather/climate issues in the Hawaii/Pacific area, especially over the islands of Hawaii. Towards this end, an operational mesoscale forecast model, the NCEP mesoscale spectral model (MSM), has been ported to the Maui High Performance Computing Center (MHPCC). This model is now run on an experimental basis in near-real time. The model and associated forecast products are described below.

## 2 MSM

HWCMO's global to regional modeling system, originally developed by NOAA's National Centers for Environmental Prediction (NOAA/NCEP), contains two components - a low-resolution global spectral model (GSM; Kalnay et al. 1996; Roads et al. 1998a) and a high-resolution regional non-hydrostatic model (MSM; Juang, 1997), which is an outgrowth of an earlier hydrostatic regional spectral model (RSM; 1996). The GSM and MSM basically use the same equations. Similar equations include the vorticity, divergence, temperature, mass and water vapor conservation equations on 28 sigma levels as well as several subsurface layers. All models also use the same comprehensive set of physical parameterizations, and the same diagnostic and output packages. The MSM (and RSM) momentum equations differ from the GSM in that the momentum equations as opposed to the vorticity/divergence equations are used. The major difference between the GSM and MSM is that high-resolution non-hydrostatic mesoscale model (MSM) uses a more complete vertical velocity equation instead of assuming the hydrostatic equation of the GSM. The MSM (and RSM) also expresses the model variables in terms of spectral perturbations (or adjustments) to the specified large-scale fields. For the MSM (and RSM) forecasts, the regional perturbations are converted to a gridpoint representation and the sum of the global and perturbation values in physical space become the output fields.

## 3 Products

The MSM is already fully operational on individual nodes at MHPCC. We use 5 nodes to generate (almost regularly) daily products for 5 model domains, including: a 10 km domain covering all of the islands of the state of Hawaii (Archipelago), Hawaii County at 4 km resolution, Maui County at 3 km resolution, Oahu County at 2 km resolution, Kauai County at 2 km resolution. We are currently displaying 24 hour forecasts (almost every day), at 3-hour increments, at http://www.mhpcc.edu/~wswx, 2m temperature, 2m relative humidity, 10 windspeed, precipitation, soil moisture, and a derived fireweather index using these variables (Roads et al. 1997). Fig. 1 shows an example of the fireweather index (shaded) and surface wind vectors for all the islands. Higher resolution forecasts are available for individual island

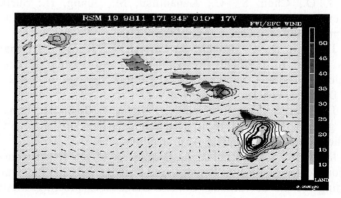

## 4  Future Developments

We expect to increase computational speed by a factor of at least 10 once the MSM code has been partially parallelized to take advantage of the multi-node structure of the IBM SP2s at MHPCC. Parallelization will make longer forecasts available at shorter wall-clock time. We also plan to port the MSM to other regions of the pacific basin and may eventually use this system to model characteristics of intense tropical storms affecting the pacific islands. An embedded boundary layer model will also eventually be developed to provide 100-meter resolution of various winds and perhaps other surface meteorological features over specific regions.

We intend to continue indefinitely our current experimental forecasts. We will thus soon build up a critical comparison between the large-scale analysis and the downscaled forecasts for all the islands. These downscaled global analyses should provide useful information for researchers attempting to better understand the relationship of small-scale island topographic controls on large-scale climate variations.

## References

1. Chen, S. -C., Roads, J.O., Juang, H. H. -M., Kanamitsu, M.: California Precipitation Simulations in the Nested Spectral Model. J. Geophys. Res. (1998) (in press, special precipitation issue)
2. Juang, H. -M. H., Kanamitsu, M.:: The NMC nested regional spectral model. Mon. Wea. Rev. **122** (1994) 3-26
3. Juang, H. -M. H.: The EMC/NCEP mesoscale spectral model: A revised version of the nonhydrostatic regional spectral model. Mon. Wea. Rev., (1997) (submitted)
4. Juang, H.-M.H., Hong, S.Y., Kanamitsu,M.: The NCEP regional spectral model : An update. Bulletin of AMS **78** (1997) 2125-2143
5. Kalnay et al.: The NMC/NCAR Reanalysis project. Bull. Amer. Met. Soc. **77** (1996) 437- 471
6. Roads, J.O., Chen, S. -C. , Fujioka, F.M. , Juang, H., Kanamitsu, M.: Global to Regional Fire Weather Forecasts. Int. Forest Fire News **17** (1997) 33-37
7. Roads, J. O., Chen, S. -C., Kanamitsu, M., Juang, H.: Surface Water Characteristics in NCEP's Reanalysis and Global Spectral Model. J. Geophys. Res.-Atmos. (1998a) (in press, special GCIP Issue)
8. Roads, J., Chen, S. -C., Ritchie, J., Fujioka, F., Juang, H., Kanamitsu, M.: ECPC's global to regional fireweather forecast system. Proceedings of International IDNDR Conference:Early Warning Systems for the Reduction of Natural Disasters, Potsdam, Germany (1998)
9. Roads, J., Chen, S. -C. , Ritchie, J. , Fujioka, F., Juang,H., Kanamitsu, M.: ECPC's global to regional forecast system. Proceedings of European Conference on Applied Climatology. Vienna, Austria, (1998c)
10. Roads, J., Chen, S.-C., Ritchie, J.: Evaluation of the Experimental Climate Prediction Center's Global to Regional and Daily to Seasonal Prediction System. Proceedings of the 23$^{rd}$ Annual Climate Diagnostics Meeting. Miami, Florida (1998d)

# GeoFEM: High Performance Parallel FEM for Solid Earth

Kazuteru GARATANI[1], Hisashi NAKAMURA[1],
Hiroshi OKUDA[2] and Genki YAGAWA[3]

[1] Research Organization for Information Science & Technology (RIST),
1-18-16 Hamamatsucho Minato-ku Tokyo, JAPAN
k-garatani@tokyo.rist.or.jp, nakamura@tokyo.rist.or.jp,
[2] Yokohama National University, 79-5 Tokiwadai Hodogaya-ku Yokohama, JAPAN
okuda@typhoon.cm.me.ynu.ac.jp
[3] University of Tokyo, 7-3-1 Hongou Bunkyo-ku, JAPAN
yagawa@garlic.q.t.u-tokyo.ac.jp

**Abstract.** The science and technology agency has begun an "Earth Simulator" project from the fiscal year of 1997, which enables the forecast of various earth phenomena through the simulation of virtual earth placed in a supercomputer. The "GeoFEM" is a parallel finite element software to be run on the "Earth Simulator" to solve problems involving the solid earth and is being developed at RIST. This project is expected to be a breakthrough in bridging the geoscience and information science fields. In this paper, we briefly describe the "GeoFEM" project, then capability for large–scale analysis is discussed and simple example analyses are shown. At this stage, the largest linear elastic problem solved by "GeoFEM" is more than 100M (100,000,000) degree of freedoms on 1,000 PEs Hitachi SR2201 at University of Tokyo.

## 1 Introduction

In order to solve global environmental problems and to take measures against the natural disasters, the science and technology agency aims to predict the global change by three fields of research, global observation, global change process research and computer simulation. The agency began a five–year project from the fiscal year of 1997 to develop an "Earth Simulator"[1], which enables the forecast of various Earth phenomena through the simulation of "virtual Earth"placed in a supercomputer. The project is considered as one of the top priorities among our national scientific agendas.

The specific research topics of the project are consist of four major themes, development of "Earth Simulator", modeling and simulation for atmospheric or oceanic field, for solid earth field and development for large–scale parallel software. The "GeoFEM"[2] which is a name of the project and software system, deals with modeling and simulation of solid earth and development of large–scale parallel software system. The GeoFEM system[3] is a parallel finite element analysis system intended for multi–physics/multi–scale problems and is being developed at RIST, or Research Institute for Information Science and Technology.

Since there are models which are not completely established in the solid earth field, the simulation must be carried out on a trial and error basis. Therefore, a joint venture with the geoscience modeling research group is crucial for the development of a targeted simulation software system. The project is expected to be a breakthrough in bridging the geoscience and information science fields.

## 2 Overview of GeoFEM Project

### 2.1 Grand Design of GeoFEM

The GeoFEM group is planning to develop the software system in the following two phases.

**(1) GeoFEM/Tiger (1997–1998)**
This phase involves the development of a multi–purpose parallel finite element software which may be applied to various scientific fields as well as becoming the basis for the "Earth Simulator" program to be developed in the next phase. A platform where various solid earth models may be read in a plug–in style, will be developed.

**(2) GeoFEM/Snake (1999–2001)**
A software system specialized in the simulation of solid earth phenomena through the use of GeoFEM/Tiger will be developed in this phase.

### 2.2 Objectives for GeoFEM/Tiger

**(1) Analysis Scale**
The linear problems of up to 100M (100,000,000) degree of freedoms (DOF) is to be solved in a practical time by current parallel computer.

**(2) Physical Phenomena**
Solid earth's physical phenomena to solve by GeoFEM/Tiger are shown at Table.1. The dynamics of the whole earth (global), which are mantle convection solved as heat–matter transfer phenomena of the viscous fluid will be simulated. In the Japanese islands scale (regional), the motion of the dislocation, buildup of tectonic stress, deformation of plates will be simulated by modeling them as elastic or viscoelastic. The simulation in the regional scale (local) will involve solving of the seismic wave propagation.

### 2.3 Developing Environment for the GeoFEM/Tiger

**(1) Hardware Environment**
"Earth Simulator" will be composed of hunreds of nodes connected by cross-bar network, each with several vectorized processors conjuncted by shared memory. As a whole system, it is believed that approximately more than 40 TFLOPS of peak processing speed will be achieved. It is also believed that about 4 terabytes of memory capacity will be supplied. However, until this "Earth Simulator" hardware is available, development of the software system will be carried

Table 1. Target of GeoFEM/Tiger

| Scale \ Phenomena | Mantle | Coupled | Plate | Wave |
|---|---|---|---|---|
| Global | ○ | ○ | ○ | |
| Regional | | | ○ | |
| Local | | | | ○ |

out on the existing hardware platforms. When "Earth Simulator" does become available for use, then the software will be improved for higher level of optimization.

**(2) Software Environment**
*Script Language for the Simulation Software*
For the main calculations, Fortran90 will be used. Control parallelization by the use of MPI as the message passing library will be implemented. However, for visualization and special maintenance of the memory, this may not be the case.
*Script Language for the Documents*
The design specifications for the source codes will be maintained in LaTeX.
*Script for the Interface between the Subsystems*
By providing the specifications for device–independent data transfer method, subsystems will be blocked from one another which will raise the level of independence at the time of the development.

**(3) Organization**
The members to develop GeoFEM system are put together at RIST. They are consist of researcher at university or institute and software engineer at private companies. Their status is endorsed by RIST and apply themselves to research and development.

## 2.4 System Composition of GeoFEM/Tiger

The system will be composed of the subsystems and integrated environment as illustrated in fig.1. Each subsystem is on the assumption that they will be linked in the memory by the device–independent interfaces. The copying of data will be avoided because of the large–scale data and pursuit of high speed computaion.

**(1) Partitioning Subsystem**
Graph partitioning is carried out based on the data from the large-scale complex CAD and mesh generator. The partitioning data, including the neighboring information, will be transferred to the succeeding subsystems. Partitioning elemental or nodal wise, and overlapped subdomains will be taken into consideration. Generally, partition with larger number of edge–cuts lead to bad convergence. In GeoFEM, multilevel methods combined with Kernighan-Lin optimization heuristics are developed as a graphical partitioner for the complicated geometries.

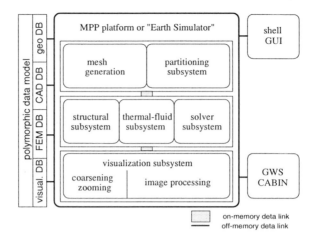

**Fig. 1.** System Configuration of GeoFEM/Tiger

**(2) Structural Analysis Subsystem**
Static or assumed time dependent algorithms will be used in the cases of linear elastic[4], viscoelastic, elastoplastic, to be able to handle various analysis conditions. Contact analysis[5] will be supported for the modeling of dislocations. Since the stiffness matrix is created locally and transferred to the solver subsystem in a form of a subroutine call, message passing is not apparent within the subsystem.

**(3) Thermal Flow Analysis Subsystem**
This subsystem will carry a function incompressible viscous flow analysis. Mantle convection is the movement of various kinds of material as a result of higher temperature in the inner part of earth. Equations which govern the process may be expressed using time, space and temperature dependent rheology or viscoelastic models. The present stage of the flow modeling "sub-system" of GeoFEM is carrying out of coupled heat and viscosity analysis by employing a parallelized FEM code. This subsystem also creates the matrix locally and transfers it to the solver, thus parallel procedure is not apparent as well.

**(4) Solver Subsystem**
In a large-scale scientific computing, linear sparse solver is one of the most time-consuming processes. In GeoFEM, various types of preconditioned iterative methods (CG, BiCGSTAB, GPBiCG, GMRES, TFQMR)[6] are implemented on massively parallel computers. It has been well-known that ILU(0) factorization is a very effective preconditioning method for an iterative solver. But it has also been well-known that this method is globally dependent data-wise and this is not the optimal methodology on parallel computers where locality is of utmost importance. In GeoFEM, "Localized" ILU(0) preconditioning method [7] has been implemented to the various types of iterative solvers[8], [9]. This method provides data locality on each processor and good parallelization effect.

## (5) Visualization Subsystem

Parallel image processing for the large-scale unstructured mesh data, including data reduction and extraction will be carried out in this subsystem[10]. Image processing calculation will be held in a same phase as the analysis for real–time visualization. As one of the image output device, CABIN, a property of the Intelligent Modeling Lab at the University of Tokyo, will be implemented[11].

The subsystems (1) through (5) will be controlled using shells. Furthermore, different types of data such as the geographical, CAD, FEM and visualization data will be be managed using a data structure model to be developed. An object-oriented system configuration will be made possible by integrating the subsystems in a device independent fashion[12].

## 3 Large Scale Analysis Feature of GeoFEM

In this section, large scale analysis feature of the GeoFEM structure sub–system is discussed.

### 3.1 Analysis Methods for Parallel FEM

For parallel FEM, the domain decomposition method (DDM) is generally used. In this method, we decompose the whole model to some subdomains to assign each PE. At GeoFEM system, the number of decomposed subdomains and PE count remain the same to keep large granularity. When solving degree of freedom (DOF), two types of DDM algorithm are taken. The one is to solve all DOF and the other is to solve boundary DOF. When we adopt the former, iterative solver such as C.G. method is usually used. The latter is still separated to two methods, those are to compose reduced matrix explicitly (substructure method, SSM) or to be converged boundary force by iterative solver. There are several kinds of DDM algorithm and each one has its merits and demerits[13],[14]. As far as large scale analyses are concerned, SSM is not suitable. GeoFEM adopts DDM solving all degrees of freedom by iterative method because of stability and applicability for nonlinear problems. In this method, proper partitioning and pre-conditioning are key technologies for stable and fast calculation. For the partitioning, Greedy or RCB method was adapted, and for the solver localized ICCG was selected in this stage.

### 3.2 Resource Estimation for Large Scale Analysis

In this section, we estimate computer resources for 100M DOF on linear elastic problem, when using the method described in the previous section. For FEM mesh, we use a simple unit cubic model as Fig.2. The model has unit length sides and it can easily change the DOF to vary division counts of each side. The memory capacity and the elapsed time problems are dominant to deal with large scale FEM analyses and most of all quantity is spent at solver. Therefore we estimate memory capacity and elapsed time required at solver.

## Memory Capacity

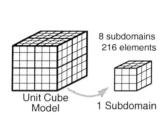

**Fig. 2.** Simplified FEM Model

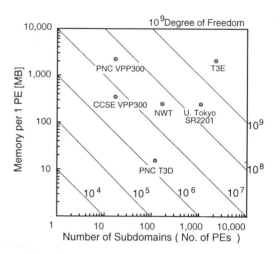

**Fig. 3.** Required Memory Estimation

At the structural analysis, memory estimation to use iterative solver is predicted by the total amount of memorizing whole stiffness matrix. The term count of the matrix is calculated by the count of connected node at each DOF. For example, term count at a DOF expressed by (*DOF per 1 node*)×(*connected node count*). When unknowns setup to displacement, then (*DOF per 1 node*) becomes 3, for the model in Fig.2; (*connected node count per 1 node*) is 27. And each term is required a double precision real and one integer type variable, it becomes 12 bytes per term. Therefore, the memory capacity to express whole stiffness matrix becomes 972×(DOF) [byte]. On taking other variables, the rough estimation is expressed below.

$$(\text{Required Memory at Solver}) = 1,000 \times (\text{DOF}) \; [\text{Byte}] \quad (1)$$

This relationship is shown in the Fig.3. In this figure, the total capacity of the memory becomes (*PE count*)×(*memory per 1PE*) considering DDM. When using SR2201 at Tokyo university which has 1,024 PE and 256MB memory per PE, we can compute 100M DOF analysis.

**Elapsed Time**

For elapsed time estimation, we measure wall clock for various DOF by single PE at SR2201. The model of calculation is represented in Fig.2 and the method for solver is ICCG. According to the parametric study, the iteration count expressed by power of DOF and the elapsed time per iteration is relative to DOF. The result is shown as the bellow equation.

$$(\text{Required Elapsed Time at Solver}) = 6.65 \times 10^{-5} \times (\text{DOF})^{1.321} \; [\text{sec.}] \quad (2)$$

1PE line of Fig.4 is plotted the upper equation. The figure includes up to 10,000 PE line and these lines are simply divided into one PE value by PE count assuming 100% parallel efficiency. As the result, when SR2201 with 1,024PE is used, we can analyze 100M DOF problem in 50 minutes. Although this estimation is one of the lower bound, however we can keep the efficiency at least dozens %, it will be analyzed within reasonable time.

### 3.3 Example Analysis

Fig.f5 shows scalability of the sample analyses on SR2201 at University of Tokyo up to 100M DOF. Those analysis fixed the nodal division count at a domain of the side to be 33, and varying PE counts 1, $2^3 = 8$, $3^3 = 27$, $4^3 = 64$, $5^3 = 125$, $6^3 = 216$, $10^3 = 1000$. For 1,000 PE analysis, number of DOF becomes $33^3 \times 3 \times 1,000 = 107,811,000$. Because of the Localized preconditioning, parallel efficiency limited about 60%. GeoFEM can show almost linear scalability.

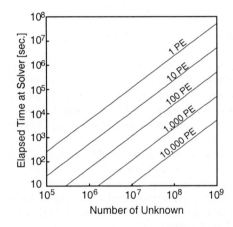

**Fig. 4.** Elapsed Time Estimation

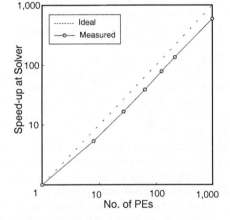

**Fig. 5.** Scalability of GeoFEM

## 4 Concluding Remarks

The overview of GeoFEM, a parallel finite element analysis system intended for the simulation of multi-physics/multi-scale problems involving the solid earth on an "Earth Simulator", and its capability for the large scale analysis were reported. The system is getting on development and maintenance aiming at publication and is already achieved 100M DOF analysis by linear elastic problem. GeoFEM/Tiger is a various objective parallel finite element software which is believed to become the basis in the parallel finite element analysis field by implementing new technologies to such factors as data structures, preconditioning under parallel environments, memory management to maximize the MPP calculations, parallel image processing of unstructured mesh data and object-oriented subsystem configuration.

## References

1. http://stagw.sta.go.jp./umi/seaearth/e9810_4.html
2. http://geofem.tokyo.rist.or.jp
3. G.Yagawa, H.Okuda and H.Nakamura : GeoFEM: Multi-Purpose Parallel FEM System for Solid Earth, Fourth World Congress on Computational Mechanics, Vol. II, (1988) pp.1048-
4. K.Garatani : GeoFEM: Multi-Purpose Parallel FEM for Solid Earth (3) Large Scale Structural Analysis for Solid Earth, Proceedings of the Conference on Computational Engineering and Science, JSCES, vol. **3** (1998)
5. M.Iizuka : GeoFEM: Multi-Purpose Parallel FEM for Solid Earth (4) Complex Earth Structure Analysis, Proceedings of the Conference on Computational Engineering and Science, JSCES, vol. **3** (1998)
6. J.J.Dongarra et al., *Templates for the Solution of Linear Systems: Building Blocks for Iterative Methods*, SIAM,1994
7. J.J.Dongarra et al. : *Solving Linear Systems on Vector and Shared Memory Computers*, SIAM,1990
8. K.Nakajima et. al. : Parallel Iterative Solvers with Localized ILU Preconditioning, HPCN'97 Proceedings, Lecture Notes in Computer Science, 1225 (1997) pp.342-350
9. K.Nakajima and H.Okuda : Parallel Iterative Solvers with Localized ILU Preconditioning for Unstructured Grids on Workstation Cluster, 4-th Japan-US Symposium on FEM in Large-Scale Computational Fluid Dynamics Proceedings (1998) pp.25-30
10. Y.Takeshima, I.Fujishiro : GeoFEM: Multi-Purpose Parallel FEM for Solid Earth (7) Unstructured Volume Visualization Approaches and Issues, Proceedings of the Conference on Computational Engineering and Science, JSCES, vol. **3** (1998)
11. I.Shirai, S.Yoshimura and G.Yagawa : Parallel Visualization of Ultra Large Analysis in Virtual Reality Environment, Proceedings of the Conference on Computational Engineering and Science, JSCES, vol. **3** (1998)
12. D.Sekita : GeoFEM: Multi-Purpose Parallel FEM for Solid Earth (6) A Strategy of the Inter/Intra System Interfaces *in Japanese*, Proceedings of the Conference on Computational Engineering and Science, JSCES, vol. **3** (1998)
13. K.Garatani, et. al. : Study on Parallelization Method of Structural-Analysis Code, HPCN'97 Proceedings, Lecture Notes in Computer Science 1225 (1997) pp.1044-1046
14. K.Garatani, H.Nakamura and G.Yagawa : Parallel Finite Element Structural Analysis Code Using DDM, HPCN'98 Proceedings, Lecture Notes in Computer Science 1401 (1998) pp.887-889

# Elastic Matching of Very Large Digital Images on High Performance Clusters

J. Modersitzki[1], W. Obelöer[2], O. Schmitt[3], and G. Lustig[2]

[1] Medical University of Lübeck, Institute of Mathematics,
Wallstraße 40, D-23560 Lübeck, Germany
modersitzki@math.mu-luebeck.de

[2] Medical University of Lübeck, Institute of Computer Engineering,
Ratzeburger Allee 160, D-23538 Lübeck, Germany
obeloeer@iti.mu-luebeck.de

[3] Medical University of Lübeck, Institute of Anatomy,
Ratzeburger Allee 160, D-23538 Lübeck, Germany
schmitt@anat.mu-luebeck.de

**Abstract.** The aim of the human neuroscanning project is to build an atlas of the human brain, based on a variety of image modalities in particular histological sections of a prepared brain. Reconstructing essential information out of deformed images is a key problem. We describe a method to correct elastic deformations. Since the method is computational expensive a parallel implementation is presented. The measurements and results shown are performed on a cluster of 48 Pentium II PCs connected via Myrinet.

## 1 Introduction

Reconstruction of deformed images is a basic problem within medical image processing (image registration). Especially, if images arise from a series of sections through a part of the human body, e.g. CT (computer tomography), MRI (magnetic resonance imaging), PET (positron emission tomography).

In this paper we concentrate on particular problems of the human neuroscanning project (HNSP) at the Medical University of Lübeck. The aim of the HNSP is to produce a three-dimensional map of a human brain based on different modalities, in particular cellular information. Here, the information is derived mainly from histological sections.

Typically, sectioning processes lead to deformed sections and consequently to deformed images. Although these deformations are small in general, they might become crucial for the reconstruction of cellular information in the human brain.

Two different approaches for correcting this kind of distortions are common. One approach is based on the idea of representing the unknown correction in terms of the coefficients of a fixed basis, such as piecewise linear functions or higher order splines. Typically these coefficients are determined by a least squares condition for some user prescribed landmarks (see, e.g., [3,4]). The second approach is based on the formulation of the problem via a non-linear partial differential equation (PDE). To solve these equations no further information on the underlying images, e.g. landmarks, is needed (see, e.g. [1, 17]).

We describe a method for correcting these kind of deformations based on a linear elasticity model leading to a non-linear PDE. This method is also used in other projects, e.g. [1], [2], [6], [7], [16].

To resolve details of the brain (e.g. neurons) very high resolution scans of the histological sections are needed. This leads to high dimensional problems. Due to memory and computational requirements of the method used a straightforward implementation can not be applied. Hence, we present a fast algorithm and its parallel implementation.

The HNSP project is described in the next section. The elasticity model is given in section 3, whereas section 4 presents our parallel implementation and performance measurements. The results and their medical discussion are given in section 5.

## 2 The Human Neuroscanning Project (HNSP)

The aim of the HNSP is the three-dimensional reconstruction of all cells of a human brain. These data should be used as the basic structure for the integration of functional data based on stochastic mapping and later on for modeling and simulation studies in such a virtual brain.

In this project a 55 year old male human post mortem brain of a voluntary donator of his corpse was fixed in a neutral buffered formaldehyde solution for 3 months. A MRI-scan of this brain was produced after fixation. Dehydration and embedding of the brain in paraffin required 3 further months. This preparatory work was followed by sectioning the brain in 20 $\mu m$ thick slices (about 7000 for this brain) by using a sliding microtome. Before each sectioning process a high resolution episcopic image ($1352 \times 1795$ pixels, 24 Bit, $7 \cdot 10^6$ Bytes) of the section plane was scanned.

After sectioning, the tissue slice was stretched in warm water at 55°C. Thereafter, it was transferred onto a microscopic slide and dried over night. The sectioning, stretching and drying processes are necessary in order to get flat tissue sections. However, this methodological steps produce non-linear deformations of each section (see Fig. 3). After drying, the sections were stained in gallocyanin chrome alum and mounted under cover-glasses in order to visualize all cells (special light microscope for analyzing large sections (LMAS)) in whole brain sections. Different neuronal entities were analyzed on different structural scales i.e. from macroscopic details down to the cellular level. In order to obtain a non-deformed stack of images a so-called elastic matching method was used. Elastic matching can be used also for multimodal matching of histological sections with non-deformed MRI-scans [14] or episcopic images. The later ones might be derived from image processing before sectioning the embedded brain. The stained sections were scanned by a transparent flat bed scanner using a resolution of 800 ppcm (or 2032 ppi) in a 8 Bit gray scale mode (size of the smallest image was $5000 \times 2000$ pixels, size of the largest image was $11000 \times 7000$ pixels (about 196 MBytes). The uncompressed amount of flat bed scanned data was approximately 700 GBytes + 40 GBytes episcopic data for one human brain.

In the following, the arbitrarily chosen sections 116 and 117 out of a total of about 7000 sections were matched. In the upper left corner of Fig. 3 the reference image (116) is shown and in the upper right corner the template image (117). In these images the left and the right hemisphere of the human brain are shown. These sections were obtained from the occipital lobe.

In this stage of the HNSP the images were scaled down from 6500 × 2300 to 512 × 512 pixels in order to keep the computation times reasonable. A straightforward rescaling based on bilinear interpolation was used.

## 3 Modeling non-linear deformations of two consecutive sections

We are looking for an elastic deformation of the template image ($T$) that simultaneously minimizes the difference between the deformed and the reference image ($R$) and the deformation energy

$$E(u,v) = \int_\Omega \frac{\lambda}{2}(u_x + v_y)^2 + \mu\left(u_x^2 + v_y^2 + \frac{1}{2}(u_y + v_x)^2\right) d(x,y),$$

where the so-called deformation field $(u,v) = (u(x,y), v(x,y))$ describes the local deformation and $\mu, \lambda$ are the so-called Lamé-constants, see, e.g., [10]. This approach enforces similarity of the images as well as connectivity of the tissue.

Applying the calculus of Euler-Lagrange we find that a minimizer is characterized by the so-called two-dimensional Navier-Lamé-equations (1), cf. e.g. [10],

$$\begin{pmatrix} f \\ g \end{pmatrix} = \begin{pmatrix} \mu(u_{xx} + u_{yy}) + (\lambda+\mu)(u_{xx} + v_{xy}) \\ \mu(v_{xx} + v_{yy}) + (\lambda+\mu)(u_{xy} + v_{yy}) \end{pmatrix} =: A \cdot \begin{pmatrix} u \\ v \end{pmatrix}. \quad (1)$$

Note, $(f,g)^\top$ (which might be viewed as a force field) depends non-linearly on the deformation, cf. eq. (2),

$$\begin{pmatrix} f \\ g \end{pmatrix} = \begin{pmatrix} \Big(T(x-u, y-v) - R(x,y)\Big) \cdot T_x(x-u, y-v) \\ \Big(T(x-u, y-v) - R(x,y)\Big) \cdot T_y(x-u, y-v) \end{pmatrix}. \quad (2)$$

An appropriate discretization of these equations finally leads to a fix-point type equation for the unknown deformation field, cf. eq (3),

$$A\big(u^{k+1}, v^{k+1}\big)^\top = \Big(f(u^k, v^k), g(u^k, v^k)\Big)^\top. \quad (3)$$

In principle, any solver can be used to compute the solution of eq. (3). However, a discretization with $m \times n$ points results in $N = 2mn$ unknowns (e.g. for 512 × 512 discretization points we end up with $N = 2^{19} = 524288$) and $A$ becomes $N \times N$. For a standard $LU$-decomposition one needs to store $\mathcal{O}(N^2)$ real numbers and approximately $\mathcal{O}(N^3)$ floating point operations. Thus, memory and computational requirements make a parallel implementation of an iterative solver for eq. (3) unavoidable.

# 4 Parallel realization and measurements

Due to a good price/performance ratio workstation clusters become an alternative to expensive dedicated parallel computers. Hence, our parallel implementation is performed on the so-called "Störtebeker Cluster" [13]. This cluster consists of 48 dual 333 MHz Pentium II nodes interconnected via Myrinet [5]. The operation system used is LINUX. For the measurements shown in Fig. 1, the parallel program uses PVM [12] as the underlying message passing system. PVM is used in order to support also different and heterogeneous platforms.

## 4.1 Parallel implementation of the elastic matching algorithm

An implementation of the parallel algorithm for $p$ processes can be divided into three parts (quantities used only locally are denoted with a subscript $_{\text{loc}}$). The repetition of part two and three is called *outer loop*. In the current implementation there is one process per node.

1. Partition the images $R$ and $T$ and the initial deformation field $(u^0, v^0)$ into $p$ stripes and distribute these stripes to $p$ processes. Every process sets $(f_{\text{loc}}^{-1}, g_{\text{loc}}^{-1}) = (0,0)$ and $k = 0$.
2. Every process applies the deformation $(u_{\text{loc}}^k, v_{\text{loc}}^k)$ to $T_{\text{loc}}$ and computes the forces $(f_{\text{loc}}^k, g_{\text{loc}}^k)$ *independently*.
   If the difference between the new and the old force field is sufficiently small, terminate.
3. Solve the linear system of equations (3) for the new deformation $(u^{k+1}, v^{k+1})$. Set $k \to k+1$ and continue with step 2.

As already pointed out, the main computational work is needed for solving eq. (3), which has to be done in any step. Here, a parallel implementation of the conjugate gradient method (CG) is used, cf. e.g. [9]. Performing this iterative scheme requires an additional so-called *inner loop*.

The basic structure of our implementation is given in Table 1. The essential computational and communication costs needed in one step of the inner CG iteration are next neighbor communication (exchange local stripes with two neighbors), two global sums (inner products $\approx 4N$ FLOPS), three local SAXPY operations (Scalar Alpha times $X$ Plus $Y$, $6N/p$ FLOPS for vectors $X, Y$ of length $N/p$), and one matrix vector multiplication (exploiting the special structure of the matrix $A$, this multiplication is $\mathcal{O}(N/p)$ FLOPS but depends on the particular discretization). Note, $N$ is the total number of unknowns (i.e. $N = 2mn$ for $m \times n$ discretization points), $p$ is the number of processes.

## 4.2 Measurements

In order to keep the measurement times reasonable, the number of steps in the outer loop as well as the maximum number of the CG steps in the inner loop is set to 50. Two measurement series with different numbers of nodes (the

**Table 1.** Principle phases of the parallel implementation of the CG-algorithm.

| (a) | computation | matrix-vector multiplication | $q_{loc} = Ap$ |
| --- | --- | --- | --- |
|     |             | inner product | $\alpha_{loc} = q_{loc}^T p_{loc}$ |
| (b) | communication | build and distribute global sum | $\alpha = \sum \alpha_{loc}$ |
| (c) | computation | two SAXPY's | $x_{loc} = x_{loc} + \alpha p_{loc}$ |
|     |             |             | $r_{loc} = r_{loc} - \alpha q_{loc}$ |
|     |             | inner product | $\beta_{loc} = r_{loc}^T r_{loc}$ |
| (d) | communication | build and distribute global sum | $\beta = \sum \beta_{loc}$ |
| (e) | computation | local vector operation | $p_{loc} = r_{loc} + \beta p_{loc}$ |
| (f) | communication | exchange local vectors | $p_{loc} \leftrightarrow$ right and left neighbor |

current implementation only supports one process per node) were performed, one matching of two 256 × 256 pixels images and a second one matching of two 512 × 512 pixels images.

The run time of the sequential version for the 256 × 256 pixels images is about 26 minutes, while the calculation of the 512 × 512 pixels images lasts about 100 minutes. Since the parallel version has a sequential part (i.e. starting the processes, initialization and distribution of the images etc.) it is clear that the speedup can not be linear (optimal). As expected, the speedup becomes better with larger problem sizes. The total run time is still very high, about 188 seconds on a 48 node cluster for the 512 × 512 pixels images.

To give inside into the program behavior three iteration steps of the CG-algorithm are shown in Figure 2. For observation and evaluation of the behavior the performance monitoring tool DELTA-T [11] is used. The *System Load* and computation/communication phases of two arbitrarily chosen processes are shown for an eight node configuration (i.e. eight processes). Figure 2 shows about 0.25 seconds of the execution time. In particular, three principle phases of the CG-algorithm as introduced in Table 1 are displayed.

The upper curve in Figure 2 shows the *System Load*. Here, 100% indicates that all eight processors are working concurrently. An average utilization of nearly 100% is reached, i.e. the parallel implementation of the CG algorithm is able to use almost the full system power. The two Gantt graphs of processes four and eight show calculation phases (black) and communication phases (white). Typically, more than 50% of the execution time of one CG step is taken by phase (a). Building the global sums (phases (b) and (d)) takes more time than exchanging the local vectors in phase (f), although much more data have to be transferred in phase (f).

If more computation nodes are used, the calculation time of phases (a), (c), and (e) is reduced. In contrast, the time for the communication phases (b) and (d)

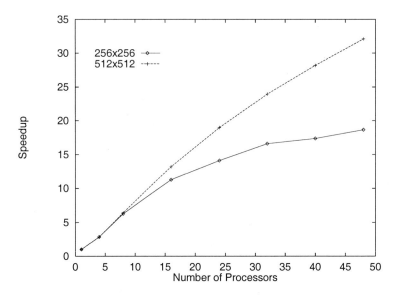

**Fig. 1.** Speedup for matches of 256 × 256 pixels images and 512 × 512 pixels images.

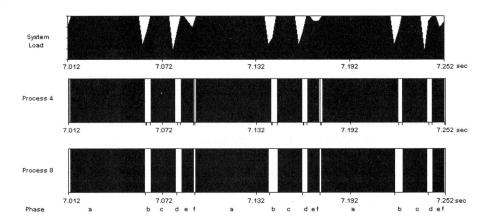

**Fig. 2.** *System Load* and computation/communication phases of the arbitrarily chosen process 4 and process 8 for an eight node configuration.

is increased. For larger numbers of nodes this leads to a lower efficiency. This might become crucial while matching two images of relatively small sizes. However, the images to be matched in the HNSP project have a high resolution and thus large sizes.

## 5 Results

Comparing the template 117 with the reference image 116 shows a difference of $||T - R||_2 \approx 10361$ in the Euclidean norm, cf. Fig. 3. After applying 500 steps of the outer loop of the elastic matching algorithm the difference has been reduced to $||T_{500} - R||_2 \approx 3123$, which is ca. 30.1%.

¿From the morphological point of view this can be considered as an adequate result. Irrespectively to the fact that there exists no other method for solving this kind of deformation problem, this technique can be considered as a sophisticated procedure to match images of tissue sections of the whole human brain.

## 6 Conclusion

The presented method allows the reconstruction of deformed images. The technique is also useful in order to align images from other modalities like PET, MRI, EEG (electroencephalography). In addition this method can be adapted easily to multimodal image matching. Moreover, this method can be extended to a variety of matching problems.

The parallel implementation makes the approach applicable and attractive for medical image processing applications, especially those from the human neuroscanning project. The parallel implementation allows the matching of images resulting from high resolution sections. Thus, linear artifacts arising from downscaling and bilinear interpolation can be reduced using large sized images in the discrete non-linear elasticity model.

¿From the medical point of view, further work has to be performed by investigating and interpreting the results. Up to now we do not know what is actually the best match in medical sense, what is undermatched and which result is overmatched (corrected template and reference are identical). ¿From the mathematical point of view a convergence proof of the overall algorithm is under work.

However, the method is a promising tool for the reconstruction process. In the future we will also parallelize the method using a multigrid solver and in particular a direct solver based on fast Fourier-type techniques (FFT) [8]. Therefore, different moduls for solving the system of linear equations with appropriate parallelization strategies have to be supplied.

Our implementation demonstrates that the elastic matching algorithm produces promising results in a reasonable amount of time on a high performance cluster system.

**Acknowledgements** We are indepted to A. Folkers and J.-M. Frahm for implementing parts of the algortithm.

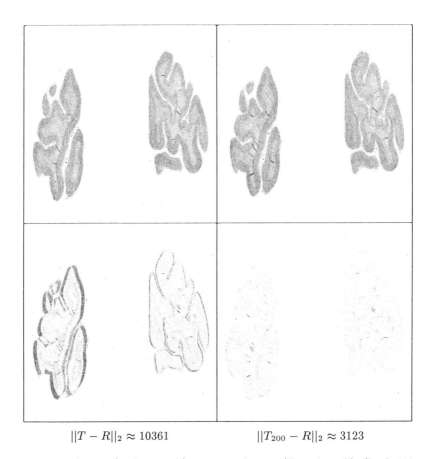

$||T - R||_2 \approx 10361$  $||T_{200} - R||_2 \approx 3123$

**Fig. 3.** Upper left: 116 (Reference $R$), upper right 117 (Template $T$), (both 512 × 512 pixels, 256 gray levels, note: differences can hardly be seen by human eyes), lower left: $|T - R|$, lower right: $|T_{500} - R|$. Here, only small differences in comparison to the first subtraction image remained, reduction: $||T_{500} - R||/||T - R|| \approx 30.1\%$.

## References

1. Amit, Y., Grenander, U., Piccioni, M.: Structural Image Restoration Through Deformable Templates, Journal of the American Statistical Association, 86(414) (1991) 376–387
2. Bajcsy, R., Kovačič, S.: Toward an Individualized Brain Atlas Elastic Matching, MS-CIS-86-71 Grasp Lap 76, Dept. of Computer and Information Science, Moore School, University of Philadelphia (1986)
3. Bookstein, F.L.: A statistical Method for Biological Shape Comparisons, J. theor. Biol. 107, (1984) 475–520
4. Bookstein, F.L.: Size and Shape Spaces for Landmark Data in Two Dimensions, Stat. Sci. 1, (1986) 181–242
5. Boden, N.J., Cohen, D., Felderman, R.E., Kulawik, A.E., Seitz, C.L., Seizovic, J.N., Su, W.-K.: Myrinet: A Gigabit-per-Second Local-Area Network. IEEE Micro 15(1) (1995) 119–128

6. Bro-Nielsen, M.: Medical Image Registration and Surgery Simulation. PhD thesis, IMM, Technical University of Denmark (1996)
7. Christensen, G.E.: Deformable Shape Models for Anatomy. PhD thesis, Sever Institute of Technology, Washington University (1994)
8. Fischer, B., Modersitzki, J.: Fast Inversion of Matrices Arising in Image Processing. Tech. Report A-98-29, Medical University of Lübeck (1998)
9. Golub, G.H., van Loan, C.F.: Matrix Computations. The Johns Hopkins University Press, Baltimore, Second edition (1989)
10. Gurtin, M.E.: An Introduction to Continuum Mechanics. Academic Press, Orlando (1981)
11. Obelöer, W., Maehle, E.: Integration of Debugging and Performance Optimization. Proc. 2nd Sino-German Workshop Advanced Parallel Processing Technologies - APPT, 17–24, Koblenz (1997)
12. Geist, A, et al.: PVM 3 Users Guide and Reference Manual. Technical Report ORNL/TM-12187, Oak Ridge National Laboratory (1993)
13. Medical University of Lübeck, Institute of Computer Engineering, Störtebeker Cluster Project. http://www.iti.mu-luebeck.de/cluster (1998)
14. Schormann, T., Henn, S., Zilles, K.: A New Approach to Fast Elastic Alignment with Applications to Human Brains. LNCS Vol. 1131 (1996) 337–342
15. Schormann, T., Dabringhaus, A., and Zilles, K.: Statistics of Deformations in Histology and Application to Improved Alignment with MRI, IEEE Transactions on Medical Imaging, 14(1) (1995) 25–35
16. Schormann, T., von Matthey, M., Dabringhaus, A., Zilles, K.: Alignment of 3-D Brain Data Sets Originating from MR and Histology, Bioimaging, Vol. 1 (1993) 119–128
17. Thirion, J.-P.: Non-Rigid Matching Using Demons, *in IEEE, Correspondence on Computer Vision and Pattern Recognition*, 1996, 245–251

# Data Intensive Distributed Computing: A Medical Application Example

Jason Lee, Brian Tierney, William Johnston
Lawrence Berkeley National Laboratory[1]
1 Cyclotron Rd.
MS: 50B-2239
Berkeley, CA 94720
{jrlee, bltierney, wejohnston}@lbl.gov

**Abstract.** Modern scientific computing involves organizing, moving, visualizing, and analyzing massive amounts of data from around the world, as well as employing large-scale computation. The distributed systems that solve large-scale problems will always involve aggregating and scheduling many resources. Data must be located and staged, cache and network capacity must be available at the same time as computing capacity, etc. Every aspect of such a system is dynamic: locating and scheduling resources, adapting running application systems to availability and congestion in the middleware and infrastructure, responding to human interaction, etc. The technologies, the middleware services, and the architectures that are used to build useful high-speed, wide area distributed systems, constitute the field of data intensive computing. This paper explores some of the history and future directions of that field, and describes a specific medical application example.

## 1.0 Introduction

The advent of shared, widely available, high-speed networks is providing the potential for new approaches to the collection, storage, and analysis of large data-objects. Two examples of large data-object environments, that despite the very different application areas have much in common, are health care imaging information systems and atomic particle accelerator detector data systems.

Health care information, especially high-volume image data used for diagnostic purposes - e.g. X-ray CT, MRI, and cardio-angiography - are increasingly collected at

---

1. The work described in this paper is supported by the U. S. Dept. of Energy, Office of Energy Research, Office of Computational and Technology Research, Mathematical, Information, and Computational Sciences and ERLTT Divisions under contract DE-AC03-76SF00098 with the University of California, and by DARPA, Information Technology Office. This report number LBNL-42690.

tertiary (centralized) facilities, and may now be routinely stored and used at locations other than the point of collection. The importance of distributed storage is that a hospital (or any other instrumentation environment) may not be the best environment in which to maintain a large-scale digital storage system, and an affordable, easily accessible, high-bandwidth network can provide location independence for such storage. The importance of remote end-user access is that the health care professionals at the referring facility (frequently remote from the tertiary imaging facility) will have ready access to not only the image analyst's reports, but the original image data itself.

This general scenario extends to other fields as well. In particular, the same basic infrastructure is required for remote access to large-scale scientific and analytical instruments, both for data handling and for direct, remote-user operation. See [8].

In this paper we describe and illustrate a set of concepts that are contributing to a generalized, high performance, distributed information infrastructure, especially as it concerns the types of large data-objects generated in the scientific and medical environments. We will describe the general issues, architecture, and some system components that are currently in use to support distributed large data-objects. We describe a health care information system that has been built, and is in prototype operation.

## 1.1 An Overall Model for Data-Intensive Computing

The concept of a high-speed distributed cache as a common element for all of the sources and sinks of data involved in high-performance data systems has proven very successful in several application areas, including the automated processing and cataloguing of real-time instrument data and the staging of data from an Massive Storage System (MSS) for high data-rate applications.

For the various data sources and sinks, the cache, which is itself a complex and widely distributed system, provides:

- a standardized approach for high data-rate interfaces;
- an "impedance" matching function (e.g., between the coarse-grained nature of parallel tape drives in the tertiary storage system and the fine-grained access of hundreds of applications);
- flexible management of on-line storage resources to support initial caching of data, processing, and interfacing to tertiary storage;
- a unit of high-speed, on-line storage that is large compared to the available disks of the computing environments, and very large (e.g., hundreds of gigabytes) compared to any single disk.

The model for data intensive computing, shown in Figure 1, includes the following:

- each application uses a standard high data-rate interface to a large, high-speed, application-oriented cache that provides semi-persistent, named datasets / objects;

- data sources deposit data in a distributed cache, and consumers take data from the cache, usually writing processed data back to the cache when the consumers are intermediate processing operations;
- metadata is typically recorded in a cataloguing system as data enters the cache, or after intermediate processing;
- a tertiary storage system manager typically migrates data to and from the cache. The cache can thus serve as a moving window on the object/dataset, since, depending on the size of the cache relative to the objects of interest, only part of the object data may be loaded in the cache - though the full objection definition is present: that is, the cache is a moving window for the off-line object/data set;

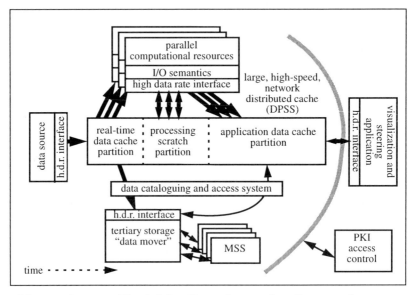

**Figure 1      Model for Data-Intensive Computing**

- the native cache access interface is at the logical block level, but client-side libraries implement various access I/O semantics - e.g., Unix I/O (upon request available data is returned; requests for data in the dataset, but not yet migrated to cache, cause the application-level read to block or be signaled);

## 1.2 The Distributed Parallel Storage Server

A key aspect of this data intensive computing environment has turned out to be a high-speed, distributed cache. LBNL designed and implemented the Distributed-Parallel Storage System (DPSS)[1] as part of the MAGIC [6] project, and as part of the U.S. Department of Energy's high-speed distributed computing program. This technology has been quite successful in providing an economical, high-performance, widely distributed, and highly scalable architecture for caching large amounts of data that can potentially be used by many different users. The DPSS serves several roles in high-performance, data-intensive computing environments. This application-oriented cache provides a standard high data rate interface for high-speed access by data sources, processing resources, mass storage systems, and user interface elements. It provides the functionality of a single very large, random access, block-oriented I/O device (i.e., a "virtual disk") with very high capacity (we anticipate a terabyte sized system for high-energy physics data) and serves to isolate the application from tertiary storage systems and instrument data sources. Many large data sets may be logically present in the cache by virtue of the block index maps being loaded even if the data is not yet available. Data blocks are declustered (dispersed in such a way that as many system elements as possible can operate simultaneously to satisfy a given request) across both disks and servers. This strategy allows a large collection of disks to seek in parallel, and all servers to send the resulting data to the application in parallel (see Figure 2). In this way processing can begin as soon as the first data blocks are generated by an instrument or migrated from tertiary storage.

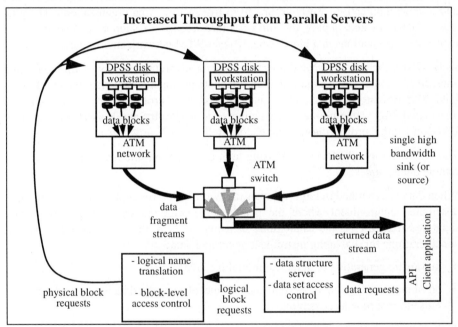

**Figure 2          DPSS Architecture**

The high performance of the DPSS - about 14 megabytes/sec of data delivered to the user application per disk server - is obtained through parallel operation of independent, network-based components. Flexible resource management - dynamically adding and deleting storage elements, partitioning the available storage, etc. - is provided by design, as are high availability and strongly bound security contexts. The scalable nature of the system is provided by many of the same design features that provide the flexible resource management (that in turn provides the capability to aggregate dispersed and independently owned storage resources into a single cache).

The DPSS provides several important and unique capabilities for a data intensive computing environment. It provides application-specific interfaces to an extremely large space of logical blocks; it offers the ability to build large, high-performance storage systems from inexpensive commodity components; and it offers the ability to increase performance by increasing the number of parallel disk servers. Various cache management policies operate on a per-data set basis to provide block aging and replacement.

## 2.0 A Medical Application Example

BAGNet was an IP over OC-3 (155 Mbit/sec) ATM, metropolitan area network testbed that operated in the San Francisco Bay Area (California) for two years starting in early 1994. The participants included government, academic, and industry computer science and telecommunications R&D groups from fifteen Bay Area organizations. The goal was to develop and deploy the infrastructure needed to support a diverse set of distributed applications in a large-scale, IP-over-ATM network environment.

In BAGNet, there were several specific projects involving subsets of the connected sites. In particular, LBNL, the Kaiser Permanente health care organization, and Philips Palo Alto Research Center collaborated to produce a prototype production, on-line, distributed, high data rate medical imaging system. (Philips and Kaiser were added to BAGNet for this project through the Pacific Bell CalREN program.)

The Kaiser project [8] focused on using high data rate, on-line instrument systems as remote data sources.

When data is generated in large volumes and with high throughput, and especially in a distributed environment where the people generating the data are geographically separated from the people cataloguing or using the data, there are several important considerations for managing instrument generated data:

- automatic generation of at least minimal metadata;
- automatic cataloguing of the data and the metadata as the data is received (or as close to real time as possible);
- transparent management of tertiary storage systems where the original data is archived;
- facilitation of cooperative research by providing specified users at local and remote sites immediate as well as long-term access to the data;
- mechanisms to incorporate the data into other databases or documents.

The WALDO (Wide-area Large-Data-Object) system was developed to provide these capabilities, especially when the data is gathered in real time from a high data rate instrument [8]. WALDO is a digital data archive that is optimized to handle real-time data. It federates textual and URL linked metadata to represent the characteristics of large data sets. Automatic cataloguing of incoming real-time data is accomplished by extracting associated metadata and converting it into text records; by generating auxiliary metadata and derived data; and by combining these into Web-based objects that include persistent references to the original data components (called large data objects, or LDOs) [9]. Tertiary storage management for the data components (i.e., the original datasets) is accomplished by using the remote program execution capability of Web servers to manage the data on mass storage systems. For subsequent use, the data components may be staged to a local disk and then returned as usual via the Web browser, or, as is the case for several of our applications, moved to a high-speed cache for access by specialized applications (e.g., the high-speed video player illustrated in the right-hand part of the right-hand panel in Figure 3). The location of the data components on tertiary storage, how to access them, and other descriptive material are all part of the LDO definition. The creation of object definitions, the inclusion of "standardized" derived-data-objects as part of the metadata, and the use of typed links in the object definition, are intended to provide a general framework for dealing with many different types of data, including, for example, abstract instrument data and multi-component multimedia programs. WALDO was used in the Kaiser project to build a medical application that automatically manages the collection, storage, cataloguing, and playback of video-angiography data$^2$ collected at a hospital remote from the referring physician.

Using a shared, metropolitan area ATM network and a high-speed distributed data handling system, video sequences are collected from the video-angiography imaging system, then processed, catalogued, stored, and made available to remote users. This permits the data to be made available in near-real time to remote clinics (see Figure 3). The LDO becomes available as soon as the catalogue entry is generated — derived data is added as the processing required to produce it completes. Whether the storage systems are local or distributed around the network is entirely a function of optimizing logistics.

In the Kaiser project, cardio-angiography data was collected directly from a Philips scanner by a computer system in the San Francisco Kaiser hospital Cardiac Catheterization Laboratory. This system is, in turn, attached to an ATM network provided by the NTON and BAGNet testbeds. When the data collection for a patient is complete (about once every 20–40 minutes), 500–1000 megabytes of digital video data is sent across the ATM network to LBNL (in Berkeley) and stored first on the DPSS distributed cache (described above), and then the WALDO object definitions are generated and made available to physicians in other Kaiser hospitals via BAGNet. Auxiliary processing and archiving to one or more mass storage systems proceeds independently. This process goes on 8–10 hours a day.

---

2. Cardio-angiography imaging involves a two plane, X-ray video imaging system that produces from several to tens of minutes of digital video sequences for each patient study for each patient session. The digital video is organized as tens of data-objects, each of which are of the order of 100 megabytes.

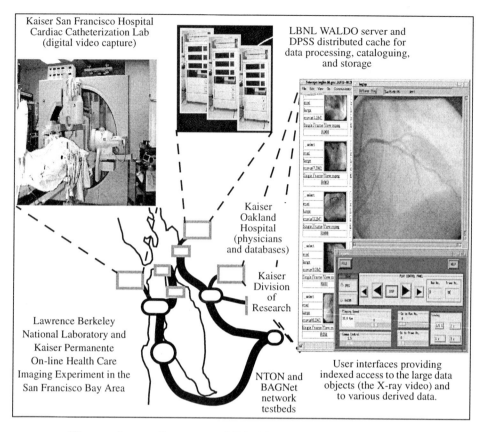

**Figure 3  Capture of Digitized Angiograms**

## 3.0 Distributed System Performance Monitoring

A central issue for using high-speed networks and widely distributed systems as the foundation of a large data-object management strategy is the performance of the system components, the transport / OS software, and the underlying network. Problems in any of these regimes will hurt a data intensive computing strategy, but such problems can usually be corrected if they can be isolated and characterized. A significant part of our work with high-speed distributed systems in MAGIC has been developing a monitoring methodology and tools to locate and characterize bottlenecks.

There are virtually no behavioral aspects of high-speed, wide area IP-over-ATM networks that can be taken for granted, even in end-to-end ATM networks. By "network" we mean the end-to-end data path from the transport API through the host network protocol (TCP/IP) software, the host network adaptors and their device drivers, the many different kinds of ATM switches and physical links, up through the corresponding software stack on the receiver. Further, the behavior of different elements at similar places in the network architecture can be quite different because they are implemented in different ways. The combination of these aspects can lead to complex and unpredictable network behavior.

We have built performance and operation monitoring into the storage system and several applications, and have designed tools and methodologies to characterize the distributed operation of the system at many levels. As requests and data enter and leave all parts of the user-level system, synchronized timestamps are logged using a common logging format. At the same time, various operating system and network parameters may be logged in the same format. This is accomplished by the Netlogger monitoring system [10], which has been used to analyze several network-generated problems that showed up in the distributed applications [11].

## 4.0 Other DPSS Applications

We have conducted a set of high-speed, network based, data intensive computing experiments between Lawrence Berkeley National Laboratory (LBNL) in Berkeley, Calif., and the Stanford Linear Accelerator (SLAC) in Palo Alto, Calif. The results of this experiment were that a sustained 57 megabytes/sec of data were delivered from datasets in the distributed cache to the remote application memory, ready for analysis algorithms to commence operation. This experiment represents an example of our data intensive computing model in operation.

The prototype application was the STAR analysis system that analyzes data from high energy physics experiments. (See [3].) A four-server DPSS located at LBNL was used as a prototype front end for a high-speed mass storage system. A 4-CPU Sun E-4000 located at SLAC was a prototype for a physics data analysis computing cluster, as shown in Figure 2. The National Transparent Optical Network testbed (NTON - see [7]) connects LBNL and SLAC and provided a five-switch, 100-km, OC-12 ATM path. All experiments were application-to-application, using TCP transport.

Multiple instances of the STAR analysis code read data from the DPSS at LBNL and moved that data into the memory of the STAF application where it was available to the analysis algorithms. This experiment resulted in a sustained data transfer rate of 57 MBytes/sec from DPSS cache to application memory. This is the equivalent of about 4.5 TeraBytes / day. The goal of the experiment was to demonstrate that high-speed mass storage systems could use distributed caches to make data available to the systems running the analysis codes. The experiment was successful, and the next steps will involve completing the mechanisms for optimizing the MSS staging patterns and completing the DPSS interface to the bit file movers that interface to the MSS tape drives.

## 5.0 Conclusions

We believe this architecture, and its integration with systems like Globus ([5], [2]), will enable the next generation of configurable, distributed, high-performance, data-intensive systems; computational steering; and integrated instrument and computational simulation. We also believe a high performance network cache system such as the DPSS will be an important component to these "computational grids" [5] and "metasystems"[4].

## 6.0 References

[1] DPSS, "The Distributed Parallel Storage System," http://www-didc.lbl.gov/DPSS/

[2] Globus, "The Globus Project," http://www.globus.org/

[3] Greiman, W., W. E. Johnston, C. McParland, D. Olson, B. Tierney, C. Tull, "High-Speed Distributed Data Handling for HENP," Computing in High Energy Physics, April, 1997. Berlin, Germany. http://www-itg.lbl.gov/STAR/

[4] Grimshaw, A., A. Ferrari, G. Lindahl, K. Holcomb, "Metasystems", Communications of the ACM, November, 1998, Volume 41, no 11

[5] Foster, I., C. Kesselman, eds., "The Grid: Blueprint for a New Computing Infrastructure," Morgan Kaufmann, publisher. August, 1998.

[6] B. Fuller and I. Richer "The MAGIC Project: From Vision to Reality," IEEE Network, May, 1996, Vol. 10, no. 3. http://www.magic.net/

[7] NTON, "National Transparent Optical Network Consortium." See http://www.ntonc.org/.

[8] Johnston, W., G. Jin, C. Larsen, J. Lee, G. Hoo, M. Thompson, B. Tierney, J. Terdiman, "Real-Time Generation and Cataloguing of Large Data-Objects in Widely Distributed Environments," International Journal of Digital Libraries - Special Issue on "Digital Libraries in Medicine". November, 1997. (Available at http://www-itg.lbl.gov/WALDO/)

[9] Thompson, M., W. Johnston, J. Guojun, J. Lee, B. Tierney, and J. F. Terdiman, "Distributed health care imaging information systems," PACS Design and Evaluation: Engineering and Clinical Issues, SPIE Medical Imaging 1997. (Available at http://www-itg.lbl.gov/Kaiser.IMG)

[10] Tierney, B., W. Johnston, B. Crowley, G. Hoo, C. Brooks, D. Gunter, "The NetLogger Methodology for High Performance Distributed Systems Performance Analysis," Seventh IEEE International Symposium on High Performance Distributed Computing, Chicago, Ill., July 28-31, 1998. Available at http://www-itg.lbl.gov/DPSS/papers.html.

[11] Tierney, B., W. Johnston, J. Lee, and G. Hoo, "Performance Analysis in High-Speed Wide Area ATM Networks: Top-to-bottom end-to-end Monitoring," IEEE Networking, May 1996. (Available at http://www-itg.lbl.gov/DPSS/papers.)

# Utilizing HPC Technology in 3D Cardiac Modeling[1]

Nikos Papazis and Dimitris Dimitrelos

Athens High Performance Computing Laboratory and
Department of Informatics, University of Athens
{papazis,ddimitr}@hpcl.uoa.gr

**Abstract.** The objective of the 3D Heartview methodology was to demonstrate the benefits of HPC technology in improving the diagnostic and clinical procedures of heart diseases. It enables an online 3D modeling of the heart structures based on 2D X-ray angiographic sequences acquired under routine clinical conditions, thus offering medical added value in the areas of operation planning, wall motion study and 3D manipulation of cardiac structures. HPC technology was utilized in the 3D modeling procedure in an effort to minimize total system response time and to result in a system applicable in every day medical practice.

## 1 Introduction

Since one century the main reason of mortality in western societies has changed from infection to cardiovascular diseases (50%) even before cancer (20%)[4],[5]. The severity of this kind of diseases has contributed to a shift in the aims of the medical imaging groups towards building systems that help standard clinical routine.

The availability and clinical relevance of 3D modeling of cardiac structures from angiographic images is already proven in a wide variety of typical applications, e.g. congenital heart defects, wall motion study and others, but the computing time required for generating a 3D model is too slow and varies between several minutes up to one hour per image. These times are unacceptable for routine applications and prevent a break-through of the method in the medical market.

The image processing system, that was developed in the context of the 3D HeartView project, needs little user interaction during modeling and works as an add-on for the rotational angiography. The software back-projects few 2D X-ray images into a 3D image-stack. It automatically creates the 3D model, visualizes and enables the calculation of the blood volume of heart chamber interior. Acceleration of the modeling process using HPC systems was the main objective of our work[1]. Keeping in mind the short term horizon of the project, we wanted to employ HPC in such a manner that would guarantee a successful outcome in respect to total system response time and fulfillment of physicians' needs.

---

[1] This work was partially funded by EC under the 3D-HeartView project (ESPRIT-24484)

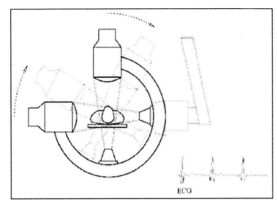

Fig. 1. Angiographic device

The rest of the paper is organized as follows: in the 2$^{nd}$ chapter the overview of the 3D HeartView methodology is presented, in the 3$^{rd}$ chapter the parallelisation strategy is analyzed and in the 4$^{th}$ chapter the results of it are presented. Finally in the 5$^{th}$ chapter some plans on the future work of our team are included.

## 2 Methodology

The standard medical practice consists of placing the patient at an angiographer. In figure 1 one is depicted. It is a machine equipped to emit x-rays and collect the images they produce. An angiographer may have one or two planes, i.e. one or two sets of x-ray source and x-ray detector devices with a fixed angle difference between them. A catheter is entered into the cardiac area of interest. At that time some contrast agent is injected in the structure. The contrast agent is a fluid that absorbs x-rays, thus making the cardiac structure apparent in the x-ray images. Rotational angiography consists of rotating the arm of the angiographer around the torso of the patient during image acquisition. The images that are produced

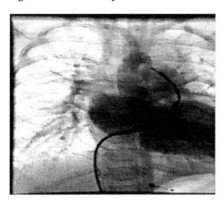

Fig. 2. 2D x-ray image

are digitized either directly by reading the digital DICOM-3 output of the angiography device, or indirectly by frame grabbing of the video output. In figure 2 a typical x-ray images is displayed. Additionally the cardiac phase is derived from the synchronously recorder ECG and is associated with each image. The digital images together with the corresponding heart phase are transferred to the HPC machine, where a 3D model of the examined heart ventricle is produced. Our 3D-modeling technique is based on backprojection [8][9], shown in figure 3, with certain enhancements for the pyramid-beam perspective[1][6]. The principle behind it is that

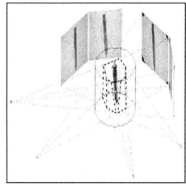

Fig. 3. Back-projection technique

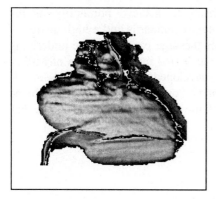

**Fig. 4.** 3D model

for every 2D x-ray image and for each pixel in that image, a virtual x-ray is back-projected to the 3D dataset. The x-ray affects the gray level value of the voxels that it intersects. The sum of the effect of all the x-rays, that intersect a given voxel, corresponds to the final gray level value of that voxel.

The cardiac phase is employed to associate images of the same cardiac time-step acquired from different view angles. The result is a volumetric dataset similar to those from CT scanners, which can be visualised and manipulated by means of ray-tracing and surface rendering techniques, using the InViVo system[1]. One is depicted in figure 4.

An advantage of the method is that it needs little user interaction during the modeling step and expects no particular target shape, therefore it is applicable for pathological deformities as well.

A disadvantage of the technique is its high computational needs which result in unacceptable for clinical practice response times, over 15 minutes per each phase. By employing HPC we reduced the modeling time from tens to some minutes, thus making the method applicable to every day clinical routine.

## 3 Parallelisation strategy

### 3.1 Hardware and Software Platform

CCi-3D by Parsytec GmBH has been chosen to be the underlying hardware platform of our system. It offers a good performance and a satisfactory ratio of performance to cost in our area of application as well as a dependable nature of functionality. It comprises of 4 dual Pentium Pro nodes, on which Windows NT operating system runs along with other hardware and software modules necessary for this kind of applications. All nodes are interconnected by a 100 Mbps Ethernet connection. The fact that the system is configured with off-the-self industrial PC technology, gives us the chance to adopt to any future changes, while keeping the overall cost of the system relatively low in the range of 18.000 to 20.000 DM. The system is divided into:
- The front-end module, which is the module the medical doctor will have in his or her office and through which he or she will have access to the graphical user interface. It consists of one dual Pentium Pro node.
- The back-end module on which the modeling algorithm will run and which is the actual parallel machine. It comprises of three dual Pentium Pro node.

The nature of our hardware and the desire to achieve portability of the implementation, compliance with the state-of-the-art standards and a high degree of efficiency led us to choose the MPI standard for message passing as the underlying software platform. In particular we have used the *WMPI* implementation of MPI, a Win32 implementation of the MPI standard, developed at the *Departamento de Endenharia Inmformatica, University of Coimbra Portugal* [2], that proved to be more reliable and of higher performance than other available Windows NT implementations at that time [7].

### 3.2 Architecture

The parallel modeler consists of two components, the *'volume-writer'* that deals with I/O and initialization operations and the *'partial-modeler'*, which carries out the parallel computation of the 3D model.

Elaborating a little on the job each component is performing, it could be stated

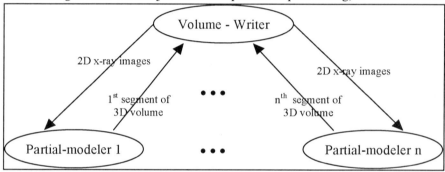

**Fig. 5.** Architectural overview of the 3D HeartView system

that after invocation the 'volume-writer' reads the 2D input x-ray images and then broadcasts the data to the instances of 'partial-modeler' over the 100Mbps Ethernet module. The architecture is displayed in figure 5. Upon reception of the data, each instance computes the appropriate segment of the 3D volume, which is uniformly distributed among them. When an instance completes its task, it passes the volume segment back to the 'volume-writer'. The 'volume-writer' upon reception of each segment forms the total 3D volume.

The algorithm makes no assumptions on the existing number of instances of the component 'partial-modeler'. Each instance first determines the number of its peers and then tries to determine which segment of the total 3D volume to compute. Thus the algorithm is scalable, in that it can deal with different hardware configurations without the need of alterations at source-level. In the case of changes in the hardware configuration the only required change in the 3D-HeartView parallel system would be in a few lines in the configuration file that WMPI uses to determine the architecture of the parallel application and consequently to spawn it[3].

## 3.3 Data partitioning

The parallelisation strategy focuses on data parallelism. We could either partition the x-ray images among the instances of the 'partial modeler' and make them compute the total 3D model, or allocate separate segments of the 3D model to each instance while providing to it the total set of 2D images. The decision was driven by the back-projection technique, in which each voxel is highly independent of others, so it can be computed in parallel. On the other hand the gray level value of each voxel is determined by pixels from every 2D image. So we had decided to opt for breaking the 3D volume among the parallel instances of 'partial-modeler', each modeling 1/n of the 3D volume using the whole set of 2D x-ray images. If we had partitioned the 2D images, we would be obliged to compute the whole 3D volume on each instance of the 'partial-modeler', thus burdening the use of memory as well as the communication between them and the 'volume-writer'. A standard case consists of having 9 2D images. Each image is approximately 257 KB, therefore the 9 images are 2,25 MB. A typical volume has a resolution of 200 units per axis and a size of 8MB. Therefore it is more efficient to partition the 3D volume.

In figure 6 the volume in a 3D axes system is depicted. The modeling technique evaluates the voxels in a manner that first accesses the voxels with the same x and y coordinate and then scans over the y-axis and last the x-axis. The latter means that for one to access all voxels having the same x- and y- coordinates, one must make leaps in the volume data structure of x*y plane slices. Therefore it had been decided that each 'partial modeler' would receive a continuous set of slices along the z-axis, instead of doing so along the x- or the y- axis, in order to make the optimum use of the cache memory.

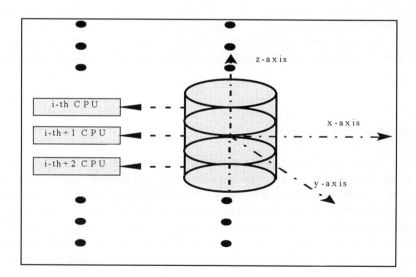

**Fig. 6.** Data partitioning

# 4 Results

## 4.1 Description of test strategy

The performance tests were conducted at the premises of the Athens High Performance Computing Laboratory and of Skoutas SA. The experiments were carried out using real patient data, as well as validation ('phantom') object data.

It is important to state that originally the 3D modeling procedure took over 1 hour to complete on an average workstation. Aggressive code optimization that was carried out during the project resulted in a decrease in the sequential modeling time to 15 -20 minutes. Thus a better basis for the parallelisation task was accomplished.

Time are measured in seconds and volume resolution in units per axis. To ensure statistical validity for each measurements 10 runs were executed. The maximum and minimum value of each set were discarded and the remainder values were kept. Each measurement was repeated 5 times and the overall mean value was computed. For each experiment a dataset of 9 2D x-ray images was used, which is an average number of frames for clinical practice.

In the context of the presented work, we were interested in developing a system that would fulfill the needs of the collaborating medical doctors. The main aspect, that needed the utilization of HPC technology, was the overall system response time, which the physician is witnessing. The latter is the main feature that would prove the proposed methodology applicable in every day clinical practice or not. Therefore speedup was calculated as the ratio of overall parallel response time, including computation as well as communication and I/O time, to sequential response time. The metric of the efficiency of the parallelisation strategy should be the decrease of this delay, rather than other features.

The current system implementing the 3D-HeartView methodology is supporting volume resolutions from 61 up to 255 units in each axis. Resolutions over 150 units per axis are the most suitable for medical applications. Consequently the measurements were done on representative volumes, i.e. 166, 200 and 255 units per axis.

## 4.2 Discussion of the results

In table 1 the achieved speedup in the total system response time in relation to the existing number of CPUs and the volume resolution is presented. Although the CCi-3D system consists of 4 nodes, only the back-end module, i.e. 6 processors, acts as the parallel machine while the front-end acts as the phycian's workstation. In principle the modeler will operate in the background on the back-end module, while the medical doctor will continue examining the original 2D x-ray images or viewing previously modeled volumes on the front-end. The first row holds the number of

|  | No CPUs | 1 | 2 | 4 | 6 |
|---|---|---|---|---|---|
| Volume |  |  |  |  |  |
| 166 |  | 1,00 | 1,72 | 3,36 | 4,40 |
| 200 |  | 1,00 | 1,74 | 3,36 | 4,40 |
| 255 |  | 1,00 | 1,70 | 3,36 | 4,38 |

**Table 1.** Speedup relative to volume resolution and number of CPUs

CPUs used in the experiments for each system configuration. The first column holds the volume resolution, for which the efficiency of our system was tested.

In figure 7 the system response time in relation to the existing number of CPUs and the volume resolution is depicted. The horizontal axis holds the number of CPUs and the vertical axis holds the system response time in seconds. There are three graphs corresponding to volume resolutions of 166, 200 and 255 units per axis.

From table 1, speedup is proven to be independent of the volume resolution computed. The amount of data increased but speedup remained almost unchanged, whereas one would expect an increase, which is the expected behavior in data parallelism. The reason for that resides on the fact that voxel are not calculated entirely independently of each other imposing restrictions on the level of exploitable parallelism, especially since we opted for partitioning the voxel space. Another cause might be the implementation of the MPI standard over Windows NT and with 100Mbs Ethernet communication. The latter will be thoroughly investigated in future work.

Another point is that speedup did not increase proportionally to the number of CPUs employed. This behavior is expected in data parallelism applications. The decrease in the computation load of each CPU was counterbalanced by an increase in the number of messages exchanged during the application.

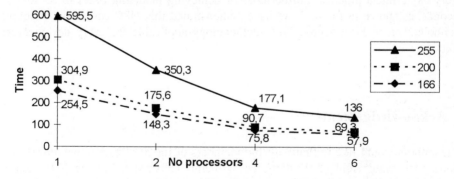

**Fig. 7.** System response time relative to volume resolution and number of CPUs

## 5 Future work

We plan to apply the proposed methodology in producing 3D models of arteries. The goal is more difficult due to the finer shape of the structures as well as their more extensive movement between cardiac cycles. The diagnosis of coronary stenosis is currently done by looking the video (or film) recording produced using X-rays angiograms and contrast agent. The degree of stenosis, as well as the degree of success after dilation, is estimated based on the experience of the supervising surgeon/cardiologist. First studies point out that the accuracy of diagnosis in qualitative manner, can be significantly improved by using 3D models of the coronary vessels. Such models would allow physicians to manipulate a spatial model of the vessel and, thus, to have a better understanding of the some times rather complicated 3D topology of the examined region.

Also we will try to make the entire process more automated, so delivering a system that can operate and be used in real time during the operation with the minimal manual intervention.

Additionally we will try to improve the parallelisation by exploiting other partitioning schemes and by utilising newer versions of the MPI implementation.

Finally we will examine methods of processing the original 2D x-ray images prior to feeding them to the modeling procedure. By enhancing the contrast of the objects from their background or segmenting the images, we expect a major improvement in the quality of the modeled 3D structures.

## 6 Conclusions

In the 3D HeartView project a system was developed that adds medical value in every day clinical practice. Furthermore by achieving modeling times of 55 to 136 seconds, as shown in figure 7, we have demonstrated that HPC can act as enabling technology in making a promising methodology applicable for every day medical practice.

## 7 Acknowledgements

The authors would like to thank all the members of the 3D-HeartView consortium, namely Dr. Hans Kehl of Westfalische Wilhelms Universitat Munster, Dr. Zenon Kyriakides, Dr. Theophilos Kolletis and Ms. Katerina Karaiskou of Onassis Cardiac Surgery Center, Dr. Georgios Sakas and Mr. Jurgen Jaeger of Zentrum fur Graphische Datenverarbeitung, Dr. Lambis Tasakos of inos GmbH, Mr. Panos Skoutas and Mr. Nikos Nychtas of Panos Skoutas S.A. and Dr. Andy Marsh of Institute of Communication and Computer Systems. Their collaboration was invaluable in making the work successful and fruitful in demonstrating that HPC can perform as an enabling technology in medical applications.

Last, but no least, we would like to thank the HPCN TTN network ans especially the INNO TTN node for their contribution and support.

## References

1. The 3DHeartView Team: The 3DHeartView project. Proceedings of HPCN'98 conference, Amsterdam, Netherlands, p1021-1023..
2. Home page of WMPI: http://dsg.dei.uc.pt/wmpi/intro.html.
3. Message Passing Interface Forum: MPI: A Message-Passing Interface Standard.
4. Dr. Kehl, HG, Jaeger, J., Kececioglu, D., Sakas, G., Gehrmann, J., Vogt, J.: Threedimensional angiography: evolution of methods and first clinical results. In Proc. of $2^{nd}$ World Congress of Pediatric Cardiology and Cardiac Surgery, Honululy, USA, 1997.
5. Dr. Kehl, HG, Jaeger, J., Kececioglu, D., Sakas, G., Rellensmann, G., Nekarda, T., Vogt, J.: Pediatric angiocardiography in three and four dimensions: Evolution of methods, validation and first clinical results. In Proceedings of the $2^{nd}$ World Congress of Pediatric Cardiology and Cardiac Surgery, Tokyo, Japan, 1998. (not printed yet).
6. Jurgen Jager : 3D HeartView. Computer Graphik Topics, Vol. 9 4/97, p6-7.
7. Home page of MPICH/NT: http://www.erc.msstate.edu/mpi/mpiNT.html.
8. Johann Radon, "áber die Bestimmung von Funktionen durch ihre Integralwerte lðngs gewisser Mannigfaltigkeiten." Ber. Verh. Sðchs. Akad. Wiss. Leipzig, Math-Nat. 69 (1917) 262-277.
9. L.A. Feldkamp, L.C. Davis, and J.W. Kress, "Practical Cone-Beam Algorithm," Optical Society of America, Vol. 1 (No. 6), pp. 612–619, 1984.

# A Parallel Algorithm for 3D Reconstruction of Angiographic Images *

R. Rivas[1], M. B. Ibáñez[2], Y. Cardinale[2] and P. Windyga[3]

[1] Facultad de Ciencias. Escuela de Computación.
Universidad Central de Venezuela. Apartado 47002, Caracas 1041-A, Venezuela
rrivas@neumann.ciens.ucv.ve
[2] Departamento de Computación y Tecnología de la Información
Universidad Simón Bolívar, Apartado 89000, Caracas 1080–A, Venezuela
{ibanez,yudith}@ldc.usb.ve
[3] School of Computer Science University of Central Florida
Orlando, FL 32816-2362
pwindyga@cs.ucf.edu

**Abstract.** Accurate diagnosis and therapeutic evaluation of coronary dysfunction is possible by tri-dimensional (3D) visualization of Coronary arteries. Reconstruction based on bi-dimensional (2D) images can be presented as a discrete optimization problem. A blind search cannot be applied, instead a Branch-and-Bound algorithm is used to explore the state space and give an intermediate result. The heuristic information used is based on knowledge based filtering in coronagraphy.
A sequential algorithm using suitable filters leads to implementations where the execution time is measured in days. In order to minimize the execution time we propose to apply parallel computing techniques.
The critical issue in parallel search algorithms is the distribution of the search space among the processors. We propose a technique to compute the total amount of work units among the processors. The technique is based on the enlargement of segments (unitary threads) representing pieces of arteries. We achieve a good load balancing and the speedup obtained is nearly optimum.

---

* This work has been supported by the project ITDC 139-82158 A2IM and Decanato de Investigación y Desarrollo USB (S1-CAI-13)

# 1   Introduction

The detection, evaluation and therapeutic follow-up of dysfunction affecting coronary arteries, are usually carried out by an imaging technique called angiography [4] (see [7]). Tri-dimensional (3D) visualization of coronary arteries can improve considerably diagnosis and maximizes patient protection by reducing exposure to radiation and contrast liquid infusion. Due to a variety of technical constrains, 3D reconstruction might be based on two orthogonal images taken simultaneously. The Fig. 1 shows such an image. These images have numerous structure superpositions as well as deformations caused by the projection process. The reconstruction based on two projections is a multiple solution problem.

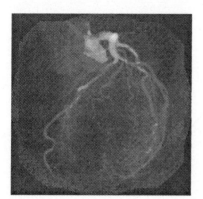 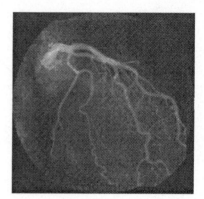

**Fig. 1.** Biplane images corresponding to left coronary artery

The GBBA group at the Simon Bolivar University [6] has developed ANIA: A tool for Angiographic Image Analysis and study. ANIA allows 2D and 3D image manipulation and it can generate the solution space for the problem of reconstruction from biplane images. From the solution space (see Fig. 2) generated by ANIA, a 3D reconstruction process must be done.

The reconstruction process involves a combinatorial generation of possible solutions (see Fig. 3), the identification of the correct solution can only be carried out using *a priori* knowledge (see [8]). This knowledge includes 3D aspects (object properties) and 2D aspects (object projection properties). None of the identified properties has enough discrimination capacity, but each one can help to eliminate a set of erroneous solutions. From the reconstruction process, it is desirable to obtain only one solution -the right one-. This research is been carried out in order to have enough discrimination conditions [10]. At the moment, it is only possible to get a reduced set of valid solutions, those which satisfy a set of pre-establish conditions. The reconstruction process using a set of conditions and its algorithm is described in Sect. 2.

---

[4] a radiography of vessels with radio opaque fluid injected

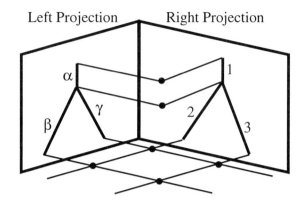

**Fig. 2.** Solution space from two 2D coronary projections

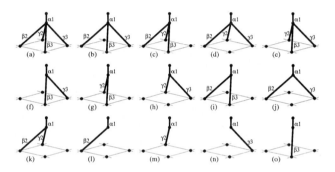

**Fig. 3.** 15 possible Partial Solution Trees for the fig 2' Solution Space

A prototype of the reconstruction process using *priori* knowledge was developed in Prolog in an early work by P. Windyga [8]. The solution space of a real coronary tree is too large [9] to be managed efficiently by sequential Prolog, thus the performance of this prototype is poor. Running on a SUN Sparc Station the Prolog's program is unable to find the set of valid solutions in one week. Based on this experience and using the set of conditions described at Sect. 2, a new sequential algorithm is implemented in C and then it is improved by including *unitary threads*. The key aspects used on this approach are given at Sect. 2.2.

The critical issue in parallel search algorithms is the distribution of the search space among the processors. There is a large amount of work in this subject [5] [4]. By increasing the length of artery segments with *unitary threads*, the search space is reduced. The reduced search space is partitioned in equal shares among the processors. Then we work with a simple load balancing technique: the static distribution of the search space (see Sect. 3). This approach normally leads to a severe load imbalance, but in our case the results are satisfactory (see Sect. 4) because the distribution does not cause high communication costs. Finally, the conclusions and future work are described at Sect. 5.

## 2 Description of the Sequential Algorithm

The process of search space generation in the 3D reconstruction of coronary arteries, begins with a *Total Solution Tree* (**TST**) containing all the possible representations of the coronary tree. From the TST a combinatorial generation of possible solutions (*Partial Solution Trees*: **PST**) is done. These solutions are validated using the following conditions.

**Ramification Condition:** The only ramification allowed is bifurcation.
**Superposition Condition:** The superposition of two PSTs segment projections onto any of the 2D plans must be short.
**Completion Condition:** Every segment at the 2D projections must have an image into the PST.
**Verticality Condition** The segments of the PSTs must have vertical orientation, that is they must going down from the main artery to the heart's inferior extreme.
**No Traversing Condition (or Vectorial Projection)** The arteries can not either traverse the heart, or going in opposite direction to the sanguine flux.

The PSTs that satisfy the conditions above described are considered the *valid solutions*.

To illustrate the properties that must be satisfied, let us examine the *PSTs* (Fig. 3) that can be built from its *TST* (Fig. 2). The ramification condition invalidates the solutions **(a), (b), (c), (d), (e)** in Fig. 3. The segments $\beta 2$ and $\gamma 2$ in Fig. 3(k) are totally superposed on the segment 2 of the right projection (Fig. 2), thus this PST is not valid. Due to the *Completion Condition* the Fig. 3(l) is not a valid solution. Note that neither the segment $\gamma$ from the left projection nor the segment 3 from the right projection are images of a PST's segment.

### 2.1 The Algorithm

In what follows we use the following terminology.

A *Total Solution Tree* $TST = <\mathcal{P}, \mathcal{E}>$ where $\mathcal{P} = \{p_1, \ldots p_m\}$ ($p_i$ are TST points) and $\mathcal{E} = \{e_1, \ldots e_n\}$ ($e_i$ are TST edges).

A *Partial Solution Tree* $PST = <\mathcal{P}', \mathcal{E}'>$ where $p'_i \in \mathcal{P}'$ and it is a copy of $p_i \in \mathcal{P}$, likewise $e'_i \in \mathcal{E}'$ and it is a copy of $e_i \in \mathcal{E}$. An edge $e = <p_k, p_l>$ in a tree is incident to the points $p_k$ and $p_l$ which are called its *endpoints*. The points in the *Partial Solution Tree* frontier are called *Active Points*: $\mathcal{A}'$.

**Algorithm**

1. The first PST is built from the *Total Solution Tree* starting with an initial point and no edges, i.e. $\mathcal{P}' = \mathcal{A}' = \{p'\}$, $\mathcal{E}' = \{\}$.
2. While there are *Active Points* ($\mathcal{A}' \neq \{\}$) do
    (a) $\forall a'_i \in \mathcal{A}'$ build $\mathcal{X}_i = \{e'_{i1}, \ldots, e'_{ir}\}$ where $e_{i1}, \ldots, e_{ir}$ are incidents to $a_i$ and none of them have an endpoint in $\mathcal{P}'$.
    (b) $\forall j_1, \ldots, j_n \in [1, 2]$ : build new PSTs using the following rules:

i. $\mathcal{E}'_{new} = \mathcal{E}' \bigcup \mathcal{X}_1^{j_1} \times \ldots \times \mathcal{X}_n^{j_n}$
ii. $\mathcal{A}'_{new}$ has the $PST_{new}$'s frontier points
iii. $\mathcal{P}'_{new}$ is the set of all the $\mathcal{E}'_{new}$'s endpoints.
Discard current PST.
(c) Validate the new edges added to the $PST_{new}$s using the *Superposition, Completion, Verticality and No Traversing Conditions*.
If any of the conditions fail, the corresponding $PST_{new}$ is discarded.

The combinatorial generation of solutions comes from the second step of the algorithm. Notice that for every *Active Point*, at most two edges are considered (*Ramification Condition*). This is a *Branch-and-Bound* algorithm and the *bound* is given by the conditions.

## 2.2 Improving the Algorithm by Using the Superposition Condition

The generation of *Partial Solution Trees* involves an exhaustive combination of the *Total Solution Tree*'s edges (see the step 2bi of the algorithm). This combination can be done in several ways (DFS [5], BFS [6]etc), depending on how the TST is visited. No matter how this process is done, a *Partial Solution Tree* may join and validate a subset of edges that other PST has already combined. This means that some work will be repeated many times. Depending on the conditions used, this situation can be partially avoided. Let us revisit the *Superposition Condition* used in our branch and bound search.

The *Superposition Condition* is useful when there is a bifurcation in the coronary tree. Let us suppose that we have a TST that has three edges $e_1 = <1,2>$ belongs to the PST been built and has two incident edges $e_2 = <2,3>$ and $e_3 = <2,4>$ that do not belong to this PST and set up the only bifurcation. The edge $e_1$ always projects its image onto a subplan -in any of the projections-, which does not intersect the one where $e_2$ projects to. This is due to the relative position of $e_1$ and $e_2$, $e_1$ is above $e_2$. The same is true for $e_1$ and $e_3$. Thus in the process of PST generation, the combinations $[e_1, e_2]$ and $[e_1, e_3]$ are always successful.

We can thus identify successful pairs of edges such as $[e_1, e_2]$ or $[e_1, e_3]$ Those are always together in a PST. A path $e_0, e_1, \ldots, e_n$ where the TST's initial point belongs to $e_0$, a TST's active point belongs to $e_n$ and $\forall i \in [0, n-1] : [e_i, e_{i+1}]$, $e_i$ and $e_{i+1}$ is a successful pair of edges, it is called an *unitary thread*. Thus, an *unitary thread* can be seen as a macro edge. The more edges an *unitary thread* has, the more combined work is saved by avoiding repeating tests among PSTs's pairs of edges.

Notice that now the sequential algorithm must consider internal points into the *unitary threads*. When two threads are combined, if their prefix are not compatible, the combination is discarded avoiding the application of the filters. Otherwise the conditions are validated and a new thread is considered.

---
[5] Depth First Search
[6] Breath First Search

## 3 Parallel Generation of Possible Solutions

The parallel generation of *Partial Solution Trees* starts by building the *unitary threads*, then the threads must be combined in all possible ways. Pairs of threads having some common points make the combination, each combination that satisfies the *conditions* is a *valid solution*.

The *unitary threads* are built before the program starts, thus it is possible to compute the amount of combinations to make before the program begins to run. In this way every process may compute the same number of combinations, therefore we may have a static load balancing. Let us suppose to have n unitary threads and p number of processors, then the number of *unitary threads* combinations that every process must made is $C$ or $C + 1$

$$C = \sum_{i=1}^{n} \binom{n}{i} \div p$$

Every processor has to do almost the same amount of combinations and hopefully the same amount of work. The variations in the workload are due to the number of *unitary threads* in the combinations, the length of the *unitary threads* and how soon the invalid solutions can be detected.

Notice that without having threads it is also possible to compute the number of combinations as

$$\sum_{i=1}^{ns} \binom{ns}{i}$$

where $ns$ is the TST's number of segments. However, if we divide the search space in the same way that we propose to do with *unitary threads*, a significant number of combinations will not include a set of segments of a valid PST. That is because the group of segments to analyze for a PST does not have any morphological relationship, in contrast to our method where every *unitary thread* has a set of segments, which belong to the same artery.

In order to determine which combinations each processor must do, the *unitary threads* are numbered from 1 to n and each processor has an identification number in $(1 \ldots p)$. Each combination will be codified as a *work unit* which is a sequence of numbers in $[1, n]$. An ordered sequence of *work units* is built following an increasing lexicographic order. The first process will do the first $C$ *work units*, the second one will perform the *work units* appearing in the positions $[C+1, 2*C]$ and so on.

The parallel algorithm work as follows. A master process generates all the *unitary threads*, and it broadcasts them to the other processors. The master process sends to the other processors the range of *work units* each one of them must generate and then combine. Once the processors have received both informations, they generate its *work units* and combine the corresponding threads following the algorithm described in section 2.

## 4  Experimental Results

We examined the performance of our application using up to eight nodes on our Parsytec Xplorer platform. Each processor node is a PowerPC 601 with 32Mbytes of RAM memory, 32K-byte cache on the processor chip for holding both instructions and data. The experiment was designed to evaluate the load balance of our system. Our data consisted of 130 segments from the Total Solution Tree, and we have 27 unitary threads.

The comparison of the parallel performance is done by measuring the runtime of the application using 1, 2, 4, 6 and 8 processors, see Fig. 4. The number of combinations made was 67 million and the runtime for one processor was 539 minutes (more than 8 hours). We also include the *ideal* time values to achieve considering the amount of processors. The small difference between both curves shows that the speedup achieved is nearly optimum.

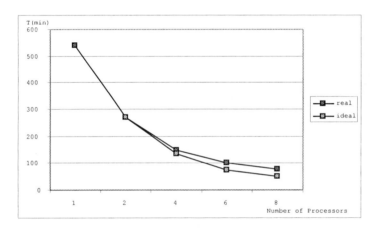

**Fig. 4.** Comparison of runtime with 1,2,4,6 and 8 processors

The Fig. 5 shows the minimum, maximum and average time taken by 1, 2, 4, 6 and 8 processors to complete the validation of its work units. Although the load balancing is not perfect, the biggest difference between the maximum and the average runtime is 13.7%. Thus this simple strategy can be a good choice for this kind of application.

With eight processors, our program can give the answer in 79 minutes. With perfect load balancing, eight processors require 67 minutes. In both cases the time is too long. In order to estimate the number of processors required to process the data in less time, we made the following experiment. We computed the time needed to calculate the 67 million combinations of our experiment in groups of five hundred thousand, then we divided in equal shares the combinations as our algorithm would. Finally we computed the time required by adding the time spent by the group of combinations chosen. Figure 6 shows that with 64

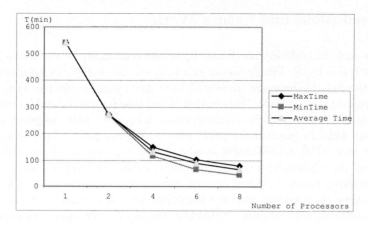

**Fig. 5.** Load balancing evaluation

processors the result can be achieved in 16 minutes which is a reasonable amount of time. This configuration is not very expensive. For better results, we can build a parallel machine with 256 processors and produce the result in only 6.2 minutes.

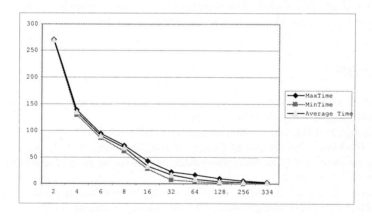

**Fig. 6.** Ideal combination distribution

## 5 Conclusions and Future Work

This is a multidisciplinary work and it will be continued by the bioengineering team (directed by P. Windyga) as much as by the computer scientists team. The runtime of the program presented here Are considerable less than the one of the original Prolog code, This makes it feasible for the GBBA researches to experiment with other discrimination conditions. The new program may also have other ways to improve its performance. We are particularly interested in looking under which conditions it is advantageous a functional parallelisation.

Due to the *Completion Condition* only a subset of *work units* are relevant. Combinations having very few threads (one or two for example) can not be a *valid solution*. However, we have not established yet which one is the safest range to consider. Once this range is established, the number of *valid solutions*, as well as the execution time, will be significantly reduced. The bioengineering group will conduct research in this direction. Once the range is known, it will be simple its consideration within the parallel program.

The parallelisation strategy chosen allows the processors to work independently once they receive the *unitary threads* as well as the range of combinations to make. It should be experimented with the definition of collaboration strategy in order to share information about incompatible combinations that some processor has already found. The question to answer is whether it exists such a mechanism to improve the performance of the program.

## References

1. I. Foster. *Designing and building parallel programs. Concepts and tools for Parallel Software Engineering.* Addison Wesley, 1995.
2. W. Group, E. Lusk, and A. Skjellum. *Using MPI.* The MIT Press, 1994.
3. R.W. Hockney. *The Science of Computer Benchmarking.* SIAM, 1996.
4. Lydia Kronsjo and Dean Shumsheruddin. *Advances in parallel algorithms.* John Wiley and sons, 1992.
5. V Kumar, A Grama, A Gupta, and G Karypis. *Introduction to parallel computing. Design and analysis of algorithms.* The Benjamin/Cummings, 1994.
6. A. La Cruz, G. Morinelli, P. Windyga, G. Bevilacqua, and J. Silva. ANIA: A tool for angiographic image analysis and study. In *XVII IEEE-EMBC Conference*, pages 381–382, Montreal, Canada, 1995. IEEE.
7. G. Passariello and F. Mora. *Imagenología Médica.* Equinoccio, 1995.
8. P. Windyga. *Evaluation et modelisation de connaissances pour la reconstruction tridimensionel du reseau vasculaire cardiaque en angiographie biplan.* PhD Thesis, Univ. Rennes France, 1994.
9. P. Windyga, G. Bevilacqua, J.L. Coatrieux, and M. Garreau. Estimation of search-space in 3D coronary artery reconstruction using angiographic biplane images. In *XVII IEEE-EMBC Conference*, pages 389–390, Montreal, Canada, 1995. IEEE.
10. P. Windyga, I. López, G. Bevilacqua, M. Garreau, and J.L. Coatrieux. Utility of 2D properties in the reconstruction of coronary arteries from biplane angiographic images. In *XVII IEEE-EMBC Conference*.

11. P. Windyga, Garreau M., Shah M., Coatrieux J.L., and LeBreton H. Three-dimensional reconstruction of the coronary arteries using a priori knowledge. In *Medical and Biological Engineering and Computing*, pages 158–164. 36, 1998.

# A Diffraction Tomography Method for Medical Imaging Implemented on High Performance Computing Environment

T. A. Maniatis, K. S. Nikita, and K. Voliotis

Department of Electrical and Computer Engineering, National Technical University of Athens, Iroon Polytechniou 9, Zografos 15780, Athens, Greece
fanis@esd.ece.ntua.gr

**Abstract.** The efficient implementation of a diffraction tomography method for medical imaging is addressed within the framework of High Performance Computing (HPC) environment. A non-linear optimization method for the solution of the inverse scattering problem is implemented on a shared memory model computer. Linear speed-up and significant reduction in the total execution time is achieved when the program is executed in parallel, enabling the feasibility of the method for realistic medical imaging applications.

## 1 Introduction

Diffraction tomography is a multiview imaging procedure where cross-sections of an object are reconstructed from scattered field measurements. The object is usually illuminated from many incident directions by a diffracting energy source, such as ultrasound or microwaves. Contrary to Computed Tomography (CT) where the laws of geometric optics are applied, in diffraction tomography the interaction between the incident field and the scattering object can only be described accurately in terms of the wave equation. Methodologies and techniques developed for the solution of the inverse scattering problem are applied in diffraction tomography. Inverse scattering is defined as the problem of finding the shape and refractive index of an inhomogeneous object from scattered field measurements performed on the object's exterior. The development of a diffraction tomography method for medical imaging is highly desirable, since it is based on the use of non-ionizing ultrasonic or electromagnetic radiation, which is considered safe at relatively low exposure levels.

Early attempts in inverse scattering relied on the linearization of the problem using the first order Born or Rytov approximation [1,2]. Despite their mathematical simplicity, these methods are useful only when the size of the scattering object is comparable to the wavelength of the incident wave and its index of refraction is not significantly different from that of the surrounding medium [3]. In order to overcome the stringent limitations of the first order, linear methods, non-linear inverse scattering methods have also been investigated [4-6]. Lately, non-perturbative methods for inverse scattering solution have been proposed, based on non-linear optimization techniques [7-9]. These methods have no theoretical restriction regarding the size or refractive index of the scattering object. However, the necessary discretization of the scattering integral equations, usually leads to an increased number of unknowns for any problem of practical importance.

Inverse scattering is a typical time consuming problem and the response time of its execution is obviously critical. A HPC architecture is an ideal platform for implementing applications that exhibit the above characteristics. Although a variety of inverse scattering methods have been reported in the literature, as far as we know none of them has addressed the problem of practical implementation of an inverse scattering method on a HPC environment.

The purpose of this work is to demonstrate the parallel execution of an inverse scattering algorithm on a HPC platform leading to the development of a realistic diffraction tomography technique for medical imaging. According to the proposed scenario, a number of different clinical sites will be connected with a HPC center where the inverse scattering algorithm will be executed. In section 2, a brief theoretical background of the proposed inverse scattering method is presented. The parallel implementation of the proposed inverse scattering algorithm on HPC architectures is discussed in section 3 along with the evaluation results. The description of the interconnection scheme between the HPC center and the clinical sites is given in section 4. Summarizing concluding remarks and a brief discussion for future work are given in the final section of this paper.

## 2   Theoretical Background

Any practical implementation of a diffraction tomography technique requires an efficient and accurate solution of the inverse scattering problem. A non-perturbative method for inverse scattering solution, based on non-linear optimization and plane wave expansion of the field inside the scatterer, has been proposed recently [10] and is briefly presented in this paragraph.

For the sake of simplicity, only two-dimensional (2D) problems will be considered. In practice, most realistic imaging problems are three-dimensional (3D) since the properties of biological bodies vary in all three spatial directions. However, if reasonable focusing of the incident wavefield can be achieved along the axis perpendicular to the plane of propagation, the contribution of out-of-plane scattering effects can be neglected without introducing significant error. Since the term "tomography" refers to a 2D imaging procedure, the 3D structure of the object under investigation can be reconstructed from a series of successive 2D cross-sectional images. A time dependence $\exp(-i\omega t)$ where $\omega$ is the angular frequency, is assumed and omitted throughout. Furthermore, we assume that the wave propagation is described by the scalar Helmholtz equation.

We consider the situation depicted in Fig. 1 where a plane monochromatic wave $\psi_0(\mathbf{r}) = \exp(ik_0\hat{\mathbf{s}}_j \cdot \mathbf{r})$ is incident on a dielectric object characterized by the complex refractive index $n_r(\mathbf{r})$. The object is embedded in a uniform background with refractive index $n_0$. The wavenumber $k(\mathbf{r})$ is related with the refractive index

$$k(\mathbf{r}) = k_0 n_r(\mathbf{r}) \tag{1}$$

where $k_0 = \omega/c_0$ is the wavenumber of the incident wave in the background medium and $c_0$ is the respective propagation velocity. The far-scattered field is measured along a circular contour that fully encloses the scattering object. The procedure is repeated for a number of $J$ different propagation directions $\hat{\mathbf{s}}_j$ and each time a new set of measurements is collected.

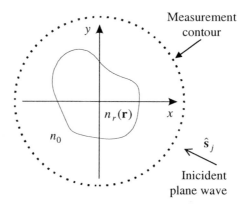

**Fig. 1.** Measurement geometry

The Lippmann-Schwinger scattering integral equation expresses the total wavefield that results from the interaction of the incident wave and the scattering object [11]

$$\psi(\mathbf{r}) = \psi_0(\mathbf{r}) + k_0^2 \int d\mathbf{r}' o(\mathbf{r}')\psi(\mathbf{r}')g(\mathbf{r}|\mathbf{r}') \qquad (2)$$

where the wavefunction $\psi(\mathbf{r})$ is some descriptor of the wave propagation and $g(\mathbf{r}|\mathbf{r}')$ is the 2D free space Green's function. The object function is defined as

$$o(\mathbf{r}) = n_r^2(\mathbf{r}) - 1 \qquad (3)$$

and characterizes the scattering properties of the object. The *scattering amplitude* is a function of the direction of observation $\hat{\mathbf{r}}_{sc}$ and is defined as

$$f_s(\hat{\mathbf{r}}_{sc}) = e^{i\frac{\pi}{4}}\sqrt{\frac{k_0^3}{8\pi}} \int e^{-ik_0\hat{\mathbf{r}}_{sc}\cdot\mathbf{r}'} o(\mathbf{r}')\psi(\mathbf{r}')d\mathbf{r}' \; . \qquad (4)$$

The scattering amplitude expresses the asymptotic behavior of the scattered field in the far region. A key point in the proposed inverse scattering method is the expression of the total wavefield in the interior of the scattering object as a superposition of plane waves in the form of a Fourier integral

$$\psi(\mathbf{r}) = \frac{1}{(2\pi)^2}\int d\mathbf{p}\, e^{i\mathbf{p}\mathbf{r}} C(\mathbf{p}) \; . \qquad (5)$$

In addition, the object function is expanded in terms of a set of $N_g$ spatially shifted and properly weighted Gaussian basis functions

$$o(\mathbf{r}) \cong \sum_{n=1}^{N_g} \alpha_n e^{-\frac{(\mathbf{r}-\mathbf{r}_n)^2}{\tau^2}} . \tag{6}$$

Using (5), the scattering integral equation (2) can be written as

$$\int d\mathbf{p} C^j(\mathbf{p}) \left[ O(\mathbf{q}-\mathbf{p}) - \frac{k_0^2}{(2\pi)^2} I(\mathbf{q},\mathbf{p}) \right] = (2\pi)^2 O(\mathbf{q}-\mathbf{k}^j) \quad j=1,\ldots,J \tag{7}$$

where $C^j(\mathbf{p})$ is the Fourier transform of the field inside the scattering object, $O(\mathbf{k})$ denotes the Fourier transform of the object function and the superscript $j$ denotes the dependence of the total wavefield on the incident plane wave, propagating in the direction of $\mathbf{k}^j = k_0 \hat{\mathbf{s}}_j$. The term $I(\mathbf{q},\mathbf{p})$ is given by the integral

$$I(\mathbf{q},\mathbf{p}) = \int d\mathbf{k} \frac{O(\mathbf{q}-\mathbf{k})O(\mathbf{k}-\mathbf{p})}{k^2 - k_0^2} . \tag{8}$$

Similarly the scattering amplitude (4) can be written as

$$f_s^j(\mathbf{k}^{sc}) = \frac{e^{i\frac{\pi}{4}}}{(2\pi)^2} \sqrt{\frac{k_0^3}{8\pi}} \int d\mathbf{p} C^j(\mathbf{p}) O(\mathbf{k}^{sc}-\mathbf{p}) . \tag{9}$$

where $\mathbf{k}^{sc} = k_0 \hat{\mathbf{r}}_{sc}$. We define the residuals in satisfying (7)

$$P^j = (2\pi)^2 O(\mathbf{q}-\mathbf{k}^j) - \int d\mathbf{p} C^j(\mathbf{p}) \left[ O(\mathbf{q}-\mathbf{p}) - \frac{k_0^2}{(2\pi)^2} I(\mathbf{q},\mathbf{p}) \right] \quad j=1,\ldots,J \tag{10}$$

and the residuals in satisfying (9)

$$R^j = f_s^j(\mathbf{k}^{sc}) - \frac{e^{i\frac{\pi}{4}}}{(2\pi)^2} \sqrt{\frac{k_0^3}{8\pi}} \int d\mathbf{p} C^j(\mathbf{p}) O(\mathbf{k}^{sc}-\mathbf{p}) \quad j=1,\ldots,J . \tag{11}$$

A cost function $Q$ is formed in terms of the residuals $P^j$ and $R^j$

$$Q(O, C^1, \ldots C^J) = \sum_{j=1}^{J} \|P^j\|_2^2 + \sum_{j=1}^{J} \|R^j\|_2^2 \tag{12}$$

where $\|\cdot\|_2$ denotes the two-norm. The inverse scattering problem is reformulated as the problem of finding the optimum values for the object function $O$ and the field inside the scatterer $C^j$, that minimize the cost function $Q$. This non-linear optimization problem is solved using the modified gradient method proposed by Kleinmann and Van den Berg [8].

## 3  Numerical Implementation on HPC platform

The numerical solution of the inverse scattering problem requires projection of the continuous models used to describe the scattering process, (7) and (9), into a finite dimensional space. Most of the methods reported in the literature employ the method of moments in order to discretize the problem [7,8]. Usually, pulse basis functions are used to approximate both the object function and the field inside the scatterer. For the accurate solution of the forward scattering problem with pulse basis functions and point matching, it has been shown [12] that the side of each elementary square cell should not exceed $0.1\lambda$, where $\lambda = 2\pi/k_0$ is the free space wavelength of the incident field. This means that a total of 100 complex variables are required for the description of the object function and 100 more for the description of the internal field, per square wavelength of the scattering object. On the other hand, for the proposed inverse scattering method, 100 complex variables per square wavelength are required for the description of the object function, plus a small number of variables (typically 12 or 24) independent of the size of the scatterer, for the description of the internal field. Nevertheless, it is evident, that for medical imaging purposes, where the size of the scatterer is several times greater than the wavelength of the incident field, the inverse scattering problem cannot be solved in any conventional computer. Therefore, the implementation of the method on a HPC platform is necessary.

Two major tasks in the proposed inverse scattering method are computationally expensive. Firstly, the calculation of the double integral terms (8) for many different pairs (**q**,**p**), can only be performed numerically. On the other hand, the iterative solution of the optimization problem using the modified gradient method, implies the explicit calculation of the cost function gradient at each step, a task that proves to be particularly time consuming. The parallelization of the algorithm concerning these two tasks is presented in the following paragraphs.

### 3.1  Parallel Numerical Calculation of the Integral Terms

The proposed inverse scattering method requires the numerical calculation of the integral terms given in (8). These integrals consist of terms that depend solely on the particular selection of the Gaussian basis functions used to describe the scatterer and the plane wave expansion of the internal field and not on the scattering object itself. Hence, they can be computed once and stored for subsequent usage. Still, their calculation involves a significant computational overhead that can be greatly reduced if these calculations are executed in parallel. The algorithm can be easily implemented in parallel on both shared and distributed memory parallel architectures. However, the calculation of each integral $I(\mathbf{q},\mathbf{p})$ for a specific pair of the vectors **q** and **p** can be performed independently from any other. Therefore, due to the inherent independent parallelism of the algorithm, it is particularly well suited to distributed memory HPC platforms with a large number of relatively fast processing elements.

The algorithm has been implemented on an Intel Paragon XP/S that is a distributed memory system with 48 nodes and on a Silicon Graphics Power Challenge that is a shared memory parallel computer with 14 nodes. These two computers are located at the High Performance Computing Center of the National Technical University of Athens. We considered a small example, where only 70 different integral terms $I(\mathbf{q},\mathbf{p})$ are calculated. This test case does not correspond to any realistic situation but rather serves the purpose of execution time assessment and is performed on the Silicon Graphics Power Challenge using various numbers of processors. Total execution time as a function of the number of processors used is given in Table 1.

**Table 1.** Total execution time for the calculation of the integrals $I(\mathbf{q},\mathbf{p})$ as a function of the number of processors used

| Number of processors | 1 | 2 | 5 | 7 | 10 | 14 |
|---|---|---|---|---|---|---|
| Execution time (sec) | 3772 | 1928 | 794 | 564 | 411 | 329 |

The speed-up for the presented test case is shown in Fig. 2 as a function of the number of processors used. Since Power Challenge is a multi-user parallel computing system that was being heavily used at the time the tests were carried out, the achieved speed-up is not 100% linear. However, a significant reduction of the total execution time is achieved.

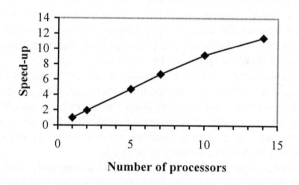

**Fig. 2.** Algorithm speed-up for the calculation of the integrals $I(\mathbf{q},\mathbf{p})$ as a function of the number of processors used

## 3.2 Parallel Implementation of the Optimization Method

The reformulation of the inverse scattering problem as a non-linear optimization one and its solution using the modified gradient method [8], requires calculation of the cost function partial derivatives with respect to the independent variables, i.e. the unknowns of the problem. However, the number of unknowns increases non-linearly with the size of the problem. Furthermore, the time required for the calculation of each specific partial derivative depends on the number of the Gaussian basis functions used to describe the scatterer. Hence, the calculation of the cost function gradient consumes most of the time within each step of the iterative optimization procedure.

In addition, each iteration of the algorithm includes an initialization phase during which special purpose matrices, whose elements are used repeatedly in the evaluation of the cost function, are calculated. This is a method of element precalculation that can speed up significantly the performance of the algorithm. Since the matrix elements to be precalculated are completely independent, their evaluation can easily be parallelized.

The total execution time of the algorithm can be approximately given by

$$T_{tot} \cong T_0 + \sum_{n=1}^{N_{iter}} T_{iter} \tag{13}$$

where $T_0$ corresponds to the initialization stage of the algorithm, and $T_{iter}$ is the time consumed for each iteration. A total of $N_{iter}$ iterations are performed. The time within each iteration can be regarded as the sum of four independent terms. Specifically,

$$T_{iter} = T_{ini} + T_{grad} + T_{CG} + T_{prep} \tag{14}$$

where $T_{ini}$ represents the initialization phase of each iteration, $T_{grad}$ is the time required for the calculation of the gradient of the cost function, $T_{CG}$ is the time required to perform a small scale minimization within each iteration using a standard conjugate gradient method and $T_{prep}$ represents the time required for the preparation of the next iteration. Since most of the time is consumed mainly for the calculation of the cost function gradient and the initialization phase of each iteration, parallel execution is considered only for these two parts of the algorithm.

The calculations involved in the evaluation of the partial derivatives and the initialization phase are completely independent therefore, they can be encapsulated on a fully scalable algorithm, independent from the underlying architecture, shared or distributed memory. Already, the proposed algorithm has been implemented and evaluated extensively on a shared memory system while its implementation on a distributed memory system is currently in progress.

Execution time assessment is achieved by performing only four iterations of the algorithm. For the test case under consideration, the object function was expanded in a set of $N_g=21$ Gaussian basis functions. Partial execution times $T_{ini}$, $T_{grad}$, $T_{CG}$ and $T_{prep}$ were recorded on the Silicon Graphics Power Challenge and are presented in Table 2. Although the parallelization performed did not affect $T_{CG}$ and $T_{prep}$, their respective values are given to stress out the fact that most of the time within each iteration is spent at the gradient calculation and the initialization phase.

**Table 2.** Partial and total execution time for the optimization method as a function of the number of processors used

| Number of processors | 1 | 2 | 4 | 8 | 14 |
|---|---|---|---|---|---|
| $T_{ini}$ (sec) | 312.75 | 157.5 | 98.25 | 83.75 | 56.25 |
| $T_{grad}$ (sec) | 242.25 | 136 | 95.5 | 80.75 | 68.25 |
| $T_{CG}$ (sec) | 2.5 | 2.5 | 2.5 | 2.5 | 2.5 |
| $T_{prep}$ (sec) | 0.5 | 0.5 | 0.5 | 0.5 | 0.5 |
| $T_{iter}$ (sec) | 558 | 296.5 | 196.75 | 167.5 | 127.5 |

If $T_0$ in (13) is ignored then we can calculate the speed-up achieved by the parallel implementation of the algorithm. Using the execution times shown in Table 2 the speed-up is plotted in Fig. 3 as a function of the number of processors used. As shown in Fig. 3 a significant speed-up is achieved as the number of processors is increased. This result demonstrates the scalability of the proposed inverse scattering method.

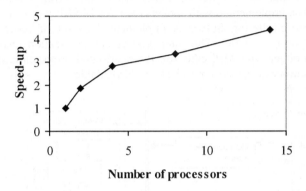

**Fig. 3.** Algorithm speed-up for the optimization method as a function of the number of processors used

## 4   Interconnection Between HPC Center and Clinical Sites

Despite the fact that diffraction tomography is still a method under investigation certain questions regarding its practical implementation at a clinical site were stimulated by the preceding presentation. Firstly, it has been made clear that any practical solution of the inverse scattering problem will be computationally intensive. The proposed non-linear optimization method, reduces the computational burden compared with similar solutions reported elsewhere [8]. However, the need for High Performance Computing power is essential for real life medical imaging applications. Since HPC facilities are not usually located at clinical sites, the need for communication between HPC site and the diagnostic center is evident. A feasible solution to this problem is the interconnection of the clinical site with the HPC center with the purpose of data exchange and remote execution of the time consuming parts of the algorithm. In fact, a HPC facility can serve many different clinical sites thus decreasing the financial cost of the proposed implementation.

According to the proposed interconnection scenario shown in Fig. 4, a number of different Diffraction Tomography Scanners interact with a single Inverse Scattering Solver, which is executed on the HPC site. Each Diffraction Tomography Scanner collects the necessary data, forms a message packet and sends it to the Inverse Scattering Solver, via a communication process (Communication Supervisor). The

Inverse Scattering Solver executes the proposed algorithm and sends back the reconstructed images. The involved communication overhead is negligible compared to the time required for the execution of inverse scattering algorithm.

The Inverse Scattering Solver is implemented on the HPC center as a parallel programming application. A detailed description of its design and implementation can be found on [13]. All the computers shown on Fig. 4 are connected on the same Ethernet backbone. The communication channels connecting the computer located on a clinical site and the parallel computer where the Inverse Scattering Solver is running, will be implemented with sockets over the TCP/IP protocol.

The Inverse Scattering Solver is implemented as set of processes running on parallel computing system. The inter-process communication channels have been implemented either with MPI calls (on the distributed programming systems) or on the shared memory of the system (when this is available).

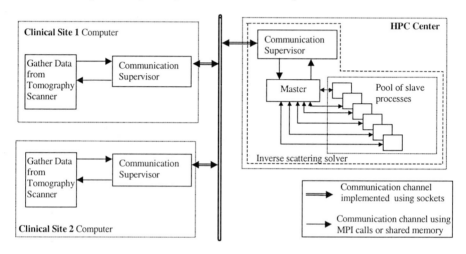

**Fig. 4.** Interconnection between clinical sites and HPC center

## 5 Conclusion

In this paper the development of a diffraction tomography method on a HPC environment has been presented. The implementation of the method for medical imaging applications relies on the efficient solution of the computationally expensive inverse scattering problem. An inverse scattering solution with no restriction regarding the size or the refractive index of the scatterer has been obtained using non-linear optimization techniques. Furthermore, a significant reduction of the total execution time of the method is achieved by parallelizing time consuming sections of the algorithm. In addition, an integrated environment is proposed, where simple workstations, such as the ones used on clinical sites, benefit from the fast execution of the inverse scattering algorithm performed on a HPC platform.

Current implementation of the algorithm has been based on the use of standard programming tools (i.e. C, TCP/IP, sockets) provided by the operating system. Even though this implementation is efficient, its porting in another parallel programming platform is not trivial. In order to ensure the portability of our application, we design a new implementation based on the use of MPI (Message Passing Interface) [14] as communication protocol of the distributed processors. MPI is a well accepted parallel programming paradigm with many implementations on various platforms.

# References

1. J. R. Shewell, and E. Wolf, "Inverse diffraction and a new reciprocity theorem," *J. Opt. Soc. Amer.*, vol. 58, pp. 1596-1603, 1968.
2. A. J. Devaney, "A filtered backpropagation algorithm for diffraction tomography," *Ultrasonic Imaging*, vol. 4, pp. 336-350, 1982.
3. M. Slaney, A. C. Kak, and L. E. Larsen, "Limitations of imaging with first order diffraction tomography," *IEEE Trans. Microwave Theory Tech.*, vol. MTT-32, pp. 860-873, 1984.
4. W. C. Chew, and Y. M. Wang, "Reconstruction of the two dimensional permittivity using the distorted Born iterative method," *IEEE Trans. Medical Imaging*, vol. 9, pp. 218-255, 1990.
5. S. Caorsi, G. L. Gragnani, and M. Pastorino, "Two-dimensional microwave imaging by a numerical inverse scattering solution," *IEEE Trans. Microwave Theory Tech.*, vol. MTT-38, pp. 981-989, 1990.
6. T. M. Habashy, M. L Oristaglio, and A. De Hoop, "Simultaneous nonlinear reconstruction of two-dimensional permittivity and conductivity," *Radio Science*, vol. 29, pp. 1101-1118, 1994.
7. D. T. Borup, S. A. Johnson, W. W. Kim and M. J. Berggren, "Nonperturbative diffraction tomography via Gauss-Newton iteration applied to the scattering integral equation," *Ultrasonic Imaging*, vol. 14, pp. 69-85, 1992.
8. R. Kleinman, and P. van den Berg, "An extended range-modified gradient technique for profile inversion," *Radio Science*, vol. 29, pp. 877-884, 1993.
9. D. Colton, and P. Monk, "A modified dual space method for solving the electromagnetic inverse scattering problem for an infinite cylinder," *Inverse Problems*, vol. 10, pp. 87-107, 1994.
10. T. A. Maniatis, K. S. Nikita and N. K. Uzunoglu, "A diffraction tomography technique using spectral domain moment method and nonlinear optimization," in *Applied Computational Electromagnetics,* N. Uzunoglu Ed., NATO – ASI Series, Berlin: Springer Verlag (in press).
11. D. Colton and R. Kress, *Inverse Acoustic and Electromagnetic Scattering Theory.* Berlin: Springer-Verlag, 1998.
12. J. H. Richmond, "Scattering by a dielectric cylinder of arbitrary cross-section shape," *IEEE Trans. Antennas Propagat.*, vol. AP-13, pp. 334-341, 1965.
13. T. A. Maniatis, *Development of Inverse Scattering Methods for Dielectric Object Imaging*. PhD Thesis, Department of Electrical and Computer Engineering, National Technical University of Athens, 1998.
14. M. Snir, S. Otto, S. Huss-Lederman, D. Walker, and J. Dongarra, *MPI-The Complete Reference.* Cambridge: M.I.T. Press, 1996.

# Track C2:

# Computational Science

# Heterogeneous Distribution of Computations While Solving Linear Algebra Problems on Networks of Heterogeneous Computers

Alexey Kalinov and Alexey Lastovetsky

Institute for System Programming, Russian Academy of Sciences
25, Bolshaya Kommunisticheskaya str., Moscow 109004, Russia
{ka,lastov}@ispras.ru

**Abstract.** The paper presents a heterogeneous distribution of computations while solving dense linear algebra problems on heterogeneous networks of computers. The distribution is based on heterogeneous block cyclic distribution which is extension of the traditional homogeneous block cyclic distribution taking into account differences in the processor performances. The mpC language, specially designed for parallel programming heterogeneous networks is briefly introduced. An mpC aplication carring out Cholesky factorization on a heterogeneous network of workstations is used to demonstrate that the heterogeneous distribution have an essential advantage over the traditional homogeneous distribution.

## 1 Introduction

Progress in network technologies is making local and even global networks of computers (in particular, networks of PCs and workstations) more and more attractive for high-performance parallel computing. While developing applications for such networks it is necessary to take into account their heterogeneity being the main peculiarity of common networks differing them from supercomputers.

The heterogeneity is displayed at least in two forms. Firstly, in the form of heterogeneity of machine arithmetics of such parallel systems. Related challenges existing in writing reliable numerical library software for heterogeneous computing environments have been analyzed in [1].

Secondly, in the form of heterogeneity of both performances of individual processors and speeds of data transfer between the processors. As a rule, to solve linear algebra problems on a heterogeneous network of computers one uses numeric software originally developed for homogeneous distributed-memory machines and later on ported to the network. As a rule, while computing on such homogeneous computer systems, a strategy of homogeneous distribution of computations over processors is used. The strategy will be referred as the HoHo strategy - "Homogeneous distribution of processes over processors - Homogeneous distribution of data over the processes", with each physical processor running one process and data being evenly partitioned among the processes.

Let us see what happens when an application, that provides a good distribution of computations and communications (due to the HoHo strategy) while running in homogeneous environments, runs on the heterogeneous network of computers. Since volumes of computations executed by different processors are approximately equal to each other, more powerful processors will wait for the weakest one at synchronization points. Therefore, the total time of computations will be determined by the time elapsed on the weakest processor. Similarly, the total time of communications will be determined by communications via the slowest communication link. So, in general, the total running time on the heterogeneous network will be close to the total running time on a homogeneous network obtained from this heterogeneous one by means of replacement of both its processors with the weakest processor and its communication links with the slowest link. In the part concerning the heterogeneity of processor performances, the statement will be experimentally corroborated in section 4.

A natural solution of this problem is heterogeneous distribution of both processes over the processors and data among the processes, taking into account differences in performances of processors and speeds of communication links. As it has been demonstrated in [2], such a distribution allows to achieve much better distribution of computations over processors of a heterogeneous computing network and, hence, to utilize its performance potential more efficiently.

Such heterogeneous distribution is a complex problem whose solution needs adequate tools. Designed specially to write efficient and portable parallel application for heterogeneous networks of computers, the mpC language [3] is just such a tool. This language is an ANSI C superset allowing to write applications adapting to differences in performances of both processors and communication links of any particular executing network. The basic idea is that an mpC application explicitly builds in run time an abstract heterogeneous computing network and distributes data, computations and communications over the network. The mpC programming system uses this information in run time to map the abstract network to any real executing network of computers in such a way that ensures efficient running of the application on the real network. More about mpC as well as the mpC free software can be found at http://www.ispras.ru/~mpc.

In the paper, we consider only the heterogeneity of processor performances. We propose a heterogeneous distribution strategy based on homogeneous distribution of involved processes over processors with each process running on a separate processor and heterogeneous block cyclic distribution of processed matrix over the processes. We investigate the strategy, named the HoHe strategy, using a typical linear algebra problem - the Cholesky factorization of square dense matrices. The problem was chosen as a well-known example of the practically important problem, whose parallel solution needs careful balancing computations and communications. Our implementation of parallel Cholesky factorization is based on the algorithm implemented in ScaLAPACK [5]. Both distribution of the involved processes and the parallel Cholesky factorization proper are performed by an mpC program calling BLAS and LAPACK functions for local computations. Note, that in this case, the parallel algorithm implemented by the mpC program for the HoHe strategy is, in general, the same as implemented by ScaLAPACK function PDPOTRF for the HoHo strategy.

Section 2 introduces the heterogeneous block cyclic matrix distribution and describes the HoHe strategy. Section 3 shortly introduces the mpC language and describes the implementation of the HoHe strategy in mpC. Section 4 gives experimental results of Cholesky factorization on a network of heterogeneous workstations using the distribution strategies.

## 2 Homogeneous distribution of processes with heterogeneous data distribution

The traditional homogeneous block cyclic data distribution [4] is determinated by grid parameters $P$ and $Q$ and block parameters $m$ and $n$. The distribution partitions the matrix into *generalized* blocks of the size $(m \cdot P)$x$(n \cdot Q)$, each of which in its turn partitioned into $(P \cdot Q)$ blocks of the same size, each going to a separate process.

Figure 1 shows an example of the homogeneous block cyclic distribution of a 12x12 matrix, block-partitioned with the block size 2x2 (m=2, n=2), over a 2x3 process grid (P=2, Q=3). In this case, generalized blocks are of size 4x6.

| | 0 | 1 | 2 | 3 | 4 | 5 | 6 | 7 | 8 | 9 | 10 | 11 |
|---|---|---|---|---|---|---|---|---|---|---|---|---|
| 0 | 0 | 0 | 1 | 1 | 2 | 2 | 0 | 0 | 1 | 1 | 2 | 2 |
| 1 | 0 | 0 | 1 | 1 | 2 | 2 | 0 | 0 | 1 | 1 | 2 | 2 |
| 2 | 3 | 3 | 4 | 4 | 5 | 5 | 3 | 3 | 4 | 4 | 5 | 5 |
| 3 | 3 | 3 | 4 | 4 | 5 | 5 | 3 | 3 | 4 | 4 | 5 | 5 |
| 4 | 0 | 0 | 1 | 1 | 2 | 2 | 0 | 0 | 1 | 1 | 2 | 2 |
| 5 | 0 | 0 | 1 | 1 | 2 | 2 | 0 | 0 | 1 | 1 | 2 | 2 |
| 6 | 3 | 3 | 4 | 4 | 5 | 5 | 3 | 3 | 4 | 4 | 5 | 5 |
| 7 | 3 | 3 | 4 | 4 | 5 | 5 | 3 | 3 | 4 | 4 | 5 | 5 |
| 8 | 0 | 0 | 1 | 1 | 2 | 2 | 0 | 0 | 1 | 1 | 2 | 2 |
| 9 | 0 | 0 | 1 | 1 | 2 | 2 | 0 | 0 | 1 | 1 | 2 | 2 |
| 10 | 3 | 3 | 4 | 4 | 5 | 5 | 3 | 3 | 4 | 4 | 5 | 5 |
| 11 | 3 | 3 | 4 | 4 | 5 | 5 | 3 | 3 | 4 | 4 | 5 | 5 |

(a) matrix distribution

| | 0 | 1 | 6 | 7 | 2 | 3 | 8 | 9 | 4 | 5 | 10 | 11 |
|---|---|---|---|---|---|---|---|---|---|---|---|---|
| 0 | 0 | 0 | 0 | 0 | 1 | 1 | 1 | 1 | 2 | 2 | 2 | 2 |
| 1 | 0 | 0 | 0 | 0 | 1 | 1 | 1 | 1 | 2 | 2 | 2 | 2 |
| 4 | 0 | 0 | 0 | 0 | 1 | 1 | 1 | 1 | 2 | 2 | 2 | 2 |
| 5 | 0 | 0 | 0 | 0 | 1 | 1 | 1 | 1 | 2 | 2 | 2 | 2 |
| 8 | 0 | 0 | 0 | 0 | 1 | 1 | 1 | 1 | 2 | 2 | 2 | 2 |
| 9 | 0 | 0 | 0 | 0 | 1 | 1 | 1 | 1 | 2 | 2 | 2 | 2 |
| 2 | 3 | 3 | 3 | 3 | 4 | 4 | 4 | 4 | 5 | 5 | 5 | 5 |
| 3 | 3 | 3 | 3 | 3 | 4 | 4 | 4 | 4 | 5 | 5 | 5 | 5 |
| 6 | 3 | 3 | 3 | 3 | 4 | 4 | 4 | 4 | 5 | 5 | 5 | 5 |
| 7 | 3 | 3 | 3 | 3 | 4 | 4 | 4 | 4 | 5 | 5 | 5 | 5 |
| 10 | 3 | 3 | 3 | 3 | 4 | 4 | 4 | 4 | 5 | 5 | 5 | 5 |
| 11 | 3 | 3 | 3 | 3 | 4 | 4 | 4 | 4 | 5 | 5 | 5 | 5 |

(b) distribution from processor point-of-view

**Fig. 1.** Example of a homogeneous block cyclic distribution of a 12x12 matrix over 2x3 process grid with the block size 2x2.

Let an executing computer system consists of a set $\mathbf{L}$ of processors and $card(\mathbf{L}) \geq P \cdot Q$. Let $P \cdot Q$ processes be distributed over $P \cdot Q$ most powerful processors in such a way, that just one process goes to each of these processors, and matrix $M$ be distributed over the processes in accordance with the heterogeneous block cyclic distribution presented below.

We also suppose that the positive real number $r_{ij}$ is associated with each of the processors and characterizes its relative performance ($i \in [0, P-1]$, $j \in [0, Q-1]$). Then, in addition to four numbers $P$, $Q$, $m$ and $n$, parametrizing the homogeneous block cyclic distribution, the heterogeneous one is parametrized by the $PxQ$ matrix $\mathbf{R} = \{r_{ij}\}$, elements of which characterize performances of the corresponding processors. Its main difference from the homogeneous distribution lies in heterogeneous data distribution inside a generalized block. Like in case of the homogeneous distribution, a generalized $(m \cdot P)x(n \cdot Q)$ block is partitioned into $(P \cdot Q)$ blocks. But in case of the heterogeneous distribution, the blocks are not of the same size, but their sizes $m_{ij}xn_{ij}$ depend on performances of processors. In the paper, we consider the simplest choice of $m_{ij}$ and $n_{ij}$ deduced from the assumption that part of matrix $M$ processed by a separate processor is proportional to its performance. That is,

$$m_{ij} \cdot n_{ij} = \frac{m \cdot P \cdot n \cdot Q \cdot r_{ij}}{\sum_{i=0}^{P-1} \sum_{j=0}^{Q-1} r_{ij}}.$$

In this case $mxn$ is the average size of the uneven blocks.

In particular, the above condition can be satisfied by the following choice of $m_{ij}$ and $n_{ij}$:

$$n_{ij} = n_j = \frac{\sum_{i=0}^{P-1} r_{ij} \cdot n \cdot Q}{\sum_{i=0}^{P-1} \sum_{j=0}^{Q-1} r_{ij}}, \qquad m_{ij} = \frac{r_{ij} \cdot m \cdot P}{\sum_{i=0}^{P-1} r_{ij}}.$$

Figure 2 shows an example of the heterogeneous block cyclic distribution of a 12x12 matrix over a 2x3 processor grid (P=2, Q=3) with the average block size 2x2 (m=2, n=2), the generalized-block size 4x6 and the following matrix of processor performances

$$\mathbf{R} = \begin{pmatrix} 6 & 4 & 2 \\ 5 & 3 & 1 \end{pmatrix}.$$

With that choice of the $m_{ij}$, a row of the distributed matrix does not have to belong to the same row of the process grid, that can lead to additional communication overheads.

## 3 Implementation of heterogeneous distribution of computations in mpC

mpC [3] is a parallel language aimed at efficiently-portable modular programming heterogeneous networks of computers. It provides facilities for specification of requirements on resources, necessary for efficient execution of the parallel application, and the mpC programming system tries to satisfy the requirements taking into account peculiarities of any particular heterogeneous network of computers.

| | 0 | 1 | 2 | 3 | 4 | 5 | 6 | 7 | 8 | 9 | 10 | 11 |
|---|---|---|---|---|---|---|---|---|---|---|---|---|
| 0 | 0 | 0 | 0 | 1 | 1 | 2 | 0 | 0 | 0 | 1 | 1 | 2 |
| 1 | 0 | 0 | 0 | 1 | 1 | 2 | 0 | 0 | 0 | 1 | 1 | 2 |
| 2 | 3 | 3 | 3 | 4 | 4 | 2 | 3 | 3 | 3 | 4 | 4 | 2 |
| 3 | 3 | 3 | 3 | 4 | 4 | 5 | 3 | 3 | 3 | 4 | 4 | 5 |
| 4 | 0 | 0 | 0 | 1 | 1 | 2 | 0 | 0 | 0 | 1 | 1 | 2 |
| 5 | 0 | 0 | 0 | 1 | 1 | 2 | 0 | 0 | 0 | 1 | 1 | 2 |
| 6 | 3 | 3 | 3 | 4 | 4 | 2 | 3 | 3 | 3 | 4 | 4 | 2 |
| 7 | 3 | 3 | 3 | 4 | 4 | 5 | 3 | 3 | 3 | 4 | 4 | 5 |
| 8 | 0 | 0 | 0 | 1 | 1 | 2 | 0 | 0 | 0 | 1 | 1 | 2 |
| 9 | 0 | 0 | 0 | 1 | 1 | 2 | 0 | 0 | 0 | 1 | 1 | 2 |
| 10 | 3 | 3 | 3 | 4 | 4 | 2 | 3 | 3 | 3 | 4 | 4 | 2 |
| 11 | 3 | 3 | 3 | 4 | 4 | 5 | 3 | 3 | 3 | 4 | 4 | 5 |

(a) matrix distribution

| | 0 | 1 | 2 | 6 | 7 | 8 |
|---|---|---|---|---|---|---|
| 0 | 0 | 0 | 0 | 0 | 0 | 0 |
| 1 | 0 | 0 | 0 | 0 | 0 | 0 |
| 4 | 0 | 0 | 0 | 0 | 0 | 0 |
| 5 | 0 | 0 | 0 | 0 | 0 | 0 |
| 8 | 0 | 0 | 0 | 0 | 0 | 0 |
| 9 | 0 | 0 | 0 | 0 | 0 | 0 |
| 2 | 3 | 3 | 3 | 3 | 3 | 3 |
| 3 | 3 | 3 | 3 | 3 | 3 | 3 |
| 6 | 3 | 3 | 3 | 3 | 3 | 3 |
| 7 | 3 | 3 | 3 | 3 | 3 | 3 |
| 10 | 3 | 3 | 3 | 3 | 3 | 3 |
| 11 | 3 | 3 | 3 | 3 | 3 | 3 |

| | 3 | 4 | 9 | 10 |
|---|---|---|---|---|
| 0 | 1 | 1 | 1 | 1 |
| 1 | 1 | 1 | 1 | 1 |
| 4 | 1 | 1 | 1 | 1 |
| 5 | 1 | 1 | 1 | 1 |
| 8 | 1 | 1 | 1 | 1 |
| 9 | 1 | 1 | 1 | 1 |
| 2 | 4 | 4 | 4 | 4 |
| 3 | 4 | 4 | 4 | 4 |
| 6 | 4 | 4 | 4 | 4 |
| 7 | 4 | 4 | 4 | 4 |
| 10 | 4 | 4 | 4 | 4 |
| 11 | 4 | 4 | 4 | 4 |

| | 5 | 11 |
|---|---|---|
| 0 | 2 | 2 |
| 1 | 2 | 2 |
| 2 | 2 | 2 |
| 4 | 2 | 2 |
| 5 | 2 | 2 |
| 6 | 2 | 2 |
| 8 | 2 | 2 |
| 9 | 2 | 2 |
| 10 | 2 | 2 |
| 3 | 5 | 5 |
| 7 | 5 | 5 |
| 11 | 5 | 5 |

(b) distribution from processor point-of-view

**Fig. 2.** Example of a heterogeneous block cyclic distribution of a 12x12 matrix over 2x3 processor grid with the average block size 2x2.

The main idea underlying mpC is that an mpC application explicitly defines a dynamic abstract heterogeneous computing network and distributes data, computations and communications over the network. The mpC programming system uses this information in run time to map the abstract computing network to any real executing network of computers in such a way that ensures efficient running of the application on this real network.

The mpC language is an ANSI C superset that introduces a new kind of managed resource, *computing space*, defined as a set of virtual processors of difference performances connected with links of different communication speeds. In run time, the virtual processors are represented by actual processes of the particular running parallel application. The programmer manages the computing space by means of creating and discarding regions of the computing space, named *network objects*, just like he manages storage creating and discarding data objects (regions of storage). At any moment of program execution, just a set of defined network objects represents the abstract computing network.

mpC application used in experiments, implements the HoHe strategy in three steps:

1. The information about performances of processors of the executing real network is updated in run time by means of execution of corresponding computations as a benchmark (namely, Cholesky factorization of a small matrix by means of the LAPACK routine `dpotf2`).
2. A network object, executing the corresponding computations, is defined in such a way that each of the $P \cdot Q$ most powerful processors of the executing network will execute only one process involved in the computations, and all the involved processes form a $P$x$Q$ process grid.

3. A driver of the Cholesky factorization reads from a file problem parameters (matrix and block sizes), forms a distributed test matrix and performs its Cholesky factorization on the PxQ process grid. Each virtual processor of the network object computes a portion of the matrix, proportional to its performance in accordance with the heterogeneous block cyclic matrix distribution described in section 2.

The above 2-dimensional strategy is a generalization of the 1-dimensional strategy based on heterogeneous data distribution and presented in [6]. The latter describes in details an mpC application carring out Cholesky factorization for that strategy.

## 4 Experimental results

We compared two distribution strategies:

- The HoHo strategy - "Homogeneous distribution of processes over processors - Homogeneous distribution of data over the processes" - the traditional distribution strategy implemented in ScaLAPACK.
- The HoHe strategy - "Homogeneous distribution of processes over processors - Heterogeneous distribution of data over the processes" implemented in mpC.

The comparison was performed for the Cholesky factorization on a network of workstations. For our experiments, we used a part of a local network consisting of 6 uniprocessor Sun workstations of different performances interconnected via 10 Mbits Ethernet. MPICH 1.0.13 was used as a particular communication platform. All workstations executed the same copy of code. Performances of the workstations, obtained by means of execution of the LAPACK routine dpotf2 performing serial Cholesky factorization, is shown in table 1.

| 1 | 2 | 3 | 4 | 5 | 6 |
|---|---|---|---|---|---|
| 190 | 190 | 190 | 190 | 100 | 280 |

**Table 1.** Performances of processors demonstrated on serial Cholesky factorization

In all our experiments we used *parallelization efficiency* as a factor characterizing how fully the parallel application utilizes the performance potential of the executing network of computers. Parallelization efficiency was defined as $S_{real}/S_{ideal}$, where $S_{real}$ was the real speedup achieved by the parallel application on the parallel system, and $S_{ideal}$ was the ideal speedup that could be achieved while parallelizing the problem. The latter was calculated as the sum of performances of processors, constituting the executing parallel system, divided by the performance of a base processor. All real speedups were calculated relative to the sequential LAPACK routine dpotf2 performing Cholesky factorization on the base processor.

Figure 3 presents the parallelization efficiency achieved by the compared distribution strategies while running on the homogeneous network 1-2-3-4 consisting of 4 workstations 1, 2, 3, and 4 with the optimal process grid (2x2) and block sizes. One can see that for large matrices the parallelization efficiency of the HoHo strategy is 8% higher than that of the HoHe strategy. Note, that the result is not due to the non-Cartesianity of the heterogeneous block cyclic matrix distribution (as shown in figure 2), since the same result was also obtained for a mpC application implementing the homogeneous block cyclic matrix distribution. So, it can be explained by more efficient implementation of Cholesky factorization in ScaLAPACK.

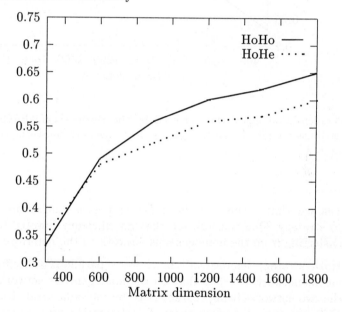

**Fig. 3.** Parallelization efficiencies achieved by the HoHo (ScaLAPACK) and HoHe (mpC) distribution strategies on the homogeneous network consisting of workstations 1, 2, 3, and 4.

Now, let us consider the heterogeneous network 1-2-5-6, consisting of workstations 1, 2, 5 and 6 and having the same total power of processors as the homogeneous network 1-2-3-4. Parallelization efficiencies achieved by the different distribution strategies for this heterogeneous network are shown in figure 4 (as before, optimal sizes of process grid and block were used).

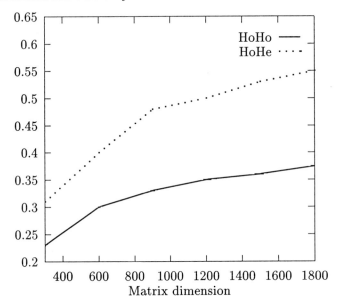

**Fig. 4.** Parallelization efficiencies achieved by the HoHo (ScaLAPACK) and HoHe (mpC) distribution strategies on the heterogeneous network consisting of workstations 1, 2, 5, and 6.

One can see, that in this case the HoHe strategy is much more efficient then the HoHo strategy. One can also see that its efficiency lowered insignificantly while transferring from the homogeneous network to the heterogeneous one.

The HoHo strategy demonstrates much worse parallelization efficiency on the heterogeneous network 1-2-5-6 than on the homogeneous network 1-2-3-4, in spite of the two networks being characterized by the same total processor performance - 760 (the sum of performances of participating processors). It conforms to the statement formulated in Introduction that the total running time (and, hence, the parallelization efficiency) provided by the HoHo strategy on a heterogeneous network is approximately the same as on a homogeneous network obtained from this heterogeneous one by means of replacement of both its processors with the weakest processor and its communication links with the slowest link. Indeed, the parallelization efficiency of the corresponding application on the network 1-2-3-4 is 1.7 times higher than on the network 1-2-5-6. At the same time, according to the above statement, this factor should be close to the ratio of performances of the weakest processors in networks 1-2-3-4 and 1-2-5-6, that is, be close to 1.9. Note, that the faster are communication links and the less powerful are processors, the higher parallelization efficiency is achieved. Therefore, the factor should be a little bit less than 1.9, since the network 1-2-5-6 has better ratio of the communication speed and the weakest processor performance than the network 1-2-3-4. Thus, we can conclude that the obtained experimental results are in good conformance with the above statement.

Figure 5 shows the parallelization efficiency achieved by the two distribution strategies on the heterogeneous network 1-2-3-4-5-6 consisting of all 6 workstations. As before, optimal sizes of process grid and block were used for each of the strategies (3x2 process grid turned out optimal for the both strategies). In this case, as before, the HoHe strategy is much more efficient then the HoHo strategy.

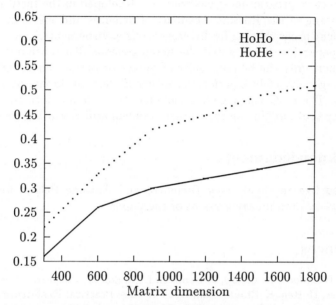

**Fig. 5.** Parallelization efficiencies achieved by the HoHo (ScaLAPACK) and HoHe (mpC) distribution strategies on the heterogeneous network consisting of workstations 1, 2, 3, 4, 5, and 6.

## 5  Conclusion

Numeric software developed for computations in homogeneous environments does not allow to utilize all performance potential of heterogeneous networks. It has been demonstrated that in this case a heterogeneous network is equal to some homogeneous network obtained from the heterogeneous one by means of replacement its processors with the weakest processor.

A natural way to answer this challenge is to develop dedicated numeric software aimed at heterogeneous environments. Such software should take into account the heterogeneity of processor performances and speeds of communication links and support heterogeneous distribution of processes, involved in computations, over processors and/or data over the processes. The distribution strategy presented in the paper is based on heterogeneous block cyclic distribution of data. It has been shown that for heterogeneous parallel environments the heterogeneous strategy is much more efficient than the traditional homogeneous strategy.

Implementation of heterogeneous parallel algorithms needs appropriate porgramming languages and tools supporting and facilitating programming heterogeneous computations. In our research, we use the mpC parallel language and its free supportive programming environment developed in the Institute for System Programming of the Russian Academy of Sciences and just aimed at portable and efficient programming for heterogeneous environments.

The paper has investigated the heterogeneous distribution strategy taking into account only the heterogeneity of processor performances. Obviously, that the heterogeneity of link performances has no less an impact on parallelization efficiency. The mpC language has some means for taking into account that heterogeneity too, but this more complex problem still is waiting its solution.

## 6 Acknowledgments

We would like to thank Jack Dongarra and Antoine Petitet for their useful comments on preliminary versions of the paper.

## References

1. L. S. Blackford, A. Cleary, J. Demmel, I. Dhillon, J. Dongarra, S. Hammarling, A. Petitet, H. Ren, K. Stanley, and R. C. Whaley Practical Experience in the Dangers of Heterogeneous Computing UT, CS-96-330, July 1996.
2. D.Arapov, A.Kalinov, A.Lastovetsky, I.Ledovskih, and T.Lewis "A Programming Environment for Heterogeneous Distributed Memory Machines", *Proceedings of the Sixth Heterogeneous Computing Workshop (HCW'97)*, IEEE Computer Society Press, Geneva, Switzerland, April 1, 1997.
3. A.Lastovetsky, The mpC Programming Language Specification. Technical Report, ISPRAS, Moscow, December 1994.
4. B.Hendrickson and D.Womble,"The Torus–wrap Mapping for Dense Matrix Calculations on Massively Parallel Computers", SIAMSSC, 15(5), 1994.
5. J. Choi, J. J. Dongarra, S. Ostrouchov, A. P. Petitet, D. W. Walker, and R. C. Whaley "The Design and Implementation of the ScaLAPACK LU, QR, and Cholesky Factorization Routines" UT, CS-94-246, September, 1994.
6. D.Arapov, A.Kalinov, A.Lastovetsky and I.Ledovskih "Experiments with mpC: Efficient Solving Regular Problems on Heterogeneous Networks of Computers via Irregularization", *Proceedings of the Fifth International Symposium on Solving Irregularly Structured Problems in Parallel (IRREGULAR'98)*, Lecture Notes in Computer Science 1457, Berkley, CA, USA, August 9-11, 1998.

# Modeling and Improving Locality for Irregular Problems: Sparse Matrix–Vector Product on Cache Memories as a Case Study

Dora Blanco Heras, Vicente Blanco Pérez,
José Carlos Cabaleiro Domínguez, and Francisco Fernández Rivera.

Dept. Electronics and Computer Science. Univ. Santiago de Compostela
Campus Sur. 15706 Santiago de Compostela. Spain.
dora,Vicente.Blanco,caba,fran@dec.usc.es

**Abstract.** *In this paper we introduce a model for representing and improving the locality of sparse matrices for irregular problems. We focus our attention on the behavior of iterative methods for the solution of sparse linear systems with irregular patterns. In particular the product of a sparse matrix by a dense vector (SpM×V) is closely examined, as this is one of the basic kernels in such codes. As a representative level of the memory hierarchy, we consider the cache memory. In our model, locality is measured taking into account pairs of rows or columns of sparse matrices. In order to evaluate this locality four functions based on two parameters called entry matches and cache line matches are introduced. Using an analogy of this problem to the Traveling Salesman Problem we have applied two algorithms in order to solve it; one based on the construction of minimum spanning trees and the other on the nearest-neighbor heuristic. These techniques were tested over a set of sparse matrices. The results were assesed through the measurement of cache misses on a standard cache memory.*

**Keywords:** sparse matrix, temporal locality, spatial locality, memory hierarchy, cache memory, iterative methods, optimization problem.

## 1 Introduction

The architectures in present machines use the concept of memory hierarchy to improve data access and consequently to reduce the increasing gap between processor and memory speeds and improving the performance in the execution of a program. Memory hierarchy typically consists of registers and caches together with primary and secondary memories in uniprocessor systems and remote memories in multiprocessor systems. The characteristics that determine the behavior of a program in a particular memory hierarchy are its temporal and spatial locality. Thus, the analysis of the data locality and the quest for its increase are fundamental questions if we are trying to improve the performance of a program [6].

It has been demonstrated that the use of memory hierarchy is effective in the execution of dense numerical codes. However, due to its indirect addressing, memory hierarchy is generally considered inefficient for irregular codes [11]. The task of evaluating and improving the locality of such codes is specially difficult on multiprocessors. On these machines there is an additional level of memory hierarchy, the one created by the memory accesses to data placed in other processors [7, 12].

In this work we focus on the behavior of iterative methods for the solution of sparse linear systems with irregular patterns. In particular we are going to closely examine the product of a sparse matrix by a dense vector ($SpM \times V$) as this is one of the basic kernels in such codes. To analyze our proposal we concentrate our study on the performance of the cache for $SpM \times V$ which depends on the values of several features arising both from the sparse matrix participating in the operation and from the particular cache memory.

## 2 Methods for Increasing the Locality of Numerical Codes

In the literature a large number of algorithms for evaluating and optimizing data locality can be found. In the case of dense codes, most approaches are based on decreasing capacity and interference misses by using *blocking* or other restructuring techniques [9]. It is well known that techniques that improve locality for a problem in an uniprocessor system will obtain an improvement when the code is executed in multiprocessor systems [2, 6, 12].

In the case of the sparse–matrix vector product ($SpM \times V$), a classic method for dealing with the problem of improving locality is applying reordering techniques like *bandwidth reduction* algorithms which derive from *Cuthill and McKee's algorithm*, and the *minimum–degree* algorithm [4].

The additional cost of memory produced by the construction and storage of the reordered matrix can be overcome in most cases. For example, when the same matrix has to be used to carry out a LU factorization and a series of $SpM \times V$, it is useful to store a reordered copy of the matrix. This reordered matrix will only be used for the $SpM \times V$. Often, such as in iterative methods [1], several products are performed sequentially. The only term that changes in the sequential products is the vector. Thus the computational cost of reordering is globally overcome.

## 3 A Qualitative Study of Locality within $SpM \times V$

Let $N$ be the number of rows or columns of a sparse matrix and $N_Z$ the total number of non–zero entries of the sparse matrix. $Cs$ is the cache size and $Ls$ is the cache line size. Figure 1(a) shows the code which corresponds to the sparse

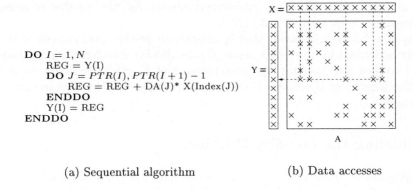

(a) Sequential algorithm  (b) Data accesses

**Fig. 1.** Algorithm for the product of a sparse matrix by a vector

matrix–vector product when the matrix is stored in *Compressed Row Storage* (CRS) format [1]. *DA*, *Index* and *PTR* are the three vectors (data, columns and row pointer) corresponding to this storage format. It provides simplicity in the access to data and flexibility to distribute data stored in this format among different processors. The accesses required by the different data structures in order to perform the product are displayed in Fig. 1(b).

Throughout this text we use the classification of cache misses proposed by Hill [8]: *compulsory misses*, *capacity misses* and *conflict misses*.

In $SpM \times V$ there are $N_Z$ references to arrays $X$, *Index* and *DA* and $N$ references to arrays $Y$ and *PTR*. The total number of references is therefore $3 \times N_Z + 2 \times N$. It can thus be verified that:

$$Total_{cachemisses} \leq (3 \times N_Z + 2 \times N)/Ls \ .$$

Arrays $Y$ and *PTR* present sequential accesses with a high spatial locality, and are not reused, so there is no temporal locality. Thus, these arrays produce mainly compulsory misses. Each of them contribute to the total number of misses with a maximum of $N/Ls$ and if, as usually, $N \ll N_Z$ then this contribution is not significant, and more important, it is almost constant. In the case of references *PTR(I)*, *PTR(I+1)*, the required data will probably remain in the cache memory between two consecutive iterations of the $I$ loop. This is because the total number of references accessed between two iterations of the $I$ loop will be at most $3 \times N$ (if row $I$ of the matrix were dense).

Arrays *DA* and *Index* present spatial locality as their elements are accessed consecutively throughout the algorithm. Their total number of references is $2 \times N_Z$. The misses produced by the accesses to these arrays are mainly compulsory although they can cause interference misses. So, the number of misses in their accesses is at least $(2 \times N_Z)/Ls$.

Array $X$ exhibits a behavior that is difficult to predict and exploit as it is indirectly addressed by the *Index* array. It may display interference misses and it must be taken into account that each element of $X$ will be reused at best $N$ times. The locality of the $SpM{\times}V$ can be improved mainly for the accesses to the vector $X$.

## 4 Modeling the Locality Problem

The locality of the $SpM{\times}V$ operation in the cache level depends among others on some features from the system like the size of the cache, the size of the cache lines, the associativity level of the cache and the replacement algorithm employed. It also depends on features from the sparse matrix like the location of the entries in each row, in terms of the number of cache lines needed to process this row or the number of cache lines of vector $X$ that are reused in the processing of the next rows.

In order to model the locality, a simple approach in which the locality is carried out over consecutive row or column pairs was used. However this model can be generalized by considering groups of rows/columns and not only two consecutive ones because there would be a potential reuse of a cache line, which stores a portion of vector $X$, in the product of two non consecutive rows of the matrix.

We propose reordering the matrix to increase the locality of the accesses to vector $X$. Two permutation vectors, $R$ and $C$, store the permutations of rows and columns. Thus:

$$\begin{aligned} AX &= b \\ (RAC)(C^{-1}X) &= Rb \end{aligned}$$

## 5 Parameters to Model Locality: Distance Functions

We will introduce certain parameters that define a magnitude we will call *distance between pairs of rows (x,y)*, denoted by $D(x,y)$, so that the smaller the value of this parameter the closer in terms of locality the two rows will be:

$$\forall \text{ rows } x, y, z. \quad D(x,y) \leq D(x,z) \Leftrightarrow locality(x,y) \geq locality(x,z) .$$

That is, if a relative order in terms of distance is established, the order in terms of locality will be the inverse.

To reflect the access locality induced in vector $X$ when computing the product of $X$ by any two rows of the sparse matrix, we consider two parameters: *entry matches* ($a_{elems}$) and *cache line matches* ($a_{lines}$) (See Fig. 2).

The number of *entry matches* between two rows of a sparse matrix is defined as the number of times that there are two non–zero elements in the same column.

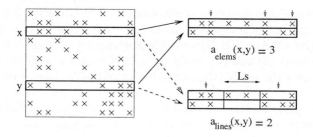

**Fig. 2.** Measurement of locality parameters

It is obvious that this situation reflects locality as the same elements of vector $X$ are reused in the product of both rows. Therefore the larger the number of *entry matches* between two rows, the larger the temporal locality shown by the accesses to vector $X$ in the product. On the other hand, each row of the matrix can be organized in $N/Ls$ blocks of size $Ls$ as shown in Fig 2. The concept of *entry matches* can be extended to *cache line matches* replacing "entries" by "blocks with at least one entry". The same concepts can be defined for columns. These parameters in terms of rows are related to the temporal locality of $X$ and, in terms of columns are related to the spatial locality of $Y$.

Without loosing generality we have considered that all the vectors will be stored starting in the first position of a cache line. This way we can avoid the sharing of cache lines between arrays and preserve the generality of our argument.

We propose several functions for defining the distance between two rows $x$ and $y$ according to the parameters we have introduced.

$$D_1(x,y) = maxa_{elems} - a_{elems}(x,y) \tag{1}$$

$$D_2(x,y) = \frac{N}{a_{lines(x,y)} + \alpha * a_{elems}(x,y) + 1} \tag{2}$$

$$D_3(x,y) = n_{lines}(x) + n_{lines}(y) - 2*a_{lines}(x,y) \tag{3}$$

$$D_4(x,y) = n_{elems}(x) + n_{elems}(y) - 2*a_{elems}(x,y) \tag{4}$$

The first of these functions establishes the distance between two rows as a difference between parameter $maxa_{elems}$ and the number of actual *entry matches* for these rows. $maxa_{elems}$ is defined as the maximum number of *entry matches* between a pair of rows of the sparse matrix. In the denominator of the second function a parameter $\alpha$ is included to weight the total number of *entry matches* between two rows with regard to the total number of *cache line matches* between them.

The third and fourth functions only differ in that one evaluates the locality between rows based on *cache line matches* and the other based on *entry matches*. $n_{elems}(x)$ is the number of entries in a row and $n_{lines}(x)$ is the number of blocks of size $Ls$ corresponding to row $x$ with at least one entry.

These functions are injective applications and they verify the triangle inequality. This inequality states that given any three vertexes of a graph $G$ ( $x$, $u$ and $v$) and given the weight $D(u,v)$. Then: $D(u,v) \leq D(u,x) + D(x,v)$. However, only functions (3) and (4) verify: $D(u,v) = 0 \Leftrightarrow x = y$. So both functions define metrics over the set $N$ (the one made up of the $N$ rows or columns of the matrix).

For a given sparse matrix the locality can be indirectly modelled by summing the distances between consecutive rows ($D_{total}$). The same could be stated for columns.

$$D_{total} = \sum_{i=0}^{N-2} D(i, i+1) .$$

In particular for the distance function (4) we have:

$$D_{total} = n_{elems}(0) + \sum_{i=1}^{N-2} 2 * n_{elems}(i) + n_{elems}(N-1) - \sum_{i=0}^{N-2} 2 * a_{elems}(i, i+1) .$$

## 6  Solving the Locality Problem

We formulate the problem of improvement of locality as a *Traveling Salesman Problem*. This is a classic combinatorial optimization problem, an NP-complete in particular. Solving it is equivalent to finding a path of the shortest length possible in a complete weighed graph [10]. In our case the function that indicates the weight of each edge satisfies the triangular inequality, so we have a symmetric TSP. The weight function reflects the *distance* between rows/columns according to the description of locality given previously.

There are different heuristics for solving it. We have opted for one based on spanning trees which is divided into two steps:

– Construction of a minimum–spanning tree of the complete graph using the *Prim algorithm* [10, 5].
– Visiting the different vertexes of the tree to establish an order of the nodes using a *depth–first search* [5].

To compare the results obtained with this algorithm we have also solved our analogy of TSP with the greedy heuristic called *nearest–neighbor algorithm* [10].

Each ordering of the graph nodes leads to a sparse matrix which is different from the original and will produce a different number of cache misses when the *SpMxV* is carried out. The minimum number of cache misses for a matrix when performing *SpMxV*, which can only be achieved with a large enough cache, will correspond to the case in which all the vectors involved in the operation are loaded once in the cache and remain until the operations are completed (infinite cache). The quality each ordering obtained can be established by comparing the number of cache misses produced with the ordering with the minimum number of cache misses for the original sparse matrix.

# 7 Results

As a test set for validating our model, we made use of several matrices with different patterns. We have selected three matrices from the *Harwell–Boeing* sparse matrix library [3] obtained from real problems and two matrices with random sparsity patterns. Fig. 3 depicts their patterns.

For solving our locality problem both, minimum–spanning trees (*Pi*) and *nearest–neighbor* heuristics (*Pi nearest*) were applied using the four distance functions introduced in Sect. 5. The notation for each of the eight resulting ways of solving includes a number which indicates the distance function used ($D_1$, $D_2$, $D_3$ or $D_4$). Thus, for each of the five matrices described above, eight different orderings were obtained.

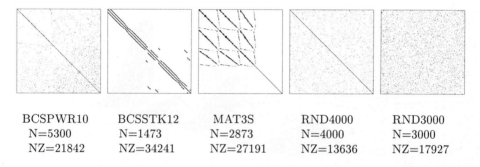

| BCSPWR10 | BCSSTK12 | MAT3S | RND4000 | RND3000 |
| N=5300 | N=1473 | N=2873 | N=4000 | N=3000 |
| NZ=21842 | NZ=34241 | NZ=27191 | NZ=13636 | NZ=17927 |

**Fig. 3.** Sparse matrices for testing

The number of cache misses was measured for each one of the matrices and varying the cache size. For all the matrices considered, when the cache size is at least *64K* words it behaves like a infinite cache (one so large it never replaces a line) and consequently there are no conflict misses nor possible improvement in the number of total misses for this specific cache configuration. Therefore, the maximum cache size tested was *32KW*. This analysis may be extrapolated to much larger memory caches and matrices. As a case study, all the measurements were carried out using a two–way set-associative cache model. The line size was 8 words, LRU as replacement algorithm and no prefetching was considered. Increasing the associativity would reduce conflict misses in the same way that increasing line size or including prefetching would reduce the compulsory misses. So, making any of the described changes on the cache organization the number of misses would be nearer the minimum. In this case, the improvement achieved by our technique would be smaller.

Figure 5 shows the total number of cache misses for matrix BCSPWR10 and the eight resulting reordered matrices. The best results are obtained for *P1* and *P4*. This behavior has also been observed for the remaining sparse matrices in the test set. For example, in Fig. 4 the behavior for the MAT3S matrix follows

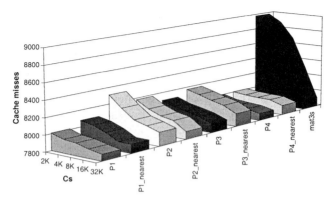

**Fig. 4.** Application of different locality optimizations to matrix MAT3S

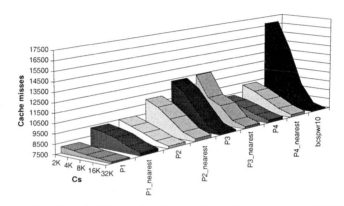

**Fig. 5.** Application of different locality optimizations to matrix BCSPWR10

the same general trend. Note that *P1* and *P4* correspond to cases for which the distance functions are constructed from the *entry matches* locality parameter which leads to closer groupings of the matrix entries than the *cache line matches* parameter.

In Figs. 6 and 7 we present the results obtained for the different matrices using algorithms *P1* and *P4*, which produce the best overall results for these matrices.

Figure 6 displays the behavior, in terms of cache misses, for the different sparse matrices when solving according to *P1*. To compare this algorithm over the different matrices the percentages of reduction of misses with respect to the original matrix were represented.

Note that the behavior, is not exactly the same for all the matrices. Some of the results, such as part of those corresponding to BCSSTK12, are below the *xy plane*. This means that in these cases no improvement was achieved with respect to the original ordering. This is because matrix BCSSTK12 presents a sparsity

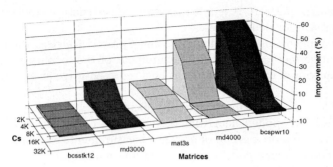

**Fig. 6.** Relative cache misses for different sparse matrices when solving $SpM{\times}V$ according to $P1$

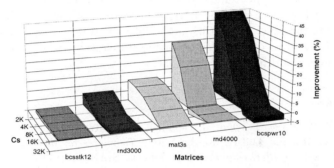

**Fig. 7.** Relative cache misses for different sparse matrices when solving $SpM{\times}V$ according to $P4$

pattern that leads to a high level of locality in the $SpM{\times}V$ product. Numerically, BCSSTK12 leads to 9348 misses for a $2K$ word cache, and the absolute minimum of cache misses for this matrix, as described in Sect. 6, is 9117. Thus, we see that for this matrix, whose N=1473, our improvement margin is no more than 231 cache misses, that is, only 2.47% of the number of misses obtained by the matrix to start with. In the case of matrix BCSPWR10 the improvement margin was 55.6% as we had 16779 misses for a $2K$ cache and the minimum is 7451 misses. Note that our reordering comes very close to this value.

The same trend as for $P1$ may be observed in Fig. 7 using the $P4$ method. In this case, the percentages of improvement are slightly smaller.

## 8  Conclusions and Future Work

In this paper a methodology for characterizing and increasing the locality of sparse matrices based on the definition of parameters associated with the order of data accesses is proposed. We achieve an improvement of locality by modifying the sparse matrix pattern. It is not necessary to restructure the code as in the classical approach for dense codes.

We have applied this methodology to the $SpM{\times}V$ which is the main kernel of most iterative methods. For the measurement of the locality we have considered the behavior of cache memories to be a representative level of the memory hierarchy.

The problem of optimizing the locality has been solved using an analogy to the TSP. We have applied heuristics based on the construction of *minimum–spanning trees* based on a group of distance measurements as a function of a set of two parameters: *entry matches* and *cache line matches*.

On applying this methodology significant improvements in our matrices were obtained. A limit to the improvement of locality may be calculated and thus it can be decided if it is advantageous or not to apply our methodology.

This analysis and improvement in locality can be extended to other numerical kernels which present different irregular accesses such as the LU, triangular solvers, etc. On the other hand, this methodology can be extrapolated to caches of a larger size and to other levels of memory hierarchy, whose impact on the total CPU time would be more important.

# References

1. R. Barret, M. Berry, T. Chan, J. Demmel, J. Donato, J. Dongarra, V. Eijkhout, R. Pozo, C. Romine, and H. van der Vorst. *Templates for the Solution of Linear Systems: Building Blocks for Iterative Methods.* SIAM Press, 1994.
2. Ken Chen. A study on the cache memory miss ratio issue in multiprocessor systems. Technical report, INRIA, Institut National De Recherche en Informatique et en Automatique, October 1990.
3. I.S. Duff, R.G. Grimes, and J.G. Lewis. User's guide for the harwell–boeing collection. Technical report, CERFACS, 1992.
4. A. George. *Direct solution of sparse positive definite systems: some basic ideas and open problems.* I. S. Duff, Academic Press, 1981.
5. Alan Gibbons. *Algorithmic Graph Theory.* Cambridge University Press, 1984.
6. J.L. Hennesy and D.A. Patterson. *Computer architectures: a quantitative approach.* Morgan Kaufman Publishers, Palo Alto, 1990.
7. Mark D. Hill and James R. Larus. Cache considerations for multiprocessors programmers. In *Communications of the ACM*, volume 33, pages 97–102. 1990.
8. M.D. Hill. *Aspects of Cache Memory and Instruction Buffer Performance.* PhD thesis, University of California, Berkeley, 1987.
9. J.J. Navarro, E. García, J.L. Larriba-Pey, and T. Juan. Block algorithms for sparse matrix computations on high performance workstations. Proc. IEEE Int'l. Conf. on Supercomputing (ICS'96), pages 301–309, 1996.
10. Gerhard Reinelt. *The Traveling Salesman. Computational Solutions for TSP applications.* Lecture Notes in Computer Science. Springer–Verlag, 1991.
11. O. Temam and W. Jalby. Characterizing the behavior of sparse algorithms on caches. Int'l Conf. on Supercomputing (ICS'92), pages 578–587, 1992.
12. Evan Torrie, M. Martonosi, C. Tseng, and M.W. Hall. Characterizing the memory behavior of compiler–parallelized applications. *IEEE Transactions on Parallel and Distributed Systems*, 7(6), December 1996.

# Parallelization of Sparse Cholesky Factorization on an SMP Cluster

Shigehisa Satoh, Kazuhiro Kusano, Yoshio Tanaka,
Motohiko Matsuda, and Mitsuhisa Sato

Tsukuba Research Center
Real World Computing Partnership
1-6-1 Takezono, Tsukuba, Ibaraki 305-0032, JAPAN

**Abstract.** In this paper, we present parallel implementations of the sparse Cholesky factorization kernel in the SPLASH-2 programs to evaluate performance of a Pentium Pro based SMP cluster. Solaris threads and remote memory operations are utilized for intranode parallelism and internode communications, respectively. Sparse Cholesky factorization is a typical irregular application with a high communication to computation ratio and no global synchronization between steps. We efficiently parallelized using asynchronous message handling instead of lock-based mutual exclusion between nodes, because synchronization between nodes reduces the performance significantly. We also found that the mapping of processes to processors on an SMP cluster affects the performance especially when the communication latency can not be hidden.

## 1 Introduction

Recent progress in microprocessors and interconnection networks motivated a trend towards high performance computing using clusters made out of commodity hardware. Symmetric multiprocessors(SMPs) are also becoming widely available. As a result, clusters of SMPs are expected to be one of the most cost-effective parallel computing platforms. In this context, we have built a PC-based SMP cluster, named COMPaS, which consists of eight quad-processor Pentium Pro SMPs connected by a Myrinet high speed network[1][2].

COMPaS(Cluster Of Multi-Processor Systems) consists of eight quad-processor Pentium Pro PC servers connected to a Myrinet switch. Each node is a Toshiba GS700 with four 200MHz Pentium Pro CPUs and has 512KB L2 caches and 512MB of main memory. The operating system on each node is Solaris 2.5.1 which supports both the Solaris thread library and the POSIX thread library. Our user-level communication layer for Myrinet, called NICAM, provides low-overhead and high-bandwidth communication. NICAM supports remote memory data transfer primitives and global synchronization primitives. Parallel programs for COMPaS can be made using Solaris threads for intranode parallelism, and remote memory operations for communications between nodes.

Our previous work based on some regular applications produced the following guidelines for effective use of COMPaS: (1) Cache-locality is crucial to attain high performance in an SMP node, because the shared-bus bandwidth is not sufficient for memory intensive programs, (2) Internode communication latency can be hidden by overlapping communication and computation. Though it is clear that synchronization between threads in the same node are faster than ones between processes running on different nodes, the performance bottlenecks for regular applications we have parallelized so far have to do with shared-bus bandwidth rather than synchronization overheads. The bus traffic can be mitigated by effective use of cache memories.

The synchronization performance is more important for irregular applications than regular applications. Sparse Cholesky factorization is one such irregular application. Irregular structure of matrices introduces irregular patterns of memory accesses and synchronization. Fortunately, a good deal of the factorization process can be performed by solving regular subproblems, i.e. computations for small dense matrices[3][4]. Blocked fan-out algorithm[3] is such an algorithm and is the basis of the Cholesky kernel in the SPLASH-2 programs[5].

The SPLASH-2 programs are a collection of parallel programs for shared address space multiprocessors. Although there are several studies which use the SPLASH-2 programs for performance evaluation, a few works are done for Cholesky on clusters. Some studies on clusters of uniprocessors use Cholesky for evaluation of software DSM systems[6][7][8]. These works show that the efficient parallel execution of Cholesky is difficult due to high synchronization overheads.

In this paper, we present parallel implementations and performance evaluation of Cholesky on an SMP cluster COMPaS. The original program is already parallelized using PARMACS macros for shared address space multiprocessors. We modify this parallel program for SMP clusters using Solaris threads and remote memory operations. The problems for such parallelization are as follow: (1) How to share data between nodes. (2) How to implement synchronization primitives. (3) How to assign processes to processors. The data sharing mechanism and the process to processor assignment policy are important for Cholesky.

There are two kinds of shared data in Cholesky: nonzero elements of the factor matrix and task queues for each of the processes. We use remote memory operations to keep these data coherent. The sharing patterns of nonzero elements and task queues are one-producer with multiple-consumers and migratory, respectively. Migratory sharing between nodes is quite expensive, because synchronization overhead between nodes is larger than one within the same node.

We have implemented two versions to share task queues, which have quite different synchronization characteristics. One is using messages which can be sent and received asynchronously with sender and receiver processes, respectively. The other is using global locks for mutual exclusion, which may induce the lock-serialization.

The mapping of processes to processors on each SMP node is also investigated, because the intranode communication is significantly faster than the internode communication.

This paper is organized as follows: Section 2 briefly describes characteristics of Cholesky. Parallel implementations of Cholesky for COMPaS is described in

1: Compute nonzero elements in the domains assigned to the process
2: Send initial tasks to each process
3: **while** the factorization is not completed **do**
4:     If the task queue of the process is empty, wait until a new task arrives
5:     Get a task from the task queue of the process
6:     Update nonzero elements in the blocks assigned to the process
7:     Send new tasks to each process
8: **done**

**Fig. 1.** Outline of the Parallel Region

section 3. Performance results and a discussion of them are presented in section 4. We conclude this work in section 5.

## 2   Characteristics of Cholesky

Cholesky is considered to be a typical irregular program, because it is characterized by a large communication to computation ratio and no global synchronization. Only the numerical factorization step of the factorization is parallelized since it is the most time-consuming step. Figure 1 shows the outline of the parallel numerical factorization step in Cholesky. In this paper, the term 'process' means a thread of control bound to each processor on an SMP cluster, not a heavy weight process.

Nonzero elements of the factor matrix are divided into domains and blocks. Each process computes the values of domains and blocks which are statically assigned to the process(lines 1 and 6 in the figure). Though elements of the domains are computed independently of other domains and blocks, elements of the blocks are dependent of some domains and blocks which may be assigned to other processes. Therefore, synchronization are required to update the values of each block.

Task queues are used for synchronization between processes. When the computation of a domain or a block is completed, its owner process notifies other processes by sending tasks to them. Each process has its own task queue. Any process can enqueue new tasks to other process's task queues, but only the owner process of the task queue can dequeue tasks from it. In figure 1, the enqueuing of tasks occurs in lines 2 and 7, and the dequeuing of tasks occurs in line 5.

There are three types of shared data in Cholesky: domains, blocks, and task queues. To parallelize Cholesky on COMPaS, we modify the original program to share these data between nodes.

Domains and blocks are data structure used to store nonzero elements of the factor matrix. Since they have the same sharing patterns, we will use the term *nonzeros* to refer to these structures. Nonzeros are assigned to owner processes statically, and can be modified by these owners only. However, values of nonzeros may be read by multiple processes. The sharing pattern of the nonzeros is called *one-producer with multiple-consumers*. A nonzero can be modified only by its owner processes. Other processes do not have access until computation of the nonzero is completed. When values of the nonzero are fixed, its owner process

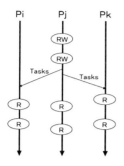
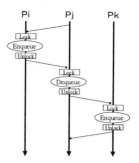

**Fig. 2.** Sharing Pattern of Nonzero Elements

**Fig. 3.** Sharing Pattern of a Task Queue

notifies other processes by sending tasks. Tasks are sent only to processes which will read the values of the nonzero. Non-owner processes can read the values of the nonzero after corresponding tasks are received. Though the owner process also reads the values, further modifications do not occur.

Task queues are used to notify the completion of the computations of each nonzero. Each process has its own task queue to receive tasks for it. The sharing pattern of the task queues is *migratory*. Task queues are consecutively accessed in read-write sequences by different processors. The owner process of the task queue only dequeues tasks from it. In contrast all the processes may enqueue tasks to it. Accesses to a task queue are guarded by mutual exclusion locks in the original program.

Figures 2 and 3 depict sharing patterns of a nonzero and a task queue, respectively. Thick arrows represent control flows of processes Pi, Pj, and Pk. The owner of the shared data is process Pj in both figures. In figure 2, the symbols R and RW in ovals denote read only access and read with modification, respectively. The nonzero is accessed by non-owner processes after corresponding tasks are received. In figure 3, the symbols Enqueue and Dequeue in ovals denote the enqueuing and the dequeuing of a task, and the symbols Lock and Unlock in rectangles denote the acquisition and the release of the mutual exclusion lock associated to the task queue. Task queues must be accessed in critical sections guarded by mutual exclusion locks.

## 3 Implementation

### 3.1 Data Sharing

For nonzeros, we do not need to share each nonzero's values until all updates for it are completed. Once the values of the nonzero are fixed, we copy them to other nodes. We inserted remote write operations to transfer nonzeros to other nodes before enqueuing tasks to notify that the values are ready to be used.

Sharing task queues is more complex. In the original program, accesses to task queues are guarded by mutual exclusion locks. The lock mechanism supported

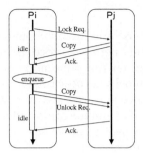 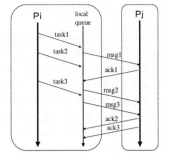

**Fig. 4.** Lock-based implementation     **Fig. 5.** Message-based implementation

by the Solaris thread library can be used for the mutual exclusion between processes in the same node. To enqueue tasks across nodes, we implemented two approaches: we call them *lock-based* and *message-based*.

The lock-based version uses remote locks, a straightforward extension of locks supported by the multithread library in the shared memory model. Figure 4 depicts how communications occur when process Pi enqueues a task to the task queue of process Pj on a different node. Pi sends a lock request message to acquire the lock for the task queue. Pj copies the latest contents of the task queue to the requester, Pi, if no process holds the task queue, otherwise, it waits until that condition holds. Pj also sends an acknowledge message to Pi. When Pi receives the acknowledge, it updates the task queue. After that, Pi copies the new contents of the task queue back to the owner process Pj, following an unlock request message. Again, Pj sends an acknowledge to Pi.

Our messages are actually remote write operations, so they are received by polling. We inserted polls either at every entry to functions which may be executed in the parallel region, or at the point waiting for tasks and messages.

Although this lock-based method is simple in the shared memory model, remote locks are blocking operations, i.e. the process stops until an acknowledge is received. It may serialize processes at critical sections. Since synchronization overheads between nodes are very expensive, the performance of the lock-based version will be poor.

The message-based implementation comes from the fact that a sender process does not have to be concerned with when the task it sent is received. In other words, we can send remote tasks asynchronously with sender processes.

Figure 5 depicts how tasks are sent from process Pi to process Pj, where they are running on different nodes. The thin vertical arrow denotes a local task queue used as a buffer for process Pi. Tasks from Pi to Pj are enqueued to this local queue by Pi. Tasks are dequeued and sent to the destination process by a polling function. This polling function performs both sending and receiving of messages, while the polling function used in the lock-based method only receives messages. Thus synchronization overheads in a message-based version are minimized.

## 3.2 Process Assignment

Processes in the original program are mapped to Solaris threads bound to each processor in COMPaS. Synchronization overheads between processes are very different, according to whether they are running on the same node or not. It is important to note that the mapping of the blocks to processes is based on a scatter decomposition, assuming that processors are arranged in a 2-D grid configuration. This mapping reduces the amount of communication and produces a regular communication pattern. Blocks must be sent to processes in the same row of processes and processes in the same column of processes.

For SMP clusters, it is preferable to map processes to processors so as to reduce internode communications rather than intranode ones. However, such a mapping requires knowledge of the communication pattern of the program.

We consider two simple mapping policies: *thread-major* mapping and *node-major* mapping. In both mappings, the first process is assigned to the first thread of the first node. In the thread-major mapping, the second process is assigned to the first thread of the second node, in other words, the node number is incremented first. The thread number is incremented first in the node-major mapping.

The thread-major mapping is well-suited for Cholesky, because threads in the same node are always associated with processes in the same row of processes. Therefore, communications between processes in the same row are performed within the same node. In the node-major mapping, some threads in the same node are associated with processes in different rows or columns.

## 4 Experimental Results and Performance Analysis

In this section, we present experimental results and discuss the performance of SMP clusters. Two input matrices from the Harwell-Boeing collection[9], namely BCSSTK15(tk15) and BCSSTK29(tk29), are used for the experiments. The size of matrix tk15 is 3948 and it contains 60882 nonzero elements. Though it is the default input matrix in the SPLASH-2 suite, it is rather small for our experiments. The size of matrix tk29 is 13992 and it contains 316740 nonzero elements. It is the largest one in the SPLASH-2 distribution.

Figures 6 and 8 show execution times of the lock-based version for matrices tk15 and tk29, respectively. Figures 7 and 9 show execution times of the message-based version for matrices tk15 and tk29, respectively. The horizontal axis denotes the number of processors, which is the product of the number of nodes times the number of threads in each node. For example, three configurations are experimented on using four processors: a quad-processor SMP, a two dual-processor SMP cluster, and a cluster of four uniprocessors.

From these experiments, we obtained the following results. The message-based version is better than the lock-based version, especially for a large number of nodes. In addition, these results show different characteristics in a trade-off between intranode parallelism and internode parallelism. In the case of the lock-based version, we obtained better performance by increasing the number of

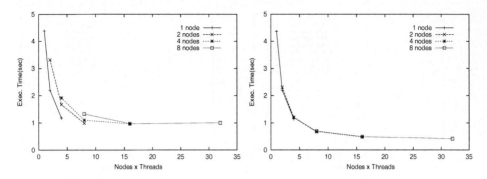

**Fig. 6.** Execution time of the lock-based version for matrix tk15

**Fig. 7.** Execution time of the message-based version for matrix tk15

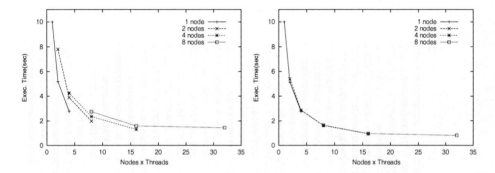

**Fig. 8.** Execution time of the lock-based version for matrix tk29

**Fig. 9.** Execution time of the message-based version for matrix tk29

processors in each node, rather than the number of nodes for a fixed number of total processors. In contrast execution times for different configurations of the same number of processors are almost the same as the message-based version.

Figures 10 and 11 show the efficiency of each implementation for both matrices respectively. In both versions, the efficiencies for matrix tk29 are better than those for matrix tk15, because matrix tk29 is larger.

We show more detail of both executions for matrix tk29 in figure 12 and figure 13, where the configuration is an eight quad-processor SMP cluster. The execution time is broken into four parts: Domain, Block, Enqueue, and TaskWait. Domain and Block are the local computation times for the domains and the blocks, respectively. Enqueue is the time for enqueuing tasks. TaskWait is task waiting time.

Since the task partitioning algorithms of each implementation are the same, local computation times are almost the same. However, the times for enqueuing are significantly different. In the case of the lock-based version, processes are serialized at the critical sections guarded by locks. Therefore, the large synchronization overhead of remote locks reduces the performance. On the other hand, the message-passing version enables hiding most of the synchronization overhead

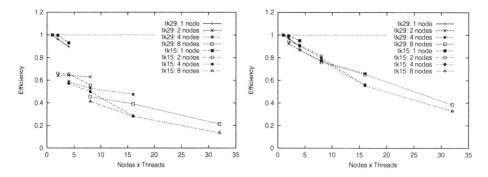

**Fig. 10.** Efficiency of the lock-based version

**Fig. 11.** Efficiency of the message-based version

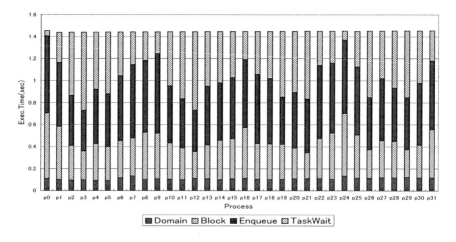

**Fig. 12.** Execution time breakdown of the lock-based version(tk29)

due to asynchronous message handling. Task waiting times of each implementation are caused mostly by inherent load imbalance. It should also be noted that the time spent for polling is less than 1% of the total execution time, for both executions. The lock-based synchronization between nodes greatly reduces overall performance for irregular applications, and asynchronous message-passing overcomes this drawback.

The number of tasks enqueued to the task queue on a different node depends on the mapping policy. Figures 14 and 15 show relative execution times of the node-major mapping for the lock-based version and the message-based version, respectively.

The assignment policy does not affect the performance of the clusters of uniprocessors. In the lock-based version, the assignment policy greatly affects the performance in the case of the clusters of quad-processor SMPs. On the other hand, the performance effect of the mapping policy is relatively small in

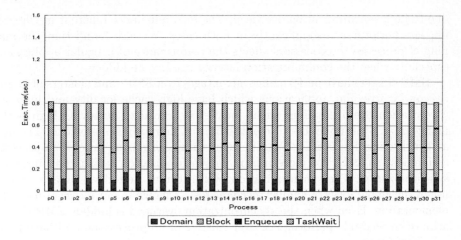

**Fig. 13.** Execution time breakdown of the message-based version(tk29)

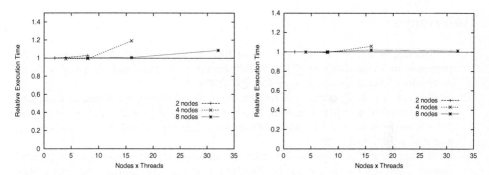

**Fig. 14.** Process assignment effect in the lock-based version

**Fig. 15.** Process assignment effect in the message-based version

the message-based version. That is not surprising, because the synchronization overhead is effectively hidden in the message-based version, due to asynchronous message handling.

## 5 Conclusions

We parallelized the sparse Cholesky factorization kernel in the SPLASH-2 programs using Solaris threads and remote memory operations, and evaluated performance on COMPaS, a Pentium Pro based SMP cluster with a Myrinet interconnection network.

The parallel implementation using asynchronous messages hides most of the synchronization overhead by overlapping computations and communications. The other implementation using global locks shows relatively low performance due to high synchronization overhead. The load imbalance inherent in Cholesky impacts performance significantly in both cases. We also found that the mapping of processes to processors affects the performance of irregular applications, especially when the communication latency can not be hidden.

SMP clusters have performance advantages over SMPs and clusters of uniprocessors for irregular applications which have high cache locality and high synchronization overhead. High cache locality is crucial to achieving high performance in an SMP node, because of the limited shared-bus bandwidth. The overall synchronization overhead can be reduced by optimizing process-to-processor assignment so as to use more intranode synchronization rather than internode synchronization. The synchronization overhead may be hidden by avoiding the use of the global locks or similar mechanisms, as shown in our message-based implementation. Even when the synchronization overhead is hidden, SMP clusters achieve comparable performance over clusters of uniprocessors, and have better cost/performance.

## References

1. Y. Tanaka, et al, COMPaS: A Pentium Pro PC-based SMP Cluster and its Experience, In *Proceedings of IPPS/SPDP workshop on Personal Computers Based Networks of Workstations*, pages 486–497, 1998.
2. Y. Tanaka, et al, Performance Improvement by Overlapping Computation and Communication on SMP Clusters, In *Proceedings of the 1998 International Conference on Parallel and Distributed Processing Techniques and Applications*, Vol.1, pages 275–282, July 1998.
3. E. Rothberg and A. Gupta, An Efficient Block-Oriented Approach To Parallel Sparse Cholesky Factorization, In *Proceedings of Supercomputing '93*, pages 503–512, November 1993.
4. A. Gupta, G. Karypis, and V. Kumar, Highly Scalable Parallel Algorithms for Sparse Matrix Factorization, IEEE Transactions on Parallel and Distributed Systems, Vol.8, No.5, pages 502–520, May 1997.
5. S. C. Woo, et al, The SPLASH-2 Programs: Characterization and Methodological Considerations, In *Proceedings of the 22nd Annual International Symposium on Computer Architecture*, pages 42–36, June 1995.
6. L. Iftode, J. P. Singh and K.Li, Understanding Application Performance on Shared Virtual Memory Systems, In *Proceedings of the 23rd Annual International Symposium on Computer Architecture*, May 1996.
7. C. Liao, et al, Monitoring Shared Virtual Memory Performance on a Myrinet-based PC Cluster, in *Proceedings of the International Conference on Supercomputing*, pages 251–258, July 1998.
8. D. J. Scales, K. Gharachorloo and C. A. Thekkath, Shasta: A Low Overhead, Software-Only Approach for Supporting Fine-Grain Shared Memory, In *Proceedings of the 7th International Conference on Architectural Support for Programming Languages and Operating Systems*, pages 174–185, October 1996.
9. I. S. Duff, R. G. Grimes and J. G. Lewis, Sparse Matrix Test Problems, In *ACM Transactions on Mathematical Software*, Vol.15, No.1, pages 1–14, March 1989.

# Scalable Parallel Sparse Factorization with Left-Right Looking Strategy on Shared Memory Multiprocessors

Olaf Schenk*[1], Klaus Gärtner[2], and Wolfgang Fichtner[1]

[1] Integrated Systems Laboratory, Swiss Federal Institute of Technology Zurich,
ETH Zurich, 8092 Zurich, Switzerland
{oschenk, fw}@iis.ee.ethz.ch
http://www.iis.ee.ethz.ch

[2] Weierstrass Institute for Applied Analysis and Stochastics,
Mohrenstr. 39, 10117 Berlin, Germany
gaertner@wias-berlin.de

**Abstract.** An efficient sparse LU factorization algorithm on popular shared memory multiprocessors is presented. Interprocess communication is critically important on these architectures - the algorithm introduces $O(n)$ synchronization events only. No global barrier is used and a completely asynchronous scheduling scheme is one central point of the implementation. The algorithm aims at optimizing the single node performance and minimizing the communication overhead. It has been successfully tested on SUN Enterprise, DEC AlphaServer, SGI Origin 2000, Cray T90, J90, and NEC SX-4 parallel computers, delivering up to 2.3 GFlop/s on an eight processor DEC AlphaServer for medium-size semiconductor device simulations and structural engineering problems.

## 1 Introduction: Semiconductor Device Simulation and Algorithms

Numerical semiconductor device and process simulation is based on the solution of a coupled system of non-linear partial differential equations (PDE's) that can be either stationary, time dependent, or even complex depending on the problem considered. The nonlinear nature of the semiconductor transport equations, with exponential relations between variables, leads to discretized equations and the sparse linear systems are typically ill-conditioned. These systems are solved by *iterative methods* or *direct methods*.

Since robust transient 2-D and 3-D simulations necessitate large computing resources, the choice of architectures, algorithms and their implementations becomes of utmost importance. Sparse direct methods are the most robust methods

---

* The work of O. Schenk was supported by a grant from the Cray Research and Development Grant Program and the Swiss Commission of Technology and Innovation under contract number 3975.1.

over a wide range of numerical properties and therefore the parallel sparse direct solver PARDISO has been integrated into complex semiconductor device and process simulation packages [4, 5].

The rapid and widespread acceptance of shared memory multiprocessor architectures, from the desktop to the high-end server, has now created a demand for efficient parallel sparse linear solvers on such shared memory architectures. The synchronization on modern multiprocessor architectures is one form of interprocess communication and can have large impact on program performance. Therefore, in section 2 we describe a method using $O(n)$ synchronization events. The experimental results are presented in Section 3. Section 4 provides a short comparison of iterative and sparse direct methods for several semiconductor device simulation examples on different architectures. Section 5 contains a summary of our results.

## 2 Parallel left-right looking strategy and asynchronous computation scheduling.

There are three levels of parallelism that can be exploited by parallel direct methods. The first one, *elimination tree parallelism* [8] is generally exploited in all parallel sparse direct packages. In practice, however, a large percentage of the computation occurs at the top levels of the tree and the sequential part of the elimination tree usually limits the efficiency. It is therefore important to consider more levels of parallelism. Another type of parallelism, called *node level parallelism*, solves in a hybrid approach at the top level of the tree the factorization of each supernode in parallel [9]. However, the size of the supernodes and the numbers of processors determine the efficiency. The third type, *pipelining parallelism*, instead, is suitable for a larger number of processors. In this case, a processor can start with the external factorization of all supernodes that are already factorized. Although a pipelining approach is difficult to realize in sparse direct solver packages, the pipelining parallelism is essential to achieve higher concurrency. Our parallel formulation of the sparse numerical factorization is based on the general framework described in [2, 10]. It successively computes fractions of supernodes called *panels*, which are introduced to increase load balancing and pipelining parallelism in the sequential part of the elimination tree.

The algorithm with reduced synchronization is shown in Figure 1. First, the scheduler is initialized with all leaves of the elimination tree (line 2). Then, a group of $p$ processes is created. Each process asks the scheduler for a new task until all panels have been factorized. The assignment of panel tasks (line 6) is completely dynamic, resulting in a significant improvement of load-balancing — if sufficiently small tasks can be assigned last. That is the reason for the introduction of the panels. We have found from experimental results that the following strategy results in panel sizes that a large enough for Level 3 BLAS and small enough to exploit load balancing and pipelining parallelism: First, if $(1 < \text{nproc} \leq 4)$ then each panel contains not more than 96 columns of the factor, if $(4 < \text{nproc})$ then the maximum panel size is reduced to 72 columns. The scheduling is dynamic but not completely independent of other processes.

```
 1:  Q ⟵ ∅
 2:  Scheduler = {initialized with leaves of the elimination tree}
 3:  for  P=1, #proc
 4:      while (Scheduler not empty)
 5:          lock I
 6:              Scheduler ⟵ Scheduler \ {J}
 7:              /* panel J is computed by process P */
 8:          unlock I
 9:          while ∃ panel K in Q  with A_{K,J} ≠ 0
10:              /* outer factorization, dependent on other processes */
11:              lock II
12:                  identify all computed descendants K of panel J
13:                  /* left looking, row traversal */
14:              unlock II
15:              A_{*,J} = A_{*,J} − A_{*,K} * A_{K,J}
16:              A_{J,*} = A_{J,*} − A_{J,K} * A_{K,*}
17:          end while
18:          /* inner factorization, independent of other processes */
19:          A_{J,J} =: L_{J,J} * U_{J,J}
20:          A_{*,J} = A_{*,J} * U_{J,J}^{-1}
21:          A_{J,*} = L_{J,J}^{-1} * A_{J,*}
22:          lock II
22:              mark parents K of panel J for pipelining of the outer
23:              factorization /* right looking, column traversal */
24:          unlock II
25:      end while
26: end for
```

**Fig. 1.** Parallel left-right looking sparse supernode block LU factorization

The first interprocess communication results from the assignment of panel J to a process — multiple execution has to be prevented by mutual exclusion. Now the process is able to compute the outer and inner factorization of panel J. In contrast to the inner factorization, which is completely independent from other processes, the outer factorization needs to be synchronized. Computing the outer factorization with a left-looking approach in line 9 (Figure 2), the process enters a second critical section and removes all panel children of panel J that are already factorized from a queue $Q$. With respect to the completely factored panels the computation is independent and the outer factorization can be performed. After having finished all the external updates of panel J, the process executes the inner factorization. The factorization of panel J terminates with a third critical section. Within this critical section all parents of panel J are informed that panel J is factorized (Figure 3: column traversal of the factor L). This is called right-looking, since the process computes the information for all panel K with K≥J. No global barrier is used within this algorithm, and the execution of different processes is completely asynchronous. Since a left-right looking strategy is being introduced, three critical sections have to be handled. In total, the number of interprocess communications is roughly $3n_p$ ($n_p$ the number of panels).

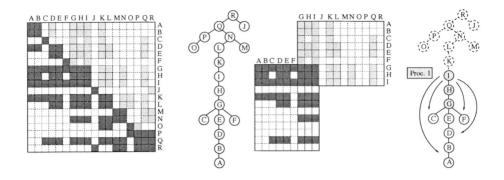

**Fig. 2.** Non-zero structure of the factors L and U, the elimination tree and the left-looking numerical factorization of supernode S(G, H, I).

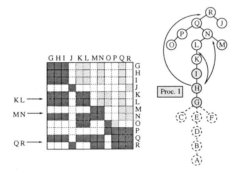

**Fig. 3.** The right-looking phase of the parallel algorithm. All panels that will be factorized by supernode S(G, H, I) are marked to exploit the pipelining parallelism.

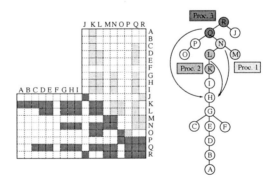

**Fig. 4.** Pipelining parallelism is essential to achieve higher parallel efficiency. Three processors can start with the external numerical factorization of the supernodes S(K, L), S(M, N), and S(Q, R) by supernode S(G, H, I).

To evaluate our left-right looking strategy, we conducted the numerical experiments on sparse matrices from semiconductor process- and device-simulations. Nested dissection was used to order these matrices [6]. Table 1 represents a set of typical matrices in semiconductor process and device simulation (systems of nonlinear partial differential equations). The purpose of the collection was to demonstrate that the approach can deliver close to peak floating point performance for practical problems. The matrices are sorted in increasing order of #flops/nnz(LU), the ratio of floating point operations (#flops) to the number of nonzeros in the factored matrix (nnz(LU)). The data re-use correlates with this ratio. Thus, we expect the performance to increase with increasing #flops/nnz(LU).

**Table 1.** Characteristics of the test matrices. All matrices are structurally symmetric. $n$ is the number of unknowns. *nnz(A)* and *nnz(LU)* defines the number of nonzeros in $A$ and $LU$. The matrices are sorted in increasing order *#flops/nnz(LU)*, the ratio of the number of floating point operations to the number of nonzeros *nnz(LU)* in the factored matrix.

| Matrix | n | nnz(A) | nnz(LU) | #flops/nnz(LU) |
|---|---|---|---|---|
| 1 3D eth-3d-eeprom | 12'002 | 630'002 | 7'052'672 | 500.43 |
| 2 3D ise-igbt-coupled | 18'668 | 412'674 | 10'755'310 | 653.82 |
| 3 3D eth-3d-eclt | 25'170 | 1'236'312 | 18'947'212 | 708.10 |
| 4 3D ise-soir-coupled | 29'907 | 2'004'323 | 30'614'337 | 925.11 |
| 5 3D eth-3d-mosfet | 31'789 | 1'633'499 | 29'582'425 | 931.30 |
| 6 3D eth-3d-eclt-big | 59'648 | 3'225'942 | 69'323'790 | 1316.31 |

## 3 Experimental results on shared memory multiprocessor systems.

In this study we consider two different shared memory architectures: parallel vector supercomputers, a 12-CPU NEC SX-4 and a 16-CPU Cray J90, and two multiprocessor servers, an 8-CPU DEC AlphaServer 8400 and an 8-CPU SGI Origin 2000. Table 2 summarizes the characteristics of the individual processors.

**Table 2.** Clock speed, on-chip and off-chip cache sizes, peak floating point and LINPACK performance of the processors.

| Processor | Clock speed | on-chip Cache | off-chip Cache | Peak Mflop/s | LINPACK Mflop/s |
|---|---|---|---|---|---|
| Alpha EV5.6 21164 | 612 MHz | 96 KB | 4 MB | 1224 | 764 |
| MIPS R10000 | 195 MHz | 32 KB | 4 MB | 390 | 344 |
| J90 | 100 MHz | - | - | 200 | 202 |
| SX-4 | 125 MHz | - | - | 2000 | 1929 |

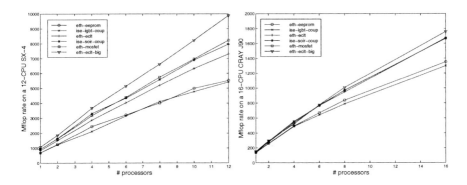

**Fig. 5.** Mflop/s on a 12-CPU NEC SX-4 and a 16-CPU Cray J90.

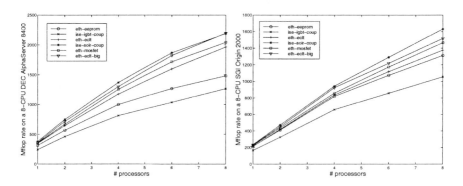

**Fig. 6.** Mflop/s on an 8-CPU DEC AlphaServer 8400 and an 8-CPU SGI Origin 2000.

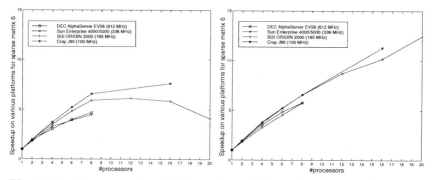

**Fig. 7.** Speedup on various platforms for a 3-D semiconductor device example with 59'648 unknowns and a left-looking approach (left) and the left-right looking (right). In contrast to the left-looking algorithm the left-right looking approach exploits pipelining parallelism. The processor first performs updates from computed descendants and finally performs any updates that may come from just-finished descendants.

For the numerical computation performed during the dense supernode numerical factorization, the single-processor vendor optimized implementation of BLAS primitives has been used. Based on the experimental results given in the Figures 5 to 7 we conclude:

- the algorithm delivers substantial speedup, even for moderate problem sizes,
- pipelining parallelism is essential to achieve higher efficiency and the new synchronization scheme has significant influence.

The speedups are computed with respect to a very efficient serial implementation of the supernode left-looking algorithm. To calibrate our speedup figures with respect to the peak performance, we compare the floating point performance of our implementation on a single processor with the single processor performance of the LINPACK benchmark in Table 2. Although the solver is designed for matrices with a very sparse structure, it delivers 160 Mflop/s on the Cray J90 or 81 % of the LINPACK performance.

Furthermore, the block numerical factorization is not designed for one special architecture like the Cray vector supercomputer. The same method achieves approximately 230 Mflop/s on a single processor SGI Origin 2000 for the sparse matrix eth-3d-eclt-big having 2% non-zero entries in the factors L and U. Comparing with the vendor optimized single-processor LINPACK test, this translates to roughly 58 % of that performance. These matrices are not unrealistically large. The largest one requires approximately 912 * 100 Mflop — hence 912 seconds on a workstation with a sustained performance of 100 Mflop/s. The details of the measurements are presented in Figure 5 to 7. Especially Figure 7 demonstrates the influence of algorithmic and hardware properties.

## 4 Comparing Algorithms and Machines for complete Semiconductor Device Simulations

For our comparison in Table 3, we used five typical semiconductor device simulation cases: one 2-D transient simulation with a hydrodynamic model and 14'662 vertices (87'732 unknowns), and four 3-D quasistationary device simulations, with 10'870, 12'699, 26'855, and 105'237 vertices (32'610, 38'097, 80'565, and 315'711 unknowns), respectively. All grids are highly irregular and the linear systems are very sparse. The irregularity of the grids leads to more complicated structures of the linear systems that must be solved during the simulation. The machines are shared memory multiprocessors and the processor characteristics are summarized in Table 2. We used an 32-processor SGI Origin 2000 with sixteen Gbytes of main memory, and an eight-processor DEC AlphaServer with two Gbytes of main memory.

Comparing the performance of two sparse direct solvers on the same machine with one single processor, we see that in all cases the new direct solver PARDISO [13] is significant faster than $SUPER_{-ISE}$ [4, 7]. The mayor bottleneck of $SUPER_{-ISE}$ is not the number of floating point operations necessary for the factorization, but rather the cost of fetching data from main memory into cache memory. Although both packages takes advantages from supernodes technique, PARDISO uses the memory hierarchy between cache and main memory significantly better. Furthermore, a nested dissection approach [6] has been integrated into PARDISO. This nested dissection fill-reducing ordering is substantially better than the multiple minimum degree algorithm for large problem sizes (up to a factor up 5 in floating point operations for larger 3-D grids). Table 3 clearly reflects the performance difference due to algorithmic improvements.

**Table 3.** Comparison of different algorithms and packages on high-end servers. Wall clock times of the solution time in hours for one complete semiconductor device simulation with the parallel DESSIS$_{-ISE}$.

|  | SLIP90$_{-ISE}$ iterative | SUPER$_{-ISE}$ direct | PARDISO direct 1 CPU | 4 CPUs |
|---|---|---|---|---|
| **2-D 14'662 vertices, six nonlinear PDE's** | | | | |
| DEC Alpha | 14.60 | 23.70 | 5.30 | 1.62 |
| SGI Origin | 22.30 | 46.70 | 9.70 | 2.81 |
| **3-D 10'870 vertices three nonlinear PDE's** | | | | |
| DEC Alpha | 1.80 | 2.30 | 0.45 | 0.12 |
| SGI Origin | 1.41 | 3.37 | 0.85 | 0.23 |
| **3-D 12'699 vertices three nonlinear PDE's** | | | | |
| DEC Alpha | 0.90 | 6.68 | 0.72 | 0.20 |
| SGI Origin | 0.90 | 5.72 | 1.00 | 0.23 |
| **3-D 26'859 vertices three nonlinear PDE's** | | | | |
| DEC Alpha | 0.92 | 45.79 | 1.33 | 0.41 |
| SGI Origin | 1.04 | 55.04 | 1.61 | 0.51 |
| **3-D 105'237 vertices three nonlinear PDE's** | | | | |
| DEC Alpha | failed | no memory | no memory | no memory |
| SGI Origin | failed | no memory | 401.1 | 16.0 on 30 CPUs |

On the other side, preconditioned sparse iterative methods play an important role in the area of semiconductor device and process simulation problems. Of the wide variety of iterative solvers available in the literature [12] it has been found that only a few are capable of solving the equations arising in semiconductor device and process simulations. This is mainly due to to the ill conditioned matrices of the coupled solution scheme. So far, BiCGstab(2) [14] with a fast and powerful incomplete LU decomposition preconditioning and reverse Cuthill-McKee ordering appears to be the best choice for solving large sparse linear systems [3, 4]. The default preconditioning in the package SLIP90$_{-ISE}$ is ILU(5,1E-2).

For increasingly difficult linear systems, the fill-in level and the threshold value for dropping a fill-in entry in the ILU decomposition is changed in five steps to ILU(20,1E-5).

However, the results for the linear solver have to be seen in the context of the complete simulation task. One often observed problem of the iterative methods used is the strong reduction of the average step size during a quasistationary or transient simulation due to convergence problems. The typical time step size for non trivial examples is in the range of 0.01 to 0.001 times that found for a direct method. Furthermore, looking at the quasistationary 3-D simulation with 105'237 vertices in Table 3 we see that the iterative solver failed during the simulation. The main memory requirement of $DESSIS_{-ISE}$ with the parallel direct solver was about 4 Gbytes on the SGI Origin 2000 for the largest example with 105'237 vertices.

## 5 Summary and outlook.

The implementation features state-of-the-art techniques [1, 10, 11]. A very high level of efficiency has been achieved on typical shared memory parallel servers and supercomputers. The proposed left-right looking supernode algorithm reduces the synchronization events required to manage the factorization to $O(n)$. This results in a parallel efficiency of approximately 75% on eight processor machines for typical real world application problems. Tangible benefits can be achieved on today's eight-processor workgroup servers. As a result, PARDISO is already in use in several commercial applications, e.g. the solver has been successfully integrated into complex semiconductor process and device simulation packages [4, 5]. It was very encouraging to observe the response of the users: the fraction of the simulation problems with enhanced complexity increased immediately.

Further improvements can be expected due to

a) better orderings,
b) assigning larger parts of the elimination tree far from the root to one process instead of scheduling the tiny leave elements,
c) introduction of the panels with respect to the update structure and size.

a) is strongly related to improved algorithms for solving the separator problem.
b) may be solved efficiently by introducing a function like $w(i)$, $i$ node number in the elimination tree, $w$ accumulated work to factorize the subtree with root $i$.
c) is purely technical but a simple way to decrease the dependence close to the root of the elimination tree.
The main advantages of direct methods are the robustness and the small constants in the non optimal complexity figure. Based on our experience with problems of the class mentioned, we expect the break even point with respect to iterative and multilevel methods for three-dimensional problems on RISC machines to occur above $10^4$ unknowns. For two-dimensional applications it will be at even larger problem sizes. However, multigrid methods may efficiently bridge the gap from $10^4$ to $10^6$ unknowns, expected in complex 3d applications. This

directs the focus of interest to improved parallel forward/backward substitution algorithms — a point that was not mentioned here at all.

## Acknowledgments.

The authors appreciate the support of following high performance benchmark groups: Silicon Graphics/Cray Research Inc., Digital Equipment Corporation, Sun Microsystems Inc and NEC. Also, we thank Esmond Ng (Oak Ridge National Laboratory) for a number of interesting discussions and hints during two visits at Zurich. Finally, we want to thank for the support within the CRAY-ETHZ SuperCluster Cooperation and the grant from the Swiss Commission of Technology and Innovation.

## References

1. C. ASHCRAFT, R. GRIMES, J. LEWIS, B. PEYTON, AND H. SIMON, *Progress in sparse matrix methods for large linear systems on vector supercomputers*, The International Journal of Supercomputer Applications, 1 (1987), pp. 10–30.
2. I. S. DUFF, *Multiprocessing a sparse matrix code on the alliant fx/8*, J. Comput. Appl. Math., 27 (1989), pp. 229–239.
3. D. FOKKEMA, *Subspace methods for linear, nonlinear, and eigen problems.*, PhD thesis, Utrecht University, 1996.
4. INTEGRATED SYSTEMS ENGINEERING AG, *DESSIS$_{ISE}$ Reference Manual*, ISE Integrated Systems Engineering AG, 1998.
5. ———, *DIOS$_{ISE}$ Reference Manual*, ISE Integrated Systems Engineering AG, 1998.
6. G. KARYPIS AND V. KUMAR, *Analysis of multilevel graph algorithms*, Tech. Rep. MN 95 - 037, University of Minnesota, Department of Computer Science, Minneapolis, MN 55455, 1995.
7. A. LIEGMANN, *Efficient Solution of Large Sparse Linear Systems*, PhD thesis, ETH Zürich, 1995.
8. J. LIU, *The role of elimination trees in sparse factorization*, SIAM Journal on Matrix Analysis & Applications, 11 (1990), pp. 134–172.
9. P. MATSTOMS, *Parallel sparse QR factorization on shared memory architectures*, Parallel Computing, 21 (1995), pp. 473–486.
10. E. NG AND B. PEYTON, *A supernodal Cholesky factorization algorithm for shared-memory multiprocessors*, SIAM Journal on Scientific Computing, 14 (1993), pp. 761–769.
11. E. ROTHBERG, *Exploiting the memory hierarchy in sequential and parallel sparse Cholesky factorization*, PhD thesis, Stanford University, 1992. STAN-CS-92-1459.
12. Y. SAAD, *Iterative Methods for Sparse Linear Systems*, PWS Publishing Company, 1996.
13. O. SCHENK, K. GÄRTNER, AND W. FICHTNER, *Efficient sparse LU factorization with left-right looking strategy on shared memory multiprocessors*, Tech. Rep. 98/40, Integrated Systems Laboratory, ETH Zurich, Swiss Fed. Inst. of Technology (ETH), Zurich, Switzerland, Submitted to BIT Numerical Mathematics, 1998.
14. G. SLEIJPEN, H. VAN DER VORST, AND D. FOKKEMA, *BiCGSTAB(l) and other hybrid Bi-CG methods*, Tech. Rep. TR Nr. 831, Department of Mathematics, University Utrecht, 1993.

# Algorithm of Two-Level Parallelization for Direct Simulation Monte Carlo of Unsteady Flows in Molecular Gasdynamics

Alexander V. Bogdanov, Igor A. Grishin, Gregory O. Khanlarov, German A. Lukianov, and Vladimir V. Zakharov

Institute for High-Performance Computing and Data Bases,
P.O. Box 71, St.Petersburg 194291, Russia
bogdanov@hm.csa.ru, {grishin, greg, monte, zvv}@fn.csa.ru
http://www.csa.ru/Inst

**Abstract.** A general scheme of two-level parallelization (TLP) has been described for direct simulation Monte Carlo of unsteady gas flows on shared memory multiprocessor supercomputers. The high efficient algorithm of parallel statistically independent runs is used on the first level. The data parallelization is employed for the second one. Two versions of TLP algorithm are elaborated with static and dynamic load balancing. The dynamic processor reallocation technique is used for the dynamic load balancing. Two gasdynamic unsteady problems were used to study speedup and efficiency of the algorithms. The conditions of the efficient usage of the algorithms have been determined.

## 1 Introduction

### 1.1 Direct Simulation Monte Carlo Method and Sequential Algorithm in Unsteady Molecular Gasdynamics

The Direct Simulation Monte Carlo (DSMC) is the simulation of real gas flows with various physical processes by means of huge number of modeling particles [1], each of which is a typical representative of great number of real gas molecules. The DSMC method conditionally divides the continuous process of particles movement and collisions into two consecutive stages (motion and collision process) at each time step $\Delta t$. The particle parameters are stored in the computer's memory. To get information about the flow field it is necessary to superimpose a grid on the computational domain. The results of simulation are averaged particles parameters in cells of the grid.

The finite memory size and computer performance make restrictions to the total number of modeling particles and cells. Macroscopic gas properties determined by particles parameters in cells at the current time step are the result of simulation. Fluctuations of averaged gas properties at single time step can be rather high owing to relatively small number of particles in cells. So, when solving steady gasdynamic problems, we have to increase the time interval of averaging (the sample size) after steady state has been achieved in order to reduce statistical error down to the required level. The averaging time step $\Delta t_{av}$ has to be much greater than the time step $\Delta t$ ($\Delta t_{av} \gg \Delta t$).

For DSMC of unsteady flows the value of averaging time step $\Delta t_{av}$ for a given problem and at the current time $t$ has to meet the following requirement: $\Delta t_{av} \ll \min t_H(x, y, z, t)$, where $t_H$ — is the characteristic time of flow parameters variation. The choice of the value of $t_H$ is determined by particular problem [2, 3]. In order to meet the condition for the averaging interval we have to carry out enough number of statistically independent calculations (runs) $n$ to get the required sample size. This leads to the increase of the total calculation time which is proportional to $n$ in the case of sequential DSMC algorithm.

The algorithm of DSMC of unsteady flows consists of two basic loops. In the first (inner) loop the single unsteady run is executed. First, we generate particles at input boundaries of the domain (subroutine **Generation**). Then we carry out simulation of particle movement, surface interaction (subroutine **Motion**) and collision process (subroutine **Interaction**) for determined number of time steps $\Delta t$. The sampling (subroutine **Sampling**) of flow macroscopic properties in cells is carried out at a given moment of unsteady process. The inner loop itself is divided into two successive steps. In the first step we sequentially carry out simulation for each of $N_p$ particles independently (**Generation, Motion, Interaction**). A special readdressing array is formed – subroutines **Enumeration, Indexing** – (it determines the mutual correspondence of particles and cells) after the first step. We have to know the location of all particles in order to fill that array. In the second step we carry out the simulation for each of $N_c$ cells independently (**Sampling**). For $t > \Delta t_s$ we accumulate statistical data of flow properties in cells.

The second (outer) loop repeats unsteady runs $n$ times to get the desired sample size. Each run is executed independently from the previous ones. To make separate unsteady runs independent it is necessary to shift a random number generator (**RNG**).

For each unsteady run three basic arrays (**P, LCR, C**) are required. The array **P** is used for storing information about particles. The array **LCR** is the readdressing array. The dimensions of these arrays are proportional to the total number of particles. The array **C** stores information about cells and macroscopic properties. The dimension of this array is proportional to the total number of cells of a computational grid. The elements of arrays **P, LCR, C** are updated at each time step $\Delta t$.

## 1.2 Parallel Statistically Independent Runs (PSIR)

The statistical independence of runs make it possible to execute them parallel. The implementation of this approach on a multiprocessor computer leads to the decrease of the number of the outer loop iterations for each processor ($n/p$ — the number of iterations for the $p$-processor computer). The data exchange between processors goes after all runs have finished and it takes infinitesimal time comparing to the overall computational time. Only one processor sequentially analyzes the results after data exchange. The range of efficient application field for this algorithm is $p \leq n$. For optimal speedup and efficiency the value of $n$ has to be multiply by $p$.

All arrays (P, LCR, C, etc.) are stored locally for each run. This algorithm can be programmed on computers with any kind of memory (shared or local). The required memory size for this algorithm is proportional to $p$.

To estimate the speedup $S_p$ and the efficiency $E_p$ of parallel algorithm with a parallel fraction of computational work $\alpha$ for the computer with $p$ processors one can use the following formulas [4]:

$$S_p(p, \alpha) = \frac{T_1}{T_p} = \frac{p}{p - \alpha(p - 1)}, \qquad (1)$$

$$E_p(p, \alpha) = \frac{S_p}{p} = \frac{1}{(1 - \alpha)p + \alpha}, \qquad (2)$$

where $T_1$ — the execution time of the sequential algorithm, $T_p$ — the execution time of a given parallel algorithm on the computer with $p$ processors. The formulas (1) and (2) are simplified because they do not take into account overheads (communication, load imbalance, synchronization, etc.).

The PSIR algorithm is coarse-grained and has high efficiency and great degree of parallelism comparing to any other parallel algorithm of DSMC of unsteady flows for the number of processor $p \leq n$. The maximum value of speedup for this algorithm can be obtained at $p = n$. The potency which gives the computer are surplus for $p > n$. Thus, the PSIR algorithm for DSMC of unsteady flows has the following condition of efficient usage: $n \geq p$. The value of parallel fraction $\alpha$ can be very high (up to $0.99 \div 0.999$) for typical problems of molecular gasdynamics [3]. The corresponding speedup is $100 \div 1000$. To get the efficiency $E_p \geq 0.5$ at $n = 100 \div 1000$ it is necessary to have $p = 100 \div 1000$ respectively. The computer architecture almost does not affect the value of $\alpha$. It is proved correct by several tests on massive-parallel supercomputers with different architectures (distributed memory Parsytec CC/16 and shared memory SPP-1600).

## 1.3 Data Parallelization of DSMC [5]

The computational time of each DSMC problem is determined by the execution time of the inner loop. The duration of this loop depends on the number of particles in the domain and the number of cells. It was stated above that the inner loop consists of two consecutive stages. The data inside each stage are independent. The elements P[k] are processed at the first stage, whereas the elements C[k] — at the second one (the elements of arrays P and C are mutually independent). Since the operations on each of these elements are independent it is possible to process them parallel. Each processor takes elements from particle array P and cell array C according to its unique ID-number, i.e. the m-th processor takes the $m$-th, $(m+p)$-th, $(m+2p)$-th, etc. elements, where "m" is the processor ID-number. Such rule of particle selection provides good load balancing because various particles require different time to process and they are located randomly in the array P .

The synchronization of processors is performed before the next loop iteration starts. Before the second stage begins it is necessary to fill the readdressing array LCR. The complete information about the array P is required for readdresing procedure. This task can not be parallelized, so it is performed by one processor. There are two synchronization points before the readdressing and after it. The reduction of the computational time is due to less number of particles and cells which have to be processed by each processor ($N_p/p$ and $N_c/p$ instead of $N_p$ and $N_c$). After the inner loop has been passed the processors also have to get synchronized.

The data from the array P is required to perform the operations on elements of array C. This data is located in the array P randomly. These arrays are stored in the shared memory in order to reduce the large data exchange between processors. The memory conflicts (several processors read the same array element) are excluded by the algorithm. The semaphore technique is used for the synchronization.

## 2 Two-Level Parallelization with Static Load Balancing

In different problems of gasdynamics the number of statistically independent runs $n$ has wide scope (from 1 to values of 1000th order and more) depending on the rate of gas properties variation and the required accuracy. It was stated above that the potency of the multiprocessor system is surplus for the application of the PSIR algorithm when the required number of statistically independent runs $n$ is less than the number of processors $p$ ($n < p$). In this case the efficient usage of computer resources of $p$-processor machine can be provided by the implementation of an algorithm of two-level parallelization (TLP algorithm). The general flowchart of TLP algorithm is shown in the fig. 1. The first level of parallelization corresponds to the PSIR algorithm, the data parallelization is employed for the second level inside each independent run. The TLP algorithm is a parallel algorithm with static load balancing.

This algorithm requires the memory size to be proportional to the number of the first level processors which compute single runs (just the same as for the PSIR algorithm). It also requires the arrays for each run to be stored in the shared memory as for the data parallelization in order to exclude great data exchange between processors.

The speedup and the efficiency of the TLP algorithm can be estimated by the following equations:

$$S_p = S_{p_1} \cdot S_{p_2} = \frac{p_1}{p_1 - \alpha_1(p_1 - 1)} \frac{p_2}{p_2 - \alpha_2(p_2 - 1)}, \quad (3)$$

$$E_p = E_{p_1} \cdot E_{p_2} = \frac{S_p}{p_1 \cdot p_2}, \quad (4)$$

where indices '1' and '2' correspond to parameters on the first level and on the second one.

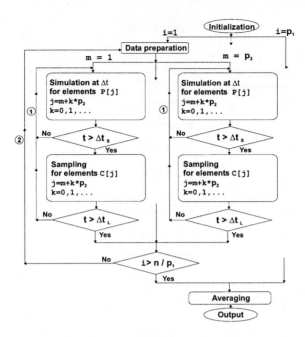

**Fig. 1.** Two-level parallelization. $\Delta t$ — time step, $\Delta t_s$ — interval between samples, $\Delta t_L$ — time of a single run, $t$ — current time, $i$ — run number (first level), $p_1$ — number of the first level processors, $m$ — the second level processor ID-number, $p_2$ — number of the second level processors, $j$ — index of array element.

The figure 2 shows the detailed flowchart of TLP algorithm for unsteady flow simulation. There are five synchronization points in the algorithm. The four of them correspond to the data parallelization. The last synchronization has to be done after termination of all runs. The synchronization is employed with the aid of the semaphore technique. In this version the iterations of the outer loop are fully distributed between the first level processors. This algorithm requires $n$ to be multiply by $p$ for uniform distribution of computer resources between single runs. In order to make the runs statistically independent we have to shift the random number generator in each run.

The HP/Convex Exemplar SPP-1600 system with 8 processors and 2Gb of memory was used for algorithm test.

To simulate operations on $p$-processor machine ($p = 1 \div 36$) we forked $p$ processes. The speedup $S_p$ was computed by the following formula: $S_p = T_1/T_p$. To get the values of $T_1$ and $T_p$ we measured the execution time of the main parent process. This process has the maximum execution time because it makes the start-up initialization before forking child processes and data processing after passing parallel computations.

To study the speedup and the efficiency of the TLP algorithm we carry out the simulation of unsteady 3-D water vapor flow in the inner atmosphere of a comet. The number of the first level processors $p_1$ was fixed and equal to 6. The

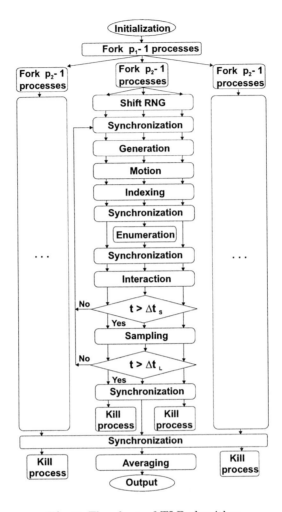

**Fig. 2.** Flowchart of TLP algorithm

number of the second level processors $p_2$ was varied from 1 to 6. The effective values of $\alpha_1$ and $\alpha_2$ (i.e. taking into account overheads) were estimated with the aid of (3) by the approximation of experimental results ($\alpha_1$ was estimated at fixed value of $p_2$, $\alpha_2$ — vice versa). We obtained the following values: $\alpha_1 = 0.998$ and $\alpha_2 = 0.97$. The figure 3 depicts the averaged experimental results for speedup and efficiency as functions of the total number of processors $p = p_1 \cdot p_2$. The same figure shows the value (marked by cross-sign) of speedup and efficiency of the PSIR algorithm (TLP algorithm turns into PSIR algorithm at $p_2 = 1$).

Thus, the TLP algorithm gives the possibility to extremely reduce the computational time required for the DSMC of unsteady flows using shared memory multiprocessor computers. The efficient usage of this algorithm is determined by

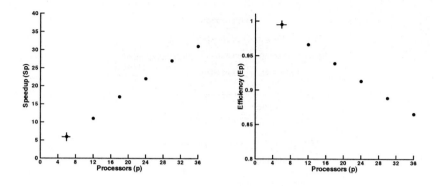

**Fig. 3.** Speedup $S_p$ (left) and efficiency $E_p$ (right) of TLP algorithm (cross — speedup of PSIR algorithm)

the condition $n < p$, and the number of processors $p$ has to be multiply by $n$ in order to provide good load balancing.

## 3 Two-Level Parallelization with Dynamic Load Balancing

The TLP algorithm with static load balancing described in section 2 has several drawbacks. It does not provide good load balancing in the following cases: first, when the ratio $p/p_1$ is not integer (part of processors are not used), second, the small values of $\alpha$ or large values of $p_2$ — in this case some processors may be idle in each run.

The increase of efficiency can be obtained by usage of dynamic load balancing with the aid of dynamic processor reallocation (DPR). The idea of the algorithm is as follows. Let us divide conditionally all available processors into two parts: leading processors $p_1$ and supporting processors which form the so called "heap" (the number of heap processors is $p_2 = p - p_1$). Each leading processor is responsible for its own run. This algorithm is similar to that of TLP but here there is no hard link of heap processors with the specific run. Each leading processor reserves the required number of heap processors before starting parallel computations (according to a special allocation algorithm). After exiting from parallel procedure the leading processor releases allocated heap processors. This algorithm makes it possible to use idle processors more efficiently, in fact this leads to execution of parallel code with the aid of more processors than in the case of TLP algorithm with static load balancing. The flowchart of TLPDPR algorithm is presented in the fig. 4.

The speedup which gives this algorithm is determined by the following basic parameters: the total available number of processors in the system $p$, the required number of independent runs $p_1 = n$ ($p_1 \ll p$), the sequential fraction of

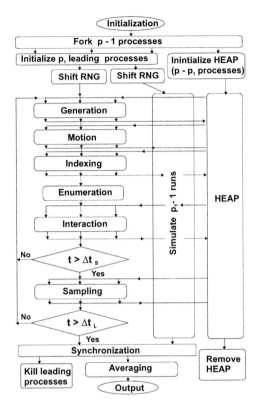

**Fig. 4.** Flowchart of TLPDPR algorithm

computational work in each run $\beta = 1 - \alpha$ and the algorithm of heap processors allocation. In this paper we use the following allocation algorithm:

$$p'_2 = (1 + \text{PRI})p_2, \quad \text{PRI} = 0 \ldots \text{PRI}^*, \quad \text{PRI}^* = \frac{\beta}{1-\beta}(p_2 - 1), \qquad (5)$$

where $p'_2$ — the actual number of the second level processors, PRI — the parameter which is estimated by simulation results of similar problems, PRI* — the estimated upper limit of the efficient range of parameter PRI. Hence, the speedup on the second level of parallelization is given by

$$S_{p_2} = \frac{1}{\beta + \frac{1-\beta}{p'_2}} \qquad (6)$$

In case of $p$ being multiply by $p_1$ and the value of PRI is equal to 0, this algorithm turns into TLP algorithm. The value of $S_{p_2}$ at PRI = PRI* gives the upper limit of speedup for a given problem.

To study the characteristics of TLPDPR algorithm we solve the problem on unsteady flow past a body. The value of sequential fraction $\beta = 0.437$, the

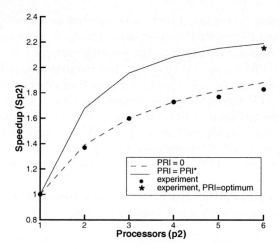

**Fig. 5.** Speedup on the second level as a function of number of second level processors $p_2$ for algorithms of TLP (PRI = 0, dashed line) and TLPDPR (PRI = PRI*, solid line), circles — experiment ($p_1 = 6$, $p = 36$, PRI = 0), asterisk — optimal value of parameter PRI

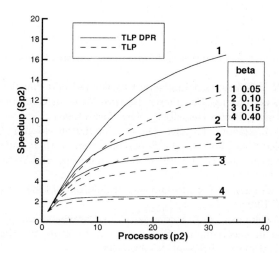

**Fig. 6.** Speedups of TLP (dashed line) and TLPDPR (solid line) algorithms as functions of number of the second level processors $p_2$ for various sequential fractions $\beta$ on the second level

number of the leading processors $p_1 = 6$. The speedup as a function of $p_2$ ($p_2 = 1 \div 6$) for PRI = 0 and PRI = PRI* is depicted in the fig. 5. The same figure shows the results of calculation for PRI = 0, the asterisk corresponds to the optimal value of parameter PRI for $p_1 = 6$ ($p = 36$). The maximum speedup $S_{p_2}$ with a given degree of parallelism ($p_1 \to \infty$), which can be estimated by the formula (6), comes to 2.3. The TLP algorithm gives the speedup ($\beta = 0.437$, $p_1 = 6$, $p = 36$) which is 80% of the maximum value. At optimum value of parameter PRI the TLPDPR algorithm gives 93% for the same case. This is equivalent to the usage of TLP algorithm on a 120-processor computer ($p = 120$, $p_1 = 6$, $p_2 = 20$). The figure 6 shows speedups of TLP and TLPDPR algorithms as functions of $p_2$ for various $\alpha_2$.

The implemented TLPDPR algorithm has the following advantages comparing to the TLP algorithm with static load balancing:

- TLPDPR algorithm makes it possible to minimize the processor idle time. It provides better load balancing;
- Better load balancing make it possible to get higher speedups under the same conditions.

The study was carried out in the Center of Supercomputing Applications (http://www.csa.ru/CSA) of the Institute for High-Performance Computing and Data Bases (http://www.csa.ru/Inst)

# References

1. G.A.Bird. Molecular Gasdynamics and Direct Simulation of Gas Flows. Clarendon Press. Oxford. 1994
2. A.V.Bogdanov, N.Y.Bykov, G.A.Lukianov. Distributed and Parallel Direct Simulation Monte Carlo of Rarefied Gas Flows. Lecture Notes in Computer Science, Vol. 1401. Springer-Verlag, Berlin Heidelberg New York (1998)
3. N.Y.Bykov, G.A.Lukianov. Parallel Direct Simulation Monte Carlo of Nonstationary Rarefied Gas Flows at the Supercomputers with Parallel Architecture. St.Petersburg. Institute for High-Performance Computing and Databases. Preprint N5-97. 1997.
4. J.M.Ortega. Introduction to Parallel and Vector Solution of Linear Systems. Plenum Press. New York. 1988
5. I.A.Grishin, V.V.Zakharov, G.A.Lukianov. Data Parallelization of Direct Simulation Monte Carlo in Gasdynamics. St.Petersburg. Institute for High-Performance Computing and Databases. Preprint N3-98. 1998.

# Parallelization of Gridless Finite-Size-Particle Plasma Simulation Codes

S. Briguglio*, G. Vlad, G. Fogaccia and B. Di Martino**

Associazione Euratom-ENEA sulla Fusione, C.R. Frascati,
C.P. 65 - I-00044 - Frascati, Rome, Italy.

**Abstract.** The main features of gridless finite-size-particle (FSP) codes are discussed, from the point of view of the performances that can be obtained, with respect both to the spatial-resolution level and the efficiency of parallel particle simulations. It is shown that such codes are particularly suited for particle-decomposition parallelization on distributed-memory architectures, as they present a strong reduction, in comparison with particle-in-cell (PIC) codes, of the memory and computational offsets related to storing and updating the replicated fluctuating-field arrays.

## 1 PIC and FSP Particle Simulation

Particle simulation codes [1] seem to be the most suited tool for the investigation of turbulent plasma behaviour. Particle simulation indeed consists in evolving, according to the equations of motion, the phase-space coordinates of a particle population in the fluctuating electromagnetic fields, selfconsistently computed, at each time step, in terms of the contribution yielded by the particles themselves (e.g., pressure perturbation), and allows one to fully retain all the relevant kinetic effects.

The equations for the fields can, in principle, be solved directly in the real space, using, e.g., finite difference methods. In many concrete situations, however, periodic spatial domains are considered. Moreover, in order to get a deeper insight in the physical mechanisms underlying the observed phenomena, the analysis is often focused on the evolution of relatively few harmonics. In such cases, it is worth solving the field equations in the Fourier space, then transforming the fields back to the real space in order to push the particles. In the present paper we want to show that, under this *few-harmonic* condition – which we assume to hold in the following –, a particular class of particle simulation codes, the *gridless finite-size-particle* (FSP) codes, is particularly suited to be efficiently parallelized on distributed-memory architectures.

The most widely used class of particle simulation codes is that of the *particle-in-cell* (PIC) codes, which compute the fluctuating fields and the required particle contribution only at the nodes of a spatial grid, then interpolating the fields

---

* briguglio@frascati.enea.it
** University of Naples "Federico II", Dipartimento di Informatica e Sistemistica, Naples, Italy

at the (continuous) particle positions in order to perform particle pushing. The introduction of a spatial grid yields a major benefit, in smoothing the short-range interactions between particles. In the absence of such a smoothing, these interactions would dominate over the long-range ones, due to the fact that, once ensured the coincidence of simulation and physical scale lengths by identifying each simulation particle with a macroparticle (i.e., a cluster of a large number of mutually non-interacting physical particles) that moves like a single physical particle [2], the density of the numerical plasma, $n_0$, is too low to satisfy the plasma condition $n_0 \lambda_D^3 \gg 1$ (with $\lambda_D$ being the Debye length). This would not correspond to what happens in a physical plasma. However, it can be easily shown that, if $L_c$ is the grid-point spacing, the condition that has to be satisfied in order to reproduce correct plasma behaviour is given by $n_0 L_c^3 \gg 1$, which replaces and relaxes the too stringent plasma condition. In an average sense, such a new condition can be written as $N_{part}/N_{cell} \gg 1$, where $N_{part}$ and $N_{cell}$ are the number of simulation particles and grid cells, respectively; it then corresponds to the requirement of disposing, for a given spatial resolution, of a number of particles large enough to accurately sample the velocity space. This requirement makes the full exploiting of parallel computers unavoidable as soon as high spatial-resolution levels have to be reached in order to investigate small-scale field fluctuations.

On the basis of these considerations, it is apparent that the main purpose of the parallelization of PIC codes is that of distributing the computational requests related to the particle population among several different processors. The computational effort related to the Fourier-space field solver is instead generally negligible and can be replicated on each processor; if this were not the case, also this portion of the simulation could be, in principle, distributed.

Two main different techniques have been developed in parallelizing PIC codes. The first approach is based on the *domain decomposition* [3]: different portions of the physical domain are assigned to different processors, together with the particles that reside on them. In this way, both the number of operations per time step and the memory space required to each computational node are reduced (in an average sense) by a factor equal, roughly speaking, to the number of processors, $n_{proc}$. Neglecting corrections due to inter-processor communication, these two quantities are indeed given by the following expressions:

$$N_{op\ d.d.}^{PIC} \approx f(N_{harm}) + \frac{1}{n_{proc}} \left( n_{FT} \times N_{harm} \times N_{cell} + n_{interp} \times N_{part} \right), \quad (1)$$

and

$$M_{d.d.}^{PIC} \approx m_{harm} \times N_{harm} + \frac{1}{n_{proc}} \left( m_{cell} \times N_{cell} + m_{part} \times N_{part} \right). \quad (2)$$

Here $N_{harm}$ is the number of Fourier harmonics retained in the simulation; $f(N_{harm})$ is an increasing function of $N_{harm}$, which corresponds to the number of operations required to update the electromagnetic fields in the Fourier space and whose precise dependence is determined by the structure of the problem and the algorithm used by the field solver; $n_{FT}$ is the number of operations needed to compute each addendum in the sum required by the Fourier transform (and anti-transform) [1] and $n_{interp}$ that of the operations required for the interpolation of the fields (and for the distribution of the particle contribution among the grid points). Moreover, $m_{harm}$, $m_{cell}$ and $m_{part}$ are the amounts of memory needed to store, respectively, a single harmonic of the complete set of Fourier-space fields, the real-space fields at each grid point and the phase-space coordinates for each particle.

The main limitation of such a method consists in the fact that inter-processor communications are required when a particle migrates from one processor to another, and this can cause the parallel version of the code to be difficult to implement and/or inefficient. Moreover, serious load-balancing problems can arise because of such migration. If such problems do not occur, however, the main advantage of the scheme emerges, corresponding to the almost linear scaling of the attainable physical-space resolution with the number of processors. The computational effort required by the replicated Fourier-space solution of the field equations – corresponding to the first terms in Eqs. (1) and (2) – is indeed negligible in comparison with the one required by the grid and particle calculations – the terms multiplied by $1/n_{proc}$.

A complementary approach to the domain decomposition is represented by the *particle decomposition* [4]. It corresponds to replicate the whole spatial domain on each processor, while distributing the particle population. No particle migrates from one processor to another, because no particle meets, in its motion, the unphysical boundaries introduced by domain decomposition, and the communication among processors is both reduced and simplified (a little amount of communication is still required, because different-processor particle contributions to pressure have to be collected and summed together, at each time step, before updating fields). Moreover, load balancing is perfectly ensured during every simulation. On the opposite side the memory request associated to the grid quantities (equilibrium and fluctuating fields), which are replicated on each processor, gives rise to a bottle-neck on the scalability of physical resolution with processors. Neglecting the corrections due to the need for communicating partial pressure data, the computational load on each processor is indeed given, in this case, by

$$N^{PIC}_{op\ p.d.} \approx f(N_{harm}) + n_{FT} \times N_{harm} \times N_{cell} + n_{interp} \times \frac{N_{part}}{n_{proc}}, \qquad (3)$$

and

$$M^{PIC}_{p.d.} \approx m_{harm} \times N_{harm} + m_{cell} \times N_{cell} + m_{part} \times \frac{N_{part}}{n_{proc}}. \qquad (4)$$

Even in the limit of infinite number of processors, in which each processor must treat a vanishingly small number of particles, the maximum resolution that can

---
[1] We refer to general situations in which the Fast Fourier Transform (FFT) algorithm is not recommended (e.g., because the Fourier space is not densely filled). Our conclusions, however, would maintain their qualitative validity even in the FFT framework.

be reached is determined by the largest size of the (replicated) grid arrays that can be stored on each node.

Although the *few-harmonic* condition does not cause the real-space resolution request to be relaxed (one is anyway interested in the high mode-number portion of the complete fluctuation spectrum), it offers a different and more convenient path toward the target of an efficient parallelization and, hence, toward the high-resolution simulations. Instead of introducing a spatial grid for the calculations of the electromagnetic fields, and interpolating such fields at each particle position to perform particle pushing, it is possible to compute the actual values of the fields at the particle position by directly transforming them from the Fourier space back to the real space. Correspondingly, particle contributions to the pressure are just Fourier-transformed and summed together, instead of being collected at the nearest grid points. In such way, the memory resources demanded by the field storage are greatly reduced, and the bottle-neck in the scalability of the resolution with $n_{proc}$ is in fact removed.

The suppression of the spatial grid, however, also eliminates the benefic smoothing of the short-range interactions between particles. To overcome such a difficulty, it is possible to resort to the FSP method [5], which consists in replacing the "singular" particles by "regular" *charge clouds* of typical size $L_s$. It is immediate to show that such a FSP ensemble behaves as a plasma under the condition $n_0 L_s^3 \gg 1$. PIC and FSP methods are expected to yield the same qualitative results if $L_c \approx L_s$.

A particle-decomposition approach to the parallelization of a gridless FSP code would then produce the following computational loads on each processor:

$$N_{op\ p.d.}^{FSP} \approx f(N_{harm}) + n_{FT} \times N_{harm} \times \frac{N_{part}}{n_{proc}}, \quad (5)$$

and

$$M_{p.d.}^{FSP} \approx m_{harm} \times N_{harm} + m_{part} \times \frac{N_{part}}{n_{proc}}. \quad (6)$$

In the case of serial simulations ($n_{proc} = 1$), considering that the quantity $f(N_{harm})$ is typically negligible in comparison with the other terms that appear in Eqs. (3) and (5), and that for PIC codes $N_{part}/N_{cell} \gg 1$, it can be easily seen that, as far as $n_{FT} \times N_{harm} > n_{interp}$ (not too few harmonics), the gridless FSP method is more expensive than the PIC one, without offering any significant advantage in terms of lower memory requests. Both the relevant terms in Eq. (3) are indeed separately smaller than the term proportional to $N_{part}$ in Eq. (5). This motivates the large diffusion of PIC codes and the poor consideration in which gridless FSP simulation has been taken until now.

In a parallel-simulation framework, however, the superiority of the FSP method is apparent. First of all, though FSP methods require back and forth Fourier transforms being executed for each particle, these calculations are distributed among different processors, differently from the replicated PIC grid calculations. From Eqs. (3) and (5), it can be seen that a gridless FSP code is expected to become more advantageous than a PIC one for

$$n_{FT} \times N_{harm} \times N_{cell} + n_{interp} \times \frac{N_{part}}{n_{proc}} \gtrsim n_{FT} \times N_{harm} \times \frac{N_{part}}{n_{proc}}, \quad (7)$$

or

$$n_{proc} \gtrsim N_{ppc}\left(1 - \frac{n_{interp}}{n_{FT} \times N_{harm}}\right), \quad (8)$$

with $N_{ppc}$ defined as $N_{ppc} \equiv N_{part}/N_{cell}$. In the FSP case, $N_{cell}$ has, by itself, no meaning. However, we can interpret such number as the number of cells that would be necessary to accurately describe the same modes we investigate by the gridless code; in other words, while, in PIC simulations, $N_{cell} \propto L_c^{-3}$, in FSP simulations we can take $N_{cell} \propto L_s^{-3}$. We can then still formally refer to $N_{ppc}$ as to an average number of particles "per cell", and look at $N_{part}$ as the product of certain levels of spatial resolution (measured by $N_{cell}$) and velocity-space one (measured by $N_{ppc}$).

Second, different from the PIC case, large efficiency can be obtained even for massively parallel simulations, as far as the number of modes, $N_{harm}$, retained in the simulation is, in spite of the high mode numbers considered (high spatial resolution), relatively small; such a *few-harmonic* condition makes indeed the terms related to $N_{harm}$ in Eqs. (5) and (6) negligible, if compared with the others and, in particular, causes the linear scaling of the maximum allowed spatial resolution with $n_{proc}$ to be recovered even for a very large number of processors. At the same time, of course, the positive feature of an automatic load balancing, typical of the particle-decomposition parallelization, is preserved in this case.

## 2 The FSP Hybrid MHD-Gyrokinetic Code

In this section we present the main features of a parallel gridless FSP version of the Hybrid MHD-Gyrokinetic Code (HMGC), a code for the investigation of Alfvénic turbulence in magnetically confined plasmas, previously implemented in a PIC version [6]. Particle-decomposition parallelization of HMGC has been presented in Ref. [4]. The code solves the coupled sets of MHD equations for the fluctuating electromagnetic fields and gyrokinetic equations of motion for collision-free energetic ions. The physical domain is represented by a three-dimensional toroidal grid, for which quasi-cylindrical coordinates are adopted (see Fig. 1): the minor-radius coordinate, $r$, and the poloidal and toroidal angles, $\vartheta$ and $\varphi$, respectively. The field solver uses finite differences in the minor radius direction and Fourier expansion in the poloidal and toroidal directions.

The FSP version of the code eliminates the grid in the $\vartheta$ and $\varphi$ directions, and introduces finite sizes for particles along the same directions. The corresponding Fourier variables are $k_\vartheta \equiv m/r$ and $k_\varphi \equiv n/R$, where $m$ and $n$ are, respectively, the poloidal and the toroidal number, and $R$ is the major-radius coordinate of the torus. The radial grid is instead maintained (so that this FSP code is not

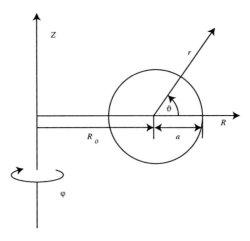

**Fig. 1.** Toroidal coordinate system $(r, \vartheta, \varphi)$ for a magnetically confined plasma with major radius $R_0$ and minor radius $a$.

a "gridless" one, in an absolute sense), because of the impossibility of adopting Fourier expansion in that direction.

The Alfvén modes, which constitute the main object of the investigation performed by HMGC, are typically characterized [7], for a given value of the toroidal number $n$, by the coupling of several poloidal harmonics, with sharper radial structure the higher the number $n$ is, and poloidal numbers such that $nq(0) - 1/2 \leq m \leq nq(a) + 1/2$ (here $q(r) \equiv rB_\varphi/R_0 B_\vartheta$ is the so-called *safety factor*, $a$ and $R_0$ are the minor and the major radius of the torus, respectively, $B_\varphi$ is the toroidal magnetic field component and $B_\vartheta$ is the poloidal one). As a consequence, both poloidal and radial resolution requests (linearly) increase with increasing $n$, as it happens – of course – for toroidal resolution. In principle, memory resources of each processor could still be saturated by the offset associated to the first term in Eq. (6). Anyway, such an offset is in fact negligible, as its mode-number dependence, for a number of retained modes proportional to $n$, is quadratic (both $m_{harm}$, which takes into account the radial dependence of each mode, and $N_{harm}$ scale linearly with $n$). In the PIC case, instead, the offset is dominated by the term $m_{cell} \times N_{cell}$, which scales as $n^3$. More precisely, with the level of precision adopted in our simulations to describe the spatial structure of the modes, the memory (in Megabytes) needed to store the field arrays for the FSP and the PIC version of HMGC, respectively, scales with the toroidal number $n$ according to the following expressions:

$$M_{field}^{PIC} \approx 0.37 n^3, \qquad (9)$$

and

$$M_{field}^{FSP} \approx 0.023 n^2. \qquad (10)$$

Here we consider linear simulations with modes characterized by a single $n$, $q(0) = 1.1$ and $q(a) = 1.9$. It can be appreciated that the PIC approach is in fact penalized both by the $n$ dependence and the proportionality coefficient. As such arrays are replicated on the different processors, the saturation of memory space (and the need for *memory paging*) is reached, in the PIC case, at lower values of $n$ and, thus, at lower space resolution. The whole memory needed to store the distributed arrays associated to particle quantities (corresponding to the last term in Eqs. (4) and (6)) scales instead, for both versions of HMGC, according to the following expression:

$$M_{part} \approx 0.23 N_{ppc} \times n^3, \tag{11}$$

with the cubic dependence on $n$ being associated to the "number of cells", $N_{cell} \equiv N_{part}/N_{ppc}$.

In Ref. [8] the PIC version of the code has been applied to the investigation of the linear stability and the nonlinear saturation of Alfvén modes in Tokamaks, in the presence of an energetic-ion population. The main results obtained in that analysis can be summarized as follows: as the energetic-ion pressure increases above a certain threshold, depending on the toroidal number $n$, the fast-growing Energetic Particle Mode (EPM) becomes unstable [7]. The EPM saturates because of a sudden displacement of a large part of the energetic particles along the minor radius. Such a displacement can, in principle, greatly enhance the energetic-particle losses, and makes the determination of the threshold for EPM destabilization an important issue in the theoretical research on reactor-relevant plasmas.

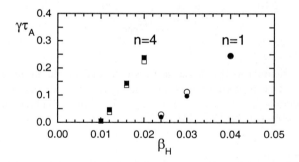

**Fig. 2.** Growth rate of Energetic Particle Modes, normalized to the inverse of the Alfvén time $\tau_A$, obtained, at different values of $n$ and $\beta_H$, by PIC (full symbols) and FSP (empty symbols) HMGC simulations.

Figure 2 shows the growth rate of the EPM's obtained by PIC and FSP H-MGC simulations at different values of $n$ and $\beta_H$ (the ratio between the energetic-particle pressure and the magnetic one). The case of a plasma with aspect ratio $R_0/a = 10$ and energetic-particle Larmor radius $\rho_H$ such that $\rho_H/a = 0.01$ is considered here. A density profile of the form $\exp\left[-\left(r^2/L_n^2\right)^{\alpha_n}\right]$ has been assumed for the energetic ions, with $a^2/L_n^2 = 2$ and $\alpha_n = 2$. Slight quantitative differences between the results obtained by the two methods can be traced back to the different algorithms adopted, but a substantial agreement is found.

## 3 Parallelization Results

The FSP HMGC has been parallelized, according to the same strategy adopted for the corresponding PIC version of the code [4], within the High Performance Fortran [9] (HPF) framework, and tested on a 16-node IBM SP parallel system, by using the IBM *xlhpf* compiler [10] (an optimized native compiler for IBM SP systems). Some obvious differences exist between the two parallel codes, related, for example, to the fact that, in the FSP case, the Fourier components of the pressure are directly calculated as a sum over the particle population, rather than transforming the corresponding sum computed at the spatial grid points [2].

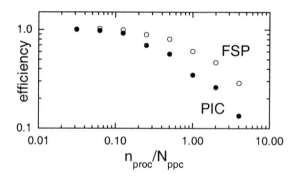

**Fig. 3.** Efficiency, defined as the speed-up factor divided by the number of processors, versus $n_{proc}/N_{ppc}$. Linear simulations retaining modes with a single toroidal number, $n = 1$, are considered. The results obtained by FSP HMGC (empty circles) are compared with those obtained by PIC HMGC (full circles).

The efficiency, defined as the speed-up factor divided by the number of processors, is plotted, in Fig. 3, versus $n_{proc}/N_{ppc}$, the inverse of the average number of particles per cell per processor. Linear simulations retaining modes with a single toroidal number, $n = 1$, are considered here. The results obtained by FSP HMGC are compared with those obtained by PIC HMGC. For such latter case, it was observed in Ref. [4] that the efficiency is almost at its ideal value as far as

---

[2] The auxiliary array, introduced in Ref. [4] with an extra dimension related to the number of processors, represents, in this case, the partial contribution (yielded by those particles that reside on a certain processor) to the Fourier transform of the pressure, rather than to the pressure itself.

$N_{ppc} \gg n_{proc}$, whereas it falls well below such a value when the average number of particles per cell seen by each processor is so small that the replicated grid calculations dominate over the distributed particle ones. It can be seen that, in the FSP simulations, the decrease in the efficiency is observed above a certain threshold in $n_{proc}/N_{ppc}$, significantly higher than the threshold found in the PIC case. This corresponds to the fact that the amount of replicated calculations is drastically reduced, and this fact allows to get high efficiency, at fixed number of particles, even at high number of processors. As a consequence, the particle-decomposition parallelization of FSP codes, different from the same parallelization of PIC codes, is suited even for massively parallel architectures.

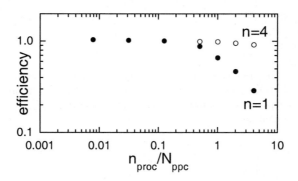

**Fig. 4.** Efficiency versus $n_{proc}/N_{ppc}$ for FSP simulations with different values of the single toroidal number retained: $n = 1$ (full circles) and $n = 4$ (empty circles). The efficiency maintains almost its ideal values up to higher values of $n_{proc}/N_{ppc}$ the higher the toroidal number $n$ is.

Figure 4 plots the efficiency values, obtained in FSP simulations with different values of $n$, versus $n_{proc}/N_{ppc}$. Note that the efficiency threshold in $n_{proc}/N_{ppc}$ comes out to increase with $n$. This fact can be easily understood, from Eq. (5), considering that the distributed part of the calculation scales with a higher power of $n$ than the replicated one does. Therefore, the two parts become comparable (with a corresponding decrease of efficiency) for higher values of $n_{proc}/N_{ppc}$ the higher the toroidal number $n$ is.

Finally, Fig. 5 shows the Central Processor Unit (CPU) *user time* required by each simulation time step versus the number of processors. Results obtained by FSP (empty symbols) and PIC (full symbols) simulations with fixed mode number, $n = 1$, and different values of $N_{ppc}$ (a), or fixed velocity-space resolution, $N_{ppc} = 4$, and different values of $n$ (b) are compared. It can be seen that the FSP approach becomes more advantageous when the number of processors exceeds a certain threshold value. In qualitative agreement with Eq. (8) and with the fact that $N_{harm} \propto n$, such a threshold comes out to increase with $N_{ppc}$ and, more weakly, with $n$.

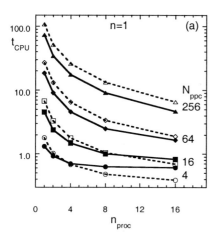 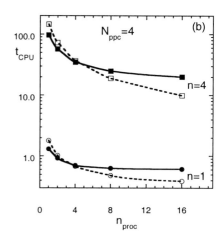

**Fig. 5.** CPU *user time* per time step versus $n_{proc}$. Results obtained by FSP (empty symbols) and PIC (full symbols) simulations with fixed mode number, $n = 1$, and different values of $N_{ppc}$ (a), or fixed velocity-space resolution, $N_{ppc} = 4$, and different values of $n$ (b) are compared.

## References

1. C. K. Birdsall and A. B. Langdon, *Plasma Physics via Computer Simulation* (McGraw-Hill, New York, 1985).
2. A. Sestero, "Basic Interactions in Real Plasmas and in the Plasmas of Numerical Experiments", Il Nuovo Cimento 9B (1972) 222-232.
3. P. C. Liewer and V. K. Decyk, "A General Concurrent Algorithm for Plasma Particle-in-Cell Codes", J. Computational Phys. 85 (1989) 302-322.
4. B. Di Martino, S. Briguglio, G. Vlad and P. Sguazzero, "Parallel Plasma Simulation in High Performance Fortran", in *High Performance Computing and Networking*, (Springer, Berlin, 1998) 203-212.
5. A. B. Langdon and C. K. Birdsall, "Theory of plasma simulation using finite-size particles", Phys. of Fluids 13 (1970) 2115-2122.
6. S. Briguglio, G. Vlad, F. Zonca, and C. Kar, "Hybrid magnetohydrodynamic-gyrokinetic simulation of toroidal Alfvén modes", Phys. Plasmas 2 (1995) 3711-3723.
7. L. Chen and F. Zonca, "Theory of Shear Alfvén Waves in Toroidal Plasmas", Physica Scripta T60 (1995) 81-90.
8. S. Briguglio, F. Zonca, and G. Vlad, "Hybrid Magnetohydrodynamic-Particle Simulation of Linear and Nonlinear Evolution of Alfvén Modes in Tokamaks", Phys. Plasmas 5 (1998) 3287-3301.
9. H. Richardson, "High Performance Fortran: history, overview and current developments", Tech. Rep. TMC-261, Thinking Machines Corporation, 1996.
10. M. Gupta, S. Midkiff, E. Schonberg, V. Seshadri, D. Shields, K.Y. Wang, W. M. Ching, T. Ngo, "A HPF Compiler for the IBM SP2", in: Proc. SuperComputing '95 (ACM, 1995).

# A Distributed Object-Oriented Method for Particle Simulations on Clusters*

Yudong Sun, Zhengyu Liang, and Cho-Li Wang

Department of Computer Science and Information Systems
The University of Hong Kong, Pokfulam Road, Hong Kong
{ydsun, zyliang, clwang}@csis.hku.hk

**Abstract.** This paper describes a distributed object-oriented method for solving $N$-body problem of particle simulations. The method allows dynamic construction of a collaborative system based on the computational requirement of an application and the available resources in the cluster. In the system, a group of objects on distributed hosts cooperate to execute the application. The method is implemented in Java and RMI. The platform-independent features of Java enable the method to support efficient distributed computing in heterogeneous environment. The performance test shows that the method can achieve good speedup and portability. The proposed method can be extended to support other scientific computing applications in distributed environment.

## 1 Introduction

*N-body problem* studies the evolution of a system of $N$ bodies (particles) under the cumulative influences on every body from all other bodies that cause the sustained body movement. It is an irregularly structured problem that exists broadly in astrophysics, plasma physics, molecular dynamics, fluid dynamics, radiosity calculations in computer graphics, and etc. [1]. The common feature of these applications is the large range of scales in the information requirements. A body in a physical domain requires progressively less information in less frequency from parts of the domain that are farther away. The domain parts and the influences from those parts are continuously changing during the evolution of the system.

Many algorithms have been proposed to solve $N$-body problem [2,3,4,5,6]. The Barnes-Hut method [2], for example, is a typical tree-based algorithm which computes the interaction between particles in space by constructing and traversing a tree which represents the interrelationship between the particles. Fig. 1 gives an example of body distribution in a two-dimensional domain and its Barnes-Hut tree representation. This is a quadtree for the bodies in 2D domain. The Barnes-Hut tree for three-dimensional domain is an octree.

---

* This research was supported by Hong Kong Research Grants Council (RGC) grant 10201696 and The University of Hong Kong CRCG grant 10200544.

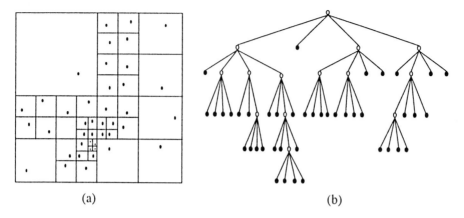

**Fig. 1.** Barnes-Hut tree in 2D domain: (a) Body distribution and domain partitioning; (b) Barnes-Hut quadtree.

Many parallel $N$-body algorithms are derived from Barnes-Hut method. Singh and et al. [7] presented a method in shared-address-space multiprocessing model. An octree is created in shared memory segment and it is globally accessible by all concurrent processes on multiple processors. Such a programming model is appropriate for shared memory systems like SMP machines. Multiple processes cooperately build a globally accessible octree and then calculate the force exerted on each body from all other bodies by tree traversal. All inter-process communication is via the shared memory. Hence the communication overhead is quite low. When solving $N$-body problem on distributed memory systems, message passing is a usual way for inter-process communication [8]. MPI and PVM are common message passing libraries.

This paper proposes a distributed object-oriented method (DOO) for solving $N$-body problems on a cluster of heterogeneous workstations and PCs. Our final goal is to extend this distributed computing framework to an Internet-based computing environment [9,10]. We choose Java and RMI (Remote Method Invocation) [11] to implement the distributed object-oriented method. With the architecture-neutral and object-oriented features, Java supports applications running in a system which is composed of different platforms. RMI provides an easy way to register remote objects to be the usable computing resources in a distributed system and allocate the computational tasks to the remote objects [11]. The mechanisms can transmit either data or methods between distributed objects and load the program codes into remote objects at runtime. Thus, the algorithm will be adaptive to the system resources and the computing requirements of an application. Java and RMI endue the distributed object method with high portability and flexibility in heterogeneous and dynamically changed environment [12].

Our method for $N$-body problem is a distributed object-oriented method rather than simply a message-passing scheme. The new method involves a group of collaborative objects. One of them plays the role of *compute coordinator*. It initiates the computing procedure by invoking remote objects on other hosts, dispatches computational tasks to them, and collects the results at the end. The registration and all interaction between those objects are through RMI interface. The distributed object-oriented method has the following advantages:

- *Platform-independent*: Code in Java and RMI is widely executable on any platform as long as JVM (Java Virtual Machine) is supported. No code modification or recompilation is needed when porting a program onto various platforms.
- *Network-computing*: DOO method is suitable for network computing on distributed platforms. It is an ideal object-oriented framework for Internet-based computation.
- *Object-migrateable*: Distributed objects can be moved from a heavily loaded machine to a less loaded one to make full use of the available computing power.
- *System-adaptive*: Remote objects can be dynamically invoked or terminated to match the system reconfiguration. Thus, the number of objects can be decreased or increased subject to the runtime configuration.
- *Method-passing*: In addition to data, method code can be transmitted from the home server to remote objects. With this capability, the new method can be automatically inserted to remote objects.

In the following sections, we illustrate the distributed object-oriented method in Section 2 and describe the DOO method for solving $N$-body problem in Section 3. The performance of running the N-body application on clusters will be discussed in Section 4. Finally Section 5 will give the conclusions about this research and briefly discuss the future work.

## 2. Collaborative System

The distributed object-oriented method is developed based on a collaborative computing model. As shown in Fig. 2, a *collaborative system* consists of a group of distributed objects on different hosts. The object can be referred to as a *compute engine*.

At the beginning, the objects start to run independently on the hosts to participate in the computing over network. The objects *register* themselves to the RMI registry. After the registration, a collaborative system is established and these objects can locate each other by looking up the registry. The distributed objects will accept the computing tasks and cooperatively process an application.

One of the distributed objects, compute coordinator, is responsible for initiating the computation. When distributed computing is demanded, it looks up the registry to find the available compute engines, and then distributes the computing tasks to these compute engines by invoking the computing method on them. The coordinator coordinates the entire computing procedure in the collaborative system and collects final results. In our $N$-body application, the compute coordinator dispatches a portion of bodies (a subset of $N$ bodies) to each engine via the RMI *interaction mechanism*. The distributed objects also interact with one another through the interaction mechanism. The interaction includes the invocation of the methods on the compute engines from the coordinator and the data exchange between the engines.

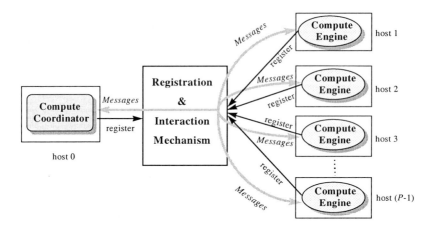

**Fig. 2.** The collaborative system on a *P*-node cluster.

The configuration of a networked system can be changed dynamically. Hosts may be turned off for maintenance or be removed from the system. New hosts can be added to the system. The workload on the hosts varies from time to time. Only idle or lightly-loaded hosts are going to accept computational tasks from other hosts. The registration mechanism enables distributed objects to freely join or disjoin a collaborative system, so that the system can be dynamically configured with the available resources. In our implementation, a class, **ComputerImpl()**, is defined to represent the compute engine. To join the collaborative system, each host creates an instance of this class.

When running the *N*-body application, a collaborative system is established. The procedure of particle simulation proceeds by iterating the computing round (see Section 3.2). It is the opportunity to reconstruct the collaborative system at the beginning of each computing round in correspondence with the current registration of the remote objects. The compute coordinator can redistribute the computing tasks to those objects. After the initiation, the coordinator takes a share of computation and works as a compute engine as well.

## 3. DOO-based *N*-body Algorithm

We developed an *N*-body algorithm based on the Barnes-Hut application in SPLASH2 suite [13]. In our algorithm, a distributed Barnes-Hut tree structure is constructed for facilitating the DOO method.

## 3.1 Distributed Barnes-Hut Tree

Let $N$ denote the total number of bodies and $P$ the number of hosts. Each host starts a compute engine. If there are four hosts, the domain in Fig. 1(a) will be partitioned into four sub-domains shown in Fig. 3(a). The domain partitioning is based on spatial locality. Thus, the number of bodies in each sub-domain is not necessarily equal. The coordinator divides $N$ bodies into $P$ portions. A portion contains the bodies in one sub-domain. Each compute engine takes one portion of bodies and builds a *local subtree* for the portion. Therefore, the global Barnes-Hut tree in Fig. 1(b) is decomposed into four subtrees shown in Fig. 3(b). The $N$-body methods in [8] and [14] also adopted distributed tree schemes.

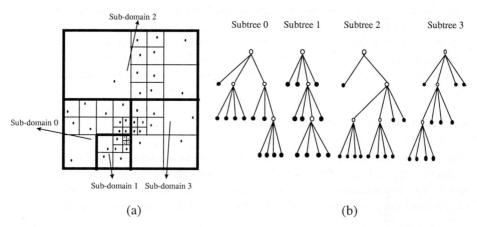

**Fig. 3.** A distributed Barnes-Hut tree in 2D for four compute engines: (a) Sub-domains after partitioning; (b) Distributed Barnes-Hut tree.

Besides its local subtree, a compute engine needs to access all remote subtrees to compute the force influences from remote bodies. To make the distributed subtrees accessible for all compute engines, each engine also creates a *partial subtree* which consists of the top levels of the local subtree and broadcasts it to all other engines. So every computer engine obtains all remote partial subtrees for its local access.

## 3.2 Computing Round

After the coordinator has assigned the portions of bodies to remote engines, the particle simulation begins and it is advanced by iterating the computing round, i.e., one step of body movement. A computing round is carried out in three phases.

1. *Subtree construction and propagation*
   Compute engines build the subtrees and broadcast the partial subtrees to all other engines. This needs an all-gathering communication. The coordinator gathers all partial subtrees and then propagates them to all compute engines by remote method calls.

2. *Force calculation*

Each engine calculates the forces on the bodies by traversing its local subtree and the received partial subtrees. The bodies that need to visit the lower levels of a subtree not existing in the particle subtree will be sent to the compute engine containing that subtree. Then the forces on these bodies is evaluated by traversing the complete subtree on that compute engine. After that, the bodies return to their original engines. Finally the forces on a body calculated both on its home engine and on remote engines are summed. The new states of a body, including its position and velocity, are determined. In this phase, all-to-all communication is performed among the compute engines.

3. *Body redistribution*

At the end of force calculation, all bodies have moved to their new positions. Some bodies may move to other sub-domains. To keep the spatial locality of the bodies, those bodies should be sent to the destination engines where their new positions locate by an all-to-all communication. In general, only a few bodies will be transmitted in this phase because of the slow pace of body movement.

## 4. Performance Tests and Analyses

The $N$-body application in distributed object-oriented method has been tested on homogeneous and heterogeneous clusters. It simulates particles' movement in 3D space. The performance has been inspected in the speedup of the application on clusters and the portability of the DOO method on heterogeneous platforms.

### 4.1 Speedup

The $N$-body application in Java and RMI has been executed on a local cluster of Pentium family PCs, including PentiumPro/233MHz, 450MHz, PentiumII/233MHz and K6-2/350MHz linked by 100Mbps Fast Ethernet, running Linux 2.0.3x.

To verify the reduction of communication traffic by the distributed tree structure, we implemented a straight-forward $N$-body method as a comparison. In this method, the compute coordinator starts a computing round by propagating $N$ bodies wholly to each compute engine, where a *complete tree* containing all bodies is built. Although each engine still computes the forces on a portion of bodies, it has to receive and store all bodies. In contrast to the *distributed tree* method in Section 3, this method can be called *complete tree* method. The force calculation for each body is local to a compute engine by traversing the complete tree. No remote tree access is required. At the end of the computing round, the compute coordinator collects the new state data of all bodies from the remote compute engines, and then broadcasts the data to all compute engines to start next computing round. There are one broadcast and one gather operation performed in each computing round. So the communication overhead is much higher than the distributed tree method.

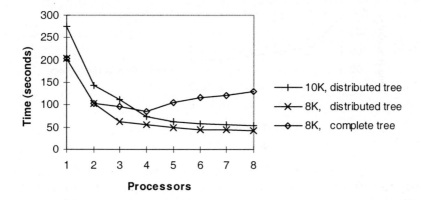

**Fig. 4.** The execution time of the two N-body methods on the homogeneous cluster of Pentium family PCs, with 8K or 10K bodies.

Fig. 4 compares the execution time of the two methods. For the distributed tree method, speedup can be observed on one to eight processors since lower communication is brought about in the broadcast of subtrees and the transmission of bodies. The distributed Barnes-Hut tree is appropriate for distributed computing. On the other hand, speedup merely occurs for smaller machine sizes in the complete tree method. When running on five processors or above, the heavy communication overhead in the broadcasting and gathering of N bodies deteriorates the performance.

### 4.2 Test in Heterogeneous Cluster

The distributed tree method has also been tested on a wider-range heterogeneous cluster of different platforms. There are totally six processors in the cluster. The processors take part in the computation in the order of two PentiumPro/450MHz PCs running Linux 2.0.36, two Sun UltraSPARC-1 workstations running Solaris 2.6 and one SGI PowerChallenge SMP (two processors are used for this computing) running IRIX 6.2. These hosts locate at different sites. Two PentiumPro's lies in the local cluster; two Sun UltraSPARC-1's belong to another distant cluster; and the SGI PowerChallenge is also a remote server. The hosts are linked together by campus network. Our code can successfully run on these platforms without any modification. Fig. 5 shows the execution time of the distributed tree method on the heterogeneous cluster.

Speedup can be attained in the heterogeneous network. It confirms that the DOO method has platform-independent features and the method supports network-computing well.

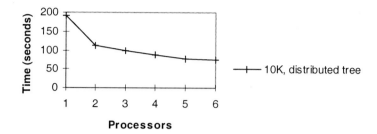

**Fig. 5.** The execution time of the distributed tree method with 10K bodies on the heterogeneous cluster. The processor order is two PentiumPro/450MHz PCs, two Sun UltraSPARC-1 workstations and two processors in SGI PowerChallenge SMP.

The execution time of the Java code is not satisfactory due to the inefficiency of Java interpreter. It is ten times slower than the implementation using C++ and MPI communication library in [15]. But Java owns attractive features of strong platform-independent portability and flexibility. We select Java and RMI to implement the method to take advantage of those features. The performance of the DOO method is expected to be enhanced with the improvement of Java packages.

## 5. Conclusions and Future Work

We have discussed a distributed object-oriented method for solving $N$-body problems on clusters. The new method is implemented in Java and RMI protocol in order to support distributed computing in heterogeneous network environment. Distributed objects can work together in the model of collaborative system. Such a system is established on the available resources in a cluster. The DOO method, with the architectural-neutral features of Java and RMI, can be considered as a promising model for distributed and parallel computing in heterogeneous environment. The model can be further expanded to a wide-area distributed computing. It also provides the opportunity to dynamically configure a computing environment whenever the computing resources have been changed. So the new method will yield high flexibility, portability and adaptability.

In our $N$-body algorithm, we proposed a distributed tree structure to reduce the communication overhead. The runtime tests show that our method is effective for solving $N$-body problems on distributed and heterogeneous systems. The DOO method is also feasible for other scientific computing applications.

In our future research, we plan to make the method adaptive to the network computing environment. Compute engines (remote objects) can join the collaborative system at runtime when they are available, and the computing tasks can be redistributed to the compute engines based on the status of the collaborative system. The method will be extended to more wide-range distributed computing.

## References

1. Singh, J. P., Hennessy, J. L., Gupta, A.: Implications of Hierarhical $N$-Body Methods for Multiprocessor Architectures. ACM Transactions on Computer Systems, Vol. 13, 2 (1995) 141-202
2. Barnes, J., Hut, P.: A Hierarchical $O$ ($N$ log $N$) Force-Calculation Algorithm. Nature, Vol. 324, 4 (1986) 446-449
3. Greengard, L., Rokhlin, V.: A Fast Algorithm for Particle Simulations. Journal of Computational Physics, Vol. 73, (1987) 325-348
4. Hernquist, L.: Hierarchical $N$-body Methods. Computer Physics Communications, Vol. 48, (1988) 107-115
5. Hu, Y. C., Johnsson, S. L., Teng, S. H.: A Data-Parallel Adaptive $N$-body Method. Proceedings of $8^{th}$ SIAM Conference on Parallel Processing for Scientific Computing (1997)
6. Salmon, J., Warren, M. S.: Parallel, Out-of-core methods for $N$-body Simulation. Proceedings of the $8^{th}$ SIAM Conference on Parallel Processing for Scientific Computing (1997), also available at http://www.cacr.caltech.edu/~johns/pubs/siam97/
7. Singh, J. P., et al.: Load Balancing and Data Locality in Adaptive Hierarchical $N$-body Methods: Barnes-Hut, Fast Multipole, and Radiosity. Journal of Parallel and Distributed Computing, Vol. 27, 2 (1995) 118-141
8. Grama, A. Y., Kumar, V., Sameh, A.: n-body Simulation Using Message Passing Parallel Computers. Proceedings of the $7^{th}$ SIAM Conference on Parallel Processing for Scientific Computing (1995) 355-360
9. Khokhar, A., Shaaban, M., Prasanna, V., Wang, Cho-Li: Heterogeneous Computing: Challenges and Opportunities. IEEE Computer magazine, Vol. 26, 6 (1993) 18-27
10. Berman, F., Wolski, R.: Scheduling from the Perspective of the Application. Proceedings of Symposium on High Performance Distributed Computing (1996), also available at http://www-cse.ucsd.edu/groups/hpcl/apples/apples.html
11. Farley, J.: Java Distributed Computing. O'Reilly & Associates Inc (1998)
12. Keren, A., Barak, A.: Adaptive Placement of Parallel Java Agents in a Scalable Computing Cluster. ACM 1998 Workshop on Java for High-Performance Network Computing (1998), also available at http://www.cs.ucsb.edu/conferences/java98/program.html
13. Woo, S. C. and et al.: The SPLASH-2 Programs: Characterization and Methodological Considerations. Proceedings of the $22^{nd}$ Annual International Symposium on Computer Architecture (1995), also available from http://www-flash.stanford.edu/SPLASH
14. Bhatt, S., Chen, M., Cowie, J., Lin, C. Y., Liu, P.: Object-Oriented Support for Adaptive Methods on Parallel Machines. Scientific Computing, Vol. 2, (1993) 179-192
15. Liu, P., Wu, J. J.: A Framework for Parallel Tree-Based Scientific Simulations. Proceedings of $26^{th}$ International Conference on Parallel Processing (1997)

# A Simple Dynamic Load-Balancing Scheme for Parallel Molecular Dynamics Simulation on Distributed Memory Machines

Naohito Sato    Jean-Marc Jézéquel

IRISA, France

**Abstract.** We propose a simple and efficient load-balancing scheme for parallel molecular dynamics simulation on distributed memory machines. It decomposes spatial domain of particles into disjoint parts, each of which corresponds with a processor and dynamically changes its shape to keep about the same number of particles throughout the simulation. In contrast to other similar schemes, ours requires no long-distance inter-processor communications but only those among adjacent processors (thus little communication overheads), whereas it still guarantees fast reduction of load-imbalance among the processors. It owes these advantages mainly to the following features: (1) The sufficiently correct *global load information* is effectively obtained with *step-wise propagation* of appropriate information via nearest neighbor communication. (2) In addition to the global load-balancing, another load-balancing procedure is also invoked on each processor without global load information in order to *suppress rapid increase or decrease of loads*. Thus, informations from remote processors can provide reliable values even after a certain period of delay. Further, we discuss how to select loads to migrate among processors so that spatial locality of the processors may be preserved. Through preliminary evaluation on an uniprocessor workstation, we have shown the scheme has strong potential for large-scale parallel molecular dynamics simulation on distributed memory machines or workstation clusters.

## 1 Introduction

One of the major aims of molecular dynamics simulation is tracing and analyzing particle interactions in a biochemical system of a huge macro-molecule, which often requires tremendous computational resources and, therefore, is considered as an important application of high-performance computing. To attain highly scalable parallel molecular dynamics simulation on distributed memory machines, it becomes crucial to reduce overheads in inter-processor communication for computing particle interaction. Since there exists strong geometric locality in particle interaction (i.e. particles interact with those at short distances), the most promising way to attain high scalability is to decompose spatial domain of particles into disjoint parts of the domain, each of which corresponds with a processor and dynamically changes its shape to keep about the same number of particles throughout the simulation. Then, in order to maintain each part of the domain in an appropriate shape, we need efficient load-balancing scheme, which may cause minimum computation overheads and minimum communication overheads for migrating loads and reducing load imbalance among the processors.

To meet these criteria, we develop a simple and efficient dynamic load-balancing scheme which requires no long-distance inter-processor communications but only those among adjacent processors (thus little communication overheads), whereas it still guarantees fast reduction of load-imbalance among the processors.

## 2 Parallel Molecular Dynamics Simulation

This section briefly surveys parallel molecular dynamics (MD) simulation and load-balancing schemes for parallel MD simulation on distributed memory machines.

### 2.1 Molecular Dynamics Simulation

First, let us consider a system of $N$ particles, denoted by $p_1, \ldots, p_N$, distributed in a bounded spatial domain. Let $m_i$, $x_i = (x_i^1, x_i^2, x_i^3)$, $v_i$, and $f_i$ denote the mass, the coordinates, the velocity of $p_i$, and the force affecting $p_i$. Then, interactions among those $N$ particles in the system can be numerically traced as follows: At each time step of simulation, (1) interaction among the particles are computed (through force accumulation) by using the predefined potentials (such as the van der Waals potential, the Coulomb potential, and the bond potentials), then (2) the coordinates and the velocity of each particle are updated by solving the motion equations. This procedure is repeated until the end of simulation. Once the total force affecting $p_i$ is accumulated, then $v_i(t)$ and $x_i(t)$ can be updated to $v_i(t + \Delta t)$ and $x_i(t + \Delta t)$ by solving $f_i(t) = m_i \dot{v}_i(t)$ and $v_i(t) = \dot{x}_i(t)$, where $\Delta t$ denotes the fixed time interval between time steps. In practice, $v_i(t + 1/2 \cdot \Delta t)$ is first computed for stability, and it is applied for computing $x_i(t + \Delta t)$ as defined in the Leapfrog method.

### 2.2 Parallel Molecular Dynamics Simulation

Molecular dynamics simulation involves inherent parallelism; that is, interaction force can be accumulated for each particle in parallel. However, the straightforward extraction of this results in poor performance due to the explosion of parallelism. Instead, we need to employ appropriate decompositions of particles onto processors so that each processor computes the same number of particle interactions. For parallel MD simulation, three major decomposition schemes have been proposed, namely (1) particle decomposition, (2) force decomposition, and (3) spatial decomposition.

The particle decomposition scheme statically decompose particles into blocks and map them onto processors [1, 6]. This is undoubtedly one of the simplest ways for guaranteeing the best load balance, though it often suffers from poor scalability due to frequent long-distance inter-processor communications in computing particle interaction.

The force decomposition scheme, in turn, decompose pairs of particles instead of particles themselves [9, 4]. It attains high-performance when the number of particles involved in the simulation does not become so large, though it still lacks enough scalability. See [9] for the detail.

In the spatial decomposition scheme, spatial domain of particles is geometrically decomposed into disjoint parts, called *subdomains*, each of which is further decomposed to small boxed regions, called *cells*. Since particles dynamically change their positions, the subdomains also exchange cells with one another and keep about the same number of particles throughout the simulation [2, 6, 3, 8]. Currently, this is the most promising decomposition scheme for attaining high performance and high scalability. (See also Figure 5.) As another spatial decomposition scheme, [7] divide the domain in an oct-tree fashion as often found in the parallel $N$-body simulation, for computing long-range interaction based on the Coulomb potential. Since high scalability plays the key role for efficient large-scale parallel molecular dynamics simulation, we develop our load-balancing scheme based on the spatial decomposition scheme.

## 2.3 Existing Load-Balancing Schemes for Parallel MD Simulation

When employing the spatial decomposition, it then becomes crucial to develop an efficient dynamic load balancing scheme to keep about the same number of particles on each processor. The efficiency of a load-balancing scheme then depends how much communication overheads can be reduced and how optimally loads can be exchanged among processors. In the rest of this section, we briefly survey some existing works for dynamic load balancing.

One of the simplest schemes is to redistribute the entire loads, i.e. cells in our case every several time step of simulation. For example, [2] employs the recursive geometric bisection along the axes for cell redistribution. However, this needs global synchronization and intensive data transfer among the processors, that results in poor scalability.

There is another sort of load-balancing scheme in which long-distance inter-processor communication is required only in collecting load information prior to load migration. In [4], load informations are propagated along linearly indexed processors by the *positional scan* operation. Each processor then determines migrating loads locally. In [6], destinations of migrating loads are selected randomly. These approaches, however, still need considerable long-distance inter-processor communications, that also result in poor scalability.

In addition, the well-known *diffusion scheme* [12] can be also applied: Each processor exchanges loads only with its neighboring processors to reduce the differences of the load sizes among them, which needs no long-distance inter-processor communication. The serious drawback of this scheme is, however, it does not guarantee optimality of load migration but often takes long time for reducing load imbalance among the processors.

# 3 Our Load-Balancing Scheme

In this section, we propose a new load-balancing scheme for parallel molecular dynamics simulation on distributed environments.

## 3.1 Suppressing Rapid Increase and Decrease of Loads

We first explain the first part of our load-balancing scheme to keep loads on each processor approximately to the same size throughout simulation as much as possible.

As explained in the previous section, the entire spatial domain is decomposed to small boxed regions, called *cells*, and they are lumped into subdomains, each of which occupies *simply-connected* area in the domain as illustrated in Figure 5 and corresponds with a distinct processor. As the hardware platform, let us assume a homogeneous 3-dimensional mesh-wise network: processors with local memories are identical with one another and connected through certain data transmission devices with the same bandwidth and latency. To identify the processors, let $N_{PE}^1, N_{PE}^2$ and $N_{PE}^3$ denote the numbers of processors along the axes respectively, and let $N_{PE}$ be $N_{PE}^1 \times N_{PE}^2 \times N_{PE}^3$. We then identify each processor as $PE_i$, where $i = (i_1, i_2, i_3)$ ranges as $0 \leq i_k < N_{PE}^k$ for $1 \leq k \leq 3$. In the same way, let $D_i$ denote the subdomain corresponding to $PE_i$. Given $N$ particles and $N_{PE}$ processors, the ideal number of particles on each processor, denoted by $\hat{N}$, is identical with $N/N_{PE}$ (i.e. $\hat{N} = N/N_{PE}$).

Clearly, $N_i$ dynamically changes its value throughout simulation: Let $N_i(t)$, instead of $N_i$, denote the number at the time $t$, and $\Delta N_i^{mot}(t)$ denote the number of particles moved into $PE_i$ as a result of particle interactions at $t$. Notice that $\Delta N_i^{mot}(t)$ is equal to the difference between the numbers of the incoming and the outgoing particles. Also, let

$\Delta N_i^{\text{mig}}(t)$ denote the number of particles to be migrated out at $t$ by our load-balancing procedure which is invoked after particle interactions are computed. Then, the following equation holds, where $\Delta t$ denote the time interval between time steps.

$$N_i(t+\Delta t) = N_i(t) + \Delta N_i(t)$$
$$\Delta N_i(t) = \Delta N_i^{\text{mot}}(t) - \Delta N_i^{\text{mig}}(t) \qquad (3.1)$$

The aim of the load-balancing is, then, to keep $N_i(t)$ close, or make it closer, to $\hat{N}$ for all $i$ through the simulation.

As a simple case, let us consider $N_i$ is approximately equal to $\hat{N}$ for all $i$ at the beginning of the simulation. Then, the aim can be restated as keeping $\Delta N_i(t)$ around 0. Since each particle changes its coordinates by a small amount of distance at a time and moves between adjacent processors, we only need to migrate particles in the converse way as those particles are moved: Let $\Delta N_{D_i \to D_{i'}}^{\text{mot}}$ denote the number of particles that have moved from $D_i$ to $D_{i'}$ prior to the load-balancing. Then, the number of particles to be migrated out of $PE_i$, denoted by $\Delta N_i^{\text{mig1}}$, can be determined as follows, where $\alpha$ is a positive constant for acceleration. Figure 1 illustrates this procedure.

$$\Delta N_i^{\text{mot}} = \sum_{i'} \left( -\Delta N_{D_i \to D_{i'}}^{\text{mot}} + \Delta N_{D_{i'} \to D_i}^{\text{mot}} \right)$$
$$\Delta N_i^{\text{mig1}} = \alpha \cdot \Delta N_i^{\text{mot}} \qquad (\alpha \le 1) \qquad (3.2)$$

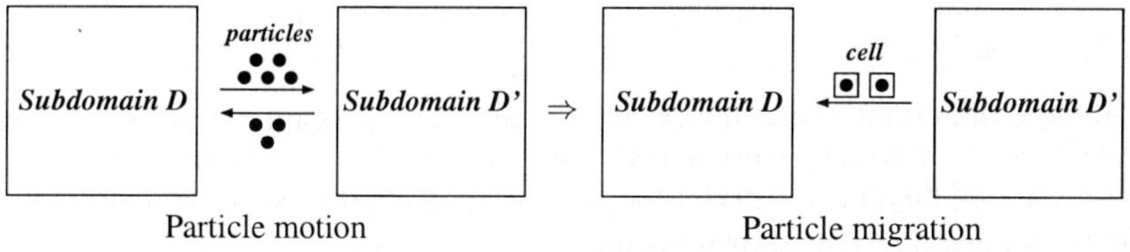

**Fig. 1.** Load balancing – 5 particles moving from $D$ to $D'$ ($\Delta N_{D \to D'}^{\text{mot}}$) and 3 particles from $D'$ to $D$ ($\Delta N_{D' \to D}^{\text{mot}}$), followed by migrating 2 particles from $D'$ to $D$ ($\Delta N_D^{\text{mig1}}$).

Clearly, this load-balancing procedure can be invoked in parallel on each processor. However, it does not always attain high-performance when invoked with $\alpha = 1$, since the procedure might migrate more particles than necessary. Further, in case a particular processor holds only a small number of particles, it may not select as many particles for migration as required by the load-balancing procedure in (3.2). Therefore, $\alpha$ is chosen to be smaller than 1 in most cases and $\Delta N_i^{\text{mig1}}$ then becomes smaller than required for perfect load-balancing. Thus, we need another mechanism to adjust the possible growing imbalance among the processors.

### 3.2 Reducing Load Imbalance among the Processors

We then explain the main part of our load-balancing scheme, which is intended to reduce load imbalance among the processors and make $N_i$ closer to $\hat{N}$ for all $i$ to recover balance.

In the same way as before, let us assume the processors and the subdomains are indexed with $(i = (i_1, i_2, i_3))$ as $PE_i, D_i$, respectively, and $PE_i$ contains $N_i(t) = N_{(i_1,i_2,i_3)}(t)$ particles at the time $t$. Then, let us define $\delta_i(t)$ as $N_i(t) - \hat{N}$, and $\sigma^k$ as the integral value of $\delta$ along the $k$-th axis ($k = 1, 2, 3$) as follows.

$$\delta_i(t) = N_i(t) - \hat{N}$$

$$\sigma^k_{(i_1,i_2,i_3)}(t) = \begin{cases} \sum_{i'_1 < i_1} \delta_{(i'_1,i_2,i_3)}(t) & (k=1) \\ \sum_{i'_2 < i_2} \delta_{(i_1,i'_2,i_3)}(t) & (k=2) \\ \sum_{i'_3 < i_3} \delta_{(i_1,i_2,i'_3)}(t) & (k=3) \end{cases} \quad (3.3)$$

Clearly, $\sigma^k$ implies the total amount of imbalance on the processors that precede $PE_i$ along the $k$-th axis. Then, the number of loads to be migrated out of $PE_i$ along the $k$-th axis to its positive direction, denoted by $\Delta_k N_i^{\text{mig2}}$, and the total number to be migrated out of $PE_i$, denoted by $\Delta N_i^{\text{mig2}}$, are defined as follows (where $\rho$ is a positive constant for acceleration). Figure 2 (2) depicts this for the 2-dimensional case.

$$\Delta_k N_i^{\text{mig2}} = \frac{\rho}{3} \cdot \left(\sigma_i^k(t) + \delta_i(t)\right) \quad (0 < \rho \leq 1)$$

$$\Delta N_i^{\text{mig2}} = \sum_{k=1,2,3} \left(-\Delta_k N_{(i'_1,i'_2,i'_3)}^{\text{mig2}} + \Delta_k N_i^{\text{mig2}}\right) \quad (3.4)$$

where $i'_k = i_k - 1, i'_m = i_m \ (m \neq k)$

The most important consequence this definition of load migration is that the load imbalance ($\delta_i$) is reduced to 0 on each processor in a *monotonous and optimal* fashion. In fact, after the load migration at each time step, $N_i(t)$ and $\sigma_i(t)$ respectively become closer to $\hat{N}$ and 0, for any $\rho \leq 1$, as shown below.

$$N_i(t) - \Delta N_i^{\text{mig2}}(t) = N_i(t) + \left(\Delta_1 N_{(i_1-1,i_2,i_3)}(t) - \Delta_1 N_i(t)\right)$$
$$+ \left(\Delta_2 N_{(i_1,i_2-1,i_3)}(t) - \Delta_2 N_i(t)\right)$$
$$+ \left(\Delta_3 N_{(i_1,i_2,i_3-1)}(t) - \Delta_3 N_i(t)\right)$$
$$= N_i(t) + \rho \cdot (-N_i(t) + \hat{N})$$
$$= (1-\rho) \cdot N_i(t) + \rho \cdot \hat{N}$$

$$\sum_{i'_k < i_k} \left(\delta_{i'}(t) - \Delta N_i^{\text{mig2}}(t)\right) = \sigma_i^k(t) - 3 \cdot \Delta_k N_{i'_{\max}}(t)$$
$$= (1-\rho) \cdot \sigma_i^k(t)$$

where $0 \leq i'_k < i_k, i'_m = i_m \ (m \neq k), i'_{\max} = (i_1, \ldots, i_k - 1, \ldots, i_3)$

As defined in (3.3), each processor needs to collect load informations through long-distance inter-processor communications prior to its load migration, that may result in poor scalability due to serious communication overheads. To eliminate those potential

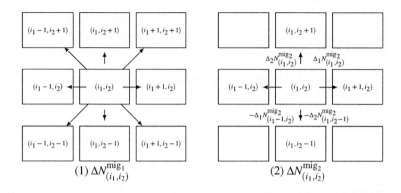

**Fig. 2.** Particle migration in $D_{(i_1,i_2)}$ — 8 and 4 adjacent processors are involved respectively in $\Delta N_i^{\text{mig}_1}$ and $\Delta N_i^{\text{mig}_2}$ (2 dimensional).

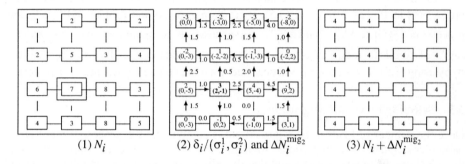

**Fig. 3.** Computing $\Delta N_i^{\text{mig}_2}$ $(N=64, N_{\text{PE}}=16, \hat{N}=4, \rho=1)$ — $\text{PE}_{(1,1)}$ for example migrates out -(-1/2)=0.5, (-1+3)/2=1, -(2/2)=-1, (2+3)/2=2.5 particles to its adjacent processors as defined in (3.4), so that the load imbalance shown in (1) becomes improved as in (3).

overheads and improve scalability, we replace (3.3) as follows.

$$\sigma^k_{(i_1,i_2,i_3)}(t) = \begin{cases} \sum_{i'_1 < i_1} c_{i_1-i'_1} \cdot \delta_{(i'_1,i_2,i_3)}(t_{i_1-i'_1}) & (k=1) \\ \sum_{i'_2 < i_2} c_{i_2-i'_2} \cdot \delta_{(i_1,i'_2,i_3)}(t_{i_2-i'_2}) & (k=2) \\ \sum_{i'_3 < i_3} c_{i_3-i'_3} \cdot \delta_{(i_1,i_2,i'_3)}(t_{i_3-i'_3}) & (k=3) \end{cases} \quad (3.5)$$

where $t_m \equiv t - m \cdot \Delta t, c_m \leq 1$

Notice that $\delta_{i'}(t - ||i - i'|| \cdot \Delta t)$ is used, in stead of $\delta_{i'}(t)$ as the load imbalance on $\text{PE}_{i'}$ which is $||i - i'||$ nodes away from $\text{PE}_i$. Thus, each processor only needs inter-processor communication with its adjacent processors (see Figure 4). Apparently, the efficiency of the load-balancing depends on the coefficients $(c_1, c_2, \ldots)$, which need to be chosen carefully for each case of simulation to attain high-performance. For example, selecting coefficients such that $|c_1| > \sum_{m>1} |c_m|$ sufficiently guarantees convergence to balanced load distribution. (As $c_{m+1}/c_m$ becomes smaller, the load-balancing then becomes closer to the one defined by the diffusion scheme discussed in the previous section.)

To complete the formal definition of our load-balancing scheme, we only need to define $\Delta N_i^{\text{mig}}$, the total number of particles to be migrated out of $\text{PE}_i$, with $\Delta N_i^{\text{mig}_1}$ and

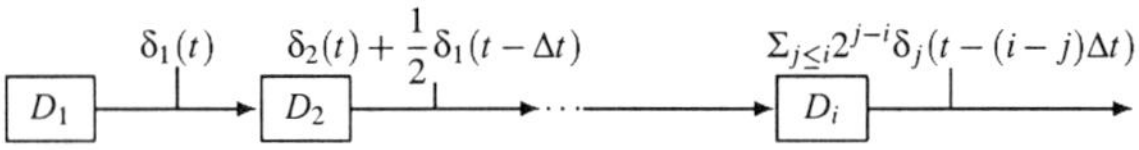

$\delta_1(t)$     $\delta_2(t) + \frac{1}{2}\delta_1(t-\Delta t)$     $\Sigma_{j\leq i} 2^{j-i}\delta_j(t-(i-j)\Delta t)$

$D_1$     $D_2$     ...     $D_i$

**Fig. 4.** Propagation of load informations (1 dimensional)

$\Delta N_i^{\text{mig}_2}$ as follows.

$$\Delta N_i^{\text{mig}} = \Delta N_i^{\text{mig}_1} + \Delta N_i^{\text{mig}_2} \qquad (3.6)$$

We are ready to present the final form of our load balancing procedure on each processor as a combination of (3.2) and (3.4): (1) On $PE_i$, $\Delta N_i^{\text{mot}}$ is computed when particles have been moved as the result of the particle interactions and $\Delta N_i^{\text{mig}_1}$ is then updated as defined in (3.2). (2) $\Delta N_i^{\text{mig}_2}$ is updated as defined in (3.4) when $\sigma^k$ ($k = 1, 2, 3$) along the axes are propagated to $PE_i$ as defined in (3.5). $\sigma$ are then updated and propagated to adjacent processors. (3) Once $\Delta N_i^{\text{mig}_1}$ and $\Delta N_i^{\text{mig}_2}$ are successfully computed, $PE_i$ then migrates out $\Delta N_i^{\text{mig}}$ of its cells for reducing its load imbalance, and it updates $N_i$ and $\delta_i$ by using (3.1) and $\sigma_i$ by using (3.5). The algorithm will be stated in a more concrete form later in this section.

### 3.3 Selecting Loads for Migration

Once the number of loads to be migrated out is computed, each processor then selects particular loads, i.e. cells of particles, to migrate out to each of its adjacent processors. In the selection, we impose a constraint that the adjacency relations among the processors should be preserved, that is, the subdomain corresponding with each processor should share its boundary with the same adjacent subdomains throughout the simulation, as illustrated in Figure 5.

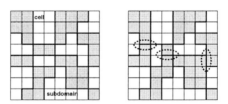

**Fig. 5.** The adjacency constraint — the right-side case breaks the constraint in three spots.

In order to satisfy this condition, we introduce a simple procedure for load selection: We always select cells along subdomain boundaries and migrate them out only when the adjacency relations between the processors sharing the boundaries will be preserved through the migration. Let us explain this in detail for the 2-dimensional case, though it can be easily extended to the 3-dimensional case. First, let us assume the adjacency condition holds on each processor. Then, the subdomain boundaries are then classified to the two distinct groups, namely (1) those shared by 2 subdomains, and (2) those shared by 4 subdomains. In the first case, a cell along the 2-processor boundary, denoted by $c$, can be safely selected for the migration across the boundary when the 8 cells around $c$ belong to the two processors sharing the boundary as specified in Figure 6 (left). In a similar manner,

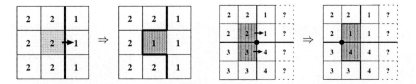

**Fig. 6.** Examples of cell migration across a 2-processor boundary (left) or a 4-processor boundary (right) — shaded cells can be migrated through the boundaries. (the numbers in the boxes denote processor indices.)

two cells sharing a 4-processor cross-point can be safely selected when the 12 cells around the two cells belong to the four processors as specified in 6 (right). Note that we can arbitrarily select cells that satisfy the above condition to sufficiently preserve the adjacency relations among the processors.

As a summary of the previous discussion, we show our load-balancing scheme in a concrete procedural form as follows.

1. Transfering $\Delta N_i^{mot}$ particles (Particle Interaction): Each processor computes $\Delta N_i^{mot}$ as well as force accumulation.
2. Computing $\Delta N_i^{mig1}$: $\Delta N_i^{mig1}(t)$ is computed by using $\Delta N_i^{mot}$ as defined in (3.2).
3. Computing $\Delta N_i^{mig2}$ and $\Delta N_i^{mig}$: As defined in (3.4), the number of particles to be migrated out to each adjacent cell is computed by using the current values of $\sigma_i$ and $\delta_i$. Also, $\Delta N_i^{mig2}(t)$ and $\Delta N_i^{mig}(t)$ are computed by using (3.4) and (3.6), respectively.
4. Updating $N_i$ and $\delta_i$: $N_i$ and $\delta_i$ are updated as shown in (3.1).
5. Updating $\sigma_i$ and migrating loads (Load Migration): $\Delta N_i^{mig}$ particles are selected as specified in Section 3.3 and migrated out to the adjacent cells. The value of $\sigma_i$ is also transmitted during the migration for the compution of (3.5).
6. Synchronization: The time step is updated by $\Delta t$ (i.e. $t \leftarrow t + \Delta t$) to proceed to the next simulation step.

## 4 Preliminary Evaluation

We have examined the feasibility and the effectiveness of our load-balancing scheme through its preliminary evaluation on an uniprocessor UltraSparc workstation. In the current prototype implementation, we depend the particle interaction part of the simulation on the AMBER [11] MD package; that is, at each time step the load-balancing procedure is invoked after the coordinates of the atoms are restored from the post-mortem output of AMBER. Figure 7 shows snapshots from the simulation of the 5159-atom acetylcholine (1ack) system. Figure 8 (left and middle) also shows the temporal changes of the normalized load sizes of the processors and the amount of inter-processor data transfer (the number of atoms), respectively. Notice that, even though the shapes of the 16 subdomains become irregular after a few steps, the amount of the inter-processor communication for force accumulation stays *almost in the same size throughout the simulation*, whereas the one for load balancing stays in the negligible size. This suggests that our scheme can attain considerable reduction of inter-processor communication in comparison with other schemes (See also the discussion in the following section).

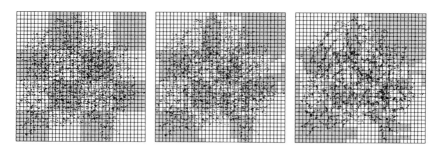

**Fig. 7.** Snapshots from the simulation of the 5159-atom system (1ack) on a 16-processor machine (at $T = 0, 1, 20$) – Each cell size is almost identical with the cutoff distance. Shades in each snapshot indicate the correspondence between the cells and the 16 processors.

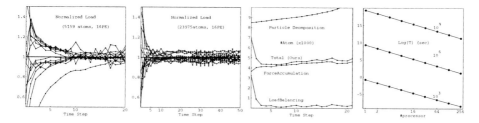

**Fig. 8.** Performance results – The first two graphs show the temporal changes of $N_i(t)/\hat{N}$ through the simulation of the 5159/23975-atom systems on the 16 processors. The third one tracks the number of atoms involved in inter-processor communication. The right-most one estimates the overall performance ($\log(T_k)$ in sec.) through 1-256 processor simulations for $10^3, 10^6, 10^9$-atom systems.

In addition, we have estimated the overall performance of the simulation with the following approximation: the computation time for the $k$-th step ($T_k$) is estimated as the sum of the four distinct parts as shown below, where $\sigma_k$ indicates the load-imbalance at the $k$-th time-step (i.e. $1 + \sigma_k = N_{\max,k}/\hat{N}$) and $\tau_j$ ($i \leq 4$) are time constants. *

$$T_k = T_{MDcomp} + T_{MDcomm} + T_{LBcomp} + T_{LBcomm}$$
$$= N_{\max,k}\tau_1 + N_{\max,k}^{2/3}\tau_2 + \sigma_k\hat{N}\log(\sigma_k\hat{N})\tau_3 + \sigma_k\hat{N}\tau_4 \quad \text{where } N_{\max,k} = (1+\sigma_k)\hat{N}$$

Figure 8 (right) shows the estimated performance for $10^3, 10^6, 10^9$-atom systems, where $\sigma$ is fixed to 0.25, $\tau_1$ and $\tau_3$ are determined to $4.1 \times 10^{-4}$ and $7.2 \times 10^{-5}$, respectively (based on the results from our UltraSparc simulation), $\tau_2$ and $\tau_4$ are deliberately chosen to $1.5 \times 10^{-4}$ and $3.2 \times 10^{-6}$, respectively (assuming a medium-speed network-connected hardware platform). This estimation strongly ensures high scalability of our scheme.

---

* $T_{MDcomp}$ includes time for computing all bonded forces. $T_{LBcomp}$ is proportional to $\sigma_k\hat{N}\log(\sigma_k\hat{N})$ since we exploit the quick sort algorithm to select migrating cells.

## 5   Discussion and Concluding Remarks

In this paper, we have proposed a new load-balancing scheme for parallel molecular dynamics simulation on distributed memory machines, and evaluated its feasibility and effectiveness through preliminary experiments on an uniprocessor workstation.

Our scheme have several advantages over other load-balancing schemes discussed in Section 2: A static load-balancing scheme such as based on particle decomposition has a serious drawback for its scalability since it needs considerable inter-processor communication for force accumulation. Dynamic load-balancing schemes, in turn, typically define how to collect load informations and how to select migrating loads: Our scheme enables them in a local and optimal fashion, that is, ours requires long-distance inter-processor communications only among adjacent processors while it still guarantees fast reduction of load-imbalance among the processors, whereas the other existing schemes lack the optimality [6], or (more often) needs rather frequent long-distance inter-processor communication [2, 4]. In fact, it is difficult to keep the amount of inter-processor communication in the same size throughout simulation with those schemes (cf. Figure 8).

Currently, we are integrating the scheme on top of the EPEE framework [5, 10] as a testbed for evaluating the practical interests of various molecular dynamics load-balancing schemes. As other possible future research directions, the following themes can be considered. (1) Applying to other practical problems: It is interesting to consider other practical applications of our scheme in addition to MD simulation. Discrete event simulation can be an interesting example. (2) Integrating with a well-founded software engineering framework (including the EPEE/UML framework): Our load-balancing scheme requires programmers to set up several parameters (including $\alpha$, $\rho$, or $c_1, c_2, \ldots$, defined in Section 3), prior to execution, that seriously affect efficiency of the load-balancing. Those parameters need to be carefully adjusted to each particular simulation, and it often becomes difficult to find optimal values for high-performance.

## References

1. Stephen E. DeBolt and Peter A. Kollman. AMBERCUBE MD. *J. of Comp. Chemistry*, 1993.
2. K. Esselink, B. Smit, and P. A. J. Hilbers. Efficient parallel implementation of molecular dynamics on a toroidal network: Multi-particle potentials. *J. of Computer Physics*, 1993.
3. R. Giles and P. Tamayo. A Parallel Scalable Approach to Short-Range Molecular Dynamics on CM-5. In *Scalable High Performance Computing Conference*, page 240. IEEE, 1992.
4. David F. Hegarty and M. T. Kechadi. Topology Preserving Dynamic Load Balancing for Parallel Molecular Simulations. In IEEE *Supercomputing*. November 1997.
5. Jean-Marc Jézéquel and Jean-Lin Pacherie. *Object-Oriented Application Frameworks*, chapter EPEE: A Framework for Supercomputing. John Wiley & Sons, New York, 1998.
6. J. Kitowski. Distributed and parallel computing of short-range molecular dynamics. *Lecture Notes in Computer Science*, 1041, 1996.
7. Kian-Tat Lim. Molecular Dynamics for Very Large Systems on Massively Parallel Computers: The MPSim Program. *J. of Comp. Chemistry*, 1997.
8. S. L. Lin, J. Mellor-Crummey, B. M. Pettitt, and G. N. Phillips Jr. Molecular Dynamics on a Distributed-Memory Multiprocessor. *J. of Comp. Chemistry*, 1992.
9. Steve Plimpton. Fast Parallel Algorithms for Short-Range Molecular Dynamics. *J. of Comp. Physics*, 1995.
10. Naohito Sato, Satoshi Matsuoka, Jean-Marc Jézéquel, and Akinori Yonezawa. A Methodology for Specifying Data Distribution using only Standard Object-Oriented Features. In *the 11th International Conference on Supercomputing*. ACM SIGARCH, July 1997.
11. P. K. Weiner and P. A. Kollman. AMBER. *J. of Comp. Chemistry*, 1981.
12. Marc H. Wellebeek-LeMair and Anthony P. Reeves. Strategies for Dynamic Load Balancing on Highly Parallel Computers. *IEEE Trans. on Parallel and Distributed Systems*, 1993.

# Resource Management for High-Performance PC Clusters[*]

Axel Keller[1], Matthias Brune[2], and Alexander Reinefeld[2]

[1] Paderborn Center for Parallel Computing
kel@uni-paderborn.de
[2] Konrad-Zuse-Zentrum für Informationstechnik Berlin
{brune, ar}@zib.de

**Abstract.** With the recent availability of cost-effective network cards for the PCI bus, researchers have been tempted to build up large compute clusters with standard PCs. Many of them are operated with workstation cluster management software in high-throughput or single user mode.
For very large clusters with more than 100 PEs, however, it becomes necessary to implement a full fledged resource management software that allows to partition the system for multi-user access.
In this paper, we present our *Computing Center Software (CCS)*, which was originally designed for managing massively parallel high-performance computers, and now adapted to modern workstation clusters. It provides
 – partitioning of exclusive and non-exclusive resources,
 – hardware-independent scheduling of interactive and batch jobs,
 – open, extensible interfaces to other resource management systems,
 – a high degree of reliability.

## 1 Introduction

Scientific computing on clusters became a very viable option in the recent years. The availability of fast and cost-effective adapter cards for short area networks, which can be plugged in into a standard PCI bus, make it possible to build up large compute clusters at modest costs.

LAN technologies like Fast Ethernet or ATM are generally not suitable for building coherent parallel systems, because they incur a considerable software overhead by protocols (e.g. the TCP/IP stack). Modern SANs (short area networks) like Myrinet [5] and SCI [11] are more suitable. By means of special hardware, they provide a data throughput of some Gigabit per second, all at a communication latency of about 10 $\mu$s.

In this paper we focus on clusters that are interconnected with *Scalable Coherent Interface* (SCI) (IEEE 1596) [11]. SCI's economical advantages (good

---
[*] The work presented in this paper was done while all three authors were at Paderborn Center for Parallel Computing, http://www.uni-paderborn.de/pc2

cost/performance ratio) mainly stem from the support of standard buses (PCI, SBus), which allows system integrators to use off-the-shelf compute nodes for building powerful systems. With the SCI hard- and software now maturing, larger systems are installed in academia and industry. Clearly, large SCI clusters with hundreds of processors can no longer be operated in single-user mode and hence there is a demand for multi user management. The CCS environment allows to operate SCI clusters in a multi-user mode just as any other high-performance computer. With CCS, SCI clusters are no longer seen as clusters but as dedicated high-performance computers.

## 1.1 Hardware Scenario: Large Dedicated SCI Systems

While CCS is also used for accessing and controlling small heterogeneous SCI clusters like the one shown in Fig. 1a, it was primarily designed for managing large dedicated compute clusters in multi-user mode. Figure 1b depicts our 32-node SCI cluster. Its 2D torus topology is made up of four vertical and eight horizontal SCI rings. Each node is equipped with two Pentium II processors at the intersection points. Due to the different physical ring lengths, the vertical and horizontal rings exhibit different communication bandwidths of 500 and 400 MBytes/s, respectively.

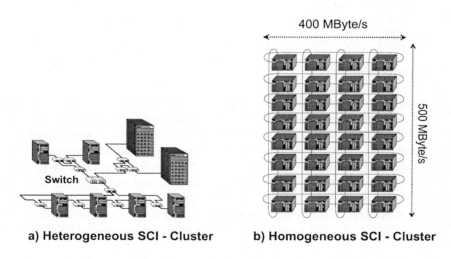

a) Heterogeneous SCI - Cluster    b) Homogeneous SCI - Cluster

Fig. 1. SCI cluster configurations at PC$^2$

Our second system has a peak performance of 86 GFlop/s. It comprises 96 nodes, each with two 450 MHz Pentium II processors and 512MB main memory. The SCI rings are routed via a 16-way SCI switch (Fig. 2). Again, this system is also operated under CCS.

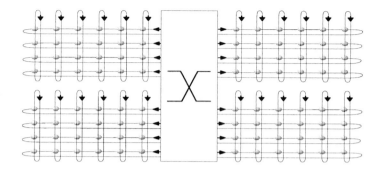

**Fig. 2.** A possible topology of a switched SCI cluster with 96 nodes

### 1.2 Software Scenario: Full HPC Environment

On the software side, our SCI clusters provide the full range of software services known from other HPC systems. This includes a spectrum of compilers (Fortran77, Fortran90, C, C++), programming interfaces (PVM, MPI, Active Messages), a parallel debugger (TotalView) and a performance monitor (Vampir). In addition, there are SCI drivers for Solaris, Linux and Windows NT. In other words, the system is capable of running multiple operating systems at the same time.

## 2 Architecture of CCS

### 2.1 Island Concept

In a computing center environment with several HPC systems, one might be tempted to operate all systems under the supervision of one central resource management system (RMS) that has different backends for the various machines. On the one hand, this approach provides a coherent user and administrator interface to all machines, but on the other hand, it is inherently vulnerable to single points of failure. Moreover, the central scheduler —and other critical software modules— might cause a performance bottleneck. We therefore introduced the *Island Concept* [13], where each machine is managed by a separate instance of the CCS software. A CCS island (Fig. 3) consists of six components:

 - The *User Interface (UI)* offers X-window or ASCII access to the machine. It encapsulates the physical and technical characteristics and it provides a homogeneous access to single or multiple systems.
 - The *Access Manager (AM)* manages the user interfaces and is responsible for authentication, authorization, and accounting.
 - The *Queue Manager (QM)* schedules the user requests onto the machine.
 - The *Machine Manager (MM)* provides an interface to the machine specific features like partitioning, job controlling, etc.

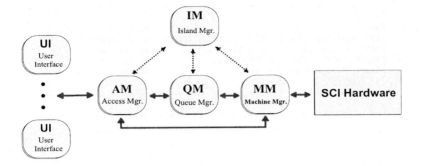

**Fig. 3.** Architecture of a CCS Island

- The *Island Manager (IM)* provides name services and watchdog functions to keep the system in a stable condition.
- The *Operator Shell (OS)* is the X-window based interface for system administrators to control CCS, e.g. by connecting to the system daemons.

## 2.2 User Interface

The *User Interface (UI)* runs in a standard UNIX shell environment like tcsh. Common UNIX mechanisms for I/O redirection, piping and shell scripts can be used. All job control signals (ctl-z, ctl-c, ...) are supported. The user shell accepts five CCS commands:

- *ccsalloc* for allocating and/or reserving resources,
- *ccsrun* for starting jobs on previously reserved resources,
- *ccskill* for resetting or killing jobs and/or for releasing resources,
- *ccsbind* for re-connecting to a lost interactive application/session,
- *ccsinfo* for getting information on the job schedule, users, job status etc.

The *Access Manager (AM)* analyzes the user requests and is responsible for authentication, authorization and accounting. CCS provides a project specific user management. Privileges can be granted to either a whole project or to specific project members, for example:

- access rights (batch, interactive, the right to reserve resources),
- allowed time of usage (day, night, weekend, ...),
- maximum number of concurrently used resources.

**The Worker Concept.** In contrast to other parallel systems, SCI clusters have nodes with full operating system functionalities. Therefore, a wide range of software packages like debuggers, performance analyzers, numerical libraries, and runtime environments are available (Sec. 1.2). Often these software packages require specific pre- and postprocessing.

For this purpose, CCS provides the so-called *worker concept*. Workers are tools to start jobs under specific run time environments. They hide specific

```
pvm,                                           #name of the worker
%CCS/bin/start_pvmJob -d -r %reqID -m %island, #run command
%CCS/bin/start_pvmJob -q -m %island,           #parse command
%root %CCS/bin/establishPVM %user,             #preprocessing
%root %CCS/bin/cleanPVM %user                  #postprocessing
```

**Fig. 4.** A worker definition for starting jobs in an PVM environment

procedures (e.g. starting of daemons, setting of environment variables, etc.) and provide a convenient way to start programs.

The behavior of a worker is defined in a configuration file (Fig. 4) by specifying five attributes:

- the name of the worker,
- the command for CCS to start the job,
- the optional parse command for detecting syntax errors,
- the optional preprocessing command (e.g. initializing a parallel file system),
- the optional postprocessing command (e.g. closing a parallel file system).

Both pre- and postprocessing can be started with either root or user privileges, controlled by a keyword. The configuration file is parsed by the user interface and can therefore be changed at run time. New workers can be plugged in without the need to change the CCS source code.

**The Virtual Terminal Concept.** With the increasing utilization of supercomputers for *interactive* use the support of remote access via WANs becomes more and more important. Unpredictable behavior and even temporary breakdowns of the network should (ideally) be hidden from the user.

In CCS, this is done by the *Execution Manager* (EM), which buffers the standard output streams (stdout, stderr). In case of a network break down, all open output streams are sent by e-mail to the user or they are written to a file. Users can re-bind to lost sessions, provided that the application is still running. CCS guarantees that no data is lost in the meantime.

### 2.3 Scheduling

In their resource requests, users must specify the expected finishing time of their jobs. Based on this information, CCS determines a fair and deterministic schedule. Both, batch and interactive requests are processed in the same scheduler queue. The request scheduling problem is modeled as an $n$-dimensional bin packing problem, where the one dimension corresponds to the continuous time flow, and the other $n-1$ dimensions represent system characteristics, such as the number of processor elements.

The integration of new schedulers is easy, because the QM provides an API to plug in new modules. This allows to adjust to specific operating modes (e.g.

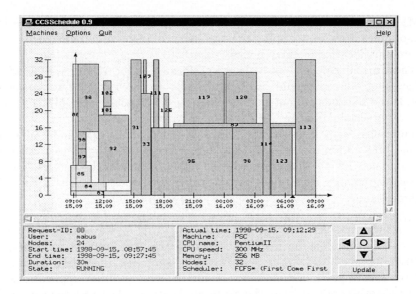

**Fig. 5.** Scheduler GUI displaying the scheduled nodes over time.

space- or time-sharing mode) by switching between several scheduling strategies at run time. This can be done by the QM itself or by the administrator.

With CCS, it is possible to *reserve resources* for a given time in the future. This is a convenient feature when planning interactive sessions or online-events. As an example, consider a user who wants to run an application on 32 nodes of the SCI cluster from 9 to 11 am at 13.07.1999. The resource allocation is done with the command: `ccsalloc -m SCI -n 32 -s 9:13.07.99 -t 2h`.

*Deadline scheduling* is another useful feature. Here, CCS guarantees the job to be completed no later than the specified time. A typical scenario for this feature is an overnight run that must be finished when the user comes back into the office next morning. Deadline scheduling gives CCS the flexibility to improve the system utilization by scheduling batch jobs at the latest possible time so that the deadline can still be met.

In our center, CCS uses an enhanced first-come-first-serve (FCFS) scheduler, which fits best to the mainly interactive request profile. Waiting times are minimized by first checking whether a new request fits into a gap of the current schedule (back-filling). Figure 5 depicts a typical schedule.

## 2.4 Topology Dependent Partitioning

From the beginning of the CCS project, we aimed on joining two conflicting goals: Maximizing the system utilization by making use of the knowledge about the system topology and maintaining a high degree of system independence for improved portability and easier adaptation to heterogeneous systems.

To deal with these two contradictory goals, we have split the scheduling process into two instances. The Queue Manager (QM) and the Machine Manager

(MM). The QM is independent of the underlying hardware architecture [8]. It has no information on mapping constraints such as the minimum cluster size or the amount/location of entry nodes.

These machine dependent tasks are performed by the MM. It verifies whether a schedule given by the QM can be mapped onto the hardware at the specified time. If the schedule cannot be mapped onto the machine, the MM returns an alternative schedule to the QM.

This separation between the hardware-independent QM and the system-specific MM allows to encapsulate system-specific mapping heuristics in small code modules. With this approach, special requests for I/O-nodes, partition shapes, or memory constraints can be taken into consideration in the verifying process.

As an example, the MM, takes system characteristics like the different speed of horizontal and vertical SCI links (see Fig. 1b), into account. With its detailed information on the machine structure the MM employs system-specific partitioning schemes. For our SCI cluster, it computes a partition according to the cost functions "minimum network interference by other applications" and "best network bandwidth for the given application". The first function prefers to use as few SCI rings as possible, thereby minimizing the number of dimension exchanges (x/y-crossovers) via other application domains, while the second function tries to map applications on single rings with maximum bandwidth.

The API of the MM allows to adapt the partitioning to arbitrary topologies, or to implement mapping modules, that are optimally tailored to the specific hardware properties.

## 2.5 Job Creation and Control

At configuration time, the QM sends the user request to the MM. The MM then allocates the compute nodes, loads and starts the application code and releases the resources after the run. Because the MM also has to verify the schedule, which is a polynomial time problem, a single MM daemon might become a computational bottleneck. We have therefore split the MM into two parts (Fig. 6), one for the machine administration and one for the job execution. Each part contains several modules and/or daemons which can run on different hosts to improve the performance.

The machine administration part consists of three separate daemons (MV, SM, CM) that execute asynchronously. A small *Dispatcher* coordinates the lower-level components.

The *Mapping Verifier (MV)* checks whether the schedule given by the QM can be realized at the specified time with the specified resources (Sec. 2.4).

The *Configuration Manager (CM)* provides the interface to the hardware. It is responsible for booting, partitioning, and shutting down the operating system software. Depending on the system's capabilities, the CM may gather subsequent requests and re-organize or combine them for improving the throughput. Additionally, the CM provides external tools with information on the allocated partition, like host names or the partition size.

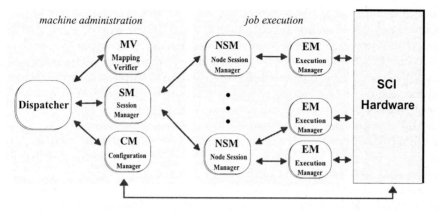

**Fig. 6.** Detailed view of the machine manager (MM)

Ideally, an RMS should provide all system features to the user, including permission to log into the owned nodes. But as a consequence, users are then able to start arbitrary processes on arbitrary nodes and the system cleanup may become difficult. In CCS, the *Node Session Manager (NSM)*, which runs on each specified entry node with root privileges, is responsible for the job controlling. At allocation time, the NSM starts an *Execution Manager (EM)* which establishes the user environment (UID, shell settings, environment variables, etc.) and starts the application. Before releasing the partition the NSM cleans up the node.

In space-sharing mode the NSM changes the passwd file to avoid concurrent logins from other users. In time-sharing mode, the NSM additionally invokes as many EMs as needed. It also gathers dynamic load data and sends it to the MM and QM, where it is used for scheduling and mapping purposes.

The *Session Manager (SM)* synchronizes the NSMs. It sets up the session, including application-specific pre- or postprocessing, and it maintains information on the status of the applications.

### 2.6 Reliability

With the transition from batch-oriented high-performance computing to interactive access, system reliability becomes an even more important issue because node breakdowns immediately influence the user's work flow. Additionally, today's parallel systems often comprise independent (workstation-like) nodes, which are more vulnerable to breakdowns than the homogeneous nodes contained in a regularly structured MPP.

An RMS must be able to detect and possibly repair breakdowns at three different levels: the computing nodes, the software daemons, and the communication network.

Many failures become only apparent when the communication behavior changes over time, or when a communication partner does not answer at all. To detect a failure and to determine its reason, the *Island Manager (IM)* maintains an up-to-date information base on the status of all system components within the island. Each CCS daemon notifies the IM when starting up or closing down, so that the IM has a consistent view on the current system status.

When a CCS daemon detects a communication problem, it closes the connection to the concerned daemon and requests the IM to re-establish the link. The IM has a number of methods to find out about the problem. Which of them to use depends on the type of target system (e.g. ping, rlogin, process state, etc.). The IM is authorized to stop erroneous daemons, to restart crashed ones, and to migrate daemons to other hosts in case of system overloads or crashes. If the IM cannot solve the problem, it sends an email with a problem report and the actions taken to the administrator.

Additionally, each CCS daemon periodically saves its state to a file for recovery purposes. At boot time the daemon reads its information and synchronizes with its communication partners. This allows to shutdown or kill CCS daemons (or even the whole island) at any given time without the risk to loose requests.

## 3 Resource and Service Description

The efficient operation of heterogeneous clusters requires a versatile resource description facility. The administrator needs it to describe type and topology of the available resources. The user needs it to specify the required system configuration for a given application. For this purpose, we developed the *Resource and Service Description RSD* [6]. RSD provides three interfaces:

- a GUI for specifying simple topologies and attributes,
- a language interface for specifying more complex and repetitive graphs (mainly intended for system administrators), and
- an API for access from within an application program.

RSD is used in nearly all CCS modules. The UI creates an RSD description out of the given parameters (or uses a given RSD description) and sends it to the QM. The QM extracts the needed information and computes a schedule. The MM verifies the schedule by mapping the user given RSD description against the static (e.g. topology) and dynamic (e.g. PE availability) information on the system resources. With RSD it is possible to build high-level tools, located on top of the CCS islands, to support setup and execution of multi-site applications running concurrently on several platforms.

## 4 Related Work

Much work has been done in the field of resource management in order to optimally utilize the costly high-performance computer systems. However, in contrast to the CCS approach described here, most of today's resource management systems are either vendor-specific or devoted to the management of LAN- or WAN-connected workstation clusters.

The *Network Queuing System NQS* [14], developed by NASA Ames for the Cray2 and Cray Y-MP, might be regarded as the ancestor of many modern queuing systems like the *Portable Batch System PBS* [3] or the Cray *Network Queuing Environment NQE* [17].

Following another path in the line of ancestors, the *IBM Load Leveler* is a direct descendant of *Condor* [15], whereas *Codine* [9] has its roots in *Condor* and *DQS*. They have been developed to support 'high-throughput computing' on UNIX workstation clusters. In contrast to high-performance computing, the goal is here to run a large number of (mostly sequential) batch jobs on workstation clusters without affecting interactive use. The *Load Sharing Facility LSF* [16] is another popular software to utilize LAN-connected workstations for high-throughput computing. For more detailed information on cluster managing software, the reader is referred to [2, 12].

These systems have been extended for supporting the coordinated execution of parallel applications, mostly based on PVM. A multitude of schemes have been devised for high-throughput computing on a somewhat larger scale, including the Iowa State University's *Batrun* [18], the Dutch *Polder* initiative [7], the *Nimrod* project [1], and the object-oriented *Legion* [10], which proved useful in a nation-wide cluster. While these schemes emphasize mostly the application support on homogeneous systems, the *AppLeS* project [4] provides application-level scheduling agents on heterogeneous systems, taking into account their actual resource performance.

## 5 Summary

The Computing Center Software (CCS) is a resource management software for the user access and system administration of dedicated high-performance systems. It has been in operation since 1992 on various massively parallel systems and workstation clusters.

With CCS, SCI clusters can be regarded as dedicated, partitionable high-performance computers that are operated in multi-user mode. CCS provides:

- different scheduling strategies known from HPC computing,
- optimal space partitioning for concurrent access by multiple users,
- a versatile resource and service description facility,
- a high degree of reliability.

The modular concept of CCS proved very useful in our adaptation. We just needed to specify the topologies of our SCI clusters with the resource and service description tool RSD [6], and we had to implement new mapping modules for the optimal partitioning of the shared SCI links.

## References

1. Abramson, D., Sosic, R., Giddy, J., Hall, B.: Nimrod: A Tool for Performing Parameterized Simulations using Distributed Workstations. 4th IEEE Symp. High Performance and Distributed Computing, August 1995.

2. Baker, M., Fox, G., Yau, H.: Cluster Computing Review. Northeast Parallel Architectures Center, Syracuse University New York, November 1995. http://www.npar.syr.edu/techreports/index.html.
3. Bayucan, A., Henderson, R., Proett, T., Tweten, D., Kelly, B.: Portable Batch System: External Reference Specification. Release 1.1.7, NASA Ames Research Center, June 1996.
4. Berman, F., Wolski, R., Figueira, S., Schopf, J., Shao, G.: Application-Level Scheduling on Distributed Heterogeneous Networks. Supercomputing, November 1996.
5. Boden, N., Cohen, D., Felderman, R.E., Kulawik, A.E., Seitz, C.L., Seizovic, J.N., Su, W.K.: *Myrinet: A Gigabit-per-Second Local Area Network.* IEEE Micro 15,1, Feb. 1995, pp. 29-36.
6. Brune, M., Gehring, J., Keller, A., Reinefeld, A.: RSD – Resource and Service Description. Intl. Symp. on High Performance Computing Systems and Applications HPCS'98, Edmonton Canada, Kluwer Academic Press, May 1998.
7. Epema, D., Livny, M., van Dantzig, R., Evers, X., Pruyne, J.: A Worldwide Flock of Condors: Load Sharing among Workstation Clusters. FGCS, Vol. 12, 1996, pp. 53–66.
8. Gehring, J., Ramme, F.: Architecture-Independent Request-Scheduling with Tight Waiting-Time Estimations. IPPS'96 Workshop on Scheduling Strategies for Parallel Processing, Hawaii, Springer LNCS 1162, 1996, pp. 41–54.
9. GENIAS Software GmbH: Codine: Computing in Distributed Networked Environments. http://www.genias.de/products/codine, January 1999.
10. Grimshaw, A., Weissman, J., West, E., Loyot, E.: Metasystems: An Approach Combining Parallel Processing and Heterogeneous Distributed Computing Systems. J. Parallel Distributed Computing, Vol. 21, 1994, pp. 257–270.
11. Hellwagner, H., Reinefeld, A. (eds.): *Scalable Coherent Interface: Technology and Applications.* Proceedings of the SCI-Europe98, Bordeaux Sept. 98. Cheshire Hensbury, 1998.
12. Jones, J., Brickell, C.: Second Evaluation of Job Queueing/Scheduling Software: Phase 1 Report. Nasa Ames Research Center, NAS Tech. Rep. NAS-97-013, June 1997.
13. Keller, A., Reinefeld, A.: CCS Resource Management in Networked HPC Systems. 7th Heterogeneous Computing Workshop HCW'98 at IPPS, Orlando Florida, IEEE Comp. Society Press, 1998, pp. 44–56.
14. Kinsbury, B.A.: The Network Queuing System. Cosmic Software, NASA Ames Research Center, 1986.
15. Litzkow, M.J., Livny, M.: Condor–A Hunter of Idle Workstations. Procs. 8th IEEE Int. Conference on Distributed Computing Systems, June 1988, pp. 104–111.
16. LSF: Product Overview. http://www.platform.com/content/products/, January 1999.
17. NQE-Administration. Cray-Soft USA, SG-2150 2.0, May 1995.
18. Tandiary, F., Kothari, S.C., Dixit, A., Anderson, E.W.: Batrun: Utilizing Idle Workstations for Large-Scale Computing. IEEE Parallel and Distributed Technics, 1996, pp. 41–48.

# The Web as a Global Computing Platform

Qusay H. Mahmoud

Etisalat College of Engineering
Emirates Telecommunications Corporation
P.O. Box 980
Sharjah, United Arab Emirates
qmahmoud@ece.ac.ae

**Abstract.** The current model of the World-Wide Web, or Web for short, has limited support for computing resources. In the current model, computing resources can be classified into two streams: server-side and client-side computing. The Common Gateway Interface (CGI) scripts are examples of server-side computing, and applets are examples of client-side computing. In this paper we point out the limitations of the Web for global distributed computing, and discuss our system that aims to add computing resources to the Web. In our proposed system clients will be able to upload code to remote, and possibly more powerful, machines where Web-based compute servers will execute the code and return the results back to clients. To help clients search for compute servers, we discuss our design of a broker system responsible for automatically allocating compute servers to clients.

## 1 Introduction

The World-Wide Web[12], or Web for short, has been very successful in what it was designed for – a network-based hypermedia information system. The current model of the Web, however, has limited support for computing. The current computing models of the Web include server-side computing using Common Gateway Interface (CGI) scripts, and client-side computing using scripted HTML files or Applets. These computing models are limited since they were designed for processing fill-out forms dynamically.

The existing computing models, however, suggest that the Web has the potential of becoming a general purpose distributed computing platform[7] in the sense that an extension to the Web's functionality to include global computing resources[8], to which clients will be able to upload code, is possible.

### 1.1 Why Global Computing?

Ted Lewis, of the Naval Postgraduate School, contends that: "The limits of parallelism seem to block further advances in processor performance beyond the next $10,000_2$ years. But a third alternative leads to the concept of an uncoordinated, globally distributed, parallel megacomputer. Such computers already exist in the form of asynchronous nodes on the Internet, but they have yet to be used to their fullest extent."[6].

In computational-intensive programs, there is always a need for more computing power and better performance. Having a global computing platform would solve many problems related to performance, fault tolerance, and resource limitation. One major advantage of such a platform would be the ability to use as many idle (and perhaps faster) machines on the Internet as possible to solve large and complex problems. An example of such a problem is the RSA-129 factoring project where more than 1600 machines were used to factor an integer of 129 digits long[10].

## 1.2 Why Web-based?

The Web, so far, has mainly been used as a global information resource system. When a user visits a Web site, generally all he[1] sees is either static or dynamic information, fill-out forms, simple applets, and animated images. The current simplified computing model of the Web allows the user to execute server-side programs using CGI scripts by calling a CGI script that will in turn call another program on the server-side. In addition, it allows the user to download mini-programs (applets), which are initially stored on the Web server-side, that will be executed on his machine. In this case, the user requests the homepage that contains the applet and the applet migrates to the client's machine and gets executed.

It is apparent that one computing model, which allows clients to upload code to the server-side for execution, is missing. In this model that we propose, the program on the client's machine is executed on a remote machine.

# 2 Limitations of the Current Web Computing Model

The file-upload feature[9], which is an extension of HTML to allow information providers to express file upload requests, can be used along with CGI to develop a simplified global Web-based distributed computing system in which clients will be able to upload their program files to remote machines for execution as shown in Fig. 1. This approach, however, has more drawbacks than advantages which are outlined in the following subsections.

## 2.1 Advantages of CGI

The advantages of using the file-upload feature and CGI to implement a simple global Web-based distributed computing system include the following:

- **Simplicity**. The main advantage of using the file-upload feature and CGI is the simplicity of implementing such a system, where only one single CGI script is needed to handle the file uploading and collecting of the output.

---

[1] "he" should be read as "he or she" throughout this paper.

- **Easiness**. From the user's point of view, it is all point-and-click. When the user is presented with an HTML form, the user needs only click on a button to get a dialog box with a view of his local disk from which he can choose the file to be uploaded for remote execution.

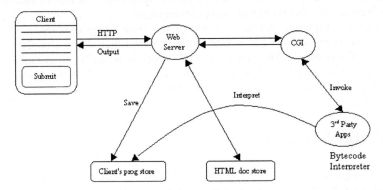

**Fig. 1.** A Simple CGI-based Computing System

### 2.2 Disadvantages of CGI

Despite the simplicity and easiness of using the file-upload feature and CGI for implementing a simple global Web-based distributed computing system, a number of disadvantages and limitations to CGI, as summarized below, would make the system unusable.

- **Incompatibility**. There are a number of Web browsers that are available to the user. Not all browsers, however, are compatible since they do not all implement the same features.
- **Space Inefficiency**. Having the whole file to be uploaded to the server's machine is a waste of space and bandwidth. The file(s) may have to be saved on the server's machine and thus taking up some valuable space.
- **Inconvenience**. If the client's program consists of multiple files, the client is required to upload all the files to the server's machine.
- **Limited I/O**. CGI was developed for the purpose of form-based information processing. In such case, once the user fills out the form and submits it, there is little, if any, interaction between the user and the script interpreting or processing the form. In real computing, however, when the program runs it may need some input from the user. Therefore, more interaction is required. This I/O limitation represents a problem when using CGI to carry out real computations.

## 3 Loading Code on the Fly

For the global computing platform to be useful, successful, efficient, and easy to adapt to, the client should not have to upload the program files to the server's machine as this is an inconvenient process. With the help of dynamic code loading, the client needs only send a URL of the location of the code to be executed and the server will in turn fetch the code from the URL and loads it dynamically (on-the-fly). Once the code has been loaded successfully, the server collects the results and sends them to the client. Dynamic class loading (via a custom class loader) is one of the innovative features built-in into the Java programming language[2].

Building a computing platform on top of the Web using dynamic code loading requires a server process (in addition to the Web server that would already be running on hosts on the Internet). The server process, which is really a sub-component of the computing platform itself, is responsible for receiving clients' requests, executing the program(s), and sending the results back to clients.

## 4 Security Considerations

Security is an important issue in every distributed system, especially when code is executing on remote machines. In general, there are two types of security problems: *nuisances* and *security breaches*. A nuisance attach simply prevents you from getting your work done. For example, client requests may overload the sever and the server's machine may crash. Security breaches on the other hand, are more serious. Your files may get deleted, for example.

In our system, the server process loads arbitrary classes into the system through a class loader mechanism. This would put the server's machine integrity at risk. This is basically due to the power of the class loader mechanism. To ensure that untrusted code cannot perform any malicious instructions, such as deleting files, the server runs in a restricted (sandbox) environment. To achieve this, we devised an extensible security model, and implemented a custom security manager.

Our extensible security model will not allow clients' code to perform any malicious actions, including: reading from or writing to files on the compute server's file system; delete files from the compute server's file system; execute any system commands (such as rm or del, ... etc); create or list files; load a new security manager; load a new class loader; and make the compute server quit. A detailed discussion of our security policy is discussed in the implementation section.

## 5 Implementation Highlights

In this section we briefly discuss the implementation of the major components of our system, including: the compute server, the class loader, the security manager, and the client.

### 5.1 Dynamic Class Loading

Class loaders are one of the cornerstones of the Java Virtual Machine (JVM) architecture. They enable the JVM to load classes without knowing anything about the underlying semantics. Classes are introduced in the Java environment when they are referenced by name in a class that is already running. The first class that gets to run is the one with the method `public static void main(String argv[])` declared in it. Once the `main` class is running, future attempts at loading classes are carried out by the class loader. The abstract class, `ClassLoader`, which is a subclass of `Object`, is contained in the *java.lang* package. Applications may inherit from the `ClassLoader` abstract class in order to extend its functionality in which the JVM dynamically loads classes. Normally, the JVM loads classes from the directory defined by the CLASSPATH environment variable on the local file system. Our compute server, however, loads classes off the network. For this, we implemented a custom class loader – `NetClassLoader`, which is capable of loading classes from remote destinations off the network.

One hidden issue when working with class loaders, that is worth discussing here, is the inability to cast an object that was created from a loaded class into its original class. The object to be returned needs to be casted. As an example, a typical usage of our `NetClassLoader` would be as follows:

```
NetClassLoader ncl = new NetClassLoader();
Object obj;
Class c;
c = ncl.loadClass("someClass");
obj = c.newInstance();
((someInterface) obj).someClassMethod();
```

Note that we cannot cast `obj` to `someClass` because only the class loader has information about the new class it has loaded. This means that a custom class loader cannot just run about any class. In fact, this is a Java limitation[2]. For example, in order for a Web browser to load an applet, the applet is implemented by inheriting from the `Applet` class. This limitation can be solved in a couple of ways: either by having a main class that each user of the system must extend (or inherit from), or by having an interface that users of the system must implement. In our case, we chose to have an interface. Our interface has the following definition:

```
public interface RemoteCompute {
   public void execute();
}
```

Now, if a user is interested in using the system, he must implement the `RemoteCompute` interface by providing an implementation for the `execute()` method. A sample program may look as follows:

```
public class MyApp implements RemoteCompute {
  public void execute() {
     System.out.println("My first remote program");
  }
}
```

---

[2] The Core Reflection API can be used to avoid this problem.

## 5.2 Custom Security Manager

The built-in Java safety features (e.g. class format verifier and automated garbage collector) ensure that the Java system is not subverted by an invalid code. These features, however, are not able to protect against malicious code. For example, imagine that a client is aware of a sensitive file (spy.txt) that exists on the compute server's file system. The client may fool the compute server in deleting that file, or even mailing it back to the client, by writing a class that may look as follows:

```
//a sample class to delete a file
public class Spy implements RemoteCompute {
  public void execute() {
    File f = new File("path to spy.txt");
    if ((f.delete() == true) {
      System.out.println("File: "+f+" is deleted");
    } else {
      System.out.println("File cannot be deleted");
    }
  }
}
```

The `SecurityManager` class, part of the *java.lang* package, provides the necessary mechanism for creating a custom security manager that defines tasks that an application can and cannot do. As an example, suppose that we wish to prevent against deleting files from the compute server's file system. The following snippet of code demonstrates how convenient it is to do so in Java.

```
private boolean checkDelete = true;
// prevent clients from deleting files
public void checkDelete(String f) {
  if (checkDelete) {
    throw new SecurityManager("Cannot delete: "+f);
  }
}
```

It is important to note that when defining a custom security manager, you must override some or all of the permission checking methods, depending on the policies enforced by the security manager. By default, all of the methods will simply throw `SecurityException`, meaning that the operation is not allowed.

To fully understand why do we need a custom security manager, the following code segment demonstrates how the Java interpreter's security manager works.

```
public boolean XX(Type arg) {
  SecurityManager sm = System.getSecurityManager();
  If(sm != null) {
    sm.checkXX(arg);
  }
}
```

This shows that when a public method call invokes the system security manager, the system determines whether the operation XX is allowed.

In the next section, we shall discuss how our security manager can be installed and used by the compute server.

## 5.3 The Compute Server

The compute server is simply a process that would be running at all times, listening to requests from clients. The server process has some form of synchronization and the ability to create new processes as needed. The Java programming language has built-in support for threads – a cheap way to create new processes that run in the same address space.

In order to achieve reliability in communication, we used TCP-based stream network connections. The `Socket` and `SocketServer` classes of the *java.net* package implement reliable connections. The main body of the compute server program looks as follows:

```
public static void main(String argv[]) {
  ComputeSecurityManager csm;
  try {
    csm = new ComputeSecurityManager();
    System.setSecurityManager(csm);
  } catch (SecurityException se) {
    se.printStackTrace();
  }
  new ComputeServer();
}
```

As you can see from the above segment of code, the first thing we do in the main program is installing our `ComputeSecurityManager` by creating an instance of it and registering it using the instruction `System.setSecurityManager(csm)`, where `csm` is an instance of `ComputeSecurityManager`. The call to the `ComputeServer()` constructor initializes the compute server to listen on an eligible port number, then it starts the thread by calling its `start()` method.

## 5.4 The Client

The client program can either be a CGI script, and applet, or even a stand-alone program that can be used from the command line. In our system we implemented the client as an applet, a CGI script, and a stand-alone program. The applet sends input to the compute server from a `TextField` component, and displays the results from the compute server in a `TextArea` component. Applets, however, can be problematic to run. The stand-alone client simply opens a connection to a compute server and sends it the URL of the code to be executed by the compute server. The compute server will then execute the code and sends the results to the client. The client will then display the results on the screen.

## 6  Searching for Compute Servers

In order for users to use our system, they need to know the URL of a compute server. This was the case with locating homepages in the early days of the Web. At present, however, search servers can be used to locate URLs of homepages. In our case, a broker can be used to dynamically match clients' requests with the available (registered) compute servers. We implemented the broker as a server process connected to a database. Each time a compute server is started it will contact the broker to register its properties (when it will be available, for how long, ... etc).

Now, whenever a client needs a compute server, it sends a request (which may include the type of the machine needed and for how long) to the broker. The broker will in turn search its database for a match. If a match is found, it will be sent to the client and then the client can either contact the compute server directly or all communications may go through the broker. We tried both approaches, but for safety and authenticated issues, we recommend the approach where all communications go through the broker.

## 7  Related Work

There has been some proposed work to take advantage of the Web and Java to implement global Web-based distributed systems. For example, Brecht et al [1] describes a system called ParaWeb that would allow users to execute serial programs on faster compute servers, or parallel programs on a variety of heterogeneous hosts. Their system includes building a parallel class library, and an extended Java runtime system so as to allow programmers to develop programs with parallelism in mind. A similar project is proposed by Fax et al [4] for high performance computing based on Web technology, where they outline a process by which to build a World-Wide Virtual Machine using Java as a candidate for Web-based computing. Their Web Virtual Machine consists of a collection of CGI that would extend the functionality of Web servers. The computation model in this system is given by user CGI processes acting as computational nodes and system CGI processes to provide the required control and management. In part, our project is similar to the above two projects since they are all using Java and the Web as the enabling technologies. The use of Java for building distributed systems that execute over the Internet has also been proposed by Chandy et al [3].

Grimshaw et al[5] proposed a global distributed parallel system, known as Legion, that aims at providing and architecture for designing and building system services. Their proposed system consists of workstations and supercomputers connected together by local networks. When a user sits at a terminal connected to Legion, he will have the illusion of a single virtual machine. It is important to note that the Web is not a component of Legion. Globus[11] is another project that is developing the fundamental technology that is needed to build computational grid, execution environments that enable an application to integrate geographically-distributed computational and information resources.

In this paper we described our system, which has distinct features from the proposed systems described above. The systems proposed do not have a dynamic compute resource allocation mechanism – broker, which is capable of satisfying clients' requests dynamically. Also, our system consists of a set of compute servers running on nodes around the Internet. The nodes act as compute servers where users will be able to contact these nodes through a broker and upload a reference (URL) of the code to be executed.

## 8  Conclusions

The current computing model of the Web was mainly designed for processing fill-out forms. Therefore, it cannot be used for real computing. In this paper, we have discussed the limitations of the Web model for global distributed computing and, our work on Web-based computing and demonstrated our idea of the Web as a Global Computing Platform. We briefly discussed our implementation of a simple Web-based distributed computing system using Java (as the architectural neutral feature of this powerful language makes it an ideal candidate for implementing such a system). We conclude that as Web-based distributed computing is hindered by administrative and architectural constraints, lots of research is needed to solve such problems by developing new models of programming and computations for use on the web.

## Acknowledgments

I am grateful to Dr. Weichang Du of the University of New Brunswick, Canada, for his supervision and guidance throughout the preparation of my thesis. Which was entitled *Design and Implementation of a Web-based Distributed Computing System*. Also, I wish to thank the anonymous referees for their comments which helped me to improve the presentation of this work.

## About the author

Qusay H. Mahmoud works for Etisalat College of Engineering, Emirates Telecommunications Corporation, Sharjah, United Arab Emirates. He holds a B.Sc. in Data Analysis and a Masters degree in Computer Science, both from the University of New Brunswick, Canada. Previously he worked for Newbridge Networks and the School of Computer Science at Carleton University, both in Canada. He is the author of over 40 technical articles on Java, and the upcoming book *Distributed Programming with Java*, Manning Publications Co.

# References

1. Brecht, T., Sandhu, H., Shan, M., Talbot, J.: ParaWeb: Towards World-Wide Supercomputing. In Proceedings of the Seventh AC SIGOPS European Workshop, Connemara, Ireland, September (1996) 181-188.
2. Campione, M., Walrath, K.: The Java Tutorial: Object-Oriented Programming for the Internet. Addison-Wesley, 1996.
3. Chandy, K.M., Dimitrov, B., Le, H., Mandleson, J., Rifkin, A., Sivilotti, P.A.G., Tanaka, W., Weisman, L.: A world-wide distributed system using Java and the internet. In Fifth IEEE International Symposium on High Performance Computing (HPC5), Syracuse, New York, August 1996.
4. Fox, G.C., Furmanski, W.: Towards Web/Java High Performance Distributed Computing – an evolving virtual machine. In Fifth IEEE International Symposium on High Performance Distributed Computing (HPDC5), Syracuse, New York, August 1996.
5. Grimshaw, A.S., Wulf, W.A., the Legion Team: The Legion Vision of a Worldwide Virtual Computer. Communications of the ACM, vol. 40, No.1, January 1997.
6. Lewis, T.: The Next $10,000_2$ Years: Part I & II. Communications of the ACM, April 1996.
7. Mahmoud, Q.H.: Design and Implementation of a Web-based Distributed Computing System. Masters Thesis, University of New Brunswick, Canada, 1997.
8. Mahmoud, Q.H.: Global Web-based Computing. A Poster Presentation at The $5^{th}$ International Conference on High-Performance Computing (HiPC'98), Chennai, India, December 1998.
9. Nebel, E., Masinter, L.: RFC1876: Form-based File Upload in HTML. Network Working Draft Document, Novemeber 1995.
10. RSA Data Security, http://www.rsa.com.
11. The Globus Project, http://www.globus.org.
12. World-Wide Web Consortium, http://www.w3.org.

# WebFlow: A Framework for Web Based Metacomputing

Tomasz Haupt, Erol Akarsu, Geoffrey Fox
{haupt,akarsu,gcf}@npac.syr.edu

Notheast Parallel Architecture Center at Syracuse University. 111 College Place
Room 3-217, Syracuse, NY 13244-4100, USA

**Abstract.** We developed a platform independent, three-tier system, called WebFlow. The visual authoring tools implemented in the front end integrated with the middle tier network of servers based on CORBA and following distributed object paradigm, facilitate seamless integration of commodity software components. We add high performance to commodity systems using GLOBUS metacomputing toolkit as the backend.

## 1 Introduction

Programming tools that are simultaneously sustainable, highly functional, robust and easy to use have been hard to come by in the HPCC arena. This is partially due to the difficulty in developing sophisticated customized systems for what is a relatively small part of the worldwide computing enterprise. Thus we have developed a new strategy - termed HPcc High Performance Commodity Computing [1] - which builds HPCC programming tools on top of the remarkable new software infrastructure being built for the commercial web and distributed object areas.

This leverage of a huge industry investment naturally delivers tools with the desired properties with the one (albeit critical) exception that high performance is not guaranteed. Our approach automatically gives the user access to the full range of commercial capabilities (e.g. databases and compute servers), pervasive access from all platforms and natural incremental enhancement as the industry software juggernaut continues to deliver software systems of rapidly increasing power.

We add high performance to commodity systems using a multi tier architecture with traditional HPCC technologies such as MPI and HPF supported as the backend of a middle tier of commodity web and object servers.

Our research addresses needs for high level programming environments and tools to support distance computing on heterogeneous distributed commodity platforms and high-speed networks, spanning across labs and facilities. More specifically, we are developing WebFlow - a scalable, high level, commodity standards based HPDC system that integrates:

- High-level front-ends for visual programming, steering, and data visualization built on top of the Web and OO commodity standards (Tier 1).

- Distributed object-based, scalable, and reusable Web server and Object broker Middleware (Tier 2)
- High Performance back-end implemented using the metacomputing toolkit of GLOBUS [2] (Tier 3)

We have demonstrated the fully functional prototype of WebFlow during Alliance'98 meeting as applied to a metacomputing application Quantum Simulations [3], and at SuperComputing '98 were we demonstrated the Landscape Management System [4] using WebFlow over a geographically distributed system.

In this paper we critically summarize our experience using WebFlow, and for the first time we present a redesigned, CORBA-based middle-tier. The paper is organized as follows. In section 2.0 we provide the WebFlow overview from the user point of view. In section 3.0 we discuss the original implementation of the middle-tier based on our experience applying the WebFlow to real applications. Section 4.0 is devote to our new design which is meant to remedy shortcomings indicated in Section 3.0. The paper is summarized in Section 5.0

## 2 Overview of WebFlow

The visual HPDC framework introduced by this project offers an intuitive Web browser based interface and a uniform point of interactive control for a variety of computational modules and applications, running at various labs on different platforms. Our technology goal is to build a high-level user friendly commodity software based visual programming and runtime environment for High Performance Distributed Computing.

WebFlow is a particular programming paradigm implemented over a virtual Web accessible metacomputer. We support many different programming models for the distributed computations: from coarse-grain dataflow to object-oriented to fine-grain data-parallel model. In the dataflow regime, a WebFlow application is given by a computational graph visually edited by end-users, using Java applets. The dataflow WebFlow modules, atomic computational units of an metacomputing application, exchange data via input and output ports, in a way similar to that used in AVS [5]. This exchange of data can realized in the following way. Whenever a module is ready to send the data (encapsulated as an object), it fires an event triggering all modules interested in receiving these data to call the corresponding method to retrieve the data. This model can be generalized to allow the module to fire arbitrary events, and add arbitrary event listeners. As the result, the module can invoke an arbitrary method of the other modules involved in the computation. The reason why we distinguish between the dataflow and object-oriented model has an historical origin. Our first WebFlow implementation [6,7] supported the dataflow model exclusively. Also, we use a different strategies to implement the front-end of the system for the dataflow and the general object-oriented models.

Nothing prohibits the user to encapsulate a data parallel application as a single WebFlow module. In this case the user is solely responsible for the interprocessor communications (we used HPF and MPI-based codes to run WebFlow modules on a multiprocessor systems[8]). More, using the DARP system[9], implemented as a WebFlow module, we were able to interactively control an HPF application at runtime, and dynamically extract the distributed data and send them to a visualization engine. This approach can be used for computational steering, runtime data analysis, debugging, and interprocessor communications on demand. Finally, we integrated two independently written applications that write checkpoint data [4]. We used WebFlow to detect the existence of the new data, and transfer them to the other application.

Modules are written by module developers, people who have only limited knowledge of the system on which the modules will run. They not need concern themselves with issues such as: allocating and running the modules on various machines, creating connections among the modules, sending and receiving data across these connections, or running several modules concurrently on one machine. The WebFlow system hides these management and coordination functions from the developers, allowing them to concentrate on the modules being developed.

An important part of our design is devoted to providing a secure and seamless access to remote resources, in particular to HPCC systems. We build the WebFlow security infrastructure on top of the SSL protocol [11]. In particular, we use GSS-API [12] based Globus GRAM to allocate HPCC resources. In this sense, the WebFlow can be regarded as a visual, high-level user interface to Globus, or conversely, we may state that we use Globus as the high performance backend of the WebFlow.

## 3  WebFlow Applications

### 3.1  Quantum Simulations (QS)

QS project is a part of the Alliance Team B and its primary purpose is demonstrate feasibility of layering WebFlow on top the Globus metacomputing toolkit. This way WebFlow serves as a job broker for Globus, while Globus (or more precisely, GRAM-keeper) takes responsibility of actual resource allocation, which includes authentication and authorization of the WebFlow user to use computational resources under Globus control.

This application can be characterized as follows. A chain of high performance applications (both commercial packages such as GAUSSIAN or GAMESS or custom developed) is run repeatedly for different data sets. Each application can be run on several different (multiprocessor) platforms, and consequently, input and output files must be moved between machines. Output files are visually inspected by the researcher; if necessary applications are rerun with modified input parameters. The output file of one application in the chain is the input of the next one, after a suitable format conversion.

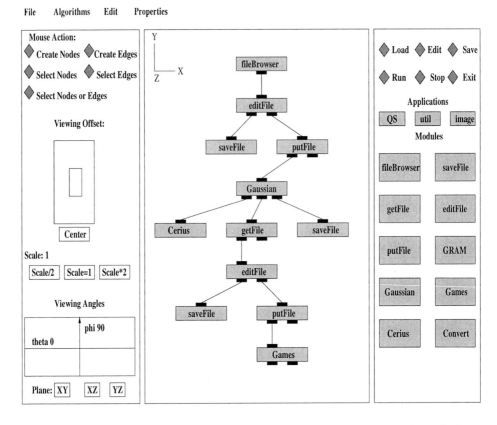

**Fig. 1.** Composing a dataflow application using the WebFlow visual editor (a Java applet)

GAUSSIAN and GAMES are run as Globus jobs on Origin2000 or Convex Exemplar at NCSA, while all file editing and format conversion a performed on the user's desktop.

For QS we are using the WebFlow editor as the front-end (c.f. fig 1). The WebFlow editor provides an intuitive environment to visually compose (click-drag-and-drop) a chain of data-flow computations from preexisting modules. In the edit mode, modules can be added to or removed from the existing network, as well as connections between the modules can be updated. Once created network can be saved (on the server side) to be restored at a later time. The workload can be distributed among several WebFlow nodes (WebFlow servers) with the interprocessor communications taken care of by the middle-tier services. More, thanks to the interface to the Globus system in the backend, execution of particular modules can be delegated to powerful HPCC systems. In the run mode, the visual representation of the metaaplication is passed to the middle-tier by sending a series of requests (module instantiation, intermodule communications) to the Session Manager.

## 3.2 Land Management System (LMS)

LMS project is sponsored by CEWES MSRC at Vicksburg, MS, under the DoD HPC Modernization Program, Programming Environment and Training (PET). The first, pilot phase of the project can be described as follows. A decision maker (the end user of the system) wants to evaluate changes in vegetation in some geographical region over a long time period caused by some short term disturbances such as a fire or human's activities. One of the critical parameters of the vegetation model is soil condition at the time of the disturbance. This in turn is dominated by rainfalls that possibly occur at that time. Consequently, the implementation of this project requires:

- Data retrieval from many different remote sources (web sites, databases)
- Data preprocessing to prune and convert the raw data to a format expected by the simulation software.
- Execution of two simulation programs: EDYS for vegetation simulation including the disturbances and CASC2D for watershed simulations during rainfalls. The latter results in generating maps of the soil condition after the rainfall. The initial conditions for CASC2D are set by EDYS just before the rainfall event, and the output of CASC2D after the event is used to update parameters of EDYS.
- Visualizations of the results.

The purpose of this project was to demonstrate feasibility of implementing a system that would allow launching and controlling the complete simulation from a networked laptop. We successfully implemented it using WebFlow with WMS and EDYS encapsulated as WebFlow modules running locally on the laptop and CASC2D executed by WebFlow on remote hosts

For this project we developed a custom front-end that allows the user to interactively select the region of interest by drawing a rectangle on a map, select the data type to be retrieved, launch WMS to preprocess the data and make visualizations, and finally launch the simulation with CASC2D running on a host of choice.

## 3.3 Shortcomings of the original WebFlow implementation

In the original prototype of WebFlow [6], the middle-tier is given by a mesh of Java enhanced Web Servers, running servlets that manage and coordinate distributed computation. This management is implemented in terms of the three servlets: Session Manager, Module Manager, and Connection Manager. These servlets are URL addressable and can offer dynamic information about their services and current state. Each of them can also communicate with each other through sockets. Servlets are persistent and application independent.

In spite of the success of the WebFlow project we see that our current implementation suffers form some limitations. Two the most obvious areas of improvement we want to achieve are fault tolerance and security. However, instead of adding complexity to already complex and to large extend custom protocol of exchanging data between the servlets based on low-level sockets, we re-implemented the WebFlow middle-tier using industry standards distributed object technologies: JavaBeans [13] and CORBA [14].

## 4 CORBA Based Implementation of WebFlow

The architecture of the new implementation of the WebFlow is shown in fig.2:

Architectureofthe WebFlow system Visual authoring tools, PSEs, or custom developed application front-ends allow the user to specify or compose her application, and they implement WebFlow interface to the middle-tier. This way a particular front-end solution comes as a "plug-in" to the system, and can be easily replaced by another one. It is our experience that different applications require different front-ends. The WebFlow API specifies how the front-end interacts with the middle-tier. It includes establishing and termination of a WebFlow

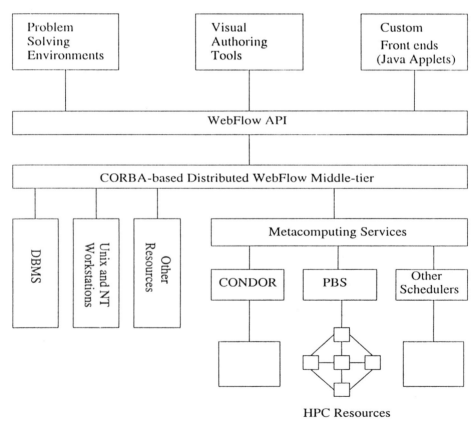

Figure 2. Architecture of the WebFlow System

session, which generally requires authentication of the user and authorization to use the resources. We base the authentication process on X.509 certificates and authorization services we model after AKENTI system [15]. We are in the early stage of implementing this functionality, and a more detailed description will be provided in a forthcoming paper [16]. Further, the WebFlow API defines how the user creates applications from predefined modules in a form of an abstract job description that specify all resources (hardware and software) needed to build and execute the meta-application. The abstract job description may request a particular resource or delegate the selection of the resources that match the application requirements to the system. The WebFlow middle tier, in turn, delegates this task to the services provided by a metacomputing toolkit such as Globus' MDS or CONDOR's matchmaker. Finally, the WebFlow API provides means to control of the modules life cycle.

The middle-tier is given by a mesh of WebFlow servers implemented as pairs of Web server and ORB. Web Servers are needed for communications with the front-end. In particular, the front-end applets are downloaded from these servers. We tested two secure (i.e., supporting SSL) Web servers: Apache [17] and Jigsaw[18]. All communications between peer WebFlow servers and the back-end modules are facilitated using CORBA. We use ORBacus [19], a free ORB implementation for this purpose, and we are testing JWORB [10], a multiprotocol server in Java that supports both http and IIOP.

Once the user is authorized to use the WebFlow resources, a session context (a container object) is created for her. Within her session context, the user creates application contexts, so that the user can run several independent applications within one WebFlow session. Within the application context, the user builds her application from preexisting modules, and registers the events and corresponding event listeners with an event adapter (which is run as a CORBA service). The event adapter maintains a translation table, and in conjunction with the CORBA name service and interface repository it allows for binding of events with methods of the target modules independently of the modules location. For modules than run on remote WebFlow servers proxy modules are automatically produced. The proxy modules make it easier to control the life cycle of the modules in the application context. In addition, they are necessary to maintain communications with the front-end controls, implemented as Java applets, of the remote module (to avoid the Java sandbox restrictions).

The WebFlow modules are CORBA objects. Typically, we implement them as the Java beans. The very fact that we are using CORBA and not Java RMI makes it possible to include objects written in languages different than Java, in particular C++.

For high performance computations the WebFlow modules serves as proxies to services provided by the metacomputing toolkits. We have demonstrated this in our implementation of the Quantum Simulations. The codes that actually run on Origin 2000 were either made available to us only in a binary form (GAUSSIAN and GAMES are commercial products), or were written in Fortran. Using them within WebFlow did not require any modifications, letting alone rewriting them in Java. We just created WebFlow proxy modules that on behalf of the user generated requests to the Globus GASS to stage the data and executables on the target machine, and retrieve the results, and requests to the Globus GRAM to allocate the resource (including authentication and authorization). Note that all GRAM requests to be specified in a low-level RSL (resource specification language) were generated by the WebFlow modules in the fly by interpreting the abstract job description which in turn was created using a visual WebFlow front-end running as an applet within a Netscape browser.

## 5  Summary

To summarize, we developed a platform independent, three-tier system: the visual authoring tools implemented in the front end integrated with the middle tier network of servers based on CORBA and following distributed object paradigm, facilitates seamless integration of commodity software components. In particular, we use the WebFlow as a high level, visual user interface for GLOBUS. This not only makes construction of a meta-application much easier task for an end user, but also allows combining this state of art HPCC environment with commercial software, including packages available only on Intel-based personal computers.

## References

1. G. Fox and W. Furmanski, "HPcc as High Performance Comodity Computing", chapter for the "Building National Grid" book by I. Foster and C. Kesselman, http://www.npac.syr.edu/users/gcf/HPcc/HPcc.html
2. I. Foster and C. Kesselman, "Globus: A Metacomputing Infrastructure Toolkit", International Journal of Super computing Applications, 1997. See also Globus Home Page: http://www.globus.org
3. Quantum Simulations , http://www.ncsa.uiuc.edu/Apps/CMP/cmp-homepage.html
4. T. Haupt, "WebFlow High-Level Programming Environment and Visual Authoring Toolkit for HPDC (desktop access to remote resources)", Supercomputing '98 technical presentation, see http://www.npac.syr.edu/users/haupt/WebFlow/papers/SC98/foils/index.htm
5. Advanced Visualization System, http://www.avs.com/
6. D. Bhatia, V. Burzevski, M. Camuseva, G. C. Fox, W. Furmanski and G. Premchandran, "WebFlow - a visual programming paradigm for Web/Java based coarse grain distributed computing", Concurrency: Practice and Experience, Vol. 9 (6), pp. 555-577, June 1997.
7. T. Haupt, E. Akarsu, G. Fox, W. Furmanski, "Web based metacomputing ", Special Issue on MetaComputing for the FGCS Int. Journal on Future Generation Computing Systems, see also http://www.npac.syr.edu/users/haupt/WebFlow/papers/FGCS/index.html
8. G. Fox, W. Furmanski and T. Haupt, SC97 handout: High Performance Commodity Computing (HPcc), http://www.npac.syr.edu/users/haupt/SC97/HPccdemos.html

9. E. Akarsu, G. Fox, T. Haupt, DARP: Data Analysis and Rapid Prototyping Environment for Distribute High Performance Computations, Home Page http://www.npac.syr.edu/projects/hpfi/
10. JWORB Project Home Page
    http://osprey7.npac.syr.edu:1998/iwt98/projects/worb, see also G. C. Fox, W. Furmanski and H. T. Ozdemir, "JWORB (Java Web Object Request Broker) based Scalable Middleware for Commodity Based HPDC", submitted to HPDC98.
11. SSL, Netscape Communications, Inc,
    http://home.netscape.com/eng/ssl3/index.html
12. RFC 1508, RFC 2078
13. Sun Microsystems, Inc., http://java.sun.com
14. CORBA - OMG Home Page http://www.omg.org
15. AKENTI home page: http://www-itg.lbl.gov/security/Akenti/homepage.html
16. Check the WebFlow home page at http://www.npac.syr.edu/users/haupt/WebFlow/demo.html

# Dynamite - Blasting Obstacles to Parallel Cluster Computing

G.D. van Albada[1], J. Clinckemaillie[2], A.H.L. Emmen[3], J. Gehring[4], O. Heinz[4], F. van der Linden[1], B.J. Overeinder[1], A. Reinefeld[5], P.M.A. Sloot[1]

[1] Department of Computer Science, Universiteit van Amsterdam, Kruislaan 403, 1098 SJ Amsterdam, The Netherlands
[2] Engineering Systems International, 20 Rue Saarinen, F-94578 Rungis SILIC 270, France
[3] Genias Benelux BV, James Stewartstraat 248, 1325 JN Almere, The Netherlands
[4] Paderborn Center for Parallel Computing, Fürstenallee 11, 33102 Paderborn, Germany
[5] Konrad-Zuse-Zentrum für Informationstechnik, Takustrasse 7, D-14195 Berlin, Germany

**Abstract.** Workstations make up a very large fraction of the total available computing capacity in many organisations. In order to use this capacity optimally, dynamic allocation of computing resources is needed. The Esprit project Dynamite addresses this load balancing problem through the migration of tasks in a dynamically linked parallel program. An important goal of the project is to accomplish this in a manner that is transparent both to the application programmer and to the user. As a test bed, the Pam-Crash software from ESI is used.

## Introduction

Workstations have become ubiquitous in many organisations. By their nature, they are often used intensively during normal working hours, and are often largely idle otherwise. They represent a huge reservoir of computing capacity that can be used much more efficiently.

Thus, we currently witness a shift of emphasis in high-performance computing from expensive, special-purpose monolithic systems to the use of clusters of workstations or PCs.

When using time-shared workstation clusters as HPC compute servers, however, one has to cope with the dynamical behaviour of the compute nodes, the network load and the application tasks. These can lead to local load imbalances, which hamper the application's execution speed and the overall system performance.

The application itself can also exhibit dynamic behaviour due to changes in the load per task (e.g. contact problems in car crash simulations). This leads to serious load imbalances, which are difficult to resolve, even on dedicated parallel platforms that offer a constant performance per node. When the node performance changes dynamically, as in workstation clusters, the situation becomes even more difficult.

Also, running a HPC task on a workstation may jeopardise its primary purpose of providing computing capacity to a particular employee.

Solving these problems requires that work somehow be migrated from one node to another. This can be done internally to the parallel application, but such an approach requires a major adaptation of each individual program. Various solutions have been developed to improve the load distribution for workstations. These range from systems that schedule parallel or sequential jobs on free workstations, such as LSF [1], via systems that can also migrate sequential jobs, such as Codine [2] and Condor [3, 4], to systems that also aim to migrate tasks in parallel jobs. MPVM/MIST [5, 6] does this for PVM based jobs, Hector [7] for MPI.

In the ESPRIT project 23499 "DYNAMITE", we develop a dynamic execution environment that handles the load balancing of parallel applications in a dynamically changing cluster environment by migrating individual tasks in a manner that is robust, efficient and transparent to the user and the application programmer. The DYNAMITE software is based on PVM 3.3.11 and is called Dynamic PVM [8] or DPVM for short. DPVM is totally transparent to the user's application: existing PVM codes need only be linked to the DPVM library. The DYNAMITE system is intended for environments requiring a relatively infrequent redistribution of workload for large applications that can run for several days. We strive for a response time of at most a few minutes and a minimal overhead, but give an absolute priority to reliability and stability.

In constructing such an environment, the following problems need to be addressed:
- migration of dynamically linked tasks,
- migration of communication endpoints,
- load monitoring,
- task (re-)allocation,
- job preparation

In this paper, we describe the ongoing work in the DYNAMITE project. The first two issues will be addressed in the next section. Subsequently we will address load monitoring, task allocation and job preparation in separate sections, before coming to our conclusions.

## Migration

Migration of tasks requires that the state of the task is captured, after which a new task is started on the target machine, initialised with the captured state. Correct migration is difficult because the interactions of the task with its environment need to be taken into account. A completely transparent migration, which cannot be detected by the task or its communication partners, is almost impossible to realise, but is not usually necessary either. We strive to migrate dynamically linked tasks with open files, communicating with other tasks solely through PVM.

For the migration of tasks with open files, we impose the additional requirement that these files can be accessed using the same path on both the source and target machine.

We have implemented the migration mechanism making use of a full checkpoint of the task. Though it requires additional communication and I/O compared to a mechanism based on a direct transfer of the task image from source to target machine, we have decided to use this approach for reasons of robustness and clarity of implementation.

Pilot versions of the checkpointer and migrator were implemented for the SUN Solaris operating system, and have been tested on OS versions 2.5.1 and 2.6 on UltraSparc workstations.

**Migration of dynamically linked code**

As stated, as the first step in the migration of a task, a checkpoint dump is made.

The checkpointing implementation used in DYNAMITE differs from existing implementations in two ways. Firstly, the checkpointing code is not linked into the program itself. Instead, it is present in the dynamic loader, a piece of code loaded before the actual program is run. The task of the dynamic loader is to load the shared libraries required by the program. Most Unix systems implement shared libraries using a dynamic loader, and have an option to specify a different loader for each program. This option is used to specify our own dynamic loader.

In DYNAMITE, this dynamic loader will perform these tasks as usual, but will also contain code to handle checkpointing signals, and to keep information on the used shared libraries. This means that it can take care of creating the checkpoint, and restoring it, using the exact same memory mappings for shared libraries. This is important, because shared libraries are normally not guaranteed to be mapped on the same memory address, which would make restarting the application impossible.

The other new aspect in the checkpointing code is the propagation of checkpointing signals. This means that the dynamic loader will, before creating the checkpoint, signal the application to allow it to save state that can possibly not be saved within the framework of the normal checkpointing procedure.

The current implementation of the checkpointer has the following limitations:
- The checkpointed task should not be multithreaded. This limitation applies to all PVM programs anyway.
- The checkpointed task should not have any files or network connections open, save for those serviced by PVM. Migration of open files will be supported in a later version.
- The checkpointer writes a full checkpoint to a file, including any mapped dynamic libraries and the complete data segment. An earlier version of DPVM used a migration approach in which most of the data segment was transferred directly from the old to the new task image through a socket. While this approach has a speed advantage, it hampers a robust implementation.

It is not necessary for the original task still to be active for the restart, as would be the case when part of the image is transferred directly from the old to the new task. The checkpoint file is an executable in its own right, and can thus be restarted in the usual way.

The job of the migrator, which is part of the DPVM library, is to start a new task on the target machine, using the checkpointed executable.

## Migration of communicating tasks

A main objective of the DPVM migration facility is transparency of the migration protocol. With respect to the task selected for migration this implies transparent suspension and resumption of execution: the task has no notion that it is migrated to another host, and the communication can be delayed without failure triggered by migration of one of the tasks. The work upon which our implementation is based is described in [8]

The first step in the migration protocol is the creation of a new process context at the destination host by sending a message to the PVM daemon (pvmd) representing that host. Next, the master pvmd updates its routing table to reflect the new location of the task. Before the task selected for migration is suspended, the communication between this task and its pvmd has to be flushed. Then the task is disconnected from its local pvmd and messages arriving for that task are refused by the task's original pvmd. The master pvmd will now broadcast the new location to all other pvmds, so that any subsequent message is directed to the task's new location.

The next phase is the actual migration of the process. The original task is checkpointed and the newly created process on the destination host is requested to restart the checkpoint.

Finally, after the checkpoint is read, the original state of the task (among which data, stack, signal mask, and registers) is restored and the task is restarted. Any message that arrived during the checkpoint/migration phase is then delivered to the restarted task.

## Packet Routing

In PVM the task identifier, task id for short, is a unique identifier that serves as the task's address and therefore may be distributed to other PVM tasks for communication purposes. For this reason, the task id must remain unchanged during the lifetime of a task, even when the task is migrated.

This has implications for the packet routing of messages. The task id contains the host identifier at which the task is enrolled and a task sequence number. This information is used by the pvmd to route packets to their destination, i.e., to the appropriate pvmd and task. When a task is migrated to another host, this routing information is not correct anymore. Therefore, an additional routing functionality must be incorporated in the pvmd routing software in order to support the migration of tasks. An important design constraint is that the routing facility must be highly efficient and should not impose additional limitations on the scalability.

To provide transparent and correct message routing with migrating tasks, the task ids must be made location independent, virtualising the task ids. This is accomplished by maintaining additional routing information tables in all pvmds. These routing tables are consulted for all inter-task communication. Upon migration of a task, first the routing table of the master pvmd is updated to reflect the change in location of the migrated task. Next, the master pvmd broadcasts the routing table change to all other pvmds, so that each routing table reflects the actual location of all migrated tasks in the system.

**Direct Connections**

The basic mode of communication in PVM is through the daemon. For reasons of efficiency, PVM allows tasks to request a direct connection to another task. This complicates the rerouting of the communication. The main problem of direct communication connections is making sure that all communication has been flushed. Simply breaking the connections may result in loss of messages and is not acceptable. Several approaches are possible. Some involve shutting down communication for the whole system temporarily, but this may cause unnecessary delay. Another approach is to leave an agent in place that takes care of the connection as long as it has not been confirmed as flushed by the other side.

A related problem occurs in the implementation of task migration in MPI. MPI is a specification that is often implemented using direct connections, as in MPICH, a popular MPI implementation. See [7] for one possible solution for the problems that occur when implementing task migration for this MPI version.

**Resource Monitoring**

Any migration decision has to be based on the information that is currently available about the cluster. This refers to the state of the hardware as well as to the runtime behaviour of the applications. The typical approach taken by most cluster management systems is to measure the load on each available host and of each application process. The busiest tasks are then moved to the least loaded nodes until a satisfactory state is achieved. This strategy has been proven well suited for running independent jobs on networks of workstations, but it performs less well for parallel applications as it completely neglects communication between interdependent tasks. This drawback is especially apparent in environments with significant performance differences between the nodes. In such scenarios, it is often the case that larger machines (typically SMPs or NUMAs with 4 to 16 processors) are assigned multiple processes. It is then desirable to have frequently communicating tasks grouped together on big machines (Figure 1).

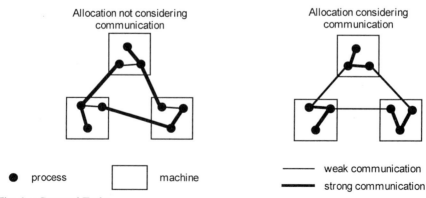

**Fig. 1.** : Grouped Tasks

In the first case, the sequential load is equally balanced but the communication is not. Therefore, the monitoring tool must also keep track of the communication between the tasks. In order to make an optimal migration decision, the following information is needed:
- available capacity on each node (CPU, memory, disk space),
- current load of each node,
- required capacity for each task,
- network connectivity and capacity,
- communication pattern for each task.

Each of these items can be measured at execution time by monitoring software, but we assume that node capacity and network properties are sufficiently stable that they can best be specified beforehand by the system administrator. Therefore, we have chosen a textual representation of the static resources (see [9] for further details).

Detailed information about the network topology can be obtained from a "Network Resource Description" file that is used for migration decisions. Tasks should preferably be migrated to nodes in the same subnet. This provides locality for the messages and prevents that a large amount of data has to be routed from one subnet to the other. If it is not possible to fulfil the requirement for locality then nodes in adjacent subnets are selected.

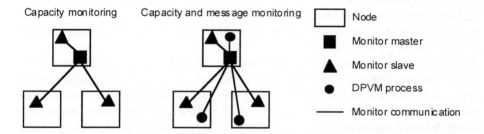

**Fig. 2.** : Capacity and Message Monitoring

Because of the assumed dynamic behaviour of the application and the system load, the other items need to be obtained by monitoring software. Information about load and capacity must be collected from all nodes of the cluster, also those where currently no task of the parallel application is running. This is accomplished by running a small monitor program (monitor slave) on each node (Figure 2).

The statistics obtained by the monitor slaves are sent to the monitor master process that is not only responsible for maintaining the whole cluster statistics but also has to make migration decisions. The information on communication patterns is obtained directly from the DPVM environment. Therefore, DPVM has been enhanced by a message monitoring thread. This thread keeps track of each message sent and received. These communication statistics are also sent to the monitor master process that is depicted in detail in Figure 3.

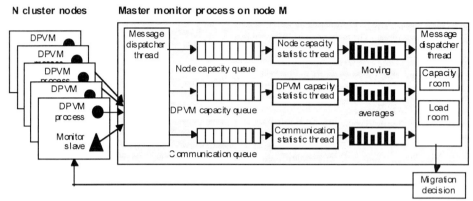

**Fig. 3.** : Architecture of the Monitor Master Process

The monitor master process consists of five threads that operate concurrently. The message dispatcher thread identifies each message received and appends it to the appropriate queue. There exist three different queues:
- a node capacity queue to store the information from the monitor slaves (CPU, memory, I/O, ...),
- a DPVM capacity queue to store the information about CPU and memory utilisation of the DPVM processes and
- a communication queue to store the information about the communication activity between the DPVM processes.

The queues act as an intermediate store because the statistics threads are only active every $j$ seconds, where $j$ can be adapted to the application monitored. Long running applications don't need a short monitoring interval and therefore the statistics need not be updated regularly. Each statistic thread maintains a ring-buffer (not shown) where the last $l$ entries are stored. Each entry in the ring-buffer corresponds to a snapshot of the monitored data at a certain point in time. It is obvious that it is not practicable to store all values since monitoring has begun. Therefore, we have chosen to implement a moving average scheme that keeps track of the last $l$ entries.

This scheme has the advantage that we can apply a recursive formula that depends only on the newest and oldest value of the ring-buffer. This speeds-up calculation of the moving averages and decreases the monitoring overhead. To allow further processing, e. g. to visualise the data sets, the statistical data is also written to disk (not shown).

## Migration Decider

The migration decider is the main part of the scheduler thread that is executed periodically by the monitor master process. Based on the monitored data, the migration decider has to judge about where and when to migrate a task from an overloaded node. Additionally the task to be moved causes some constraints on the migration decision. Therefore, the master load monitor has to supply some normalised values about the attributes CPU, memory, and disk swap space of each node and additionally the available network capacity.

The increasing interest in distributed computing has lead to intensive scientific research in load balancing schemes for distributed memory systems [2, 5, 10, 11, 12, 13, 14, 15]. Because not every load-balancing scheme is applicable to every application, the migration decider has been designed in a flexible manner to support a broad range of applications. For the first prototype we have implemented a straightforward solution with a greedy-like algorithm and constraints lists.

We call $c_{i,j}$ the available capacity of the attribute $i$ of the node $j$. In conjunction with priority coefficients $k_i$ for each attribute we are able to calculate the local available capacity $C_j$ of the node $j$ which is given by (1)

$$C_j = \sum_i k_i \times c_{i,j} \qquad (1)$$

Using the priority coefficients $k_i$ we can adapt the load-balancing scheme to the needs of different applications. Applications with a high demand in CPU and memory capacity like Pam-Crash will use a high value for these priority coefficients. All $C_j$ will then be sorted. Sorting $C_j$ in ascending order provides us a data-set which comprises the *capacity room C*. Sorting $C_j$ in descending order provides the data-set for the *load room L*. Each of these data sets are managed as priority queues (heaps) as indicated by Figure 4.

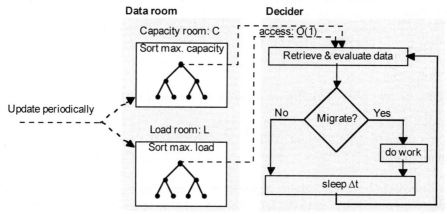

**Fig. 4.** : Architecture of the Migration Decider

The migration decider only looks at the first element of each heap. The first element of the *capacity room C* represents the node with the highest available capacity. Whereas the first element of the *load room L* represents the most heavily burdened node. A migration will be triggered, if the following conditions are met:
$L_1 > T$ and $C_1 > 1 - T$ with $i, j$ in $\{1, ..., n\}$ and $n = $ *number of nodes,* where $T$ denotes an application specific threshold level for the task migration. By using heaps for the data management, the migration decider task is able to retrieve the essential information with minimum effort $O(1)$. Additionally, updating elements in the data room can be done with $O(n * \log n)$. Although there exist other schemes with faster access to the data elements (e. g. linked lists) if only a few number of tasks have to be

considered but by using heaps we are not limited to support only a small number of tasks.

As illustrated in Figure 5, the algorithm of the decider is straightforward. The function `CheckForMigration` will be called periodically to check if the load index of the most loaded node is higher than a user defined threshold level and furthermore if a node exists which has enough remaining capacity (migration mapping). When the decision for migration is taken, the tasks are moved from the 'overloaded' node to the node with the best capacity left. Thereafter both data rooms are reordered by setting the load and capacity indices of the corresponding nodes to default values and by re-sorting the data heaps. By using a recursive algorithm, the whole migration is done in one global step. As a result, the application uses the whole workstation cluster efficiently and expensive compute time is not wasted migrating single tasks one at a time.

```
CheckForMigration () {
    /* Will be triggered at least every t seconds */
    if (GetMaxLoadFromListOfLoadedNodes() <= Threshold) return;
    if (GetBestCapacity() > (1.0 - Threshold)) {
        /* there exists a node which is less burdened;
           do the migration stuff */
        DoMigrationStuff();
        UpdateLoadRoom();       /* effort: O(n * log n) */
        UpdateCapacityRoom();   /* effort: O(n * log n) */
        CheckForMigration();    /* do the recursion */
    }
}
```

**Fig. 5.** : Pseudo-Code of the Recursive Algorithm for the Decider Module


**Job preparation**

As is the case for every parallel application, an application using the DYNAMITE environment must be split into separate tasks. These tasks must be started on the nodes of the assigned cluster. Usually, in FEM applications, such as Pam-Crash [16], and many others, this is accomplished by partitioning the problem data over the available nodes in proportion to the capacity of a node. This will result in a tight fit, which is fine if there are no variations in load or capacity. For DYNAMITE we are considering two other approaches:
1. Sparse decomposition. When the aim is to allow any one workstation from a pool of (equal) workstations to be temporarily used for other purposes, the task should be split into fewer subtasks than the number of available nodes. In this way, flexibility is gained at a cost in performance.
2. Redundant decomposition. When the aim is to allow for the redistribution of work in an application that produces a dynamically changing load, it may be preferable

to split the data so that every workstation gets more than one partition. In this way load can easily be shifted, albeit at a cost in communication efficiency.

Beside this additional choice in the partitioning, running a task under DYNAMITE also requires the monitoring tasks to be started together with the DPVM system. Though this need not require any additional effort on the side of the user, we will provide a simple GUI to assist the user in starting his DYNAMITE empowered application.

## Conclusions

DYNAMITE will provide the application developer with a robust tool that makes it possible to respond flexibly to dynamic changes in the available system capacity and application workload. The DYNAMITE system will migrate (dynamically linked) tasks from a parallel program when necessary. The overhead involved will be very small compared to the possible cost of a load imbalance. The system structure is modular so that it can easily be adapted to specific application requirements. In the development phase this modularity will be used for experimentation with various migration policies.

## References

[1] S. Zhou, X. Zheng, J. Wang and P. Delisle, *Utopia: A load sharing facility for large heterogeneous distributed computer systems*, Software – Practice and Experience, v. 23, n. 12, pp. 1305–1336, 1993

[2] http://www.genias.de/products/codine

[3] J. Pruyne and M. Livny, *Managing Checkpoints for Parallel Programs* - Poc. IPPS Second Workshop on Job Scheduling Strategies for Parallel Processing, 1996

[4] M. Litzkow, T. Tannenbaum, J. Basney, and M. Livny, *Checkpoint and Migration of Unix Processes in the Condor Distributed Processing System* - Technical Report 1346, University of Wisconsin, WI, USA, 1997

[5] J. Casas, D.L. Clark, R. Konoru, S.W. Otto, R.M. Prouty and J. Walpole, *MPVM: A migration transparent version of PVM*, Usenix Computer Systems, v. 8, n. 2, Spring, pp. 171–216, 1995

[6] J. Casas, D. Clark, P. Galbiati, R. Konuru, S.Otto, R. Prouty and J. Walpole, *MIST: PVM with Transparant Migration and Checkpointing*, Third Annual PVM Users' Group Meeting, Pittsburgh, PA, 1995

[7] J. Robinson, S.H. Russ, B. Flachs, B. Heckel, A Task Migration Implementation of the Message-Passing Interface. Proceedings of the 5th IEEE international symposium on high performance distributed computing, pp. 61-68, 1996

[8] B.J. Overeinder, P.M.A. Sloot, R.N. Heederik, L.O. Hertzberger, *A dynamic load balancing system for parallel cluster computing*, Future Generation Computer Systems 12, pp. 101-115, 1996

[9] Matthias Brune, Jörn Gehring and Alexander Reinefeld, *Heterogeneous Message Passing and a Link to Resource Management,* Journal on Supercomputing, Vol. 11, Kluwer, Boston, pp. 355–369, 1997, http://www.uni-paderborn.de/pc2/services/public/1997/97012.ps.Z

[10] F. Bonomi and A. Kumar, *Adaptive optimal load balancing in a nonhomogeneous multi-server system with a central job scheduler,* IEEE Trans. on Computers, v. 39, n. 10, pp. 1232–1250, 1990

[11] J. Casas, R. Konoru, S.W. Otto, R. Prouty and J. Walpole, *Adaptive load migration systems for PVM,* Proceeedings of Supercomputing '94, Washington DC, pp. 390–399, 1994

[12] M. Hamdi and C.K. Lee, *Dynamic load balancing of data parallel applications on a distributed network,* Proceedings of 1995 International Conference on Super-computing, Barcelona, pp.170–179, 1995

[13] R. von Hanxleden and L.R. Scott, *Load balancing on message passing architectures,* Journal of Parallel and Distributed Computing, v. 13, pp. 312–324, 1991

[14] R. Diekmann, B. Monien and R. Preis, *Load Balancing Strategies for Distributed Memory Machines,* Parallel and Distributed Processing for Computational Mechanics: Systems and Tools, B.H.V. Topping (ed.), Saxe-Coburg, 1998

[15] T. Decker, M. Fischer, R. Lüling and S. Tschöke, *A Distributed Load Balancing Algorithm for Heterogeneous Parallel Computing Systems,* Proceedings of the 1998 International Conference on Parallel and Distributed Processing Techniques and Applications (PDPTA'98), H. R. Arabnia (ed.), CSREA Press, Volume II, pp. 933–940, 1998

[16] http://www.esi.fr/products/crash/index.html

# Iterative Momentum Relaxation for Fast Lattice-Boltzmann Simulations

D. Kandhai[1], A. Koponen[2], A. Hoekstra[1], and P.M.A. Sloot[1]

[1] Department of Mathematics, Computer Science, Physics and Astronomy,
University of Amsterdam,
Kruislaan 403 NL-1098 SJ Amsterdam,
Netherlands
email:{kandhai, alfons, sloot}@wins.uva.nl

[2] Department of Physics,
University of Jyväskylä,
P.O. Box 35, FIN-40351 Jyväskylä,
Finland
email:{antti.koponen}@phys.jyu.fi

**Abstract.** Lattice-Boltzmann simulations are often used for studying steady-state hydrodynamics. In these simulations, however, the complete time evolution starting from some initial condition is redundantly computed due to the transient nature of the scheme. In this article we present a refinement of body-force driven lattice-Boltzmann simulations that may reduce the simulation time significantly. This new technique is based on an iterative adjustment of the local body-force and is validated on a realistic test case, namely fluid flow in a static mixer reactor.

## 1 Introduction

The lattice-Boltzmann method (LBM) is a mesoscopic approach based on the kinetic Boltzmann equation for simulating fluid flow [1, 2, 3, 4]. In this method fluid is modeled by particles moving on a regular lattice. At each time step particles propagate to neighboring lattice points and re-distribute their velocities in a local collision phase. This inherent spatial and temporal locality of the update rules makes this method ideal for parallel computing [5]. During the last years, LBM has been successfully used for simulating many complex fluid-dynamical problems, such as suspension flows, multi-phase flows, and fluid flow in porous media. All these problems are quite difficult to simulate by conventional methods [3, 6, 7, 8].

However, as most numerical algorithms, the standard lattice-Boltzmann scheme also has its shortcomings. For instance, in a recently performed comparative study between the finite element and the lattice-Boltzmann method for simulating steady-state fluid flow in a SMRX static mixer reactor, it became evident that the computational time (on a sequential machine) required by the lattice-Boltzmann method was higher than that of the finite element method for obtaining the same level of accuracy. The memory requirements on the other hand were lower for the lattice-Boltzmann simulations (details can be found in Ref. [9]). It can be argued that the longer computational time of LBM is a direct consequence of the transient nature of this scheme. In this article, we will present a new technique, namely the Iterative Momentum Relaxation technique (IMR), which can significantly reduce the saturation time. In this technique the body force which is often used to drive a flow in lattice-Boltzmann simulations, is adjusted dynamically by calculating the average loss of momentum due to viscous forces.

In section II we first review the basics of the lattice-Boltzmann method and the IMR technique. In section III we discuss a benchmark application, namely fluid flow in the SMRX reactor, and finally we present the results obtained with the IMR technique.

## 2 Simulation method

### 2.1 The lattice-Boltzmann BGK method

Basically, the time evolution of the lattice-Boltzmann model consists of a propagation phase, where particles move along lattice bonds from a lattice node to one of its neighbors, and a collision phase with a local redistribution of the particle densities subject to conservation of mass and momentum. The simplest and currently the most widely used lattice-Boltzmann model is the so-called lattice-BGK (Bhatnagar-Gross-Krook) model. Here the collision operator is based on a single-time relaxation to the local equilibrium distribution [2, 13].

The time evolution of the lattice-BGK model is given by [13]

$$f_i(\mathbf{r} + \mathbf{c}_i, t+1) = f_i(\mathbf{r}, t) + \frac{1}{\tau}(f_i^{(0)}(\mathbf{r}, t) - f_i(\mathbf{r}, t)), \qquad (1)$$

where $f_i(\mathbf{r}, t)$ is the density of particles moving in the $\mathbf{c}_i$ direction, $\tau$ is the BGK relaxation parameter, $f_i^{(0)}(\mathbf{r}, t)$ is the equilibrium distribution function towards which the particle population is relaxed. The hydrodynamic fields, such as the density $\rho$ and the velocity $\mathbf{v}$, are obtained from moments of the discrete velocity distribution $f_i(\mathbf{r}, t)$ (here $N$ is the number of links per lattice point):

$$\rho(\mathbf{r}, t) = \sum_{i=0}^{N} f_i(\mathbf{r}, t) \qquad \text{and} \qquad \mathbf{v}(\mathbf{r}, t) = \frac{\sum_{i=0}^{N} f_i(\mathbf{r}, t)\mathbf{c}_i}{\rho(\mathbf{r}, t)}, \qquad (2)$$

and a common choice for the equilibrium distribution function is [13],

$$f_i^{(0)} = t_i \rho (1 + \frac{1}{c_s^2}(\mathbf{c}_i \cdot \mathbf{v}) + \frac{1}{2c_s^4}(\mathbf{c}_i \cdot \mathbf{v})^2 - \frac{1}{2c_s^2}v^2), \qquad (3)$$

where $t_i$ is a weight factor depending on the length of the vector $\mathbf{c}_i$, and $c_s$ is the speed of sound. The lattice-Boltzmann model presented here yields the correct hydrodynamic behavior for an incompressible fluid in the limit of low Mach and Knudsen numbers [13].

Beside the computational kernel described above, flow simulations require a consistent set of boundary conditions for the solid walls and the in- and outlets. In Lattice-Boltzmann simulations solid walls are often imposed by using the bounce-back method, while inlet and outlets can be implemented by using pressure/velocity boundaries or body-forces[16, 17]. In the case of pressure/velocity boundaries the particle densities $f_i$ at the inlet and outlet are chosen such that they yield some consistent values for the velocity or pressure. In the body-force approach, which is somewhat restricted to problems with a periodic geometry, the flow is driven by adding a fixed amount of momentum along the flow direction at each lattice point. The overall effect is that a pressure gradient is imposed between the inlet and outlet. For low Reynolds number flows, it has been shown for several benchmark problems that in the stationary state the hydrodynamic behavior of both the body-force and pressure/velocity boundaries are similar[17].

## 2.2 The Iterative Momentum Relaxation (IMR) Technique

As stated in our previous section, lattice-Boltzmann flow simulations are often driven by a body force. According to Newton's second law, the net force acting on the fluid phase during the simulation is equal to the rate of change of the total momentum,

$$\frac{d\mathbf{P}(t)}{dt} = \mathbf{Q} - \mathbf{T}(t), \qquad (4)$$

where $\mathbf{P}(t)$ is the total momentum, $\mathbf{Q}$ is the total body-force and $\mathbf{T}(t)$ is the total viscous friction force due to the obstacles. In standard lattice-Boltzmann simulations the body-force is kept constant during the simulation, while the friction force depends on the velocity field and the geometry of the problem. A steady-state solution is reached when the total body force $\mathbf{Q}$ acting on the fluid is completely cancelled by the viscous friction force $\mathbf{T}$ due to the walls and obstacles.

The main idea of the IMR technique is to reduce the saturation time by adjusting the applied body force during the iteration depending on the change of fluid momentum at the iteration step considered. For some fixed amount of iteration steps (considered as a time interval in IMR) the momentum loss is computed and used to calculate the friction force acting on the fluid during that time interval as follows,

$$\mathbf{T}(t) = \mathbf{Q}(t) - \frac{d\mathbf{P}(t)}{dt}. \qquad (5)$$

The body-force for the next time interval is then set equal to this guess. Notice that in this formulation, the body-force is no longer constant. Moreover, this strategy does not influence the explicit character of the Lattice-Boltzmann algorithm and thus its efficient and easy parallelization.

In summary, the IMR technique can be described by the following algorithm. First a flow is initialized. After every $t_{step}$ time steps, the following iterative procedure (where $k$ denotes the iteration counter of the IMR-loop) is repeated:

1. Calculate the momentum change $(\Delta P)_k$ of the fluid phase in the direction of the body force during the next time step.
2. Calculate the average momentum loss $T_k = Q_k - (\Delta P)_k$ ($Q_k$ is the total body force at the iteration step $k$) of the fluid due to the viscous forces during this time step.
3. Choose a new body force as $Q_{k+1} = T_k$.

The new body force $Q_{k+1}$ accelerates the fluid during $t_{step}$ time steps before returning to step 1. The simulation is carried out until the body force $Q$ and the total momentum reaches an acceptable degree of convergence. This is similar to the heuristical approach for the convergence criteria used in standard lattice-Boltzmann simulations.

## 3 Simulation results

To validate the IMR technique, we have simulated fluid flow in an SMRX static mixer reactor (see Ref. [9] for details). The SMRX static mixer reactor is a technology introduced 15 – 20 years ago, which has gained more and more in popularity within the chemical industry over recent years. It is a plug-flow type reactor filled with a series of SMRX static mixer elements (see Figure 1) turned at 90 degrees with respect to each other.

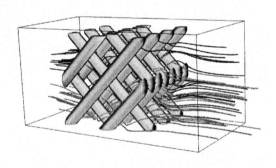

**Fig. 1.** The SMRX static mixer element. The reactor consists of an SMRX element placed in a rectangular duct. The inlet and outlet sections are of the same size as the element itself. The flow is from left to right. The streamlines illustrate the mixing process along the reactor.

The mixer element consists of specially designed stationary obstacles which promotes mixing of fluid flowing through it. Its mixing mechanism relies on splitting, stretching, reordering and recombination of the incoming fluid streams. In this communication we focus on only one SMRX element. Due to usually rather complex flows and geometries, only few 3D numerical simulations of flow through static mixers were performed in the past [15]. We have taken this application as a benchmark, since it is one of the very few cases of fluid flow in complex geometries with well documented results from traditional numerical methods and experimental data.

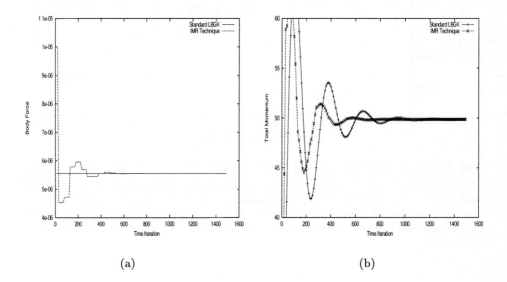

(a)        (b)

**Fig. 2.** Left: Body-force as a function of the time iteration. Right: the time evolution of the total x-momentum along the SMRX reactor for the standard LBGK algorithm and the IMR technique is shown. $\tau = 1$ and the element dimensions are $56 \times 56 \times 56$ lattice points.

The time evolution of the body-force and the total momentum along the flow direction for the lattice-Boltzmann simulations with a constant body-force and the IMR technique is shown in Fig. 2. These simulations were performed for an element discretization of $56 \times 56 \times 56$ lattice points and the relaxation parameter $\tau$ was equal to 1. It is clear that the damping of the oscillatory behavior of the momentum is enhanced by the IMR technique. This is a result of the feedback of the flow field on the body-force. Moreover, both approaches clearly converge to the same value for the total momentum.

The time evolution of the relative difference in the total momentum along the flow direction, $\frac{|\Delta P_x|}{|P_x|}(t)$ ($\Delta P_x$ is computed between two results of two successive IMR trials), for the standard LBGK algorithm and the IMR technique is shown in Fig. 3. From this figure, it is evident that with the IMR method the relative difference converges faster to some level of tolerance.

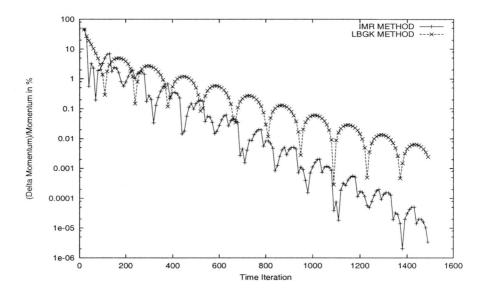

**Fig. 3.** The relative difference (in %) in the total momentum along the flow direction, $\frac{|\Delta P_x|}{|P_x|}(t)$, as a function of time, for the standard LBGK algorithm and the IMR technique. $\tau = 1$ and the element dimensions are $56 \times 56 \times 56$ lattice points. In both cases the oscillatory behavior is due to a non-zero initial velocity field. In the case of the IMR technique more oscillations are present due to the iterative refinement of the body-force.

In Fig. 4 we show the relative difference in the mean velocity along the reactor (in %) for different time-steps, in the case of the standard LBGK method (on the left) and the IMR technique (on the right). As reference data we have used the simulation results obtained after 1500 time-steps, as then the simulations were completely saturated in both cases. With the IMR technique 1% accuracy in the velocity and the pressure fields compared to the reference data, was already reached after 550 time steps, whereas the constant body-force method required around 1000 time steps to reach a similar accuracy. Moreover, the steady state solution of both approaches are very close to each other (data not shown). The relative difference in the mean velocity along the reactor, between the stationary state of both approaches, is smaller then 0.07%. In Ref. [9] we have shown in detail that the results of the standard LBGK method are also in good agreement with Finite Element calculations and experimental data. Thus we can conclude that the results obtained by the IMR technique are also consistent with experimental data.

In this test case we have used $t_{step} = 50$. Tests with some other values of $t_{step}$ did not show significant improvements in the overall benefit gained by the IMR technique. Similar speed up results were also found for other Reynolds numbers provided that the flow is laminar. More detailed investigation and application of the IMR technique to other problems will be reported elsewhere[17].

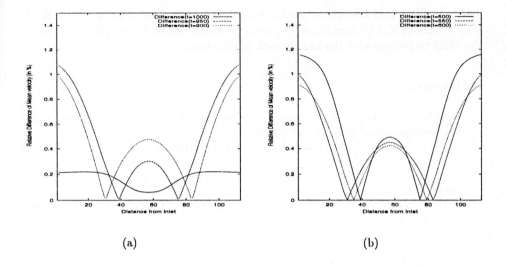

**Fig. 4.** On the left the relative difference in the mean velocity (in %) of the standard LBGK method is shown for $t = 900$, $t = 950$ and $t = 1000$. On the right the relative difference in the mean velocity (in %) of the IMR technique is shown for $t = 500$, $t = 550$ and $t = 600$. In both cases the relative difference is computed with respect to the simulation result at $t = 1500$ time-steps (simulation is then completely saturated) and the mean velocity is computed at different cross-sections along the reactor. $\tau = 1$ and the element dimensions are $56 \times 56 \times 56$ lattice points. The relative error is higher at the inlet and outlet, because the mean-velocity is smaller at those locations.

## 4 Conclusions

In many lattice-Boltzmann simulations, the complete time evolution of the system is computed with a constant body force starting from some initial velocity and pressure fields. The number of time steps which is required to reach the steady state can then be very large in some cases. We presented a new technique for reducing the number of time steps that is needed to reach the steady state for body-force driven flows. This strategy does not influence the explicit character of the Lattice-Boltzmann algorithm and thus its efficient and easy parallelization. We conclude that at least in problems involving laminar flow, the IMR technique can be very efficient in decreasing the number of time steps needed to reach the steady state.

## Acknowledgments

This work was partly carried out within the MPR (Massive Parallel Computing) project "Many Particle Systems" funded by the Dutch foundation for basic research. We would also like to thank Robert Belleman, David Vidal, Huub Hoefsloot, Markku Kataja and Jussi Timonen for many useful discussions concerning the IMR technique and the benchmark application.

## References

1. R. Benzi, S. Succi and M. Vergassola, The lattice-Boltzmann equation - theory and applications, *Phys. Rep.*, **3**, 145 (1992).
2. S. Chen, Z. Wang, X. Shan and G. Doolen, Lattice-Boltzmann computational fluid dynamics in three dimensions, *J. of Stat. Phys.*, **68**, 379 (1992).
3. D. H. Rothman and S. Zaleski, *Lattice gas cellular automata*, Cambridge University Press, (1997).
4. S. Chen and G.D. Doolen, Lattice Boltzmann method for fluid flows, *Annu. Rev. Fluid Mech.*, **30**, 329 (1998).
5. D. Kandhai, A. Koponen, A. Hoekstra, M. Kataja, J. Timonen and P.M.A. Sloot, Lattice-Boltzmann hydrodynamics on parallel systems, *Comp. Phys. Commun.*, **111**, 14 (1998).
6. A. Koponen, D. Kandhai, E. Hellén, M. Alava, A. Hoekstra, M. Kataja, K. Niskanen, P. Sloot and J. Timonen, Permeability of three-dimensional random fiber webs, *Phys. Rev. Lett.*, **80**, 716 (1998).
7. D.S. Clague, B.D. Kandhai, R. Zhang and P.M.A. Sloot, On the hydraulic permeability of (un)bounded fibrous media using the Lattice-Boltzmann method, *Submitted*.
8. J.A. Kaandorp, C. Lowe, D. Frenkel and P.M.A. Sloot, Effect of nutrient diffusion and flow on coral morphology, *Phys. Rev. Lett.*, **77**, 2328 (1996).
9. D. Kandhai, D. Vidal, A. Hoekstra, H. Hoefsloot, P. Iedema and P.M.A. Sloot, Lattice-Boltzmann and finite element simulations of fluid flow in a SMRX mixer, *Int. J. Num. Meth. Fluids*, accepted for publication.
10. G. McNamara and G. Zanetti, Use of the Boltzmann equation to simulate lattice-gas automata, *Phys. Rev. Lett.*, **61**, 2332 (1988).
11. F.J. Higuera and J. Jemenez, Boltzmann approach to lattice gas simulations, *Europhys. Lett.*, **7**, 663 (1989).
12. F.J. Higuera, S. Succi and R. Benzi, Lattice gas-dynamics with enhanced collisions, *Europhys. Lett.*, **9**, 345 (1989).
13. Y.H. Qian, D. d'Humieres and P. Lallemand, Lattice BGK models for Navier-Stokes equation, *Europhys. Lett.*, **17**, 479 (1992).
14. A. J. C. Ladd, Numerical simulations of particulate suspensions via a discretized Boltzmann equation. Part 1. theoretical foundation, *J. Fluid Mech.* **271**, 285 (1994); A. J. C. Ladd, Numerical simulations of particulate suspensions via a discretized Boltzmann equation. Part 2. numerical results, *J. Fluid Mech.* **271**, 311 (1994).
15. E.S. Mickaily-Huber, F. Bertrand, P. Tanguy, T. Meyer, Albert Renken, Franz S. Rys and Marc Wehrli, Numerical simulations of mixing in an SMRX static mixer, *The Chem. Eng. J.*, **63**, 117-126 (1996).
16. S. Chen, D. Martinez and R. Mei, On boundary conditions in lattice Boltzmann methods, *Phys. Fluids*, **8**, 2527 (1996).
17. D. Kandhai, A. Koponen, A. Hoekstra, M. Kataja, J. Timonen and P.M.A. Sloot, Implementation Aspects of 3D lattice-BGK: Boundaries, Accuracy and a Fast Relaxation Method, *Submitted*.

# Lattice Gas:
# An Efficient and Reusable Parallel Library
# Based on a Graph Partitioning Technique

Alexandre Dupuis and Bastien Chopard

University of Geneva, CUI, 24, rue Général-Dufour, CH - 1211 Geneva, Switzerland
[Alexandre.Dupuis|Bastien.Chopard]@cui.unige.ch,
WWW home pages: http://cuiwww.unige.ch/~[dupuis|chopard]/

**Abstract.** We present a parallel library which can be used for any lattice gas (LG) application. A highly reusable implementation, as well as a general parallelization scheme based on graph partitioning techniques are developed. We show that the performance we obtain with our approach compares favorably with the plain, classical implementation of LG models on regular domains whereas on irregular domains, it can even be better. We propose a theoretical expression for the execution time and we validate our analysis in the case of the problem of wave propagation in urban areas.

## 1 Introduction

The lattice gas (LG) technique is a fast developing numerical tool aimed at simulating, on a computer, various physical phenomena such as complex fluid flows, reaction-diffusion systems or wave propagation processes [1,2]. In the present terminology, LG systems consist of the cellular automata models known as lattice gas automata and lattice Boltzmann (LB) methods that are now viewed as serious competitors to traditional CFD techniques.

From a computational point of view, a LG model is defined on a regular $d$-dimensional lattice, with $z$ main directions labeled $v_i$, $i = 1, ..., z$. For instance, on a 2D square lattice, $z = 4$ and the $v_i$'s correspond to unity vectors pointing north, west, south and east respectively. On each lattice site $r$ are attached $z$ variables $f_i$ (boolean, integer or real), which travel in lattice direction $v_i$. The system evolves in discrete time steps $t = 0, 1, ...$ . At each time step, all the $f_i$'s entering the same site interact according to some collision operator $\Omega$ whose effect is to produce new values for the $f_i$ before they propagate to the nearest neighbor sites. The collision term $\Omega$ is chosen according to the physical system under study.

The dynamics of a LG model can thus be expressed as

$$f_i(t+1, r+v_i) = f_i(t, r) + \Omega_i(f(r, t)) \tag{1}$$

The computation takes place on a spatial domain $D$ which can be very sparse (or irregular) depending on the boundary condition which is used. This is the case, in particular for a flow in porous media, or for wave propagation in disordered systems.

The time evolution in equation (1) is usually split into two phases: (i) a collision step in which $\Omega_i(f)$ is computed in parallel at all sites $r$; (ii) a propagation step in which all $f_i$'s are moved to the appropriate neighbors on the lattice. Phase (ii) can be performed in parallel and requires inter-processor communications.

In the LG community performance issues are paid more attention than software engineering concepts such as reusability or clarity. Adding too many software layers in the computer implementation can be very slow and consequently not adapted. On the other hand, in view of producing a general library for LG models, some reusability is highly desirable in order to be able to change the lattice topology or the collision term. In such a library one also has to deal in a generic way with boundary conditions, i.e. with computational domains $D$ which can be quite irregular.

To illustrate these two trends, see for example [3] and [4]. These two implementations have both a quality. The first one is efficiently coded while the second one is much more reusable and elegant, but rather slow.

Due to the success of the LG approach and the large range of possible applications, there is a growing demand for an efficient and reusable tool, saving developing time without degrading too much the CPU performance. An interesting attempt is realized in [5] for cellular automata. A working environment and a specific language that extends C are provided. This system is however too general to ensure optimal performance for LG models.

In this paper we propose a parallel object oriented library for LG applications. It uses a graph partitioning technique to decompose the domain across several processors. Any lattice topology and collision rule can be implemented with the same software structure and any irregular domain can be treated efficiently. We also show that, when the domain is sparse enough, our approach is faster than the direct LG implementation which extends the computation to a rectangular domain and masks the site that should not be present.

## 2 Library implementation

### 2.1 Motivation

In general a 2D LG application with any neighbor topology is represented as a 2D array (called a matrix in what follows) in the computer. The main advantage of this data structure is the intrinsic ease to access any site and its neighbors. Therefore, fast memory access is possible without indirection. On the other hand, a matrix requires us to keep intact the space even if it presents some "holes" such as obstacles in the simulation of porous media flows, or buildings in a wave propagation simulation in a city. In what follows, we shall call this approach the classical method.

Here we propose to represent the lattice with an object oriented approach. We consider a vector of cells instead of a matrix. Every cell knows its neighborhood through a list of indexes relative to the vector. This vector can be seen as a flattened matrix where we have added the topological information.

With this representation the unused areas (holes) are no longer included in the computation and domains of any shape can be treated equally well. Also, it is easy to remove an arbitrary number of cells, move a block of cells of any shape to another location or to another processor. Finally a lattice traversal is straightforwardly executed.

## 2.2 The Object Oriented Model

In our Object Oriented Model (OOM), there are three main entities: the cell, the lattice and the (physical) model. A cell is an entity which contains the $f_i$'s, the neighborhood information and methods. For polymorphic reasons, an abstract virtual class called VCell is defined. Every usable cell has then to inherit from VCell and it will be of type VCell as well as of its own type. A lattice is a vector of cells. In order to regroup common methods, an abstract virtual lattice called VLattice is defined. Every lattice inherits from it and has to define the abstract methods specified in VLattice. This hierarchy is depicted in figure 1.

With these two building blocks, we can construct a model, i.e. any lattice gas application. A model uses a specific lattice by inheriting from it and implements the abstract methods.

## 2.3 Parallelization

**Domain decomposition** We want to provide a general domain decomposition applicable onto every lattice (i.e. onto a VLattice). We choose to divide the space into $k$ subsets. To ensure load balancing in a simple way, we can set $k = \alpha p$, where $p$ is the number of processors and $\alpha$ is a positive constant. The subsets are then distributed randomly to the processors in a round robin way. After this distribution every processor will then have $\alpha$ subsets, corresponding to all regions of the domain. However, this technique is not scalable and, if $p$ becomes large, one has to chose $\alpha = 1$ to ensure locality of the communications.

In most current LG implementations, the subsets are rectangular areas (e.g. bands or squares). From a compactness point of view (i.e. communication to computation ratio) this is not the best choice especially when the domain is sparse.

The general decomposition problem can be formulated as follows: find $k$ subsets such that the number of cells with a neighbor in a different subset is minimized. This is nothing else than the graph partitioning problem that we shall solve for the domain decomposition of the OOM.

**Graph partitioning problem** The graph partitioning problem (GPP) amounts to looking for $k$ subsets of a graph where the number of cut edges is minimized.

More formally, let $G = (V, E)$ be a graph where $V$ is a set of $n$ nodes and $E$ is a set of $e$ edges. Let $\Omega = (\omega_{ij})$ with $i, j = 1, \cdots, n$ be a connectivity matrix

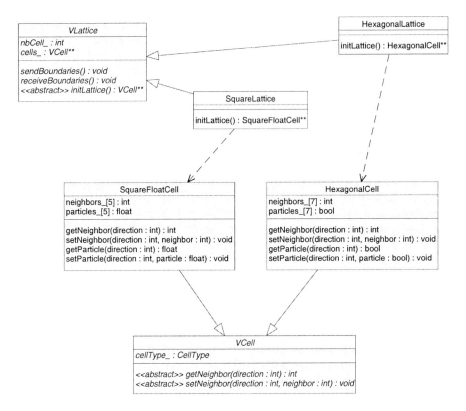

**Fig. 1.** The hierarchy of the Object Oriented Model (OOM). The attributes and methods depicted are only the basic ones (more can be added). The upper part of each box is the name of the class, the middle part contains the attributes and the lower part is used for the methods of the class. A solid arrow indicates an inheritance relation and a dashed arrow indicates an utilization relation.

describing $E$. Let $k$ be a positive integer. A $k - way$ partition of $G$ is a set of disjunct node subsets $s_1, \cdots, s_k$ such that $\bigcup_{i=1}^{k} s_i = G$.

For any $v \in V$, we define the function $\sigma$ as: $\sigma(v) = b$, if $v \in s_b$. The cost of a partition is defined as

$$C = \sum_{\forall ij, \sigma(i) \neq \sigma(j)} \omega_{ij}$$

The GPP consists in minimizing $C$, for a given $k$. This problem is known to be NP-Complete [6]. Indeed, a combinatorial calculation already shows that for a small graph of 40 vertices there are $10^{20}$ 4-way possible partitions.

Therefore, to solve this problem one has to find relevant heuristics. After the famous Kernighan-Lin algorithm [7] in 1970 many interesting approaches have been proposed such as greedy heuristics [8–10], genetic algorithms [11, 12], simulated annealing techniques [13], spectral partitioning methods [14] or multilevel algorithms [15].

Two important criteria of such methods are the cost of the partition computation and the quality of the partition. The multilevel algorithms seem to well satisfy these criteria [15, 16].

There are several available libraries solving the GPP. The most famous ones are Metis [17, 18], Chaco [19], Party [20] and Scotch [21]. For our case, the Metis library which implements multilevel algorithms, turned out to be the best one. Indeed, our tests show that Metis is faster and that its partitions are well adapted to our needs. To illustrate the use of the Metis library, we present in figure 2 a 20-way partition of an irregular domain.

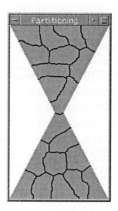

**Fig. 2.** The 20-way partition obtained with the use of the Metis library on a single processor of an IBM SP2 in 2.1 seconds. This is an illustration of the graph partitioning problem applied to an irregular domain composed of 48241 vertices and 143278 edges. The subset boundaries are depicted in black while the subset are shown in gray. The white areas do not belong to the lattice.

**Communications** In the classical approach, where the domain $D$ is represented as a $d$-dimensional array and the irregularities are implemented by masking the corresponding sites, interprocessor communications are easily performed by exchanging regular array section between adjacent processors.

In the OOM library, the communication step consists of building a communication buffer for every processor belonging to the neighborhood of each subset boundary.

We use asynchronous, non-blocking communications in which communications overlap with the computation. To this end, one proceeds as follows: (i) the boundary cells are sent out; (ii) the inside part of each subdomain is updated; (iii) the neighbor cells are received from the remote processors and the subset boundaries updated. In most LG applications, the communications are totally overlapped with the collision step.

## 3 Parallel performances

### 3.1 Outline

In order to compare the OOM with the classical method (full array, block partitioning) we consider the run time complexity of both methods. For the sake of the comparison, we assume a 2D domain. For the classical method we also assume that the domain is decomposed into bands. The communications are then composed of the lines between each band. Non-blocking message passing primitives are also used to overlap communications with computations. In this method, even if the domain is sparse, every band has the same size. This implies a surplus of time for the domain traversal as well as data to communication.

### 3.2 Theoretical models of performances

To evaluate the execution time of the three phases: collision, in-processor propagation and inter-processor propagation (communication), we introduce the following notation: $n^2$ is the number of sites, $p$ is the number of processors and $\delta$ is a real value $\in [0; 1]$ expressing the domain irregularity, i.e. the fraction of the $n^2$ sites which actually belong to the domain. A theoretically execution time for the three stages is given below. We start with the case of OOM.

**The OOM library:**

- **Collision:** $\frac{n^2}{p}(1-\delta)$ sites are computed in a $T_{comp1}$ time.
- **Propagation:** $\frac{n^2}{p}(1-\delta)$ sites are shifted in a $T_{shift1}$ time.
- **Communication:** The length of the subset boundary can be estimated as $\alpha\sqrt{\frac{n^2}{p}(1-\delta)}$ where $\alpha = 4$ if the subset is a square and $\alpha = 2\sqrt{\pi}$ if the subset is a disk. We suppose that there are $(p-1)$ buffers to build and that the cell distribution among the processors is uniform. Then the time to build

the messages is $\alpha\sqrt{\frac{n^2}{p}(1-\delta)}T_{pack}$. The non-blocking communications are started up in a time $T_{startup1}$ and executed in a time $\alpha\sqrt{\frac{n^2}{p}(1-\delta)}T_{comm}$.

Thus, the total theoretical execution time can be written as

$$T_{OOM}(p,n,\delta) = \frac{n^2}{p}(1-\delta)T_1 + \alpha\sqrt{\frac{n^2}{p}(1-\delta)}T_{pack}$$

$$+ \max(0, \alpha\sqrt{\frac{n^2}{p}(1-\delta)}T_{comm} - T_{col1}) + T_{startup1} \qquad (2)$$

where $T_1 = T_{comp1} + T_{shift1}$ and $T_{col1} = \frac{n^2}{p}(1-\delta)T_{comp1}$.

**Classical method** The theoretical execution time for the three stages can be expressed as

- **Collision:** $(1-\delta)\frac{n^2}{p}$ sites are computed in a $T_{comp2}$ time and $\delta\frac{n^2}{p}$ are only visited in a $T_{visit}$ time.
- **Propagation:** $(1-\delta)\frac{n^2}{p}$ sites are shifted in a $T_{shift2}$ time and $\delta\frac{n^2}{p}$ are only visited in a $T_{visit}$ time.
- **Communication:** the non-blocking communications are started up in a time $T_{startup}$ and executed in a time $2nT_{comm}$ as the length of the band boundary is equal to $n$.

The execution time for this classical method can be summarized as

$$T_{classic}(p,n,\delta) = \frac{n^2}{p}(T_2 - \delta(T_2 + 2T_{visit}))$$

$$+ \max(0, 2nT_{comm} - T_{col2}) + T_{startup} \qquad (3)$$

where $T_2 = T_{comp2} + T_{shift2}$ and $T_{col2} = \frac{n^2}{p}(1-\delta)T_{comp2} + \delta\frac{n^2}{p}T_{visit}$.

## 3.3 Measured performance

**A typical application** To validate the theoretical model of performances, we choose the so-called ParFlow application [22, 2] (wave propagation in a city) which requires a square lattice and a simple collision operator: one has typically

$$f_i(t+1, \mathbf{r}+\mathbf{v}_i) = \frac{\mu}{2}\sum_{j=1}^{4} f_j - f_{i+2}$$

where $\mu$ is an attenuation coefficient (for example $\mu = 1.0$ means that there is no obstacle while $\mu = 0.0$ means that there is a building). If the wave is reflected on the building walls, it is unnecessary to compute the wave inside the buildings. $\delta$ corresponds to the ratio between the total building area and the total area.

**Data fits and discussion** Figure 3 shows the actual performance of the two implementations of the ParFlow method. For $\delta = 0$, the plain method is twice faster than the OOM implementation. However, as the irregularity of the domain increases, the two execution times intersect and OOM becomes faster. The value of $\delta$ for which this happens depends on the application. As shown here for the ParFlow application, the value of $\delta$ weakly depends on $p$. A more detailed analysis is under study.

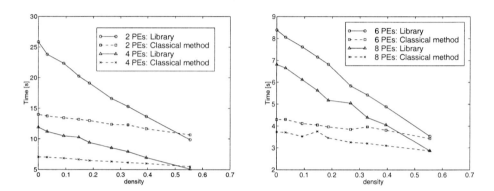

**Fig. 3.** Execution time as a function of the building density, for several number of processors, on a 2D array of size $n = 512$, for 100 iterations. Solid lines show the performance of OOM, whereas dashed lines corresponds to the plain, classical approach.

The performance model proposed in the previous section can be validated with the present time measurements. Let us rewrite (2) and (3) for given values of $n$ and $p$. For $n$ large enough, the collision time is greater than the communication time and the last term of (2) and (3) vanishes. Thus

$$T_{library}(\delta) = c_1(\sqrt{1-\delta})^2 + c_2(\sqrt{1-\delta}) + c_3 \qquad (4)$$
$$T_{classic}(\delta) = c_4\delta + c_5 \qquad (5)$$

with $c_1 = \frac{n^2}{p}T_1$, $c_2 = \alpha\sqrt{\frac{n^2}{p}}T_{pack}$, $c_3 = T_{startup1}$, $c_4 = -\frac{n^2}{p}(T_2 + 2T_{visit})$ and $c_5 = T_{startup2} + \frac{n^2}{p}T_2$.

Using these equations, the values of the $c_i$ can be extracted from the measured performance trough a mean-square data fit. Figure 4 shows the prediction of (4) and (5) as a function of the building densities $\delta$, for various number of processors $p$. We see that the theoretical curves fit pretty well the measured times.

## 4 Conclusion

This paper presents an elegant and efficient parallel implementation of a library which can be used for any lattice gas applications. Its hierarchical modeling of

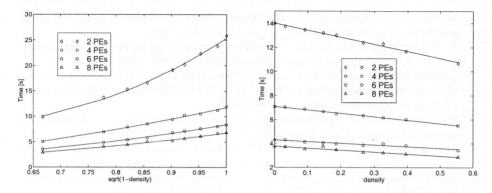

**Fig. 4.** Fit (solid line) of the measured execution times (isolated points) for various number of processors and various building densities according to equations (4) and (5). The left hand side shows the OOM times, while the execution times of the classical method are depicted on the right.

the objects implies a high reusability. A general parallelization scheme, based on a graph partitioning method is proposed.

The performance of our library is described by a general and theoretical model giving the parallel execution time in terms of the number of processors, problem size and irregularity degree.

We have validated our analysis by considering a specific application consisting of simulating wave propagation in a city. We have shown that the performance offered by our flexible and reusable implementation is at worst a factor two slower than the classical approach. When the domain irregularity is large enough, the OOM library may even surpass it.

## Acknowledgment

This research is granted by the Swiss National Science Foundation.

## References

1. D. Rothman and S. Zaleski. *Lattice-Gas Cellular Automata: Simple Models of Complex Hydrodynamics.* Collection Aléa. Cambridge University Press, 1997.
2. B. Chopard and M. Droz. *Cellular Automata Modeling of Physical Systems.* Cambridge University Press, 1998.
3. Pascal O. Luthi. *Lattice Wave Automata.* PhD thesis, University of Geneva, 1998.
4. Frédéric Guidec, Patrice Calégari, and Pierre Kuonen. Parallel irregular software for wave propagation simulation. In *HPCN'97 High-Parallel Computing and Networking*, Lecture Notes in Computer Science, pages 84–94. Springer-Verlag, 1997.
5. S. Di Gregorio, R. Ringo, W. Spataro, Giandomenico Spezzano, and Domenico Talia. A parellel cellular environment for high performance scientific computing. In H. Liddell at al., editor, *HPCN'96 High-Peformance Computing and Networking*, pages 514–521, Berlin, 1996. Springer-Verlag.

6. M. Garey, D. Johnson, and L. Stockmeyer. Some simplified NP-complete graph problems. *Theoritical Computer Science*, 1:237–267, 1976.
7. B. W. Kernighan and S. Lin. An efficient heuristic procedure for partitioning graphs. *The Bell system technical journal*, 49(1):291–307, 1970.
8. Roberto Battiti, Alan Bertossi, and R. Rizzi. Randomized greedy algorithms for the hypergraph partitioning problem. In *DIMACS Workshop on Randomization Methods in Algorithm Design*, October 1997.
9. Roberto Battiti and Alan Bertossi. Greedy, prohibition, and reactive heuristics for graph partitioning. *IEEE Transactions on Computers*, to appear.
10. M.Laguna, T.A. feo, and H.C. Elrod. A greedy randomized adaptative search procedure for the two-partition problem. *Operations Research*, 42:667–687, 1994.
11. Thang Nguyen Bui and Byung Ro Moon. Genetic algorithm and graph partitioning. *IEEE Transactions on Computers*, 45(7):841–855, July 1996.
12. Gregor von Laszewski and Heinz Mühlenbein. Partitioning a graph with a parallel genetic algorithm. In *Parallel problem solving from nature*, pages 165–169, 1991.
13. D.S. Johnson, C.R. Aragon, L.A. McGeoch, and C. Schevon. Optimization by simulated annealing: An experimental evaluation. *Operations Research*, 37:865–892, 1989.
14. Alex Pothen, H.D. Simon, Lien Wang, and Stephen T. Bernard. Towards a fast implementation of spectral nested dissection. In *Supercomputing '92*, pages 42–51, 1992.
15. George Karypis and Vipin Kumar. A fast and highly quality multilevel scheme for partitioning irregular graphs. Technical Report 95-035, Departement of Computer Science, University of Minnesota, 1995.
16. Robert Leland and Bruce Hendrickson. An empirical study of static load balancing algorithms. In *Scalable High-Performance Computing Conference (SHPCC'94)*, pages 682–685, 1994.
17. http://www-users.cs.umn.edu/~karypis/metis/.
18. George Karypis and Vipin Kumar. *A Software Package for Partitioning Unstructured Graphs, Partitioning Meshes, and Computing Fill-Reducing Orderings of Sparse Matrices*, November 1997.
19. ttp://www.cs.sandia.gov/CRF/chac.html.
20. http://www.uni-paderborn.de/cs/robsy/party.html.
21. http://www.labri.u-bordeaux.fr/Equipe/ALiENor/membre/pelegrin/scotch/.
22. B. Chopard, P.O. Luthi, and Jean-Frédéric Wagen. A lattice boltzmann method for wave propagation in urban microcells. *IEE Proceedings - Microwaves, Antennas and Propagation*, 144:251–255, 1997.

# Algorithms of Parallel Realisation of the PIC Method with Assembly Technology

M.A.Kraeva, V.E.Malyshkin

Supercomputer Software Department
Institute of Computational Mathematics and
Mathematical Geophysics, Russian Academy of Sciences
6, pr. Lavrentieva, 630090, Novosibirsk, Russia
{ kraeva, malysh }@ ssd.sscc.ru

**Abstract.** The main idea of the Assembly Technology and its application to parallelisation of the Particle-In-Cell (PIC) method is considered. The algorithms of the PIC method realisation for multicomputers are presented. Dynamic load balancing for the PIC method realisation is discussed. The PIC realisation with the assembly technology is based on construction of a fragmented parallel program which is able to send its fragments for execution in underloaded processor nodes of multicomputer. Assignment of a fragment for execution on a processor element is done dynamically in the course of execution. This is the basis of the dynamic load balancing algorithm.

## 1. Introduction

The PIC method is a challenge for application of parallel computing technologies. Generally, the direct modelling of the physical phenomenon on the basis of description of the phenomenon behaviour in the local area instead of solution of a very complex system of differential equations in the whole multidimensional space (usually more than 3 dimensions) is now a very promising direction of investigation. Such modelling demands realisation of numerical methods with irregularity and even a dynamically changing irregularity of data structure. These methods are very difficult for parallelisation (adoptive mesh, variable time step, particles, etc.) and high performance realisation..

The Assembly Technology (AT) has been specially developed for the realisation of numerical algorithms on multicomputers [1-3]. Fragmentation and dynamic load balancing are the key features of the AT.

## 2. The PIC method

The PIC methods have numerous applications in the simulation of plasma by modelling of plasma as a set of a huge number of test particles and following the evolution of orbits of individual test particles in the self-consistent electromagnetic field. A real physical space is represented by a model of simulation domain (space of modelling - SM). The whole SM is divided by the rectangular grid into cells, upon which the electric $E$ and magnetic $B$ fields are discretised. Values of the fields at any point inside the cell are calculated by interpolation.

Each charged particle is characterised by its co-ordinates and velocities. Instead of solution of equations in the 6D space of co-ordinates and velocities, the dynamics of the system is followed by integrating the equation of motion of every particle in a series of discrete time steps. A particle motion is driven by the Lorentz force, which depends on the electric and magnetic fields. Thus, at the *first phase* of each step for every particle, the Lorentz force is calculated from the values of electromagnetic fields at the nearest grid points (only the values of electrical and magnetic fields inside the cell, where the particle is located, are used for calculations). It means that at any moment of modelling a particle belongs to a certain cell. At the *second phase*, the new values of velocities and co-ordinates of all the particles are calculated. So, a particle can fly from one cell to another. The sizes of a time step and of a cell are chosen in such a manner that a particle cannot fly farther than into the adjacent cell at one step of modelling. Particle motion generates the current density $U$, which is also discretised upon the rectangular grid (*the third phase*). At the last phase, the values of the electric $E$ and the magnetic $B$ fields inside SM are recalculated from the current density $U$ by solving Maxwell's equations. Only the values of the electrical and the magnetic fields inside the cell and its adjacent cells are involved in computations. The number of time steps is defined by a physical experiment.

Let us notice that none of the particles affects another particle. Hence, the calculations at the first three phases are done independently.

The basic algorithm for this electromagnetic PIC code is as follows:

Set the initial values of the variables describing the particles and the fields; t=0.
Time loop {
   t:=t+Δt;
   For each particle {
      1. Interpolate the electromagnetic field to the particle position to obtain the force affecting the particle (gather phase);
      2. Calculate the new coordinates and velocity of the particle;
      3. Calculate the charge carried by a particle to the cell vertices within the time step to obtain the current charge and density (scatter phase) }
   4. Solve Maxwell's equations to update the electromagnetic field }

A more detailed description of the PIC method can be found in [4,5].

## 3. Problems of the PIC realisation

Plasma simulation requires high performance and the large volume of memory. The grid may comprise several hundred thousand points. To study the problem of plasma cloud expansion in magnetized background with non-uniform magnetic field, we used the 60x60x100 grid. 14 million particles were used for the modeling. This motivates desirability to implement the PIC codes on the powerful supercomputers.

On the other hand, the PIC algorithm has the great possibility for the parallelisation, because all the particles move independently. The volume of computations at the first three phases of each time step is proportional to the number of particles. About 90% of multicomputer resources are spent for the particle processing. Thus, to implement the PIC code on multicomputer, the equal numbers of particles should be assigned to each processor element (PE). However on the MIMD distributed memory multicomputers, the performance characteristics of the PIC code crucially depend on how the grid and particles are distributed among the PEs. In order to decrease the communication overheads at the first and the third phases of a time step it is required that the PE contains both cells (values of the electromagnetic fields at the grid points) and the particles located inside them. Unfortunately, in the course of modelling, some particles might fly from one cell to another. To satisfy the previous requirement, two basic decompositions can be used.

In the so-called Lagrangian decomposition the equal number of particles are assigned to each PE with no regard for their position in the SM. In this case the values of the electromagnetic fields, the current charge and density at all the grid points should be copied in the every PE. Otherwise, the communication overheads at the first and the third phases will decrease the effectiveness of parallelisation. Disadvantages of the Lagrangian decomposition are the following:

- strict memory requirements;
- communication overheads at the forth phase (to update the current charge and density).

•In the Eulerian decomposition, each PE contains a fixed rectangular subdomain, including electromagnetic fields at the corresponding grid points and particles in the corresponding cells. If a particle leaves its subdomain and flies to another subdomain in the course of modelling, then this particle should be transferred to the PE containing this latter subdomain. Thus, even with the equal initial workload of Pes, in several steps of modelling, some PEs might contain more particles than the others. This results in the load imbalance. The character of the particles motion does not fully depend on equations, but also on initial particles distribution. For example, if all the active particles are concentrated inside a single cell that might initiate explosion of plasma cloud.

It is clear that the parallel implementation of the PIC method on the distributed memory multicomputers demands the dynamic load balancing.

## 4. Assembly Technology

The Assembly Technology (AT) was applied to the PIC method parallelisation. Numerical methods, generally, and the PIC method, in particular, are very suitable for the application of the AT.

The Assembly Technology includes different aspects of solving an application problem on multicomputers. Here, only the fragmentation aspect will be discussed. In accordance with the AT the description of an application problem should be divided into a system of reasonably small fragments (fragments of different size for solution of different problems), representing the realisation entities of a problem description. The program, which realises a fragment (P_fragment) contains both data and the code. In other words, a program, realising an application problem, is assembled out of small P_fragments of computations connected through variables for data transfer.

The fragmented structure of an application parallel program is kept in the executable code and provides the possibility for organisation of flexible and high performance execution of a program. The general idea is the following. The assembled program is represented in a programming language, and is implemented on multicomputer as a system of executable P_fragments. If these fragments are small enough, then initially, for each PE of multicomputer, the equal workload is assembled out of the P_fragments. After that the program of a PE is executed, looping over all the P_fragments, assigned to this PE.

If the workload of a PEs changes in the course of computation, and, at least, one of PE becomes overloaded, then a part of P_fragments, that were assigned for execution in the overloaded PE, should fly to the less loaded PEs equalizing the workload of multicomputer PEs [6]. The dynamic load balancing of multicomputer is based on such a fragmentation.

## 5. Parallelisation of the PIC method

Now let us consider the assembly approach to the PIC method parallelisation in order to demonstrate the application of AT.

A layer of the SM was chosen as a minimal fragment for implementation of PIC on the line of PE. For implementation on the grid of PE, a column was chosen as minimal fragment. These fragments consist of the adjacent cells of SM (Figs . 1, 2). Such fragments contain both data (particles inside the cells of a fragment and values of electromagnetic fields at their grid points) and the procedures which operate with these data. When a particle moves from a cell of one fragment to a cell of another fragment, it should be removed from the first fragment and added to another one. Thus, we can say that AT uses the Eulerian decomposition. In such a way, the three-dimensional simulation domain is initially partitioned into N blocks (where N is the number of PEs), which contain approximately the same number of particles (Figs . 1, 2). When the load imbalance is crucial, some layers (columns) of an overloaded block are transferred to another less loaded block (from one PE to another). In the course of modelling, the adjacent blocks are located in the linked PEs. Therefore, the adjacent cells are located in the same or in the linked PEs. It is important for the second phase, at which some particles can fly from one cell into another and for the fourth phase, when for recalculation of values of electromagnetic fields in a certain cell, values in the adjacent cells are also used.

Let us consider the case of the grid of PEs in more detail. Let the number of PEs be equal to $l \times m$. Then SM is divided into $l$ blocks orthogonal to some axis. Each block consists of several adjacent layers and contains about *NP/l* particles. The Block_i is assigned for processing to the i-th row of the 2-D grid (Fig. 2). Blocks are formed as to provide an equal total workload of every PEs row of the processor grid. Then every block_i is divided into *m* sub-blocks *block_ij*, which are distributed for processing among *m* PEs of the row. These sub-blocks are composed as to provide an equal workload of every PE of the row. If overloading of at least one PE occurs in the course of modelling, this PE is able to recognise it at the moment when the number of particles substantially exceeds *NP/(l×m)*. Then this PE initiates the rebalancing procedure.

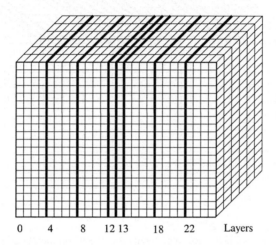

**Fig. 1.** Decomposition of SM for implementation of PIC on the line of PEs

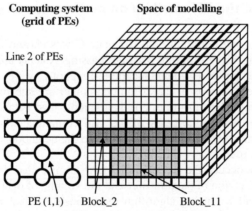

**Fig. 2.** Decomposition of SM for implementation of PIC on the 2D grid of PEs

In the case of the hypercube communication topology the 2D grid of PEs can be mapped into the hypercube keeping the neibourhood of PEs. Thus, the same technique as in the case of the grid can be used.

If the number of layers $k \approx N$ (or $k \approx l$ in the case of the grid of PEs), it is difficult to divide the SM into blocks with the equal number of particles. Also, if particles are concentrated inside a single cell, it is impossible to divide SM into equal subdomains.

In order to achieve the better load balance, the following modified domain decomposition is used. The adjacent layers are copied in, at least, 2 PEs or more if necessary (Fig. 3) – these are *virtual layers*. The set of particles in such layers is distributed among all these PEs. During the load balancing at first particles are redistributed among PEs, and only if necessary, the layers are also redistributed.

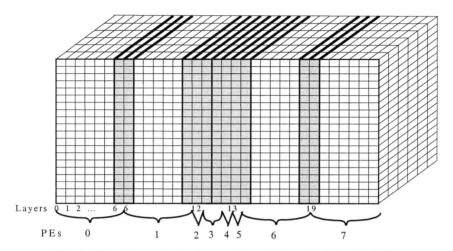

**Fig. 3.** Virtual layers for implementation of PIC on the 2D grid of PEs

## 6. Realisation of the PIC method on multicomputers

In order to provide good portability the language C was chosen for the parallel PIC code implementation. Since different message passing systems are used for different multicomputers, the special communication functions were designed. In the compilation time the functions of target multicomputer are substituted. In particular, the MPI functions might be substitute. For the dynamic load balancing of PIC several algorithms were realised. In the cases of the grid communication structure and the virtual fragments (when particles can fly not only to the neighbouring PEs) the special tracing functions are used.

If load imbalance occurs, the procedure BALANCE is called. In this procedure the decision about the direction of data transfer and the volume of data to be transferred is taken. For every load balancing algorithm there exists its own realising procedure BALANCE. The procedure TRANSFER is used for the data transfer. There are two realisations of this procedure: for the line of PEs and for the grid of PEs. The procedure is the same for any load balancing algorithm on the line of PEs. The parameters of the procedure are the number of particles to be exchanged and the direction of data transfer.

Let us consider algorithms of the dynamic load balancing of PIC. All the PEs are numerated. In the case of the line of PE, each PE has number $i$, where $0 \leq i <$ *number of PEs*. In the case of the $(l*m)$ grid of PEs, the number of PE is the pair $(i,j)$, where $0 \leq i < l$, $0 \leq j < m$. In the same way, layers and columns of SM are numerated.

## 6.1 The centralised load balancing algorithm on the line of PEs

Let *NP* particles be used for modelling. Then the SM is divided into *N* blocks (where *N* is the number of PEs), each block consisting of several adjacent layers and containing about *NP/N* particles (Fig. 1). The layers are enumarated by the integer numbers $0, 2, ..., k-1$, *k* is the number of layers of the SM. If the overload of at least one PE occurs in the course of modelling, this PE is able to recognise it at the moment when the number of particles substantially exceeds *ANP=NP/N*. Then this PE initiates the rebalancing procedure. All the PEs run BALANCE. PE*0* requires the number of particles inside every *j*-th layer of SM located in PE*i*, from every PE*i*, $i=1,...,N-1$, puts this information into *A[j]* (the size of *A* is equal to *k*). In such a way *A[i]* contains the number of particles inside the *i*-th layer. Then the array *A* is broadcasted to all the PEs. Each PE builds the array *S* (the size of *S* is equal to *N*) as shown in Fig. 4. In such a way, *S[i]* contains the minimal number of all the layers located inside the PE*i*. For the case in Fig.1, S={0,4,8,12,13,14,18,22}.

After that each PE runs the procedure TRANSFER. In this algorithm, contrary to the case with virtual layers, the first parameter of the procedure TRANSFER is equal to the number of particles in the layers to be transferred. By calling this procedure, the PEs realise the layers redistribution.

This load balancing algorithm is suitable for multicomputers with the small number of PEs (about 16 or so), because it requires the global data exchange.

## 6.2 The centralised algorithm with virtual layers

The idea of virtual layers was discussed in the previous paragraph. The algorithm of the load balancing is shown in Fig. 5. In the procedure TRANSFER, at first particles are transferred among PEs, and only if necessary the layers are also redistributed. In this algorithm, contrary to the algorithm in 6.1, the first parameter of the procedure TRANSFER could be unequal to the number of particles in layers to be transferred. If after the particle transfer no particle remains in the layer, this layer is removed from PE. The last layer from which particles were sent to the neighbouring PE becomes the virtual layer for these two PEs. Therefore, if the number of layers $k \approx N$ (N is the number of PEs), the workload can be rebalanced without the layers transfer, just the particles from virtual layers can be moved from one PE to another.

## 6.3 The centralised load balancing algorithm on the grid of PEs

If overload of at least one PE occurs in the course of modelling, this PE is able to recognise it at the moment when the number of particles substantially exceeds NP/(l×m) (more than the initially chosen ε). Then this PE initiates the rebalancing procedure. All the PEs run BALANCE. PEs *(i,0)* of every row require the number of particles in every column of SM from the other PEs of the *i*-th row, put this information into the array A and broadcast it back to all the PEs of the *i*-th row. Then all the PEs *(i,0)* broadcast the collected information to each other and make a decision about the necessity to redistribute layers among the rows of PEs. After that PEs of every row make a decision which columns should be transferred among the PEs of the row. Thus, the algorithm in Fig. 4 is implemented twice in the course of the procedure BALANCE execution. After that each PE runs the procedure TRANSFER. In such a way, first the data transfer is planned, then the data transfer is realised.

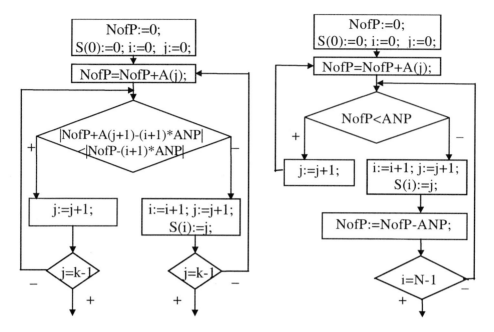

**Fig. 4.** Centralised load balancing algorithm on the line of PEs

**Fig. 5.** Centralised load balancing algorithm with virtual layers

Let us note that in the case of the grid of PEs the data might be transferred not only between the neighbouring PEs. It is because blocks Block_i, $i=0,...,l-1$, are assigned for processing on the i-th rows of the 2D grid independently (Fig. 6).

### 6.4 The diffusive load balancing algorithm

This algorithm is suitable for large multicomputers (with the number of PEs more than 16). This is the decentralised algorithm. The decision is taken on the information about load balancing in the local area of PEs. Thus, this algorithm gives the worse quality of rebalance than the centralised one, but it does not require the global data exchange. To use this algorithm, the modified domain decomposition (with virtual layers) should be chosen.

Another advantage of this algorithm is the fact that it can be used for solving problems with a non-constant number of particles, as it does not make use of the knowledge about the amount of model particles in the whole SM.

The name " diffusive " means that particles are transferred between the linked PEs in several steps, so, in k steps of balancing the particles from PE$i$ could fly into PE$i+k$ and PE$i-k$. In the procedure BALANCE, the *basic diffusive algorithm* with the size of the local area equal to 2, is used [7]. It is possible to choose the number of steps of diffusive algorithm - step_d. At the odd steps, both processors in every couple of PEs $2i$ and $2i+1$ (where $0 \leq i \leq N/2-1$, $N$ is the number of PEs), exchange information about the number of particles and run the procedure TRANSFER to exchange particles and to provide the equal workload of this couple of PEs (Fig. 7). At the even steps, both processors in every couple of PEs *2i-1* and *2i* (where $1 \leq i \leq (N-1)/2$), exchange information about the number of particles and run the procedure TRANSFER to exchange particles and to provide the equal workload of this couple of PEs. Thus, in the procedure BALANCE, the procedure TRANSFER is called step_d times.

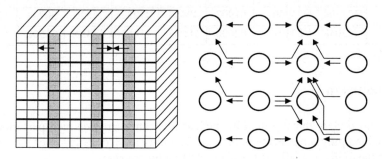

**Fig. 6.** Direction of data transfer for implementation of PIC on the 2D grid

**Fig. 7.** Local areas for the basic diffusive algorithm

### 6.5 Imbalance Threshold

To call the procedure BALANCE, the PEs need to recognise that the imbalance has occurred. In the case the number of particles is changed in the course of modelling diffusive load balancing algorithm is used and the procedure BALANCE is called after each time step of modelling.

If centralised algorithms are used for the dynamic load balancing, the PEs exchange information about load balancing, so every PE has information about the number of particles in all the PEs. In every PE, the difference *mnp-NP/N* (where *mnp* is the maximum number of particles in PE, *NP* is the total number of particles, *N* is the number of PEs) is calculated. If the difference is more than the threshold Th, the procedure BALANCE is called. The threshold can be a constant, chosen in advance, or an adaptive number. In the latter case, initially Th=0. In the course of modelling, the time *t_part*, which is required to implement steps 1)-3) of the PIC algorithm for one particle, is calculated. After every BALANCE call the time of balancing t_bal is calculated. Th is assigned to be equal to t_bal/t_part (how many particles could be processed for the same time as one balancing requires). After each next step of the PIC algorithm, Th is decreased by *mnp-NP/N*. When the value of Th is negative, BALANCE is called.

### 6.6 Comparison of effectiveness for different load balance algorithms

Let us compare the described algorithms for the problem of modelling of cloud plasma explosion in magnetised background. The grid size is equal to 24x24x36 points. The number of particles is equal to 800000 (240128 cloud particles and 559872 background particles). In Tab.1, the total time of program execution is shown.

**Table 1.** Comparison of effectiveness of the load balance algorithms. WB – without balancing, CAVL – centralised algorithm with virtual layers, DA – diffusive algorithm

| N of PE | 2 PE | 3PE | 4PE | 5PE | 6PE | 7PE | 8PE | 20PE |
|---|---|---|---|---|---|---|---|---|
| WB | 121768 | 107878 | 68801 | 63481 | 65332 | 47470 | 46716 | 24687 |
| CAVL | 122294 | 81900 | 61902 | 49453 | 41223 | 35826 | 30744 | 13227 |
| DA | 122048 | 83191 | 62606 | 50083 | 41465 | 34451 | 31508 | 13353 |

## 7. Conclusion

The described algorithms of the PIC method realization provide high performance of the assembled program execution, its high flexibility in reconstruction of the code and dynamic tunability to available resources of a multicomputer.

High performance of the program execution provides the modelling of big size problems such as the study of a cloud plasma explosion in magnetised background, modelling of interaction of a laser impulse with plasma. For the first problem with a non-uniform magnetic field, the 60x60x100 grid and 14 million particles were used for modelling. Process of modelling made more than 2000 time steps.

Flexibility of program construction with the assembly technology provides assembling a certain program for solution of a certain problem, but not a general program. This also improves the performance of an assembled program.

We apply the assembly technology to realisation of different numerical methods and hope to create the general tool to support realisation of mathematical approximating models.

## References

1. V.A.Valkovskii, V.E.Malyshkin.Synthesis of Parallel Programs and Systems on the Basis of Computational Models.// Nauka, Novosibirsk, 1988. (In Russian)
2. V.Malyshkin. Functionality in ASSY System and Language of Functional Programming.// Proc. "The First Aizu International Symposium on Parallel Algorithms/Architecture Synthesis", 1995, Aizu-Wakamatsu, Fukushima, Japan. IEEE Comp. Soc. Press, Los Alamitos, California. P. 92-97.
3. M.Kraeva, V.Malyshkin. Implementation of PIC Method on MIMD Multicomputers with Assembly Technology.// Proc. of HPCN Europe 1997 (High Performance Computing and Networking) Int. Conference, LNCS, Vol.1255, Springer Verlag, 1997. pp. 541-549.
4. Hockney R, Eastwood J. Computer Simulation Using Particles.// McGraw-Hill Inc. 1981.
5. Berezin Yu.A., Vshivkov V.A. The method of particles in rarefied plasma dynamic.// Novosibirsk: Nauka, 1980 (in Russian)
6. M.Kraeva, V.Malyshkin. Dynamic load balancing algorithms for implementation of PIC method on MIMD multicomputers.// Programmirovanie, N 1, 1999 (accepted for publication, in Russian)
7. A.Corradi, L.Leonardi, F.Zambonelli Performance Comparison of Load Balancing Policies based on a Diffusion Scheme// Proc. of the Euro-Par'97 LNCS Vol. 1300, p.882-886.

# Computational Aspects of Multi-Species Lattice-Gas Automata*

D. Dubbeldam, A.G. Hoekstra and P.M.A. Sloot

University of Amsterdam,
Faculty for Mathematics, Computer Science, Physics, and Astronomy,
Kruislaan 403, 1098 SJ Amsterdam,
The Netherlands,
Tel +31 20 525 7563, Fax +31 20 525 7490,
http://www.wins.uva.nl/research/pscs/,
Email:{dubbelda, alfons, sloot}@wins.uva.nl

**Abstract.** We present computational aspects of a parallel implementation of a multi-species thermal lattice gas. This model, which can be used to simulate reaction-diffusion phenomena in a mixture of different fluids, is analyzed for a fluid system at global equilibrium. Large system sizes combined with long-time simulation makes parallelization a necessity. We show that the model can be easily parallelized, and possesses good scalability. Profiling information shows the random number generator has become a bottleneck. The model can be statistically analyzed by calculating the dynamic structure factor $S(k,\omega)$. As an illustration, we measure $S(k,\omega)$ for a one-component system, and extract the values of transport coefficients from the spectra. Finally, $S(k,\omega)$ is shown for a two-component thermal model, where the central peak is more complicated, due to the coupled entropy-concentration fluctuations.

## 1 Introduction

Lattice-gas automata (LGA) are a relative novel method to simulate the hydrodynamics of incompressible fluids [1]. The flow is modeled by particles which reside on nodes of a regular lattice. The extremely simplified dynamics consists of a streaming step where all particles move to a neighboring lattice site in the direction of its velocity, followed by a collision step, where different particles arriving at the same node interact and possibly change their velocity according to collision rules. The main features of the model are exact conservation laws, unconditional stability, a large number of degrees of freedom, intrinsic spontaneous fluctuations, low memory consumption, and the inherent spatial locality of the update rules, making it ideal for parallel processing.

Different LGA models exist, both in two and three dimensions, where the models differ in the number of used velocities and the exact definition of the collision rules (see [2–4]). The basic LGA model proposed by Frisch, Hasslacher,

---

* presenting author: D. Dubbeldam

and Pomeau (the FHP I model) is a two dimensional model, based on a triangular grid, where up to six particles (hexagonal symmetry) may reside at any of the sites [5]. The model was extended with rest-particles (FHP II model) and the maximization of the number of collision rules (FHP III model), resulting in a higher maximum Reynolds number. The model of Grosfills, Boon, and Lallemand (GBL-model) introduced non-trivial energy conservation to the LGA's, allowing the simulation of temperature, temperature gradients, and heat conduction [6,7].

The models can be extended with multiple species, where each species is tagged with a different color, but differs in no other way [1]. In this paper we focus on such an extended LGA model, which can be used to simulate reaction-diffusion phenomena in a mixture of different fluids. The only four properties that are conserved in this model are mass, momentum, energy, and color, in case there is only diffusion and no reaction.

The Boolean microscopic nature, combined with stochastic micro-dynamics, results in intrinsic spontaneous fluctuations in LGA [6–8]. Such fluctuations can be described by the dynamic structure factor $S(k,\omega)$, the space and time Fourier transform of the density autocorrelation function [9]. The fluctuations extend over a broad range of wavenumbers $\mathbf{k}$ and frequencies $\omega$. $S(k,\omega)$ of real fluids can be measured by light-scattering experiments, where the measured quantity is the power spectrum of density fluctuations. In the hydrodynamic limit, one observes two Doppler-shifted Brillouin peaks and a central Rayleigh peak. Such spectra are also observed in GBL [6,7]. A measured spectrum from a GBL simulation contains transport coefficient information. The Brillouin peaks correspond to the sound modes, the Rayleigh peak corresponds to energy density fluctuations and diffusion. We are interested in fluctuations in two-species GBL models, and therefore have extended our multiple-species GBL with routines to measure $S(k,\omega)$. The main goal in the present work is to discuss the computational aspects of this model.

## 2 The GBL model

The two dimensional GBL lattice gas model is based on a triangular lattice with hexagonal symmetry for isotropy reasons. The particles have unitary mass with no spatial extension. The model evolves according to the following dynamical rule:

$$n_i(\mathbf{x} + \mathbf{c_i}, t+1) = n_i(\mathbf{x}, t) + \Delta(\mathbf{n}(\mathbf{x}, t)) \qquad i = 1, \ldots, 19 \qquad (1)$$

where the $n_i$ are Boolean variables representing the presence/absence of a particle at site $x$ at time $t$, and $\Delta(\mathbf{n}(\mathbf{x},t))$ is the collision operator, and $\mathbf{c_i}$ is the velocity vector. The exclusion principle (no more than one particle is allowed at a given time in a certain channel) garantues the convenient specification of a state of a node as a 19-bits integer.

The introduction of temperature requires a multiple-speed model (ideally a velocity distribution). The GBL-model has four different speeds, $0, 1, \sqrt{3}$, and 2, which corresponds to a kinetic energy of $\frac{1}{2}mv^2 = 0, \frac{1}{2}, \frac{3}{2}$, and 2, respectively.

The model has one rest particle, 6 velocities of speed 1, 6 velocities of speed $\sqrt{3}$, and 6 velocities of speed 2. The mapping of these velocities to bits are as shown in figure 1(a). As an example, in the streaming step bit 7 will be propagated two lattice nodes to the right.

The collisions redistribute mass, momentum, and energy among the 19 channels of each node at every time step such that mass, momentum, and energy are conserved. The collision outcome is chosen randomly among all states belonging to the same class as the input state (including the input state itself). This procedure is equivalent to the *random algorithm* [10]

Our implementation of the GBL model is such that it contains all other known models based on a triangular grid, FHP I (bit 0, ... ,5), FHP II/III (bit 0, ... ,6) etc. Hence, to simulate another lattice gas model we stream only a subset of the 19 bits, and we use a different collision lookup-table. To simulate a square grid, the streaming has to be slightly adjusted.

## 3 Implementation Aspects

*The triangular grid* To denote each node in a set of integer coordinates $(x, y)$ we multiply the coordinates by the factor $(2, \frac{2}{\sqrt{3}})$. Thus, each node of a triangular lattice can be uniquely mapped to a node of a square lattice (figure 1). Note that when working with rectangular space, the lengths along the vertical axes should be multiplied by $\frac{\sqrt{3}}{2}$. To avoid an awkward diamond shaped grid, the streaming step is different for even and odd parity of the lattice (see figure 1(b) and 1(c)). The conversion of a rectangular shaped triangular lattice now remains rectangular.

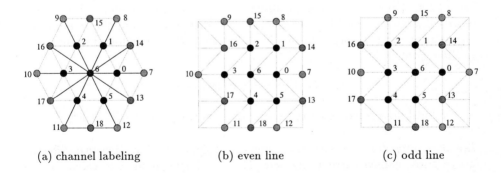

(a) channel labeling     (b) even line     (c) odd line

**Fig. 1.** In (a) the channel labeling per lattice node is shown. The conversion from the triangular lattice to the square lattice is based on a the parity of the lattice. On even lines we propagate using (b), for odd lines we propagate using (c).

*Parallelization* Parallelization of grid based algorithms like the lattice-gas automata is done by means of a data decomposition strategy, where the computational domain is decomposed into sub-domains. Each processor performs computations on a certain sub-domain and exchanges information with other nodes in order to resolve dependencies. For that purpose we use a ghost-boundary of two lattice nodes (the GBL model has a highest velocity of 2, i.e. each lattice site streams its particles of velocity 2 to its next-nearest neighbors). The Fast Fourier Transform (FFT) used to measure the $S(k,\omega)$ is a parallel version of FFTW [11] for in-place, multi-dimensional transforms on machines with MPI [12]. It has the requirement that the decomposition is a slice-decomposition (equal sub-volumes in one dimension). In this paper we restrict ourselfs to the dynamical structure factor for a fluid system at global equilibrium. We have periodic - or no-slip boundary conditions and no obstacles in the fluid. Hence, we have near-optimal load-balancing. If the lattice grid is not rectangular and/or contains obstacles in the fluid the decomposition can be done with the ORB-method [13, 14]. In this case, the parallel FFT routine needs to be adjusted.

*Initialization* The GBL model can be shown to be free of known spurious invariants, which have plagued other LGA's. These models must be correctly initialized to avoid such invariants. The best known spurious invariant is the conservation of total transverse momentum on even and odd lines every two time steps. This spurious invariant can be eliminated by choosing initial conditions such that the total transverse momentum is zero [8]. We accomplish this by initializing the density only pairs-wise with opposite velocities.

*Data structures* The GBL model uses 19 velocities and can be conveniently represented as a 19-bit integer. The complete lattice is allocated as a single array of a structure containing $n + 1$ longs (a long is 4 bytes), where $n$ is the number of species (colors). One state denotes the presence/absence of the particle, and the $n$ longs denotes the presence/absence of the particular colors. This generic approach allows a variable number of species (decided at compile-time) at the cost of using more memory. An index to a lattice node can be found in the array by the relation:

$$\text{lattice}(x,y) = x + y \times \text{lattice width} \qquad (2)$$

i.e. the local data is expected to be stored in row-major order (C order).

*The streaming-step* The streaming step is accomplished by using two different lattice grids, named current and new. The new lattice is used to calculate the new state at the next time-step from the values of the current lattice. After updating the lattice, we swap the pointers to the lattices and the new lattice becomes the current lattice. If memory size is a problem it would be possible to do the streaming step in-place, at the cost of accessing each lattice point several times instead of just once. Since our models are all two-dimensional and using domain decomposition, memory should not be a real problem here.

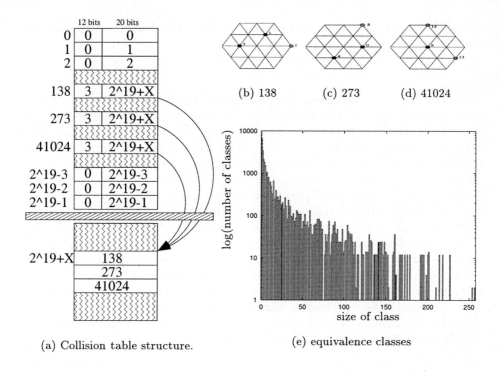

**Fig. 2.** Collision table. The first $2^{19}$ indices are divided into 12 and 20 bits. The left 12 bits denotes the number of collision outcomes, the right most 20 bits denotes a index-number from where the collision outcomes are stored in the table. These collision outcomes start from index $2^{19}$. The figure shows an example for an equivalence class of size 3. The three configurations are shown in (b),(c), and (d); (e) shows the histogram of the 29925 equivalence classes of the GBL model (largest class 257).

*Collision tables* The collision step consists of updating the lattice node with its collision outcome. In the 6 and 7 bits models, we use a collision table of words (16 bits) of size $2^6$ and $2^7$ respectively. In these models we have at most two collision outcomes, hence we store two 8 bits outcomes in each table entry and choose one randomly. If there is only one outcome, both values are equal.

The models which use more than 7 bits are implemented as equivalence classes. An equivalence class is formed by all states having the same mass, momentum, and (for thermal models) energy. For the 19-bits model this means a collision table of $2^{19}$ indices, followed by the equivalence classes, see figure 2(a). Every index of an element in a class points to the start of the class (138, 273 and 41024 all point to $2^{19} + X$). The left 12 bits are used to indicate the number of collision outcomes. If the number is zero, the outcome is equal to the input state. Otherwise the value of the right 20 bits is an index pointing to the first outcome possibility, followed by the other possible outcomes, of which we choose one at random. It is clear that the input state is also among them (meaning no collision), but since most classes are quite large this has little influence (see figure 2(e)).

*Color redistribution (diffusion)* The problem at hand is how to distribute several color particles random over the collision outcome. We start with the collision outcome in which we find $n$ ones, denoting the presence of the particles. This value is being used for the distribution of the colors. Now we have the first $m$ particles of a certain color. We start from right to left, look for a one in the collision outcome, and set the particle at this color if a random number between 0 and 1 is smaller than $m$ divided by $n$. If this is indeed the case we set the bit from 1 to 0 (the original collision outcome was stored for further use), and we continue with $n-1$ and $m-1$ (we now have to distribute $m-1$ over $n-1$). If the random number was larger than or equal to $m$ divided by $n$ we do not set the particle color and continue with $n-1$ and $m$ (we now have to distribute $m$ over $n-1$). Note that the procedure is correct for $m \geq n$, since than the particles will always be set to the color. The other colors are treated similarly, with one change: we now have to distribute $r$ color particles over $n-m$ ones. Note that the number of random generated numbers depends on the number of ones in the collision outcome, i.e. the density.

*Calculating $S(k,\omega)$* The dynamic structure factor $S(k,\omega)$ is treated as a spectral function (a function of $\omega$ at fixed values of the wave number $k$). Calculation of $S(k,\omega)$ first requires a spatial two dimensional Fourier transform of the density, which is done at simulation time, followed by a one-dimensional Fourier transform in time of the spatial Fourier transformed data, done after simulation time.

The spatial transform is performed every time step, and the wave numbers we are interested in are stored. The spatial transform results in a two-dimensional array, scattered over the processors. The first row of the array of the first process contains the wavenumbers $k_x = $ index $\times \frac{2\pi}{L_x}$, the first column of the array (scattered over the processors) represents the wave numbers $k_y = $ index $\times \frac{2\pi}{L_y} \times \frac{2}{\sqrt{3}}$, where $L_x$ is the lattice size in $x$-direction, and $L_y$ is the lattice size in the $y$-direction.

After the simulation we Fast Fourier Transform the stored wave numbers in time. To reduce the initially large variance we apply the technique discussed in [15]. We partition the data into $K$ segments each of 16384 consecutive data points. Each segment is separately FFT'd to produce an estimate. We let the segments overlap by one half their length. Finally, the $K$ estimates are averaged at each frequency, reducing the initial variance by $\sqrt{K}$. Data windowing is used to reduce "leakage". The $S(k,\omega)$ is scaled as $\int_{-\pi}^{\pi} S(\mathbf{k},\omega)\,d\omega = 2\pi S(\mathbf{k})$ [9]. Thus, the area under $S(k,\omega)/S(k)$ is $2\pi$ and the spectra shown in the next section are scaled accordingly.

# 4 Computational aspects

## 4.1 Profile analysis

A (sequential) profile of the code can be used to provide inside in the fraction of the execution time spent in a function, divided in four categories: collision step (13%), streaming step (6%), the Fast Fourier Transform (18%), and color redistribution (63%) (for a simulation of the GBL model at lattice size $512 \times 512$, reduced density 0.3, periodic boundary conditions in both directions, 50% red particles, 50% blue particles). We note the large fraction of the execution time spent in the color redistribution step, which consists almost entirely of random number generation.

The random number generator used is the *ran2* recommended in Numerical Recepices [15]. During this particular simulation we generate more than $10^{13}$ random numbers. This means that to prevent correlations we have to use a random number generator with a large period (the cycle of *ran2* is $10^{18}$).

A solution for the low $b$-bit two-species LGA models is to construct one collision table of size $2^{2b}$ where the configuration consists of both place *and* color information. However, for multiple species and/or the GBL model this would result in a much too large collision table. Further improvements in speed should thus focus on a faster random number generator. A first optimization is the use of antithetic variables, i.e. if $u$ is a random number (uniform $[0, \ldots, 1]$) then so is $1 - u$ [16].

## 4.2 Scalability results

To reduce the initially large variance in a spectral measurement, a large number of time steps is needed. For measurements at small values of **k** we have to use large lattice sizes. Parallel computing is exploited here to facilitate efficient simulation. The two factors controlling the efficiency of parallelization are the ratio between the communication time and calculation time, and the balance of workload among the processors. Since we have near-optimal load-balancing (square lattice grid and absence of obstacles in the fluid) only the first factor applies here. To measure scalability we have performed measurements of the execution time per iteration for a different number of lattice sizes at various number of processors. The LGA simulation data is a two-species GBL simulation of a fluid at global equilibrium, reduced density 0.3, periodic boundary conditions in both directions, 50% red particles, 50% blue particles. A spatial Fast Fourier Transformation is performed every time-step.

The results of the scalability measurements are shown in figure 3. For small lattice size the scalability is bad, due to the large communication overheads. However, for a more realistic choice of lattice size we see an almost perfect scaling.

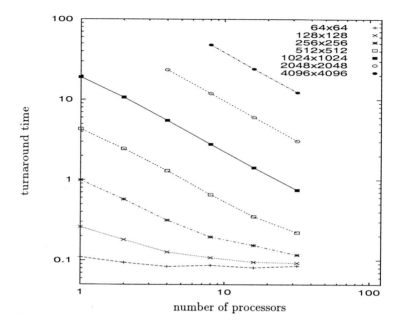

**Fig. 3.** The execution time in seconds per iteration for different lattice sizes (two-species GBL simulation of a fluid at global equilibrium, reduced density 0.3, periodic boundary conditions in both directions, 50% red particles, a spatial Fast Fourier Transformation is performed every time-step).

## 5 Case study

### 5.1 Transport coefficient measurements

The dynamic structure factor $S(k,\omega)$ for real fluids in the Landau-Placzek approximation (small values of $k$ and $\omega$) reads [9]:

$$\frac{S(\mathbf{k},\omega)}{S(\mathbf{k})} = \left(\frac{\gamma-1}{\gamma}\right)\frac{2\chi k^2}{\omega^2 + (\chi k^2)^2} + \frac{1}{\gamma}\sum_{\pm}\frac{\Gamma k^2}{(\omega \pm c_s k)^2 + (\Gamma k^2)^2} \\ + \frac{1}{\gamma}[\Gamma + (\gamma-1)\chi]\frac{k}{c_s}\sum_{\pm}\frac{c_s k \pm \omega}{(\omega \pm c_s k)^2 + (\Gamma k^2)^2} \qquad (3)$$

where $S(k)$ is the static structure factor. Here $\chi$ is the thermal diffusivity, $\Gamma = \frac{1}{2}[\nu + (\gamma-1)\chi]$ the sound damping, where $\nu$ is the longitudinal viscosity; $c_s$ is the adiabatic sound velocity; and $\gamma$ is the ratio of specific heats. A typical spectrum as found in real fluids consists of three spectral lines, a central peak (Rayleigh peak), which arises from fluctuations at constant pressure and corresponds to the thermal diffusivity mode, and two shifted peaks (Brillouin peaks), which arise from fluctuations at constant entropy and correspond to the acoustic modes.

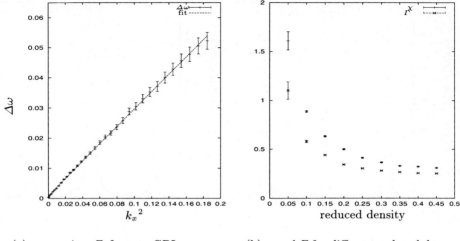

(a) measuring $\Gamma$ from a GBL spectrum.

(b) $\chi$ and $\Gamma$ for different reduced densities.

**Fig. 4.** Figure (a) plots $\Delta\omega$ of the Brillouin peaks (with error bars) as a function of $k^2$. This provides an experimental measurement of $\Gamma 0.288 \pm 0.001$ (value of the slope) at a reduced density of 0.3. Figure (b) shows the measurements (with error bars) of $\chi$ and $\Gamma$ for different reduced densities.

Equation (3) for $S(k,\omega)$ is known as the Landau-Placzek approximation and also holds for the GBL model in the limit for small $k$ and small $\omega$ [6,7]. A measured spectrum from a GBL simulation contains transport coefficient information. We can extract the sound damping $\Gamma$ as the half-width of the Brillouin peaks: $\Delta\omega = \Gamma k^2$. We accomplish this by plotting $\Delta\omega$ as a function of $k^2$. A least-square fit provides the value of the slope, which is an experimental value of $\Gamma$. Figure 4(a) shows we find $\Gamma = 0.288 \pm 0.001$ for a reduced density of 0.3. The thermal diffusivity coefficient $\chi$ is obtained as the half-width of the central peak: $\Delta\omega = \chi k^2$. The experimental determination of the position of the Brillouin peaks is a measurement of the adiabatic sound velocity $c_s$. The ratio of the specific heats $\gamma$ is obtained as the ratio of the integrated intensity of the Rayleigh peak to those of the Brillouin peaks. Our experimental measurements of the transport coefficients from a GBL simulation are in agreement with other results [7]: $c_s = 1.275 \pm 0.001$, and $\gamma = 1.32 \pm 0.02$, and the dependence of $\Gamma, \chi$ on the reduced density is shown in figure 4(b).

## 5.2 The spectra of a two component thermal lattice gas

The $S(k,\omega)$ for a two-component system has a spectrum structure where it is difficult to separate the contributions from concentration fluctuations and

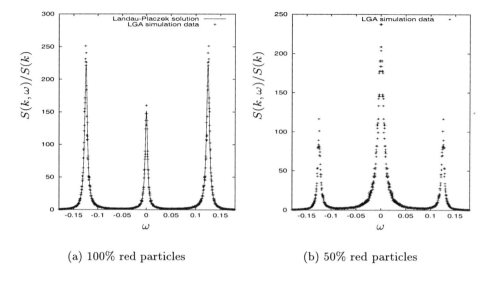

(a) 100% red particles  (b) 50% red particles

**Fig. 5.** The spectral function $S(k,\omega)$ for a (a) 100% and (b) 50% red GBL lattice gas ($k_x = 8 \times \frac{2\pi}{512}$, reduced density 0.3), data obtained from 1000000 timesteps.

entropy fluctuations [9]. The two contributions are independent if one of the two components is in trace amounts, but in general the two are not decoupled.

Figure 5(a) shows $S(k,\omega)$ for the standard GBL model. The simulation data obtained from the lattice gas simulation is fitted with the analytical solution, providing the transport coefficients in lattice units. Figure 5(b) shows $S(k,\omega)$ for the two-species GBL model. We notice a significant contribution from concentrations fluctuations to the amplitude of the Rayleigh peak.

## 6 Concluding comments

The parallel implementation of a lattice gas automata is straight forward, due to the inherent spatial locality. The GBL model uses velocities of two lattice spacing, hence we now have to use ghost-layers consisting of two lattice nodes for the implementation of the boundary conditions. The implementation of the collision step can simply be done using a collision table, where we obtain the collision outcome in at most two memory references and one random number generation.

Our extension of the model with multi-species, where the number of species is decided at compile time, requires a large amount of random numbers. The most widely used generators (like *random* and *drand48*) are not adequate. We recommend the use of the random number generator *ran2* as found in numerical recipes. A profile shows the random generator has become a large bottleneck. A first optimization is the use of antithetic variables.

The computation of the dynamic structure factor $S(k,\omega)$ requires a spatial Fourier transform at every time step. The parallel FFT expects a slice-decomposition of the lattice grid, but has otherwise no restrictions to the implementation. As an illustration we have shown how to calculate two transport coefficients, $\Gamma$ (sound damping) and $\chi$ (thermal diffusivity), from the Rayleigh-Brillouin spectrum. These measurements were performed for different reduced densities. Finally, we presented a new result, namely the spectrum of a two-species GBL LGA. The spectrum clearly shows the added contribution of concentration fluctuations to the Rayleigh peak. It is in general not easily possible to derive the independent values of $\chi$ and $D$ (the diffusion coefficient) from this spectrum.

# References

1. D H. Rothman and S. Zaleski. *Lattice gas cellular automata*. Cambridge University Press, 1997.
2. D. d'Humières, P. Lallemand, and U. Frisch. Lattice gas models for 3-d hydrodynamics. *EuroPhys. Lett.*, 2:291–297, 1986.
3. K. Diemer, K. Hunt, S. Chen, T. Shimomura, and G. Doolen. Velocity dependence of reynolds numbers for several lattice gas models. In Doolen, editor, *Lattice Gas Methods for Partial Differential Equations*, pages 137–178. Addison-Wesley, 1990.
4. U. Frish, D. d'Humières, B. Hasslacher P. Lallemand, Y. Pomeau, and J.P. Rivet. Lattice gas hydrodynamics in two and three dimensions. *Complex systems*, 1:649–707, 1987.
5. U. Frish, B. Hasslacher, and Y. Pomeau. Lattice-gas automata for the navier-stokes equation. *Physical Review Letters*, 56:1505, April 1986.
6. P. grosfils, J.P. Boon, and P. Lallemand. Spontaneous fluctuation correlations in thermal lattice-gas automata. *Physical Review Letters*, 68:1077–1080, 1992.
7. P. grosfils, J.P. Boon, R. Brito, and M.H. Ernst. Statistical hydrodynamics of lattice-gas automata. *Physical Review E*, 48:2655–2668, 1993.
8. D. Hanon and J.P. Boon. Diffusion and correlations in lattice-gas automata. *Physical Review E*, 56:6331–6339, 1997.
9. J.P. Boon and S. Yip. *Molecular Hydrodynamics*. McGraw-Hill, New York, 1980.
10. M. Hénon. Isometric collision rules for the four-dimensional fchc lattice gas. *Complex Systems 1*, pages 475–494, 1987.
11. Copyright (C) 1998 Massachusetts Institute of Technology. The fastest fourier transform in the west. *http://theory.lcs.mit.edu*.
12. University of Chicago and Mississippi State University. Mpich. *http://www.mcs.anl.gov/mpich/*.
13. D. Kandhai, A. Koponen, A. Hoekstra, M. Kataja, J. Timonen, and P.M.A. Sloot. Lattice-boltzmann hydrodynamics on parallel systems. *Computer Physics Communications*, 111:14–26, 1998.
14. D. Kandhai, D. Dubbeldam, A.G. Hoekstra, and P.M.A Sloot. Parallel lattice-boltzmann simulation of fluid flow in centrifugal elutriation chambers. In *Lecture Notes in Computer Science*, number 1401, pages 173–182. Springer-Verlag, 1998.
15. W.H. Press, B.P. Flannery, S.A. Teukolsky, and W.T. Vetterling. *Numerical Recipes in C*. Cambridge University Press, 1988.
16. S.M. Ross. *A course in Simulation*. Maxwell Macmillian, New York, 1991.

# Towards a Scalable Metacomputing Storage Service

Craig J. Patten, K.A. Hawick and J.F. Hercus

Advanced Computational Systems Cooperative Research Centre
Department of Computer Science, University of Adelaide,
Adelaide, SA 5005, Australia
Email: {cjp,khawick,james}@cs.adelaide.edu.au

**Abstract.** We describe a prototypical storage service through which we are addressing some of the open storage issues in wide-area distributed high-performance computing. We discuss some of the relevant topics such as latency-tolerance, hierarchical storage integration, and legacy and commercial application support. Existing high-performance computing environments are either ad-hoc or focus narrowly on the simple client-server case. The storage service which we are developing as part of the DISCWorld metacomputing infrastructure will provide high-performance access to a global "cloud" of storage resources in a manner which is scalable, secure, adaptive and portable, requiring no application or operating system modifications. Our system design provides flexible, modular and user-extensible access to arbitrary storage mechanisms and on-demand data generation and transformations. We describe our current prototype's status, some performance analysis, other related research and our future plans for the system.

## 1 Introduction

Wide-area distributed high-performance computing, or metacomputing [4], is becoming an increasingly active research field. Relatively recent advances in high-bandwidth wide-area network technology have released the centralised supercomputer's hold on high-performance computing, and provided for greater integration of geographically distributed high-performance computing, communications, storage and visualisation resources.

These metacomputing systems naturally present many interesting challenges in the areas of storage and I/O. Most existing systems for handling data storage and retrieval across networks were designed for different workloads and operational environments, and did not address many of the issues which are unique to wide-area distributed computing. Within the DISCWorld [10, 11] metacomputing infrastructure project, we are developing a storage system to address some of these issues.

This system, the DISCWorld Storage Service, has been designed to provide a standard file system interface to a global "cloud" of storage resources. The defining attribute of the system is its sole purpose of providing a storage service for applications in a wide-area metacomputing environment; we are not developing a general distributed file system.

Most existing technology in the area of distributed storage and I/O is designed to address different problems to those encountered in metacomputing systems; this is our rationale for working towards a scalable metacomputing storage service. Issues such as separate administrative zones, large network latencies, non-uniform network topologies, highly heterogeneous systems, security and the need to support custom as well as legacy and commercial applications all present problems which cannot be suitably addressed using existing systems.

Our aim is not to produce a system which attempts to solve all of these problems. Rather, we are concentrating on providing a storage service which is decentralised and scalable, portable *without* requiring application or operating system modifications or additions, capable of handling various wide-area networking issues such as high-latency links and flexible and adaptive in its use of communications and storage resources.

Our main driving applications for the system centre around Geographical Information Systems (GIS), satellite imagery archives [9] and other scientific applications exhibiting large demands on the underlying storage infrastructure. Locally we store an up-to-date GMS5 [8] satellite imagery archive. We have built some example driver applications on top of this archive, such as the ERIC [12] satellite image browser.

In Section 2 we discuss in more detail the various issues relevant to storage and I/O in metacomputing systems. In Section 3 we describe the architecture and current prototype implementation of the DISCWorld Storage Service, and in Section 5 we discuss various performance issues, and preliminary results. Finally we conclude with our future plans for DSS.

## 2 Metacomputing Storage

Storage and I/O in metacomputing systems present different challenges than those addressed by traditional distributed/network storage technology. The magnitudes of both the workloads and wide-area network latencies are two of the most obvious. Existing distributed file systems are predominately geared towards "everyday" data usage across relatively local areas, where network latency is not a bottleneck, and the file size distribution is skewed much lower than in high-performance scientific computing. Existing mechanisms for remote data access in metacomputing systems, whilst generally designed for bulk data transfer over high-latency networks, are however less sophisticated than their distributed file system counterparts. We discuss some existing systems in Section 4.

Besides latency, other network issues also impinge on the design of any distributed storage system across the wide area. An increased level of heterogeneity in network and storage technology, higher network topology complexity and greater reliability problems all present challenges rarely seen in local area storage management. Across the wide-area, computers are now often reachable through multiple different networks, sometimes possessing multiple network interfaces with different performance characteristics. Wide-area high-performance network management itself offers reliability and robustness problems to be addressed.

Administrative and security issues become increasingly complex in wide-area environments. Computers across these networks belong to different institutions and companies, and reside in many different countries. Global user and host authentication and differing security requirements and infrastructures present challenges to developers of metacomputing systems and applications.

As these networks spread over wider areas and connect greater numbers of computers, the collective level of heterogeneity in the system increases. Portability then becomes a more important attribute of any storage system, if it is to operate over the wide area. It is therefore important for a storage system to provide access to applications without requiring access to source code or custom operating system modifications. For example, commercial or legacy applications where source code is unavailable. Any distributed storage system which requires source code modifications, applications to be relinked against custom libraries, or custom operating system modules or modifications, consequently renders an immense body of existing and future software and platforms as incompatible.

Recently emerging in wide-area distributed computing is the concept of remotely "leasing" access to high-performance computing resources [5]. In a storage context, this raises a number of technical issues such as data security, access methods for leased storage resources and associated electronic commerce mechanisms and policies.

In a distributed computing system where the paradigm extends beyond the simple client-server case, bulk data transfer becomes more complex, in that the source, destination and initiatior of such transfers may each comprise of multiple, geographically distributed hosts. A user or application on a mobile computer with poor connectivity may wish to transparently transfer data between distributed storage resources possessing high-performance connectivity without involving their poorly-connected host in the actual transfer. This example can be extended to the cases of arbitrary parallel data transfers such as gather/scatter and on-demand bulk data transformations between distributed storage resources. Laterally extending the data replication model to enable intermediate proxy and nearby caching and prefetching nodes also holds great potential for distributed storage performance. Existing distributed file systems and metacomputing remote data access mechanisms cannot provide the potential performance benefits available from such flexible and integrated use of storage resources in such an infrastructure.

The storage service we describe provides a framework for addressing these limitations and has the potential to drastically improve wide area data transfer performance.

## 3 DISCWorld Storage Service (DSS) Architecture

The DISCWorld Storage Service (DSS) architecture is illustrated in Figure 1. Each node in the metacomputing system may support a DSS daemon that emulates a Network File System [15, 18, 19] (NFS) server called the DISCWorld File System, DWorFS [13]. The NFS interface was chosen to allow commercial and legacy applications access to the DSS architecture without requiring any source or runtime modifications. All that is required is that the systems running the applications can NFS mount the file system exported by DWorFS. Other systems which use user-level NFS daemons include the Cryptographic File System [1], the Semantic File System [7] and ATTIC [4].

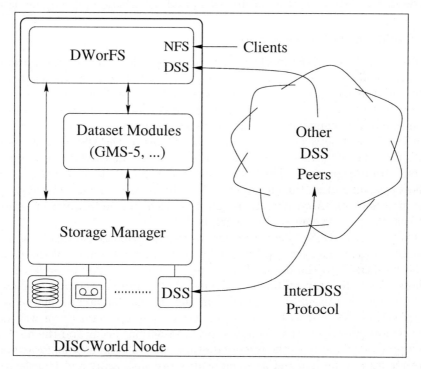

**Fig. 1.** Overview of the DISCWorld Storage Service architecture. Each node executes a DSS daemon, controlling local storage resources and brokering access to those resources for NFS clients on their local network(s) and remote DSS peers.

To emulate the file systems presented by the DWorFS layer, the DSS architecture uses dynamically-loaded modules. Each module is responsible for presenting its underlying storage architecture as if it were a real file system, regardless of its internal structure.

Consider Figure 1. The DISCWorld node on the left executes a DSS daemon, which consists of the front-end DWorFS NFS interface, the DSS Storage Manager and underlying dataset and storage modules. NFS requests from local clients are received at the DWorFS layer, and passed down to dataset-specific modules, which process the request accessing whatever storage resources are available through the Storage Manager. The Storage Manager communicates with storage modules which control access to local storage resources and remote DSS nodes through the DSS module and protocol. Incoming requests from other DSS nodes are received by the DWorFS layer and processed similarly.

The major strength of the DSS architecture is that the dynamically-loaded modules can be of arbitrary complexity. Therefore a DSS implementation is able to support arbitrary storage architectures, some of which may be virtual rather than physical in nature. It is also able to provide transparent access to distributed resources via the DWorFS layer without imposing any restrictions on the underlying networking protocols and infrastructure that may be used.

### 3.1 Wide Area Network Support

As discussed below, NFS performance characteristics may be appropriate to the high-bandwidth, low-latency behaviour of local area networks but perform badly with the high latency nature of wide area networks. This is especially true if large bulk data transfers are required. To address this issue the DSS architecture assumes that a dynamically-loaded module will be available to implement a bulk transfer protocol between DSS daemons. In Figure 1, the DSS storage module fulfills this role and is able to provide direct access from local dataset modules to the bulk transfer protocol. The bulk transfer protocol can then be used for tasks such as streaming data from a tape silo directly to DSS daemons elsewhere on a wide area network. Alternatively, it could be used to achieve efficient transfers between file systems emulated by local modules.

The design of DSS differs from most other distributed storage systems in that the storage manager can be configured to take advantage of the concept of the "distance" to the data. Most systems which incorporate prefetching or caching, generally stay within the client-server paradigm. Data is prefetched from the server, cached at some level in the storage hierarchies of the server and/or client. When distributed storage systems accept data to be written to some remote host, the usual request-reply transactions occur and the data generally proceeds directly to the remote destination. DSS will enable, for example, write requests to be buffered at some nearby fast disk array and transfer them to their final destination in bulk. We are also looking at the capability of arbitrarily configuring a DSS node to utilise other DSS nodes for proxy caching and prefetching.

### 3.2 GMS5 Satellite Imagery Access

To demonstrate the utility of this architecture we have implemented a module to interface to our GMS5 satellite image repository. Assuming that the /dss directory on a system is the local mount-point for the DSS, an application could attempt to open the file

```
/dss/GMS5/vis/98/02/02/05/32/250x200+950+1700.hdf
```

This prompts the GMS5 module to access its image repository, check for the existence of the named image, and if successful, present, through the DWorFS

NFS layer, a file handle for the requested "file". In the above example, a 250x200 portion of the Hierarchical Data Format (HDF) GMS5 satellite image for 2/2/1998 0532 UTC, cropped from coordinates (950,1700), generated on-demand. The application is shielded from the underlying organisation of the data, which in our archive is a collection of compressed HDF files, but which could be, for example, a database or distributed persistent object store.

### 3.3 Prototype Hierarchical File System

Robotic tape silos form the lowest level of many large scale on-line and near-line data storage and backup systems. They are used to provide economic solutions to the storage of large data sets which can not realistically be stored on disk. However, their integration into the storage hierarchy is anything but ubiquitous or standardised and invariably is proprietary and expensive, application and data specific, or involves human operator intervention.

The DSS allows the integration of hierarchical storage into the system through the use of the customisable storage modules. The module for a tape silo will present the same interface as any other storage module. Its internal behavior will be different, to reflect the different access properties, namely high latency, of the underlying storage resource. Our prototype hierarchical storage module, currently a simple proof-of-concept at this point in time, provides applications with read/write access to files stored within a tape silo.

The very large latencies of tape storage systems are the most important factor to be considered in the development of the module for them. The issues involved are similar to those involved for storage systems operating over a high latency, average bandwidth network, although the latencies are much larger. In our project we use a StorageTek TimberWolf 9740 tape silo containing two Redwood SD-3 drives. The average latency for access to data stored in this system is approx. 75 seconds [17]. This figure is dominated by the tape seek time which can be reduced by using lower capacity tapes (this figure assumes a 50GB tape). However, regardless of the size of tape used, the latency dominates the access times for even very large (larger than 100MB) data objects.

As an example, consider an archive of large medical or scientific images stored on a tape silo. To overcome the latency problems of the silo, a local high-speed disk array is used as a cache for the silo, to buffer writes to the store and provide space for frequently used data and prefetching. This hierarchical storage resource can be readily managed by attaching appropriate dataset and storage modules to DSS.

## 4 Related Work

A large body of work exists in the field of distributed file systems, and the research effort into metacomputing, and metacomputing storage mechanisms, is growing. The Andrew File System [16] (AFS) and NFS have proven highly successful for remote data access in predominately workstation environments, however these and other existing distributed file systems are not directly applicable in a metacomputing environment. An indepth study of all existing technology and its relation to our work is beyond the scope of this paper, however we briefly discuss a few systems with similar goals to our own.

As part of the Globus [6] metacomputing infrastructure, Remote I/O (RIO) provides remote data access, including some collective operations, through a latency-tolerant protocol. It is accessed through an Abstract Data I/O (ADIO) programming API, and is used by systems such as a message-passing interface implementation. Also part of Globus, Global Access to Secondary Storage (GASS) provides a set of replacement I/O calls such as read() and write() on Unix systems, which can communicate with remote HTTP, FTP and GASS servers, and perform some rudimentary caching. Applications wishing to use this functionality must be recompiled with these custom routines.

WebFS, part of the WebOS [23] system, provides access to the HTTP namespace through a standard file system interface transparent to user applications. However, HTTP currently has limited semantics in an I/O context and is only slightly more sophisticated than anonymous FTP. Their current system is specific to the Solaris OS, using a kernel module to implement the functionality.

WebNFS is an extension to the standard NFSv3 protocol by Sun, to provide access to NFS servers from within web-enabled appliations, and to improve the handling of latency in the NFS protocol. Through using a mechanism called Multi-Component Lookup (MCL), where lookup requests can contain entire pathnames rather than just single pathname components, WebNFS removes some of the inherent latency-intolerance in the NFS protocol.

## 5 Performance Discussion

The Sun Network File System (NFS) is the most ubiquitous existing distributed file system implementation. NFS is generally implemented with a relatively small block transfer size such as 8192 bytes (8KB). Under NFS Version 2 (NFSv2) this is the specified maximum, whereas the NFS Version 3 (NFSv3) protocol specification defines no limit; clients can request a desired read and write block transfer size, which the server has the choice of honouring. For example, our investigations show that Digital UNIX 4.0D has maximum read/write block sizes of 8KB, and Solaris 2.6 implements a maximum (and default) read/write block size of 32KB. The Andrew File System (AFS) allows block sizes of 64KB.

An NFS read request from an application program will typically be implemented as a sequence of requests for 8KB data blocks, each incurring a latency or overhead cost as well as the bandwidth cost to transfer the data. This is unacceptably slow for transfers across wide area networks where the latency of a single request may be of the order of tens or hundreds of milliseconds or more. Standard NFS file/directory lookup requests, which issue only one pathname component per request, to provide internal field separator independence, are highly intolerant of latency. WebNFS does address this issue through using Multi-Component Lookup (MCL), however it is yet to become widespread functionality.

Performance tests using the Internet Protocol over available wide-area networks have indicated approximate round-trip times (RTT) of 20ms between the Australian cities of Adelaide and Canberra, 200ms across a dedicated research link between Canberra and Japan, and approximately half a second across conventional Australia-USA-Japan internet links.

Consider the cost to transfer a 5MB file using conventional NFS, with an 8KB transfer block size. This requires some 640 request/reply pairs, incurring a latency cost of approximately 12.8 seconds for the Adelaide-Canberra link and 128 seconds between Canberra and Japan. It is likely that as faster networks become available this will become the dominant cost, if it is not already. The dedicated research link available to us between Canberra and Japan has a bandwidth of 1Mbit/s (128KB/s), thus contributing only 40 seconds to the transfer time. Hence, the latency overhead comprises 128/(40+128) or approximately 81%. This figure will surely worsen as more bandwidth becomes available, whereas the latency is fixed by speed of light limitations. We have a dedicated link of 10Mbit/s between Adelaide and Canberra, yielding a transfer time component of 4s, which is likewise smaller than the overhead component. At 155Mbit/s, a rate which our link is capable of being configured for, the bandwidth component is almost negligible compared to the latency overhead.

This analysis shows the critical importance of a bulk data transfer mechanism that uses higher transfer block sizes and a latency-tolerant protocol. Existing implementations of NFS are not optimised for wide area use. The limited amount of block size tuning which NFSv3 implementations allow, offers only a factor of two or four improvement. At least an order of magnitude improvement is desirable.

Performance gains can also be achieved by incorporating prefetching and pipelining transfer mechanisms into a wide area bulk data transfer system. Whether and how to support prefetching and caching in NFS is up to the implementors. However, existing implementations are not aggressive enough to effect a large performance improvement for the bulk data transfers used in wide-area metacomputing, as they are geared towards different I/O workloads. Typically, when an NFS client issues a read request, the requested block and perhaps some of the following blocks are read by the server. For large transfers across wide-area networks, better performance could be achieved by this request being used as a hint to start the transfer of larger portions of the file to some "closer" storage resource in anticipation of future requests, for example a fast disk array on a nearby LAN, or a local cache disk.

Our current implementation of DSS is still in its early prototype stage, and we do not yet possess detailed performance results. However, we have performed some preliminary experiments using NFSv3 (8KB blocks) across the loopback interface on Digital UNIX hosts, as applications could well be expected to perform if executing on a DISCWorld Storage Service node. Results indicate that on systems where the local file system read and write bandwidth was approximately 4.4MB/s, the read performance dropped to 4.0MB/s, and write performance dropped to 1.5MB/s. Whilst the write performance is not especially encouraging, we believe the performance tradeoff for the sake of portability is, at least for now, worth making.

A more detailed performance evaluation of using the DWorFS NFS layer and a dynamically-loaded dataset module to provide access to a persistent object store has been undertaken by Brown [2]. The untuned user-space DWorFS implementation used for that study shows performance which suffers relative to the Solaris 2.5 NFS server due to the blocksize and synchronous write limitations of NFSv2, used by DWorFS, but in some write-heavy operations outperforms the Solaris server when the module minimises stable store write activity.

## 6 Conclusions and Future Directions

We have described a distributed storage service through which we are investigating the storage and I/O issues in metacomputing systems. The key attributes of this system are its decentralised design, the transparent interface for user applications, its extensibility in dataset and storage management, and its ability to take advantage of arbitrary storage resources in a flexible fashion.

Future work obviously consists of a complete DSS implementation and detailed performance analysis thereof. Using the system, we also intend to investigate the areas of data access pattern analysis and prediction, and compression and data layout issues. Security is of course also an area to be addressed; there are however other research efforts focussing more intensely on this aspect of distributed storage, and we will likely draw on their results and experiences rather than concentrate on this topic ourselves.

Our initial experience and performance results gained from implementing the upper layers of the DISCWorld Storage Service have shown that the design ideas described in this paper are feasible to implement. When we have completed the full implementation we will be able to investigate typical access behaviours, with a view to verifying our approximate predictions and design assumptions.

## 7 Acknowledgements

This work is being carried out as part of the Distributed High Performance Computing Infrastructure (DHPC-I) project of the Research Data Networks (RDN) Cooperative Research Center and is managed under the On-Line Data Archives Program of the Advanced Computational Systems (ACSys) CRC. RDN and ACSys are established under the Australian Government's CRC Program. Thanks to F.A.Vaughan for his encouragement in developing the system described here.

## References

1. Matt Blaze, A Cryptographic File System for Unix, *Proc. First ACM Conference on Communications and Computing Security*, Fairfax, Virginia, November 1993.
2. A. L. Brown, Utilising NFS to Expose Persistent Object Store I/O, *Proc. Sixth IDEA Workshop*, Rutherglen, Australia, January 1998.

3. Vincent Cate and Thomas Gross, Combining the Concepts of Compression and Caching for a Two-Level Filesystem, *Proc. Fourth ASPLOS*, Santa Clara, April 1992, pp. 200-211.
4. C. Catlett, and L. Smarr, Metacomputing, *Comm. ACM*, 35 (1992), pp. 44-52.
5. B. Christiansen, P. Cappello, M. F. Ionescu, M. O. Neary, K. E. Schauser, and D. Wu, Javelin: Internet-Based Parallel Computing Using Java, *Proc. ACM Workshop on Java for Science and Engineering Computation*, June 1997.
6. I.Foster, C.Kesselman, Globus: A Metacomputing Infrastructure Toolkit, *Intl. Journal of Supercomputer Applications*, 11(2):115-128, 1997.
7. David K. Gifford, Pierre Jouvelot, Mark A. Sheldon, James W. O'Toole, Jr., Semantic File Systems, *Proc. Thirteenth Symposium on Operating Systems Principles*, 1991.
8. Japanese Meteorological Satellite Center, The GMS User's Guide, 2nd Ed., 1989.
9. K.A.Hawick, H.A.James, K.J.Maciunas, F.A.Vaughan, A.L.Wendelborn, M.Buchhorn, M.Rezny, S.R.Taylor and M.D.Wilson, Geographic Information Systems Applications on an ATM-Based Distributed High Performance Computing System, *Proc. HPCN Europe '97*, Vienna, April 1997.
10. K.A.Hawick, P.D.Coddington, D.A.Grove, J.F.Hercus, H.A.James, K.E.Kerry, J.A.Mathew, C.J.Patten, A.J.Silis, F.A.Vaughan, DISCWorld: An Environment for Service-Based Metacomputing. *Future Generations of Computer Science Special Issue on Metacomputing*. Also DHPC Technical Report DHPC-042, April 1998.
11. K.A.Hawick, H.A.James, Craig J. Patten and F.A.Vaughan, DISCWorld: A Distributed High Performance Computing Environment, *Proc. HPCN Europe '98*, Amsterdam, April 1998.
12. H.A.James and K.A.Hawick, A Web-based Interface for On-Demand Processing of Satellite Imagery Archives, *Proc. Australian Computer Science Conference (ACSC) '98*, Perth, February 1998.
13. Craig J. Patten, F.A.Vaughan, K.A.Hawick and A.L.Brown, DWorFS: File System Support for Legacy Applications in DISCWorld, *Proc. Fifth IDEA Workshop*, Fremantle, February 1998.
14. Brian Pawlowski, Chet Juszczak, Peter Staubach, Carl Smith, Diane Lebel, David Hitz, NFS Version 3 Design and Implementation, *Proc. USENIX 1994 Summer Conference*, June 1994.
15. R. Sandberg, D. Goldberg, S. Kleiman, D. Walsh, and B. Lyon, Design and Implementation of the Sun Network Filesystem, *Proc. USENIX 1985 Summer Conference*, pp. 119-130, 1985.
16. M.Satyanaranan, J.H.Howard, D.N.Nichols, R.N.Sidebotham, A.Z.Spector, and M.J.West, The ITC Distributed File System: Principles and Design, *Proc. Tenth Symposium on Operating Systems Principles*, pp. 35-50, 1985.
17. StorageTek TimberWolf 9740 Specifications, URL http://www.stortek.com/StorageTek/hardware/tape/9740.
18. Sun Microsystems, NFS V2 Specification, RFC 1094, March 1989.
19. Sun Microsystems, NFS V3 Specification, RFC 1813, June 1995.
20. Sun Microsystems, WebNFS Client Specification, RFC 2054, October 1996.
21. Sun Microsystems, WebNFS Server Specification, RFC 2055, October 1996.
22. Sun Microsystems, WebNFS, April 1997.
23. Amin Vahdat, Thomas Anderson, Michael Dahlin, Eshwar Belani, David Culler, Paul Eastham, and Chad Yoshikawa, WebOS: Operating System Services for Wide Area Applications, *Proc. Seventh HPDC Conference*, Chicago, July 1998.

# Dynamic Visualization of Computations on the Internet

Jie Li and Elise de Doncker *

Western Michigan University, Kalamazoo MI 49008, USA,
{jli,elise}@cs.wmich.edu

**Abstract.** We present the design of a visualization tool which can be incorporated in parallel/ distributed code to track computational parameters with respect to convergence and the distribution of the work remaining for the parallel processes. We describe its functionality, algorithms and implementation in Java, as it is applied to a package for multivariate integration. Implementation aspects include message passing and control flow in its interaction with the application code. The approach is mainly asynchronous and based on multi-threading.

## 1 Introduction

Software implementing scientific algorithms is often complex especially for users from different disciplines. For a user it is thus helpful to view the behavior of the algorithm for the problem at hand, to verify that the problem is posed appropriately (and the inputs supplied accordingly), for help in debugging and for assessing the algorithm's efficiency (or its lack of it). This is particularly useful when a software package supplies what seems like more than one viable method to accomplish the same task. Furthermore, the user's faith in a method may be really taunted if the problem is solved on parallel processors.

On the other side, a tool for visualization of the computations helps the developers with debugging and with the cycle of performance evaluation and improvements. In this paper we will discuss the development and functionality of a tool for the visualization of computations on a distributed memory system. The framework is generally that of adaptive algorithms where tasks are selected and partitioned, their subtasks evaluated and inserted into a task pool. Our application is to a package (ParInt) which handles the computation of multivariate integrals by adaptive methods [1]. Other applications that fit a task pool model are, for example, some radiosity algorithms for image rendering, and some branch-and-bound methods.

The visualization component is invoked from a Graphical User Interface (GUI) implemented as a Java applet. The latter allows running the computational software over the Internet by a user at a remote site. In this paper we will describe the design of the visualization tool, its incorporation into the

---

* Research supported in part by the National Science Foundation under grant CCR-9405377.

GUI applet, its interaction with the Java server and its functions called by the back-end software for collecting/ sampling of the data.

## 2 Computational characteristics

The terms **automatic**, **adaptive** and **distributed** characterize the algorithms supported by this type of visualization tool. The term *automatic* refers to the fact that the results are required to within a user-specified accuracy or error tolerance $\epsilon$. The latter may measure an absolute or a relative error or a mixture of both. The algorithm performs the computation of its approximation $A$ to the (unknown) solution $S$, along with an error estimate $E$ pertaining to $A$, until it finds that $E \leq \epsilon$. The estimated error $E$ is intended to bound the actual absolute error $||S - A||$.

The term *adaptive* indicates that the course of the computation depends on the problem instance at hand. This may be achieved by partitioning the domain of the problem. A fixed formula is used over the original domain; if the error estimate is not satisfactory the domain is subdivided and the local formula applied over each part. From then on the region (task) with the largest error estimate is selected for further subdivision at each step. As a result, successive subdivisions are concentrated in areas where the estimated error is high, i.e., where the problem is difficult. The priority queue is implemented by a heap on the subregions, keyed with the subregion estimated error.

The automatic adaptive algorithm is *distributed* by partitioning the original domain among $p$ available (worker) processes. A distributed global heap version may consider the next $p$ regions of highest estimated error at each step. Alternatively, each worker process keeps a local heap [2]. Since the local heaps at different processes generally differ in size depending on the behavior of the problem on the originally assigned regions, a load balancing mechanism is in place to send work from busy to less busy processes.

The visualization tool displays dynamic information on: 1) the regions with error estimates larger than a given threshold, for each worker process; 2) the newly generated error estimates (over a set time window) for each worker process; and 3) the distribution of the regions over the worker processes at a given time. Type 1) gives information on the load of each process in terms of the number of tasks of which the error estimate exceeds the given threshold; 2) on the recent changes of load for each process; and 3) on the distribution of error and load over the processes.

## 3 System architecture and design

The major tasks of the visualization system are partitioned into three parts:

1. data sampling and collecting: done by the back-end computational code, which incorporates the function calls to generate the information needed for visualization;

2. data visualization: to convert the raw data for display (into easily understood representations such as graphs, histograms ...);
3. communication among the different parts: the back-end program, the Java applet (GUI) and the Java server.

The architecture of the system as applied to ParInt is depicted in Figure 1. Initially, the user starts the Java applet from a web browser and selects the

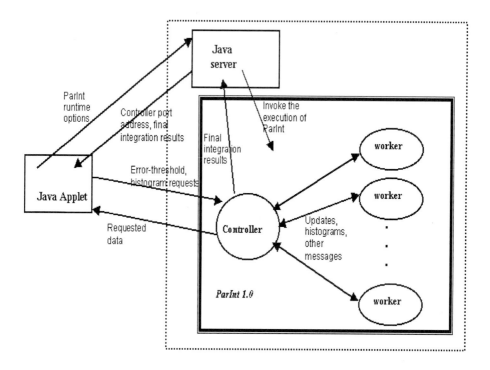

**Fig. 1.** Architecture

desired (ParInt) runtime options. Visualization can be turned on or off. A connection is established with the Java server at the ParInt site. The Java server then invokes the ParInt program according to the given runtime options. ParInt starts its parallel execution by invoking the requested number of processes, one of which acts as a controller. If the visualization option is turned on, the controller acts as a stream server listening on a specific port number assigned by the Java server. The Java server sends the address (socket port number) of the controller to the applet and the applet connects with the controller. During the computation, the applet sends visualization requests to the controller; the worker processes send the visualization data to the controller and the controller sends the necessary data to the applet. The applet displays the data in the requested form. At the end of the computation, the Java server collects the final results from the controller and sends them to the applet, where they are displayed for the user.

Regarding the interface with the back-end program (ParInt), the main design goals were to minimize the modification of the original back-end code and the extra communication and computation overhead involved in the visualization. The latter involves the additional information passed between the controller and workers as well as the computation required to process the visualization data.

Regarding the types of display in the Java part, we chose to represent the error distribution by a histogram of the *number of regions (tasks)* vs. *error level*. A time-series diagram is used to view the behavior of the estimated error and the load (number of error estimates above a given threshold) for each process. The relative independence of the visualization code from the rest of the applet has been an important design issue, especially since the applet is still under development.

The underlying communication interface between the three parts was realized using sockets, since they provide an easy way to connect C and Java programs.

## 4 Implementation

In this section we outline the incorporation of the visualization tool into the ParInt code (for multivariate integration) written in C and layered over MPI. We describe the message passing required for the visualization functions in the controller and worker algorithms of the computational code. We also give the the data retrieving algorithms for the time series diagrams representing processor loads (number of tasks) and errors vs. time, and for the histograms representing the number of tasks vs. error level. The organization of the visualization code is also outlined.

**Message passing and control flow.** If visualization is turned on, the controller will first create a socket and listen on the provided port number, until there is a connection request from the applet. During the computation, the controller probes to receive new error thresholds or histogram requests from the Java applet, in order to forward them to the workers. The controller also probes for messages from the workers with error estimates, loads or histogram data in each iteration of the computation loop; upon arrival these are sent to the applet.

Within each iteration of their major computation loop, the workers probe for new error thresholds or histogram requests. If an error threshold is received, then the subsequent counting of the regions will be based on this value, until a newer one arrives. The error estimates and corresponding number of regions are sent to the controller, embedded in an update message. If a histogram request is received, the worker sends the data to the controller including the worker id, the histogram array, the minimum and maximum error level (for scaling along the horizontal axis) and a time stamp.

In order not to slow down the computation significantly, the communications are performed asynchronously, apart from those of the initial error thresholds.

**Data retrieving algorithms.** The local error estimates required for display are provided by the back-end code as they are used to key the local heap of their process. Thus, additional work is only required to count the regions with error above the specified threshold (constituting the load) and to calculate the histogram of the number of regions per error level. These data are obtained by walking through the heap. The heap traversal is performed iteratively (as opposed to recursively) to minimize computational overhead (cf. Figure 2 and Figure 3 with pseudocode algorithm representations for the load and for the histogram, respectively). The histogram is scaled and divided into error levels of

```
Count = 0; // number of regions
Bool ParentOK; // parent does/ does not exceed threshold

P points to root of heap;
While(true) {
        If P's region-error exceeds threshold {
                ParentOK = true;
                Count++;
        }
        If ParentOK and has left child then
                P points to left child;
        Else if ParentOK and has right child then
                P points to right child;
        Else { // no children or Parent's region-error too small
                Do {
                        Walk up through heap;
                } until unvisited right sibling or root is reached;
                If root was reached then
                        Return Count;
                P points to unvisited right sibling;
        }
}
```

**Fig. 2.** Algorithm for load

equal width. The algorithm counts how many regions of the heap fall into each error level.

**Code Organization.** To minimize the changes to the back-end code, all major visualization functions are confined to separate files (visual-ctrl.c for the controller and visual-wrkr.c for the worker; header file visual.h was created to contain the new prototypes). The main changes in the ParInt controller and worker functions are done by calls to the visualization functions only; which is

```
            P points to root of heap;
            Divide error levels evenly between low_exp and high_exp;
            While(true) { // associate region pointed at by P with appropriate error level
                    If P has left child then
                            P points to left child;
                    Else if P has right child then
                            P points to right child
                    Else { // no children or Parent's region-error too small
                            Do {
                                    Walk up through heap;
                            } until unvisited right sibling or root is reached;
                            If root was reached then // finished traversal
                                    Return;
                            P points to unvisited right sibling;
                    }
            }
```

**Fig. 3.** Algorithm for histogram

made possible by the modular design. For maximum modularity, file scope static variables and functions are used in visual-ctrl.c and visual-wrkr.c.

### 4.1 The Java applet

**Visualization parameters.** The user may specify parameters for the three types of diagrams drawn for this application.

- The local estimated error and load vs. time for each process shows the $N$ most recent samples and is updated dynamically. The user may select the time interval to refresh the diagram and the number of samples $N$. Sample diagrams depicting local estimated errors and loads of four worker processes are given in Figure 4. For the sample problem at hand, the graphs show that worker 0 has most of the work initially, but the other workers receive work subsequently through load balancing.
- The global estimated error and number of regions diagram shows the current values vs. time and is updated dynamically. The user specifies the time interval for updating.
- The histogram of the number of regions per error level can be queried by the user at any time during the computation for any worker. The user can also "zoom in" on parts of the histogram by specifying the new error range. A sample is given in Figure 5.

In summary, the following options can be set via the applet to control the visualization functions: to turn visualization on/off; whether or not to display grids on the diagrams; to set the number of samples displayed in the time series diagrams; to specify the time interval for updating the dynamic

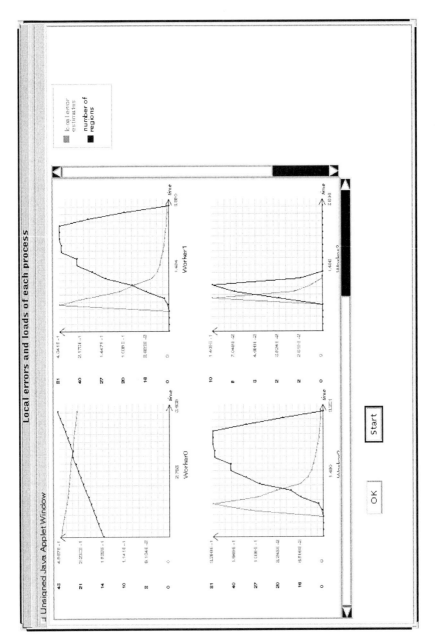

Fig. 4. Local estimated errors and loads

diagrams; and to set the error threshold. To this end, six windows can be opened: the visual options dialog; the local error estimates and load window; the global error estimates and number of regions window; the process id dialog; the histogram window; and the ZoomIn dialog, to focus on a histogram slice by specifying the new error range.

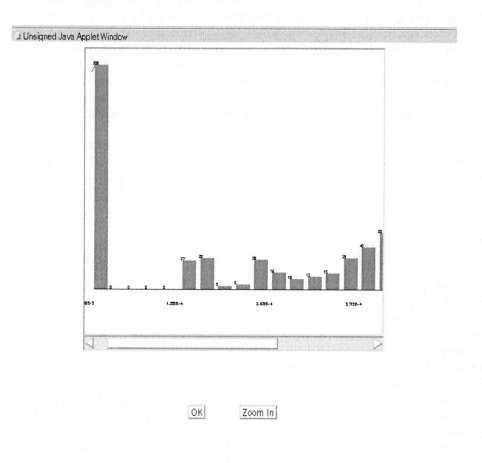

**Fig. 5.** Histogram

**Synchronization and mutual exclusion.** In Java, the Socket read and write primitives are all blocked primitives, which means that the thread will block on read or write if it is not ready to receive or send the data. We use multiple threads to resolve this problem since we need to receive data continuously in order to display the newest data dynamically.

When the connection with the ParInt controller is established, a thread listener is started which is dedicated to receiving data from the ParInt controller. A second thread is used to await the final results from the Java server. After the results arrive they are displayed. These threads run concurrently with the applet to alleviate blocking. There are two cases which require a synchronization of threads. 1) When the user requests a histogram, the query is sent to the controller immediately. Since all the data are received by the listener thread, a problem exists with respect to determining the arrival of the histogram. Therefore, the communication of the histogram needs to be synchronized. 2) The listener thread updates the data each time it receives an update from the controller. On the other hand, the drawing threads read the data at fixed time intervals. Updating the data while a diagram is being read and drawn would lead to inconsistent results; the updating and reading operations are mutually exclusive.

A semaphore class is used to resolve these problems [3]. In the first case, a P() at the GUI action handling thread operation is used to wait for the result; a V() operation is used by the listener to acknowledge the arrival of expected data and release the blocking thread. In the second case, pairs of P() and V() are used at the reading and writing ends to implement mutual exclusion.

## 5 Conclusion

We presented a dynamic visualization tool which can be incorporated into parallel/ distributed computational code to view information regarding the convergence of the computation and the partitioning of the work remaining for the parallel processes. For the algorithm or software designer the visualization tool may serve to assess the efficacy of the computational methods such as a load balancing strategy, or to tweak implementation parameters. For the user it provides an indication that the problem is solved efficiently or, otherwise, that it may need to be posed differently. In the context of multivariate integration, for example, a singularity of the integrand function interior to a subregion will escalate the number of subdivisions around the singularity and the load of the affected process. In that case, the user would generally benefit from pre-partitioning the domain so that the singularity is on or along an edge if possible. The interface of the visualization system with the application back-end code is relatively simple in view of the modularity of the design. Our use of Java for the implementation enhances its portability.

By way of comparison, some visualization systems of this nature are very application oriented. For example, the CAST project focused on the visualization of ocean data and delivered the ENVIS ENvironment VISualization system (http://www.cast.msstate.edu/CompletedProj/ENVIS.html). The goal of the SMART project is to design a Parallel Volume Visualization toolkit (PVV), with applications such as the modeling of pollution dispersion and the visualization of medical data (www.lsi.usp.br/smart). Others are targeted to certain parallel architectures or systems. The ParVis project targets massively parallel computer architectures (www.rigi.csc.uvic.ca/anu/parvis). The Distributed Array Query and Visualization (DAQV) system enables external clients to access and interact with distributed arrays. It is oriented toward Fortran applications implemented as SPMD processes which may communicate via a message passing system (www.cs.uoregon.edu/ hacks/research/daqv). Yet another class of visualization systems allow for process event tracking, for example to visualize sends and receives (as supported by XPVM/ PGPVM; and MPE). Exploratory Visualization (swarm.cs.wustl.edu/projects/explore) enables dynamic and animated process event monitoring.

**Acknowledgment.** We thank Srinivas Hasti and Rodger Zanny for discussions regarding the Java server and the computational back-end code of ParInt.

# References

1. E. DE DONCKER, A. GUPTA, J. BALL, P. EALY, AND A. GENZ, *Parint: A software package for parallel integration*, in $10^{th}$ ACM International Conference on Supercomputing, Kluwer Academic Publishers, 1996, pp. 149–156.
2. E. DE DONCKER, A. GUPTA, AND P. EALY, *Two methods for load balanced distributed integration*, in Lecture Notes in Computer Science, vol. 1067, Springer-Verlag, 1996, pp. 562–570.
3. E. W. DIJKSTRA, *Co-operating sequential processes*, in Programming Languages, F. Genuys, ed., Academic Press, New York, 1965, pp. 43–112.

# A Flexible Security System for Metacomputing Environments*

Adam Ferrari, Frederick Knabe, Marty Humphrey,
Steve Chapin, and Andrew Grimshaw

University of Virginia
Department of Computer Science

**Abstract.** A metacomputing environment is a collection of geographically distributed resources (people, computers, devices, databases) connected by one or more high-speed networks, and potentially spanning multiple administrative domains. Security is an essential part of metasystem design—high-level resources and services defined by the metacomputer must be protected from one another and from corrupted underlying resources, and underlying resources must minimize their vulnerability to attacks from the metacomputer level. We present the Legion security architecture, a flexible, adaptable framework for solving the metacomputing security problem. We demonstrate that this framework is sufficiently flexible to implement a wide range of security mechanisms and high-level policies.

## 1 Introduction

Legion [5, 6] is a distributed computing platform for combining very large collections of independently administered machines into single, coherent environments. Like a traditional operating system, Legion provides convenient user abstractions, services, and policy enforcement mechanisms over a diverse set of lower-level resources. The difference is that in Legion, these resources may consist of thousands of heterogeneous processors, storage systems, databases, legacy codes, and user objects, all distributed over wide-area networks spanning multiple administrative domains. Legion provides the means to pull these scattered components together into a single, object-based *metacomputer* that accommodates high degrees of flexibility and site autonomy.

Security is an essential part of the Legion design. In a metacomputing environment, the security problem can be divided into two main concerns: (1) protecting the metacomputer's high-level resources, services, and users from each other and from corrupted underlying resources, and (2) preserving the security policies of the underlying resources that form the foundation of the metacomputer and minimizing their vulnerability to attacks from the metacomputer level. For example, restricting who is able to configure a metacomputer-wide scheduling service would fall in the first category. Its solution requires metacomputer-specific definitions of identity, authorization, and access control. Meanwhile, enforcing a policy that permits only those metacomputer

---

* This work was funded by DARPA contract N66001-96-C-8527, DOE grant DE-FD02-96ER25290, DOE contract Sandia LD-9391, and DOE D459000-16-3C

users who have local accounts to run jobs on a given host falls in the second category. Its solution might require a map between local identities and verifiable metacomputer identities.

To satisfy users and administrators, a full security solution must address and reconcile both of these security concerns. Users must have confidence that the data and computations they create within the metacomputer are adequately protected. Administrators need assurances that by adding their resources to a metacomputer (and thus making those resources more accessible and valuable to users), they are not also introducing unreasonable security vulnerabilities into their systems.

Attempting to incorporate security as an add-on late in the implementation process has been problematic in a number of first-generation metacomputing systems such as PVM, MPI, and Mentat. To avoid this pitfall, the Legion group has addressed security issues since the earliest design phases [10]. Our metacomputing security model has three interrelated design goals: flexibility, autonomy, and breadth. *Flexibility* demands that the framework be adaptable to many different security policies and allow multiple policies to coexist. *Autonomy* is essential so that organizations and users within a metacomputing environment can select and enforce their desired security policies independently. Finally, *breadth* refers to the ability of the metacomputer's architectural framework to enable a rich set of security policy features.

These goals are strongly driven by our view that a fundamental capability of a metacomputer is its ability to scale over and across multiple trust domains. A Legion "system" is really a federation of meta- and lower-level resources from multiple domains, each with its own separately evaluated and enforced security policies. As such, there is no central kernel or trusted code base that can monitor and control all interactions between users and resources. Nor is there the concept of a superuser—no one person or entity controls all of the resources in a Legion system.

If it is to satisfy a broad range of security needs, our architecture must allow the implementation of a number of different security features. These include

– Isolation
– Access control for resources
– Identity of principals
– Detection and recovery
– Communication privacy and integrity
– Integration with standard mechanisms

The first point, isolation, refers to the ability of components in the metacomputer to insulate themselves from security breaches in other parts of the system. This feature is particularly important in large Legion networks, where we must generally assume that at least some underlying hosts have been compromised or may even be malicious.

In this paper we elaborate a metacomputing architecture based on our design goals that addresses both parts of the metacomputing security problem. In our discussion, we present examples of mechanisms we have designed or implemented within the architecture that enable a number of useful security policies, and provide examples of those policies.

## 2 Architectural Support for Security

Legion is composed of independent, active objects. All entities of interest within the system—processing resources, storage, users, etc.—are represented by objects [7]. Legion objects communicate via asynchronous method calls supported by an underlying message passing system. Each method call contains actual parameters and an optional set of *implicit parameters,* metadata that is available to called objects. Objects are instances of classes that define their interface, which is required to be a superset of a minimal *object-mandatory* interface. Object-mandatory methods include functions such as an interface query and methods to implement object persistence.

Legion objects are persistent, and are defined to be in one of two states: *active* or *inert*. When an object is active, it is hosted within its own running process and can service method calls. When an object is inert, its state (called its Object Persistent Representation, or OPR) is preserved on a storage device managed within the system. Objects implement internal methods to store and recover their dynamic state.

**Legion Runtime Library** The implementation of Legion objects is supported by a *Legion Runtime Library* (LRTL) interface. The LRTL defines the interfaces to services such as message passing, object control (e.g., creation, location, deletion), and other basic required mechanisms.

A critical element of the LRTL is its flexible, configurable protocol stack [9]. All of the processing performed in the construction of method calls at the sender and in handling them at the recipient is configured using a flexible, event-based model. This feature makes it especially convenient for tool builders to provide drop-in protocol layers for Legion objects. For example, adding message privacy through a cryptographic protocol is simply a matter of registering the appropriate message processing event handlers into the Legion protocol stack—the added service is transparent to the application developer.

**Core Objects** Within the Legion object model we define the interfaces to a set of basic classes that are fundamental to the operation of the system and that support the implementation of the object model itself.

*Host Objects* in Legion represent processing resources. When a Legion object is activated, it is a Host Object that actually creates a process to contain the newly activated object.[1] The Host Object thus controls access to its processing resource and can enforce local policies, e.g., ensuring that a user does not consume more processing time than allotted.

*Vault Objects* in Legion represent stable storage available within the system for containing OPRs. Just as Host Objects are the managers of active Legion objects, Vault Objects are the managers of inert Legion objects. For example, Vaults are the point of access control to storage resources, and can enforce policies such as file system allocations.

Hosts and Vaults provide the system with interfaces to processing and storage resources. The use of these interfaces is encapsulated by *Class Manager Objects*.[2] Class

---

[1] In some environments, the Host Object may enter the object as a new job to run in a queue management system, but this difference is transparent to the rest of Legion.

[2] In many of the cited Legion references, Class Manager Objects are referred to simply as "Class Objects."

Managers are responsible for managing the placement, activation, and deactivation of a set of objects, or *instances,* of a given class. They provide a central mechanism for specifying policy for a set of like objects. Policies set by the Class Manager include defining which implementations are valid for instances, which hosts are suitable for execution of instances, which users may create new instances, and so on. In addition to setting policy for instances, Class Managers serve as location authorities for instances, supporting the binding of object ids to low-level object addresses (typically an IP address plus port number).

A critical aspect of the Legion core object classes is that they define interfaces, not implementations. The Legion software distribution provides a number of default reference implementations of each core object type, but the model explicitly enables and encourages the configuration, extension, and even replacement of local core object implementations to suit site- and user-specific requirements. For example, by replacing the implementation of the Host Object, a site can define arbitrary mechanisms and policies for the usage of their computational resources.

## 3  Security Features in Legion

The Legion architecture is the critical foundation for satisfying the flexibility, autonomy, and breadth goals for our metacomputing security model. We now consider how those goals are met in the current system implementation.

**Identity** In Legion, every object is identified by a unique, location-independent Legion Object Identifier, or LOID. LOIDs consist of a variable number of binary fields. As a default Legion security practice, we use one of the LOID fields to store an X.509 certificate including (at a minimum) an RSA public key. By including an object's public key in its LOID, we make it easy for other objects to encrypt communications to that object or to verify messages signed by it. Objects can just extract the key from the LOID, rather than looking it up in some separate database, which eliminates some kinds of public key tampering.

Users in Legion also have LOIDs. A user creates his own LOID, which is then registered with the system and entered in appropriate system groups and access control lists by resource providers. When an object makes a call on behalf of the user, the user's LOID and associated credentials provide the basis for authentication and authorization. The ownership of a user's LOID resides in the user's unique knowledge of the private key that is paired with it. The private key is kept encrypted on disk, on a smart card, or in some other safe place.

For a resource, the essential step in deciding whether to grant an access request is to determine the identity of the caller. If a user communicates directly with the target object, he can establish his identity relatively easily with an authentication protocol. In a distributed object system, however, the user typically accesses resources indirectly, and objects need to be able to perform actions on his behalf. To transfer the user's identity in Legion, we issue *credentials* to objects. A credential is a list of rights granted by the credential's maker, normally the user or his proxy. A credential is passed through call chains, and is presented to a resource to gain access. The resource checks the rights in the credential and who the maker is, and uses that information in deciding to grant access.

There are two main types of credentials in Legion: *delegated credentials* and *bearer credentials*. A delegated credential specifies exactly who is granted the listed rights, whereas simple possession of a bearer credential grants the rights listed within it. A credential specifies the period for which it is valid, who is allowed to use the credential, and which method calls it can be used for. The credential also includes the identity and digital signature of its maker.

Tools or commands directly executed by the user create the credentials they need to carry out their actions. The credentials are made as specific as possible to avoid unnecessary dispersion of authority. Short timeouts in credentials, coupled with user-specific *Refresh Objects* that can revalidate expired credentials, permit a variety of recovery tactics if a credential or user key is stolen. Additional details concerning credentials and credential refresh can be found in [2].

**Access Control** In Legion, access is defined as the ability to call a method on an object. The object may represent a file, a Legion service, a device, or any other resource. Access control is not centralized in any one part of the Legion system. Each object is responsible for enforcing its own access control policy. It *may* collaborate with other objects in making an access decision, and indeed, this allows an administrator to control policy for multiple objects from a single point. The Legion architecture does not require this, however.

The general model for access control is that each method call received at an object passes through a *MayI* layer before being serviced. MayI is defined on a per-object basis, and is specified as an event in the configurable LRTL protocol stack [9]. MayI decides whether to grant access according to whatever policy it implements. If access is denied, the object will respond with an appropriate security exception.

MayI can be implemented in multiple ways. The default LRTL MayI implementation is based on access control lists and credential checking. In this MayI, *allow* and *deny* access control lists containing user and group LOIDs can be specified for each method in an object. When a method call is received, the credentials it carries are checked by MayI and compared against the access control lists. Multiple credentials can be carried in a call; checking continues until one provides access.

The form of access control provided by the default MayI is sufficient for some kinds of objects, such as file objects, but not for others. The LRTL configurable, event-based protocol stack makes it easy to replace or supplement the default MayI with extra functionality. Furthermore, the default MayI itself is relatively simple to modify if, for example, new forms of credentials or different kinds of access control lists must be supported. With the Legion security architecture, these types of changes can be made on a local basis without affecting other parts of a Legion system.

**Communication Privacy and Integrity** Encryption and integrity services are provided at the level of Legion messages. When a Legion message is prepared for sending, an event handler that implements a message security layer is triggered. This layer inspects the implicit parameters accompanying a message to determine which security functions to apply. In the current LRTL, a message may be sent with no security, in *private mode*, or in *protected mode*. In both private and protected modes, certain key elements of a message (e.g., any contained credentials) are encrypted using the public key of the recipient. The functional difference between the two modes is in how the rest of the message is treated. In private mode it is encrypted, whereas in protected mode only a digest is generated to provide an integrity guarantee. Unless private mode is already on, protected mode is selected automatically if a message contains credentials. This is a failsafe measure to prevent credentials from being transmitted in the clear. Details of the encryption mechanisms can be found in [2].

Because the mode in use is stored in implicit parameters, it propagates through call chains. For example, a user can select private mode when calling an object. All subsequent calls made by objects on behalf of the user will also use private mode. The default security layer does not provide mutual authentication. The sender can be assured of the identity of the recipient, because only the desired recipient can read the encrypted parts of the message. The recipient usually doesn't care who the actual sender is; its decisions are based solely on the credentials that arrived in the message.

**Object Management and Isolation** The management of active and inert objects by Legion core objects is an important point of local security mechanism and policy in Legion. Fundamentally, Legion software runs on existing operating systems with their own security policies. It is therefore critical that the implementation of the Legion object model ensure that extra-Legion mechanisms cannot be used to subvert higher-level security mechanisms. Similarly, it is important to ensure that Legion does not break local security policies at a site. A local system administrator is generally concerned with who can create processes on his system via Legion, what those processes can do, and who pays for their resource use. On Legion's side, there is a need to prevent user objects from interfering with one another or with core system objects (e.g., Hosts and Vaults), and to maintain the privacy of persistent state (OPRs). The latter is particularly significant because objects store their private keys in their OPRs.

The needs of Legion are common to any multi-user operating system, and our approach to providing them is to leverage off of existing operating system services. Our general strategy for isolating objects from one another in the default Legion implementation is to use separate accounts to execute different user objects. Similarly, we use local accounts and storage system protections to protect OPRs.

Accounts that can be used for these purposes fall into two categories. For those Legion users who happen to have accounts on the local system, processes and storage that represent the user's objects can be owned by the user's local account. For other users, we support the use of a pool of generic accounts that are designated for Legion use. The generic accounts usually have minimal permissions (e.g., no home directory, no group memberships, etc.). The local Host and Vault Objects use their own dedicated local accounts to ensure isolation from other user objects.

We encapsulate the privileged operations necessary for this policy in a *Process Control Daemon* (PCD) that executes on the host, providing services to the Host and Vault in a controlled fashion. The PCD is a small, easily vetted program that runs with root permissions. It is configured only to allow access by the user account on which the Host and Vault Objects are running. Its key functions are recursive change ownership of a directory, process creation under a designated account, and process termination. The PCD limits the user-ids to which these operations can be applied to a set configured by the local system administrator. The set includes the generic Legion accounts and potentially the accounts of local Legion users.

Alternatively, a local site policy may require that Kerberos be used to authenticate access to all local user accounts. Depending on the local Kerberos configuration, the Host Object can use forwarded Kerberos credentials, entries in users' Kerberos authorization files, or callbacks to user credential proxies to start objects on the appropriate accounts. The point is not how this is done, but that it can be done: Legion can adapt to a large range of security standards as necessary.

## 4 Policy Examples

Although Legion's flexibility allows the implementation of a wide variety of security mechanisms, application developers and site administrators typically have higher-level policy specifications in mind when using software. The particular underlying mechanisms are less important, as long as the user can be assured that high-level policy requirements are being met. In this section, we consider illustrative examples of how the Legion system architecture and existing Legion tools can be organized to meet sample site and application policies.

**Site Isolation** A Legion system can consist of multiple domains, each possibly in a different organization or trust domain. System administrators contributing resources to a larger metasystem typically require certain site-isolation properties. For example, consider a site that makes resources available to Legion, and is managed by a given local Legion administrator, who we will call *Admin*. A reasonable policy is that no matter how subverted any external sites in the Legion system might be, no intruder can invoke methods on local Legion resources as Admin. Such a policy is clearly desirable since Admin is likely to have administrative control over critical local resources: who can use which machine, and for how long; who can access which locally stored OPRs; etc. The ability to invoke methods as Admin is tantamount to complete control of the local Legion software.

The desired isolation policy can be achieved through a number of straightforward safeguards enabled by the Legion framework. First and foremost, all of the core objects managing the local site should be started and configured by Admin. This isolated domain startup avoids any external trust dependencies on outside systems. However, to achieve the desired functionality of a metacomputer, the local domain will be connected to some set of external Legion domains. After this link to the external (and untrusted) system is made, Admin must ensure that no messages containing his credentials are sent to off-site objects, as a subverted or malicious external site could then use Admin's credentials to break the isolation policy. However, simply stating that Admin should not pass credentials off-site is not good enough—Admin might make a simple mistake that could break the policy, so we would like automated enforcement of this safety measure. Such automated enforcement is easy in Legion: Admin simply uses a version of the LRTL in which the protocol stack is configured with an extra event handler for the message-send event. If a message is inadvertently directed off-site while containing Admin credentials, the message is blocked and the event handler raises an exception. With this simple modification to Admin's Legion environment, he can be assured that his credentials will not be dispersed to untrustworthy off-site objects.

Ensuring that Admin does not communicate with off-site objects has a desirable secondary effect. Since Admin cannot communicate with external, untrustworthy sites, he cannot place critical objects on resources at these sites. This benefit extends to an array of potentially critical, but not necessarily obvious, resources. For example, suppose Admin maintains a local Group Object listing the set of users that are allowed to start objects on local resources. If this object were allowed to execute on an untrustworthy site, its contents could be modified by a malicious resource owner, and local site-resource usage policy could be broken.

The two mechanisms described above, in combination with carefully configured access control for local core objects such as Hosts, Vaults, and critical Class Managers, ensure that the desired isolation policy will be met. Off-site objects will neither be able to generate nor steal local Admin's credentials. External callers will be prevented from invoking unauthorized methods on local critical resources, ensuring that local access control is not tampered with, local resource usage policies are not modified, and that security failures in other domains do not have serious consequences for the local site.

**Site-Wide Required Access Control** The Legion access control model as presented in Section 2 is based on the assumption that users will configure access control for their own objects. This concept adds a powerful level of flexibility to the system—for example, it makes arbitrary resource access policies possible. However, on first examination it appears to relinquish the ability for a system administrator to set site-wide policies about access control for user objects. For example, the default Legion access control configuration does not grant the administrator for a Legion domain access to other users' objects within the domain—there is no root user who can read any file or use any program in the domain. Such lack of ability to configure global, site-wide, mandatory access control policies may be unacceptable at some sites. However, the flexibility of the Legion architecture allows us to address this issue in a straightforward fashion using existing tools.

As an example of a site-wide access control policy, we consider the problem of prohibiting access to files by outside users. The Legion system defines a basic File Object that can be used to represent a file in the system. Access control for the normal Legion File Object is based on the default Legion MayI mechanism, which places no restrictions on what LOIDs (i.e., what users) may be placed on access control lists. To enforce the policy that files may not be accessed by outside users, we effectively want a way to control which LOIDs may be placed on the ACL for local file objects. We can achieve this policy using the power of local Host Objects to control access to local resources. The Host Objects at the site (which are owned and controlled by the local administrator) are a point of resource access policy—they define which types of objects may run at the site. Using this feature, the site administrator can strictly limit the classes of objects that may run at the site. In particular, the allowable set of classes can be limited to those that are approved by the system administrator. The list of allowable classes can be configured to only include file objects with an alternate MayI layer—an extended version of the default ACL mechanism that also verifies that allowed LOIDs are in a well-known group containing only the local site users. Given this simple configuration, the site administrator can ensure that files are not inadvertently exported to outside users through Legion. Furthermore, this approach generalizes to other site-wide access control restrictions, and other similar site-wide policy enforcement problems.

**Firewalls** Firewalls are a simple fact of life at many security-conscious institutions. While firewalls are not addressed explicitly in the Legion model, the Legion architecture is sufficiently flexible to accommodate firewalls with ease. As is typical in firewall situations, a proxy on the firewall host is the natural solution. However, the ability to use custom versions of the Legion core objects, and the flexible protocol stack model of the LRTL, allow proxy-based solutions to be employed in Legion in an especially straightforward, user-transparent way.

Objects started on hosts behind a firewall automatically have a Proxy Object on the firewall host assigned to them by their Host Object (in some cases, each user might desire their own proxy object; in other cases, a shared proxy object is acceptable; either model is simple to support). The object address for a newly activated object behind the firewall that is reported to the object's Class Manager is actually the address for the Proxy Object—when callers of the object bind its LOID to an object address, they will be given the address of the Proxy Object. The Proxy Object then acts as a simple reflector, forwarding any received messages to their intended destinations behind the firewall. Use of the Proxy Object to forward outbound messages from callers behind the firewall is automated by a transparent add-in event handler in the LRTL protocol stack.

**Resource Selection Policy** In principle, a user of a metacomputer shouldn't need to care which resources are used to execute his jobs. In practice, however, the trustworthiness of the resources that are selected for certain applications is of critical interest to the user. Policies regarding which resources may be used to execute objects are logically localized within the Class Managers of a user's object classes. In principle, any site selection policy can be encoded in a user's Class Manager Objects, giving the user total control over the selection and use of trustworthy sites.

Although this problem is solved cleanly at the architectural level in Legion, we deemed this issue of site selection for application users important enough to warrant special features in the default Class Manager Object reference implementations. All default Class Managers in Legion check for certain implicit parameters that can be used to limit resource selection. By setting these implicit parameters in his Legion environment (using a provided tool), the user can configure a resource selection policy that will propagate to all "create instance" methods called on Class Manager objects on behalf of the user. Of course, the architectural principle that users can encode any resource selection policy they wish in their own Class Manager implementations still holds; in fact, a convenient model for such customization is supported by the default Class Manager's ability to be configured to use an external *Scheduler Object* with a well-known interface. However, in the common case, where a user can generate a list of sites that he deems trustworthy and indicate this in his environment, the default implementation provides the mechanism to implement an effective resource selection policy.

## 5 Related Work

Two projects that incorporate security into large-scale distributed computing platforms are Globus and WebOS. Globus [3] is a "bag of services" model for metacomputing, in contrast to Legion's integrated environment approach. The Globus Security Infrastructure is a single sign-on authentication system that is deployed at each site in a Globus network. Different underlying authentication protocols such as Kerberos and SSL may be plugged into the infrastructure via GSS-API modules. A local Globus site uses the authentication information it receives in a request to make authorization decisions; it can also call back to a user's proxy to confirm the request.

The Globus Security Infrastructure essentially focuses on one component of the overall metacomputing security problem. Legion, with its "network operating system" perspective, addresses broader issues that allow the development of sophisticated security policies to manage metacomputer resources, as described in Section 4. The Legion architecture fundamentally permits greater autonomy and flexibility in the choice of security technologies and approaches. Globus does address an important part of the metacomputing security puzzle, however, and it could be chosen as an alternate mechanism to the current RSA approach for implementing identity and integrity in a Legion system.

CRISIS [1] is the security architecture for WebOS. WebOS is fundamentally different from Legion in terms of the basic services provided. WebOS provides a single, traditional file system and a fixed interface for authenticated remote process creation. CRISIS defines careful, effective security policies for these basic services. However, the CRISIS solution does not provide a means for easily developing security policies for new mechanisms as they are added to WebOS, nor does it provide a means for modifying the security policies supported for the existing services.

Two other projects related to security efforts in Legion, although not with the focus on metasystems, are Java and CORBA. The computational model of Java [4] (JDK 1.2) requires identity and authentication in order to execute digitally signed code downloaded from a remote site. The JDK provides per-class (or per-application) protection domains. However, it differs significantly from Legion in its lack of support for per-site security mechanisms, delegation, and user authentication.

The security model of CORBA [8] encompasses identification and authentication, authorization and access control, auditing, security of communication, non-repudiation, and security information administration. Typically, an ORB vendor implements CORBA security using existing technology such as GSS-API, Kerberos, and SESAME. Many of the goals of the CORBA security model are similar to the goals of the Legion security model, including simplicity, scalability, usability, and flexibility. However, CORBA is not a metacomputing system—it does not construct an operating system-like environment using underlying distributed resources. Given this fundamental difference in target use, CORBA does not address the metacomputing security problem.

## 6 Conclusions

We have presented the basic security architecture of the Legion system, and we have demonstrated that our design is sufficiently flexible to accommodate a wide variety of security-related mechanisms. This flexibility is critical to the successful deployment and use of metacomputing software. One-size-fits-all software dictated by a single group will never satisfy the requirements of the wide range of users and resource providers in a large-scale, cross-domain environment. We have also demonstrated that flexibility does not come at the price of complete lack of control. Within the flexible Legion framework, we showed how a number of important site-wide and application-wide security policies could be achieved. Naturally, the set of policies presented is only a small fraction of the policies that will be needed across the complete Legion environment.

The Legion system, including the security features described here, is currently publicly available. It is widely deployed on hundreds of machines at dozens of sites spanning multiple trust domains. Key portions of the software, such as the PCD described in Section 2, have been vetted and approved by system administrators at sites such as the San Diego Supercomputing Center and the US Naval Oceanographic Office (NAVO). In the future, we plan to continue deployment of Legion, developing additional mechanism and adapting to new site-local policies as required. We are also in the process of measuring the performance impact of key Legion security mechanisms.

## References

1. E. Belani, A. Vahdat, T. Anderson, and M. Dahlin. CRISIS: A wide area security architecture. In *Seventh USENIX Security Symposium*, Jan. 1998.
2. A. Ferrari, F. Knabe, M. Humphrey, S. Chapin, and A. Grimshaw. A flexible security system for metacomputing environments. Technical Report CS-98-36, Department of Computer Science, University of Virginia, Charlottesville, Virginia, Dec. 1998.
3. I. Foster, C. Kesselman, G. Tsudik, and S. Tuecke. A security architecture for computational grids. In *Fifth ACM Conference on Computers and Communications Security*, Nov. 1998.
4. L. Gong, M. Mueller, H. Prafullchandra, and R. Schemers. Going beyond the sandbox: An overview of the new security architecture in the Java development kit 1.2. In *USENIX Symposium on Internet Technologies and Systems*, pages 103–112, Dec. 1997.
5. A. S. Grimshaw and W. A. Wulf. Legion: A view from 50,000 feet. In *Fifth IEEE Symposium on High Performance Distributed Computing*, Aug. 1996.
6. A. S. Grimshaw and W. A. Wulf. The Legion vision of a worldwide virtual computer. *Communications of the ACM*, 40(1):39–45, Jan. 1997.
7. M. Lewis and A. Grimshaw. The core Legion object model. In *Fifth IEEE Symposium on High Performance Distributed Computing*, Aug. 1996.
8. Object Management Group. CORBAservices: Common object services specification, security service specification. Version 97-12-12, 1998.
9. C. Viles, M. Lewis, A. Ferrari, A. Nguyen-Tuong, and A. Grimshaw. Enabling flexibility in the legion run-time library. In *International Conference on Parallel and Distributed Processing Techniques and Applications*, pages 265–274, June 1997.
10. W. A. Wulf, C. Wang, and D. Kienzle. A new model of security for distributed systems. Technical Report CS-95-34, Department of Computer Science, University of Virginia, Charlottesville, Virginia, Aug. 1995.

# Computational Experiments Using Distributed Tools in a Web-Based Electronic Notebook Environment

Allen D. Malony and Jenifer L. Skidmore and Matthew J. Sottile

Department of Computer and Information Science, University of Oregon, Eugene, Oregon 97403-1202

**Abstract.** Computational environments used by scientists should provide high-level support for scientific processes that involve the integrated and systematic use of familiar abstractions from a laboratory setting, including notebooks, instruments, experiments, and analysis tools. However, doing so while hiding the complexities of the underlying computational platform is a challenge. ViNE is a web-based electronic notebook that implements a high-level interface for applying computational tools in scientific experiments in a location- and platform-independent manner. Using ViNE, a scientist can specify data and tools, and construct experiments that apply them in well-defined procedures. ViNE's implementation of the experiment abstraction offers the scientist easy-to-understand framework for building scientific processes. This paper discusses how ViNE implements computational experiments in distributed, heterogeneous computing environments.

## 1 Introduction

The increasing application of high-performance computing, storage, and graphics systems to scientific problem solving coupled with the ever-widening access to and interaction with scientific tools and knowledge made possible by high-performance networking, offers the potential to remove physical constraints of space and time in scientific investigation. However, scientists routinely employ *processes* in their work that specify many aspects of experimental procedures, including measurement protocols, sequenced data analyses, process monitoring, and results presentation. These processes provide the framework for scientific study and are commonly reflected in the physical operations of the traditional laboratory environment. Scientists also follow processes in computational environments that include the use of simulation programs, databases, analysis and visualization tools, networked computing resources, and web-based documentation [2, 3]. However, whereas these computational "instruments" are highly advanced, the support for the computational science processes is often simplistic and limited. One reason is that the processes are naturally defined with respect to the scientific domain of abstraction and these abstractions are not easily mapped to the system-level complexities of the computational environment and infrastructure. In this paper, we discuss a web-based framework called *ViNE* designed to support computational scientific processes in a high-performance distributed environment.

The Virtual Notebook Environment (*ViNE*) is a platform-independent system that manages a range of scientific activities across distributed, heterogeneous computing platforms [10]. On the one hand, ViNE provides the web-based equivalent of common paper-based lab notebooks with additional features for notebook sharing, security, and collaboration. Other electronic notebook systems offer similar capabilities [4, 5, 6, 8] in addition to features such as automatic data acquisition [9] and image annotation [5]. On the other hand, ViNE is unique in its ability to represent, manage, and execute *computational experiments*; i.e., support scientific processes that involve the use of data, tools, and programs throughout the scientist's computing repetoire. ViNE hides system-level complexities, allowing scientists to specify and launch computational experiments using a visual specification language. The scientist is freed from concerns about inter-tool connectivity, data distribution, data management, and machine idiosyncracies. Experiment specification is abstracted in the notebook interface and results from experiments can be dynamically linked back into the notebook content.

We focus here specifically on ViNE's support for computational experiments. In particular, we describe how ViNE represents experimental processes in the notebook and executes the experiments in a distributed, heterogeneous system. In section §2, we give a high-level view of ViNE's system architecture. Section §3 briefly describes some of the components that make up the ViNE system. The framework for computational experiments is discussed in detail in Section §4. We conclude the paper with our future plans.

## 2 ViNE Architecture

The Virtual Notebook Environment provides functionality through five components: *browse, administration, data organization, tool organization,* and *experiment* [10]. The administration and browse components provide the basic notebook functions which include managing notebook structure (pages, chapters, content), security parameters, and navigation. Extended functionality such as describing data, tools, and analysis tasks are provided by the data organization, tool organization and experiment functional components, respectively. These five components work together to access, store and manipulate notebooks, data, and tools on a given machine.

ViNE is constructed to operate in a distributed, heterogeneous system, and its components are spread across the machines used in the computational environment. A *leaf* is the abstract entity that contains the functional components, notebooks, data, and tools. In addition, a leaf contains a communication server, called a *stem*, and a webserver. The stem is responsible for sending and receiving URL formatted messages that encode ViNE's commands and protocols between the leaves. The webserver provides the notebook's user interface.

A leaf can contain all five components, providing full ViNE functionality for the user. It is also possible to have a more restricted leaf where only the notebook administration and browsing components are available. This type of leaf is a notebook server that only provides the basic notebook functions and does not have any extended functionality. In addition, there are special leaves that do not contain any components, but do have stems and web servers. These leaves can be used to provide access to data stored in large data repositories or act as computation servers running specialized tasks, offering users with a wider spectrum of tools. These leaves can be thought of as data and computation servers that do not house notebooks. The leaves can be configured and located in a way that best fits the distributed system.

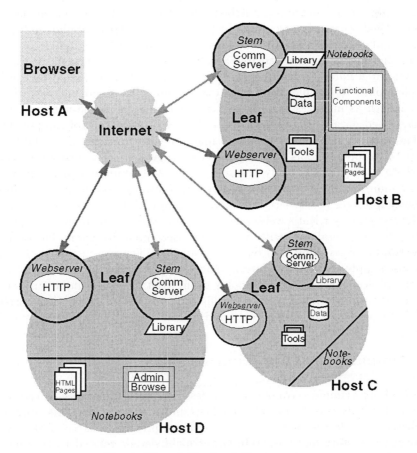

**Fig. 1.** Virtual Notebook Environment

Together, the leaves distributed across the computational platforms define the notebook environment. An example of a notebook environment can be seen in Figure 1. It contains a leaf that provides complete functionality (Host B), a computational and data server (Host C), and a notebook server (Host D). ViNE's architecture consists of one or more notebooks environments.

## 3 ViNE Components

The *data organization, tool organization,* and *experiment* components work together to provide the extended functionality of the notebook. The data and tool organization components implement the management facilities for a notebook's resources. Once data and tools have been described in the notebook, the user is able to combine their data and tools to perform analysis tasks, which we call *experiments*; see section §4.

The data component allows notebooks to reference data throughout the ViNE system. ViNE currently recognizes just two types of data, tabular and general data. Tabular data is a matrix of values, fully described by its dimensions. Both tabular and general data are treated as a flat files. A general data file is described by giving it a name, providing the name of the host it is located on, and the complete path to the file. Data that can be interpreted and understood when viewed, such as matrices and images, can be displayed on the notebook pages. All described data is available for use in the experiment control component.

The tool organization component maintains descriptions of available tools and their wrappers. Tools are categorized as either general or custom depending on whether they can be fully controlled from the command line (general) or require more complex interaction and control (custom). Wrappers are Common Gateway Interface (CGI) scripts that run a tool with the given parameters and can be accessed by a leaf's webserver. ViNE provides wrappers for general tools. To register a general tool with ViNE, the user enters its name, location, and a description of its input and output parameters. For a a custom tool, he/she must also supply a wrapper to interface between ViNE and the tool. The wrapper is responsible for accepting a complex string containing the input and output information, translating it for the specific tool, and executing the tool correctly. In either case, once the tool is registered, it can be used in experiments.

## 4 Computational Experiments

The ViNE experiment component provides users with the ability to create and execute distributed tasks for data analysis and manipulation. This component was designed with the target user in mind and provides a layer of abstraction between the system and user. This layer hides details related to the function and implementation of the underlying data transport and control mechanisms, leaving the user to concentrate only on the task they wish to accomplish. The experiment component is divided into two sub-components that together provide all necessary functions - the *experiment builder* and the *experiment controller*.

Before discussing the software architecture, it will be necessary to introduce the representation of experiments within the environment. In general, an *experiment* within ViNE is a sequence of data transfers and operations. An experiment is represented as a directed acyclic graph of computational nodes and data-flow connections. This representation is advantageous to the implementation for many reasons. For constructing experiments, one can easily construct the data-flow graph within a visual application. Well known algorithms for proving certain characteristics about graphs, such as cycle elimination, allow for simple error-detection at construction time to avoid potential problems at execution. Finally, the use of a graph representation in the controller provides the capability to easily describe parallel task execution within a single experiment.

One of the important design considerations within experiments was defining the input/output characteristics of the nodes and data-flow connections. As stated above, when a tool is introduced into ViNE the user must describe the I/O characteristics of the wrapper. Since experiments also involve data files and previously constructed experiments, the I/O treatment of such entities also must be defined. It is assumed that a data files is available for either reading or writing at any given moment. In order to protect data integrity, data files are allowed a single input during an experiment. Since read operations cannot damage data and data is assumed to be static once created, an arbitrary number of entities can receive input from a given data file. The I/O characteristics of existing experiments are defined so that any experiment may be used within a new experiment. More explanation of experiment construction and execution is necessary to describe this fully and is given in the experiment control section below.

### 4.1 Experiment Builder

Construction of experiments is accomplished using a Java [1] applet executed within any Java-enabled web browser. The builder provides a visual environment in which users may choose data, tools, and experiments contained within their notebook to create new experiments. Entities are chosen and placed into a canvas on which the user creates connections between the entities to represent data-flow within the experiment. Tools within an experiment may be configured within the applet by entering information into a configuration frame which appears when the tools is added or selected. These configuration frames use information from the tool description component to allow the user to set parameters for the tool execution. Tools which do not utilize the general wrapper architecture may provide a custom frame which is dynamically loaded into the applet to allow user configuration of the custom wrapper.

The custom tool configuration frames and other portions of the builder rely on its object-oriented design and implementation. The representation of an experiment within the builder is as a set of objects representing each node and connection. The node classes all derive from a generic parent containing the basic information shared by all node types, while providing an abstract API which each specific sub-class must provide regardless of the implementation. By using an approach such as this, the generic API provided to entities allows classes such as the configuration frames to be rapidly created and integrated into the basic system with no modification.

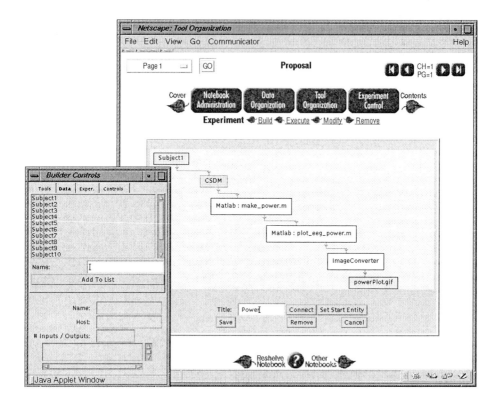

**Fig. 2.** Experiment Builder

In addition to the classes which represent the experiment itself, the experiment builder contains classes which perform experiment verification and routine GUI tasks. The GUI classes do not deserve much attention and simply provide a flexible canvas for drawing experiments while providing access to configuration information. The verification classes are able to rapidly test the experiment during construction to ensure that the resulting graph conforms to certain basic requirements. With the addition of data-flow connections, the verification class notifies the user and prevents the creation of dangerous cycles or impossible input/output combinations (such as an inappropriate number of input or output streams.) Database access scripts, implemented using Perl, provide a means for querying and modifying the database's existing experiment, tool, and data information. A simple CGI-based communication scheme to these backend database access scripts, in combination with the applet components, provide all required functionality for experiment creation.

## 4.2 Experiment Controller

Like the experiment builder, the ViNE *experiment controller* application is also implemented using Java for a variety of reasons. An object based design was used to implement the controller, taking advantage of Java thread, networking, and I/O constructs available within the default library. The controller utilizes the Java-based stems to perform actions such as locating data and tools, retrieving security information, and transferring data between leaves. At execution time the controller is invoked on a single leaf and manages data and tool invocation for the duration of the experiment.

Before execution the controller reads the description of the data flow graph as constructed in the builder applet. Since this graph was checked for errors during construction, this process is not required within the controller. The data flow graph is reconstructed as a network of objects within the controller, with each object containing required dependency information so that the correct order of operations will occur. Synchronization is accomplished by implementing each node as a thread object and using built in thread methods for monitoring dependency states. Reconstructing the experiment graph involves two steps - first, each node is created with all information relevant to finding and executing or moving the appropriate entity. If the node represents an existing experiment, the representative node is substituted by the entire experiment graph.

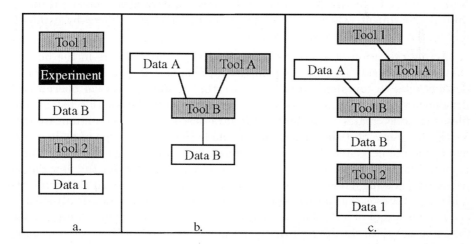

**Fig. 3.** Experiment Substitution : Experiment (a) contains inlined experiment (b), resulting in the actual executed experiment (c).

Once the nodes exist as threads, the connection information is read and each connection is established. Connections are established by giving each node with an input a reference to the node providing the input data. Nodes which are referenced as dependencies also are given a count of nodes who require their output. This is used later in managing execution workspaces. This node is then monitored by the receiving node in so that execution begins only when the output data is ready. Once all nodes and connections are created, all nodes are given a reference to a single semaphore thread and each thread is started. Once all threads are started, each attempts to **join** the semaphore which ceases execution when the last node is started. This causes all threads to start simultaneously. Each thread then attempts to **join** all threads it requires as input. Threads which have no dependency list start immediately while all others enter a waiting state. The experiment is completed when all node threads have terminated.

Experiment substitution for inlined experiments, as mentioned above, involves a more complicated I/O treatment than data or tools. Input to an inlined experiment depends on which nodes within the inlined experiment have no dependencies. If a tool occurs in this set of nodes with no dependencies, any input given to the experiment is passed to these tools regardless of whether or not it is used. If data files occur as nodes without dependencies, then incoming data is ignored for these particular nodes. Output from an inlined experiment is somewhat ambiguous to define. Since experiments are not required to have single input/single output behavior, an experiment may produce many data files as the final step of execution. Deciding which of these to pass along to the following nodes in the new experiment is difficult, and for obvious reasons such as performance and space constraints, all of the output data cannot simply be given as input leaving the receiver to decide which to use. For these reasons, output from an inlined experiment is assumed to arrive in data files defined within the inlined experiment, which can be utilized in the new experiment by explicitly adding the required data files as nodes.

Actual execution of a node is a simple process which occurs in the body of the thread objects. The first step in the lifetime of a thread is to wait for its dependencies to complete execution. Once this occurs, the thread queries each dependency thread for a list of available output data. Given this list of data files, the node creates an instance of a Java-based client to communicate with ViNEs Java servers. Using this client, the node checks what type of node it is to determine the next step. If the node is a data file, input from the previous node is moved into the notebook and a database entry is created for the new data if necessary (existing data of a given logical name is overwritten). At this point, the node terminates execution and threads that depend on it automatically start. If the node is a tool object, the execution process is somewhat more complex. The first step is to use the Java client to create a workspace on the host containing the tool, into which all input and output data is temporarily stored. The Java client is then given the set of inputs which are moved from the dependency nodes to the current workspace. Once all data is moved, the dependency nodes are notified and are allowed to remove their workspaces if all nodes which depend upon them have sent notification.

When all input data is within the workspace, the node then copies any additional files required for tool execution to the workspace. This is required for tools such as MatLab [7], in which programs may require additional modules to exist within the workspace where the execution occurs. As soon as all data

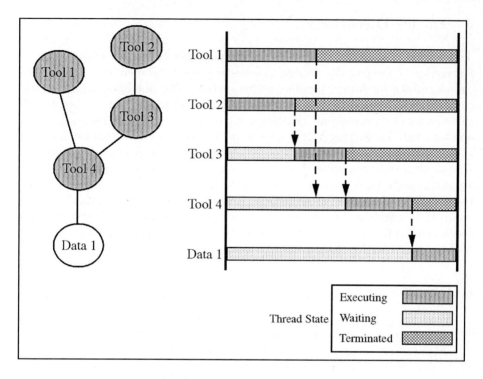

**Fig. 4.** Thread Lifetime

and other required files are completely moved, the client is used to invoke the CGI-based tool wrapper. When the tool has completed execution, the wrapper returns output information to the client, which then passes it to the node thread for further processing. The node uses this output information, along with a directory listing of the workspace to build the list of output files created during runtime. This list is then made available for other nodes which depend on this node, and execution of the thread terminates.

Experiment output is placed within the notebook of the user for future use and analysis. If, for example, a visualization is created, the user may then place this within a notebook page for later browsing. Status information of experiments is also stored within the notebook so users can monitor running experiments or analyze the behavior of previously executed experiments. Since experiments are run within the experiment controller, an arbitrary number of simultaneous experiments may execute, constrained only by system resources. Each executing experiment involves starting a new instance of the controller software, while all using a common Java server. By using a common Java server to which all clients connect for data movement and workspace creation, collisions of workspaces and intermediate file naming is prevented.

## 5 Future Directions

The current version of ViNE is a prototype being tested by scientists at the University of Oregon [2]. During this testing, we have identified a number of enhancements for future versions of the system. The notebook administration component will be extended with a framework for integrating editors appropriate to a variety of data content, allowing users to directly manipulate images, data, and text without leaving ViNE. Currently, we use a persistent-server model for leaf process scheduling but to support systems with low resources, we will provide on-demand processing. For the scientist with very large amounts of data, we will provide more sophisticated ways to organize it, and we need to be able to link directly to existing databases. Adding support for automated maintenance of the tools' configurations described in the notebook is another area of future development. Finally, we will enhance the experiment component, adding facilities for experiment monitoring, steering, and analysis.

## References

1. K. Arnold and J. Gosling, *The Java Programming Language*, Addison-Welsey, 1996.
2. Brain Electrophysiology Laboratory.
   http://hebb.uoregon.edu/brainlab/belHome.html.
3. J. Cuny et. al. Building Domain-Specific Environments for Computational Science: A Case Study in Seismic Tomography, *International Journal of Supercomputing Applications and High Performance Computing*, Vol. 11, No. 3, pp. 179-196, 1997.
4. D. Edelson, R. Pea, and L. Gomez, The Collaboratory Notebook, *Communications of the ACM*, Vol. 39, No. 4, April 1996, pp. 32–33.
5. A. Geist and N. Nachtigal, Oak Ridge National Laboratory Electronic Notebook Project. http://www.epm.ornl.gov/geist/java/applets/enote/.
6. J. Hong et al., Personal Electronic Notebook with Sharing, *Proceedings of the Fourth Workshop on Enabling Technology: Infrastructure for Collaborative Enterprises*, April 1995.
7. The Math Works, *MATLAB: High Performance Numeric Computation and Visualization Software Reference Guide* Natick, MA. 1992.
8. J. Myers et al., Electronic Laboratory Notebooks for Collaborative Research, *Proceedings of the IEEE Fifth Workshop on Enabling Technology: Infrastructure for Collaborative Enterprises*, June 1996.
9. B. Rex and D. St. Pierre, PNNL NMR Electronic Logbook, *Proceedings of the IEEE Fifth Workshop on Enabling Technology: Infrastructure for Collaborative Enterprises*, June 1996.
10. J. Skidmore, et al., A Prototype Notebook-Based Environment for Computational Tools, *Proceedings of SC98*, November 1998.

# A Parallel/Distributed Architecture for Hierarchically Heterogeneous Web-Based Cooperative Applications

Thomas L. Casavant, Todd E. Scheetz, Terry A. Braun,
Kyle J. Munn, and Sureshkumar Kaliannan

Parallel Processing Laboratory,
Dept. of Electrical and Computer Engineering,
and the Genetics Program
University of Iowa
Iowa City, IA 52242
U.S.A.
genomap@eng.uiowa.edu

**Abstract.** A new class of applications is described which requires cooperation among diverse users in multiple data and problem instance domains. The hierarchy of parallelism includes heterogeneity within a single instance of the problem, homogeneity among subsets of users within a problem domain, and multiple problem domains which share computational resources – software and hardware. The core of the architecture is a socket-server which registers clients and servers (both statically and dynamically), and assures isolation of users in separate problem domains. The users all see the system as a set of functions accessible via the WWW. The particular problem of genetic linkage analysis is used as a case study to illustrate and implement the architecture. GenoMap, the first implementation of this system is being deployed for several groups of cooperating users at multiple institutions in a study to isolate the genomic locus of the controlling gene(s) in several diseases including autism. More than 400 genetic markers are being analyzed from more than 300 individuals in this study. The users span geneticists, clinical physicians, statisticians, disease specialists, laboratory technicians, and computer scientists/engineers.

## 1 Introduction

This paper describes a general software architecture for a class of applications made possible by the WWW. The motivation for our study is a medical research application in the Human Genome Project (HGP) – Genetic Linkage Analysis. Genetic Linkage Analysis is the primary method used to determine the association between a genetically-linked trait (of interest to us are disease traits such as cancer, autism, etc.) and a locus of the human genome (a search space of some three billion base pairs). This particular problem requires cooperation among a diverse collection of researchers including geneticists, clinical physicians, statisticians, disease specialists, laboratory technicians, and computer scientists/engineers. The scale of this problem would clearly indicate the need for large scale informatics support. Historically, however, relatively rare, simple traits have been studied and ad hoc methods for data storage and analysis have been employed successfully to isolate only hundreds of disease genes (from a set of approximately 100,000 candidates) to date. The experience gained in these studies, coupled with the application of distributed/parallel computing methods, is making possible the systematic approach of the isolation of all the disease genes in the human genome. Below, we characterize the primary parallel/distributed characteristics of this class of applications. Sections 2 and 3 describe GenoMap and its various components. The current implementation of GenoMap is briefly discussed in Section 4, for more information the reader is directed to [1]. To date, the first internal release of GenoMap has been in use for approximately 2 years and is presently being employed in the analysis of more than 400 genetic loci with data describing disease traits from more than 300 individuals with respect to the disease autism.

The class of applications of interest have parallelism present at three distinct levels. We describe them in a hierarchy, and also distinguish between multiplicity of users (sets of clinicians) and multiplicity of problems (diseases).

**Level 1.** Heterogeneous parallelism within a single problem instance. These components represent functional parallelism.
**Level 2.** Multiple instances of each functional component referred to in level 1.
**Level 3.** Multiple instances of the entire problem class, potentially with multiple instances of each functional component in level 1.

Level 1 illustrates a key attribute of an application of this class. In the genetics linkage setting, this refers to the need to support the computational demands of each of the diverse members of the team of users – physicians, statisticians, laboratory technicians, etc. Each user needs to see the system through the "window" of their part of the collaboration. In level 2, we are describing the oft-present attribute of scale. As a problem instance becomes large, not only do computational demands grow, but so do the number of project personnel and their geographical dispersion. In the case of our autism study, the scale of the project dictated that distinct groups (one at the University of Iowa, one at Tufts University, and one at Vanderbilt University) cooperate in the labor-intensive gathering of phenotypes (patient status w.r.t. autism), genotypes (the states of the 400 different genome loci), and conducting the many analyses required to determine which (if any) of the loci were primary candidates for autism. At the third level of the hierarchy, it is desirable to manage the computational resources, in addition to the data resources, in a cooperative way. This involves the sharing of software and hardware. Users in many problems domains (such as medicine and biology) do not have easy access to the kind of support for cooperative computing as dictated in this class of applications. Thus, the solution to such a problem requires the ability to support sharing of data and software, while assuring secure isolation of data from multiple distinct instances of the problem.

Clearly, in handling information regarding genetic traits, and diseases, the need for such security is clear.

We summarize the requirements of our solution as follows:

- The solution must support seamless sharing and cooperation of users at levels 1 and 2.
- The solution must assure that parallelism at level 3 is supported with secure isolation of distributed components.

GenoMap, a distributed, Web-based system for the support of genetic linkage analysis, illustrates the class of applications, and the key elements of our approach to solving the problem.

## 2 The GenoMap System from the User View

GenoMap is a suite of independent, yet inter-related tools primarily developed in Java. These tools manipulate data from a domain-specific networked database, allowing sharing of information among multiple distributed clients without replication and coherency problems. The main goal of GenoMap is to provide a portable, intuitive interface for managing the information associated with the gene location/discovery process. The following table briefly describes the various services provided by GenoMap. A more complete description of each these components appears elsewhere [1], [2].

Table 1. GenoMap Components

| Name of Software Component | Description |
|---|---|
| Subject Log | Recording and searching for subject specific information. |
| Marker Log | Recording and searching for marker specific information. |
| Trait Log | Recording and searching for trait specific information. |
| Linkage Experiment Editor | Creating, viewing, and updating linkage experiments |
| Genotyping Assistant | Creating, viewing, and updating genotyping experiments |
| GenoScape | Viewing and genotyping of genotyping gel images |
| GenoScape Launcher (Client/Server) | Front-end interface to genotyping tool |
| Verification (Client/Server) | Interface to display all recorded genotypes together |
| Linkage Analysis (Client/Server) | Interface to multiple linkage analysis packages |
| Socket Server | Manages communication between clients and servers, plus handles load-balancing |

## 3 GenoMap Architecture

GenoMap is a large-scale, distributed, heterogeneous, client-server application to support the systematic exploration of the genome to narrow, and ultimately identify, the locus of a particular gene (or set of genes) involved in a disease or trait. In contrast to many applications developed in support of the Human Genome Project (HGP) [5] to date, GenoMap does not involve gathering or analysis of any DNA sequence data. Rather, the fundamental informational components of this *functional genomics* [8] application are 1) familial relationships (pedigree information), 2) clinical observations of disease, or trait (phenotype), 3) sets of known polymorphic genome loci (genetic markers), and 4) information about the state of candidate loci for the individuals being studied (genotypes). There are two primary ramifications of these characteristics that distinguish this problem from most existing network-based genome analysis applications.

1. The need for privacy regarding pedigrees, genotypes, and phenotypes, and
2. the need to support a diverse collection of cooperating individuals in the gene isolation process.

The first implication requires protection of the data being used to perform analyses, and the second naturally suggests a heterogeneous, distributed solution. However, if these are the primary requirements of the solution, then they present an immediate conflict. Distribution of data to heterogeneous computation sites inherently involves taking risks with the security of data. The approach taken by GenoMap provides for security by:

- Verifying identities of individuals in the granting of *access* to sensitive data through the network:
    - Passwords protect access to the the tools themselves.
    - Database contents can only be retrieved in the context of a single functional tool of the system. Thus users are prevented from making "custom" queries to extract information that would compromise individual privacy.
    - The database structure itself hinders the association of individual identities with their clinical, or genetic information.
- Databases containing sensitive data can be kept localized with the computations accessing them, thus allowing local administrative control, and allowing physical restrictions to be placed on the set of IP addresses allowed to access a particular database.

Most of GenoMap has been implemented in Java with a socket-oriented, client-server design employing recent applet security features [3]. The gene identification application supported by GenoMap is characterized by a two stage process of data gathering and verification, followed by an analysis phase known as *genetic linkage analysis* [7], [9]. Java applets provide interfaces for specifying *linkage experiments*, support for management of the data collection and verification process, and interacting with statistical linkage analysis packages [4]. One of the data collection tools is a large C/X-windows application – GenoScape [2], for scoring gels with repeat markers [6], and many of the analysis packages are pre-existing and run in a UNIX environment (originally written in various languages – Fortran, Pascal, etc.). However, to support both security, interoperability, and sharing of the software among multiple laboratories and projects, a novel component of our system is a *socket server process* (SSP) that provides a naming service to the applets and applications, controls access to applications and data, and provides a load balancing function.

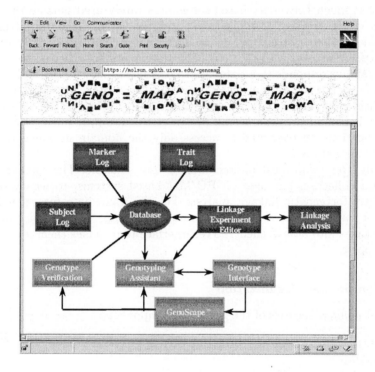

**Fig. 1.** User's WWW View of GenoMap

Figure 1 shows the main interface window that provides not only the common user view of the GenoMap system, but also serves as a central starting point for launching all GenoMap functions, and validating user's identity.

### 3.1 Requirements

The GenoMap system is designed to support a diverse collection of users in a wide-area network environment. The data objects being managed are of the most sensitive nature – usually identifying family relationships among members, some of whom may carry stigmatizing genetic diseases.

An additional requirement is that GenoMap be portable and sharable among multiple research laboratories. Due to the difficulties and costs associated with updating and distributing copies of software, we have decided that GenoMap must be a web-based system – i.e., whenever possible, the only static copies of GenoMap applets and applications will be stored on the web server at the University of Iowa. While appearing to be a potential bottleneck in several ways, in fact, this allows our group to make full use of the system locally, while not having to expend efforts supporting users outside of our local site with updates, releases, and copies of documentation. In fact, these tasks represent the primary reason that much University-based software is rarely used outside the domain in which it was created. In order to support the use of GenoMap at multiple, geographically dispersed sites we define the notion of a *database domain* or simply *domain*. Users register with the GenoMap system and are assigned to a domain. The domain corresponds to a database that is most likely stored and served on a machine within the administrative domain of the user. This greatly reduces the chances that security violations with respect to privacy of data will occur. However, it complicates the overall design of GenoMap, requiring that applets and applications be restricted to access only the domain in which they were originally instantiated.

Finally, GenoMap must interact with extant software. The software ranges from data collection packages on PC/Mac-based systems, to legacy codes in FORTRAN for genetic linkage analysis. For the immediate future, it is a requirement for GenoMap to directly be able to "wrap" the interface details of such packages into a Java client/server interface that makes them appear to be a Java applet.

### 3.2 Component Organization

A typical *domain* consists of the following components:

***Web Server*** It hosts all the User Interface applets for the services offered by GenoMap.

***Services*** A service in GenoMap consists of two or more components/resources; A client/applet which implements a User Interface, and a backend server which does all the computations, implements all file accesses, or is a database which stores all the information required for the service. Not all services require a backend server. Each user has privileges which determine which service she/he is allowed to access.

***Server*** A server is an application that performs computations, or stores/retrieves data on behalf of a client. A database is a type of server.

***Clients*** Clients are the software/applets that allow access to the servers through WWW browsers.

***Socket Server*** The socket server primarily performs the following four functions:

   **1. Storage of Information** The socket server keeps track of where the servers are running and to which port each is listening for client requests.

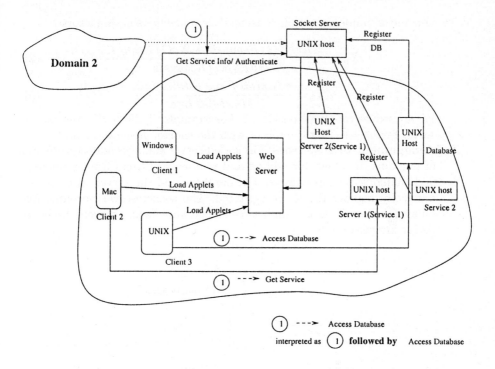

**Fig. 2.** Domain Components

It stores the database location and maintains the list of registered domains and the users in each domain along with their privileges.

2. **Load Balancing** The goal of load balancing in GenoMap is to evenly distribute the requests for services among the multiple server instances. This is achieved by simply providing the appropriate server location information to client applets. The socket server selects the appropriate server depending on the following parameters.
   - Number of client requests currently being serviced by each server.
   - Individual server parameters, such as the maximum number of requests the server can handle simultaneously/concurrently.
   - Idle time since last serviced request.

   A LoadValue is calculated using the following equation

$$LV = \sum_{i=1}^{n} W_i X_i \qquad (1)$$

where

$n$ = number of factors used to calculate LoadValue(LV)
$W_i$ = Weight(importance) assigned to that factor
$X_i$ = Normalized value of the parameter

Our initial implementation considers the following simple factors:

$$X_1 = \frac{MaxNoOfRequests - NoOfRequests}{MaxNoOfRequests} \tag{2}$$

$$X_2 = \frac{MaxIdleTime - IdleTime}{MaxIdleTime} \tag{3}$$

$W_i$ are values ranging from $0\ldots 1$. For example, if $W_1 > W_2$, the socket server favors selecting the server with the shorter *Request* queue.

3. **Management of resources** The socket server manages addition/deletion of domains. It allows for adding and removing of users from a domain and changing their privileges. The database and the servers can be dynamically registered and unregistered. The socket server accounts for failures during a transaction between client applets and servers. Refer to [1] for further details.

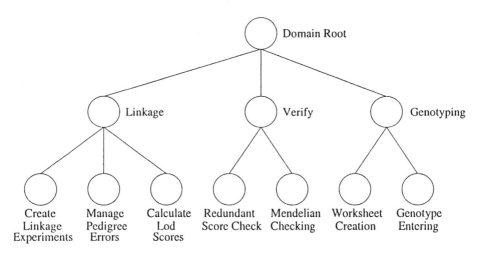

**Fig. 3.** Authentication Hierarchy

4. **Authentication** GenoMap uses an authentication scheme based on naming (Figure 3). Each domain is divided into logical sub-domains corresponding to the services offered in that domain. Each service is further divided into sub-domains giving finer control of the service. Each user is placed in one or more nodes. The user's access rights are assigned according to their position in the hierarchy. The closer the user is to the root, the more rights he/she has.

### 3.3 Scenario

Domain creation involves setting up the database and registering it with the socket server, starting the servers (multiple instances to increase the availability) if required, and creating a user group with appropriate privileges. Once created the users can access the services using the client applets. First, the client authenticates itself with the socket server and then retrieves the server (hostname, port, etc.) and database information. The client may have to provide further authentication before being allowed to access the database's information.

Adding a new service is very simple and straight forward. All one has to do is to implement the User Interface and the Server (if required); register the service with the socket server and update the privileges for all users who require access to the new service.

### 3.4 Virtual Domain

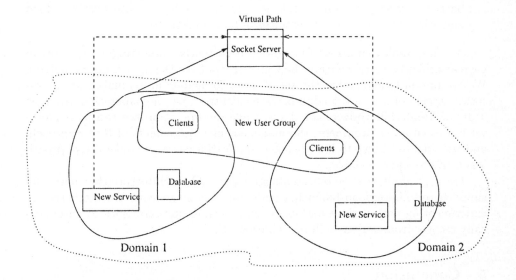

**Fig. 4.** Illustration of Virtual Domains

A Virtual domain (Figure 4) is formed by combining two or more existing domains to offer a new service which utilizes the resources of participating domains. A new User Group is formed that has rights to access the services. The backend servers/clients can access the resources of the other domain only after proper authentication from the socket server.

## 4  Implementation and Usage to Date

Apart from specific implementation details, GenoMap is a networked application consisting of a set of interacting 1) Java applets and applications, 2) Java applets/applications and C/X-windows user interfaces, 3) Java and C applications that interact with local files systems, and 4) Java applets that interact with a database server and Java and C applications. Most of the clients are Java applets which provide various functionalities such as verification of genotypes to users. To meet the basic requirements outlined in the previous section, there are two different design approaches to this problem.

1. Clients contact their paired servers individually and registration is done on an applet/server pair basis. Servers then use "well-known" socket addresses to communicate with their client applet(s). This approach is the basis for our first implementation of GenoMap.
2. A separate Socket Server process is implemented and registered with the Web Server. The socket server authenticates valid users of GenoMap, passes the location information for the applet clients needing to contact various backend servers and the Database. The Socket Server also performs load balancing among multiple instances of backend servers.

Our first implementation of GenoMap is currently supporting a large autism gene identification study within the University of Iowa Department of Pediatrics. We are presently expanding the set of users to include individuals and laboratories outside of our direct administrative control (e.g., a cooperating group at Tufts or Vanderbilt University). Thus, while the first approach can be made to work in such an environment, the management requirements of that approach, and the limited ability to load-balance application activity, make the approach less attractive and scalable.

The second alternative represents an expandable architecture that naturally supports automatic load-balancing and ease of management. We are currently experimenting with various load-balancing schemes and strategies to automatically recover from various failure scenarios [1].

## 5 Conclusion

The conflicting requirements of security and heterogeneity are at the heart of the approach taken in GenoMap. Our approach provides for security by verifying identities of individuals in the granting of access to sensitive data through the network. The database itself is physically partitionable across lab boundaries, and access to a database is encapsulated within well-defined APIs.

GenoMap has been implemented primarily in Java with a socket-oriented, client-server design employing recent applet security features. GenoMap is currently being used in the collection of data for, and analysis of, a number of relatively small gene identification studies. It is also being used in one large genome-wide screening for the locus of the gene(s) involved in autism. The former shows its usefulness in supporting users who want to employ the analysis features of GenoMap, while not needing the large-scale data collection and management facilities, while the latter shows the usefulness in managing gene identification studies that would have been unmanageable without such a system.

# References

[1] T. L. Casavant, K. J. Munn, T. A. Braun, T. E. Scheetz, S. Kaliannan, "An illustration of a Parallel/Distributed Archtecture for A hierachially Heterogenous Web-Based Cooperative Applications", University of Iowa, Technical Report TR-ECE-981213,
http://www.eng.uiowa.edu/~ tomc/papers/genomapLong.tar.gz

[2] T. E. Scheetz, K. J. Munn, T. A. Braun, V. Sheffield, E. M. Stone, T. L. Casavant, "GenoMap: Adistributed system for unifying genotyping and genetic linkage analysis", Parallel Computing 24 (1998), pp. 1567-1592

[3] G. Cornell and C. S. Horstmann, *Core Java*, Prentice Hall, Upper Saddle River, New Jersey, 1997.

[4] R. W. Cottingham and R. M. Idury, "Faster Sequential Genetic Linkage Computations," *American Journal of Human Genetics*, 53:252-263, 1993.

[5] Deparment of Energy, "Five Years of Progress in the Human Genome Project," *Human Genome News*, Volume 7, Numbers 3-4, September-December 1995. Available via the WWW from www.ornl.gov in TechResources/Human_Genome/publicat/hgn/v7n3/04progre.html (September, 1997).

[6] J. F. Gusella and N. S. Wexler, "A polymorphic DNA marker genetically linked to Huntington's disease," *Nature*, Volume 306, November 17, 1983.

[7] J. Lalonel, R. White, "Analysis of Genetic Linkage," Emery & Rimoin's Principles and Practice of Medical Genetics, pp. 111-125, 1996.

[8] R. L. Mynatt, R. J. Miltenberger, M. L. Klebig, L. L. Keifer, J-H Kim, M. B. Zemel, J. E. Wilkinson, W. O. Wilkison, and R. P. Woychik. "Analysis of the function of the agouti gene in obesity and diabetes," *Proceedings: International Business Communications 2nd Annual International Symposium: Obesity, Advances in Understanding and Treatment*, In press.

[9] J. Ott, *Analysis of Human Genetic Linkage*, Johns Hopkins University Press, Baltimore, 1991, pp. 108-141.

# Effective Dynamic Load Balancing of the UKMO Tracer Advection Routines

Douglas Andrew Smith

Edinburgh Parallel Computing Centre, University of Edinburgh, Edinburgh

**Abstract.** The United Kingdom Meteorological Office's routine which models the advection of tracers in the atmosphere suffers from a large load imbalance which can be as large as 30:1 in global forecasting runs. By dynamically changing the data distribution we have reduced the load imbalance. Our method produces approximately a factor of two speed up in the imbalanced section of the code during low resolution climate runs.

## 1 Introduction

The United Kingdom Meteorological Office (UKMO) has its own forecasting code known as the Unified Model. It is used both to predict the coming week's weather and the climate in hundreds of years time. Obviously predicting yesterday's weather today is of no use to anyone and for this reason UKMO uses one of the most powerful supercomputers in the world, an 880 processor Cray T3E.

Edinburgh Parallel Computing Centre (EPCC) specialises in Novel and Advanced Computing. For the past 9 months EPCC has been contracted to carry out single processor optimisation and parallel optimisation on parts of the Unified Model. We have carried out single processor optimisation on the sections of the Unified Model dealing with calculating the effects of solar radiation on the atmosphere and with calculating boundary layer effects. We are also in the process of parallelising the code that models the oceans surrounding the United Kingdom.

This paper describes the work carried out by EPCC to perform parallel load balancing of the routines used to calculate the advection of Tracer Variables in the Unified Model.

## 2 Tracer Advection

When pollutants or particulates are released into the atmosphere they are moved around by the process of advection. The UKMO tracer advection scheme aims to model this process to allow the concentration of pollutants to be calculated so that their effects on, for example, the climate can be calculated.

The equation describing the advection is

$$\frac{\partial C}{\partial t} + u\frac{\partial C}{\partial x} + v\frac{\partial C}{\partial y} + \dot{\eta}\frac{\partial C}{\partial \eta} = 0 \tag{1}$$

where $C$ is the concentration of the tracer being advected, $x$ and $y$ are the standard Cartesian coordinates and $\eta$ is the vertical coordinate. $u$ and $v$ are the wind speeds in the East-West and North-South directions respectively which are advecting the tracer.

The tracer advection routines are run in two types of model. Firstly there is the global model where the world's weather and climate is modelled. There is also the Limited Area Model where only a part of the globe is modelled, usually at a higher resolution. Both cases use nearly the same code, differing only in how the edges of the grid are treated.

## 3  Von Neumann Stability Analysis

In the Unified Model, equation 1 is discretised using finite differences. The effect of the advection at each timestep is calculated in 3 stages. Firstly the tracer is advected in the East-West direction, followed by the North-South direction and then finally it is advected vertically. It is the advection in the East-West direction that is by far the most load-imbalanced and it is here that we have carried out our optimisations.

Discretising equation 1 will lead to instabilities if the discretised timestep is too large. Instabilities mean that some of the eigenmodes of the difference equation may grow exponentially.

In order to avoid this problem the Courant-Friedrichs-Lewy stability criterion must be obeyed. The stability condition is that

$$\frac{|u|\Delta t}{\Delta x} \leq 1 \tag{2}$$

where $|u|$ is the wind speed that is advecting the tracer, $\Delta t$ is the discretised timestep and $\Delta x$ is the spatial lattice spacing. The value of the left hand side of equation 2 is known as the Courant Number.

UKMO uses a regular latitude/longitude grid in the Unified model. For the global model this means that the spherical world is mapped on to a rectangular grid. Thus each row of constant latitude has the same number of points as any other but the physical spacing between the points decreases from the equator towards the poles.

The Limited Area Model uses a rotated grid with the model area centred on the equator. The variation in Courant number is very much less as the variation in latitude is not as great and load balancing is not really an issue here. From now on we will only discuss the global model.

For the East-West advection the Courant number at each point is given by

$$\nu = \frac{u\Delta t}{na\Delta\lambda cos(\phi)} \tag{3}$$

where $u$ is the windspeed, $a$ the radius of the Earth, $\lambda$ the longitude and $\phi$ the latitude. If the windspeeds across the world were relatively uniform then it is obvious that the Courant number will, in general, increase towards the poles and the advection may become unstable near the poles.

In the Unified Model, the condition for stability is that

$$\nu < \frac{1}{4} \qquad (4)$$

Thus in order for this condition to be obeyed, rather than carry out one sweep with timestep $\Delta t$, $n$ sweeps are carried out with a timestep of $\frac{\Delta t}{n}$ where $n$ is chosen so that the stability condition is obeyed.

It is also a requirement that the number of sweeps increases monotonically from the equator to the poles.

## 4 The original code

The original tracer advection code used a 2-d grid decomposition with each processor getting $n$ row segments (see figure 1). The tracer advection code advects a single vertical level of the data at a time; N_LEVELS vertical levels in the model means N_LEVELS calls to the tracer advection routine.

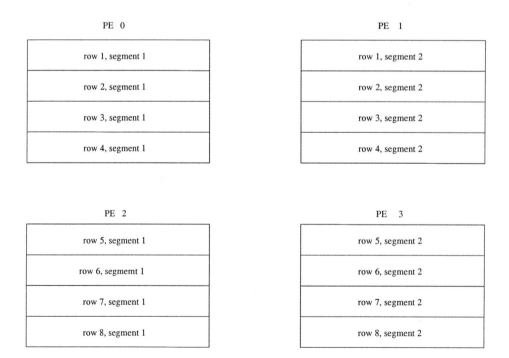

**Fig. 1.** Original data decomposition for Unified Model showing allocation of row segments to different processors

The tracer advection routines involve finite differences between neighbouring points. With the above decomposition this means we need communications between neighbouring processors in the East-West direction for the East-West advection. Each processor therefore needs a 1-d halo surrounding its data. The finite differences have to be calculated after each sweep and therefore the haloes must also be updated after each sweep.

Pseudo-code for the original tracer advection code is given below. MAX_SWEEP is the maximum number sweeps that have to be carried out on a row and is set globally across all the processors.

```
DO SWEEP=1,MAX_SWEEP

    FIND ROWS WHERE N_SWEEP(ROW).ge.SWEEP

    DO ROWS=FIRST_ROW_TO_BE_UPDATED,
                LAST_ROW_TO_BE_UPDATED
        CARRY OUT ADVECTION SWEEP
    END DO

    UPDATE HALOES

END DO
```

Specifying that the number of sweeps should increase monotonically towards the poles means that at each advection sweep a contiguous block of rows is advected. This was presumably to allow more efficient use of the vector machine that this code was originally run on by giving it vectors as long as possible to process.

The updating of the haloes after each sweep is carried out by the routine **SWAPBOUNDS**. It is a synchronising routine and has to be called by all of the processors. If a processor has no work to do, it will sit in **SWAPBOUNDS** until all the other processors get there and then update its haloes even if they haven't changed since the last call.

The number of sweeps to be carried out on each row of the grid increases towards the poles and therefore the processors nearer to the poles will have more work to do. In the global climate model the number of sweeps at the poles can be as many as 60-70 while the number of sweeps at the equator can be as little as 1 and the resulting load imbalance between processors can be as much as 30 to 1. Because of the size of the load imbalance the **SWAPBOUNDS** routine sits at the top of the profile of this code.

## 5 The new code

### 5.1 The new data distribution

As noted above the Unified Model uses a 2-dimensional logical processor grid to decompose the data. For the East-West advection this is not the ideal data distribution to have. The East-West advection works on a 1-d strip, a row, of the data. Using a 2-d decomposition, with only partial rows on each processor means that after every advection step the haloes of that row must be updated. If, however, we had 1 full row of the data wholly contained on one processor then all the sweeps of the advection could be carried out on this row without any need for communications between processors, We therefore dynamically redistribute the data into a 1-d decomposition for the East-West tracer advection calculation and then distribute the data back into the 2-d decomposition afterwards.

Each row has its own particular number of sweeps that need to be carried out on it. If we now assign full rows to each processor we can assign the rows so that the total number of sweeps to be carried out by each processor is approximately the same.

Figure 2 shows the new data distribution scheme for the tracer advection code. A similar data redistribution scheme is used in the Fourier Filtering routines of the Unified Model.

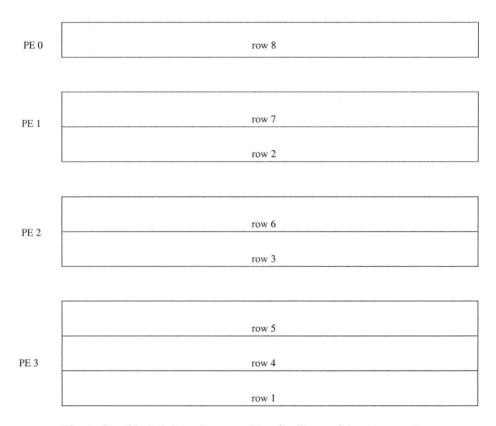

**Fig. 2.** Possible 1-d data decomposition for Tracer Advection routines

## 5.2 Redistributing the data

The Unified Model performs interprocessor communication using the GCOM libraries written by SINTEF. These are a set of wrappers that can be compiled to allow the user to access the Cray Shared memory libraries, the fastest communications libraries on the T3E, or MPI making the code portable.

Our redistribution was carried out using the routine **GCG_RALLTOALLE** which carries out an all-to-all redistribution of the data. This routine needs to be passed two lookup tables, one send table and one receive table which contain maps describing the requested data movements to and from each processor. We have a separate lookup table for each level to allow for different distributions of rows.

Each processor can send and receive up to MAX_MESG messages where MAX_MESG is a parameter that can be set arbitrarily at compile time. This means that each processor can only be allocated up to $\frac{MAX\_MESG}{nproc\_x}$ rows before the lookup table is overfilled. If a processor is allocated more rows than this then the code will abort.

## 5.3 The load balancing algorithm

Having decided to give each processor full rows of data we now have to allocate the rows to each processor. Obviously we want the rows to be allocated to each processor so that the load balance is the best possible for this dataset.

Our load balancing algorithm works as follows. Firstly we sort the rows so that they are ordered into decreasing number of sweeps. Then we loop over the rows in decreasing order allocating each row to the processor that currently has to carry out the fewest number of sweeps.

There is one slight modification to this algorithm. The lookup tables are only of finite size and it is possible that for a really unbalanced data set one processor may start to get more rows than there is room for in the lookup table. The algorithm accounts for this by keeping count of the number of rows allocated to each processor. The number of rows that each processor can receive is given by MAX_ROWS defined by $MAX\_ROWS = \frac{MAX\_MESG}{nproc\_x}$. If the processor with the least amount of work is already receiving MAX_ROWS rows then there is no room for any more rows. In this case the row is allocated to the least loaded processor which has room for the row.

# 6 Performance gains

UKMO provided EPCC with a standard resolution global climate model model. This uses a grid of 96 Columns (East-West) by 73 rows (North -South) by 19 levels. All of the rows except the polar rows take part in the East-West advection giving 71 rows to be distributed across all the processors in a load balanced way.

We have timed both the original code and the new load-balanced code on 8 and 16 processors. Table 1 shows the wall clock times of the original code and the new load balanced code for the East-West sweeps.

**Table 1.** Gains for tracer advection code

| proc_x | proc_y | orig. time(s) | new time(s) | %saving |
|--------|--------|---------------|-------------|---------|
| 2 | 4 | 1.48 | 0.68 | 54 |
| 4 | 2 | 1.33 | 0.67 | 49 |
| 4 | 4 | 0.88 | 0.43 | 51 |

The new load balanced code has given a factor of approximately two in the performance of this routine. We note that the results produced by the load balanced code are bit identical to the results produced by the original code.

## 7 Further optimisations

We noted before that the tracer advection routines calculate the advection for each level at a time. There are 71 rows to be advected in the East-West direction and these have to be distributed over a number of processors. There will probably still be a residual load imbalance as there is not enough granularity amongst the rows to fully balance the load. If we were to change the call interface so that the advection took place for all the rows on all the levels in a single call to the routine, we could use all these rows in the load balancing algorithm leading to an even better load balance.

## 8 Summary

The United Kingdom Meteorological Offices tracer advection routines suffer from poor load-balancing during the calculation of advection in the East-West directions. By dynamically changing the grid decomposition from a 2-dimensional one to a 1-dimensional one and allocating the rows using a simple load balancing algorithm we have produced an improvement of a factor of two in the performance of this piece of code.

Any computational scheme that suffers from CFL instability, for instance any difference scheme with a fixed timestep and variable spatial grid sizes, could benefit from dynamic load balancing of the kind described here.

## Acknowledgements

This work was funded by the United Kingdom Meteorological Office. The author would like to thank Steve Booth, Rob Baxter and Paul Burton for useful discussions.

# Dynamic Load Balancing in Parallel Finite Element Simulations

Arjen Schoneveld[1]*, Martin Lees[2], Erwan Karyadi[2] and Peter M.A. Sloot[1]

[1] Parallel Scientific Computing and Simulation Group
Faculty of Mathematics, Computer Science, Physics and Astronomy
University of Amsterdam
Kruislaan 403, 1098 SJ Amsterdam
The Netherlands

[2] The MacNeal-Schwendler Corporation
Groningenweg 6
2803 PV Gouda
The Netherlands

**Abstract.** In this paper we introduce a new method for parallelizing Finite Element simulations enabling the use of dynamic load balancing. A physical space partitioning is obtained by dividing the bounding cube into a large number of sub cubes. The cube mesh together with a workload attribute assigned to each cube is used to present an abstract view of the simulation. Based on this abstract view a dynamic load balancing process decides on a possible local repartitioning of the mesh. The dynamic load balancing process itself is diffusion based, that is cubes are migrated between neighboring partitions. A parallel simulation framework (P-CAM) is used to implement the dynamic load balancer.

## 1 Introduction

Most existing partitioning methods for parallel Finite Element simulations are based on examining the connectivity of the element mesh and dividing the mesh in such a way that the number of cut edges is minimized (see e.g. [8,7]). These methods assume data flow only between adjacent elements in the mesh. For different reasons this is not applicable to the explicit finite element solvers of MSC/DYTRAN [1]. Specifically, such explicit solvers can operate on several meshes at once. Data can flow between two meshes that are not even related by connectivity due to *contact* and *coupling*. MSC/DYTRAN contains two finite elements solvers: Lagrangian (structural) and Eulerian (fluid). Beside these two main solvers, which are entirely separate, an algorithm for structure/structure interaction (contact) and fluid/structure interaction (coupling) is also provided. Two different meshes are used to represent contact and coupling between elements. Both reside in the same physical space. These interactions can not be derived from the connectivity of the mesh and have to be determined by a spatial

---
* presenting author

search. Because of this we have decided to partition the model in physical space to ensure that data flow can be localized to single CPUs or nearest neighbor CPUs. Physical partitioning is realized by superimposing a coarse imaginary 3D mesh of 512 cubes over the model meshes. Each cube can contain entities from various meshes. Each cube has a property called workload, which is the amount of CPU time required to process all the elements in the cube. The workload per cube varies over time as the cube's content changes in quantity or character. The physical space is divided into 512 ($8^3$) cubes in order to create a *redundant decomposition* [3], resulting in many more cubes than CPUs. A redundant decomposition is necessary to enable the dynamic load balancing process to shift cubes between processors in order to resolve an imbalance situation.

In this paper we will introduce a dynamic load balancing strategy for the Euler solver of the MSC/DYTRAN program. In section 2 we will discuss a framework for parallel simulations, P-CAM, in which we will embed a dynamic load balancing module. Different load balancing strategies applied to a finite element simulation of an underwater explosion will be discussed in section 3. Finally we will discuss the results of the experiments and give some conclusions in section 4.

## 2  A Framework for parallel simulations

The concept of Dynamical Complex Systems (DCS) [9], a set of interconnected virtual particles which evolve through time using an execution model, is implemented in a Parallel Cellular Automata Modeling environment called P-CAM [6]. The definition of a virtual particle can be completely arbitrary, as long as some basic computation is carried out by this particle [9, 4]. The interconnection structure between particles can be anything from a regular grid to randomly connected graphs. During the evolution of the system, particles and connections can be created or annihilated. To support parallelism, virtual particles can be allocated to virtual processors. In addition, the allocation is allowed to be dynamic, that is, a virtual particle may be re-allocated to another virtual processor.

The basic atomic unit of computation in P-CAM is the cell or virtual particle, where atomic denotes the computational atomicity of the cell. Each cell can be connected to any other cell. Initially the cells are partitioned in an arbitrary manner over the available processors. In order to support transparent re-assignment of cells to processors, P-CAM supports the migration of an arbitrary set of cells to an arbitrary processor. Essentially, cell migration involves packing the cell (and its state), sending it to the designated processor and notifying all processors that have connections to it. In other words, the parallel task graph (the cells and their interconnections) can be dynamically re-mapped to processors.

In this paper we will apply the functionality of the framework to implement a dynamic load balancing system for an explicit finite element solver. The partitioning in cubes, of the physical domain in which the finite element meshes are located, is used to represent the cells (the cubes) and their interconnection (a

3D grid). Because DYTRAN is a existing software package containing a huge amount of Fortran code we have chosen to separate the simulation code from the dynamic load balancing process: the simulation is abstracted to *workload* values assigned to cubes. During the simulation, DYTRAN periodically updates the workload values to the dynamic load balancing process, which in turn decides if and how to migrate the "physical" cubes, corresponding to simulation sub tasks.

**Dynamic Load Balancing**

A serious problem in many simulation applications is the fact that the amount of workload associated with the parallel processes can be subject to change during program execution. This change is likely to induce imbalance in the workload distribution and, hence, requires some action, in order to rebalance the load over the processes [5].

Hence, we need a dynamic load balancing method to assure that the allocation of atomic processes will only vary "gradually". Small variations in workload distribution should only induce small variations in the process allocations. That is, subsequent partitions should "look alike". Secondly, we need a method which is not expensive to perform, and as such, is not a seriously hampering factor when it comes to real time execution.

In this section, we will formally introduce the load balancing problem. Next, we discuss two state-of-the-art methods to solve the load balancing problem. The input that both methods obtain is the workload per parallel process, and the process connectivity. With this information a "workflow" pattern can be computed. This workflow denotes how much work has to be migrated between every connected pair of processes in order to realize load balancing.

For the purpose of load balancing, we have to take focus on a subset of its attributes: We are only interested in the workload associated with the *atomic* processes (or cells), and their connectivity.

We assume that we have a number of parallel processes, each of which can be composed of zero or more *atomic* processes (or cells if you like). At this point we make no explicit assumption whether the processes are allocated on parallel *processors*. They may either be executed concurrently on a single multi-tasking processor, or in parallel on a distributed memory computer system (or some combination of both).

We view the parallel simulation program as a graph $H = (T, E)$, with $T$ the set of parallel processes, and $E$ the set of interprocess links. If we consider $P$ to be the set of *atomic* processes to be distributed among the parallel processes we can define the following:

- A mapping is a function $\pi : P \to T$, assigning each *atomic* process to a parallel process.
- The *weight* of process $i$ relative to a mapping $\pi$ is defined as

$$weight_\pi(i) = \sum_{\mu \in P : \pi(\mu) = i}^{|P|} comp(\mu), \qquad (1)$$

where $comp(\mu)$ denotes the abstract computational work of the *atomic* process $\mu$, which, for instance, can be expressed in *flop*. The actual execution time associated with the computational work $comp(\mu)$ can be calculated when the CPU speed ($CPUspeed$), which is for instance expressed in units *flop/s*, is known. For the moment we assume that we allocate each parallel process to a unique processor, and that the CPUs are equally fast. Hence, we may replace the weight $weight_\pi(i)$ by the "time complexity", $t_\pi(i)$, which is expressed in seconds.

$$t_\pi(i) = weight_\pi(i)/CPUspeed, \qquad (2)$$

– The global cost function $\Gamma(\pi)$, is derived from the variance of the load:

$$\Gamma(\pi) = \sum_{i \in T} (\bar{l} - t_\pi(i))^2 , \qquad (3)$$

where

$$\bar{l} = \frac{\sum_{i \in T} t_\pi(i)}{|T|} , \qquad (4)$$

that is, the average execution time per parallel process.

Consequently, the load balancing problem can be defined as follows:
*find a mapping $\pi$ which minimizes the global cost function $\Gamma(\pi)$.*

Cybenko [2] has shown that the global minimum for Eq.(3) can be reached by parallel optimization of the local load, that is, by locally optimizing:

$$\Gamma_i(\pi) = \sum_{(i,j) \in E} (t_\pi(i) - t_\pi(j))^2, \; \forall i, j \in T, \qquad (5)$$

under the constraints that the graph is connected and not bipartite.

In [6], a *Poisson iteration* method to determine the workflow is discussed in detail. The solution of this iteration method is a $|T| \times |T|$ matrix describing the amount of work which flows from parallel process $i$ to $j$. Note that the workflow is zero if $(i, j) \notin E$.

**Cell Selection** After the workflow has been determined, the next question is, given the workflow, how to select cells from the overloaded processes that are most suited for migration to under-loaded processes. In the following, two different heuristic methods are proposed as a solution to this problem. At this point we can use the advantage of the cell based application framework. It allows us to define a clean and simple interface to the cell selection methods, which makes it easy to experiment with the different strategies. No alterations to the kernel code are necessary.

In the cell selection heuristics we migrate cells between processes that lie on (or close to) the inter process boundaries. This idea, amongst others, finds support in the work by Walshaw *et al.* [10], where comparable assumptions are made for selecting "migratable elements" for dynamic load balancing in finite element meshes.

In addition to the connectivity of the cells we can associate a spatial coordinate when cells correspond to a process that is simulated in Euclidean space. At this point, a cell selection strategy might incorporate this *a priori* knowledge and take advantage of it.

To explain the functionality of each of the two selection methods below, we use a simple task graph depicted in Fig. 1.

The cell oriented view maintained by the kernel allows for easy processing of a number of key characteristics of all domains. Amongst others, for each process, immigrant and emigrant lists are kept up to date specifying which local cells lie on inter process boundaries. As we will see below, this is especially convenient for parallel processes in which we select cells to resolve the workflow.

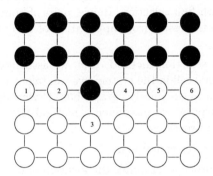

**Fig. 1.** Two processes $i$ (white cells) and $j$ (black cells), each consisting of a number of computational cells.

Consider two parallel processes $i$ and $j$, depicted in Fig.1, with process $j$ consisting of the black cells, and process $i$ of the white cells. We assume that each cell has an equal workload. Furthermore, $max\_deg$ is the maximum number of cells that one cell can be connected to (i.e. $max\_deg = 4$ in this case.). Clearly, the example system is out of balance. $j$ has 13 black cells, vs. 17 white cells on $i$.

Without too much effort we can imagine that a workflow, of 2 cells, will be directed from process $i$ to process $j$ (in general the size of the workflow will have to be provided by any of the above workflow algorithms of the previous section. For this simple example we can solve it by hand).

Note that in Fig.1 only the cells on the boundary in process $i$ are numbered.

The first cell selection heuristic utilizes purely the connectivity information of the cells. The method, *maxcon*, operates as follows:

(1) For each boundary cell determine "how strong" it is connected to its local process, and how strong it is "pulled" by one or more external domains.

For instance, in Fig.1, cell 1 has two connections with process $i$, whereas it only has one connection to process $j$. On the other hand, cell 2 has two connections to both process $i$ and process $j$, and in that sense no specific stronger "desire" to be on any of the two processes. Cell 3 is connected quite strongly to process $i$, since it has three connections to this process, opposed to one connection to a cell on process $j$. In this way we can "categorize" the boundary cells, from weakly connected (no connections with any cell on the host party; note that it is, in principle, possible that a cell is completely surrounded by "alien" cells), to strongly connected ($max\_deg$ - 1 connections with the host process, not $max\_deg$, since any cell with $max\_deg$ neighboring cells residing on the same host process is an *internal* cell and, hence, cannot be present in the BoundaryList.).

(2) After this categorization a breadth-first search is applied. The first cell is selected from the list of weakly connected cells.

In the example this means that cells 2 and 4 are first candidates as "starting cells" to start a breadth-first selection. Hence, pairs of boundary cells with either cell 2 or cell 4 will be the result from application of this algorithm.

In the second cell selection method, *com*, we associate a set of coordinates with each computational cell in the system, we can take advantage of this a priori knowledge, in addition to the purely "graph-based" view applied in the previous methods. For this purpose, we introduce a method that utilizes the center of mass (com) of parallel processes.

The algorithm operates as follows:

(1) Each parallel process computes the location of its private center of mass, **com**, in the simulation space as follows:

$$com = \frac{\sum_{i=1}^{N} m_i x_i}{N}, \qquad (6)$$

where the summation runs over $N$ local cells with coordinates $x_i$ and masses $m_i$. In our case we can take the cell-mass as unity.

(2) Each process notifies its neighboring processes about the value of its local **com**, and stores the centers of masses that it receives from its neighbors.

(3) For each local cell it is determined to which neighboring center of mass, besides that of the local process, it is closest. And for each neighbor process a list is created which contains the IDs of the "closest" cells.

(4) Each of these lists are sorted in ascending order. That is, the cells closest to a neighbor process are first in the list.

After these steps, essentially the same algorithm as in the other methods is applied. If a workflow from one process to another is required, we simply take so many subsequent cells from the List that was sorted on distance, that the corresponding work amount satisfies the required flow.

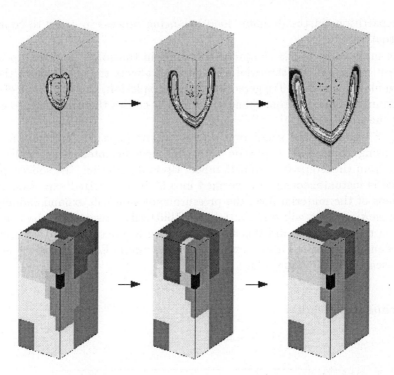

**Fig. 2.** Under water explosion simulation of pressure fringes and the current cube partitions, with $P = 8$, MAX_IMB=7%, com cell selection and DLB_FREQ=25. The size of the physical domain is 1.5x1.5x1.0m, with 160000 elements and the specific internal energy of the explosion is 5.706E+6 Joule. From left to right the iterations 231, 430 and 625 are displayed for the 1/4 model.

## 3 Case study: Finite Element Simulation of Underwater Explosion

We will use a finite element simulation of an underwater explosion as our study object. Because there are two materials in Euler domain (gas explosion and water) we use multimaterial elements. The explosive is placed in the middle of the bucket and resides at the corner of two symmetry planes about 22 cm below the water surface. The region where the explosive is initially located is meshed very fine to be able to accurately compute the detonation process.

The computational mesh remains fixed in space and in time and the material (water, gas) can flow through the mesh from element to element. Due to the asymmetrical nature of the reduced model (only a quarter of the original physical model is used in the simulation) and the propagation of a shock wave through the water upon detonation, we can expect an imbalance situation when the model is initially partitioned in equally sized volumes (ORB partitioning). Therefore we need repartitions in the dynamic load balancing process in order to cope with this situation.

The major part of the computational time in the solver is taken up by the transport process of multimatrial elements and affects the workload in the current simulation process. Triggered by the gas explosion, the transport of mass, momentum and energy from element to element causes the workloads to change during the simulation.

Fig. 2 visualizes the simulation process of the pressure distribution in the model using "pentolite" explosion with SIE (specific internal energy)= 5.706 MJoule from timestep=0.15 to 0.45 mSec. Upon detonation, the solid explosive material is instanteneously transformed into high pressurized explosion gasses. By means of the material flow, the pressure explosion will expand radially and after some timesteps will distribute like a "butterfly" away from the charge as shown in the figure. We did this simulation with 8 processors and the dynamic load balancing process is also shown in the same figure. Each processor partition is represented by a unique color.

**Experimental Results**

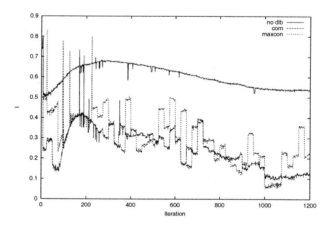

**Fig. 3.** The evolution of the imbalance parameter for three load balancing strategies: no dynamic load balancing (No dlb), dynamic load balancing with com cell selection (com) and with maxcon cell selection (maxcon)

We did an extensive series of experiments on a IBM SP2 cluster using 4, 8 and 16 processors. The communication primitives for DYTRAN have been implemented using the MPI message passing library for Fortran. The dynamic load balancing or partitioning process runs on 1 processor. Prior to repartitioning all cube loads are collected on processor 0 by a global communication step. Each parallel simulation is initially decomposed using an orthogonal recursive bisection (ORB) of the cubes. After the initial decomposition one out of the three available load balancing methods is chosen: no dynamic load balancing (*no dlb*), dynamic load balancing with *maxcon* cell selection and dynamic load balancing with *com* cell selection. Two important parameters for dynamic load balancing are the frequency (DLB_FREQ) of calling the partitioner and the tolerated level of load imbalance (MAX_IMB), where load imbalance $\mathcal{I}$ is defined as a function of the average load, $\bar{l}$ (Eq. 4), and the maximal load $l_{max}$:

$$\mathcal{I} = \frac{l_{max} - \bar{l}}{l_{max}} \qquad (7)$$

These two parameters have been chosen such that they are optimal for this specific simulation: DLB_FREQ=25 and MAX_IMB=7%. The workload for each cube that is supplied to the partitioner is averaged over the last 25 iterations to reduce the noise level and to ignore spurious dynamics. Note that decreasing DLB_FREQ increases the communication overhead due to the global communication step that is involved in collecting the individual cube workloads.

In Fig. 3 the evolution of the imbalance, $\mathcal{I}$, on an 8 processor partition is given for each of the three load balancing strategies. The load is defined as the time to complete a complete iteration of the Euler kernel, that is without communication times. Note that including the communication time in such a synchronized simulation would result in $\mathcal{I}$ being approximately zero. The spikes in the plots are due to unknown delays in the timing measurements. The block wall character of the dlb strategies is caused by the value of DLB_FREQ. The width of the blocks is exactly 25 iterations.

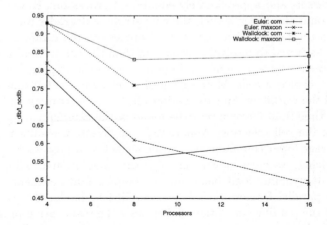

**Fig. 4.** Dynamic load balancing efficiency ($t_{dlb}/t_{nodlb}$) of the Euler kernel and the Wall clock time for the two load dynamic balancing strategies: com and maxcon, for 4, 8 and 16 processors

To quantify the efficiency of the dynamic load balancing process we plot the dlb time $t_{dlb}$ versus the nodlb time $t_{nodlb}$. In Fig. 4, $\frac{t_{dlb}}{t_{nodlb}}$ for the Euler kernel and the wallclock time is depicted for the maxcon and com strategies on 4, 8 an 16 processors.

## 4 Discussion and conclusion

In this paper we have described a physical domain partitioning approach to the dynamic load balancing of a parallel finite element simulation. A parallel simulation framework, P-CAM, has been used to implement a separate dynamic load balancing module. The simulation model is decomposed into 512 cubes, where each cube contains a part of the finite element mesh. Associated with each cube is a workload, quantifying the total CPU time to finish a single iteration of the Euler solver. The 3D cube mesh and its corresponding workloads are used by the dynamic load balancing module to decide whether to repartition and how to resolve the load imbalance by reallocating cubes to processors. Using an underwater explosion simulation, the computational performance of a statically decomposed parallel process is compared with a dynamically load balanced version. From the experimental results presented in Figs. 3 and 4 it is clearly demonstrated that dynamic load balancing pays off in terms of turn around time. From Fig. 3 a clear reduction of the load-imbalance $\mathcal{I}$ in the Euler solver can be observed for different cell selection strategies compared to static decomposition. Also from this figure it can be observed that the com selection strategy seems slightly better in terms of load-imbalance than the maxcon selection strategy. This observation is supported by the results of Fig. 4 where the efficiency of the dlb strategies is compared to no dlb, for 4, 8 and 16 processors. The wall clock time efficiency of the com strategy is better than maxcon for 8 and 16 processors and approximately equal for 4 processors. However, just taking into account the Euler kernel, the com strategy degrades for 16 processors compared to the maxcon strategy. For 16 processors the center of mass based strategy is less suited to resolve the workflow than the connection based scheme. This is not caused by some inherent imperfection of the strategy, but merely by an "unlucky" choice of cells. More specifically, cells are selected that result in an overshoot of the workflow. An explanation why the com strategy still has a lower turnaround time than maxcon must be found in the communication pattern resulting from the cell selection. Apparently com selection results in allocations with on average lower communication volumes between processors.

Summarizing, we have shown that physical domain partitioning is a viable route towards dynamic load balancing of complex finite element simulations. A re-usable parallel framework has been applied to model the computational dynamics of the parallel simulation and to steer the reallocation process of simulation tasks. However, many open issues remain in this work. The granularity of the sub-cubes is a very important question. Large cubes induce less overhead in migration and communication but also less room for balancing the workload. Small cubes result in a lot of overhead and great flexibility in dynamic load balancing. The scale of the spatial events in the specific simulation plays in important role in choosing the correct cube size, small spatial effects require a small granularity in the decomposed physical domain. Other unsolved issues are the frequency of calling the partitioning process and the tolerated level of imbalance. Ideally one would like to reside to a kind of adaptive parameter tuning of these quantities.

The development of parallel DYTRAN is still continuing at MSC. In our case study we have left out the Lagrange part of the original simulation model (a rigid boat) considering that the same simulation with Lagrange/Euler coupling is still part of ongoing study.

# References

1. The MacNeal-Schwendler Corporation. Msc/dytran version 4.0 user's manual.
2. G. Cybenko. Dynamic load balancing for distributed memory multiprocessors. *Journal of Parallel and Distributed Computing*, 7:279–301, 1989.
3. J.F. de Ronde, A. Schoneveld, and P.M.A. Sloot. Load balancing by redundant decomposition and mapping. *Future Generation Computer Systems*, 12(5):391–406, 1997.
4. W. Dzwinel. Virtual particles and search for global optimum. *Future Generation Computer Systems*, 12(5):371–389, 1996.
5. Benno J. Overeinder and Peter M. A. Sloot. Breaking the curse of dynamics by task migration: Pilot experiments in the polder metacomputer. In Marian Bubak, Jack Dongarra, and Jerzy Waśniewky, editors, *Recent Advances in Parallel Virtual Machine and Message Passing Interface*, volume 1332 of *Lecture Notes in Computer Science*, pages 194–207, 1997.
6. A. Schoneveld and J.F. de Ronde. P-cam: A framework for parallel complex systems simulations. to appear in FGCS (special issue on Cellular Automata).
7. A. Schoneveld, J.F. de Ronde, P.M.A. Sloot, and J.A. Kaandorp. A parallel cellular genetic algorithm used in finite element simulation. In H-.M. Voigt, W. Ebeling, I. Rechenberg, and H-.P. Schwefel, editors, *Parallel Problem Solving from Nature (PPSN IV)*, pages 533–542, 1996.
8. H. D. Simon. Partitioning of unstructured problems for parallel processing. *Computing Systems in Engineering*, 2(2/3):135–148, 1991.
9. P.M.A. Sloot, A. Schoneveld, J.F. de Ronde, and J.A. Kaandorp. Large-scale simulations of complex systems, part i: conceptual framework. Technical Report Working Paper 97-07-070, Santa Fe Institute, 1997.
10. C. Walshaw, M. Cross, and M.G. Everett. A localised algorithm for optimising unstructured mesh partitions. *International journal of supercomputer applications*, 4(9):280–295, 1995.

# A Load Balancing Routine for the NAG Parallel Library

Rupert W. Ford[1] and Michael O'Brien[2]

[1] Centre for Novel Computing,
Department of Computer Science,
The University of Manchester,
Manchester M13 9PL, U.K.
rupert@cs.man.ac.uk
http://www.cs.man.ac.uk/cnc
[2] Military Aircraft and Aerostructures,
British Aerospace,
Warton Aerodrome,
Lancashire PR4 1AX, U.K.
Michael.OBrien@bae.co.uk

**Abstract.** This paper describes a load balance routine which has been developed for the NAG Parallel Library. This routine is designed for load balance problems where each task can be computed independently and allows the user to choose from a number of different load balance strategies. The benefits of this routine are discussed in terms of both performance and ease of use, and results are presented for a production RCS prediction code on a Cray T3D and a SGI Origin 2000.

## 1 Introduction

The load balance routine described in this paper has been developed for inclusion in the NAG Parallel Library [7]. This library is a collection of portable, memory scalable, parallel Fortran 77 routines for the solution of numerical and statistical problems.

This work forms part of the ESPRIT Framework IV project P20018 PINEAPL (Parallel Industrial Numerical Applications and Portable Libraries). The aim of the project is to develop an application-driven, general-purpose library of parallel numerical software to significantly extend the scope of the NAG parallel library. The project (coordinated by NAG) is driven by applications from four industrial end users, representing the needs of the numerical scientific and engineering market. In this project parallel library experts are paired with end users; in the University of Manchester's case the end user is British Aerospace (BAe).

One of BAe's applications ("System AB3") involves the prediction of the radar cross section (RCS) of an aircraft's air intake duct. The particular technique requires raytracing to calculate the RCS of an arbitrary shaped duct. A ray tracer developed at the University of Manchester has been integrated into BAe's "System AB3".

Parallelism can be naturally exploited at the level of rays as each ray can be calculated independently. Note, the geometry is simple enough to allow its replication on each processor. However, although rays can be independently calculated, their computational cost will vary significantly, depending on the path a ray takes. This means that a static equal allocation of rays to processors will not necessarily give a load balanced solution.

The load balance routine (called Y01CAFP) was developed to solve the above and similar load balance problems. Y01CAFP is therefore designed to minimise the elapsed time for $n$ independent tasks, where $n$ is fixed and known, running on $p$ processors. The solution to such a problem is often termed task farming, as tasks may be sent (farmed out) to other processors. It is primarily designed for problems where $n \gg p$ and the time for each task is variable and unknown, however it can be of benefit for problems where the time for each task is known but the distribution of tasks is not regular. It is also useful for distributing tasks when all data is held on the root (master) processor.

The next section summarises the design philosophy of the NAG Parallel library and describes the main features of its implementation. This allows a detailed description of Y01CAFP in the subsequent section. Section 4 discusses the BAe RCS application and the test case used for evaluation. Section 5 presents the results of running the test case on a Cray T3D and a SGI Origin 2000 and finally, Section 6 gives our conclusions.

## 2 NAG Parallel Library

The routine described in this paper has been developed for inclusion in the NAG Parallel Library [7]. This library is a collection of parallel Fortran 77 routines for the solution of numerical and statistical problems. The library is divided into chapters, each devoted to a branch of numerical analysis or statistics. The library is primarily intended for distributed memory parallel machines, including networks and clusters, although it can readily be used on shared memory parallel systems that implement PVM [6] or MPI [9]. The library supports parallelism and memory scalability, and has been designed to be portable across a wide range of parallel machines. The library assumes a Single Program Multiple Data (SPMD) model of parallelism in which a single instance of the user's program executes on each of the logical processors.

The NAG Parallel Library uses the Basic Linear Algebra Communication Subprograms (BLACS) [5] for the majority of the communication within the library. Implementations of the BLACS, available in both PVM and MPI, provide a higher level communication interface. However, there are a number of facilities that are not available in the BLACS, such as sending multiple data types in one message (multiple messages must be sent) and non-blocking sends and receives. There is, therefore, a clear trade-off between code portability (plus ease of maintenance) and performance. As performance is crucial in load balancing, much of the communication is written in PVM and MPI.

The library is designed to minimise the user's concern with use of the BLACS, PVM or MPI, and present a higher level interface using library calls. Task spawning and the definition of a logical processor grid and its context is handled by the parallel library routine Z01AAFP. On completion the library routine Z01ABFP is called to undefine the grid and context. The routines Z01AAFP and Z01ABFP can be considered as left and right braces, respectively, around the parallel code.

## 3 Load Balance Routine (Y01CAFP)

### 3.1 Code integration

Y01CAFP assumes the problem requiring load balancing is written in the form

```
DO I=1,NLOCAL
  CALL TASK(I)
END DO
```

where all data is passed into TASK through COMMON, the index I distinguishes the actual task to be performed and all tasks are independent. The routine then replaces the above code fragment. The user must therefore modify the program so that it conforms to this specification.

The user must also supply a routine (whose specification is defined in the documentation) which will pack or unpack the data required to compute a task for a range of contiguous indices. Y01CAFP will call this routine to pack and unpack data into the appropriate indices. Y01CAFP supplies pack (NAGPACK) and unpack (NAGUNPACK) routines to facilitate this task. These routines are wrappers around the PVM and MPI versions and have a similar syntax.

As well as specifying how many tasks are on a particular processor (NLOCAL) the user must also specify the maximum number of tasks that *could* be computed on that processor (NMAX). The load balancer will use the space between NLOCAL and NMAX to perform any required remote computation. NMAX must be large enough to allow any required remote computation to take place. The actual amount is defined in the user documentation and the program will give an error if NMAX is not large enough.

In the PVM implementation Y01CAFP makes use of the system buffers to buffer data. As MPI does not support system buffers the user must supply a buffer large enough to send and receive the largest message.

### 3.2 Load Balancing options

Y01CAFP allows the user to select one of four different load balancing strategies, 'ASIS', 'BLOCK', 'CYCLIC' and 'GRAB'. In addition 'CYCLIC' and 'GRAB' have a block size (BSIZE) which is set by the user. If all of the data is initially on the root processor (NLOCAL=0 on all other processors) then the master/slave (MASLV) option can be set. This option dedicates the root processor to communication. In this case Y01CAFP is effectively parallelising the application.

Note that, changing the load balancing options in Y01CAFP will not affect the results, only the load balance and therefore, solution time. Y01CAFP accepts any initial data distribution and the final distribution will be the same as the initial distribution. Y01CAFP provides an indication of how it has performed through the TINFO and NINFO arrays. These arrays give timing and counting information respectively.

**ASIS:** 'ASIS' performs no load balancing. It is useful to test the correct working of the code when the load balancing routine is first used. It can also be used to determine the load imbalance inherent in the problem using the NINFO and TINFO output arrays and gives a non-load balanced timing result allowing comparison with any load balanced results.

**BLOCK:** 'BLOCK' should be used when the computational costs of the tasks are the same but their distribution across processors is irregular. Note, if the distribution were regular in this case, the problem would already be load balanced.

The implementation of 'BLOCK' takes a given task distribution and redistributes it so that each processor has no more than $\lceil n/p \rceil$ tasks. It attempts to minimise the number of messages sent by a combination of sending tasks from the most loaded processor to the least loaded processor and looking for pairs of equally overloaded and underloaded processors [10].

**CYCLIC:** 'CYCLIC' should be used when computational costs for successive tasks (in iteration space) are similar, but the load varies over many iterations. The distributed task indices are treated as a single global index ordered by processor identifier.

The implementation of 'CYCLIC' firstly computes the global iteration space. Secondly, processors send all local tasks which require redistribution. Note, if a processor needs to send more than one block to the *same* remote processor, it does so in separate messages. Thirdly, processors compute any local tasks, and finally, processors compute any remote tasks and return the results.

**GRAB:** 'GRAB' should be used when the computational costs of the tasks are unknown. In this case a regular distribution of tasks (as given by the two previous strategies) may result in load imbalance. With this option each processor performs its own computation then steals 'BSIZE' tasks from any processors which are still computing until all work has completed.

The implementation of 'GRAB' checks for any task requests after computing each local task (of size 'BSIZE'). If it receives a request and has more than 'BSIZE' tasks remaining it sends these to the requesting processor, otherwise it sends a negative acknowledgement (NACK). When a processor has finished its own tasks it requests each processor in turn for work. It completes when it has received NACK's from all other processors and has sent NACK's to all other processors.

**MASLV:** The 'MASLV' option is only relevant when all tasks are on the root processor i.e: NLOCAL= 0 on all processors except root. If, in this case, MASLV is .TRUE. the root processor does not take part in any computation, it is used purely for communication. This option is useful when the cost of communication is high enough to significantly slow the root processor. For example, if the communication costs of sending and receiving data were equal to the computation costs the root processor would take approximately twice as long as the other processors. This effect increases with the number of processors, the amount of data transfered and the speed of the processor. It decreases with the speed of the network. Output from NINFO and TINFO, helps the user these effects.

## 4 RCS Example

### 4.1 Description

The purpose of this application is to predict the radar cross section (RCS) of an aircraft's air intake duct. Ducts are particularly important as they act as a waveguide propagating energy (Electro-magnetic (EM) waves) back in the receiver direction. Therefore a large portion of an aircrafts RCS is due to duct reflection. Ray tracing techniques [2] are useful for RCS analysis as they allow the realistic modelling of physical systems with arbitrary shaped ducts and different absorption characteristics [1].

Manchester University has developed a ray tracer for inclusion into BAe's "System AB3". This code uses raytracing to calculate the RCS of an arbitrary shaped duct. The duct geometry is designed using the CAD package CATIA whose surfaces are output as parametric bi-cubic patches [2] in PATRAN [3].

A user generated "ASPECT" file controls the position, direction, frequency, angle and polarisation of the initial EM rays. Rays are exclusively directed inside the duct as this is the area of interest. These rays are then ray traced by AB3. At each ray/surface intersection the EM characteristics of the ray are modified based on the intersected surface characteristics. The rays are terminated either when they emerge from the duct or when their energy falls below an appropriate threshold. AB3 integrates the emerging rays to obtain the RCS.

### 4.2 Integration into the NAG Parallel library

In AB3, a set of rays, whose starting points are arranged in a two dimensional grid, are traced into the duct. This was implemented as a double loop over the rays initial coordinates. To make the AB3 code conform to the load balance specification this double loop had to be changed to single index. The data was already passed into the routine using common. The packing and unpacking routine was simple to implement as all data was dependent on the index.

The initial implementation added the NAG begin parallel (Z01AAFP) and end parallel (Z01ABFP) calls around the code. All non-root processors then skipped the initialisation and waited in Y01CAFP while the root processor set up all the data. In this case the load balancer distributes the work from the root to the remaining processors and acts as if it is parallelising the code.

Whilst this version was a useful starting point, the memory requirements of the root processor meant that this version would not scale to large problem sizes.

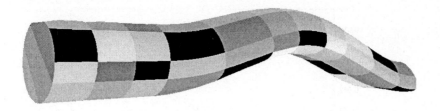

**Fig. 1.** External view of a de-classified BAe duct

To remove this limitation the data was pre-distributed amongst the processors. This was done in two ways. The first (termed block) assigned the first $\lceil n/p \rceil$ rays to the first processor, the next $\lceil n/p \rceil$ rays to the second processor and so on. The second (termed cyclic) assigned the first ray to the first processor, the second ray to the second processor and so on, wrapping round to the first processor after the last processor.

The final RCS prediction code section (integrating the emerging rays to obtain the RCS) has not been parallelised. The results from each processor are sent to the root processor and it performs the computation. This section is included in the timing results presented and for large numbers of processors becomes an important factor.

### 4.3 Test case

The test case used in this paper is a duct which has the complexity of real ducts currently in use and/or being developed by BAe, but has been modified so that it can be de-classified. The *external* visual ray tracing of this duct has been performed using the ray tracer developed at the University of Manchester (which is a modification of krt[4] to allow bi-cubic patch intersection). This is also the ray tracer which has been modified to form part of "System AB3". The patches have been artificially shaded to highlight them, see Figure 1.

## 5 Results

In all versions described in this section the grab option uses a block size of $l/(5p^2)$, where $p$ is the number of processors and $l$ is the total number of rays. Smaller block sizes were investigated but this made no difference to the performance. In this section the problem sizes are given in terms of the ray density which is the number of rays per wavelength. In the example code the wavelength is 3cm and the frequency is approximately 10GHz. The total number of rays is also given for reference. At the time the following results were taken the TINFO and NINFO performance analysis arrays and the CYCLIC option described in Section 3.2 were not implemented.

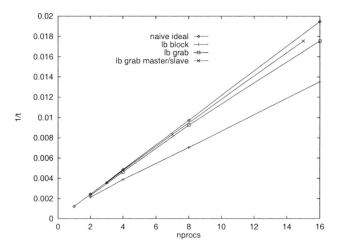

**Fig. 2.** SGI O2000, initial distribution all on root, ray density 13 (106,912 rays).

In Figure 2 the initial data is stored on the root processor. The reciprocal of wall-clock time is given on the y-axis giving the equivalent of a speedup graph without normalising the time taken. The 'naive ideal' line is simply the sequential time divided by the number of processors. The load balance block option suffers from load imbalance which is improved by the load balance grab option. In these cases the root processor is both computing its own work and sending and receiving work to and from the other processors. To determine this performance penalty the grab option is repeated with the master/slave option set to true. Note, to show this overhead the root processor is not included in nprocs which in this case is the number of computing processors. This shows that much of the remaining difference from the ideal line is due to this overhead.

The data was then pre-distributed in equal sized blocks across the processors. With the load balance block option set, the result is identical to the block option in Figure 2 and is therefore not presented. This result shows that the load balance block option is efficient when the data is all on one processor. Note, in this pre-distributed case Y01CAFP does not have to perform any data re-distribution. The load balance grab option is given in Figure 3 and it performs as well as Y01CAFP with grab and master/slave options (originally shown in Figure 2) which is also displayed as a reference. This shows that pre-distributing the data removes the data transfer bottleneck from the root processor.

The data was then pre-distributed in a cyclic manner. Figure 3 shows that with the load balance block option the performance is as good as the other two options. The load balance block option actually performs no data re-distribution in this case. This means that for this problem pre-distributing the rays in a cyclic manner gives very good load balance. The load balance grab option gives no further improvement and is therefore not included.

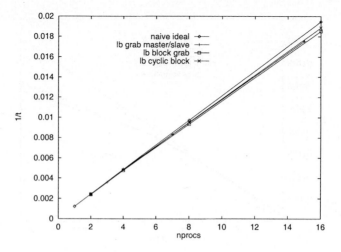

**Fig. 3.** SGI O2000, initial distribution block and cyclic, ray density 13 (106,912 rays).

Figure 4 presents results for the same test case scaling up to a much larger number of processors on a Cray T3D. The initial block distribution with the load balance block option performs worst due to load imbalance. Changing this option to grab brings the performance close to that for the initial cyclic distribution with the load balance block option. The initial cyclic distribution with the load balance grab option performs the best by a small margin. The performance improvement falls off primarily due to the sequential fraction mentioned in Section 4.2.

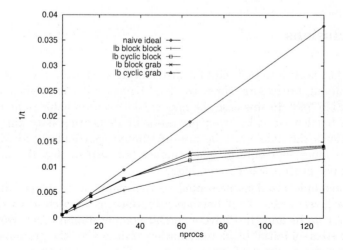

**Fig. 4.** Cray T3D, ray density 13 (106,912 rays).

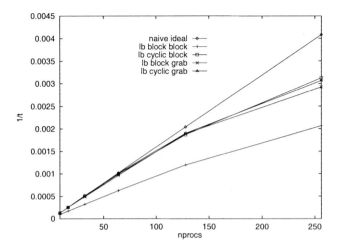

**Fig. 5.** Cray T3D, ray density 50 (1,586,356 rays).

Figure 5 presents results for the same options as the previous figure with a much greater ray density of 50 (1,586,356 rays). The trends are very similar however the larger problem size is more scalable. At 256 processors the cyclic pre-distribution actually performs slightly worse with the load balance cyclic option than the load balance block option.

Figure 6 again presents results for the same options as Figure 4 for a ray density of 200 (25,370,577 rays). At this density the problem will only run on 128 or more processors due to memory limitations. In this example all options scale linearly except the initial block distribution with load balance block option which suffers from load imbalance.

## 6 Conclusions

Y01CAFP has proven to be useful for distributing work from the root processor to the remaining processors. Note, the load balancer is effectively parallelising the application here. In this case the master/slave option helps reduce the communication bottleneck at the root processor by dedicating it to this task. For larger problems the data must be pre-distributed (particularly for distributed memory machines) not only for performance but also so that the memory requirements per processor is not too high.

In the example problem presented in this paper a cyclic pre-distribution of the data gives a near load balanced solution (as rays close to each other follow similar paths and thus have a similar computational cost). However, for all problem sizes an initial block distribution of data with the grab load balance option gives very similar performance results. This suggests that for a different dataset, or different problem entirely, where an initial cyclic distribution is not

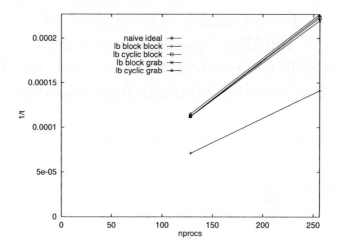

**Fig. 6.** Cray T3D, ray density 200 (25,370,577 rays).

feasible or does not give a load balanced solution an initial block distribution of data with the grab load balance option will load balance the problem.

The initial integration of the load balancer into the RCS code was relatively simple and BAe are now using the NAG parallel library and the Y01CAFP load balance routine in production runs with much greater ray densities and more realistic geometries than were previously possible.

In summary Y01CAFP has proven to be a very powerful, flexible and useful load balancing routine.

## References

1. Ling H. et al, *Shooting and Bouncing Rays: Calculating the RCS of an Arbitrarily Shaped Cavity*, IEEE Transactions on Antennas and Propagation, Vol. 37, No. 2, February 1989, pp. 194-204
2. Watt A., *Fundamentals of Three Dimensional Computer Graphics*, Addison Wesley, 1989.
3. PATRAN Plus User Manual.
4. Keates M., Hubbold J., *Accelerated Ray Tracing on the KSR1* UMCS-94-2-2
5. J. Dongarra and R. C. Whaley, (1997) *A User's Guide to the BLACS v1.1*, Technical Report CS-95-281, University of Tennessee, Knoxville, Tennessee.
6. A. Geist, A. Beguelin, J. Dongarra, R. Manchek, W. Jiang, and V. Sunderam, (1994), *PVM: A Users' Guide and Tutorial for Networked Parallel Computing*, The MIT Press, Cambridge, Massachusetts.
7. N.A.G., (1997) *N.A.G. Parallel Library Manual, Release 2*, N.A.G. Ltd., Oxford.
8. N.A.G., (1997) *N.A.G. Fortran Library Manual, Mark 17*, N.A.G. Ltd., Oxford.
9. M.Snir, S.Otto, S.Huss-Lederman, D.Walker and J.Dongarra, (1996) *MPI: The Complete Reference*, The MIT Press, Cambridge, Massachusetts.
10. R.Ford (1998) *A Message Minimisation Algorithm* CNC Technical Report, Department Of Computer Science, The University of Manchester, Manchester, U.K.

# Performance Assessment of Parallel Spectral Analysis: Towards a Practical Performance Model for Parallel Medical Applications

F. Munz[1,2], T. Ludwig[2], S. Ziegler[1], P. Bartenstein[1], M. Schwaiger[1], and A. Bode[2]

[1] Nuklearmedizinische Klinik und Poliklinik des Klinikums rechts der Isar
[2] Lehrstuhl für Rechnertechnik und Rechnerorganisation
Technische Universität München (TUM)

email: Munz@Informatik.TU-Muenchen.DE

**Abstract.** We present a parallel, medical application for the analysis of dynamic positron emission tomography (PET) images together with a practical performance model. The parallel application improves the diagnosis for a patient (e. g. in epilepsy surgery) because it enables the fast computation of parametric images on a pixel level in contrast to the traditionally used region of interest (ROI) approach. We derive a simple performance model from the application context and demonstrate the accuracy of the model to predict the runtime of the application on a NOW. The model is used to determine an optimal value for the length of the messages with regard to the per message overhead and the load imbalance. (**Keywords**: positron emission tomography, parallel kinetic modeling, PVM, practical performance prediction, network of workstations, medical application)

## 1 Introduction

During the last decade positron emission tomography (PET) has developed as a medical imaging modality with unique sensitivity. PET enables the regional brain function to be assayed in a fully quantitative and non-invasive way, whereas traditional tomographic technologies (e.g. computer tomography) are only capable of imaging the morphological structure of the body. In this paper we present a parallel implementation of spectral analysis, a technique which is used to extract physiological parameters of dynamic PET scans.

Although we have previously shown that computationally intensive functional imaging algorithms such as iterative image reconstruction from projections can expect a good performance improvement even on cluster of workstations [11], and their implementation is rather straightforward [2], the number of routinely used parallel applications in medicine is still very low. Often inferior algorithms are used where better, but computationally more intensive algorithms are available. We believe, that besides the increased complexity in the implementation phase, the main reason for this is the missing predictability of the application performance of a parallel program and the trial and error approach to determine optimal parameter settings. Therefore we derive an analytical performance model based on architectural parameters of the interconnection network and the application context.

The remainder of this paper is structured as follows. In Sect. 2, we describe the medical background which is important to understand the context of our application. Section 3 describes the parallel implementation of the algorithm on a cluster of workstations. We emphasize the importance of an adequate performance model to determine the optimal mapping of data on processors. In Sect. 4, we examine the most popular abstract machine models and evaluate their applicability in regard to our hardware and software environment. We then derive an explicit formula to predict the performance of our application. In Sect. 5, we demonstrate the accuracy and usability of our model for the determination of optimal parameter settings. Finally, Sect. 6, summarizes the main points of the paper.

## 2 Kinetic Modeling

To acquire a PET scan, a physician injects a small amount of a radioactively labeled substance (tracer) with known physiological properties such as water, sugar or molecules binding to a particular type of neuronal receptor. A PET camera measures the decay of the radioisotopes in circumferential arrays of scintillation detectors. The reconstruction of the measured projection data is a complex task and defines a field of intensive research [11].

An important aspect of PET is the ability to monitor the accumulation or disappearance of the biological tracer over time. During a dynamic PET study datasets are acquired at 25 points of time $t_j$, consisting of 47 slices with 128 x 128 pixels each. Fig. 1 shows plane number 15 of six out of 25 frames.

**Fig. 1.** Frame number 1,5,10,15,20 and 25 of a dynamic PET scan.

Traditionally, compartmental models, formulated by differential equations, are used to relate the time course of the supply in arterial plasma $C_m(t)$ to the concentration of the tracer in cerebral tissue $C_{tiss}(x, y, z, t_j)$. Fitting the free variables of the model to the measured data yields quantitative physiological parameters such as blood flow or the density of opiate receptors per ml tissue.

However, compartmental models have three major disadvantages: they normally do not work on a pixel level because of numerical instabilities and noise, they require a priori assumptions of tracer kinetics and they involve systems of differential equation which can be difficult to solve.

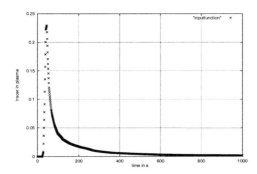

**Fig. 2.** Arterial input function of a 24 year old male volunteer.

To overcome these shortcomings CUNNINGHAM proposed a general modeling approach that allows the determination of the unit impulse response functions (IRF) of the tracer kinetics at a pixel level [5]. In the spectral analysis method the PET tissue time activity curve (TAC) is modeled as a linear combination of basis functions each of which is a single exponential in time convolved with the arterial input function $C_m(t)$:

$$C_{tiss}(t_j) = \sum_{i=1}^{i=k} \alpha_i C_m(t) \otimes e^{-\beta_i t}, \; \alpha_i >= 0, \; \lambda <= \beta_i <= 1 \qquad (1)$$

$k$ is the number of basis functions, usually 64. The $\beta$ values are chosen to cover the spectrum of kinetic behavior, from the physical decay constant $\lambda$, the slowest possible clearance, to the fastest dynamic (e.g. $1s^{-1}$). We determine the $\alpha_i$ values, which best fit the measured data given the predetermined and fixed values of $\beta_i$ with the non-negative least squares (NNLS) algorithm described by LAWSON and HANSON [10]. Note, that this has to be done for $128 \cdot 128 \cdot 47 = 770048$ TACs.

## 3 Parallel Implementation

### 3.1 Target Architecture

Our target architecture is a cluster of 20 Sun 60 workstations with a 300 MHz Ultra SPARC II CPU. Each workstation is equipped with 384 Mbyte of RAM.

All workstations are running Solaris 2.6. A 3COM 3300 fast Ethernet switch interconnects the workstations with 100 Mbit/s. Special care was taken to avoid any kind of external load on the machines and all benchmarks and other measurements were carried out at night and/or during the weekend.

### 3.2 Implementation of the Application

The distributed application is build on top of the PVM 3.4 message passing library. We use a simple heuristic to distribute the computations across the nodes of the cluster. In the following we assume to use *proc* nodes out of 20 available nodes. First, PVM is started on all involved nodes and a worker process for the calculation of the parametric images is assigned to every pvmd. Setup values are broadcasted to every worker process to initialize the numerical routines. Then $gr$ TACs are sent to every node. Whenever a node is finished it returns the $m$ physiological parameters for every TAC, and another set of $gr$ TACs are assigned to that node. This is done to keep all nodes constantly busy and to minimize load imbalance.

Although the flow of control and the communication properties of this data distribution scheme is quite simple, there are some open questions which cannot be answered easily. Predictions of possible speedups or the time to run the application on *proc* nodes are difficult to achieve and will depend on the granularity $gr$ of the algorithm. Assuming a fixed number of nodes, we will expect more per-message overhead for a smaller $gr$, but an increased load imbalance for a larger $gr$. The optimum value $gr_{opt}$, could of course be determined by measuring the runtime for different $gr$ values, but as $gr_{opt}$ is a function of *proc* and the underlying hardware, this proceeding seems to be rather tedious.

## 4 Performance Model

### 4.1 Abstract Machine Models

The predominant theoretical model of parallel computation certainly is the Parallel Random Access Machine (PRAM)[6]. Many algorithm designers in theoretical computer science use the PRAM model due to its simple and clean semantic attributes mainly for proving complexity bounds and a gross classification of algorithms. However, PRAM is of limited practical use as it assumes synchrony at an instruction level, interprocessor communication with zero overhead, zero latency and infinite bandwidth. For that reason application programmers cannot rely on the PRAM big-O notations of complexity bounds (which may hide a constant factor of any size) when they implement a particular algorithm on a real parallel machine [9]. For instance, we would expect linear speedup for our application under PRAM which is certainly not the case on a network of workstations. Actually, the opposite could be true on a real parallel machine. If we assume a very short computation time, then the distributed calculation of functional images might take even longer than running the application on a single node.

To avoid such loopholes, a large amount of effort has been put into the development of more sophisticated models keeping in mind the balance between accuracy and simplicity. VALIANT proposed the Bulk Synchronous Parallel (BSP) model to bridge the gap from theory to practice [14]. Under BSP the execution of a parallel program is divided into supersteps. Supersteps consist of $h$-relations (communication abstractions of at most $h$ words) which take $g \cdot h$ cycles and computation cycles of length $w$, where $w$ is the maximum number of cycles spent for computation by any processor. At the end of each superstep synchronization is forced by a barrier operation with overhead $l$. Opposite to the PRAM model, communication and synchronization overhead is modeled in BSP. HWANG and XU introduced the Phase Parallel Model (PPM), which is even more comprehensive because it includes parallelism overhead [9]. The PPM distinguishes among three non-overlapping phases: In the parallelism phase $T_{par}$ process creation is accomplished, during the computation phase $T_{comp}$ processors execute instructions on local memory, and communication as well as synchronization is done in the interaction phase $T_{interact}$. Under PPM the total execution time of a superstep is modeled as follows:

$$T_n = T_{comp} + T_{par} + T_{interact} \qquad (2)$$

The LogP model, published by CULLER in 1993, exposes more architectural parameters. Based on the four parameters $L$ (network latency), $o$ (processor overhead), $g$ (message gap) and $P$ (processors), CULLER describes a microbenchmark signature to determine the receive overhead $o_r$ and the minimum time between two consecutive messages $g$ [4]. As shown in Fig. 3, modeling the startup-time $t_0$ as three terms $t_0 = o_s + L + o_r$ invites the algorithm designer to hide latency $L$ by overlapping communication and computation. The LogP model is based on an Active Message (AM) style of communication. AM provides low overhead, communication primitives for short messages which resembles a kind of lightweight RPC. Finally, ALEXANDROV published LogGP, an extension to LogP which also incorporates long messages[1]. RUGINA recently presented an algorithm that uses the communication pattern of a parallel program to simulate the execution under the LogGP model, mentioning that explicit formulas for the runtime of data parallel applications even when all processors get equal shares of data are rather complicated [12].

### 4.2 Modeling Image Analysis Algorithms

The various machine models, which we described in the previous Sect., only represent a first step in understanding the performance of a parallel application as they model only the underlying hardware. All models require as an input the communication properties of the algorithm being analyzed [13]. Although it is usually inherently more difficult, and sometimes even impossible, to describe these properties for a particular algorithm, many image analysis applications represent a class where this is feasible. For benchmark routines (e.g. SPLASH 2) and numerical algorithms (e.g. FFT and LU decomposition) researchers have

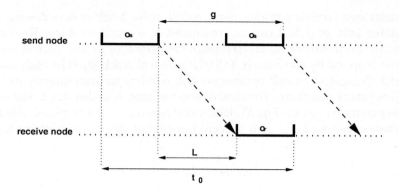

**Fig. 3.** LogP parameters and their relation to $t_0$

tried to investigate the communication behavior. In the following Sect. we will explain a more pragmatic approach that is well suited for image analysis applications with a high degree of inherent data parallelism. We develop a formula for the runtime of the program from the application context and then show the similarity to the PPM model by rearranging the equations.

After the parallel virtual machine is set up, the worker processes are started and initialized. Then *proc* send primitives are necessary to transmit $gr$ TACs to all waiting worker tasks. The time that is spent until the last worker returns its results is $(s_{gr} + comp_{gr} + r_{gr})$, where $s_{gr}$ denotes the time to send $gr$ TACs to the worker, $comp_{gr}$ denotes the time to calculate the physiological parameters for $gr$ TACs and $r(gr)$ is the time to send back $m$ physiological parameters to the master. During this time the results from the other workers were received already, and according to the heuristic described in 3.2 new sets of TACs were assigned to the idle worker processes. This will be repeated until all TACs are sent out, then the remaining *proc* results will be received. This leads us to an estimation for the time that is needed for interaction and computation:

$$T_{comp} + T_{interact} = (s_{gr} + comp_{gr} + r_{gr})\frac{no\_messages}{proc} \qquad (3)$$

$$= \frac{time\_tac \cdot no\_tacs}{proc} + (r_{gr} + s_{gr})\frac{no\_tacs}{proc \cdot gr} \qquad (4)$$

Equation 4 reveals two terms. The computation time $T_{comp}$ is reciprocal to the number of processors *proc* but independent of the granularity $gr$, whereas the interaction time depends on the granularity as well on the number of available processors. For load imbalance $T_{imbal}$, which occurs for example when $no\_messages = no\_tacs/no\_proc$ modulus *proc* is not equal to zero we add another $T_{imbal} = gr \cdot time\_tac$ to the execution time per plane.

An accurate model of the interaction time $T_{interact}$ requires some more considerations. The parameter $s_{gr}$ represents the time to transfer $gr$ TACs from the master to a receiving worker node, which under LogP is $o_s + L + o_r$, taken for granted that $gr$ TACs can be considered as a small message. However, in our cluster environment $o_s$ is very large compared to $L$ due to the software overhead imposed by the Solaris TCP/IP protocol stack [3]. The high software overhead obscures the LogP parameters and restricts us from directly using the microbenchmark signature. For that reason we lump together $o_s$, $L$ and $o_r$ into a single parameter $t_0$ (see Fig. 3). HOCKNEY proposes a linear model, where the interaction overhead $T_{interact}$ is a linear function of the message length $n$ [8]:

$$T_{interact} = t_0 + \frac{m}{r_\infty} \qquad (5)$$

Here the startup time is denoted by $t_0$ and the asymptotic bandwidth by $r_\infty$. HOCKNEY presents two more interaction parameters in his report. The half-peak length $n_{1/2} = t_0 \cdot r_\infty$ characterizes the message length when half of the peak bandwidth is achieved and $\pi_0 = 1/t_0$ denotes the small message performance. The parameters $t_0$ and $r_\infty$ can be obtained by executing a simple benchmark program: The master process sends a message of length n to a worker process. The worker process echos back the same messages immediately. The round trip delay for the ping-pong benchmark is measured at the master process for increasing values of $n$ and divided by two. Fitting the obtained data to (5) yields $t_0$ as the intersection with the y axis and $r_\infty$ as the slope of the line [7]. Figure 4 illustrates the measured times for the ping-pong benchmark on our Sun cluster together with the line of best fit.

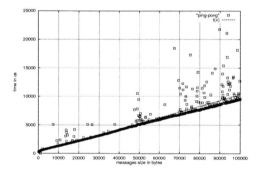

**Fig. 4.** Times for ping-pong benchmark in $\mu$sec. The benchmark was repeated 20 times and averaged to get rid of cache misses or buffer allocation effects.

HWANG and XU presented the values for the SP-2 multicomputer, which are appended to table 4.2 for comparative purposes.

We also investigated the parallelism overhead $T_{par}$ which occurs when the master process starts $proc$ worker processes. $T_{par}$ is linear in the number of

**Table 1.** Comparison of Hockney's benchmark parameters for SP-2 [9] and local Sun cluster.

|             | $t_0$      | $r_\infty$     | $\pi_0$    | $n_{1/2}$  |
|-------------|------------|----------------|------------|------------|
| Sun Cluster | 301 $\mu$s | 10.03 MByte/s  | 3.32 kHz   | 3019 bytes |
| IBM SP-2    | 46 $\mu$s  | 28.57 MByte/s  | 21.74 kHz  | 1314 bytes |

processors started. Curve fitting the measured startup times to a straight line, as shown in Fig. 5 gives us (6):

$$T_{par} = 0.0242s + 0.0193 \frac{s}{proc} \quad (6)$$

Equation 6 assumes that PVM is running on all nodes already, which is usually the case in our cluster for batch processing. For registering the program to PVM and controlling that a pvmd is running on every node we have to add another 0.09 s. This time can be significantly larger if something goes wrong (e.g. if names cannot immediately be resolved because of nameserver problems).

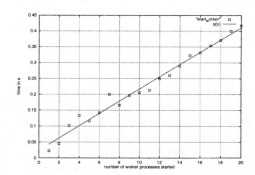

**Fig. 5.** Time in seconds to start *proc* worker processes, when pvmd is running on all nodes already.

The calculation for a single TAC takes $time\_tac = 1230\mu s$ on average with a standard deviation of 371. The large deviation is due to application specific reasons. We believe that it cannot be modeled adequately. In addition there is a sequential part of the program $T_{seq}$ that cannot be parallelized and consumes about 0.08 s.

## 5 Results

Putting the terms of Sect. 4.2 together leads us to the following performance model for parallel spectral analysis:

$$T(proc, gr) = T_{seq} + T_{par}(proc) \\ + no\_planes(T_{comp}(proc) + T_{imbal}(gr) + T_{interact}(proc, gr)) \quad (7)$$

We first use the model from (7) to predict the runtime for a fixed granularity $gr = 128$. As Fig. 6 (left) shows, the runtime of the program is predicted exactly over the whole range of available processors. Note, that the model is not fitted to the measured data — it is purely based on the network parameters $t_0$, $r_\infty$ and the application context, in particular $time\_tac$.

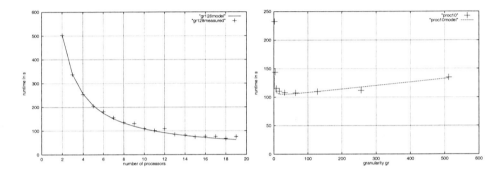

**Fig. 6.** Comparison of predicted run time and measured run time for $gr = 128$ (left) and $proc = 10$ (right).

We then apply the performance model to predict the optimum message length $gr$ for a fixed number of processors. The results in Fig. 6 (right) show that the model accurately predicts the shape of the curve and allows a good estimation of the optimal $gr$ value. Only for very small values of $gr$ the model tends to underestimate the runtime. As our model incorporates the per message overhead, we assume that this effect occurs because the processor becomes the bottleneck as the messages get smaller and the message frequency increases.

## 6 Summary

We presented a medical application for the kinetic analysis of functional images together with a pragmatic and accurate performance model. To our knowledge, this is the first performance model for a parallel medical application. The performance model is based on three basic parameters ($t_0, r_\infty$ and $time\_tac$) and predicts the runtime for an arbitrary number of processors and message granularity reasonable well. We believe that the model makes the trial and error approach to optimize application parameters obsolete. In the future we will use this model to quantify the benefit of implementing the application on a cluster with a SCI interconnect and on a massively parallel processor (IBM SP-2).

## Acknowledgment

We would like to thank Vin Cunningham and Terry Jones for sustaining our work on spectral analysis. Michael Herrman provided help, troubleshooting and useful information about the Sun cluster at the TUM. Thanks to Alex Drzezga for the PET scans. This work is part of an ongoing collaboration between the Lehrstuhl für Rechnertechnik und Rechnerorganisation (LRR-TUM) and the Klinik und Poliklinik für Nuklearmedizin rechts der Isar. This study was partly supported by the Deutsche Forschungsgemeinschaft (SFB 462, Sensomotorik).

## References

1. ALEXANDROV, A., IONESCU, M., SCHAUSER, K., AND SCHEIMAN, C. LogGP: Incorporating Long Messages into the LogP Model. *Proc. of the 7th Annual ACM Symp. on Parallel Algorithms and Architectures* (1995), 95–105.
2. ARABNIA, H. High-Performance Computing and Applications in Image Processing and Computer Vision. In *High Performance Computing* (1997), C. Polychronopoulos, K. Joe, K. Araki, and M. Amamiya, Eds., vol. 1336 of *Lecture Notes in Computer Science*, Springer-Verlag, p. 72.
3. CLARK, D., JACOBSON, V., ROMKEY, J., AND SALWEN, H. An Analysis of TCP Processing Overhead. *IEEE Communications Magazine* (June 1989), 23–29.
4. CULLER, D., KARPAND, R., PATTERSON, D., SAHAY, A., SCHAUSSER, K., SANTOS, E., SUBRAMONIAN, R., AND VON EICKEN, T. LogP: Towards a Realistic Model of Parallel Computation. In *Proc. ACM Symp. on Principles and Practice of Parallel Programming* (May 1993).
5. CUNNINGHAM, V. J., AND JONES, T. Spectral Analysis of Dynamic PET Studies. *Journal of Cerebral Blood Flow and Metabolism 13* (1993), 15–23.
6. FORTUNE, S., AND WYLLIE, J. Parallelism in Random Access Machines. In *Proceedings of the Tenth ACM Symposium Thery of Computing* (May 1978).
7. HOCKNEY, R. Performance Parameters and Bechmarking of Supercomputers. *Parallel Computing 17* (1991), 1111–1130.
8. HOCKNEY, R. The Communication Challenge for MPPs: Intel Paragon and Meiko CS-2. *Parallel Computing 20* (1994), 389–309.
9. HWANG, K., AND XU, Z. *Scalable Parallel Computing*. Mc Graw-Hill, 1998.
10. LAWSON, C. L., AND HANSON, R. J. *Solving Least Squares Problems*. Prentice Hall Series in Automatic Computation. Prentice-Hall, Englewood Cliffs, NJ, 1974.
11. MUNZ, F., STEPHAN, T., MAIER, U., LUDWIG, T., BODE, A., ZIEGLER, S., NEKOLLA, S., BARTENSTEIN, P., AND SCHWAIGER, M. NOW Based Parallel Reconstruction of Functional Images. In *Proceedings of the First Merged International Parallel Processing Symposium and Symposium on Parallel and Distributed Computing* (Los Alamitos, California, USA, April 1998), B. Werner, Ed., IEEE Computer Society Technical Committee on Parallel Processing, pp. 210–214.
12. RUGINA, R., AND SCHAUSER, K. E. Predicting the Running Times of Parallel Programs by Simulation. In *Proceedings of the 12th International Parallel Processing Symposium and 9th Symposium on Parallel and Distributed Processing, Orlando, FL* (April 1998).
13. SINGH, J. P., ROTHBERG, E., , AND GUPTA, A. Modelling Communication in Parallel Algorithms: A Fruitful Interaction between Theory and Systems? *Proc. of the 10th Annual ACM Symposium on Parallel Algorithms and Architectures* (1994).
14. VALIANT, L. A bridging model for parallel computation. *Comm. of ACM 33*, 8 (1990), 103–111.

# Parallel Algorithm and Processor Selection Based on Fuzzy Logic

Shuling Yu, Mark Clement, Quinn Snell, and Bryan Morse

Computer Science Department
Brigham Young University
Provo, Utah 84602-6576
(801) 378-7608
{shulingy, clement, snell, morse }@cs.byu.edu
http://ccc.cs.byu.edu

**Abstract.** The face of parallel computing has changed in the last few years as high performance clusters of workstations are being used in conjunction with supercomputers to solve demanding computational problems. In order for a user to effectively run an application on both tightly coupled and network based clusters, he must often use different algorithms that are suited to the network available on the computing platform. An application may also be able to effectively utilize a different number of processing nodes with a particular algorithm and processor configuration. It is difficult for a user to determine which set of parameters to select in order to customize the application for an available computing environment. The principal aim of this research is to show that fuzzy logic can be used to select the most efficient algorithm and an optimal number of processors for a parallel application. In this paper we examine three algorithms for image convolution which each have advantages depending on the available architecture and problem size. A fuzzy logic technique is developed which is able to make effective selections, freeing the user from an otherwise daunting task. The fuzzy logic selection system is easy to set up and these results can be extended to additional applications.
**Keywords: algorithm selection, parallel algorithms, image convolution, fuzzy logic.**

## 1 Introduction

Several techniques have been used to increase the speed of parallel computations at the algorithmic and architectural level [3, 11]. Algorithms can be selected which perform fewer communications in loosely coupled configurations at the cost of increased computations. When high speed networks are available, these algorithms should be replaced by more efficient schemes that may perform more communications. In order to keep the grain size large enough to achieve acceptable performance, the number of processors should be limited by the size of the problem to be solved. Choosing the correct parallel algorithm and the optimal number of processors can have a significant effect on the performance of certain parallel applications. The exact effect of these factors is uncertain and difficult to represent with a closed formula. In this paper, we propose a new optimization method based on fuzzy logic to automatically select the appropriate parallel algorithm and parallel scope for a specific problem on a given hardware platform.

Fuzzy logic is multi-valued logic. Whereas Boolean logic has just two values 0 and 1, fuzzy values vary from an minimum to a maximum as a function of the input [14]. Fuzzy logic matches the approximate nature of human reasoning fairly well in many cases. Fuzzy set theory provides a mechanism for carrying out approximate reasoning processes when available information is uncertain, incomplete, imprecise or vague [5].

In the fields of network management and performance prediction, fuzzy models have been applied and/or proposed in a number of areas. Ficili and Panno [2] discuss the use of fuzzy approaches in the policing of packetized voice sources. Man and Isak [9] propose a new fuzzy clustering algorithm to detect and characterize ring-shaped clusters and combinations of ring-shaped and compact spherical clusters. Holtzmann [7] discusses the use of the fuzzy set theory in handling an uncertain service time of a queuing system and forecasting new service in broadband traffic. Schopf and Berman [10] propose a method to use a set of possible values and their probabilities (called stochastic values) as parameters to adaptable performance prediction models on production distribution systems. But the computation of stochastic values is much more complex than fuzzy logic computations. This research applies the principles of fuzzy logic in a new way to select the appropriate parallel algorithm and number of processors for parallel applications.

## 2 Parallel Image Convolution Algorithms

Image processing algorithms are inherently parallel in nature [13, 8]. The convolution operation arises in variety of ways with images [1, 6, 12]. Digital filtering, image smoothing, image sharpening and other image processing operations all involve convolution. The result of convolution (computed at a pixel level) is dependent only on the values of its neighboring points. These operations can be efficiently mapped to the parallel arena. In this section, we discuss the execution time for three parallel convolution algorithms. The models and their performance were described in more detail in prior work [15]. The algorithms include:

- Conventional parallel convolution. Each processor exchanges border data at every iteration.
- Redundant Boundary Computation (RBC) convolution. The edge data of each block is distributed to overlap with its neighboring block, so there are no communications between iterations. Processors perform the same computations for the boundary elements so the total number of computations performed by this algorithm can be much larger than the conventional algorithm for large kernel sizes. When communications are expensive compared to computation time, this algorithm can improve the overall speed of the convolution.

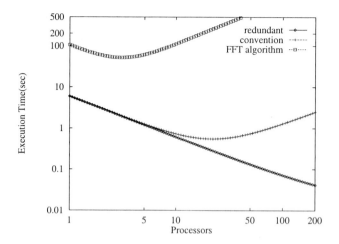

**Fig. 1.** Execution time for the three algorithms with $N = 1024, m = 3 and K = 3$ on Ethernet-connected networks

- (Fast Fourier Transform) FFT convolution. The image and its kernel are transformed into the frequency domain, then pointwise multiplication is performed, the result of convolution is obtained as the inverse FFT of the product. Parallel algorithms are used to compute the FFT and inverse FFT.

Figures 1 and 2 show the execution time for the three algorithm when a 1024 × 1024 image is used on a shared media network.

In analyzing the algorithms, $N$ defines the size of the problem. This research assumes that images are square with $N \times N$ pixels. Certain problems require convolution to be run for several iterations. We will use $m$ to denote the number of iterations for the given problem. The size of the kernel used for convolution will be represented using the variable $K$. From Figure 1, we can see that the RBC algorithm has better scalability and execution time than the other two algorithms when the kernel size and the number of iterations are small. Figure 2 shows that the FFT algorithm has poor scalability due to sever bandwidth competition. It has better execution time only if the processor count is small. The conventional algorithm has better execution time than the RBC algorithm when the kernel size and number of iterations are larger. Since no algorithm is best in all cases the user must select the best algorithm based on specific parameters. Since these parameters are difficult to obtain accurately, fuzzy logic can be effectively applied to make the process easier.

Experimental data was generated using the Parallel Virtual Machine (PVM) libraries[4] on a network of Hewlett Packard workstations connected with Ethernet in order to train the fuzzy system. In order to simplify the comparisons, slave execution time was used since data distribution and collection time was the same for all three algorithms. Table 1 summarizes experimental data[15]. For small kernels (with size less than 51), the conventional algorithm and the

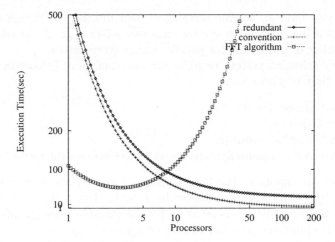

**Fig. 2.** Execution times for the three algorithms with $N = 1024, m = 21 and K = 6$ on Ethernet-connected networks

| Image | Kernel | Iteration | Algorithm | Proc. |
|---|---|---|---|---|
| 512 | 3 | 3 | RBC | 2 |
| 512 | 9 | 6 | Conventional | 8 |
| 512 | 51 | 3 | FFT | 8 |
| 1024 | 3 | 3 | RBC | 4 |
| 1024 | 9 | 6 | Conventional | 8 |

**Table 1.** The preferred algorithm and number of processors for different situations as measured on Ethernet-connected networks with PVM

RBC algorithm exhibit better performance. The RBC algorithm is a better selection for situations with small kernel sizes and a small number of iterations. Note that the number of processors which can be effectively is limited by the problem size. The 1024 size image can make use of additional processors and this number increases with larger problems. The FFT algorithm can achieve better performance only when the kernel size is large. Because the Fourier Transform uses complex calculations, the overhead of data type transformation, complex calculation, and the use of larger memory space reduce the benefits of the FFT algorithm.

## 3  Fuzzy Selection System

Parallel convolution is too complex to model precisely using analytical methods. The actual congestion present in the network and the effects of transient loads on processors make it difficult to select correct runtime parameters. Fuzzy logic can be used to solve problems where the results are derived from uncertain and incomplete information. In this section, we will discuss the design and implementation of a fuzzy selection system that can select a good algorithm and an efficient number of processors for parallel image convolution.

The fuzzy selection system requires the abstraction of factors in fuzzy sets. The factors that we chose are:

- The input image size $N$
- The kernel size $K$
- The number of iterations $m$
- The *ratio* of the communication speed to the processing speed

These factors are inputs to the fuzzy system. The outputs are the selected algorithm and the number of processors. In order to design the fuzzy system, we first specify the fuzzy sets associated with each variable, then specify the shape of the membership functions and finally, set up the fuzzy rules.

We use three fuzzy sets for the input variables ["large" (L), "medium" (M), and "small" (S)], three fuzzy sets for the algorithm output ["Conventional" (C), "Redundant" (R) and "FFT" (F)], and five fuzzy sets for the number of processors output ["Large Large" (LL), "Small Large" (SL), "Medium" (M), " Large Small" (LS), and "Small Small" (SS)].

Fuzzy sets are determined by their membership functions. Fuzzy membership functions are the mechanism through which the fuzzy system interfaces with the inputs and outputs. We use trapezoidal functions as membership functions. There are three trapezoidal types: left, regular, and right (as shown in Figure 3). The triangle membership function is used for the **medium** classification, while left and right trapezoids are used for small and large respectively (see Figure 4). The shape of the trapezoidal membership function is completely determined by the points which define its linear segments. For example, a regular trapezoid, as shown in Figure 3, has three linear segments defined by the four points a,b,c, and d. The input membership functions are shown in Figure 4 (we use the same membership functions for all fuzzy inputs, but this could be changed if some input had a different behavior). The corresponding point values which define the shape are shown in Table 2. The number of processors output for each algorithm is shown in Table 3.

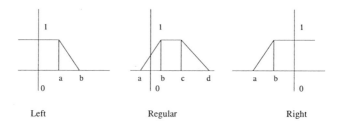

**Fig. 3.** Three types of trapezoids

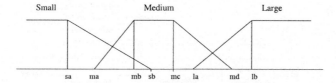

**Fig. 4.** Membership functions for image, kernel, iteration and ratio inputs. The values from Table 2 are used to customize the membership function for each input.

|           | sa  | sb  | ma  | mb  | mc  | md  | la  | lb   |
|-----------|-----|-----|-----|-----|-----|-----|-----|------|
| Image     | 100 | 250 | 200 | 300 | 500 | 800 | 700 | 1200 |
| Kernel    | 3   | 7   | 5   | 9   | 21  | 41  | 31  | 51   |
| Iteration | 1   | 5   | 2   | 7   | 10  | 15  | 12  | 20   |
| Ratio     | 1   | 8   | 5   | 10  | 50  | 100 | 70  | 120  |

**Table 2.** Membership function values for inputs. These values are used to define the trapezoids in Figure 4

| Algorithm and Number of Processors ||||
|----|--------|---------|--------|
|    | RBC(1) | CONV(2) | FFT(3) |
| SS | 1      | 2       | 2      |
| LS | 4      | 5       | 5      |
| M  | 8      | 10      | 10     |
| SL | 16     | 20      | 20     |
| LL | 32     | 32      | 32     |

**Table 3.** Action values for the number of processors output for each algorithm. These values were derived from the experimental data described earlier.

The fuzzy system has two stages. First it selects the appropriate algorithm with the fuzzy set's membership functions and fuzzy rules associated with algorithm selection. Then it selects the optimal number of processors using the algorithm which has been selected. The system is shown in Figure 5. Fuzzy systems represent well-defined deterministic functions. The output of a fuzzy system is precise, not fuzzy. There are several approaches for obtaining "defuzzification" output from a fuzzy system. This research uses a "Max of Mins" method that is not as computationally demanding as other techniques.

We store the rules in a "virtual" multidimensional array where each rule carries with it index information that specifies its location. With this approach we can store array entries that are actually used, not the full array. The order in which the rules are stored is not significant, they are stored consecutively.

Table 4 represents the rules used for selecting the appropriate algorithm. The "-1" indicates that the variable does not effect the rule. Note that for a Large (L) kernel size, the FFT algorithm is selected, ignoring the other factors as shown in line 1. Lines 9 and 10 show that for a large image size, a medium kernel size and

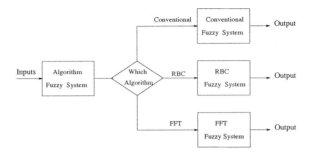

**Fig. 5.** Logical organization of the fuzzy selection system

| | Image Size | Kernel Size | Iteration | Ratio | algorithm |
|---|---|---|---|---|---|
| 1 | -1 | L | -1 | -1 | FFT |
| 2 | -1 | S | S | -1 | RBC |
| 3 | -1 | M | S | -1 | CONV |
| 4 | -1 | S | M | -1 | CONV |
| 5 | -1 | S | L | L | CONV |
| 6 | -1 | S | L | M | RBC |
| 7 | -1 | S | L | S | RBC |
| 8 | L | M | L | L | CONV |
| 9 | L | M | L | M | CONV |
| 10 | L | M | L | S | RBC |
| 11 | M | M | L | -1 | FFT |
| 12 | S | M | L | -1 | FFT |
| 13 | L | M | M | -1 | CONV |
| 14 | M | M | M | -1 | CONV |
| 15 | S | M | M | -1 | FFT |

**Table 4.** Fuzzy rules for algorithm selection

a large number of iterations, the communication speed is used to select either the RBC or CONV algorithms. The RBC algorithm is selected if the processing speed is much faster than the communication speed.

Figure 5 represents the rules used for the RBC algorithm in determining the number of processors to use. Similar tables were developed for the other two algorithms. Note that for a large image size, a small kernel size and a large number of iterations (lines 4 and 7), the number of processors is dependent on the ratio of communication to processor speed. if the communications are slow (line 7), then a smaller number of processors will be used to avoid congestion in the network.

|   | Image Size | Kernel Size | Iteration | Ratio | Processors |
|---|---|---|---|---|---|
| 1 | L | S | S | -1 | M |
| 2 | M | S | S | -1 | LS |
| 3 | S | S | S | -1 | SS |
| 4 | L | S | L | M | LL |
| 5 | M | S | L | M | SL |
| 6 | S | S | L | M | M |
| 7 | L | S | L | S | SL |
| 8 | M | S | L | S | M |
| 9 | S | S | L | S | LS |
| 10 | L | M | L | S | LL |

**Table 5.** Fuzzy rules for selection of the number of processors for the RBC algorithm

## 4 Experimental Results

The experimental results obtained from the fuzzy system in different situations are shown in Table 6. Referring to the table, we see that the selection by the fuzzy system corresponds to the experimental results. The RBC algorithm achieves a lower execution time when the kernel size (3) and number of iterations (3) are small. The communication speed does not effect its performance. The conventional algorithm has the lowest execution time when the kernel size (9) and number of iterations (6) is moderate. The preferred number of processors increases when the ratio of communication speed to computation speed increases. Future work will validate the results on other networks and to plot results for other intermediate values. When the kernel size and number of iterations is very large, the FFT algorithm exhibits the lowest execution time. These results clearly demonstrate the feasibility of the fuzzy selection system. The fuzzy selection system is flexible and easy to implement and has the ability to handle FAM matrices of arbitrary dimension. These results can be extended to other applications through altering the membership functions and/or the fuzzy rules.

## 5 Conclusions

Parallel computing is an effective way to deal with computationally intensive problems such as convolution. The appropriate parallel algorithm and number of processors must be chosen carefully in order to minimize the execution time. The performance of parallel applications is affected by many factors, such as problem size, algorithm complexity, networks features and processor performance. These factors are uncertain and often difficult to specify exactly.

In this paper, a fuzzy logic method is developed to deal with this problem. The selection method abstracts the features of the application in fuzzy sets. Then a fuzzy associative memory (FAM) transforms these sets into weights. The set of weights is then defuzzified to generate the result of the selection. Parallel image

| Image | Kernel | Iteration | ratio | Selected Algorithm | Number of Processors | Experiment Execution Time Conventional | RBC | FFT |
|---|---|---|---|---|---|---|---|---|
| 512 | 3 | 3 | 0 | RBC | 2 | 2.8 | 2.48 | 39.5 |
| 512 | 9 | 6 | 0 | Conventional | 8 | 4.9 | 6.6 | 18.8 |
| 512 | 51 | 3 | 0 | FFT | 9 | 44.5 | 65.3 | 18.0 |
| 1024 | 3 | 3 | 0 | RBC | 4 | 10.2 | 10.0 | 200.0 |
| 1024 | 9 | 6 | 0 | Conventional | 11 | 17.5 | 23.3 | 100.5 |
| 512 | 3 | 3 | 100 | RBC | 2 | | | |
| 512 | 9 | 6 | 100 | Conventional | 11 | | | |
| 512 | 51 | 3 | 100 | FFT | 11 | | | |
| 1024 | 3 | 3 | 100 | RBC | 4 | | | |
| 1024 | 9 | 6 | 100 | Conventional | 19 | | | |

Table 6. Results of the fuzzy selection system

convolution is used as an example to illustrate how a fuzzy logic selection system is developed. Three parallel convolution algorithms are implemented on HP workstation cluster and their performance is evaluated. A fuzzy logic selection system is shown to be effective in automatically selecting the correct algorithm and number of processors for a certain problem and system environment.

In order for parallel processing to achieve widespread use, the user must be able to achieve reasonable performance without an extraordinary amount of analysis or examination of the system. Many researchers have suggested using machines on the internet for parallel computations. Determining system parameters for a collection of these machines will be even more difficult than it is when using a cluster of homogenous workstations. Fuzzy logic methods, such as those described here, can allow internet based environments to be exploited by users when exact system parameters are unknown.

## References

[1] R. N. Bracewell. *Two-demensional Imaging*. Prentice-Hall Inc., 1995.

[2] G. Ficili and D. Panno. Performance Analysis of a Fuzzy System in the Policing of Packetized Voice Sources. In *Broadband Communications'96 - Global infrastructure for the information age Proceedings of the Internation*, pages 211–222, 1996.

[3] R. D. W. G. C. Fox and P. C. Messina. *Parallel Computing Works!* Morgan Kaufmann Publishers Inc., 1994.

[4] A. Geist, A. Beguelin, J. Dongarra, W. Jiang, R. Manchek, , and V. Sunderam. *PVM: Parallel Virtual Machine–A User's Guide and Tutorial for Networked Computing*. MIT Press, 1994.

[5] R. T. H. T. Nguyen, M. Sugeno and R. R. Yager. *Theoretical Aspects of Fuzzy Control*. John Wiley and Sons, Inc, 1993.

[6] R. Haralick and L. Shapiro. *Computer and Robot Vision*. Addison Wesley Publishing Company, 1992.

[7] J. M. Holtzmann. Coping with Broadband Traffic Uncertainties: Statistical Uncertainty, Fuzziness, Neural Networks. In *IEEE Workshop on Computer Communications*, Dana Pt, California, Oct. 1989.

[8] S. Levialdi. *Integrated Technology for Parallel Image Processing*. Academic Press, INC., 1985.

[9] Y. Man and I. Gath. Detection and Separation of Ring-Shaped Clusters Using Fuzzy Clustering. *IEEE*, 1994.

[10] J. M. Schopf. Performance Predication in Production Enviornments. In *University of California*, 1997.

[11] C. L. Seitz. *Resources in Parallel and Concurrent Systems*. ACM Press, 1991.

[12] J. Tenber. *Digital Image Processing*. Prentice-Hall Inc., 1991.

[13] D. S. R. W. E. Alexander and C. S. G. Jr. Parallel Image Processing with the Block Data Parallel Architecture. *Proceeding-of-the-IEEE*, July 1996.

[14] S. T. Welstead. *Neural Network and Fuzzy Logical Application in C/C++*. John Wiley and Sons INC., 1994.

[15] S. Yu. *Algorithm Selection For Parallel Image Convolution*. Brigham Young University. Master Thesis, 1998.

# Recurrent Neural Network Approach for Partitioning Irregular Graphs

M-Tahar Kechadi

Parallel Computational Research Group,
Department of Computer Science
University College Dublin
Belfield, Dublin 4, Ireland.
e-mail: tahar.kechadi@ucd.ie

**Abstract.** This paper is concerned with utilizing a neural network approach to solve the $k$-way partitioning problem. The $k$-way partitioning is modeled as a constraint satisfaction problem with linear inequalities and binary variables. A new recurrent neural network architecture is proposed for $k$-way partitioning. This network is based on an energy function that controls the competition between the partition's external cost and the penalty function. This method is implemented and compared to other global search techniques such as simulated annealing and genetic algorithms. It is shown that it converges better than these techniques.

## 1 Introduction

$K$-way partitioning, known also as graph partitioning problem, consists of dividing the vertices of a given graph into $k$ disjoint subsets of fixed size in such a way as to minimize the connections between subsets. This problem arises in many areas of computation including applications in fluid dynamics, mechanics, task scheduling, large sparse matrices [15] and VLSI design [17]. In VLSI design, communication rates between components have to be optimized in order to reduce communication delays. In scientific computing the efficient implementation of many parallel applications usually requires the solution to $k$-way partitioning of their task graphs. The graph vertices represent the computational tasks and the edges represent the data dependencies between tasks. The problem is to map the vertices (computational tasks) of a graph which represents the user application onto $k$ processor-nodes such that i) the communication cost is minimized and ii) the work-load is equalized among the processor-nodes.

Many techniques have been proposed and tested for different applications, including simulated annealing methods [8, 9], genetic algorithms [2], spectral partitioning algorithms [5, 15, 16], geometric partitioning [14] and multilevel graph partitioning [6]. Since the $k$-way partitioning is NP-complete, all these methods find reasonably good solutions. The efficiency of any algorithm can be characterized by its computational complexity and how good a partitioning it achieves. Any efforts to improve one of those characteristics usually results in the other becoming worse.

The Kernighan-Lin algorithm [13] gives good solutions at a local level and is much faster than simulated annealing [15]. However it depends on the initial partition and does not fully address the global problem. The class of spectral partitioning techniques usually produces good solutions for a variety of applications but they are expensive in time [5, 11]. These techniques can be improved upon by using the multilevel partition algorithm [1] without any loss of the quality of partitioning for some applications. The class of geometric graph partitioning methods can be used when geometric information is available. They are much faster but usually obtain worse solutions than spectral partitioning methods. Multiple trials are often required to produce solutions as good as those obtained by spectral partitioning techniques [10]. Multilevel graph partitioning reduces the size of the initial graph by contracting vertices and edges. The algorithm consists of partitioning the contracted graph and then uncontracting it to construct a partition for the initial graph. This class of partitioning seems to be better than previous ones [11, 10].

Recently we developed a new algorithm, called SLM, based on legal migrations and an energy function [12]. A migration is legal if the energy of the system decreases. The aim of this algorithm is to use simple legal migrations (simple vertex move) to reach local minimum, and then use a legal migration (subset move) to leave that local minimum. An algorithm for finding a subset candidate to move is also described. Its efficiency depends on the connectivity of a graph.

This paper deals with another technique, a recurrent neural network (RNN). The use of neural networks for solving an optimization problem is not new. Hopfield and Tank [7] had already suggested effective and fast neural networks for some optimization problems. The optimization problem is represented by an energy function that can be iteratively minimized by the dynamics of a recursive neural network. Because of the NP-complete nature of the *k-way* partitioning problem, in general, it is not possible to force the network to converge to the optimal solution. Therefore, similarly to heuristic methods, there is no guarantee that the network will find the optimum solution. Most artificial neural networks applied to optimization problems are local optimization techniques. This means that when the system is in a local minimum the network is not able to make a move to another solution better than the current one.

The technique studied in this paper is an RNN method based on a sophisticated cost function. The cost function is designed in such a way as to avoid local search. This is done by adding a new term to the cost function called internal energy of the system. When the internal energy is high the system becomes unstable and when it tends to zero the system will converge to its stable state which constitutes the final solution.

## 2   Neural network Method

The artificial neural network (ANN) technique in the domain of optimization problems brings something new in terms of searching for an optimal solution or at least for a reasonably good solution. Instead of fully or partly exploring the different possible configurations as in heuristic methods, it feels its way in a fuzzy manner toward good solutions. This process is done in a way that allows for a statistical interpretation of the results. The ANN approach is more promising for solving such problems.

Let $G = (V, E)$ be an undirected weighted graph, where $V$ is the set of $n = |V|$ vertices and $E$ is the set of $l = |E|$ edges. We assume that both vertices and edges are weighted. A $k$-way partition is a partitioning of the vertex set $V$ into $k$ disjoint subsets. These subsets are called nodes and denoted by $N_{i, i=1,\cdots,k}$. The $k$-way partitioning problem is the problem of finding a $k$-way partition with the minimum external cost and the difference between the weights of nodes is also minimized. The external cost (resp. internal cost) of a subset is defined as the sum of the weights of external edges (resp. internal edges).

$K$-way partitioning is viewed as a classical constrained optimization problem.

$$\min E(x)$$
$$\text{subject to } \begin{cases} g_i(x) = 0 & i = 1, \cdots I \\ x \in \{0, 1\}^{k \times n} \end{cases} \qquad (1)$$

where $g_i(x)$ are linear functions, called residuals, and $E(x)$ is the cost function of the system. This is a standard approach in the design of ANNs [7,3]. It is well known that the dynamics of the network is governed by an appropriate cost function. For some classical optimization problems neural network approaches perform badly in the situation where a cost function is competing with hard constraints. Therefore the choice of an appropriate cost function is a prerequisite in the design of an ANN to an optimization problem. The choice of a cost function has to fulfill the following two criteria:

- The lowest energy state corresponds to the optimal solution.
- The problem constraints are not very "hard" in order to let the network explore the space of solutions with more freedom.

## 3 The Cost Function

Let $x$ be the current partition. Based on the criteria defined above, $k$-way partitioning can be mapped onto a cost function of the form:

$$E(x) = f(x) + P(x) \qquad (2)$$

where $f(x)$ represents the external cost of the partition $x$ and $P(x)$ is a penalty function that controls the degree of satisfiability (feasibility) of the residuals $g_i(x)$. The performance of the neural network will depend on the competition between $f(x)$ and $P(x)$. We will see later, while defining the penalty function, how to control this competition and guide the network to find a good solution.

## 3.1 External cost function

Let $w_{ij}$ denote the weight of the edge $(v_i, v_j)$ and $x_i$ be a binary k-vector ($k$ is the partition size) associated to a vertex $v_i$ and defined by

$$x_{ij} = \begin{cases} 1 & \text{if } v_i \in N_j \\ 0 & \text{otherwise} \end{cases} \qquad (3)$$

Note that the partition $x$ is given by the $n$ k-vectors; $x = (x_1, x_2, \cdots, x_n)$. The function $f(x)$ which expresses the total external cost of a partition $x$ is given by

$$f(x) = \frac{1}{2} \sum_{i,j}^{n} w_{ij} x_i (u_k - x_j) \qquad (4)$$

where $u_k$ is a vector of $k$ components consisting of all ones; $u_k = (1, 1, \cdots, 1_k)$, and is called a *unit vector*.

## 3.2 Penalty Function

In the constrained optimization problem the penalty function expresses the constraints of the system in such a way that the cost will be high for the infeasible solutions. A penalty term is associated to each residual in the system. This term can be governed only by a positive scaling factor $\alpha$ like $\alpha g_i(x)$ or can be represented by a quadratic form. In this paper we choose a quadratic form to build the penalty function and this is given by

$$P(x) = \frac{\alpha}{2} \left[ \left( \sum_{i=1}^{n} x_i - m \right)^2 + \left( n - \sum_{i=1}^{n} x_i^2 \right) \right] \qquad (5)$$

where each component of the vector $m$ expresses the size of the corresponding subset $N_i$ and $\alpha$ is a scaling factor called penalty parameter. The cost function is the sum of the external cost and the penalty function given by the equations (4) and (5) respectively.

$$E(x) = \frac{1}{2} \sum_{i,j}^{n} w_{ij} x_i (u_k - x_j) + \frac{\alpha}{2} \left[ \left( \sum_{i=1}^{n} x_i - m \right)^2 + \left( n - \sum_{i=1}^{n} x_i^2 \right) \right] \qquad (6)$$

Usually the penalty function, as formulated above, leads the dynamics of the system into a local minimum depending on the initial conditions. When a local minimum is reached, the system is trapped in that state. In order to force the system to leave the current local minimum an additional penalty term is introduced based on the barrier function methods [18]. We call this penalty term the *internal energy* of the system.

The internal energy function is designed in such a way that its value is small at points far from the constraints boundaries. As the system approaches the constraint boundaries its internal energy increases to infinity. Based on this idea, the internal energy of the system can be expressed by

$$E_{int}(x) = -\frac{1}{\gamma} \ln \left( \left| \sum_{i=1}^{n} x_i - m \right| u_k \right) \tag{7}$$

where $\gamma > 0$ is a parameter of the system used to control the internal energy of the system. The absolute value of a vector is defined as the absolute value of its elements. The factor $1/\gamma$ is the temperature of the system. By adding $E_{int}(x)$ to the penalty function the cost function given by the equation (6) becomes

$$E(x) = \frac{1}{2} \sum_{i,j}^{n} w_{ij} x_i (u_k - x_j) +$$

$$\frac{\alpha}{2} \left[ \left( \sum_{i=1}^{n} x_i - m \right)^2 + \left( n - \sum_{i=1}^{n} x_i^2 \right) - \frac{\delta_0}{\gamma} \ln \left( \left| \sum_{i=1}^{n} x_i - m \right| u_k \right) \right] \tag{8}$$

where

$$\delta_0 = \begin{cases} 0 & \text{if } \sum_{i=1}^{n} x_i - m = 0 \\ 1 & \text{otherwise} \end{cases} \tag{9}$$

The internal energy function has only one minimizer called the analytic center [4]. Intuitively the internal energy expresses a distance to the constraint boundaries of the system. The neural network will attempt to bring the system to a state where the solution at that state is close to the analytic center of the problem. By cooling the system (increasing $\gamma$) the barrier vanishes with time and then allows the system to stabilize near the constraints boundaries.

## 4 The Network Architecture

The cost function defined above is optimized iteratively. A RNN is then defined such that the solution of the problem is given by the neuron outputs at each time step. The final solution is found when the RNN has converged and the system is cooled down.

Let $Out_j$ be the output of the neuron $j$. The transfer function that maps the inputs to outputs is commonly a sigmoïd function in the interval $[0, 1]$.

$$Out_j = \frac{1}{1 + e^{-int_j/\rho}} \tag{10}$$

where $\rho > 0$ controls the gain of the activation function. Notice that when $\rho$ tends to zero the neural network outputs are booleans. This feature allows the network to work in a continuous model to produce boolean outputs.

## 5 Equations of motion

The new inputs and outputs of each neuron are derived from the cost function. The cost function behaves as the error made on the neural network outputs: The smaller the value of the cost function the better the solution given by the network outputs. The discrete equations of the inputs are given by

$$int_j^{t+\Delta t} = int_j^t - \Delta_j E \Delta t \tag{11}$$

where

$$\Delta_j E = \frac{\partial E(x)}{\partial x_j} = \sum_{i \neq j}^{n}(w_{ij} - x_i(1 - 2w_{ij})) - (m + \theta(x)) \tag{12}$$

and $\theta(x)$ is defined by

$$\theta(x) = \begin{cases} 0 & \text{if } \sum_{i=1}^{n} x_i - m = 0 \\ \frac{u_k}{2\gamma(|\sum_{i=1}^{n} x_i - m|u_k)} & \text{otherwise} \end{cases} \tag{13}$$

## 6 The Network Dynamics

The equations of motion as derived above describe the asynchronous update model; the neurons are updated one by one. The neuron to be updated is randomly chosen leaving the others for future iterations. However, in order to improve network convergence, at each iteration all the outputs of the neurons are computed and we choose at random one neuron among those that have had their outputs changed at the previous iteration. The initial solution of the system is considered to be continuous (with $0 \leq x_{ij} \leq 1$) because the network is supposed to work in a continuous model.

This optimization model has the same behavior as the simulated annealing model where the temperature is represented by the internal energy of the system. When the internal energy is high the system has more freedom to escape local minima. When the internal energy is small the system converges to a minimum. To cool down the system (to reduce its internal energy) the coefficient $\gamma$ is incremented after each iteration. The convergence speed is defined by the function used to increment $\gamma$. The convergence is achieved when the internal energy approaching zero as shown in the following:

$$\frac{d\tilde{E}}{dt} = \sum_{j=1}^{n} \frac{\partial \tilde{E}}{\partial x_j} \frac{dx_j}{dt} = \sum_{j=1}^{n} \left(\frac{-\partial int_j}{dt}\right) \frac{dx_j}{dt} = -\sum_{j=1}^{n} \frac{dx_j}{dint_j} \left(\frac{dint_j}{dt}\right)^2 \tag{14}$$

Note that $\tilde{E}$ is the total cost function without the internal energy term. As the activation function of the outputs given by the equation (10) is monotonic increasing, we get:

$$\frac{d\tilde{E}}{dt} \leq 0 \tag{15}$$

This means that $\tilde{E}(x)$ is an energy function of the system to be optimized.

The activation function of the inputs is given by the equations of motion (11) and (13). The network consists of three layers of processing units. The first layer calculates the residuals $g_i(x)$ and the errors $\partial P(x)/\partial x_i, (i = 1 \cdots n)$. The second layer combines the errors for each neuron and computes the adjusted new inputs. In the last layer the new outputs are calculated by integrating in time the input values. The recurrent neural network implementing these three layers is given in figure (1).

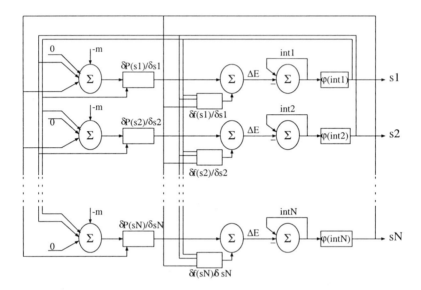

**Fig. 1.** Recurrent neural network architecture using penalty function approach.

This method is implemented in C++ language. The results are obtained on SGI O2 workstation. Geometric graphs are used to study the profile and the performance of this method. In this paper the performance of the RNN method is only compared to that of simulated annealing and genetic algorithm. These techniques are chosen because they perform global search and the simulated annealing has the same general behavior as the RNN method. Figure (2) depicts the general profile of the RNN technique. The data of figure (2) are the average values of 1000 different geometric graphs with $n = 500$ and $l = 79127$.

At high temperature, the system is in unstable states and most of the solutions are not feasible. Starting the optimization with initial temperature around 50 helps the system to converge quickly and also reduces the number of unfeasible solutions. Figures (3) and (4) show that RNN method performs better than both simulated annealing and genetic algorithm. The result comparison is much more easier with simulated annealing because of their similar general behavior. The system parameters with genetic algorithm are completely different

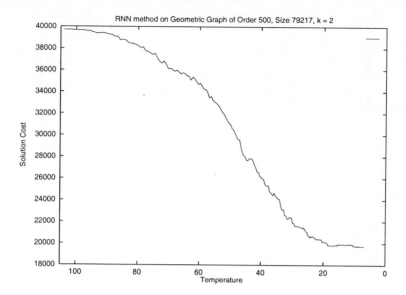

**Fig. 2.** Typical profile of RNN technique.

from those of RNN method. The genetic algorithm curve in figure (4) is generated from the best parameter configuration among all the performed runs. The implemented genetic algorithm and simulated annealing are described in [2] and [8] respectively.

## 7 Conclusion

This paper discussed the recurrent neural network technique for solving $k$-way partitioning of unstructured graphs. The ANN approach applied to unconstrained optimization problem is reviewed and original cost function in the case of $k$-way partitioning is presented. The cost function is based on the internal energy of the system which allows the system to be able to explore the whole space of solutions and also to be able to escape local minima. One of the feature of this technique is that it performs 0/1 constrained optimization problem in continuous model. Experimental results show that the technique performs well and its convergence speed is better than global search techniques like simulated annealing and genetic algorithm. We still need to study the performance of RNN technique with different initial conditions and explore the space of its parameters. The study of its scalability and comparison with other methods will also be part of our future work.

## References

1. S.T. Barnard and H.D. Simon. A Fast Multilevel Implementation of Recursive Spectral Bisection for Partitioning Unstructured Problems. In *Proceedings of the*

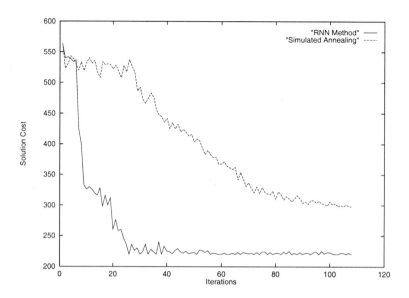

**Fig. 3.** RNN technique vs. Simulated annealing technique.

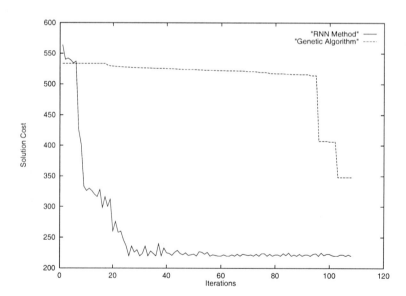

**Fig. 4.** RNN technique vs. genetic algorithm technique.

*6th SIAM Conference on Parallel Processing for Scientific Computing*, pages 711–718, 1993.
2. T.N. Bui and B.R. Moon. Genetic Algorithm and Graph Partitioning. *IEEE Transactions on Computers*, 45(7):841–855, July 1996.
3. A. Cichocki and A. Bargiela. Neural networks for solving linear inequality systems. *Parallel Computing*, 22(11):1455–1475, January 1997.
4. C.C. Gonzaga. Path-following methods for linear programming. *SIAM Review*, 32(2):167–224, .
5. B. Hendrickson and R. Leland. An Improved Spectral Graph partitioning Algorithm for mapping parallel Computations. Technical Report SAND92-1460, Sandia National labs, Albuquerque, NM., 1992.
6. B. Hendrickson and R. Leland. A Multilevel Algorithm for Partitioning Graphs. Technical Report SAND93-1301, Sandia National labs, Albuquerque, NM., 1993.
7. J.J. Hpfield and D.W. Tank. Neural computation of decisions in optimization problems. *Biological Cybernitics*, 52:141–152, 1985.
8. D.S. Johnson, C.R. Aragon, L.A. Mcgeoch, and C. Schevon. Optimization by Simulated Annealing: an Experimental Evaluation; Part I, Graph Partitioning. *Operations Research*, 37(6):865–892, November-December 1989.
9. D.S. Johnson, C.R. Aragon, L.A. Mcgeoch, and C. Schevon. Optimization by Simulated Annealing: an Experimental Evaluation; Part II, Graph Coloring and Number Partitioning. *Operations Research*, 39(3):378–406, May-June 1991.
10. G. Karypis and V. Kumar. A Fast and High Quality Multilevel Scheme for Partitioning Irregular Graphs. Technical Report TR 95-035, Department of Computer Science, University of Minnesota, July 1995.
11. G. Karypis and V. Kumar. Parallel Multilevel Graph Partitioning. Technical Report TR 95-036, Department of Computer Science, University of Minnesota, June 1995.
12. M-Tahar Kechadi and D.F. Hegarty. A parallel technique for graph partitioning problems. In *Proceedings of The International Confrence and Exhibition on High-Performance Computing and Networking (HPCN)*, pages 449–457, Amsterdam, Netherlands,, April 20-23 1998. Springer.
13. B. Kernighan and S. Lin. An Efficient Heuristic Procedure for Partitioning Graphs. *Bell Syst. Tech. Journal*, 29:291–307, February 1970.
14. G.L. Miller, S-H. Teng, and S.A. Vavasis. A Unified Geometric Approach to Graph Separators. In *Proceedings of the 31st Annual Symposium on Foundations of Computer Science*, pages 538–547, 1991.
15. A. Pothen, H.D. Simon, and K-P. Liou. Partitioning Sparse Matrices with Eigenvectors of Graphs. *SIAM Journal of Matrix Analysis and Applications*, 11(3):430–452, July 1990.
16. A. Pothen, H.D. Simon, L. Wang, and S.T. Barnard. Towards a Fast Implementation of Spectral Nested Disection. In *Proceedings of Supercomputing'92*, pages 42–51, 1992.
17. K. Shahookar and P. Mazumder. VLSI Cell Placement techniques. *ACM Computing Surveys*, 23(2):143–220, June 1991.
18. G.N. Vanderplaats. *Numerical Optimization Techniques for Engineering Design.* McGraw-Hill, New-York, 1984.

# Track C3:

# Computer Science

# JIAJIA: A Software DSM System Based on a New Cache Coherence Protocol[1]

Weiwu Hu   Weisong Shi   and   Zhimin Tang

Institute of Computing Technology

Chinese Academy of Sciences, Beijing 100080

E-mail: {hww,wsshi,tang}@water.chpc.ict.ac.cn

**Abstract.** This paper describes design and evaluation of a software distributed shared memory (DSM) system called JIAJIA. JIAJIA is a home-based software DSM system in which physical memories of multiple computers are combined to form a larger shared space. It implements the lock-based cache coherence protocol which totally eliminates directory and maintains coherence through accessing write notices kept on the lock. Our experiments with some widely accepted DSM benchmarks such as SPLASH2 program suite and NAS Parallel Benchmarks indicate that, compared to recent software DSMs such as CVM, higher performance is achieved by JIAJIA. Besides, JIAJIA can solve large problems that cannot be solved by other software DSMs due to memory size limitation.

## 1 Introduction

Over the past decade, software Distributed Shared Memory (DSM) systems have been extensively studied to provide a good compromise between programmability of shared memory multiprocessors and hardware simplicity of message passing multicomputers. Many software DSM systems, such as Ivy[12], Midway[2], Munin[4], TreadMarks[9], and CVM[10] have been implemented on the top of message passing hardware or on network of workstations. However, software DSMs suffer from the high communication and coherence-induced overheads caused by the high level of implementation and large granularity of coherence. Many techniques, such as multiple-writer protocol[4], lazy release consistency[8], and hardware support[3], have been proposed to reduce false sharing, minimize remote communication, and hide communication latency.

This paper introduces our design and evaluation of a software DSM system called JIAJIA. JIAJIA is characterized by its lock-based cache coherence protocol[6] for scope consistency (ScC)[7]. The protocol is lock-based because it totally eliminates directory and all coherence related actions are taken through write notices kept on the lock. Compared to the directory-based protocol, the lock-based protocol is simpler and consequently more efficient and scalable.

Another distinguishing feature of JIAJIA is that it combines physical memories of multiple computers to form a larger shared space. In other recent software DSMs such as TreadMarks and CVM, the shared address space is limited by the local memory size of one machine. In JIAJIA, each shared page has a home node

---

[1] The work of this paper is supported partly by National Climbing Program of China, Natural Science Foundation of China, and President Young Creation Foundation of the Chinese Academy of Sciences.

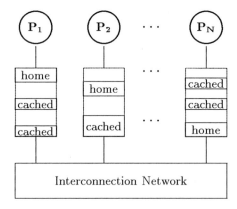

**Fig. 1.** Memory Organization of JIAJIA

and homes of shared pages are distributed across all hosts. With this memory organization, the size of shared space can be as large as the sum of each machine's local memories. Since each page has a home, JIAJIA totally eliminates the complexity of local diffs keeping, garbage collection, local address to global address translation, and timestamp maintenance.

Performance measurements with some widely accepted DSM benchmarks such as SPLASH2 program suite and NAS Parallel Benchmarks indicate that, compared to recent software DSM systems such as CVM, higher speedup is reached by JIAJIA. Besides, JIAJIA can solve large problems that cannot be solved by other software DSMs due to memory size limitation.

The rest of this paper is organized as follows: Section 2 introduces the memory organization of JIAJIA. Section 3 describes the lock-based cache coherence protocol implemented in JIAJIA. Section 4 gives JIAJIA's application programming interface. Section 5 presents our evaluation results and analysis. The work of this paper is summarized in Section 6.

## 2 Memory Organization

Figure 1 shows JIAJIA's organization of the shared memory. Unlike other Software DSMs that adopt the COMA-like memory architecture, JIAJIA organizes the shared memory in a NUMA-like way. In JIAJIA, each shared page has a home node and homes of shared pages are distributed across all nodes. References to local part of shared memory always hit locally. References to remote shared pages cause these pages to be fetched from its home and cached locally. When cached, the remote page is kept at the same user space address as that in its home node. In this way, shared address of a page is identical in all processors, no address translation is required on a remote access.

In JIAJIA, shared pages are allocated with the mmap() system call. Each shared page has a fixed global address which determines the home of the page. Initially, a page is mapped to its global address only by its home processor. Reference to a non-home page causes the delivery of a SIGSEGV signal. The SIGSEGV handler then maps the fault page to the global address of the page in

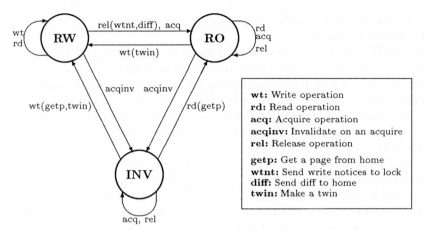

**Fig. 2.** State Transition of the Protocol

local address space. Since the total shared memory allocated may be larger than the physical memory of one host, mapping too many remote pages will break the system down. To avoid this, each host maintains a "cache" data structure to record all locally kept non-home pages. Any locally kept remote page must find a position in the local cache. If the number of locally kept remote pages is larger than the maximum number allowed, some aged cache pages must be replaced (unmapped) to make room for the new page.

With the above memory organization, JIAJIA is able to support shared memory that is larger than physical memory of one machine. In other software DSMs such as TreadMarks and CVM, the shared space is limited by the physical memory of one machine because no cache replacing mechanism is implemented and hence each host should have the capability of holding all shared pages. Besides, in these systems, each processor maintains a large local page table to keep directory information (to locate a page on a page fault), twins, diffs, protect states, and local and global addresses of all pages. The size of this page table scales linearly with the number of shared pages while the local memory of each processor does not scale as well. In JIAJIA, homes of shared pages are distributed across all processors and hence total shared pages allocated are not limited by the physical memory of a machine. Instead of keep information of all shared pages, the page table of JIAJIA contains only information about its "cached" pages. For each cached page, the page table only keeps its address, protect state, and a twin for writable pages[2].

---

[2] In the recent version of JIAJIA, there is also a page table of all shared pages. However, each item of the page table requires only six bytes.

## 3 Lock-Based Cache Coherence Protocol
### 3.1 Basic Protocol
Based on the observation that the benefit of complex design of a software DSM system may be offset by the system overhead caused by the complexity, the cache coherence protocol of JIAJIA is designed as simple as possible. Instead of supporting multiple consistency models and adaptive write propagation strategies as some previous DSMs did, JIAJIA takes a fixed memory consistency model and fixed write propagation (write-invalidate) strategy. Multiple-writer technique is employed to reduce false sharing.

JIAJIA implements the scope consistency (ScC)[7] which is even lazier than lazy release consistency (LRC)[9]. Adopting the ScC greatly simplifies the lock-based cache coherence protocol of JIAJIA. In TreadMarks which implements LRC, complex data structures such as *intervals* and *vector timestamp* is employed to record the "happen-before-1" relation of memory accesses among different processors. In JIAJIA, ScC does not require to implement the complete "happen-before-1" relation. Instead, it only requires previous intervals related to the acquired lock to be visible to the acquiring processor on an acquire.

The distinguishing characteristic of JIAJIA's cache coherence protocol is its lock-based feature, i.e., it does not rely on any directory information to maintain coherence. Instead, coherence is maintained through writing and reading *write-notices* on the lock. The lock-based protocol can be summarized as follows.

In the protocol, each page has a home and can be cached by a non-home processor in one of three states: Invalid (INV), Read-Only (RO), and Read-Write (RW). Since the multiple-writer technique is assumed, a page may be cached by different processors in different states concurrently. As a special kind of shared object, each lock also has a home node.

On a release, the releaser performs a comparison of all cached pages written in this critical section with their twins to get diffs of this critical section. These diffs are then sent to their associated homes. At the same time, a release message is sent to the associated lock manager to release the lock. Besides, the releaser piggybacks write-notices of the critical section on the release message notifying the modifications in the critical section.

On an acquire, the acquiring processor sends an acquiring request to the lock manager. The requesting processor is then stalled until it is granted the lock. When granting the lock, the lock manager piggybacks write-notices associate with this lock on the granting message. After the acquiring processor receives this granting message, it invalidates all cached pages that are notified as obsolete by the associated write-notices.

A barrier can be viewed as a combination of a lock and an unlock. Arriving at a barrier ends an old "critical section", while leaving a barrier begins a new one. In this way, two barriers enclose a critical section. On a barrier, all write notices of all locks are cleared.

On a read miss, the associated memory page is fetched from the home in RO state in the local memory.

On a write miss, if the written page is not presented or is in INV state in the local memory, it is fetched from the home in RW state. If the written page

Table 1. Message Costs of Shared Memory Operations

| Protocols | Access Miss | Lock | Unlock | Barrier |
|---|---|---|---|---|
| LRC | 2m | 3 | 0 | 2(n-1) |
| lock-based | 0 or 2 | 2 | f+1 | 2(n-1)+F |

m = # concurrent last modifiers for the missing page
f = # messages to send diffs
n = # processors in the system
$F = \sum_{i=1}^{n}$(# messages to send diffs by processor $i$)

is in RO state in the local memory, the state is turned into RW. A write-notice is recorded about this page and a twin of this page is created before writing.

Figure 2 shows the state transition of the above protocol.

### 3.2 Advantages and Disadvantages

Table 1 shows the number of messages sent on an ordinary access miss, a lock, an unlock, or a barrier in the lazy release protocol and the lock-based protocol. It can be seen from Table 1 that, compared to the LRC, our protocol has less message cost on both ordinary accesses or lock, but requires to write diffs back to home of associated pages on a release or a barrier. Besides, the lock-based protocol is free from the overhead of maintaining the directory.

### 3.3 Optimization of the Protocol

The above description of the lock-based cache coherence protocol does not include any optimization. We made the following optimizations to the above basic protocol.

- In the basic protocol, a cached page is invalidated on an acquire (or barrier) if there is a write notice in the lock indicating that this page has been modified. However, if the modification is made by the acquiring processor itself, and the page has not been modified by any other processors, then the invalidation is unnecessary since the modification has already been visible to the acquiring processor. With this optimization, a processor can retain the access right to pages modified by itself on passing an acquire (or barrier).
- Another improvement to the basic protocol is called the *read notice* technique. In the basic protocol, write notices produced in a critical section is sent to the lock on a barrier. However, if a page is modified only by its home node, and there is no other processors have read the page since last barrier, then it is unnecessary to send the associated write notice to the lock on the barrier. To do this, a *read notice* is recorded in the home of a page any time a remote get page request is received.
- A method similar to *incarnation number* introduced in Midway[2] and ScC[7] is adopted to eliminate unnecessary invalidation on locks. With this method, each lock is associated with an incarnation number which is incremented when the lock is transferred. Besides, each processor maintains a local incarnation number for each lock. When a write notice is recorded in the lock, the current incarnation number of the lock is recorded as well. On a lock

acquire, the acquiring processor includes its incarnation number of the lock in the acquiring message. On a lock grant, only write notices which have an incarnation number larger than that in the request is sent to the acquiring processor. The acquiring processor then sets its local incarnation number of the lock to the lock's current incarnation number and invalidates cached pages according to received write notices as normal.
- A cache only write detection (CO-WD) scheme is proposed to reduce the number of page faults. Normally, in a home-based software DSM, both home and cache pages are write-protected at the beginning of an interval to detect writes of the interval. The CO-WD scheme does not write-protect home pages but invalidates all cached pages at the beginning of an interval as if all home pages have been modified previously. As a result, pages faults caused by writing home pages are eliminated at the cost of some extra cache miss.

## 4 Programming Interface

The API of JIAJIA is similar to that of other software DSM systems. It provides six basic calls: jia_init(argc, argv), jia_alloc(size), jia_lock(lockid), jia_unlock(lockid), jia_barrier(), and jia_exit() to the programmer. Besides, JIAJIA offers some subsidiary calls: jia_setcv(condvar), jia_resetcv (condvar), and jia_waitcv(condvar) to provide the conditional variable synchronization method, jia_clock() to return the elapsed time since the start of application in seconds in float type, and jia_error(char *str) to print out the error string str and shut down all processes.

JIAJIA provides two variables jiapid and jiahosts to the programmer. They specify the host identification number and the total number of hosts of a parallel program.

JIAJIA looks for a configuration file called .jiahosts in the directory where the application runs. This file contains a list of hosts to run the applications, one per line. Each line contains 3 entries: *machine name, user account,* and *password*. The first line of .jiahosts should be the master on which the program is started.

A distinguishing feature of JIAJIA's API is that it allows the programmer to flexibly control the initial distribution of homes of shared locations. The basic shared memory allocation function in JIAJIA is jia_alloc3(size, blocksize, starthost) which allocates size bytes cyclically across all hosts, each time blocksize bytes. The starthost parameter specifies the host from which the allocation starts. The simple call jia_alloc(size) equals jia_alloc3(size, Pagesize, 0).

Another interesting character of JIAJIA's API is that it also provides some MPI-like message passing calls: jia_send(buf, len, topid, tag), jia_recv (buf, len, frompid, tag), jia_bcast(buf, len, root), and jia_reduce( sndbuf, rcvbuf, count, operation, root). These message passing calls allows the programmer to write message passing program for some modules (or just port existing message passing modules) and write shared memory program for other modules. Besides, using some message passing primitives in shared memory programs helps to improve performance in many cases.

Table 2. Characteristics of Benchmarks and Execution Results

| Appl. | Size | Shared Memory | Barr. # | Lock # | Seq. Time | 8-proc. Time | | Speedup | |
|---|---|---|---|---|---|---|---|---|---|
| | | | | | | JIAJIA | CVM | JIAJIA | CVM |
| Water | 1728 mole. | 484KB | 35 | 520 | 178.00 | 26.47 | 39.79 | 6.72 | 4.47 |
| Barnes | 16384 | 1636KB | 28 | 64 | 413.24 | 64.75 | 66.02 | 6.38 | 6.26 |
| LU | 2048 × 2048 | 32MB | 128 | 0 | 84.86 | 25.04 | 35.21 | 3.39 | 2.41 |
| LU | 8192 × 8192 | 512MB | 512 | 0 | 5464.80* | 909.39 | — | 6.01 | — |
| EP | $2^{24}$ | 4KB | 1 | 8 | 49.69 | 6.30 | 6.30 | 7.89 | 7.89 |
| IS | $2^{24}$ | 4KB | 30 | 80 | 30.10 | 4.84 | 4.59 | 6.22 | 6.56 |
| SOR | 2048 × 2048 | 16MB | 200 | 0 | 68.44 | 11.45 | 15.25 | 5.98 | 4.49 |
| SOR | 8192 × 8192 | 256MB | 200 | 0 | 1235.76** | 166.20 | — | 7.44 | — |
| TSP | 20 cities | 788KB | 0 | 1121 | 175.36 | 33.25 | 76.20 | 5.27 | 2.30 |

*: Estimated as eight times of 4096 × 4096 LU sequential time (683.10 seconds), sequential time of 8192 × 8192 LU is not available due to memory size limitation.
**: Estimated as four times of 4096 × 4096 SOR sequential time (308.94 seconds), sequential time of 8192 × 8192 SOR is not available due to memory size limitation.

## 5 Performance Evaluation and Analysis

The evaluation is done in the Dawning-1000A parallel machine developed by the National Center of Intelligent Computing Systems. The machine has eight nodes each with a PowerPC 604 processor and 256MB memory. These nodes are connected through a 100Mbps switch Ethernet. In the test, all libraries and applications are compiled by gcc with the -O2 optimization option. To compare the performance of JIAJIA with other software DSMs, same applications are also run under CVM[10] which is another software DSM system developed in Maryland University.

We port some widely accepted DSM benchmarks to evaluate the performance of JIAJIA. This paper shows results of seven applications, include Water, Barnes, and LU from SPLASH and SPLASH2[14, 15], EP and IS from NAS Parallel Benchmarks[1], and SOR and TSP from Rice University[13]. Memory reference characteristics of these applications are shown in the left part of Table 2.

The right part of Table 2 shows sequential and eight-processor parallel run time of the benchmarks. As indicated in the table, the 8192 × 8192 LU and SOR cannot be run on single machine due to memory size limitation and the corresponding sequential run times are estimated values. Table 3 shows some runtime statistics of JIAJIA and CVM, include message counts, message amounts, SIGSEGV signal counts, and remote access counts.

It can be seen from Table 2 that, for most tested applications, JIAJIA achieves satisfied performance and speedup.

Both Water and Barnes are N-body problems and are characterized with tight sharing. As has been stated, the lock-based protocol of JIAJIA takes all coherence related actions in synchronization points and has least message overheads in ordinary write and read misses. In Water and Barnes, the number of shared pages is small, and this small number of pages are referenced frequently by multiple processors. Hence, the overhead at synchronization point is not high, and page faults caused by ordinary write and read operations introduce least messages in JIAJIA. As a result, compared to CVM, JIAJIA achieves a higher speedup in Water and Barnes. Statistics in Table 3 show that CVM introduces

**Table 3.** Eight-Processor Execution Statistics

| Appl. | Messages | | Msg. amt.(KB) | | SIGSEGVs | | Remote accesses | |
|---|---|---|---|---|---|---|---|---|
| | JIAJIA | CVM | JIAJIA | CVM | JIAJIA | CVM | JIAJIA | CVM |
| Water | 10828 | 160134 | 16850 | 72408 | 4892 | 30899 | 2847 | 19416 |
| Barnes | 37018 | 114284 | 80569 | 64960 | 34144 | 37370 | 17844 | 18193 |
| LU2048 | 25992 | 49283 | 99950 | 203604 | 32663 | 23998 | 12072 | 11874 |
| LU8192 | 386664 | — | 1569946 | — | 1108875 | — | 189720 | — |
| EP | 77 | 105 | 60 | 65 | 22 | 23 | 14 | 14 |
| IS | 1050 | 984 | 896 | 2710 | 230 | 240 | 140 | 150 |
| SOR2048 | 8412 | 11288 | 11763 | 12662 | 2800 | 9650 | 2800 | 2786 |
| SOR8192 | 8413 | — | 46135 | — | 2800 | — | 2800 | — |
| TSP | 19773 | 15773 | 24265 | 4948 | 8312 | 8773 | 5580 | 6069 |

much more remote misses and consequently communication than JIAJIA in Water, while JIAJIA and CVM cause similar number of remote misses in Barnes. Table 3 also shows that, with the same number of remote misses, CVM causes more messages but less message amounts than JIAJIA. This is because compared to JIAJIA, CVM requires more messages to locate the owner and collect diffs of a page, but transfers only diffs instead of the whole page (as in JIAJIA) on a remote miss.

Computation-to-communication ratios of both LU and SOR are $O(N)$ where $N$ is the problem size. As a result, speedups of both LU and SOR are acceptable and scale with the problem size. Frequent inter-processor synchronization contributes the main reason for the moderate speedup of LU and SOR in the $2048 \times 2048$ cases, because computation steps of LU and SOR are separated by barriers. Besides, in LU, only the updating trailing submatrix computation phase of each step is fully parallelized. In JIAJIA, matrices are initially distributed across processors in the way such that each processor keeps in its home the data it processes (writes). As a result, no diffs are produced and propagated in JIAJIA since all modifications are made directly to home pages[16]. This contributes an important reason for the superior performance of JIAJIA to CVM. Another reason for the performance difference between JIAJIA and CVM is that CVM requires a cold startup time to distribute the matrix across processors (though the -m2 CVM command line option is used in SOR). Statistics in Table 3 show that, CVM causes more messages than JIAJIA in LU and SOR though remote misses are similar in JIAJIA and CVM.

Both EP and IS have separate computation and communication phases, i.e., computation of EP and IS happen locally and communication happens only at the end of computation. EP achieves a speedup of 8 in both JIAJIA and CVM because the communication and computation ratio of EP is low. In IS, the most time-consuming computation is for each processor to count its local part of keys, while summing the counting results in the local buckets up into the shared bucket constitutes the communication work. We keep the number of buckets at 1024 which makes the communication amount relatively small compared to the computation work of counting $2^{24}$ keys. As a result, an acceptable speedup is achieved in both JIAJIA and CVM. CVM slightly outperforms JIAJIA in IS because JIAJIA has more overhead than CVM on a release. In JIAJIA, diffs have to be sent to its home before the release message is sent to the lock, while CVM keeps diffs locally. As a result, when summing up values of private buckets into the shared bucket, JIAJIA takes more time for each processor to enter and leave the critical section sequentially.

TSP specializes in that all inter-processor synchronization are taken through locks. In TSP, each processor frequently reads from and writes to the pool of tours and the priority queue, causing tight sharing of pages (a page can store 27 paths in TSP[13]). The result of Table 2 and Table 3 seems contradicting: JIAJIA causes more message amounts than CVM but performs better in TSP. Further experiments show that the single processor execution time of CVM is 478.07 seconds, much larger than sequential time. When compared to its own single processor time, CVM achieves a better speedup (6.29). The reason for the more message amounts of JIAJIA than CVM lies on the different memory organization and coherence protocol in CVM and in JIAJIA. On a page fault, CVM only transfers diffs which is rather small in TSP, while JIAJIA fetches the whole page from its home. Table 3 shows that JIAJIA has similar number of messages with CVM but four times more message amounts than CVM.

It can also be seen from Table 2 that LU-8192 and SOR-8192 which cannot run on CVM due to memory size limitation can be run in JIAJIA with multiple processors.

## 6 Conclusions and Future Work

The above evaluation and analysis can be summarized as follows.
- With its NUMA-like memory organization scheme, JIAJIA can combine memories of multiple processor to form a large shared memory.
- JIAJIA achieves better performance than CVM for most tested applications. This is mainly because the simplicity of the lock-based cache coherence protocol entails little system overheads on ordinary access miss. Other factors, such as the home-based memory organization which requires no diffs generating for home pages, the uniformed local and global addresses mapping of JIAJIA, and the flexible API which allow the programmer to control initial distribution of shared data, also help to improve performance.
- The lazy release protocol of CVM has less overhead than the lock-based coherence protocol of JIAJIA on release operation. Besides, JIAJIA needs to take a full page from home on a page fault, while CVM only takes diffs of the fault page and helps to reduces communication amount.

Our future work include improving JIAJIA Software DSM system, optimizing the lock-based cache coherence, and developing a hardware prototype to implement the protocol. Further information about JIAJIA is available at http://www.ict.ac.cn/chpc/index.html.

## References

1. D. Bailey, J. Barton, T. Lasinski, and H. Simon, "The NAS Parallel Benchmarks", *Technical Report 103863*, NASA, July 1993.
2. B. Bershad, M. Zekauskas and W. Sawdon, "The Midway Distributed Shared Memory System", in *Proc. of the 38th IEEE Int'l CompCon Conf.*, pp. 528–537, Feb. 1993.

3. R. Bianchini, L. Kontothanassis, R. Pinto, M. Maria, M. Abud, and C. Amorim. "Hiding Communication Latency and Coherence Overhead in Software DSMs", in *the 6th Int'l Conf. on Architectural Support for Programming Languages and Operating Systems*, Oct. 1996.
4. J. Carter, J. Bennet, and W. Zwaenepoel, "Implementation and Performance of Munin", in *Proc. of the 13th ACM Sym. on Operating Systems Principles*, pp. 152–164, Oct. 1991.
5. K. Gharachorloo, D. Lenoski, J. Laudon, P. Gibbons, A. Gupta, and J. Hennessy, "Memory Consistency and Event Ordering in Scalable Shared Memory Multiprocessors", in *Proc. of ISCA'90*, pp. 15–26, May 1990.
6. W. Hu, W. Shi, Z. Tang, and M. Li, "A Lock-based Cache Coherence Protocol for Scope Consistency", *Journal of Computer Science and Technology*, Vol. 13, No. 2, pp. 97–109, March 1998.
7. L. Iftode, J. Singh and K. Li, "Scope Consistency: A Bridge Between Release Consistency and Entry Consistency", in *Proc. of the 8th Annual ACM Sym. on Parallel Algorithms and Architectures*, June 1996.
8. P. Keleher, A. Cox, and W. Zwaenepoel, "Lazy Release Consistency for Software Distributed Shared Memory", in *Proc. of ISCA'92*, pp. 13–21, 1992.
9. P. Keleher, S. Dwarkadas, A. Cox, and W. Zwaenepoel, "TreadMarks Distributed Shared Memory on Standard Workstations and Operating Systems", in *Proc. of the 1994 Winter Usenix Conf.*, pp. 115–131, Jan. 1994.
10. P. Keleher, "The Relative Importance of Concurrent Writers and Weak Consistency Models", in *Proc. of the 16th Int'l Conf. on Distributed Computing Systems*, pp. 91–98, May 1996.
11. D. Lenoski, J. Laudon, K. Gharachorloo, P. Gibbons, A. Gupta, and J. Hennessy, "The Directory-Based Cache Coherence Protocol for the DASH Multiprocessors", in *Proc. of ISCA'90*, pp. 148–158, June 1990.
12. K. Li, "IVY: A Shared Virtual Memory System for Parallel Computing", in *Proc. of ICPP'88*, Vol. 2, pp. 94–101, Aug. 1988.
13. H. Lu, S. Dwarkadas, A. Cox, and W. Zwaenepoel, "Quantifying the Performance Differences Between PVM and TreadMarks", *Journal of Parallel and Distributed Computing*, Vol. 43, No. 2, pp. 65–78, June 1997.
14. J. Singh, W. Weber, and A. Gupta, "SPLASH: Stanford Parallel Applications for Shared Memory", *Computer Architecture News*, 20(1):5–44, Mar. 1992.
15. S. Woo, M. Ohara, E. Torrie, J. Singh, and A. Gupta, "The SPLASH-2 Programs: Characterization and Methodological Considerations", in *Proc. of ISCA'95*, pp. 24–36, 1995.
16. Y. Zhou, L. Iftode and K. Li, "Performance Evaluation of Two Home-based Lazy Release Consistency Protocols for Shared Virtual Memory Systems", in *Proc. of the 2nd USENIX Sym. on Operating System Design and Implementation*, Seattle, Oct. 1996.

# Efficient Analytical Modelling of Multi-Level Set-Associative Caches[*]

John S. Harper[**], Darren J. Kerbyson, and Graham R. Nudd

High Performance Systems Group,
University of Warwick, Coventry CV4 7AL, UK.

**Abstract.** The time a program takes to execute is significantly affected by the efficiency with which it utilises cache memory. Moreover the cache miss behaviour of a program is highly unstable, in that small changes to input parameters can cause large changes in the number of misses. In this paper we describe novel analytical methods of predicting the cache miss ratio of numerical programs, for sequential hierarchies of set-associative caches. The methods are demonstrated to be applicable to most loop nests. They are also shown to be highly accurate, yet able to be evaluated orders of magnitude faster than a comparable simulation.

## 1 Introduction

As the gulf between processor and memory speeds continues to widen, caches become increasingly important as a method of hiding memory latency. Cache performance is one of the most important factors affecting program execution time, especially for large scientific codes where memory access patterns are often regular, and in some cases subject to pathological cache behaviour. Thus cache behaviour is a critical component when modelling any high performance system. This paper describes the novel cache modelling techniques that are used by the PACE performance analysis toolset [8]. PACE allows parallel programs to be modelled analytically, giving rapid, but accurate, predictions.

The PACE toolset implements a layered approach to modelling the performance of parallel systems. The core of the system is a language allowing developers to describe the performance aspects of their applications. This description is compiled and evaluated to provide detailed performance predictions about the application. Predictions are usually observed to be within twenty percent of actual run-times. Due to the analytical nature of the methodology, predictions are made very rapidly, allowing performance prediction to be employed in unusual areas, for example run-time scheduling and program tuning [6].

Analytical techniques of predicting cache behaviour are an active area of research, due to the high cost of modelling caches through simulation. Existing analytical models can be broadly divided into two areas: the first includes

---

[*] This work is funded in part by DARPA contract N66001-97-C-8530, awarded under the Performance Technology Initiative administered by NOSC.
[**] john@dcs.warwick.ac.uk

those that encapsulate the memory workload through a number of constant parameters—usually obtained through the study of memory access traces—for example [1, 10]. The second category includes models that analyse the high-level structure of the program, generally concentrating on loop constructs and the array references nested within these loops, for example [7, 12, 3, 11, 4].

For performance prediction purposes, where the input to the cache model is a representation of the program source code, the second type of analytical model is the most suitable. Unfortunately, all existing models have limitations preventing their use in every scenario. None consider both direct-mapped and set-associative caches, for all levels of the memory hierarchy, or with enough flexibility to express the majority of loop nests found in scientific codes.

The contribution of this paper is to present a set of models addressing these limitations. Building on aspects of the direct-mapped model of Temam et al. [11], novel methods are described that encompass set-associative caches and multiple levels of caching; these are the most common cache architectures, thus models are very valuable. The applicability of the methods is such that a wide range of numerical codes may be evaluated. Accuracy is shown to be high, comparable with simulation, for both primary and external caches, yet predictions are made orders of magnitude faster than simulations.

The following section describes the input to the models, and the constraints on the problems that may be addressed. Sec. 3 introduces the fundamental operation of the models—forming the *cache footprint* of a set of array references over a subset of loop iterations. Using this concept, Sec. 4 describes how the number of cache misses is predicted, for both primary and external caches. Finally, Sec. 5 shows example predictions made by the models for several loop nests.

## 2 Description of Memory Reference Behaviour

Within the PACE toolset, performance models are made by breaking an application into two conceptual areas: computational workloads and parallelisation strategies. Computational sub-tasks use a notation called *Software Execution Graphs (SEGs)* to describe the control flow and workload of the task being modelled. A SEG is a graph with nodes representing either computational basic-blocks or methods of changing control flow (i.e. loop structures and conditional statements). Computation nodes may be described in a number of ways, ranging from a benchmarked timing, to a list of the abstract operations forming the block [9]. These computational sub-tasks are combined with parallel constructs, such as message passing primitives, to form a model of the entire program [8]. The full extent of this description is beyond the scope of this paper.

Considering only sequential pieces of computation, there is a one-to-one mapping between loop and conditional statements in the program source code, and loop and conditional nodes in the SEG representation. Annotating this graph with the array references contained in the program (with the array indices dependent on the indices of the loop statements) combined with details of the array characteristics (base-address and dimensions) gives all necessary infor-

```
                                    compute <is clc, SISL>;
                                    loop i (<is clc, LFOR>, 0, N-1) {
  for (i = 0; i < N-1; i++) {         compute <is clc, SISL>;
    for (j = 0; j < N-1; j++) {       loop j (<is clc, LFOR>, 0, N-1) {
      X[i][j] = X[i+1][j]               compute <is clc, TDSL, 5*ADSL>;
           + Y[i][j] + Y[i][j+1]        ref X[i+1,j], Y[i,j], Y[i,j+1],
           + Z[j][i] + Z[j][i+1]            Z[j,i], Z[j,i+1], X[i,j]:w;
    }                                 compute <is clc, INLL, CMLL>;
  }                                 }
                                    compute <is clc, INLL, CMLL>;
                                  }
          (a) C code
                                          (b) PACE description
```

**Fig. 1.** Example loop nest

mation for making predictions of cache behaviour. Fig. 1 shows an example computational task; the compute statements describe the atomic language operations, and therefore do not affect the cache behaviour. There is a direct link between loops and array reference in the source code, and equivalent details in the performance model. The array characteristics (the base addresses, type, and dimensions) are defined elsewhere in the performance model.

There are few constraints on the problems involved, the loop bounds must be known at evaluation-time, and constant through the inner iterations of the loop. Loop bounds (and array indices) are automatically normalised; array references must be of the form:

$$X[\alpha_{11}j_{\gamma_{11}} + \alpha_{12}j_{\gamma_{12}} + \cdots + \beta_1, \ldots], \tag{1}$$

where $X$ is the name of the array, and $\alpha_{ik}$, $\gamma_{ik}$, and $\beta_i$ are constants, with $i$ ranging from one to the number of dimensions in the array, and $k$ ranging from one upwards. The variables $j_1 \ldots j_n$ represent the loop indices of the loops above the reference (i.e. the reference is in the body of loop $n$). The order of the indices may be either Fortran or C style (i.e. column-first or row-first).

Table 1 presents an analysis of all loop nests in the serial versions of the *NAS Parallel Benchmarks (NPB)*, a set of numerical benchmark codes consisting of around 15000 lines of Fortran [2]. The method of analysis was to consider separately each loop nest that accesses arrays, noting which are suitable for modelling in PACE. Where the nest occurs in a subroutine, all parameters to the subroutine are assumed to be known; any global variables that are only set at initialisation are also treated as constants. There are several reasons why a loop may not be modelled: it contains a non-trivial function call (PACE expands calls inline, so the normal model constraints apply; also allowed are calls to FPU operations); the loop body has non-local entrances or exits; the iteration

| NPB program | total | | unsuitable nests | | | | total suitable | |
|---|---|---|---|---|---|---|---|---|
| | nests | sub-nests | Func. Body | Iter. Ref. | | Cond. | nests | sub-nests |
| BT | 66 | 220 | 22 | | | | 46 (70%) | 169 (77%) |
| CG | 30 | 41 | 1 | 8 | 8 | | 12 (40%) | 22 (54%) |
| EP | 5 | 7 | 2 1 | | 1 | | 3 (60%) | 3 (43%) |
| FT | 17 | 40 | 1 1 | 1 | 1 | | 13 (76%) | 34 (85%) |
| LU | 48 | 156 | | | | | 48 (100%) | 156 (100%) |
| MG | 39 | 89 | 4 2 | 2 | | 2 | 32 (82%) | 76 (85%) |
| SP | 91 | 310 | | | | | 91 (100%) | 310 (100%) |
| total | 296 | 863 | 30 4 | 9 | 12 | 2 | 245 (82%) | 770 (89%) |

**Table 1.** Analysis of loop nests in NPB suitable for modelling

bounds are unsuitable; an array reference can't be expressed; there are conditional statements that can't be modelled. The results are classified by "nests" and "sub-nests"; a "nest" is a top-level loop nest, a "sub-nest" is a loop nest at any level, including those contained within other sub-nests.

The results are encouraging, with 82% of all top-level loop nests being suitable for modelling (this compares favourably with other cache models [4]). Of the two benchmarks with less than 70% loops suitable, the 'CG' benchmark uses several sparse matrices, and therefore certain array references use data dependent indices; the 'EP' benchmark is perhaps too small to give a meaningful analysis. Function calls are the most common cause of loops that can't be modelled, these are mostly calls to the random number generator used to initialise some arrays, in real codes this is unlikely to be a problem.

## 3 Footprint Modelling

The methodology of this paper relies on one main procedure to predict cache interference (this includes both conflict and capacity effects), for both temporal and spatial reuse. Given a group of array references that only differ in their constant terms (the $\beta_k$ terms of (1)) or the array they access, the task is to identify their cache footprint across a subset of all loop iterations.

The footprint contains all distinct cache locations accessed by the references over the specified loop iterations. It is represented as a two-dimensional structure, one dimension representing the sets in the cache, the other the number of data items stored in each set. Calculating a cache footprint is a three-step process:

**1. Identify the *memory* footprint of one array reference.** This models all memory locations accessed by the reference over the loop iterations, and is normally a number of contiguous memory regions, spaced at constant intervals. In rare cases, a more complex structure will be identified, involving several dif-

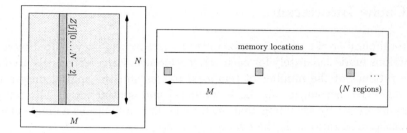

**Fig. 2.** Array and memory footprints of $Z[j][i]$ for one iteration of loop $i$.

ferent sizes of interval. Fig. 2 shows the memory footprint of reference $Z[j][i]$ (from Fig. 1(a)) for a single iteration of loop $i$.

**2. Map the memory footprint into the cache.** The memory address space is theoretically infinite, but the cache has only a finite number of locations. If a cache has capacity $\mathcal{C}$, and each set in this cache contains $a$ lines (the associativity), then the cache contains $\mathcal{C}/a$ unique locations. Each memory address $x$ maps to cache location $x \bmod (\mathcal{C}/a)$. We use a modified form of the Euclidean algorithm of Temam et al. [11] to map the memory footprint into the cache address space. This gives the cache locations containing data, and the average number of items stored by each such location.

**3. Combine the cache footprints of the individual array references.** Since the array references being considered have similar access patterns, their cache footprints also have the same pattern, but are "rotated" in the cache. Due to this common form, it is possible to combine copies of the footprint calculated in step (2), one per reference, such that only distinct data items are included.

The end result is a sequence of two-dimensional regions representing the cache locations accessed by a group of similar references for a particular subset of the loop iterations, and the average number of distinct data items stored in each location [5]. Fig. 3 shows an example footprint, from the 2D Jacobi loop nest shown in Sec. 5 (Fig. 4(b)).

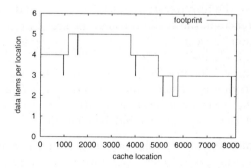

**Fig. 3.** Example cache footprint from Jacobi kernel ($\mathcal{C}/a = 8192$)

## 4 Cache Interference

The total number of cache misses incurred by a loop nest is evaluated by adding predictions made separately for each array reference. Each reference's prediction is the product of the number of temporal misses and the spatial miss ratio of that reference, accounting for both temporal and spatial reuse. The temporal misses are found by assuming that each cache line only stores a single element; the spatial miss ratio adds the effects of line size.

For each reference $R$, we say that its reuse is due to a *reuse dependence*, a reference $\vec{R}$ that accesses data items that will subsequently be re-accessed by $R$. If $\vec{R} = R$ there is a *self-dependence*, otherwise there is a *group-dependence*. For any reuse dependence $\vec{R}$, two parameters can be identified. Firstly, the reuse occurs on a loop at depth $l$ in the nest; secondly, there are $\delta$ iterations of loop $l$ between $\vec{R}$ accessing a data element and $R$ reusing it.

Consider the group-dependence between $R = Z[j][i]$ and $\vec{R} = Z[j][i+1]$ in Fig. 1(a). Here $R$ re-accesses the column $Z[0 \ldots N-2][i]$ of the matrix, accessed by $\vec{R}$ in the previous iteration of loop $i$. Thus for this dependence, $l = 1$ ($i$ is the outermost loop), and $\delta = 1$.

The miss prediction for any dependence is the sum of two terms: the *compulsory* misses, related to the number of distinct elements accessed by $R$ on the first $\delta$ iterations of loop $l$; and the *interference* misses, related to how the data accessed by any $\delta$ iterations of loop $l$ maps into the cache, and how it affects the data being reused by $R$ at that point. Both spatial and temporal dependences can be evaluated in this manner.

The array references in the loop nest are partitioned into groups, firstly by the loop body they occur in, and secondly by their access pattern, so that each group contains references from a single loop body, with only their $\beta_k$ terms differing. Each group is then considered as an atomic source of interference on each dependence, the overall interference prediction being found by a "union" of the probabilities of each group causing interference, assuming independence between groups.

Interference from a group of references on a reference $R$ is found by identifying two cache footprints (as in the previous section): the first containing the data accessed by $\vec{R}$ during the first iteration of loop $l$; the second containing the data accessed by all references in the group through the first $\delta$ iterations of loop $l$. The size of the intersection between these footprints gives the amount of interference occurring. Two methods may be used to evaluate the intersection; if $R$ is a member of the group of references being considered, then the two footprints are compared exactly, region-by-region. Alternatively, when $R$ is not a member of the group, there is no basis for accurate comparison, and a statistical method is used, assuming that the footprints are independent [5].

**Primary cache interference.** Evaluating interference in the primary cache is relatively straightforward, it is generally known that every array reference always accesses the primary cache. This is contrary to the external cache where accesses are generally linked to primary cache misses. If the primary cache has associativity $a_1$, then interference occurs wherever the average number of items per location, in the intersection of the two footprints, is greater than $a_1$. Since the average items in a cache location may be fractional, the probability of interference is found by subtracting $a_1$ from this value, but truncating the minimum and maximum probabilities at zero and one respectively, for natural reasons.

**External cache interference.** Detecting interference in an external cache is not so simple, since the stream of accesses is not known *a priori*, but depends upon the behaviour of the previous cache. However, by pretending that the external cache has the same number of cache locations as the primary cache, but with much higher associativity, it is possible to use the footprints of the primary cache to predict interference in the external cache. If the primary cache has capacity $C_1$ and associativity $a_1$, it contains $C_1/a_1$ locations. If an external cache at level $x$ in the hierarchy has capacity $C_x$, then it too can be thought of as having $C_1/a_1$ locations, by letting its *pseudo-associativity* $a'_x$ be equal to:

$$a'_x = \frac{a_1 C_x}{C_1}.$$

This allows the primary cache footprints to be applied to any external cache; interference is predicted in a similar manner as in the primary cache, but occurs where the number of data items per cache location is greater than the *pseudo-associativity* $a'_x$. It is trivial to show that it is not possible to predict interference in level $x$ where the data would not actually miss in level $x-1$, provided each cache at level $x$ has $C_x > C_{x-1}$, which should always be the case.

As in the primary cache, the probability of a footprint region causing interference is dependent on the number of data items per location; but in an external cache interference begins at $a'_x$ items, and peaks at $a'_x + a'_x/a_x$, or $3a'_x$, depending whether an identity, or random, virtual-to-physical mapping is being modelled. Also, in a random mapping all interference is evaluated using the statistical intersection method.

## 5 Example Results

The models described in this paper have been implemented (about 6000 lines of C code), as part of the PACE environment. All architecture parameters are defined to match the particular hardware system being examined. To demonstrate the validity of the models, predictions made for several example loop nests are shown. The loops include the stencil operation previously shown (Fig. 1(a)), an unblocked matrix-multiply (Fig. 4(a)) and a two-dimensional Jacobi loop (Fig. 4(b)). These examples were chosen for their markedly different behaviour. Fig. 5 shows predicted miss-ratios for primary and secondary caches of varying associativity, plotted against matrix size; also plotted on each figure is the difference between the predicted and simulated values, the closer this is to the zero-axis, the higher the accuracy. The secondary cache is assumed to be physically addressed with a random mapping between virtual and physical addresses.

```
                                    DO J = 1, N-2
                                      DO I = 1, N-2
                                        VXN(I,J) = (c0*VX0(I,J)+dty2
   DO I = 0, N-1                          * (VX0(I-1,J)+VX0(I+1,J))
     DO J = 0, N-1                        + dtx2*(VX0(I,J+1)+VX0(I,J-1))
       Z(J, I) = 0.0                      - dtx*(P0(I,J)-P0(I,J-1))-c1)
       DO K = 0, N-1                      * IVX(I,J)
         Z(J, I) = Z(J, I)             VYN(I,J) = (c0*VY0(I,J)+dty2
                 + X(K, I) * Y(J, K)      * (VY0(I-1,J)+VY0(I+1,J))
       ENDDO                              + dtx2*(VY0(I,J+1)+VY0(I,J-1))
     ENDDO                                - dty*(P0(I-1,J)-P0(I,J))-c2)
   ENDDO                                  * IVY(I,J)
                                      ENDDO
                                    ENDDO
      (a) Matrix multiply
                                                    (b) 2D Jacobi
```

**Fig. 4.** Example loop nests

All miss ratios are global values, the number of misses at that level, divided by the total number of array accesses.

As demonstrated by Fig. 5, the accuracy of the predictions is good, both for the primary and secondary caches. Perhaps the most striking feature is just how ineffective set-associative caches are at absorbing cache interference—even with $a_1 = 4$, the primary cache miss ratio is still very varied, small changes in inputs leading to vast differences in miss ratio. The models mostly pick up these differences very well, with a slight exception being some of the periodic peaks in the primary jacobi plot, which are either under- ($a_1 = 1$, $a_1 = 2$) or over- ($a_1 = 4$) predicted. The cause of these peaks is spatial cross interference between similar references ("ping-pong" interference), and may suggest that this area of the model requires refinement.

## 6 Summary

A method of describing numerical loop nests has been described suitable for performance modelling purposes. The expressiveness of this method has been shown to allow description of the majority of loop nests in a set of typical numerical programs (82% of NPB nests). Working from these descriptions and details of the data structures being referenced, a method of modelling the cache footprint of a group of similar references has been described. Such cache footprints give a detailed analytical description of how any data items accessed map into the cache, thus allowing the level of cache interference to be evaluated.

Models of set-associative cache interference have been described, both for primary and external caches, architectures that have not previously been analysed in this way. The models use cache footprints defined over specific loop iterations to gauge their effect on any potential data reuse, for both spatial and temporal reuse. It has been shown how the footprints of the primary cache behaviour may

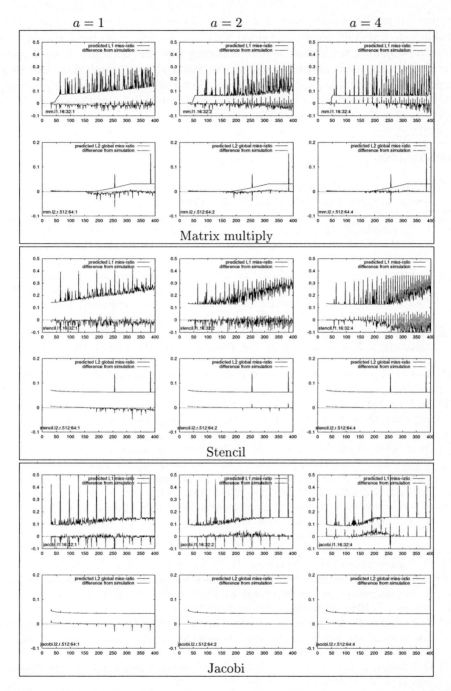

**Fig. 5.** Predicted miss ratios and errors for $\mathcal{C}_1 = 16K$, $\mathcal{L}_1 = 32$, $a_1 = \{1, 2, 4\}$; and $\mathcal{C}_2 = 512K$, $\mathcal{L}_2 = 64$, $a_2 = \{1, 2, 4\}$ ($a_1 = 1$) configurations.

be used to predict interference in all external caches, thus reducing the cost of evaluation. Example results obtained with the models have been shown, and compared against simulation results, showing that the accuracy is very good. Predictions are made orders of magnitude faster than a simulation, and most importantly, prediction rate does not depend on the number of memory accesses as with simulation (i.e. the array *dimensions*), but mainly on the static size of the loop nest (i.e. the number of array *references*).

## References

[1] A. Agarwal, M. Horowitz, and J. Hennessy. An analytical cache model. *ACM Trans. Comput. Syst.*, 7(2):184–215, May 1989.
[2] D. H. Bailey, E. Barszcz, J. T. Barton, D. S. Browning, R. L. Carter, D. Dagum, R. A. Fatoohi, P. O. Frederickson, T. A. Lasinski, R. S. Schreiber, H. D. Simon, V. Venkatakrishnan, and S. K. Weeratunga. The NAS Parallel Benchmarks. *The International Journal of Supercomputer Applications*, 5(3):63–73, Fall 1991.
[3] J. Ferrante, V. Sarkar, and W. Thrash. On estimating and enhancing cache effectiveness. In U. Banerjee, D. Gelernter, A. Nicolau, and D. Padua, editors, *Proceedings of Languages and Compilers for Parallel Computing*, volume 589 of *LNCS*, pages 328–343, Berlin, Germany, Aug. 1992. Springer.
[4] S. Ghosh, M. Martonosi, and S. Malik. Cache miss equations: An analytical representation of cache misses. In *Proceedings of the 11th ACM International Conference on Supercomputing*, Vienna, Austria, July 1997.
[5] J. S. Harper, D. J. Kerbyson, and G. R. Nudd. Analytical modeling of set-associative cache behavior. To appear in *IEEE Transactions on Computers*.
[6] D. J. Kerbyson, E. Papaefstathiou, and G. R. Nudd. Application execution steering using on-the-fly performance prediction. In *High-Performance Computing and Networking*, volume 1401 of *LNCS*, pages 718–727. Springer, 1998.
[7] M. S. Lam, E. E. Rothberg, and M. E. Wolf. The cache performance and optimizations of blocked algorithms. In *Proceedings of the Fourth International Conference on Architectural Support for Programming Languages and Operating Systems*, pages 63–74, Santa Clara, California, 1991.
[8] G. R. Nudd, D. J. Kerbyson, E. Papaefstathiou, J. S. Harper, S. C. Perry, and D. V. Wilcox. PACE: A toolset for the performance prediction of parallel and distributed systems. *Journal of High Performance and Scientific Applications*, 1999.
[9] E. Papaefstathiou, D. J. Kerbyson, G. R. Nudd, D. V. Wilcox, J. S. Harper, and S. C. Perry. A common workload interface for performance prediction of high performance systems. In *Workshop on Performance Analysis and its Impact on Design (PAID98)*, Barcelona, Spain, June 1998.
[10] J. P. Singh, H. S. Stone, and D. F. Thiebaut. A model of workloads and its use in miss-rate prediction for fully associative caches. *IEEE Trans. Comput.*, 41(7):811–825, July 1992.
[11] O. Temam, C. Fricker, and W. Jalby. Cache interference phenomena. In *Proceedings of ACM SIGMETRICS*, pages 261–271, 1994.
[12] M. E. Wolf and M. S. Lam. A data locality optimizing algorithm. In *Proceedings of the SIGPLAN '91 Conference on Programming Language Design and Implementation*, volume 26, pages 30–44, June 1991.

# Buffer Management in Wormhole-Routed Torus Multicomputer Networks

Kamala Kotapati[1] and Sivarama P. Dandamudi[2]

[1] Newbridge Networks, Kanata, Canada, kkotapat@ca.newbridge.com
[2] Center for Parallel and Distributed Computing, School of Computer Science, Carleton University, Ottawa, Canada, sivarama@scs.carleton.ca

**Abstract.** This paper focuses on buffer management issues in wormhole-routed torus multicomputer networks. The commonly used buffer organizations are the centralized and dedicated buffer organizations. The results presented in this paper indicate that the dedicated buffer organization provides better performance than the centralized organization under uniform traffic. Under high hot-spot traffic, the centralized organization outperforms the dedicated. The hybrid buffer organization proposed in this paper inherits the merits of the centralized and distributed organizations. The results presented suggest that the hybrid buffer organization can be designed to configure dynamically from the dedicated to centralized depending on the traffic conditions.

## 1 Introduction

In distributed-memory multicomputers, each node consists of a processor and its local memory. Nodes communicate by explicitly passing messages through a processors-to-processor interconnection network. Torus interconnection network, which is considered here, is widely used in current multicomputer systems.

Wormhole routing is used in current multicomputer systems and is the focus of this paper. In wormhole routing, a message (packet) is divided into a number of *flits* for transmission [7]. Normally, the bits constituting a flit are transmitted in parallel between two routers. The header flit (or flits) of the message governs the route. As the header advances along a specified route, the remaining flits follow it in a pipelined fashion. If the header flit is blocked, the flow control within the network blocks the trailing flits and these flits remain in flit buffers along the established route. Once a channel has been acquired by a packet, it is reserved for the packet. The channel is released when the last, or tail, flit has been transmitted on the channel. Flits passing between two adjacent nodes use a handshaking protocol.

Wormhole routing is a popular switching technique in the new generation multicomputer systems as wormhole-routed networks provide, in the absence of network contention, network latencies that are relatively insensitive to path length. For example, the Cray T3D and T3E systems use wormhole routing in a 3D torus network.

Wormhole-routed networks are susceptible to deadlocks. To avoid deadlocks and to improve performance, virtual channels have been introduced [3]. A virtual channel consists of a buffer that can hold one or more flits of a message and associated state information [4]. Several virtual channels may share the bandwidth of a single physical channel. Virtual channels separate the allocation of buffers from allocation of channels by providing multiple buffers for each channel in the network. Adding virtual channel flow control to a network makes more effective use of both resources by decoupling their allocation. The only expense is a small amount of additional control logic. A more comprehensive treatment of the topic is given in [5].

Buffer management is an important performance issue in wormhole-routed networks. Two basic buffer organizations are the centralized and dedicated organizations. Each organization has its advantages and disadvantages. Here we propose a hybrid buffer organization that inherits the merits of the two basic buffer organizations. Performance evaluation results presented here and elsewhere [6] suggest that the hybrid organization can be designed to configure dynamically from the dedicated to centralized depending on the traffic conditions.

## 2 The Negative-Hop Algorithm

Performance comparison of three algorithms—negative-hop (NHop), e-cube, and *-channel—indicates that the NHop performs better than the e-cube and *-channel algorithms in torus networks [2]. This is the rationale for using the NHop algorithm in this paper.

The NHop algorithm was proposed in [2]. The network is partitioned such that there are no cycles in any partition. Each partition is assigned a unique number. A negative hop is a hop that takes a message from a node in a higher numbered partition to a node in a lower numbered partition. In torus networks, two partitions are used to implement the NHop algorithm.

In the NHop algorithm, a message can use any hop that takes it closer to its destination. A message that has taken $i$ negative hops uses a $c_i$ virtual channel for its next hop. Direct deadlocks cannot occur with wormhole switching, since messages exchanged between neighbors use distinct physical channels. To avoid deadlocks, the NHop algorithm requires at most $1 + (\frac{H_I(C-1)}{C})$ virtual channels, where $H_I$ is the maximum number of inter-partition hops a message can take and $C$ is the number of distinct partitions.

A problem with this basic NHop algorithm is that, for many networks, it may require too many virtual channels [1]. The channel requirements can be reduced using improved negative hop schemes (INHop), which are based on the negative hop scheme. The basic technique of INHop algorithm is discussed next.

The NHop algorithm can be improved by giving more choice of virtual channels for messages in higher classes. A message with virtual channel class $i \geq 0$ may use any virtual channel of classes 0,...,$i$. For example, if a message of class 2 does not find a virtual channel of class 2 in the path to its next host, the message selects any free virtual channel in classes 0 and 1 that is in its path, relabels it as 2 and uses it. A virtual channel relabeled by a message of higher class number returns to its original class after the message has relinquished it. A blocked message, however, can only wait for a virtual channel of its class.

Deadlocks cannot occur since each blocked message waits for virtual channels as per the original algorithm. Starvation may be avoided by ensuring that a virtual channel is relabeled to a higher class only when there are no messages of its class waiting for it. For example, if a message is eligible for class $i$ virtual channel and there is no virtual channel of that class available, the message can take any virtual channel of class 0 to $i$. The precondition to acquire such lower class (say $p$) is possible only if there is no other message waiting for the virtual channel of class $p$. Using *ranges* of classes to select virtual channels gives priority to messages that have already used many virtual channels. In the remainder of the paper NHop refers to the improved NHop algorithm.

## 3 Buffer Management

Nodes require buffering due to conflicts that arise when several messages simultaneously require the use of the same link. A scheme is necessary to allocate buffers of a node among the virtual channels of the node.

There are two common solutions to this situation [9]: dedicated and centralized. These two schemes offer an increased degree of buffer sharing and increased channel utilization but at the cost of increased complexity in the control circuitry. They are distinguished by the restrictions on the placement of each channel's messages. In the dedicated scheme, there is minimal sharing of buffers and each channel has a set of dedicated buffers, i.e., its own FIFO queue. In the centralized scheme, there is maximal sharing of buffers and each node has a centralized pool of buffers that all channels can share.

The centralized scheme offers the most sharing at the cost of additional control circuitry. In addition, this scheme could also suffer from *buffer hogging*. Buffer hogging occurs when one output port becomes congested and uses a disproportionately large portion of the buffer pool, impeding traffic on other ports. Under these circumstances, better performance is obtained if the degree of buffer sharing is limited, and each port is limited to some maximum number buffers. Note that buffer assignment problem does not exist in the dedicated buffers scheme.

### 3.1 Buffer Organizations

The objective of this paper is to study the performance of the torus network under different buffer organizations. Three buffer organizations, briefly described next, are chosen for conducting this study. See [6] for a more detailed discussion.

**Dedicated buffer organization** In this organization, each virtual channel leading into a router has its own flit buffer. That is, $m = pv$ where $m$ is the number of flit buffers used, $p$ is the number of incoming physical channels, and $v$ is the number of virtual channels per physical channel.

In this organization, no delay is incurred by messages to acquire a buffer. The major components of delay are the flow control from incoming physical channels to flit buffers, crossbar delay from flit buffers to the outputs of crossbar, and the virtual channel controller delay from the outputs of the crossbar to outgoing physical channels. For the header flit, header decode and update as well as channel selection are the additional costs.

**Centralized buffer organization** In this organization, all the buffers are shared by all virtual channels. Each buffer is assigned a class. The virtual channel goes through a crossbar before accessing its exclusive buffer. Once a flit buffer is allocated to a virtual channel, it remains associated with that virtual channel until it is released.

The header decode, update and channel selection are similar in both dedicated and centralized organizations. In the centralized organization, however, when a header flit arrives, it is allocated a central buffer by establishing a connection through a crossbar. Once the connection is established, the allocated central flit buffer acts as a dedicated flit buffer to that virtual channel, and the transit of data flits is similar to that of the dedicated flit buffer implementation.

**Hybrid buffer organization** The dedicated and centralized buffer organizations have their advantages and disadvantages. The hybrid buffer organization proposed here inherits the merits of the two organizations. In this organization, each channel is assigned a particular number of dedicated buffers. So, in order to use these buffers the virtual channels need not go through the crossbar. The remaining buffers are organized centrally. If $p$ is the number of physical channels per node and $l$ is the number of dedicated buffers assigned to each physical channel, the remaining $(b - pl)$ buffers are maintained as a central pool of buffers, where $b$ is the total number of buffers.

The header decode and update and channel selection are same as in the other two organizations. The flow control is done partly as in the dedicated buffer organization and partly as in the centralized buffer organization, depending on the type of the virtual channel (dedicated or centralized). Messages that could get a dedicated buffer virtual channel suffer no extra delay at the router. If a message is taking a centralized buffer virtual channel, it suffers more delay as explained in the centralized buffer organization.

## 4 System Model

A simulator has been developed to compare the performance of three buffer organizations. The simulator has the flexibility to support various topologies, routing algorithms and traffic patterns. This section describes the parameters of the simulation model and their default values.

In our experiments, we have used a 2-dimensional 256-node (16,2) torus network. The number of classes is fixed at 9. The number of virtual channels per physical channel is fixed at 9 in the dedicated organization and 18 in the other two buffer organizations. The virtual channels of the NHop algorithm are divided uniformly among the nine different classes. Multiple virtual channels mapped to a physical channel share its bandwidth in time-multiplexed manner. The total number of buffers per node is maintained at 36 for all three buffer organizations. In the hybrid organization, the number of dedicated buffers is set to 16. The remaining 20 buffers are held in a central pool. The body flit transmission time is 1 clock cycle and the header flit transmission time is 2 clock cycles if a centralized buffer is used and 1 clock cycle if a dedicated buffer is used.

A processor generates messages with a mean message generation time of $T$. The messages have a fixed length of 20 flits.

In wormhole routing, bubbles could be introduced, especially at low traffic, in transmission of consecutive flits of message because of asynchronous pipelining. To reduce these bubbles, the depth of buffers is fixed to 4, i.e., each buffer can hold four flits of the same message. Whenever, a buffer has space for one or more flits, next data flit is sent from the previous router in the path.

**Traffic Patterns** Two traffic patterns are considered in this paper.

**Uniform** Uniform traffic is widely used in simulation studies and serves as a benchmark traffic pattern. In this traffic pattern, a node sends messages to any other node in the network with an equal probability. Uniform traffic could be representative of the traffic generated in massively parallel computations in which array data are distributed among the nodes using hashing techniques.

**Hot-spot** In hot-spot traffic a single node (hot node) receives a specified fraction of the total messages generated in the network. More specifically, given that $N$ is the total number of nodes in the network, $m$ is the number of messages generated per node per a clock cycle ($0 \leq m \leq 1$), $h$ is the fraction of messages directed at the hot-spot, each node generates messages directed to hot-spot at the total rate of $mh$ [8]. The effective number of messages to hot-spot are $m(1-h) + mhN$.

## 5 Simulation Results

This section presents the results of the simulation experiments conducted to evaluate the performance of the centralized, dedicated, and hybrid buffer organizations using the NHop algorithm. For each simulation run, thirty one batches of 2000 messages were used. First batch results were discarded to allow the network to reach steady state. It was observed that the network reaches steady state after the first batch. This strategy has produced 95% level confidence intervals that are within 5% of the mean value reported unless the network load is very high. Average response time is used as the performance metric, which represents the average delay encountered by a message. Due to space restrictions, only a sample of the results is presented here. Complete results are available in [6].

### 5.1 Uniform Traffic Results

**Base class results** This experiment is aimed at establishing a base set of results. From Figure 1 it is clear that the dedicated buffer organization has better

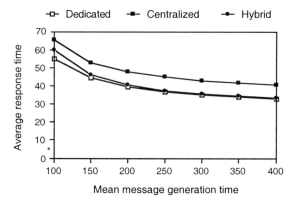

**Fig. 1.** Performance under uniform traffic with exponential inter-message generation times. All other parameters are set to their default values given in Section 4.

performance than the centralized buffer organization. Performance of the hybrid organization closely follows the dedicated organization. Better performance of the dedicated organization is due to the fact that no additional delay is involved in acquiring a buffer.

When the traffic is uniformly distributed, the necessity to increase the number of active virtual channels in one direction by shifting the buffers from under utilized links does not arise. In this scenario, the centralized organization incurs additional overhead to acquire buffers from the central pool without any gain in performance. It can be observed from the graph that the difference in the message delay between the centralized and dedicated organizations is approximately equal to the average path length. This difference can be attributed to the extra clock cycle spent by the header flit to acquire a buffer at each node on its path.

At low message traffic (i.e., at higher inter-message generation times), performance of the hybrid organization is similar to that of the dedicated as there may not be any need to acquire buffers from the central pool. As the message generation rate increases (i.e., smaller inter-message generation times), the need to acquire central buffers increases. This increases the difference in performance between the two organizations.

**Hyper-exponential message generation** In this experiment, inter-message generation times are hyper-exponentially distributed with a coefficient of variation (CV) of 4. That is, the messages are generated in a clustered manner. This might represent the communication pattern when a processor is synchronizing with other processors.

It can be seen from Figure 2 that the hybrid buffer organization has better performance than the other two organizations for $T \geq 220$. For $T < 220$, it has slightly worse performance than the dedicated buffer organization. This could be attributed to its partial capability to respond to the dynamic conditions in the

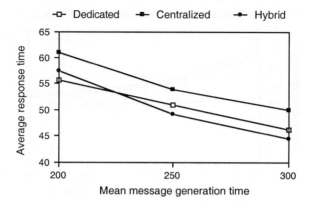

**Fig. 2.** Performance under uniform traffic with hyper-exponential inter-message generation times with a CV of 4. All other parameters are set to their default values given in Section 4.

network. As in the previous results centralized has the worst performance. Furthermore, clustered nature of message generation results in every organization having a higher latency when compared to the previous results.

**Variable message length** In this experiment the length of the messages generated by each node in the network is varied. To be specific, for a given mean message length, which is 20 flits in our experiments, nodes generate messages whose lengths are hyper-exponentially distributed with a CV of 4. Therefore, a node generates several short messages and few very long messages. This scenario is similar to applications such as video transmission where several short messages are exchanged by the nodes before the actual long video data are transmitted.

The results of the experiments are given in Figure 3. In contrast to the previous results, these results indicate that the centralized organization provides better performance than the other two organizations. The main reason for this behavior is that long messages hold the resources for a long time, which leads to conditions somewhat similar to the hot-spot behavior (discussed next). The centralized organization could respond to such dynamic conditions in a better way by shifting buffer based on the demand as it has a central pool of buffers.

## 5.2 Hot-Spot Traffic

The results for the hot-spot traffic are presented in three plots in order to provide an understanding of the underlying behavior. First plot gives the average response time for only the regular (i.e., non-hot-spot) messages. The second plot presents the results for the hot-spot messages only. The third plot gives the average response time results for all messages (i.e., regular and hot-spot).

**Regular delay** The delay involved in delivering regular messages is shown in Figure 4. It can be seen that the dedicated and hybrid organizations provide

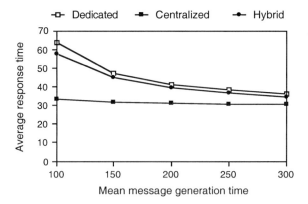

**Fig. 3.** Performance under uniform traffic with exponential message generation times and hyper-exponential message lengths with a mean of 20 flits and CV of 4. All other parameters are set to their default values given in Section 4.

superior performance when the hot-spot rate $h$ is less than about 9%. However, as the hot-spot rate increases beyond this, the centralized organization is better because of its ability to allocate buffers from the central pool on a demand-driven basis. It should be pointed out that, in practice, hot-spot rate of 10%, for example, is considered unreasonably high.

**Hot delay** The delay suffered by the hot-messages as a function of hot-spot rate $h$ is given in Figure 5. The data presented in this figure show that, when $h \leq 5$ the centralized organization performs worse than the other two organizations. At higher hot-spot rates, for reasons discussed before, the centralized organization provides substantially better performance than the other two buffer organizations.

**Total delay** The total delay suffered by the messages in three different organizations is given in Figure 6. As the total delay is the weighted average delay and the fraction of hot-spot messages is small, the total delay behavior is similar to that of the regular delay shown in Figure 4.

Simulation results presented in [6] indicate that these observations are true even when the inter-message generation times and message lengths are hyper-exponentially distributed.

It should be noted that the performance is sensitive to the number of buffers available. If the number of dedicated buffers is small, the centralized organization may perform better even for smaller hot-spot rates.

## 6 Conclusions

The dedicated buffer organization is better than the centralized organization for the uniform traffic in all scenarios except when the message traffic consists of a mix of small and large messages. In this case, the centralized organization is

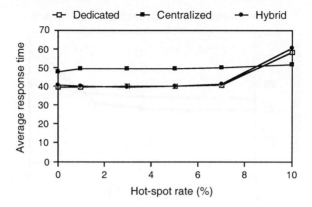

**Fig. 4.** Performance under hot-spot traffic: Regular message delay when the mean message generation time fixed at 200 (Other parameters are set to their default values.)

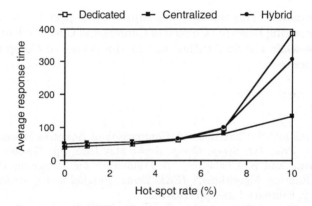

**Fig. 5.** Performance under hot-spot traffic: Hot message delay when the mean message generation time fixed at 200 (Other parameters are set to their default values.)

better. The centralized buffer organization is also better under high hot-spot traffic. The dedicated organization, however, provides better performance even for hot-spot traffic if the hot-spot rate is low.

The important conclusion from the data presented here and elsewhere [6] is that no single buffer organization is universally superior for all workloads. Ideally what is required is a buffer organization that changes from the centralized to distributed and vice versa depending on the traffic conditions. The proposed hybrid organization provides such a mechanism to dynamically adopt the buffer organization to the traffic conditions. Further work is needed in the area of detecting traffic conditions and configuring the hybrid organization appropriately.

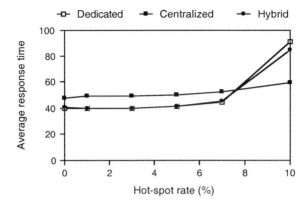

**Fig. 6.** Performance under hot-spot traffic: Total delay when the mean message generation time fixed at 200 (Other parameters are set to their default values.)

**Acknowledgements** This research was supported in part by the Natural Sciences and Engineering Research Council of Canada and Carleton University. This work was done while the first author was at the School of Computer Science, Carleton University.

## References

1. R.V. Boppana and S. Chalasani, "A Comparison of Adaptive Wormhole Routing Algorithms," *Proc. Int. Symp. Comp. Arch.*, May 1993, pp. 351-360.
2. R.V. Boppana and S. Chalasani, "A Framework for Designing Deadlock-Free Wormhole Routing Algorithms," *IEEE Trans. Parallel and Distributed Systems*, Vol. 7, No. 2, February 1996, pp. 169-183.
3. W.J. Dally, "Virtual-Channel Flow Control," *IEEE Trans. on Parallel and Distributed Systems*, Vol. 3, No. 2, March 1992, pp. 194-205.
4. W.J. Dally and H. Aoki, "Deadlock-free Adaptive Routing in Multicomputer Networks using Virtual Channels," *IEEE Transactions on Parallel and Distributed Systems*, Vol. 4, No. 4, April 1993, pp. 466-475.
5. J. Duato, S. Yalamanchili, and L. Ni, *Interconnection Networks: An Engineering Approach,* IEEE Computer Society Press, 1997. (More comprehensive set of references are available in this book.)
6. K. Kotapati, *The Buffer Management in Wormhole-Routed Torus Multicomputer Networks*, MCS thesis, School of Computer Science, Carleton University, 1997.
7. L.M. Ni and P.K. McKinley, "A Survey of Wormhole Routing Techniques in Direct Networks," *IEEE Computer*, February 1993, pp. 62–76.
8. G. F. Pfister and V. A. Norton, "Hot-Spot Contention and Combining in Multistage Interconnection Networks," *IEEE Trans. Computers*, Vol. 34, No. 10, October 1985, pp. 943–948.
9. D.A. Reed and R.M. Fujimoto, *Multicomputer Networks: Message Based Parallel Processing,* MIT Press, Cambridge, Mass., 1987.

# Performance Analysis of Broadcast in Synchronized Multihop Wireless Networks

Kuang-Hung Pan[1], Hsiao-Kuang Wu[2], Rung-Ji Shang[3], and Feipei Lai[1,3]

[1]Dept. of Electrical Eng. Room 342, National Taiwan University, Taipei, Taiwan
kpan@archi.csie.ntu.edu.tw, flai@cc.ee.ntu.edu.tw
[2]Dept. of Computer Sci. & Infor. Eng., National Central University, ChungLi, Taiwan
hsiao@csie.ncu.edu.tw
[3]Dept. of Comp. Sci. & Infor. Eng. Room 419, National Taiwan University, Taipei, Taiwan
shang@orchid.ee.ntu.edu.tw

**Abstract.** Analyzing the performance of broadcast in wireless networks is imperative because of the importance of broadcast and the difference between wireless communications and wired communications. In a practical system, the transmitted packets might arrive at the receiver later than desired. Thus, synchronization is an important issue to be analyzed. In this paper, we analyze the performance of broadcast with consideration of such imperfect synchronization. We find that randomized distributed broadcast algorithm provides guaranteed performance even with imperfect synchronization.

## 1 Introduction

Broadcast is an important task in distributed computing and systems. It can be used for disseminating information among a set of receivers [1], and exchanging messages in distributed computing [2]. In a multihop wireless network, there is not a base station that can communicate with all users directly, and not all users can receive the messages from all other users directly [1]. Therefore, it is not so simple to solve the broadcast problem in a multihop wireless network. Modern applications of such multihop networks include disaster rescue and ad-hoc personal communications services [3-6].

In previous work, Chlamtac and Weistein proposed a centralized broadcast algorithm with time complexity $O(Rlog^2V)$, where $R$ is the radius of the network, and $V$ is the number of mobile users in the network [7]. A broadcast protocol based on multi-cluster architecture [8] was proposed in [9]. In [10], it was proved that any broadcast algorithm requires $\Omega(log^2V)$ time slots to finish for a radius-2 network. In [11], it was proved that any distributed broadcast algorithm requires $\Omega(RlogV)$ time slots to finish. Bar-Yehuda, Goldreich, and Itai in [12] proposed a distributed randomized algorithm, which needs only $O((R+logV/\varepsilon)logV)$ time slots to finish with probability $1-\varepsilon$. This algorithm is exponentially superior to any distributed deterministic algorithm, which takes $\Theta(V)$ time slots even for a radius-3 network. Based on this randomized algorithm, a routing and multiple-broadcast algorithm was proposed in [13]. In [14], an $\Omega(D\log(V/R))$ lower bound for randomized broadcast algorithm was provided by Kushilevitz and Mansour.

Besides, more considerations should be taken to better describe the real environment in wireless networks. In the previous work [15], the channel reliability was introduced to take into account the variations in power levels [16]. In this paper, we analyze the performance of broadcast with considerations of imperfect synchronization. Synchronized (slotted) protocols often yield better performance than unslotted protocols, such as the advantages of slotted ALOHA over unslotted ALOHA [1]. However, the performance of synchronized protocols is often degraded by imperfect synchronization. If the synchronization is not perfect so that the delay in receiving a packet can be comparable to slot duration (a slot is about of the same duration of a packet), the packets transmitted in a slot might collide with those transmitted in the next slot. In multihop wireless networks, such imperfect synchronization might come from propagation delay and imperfect clock synchronization due to the difficulty of synchronizing all processors in a distributed multihop network during the setup of the network.

In this paper we analyze the performance of broadcast with slot level imperfect synchronization in multi-hop wireless networks. The rest of the paper is organized as follows. Section II provides the system model and states the broadcast algorithms. Section III analyzes the performance of the broadcast algorithm. Section IV gives numerical examples and discussions. Conclusions are provided in section V.

## 2 Preliminaries

The network is represented by an undirected graph $G$, in which a node denotes a mobile user and an edge between two nodes means that these two nodes are connected. The degree of a node is the number of neighbors this node has. The degree $D$ of the network is the maximum degree of the nodes. The radius $R$ of the network is the maximum number of edges in the shortest path between any two nodes. The probability that a processor is so busy with other tasks that it temporarily does not execute the algorithm is also included. This may result from some emergent event that a processor has to deal with, or power-saving considerations with lightweight processors. Allowing the processors not to carry out broadcast temporarily thus enables the processors to deal with other urgent task or to save power. All processors are assumed to have the same characteristics, and are dedicated to executing the algorithm in a time slot with the same probability $p_r$.

The goal of the broadcast algorithm is to carry the packet to be broadcast to as many users as possible. Each user in the network executes the broadcast algorithm.

Centralized and deterministic broadcast algorithms were shown that they do not perform well in multihop wireless networks. Instead, distributed and randomized [17] broadcast algorithms were shown to be able to fulfill the task in bounded time [12,15]. In this paper, the randomized distributed algorithm in [12] is adopted for the performance analysis of distributed randomized algorithm. In this algorithm, any user receiving the message to be broadcast (denoted as $m$) in this time slot will broadcast it in the next time slot. To prevent the situation that too many nodes are transmitting and collisions of packets are too frequent, the algorithm uses a procedure called "Decay". In "Decay", a processor that is transmitting at this time-slot may stop transmitting at the next time slot with a specified probability $1-q$, where $q$ is known as the retransmission probability. This process repeats for $\lceil \tau \rceil$ time slots.

```
Procedure Decay(⌈τ⌉, m);
repeat at most ⌈τ⌉ times (at least once)
    transmit m to all neighbors;
    exit with probability 1-q;
```

The whole randomized algorithm uses $\lceil \tau \rceil$ as the number of time slots in a stage, i.e., the transmitting processors restart executing "Decay" every $\lceil \tau \rceil$ time slot [12]. In time slot 0, a processor, called the source, generates and broadcasts the message $m$ to other users. Then every processor executes the following broadcast procedure.

```
Procedure Broadcast;
Wait until receiving the message m;
do T times
    Wait until ( Time mod ⌈τ⌉) =0;
    Decay(⌈τ⌉, m);
end do.
```

For the situation where a sequence of packets has to be broadcast, the above algorithm can serve as one step in a comprehensive algorithm, in which the packets are broadcast in an ordered sequence.

## 3 Performance Analysis of Broadcast Algorithms

### 3.1 Probabilities of Successful Reception

Synchronization can refer to that at carrier phase, bit, word [18], algorithm [2,17], etc. In this paper, we are considering synchronization at slot level. If the time lag between simultaneously received packets is small enough in comparison with the duration of a packet (slot), the delay can be compensated by the guard period and this packet will not seriously damage the reception in the next slot. Otherwise, the delay cannot be compensated by the guard period and this packet may seriously damage the reception at the next slot.

In this paper, Rayleigh fading is assumed [16]. Equivalently, their in-phase and quadrature-phase components are Gaussian distributed [16,18]. Incoherent binary FSK is used for modulation and demodulation [16,18]. Assume the average energy per bit [18] of each received packet is $E_b$. The power level of the Gaussian noise at the receiver is $N_0$. We assume the fading is slow. Thus, if one bit is not received successfully, it probably means the fading or interference is severe for the whole packet containing the bit, and the packet is unlikely to be received successfully. The probability of bit error for incoherent binary FSK in a Rayleigh fading channel is $P_b=1/(2+z)$, where $z$ is the average signal-to-noise ratio [18]. With Rayleigh fading, the power level of the sum of the interfering signals can be calculated as $(n-1)E_b$ [18]. Thus, the average signal-to-noise ratio is $z=E_b/(N_0+(n-1)E_b)$

## 3.2 Performance analysis of broadcast with imperfect synchronization

Because the packets transmitted in a time slot might arrive in the next time slot, we have to take into account the number of transmissions in the previous slot to calculate the probability of successful reception. Consider the situation where there are $n$ neighbors trying to broadcast the message $m$ to user $i$ at the $0^{th}$ time slot and there were $n^-$ transmissions at the $(-1)$-th time slot. Let $f(n,n^-,t)$ denote the probability that user $i$ receives $m$ before time $t$ in this situation. Since the upper bound of $n$ is $D$, $\lceil \tau \rceil$ should be chosen a minimum value such that $f(D,0,\lceil \tau \rceil) \geq P_s$, where $P_s$ is the required probability of success in a stage. Therefore, the difference in the performances of synchronized and non-synchronized reception can be known by evaluating $f(D,0,t)$.

The probability that the packets transmitted at a time slot collide and destroy the reception at the next slot due to propagation delay or imperfect clock synchronization is denoted as $p_d$. The recurrence relation of $f(n,n^-,t)$ is

$$f(n,n^-,t)=(1-p_d)^{n^-} P_{suc}[n]+(1-(1-p_d)^{n^-} P_{suc}[n])\sum_{j=0}^{n}C_{n,j}\, q^j(1-q)^{n-j}(f(j,n,t-1)), \quad (1)$$

where $(1-p_d)^{n^-} P_{suc}[n]$ represents the probability that user $i$ receives $m$ successfully in the 0-th time slot, and in the term $(1-(1-p_d)^{n^-} P_{suc}[n])\sum_{j=0}^{n}C_{n,j}\, q^j(1-q)^{n-j}f(j,n,t-1)$, $(1-(1-p_d)^{n^-} P_{suc}[n])$ is the probability that $i$ did not receive $m$ successfully in the 0-th time slot, and $C_{n,j}q^j(1-q)^{n-j}$ represents the probability of occurrence that $j$ out of $n$ neighbors keep transmitting in the next time slot, which is multiplied by $f(j,n,t-1)$, the probability of successful reception with $n$ neighbors transmitting in the previous slot for each case. Assume the probability that a processor is taking a rest to be $p_r$, the probability that a user receives a packet successfully is

$$P_{suc}[n] = (1-p_r)\sum_{j=0}^{n}C_{n,j}\,(1-p_r)^j p_r^{(n-j)} P_{rec}[j], \quad (2)$$

where $(1-p_r)$ is the probability that the receiving node is not taking a rest, $C_{n,j}(1-p_r)^j p_r^{(n-j)}$ represents the probability that $j$ neighbors are not taking a rest and are transmitting the packet in this time slot, and $P_{rec}[j]$ is the probability that a packet can be received correctly in a collision of $j$ packets. $P_{rec}[j]$ can be approximately obtained from the analysis of $P_b$ in the previous subsection.

Besides the numerical approximate solution we will use in the next section, we also solve the recurrence relation by the analytical approach modified from [12]. We derive the following theorems, which provide an upper bound on the time complexity of this algorithm. This bound is not meant to be tight. Rather, this bound shows that the performance of this algorithm is still satisfactory in spite of imperfect synchronization.

The recurrence relation of the bound on $f(n,n^-,t)$ is obtained as follows.

for $n=1$, $f(1,0,t)=1$, $f(1,n^-,t)=(1-p_d)^{n^-} +(1-(1-p_d)^{n^-})qf(1,1,t-1)$,
for $n \geq 2$,

$$f(n,n^-,t) \geq \sum_{j=0}^{n} C_{n,j} q^j (1-q)^{n-j} f(j,n,t-1), \qquad \forall n^- \in N \qquad (3)$$

where $C_{n,j} q^j (1-q)^{n-j}$ represents the probability that $j$ neighbors keep transmitting in the next time slot, which is multiplied by $f(j,0,t-1)$, the probability of successful reception for each case, and $\sum_{j=0}^{n} C_{n,j} q^j (1-q)^{n-j} f(j,n,t-1)$ is hence the probability of successful reception after the fist time slot.

*Theorem 1*

$$f(n,n^-,\infty) \geq \frac{(1-p_d)q}{1-p_d q} \frac{2q(1-q)}{1-q^2} \qquad \text{for } n \geq 2, \forall n^- \in N \qquad (4)$$

*Proof:*

Let $t \to \infty$ and $n^- = \infty$ in (1) for $n \geq 2$, we obtain

$f(n,\infty,\infty) = \sum_{j=0}^{n} C_{n,j} q^j (1-q)^{n-j} f(j,n,\infty)$. Obviously, $f(0,n^-,\infty) = 0$. Also, $f(1,0,\infty) = 1$, and

$f(1,n^-,\infty) = (1-p_d)^{n^-} + (1-(1-p_d)^{n^-}) qf(1,1,\infty)$,
for $n^- = 1$, $f(1,1,\infty) = (1-p_d) + (1-(1-p_d))qf(1,1,\infty)$,
$\Rightarrow f(1,1,\infty) = (1-p_d)/(1-(1-(1-p_d))q) = (1-p_d)/(1-p_d q)$.
Also, $f(1,n^-,\infty) \geq f(1,\infty,\infty)$ because the minimum of $f(1,n^-,\infty)$ is achieved when $n^- \to \infty$, i.e., the interference from the packets in the previous slot is maximum.
Therefore, $f(1,n^-,\infty) \geq q(1-p_d)/(1-p_d q)$.

The lower bound of $f(n,\infty,\infty)$ is obtained as follows.

$$f(n,\infty,\infty) = \sum_{j=0}^{n} C_{n,j} q^j (1-q)^{n-j} f(j,n,\infty),$$

for $n=2$, $f(2,\infty,\infty) = \sum_{j=1}^{2} C_{n,j} q^j (1-q)^{n-j} f(j,2,\infty) \geq 2q(1-q)q(1-p_d)/(1-p_d q) + q^2 f(2,\infty,\infty)$

so $f(2,\infty,\infty) \geq \dfrac{(1-p_d)q}{1-p_d q} \dfrac{2q(1-q)}{1-q^2}$.

Similarly, we can use mathematical induction to find the lower bound of $f(n,\infty,\infty)$:

assume $f(j,n,\infty) \geq f(j,\infty,\infty) \geq \dfrac{(1-p_d)q}{1-p_d q} \dfrac{2q(1-q)}{1-q^2}$, for $2 \leq j \leq n-1$,

$$f(n,\infty,\infty) = \sum_{j=0}^{n} C_{n,j} q^j (1-q)^{n-j} f(j,n,\infty)$$

$\Rightarrow (1-q^n) f(n,\infty,\infty) \geq \sum_{j=1}^{n-1} C_{n,j} q^j (1-q)^{n-j} f(j,n,\infty)$

$\Rightarrow (1-q^n) f(n,\infty,\infty) \geq q(1-q)^{n-1} q(1-p_d)/(1-p_d q) + \dfrac{(1-p_d)q}{1-p_d q} \dfrac{2q(1-q)}{1-q^2} \sum_{j=2}^{n-1} C_{n,j} q^j (1-q)^{n-j}$

$$\Rightarrow f(n,n^-,\infty) \geq f(n,\infty,\infty)$$

$$\geq \frac{n(1-q)^{n-1}q}{1-q^n}\frac{(1-p_d)q}{1-p_dq} + \frac{1}{1-q^n}\frac{2q(1-q)}{1-q^2}\frac{(1-p_d)q}{1-p_dq}\left((1-q)+q)^n - q^n - (1-q)^n - n(1-q)^{n-1}q\right)$$

$$= \frac{(1-p_d)q}{1-p_dq}\frac{2q(1-q)}{1-q^2}\frac{1}{1-q^n}\left(1-q^n-(1-q)^n + n(1-q)^{n-1}q\left(\frac{1-q^2}{2q(1-q)}-1\right)\right)$$

$$= \frac{(1-p_d)q}{1-p_dq}\frac{2q(1-q)}{1-q^2}\frac{1}{1-q^n}\left(1-q^n-(1-q)^n + \frac{n}{2}(1-q)^n\right)$$

$$\geq \frac{(1-p_d)q}{1-p_dq}\frac{2q(1-q)}{1-q^2}, \text{ for } n\geq 2. \qquad \#$$

*Theorem 1* is used to derive *Theorem 2*, which gives a lower bound on $\tau$.

*Theorem 2*

$$f(n,0,t) > \frac{3}{2}\frac{(1-p_d)q}{1-p_dq}\frac{q(1-q)}{1-q^2}$$

for $n\geq 2$ and $t \geq \left(\frac{\log_2 n}{\log_2(1/q)}\right) + \left(\frac{1}{\log_2(1/q)}\right)\log_2\left(\frac{(1-p_dq)2(1-q^2)}{(1-p_d)q(1-q)}\right)$.

*Proof*:

$f(n,0,t) \geq Prob$(the user eventually has received the message $m \wedge$ at least one transmitter keeps transmitting at time $= t$)

$$\geq f(n,0,\infty) - nq^t > \frac{(1-p_d)q}{1-p_dq}\frac{2q(1-q)}{1-q^2} - nq^t$$

$$> \frac{3}{2}\frac{(1-p_d)q}{1-p_dq}\frac{q(1-q)}{1-q^2}, \text{ for } nq^t \leq \frac{1}{2}\frac{(1-p_d)q}{1-p_dq}\frac{q(1-q)}{1-q^2}$$

i.e, $t\geq \log_q\left(\frac{1}{n}\frac{(1-p_d)q}{1-p_dq}\frac{q(1-q)}{2(1-q^2)}\right) = \left(\frac{1}{\log_2 q}\right)\log_2\left(\frac{1}{n}\frac{1-p_d}{1-p_dq}\frac{q(1-q)}{2(1-q^2)}\right)$

$$= \left(\frac{\log_2 n}{\log_2(1/q)}\right) + \left(\frac{1}{\log_2(1/q)}\right)\log_2\left(\frac{(1-p_dq)2(1-q^2)}{(1-p_d)q(1-q)}\right). \qquad \#$$

From *Theorem* 2, the time needed to reach $P_s$ is logarithmically increasing with $n$. By *Procedure Broadcast*, the randomized algorithm uses $\lceil \tau \rceil$ as the number of time slots in a stage. From *Theorem 2*, $\tau \leq \left(\frac{\log_2 D}{\log_2(1/q)}\right) + \left(\frac{1}{\log_2(1/q)}\right)\log_2\left(\frac{(1-p_dq)2(1-q^2)}{(1-p_d)q(1-q)}\right)$, if we choose $P_s = \frac{3}{2}\frac{(1-p_d)q}{1-p_dq}\frac{q(1-q)}{1-q^2}$ and the upper bound of $n$ is $D$. The upper bound on $\tau$ is thus established.

The time slots required by the broadcast algorithm can be obtained as follows. Let $T_R$ be the time a user receives the message and $C(\delta) = \sqrt{\log_2(V/\delta)}$. For *perfect synchronization* [12], if we choose $P_s = 0.5$ and $\lceil \tau \rceil = 2\lceil \log D \rceil$, we have $Prob(T_R > \lceil \tau \rceil N(\delta)) < \delta/V$, where $N(\delta) = 2R + 5C(\delta)Max(C(\delta), \sqrt{R})$, and

$$Prob(Max(T_R) > \lceil \tau \rceil N(\delta)) < \delta. \tag{5}$$

For *imperfect synchronization*, if we choose $P_s = \dfrac{3}{2}\dfrac{(1-p_d)q}{1-p_d q}\dfrac{q(1-q)}{1-q^2}$, and $\lceil \tau \rceil =$

$\left(\dfrac{\log_2 D}{\log_2(1/q)}\right) + \left(\dfrac{1}{\log_2(1/q)}\right)\log_2\left(\dfrac{(1-p_d q)2(1-q^2)}{(1-p_d)q(1-q)}\right)$, we have $Prob(T_R > \lceil \tau \rceil N(\delta)) < \delta/V$ by

Theorem 3 in [15], where $N(\delta) = R/P_s + (5/2P_s)C(\delta)Max(C(\delta),\sqrt{R})$, and

$$Prob(Max(T_R) > \lceil \tau \rceil N(\delta)) < \delta \tag{6}$$

by Theorem 4 in [15]

## 4 Numerical Examples and Discussions

The physical environment and the design of operating power determine the network topology and the parameters $R$ and $D$. It can be seen that minimizing the time complexity requires minimizing both $R$ and $D$. However, there is a tradeoff between $D$ and $R$. For example, if the transmitters increase their power levels, $R$ could decrease and $D$ could increase. Therefore, evaluating the performance of the broadcast algorithm is very helpful in saving not only time but also power. How to decide an optimum value depends on the values of $f(n,0,t)$. In this section, we evaluate equation (1) using recursive method to give numerical examples of $f(n,0,t)$ and $\lceil \tau \rceil$.

The influences of different parameters on $f(n,0,t)$ are illustrated in Fig. 1, where $z_0 = E_b/N_0 = 5, 15, 25$(dB) and $p_r = 0.06$. It can be seen that $f(n,0,t)$ decreases with $p_d$ increasing. Again, the curves show that the smaller $q$ is, the faster $f(n,0,t)$ grows with $t$, and the asymptotic value of $f(n,0,t)$ is smaller. The larger $p_d$ is, the smaller $f(n,0,t)$ is. Raising the power to increase the signal-to-noise ratio is effective only when the transmitter signal power level is really too low. When collisions rather than noise limit the performance, raising the transmitter signal power level is not effective.

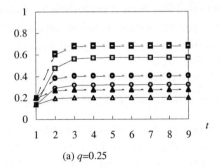

(a) $q=0.25$

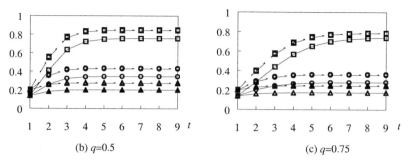

(b) $q=0.5$            (c) $q=0.75$

**Fig. 1** Figure of $f(n,0,t)$, with $n=5$, $p_r=0.06$.

Fig. 2 illustrates the relation between $f(n,0,t)$ and $p_d$. It is verified again that $f(n,0,t)$ degrades with $p_d$. When $p_d$ increases toward 1, $f(n,0,t)$ decreases and approaches the probability that only the reception in the first slot (without interference from the previous slot) is successful. Again, after a threshold, more increase in the transmitter power is useless.

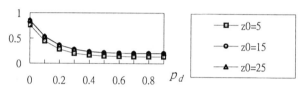

**Fig. 2** The relation between $f(n,0,t)$ and $p_d$. Assume $n=5$, $p_r=0.06$, and $t=9$ (slots).

### 4.2 Optimization of the network topology

Consider the network topology of the three examples in Fig. 3: (a) $D=2$, $R=8$; (b) $D=3$, $R=4$; (c) $D=8$, $R=2$. Let $N=9$ and $\delta=0.01$, then $\log_2(N/\delta)=9.8$, $C(\delta)=\sqrt{\log_2(V/\delta)}=3.13$, $\text{Max}(\sqrt{R},C(\delta))=3.13$. For perfect synchronization, $N(\delta)=2*R+5*9.8=2R+49$, the time slots needed to carry out broadcast can be obtained from (5) as 130 for Fig. 3(a), 228 for Fig. 3(b), and 330 for Fig. 3(c).

For imperfect synchronization, the time slots required can be obtained from (6). We choose $P_s = \frac{3}{2}\frac{(1-p_d)q}{1-p_d q}\frac{q(1-q)}{1-q^2}$ and $q=0.5$, then $N(\delta)=R/P_s+24.5/P_s$. (i) Let $p_d=0.1$, we have $P_s=0.24$, $\lceil \tau \rceil = \lceil \log_2 D+2.66 \rceil$, and the time slots needed to carry out broadcast are 544 for Fig. 3(a), 595 for Fig. 3(b), and 690 for Fig. 3(c). (ii) Let $p_d=0.2$, we have $P_s=0.22$, $\lceil \tau \rceil = \lceil \log_2 D+2.75 \rceil$, and the time slots needed to carry out broadcast are 592 for Fig. 3(a), 648 for Fig. 3(b), and 750 for Fig. 3(c).

From this example, we can see that the smaller the $D$, the less the time needed. The radius $R$ may be allowed to be larger if $D$ can be reduced. This is because the occurrence of collisions is the main obstacle of broadcast. And the collision due to the transmission in the previous slot seriously damages the performance.

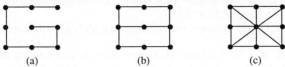

**Fig. 3** Logical graph of networks with 9 users

## 5 Conclusions

Centralized algorithms and deterministic algorithms are even more difficult to carry out with packet (slot) level non-synchronization because trying to control everything according to a perfect global clock is not practical. The distributed randomized algorithm proposed in [12] is more practical than centralized and deterministic algorithms. We generalize the results in [12] to more general situation to include imperfect synchronization. By the theorems obtained in this paper, this algorithm is guaranteed to fulfill the broadcast task in reasonable time. From *Theorem 2*, The time complexity grows only logarithmically with the degree $D$ of the network.

We also found that the number of time slots needed grows with the probability of being non-synchronized. Besides, raising the transmitter power is not effective, and trying to reduce the non-synchronization at packet (slot) level will be more effective. Also, increasing retransmission probability $q$ can increase the reachable probability in a stage. Meanwhile, increasing $q$ also decreases the rate $f(n,n^-,t)$ grows with $t$. Therefore, $q$ can be chosen a proper value to obtain the optimized performance according to the given environment. In addition, how to optimize the network topology to reduce the time needed for broadcast is also examined.

For future research, more accurate calculation of the probability of successful reception of a packet in collision with more practical and accurate considerations can be analyzed. For instance, the probability of successful reception of a packet should be calculated according to the coding techniques used and the statistics of the interference. The relation between successful reception at packet level and signaling level can be further investigated. We might also investigate the use of diversity to distinguish packets from different users to increase the probability of successful reception [19]. Also, the performance of multiple-messages broadcast, routing [14,20], and multicast can be investigated. In addition, other possible methods of broadcast in multi-hop wireless networks can also be studied. For example, how to carry out broadcast in a completely asynchronous (i.e., unslotted) manner and dynamic broadcasting [21] can be studied.

## References

1. D. P. Bertsekas and R. G. Gallager, *Data Networks*, $2^{nd}$ edition, Englewood Cliffs, N.J. : Prentice Hall,1992.

2. D. P. Bertsekas and J. N. Tsitsiklis, *Parallel and Distributed Computation: Numerical Methods*, Englewood Cliffs, N.J. : Prentice Hall, 1989.
3. T.-W. Chen, J.T. Tsai, and M. Gerla, "QoS routing performance in multihop, multimedia, wireless networks," in *Proc. ICUPC'97*.
4. C.-C. Chiang and M. Gerla, "Routing and multicast in multihop, mobile wireless networks," in *Proc. ICUPC'97*.
5. C.-C. Chiang, H.-K. Wu, W. Liu, and M. Gerla, "Routing in Clustered Multihop Mobile Wireless Networks with Fading Channel," in *Proc. IEEE Singapore International Conference on Networks (SICON'97)*.
6. C. R. Lin and M. Gerla, "MACA/PR: An Asynchronous Multimedia Multihop Wireless Network," in *Proc. INFOCOM '97*.
7. I. Chlamatic and O. Weistein, "The Wave Expansion Approach to Broadcasting in Multihop Radio Networks," *IEEE Trans. Commun.*, Vol. 39, No. 3, Mar. 1991, pp. 426-433.
8. M. Gerla and J. T. Tsai, "Multiculster, mobile, multimedia radio networks," *Wireless Networks*, vol. 1, pp. 255-265, 1995.
9. E. Pagani and G.P. Rossi, "Reliable broadcast in mobile multihop packet networks," in *Proc. MOBICOM'97*, pp. 34-42.
10. N. Alon, A. Bar-Noy, N. Linial, and D. Peleg, "A Lower Bound for Radio Broadcast," *Journal of Computer and System Sciences*, 43, 1991, 290-298.
11. D. Bruschi and M. Del Pinto, "Lower bounds for the broadcast problem in mobile radio networks," *Distributed Computing*, 1997 10, 129-135.
12. R. Bar-Yehuda, O. Goldreich, and A. Itai, "On the Time-Complexity of Broadcast in Multi-hop Radio Networks: an Exponential Gap between Determinism and Randomization," *Journal of Computer and System Sciences*, 45, 1992, 104-126.
13. R. Bar-Yehuda, A. Israeli, and A. Itai, "Multiple communication in multi-hop radio networks," *SIAM J. Comput.*, 22, 1993, pp. 875-887.
14. E. Kushilevitz and Y. Mansour, "An $\Omega(D \log(N/D))$ lower bound for broadcast in radio networks," *SIAM J. Comput.*, 27, 1998, pp. 702-712.
15. K.-H. Pan, H.-K. Wu, R.-J. Shang, and F. Lai, "Performance analysis of broadcast in wireless networks with channel reliability," in *Proc. the $2^{nd}$ IASTED International Conference on Parallel and Distributed Computing and Networks (PDCN'98)*, pp. Brisbane, Australia, Dec. 14-16, 1998.
16. T. S. Rappaport, *Wireless Communications: Principles and Practice*, Prentice Hall PTR, New Jersey, 1996.
17. R. Motwani and P. Raghavan, "Randomized Algorithms," Cambridge, New York, 1995.
18. R. E. Ziemer and W. H. Tranter, *Principles of Communications*, Boston, MA: Houghton Mifflin, 1995, $4^{th}$ edition.
19. G. W. Wornell and M. D. Trott, "Efficient signal processing techniques for exploiting transmit antenna diversity on fading channels," *IEEE Trans. Signal Processing*, Vol. 45, No. 1, pp. 191-205, Jan. 1997.
20. G. D. Stamoulis and J. N. Tsitsiklis, "Efficient Routing Schemes for Multiple Broadcasts in Hypercubes", *IEEE Trans. Paral. Distri. Syst.*, Vol. 4, No. 7, pp. 725-739, July 1993,.
21. E. A. Varvarigos and D. P. Bertsekas, "Dynamic Broadcasting in Parallel Computing," *IEEE Trans. Paral. Distri. Syst.*, Vol. 6, No. 2, pp. 120-131, Feb. 1995.

# EARL - A Programmable and Extensible Toolkit for Analyzing Event Traces of Message Passing Programs -

Felix Wolf[1] and Bernd Mohr[2]

[1] TU Darmstadt, Graduiertenkolleg "Infrastruktur für den elektronischen Markt",
64283 Darmstadt, Germany
fwolf@rbg.informatik.tu-darmstadt.de
[2] Forschungszentrum Jülich GmbH, ZAM, 52425 Jülich, Germany
b.mohr@fz-juelich.de

**Abstract.** This paper describes a new meta-tool named EARL which consists of a new high-level trace analysis language and its interpreter which allows to easily construct new trace analysis tools. Because of its programmability and flexibility, EARL can be used for a wide range of event trace analysis tasks. It is especially well-suited for automatic and for application or domain specific trace analysis and program validation. We describe the abstract view on an event trace the EARL interpreter provides to the user, and give an overview about the EARL language. Finally, a set of EARL script examples are used to demonstrate the features of EARL.

## 1 Motivation

Using event tracing to analyze the behavior of parallel and distributed applications is a well accepted technique. In addition, there are a multitude of powerful, graphical event trace analysis tools (e.g., AIMS[9], Paradyn[8], Pablo[6], SIMPLE[5], VAMPIR[3], Upshot[4], and many more). However, all of them have one or more of the following shortcomings:

1. The biggest problem (especially with graphical tools) is that event traces generated on today's large and fast machines are getting very big. Either the tools show the recorded behavior by displaying an animation or they read in the whole trace at once, display it, and then allow the user to zoom in and out. If analyzing very large traces, a user looking for problems/bottlenecks would either have to watch or zoom in and out for a long time. In the second case, it is possible that the trace is too large to be read in in total. In either case, the user never knows whether he missed something because he didn't look carefully enough or zoomed in at the wrong places. Clearly, a more "automatic" way of analyzing large traces is needed.
2. Although the tools provide a large number of graphical views the right one needed for the application might not be available. Typically, many of these

tools cannot be used if the analysis has to be carried out in a domain or application specific way not covered by the graphical displays provided with the tools. Also, many cannot handle user-defined events in a useful way. A more flexible and easily programmable tool is needed. Such a tool would also allow a tool expert to explore new ideas for trace analysis tools or to quickly implement custom-made tools for ordinary users if needed.
3. Traditional performance analysis tools have very little or no support for conducting experiments (i.e., repeated measurements with varying processor numbers or input data sets). Often, they cannot analyze more than one event trace at the same time (e.g., to support trace comparisons).

We therefore designed and implemented a new meta-tool called EARL (Event Analysis and Recognition Language). EARL is actually a new high-level trace analysis language which allows to easily construct new trace analysis tools by writing scripts in the EARL language. These are then executed by the EARL interpreter. Section 2 describes the EARL language and the implementation of the interpreter in more detail. Two longer EARL script examples in Section 3 show how easy it is to use EARL to implement new trace analysis tools. Section 4 discusses related work and Section 5 concludes the paper and describes the enhancements to EARL we hope to implement in the future.

## 2 The EARL Toolkit

In order to achieve the highest degree of flexibility and programmability for analyzing event traces, we designed and implemented a new high-level trace analysis language. Although EARL is designed to be a generic event trace analysis tool, the current prototype (which is described in this paper) concentrates on the analysis of event traces generated from message passing programs. This is not really a restriction as most uses of event tracing are in the field of parallel programming on distributed memory machines (which almost all use a one or two sided message passing scheme for communication). In addition, analysis of message passing traces is well understood and therefore allowed us to provide high-level, well known abstractions as the programming interface to an EARL user.

### 2.1 Abstract View on an Event Trace

Much of the power of EARL comes through its very high-level abstraction of an event trace which allows a programmer to concentrate on the trace analysis and let EARL take care of the different trace formats and their encoding of functions and event types, of input handling and buffering, and of keeping track of message queues and call stacks.

An EARL programmer can view an event trace as a sequence of *events*. The events are sorted according to their timestamp and numbered starting at one. There are different *event types*. EARL defines four predefined event types: entering (named **enter**) and leaving (**exit**) a region, and sending (**send**) and receiving (**recv**) a message. There may be more event types defined depending on the underlying trace format. A *region* is a named section of the traced program (e.g., it could be a loop or basic block, but mostly it is a function or subroutine). If supported by the trace format, regions may be organized in *groups* (e.g., user or system functions).

An event type is represented by a n-tuple of *attributes*. An event (instance) is defined by a corresponding n-tuple of values assigned to these attributes. The number of attributes depends on the type of the event. However, all event types have the following attributes in common:

**num:** The number of the event.
**node:** The location (cpu, pe, or node) where the event happened. Nodes are numbered for 0 to n-1 where n is the total number of locations used by the parallel program.
**time:** The timestamp of the event as a floating point value in seconds.
**type:** The event type is explicitly given as a attribute value.
**enterptr:** The number of the **enter** event which determines the region in which the event happened. For **exit** events, this means that their **enterptr** refers to the matching **enter**.

The **enter** and **exit** event types have additional **region** and **group** attributes specifying the name of the region entered or left and its group, and **send** and **recv** have attributes describing the destination (**dest**), source (**src**), tag (**tag**), length (**len**), and communicator (**com**) of the message. In addition, the **recv** event type has a **sendptr** attribute pointing to the corresponding **send** event.

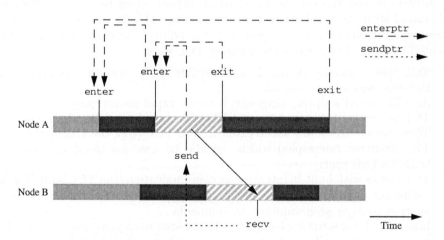

**Fig. 1.** References provided by **sendptr** and **enterptr**

In addition to the basic event trace model, EARL provides the concepts of *regions* and *messages*. These are defined as pairs of matching events: **enter/exit** or **send/recv** respectively. In the EARL language, these concepts are supported in the form of the `enterptr` and `sendptr` which are calculated automatically by EARL based on the information available in the event trace and the semantic knowledge about the message passing programming model. For each position in the event trace, EARL also defines a *region stack* per node and a *message queue* implemented as lists of `enter` and `send` events which define the regions entered and messages not yet received at that time. In addition to query theses structures directly, it is possible to navigate step-by-step through the region stack using the `enterptr` attribute or to trace back messages by following the `sendptr` attribute (see Figure 1).

All these facilities together allow to easily process complex event patterns made out of regions and messages. This is demonstrated by the examples in Section 3.

## 2.2 Implementation Notes

The EARL interpreter reads and decodes the underlying trace format and maps it automatically to the EARL event types and attributes. This allows the programmers to write their trace analysis scripts independent from the format of the event trace and of the encoding of event types and function/region names. EARL only requires that the event trace recording system uses global timestamps (by means of hard- or software synchronisation) and that the recording of each message transfer is complete (i.e., all of its send and receive events are recorded or none; otherwise the matching of `send` and `recv` would fail resulting in incorrect `sendptr` attributes). Currently, EARL supports the VAMPIR[3] and ALOG[4] trace formats.

Instead of re-inventing the wheel when implementing the EARL language, we started with the well known scripting language TCL[2] and extended it with commands for event trace and event record handling. The extensions are implemented in C++. The reasons for choosing TCL were:

- TCL was originally designed as extensible tool command language, and therefore was easy to extend.
- As a high-level scripting language, it allows rapid prototyping.
- TCL is very portable (it runs on UNIX, Macintosh, and Windows systems)
- There are already many useful extensions to TCL (e.g., the graphical toolkit TK, piecharts, bargraphs) which can also be used for the development of trace analysis tools.
- TCL comes with built-in interprocess communication (the TCL built-in `send` command and sockets) which makes it easy to integrate it with other trace analysis tools or programming environments.
- Like other Unix scripting languages, TCL allows to execute and control other processes very easily making it very suitable to implement trace analysis experimentation tools.

To improve efficiency, EARL automatically caches the most recently processed events in the *history buffer* and stores important trace state information (including the region stacks and the message queue) at fixed intervals in so-called *bookmarks* to speed-up random access to events.

## 2.3 List of EARL Commands

This section gives an overview of the new commands we added to TCL in order to allow high-level, portable, and efficient event trace analysis. The EARL extensions follow the object-oriented style which is also used in TK: the command to open a event trace returns a *trace object handle* which is automatically registered as a new TCL command. The other EARL functions are implemented as methods of the trace object. EARL supports the following functionality:

**Trace handling:** The `earl open` command takes the filename of the event trace as an argument and returns a handle to it. It has optional switches to pass the trace format (`-format`), the size of the history buffer (`-hist`), and the distance between bookmarks (`-mark`). The `close` method closes the event trace and releases all related resources.

**Event access:** EARL provides two methods for accessing events: `set` fills an specified associative array `arr` in a way that $\text{arr}(\text{attr}_i) == \text{value}_i$, while the `get` method returns an event as a list $\{\text{attr}_1 \text{ value}_1 \text{ attr}_2 \text{ value}_2 \ldots \text{attr}_n \text{ value}_n\}$. Both take the number of the event to process as argument. In addition, they set the *current event pointer* to the processed event unless the optional switch `-fetchonly` is used. Both methods come in three flavors: the user can pass the number of the event to process, or move through the trace sequentially forward or backward (relative to the current event pointer) by using the additional methods `setnext` and `getnext` or `setprev` and `getprev`.

**State access:** EARL automatically keeps track of the state of the region stacks and the message queue for the current event. The `stack` method returns the stack of a specified node as a list of either the region names (`-sym` switch) or the event number of the corresponding `enter` events (default). The `queue` method returns the message queue as a list of event numbers which point to the corresponding `send` events.

**General information access:** The `info` method gives access to general information about the event trace. It allows to get a list of all defined event types (`eventtypes`), a list of all attributes for a specified type (`attributes`), the filename (`filename`) and format (`format`) of the event trace, the number of nodes used in the parallel application (`nodecount`), and a list of all defined regions (`regions`) and groups (`groups`).

**Statistics:** Event trace analysis often involves keeping statistics of a large number of values like the execution times of a region or the transfer rates of messages. EARL supports this by providing *statistic objects*. The command `earl stat` creates a new statistic object and returns a handle to it. The method `addval` adds a new value to the data set. At any point, the user can ask for the number of values in the data set (method `count`), the minimum (`min`), maximum (`max`), mean (`mean`), median (`med`), sum (`sum`), variance (`var`), and the 25% and 75% quantiles (`q25`, `q75`). The quantiles (`med, q25, q75`) are actually estimates computed with the P2 algorithm[10] which makes it unnessary to store the complete dataset. Finally, there are methods to `reset` or `delete` statistic objects.

# 3 EARL Script Examples

This section describes two EARL script examples. Each of them is generic in the sense that it can be used with any message passing trace supported by EARL. Although simple (all are around 20 lines of code) they perform quite complex calculations. The simplicity comes from the abstractions defined in the EARL event trace model and the high-level nature of the TCL scripting language.

## 3.1 Example 1: Compute Wasted Time of MPI_Recv

The first example demonstrates the capabilities of EARL for solving nonstandard problems, especially recognizing complex events patterns. Consider the following: For a set of event traces from a parallel MPI program, determine the time which is wasted when a MPI_Recv is posted before the corresponding MPI_Send was executed (see Figure 2).

Here is the complete EARL script code[1]:

```
 1:   #!/usr/local/bin/earl
 2:   foreach arg $argv {
 3:     set t [earl open $arg]
 4:     set sum_wasted 0
 5:     while {[$t setnext curr] != -1} {
 6:       if {$curr(type) == "recv"} {
 7:         $t set recv_start $curr(enterptr) -fetchonly
 8:         if {$recv_start(region) != "MPI_Recv"} continue
 9:         $t set send $curr(sendptr) -fetchonly
10:         $t set send_start $send(enterptr) -fetchonly
11:         if {$send_start(region) != "MPI_Send"} continue
12:         set wasted [expr $send_start(time)-$recv_start(time)]
13:         if {$wasted>0} {
14:           set sum_wasted [expr $sum_wasted+$wasted]
15:         }
16:       }
17:     }
18:     puts "[$t info filename]:  $sum_wasted seconds wasted."
19:     $t close
20:   }
```

Line 1 is a special comment which tells a Unix system which command to use to execute the following script file. Line 2 loops through a set of trace files specified as command line arguments. Line 3 opens the trace file which is specified by the current command line parameter arg and stores the handle in variable t. The while in line 5 steps sequentially through the event trace setting the array curr to the next event. If we find a recv event (line 6), we fill the array

---

[1] The line numbers are not part of the source code.

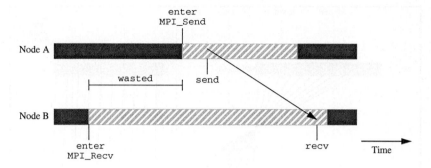

**Fig. 2.** Wasted Time in Message Passing Programs

recv_start with the enter event of the enclosing region (line 7). If the enclosing region is not MPI_Recv (the message could have been sent from another routine, e.g., MPI_Broadcast), we skip the rest of the loop and continue the search (line 8). Next, we set array send to the corresponding send event (line 9), and again check whether it originated from a MPI_Send (lines 10 and 11). We compute the difference between the begin of MPI_send and MPI_Recv (line 12) and add it to the variable sum_wasted if MPI_Recv executed before MPI_Send (line 14). Finally, we print the result (line 18) and close the trace (line 19).

### 3.2 Example 2: Passing Messages Out of Order

The second example demonstrates how EARL can be used to find programming errors in message passing programs. The example is taken from the Grindstone test suite for parallel performance tools[12] and highlights the problem of passing messages "out-of-order". This problem could arise if one process is expecting messages in a certain order, but another process is sending messages which are not in the expected order. In Figure 3, an extreme example is shown: in the first part of the program, Node 1 is processing incoming messages in the opposite order they were sent from Node 0. Processing them in the order they were sent would not only speed-up the program but also requires much less buffer space for storing unprocessed messages.

The EARL code for this example is trivial:

```
 1:   #!/usr/local/bin/earl
 2:   set t [earl open [lindex $argv 0]]
 3:   while {[$t setnext curr] != -1} {
 4:     if {$curr(type) == "recv"} {
 5:       foreach send [$t queue $curr(node) $curr(src)] {
 6:         if {$send < $curr(sendptr)} {
 7:           puts "Received message in wrong order:"
 8:           puts "  on node $curr(node) at $curr(time)"
 9:           puts "  call stack: [$t stack $curr(node) -sym]"
10:           break
```

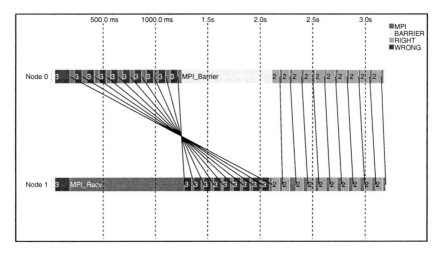

**Fig. 3.** Passing Messages Out of Order

```
11:            }
12:        }
13:    }
14: }
15: $t close
```

We open the trace file specified as first command line parameter (line 2) and sequentially loop through the events of the trace (line 3). If we find a `recv` event (line 4), we check all messages still in the message queue sent to the current node from the same source node as the current message (line 5), whether the corresponding `send` event happened before the `send` event of the current message (line 6). As we only have to determine the order of these events in time and EARL provides references to other events in the form of event numbers, the necessary comparison can be done by comparing the references since the event numbers are assigned in chronological order. If we find such an outstanding message, an error message is printed (line 7 to 9).

## 4 Related Work

EARL is certainly not the first programmable tool for event trace analysis:

- EDL[1] was one of the first trace analysis tool which was programmable. EDL allows to define custom hierarchies of events based on regular event expressions. The expressions were translated into an automaton which tried to locate the defined events in an event trace. While working well for sequential programs, parallel programs required the use of a special interleaving operator. The use of this operator results in huge automatons so the tool can only be used with small short parallel programs or simple parallel patterns only.

- SIMPLE[5] is an environment for event trace analysis. It includes a large set of tools for specific tasks, each of which defines its own command language for adapting it to specific trace formats or application areas. Although very powerful, this makes the usage of SIMPLE very complex. Also, it is not possible to combine the tasks of the different tools, so that in the worst case, each task requires the processing of the whole event trace. EARL on the other side, allows to combine different scripts, so that an event trace must be read only once.
- Pablo[6] is a very powerful, programmable, graphical event trace analysis tool. Pablo can be programmed by arranging predefined modules for trace input, event record processing, and visualization in a configuration window and connecting them. The modules are highly configurable. It is simple to use as long as the desired analysis matches the intended use of the predefined modules otherwise the graphical programming can be cumbersome or difficult. It cannot easily be extended because the implementation of new (user-defined) modules is quite complex.

Another solution to the shortcomings listed in the motivation is to use no tracing at all. Paradyn[8] is a tool for on-line performance analysis and optimization of parallel programs. It automatically tries to locate performance bottlenecks. Measurement overhead is kept low by dynamic and selective instrumentation. It is also one of the few tools which support experiment management[11].

## 5 Conclusion and Future Work

We just completed our first prototype of EARL. Early experiments show that although simple in design, EARL is a powerful and easy to use meta-tool for experts[2] to implement generic or custom-made program or application domain specific event trace analysis tools. Because of its programmability and flexibility, EARL can be used for a wide range of event trace analysis tasks:

- calculation of performance indices and trace statistics of all kinds
- finding all locations of possible bottlenecks (which then can be analyzed with traditional graphical trace analysis tools if necessary)
- performance visualization and animation (as far as TK or other TCL graphics extensions are suitable for this task)
- experiment management where within an experiment the instrumentation of the programs and generation of traces is based on results calculated from earlier runs. This allows to implement automatic program optimization tools.
- application or domain specific versions of these tasks

In the next months, we want to implement a library of useful generic EARL scripts and subroutines which then can be used by programmers to analyze their parallel applications. We also hope to implement additional decoder modules for

---

[2] especially if they know Tcl :-)

other trace formats (e.g., PICL[7] or SDDF[6]) and to add more direct support for traces generated by programs based on other programming paradigms than message passing.

In addition, we recently started a new project to design and implement an environment for the automatic detection of standard bottlenecks in parallel or distributed applications called KOJAK (Kit for Objective Judgement and Automatic Knowledge-based detection of bottlenecks)[13]. In this project, we plan to explore different ways of representing and locating bottlenecks. Here, we plan to use EARL in order to easily implement and evaluate the different methods.

## References

1. P. Bates, Debugging Programs in a Distributed System Environment, Ph.D. Thesis, University of Massachusetts, February 1986.
2. J. Ousterhout, Tcl and the Tk Toolkit, Addison-Wesley, 1994.
3. A. Arnold, U. Detert, and W.E. Nagel, Performance Optimization of Parallel Programs: Tracing, Zooming, Understanding, in: R. Winget and K. Winget, editors, Proc. Cray User Group Meeting Spring 1995, pages 252–258, Denver, CO, 1995.
4. V. Herrarte and E. Lusk, Studying Parallel Program Behavior with Upshot, Technical Report ANL-91/15, Mathematics and Computer Science Division, Argonne National Laboratory, August 1991.
5. B. Mohr, Standardization of Event Traces Considered Harmful or Is an Implementation of Object-Independent Event Trace Monitoring and Analysis Systems Possible? in: J.J. Dongarra and B. Tourancheau, editors, Proc. CNRS-NSF Workshop on Environments and Tools For Parallel Scientific Computing, volume 6 of Advances in Parallel Computing, pages 103–124, Elsevier, September 1992.
6. Reed, D.A. and Olson, R.D. and Aydt, R.A. and Madhyasta, T.M. and Birkett, T. and Jensen, D.W. and Nazief, A.A. and Totty, B.K., Scalable Performance Environments for Parallel Systems, in: Proc. 6th Distributed Memory Computing Conference, pages 562–569, IEEE Computer Society Press, 1991.
7. G.A. Geist, M.T. Heath, B.W. Peyton, and P.H. Worley, PICL: A Portable Instrumented Communication Library, Technical Report ORNL/TM-11130, Oak Ridge National Laboratory, Tennessee, July 1990.
8. B.P. Miller, M.D. Callaghan, J.M. Cargille, J.K. Hollingsworth, R.B. Irvin, K. Kunchithapadam, K.L. Karavanic, and T. Newhall, The Paradyn Parallel Performance Measurement Tools, IEEE Computer 28(11), November 1995.
9. J. C. Yan, S. R. Sarukkai, and P. Mehra, Performance Measurement, Visualization and Modeling of Parallel and Distributed Programs using the AIMS Toolkit, Software Practice & Experience, Vol. 25, No. 4, pages 429–461, April 1995.
10. R. Jain, I. Chlamtac, The P2 Algorithm for Dynamic Calculation of Quantiles and Histograms Without Storing Observations, in: Communcations of the ACM, Vol. 28, No. 10, Oct 1985.
11. K.L. Karavanic, B.P. Miller, Experiment Management Support for Performance Tuning, in: Proc. Supercomputing'97, San Jose, Nov 1997.
12. J.K. Hollingsworth, M. Steele, Grindstone: A Test Suite for Parallel Performance Tools, Computer Science Technical Report CS-TR-3703, Univ. of Maryland, 1996.
13. M. Gerndt, B. Mohr, F.Wolf, M. Pantano, Performance Analysis on CRAY T3E, Euromicro Workshop on Parallel and Distributed Processing (PDP '99), IEEE Computer Society, pages 241–248, 1999.

# XSIL: Extensible Scientific Interchange Language

Kent Blackburn[1], Albert Lazzarini[1]
Tom Prince[1,2], Roy Williams[2]

[1]LIGO Laboratory, California Institute of Technology 18-34, Pasadena, CA 91125, USA
{kent,lazz}@ligo.caltech.edu

[2]Center for Advanced Computing Research, Caltech 158-79, Pasadena, CA 91125, USA
{prince, roy}@cacr.caltech.edu

**Abstract.** We motivate and define the XSIL language as a flexible, hierarchical, extensible transport language for scientific data objects. The entire object may be represented in the file, or there may be metadata in the XSIL file, with a powerful, fault-tolerant linking mechanism to external data. The language is based on XML, and is designed not only for parsing and processing by machines, but also for presentation to humans through web browsers and web-database technology. There is a natural mapping between the elements of the XSIL language and the object model into which they are translated by the parser. As well as common objects (Parameter, Array, Time, Table), we have extended XSIL to include the IGWDFrame, used by gravitational-wave observatories.

## 1 Introduction

The Extensible Scientific Interchange Language (XSIL) is designed to represent collections of common scientific data objects. There are constructors for objects such as Parameters, Arrays, Tables, and other types, as well as support for binary objects, both MIME-typed and untyped. There is also a container object that may contain these other types as well as other containers, so that an XSIL file can be hierarchical. For each object, the entire object with all its data may be represented in the XSIL file. Alternatively the XSIL file may only contain metadata, with the bulk binary data external: there is a flexible Stream object that may contain URL-type links to external data.

XSIL is based on the XML language[1], an industry standard for which a large amount of software is available in the form of editors and parsers, as well as the familiar Microsoft and Netscape browsers. The aim has been to keep the language simple, intuitive, and easy to create and use, using a text editor, an XML editor, or from a program. The XSIL file can be used by either a human or a computer: it can be displayed, edited, summarized, sorted, or printed for human consumption; or it can be parsed as machine input with a number of methods and tools, as outlined below.

XSIL is a way to represent collections of scientific data objects—small objects with data explicitly contained in the file, and large objects represented by the salient metadata and references to binary files elsewhere. We intend the XSIL format to be used:

- As a flexible and general transport format between disparate applications in a distributed archiving and computing system; a text-based object serialization that can be handled by common tools, or
- As documentation mechanism for collections of data resulting from experiments or simulations; with all the parameters, structure, filenames and other information needed to keep a complete scientific record.
- As an "ultra-light" data format: a user can, if he wants, simply read the markup, then delete all except the actual data, or all except for the filenames where the data may be found.

We should note that an analogous format[2], WDDX, is under development as part of Allaire's Cold Fusion product line.

## 1.1 Applications

The LIGO project[3] (Laser Interferometric Gravitational wave Observatory) is a large, federally funded physics experiment that will produce several megabytes per second, 24 hours per day, 7 days per week. This data will be processed, looking for matches with astrophysically significant events, for example coalescence of neutron stars and black holes. The data will also be processed and distributed in other ways, and will be supplemented by instrument status data, candidate events from the pattern matching, and other data. While a format has been fixed for the raw data, in collaboration with the French/Italian VIRGO observatory[4], there is a need for a more flexible, more generic format for many of the other datasets, which motivated the design of XSIL.

Other projects at Caltech and elsewhere which may benefit from XSIL include

- Digital Puglia Synthetic Aperture Radar Atlas[5], an archiving and processing facility for knowledge-discovery in remote-sensing databases,
- Digital Sky, a prototype confederation of astronomical surveys,
- Center for Simulation of the Dynamic Response of Materials[6], a multidisciplinary consortium at Caltech for simulations at multiple scales.
- Interferometric SAR Library, a facility to improve the usability of this promising technology.

## 2 Presentation and Content

The syntax of XSIL is based on XML[1] (eXtensible Markup Language), now an industry standard for representing structured textual documents. XML combines the popularity of HTML in the wide Internet community with the battle-hardened power of SGML in the library community. Every XSIL file is an XML file; some references to XML are at the end of this document. Here is a small example of an XSIL file:

```
<?xml version="1.0"?>
<!DOCTYPE XSIL SYSTEM "XSIL.dtd">
<XSIL>
    <Comment>Five Measurements of voltage</Comment>
    <Param Name="Gain" Unit="milliVolt">1.453</Param>
```

```
    <Array>
        <Dim>5</Dim>
        <Stream><Metalink Format="Text" Delimiter=" \n"/>
            1.28374 1.23453 1.94847 2.148474 2.39484
        </Stream>
    </Array>
</XSIL>
```

The first line is like a "magic number" which must be the first line of any XML file. The second line says that this XML file is of a particular kind, an XSIL file. The <Comment> element is supposed to contain text that appears in presentation of XSIL, but does not play a part when the computer parses the file. This is followed by a <Param> object and and <Array> object. The <Param> object is an entry in a list of keyword-value pairs — there may also be units and comments associated with the <Param>. The <Array> has one dimension, with five elements, as indicated by the <Dim> element, and the array is then expressed explicitly in ASCII text through the <Stream> tag.

## 2.1 Presentation

One of the advantages of basing the XSIL format on a standard language such as XML, is that standard desktop tools can be used to view and edit the file. Figure 1 shows how

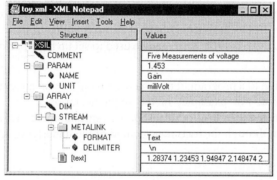

**Fig. 1**: The file above rendered by an XML-editor.

this file looks when viewed with an XML editor, in this case XML Notepad from Microsoft. There can also be customized presentation of XML and XSIL files in the browser through style sheets. Large numbers of XML tools are coming to market, for browsing, editing, sorting, and converting XML files. Some examples are shown for a more sophisticated XSIL example later in this document.

## 2.2 Parsing XSIL

In addition to providing a document that can be browsed by a human, an XSIL file is designed to allow computers to "understand" its content. By this, we mean that the structure defined by the XSIL file is mapped into objects on the program side—that is, an API. The current collection of XML parsing implementations are based either on:

- Event based parsing: The user associates handlers with tag names, then hands control to the parser.

- Document Object Model[7,8]: The parser returns a document object which the user can interrogate, a fully-formed tree with the tags and text from the XML attached.

In both cases, access to the attributes of an element is generally through an associative lookup: given an attribute name, there may be a corresponding value, but it may be that the list of attribute names cannot be retreived.

## 3 The XSIL Language

### 3.1 Objects and Streams

In the XSIL language, there are *objects* and *streams*. An object defined in the XSIL API implements the XSILObject interface, which has the method readStream(), as well as other methods to get the name and type of an object. While XSIL itself is designed for flexibility rather than speed, it is expected that streams may be implemented with an opposite sensibility: as powerful, high-bandwidth, possibly parallel, data streams. A stream is defined in the XSIL file with a <Stream> element, which may contain data explicitly, or may be a link to data stored elsewhere through a URL-like syntax.

Streams provide a separation between control and data; XSIL is the control and its Stream carries the data. Just as the anchor chain of a ship is pulled to shore with a small rope, so the flexible, low-bandwidth XSIL can be used to set up high-performance data channels.

When an object is read in to a program, it may have an input stream attached to it, from which it may read data. For example, in the implementation of the Array class, dimensions of the Array are used to allocate an appropriate amount of memory, then the data can then be read in through the readStream method of the underlying XSILObject.

### 3.2 XML

XML is much more than "glorified HTML", but rather it is a "language for creating languages". It is a hierarchical structure of elements that may contain other elements. An *element* generally consists of a *start tag*, a *body*, and an *end tag*, for example: <Fruit>Banana</Fruit>, where start and end tags are distinguished by the presence of a slash. An element may be *empty*, meaning that there is only a single tag, with no body, for example <EmptyElement/>; note the position of the slash. Elements may contain *attributes*, for example: <Fruit color="yellow">. We should point out that XML is case-sensitive, so that <Apple>, <apple> and <APPLE> are all different tags.

Finally we distinguish *presentation* and *parsing:* When an XML document is presented, the element structure is used to generate formatting information so that a human can visualize its content—conversion to HTML, TeX, VRML, etc. When it is parsed, a computer reads the file and the elements are converted to objects that are returned from the API.

## 3.3 The Container Object

In the XSIL language, there is a generic <XSIL> element which can contain other elements, including other <XSIL> elements, thus inducing a hierarchy. Each of these may have a Name attribute, to provide hierarchical naming that is visible from the API. Furthermore, the XSIL file must be enclosed in <XSIL>...</XSIL> tags, so that when the file is parsed, it is always a single XSIL object that is passed back. Here is an XSIL fragment that consists of a container hierarchy with an array at the second level and a parameter at the third level:

```
<XSIL Name="Fruit">
    <XSIL Name="YellowFruit">
        <Array><Dim>7</Dim></Array>
        <XSIL Name="Banana">
            <Param Name="Inductance">1.34</Param>
        </XSIL>
    </XSIL>
</XSIL>
```

## 3.4 Object Representation in XSIL

In general, we would like to attach data and metadata to the branches and leaves of the tree of containers, presumably in the form of objects. We may point out here that the terms "object" and "element" have similar meanings here: except that "object" implies a program perspective (the API), whereas "element" implies the document perspective (XML). Thus in this paper these points of view may be somewhat loosely interchanged.

The basic data element of XSIL is simply <Object>, corresponding to the base-class XSILObject in the API. All XSIL elements may contain the following elements:

- <Comment>: this text is not parsed, it is presumably natural language.
- <Param>: to define associations between names and values that are specific to the containing object, and that are not accessible to other objects.
- <Stream>: a definition of the input data stream that the object may draw upon from the API.

Also, most XSIL objects can have certain attributes:

- Name: A string representing the name of the object; defaults to the null string.
- Type: The type of the object or of the relevant primitive type: defaults to "double"
- Unit: A string representing the physical units associated with a number or parameter, for example "Hz" or "km"; defaults to the null string.

In addition to these common properties, perhaps the most important aspect of the generic object is that it may have access to a Stream object, if one has been defined. A generic object might be used to reference a binary file, with the assumption that the meaning is in the metadata or in natural-language text (such behavior is condoned, not encouraged). The generic object may refer to a MIME-typed object, in which case it is expected that the information about interpreting the stream is contained in the MIME-type, which is available at the beginning of the stream.

For more specific objects, we expect the tags of the XSIL definition to provide enough metadata to the implementation that it knows how to read the associated stream.

### 3.5 Some XSIL Objects

XSIL is an extensible language, meaning that users can implement their own objects or subclass the existing objects. There are mechanisms in XSIL for this through addition of extra handlers in the parsing API, through addition of extra sections in the DTD for syntax checking (see below), and though addition of extra parts to the style language for presentation. In this list are some common scientific objects that many applications can use, such as Parameter, Time, Array, Table. We will extend XSIL with a format that may be less popular: expressing the metadata from the IGWD-Frame file, which is a standard file format for recording data from Gravitational-Wave detectors.

#### 3.5.1 Parameter

A parameter in XSIL is an association between a name and a value, perhaps with additional attributes such as Unit and Type. For example:

<Param Name="Fruit_Mass" Unit="kg">0.387</Param>

As may be obvious, the meaning here is "Fruit_mass = 0.387 kg", which is the kind of thing usually found in "parameter files" or "header files" in scientific computing. At the API level, there is a dictionary of these parameters available, perhaps with unit conversion, allowing an easy lookup of critical parameters.

All XSIL objects may contain Parameters, as well as Comments. Thus the Parameter object is different from other objects: while a container may contain a Table element, which in turn may contain Parameters, but the Table may not contain other Tables.

#### 3.5.2 Time

In the LIGO experiment, as with many other experiments in physical science, it is critical that timing information be not only accurate, but also easy to understand. The <Time> element in XSIL can represent either "natural" time (ISO-8601 standard, YYYY-MM-DD HH:MM:SS.mmmuuunnn), or GPS time, or "Unix time" (seconds since 1/1/1970). The different formats are differentiated by the Type attribute in the tag:

<Time Type="ISO-8601">1998-11-08 17:40:00.032</Time>

#### 3.5.3 Array

An array is a collection of numbers (or other primitive type) referenced by subscripts, which is a list of integers whose maximum values are given by the list of Dimensions of the Array. This definition is very close conceptually to a Fortran or C array, with the Type attribute of the <Array> tag specifying which primitive type is contained in the array (float, int, etc.). The XSIL element specifies the dimensions of the array, but it does not specify the subscript ranges. For a dimension of 5, a Fortran binding of the API would label subscripts from 1 to 5, but a C binding would have subscripts from 0 to 4.

As with other XSIL objects, the Array tag may have Name and Type attributes, and the

Array element may contain Comment, Parameter, and Stream elements. The only element specific to this class is <Dim>, for example:

```
<Array Type="int">
    <Dim>5</Dim>
    <Dim>3</Dim>
</Array>
```

which specifies a 5x3 array of integers, with the last dimension changing fastest. The presumption is that 15 integers may be read from the Stream associated with this Array.

### 3.5.4   Table

A table is an unordered set of *records*, each of the same format, where a record is an ordered list of values. The contents of a record are defined by column headings, each of which may have a unit and a type. This definition of a table should be thought of as similar to the table object that is found in a relational database; we should point out that this is *not* the complex and exotic typographical beast of TeX or HTML.

The only tag specific to the Table object is <Column>, which specifies the name, type, and possibly units associated with one of the columns of the table. It can be thought of as the heading of a column in a table. In Figure 2 is shown a small table definition, together with a presentation of the table in HTML.

```
<Table>
    <Column Name="Fruit"/>
    <Column Name="Color"/>
    <Column Name="Mass" Unit="kg"/>
    <Stream>
        <Metalink Format="Text" Delimiter=",\n"/>
        Banana,Yellow,0.43
        Cherry,Red,0.01
    </Stream>
</Table>
```

**Fig. 2**: An XSIL fragment expressing a Table of Fruit. To the right is a presentation of the table using HTML viewed in a browser: because XSIL is an XML dialect, such translations can be done with great facility.

### 3.5.5   IGWD Frame

We are implementing an XML definition of the metadata found in a IGWD Frame[9], the principle data object of the LIGO[3] (USA) and VIRGO[4] (France/Italy) gravitational-wave observatories. The idea is that an XSIL file can contain the metadata for these objects, such as cataloguing information, perhaps with embedded content summaries. The Frame object in XSIL will also include a Stream that provides access to the actual binary file. The metadata includes timing information, the observatory at which the data was taken, natural language "history" records, together with the list of data channels that are recorded in the file. Each named data channel represents a stream of data from a particular instrument at the observatory, and has a status, data rate, gain, offset, etc.

When the parser sees a <IGWDFrame> element, it should pass control to a Frame reader module, which assumes the existence of an XSIL Stream object from which it can read the data. For more detail, see section 7.

## 4 Streams

When an XSIL element is parsed, the implementation of the resulting object may use the readStream() method of the object, for example an Array object would first read the type and dimensionality of the Array, then proceed to read in the data from its Stream. Thus a Stream may be thought of as a data socket, together with metadata about the link: data encoding, recommended timeout values, permission information, delimiter characters, and so on. The data may be contained in the XSIL file directly, either as readable text or as binary that is encoded to text by uuencode or base64.

### 4.1 A Stack of Streams

As the file is parsed, a stack of stream objects is created, with a new Stream pushed on the stack whenever such an element is encountered, and the stack is popped from the stcak and closed when certain end-tags are encountered. When an element is parsed into an object, it is given access to the Stream which is at the top of the stack. Many objects may then be read serially from the same Stream. For example:

```
<Stream Name="Yangtze">...</Stream>
<XSIL>
    <Array Name="Panda">....</Array>
    <Stream Name="Thames">.....</Stream>
    <Array Name="Quince">....</Array>
    <Array Name="Pumpkin">....
        <Stream Name="Hudson">....</Stream>
    </Array>
</XSIL>
<Array Name="Bamboo">....</Array>
```

Here the Array named "Panda" is read from the only open Stream, "Yangtze". A new stream, "Thames" is then pushed on to the stack, from which the Array "Quince" is read. The Array "Pumpkin" comes with its own Stream, called "Hudson", which is closed and popped as soon as the Array has been read. When the final Array in the example is reached, "Bamboo", only one Stream is still on the stack, which is "Yangtze", from which it is read.

### 4.2 Stream Element

A Stream element may contain actual data or a link to the data.

#### 4.2.1 <Data>: Explicit Data

If the data is present explicitly, it is assumed to be unparsed character data; either as delimited text (as specified in the Delimiter attribute of the Metalink tag); alternatively it may binary encoded as text in one of various ways.

### 4.2.2 &lt;Link&gt;: External Data

Another way to specify a data stream is by an external reference, which is done in XSIL with a &lt;Link&gt; element. The content of the Link element has a URL-like syntax:

protocol://hostname:port/filename

where the data may be on a local file (file), on a web- or ftp server (http, ftp), or other idiosyncratic words like tape. From the API perspective, the protocol list can be extended by suitable handlers.

### 4.2.3 Metalink

There may be also be a &lt;Metalink&gt; element associated with the &lt;Stream&gt; or with &lt;Link&gt; elements. This empty element provides metadata about the link itself: how binary has been encoded to text, the delimiter character between numbers in a text stream, the recommended timeout when accessing this data, access protocols, access restriction information, etc. etc.

### 4.2.4 Fault Tolerance

There may be multiple &lt;Link&gt; elements for a given Stream, with the assumption being that any of the links can supply the data. In this way we can provide transparent fault-tolerance for access to the data: one link may be to volatile disk (with a small timeout), and another may link to an archival tape-robot, with a long timeout. The data may also be on a tape offline if both of these fail.

## 5 An Example XSIL File

To illustrate some of the capabilities of the XSIL format, we have created a fairly comprehensive example, which is page from an "eletronic logbook" for computational scientists:

```
<?xml version="1.0"?>
<!DOCTYPE XSIL SYSTEM "XSIL.dtd">
<XSIL Name="Sample XSIL File">
    <Comment>LIGO power spectrum of 32 magnetometers</Comment>
    <Comment>Created by Jill and Sue</Comment>
    <Param Name="LIGOType">Power Spectrum</Param>
    <Time Name="StartTime" Unit="GPS">609847463.78237325</Time>
    <Param Name="FreqSamp" Unit="Hz">
        <Comment>This is the sampling frequency</Comment>
        1024
    </Param>

    <Stream>
<!-- Open a Stream: this data is at Cacr, Hanford, and on a tape in the fireproof vault. The implication is to open ONE of them -->
        <Link><Metalink Format="bigend" Timeout="600" Protocol="DCE"/>
            file://hpss.cacr.caltech.edu/magval_09_25_97.bin
        </Link>
        <Link><Metalink Format="base64"/>
            file://hanford.ligo.caltech.edu/magval_09_25_97.base64
        </Link>
        <Link>tape://347846-6/756473</Link>
```

```
        </Stream>
<!-- Here is an Array object, the data comes from the stream above -->
    <Array Name="Magread" Type="double">
        <Comment>Magnetometer Readings from Sept. 19</Comment>
        <Param Name="Gain">40.76</Param>
        <Dim>1024</Dim>
        <Dim>32</Dim>
    </Array>
<!-- This Array has its own data -->
    <Array Name="Magcal" Type="Float">
        <Comment>Magnetometer Calibration from Sept. 19</Comment>
        <Dim>32</Dim>
        <Stream><Metalink Format="Text" Delimiter=" \n"/>
1.28374 1.23453 1.94847 2.148474 2.39484 2.84746 3.10928 4.92827
5.28374 5.23453 5.94847 6.148474 6.39484 6.84746 7.10928 8.92827
9.28374 9.23453 9.94847 10.18474 10.3984 10.8446 11.1928 12.9827
13.2874 13.2453 13.9847 14.18474 14.3984 14.8446 15.1928 16.9827
        </Stream>
    </Array>
<!-- Nested containers -->
    <XSIL Name="Magnet parameters">
        <Comment>This is the magnetic value and offset</Comment>
        <Param Name="Magname">BerthaSQUID</Param>
        <Array Name="Magvalue" Type="Complex">
            <Dim>32</Dim>
            <Stream><Link><Metalink Format="bigend"/>
                file://hpss.cacr.caltech.edu/magval.bin
            </Link></Stream>
        </Array>
    </XSIL>
<!-- A table expressed in XSIL -->
    <Table>
        <Column Name="ChannelName" Type="String"/>
        <Column Name="Site"        Type="Float"   Unit="meter"/>
        <Column Name="Clock"       Type="Float"   Unit="hour"/>
        <Column Name="Description" Type="String"/>
        <Stream><Metalink Format="Text" Delimiter=",\n"/>
BOX_01_09, 2770, 3, Temperature for the apple
BOX_01_17, 3880, 6, Pressure inside the banana
BOX_01_23, 3990, 8, Pressure in the Banana Cryopump
        </Stream>
    </Table>
<!-- Finally a generic XSIL object. Here Type means MIME-Type -->
    <Object Name="ExcelChannels" Type="application/vnd.ms-excel">
        <Comment>Same table in Excel format</Comment>
        <Stream><Link>
        http://www.ligo.caltech.edu/bertha/channels.xls
        </Link></Stream>
    </Object>
</XSIL>
```

## 5.1 Attributes and Elements

In many ways attributes and elements are treated equally by presentation and parser mechanisms. The major difference seems to be that whereas attributes are simpler—only a string is returned, but elements can be arbitrarily extended. We shall provide element-based versions of the Name, Type, and Unit attributes.to allow their extendability.

## 6 DTD Definition of XSIL

An XML file may be associated with a Document Type Definition (DTD) which defines the allowed tag names in the document, and how these fit together: which elements may contain which other elements, and how many of each element there may be. For example, an <Array> element may have multiple <Dim> element to specify its dimensionality, but it cannot contain <Table> or <Object> elements.

```
<!ELEMENT XSIL((XSIL|Comment|Object|Param|Stream|Array|Table)*)>
<!ATTLIST XSIL Name CDATA "" Type CDATA "">

<!ELEMENT Comment(PCDATA)>

<!ELEMENT Object(Comment | Stream)*>
<!ATTLIST Object Name CDATA "" Type    CDATA      ""  >

<!ELEMENT Param((PCDATA | Comment)*)>
<!ATTLIST Param Name CDATA "" Unit    CDATA "" >

<!ELEMENT Stream((PCDATA | Link | Metalink)*)>
<!ELEMENT Link((PCDATA | Metalink)*)>
<!ELEMENT Metalink(#PCDATA)>
<!ATTLIST Metalink
    Format CDATA "" Timeout CDATA "" Delimiter CDATA "" >

<!ELEMENT Array((Dim | Param | Stream | Comment)*)>
<!ATTLIST Array Name CDATA "" Type CDATA "">
<!ELEMENT Dim(PCDATA)>
<!ATTLIST Dim Name CDATA "">

<!ELEMENT Table((Column | Param | Stream | Comment)*)>
<!ELEMENT Column EMPTY>
<!ATTLIST Column Name CDATA "" Type CDATA "" Unit CDATA "">
```

When a parser sees a document that is supposedly XSIL, it can first use the DTD to check that it meets the specification, that some object does not have two names, that an Array does not contain a Table. This part of the validation can be done with any validating XML parser. After this the parser can check XSIL-specific validation, such as making sure that the dimension of an array is integer.

A DTD may also be thought of as a collaboration mechanism. The members of the collaboration jointly agree on a DTD, then each side can implement against this specification—this is analogous to the collaboration mechanism with object-orented programming, where there is joint agreement on objects and their methods, followed by independent implementation.

## 7 Extending XSIL

After years of developement, object-oriented languages such as C++ and Java are very good at defining inheritance; unfortunately, XML has not come so far, and the DTD mechanism is not structured for element inheritance and subclassing. Other specifications of document type are under discussion[10,11].

As an illustration, let us consider how XSIL will be extended to include the IGWD

Frame. In the Document Object Model, a document object has been passed back through the API representing the XSIL file, and an element has been found whose name is <IGWDFrame>. Control is then passed to an XSIL-Frame object whose function it is to interface between the Frame and the XSIL. This object reads in the metadata associated with the Frame—provenance, timing, size, channel list, etc.—then returns the XSIL-Frame object to the application. The application may use the XSIL-Frame object as metadata, to present, to choose, to update the catalogue, or it may actually open the Frame file and read in all of its data. From the XSIL Stream object can be extracted a C++ stream or a C file descriptor, which can be passed to the existing Frame library for reading.

## 8 References

[1] XML resources:
http://www.xml.com

[2] Web Distributed Data Exchange (Allaire Corp.)
http://www.allaire.com/developer/wddx/

[3] LIGO (Laser Interferometric Gravitational wave Observatory)
http://www.ligo/caltech.edu/

[4] VIRGO Gravitational wave Observatory
http://www.pg.infn.it/virgo

[5] G. Aloisio, M. Cafaro, R. Williams, Digital Puglia Synthetic Aperture Radar Atlas, Proceedings of HPCN99, Amsterdam, April 12-16

[6] A Center for the Dynamic Response of Materials,
http://www.cacr.caltech.edu/ASAP/

[7] Document Object Model Specification (W3C)
http://www.w3.org/TR/WD-DOM/level-one-xml

[8] Document Object Model Resources
http://www.xml.com/xml/pub/DOM

[9] IGWD-Frame Class Library
http://www.ligo.caltech.edu/~wmajid/fcl/index.html

[10] Document Content Description for XML
http://w3c.org/TR/NOTE-dcd

[11] XML-Data, a Schema Definition Language for XML written in XML
http://www.w3.org/TR/1998/NOTE-XML-data/

# ForkLight: A Control–Synchronous Parallel Programming Language

Christoph W. Keßler and Helmut Seidl

FB IV - Informatik, Universität Trier, 54286 Trier, Germany
e-mail: {kessler,seidl}@psi.uni-trier.de

**Abstract.** ForkLight is an imperative, task-parallel programming language for massively parallel shared memory machines. It is based on ANSI C, follows the SPMD model of parallel program execution, provides a sequentially consistent shared memory, and supports dynamically nested parallelism. While no assumptions are made on uniformity of memory access time or instruction–level synchronicity of the underlying hardware, ForkLight offers a simple but powerful mechanism for coordination of parallel processes in the tradition and notation of PRAM algorithms: Beyond its asynchronous default execution mode, ForkLight offers a mode for control–synchronous execution that relates the program's block structure to parallel control flow.

We give a scheme for compiling ForkLight to C with calls to a very small set of basic shared memory access operations like atomic *fetch&add*. This yields portability across parallel architectures and exploits the local optimizations of their native C compilers. Our implementation is publically available; performance results are reported. We also discuss translation to OpenMP.

**1. INTRODUCTION.** Parallel processing offers an attractive way to increase computer performance. MIMD architectures, where program control is individual to each processor, offer high flexibility in programming, but devising, implementing and debugging parallel programs is difficult. Even more difficulties arise if the programmer has to care about explicit data distribution to achieve reasonable performance. Automatic parallelization and automatic data distribution techniques are still limited to rather regular programs. Addressing the latter problem, parallel computer manufacturers and research groups recently devised several types of massively parallel (or at least scalable) machines providing a shared memory. Some of these still require program tuning for locality to perform efficiently, e.g. Stanford DASH [18], while others use multithreading to hide the memory access latency and high–bandwidth memory networks, and thus become more or less independent of locality issues, e.g. Tera MTA [2].

So far there is only one massively parallel shared memory architecture that offers uniform memory access time (UMA), the SB-PRAM [1]. Due to a common clock all processors of the SB-PRAM work synchronously, i.e. they start (and complete) execution of an instruction simultaneously. This synchronicity makes parallel programming very easy, as it leads to deterministic parallel program execution; it is particularly well suited for the implementation of synchronous algorithms using e.g. fine–grained pipelines or data–parallel operations. Furthermore, such synchronous MIMD machines with UMA shared memory are very popular in theoretical computer science, where they are known as PRAMs (*P*arallel *R*andom *A*ccess *M*achines).

On the other hand, no massively parallel MIMD machine that is commercially available today is UMA or synchronous in this sense. Rather, the following features are common: The user sees a large amount of threads (due to scalable architecture and multithreading) and a monolithic shared memory (due to a hidden network). There is no common clock. The memory access time is non–uniform, but more or less independent of locality (due to multithreading). Program execution is asynchronous (due to the

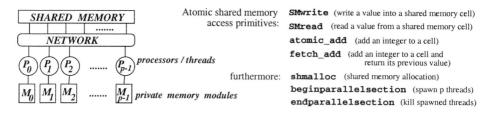

**Fig. 1. Asynchronous PRAM model with shared memory access operations.** The seven operations on the right hand side are sufficient to handle shared memory parallelism as generated by the ForkLight compiler. The processors' native load/store operations are used for accessing private memory. Other thread handling functions like inspecting the thread ID or the number of threads are implemented by the compiler's run–time library.

previous two items, and because of system features like virtual memory, caching, and I/O). Also, there is efficient hardware support for atomic *fetch&op* instructions. — A typical representative of this class of parallel machines is the Tera MTA [2].

In order to abstract from particular features and to enhance portability of parallel algorithms and programs, one often uses a *programming model* that describes the most important architecture properties. Suitable parameterization allows for straightforward estimates of run times; such estimations are the more accurate, the more the particular parallel hardware fits the model used. In our case, the programming model is the Asynchronous PRAM introduced in the parallel theory community in 1989 [13, 9, 10]. An *Asynchronous PRAM* (see Fig. 1) is a MIMD parallel computer with a sequentially consistent shared memory. Each processor runs with its own clock. No assumptions are made on uniformity of shared memory access times. Thus, much more than for a true PRAM, the programmer must explicitly take care of avoiding race conditions (nondeterminism) when accessing shared memory locations or shared resources (screen, shared files) concurrently. We add to this model some atomic *fetch&op* instructions like fetch_add and atomic update instructions like atomic_add, which are required for basic synchronization mechanisms. This is not an inadmissible modification of the original Asynchronous PRAM model, as software realizations for these primitives do exist (but incur significant overhead [23]). In short, this programming model is closely related to the popular PRAM and BSP models but offers, in our view, a good compromise, as it is closer to real parallel machines than PRAMs and easier to program than BSP.

The PRAM programming model, as supported e.g. by **Fork95** [15], *ll* [19], and *Modula-2\** [21], offers deterministic write conflict resolution and operator–level synchronous execution: there are no race conditions at all, data dependencies need not be protected by locks or barriers. Unfortunately, this ideal parallel programming model leads to very inefficient code when compiled to asynchronous machines, in particular if the compiler fails to analyze data dependencies in irregular computations and resorts to worst-case assumptions. Hence, programming languages especially designed for "true" PRAMs, cannot directly be used for Asynchronous PRAMs.

In this paper we propose **ForkLight**, a task–parallel programming language for the Asynchronous PRAM model. It retains a considerable part of the programming comfort known from **Fork95** but drops the requirement for exactly synchronous execution. Rather, synchronicity is relaxed to the basic block level (*control–synchronicity*).

The *fork-join model* of parallel execution, adopted e.g. by ParC [3], Tera-C [6], or Cilk [17], corresponds to a tree of processes. Program execution starts with a sequential

process, and any process can spawn arbitrary numbers of child processes at any time. While the fork-join model directly supports nested parallelism, the necessary scheduling of processes requires substantial support by a runtime system and incurs overhead.
— In contrast, in the *SPMD model* of parallel execution, all $p$ available processors (or threads) are running just from the beginning of main execution; they are to be distributed explicitly over the available tasks, e.g. parallel loop iterations. Given fixed sized machines, SPMD seems better suited to exploit the processor resources economically.

ForkLight follows the SPMD model. Coordination is provided e.g. by composing the threads into groups. The threads of a group can allocate group–local shared variables and objects. In order to adapt to finer levels of nested parallelism, a group can be (recursively) subdivided into subgroups. In this way, ForkLight supports a parallel recursive divide–and–conquer style as suggested e.g. in [8], as well as data parallelism, task farming, and other parallel algorithmic paradigms. In contrast, most other languages adopting the SPMD model, like Denelcor HEP Fortran [14], EPEX/Fortran [12], PCP [4], Split-C [11], AC [7], support only one level of parallelism and one global name space; only PCP has a hierarchical group concept similar to that of ForkLight. Moreover, in ForkLight, control–synchronous execution can locally be relaxed towards totally asynchronous computation as desired by the programmer, e.g. for efficiency reasons.

For the compilation of ForkLight discussed in Section 3, we only assume a shared memory and efficient support of atomic increment and atomic *fetch&increment* operations (see Fig. 1). These are powerful enough to serve as the basic component of simple locking/unlocking and barrier mechanisms and to enhance e.g. the management of parallel queues or self–scheduling parallel loops [22] and occur in several routines of the ForkLight standard library. A source–to–source compiler for ForkLight has been implemented based on the methods given in Section 3. It generates C source code plus calls to the routines listed in Fig. 1 that are currently implemented by calls to the shared-memory P4 routines [5], which provides platform independence. We also give an implementation scheme and optimization for OpenMP [20]. We report results on Solaris multiprocessor workstations with P4 support and on the SB-PRAM. Our implementation is available at http://www.informatik.uni-trier.de/~kessler/forklight

## 2. LANGUAGE DESCRIPTION.
ForkLight extends C by constructs for group handling, declaration and allocation of shared variables, and relaxing control–synchronicity.
**Shared and private variables.** Variables are classified either as *private* (this is the default) or as *shared*; the latter are to be declared with the storage class qualifier sh. Here "shared" always relates to the thread group (see later) that executes that variable's declaration. Private variables exist once for each thread, whereas shared variables exist only once in the shared memory subspace of the thread group that declared them.

The *total number of started threads* is accessible through the constant shared variable __P__, the *physical thread ID* through the function _PhysThreadId().
**Execution modes and regions.** There are two different program execution modes that are statically associated with source code regions: control–synchronous mode in control–synchronous regions, and asynchronous mode in asynchronous regions.

In *control–synchronous mode* ForkLight maintains the invariant that *all threads belonging to the same (active) group work on the same basic block*. Subgroup creation and implicit barriers occur only in control–synchronous mode.

In *asynchronous mode*, control–synchronicity is not enforced. The group structure is frozen; shared variables and automatic shared heap objects cannot be allocated.

There are no implicit synchronization points. Synchronization of the current group can, though, be explicitly enforced by a barrier() call or by a *barrier statement* denoted by a sequence of at least three ='s to optically emphasize it in the program code.

Functions are classified as control–synchronous (to be declared with type qualifier csync) or asynchronous (this is the default). main() is asynchronous by default. A control–synchronous function is a control–synchronous region, except for (blocks of) statements explicitly marked as asynchronous by async or as sequential by seq. An asynchronous function is an asynchronous region, except for statements explicitly marked as control–synchronous by start or join. async <stmt> causes the processors to execute <stmt> in asynchronous mode. In other words, the entire <stmt> (which may contain loops, conditions, or calls to asynchronous functions) is considered to be part of the "basic" (w.r.t. control–synchronicity) block containing this async. There is no implicit barrier at the end of <stmt>. If the programmer desires one, (s)he may use an explicit barrier. seq <stmt> causes <stmt> to be executed by exactly one thread of the current group; the others skip <stmt>. There is no implicit barrier at the end of <stmt>.

Asynchronous functions and async blocks are executed in asynchronous mode, except for start <stmt> statements that are only permitted in asynchronous mode. start switches to control–synchronous mode for its body <stmt>. It causes all __P__ threads to barrier–synchronize, form a group, and execute <stmt> simultaneously and in control–synchronous mode, with unique thread IDs $ numbered from 0 to __P__ $-1$. A generalization of start, the join statement, allows to more flexibly collect a variable amount of threads over a specified time or event interval and make them execute a control–synchronous statement [16].

In order to maintain the static classification of code, within asynchronous regions only async functions can be called. In the other way, calling an async function from a control–synchronous region is always possible and results in an implicit entering of the asynchronous mode. Shared local variables can only be declared / allocated within control–synchronous regions. In particular, asynchronous functions must not allocate shared local variables. All formal parameters must be private. There is an implicit group–wide barrier synchronization point at entry to control–synchronous functions.

**Nested parallelism and groups.** ForkLight programs are executed by *groups* of threads, rather than by individual threads. When program execution starts, there is just one group, the *root group*, that contains all available threads. Groups may be recursively subdivided. Thus, at any point of program execution, all presently existing groups form a tree–like *group hierarchy*. Only the *leaf groups* of the group hierarchy are active. Subgroups of a group can be distinguished by their *group ID*. A thread can access its current group's ID through the constant @. The subgroup ID @ can be preset by the programmer at the fork statement. join and start set @ to 0. The group–relative thread ID $, a private constant variable, is automatically computed whenever a new subgroup is created, by renumbering the subgroup's $p$ threads from 0 to $p-1$. $ and @ are automatically saved when splitting the current group, and restored when reactivating it. All threads of a group have access to a common shared address subspace. allocating them. A thread can inspect the number of threads in its current group by the function groupsize().

At entry to a control–synchronous region (i.e., a join or start body), the threads form one single thread group. However, without special handling control flow could diverge for different threads at conditional branches such as if statements, switch statements, and loops. Only in special cases it can be statically determined that all threads are going to take the same branch of control flow. Otherwise, control–synchronicity could be lost. To avoid this, **ForkLight** guarantees control–synchronicity by suitable automatic group splitting. Nevertheless, the programmer may know in some situations that such a splitting is unnecessary. For these cases, (s)he can specify this explicitly:

We consider an expression to be *stable* if it is guaranteed to evaluate to the same value on each thread of the group for all possible program executions, and *unstable* otherwise. An expression containing private variables (e.g., $) is generally assumed to be unstable. But even an expression $e$ containing only shared variables may be also unstable: Since $e$ is evaluated asynchronously by the threads, it may happen that a shared variable occurring in $e$ is modified (maybe as a side effect in $e$, or by a thread outside the current group) such that some threads of the group (the "faster" ones) use the old value of that variable while others use the newer one, which may yield different values of $e$ for different threads of the same group. Technically, the compiler defines a conservative, statically computable subset of the *stable expressions* as follows:

(1) A (shared) constant is a stable expression. (This also includes @ and shared constant pointers, e.g. arrays.)

(2) The pseudocast stable(e) is stable for any expression $e$ (see below).

(3) If expressions $e_1$, $e_2$ and $e_3$ are stable, then also the expressions $e_1 \oplus e_2$ for $\oplus \in \{+, -, *, /, \%, \&, |, \hat{}, \&\&, ||\}$, $\ominus e_1$ for $\ominus \in \{-, \sim, !\}$, $e_1[e_2]$, $e_1.field$, $e_1 \rightarrow field$, $*e_1$, and $e_1?e_2:e_3$ are stable.

(4) All other expressions are conservatively regarded as unstable.

Conditional branches with a stable condition expression do not affect control–synchronicity. Branches in control–synchronous mode with unstable conditions lead to automatic splitting of the current group into subgroups — one for each possible branch target. Control synchronicity is then only maintained within each subgroup. Where control flow reunifies again, the subgroups cease to exist, and the previous group is restored. There is no implicit barrier at this program point. (A rule of thumb: Implicit barriers are, in control–synchronous mode, generated only at branches of control flow, not at reunifications.) For an unstable two–sided if statement, for instance, two subgroups are created. The processors that evaluate their condition to true join the first, the others join the second subgroup. The branches associated with these subgroups are executed concurrently. For a loop with an unstable exit condition, one subgroup is created that contains the iterating processors.

Nevertheless, sometimes unstable branch conditions in control–synchronous mode may be tolerable without automatic group splitting, for performance tuning purposes. The pseudocast stable (<expr>) causes the compiler to treat expression <expr> as stable; the compiler then assumes that the programmer knows that possible unstability of <expr> will not be a problem in this context, for instance because (s)he knows that all processors of the group will take the same branch.

Splitting a group into subgroups can also be done explicitly. Executing

fork ( $e_1$; @=$e_2$ ) <stmt>

means the following: First, each thread of the group evaluates the stable expression $e_1$ to the number of subgroups to be created, say $g$. Then the current leaf group is deactivated and $g$ subgroups $g_0, ..., g_{g-1}$ are created. The group ID of $g_i$ is set to $i$. Evaluating expression $e_2$ (which is typically unstable), each thread determines the index $i$ of the newly created leaf group $g_i$ it will become member of. If the value of $e_2$ is outside the range $0, ..., g-1$ of subgroup IDs, the thread does not join a subgroup and skips <stmt>. The IDs $ of the threads are renumbered consecutively within each subgroup from 0 to the subgroup size minus one. Each subgroup gets its own shared memory subspace, thus shared variables and heap objects can be allocated locally to the subgroup. Now, each subgroup $g_i$ executes <stmt>. When a subgroup finishes execution, it ceases to exist, and its parent group is reactivated as the current leaf group. Unless the programmer writes an explicit barrier statement, the processors may immediately continue with the following code. Note that empty subgroups (with no threads) are possible; an empty subgroup's work is immediately finished, though.

**Pointers and heaps.** Since the private address subspaces are not embedded into the global shared memory but addressed independently, shared pointer variables must not point to private objects. As it is, in general, not possible to statically verify whether the pointee is private or shared, dereferencing a shared pointer containing a private address will lead to a run time error. Nevertheless any pointer may point to a shared object.

ForkLight supplies three kinds of heaps. First, there is the usual, private heap for each thread with the (asynchronous) functions malloc() and free() known from C. Second, there is a global, permanent shared heap with the asynchronous functions shmalloc() and shfree(). Finally, there is one automatic shared heap for each group that is intended to provide fast temporary storage blocks local to a group. Consequently, the life range of objects allocated on the automatic shared heap by the control–synchronous shalloc() function is limited by the life range of the allocating group.

Pointers to functions are also supported. In control–synchronous mode, dereferencing a pointer to a control–synchronous function is only legal if it is stable.

**Standard atomic operations.** Atomic *fetch&op* operations, also known as *multiprefix* computations when applied in a synchronous context with priority resolution of concurrent write accesses to the same memory location, are available as standard functions called fetch_add, fetch_max, fetch_and and fetch_or, to give the programmer direct access to these powerful operators. They can be used in control–synchronous as well as in asynchronous mode. Note that the order of execution for concurrent execution of several, say, fetch_add operations to the same shared memory location is not determined in **ForkLight**. For instance, computing the size $p$ of the current group and new thread ID variables myrank consecutively numbered $0,1,...,p-1$ may be done by this code, where the function-local integer variable p is shared by all threads of the current group.

```
csync void foo( void ) {
  int myrank;
  sh int p = 0;
  ===== //guarantees p is init.to 0
  myrank = fetch_add( &p, 1 );
  ===== //guarantees p is groupsize
  ...
}
```

The atomicity of these operators is very useful to access semaphores in asynchronous mode, e.g., simple locks that sequentialize access to some shared resource where necessary. In its standard library, **ForkLight** offers simple locks, fair locks, and reader–writer locks, and further atomic memory operations void atomic_op().

## 3. COMPILATION.

The ForkLight implementation uses its own shared memory management using a sufficiently large slice of shared memory. To the bottom of this shared memory slice we map the shared global initialized resp. non–initialized variables. In the remainder of this shared memory part we arrange a shared stack and an automatic shared heap (see figure). Group splitting operations also cause splitting of the remaining shared stack space, creating an own shared stack and automatic heap for each subgroup; (i.e., a "cactus" stack and heap). The shared stack pointer sps and the

**Fig. 2.** Barrier synchronization pseudocode and shared group frame

$R_{csc} \leftarrow$ csc;  $R_{next} \leftarrow R_{csc} + 1$;
**if** $(R_{next} > 2)$   $R_{next} \leftarrow 0$   // wrap–around
atomic_add( gps+$R_{next}$, 1);
atomic_add( gps+$R_{csc}$, -1);
**while** (SMread($sc[R_{csc}]) \neq 0$) ;   // wait
csc$\leftarrow R_{next}$;

|  | ↑ |
|---|---|
|  | group–local shared var's |
|  | $sc[2]$ |
|  | $sc[1]$ |
| gps→ | $sc[0]$ |

automatic shared heap pointer eps may be permanently kept in registers on each thread. A shared stack or automatic heap overflow occurs if sps crosses eps. Another shared memory slice is allocated to install the global shared heap. Initially, the thread on which the user has started the program executes the startup code, initializes the shared memory, and spawns the other threads as requested by the user. All these threads start execution of the program in asynchronous mode by calling main(). A private stack and heap are maintained in each thread's private memory by the native C compiler.

**Group frames and group–wide barrier synchronization.** For each group the compiler keeps a shared and private group frames. The *shared group frame* (see Fig. 2) is allocated on the group's shared stack. It contains the shared variables local to this group and *three synchronization cells* $sc[0]$, $sc[1]$, $sc[2]$. Each thread holds a register gps pointing to its current group's shared group frame, and a private counter csc indexing the *current synchronization cell*. When a new group is created, csc is initialized to 0, $sc[0]$ is initialized to the total number of threads in the new group, and $sc[1]$ and $sc[2]$ are initialized to 0. If no thread of the group is currently at a barrier synchronization point, the current synchronization cell $sc[\text{csc}]$ contains just the number of threads in this group.

At a group–wide barrier synchronization, each thread atomically increments the next synchronization cell by 1, then atomically decrements the current synchronization cell by 1, and waits until it sees a zero in the current synchronization cell, see Fig. 2. The algorithm guarantees that all threads have reached the barrier when a zero appears in the current synchronization cell. Only then they are allowed to proceed. At this point of time, though, the next current synchronization cell, $sc[R_{next}]$, already contains the total number of threads, i.e. is properly initialized for the following barrier synchronization. Once $sc[R_{csc}]$ is 0, all threads of the group are guaranteed to see this, as this value remains unchanged at least until after the following synchronization point. The run time is, for most shared memory systems, dominated by the group–wide atomic_add and SMread accesses to shared memory, while all other operations are local.

Each shared group frame is complemented by a *private group frame* containing the private group information, pointed to by register gpp. It contains the current values of the group ID @, the group–relative thread ID $, and the current synchronization cell index csc. @ needs not be stored on the shared group frame since it is read–only. Also, the parent group's shared stack pointer sps, group–local heap pointer eps, and group frame pointer gps are stored in the private group frame, together with the parent group's gpp, thus the parent group can easily be restored when being reactivated.

**Translation of a function call.** Asynchronous functions are just compiled as known from sequential programming, as no care has to be taken for synchronicity. A control–synchronous function with shared local variables needs to allocate a shared group frame. As these variables should be accessed only after all threads have entered the function, there is an implicit group–wide barrier at entry to a control–synchronous function.

**Translation of the fork statement.** A straightforward implementation assumes that all $k$ subgroups will exist and distributes shared stack space equally among these. For fork ( $k$; @=e ) <stmt> the following code is generated:

(1) $R_k \leftarrow$ eval($k$); $R_@ \leftarrow$ eval($e$); slice $\leftarrow \lfloor$(eps-sps)$/R_k\rfloor$;
(2) **if** $(0 \leq R_@ < R_k)$ { sc $\leftarrow$ sps+$R_@$ * slice; SMwrite( sc, 0 ); }
(3) barrier local to the (parent) group       // necessary to guarantee a zero in sc[0]
(4) **if** $(0 \leq R_@ < R_k)$ { $R_\$ \leftarrow$ fetch_add(sc,1); allocate a private group frame pfr
    and store there gps, eps, sps, gpp; init. new csc field to 0, @ field to $R_@$, $ field to $R_\$$
    **if** $(R_\$ = 0)$ { sc[1]$\leftarrow$ 0; sc[2]$\leftarrow$ 0; }                              }
(5) barrier local to the (parent) group       // guarantees final subgroup sizes in sc[0]
(6) **if** $(0 \leq R_@ < R_k)$              // enter subgroup $R_@$              otherwise skip
    { gps$\leftarrow$ sc; gpp$\leftarrow$ pfr; sps$\leftarrow$ gps+3+#sh.locals; eps$\leftarrow$ gps+slice; } **else** goto (9)
(7) code for <stmt>
(8) atomic_add( gps+csc, -1);       // cancel membership in the subgroup
    leave the subgroup by restoring gps, sps, eps, gpp from the private group frame
(9) (next statement)

The overhead of the above implementation mainly consists of the parallel time for two group–wide barriers, one subgroup–wide concurrent SMwrite, and one subgroup–wide fetch_add operation. Also, there are two exclusive SMwrite accesses to shared memory locations. The few private operations can be ignored, since their cost is usually much lower than shared memory accesses. There are some *optimizations:* (1) group splitting and barriers can be skipped if the current group consists of only one thread. (2) Some of the subgroups may be empty. In that case, space fragmentation can be reduced by splitting the parent group's stack space in only that many parts as there are different values of $R_@$. (3) Not all group splitting operations require the full generality of the fork construct. Splitting into equally (or weighted) sized subgroups, as in PCP [4], can be implemented with only one barrier and without the fetch_add call, as the new subgroup sizes and ranks can be computed directly from locally available information.

**Accessing local shared variables.** Subgroup creation within the same function can be statically nested. The compiler determines the group nesting depth of the declaration of each variable x (see example on the right), call it $gd(x)$, as well as that of each use of x, call it $gu(x)$. For each use of x, the compiler computes the difference $d = gu(x) - gd(x)$ and, where $d > 0$, inserts code to follow the chain of gps pointers $d$ times upwards the group tree, in order to arrive at the group frame containing x. Because $d$ is typically quite small, this loop is completely unrolled. Note that all these read accesses but the last one are private memory accesses.

```
csync void foo()
{fork(...) {
   sh int x=@; //decl
   ...
   fork(...)
   ... = x;   //use
}}
```

**Translation to OpenMP.** OpenMP [20] is a shared–memory parallel application programming interface for Fortran and C/C++, consisting of a set of compiler directives and several run time library functions. Although OpenMP is, in principle, task-parallel, it is more tailored towards dataparallel programming. Memory consistency must be enforced by the programmer by explicit flush() operations. OpenMP follows the fork-join execution model. Nevertheless, nested parallelism is only an optional feature of the implementations. Varying defaults for declarations of shared and private variables make OpenMP programs quite hard to read. Currently some OpenMP implementations are available for Fortran but not yet for C/C++.

A transcription of the existing **ForkLight** back-end to OpenMP is straightforward: begin- and endparallelsection() correspond to a omp parallel directive at the top level of the program. The shared stack and heap are simulated by two large arrays declared `shared volatile` at this directive, `SMread` and `SMwrite` become accesses of these arrays. Explicit `flushing` after `SMwrites` is not necessary for `volatile` shared variables. `omp barrier` and other synchronization primitives of OpenMP cannot be used for **ForkLight** because they are not applicable to nested SPMD computations; thus we will use our own implementation for the synchronization routines. The `atomic` directive of OpenMP is applicable to increment and decrement operators, but `fetch_add` has to be expressed using the sequentializing `critical` directive. In order to increase scalability, we propose *hashing of critical section addresses*: The `omp critical` directive optionally takes a compile–time constant name (string) as a parameter. `critical` sections with different names may be executed concurrently, while entry to all `critical` sections with the same name is guarded by the same mutual exclusion lock. For our implementation of atomic memory operations like `fetch_add` we use a static finite set of locks and distribute the shared memory accesses among these by a hash function.

**Performance Results.** We have implemented the compiler for the two parallel platforms that are currently accessible to us and that are still supported by P4: multiprocessor Solaris workstations and the SB-PRAM. These two completely different types of architecture represent two quite extremal points in the spectrum of shared–memory architectures regarding execution of P4 / **ForkLight** programs.

On a loaded four–processor Solaris 2.5 workstation, where atomic memory access is sequentialized, we observed good speedup for well–parallelizable problems like pi–calculation or matrix product but only modest or no speedup for problems that require frequent synchronization.

| problem | size | time [s] with # threads | | | |
|---|---|---|---|---|---|
| | | 1 | 2 | 3 | 4 |
| Pi-calculation | 5000000 | 22.61 | 14.55 | 11.37 | 7.86 |
| matrix product | 200 x 200 | 41.08 | 24.16 | 19.22 | 14.49 |
| par. mergesort | 48000 | 4.39 | 3.58 | n.a. | 2.33 |
| par. quicksort | 120000 | 14.50 | 9.83 | n.a. | 6.35 |

On the SB-PRAM we exploited its native `fetch_add` and `atomic_add` operators which do, differently from standard P4, not lead to sequentialization. For the SB-PRAM prototype at Saarbrücken we obtained these results.

| problem | size | time [ms] with #threads | | | | | | |
|---|---|---|---|---|---|---|---|---|
| | | 1 | 2 | 4 | 8 | 16 | 32 | 64 |
| par. mergesort | 1000 | 573 | 373 | 232 | 142 | 88 | 57 | 39 |
| par. mergesort | 10000 | 2892 | 4693 | 3865 | 1571 | 896 | 509 | 290 |
| par. quicksort | 1000 | 1519 | 807 | 454 | 259 | 172 | 145 | 130 |

We observe: (1) Efficient support for non-sequentializing atomic `fetch_add` and `atomic_add`, as in SB-PRAM or Tera MTA, is essential when running **ForkLight** programs with large numbers of threads. Executables relying only on pure P4 suffer from serialization and locking/unlocking overhead and are thus not scalable to large numbers of threads. — (2) On a non–dedicated, loaded multiuser / multitasking machine like our Solaris multiprocessor workstation, parallel speedup suffers from poor load balancing due to stochastic delaying effects: the processors are unsymmetrically delayed by other users' processes, and at barriers these delays accumulate. — (3) Even when running several P4 processes on a single processor, performance could be much better for $p > 1$ if the **ForkLight** run time system had complete control over context switching for its own threads. Otherwise, much time is lost spinning on barriers to fill the time slice assigned by an OS scheduler that is unaware of the parallel application. This is an obvious weakness of P4. — (4) Explicit load balancing in an SPMD application may be problematic in particular for small machine sizes (quicksort), or when the hardware scheduler does not follow the intentions of the user, e.g. when the scheduler maps several P4 threads to a processor where only one thread was intended for. — (5) Where these requirements are met, our prototype implementation achieves acceptable speedups and performance scales quite well even for rather small problem sizes.

**4. CONCLUSION.** From a software engineering point of view, control–synchronous execution is an important guidance to the programmer since it transparently relates the program block structure to the parallel control flow. By the support of nested parallelism in **ForkLight**, programming is as comfortable as in fork-join languages.

## References

1. F. Abolhassan, R. Drefenstedt, J. Keller, W. J. Paul, D. Scheerer. On the physical design of PRAMs. *Computer Journal*, 36(8):756–762, Dec. 1993.
2. R. Alverson, D. Callahan, D. Cummings, B. Koblenz, A. Porterfield, B. Smith. The Tera Computer System. *Proc. 4th ACM Int. Conf. on Supercomputing*, pp. 1–6, 1990.
3. Y. Ben-Asher, D. Feitelson, L. Rudolph. ParC — An Extension of C for Shared Memory Parallel Processing. *Software – Practice and Experience*, 26(5):581–612, May 1996.
4. E. D. Brooks III, B. C. Gorda, K. H. Warren. The Parallel C Preprocessor. *Scientific Programming*, 1(1):79–89, 1992.
5. R. Butler, E. Lusk. Monitors, Messages, and Clusters: The P4 Parallel Programming System. *Parallel Computing*, 20(4):547–564, April 1994.
6. D. Callahan, B. Smith. A Future–based Parallel Language for a General–Purpose Highly-parallel Computer. Tera Computer Company, http://www.tera.com, 1990.
7. W. W. Carlson, J. M. Draper. Distributed Data Access in AC. In *Proc. ACM SIGPLAN Symp. on Principles and Practices of Parallel Programming*, pages 39–47. ACM Press, 1995.
8. M. I. Cole. *Algorithmic Skeletons: Structured Management of Parallel Computation*. Pitman and MIT Press, 1989.
9. R. Cole, O. Zajicek. The APRAM: Incorporating Asynchrony into the PRAM model. *Proc. 1st Ann. ACM Symp. on Par. Algorithms and Architectures*, pp. 169–178, 1989.
10. R. Cole, O. Zajicek. The Expected Advantage of Asynchrony. *JCSS* 51:286–300, 1995.
11. D. E. Culler, A. Dusseau, S. C. Goldstein, A. Krishnamurthy, S. Lumetta, T. von Eicken, K. Yelick. Parallel Programming in Split-C. *Proc. Supercomputing'93*, Nov. 1993.
12. F. Darema, D. George, V. Norton, G. Pfister. A single-program-multiple-data computational model for EPEX/FORTRAN. *Parallel Computing*, 7:11–24, 1988.
13. P. B. Gibbons. A More Practical PRAM model. In *Proc. 1st Annual ACM Symposium on Parallel Algorithms and Architectures*, pages 158–168, 1989.
14. H. F. Jordan. Structuring parallel algorithms in an MIMD, shared memory environment. *Parallel Computing*, 3:93–110, 1986.
15. C. W. Keßler, H. Seidl. The Fork95 Parallel Programming Language: Design, Implementation, Application. *Int. Journal of Parallel Programming*, 25(1):17–50, Feb. 1997.
16. C. W. Keßler, H. Seidl. ForkLight: A Control–Synchronous Parallel Programming Language. Tech. Report 98-13, Univ. Trier, FB IV–Informatik, 54286 Trier, Germany, 1998.
17. C. E. Leiserson. Programming Irregular Parallel Appplications in Cilk. *Proc. IRREGULAR'97*, pp. 61–71. Springer LNCS 1253, 1997.
18. D. Lenoski, J. Laudon, K. Gharachorloo, W.-D. Weber, A. Gupta, J. Hennesy, M. Horowitz, M. S. Lam. The Stanford DASH multiprocessor. *IEEE Computer*, 25(3):63–79, 1992.
19. C. León, F. Sande, C. Rodríguez, F. García. A PRAM Oriented Language. *EUROMICRO Wksh. on Par. and Distr. Processing*, pp. 182–191. IEEE CS Press, 1995.
20. OpenMP ARB. OpenMP White Paper, http://www.openmp.org/, 1997.
21. M. Philippsen, W. F. Tichy. Compiling for Massively Parallel Machines. In *Code Generation – Concepts, Tools, Techniques*, pp. 92–111. Springer Workshops in Computing, 1991.
22. J. M. Wilson. *Operating System Data Structures for Shared-Memory MIMD Machines with Fetch–and–Add*. PhD thesis, New York University, 1988.
23. X. Zhang, Y. Yan, R. Castaneda. Evaluating and Designing Software Mutual Exclusion Algorithms on Shared-Memory Multiprocessors. *IEEE Par. & Distr. Techn.*, 4:25–42, 1996.

# HPF Parallelization of a Molecular Dynamics Code: Strategies and Performances

B.Di Martino[1], M.Celino[2], V.Rosato[2]

[1] Dipartimento di Ingegneria dell'Informazione
Second University of Naples, Napoli, Italy ***
[2] ENEA - HPCN Project, C.R.Casaccia
CP 2400, 00100 Roma AD, Italy [†]

**Abstract.** The High Performance Fortran (HPF) environment has been used to efficiently parallelize a Molecular Dynamics (MD) code. The MD code is a Tight-Binding code, properly specialized for semiconductor materials and characterized by inhomogeneous data distribution. Furthermore, the electronic properties are taken into account during the atomic dynamics: for this reason a large sparse matrix is built and diagonalized to compute the whole body of its eigenvalues and eigenvectors at each time step. For the diagonalization task, a parallel mathematical routine implemented. The strategy of parallelization, the integration of the mathematical routine in the HPF code are described and discussed. All the benchmark are performed on IBM SP architecture.

## 1 Introduction

Computer simulations at the atomic scale are among the most powerful theoretical tools used to investigate the properties of matter, from complex molecules to semiconductors. Materials Science has largely benefited of the enormous progress achieved in the last twenty years in this area which has let available a large set of models based on classical [1] up to fully ab-initio or quantum mechanical [2] representations. The Tight Binding (TB) model [3,4] has recently called the attention as being able to perform simulations of semiconductors materials combining high accuracy and a lower computational cost with respect to fully quantum mechanical (ab-initio) schemes.

Although being much simpler than ab-initio approaches, the computational complexity of TB algorithms is still considerable when used in combination to a Molecular Dynamics (MD) technique (TBMD) because it requires the repeated diagonalization of a large and sparse matrix. It is well known that the computational complexity of the matrix diagonalization scales with $N^3$ ($O(N^3)$), where $N$ is the number of atoms of the system. As a consequence, the practical size of the simulated systems cannot exceed the limit of $N = 200 - 300$ atoms on a powerful workstation. Further difficulties arise with the memory occupancy as

---

*** Email: dimartin@grid.unina.it
[†] Email: celino,rosato@casaccia.enea.it

the size of the matrix involved in TB calculations is of the order $kN * kN$ (with $k \simeq 10$).

This limit can be overcome by porting the TBMD code on parallel computers. The aim of the present study is to realize a TBMD code which allows the study of systems constituted by a number of atoms of about $10^3$ in the $O(N^3)$ formulation. In this work we report the parallelization of a $O(N^3)$ TBMD code, performed on a distributed memory MIMD computer (IBM SP), by using HPF (High Performance Fortran) and a set of mathematical libraries, contained in the package PESSL (Parallel Engineering and Scientific Subroutine Libraries) [5], suitably parallelized to fit the architectural features of that computer.

## 2  The TB mathematical formalism

The goal of a generic MD code is to generate time trajectories of a system constituted by $N$ atoms by solving the equations of motion depending on the forces acting on each atom. The forces can be calculated in a classical or in a quantum mechanics representation and they are the main computational task in both cases. Whereas in the classical representation force evaluation is essentially related to the calculation of the interparticle distances, in the TB representation there is a further computational complexity given by the diagonalization of the TB matrix (whose construction is related to the computation of the interparticle distances). The computational cost is, then, $O(N^2)$ scaling, in the case of classical regime and $O(N^3)$ in the TB case.

The TB formulation is based on the adiabatic approximation of the Hamiltonian $H_{tot}$ of a system of atoms and electrons in a solid [3]

$$H_{tot} = T_i + T_e + U_{ee} + U_{ei} + U_{ii} \tag{1}$$

where $T_i$ and $T_e$ are the kinetic energy of atoms and electrons, $U_{ee}$, $U_{ei}$, $U_{ii}$ are the electron-electron, electron-atom and atom-atom interactions, respectively.

Referring to the theory of one electron moving in the presence of the average field due to the other valence electrons and atoms, the reduced one-electron Hamiltonian can be written

$$h = T_e + U_{ee} + U_{ei} \tag{2}$$

giving the eigenvalues (energy levels) $\epsilon_n$ and the eigenfunctions $|\Psi_n>$, where $n$ is the rank of the matrix $h$. In the TB scheme, the eigenfunctions are represented as linear combinations of atomic orbitals $|\phi_{l\alpha}>$

$$|\Psi_n> = \sum_{l\alpha} c_{l\alpha}^n |\phi_{l\alpha}> \tag{3}$$

where $l$ is the quantum number indexing the orbital and $\alpha$ labels the atom. The expansion coefficients $c_{l\alpha}^n$ represent the occupancy of the $l$-th orbital located on the $\alpha$-th atom.

In the present TB scheme, the elements of the $h$ matrix, $< \phi_{l'\beta}|h|\phi_{l\alpha} >$, connecting $\alpha$ and $\beta$ nearest neighbours atoms are constituted by the product of two contributions

$$< \phi_{l'\alpha}|h|\phi_{l\beta} >= a_{l'l} f(r_{\alpha\beta}) \qquad (4)$$

where $a_{l'l}$ are parameters fitted on ab-initio or experimental results, and $f(r_{\alpha\beta})$ is a scaling function dependent on the particle distance $r_{\alpha\beta}$. As a further approximation, a minimal basis set, i.e. the minimal set of electronic orbitals for each atom is usually adopted: four basis functions ($s$, $p_x$, $p_y$, $p_z$) per atom are known to be sufficient for a satisfactory description of the valence bands in the case of elemental semiconductors (silicon and carbon). In order to obtain the eigenvalues (the single-particle energies) $\epsilon_n$ and the eigenvectors $c_{l\alpha}^n$ it is necessary to solve the secular problem at each MD time-step. This implies repeated diagonalization of the matrix $h$, which introduces the $O(N^3)$ scaling. The rank of the matrix is determined by $n = N*k$ where $N$ is the number of atoms in the simulated systems and $k$ is the dimension of the basis set ($k = 4$ in the simplest case of elemental semiconductors). Once the eigenvalues and the eigenvectors are known, the attractive potential energy can be computed and summed to the repulsive part $U_{rep}$ derived from a many-body approach [6]:

$$E_{tot} = \sum_n \epsilon_n f(\epsilon_n, T) + U_{rep} \qquad (5)$$

where $U_{rep}$ is of the form:

$$U_{rep} = \sum_{\alpha > \beta} \Phi(r_{\alpha\beta}) \qquad (6)$$

$U_{rep}$ takes also into account the overlap interaction originated by the non-orthogonality of the basis orbitals and the possible charge transfer.

The Hamiltonian of eq.(1) refers only to the atomic coordinates describing the internal degrees of freedom of the system. In order to simulate the interaction with the surrounding, suitable MD schemes have been devised to account for a coupling between the internal degrees and the external degrees of freedom (thermal bath and possibility of imposing to the system an external stress). Introducing further degrees of freedom simulates the coupling. The MD simulation in the isothermal-isobaric ensemble requires the presence of 10 extra variables which are coupled to the internal degrees of freedom via suitable parameters which can be adjusted to ensure the fastest convergence to equilibrium (Parrinello-Rahman and Nosé algorithm [7]).

After the calculation of the force on each atom, the equations of motion can be integrated by using a finite difference scheme. In this code, a sixth-order predictor-corrector scheme (VI order Gear algorithm) [8] has been implemented in view of its good accuracy.

## 3  The sequential TBMD computation

The goal of a generic MD code is to generate time trajectories of a system constituted by $N$ atoms by solving the equations of motion depending on the forces acting on each atom. The forces can be calculated in a quantum mechanics representation and they are the main computational task.

The layout of the generic iteration step of the sequential TBMD code consists in the following steps: (1) Predictor part of the integration of the dynamical equations for the evolution of the $3N+10$ degrees of freedom; (2) Computation of the interparticle distances: this is used to fill the matrix of nearest neighbors and the array to store the number of nearest neighbors of each atom; (3) Computation of the matrix elements; (4) Diagonalization of the real skew matrix $h$ for the computation of the spectrum of eigenvectors and eigenvalues. This is the most computational intensive part of the code; (5) Computation of the attractive part of the atomic forces (Feynman routine) by using the eigenvalues and the eigenvectors computed in the previous step; (6) Computation of the repulsive part of the atomic forces: this part of the forces depends only by the distances among the atoms: the nearest neighbors matrix is thus used; (7) Corrector part of the integration of the dynamical equations for the evolution of the $3N + 10$ degrees of freedom; (8) All the physical quantities are evaluated and stored on disk.

The predictor-corrector scheme requires the knowledge of the atomic coordinates till the fifth order derivative: the whole atomic system is thus described by 3*6 arrays of length $N$ for the coordinates and by further 6 arrays of the same length to store the forces (3 for the attractive part and 3 for the repulsive part). A $N * N$ matrix is requested to store the interparticle distances. For minimizing the computation of the TB $h$ matrix, the components of the forces and of the distances are also stored in 9 arrays of dimension $N$. The $h$ matrix of dimensions at least $4N * 4N$ must be allocated.

## 4 Code parallelization by using HPF and PESSL

To reach the target of the full parallelization of the TBMD code, we have chosen the approach of the utilization of very high level frameworks: HPF parallel language and PESSL mathematical library. High Performance Fortran (HPF) [9] is a high level language that simplify the programming and the porting of codes on distributed or shared memory parallel computers.

Essentially HPF [10] provides a standardized set of extensions to Fortran 90 and Fortran directives to enable the execution of a sequential program on parallel computer systems. Its main feature is to eliminates the complex, error prone task of explicitly programming how, where and when to pass messages between processors on distributed memory machines. HPF allows for testing the efficiency attained with different data mappings by simply changing the directives in the program instead of recoding a message-passing program. The writing of efficient HPF programs is not, however, a trivial task. Indeed, the high-level nature of the language prevents the user from clearly understanding the behavior of the parallel code being produced by the compiler, and that can frequently lead to unefficient codes. Anyway, HPF is broadly recognized as being the tool of choice for porting big legacy sequential codes to parallel architectures, when the computation exhibits regular data-parallel behavior, and it can be effective even when the computation presents characteristics of irregularity (see, e.g., [11]).

The different tasks in which the computation is decomposed, described in section (3), are all amenable to be parallelized in the data parallel paradigm. The main issue which arises in depicting the parallelization strategy is the selection of the optimal layout for the data structures, most of them involved in more than one (or all) computational tasks.

Step (1) and (7) (the predictor and corrector steps) involve sweeps over the following monodimensional arrays: the particles coordinates, their derivatives up to the fifth order and the total forces acting on each particle. The do loops implementing those sweeps do not present loop-carried dependences; thus their iterations can be distributed among the processors using the alignment and the cyclic distribution.

The step (2) computes the distances among the atoms and it involves the construction of the "Verlet lists" together with the corresponding filling marker vectors, which keep track of the symmetry and the sparsity of the values stored and are consequently used to reduce the amount of computations when accessing the arrays. The optimal layout for these data structures is their alignment with the arrays of coordinates, and thus their distribution is cyclic over the first coordinate. The iterations of the double loop nests, which perform the computation of distances and Verlet lists, would be distributed accordingly. Unfortunately, this data layout is in conflict with the optimal data layout for the last step of step (5). This step updates the data structures (monodimensional arrays) representing the attractive part of the forces, and uses auxiliar twodimensional arrays. The elements of these arrays are computed by sweeping over the twodimensional distance arrays and, for each distance $(i,j)$, by reducing the $4i-th$ to $(4i+4)-th$ and $4j-th$ to $(4j+4)-th$ rows of the Hamiltonian matrix. Thus, for each sweep of a row of the distance matrix $d$, there is a sweep of the whole matrix $h$. If we follow the data layout prescribed above for the $d$ matrix (i.e. cyclic distribution of the rows over the processors) we should replicate the whole matrix $h$ over the processors, to reduce the amount of communications. This is unfeasible, as the $h$ is the largest structure of the whole data set, being 16 times larger (in the case of a minimal basis set) than the $d$ matrix.

The only way to cope with this constraint is to give up on distributing the distance matrix $d$ (thus replicating it over the processors). The hamiltonian $h$ is instead distributed over the second dimension (i.e. distribute its columns). The sweep over the elements of the $d$ matrix is thus replicated, but the reduction of the 8 rows of $h$, for each distance $(i,j)$, can be performed in parallel: each processor sweeps the portion of rows of $h$ assigned to it, and performs a partial reduction of those elements. The results of the partial reduction operated from each processor are then composed and the result of the final reduction is then broadcasted. The amount of distribution of the iterations of the overall loop nest (and thus the degree of parallelization) remains the same as in the case of the alternative data layout (distributed distances, and replicated hamiltonian).

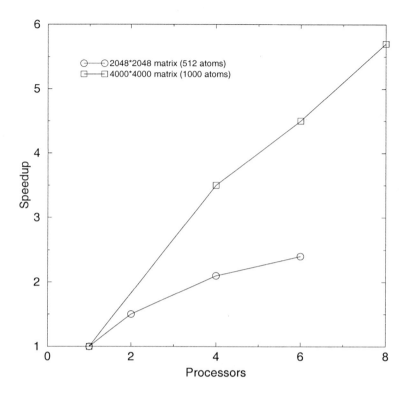

**Fig. 1.** Speedup obtained using HPF+PESSL parallel Molecular Dynamics code on the IBM SP located in IBM Centre of Poughkeepsie (USA). It is clear the good speedup obtained till 8 nodes.

In step (3) and (4) the computation and diagonalization of the Hamiltonian twodimensional matrix h(n,n), and computation of its eigenvalues in a monodimensional array are performed. The PESSL routine SYEVX is called for this task. The parameters in the tuning process are those: (a) to control the absolute tolerance on the eigenvalues, which has been tuned to obtain the minimum of tolerance to improve the efficiency of the algorithm for the orthogonalization of the eigenvectors; (b) to balance the workspace and the memory requested for the eigenvectors calculations.

The use of the PESSL routine introduces extra inter-node communications because the SYEVX routine needs the matrix to be distributed in both dimensions as cyclic. We thus need to redistribute the hamiltonian matrix h during this step, and to restore its distribution over the columns as needed in the Feynman routine.

## 5  Results and discussions

To validate the porting and the data distribution adopted we have performed several runs on two IBM SP2 platforms: the first one is located in ENEA Research Centre in Frascati (Italy) and the other in the IBM Centre of Poughkeepsie (USA). The first has 390-series processors with 66 MHz and 128 MB RAM, while the second has 397-series with 160 MHz and 256 MB RAM. The runs have been performed on 2, 4, 6 and 8 processors. A clear speedup till 8 nodes is shown both using the SP2 at the ENEA Centre and the one at the IBM Centre (see Figure (1)).

We can observe: (1) The approach described in the last section consists in distributing all the atomic coordinates obtaining in the first and in the last phase a speedup equal to the number of the processors; (2) The Feynman routine, even if the atomic coordinates are distributed all over the processors thus requiring a communication among all the processors, in the case of 640 atoms, shows a clear speedup of 2 going from 4 to 8 processors; (3) Also the SYEVX parallel routine has a good speedup till 8 nodes allowing its efficient implementation and its use at each time step.

At this point we must underline that the good performance is obtained even if we use a mathematical library that does not utilize the sparsity of the $h$ matrix to improve the performance of the inter-node communications during the diagonalization process.

Furthermore we can conclude that: (1) even if HPF is not completely suitable for irregular problems, like MD simulations, we have obtained good speedup and efficiency; (2) even if different data distributions (one for the matrix diagonalization routine and the other for the MD routines) are needed, the overall performance are good because the parallelization strategy adopted can minimize the data communications.

Using this approach, we have reached the target of simulations with a number of atoms of order $10^3$: the parallel code performs the MD simulation in a reasonable amount of wall clock time and the RAM occupancy of the whole code can be managed.

## 6  Acknowledgments

We are very grateful to M.Briscolini and S.Filippone (IBM Italia) for their kind help in the use of the PESSL mathematical libraries and to run the benchmarks on the IBM SP2 located in the IBM Centre of Poughkeepsie.

We would like also to acknowledge L.Colombo (University of Milan) for useful discussions on the implementation of the TB algorithm.

## References

1. M.P.Allen and D.J.Tildesley, "Computer simulation of liquids", Clarendon Press, Oxford, 1987.

2. R.Car and M.Parrinello, Phys.Rev.Lett. **55** (1985) 2471.
3. L.Colombo in: Annual Review of Computational Physics IV, edited by D.Stauffer (World Scientific, Singapore, 1996) p.147.
4. C.Z.Wang, K.M.Ho, "Tight Binding Molecular Dynamics studies of covalent systems" in "New methods in Computational Quantum Mechanics". Ed. I.Prigogine, S.A.Rice, vol.XCIII in Advances in Chemical Physics, John Wiley & Sons, 1996.
5. 'Parallel Engineering and Scientific Subroutine - Guide and Reference, Release 2', IBM (1996).
6. I.Kwon, R.Biswas, C.Z.Wang, K.M.Ho, C.M.Soukoulis, Phys.Rev.B **49** 7242 (1994).
7. M.Parrinello and A.Rahman, J.Chem.Phys. **72**, 2662 (1982); S.Nosé and M.L.Klein, Molecular Physics, **50** 1055-1076 (1983).
8. C.W.Gear, "The numerical integration of ordinary differential equations of various orders". Report ANL 7126, Argonne National Laboratory (1966); C.W.Gear, "Numerical initial value problems in ordinary differential equations". Prentice-Hall, Englewood Cliffs, NJ (1971).
9. High Performance Fortran Forum, "High Performance Fortran Language Specification", Version 1.1, 1994. High Performance Fortran Forum, "High Performance Fortran Language Specification", *Scientific Programming*, **2**(1-2) (1993) 1-170.
10. F.Darema et al., "A single-program-multiple-data computational model for EPEX/FORTRAN", Parallel Computing, **7**, (1988) 11-24.
11. B.Di Martino, S.Briguglio, G.Vlad and P.Sguazzero, Lecture Notes in Computer Science n. 1401, Springer-Verlag, April 1998 (Proc. of *High Performance Computing and Networking '98* Conference, Amsterdam (Nl), Apr. 1998).

# Design of High-Performance C++ Package for Handling of Multidimensional Histograms

Marian Bubak[1,2], Jakub T. Mościcki[1], Jamie Shiers[3]

[1] Institute of Computer Science, AGH, al. Mickiewicza 30, 30-059, Kraków, Poland
[2] Academic Computer Centre – CYFRONET, Nawojki 11, 30-950 Kraków, Poland
[3] IT/ASD, CERN 1211 Geneva 23, Switzerland

*email:* bubak@uci.agh.edu.pl, moscicki@student.uci.agh.edu.pl, Jamie.Shiers@cern.ch
*phone:* (+48 12) 617 39 64, +41 22 767 4928
*fax:* (+48 12) 633 80 54, +41 22 767 8630

**Abstract.** This paper evaluates different object-oriented models and design techniques for development of high-performance numerical C++ software to handle multidimensional data. Two fundamental approaches are considered: the classical inheritance driven class hierarchy and the static polymorphism techniques. The inheritance driven approach exhibits intrinsic drawbacks, whilst static polymorphism techniques based on traits may significantly increase the efficiency and extensibility of the design. In the paper, a special attention is given to the analysis of efficiency and extensibility of the proposed solutions in ODBMS environment. As the result of this reaserch the HistOOgrams Package has been developed at CERN as a part of LHC++ environment. The results of this study may be helpful in object-oriented design of other numerical packages.

**Keywords**: multidimensional data analysis, static polymorphism, class inheritance, excessive deferment

## 1 Introduction

Large scale experiments in particle physics require powerful computing environments. Large Hadron Collider (LHC) in CERN, scheduled to start off in 2005, is expected to generate about 5 PB of data a year (160 MB/s). Storage systems used nowadays for existing experiments are being revised to scale up to the predicted data throughput. The transformation to new storage system architecture is based on high-performance ODBMS (Object Database Management System) multi-petabyte data-handling environment and is followed by a new object-oriented software for data analysis. CERN LHC++ project [1] delivers solutions for large scale experiments, primarily for the Large Hadron Collider, in the ODMG-compliant [2] ODBMS environment. Histogram-based analysis of multidimensional data is an important part of particle physics data analysis process and therefore a distinct software component for multidimensional histogram handling has been included into the LHC++ project.

Recently, there have been many trends to exploit advanced features of C++ standard [3], e.g., STL [4], in the field of scientific high-performance computing. The research focussed mainly on static polymorphism (e.g., [5–8]) indicated that near-optimal, comparable to FORTRAN performance is attainable for C++ numerical applications. This paper shows how similar techniques may be successfully applied for handling multidimensional data in numerical applications.

We have investigated the impact of multidimensionality on the design of the HistOOgrams Package – the LHC++ component for histogram handling. The efficiency and extensibility of considered design techniques as well as conformity to emerging ANSI/ISO C++ and ODMG standards were of primary importance. The discussion starts with a brief presentation of the user requirements and the key results of object-oriented analysis (OOA) phase. In the subsequent sections an inheritance-driven design and template-based solution are presented. The paper is summarized with the performance benchmarks.

## 2  OOA Model of Multidimensional Histograms

The main task of the HistOOgrams Package is creation, filling and retrieving contents of histograms for batch-job data processing. 1- and 2-D Cartesian histograms (i.e. bins subsets of $R^n$) are of special concern because they cover over 90% of typical data analysis runs and they require the best optimisation. Additional requirements are extensibility (arbitrary partitions of multidimensional problem spaces) and development of an easy to use, type-safe interface.

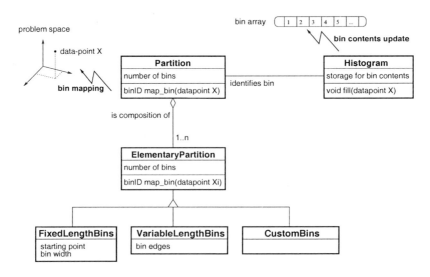

**Fig. 1.** Simplified OOA diagram of multidimensional histogram handling.

The key results of object-oriented analysis are presented in Fig. 1. Histogram filling is a crucial operation and has a significant impact on the design. Typically, particle physics data analysis batch jobs perform an iterative histogram filling with the particle detector event data. Upon Histogram::fill(X) request from the client application loop, a data-point X is mapped into the identifier (binID) of bin corresponding to X which is then used for the bin array lookup and update of relevant bin contents. This functional decoupling of the histogram filling process into *bin mapping* and *bin contents update* is reflected in the OOA model: the histogram maintains a private bin array for bin contents store and update, whilst Partition object associated with Histogram is responsible for bin mapping.

The essential item for the analysis of the impact of multidimensionality on the OO design is a $n$−dimensional vector $X$: $\{X_1 x_1; \cdots; X_i x_i; \cdots; X_n x_n\}$. It has three key properties: the dimension $(n)$, type $(X_1, \cdots, X_n)$, and composition (homogeneous or heterogeneous; homogeneous composition means that $X_1, \cdots, X_n$ are the same types). Partition object aggregates $n$ ElementaryPartition objects – each corresponding to one dimension of the problem space. ElementaryPartition provides partial mapping for $x_i$ element of data-point $X$ what, in turn, enables to compute binID. The partitioning is defined in classes derived from ElementaryPartition.

The main design problem is to enable an efficient implementation of the data-point $X$ representation in the interfaces of partition and histogram classes while taking into account the user requirements. The heterogeneous space is especially difficult to handle because programming languages usually have no support for heterogeneous collections.

## 3 A Straightforward Design – Inheritance Driven Class Hierarchy

First, we consider a straightforward design and implementation of the object model introduced in the previous section. In this approach, the concepts from OOA model are almost directly transferred into the design, and the partition classes directly reflect the OOA model. It is necessary to notice that we restrict the desing to the Cartesian problem spaces with homogeneous composition. For the clarity of presentation, we confine the discussion to 2D-real histogram with equidistant bins (fixed length bins). However, it may be easily extended on any other types of histograms and problem spaces.

The root classes Histogram and Partition (Fig. 2) serve as a base for all operations and attributes which are not dependent on any specific problem space and which are omitted for clarity. Thus, the base classes do not contain references to X in their interfaces. The lower are classes in the inheritance hierarchy the more specific problem spaces are handled.

CSpaceHistogram and CSpacePartition are related to a general Cartesian problem space ($n$−dimensional homogeneous composition of $R$) and they correspond to Partition and Histogram classes in OOA diagram (Fig. 1). The generalised data-point is an array of doubles, $X$ = double[0..n-1]. CSpaceHisto2D is related to 2-dimensional Cartesian problem space; $X$ is a pair of doubles $x$,

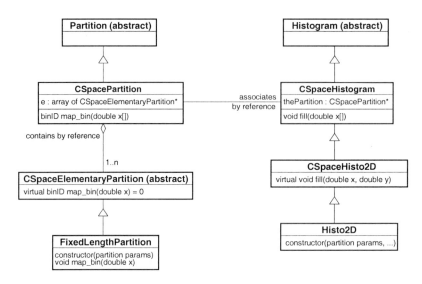

**Fig. 2.** Straightforward design - OOA directly transfered to OOD.

$y$. CSpaceHisto2D is able to handle any 2-D combination of CSpaceElementary-Partitions.

On the other hand Histo2D is dedicated to one specific combination: two FixedLengthPartitions. Hence, it is possible to isolate the user from using partition classes: Histo2D may create appropriate partition objects automatically. The parameters for elementary partitions are specified in Histo2D constructor.

The efficiency analysis concerns two issues: histogram creation and filling. These two operations are critical as they may be performed many thousand or even million times per one analysis task. The dynamic model of histogram creation is similar to the dynamic model of histogram filling and therefore we focus the discussion on the latter (Fig. 3).

Upon client call Histo2D::fill(x,y) the coordinates x, y are packed into a temporary array {x,y} which is then processed by CSpaceHistogram::fill. The array is sent to CSpacePartition::map_bin which iterates over its elementary partitions with a query for partial mapping (virtual FixedLenghtPartition::map_bin). The *transient* component of efficiency involves 3 virtual calls per one filling operation.

In the ODBMS implementation there is an additional call overhead due to object persistence. Persistent objects are created and accessed in the persistent address space of the ODBMS. The member function call of a persistent object is slower than the C++ function call of a transient object as there is an additional indirection level of Object Identifier (OID).

In the straightforward design all partition and histogram objects are distinct ODBMS entities with OID access and histogram → partition association and partition → elementary partition aggregation are implemented *by reference*. The *persistent* component of efficiency involves 4 OID calls per one filling operation.

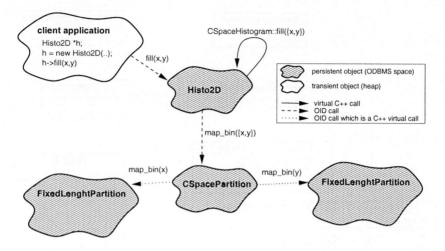

**Fig. 3.** Message path and dynamic model for Histo2D filling operation.

The message path for handling critical operations is long and inefficient for two reasons: simple case of 2D histogram of doubles is handled with general mechanisms of CSpaceXXX classes, what implies virtual calls and inefficient data handling, and the overall structure of auxiliary partition objects is too complex, which in turn, has a crucial impact in the ODBMS environment and results in slow OID access.

## 4 Refined Class Hierarchy

The efficiency analysis presented in the previous section shows drawbacks of the direct transfer from OOA to OOD. They result from tight coupling of the general $n$−dimensional and specific 2D histograms. The goal of the refinement is to shorten the message path for filling and creation operations by bypassing CSpaceXXX classes, i.e., by accessing FixedLengthPartition directly from Histo2D. In this way the histogram → partition association is logically shifted down the hierarchy by one level (see Fig. 4).

CSpacePartition2D is a new abstract class which aggregates common properties and operations of 2D Cartesian partitions and fixes the interface for the pure virtual map_bin method. There are two kinds of partition classes: abstract ones like CSpacePartition2D, which are used to interface with histogram classes and concrete ones like FixedLengthPartitio2D, which are used to build the partition hierarchy.

The message path is shorter by one call compared to the previous design and it does not go through the slow processing in CSpaceXXX classes. There are two virtual calls (CSpaceHisto2D::fill and CSpacePartition2D::map_bin) per one filling. FixedLengthPartition1D objects are aggregated in FixedLengthPartitio2D object *by value* and therefore the virtual call overhead is reduced. The number of

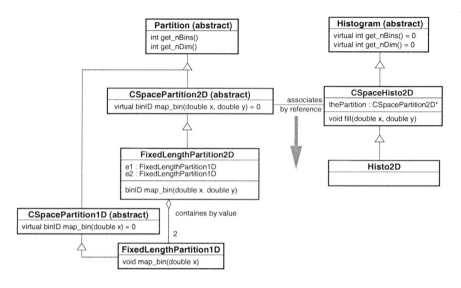

**Fig. 4.** Refined design for inheritance-based histogram handling.

OID calls per fill operation is reduced to 2 as only two separate ODBMS entities are now considered: Histo2D and FixedLenghtPartition2D objects.

Virtual calls could be eliminated completely by bypassing the level of abstract classes with a pointer to FixedLengthPartition2D embedded directly in the Histo2D (i.e., the association link shifted down the hierarchy as marked with the bold arrow on the diagram). Then it would be possible to have *by value* association and to omit one OID call.

Unfortunately, undesirable effect of *excessive deferment* occurs when the association link is shifted too far to the bottom of the hierarchy. To explain this effect the interface of Histogram class should be considered.

In the interface of Histogram class there are several methods which are mirrored from corresponding partition classes by message delegation, e.g., Histogram::get_nBins (number of bins) and Histogram::get_nDim (dimensionality of the problem space). If the association link is not contained in Histogram class, then these methods must be declared as pure virtual ones. The same apply to methods in the CSpaceHisto2D which depend on the interface of CSpacePartition2D. The definition of all these pure virtual methods is deferred to the class which contains the association link. The lower the association link is implemented in the class hierarchy, the more method definitions must be deferred. In this way one may get to the point where the deferment is too excessive and the profits of object-oriented approach are lost: leaf classes are unnecessarily forced to provide identical definitions of many deferred methods which could be provided only once in the base class. In C++, a way out is to place the association link at the top level and to use downcasting at the lower ones. However, this technique requires Runtime Type Identification (RTTI) and reduces both efficiency and clarity of the design.

## 5 Static Polymorphism on Stage

Let us consider a histogram class parameterised with partition type (Fig. 5). In this situation, excessive deferment disappears: Histogram<PARTITION> class can safely fix the definition of get_nBins. Parameterisation enables to *declare* a histogram → partition association link in the base class and *implement* it as low in the hierarchy as it is required.

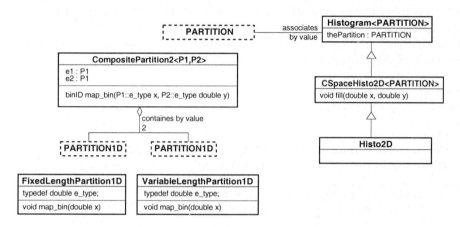

**Fig. 5.** Overview diagram of the design based on static polymorphism

The important change concerns the partition classes. Instead of creating a complete hierarchy of 1D, 2D, 3D, ... partitions, we directly create only basic 1D partitions (building blocks) and we have a couple of auxiliary classes (composition tools) which are able to generate automatically partition classes of higher dimensions. To get a new class, we instantiate a composition tool with the chosen building blocks. This scheme offers a great deal of flexibility: we can easily create higher-dimension partitions, mix partitions of different types, and finally, we can extend the system in the simple, well-defined manner by adding new building blocks.

Building blocks are 1D partition classes like FixedLengthPartition1D and VariableLengthPartition1D. In general, a building block partition P must provide e_type – the specification of the elementary type of partition. In this way, partition P declares to provide map_bin(P::e_type x) method. This technique is borrowed from the traits [9]. The typical declaration of P is following:

```
class FixedLengthPartition1D
{ ...
  typedef double e_type;
  binID map_bin(double x);
  ... };
```

An example of the composition tool is the class CompositePartition2<P1,P2> which merges two 1D partitions P1, P2 into a single composite 2D partition. An extension of the partition classes with a 2D fixed length partition may be easily obtained with:

```
typedef CompositePartition2<FixedLenghtPartition1D,FixedLengthPartition1D>
    FixedLengthPartition2D;
```

FixedLenghtPartition2D class can be used to instantiate CSpaceHisto2D<P> which is inherited by Histo2D class (see Fig. 5). Histo2D uses partition classes transparently to the client code. This design offers also flexibility for developers: it is relatively easy to create new composition tool classes. Let us consider the example of the CompositeParititon2 class:

```
template <class P1, class P2>
class CompositePartition2<P1,P2>
{
  P1 p1; P2 p2;
  CompositePartition2( P1 tp1, P2 tp2 ) : p1(tp1), p2(tp2) {}
  binID map_bin(P1::e_type x, P2::e_type y)
    {
       binID i1 = p1.map_bin(x), i2 = p2.map_bin(y);
       return composition_mapping(i1,i2);
    }
  ...
};
```

CompositePartition2 delegates the bin mapping to the lower order partitions P1, P2 and calculates the linear binID based on partition binIDs i1, i2. The global function composition_mapping performs simple linearisation of 2D indices. Such an approach enables to achieve high level of static polymorphism-based code reuse.

This design provides also the possibility of inheritance-based code reuse for the partition and histogram classes. So, for example, similarly to the previous designs, it is possible to have a Partition class that captures common, dimension-independent properties and is inherited by the building blocks and composition tools. Such a structuring does not degrade efficiency of critical operations, and moreover, it offers the flexibility of dynamic polymorphism and inheritance. Within the same design the *hybrid approach* involves two orthogonal structures: the static polymorphism with heavily inlined *compositional* reuse for robust handling of the critical operations of multidimensional-interface and the dynamic polymorphism (virtual functions) with the inheritance-based reuse for the rest of the interface.

Parameterisation of the histogram classes enables to achieve very short message paths for the critical operations. Moreover, the association link may be implemented by an embedded (aggregated *by value*) partition object. Virtual calls are eliminated and inlining is possible. Since a partition is embedded by value in a histogram, the number of OID accesses is reduced to one per filling operation. This also implies faster creation since only one database entity (Histogram) is created.

## 6 Implementation and Benchmarks

The HistOOgrams Package was implemented on all major Unix platforms and Windows NT. The design strategy aimed to obtain a single, multi-platform source code. The Package has no provisions for graphical display nor for explicit persistence. It was implemented on top of Objectivity/DB version 4 [10] and is a lightweight collection of classes (with about 5,000 lines of source code).

A suit of benchmarks has been performed on HP 735/125 with HP-UX 10.20 and on Sun SPARCserver 20/152 with Solaris 2.5.1. The basic test consisted of filling different types of 1D and 2D histograms 500,000 times per run. The results of tests in persistent environment were corrected to compensate the ODBMS implicit I/O time which is not present in the transient tests. We observed different performance in the transient and persistent versions of the straightforward design; the transient version was 7 to 12 times faster than the persistent one. However, this difference is much smaller for the design which involves static polymorphism (from cca. 1.3 to 1.5 in favour of the transient version).

The direct comparison between three designs under discussion shows that the first refinement of straightforward design leads to speed-up 5-7 times for the persistent version, whilst the second refinement (involving the static polymorphism) gives an additional improvement (relative to the first refinement) of factor 2. For the transient version, the differences between these designs are smaller and the ratio between the worst and the best design is about 3.

The benchmark results confirm the analysis of efficiency presented in the previous sections. The performance issues are of special concern in the ODBMS environment: performance improvements of the subsequent designs are significantly larger for the persistent tests.

## 7 Summary

We have discussed and evaluated the application of the object-oriented techniques for development of the efficient numerical library to handle inherently multidimensional data.

It was shown that a design which directly reflects the OOA does not necessarily lead to an efficient solution. Some intrinsic disadvantages of the classical OO techniques heavily based on dynamic polymorphism and inheritance were investigated. Few of them may be partially overcome in C++ language. On the other hand, static polymorphism offers a comprehensive solution: simple, highly efficient and flexible. The design based on the static polymorphism applies a relatively small subset of ANSI/ISO C++ extensions for templates what is important for multi-platform targeted projects. The techniques discussed in the paper were tested in the ODMG-compliant ODBMS environment.

Design techniques and efficiency analysis presented in this paper may also be useful for development of other numerical packages.

**Acknowledgments.** This research was done in the framework of the CERN Technical Student Programme in 1997. JTM thanks people from the LHC++ and RD45 teams for enlightening discussions on the design of the HistOOgrams Package. Special thanks go to Olivier Couet, Dirk Duellmann and Dino Ferrero Merlino for their contribution and many helpful comments and suggestions. This research was partially supported by the AGH grant.

# References

1. Status Report of the LHC++ Project.
   http://wwwinfo.cern.ch/asd/lhc++/sr98/index.html
2. Object Database Management Group: The Object Database Standard: ODMG-93, Release 1.2. Morgan Kaufmann Publishers, ISBN 0-201-63398-1
3. ANSI/ISO C++ Standard. http://webstore.ansi.org/AnsiDocStore
4. Musser, D.R., Saini, A.: STL Tutorial and Reference Guide: C++ Programming with the Standard Template Library. Addison Wesley, ISBN 0-201-63398-1.
5. High Performance C++. http://www.extreme.indiana.edu/hpc
6. Blitz++ - a C++ library for scientific computing.
   http://monet.uwaterloo.ca/blitz/
7. Template Numerical Toolkit (TNT). http://math.nist.gov/tnt/
8. Veldhuizen, T.L., Jernigan, M.E.: Will C++ be faster than Fortran? In: Proceedings of the 1st International Scientific Computing in Object Oriented Parallel Environments (ISCOPE'97).
9. Myers, N.C.: Traits: a new and useful template technique. In: C++ Report, June 1995
10. Objectivity/DB, Objectivity Inc., www.objectivity.com

# Dynamic Remote Memory Acquiring for Parallel Data Mining on PC Cluster: Preliminary Performance Results

Masato OGUCHI[1,2] and Masaru KITSUREGAWA[1]

[1] Institute of Industrial Science, The University of Tokyo
7-22-1 Roppongi, Minato-ku Tokyo 106-8558, Japan
[2] Informatik4, Aachen University of Technology
Ahornstr.55, D-52056 Aachen, Germany
oguchi@i4.informatik.rwth-aachen.de

**Abstract.** Recently data intensive applications such as data mining and data warehousing have been focused as one of the most important applications for high performance computing. As a platform, PC/WS cluster is a promising candidate for future high performance computers, from the viewpoint of good scalability and cost performance ratio. We have developed a large scale ATM connected PC cluster until now, and implemented several database applications, including parallel data mining, to evaluate their performance and the feasibility of such applications over PC clusters.

Different from the conventional scientific calculations, association rule mining, which is one of the most well known problems in data mining, has a peculiar feature for its usage of main memory. It needs a lot of small data on main memory at each node, and the number of those areas suddenly swells to be enormous during the execution. Thus the requirement of memory space changes dynamically and becomes tremendously large. As a result, contents of memory must be swapped out if the requirement exceeds the real memory size. However, because the size of each data is rather small and all the areas are accessed almost at random, swapping out to a storage device is expected to degrade the performance dramatically in this case. Therefore some other methods must be introduced to perform large scale data mining on PC clusters, which may require huge memory dynamically.

We are investigating the feasibility of using available idle nodes' memory as a swap area when some nodes need to swap out its real memory contents, during the execution of parallel data mining on PC clusters. Idle nodes are expected to exist in large scale clusters. In this paper, we report our preliminary results in which application executing nodes acquire extra-memory dynamically from several available idle nodes through ATM network. The experimental result on the PC cluster shows this method is expected to be considerably better than using hard disks as a swapping device.

# 1 Introduction

Data mining has attracted a lot of attention recently from both the research and commercial community, for finding interesting trends hidden in large transaction logs. Because of the progress of bar-code technology, large transaction processing system logs have been accumulated. Such data was just archived and not used efficiently until recently. The advance of microprocessor and secondary storage technologies allows us to analyze the vast amount of transaction log data to extract interesting customer behaviors. For very large mining operations, however, parallel processing is required to supply the necessary computational power.

PC/WS(Personal Computer/Workstation) clusters have become a hot research topic in the field of parallel and distributed computing. They are regarded to play an important role as large scale parallel computers in the next generation, from the viewpoint of good scalability and cost performance ratio. Various research projects to develop and examine PC/WS clusters have been performed until now[1][2][3][4][5]. However, most of them only measured basic characteristics of PCs and networks, and/or some small benchmark programs were examined. We believe that data intensive applications such as data mining and ad-hoc query processing in databases are quite important for future high performance computers, in addition to the conventional scientific applications. We have developed a pilot system of ATM connected PC cluster consists of 100 Pentium Pro PCs, and evaluated it with database applications including the TPC-D benchmark and a data mining application[6][7].

One typical problem in data mining is mining of association rules from transaction log data. Different from the conventional scientific calculations, association rule mining has a peculiar feature for its usage of main memory. It needs a lot of small data on main memory at each node, and the number of those areas suddenly increases to be enormous during a certain step of the execution, so that the requirement of memory becomes tremendously large. In this case, part of memory contents must be swapped out if the requirement exceeds the real memory size.

Because the size of each data is rather small and all the areas are accessed almost at random, however, swapping out to a secondary storage system should cause terrible performance degradation. Hence we are investigating the feasibility of using available idle nodes' memory as a swap area. Since idle nodes are expected to exist in large scale clusters, we can make use of them. In this paper, we report our preliminary experimental results in which application executing nodes acquire extra-memory dynamically from several idle nodes on the ATM connected PC cluster. The rest of paper is organized as follows: In Section 2, data mining application and its parallelization are explained. An overview of our PC cluster is shown and implementation of parallelized data mining on PC cluster is related in Section 3. In Section 4, the method of dynamic remote memory acquiring for parallel data mining is explained, and performance results of the experiments are shown and analyzed in Section 5. Finally some concluding remarks are stated in Section 6.

## 2 Parallel data mining application

### 2.1 Association rule mining

Data mining has been focused as an important application in the field of high performance computing. Data mining is a method of the efficient discovery of useful information such as rules and previously unknown patterns existing among data items embedded in large databases, which allows more effective utilization of existing data. One of the most well known problems in data mining is mining of association rules from a database, so called "basket analysis"[8]. Basket type transactions typically consist of transaction id and items bought per-transaction. An example of the association rule is "if customers buy A and B then 90% of them also buy C". The most well known algorithm for association rule mining is Apriori algorithm proposed by R. Agrawal of IBM Almaden Research[9].

### 2.2 Parallelization of the application

In order to improve the quality of the rule, we have to analyze very large amounts of transaction data, which requires considerably long computation time. We have studied several parallel algorithms for mining association rules until now[10], based on Apriori. One of these algorithms, called HPA(Hash Partitioned Apriori), is implemented and evaluated on the PC cluster.

Apriori first generates candidate itemsets, then scans the transaction database to determine whether the candidates satisfy the user specified minimum support. At first pass (pass 1), a support for each item is counted by scanning the transaction database, and all items which satisfy the minimum support are picked out. These items are called large 1-itemsets. In the second pass (pass 2), 2-itemsets (length 2) are generated using the large 1-itemsets. These 2-itemsets are called candidate 2-itemsets. Then supports for the candidate 2-itemsets are counted by scanning the transaction database, and large 2-itemsets which satisfy the minimum support are determined. This repeating procedure terminates when large itemset or candidate itemset becomes empty. Association rules which satisfy user specified minimum confidence can be derived from these large itemsets.

HPA partitions the candidate itemsets among processors using a hash function as in the hash join in relational database. HPA effectively utilizes the whole memory space of all the processors. Hence it works well for large scale data mining. Each step of the algorithm is as follows:

1. Generate candidate $k$-itemsets:
   All processors have large $(k-1)$-itemsets on their memory when pass $k$ starts. Each processor generates candidate $k$-itemsets using large $(k-1)$-itemsets, applies a hash function and determines a destination processor ID. If the ID is its own, inserts it into the hash table, otherwise discards.
2. Scans the transaction database and counts the support value:
   Each processor reads the transaction database from its local disk. Generates $k$-itemset from that transaction and applies the same hash function used in

phase 1. Determines the destination processor ID and sends the $k$-itemset to it.
When a processor receives these itemsets, it searches the hash table for a match, and increments the match count.
3. Determine large $k$-itemsets:
Each processor checks all the itemsets it has and determines large itemsets locally, then broadcasts them to the other processors. When this phase is finished at all processors, large itemsets are determined globally. The algorithm terminates if no large itemset is obtained.

It is known that candidate itemsets in pass 2 is extremely larger than other passes. This relatively frequently happens in association rule mining.

## 3 Our PC cluster pilot system

### 3.1 An overview of the cluster

In our PC cluster, 100 PCs are connected with an ATM switch. 200MHz Pentium Pro PCs are used as a node of the cluster. Each node consists of components shown in Table 1.

Table 1. Each node of PC cluster

| CPU | Intel 200MHz Pentium Pro |
|---|---|
| Chipset | Intel 440FX |
| Main memory | 64Mbytes |
| Disk drive | 2.5Gbytes IDE hard disk |
| OS | Solaris2.5.1 for x86 |
| ATM NIC | Interphase 5515 PCI ATM Adapter |

All nodes of the cluster are connected with a 155Mbps ATM LAN as well as Ethernet. We use RFC-1483 PVC driver, which supports LLC/SNAP encapsulation for IP over ATM[11][12]. Only UBR traffic class is supported in this driver. TCP/IP over ATM is used as communication protocols.

HITACHI's AN1000-20 is used as an ATM switch. Since this switch has 128 port 155Mbps UTP-5, all nodes can be connected directly with each other, composing a star topology rather than a cascade configuration. An overview of the PC cluster is shown in Figure 1. On this PC cluster, we achieved about 120Mbps throughput in the case of point-to-point communication, even with TCP/IP over ATM protocol[7].

### 3.2 Implementation of parallelized data mining on PC cluster

HPA program has been implemented on the PC cluster pilot system. Each node of the cluster has a transaction data file on its own hard disk. At each node, two processes are created and executed: One process makes candidate itemsets from previous large itemsets, and sends them to the other processes, which put the data into a hash table. Also in the data counting phase, one process generates itemsets by scanning the transaction data file, and sends them to the other processes on the node decided by the hash function, which checks and increments its hash table value appropriately.

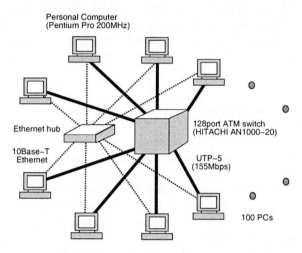

**Fig. 1.** An overview of the PC cluster

Solaris TLI (Transport Layer Interface) system calls are used for the interprocess communication. All processes are connected with each other by TLI transport endpoints thus forming mesh topology. /dev/tcp is used as a transport layer protocol, which is two-way connection based byte stream. On the ATM level, PVC (Permanent Virtual Channel) switching is used since data is transferred continuously among all processes.

Transaction data is produced using data generation program developed by Agrawal, designating some parameters such as the number of transaction, the number of different items, and so on. The produced data is divided by the number of nodes, and copied to each node's hard disk.

During the execution of HPA, itemsets are kept on memory as listed structures which are classified by a hash function. That is to say, all itemsets having the same hash valued are assigned to the same hash line on the same node, and their structures are connected with each other to form a list.

## 4 Dynamic remote memory acquiring for parallel data mining

### 4.1 Background of the experiment

As was mentioned in Section 2, the number of candidate itemsets in pass 2 is extremely larger than other passes in association rule mining. This depends sensitively on user specified conditions such as minimum support value, and it is difficult to predict how large the number is before execution. Therefore, it may happens that the number of candidate itemset suddenly swells in this step so that the requirement of memory becomes tremendously large. When the requirement is over the real memory size, part of memory contents must be swapped out. However, because the size of each data is rather small and all the data is accessed almost at random, swapping out to a storage device is considered to degrade the performance dramatically in this case. In the rest of this paper, we will discuss a method in which available idle node's memory is used as a swap area, in the case of huge memory being dynamically required during the execution of parallel data mining.

### 4.2 Mechanism of dynamic remote memory acquiring

In this experiment, we set a limit value for memory usage of candidate itemsets at each node. When the amount of memory usage is over this value during the execution of HPA program, part of contents are swapped out to available idle nodes' memory. Although idle nodes should be found dynamically in a real system, we decided them statically for the sake of experiments, and call them "memory servers". That is, the application execution node acquires memory area dynamically from one of memory servers when it is needed. The number of memory servers can be appropriately changed from one to many in the experiment. The unit of swapping operation is a hash line, which is a listed structures as was explained in Section 3. The destination memory server for each hash line is also decided by the hashing.

When the number of candidate itemsets becomes extremely large in pass 2 and the amount of memory usage exceeds a specified value, the node sends out some of its memory contents to a destination memory server. The swapped out hash line is selected by LRU manner. At the memory server, the received contents are allocated and written on its main memory.

On the other hand, a pagefault occurs when an application execution node accesses to contents which have been swapped out. In this case, the node is able to know which memory server stores the contents by calculating the hash value. It sends a request to this memory server, and the memory server sends back the requested hash line. After the application execution node receives the contents, they are allocated and written on main memory again, then the execution of application resumes. Replacements of data are decided by LRU manner.

The basic behavior of this experiment has something in common with distributed shared memory systems (See [13], for example). If data structures insides applications are considered in distributed shared memory, almost the same effect can be expected. Thus our mechanism might be regarded as equivalent with an optimized case of distributed shared memory for a particular application.

# 5 Performance results of dynamic remote memory acquiring

## 5.1 Implementation of experimental system on PC cluster

The parameters used in the experiment are as follows: The number of transactions is 1,000,000, the number of different items is 5000, and the minimum support is 0.1%. The size of the transaction data is about 80Mbytes in total. The message block size is set to be 4Kbytes, and the disk I/O block size is 64Kbytes, which are suitable values for our cluster[14].

The number of application execution node is 8 in this evaluation. The number of memory server is changed from 1 to 16. The number of hash line for candidate itemsets is 800,000 in total, hence about 100,000 hash lines are assigned to each node during the execution. The unit of swapping operation is a hash line, which can be contained in one message block.

With the above conditions, the number of candidate itemsets in pass 2 becomes 4,871,881 in total. These candidate 2-itemsets are assigned to each node using a hash function. The numbers of candidate 2-itemsets at each node are shown in Table 2. Although the itemsets are assigned using a hash function, the numbers at each node are not completely equal. This frequently happens in the execution of HPA, because a kind of skew usually exists in transaction data in association rule mining. We have also developed a method to treat it, which can be found in another literature[15].

Since each candidate itemset occupies 24bytes in total(structure area + data area), approximately 14-15Mbytes of memory is filled with these candidate itemsets at each node.

| node 1 | node 2 | node 3 | node 4 | node 5 | node 6 | node 7 | node 8 |
|--------|--------|--------|--------|--------|--------|--------|--------|
| 602559 | 641243 | 582149 | 614412 | 604851 | 596359 | 622679 | 607629 |

**Table 2.** The numbers of candidate 2-itemsets at each node

## 5.2 Performance results and their analysis

The execution time of pass 2 of HPA program, when the number of memory servers changes from 1 to 16, is shown in Figure 2. In this figure, the result of 5 different cases are shown. The upper 4 lines are the cases of memory usage for candidate itemsets being limited as 12[MB], 13[MB], 14[MB], and 15[MB], respectively. The lowest line is the case with no memory usage limit.

The execution time becomes longer as the memory usage limit size becomes small, since the number of swap out increases in such cases.

When the number of memory server is small, the execution time is quite long especially when the memory usage limit size is smaller. Apparently memory server(s) become bottleneck in these cases. This bottleneck is resolved when the number of memory server is 8 - 16 in this experiment.

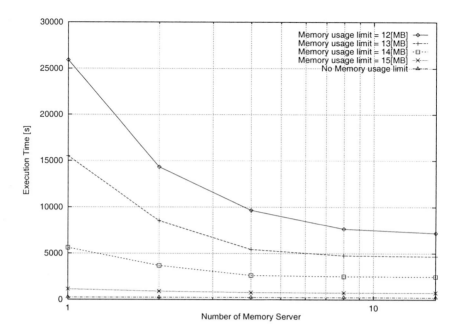

**Fig. 2.** Execution time of HPA program (pass 2)

| Usage limit | node 1 | node 2 | node 3 | node 4 | node 5 | node 6 | node 7 | node 8 |
|---|---|---|---|---|---|---|---|---|
| 12[MB] | 1606258 | 2925254 | 1306521 | 2361756 | 1671840 | 1723410 | 2166277 | 2545003 |
| 13[MB] | 885798 | 1896226 | 593000 | 1374688 | 932374 | 896150 | 1326941 | 1375398 |
| 14[MB] | 254094 | 1003757 | | 512984 | 286945 | 191102 | 601657 | 407628 |
| 15[MB] | | 268039 | | | | | | |

**Table 3.** The numbers of pagefaults when memory usage limit is changed

When memory usage is limited, the execution time is quite longer than the no memory limit case. This is because the number of swap out is extremely large. In Table 3 the numbers of pagefaults at each node are shown. Pagefaults occurred at only one node when the memory usage limit is 15[MB], and at seven nodes when the memory usage limit is 14[MB]. We can calculate the execution time of each pagefault as follows: We will focus on the case when the number of server nodes is 16, in which memory servers are not considered to be bottleneck. In the case of memory usage limit being 13[MB], for example, the execution time of the program is 4674.0[s], and the difference of the execution time between the no memory limit case is 4427.0[s]. The total execution time is decided by the busiest node of swapping operation. In this case the maximum number of pagefaults is 1896226. Thus the execution time of each pagefault can be obtained as dividing

| Usage limit | Exec. time | Diff. of Exec. time | Max # of pagefaults | Exec. time of pagefault |
|---|---|---|---|---|
| 12[MB] | 7183.1 [s] | 6936.1 [s] | 2925243 | 2.37 [ms] |
| 13[MB] | 4674.0 [s] | 4427.0 [s] | 1896226 | 2.33 [ms] |
| 14[MB] | 2489.7 [s] | 2242.7 [s] | 1003757 | 2.22 [ms] |
| 15[MB] | 757.3 [s] | 510.3 [s] | 268093 | 1.90 [ms] |

**Table 4.** The execution time of each pagefault (The number of server nodes is 8)

the difference of the execution time by the maximum number of pagefaults, 2.33[ms] in this example. The rest cases are also calculated in Table 4.

The execution time of each pagefault consists of round trip delay time, data transmission time, and memory allocation and/or search time at memory servers. The point-to-point round trip time on our PC cluster is approximately 0.5[ms] as mentioned in [14]. Since the point-to-point throughput is about 120[Mbps] on our cluster, and each pagefault data is contained in one message block (4Kbytes), the data transmission time can be calculated from these values as approximately 0.3[ms]. The rest of time is considered to be the memory servers operation time.

We can compare this pagefault execution time with the access time of hard disks. According to a state-of-art specification of SCSI hard disks, HITACHI DK319H 7,200[rpm] disks for example, the average seek time for read is about 7.5[ms] and the average rotation waiting time is about 4[ms]. In the case of latest fast hard disks such as HITACHI DK3E1T 12,000[rpm] disks, the average seek time for read is about 5[ms] and the average rotation waiting time is about 2.5[ms]. Therefore, it takes at least 11.5[ms] to read data from 7,200[rpm] hard disks, and 7.5[ms] in the case of 12,000[rpm] hard disks. Compared these values with the results of the experiment, we can conclude that performance of the application must become terrible if we use a secondary storage system as a swapping device.

# 6 Conclusion

In this paper, our preliminary experimental results are shown and discussed in which application executing nodes acquire extra-memory dynamically from available idle nodes on ATM connected PC cluster. Because the number of pagefaults during the execution is enormous, the execution time of each pagefault is crucial for total performance. The experimental result shows this method is expected to be considerably better than using hard disks as a swapping device. We are planning to perform the experiment using hard disks as a swapping device, for the sake of comparison.

## Acknowledgment

This project is partly supported by JSPS (Japan Society for the Promotion of Science) RFTF Program and NEDO (New Energy and Industrial Technology Development Organization). We would thank to Prof. Spaniol in Aachen University of Technology for supporting our research.

## References

1. Huang, C., McKinley, P. K.: Communication Issues in Parallel Computing Across ATM Networks. IEEE Parallel and Distributed Technology. **2**, 4 (1994) 73–86
2. Carter, R., Laroco, J.: Commodity Clusters: Performance Comparison Between PC's and Workstations. Proceedings of the Fifth IEEE International Symposium on High Performance Distributed Computing. (1996) 292–304
3. Culler, D. E. , Dusseau, A. A., Dusseau, R. A., Chun, B., Lumetta, S., Mainwaring, A., Martin, R., Yoshikawa, C., Wong, F.: Parallel Computing on the Berkeley NOW. Proceedings of the JSPP'97. (1997) 237–247
4. Barak, A., La'adan, O.: Performance of the MOSIX Parallel System for a Cluster of PC's. Proceedings of the HPCN Europe 1997. (1997) 624–635
5. Tezuka, H., Hori, A., Ishikawa, Y., Sato, M.: PM: An Operating System Coordinated High Performance Communication Library. Proceedings of the HPCN Europe 1997. (1997) 708–717
6. Tamura, T., Oguchi, M., Kitsuregawa, M.: Parallel Database Processing on a 100 Node PC Cluster: Cases for Decision Support Query Processing and Data Mining. Proceedings of SC97: High Performance Networking and Computing. (1997)
7. Oguchi, M., Shintani, T., Tamura, T., Kitsuregawa, M.: Optimizing Protocol Parameters to Large Scale PC Cluster and Evaluation of its Effectiveness with Parallel Data Mining. Proceedings of the Seventh IEEE International Symposium on High Performance Distributed Computing. (1998) 34–41
8. Fayyad, U. M., Shapiro, G. P., Smyth, P., Uthurusamy, R.: Advances in Knowledge Discovery and Data Mining. The MIT Press. (1996)
9. Agrawal, R., Imielinski, T., Swami, A.: Mining Association Rules between Sets of Items in Large Databases. Proceedings of the 1993 ACM SIGMOD International Conference on Management of Data (1993) 207–216
10. Shintani, T., Kitsuregawa, M.: Hash Based Parallel Algorithms for Mining Association Rules. Proceedings of the Fourth IEEE International Conference on Parallel and Distributed Information Systems (1996) 19–30
11. Heinanen, J.: Multiprotocol Encapsulation over ATM Adaptation Layer 5. RFC1483 (1993)
12. Laubach, M.: Classical IP and ARP over ATM. RFC1577 (1994)
13. Amza, C., Cox, A. L., Dwarkadas, S., Keleher, P., Lu, H., Rajamony, R., Yu, W., Zwaenepoel, W.: TreadMarks: Shared Memory Computing on Networks of Workstations. IEEE Computer. **29**, 2 (1996) 18–28
14. Oguchi, M., Shintani, T., Tamura, T., Kitsuregawa, M.: Characteristics of a Parallel Data Mining Application Implemented on an ATM Connected PC Cluster. Proceedings of the HPCN Europe 1997. (1997) 303–317
15. Shintani, T., Kitsuregawa, M.: Parallel Mining Algorithms for Generalized Association Rules with Classification Hierarchy. Proceedings of the 1998 ACM SIGMOD International Conference on Management of Data (1998) 25–36

# The Digital Puglia Project: An Active Digital Library of Remote Sensing Data

Giovanni Aloisio[1], Massimo Cafaro[1], Roy Williams[2]

[1] Facoltà di Ingegneria, Università degli Studi di Lecce
Via per Monteroni, 73100 Lecce, Italy
{aloisio,cafaro}@sara.unile.it
[2] Center for Advanced Computing Research
California Institute of Technology, Pasadena, California, USA
roy@cacr.caltech.edu

**Abstract.** The growing need of software infrastructure able to create, maintain and ease the evolution of scientific data, promotes the development of digital libraries in order to provide the user with fast and reliable access to data. In a world that is rapidly changing, the standard view of a digital library as a data repository specialized to a community of users and provided with some search tools is no longer tenable. To be effective, a digital library should be an active digital library, meaning that users can process available data not just to retrieve a particular piece of information, but to infer new knowledge about the data at hand. Digital Puglia is a new project, conceived to emphasize not only retrieval of data to the client's workstation, but also customized processing of the data. Such processing tasks may include data mining, filtering and knowledge discovery in huge databases, compute-intensive image processing (such as principal component analysis, supervised classification, or pattern matching) and on demand computing sessions. We describe the issues, the requirements and the underlying technologies of the Digital Puglia Project, whose final goal is to build a high performance distributed and active digital library of remote sensing data.

## 1 Introduction

The last three years have seen the transition from isolated personal computers to universal Internet connectivity; now in Europe and the USA we see computing centers joining into ever-larger geographically distributed collaborations. Taking these facts together, it is clear that supercomputers and data archives will soon become part of a global data and computing fabric. Some users will not be aware of the architecture or the location of the machines executing their jobs, and yet other users may construct and schedule a complex, distributed, metacomputer with heterogeneous computers and data resources providing services to some central objective.

In the paper we show how such a fabric can be exploited to create an "active digital library" of remote-sensing data. The term "library" implies that data is organized for easy access by a community of users. The term "active" implies that the library provides computing services in addition to data-retrieval services, so that users can initiate computing jobs on remote supercomputers for processing, mining, and filtering of the data in the library.

In many cases, data must be processed by a supercomputer before a human can extract any knowledge from it; only with filtering, mining, and visualization algorithms does it expose its knowledge content. A good example of such data is Synthetic Aperture Radar (SAR) images of the surface of the Earth. In this case, sometimes the user does not need just the delivering of the ground image selected from the library, but she needs to start on this image compute-intensive image post-processing (such as mosaicking, registration, interpolation, rotation, GIS integration, or tasks such as Principal Component Analysis, Singular Value Decomposition, Maximum Likelihood Classification or fusion with other data, such as a digital elevation model). Furthermore, "on-demand computing" sessions could be also required for raw-data that have not yet been processed.

In the following, we present the *Digital Puglia* project we conceived to create an "Active Digital Library" (ADL) of remote-sensing data.

## 2 The Digital Puglia Project

Our ADL prototype refers to the Puglia region of southern Italy and is built using SAR raw data provided by the Italian Space Agency. The Digital Puglia ADL is based on three joined projects:

- the SARA Digital Library,
- the Globus Metacomputing Toolkit,
- the SAR processing on Wyglaf.

**SARA** (Synthetic Aperture Radar Atlas) [1,2] is a web-based digital library that has been running at Lecce, Caltech, and the San Diego Supercomputer Center for over a year. Data is replicated on multiple servers to provide fault-tolerance and also to minimize the distance between client and server. SARA already allows clients to download SAR images from the public domain SIRC dataset. A client navigates web pages containing Java applets that implement a GUI (Graphical User Interface) showing a map of the world. Clicking on the map zooms in on a part of the world until the user can see the coverage of the atlas in terms of the SAR images, which are perhaps 50km in size. Chosen subsets of the image can then be downloaded in any of a variety of formats.

The **Globus** project [3] is developing the basic infrastructure required to support computations that integrate geographically distributed computational and information resources. Such computations may link tens or hundreds of resources located in multiple administrative domains and connected using networks of widely varying capabilities. Existing systems have only limited abilities for identifying and integrating new resources, and lack scalable mechanisms for authentication and privacy. The Globus project is building a parallel programming environment that supports the dynamic identification and composition of resources available on national-scale internets, and mechanisms for authentication, authorization, and delegation of trust within environments of this scale.

The **SAR processing on Wyglaf** project [4] is developing a parallel SAR processor on a Beowulf cluster of PCs (Wyglaf) available at the HPC Laboratory of the University of Lecce. Goal of the project is also to provide on-demand SAR processing capabilities.

A primary objective of the Digital Puglia project is to use heterogeneous distributed computing resources for an on-demand SAR processing service. The client can order processing of a dataset through the Internet, then retrieve the resulting multichannel images when processing is complete.

Joining SARA and Globus allows servicing multiple user requests for on-demand SAR processing. A collection of computing resources, such as the Wyglaf cluster at Lecce or the resources of the Center for Advanced Computing Research at Caltech in California, can be exploited to fulfill the user requirements in terms of time constraints and geographic location. Thus, depending on where the query originates and how stringent are the temporal constraints, a suitable ensemble of computing facilities will be selected by a simple click in the SARA GUI exploiting the Globus Toolkit.

In the rest of the paper we will concentrate on the SARA project and its new features. In fact, the SARA architecture (the kernel of the Digital Puglia Project) was changed with respect to the first version to meet new requirements (such as on-demand SAR processing) and to adapt its structure to the new emerging technologies (such as XML). In the following we refer to Digital Puglia as DPSARA.

## 3 Technologies Underlying the ADL Implementation

In this section, we briefly make explicit our digital library requirements. In our ADL:

- contents should be accessible on the web;
- users should be allowed to do different operations on the basis of their authorizations;
- a structured data base of SAR images must be provided and data should be searched for by means of complex queries;
- operations include, but are not limited to, retrieval of image data, on demand processing of raw data for authorized users, on demand postprocessing such as Principal Component Analysis, supervised or unsupervised classification, multitemporal image production and so on.

To meet these requirements we advise the use of the following technologies:

- a Java enabled web server which allows servlet execution (such as Sun's Java Web Server or Netscape's Enterprise Web Server);
- Java applets on the client side and Java servlets on the server side. The applets will provide the necessary interaction between the user and the library, and the servlets will be used to provide the same functionality as CGI together with the new possibility to establish interactive sessions no more stateless;

- a database supporting SQL and Sun's JDBC (Java DataBase Connectivity) which will be used as the communication protocol between servlets and the database;
- a GIS software to plot and annotate geographic maps;
- a Beowulf cluster for parallel computing on commodity hardware following the recent trend in this area;
- the MPI standard for message passing;
- HTML and XML.

The former technologies have proven themselves to be reliable and useful in the construction of distributed web-based applications. XML has gained in the last few months considerable attention, and we briefly describe its usefulness in the context of Digital Puglia.

## 4 XML, an emerging technology on the web

In the past the only way to exchange data between different software was the use of available import/export filters. XML has the potential to become a universal platform for data exchange and we adopt it as a standard to communicate between the different services provided by Digital Puglia.

XML [5] is a proper subset of SGML, intended to port on the web the essential features and the inherent power of SGML avoiding its well known complexity.

XML allows the creation of customized markup languages (i.e. an XML document resembles an HTML document, but the set of tags is not fixed and can be locally defined and extended). An XML file is just a plain ASCII file, thus can be generated using a common text editor. Moreover, it can be checked for well formedness and validity against its Document Type Definition (DTD), which is a subsidiary document that specifies allowed syntax. The check can be performed by a parser which can be written in C or Java.

XML thus makes interfaces transparent and easy to use. Collaborative working on a project becomes a simple matter, since once an agreement is reached in a group about what element types can be used, and the relevant DTD is written, then the people involved in a project can develop its part of the project referring to the defined DTD, aided by a validating parser to check for mistakes.

These features allow:

- intelligent search of a particular piece of information;
- intelligent check and validation of data structure;
- performing complex queries on the data;
- linking different types of information in a richer way with respect to HTML;
- creation of standards of XML element types for industry or specialized communities.

# 5 DPSARA metadata and XML representation

In the following we present the metadata describing our remote sensing images with reference to SARA, an XML DTD (Document Type Definition) for them, and an example image track described in XML using the DTD to check and validate the XML. The entity relationship model for the metadata is shown in Fig.1.

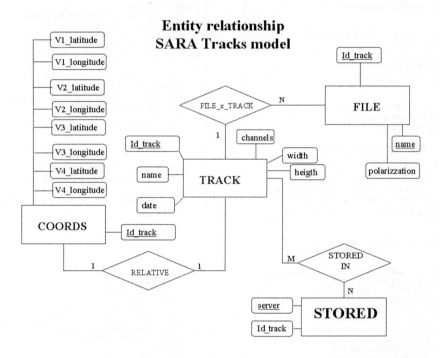

**Fig. 1.** Entity Relationship model for track metadata

The database consists of four tables. The Track table houses information about the track image such as its name, date of acquisition, unique id, width, height, and number of channels.

The Coords table contains the coordinates of latitude and longitude of the four vertex surrounding the image. In the File table the filenames of the files constituent the image are reported, and finally the Stored table contains the information about where the image is actually stored, that is, one of the web server which compose this distributed digital library.

A DTD for this metadata is presented in Fig.2, while an example of XML file produced according to the DTD is reported in Fig.3.

```
<?XML version='1.0'?>
<!-- DTD for SARA tracks -->
<!—Digital Puglia Project -->

<!ELEMENT SARAMETADATA
(SARAQUERYRESULT*)>
<!ELEMENT SARAQUERYRESULT
(SARATRACK,SARACOORDS,SARAFILES,
SARASTORED+)*>

<!ELEMENT SARATRACK
(NAME,TRACKDATE,WIDTH,HEIGHT,
CHANNELS)>

<!ELEMENT NAME (#PCDATA)>
<!ELEMENT TRACKDATE (#PCDATA)>
<!ELEMENT WIDTH (#PCDATA)>
<!ELEMENT HEIGHT (#PCDATA)>
<!ELEMENT CHANNELS (#PCDATA)>
<!ATTLIST SARATRACK IDTRACK ID
#REQUIRED>                         1
```

```
<!ELEMENT SARACOORDS
(SARACOORD+)>
     <!ELEMENT SARACOORD (LAT,LON)>
     <!ELEMENT LAT (#PCDATA)>
     <!ELEMENT LON (#PCDATA)>
     <!ATTLIST SARACOORD IDTRACK IDREF
#REQUIRED>

     <!ELEMENT SARAFILES (SARAFILE+)>
     <!ELEMENT SARAFILE (POLARIZATION)>
     <!ELEMENT POLARIZATION (#PCDATA)>
     <!ATTLIST SARAFILE NAME ID
#REQUIRED>
     <!ATTLIST SARAFILE IDTRACK IDREF
#REQUIRED>


     <!ELEMENT SARASTORED EMPTY>
     <!ATTLIST SARASTORED SERVER ID
#REQUIRED>
     <!ATTLIST SARASTORED IDTRACK IDREF
#REQUIRED>                           2
```

**Fig. 2.** Sara Track DTD

```
<?XML VERSION='1.0'?>
     <!DOCTYPE  SARAMETADATA  SYSTEM
"SaraTrack.dtd">
     <SARAMETADATA><SARAQUERYRESULT>
     <SARATRACK IDTRACK = "11829">
     <NAME>Sena Madureira, Brazil</NAME>
     <TRACKDATE>04-16-1994</TRACKDATE>
     <WIDTH>3624</WIDTH>
     <HEIGHT>7995</HEIGHT>
     <CHANNELS>4</CHANNELS>
     </SARATRACK>
     <SARACOORDS>
     <SARACOORD>
     <LAT>297.575</LAT>
     <LON>-18.588</LON>
     </SARACOORD>
     <SARACOORD>
     <LAT>297.206</LAT>
     <LON>-18.798</LON>
     </SARACOORD>
     <SARACOORD>
     <LAT>297.696</LAT>
     <LON>-19.574</LON>
     </SARACOORD>
     <SARACOORD>                      1
```

```
     <LAT>298.066</LAT>
     <LON>-19.363</LON>
     </SARACOORD>
     </SARACOORDS>
     <SARAFILES><SARAFILE      NAME    =
"pr11829_byt_hh" IDTRACK = "11829">
     <POLARIZATION>LHH</POLARIZATION>
     </SARAFILE>
     <SARAFILE   NAME    =   "pr11829_byt_hv"
IDTRACK = "11829">
     <POLARIZATION>LHV</POLARIZATION>
     </SARAFILE>
     <SARAFILE   NAME    =   "pr11830_byt_hh"
IDTRACK = "11829">
     <POLARIZATION>CHH</POLARIZATION>
     </SARAFILE>
     <SARAFILE   NAME    =   "pr11830_byt_hv"
IDTRACK = "11829">
     <POLARIZATION>CHV</POLARIZATION>
     </SARAFILE>
     </SARAFILES>
     <SARASTORED    SERVER   =   "CACR_IIPSS"
IDTRACK = "11829"/>
     </SARAQUERYRESULT>
     </SARAMETADATA>                   2
```

**Fig. 3.** An example XML file

# 6 Integration of XML-based services in SARA

XML was adopted to exchange metadata and control information between the various services composing SARA.

We use a three-tier architecture as in fig. 4 to clearly define what service does what, allowing for a more flexible separation of scope and to prepare for future integration with other services. The client tier can be a customized software or a web browser which interact with the middle tier generating queries and commands.

The middle tier is responsible of the management and execution of queries and commands received from the client layer. This can be done by using a number of different services, which can interact to provide the answer requested or execute the command issued.

The backend tier comprises the database management system, the data, a GIS and the computing resources.

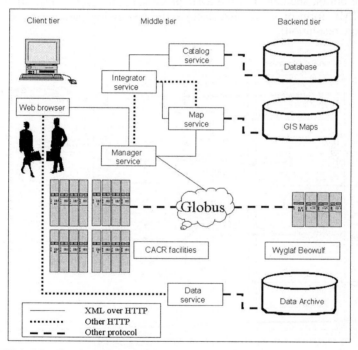

**Fig. 4.** Digital Puglia Three-Tier Architecture

More details on the use of XML for describing scientific data can be found in [6], and some insights about the new XML-Architecture of SARA are reported in [7].

## 7 On-Demand SAR processing

The "Active Digital Library" of remote-sensing data provided by the *Digital Puglia* project was conceived to allow not only retrieval of data to the client's workstation, but also customized processing of the data. Besides the usual compute-intensive image post-processing, "on-demand computing" sessions could be required for raw-data that have not yet been processed. In the case of on-demand SAR processing, the data archive must be connected to a powerful compute server at high bandwidth (controlled by a client who may be connected at low bandwidth). This means that a completely automated SAR processor must be designed and the user interface should allow control of some parameters of the SAR processing software.

The architectural solution we adopted to implement on-demand SAR processing to be integrated in the SARA interface was designed to be platform-independent.

The user (with basic password authentication) can choose on a clickable map of the world the region of interest where the SAR raw-data are available, then with a click on a simple button she can start the SAR processor. The user is also allowed to set (from a HTML form or a Java interface) some processing parameters (such as the needed image resolution).

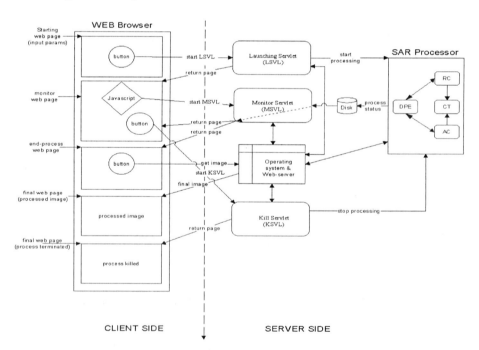

**Fig. 5.** On-Demand architecture

At this point a new window (monitor) is sent back to the user, showing how the processing is done. When the whole processing terminates, the final SAR image can be obtained by clicking on a button.

For the implementation we use three servlets and two JavaScript Routines. The processing is achieved by a sequential SAR processor until we complete the porting on Wyglaf.

In Fig.5, the architecture of the on-demand SAR processing system is showed. After chosing the track to be processed and setting some processing parameters, the SAR processing can be activated by a click on a button which starts the launching servlet (LSVL) on the server side.

The LSVL:
- controls the availability of computational resource;
- starts the SAR processor;
- immediately returns back to the user a web page, which reports the information on the state of the running process.

To allow the process monitoring a second servlet, the monitor servlet (MSVL) is used. The MSVL is called every 5 seconds by a Javascript routine, started inside of the first web page sent back from the LSVL.

The MSVL:
- controls that processing is going on;
- returns to the user another web page which contains the new output state of SAR processor (this page will active another monitor in a cyclic way).

It should be noted that the running SAR processor writes on file the intermediate messages about the state of the process, and this file is used by the MSVL to read the process information to be sent toward the client side.

The process ends when SAR processing is terminated unless the user interrupts it clicking a button which calls a kill servlet (KSVL): in this case the MSVL informs the user that the processing is terminated, giving the possibility to visualize the final processed image through a simple click on a button on the returned web page.

At this point the image is sent back to the user in a new browser window.

## Conclusions

The Digital Puglia Project is a high performance distributed Active Digital Library of remote sensing data. The library allows not only retrieval of data to the client's workstation, but also customized processing of data. The kernel of the library is the SARA architecture that was modified with respect to its first implementation to meet new requirements and to exploit new emerging technologies, like XML.

The issue and the underlying technologies of the Digital Puglia Project were presented.

## References

1. R. Williams, G. Aloisio, M. Cafaro, G. Kremenek, P. Messina, J. Patton, M. Wan,"SARA: The Synthetic Aperture Radar Atlas", http://www.cacr.caltech.edu/sara/
2. G. Aloisio, M. Cafaro, P. Messina, R. Williams, "A Distributed Web-Based Metacomputing Environment", Proc. HPCN Europe 1997, Vienna, Austria, Lecture Notes In Computer Science, Springer-Verlag, n.1225, 480-486, 1997.
3. I. Foster and C. Kesselman, "The Globus Project", http://www.globus.org
4. G. Aloisio, M. Cafaro, "The Wyglaf Beowulf machine", http://sara.unile.it/~cafaro/wyglaf.html
5. "www.xml.com"
6. R. Williams,, J. Bunn, R. Moore, and J.C.T. Pool, "Interfaces to Scientific Data Archives", Report of a Workshop sponsored by the National Science Foundation, May 1998, http://www.cacr.caltech.edu/isda
7. G. Aloisio, G. Milillo, R.D. Williams, "An XML Architecture for High Performance Web-Based Analysis of Remote Sensing Archives", to appear on FGCS Int. Journal, North Holland

## Acknowledgments

The work has been supported by the Italian Space Agency grant ASI ARS-96-118 and by the Center for Advanced Computing Research (CACR) at Caltech.

# An Architecture for Distributed Enterprise Data Mining

J. Chattratichat, J. Darlington, Y. Guo, S. Hedvall, M. Köhler, and J. Syed

Data Mining Group
Imperial College Parallel Computing Centre
180 Queen's Gate, London SW7 2BZ, UK
{jc8, jd, yg, dsh1, mk, jas5}@doc.ic.ac.uk

**Abstract.** The requirements for data mining systems for large organisations and enterprises range from logical and physical distribution of large data and heterogeneous computational resources to the general need for high performance at a level that is sufficient for interactive work. This work categorises the requirements and describes the *Kensington* software architecture that addresses these demands. The system is capable of transparently supporting parallel computation at two levels, and we describe a configuration for trans-atlantic distributed parallel data mining that was demonstrated at the recent Supercomputing conference.

## 1 Introduction

*Data Mining*, or *Knowledge Discovery in Databases* is concerned with extracting useful and new information from data, and provides the basis for leveraging the investments in data assets. It combines the fields of databases and data warehousing with algorithms from machine learning and methods from statistics to gain insight in hidden structures within the data. In order to apply the knowledge from the Data Mining process, the results need to be analysed, often with the help of visualisation tools, as well as integrated into the business process.

Data mining systems for enterprises and large organisations have to overcome unique challenges. They need to combine access to diverse and distributed data sources with the large computational power required for many mining tasks. The data mining process, as perceived by the analysts, knowledge workers and end-users of the discovered knowledge, is an interactive one that functions best when a high degree of interactivity is available. The analyses are usually refined during several iterations through the cycle of data selection, pre-processing, model building and model analysis. The best results are usually achieved by combining models from different techniques, which calls for a wide variety of integrated tools within the system, as well as openness for future extensions.

In large organisations, data from numerous sources needs to be accessed and combined to provide comprehensive analyses, and work groups of analysts require access to the same data and results. For this purpose the existing networking infrastructure, typically based on Internet technology, is to be re-used.

Confidentiality becomes a key issue, and the system architecture needs to provide security features at all levels of access. The different needs of enterprises require that a system offers a wide range of configuration options, so that it is possible to scale applications from a few client workstations to high-performance server machines.

In this article we will discuss the implications of the above requirements, focusing on the Kensington solution that employs Internet and distributed component technologies for deployment on high-performance servers such as distributed memory and shared-memory parallel machines. The next section will discuss the key functional requirements that have been outlined so far. The following chapter outlines the design and implementation of the *Kensington* enterprise data mining system, in particular Java- and CORBA-based networking and component technology. We then describe a scenario for distributed data mining that was demonstrated at SuperComputing'98 as part of the award-winning Terabyte Challenge. The final section concludes and outlines future trends in the field.

## 2 Enterprise Data Mining Requirements

Data mining system architectures for enterprises have to meet a range of demands from the field of data analysis and the additional needs that arise when handling large amounts of data inside an organisation. Modern data mining applications are expected to provide a high degree of integration while retaining flexibility. In this way they can efficiently support different types of analyses over the organisation's data. Data mining is understood to be an iterative process for the analyst [FPSS96], especially in the initial exploratory phases of the analytical task. Therefore, a high degree of interactivity is required, often combined with the need for visualisation of the data and the analytical results.

The field of data mining is developing rapidly, and the methods applied in a tool today may be superseded by more advanced algorithms in the near future. Furthermore, the convergence with statistical methods has only just started, and will grow in pace over the next few years. The need for enhancement of the existing tool set has to be reflected by a software architecture that enables the straightforward integration of new analytical components. In a similar vein, the results from the analytical functions need to be presented in portable formats, as most analysts will want to use different specialist packages to further refine or report the results.

In large organisations, the amount and the distribution of the data become an additional challenge. The size of the data may make it impractical to move it between sites for individual analytical tasks. Instead, data mining operations are required to execute "close to the database". In the absence of dedicated support for data mining and other analytical algorithms in the database management systems, this can be achieved by setting up high-performance servers in close proximity to the databases. The overall data mining system will then have to manage the distributed execution of the analytical tasks and the combination of the partial results into a meaningful total. Also, this approach can sometimes benefit from the improved generalisation power that often occurs when combining analytical models [CS96].

Three kinds of scalability requirements arise in enterprise environments; data sets can be *very large*, there may be *many sites* on which data is accumulated, and there may exist *many users* who need access to the data and the analytical results. An enterprise data mining architecture should be flexible enough to scale well in all these cases. This will require access to high-performance analytical servers, the ability to distribute the application and the capability to provide multiple access points.

## 2.1 Vertical structure of Enterprise Data Mining

Any architecture that fulfills these requirements is likely to be a *three-tier client/ server architecture*. The functional specification for the three tiers includes the client, application server and third-tier servers. The client handles interactive creation of data mining tasks, visualisation of data and models, and sampling of the data. The application server authenticates users, provides persistent storage and access control for objects, and controls task execution and data management. The third-tier servers provide both database and high-performance data mining services.

## 2.2 Horizontal structure of Enterprise Data Mining

*Software Component Architectures* have been developed to facilitate the creation of flexible distributed systems. Software components are modular building blocks that developers can combine into ensembles. The implementation of the components is encapsulated and exposed only through an application programming interface (API). New functions can be integrated into the application by embedding them into new components and attaching them to the existing system. The flexibility and extensibility required by enterprise data mining environments makes the use of software components mandatory.

A variety of component architectures are available, most notably the Java Enterprise Architecture, CORBA and DCOM. The main differences between these proponents are the range of hardware and operating systems supported, and the options for component deployment.

## 2.3 The role of high-performance systems

High-performance machines are most likely to be employed in the third tier. Parallel systems have been shown to provide appreciable speed-up for data mining tasks [CDG$^+$97], and they have been in use for parallel database machines for some time.

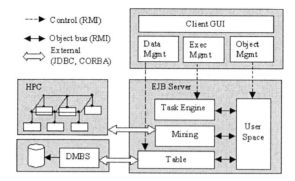

**Fig. 1.** Kensington Client and Server Components

## 3 Kensington Data Mining system

The Kensington Enterprise Data Mining system (see Figure 1) has been designed using the Enterprise JavaBeans component architecture and has been implemented in Java. It integrates parallel data mining functions that have been written in C and MPI via a CORBA interface. Databases anywhere on the Internet can be accessed via a JDBC connection. The client is built as a highly interactive Java application using JavaBeans.

### 3.1 Application server

**Design** The application logic consists of four Enterprise JavaBeans (EJB) classes, which execute in an EJB server environment. The EJB server is a generic implementation of Sun's EJB 1.0 standard, with additional capabilities for handling user login and authentication. The EJB server provides container functions for all EJB classes according to deployment options that are set as part of the application design. Each EJB class provides services for one aspect of the application, namely user object management, task execution, mining component management and database access and data storage. The EJB server provides *container-based* persistence, but allows EJBeans to manage their persistence themselves (*bean-managed* persistence).

A key advantage of the EJB architecture is the ability to run multiple cooperating servers at different sites. The Kensington EJB server provides two global services for such a set-up: user authentication and secure communication between EJB server incarnations. EJBean classes can then be flexibly deployed across these servers. An example of a distributed configuration is described in section 4. The flexible deployment options enable the use of multiple points of entry for user sessions, thereby providing scalability in the number of supported clients.

**Implementation** The four EJB classes that constitute the application server functionality are:

**UserSpace EJB** manages persistent objects for the user; the functionality of this EJBean includes a directory structure for the objects with user and group access control. The UserSpace EJB manages the persistence of user's data mining tasks and results.

**TaskEngine EJB** is the contact point for the client to control the execution of data mining tasks, including preprocessing, model generation and model evaluation; the execution can be distributed between different sites; results are returned to the client.

**Mining EJB** handles the interface with the analytical components; high-performance parallel components are accessed via a range of protocols, including CORBA and JNI.

**Table EJB** is responsible for the import and storage of data bases, via JDBC; it supports browsing the external databases from the client. As of this writing, the Table EJB utilises the container-based persistence for the user's datasets, but a bean-managed version is under construction for improved performance.

The EJB server implemented for Kensington provides all the container functions specified by the EJB-1.0 standard, functions for container-managed persistence from the EJB-2.0 standard and additional services for user session management. The latter have been designed to provide system-wide authentication for multiple cooperating EJB servers that can be distributed. In this scenario a common JNDI service (Java Networking and Directory Interface) provides global naming services. In addition, the EJB server facilitates the sharing, or *pooling* of critical resources such as network connections to databases and CORBA connections to external components.

All EJB classes can be flexibly configured within any number of EJB servers under a common JNDI umbrella. This implies that *a single application server* configuration can contain either multiple points of login for user sessions, or multiple sites for storing data, or multiple sites for mining data. A single user's or group's data can be distributed across all the sites that contain a Table EJBean, and the user can initiate a mining task from any of the login sites to execute mining functions on any of the EJB server sites that contain a Mining EJBean. Of course, all EJBean classes can also be configured to run within a single EJB server.

The services provided by the EJB classes are exposed to the client via control interfaces (cf. the *Control* connections in Fig. 1), which are realised in RMI over secure sockets (SSL). All interaction between EJB classes uses the same interfaces, thus allowing for transparently distributed instances of the EJBeans. The UserSpace EJBean serves as repository for persistent objects which interacts with the other EJB classes via an interface that functions as an *object bus*.

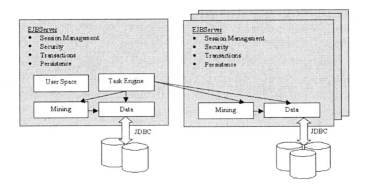

**Fig. 2.** A Scenario with Distributed Kensington EJB Servers

External interfaces have been defined for the Table and Mining EJBean. They are realised via JDBC for database access and CORBA, respectively, thus allowing maximal portability and adaptability.

### 3.2 Client

**Design** The main purpose of the client is to provide an interactive GUI with two main capabilities: interactive visual programming of data mining tasks, and three-dimensional visualisation of data and analytical models. The client/server connection makes conservative assumptions about the performance of network connections and tries to reduce the amount of data transferred between the two tiers. Kensington uses a sampling method to provide the user with enough data for interactive pre-processing and preliminary analysis without overloading either the network, or the memory of the client machine. However, the full data set can be accessed from the client on demand. The data mining tasks programmed by the users generally execute over the full data set, thus providing the most comprehensive analysis possible.

The client offers a browser for inspecting the contents of remote databases, prior to import into the Kensington server. The user only needs read access permissions to the database and tables in question. Once entered, the details of the connection are stored as *database bookmark* for future use. It is possible to reduce the data imported from the source data table by interactively formulating selection queries. Upon completion of the selection, the Table EJBean of the attached EJB server imports the data into the Kensington EJB server.

**Implementation** The client is a Java application that is connected to one or more EJB servers via the RMI protocol. The graphical user interface has been implemented with JavaBeans to allow flexible extension of the visual programming framework. Every component in the user's data mining task is represented by a JavaBean element. The data mining process consists of data sources (tables), pre-processing operations, analytical functions (data mining and statistical methods) and application nodes (model assessors and predictor nodes), cf. Fig. 4. When the user connects task nodes, the client checks the metadata definitions of the data set to check the semantic conditions of the operations.

The GUI on the client interacts with the EJB server via three components. The *Data Manager* is responsible for accessing the services of the Table EJBean and provides the elementary functions for browsing remote data bases. The *Exec Manager* is invoked on the client when the user requests the execution of a data mining task. The *Object Manager* provides a view of the user's objects held on the Kensington server. These components reflect the functions of the EJBeans on the server, while also encapsulate caching and network connection management.

### 3.3 Third-tier components: high-performance mining

**Design** The analytical components of Kensington are parallel programs that have been encapsulated as objects via the Mining EJB class. This class controls data conversion and parameter passing to each of the algorithms. Any component that executes on dedicated separate systems, such as parallel machines, can be integrated via CORBA, RMI (Java Remote Method Invocation) or JNI (Java Native Interface). The interfaces allow the introduction of further mining components with very little overhead, providing flexibility similar to that in the client. The predictive models that are computed by the data mining modules are internally represented in the emerging PMML (Predictive Model Markup Language) standard, which is developed within the Terabyte Challenge project, cf. section 4.

**Implementation** The current range of analytical functions includes classification (C4.5 [Qui93], backpropagation neural networks [Pen97] and naive Bayesian methods), clustering (self-organising maps [Rue97] and k-means methods), association rule discovery (a-priori method [Toi96]). In addition, an interface for calling functions in an S-plus statistical server has been created. This allows feeding any data from Kensington to any function available in S-plus, with results reported on the client.

A CORBA interface was designed to handle the transfer of data and control between the Mining EJBean and the external mining components. This allows the immediate integration of CORBA-capable systems. Furthermore, any data mining application can be *wrapped* by an object layer that provides the interface methods and translates the data and control parameters [CDG+98]. The open interface of the Mining EJBean allows quick migration of separately programmed components into Kensington, thus providing rapid access to the other features of the system for the data mining developer.

## 4 Distributed Parallel Data Mining for the Terabyte Challenge

The Terabyte Challenge project explores the technology to tackle data analysis projects of the largest imaginable size. The scope of the project encompasses

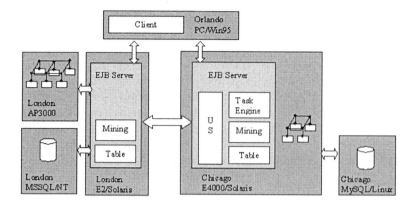

**Fig. 3.** The Trans-Atlantic Kensington Configuration for Supercomputing'98

distributed and parallel computing software technology. The Kensington system was demonstrated in a distributed setting with two host sites and a client running at the SuperComputing'98 site. Each host site consisted of a machine that ran an instance of the Kensington EJBserver. One of the sites, at the University of Illinois in Chicago, was equipped with a full complement of EJBeans and acted as the coordinating site for the whole installation. The second site, at Imperial College, ran an EJB server with only Table and Mining EJBeans (see Figure 3). The network connections were standard IP connections over Ethernet.

The client was configured with access to both server sites, which enabled the users to access and mine data concurrently across the network. It is important to note that the coordinating EJB server, and in particular the TaskEngine EJBean, act as arbitrator for control purposes only: the data that was distributed over databases at the host sites was loaded and mined locally.

In a typical client session (Fig. 4), the user browses remote databases until the required data has been identified. The browsing and loading is handled by a Table EJBean that is associated with each database at request time[1]. The user then proceeds to edit a data mining task on the client. A mining task starts with a data source, may contain some preprocessing operations and terminates with an analytical function. Separate tasks can be defined for any data source that has been loaded, including data loaded from different databases and held at different EJB server sites. Upon request, tasks are handed to the TaskEngine on the coordinating EJB server. It selects executing Mining EJBeans according to a resource allocation policy[2] and initiates the computations. Results are sent back to the coordinating TaskEngine which forwards them to the client.

---

[1] For the SuperComputing demonstration, a static mapping from databases to EJB servers was employed. Dynamic mappings are possible as well.

[2] The default policy is to execute mining operations in the EJB server that contains the data.

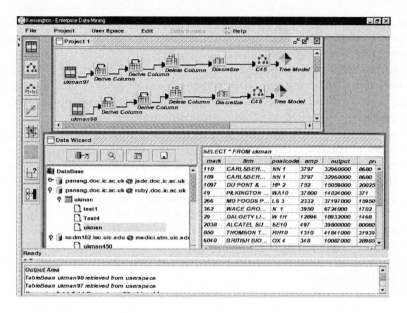

**Fig. 4.** Mining Data simultaneously in the UK and US

This configuration effectively provides parallelism at two levels. The mining tasks are parallel components that run on two different multi-processor architectures (distributed and shared memory). The overall mining task consists of mulitple mining activities which are scheduled to happen concurrently on the two different machines.

## 5  Conclusion and outlook

The Kensington system architecture has been designed to address the demands of enterprise data mining. Its flexibility results from the use of distributed components, built with Java Enterprise Beans technology. The EJB server developed within the project provides global binding, authentication and security services. The deployment of components for the application can be customised to match the computational and data resources at hand. The Kensington server has been built entirely in Java, which provides the additional benefit of platform independence.

The interfaces between components and to the environment are open to facilitate the integration of new components at all levels, including analytical functions (data mining and statistics), data pre-processing operations and visualisation and report JavaBeans. A CORBA interface provides connections to high-performance data mining services, and an interface to the S-plus package integrates numerous statistical functions.

All network connections can operate over standard TCP/IP Internet or Intranets or virtual private networks, and the system incorporates connection management and secure communication.

Data mining architectures have rapidly evolved from stand-alone applications over flat files to systems with integrated data management. Systems such as Kensington belong to the class of third generation architectures which are characterised by distribution of data and computation.

Emerging technologies for *dynamic networking*, such as Sun's Jini, will provide a layer above the current interfaces for distributed components. The main addition is the capablity to dynamically add computational resources to a network, thus enabling truly *networked applications*. These capabilities, combined with the mobility of code that has been pioneered by the Java language, will allow future enterprise data mining systems to deliver comparatively compact analytical functions to any database site on the network. The full performance potential will be leveraged when the integration of Java into the database management systems allows the mobile mining components to execute within the databases themselves.

# References

[CDG+97] J. Chattratichat, J. Darlington, M. Ghanem, Y. Guo, H. Hüning, M. Köhler, J. Sutiwaraphun, H. W. To, and D. Yang. Large scale data mining: Challenges and responses. In *Proceedings of Third International Conference on Knowledge Discovery and Data Mining*, pages 143–146, 1997.

[CDG+98] J. Chattratichat, J. Darlington, Y. Guo, S. Hedvall, M. Köhler, A. Saleem, J. Sutiwaraphun, and D. Yang. A software architecture for deploying high-performance solutions on the internet. In *High-Performance Computing and Networking*, 1998.

[CS96] P. K. Chan and S. J. Stolfo. Sharing learned models among remote database partitions by local meta-learning. In E. Simoudis, J. Han, and U. Fayyad, editors, *The Second International Conference on Knowledge Discovery and Data Mining*, pages 2–7. AAAI Press, 1996.

[FPSS96] U. M. Fayyad, G. Piatetsky-Shapiro, and P. Smyth. From data mining to knowledge discovery: An overview. In U. M. Fayyad, G. Piatetesky-Shapiro, P. Smyth, and R. Uthurusamy, editors, *Advances in Knowledge Discovery and Data Mining*. MIT Press, 1996.

[Pen97] Wanida Pensuwon. Parallel neural networks. Msc. thesis, Imperial College, University of London, 1997.

[Qui93] J. Ross Quinlan. *C4.5: Programs for Machine Learning*. Morgan Kaufman Publishers, 1993.

[Rue97] Stefan Rueger. Parallel self-organising maps. In *Proceedings of the Seventh Parallel Computing Workshop, Australian National University, Canberra, September 25–26*, 1997.

[Toi96] Hannu Toivonen. *Discovery of Frequent Patterns in Large Data Collections*. PhD thesis, Department of Computer Science, University of Finland, 1996.

# Representatives Selection in Multicast Group

Shusheng Li        A.L.Ananda

Center of Internet Research, School of Computing
National University of Singapore
lishushe@comp.nus.edu.sg, ananda@comp.nus.edu.sg

**Abstract.** *An usual problem in multicast is feedback implosion [Dan89] when a large number of receivers synchronously send feedbacks to the sender. For avoiding it, this paper provides a distributed mechanism to build a representatives set in the group according to the network traffic and group member distribution. They may be used to reflect the situation of entire group.*
*This mechanism has several major features: it seldom relies on the source, the selection of representatives is irrelevant with the source. Each member in the group decides its state by itself according to the situation around it, and doesn't need the overall group information. It is scalable and robust in adaption to network traffic. Simulations show the selection mechanism is very effective and fast to reach a stable state.*

## 1 Introduction

Representatives are used by many existing multicast protocols like RMTP [LP96], MESH [LDW97], MTP [SAK92]. But there still isn't a good representative selection solution. Especially for many-to-many multicast service, a independent representative selection mechanism which don't rely on a specified sender's multicast tree is needed. The RMTP proposed by Paul [LP96] provide a unchangeable hierarchical structure in which receivers are grouped into local regions or domains and in each domain there is a special receiver called a *Designated Receiver* which is responsible for sending acknowledgments periodically to the sender, it processes acknowledgments from receivers in its domain and retransmit lost packets to the corresponding receivers. Although it provides a fast feedback and local error recovery, it need the users to be aware of some information about the approximate location of receivers and network topography. It is often impossible to an ordinary user. MESH [LDW97] partitions receivers set along the boundaries representing a high performance network domain within the multicast network. By electing an active receiver in each subgroup, it achieves local recovery and feedback suppression. However, it doesn't address the algorithms for effective subgrouping and electing the active receivers.

In this paper, we will provide a flexible self-organizing mechanism according to the network traffic and topography. We assume there are three kind of member: sender, representative and ordinary member in a group. The representatives are selected among all receivers by a scalable distributed algorithm take responsibility to collect receiving information from receivers around them and reflect them to the sender. Every receiver will decide its state, identity in the group and where it send feedback to according to the neighborhood situation and network traffic around itself.

## 2 Construction of Neighbor List

For rapid loss recovery and impressing the exceeded feedback, at first every receiver in the group must make decision independently to change state according to knowledges which it can collect from its neighborhood. The effective neighbor recognition method is the base of the self-organizing algorithm.

One receiver recognize the neighbor situation by *Neighbor Inquiry* and *Neighbor Probe* message which will be described later, and then rank those neighbors around itself with a *preference weight*. Every receiver maintains a neighbor list to record its loss rate and the distance between them.

One neighbor item in the neighbour list is described as follow:

| |
|---|
| Round trip time to the neighbor |
| Loss rate of the neighbor |
| Preference weight |
| Whether it suppress me |
| Is the neighbor a representative |
| Its feedback no |
| ...... |

**Table 1.** Neighbour list content

The receiver ranks the neighbors around it by these parameters. When a receiver first join the group, it will use an expanding ring search (ERS) technique to look for potential neighbors. As soon as other member receives this probe, it will include its loss rate and time stamp into a reply packet and send it to the receiver by unicast. Then the receiver will use below equations to evaluate this node and build a neighbor item in the neighbor list.

### 2.1 Selection of Neighbors

A receiver doesn't accept all other receivers in the same local area to be its neighbor. The neighbour number in the neighbour list is limited by a parameter called *neighbour scope*. We provide an selection mechanism to accept neighbors.

Selection of neighbors will consider two factors:

– The loss rate of the neighbor. The receiver prefers reliable member to be its neighbor.
– The distance between the receiver and the neighbor. The receiver prefers nearer member to be its neighbors for quick loss recovery.

## 2.2 Distance Estimation

Usually TTL value in the IP packet header is used to estimate the distance between two nodes on internet. Like in a traditional IP router, TTL is also decreased by one at each multicast router. But some experiments in [XMZY97] shows that the TTL values between two hosts are usually not symmetric. That host A can reach host B with some TTL value doesn't imply that host B can reach host A with the same TTL. A receiver may be not able to get right estimation for the distance from the TTL value within the packet. Moreover, although hop count to some extent reflects the distance between two hosts, it is usually static and can't reflect dynamical network traffic changes, and the current UNIX socket API makes it difficult for a user level application to get access to the TTL information. For these reasons, we use the round trip time (RTT) to estimate the distance among receivers and calculate the preference weight in that it can provide dynamic information on network changes.

To evaluate the distance between a neighbor node and the receiver, One the receiver need to sends a probe packet to the neighbor with timestamp ($t_1$). The neighbor records the time it receives the message ($t_2$) and includes this value in its reply to the receiver node, which it will also timestamp ($t_3$). The receiver node records the time at which it receives the neighbor's reply ($t_4$). Then the distance will be:

$$distance_{(neighbor, loser)} = \frac{(t_2 - t_1 + t_4 - t_3)}{2} \quad (1)$$

## 2.3 Calculating Preference Weight

The preference weight is used to indicate the extension that a neighbor can be chosen to be a request objective when a loss is detected. As above stated, The near and reliable neighbors are more prefered to be request objectives. So the calculation of preference weight is based on these two factors.

We use below formula to evaluate the loss rate:

$$Loss(N) = 1 - \frac{(the\ number\ of\ successfully\ received)}{the\ number\ of\ the\ totally\ sent} \quad (2)$$

Then the expected time of receiving a correct packet from this neighbor with above parameters.

$$E(Time) = 2 \times distance_{(nr)} \times (1 - l) + Time_{loss} \times l \quad (3)$$

$distance_{nr}$ is the distance between the neighbor and the receiver. l is the loss rate calculated by above formula. $Time_{loss}$ is the time needed for loss recovery when this neighbour fail to provide a recovery.

According to the equation 3, we make preference weight as:

$$Weight = distance_{(nr)} + (\alpha \times distance_{(nnr)}) \times Loss \quad (4)$$

$distance_{nnr}$ is the maximum distance(RTT) in the neighbor list. $\alpha$ is a scale variable to adjust the seriousness of loss rate. Receivers arrange their neighbours in the neighbour list by this preference weight. The lower preference weight is, the more prefered a neighbour is to be a request object once a loss is detected.

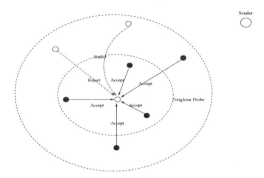

**Fig. 1.** Neighbour Selection

## 3 Organization of Neighbor Nodes

### 3.1 Neighbor Probe

*Neighbor Probe* is an extending scoped multicast message used by receivers to find new members who have just joined the group or by a new member which just join the group to recognize the environment around it. It is sent out in a fixed *Probe Interval*. The receiver which receives this probe message will reply a *Neighbour Probe Reply* message with a time stamp and its loss rate. Once a new receiver is found to join the multicast group, the receivers around it will estimate it performance by information included in the *Neighbour Probe Reply* by the new member, and then decide whether add it into the neighbour list.

The member who already exists can also reply a *Neighbour Probe* to let the prober re-estimate its performance. Once it is qualified, it can be added to the neighbor list of the prober.

But for avoiding frequent change in the content of neighbor list which may cause a decrease on system performance, the *Probe Interval* often is a rather long interval.

### 3.2 Neighbour Inquiry

As soon as, a receiver build up its neighbor list by extending scope probe. There are then periodical unicast communications among these receivers. Each receiver exchange information of loss rate and network traffic with its neighbors by an unicast message called *Neighbour Inquiry* in a fixed Inquiry Interval. It is an uni-direction inquiry, but not bi-directed. That a receiver inquires a neighbor in its neighbor list doesn't mean that the neighbor also inquiries it back in turn. The inquirer is called the *follower* of the receiver who is inquired. The receiver who is inquired by its follower must also issue a *Neighbor Inquiry Reply* to the follower to inform it current situation information. By periodically checking the inquiry reply from the neighbors, one receiver may keep updating the neighbors' performance and re-rank them in the neighbor list by recalculating their preference weight.

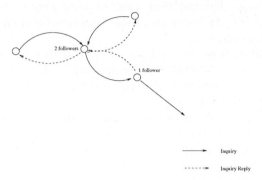

**Fig. 2.** Receivers exchange inquiry messages

### 3.3 Adjustment in Neighbour list

When network traffic changes, The original performance of some neighbour in the neighbour list may also change. So the preference weight of one neighbour should change and dynamically reflect network traffic.
There are some factors effecting the preference weight.

1. One neighbour is experiencing a congestion to the receiver.
2. One neighbour is experiencing a congestion to the source.
3. One neighbour loses connection.
4. A new receiver joins the group

By include the time stamp in the inquiry message, one receiver can check the traffic status between it and a neighbor. In case of one neighbor experiencing a congestion with the receiver, the round trip time($distance_{nr}$ in the equation 4) will be effected, then increase, then the preference weight of the neighbor will increase so that receiver may rearrange its position in the neighbor list.

In case of one neighbor experiencing a congestion with the source. the loss rate of the neighbor will increase so that increase the preference weight in the equation 4

For a flexible management on the members in the group, we allow members suddenly join and leave the group. So we set a timer in every receiver for detecting losing connection of neighbors in the neighbor list. Once one neighbor doesn't reply a inquiry in an interval, the neighbor will be considered as an unavailable member.

The new member can by found by periodical extending multicast message *Neighbor Probe*. Once it appears and responses the probe, it will be found by others and be considered to be a neighbor of others. The new member also send out a *Neighbor Probe* message with extending TTL scope for searching neighbors once it join the group. The qualified members then may be added to this new member's neighbor list. At meantime, those who receives its probe message will also consider it as a follower.

The receiver will change the sequence in the neighbor list according the new information come from the probe or inquiry reply. According to the section 2.3, the sequence in the neighbor list will directly influence the error recovery effectivity. Updating the sequence in the neighbor list will keep receivers always get a quick loss recovery. When the performance of a neighbor isn't qualified or the connection is not available, it may be deleted from the neighbor list. Once the number of neighbors in the list is lower than a *MINIMUM NEIGHBOR NUMBER*, the receiver will initiate new *Neighbor Probe* with expanding scope multicast to add new neighbors to the neighbor list.

## 4 Representative Selection

### 4.1 Inquiry Set

Every receiver maintain a *Inquiry Set* which is filled with the description of *followers*(defined in section 3.2)which send *Neighbour Inquiry* to it. Receivers will record the sum of *followers* which send *Neighbour Inquiry* to it. This sum is called *inquired number*. Another parameter is called *preference factor* which is included in a *Neighbour Inquiry* message. It indicates the receiver's position in the *follower's* neighbour list. It is used when two members compete to be a representative. As described in section 4.3, when two member have the same *inquired number*, they compare their *feedback rate*. The *feedback rate* is the sum of *preference factor* of all the *followers*. Because we queue the neighbours in the neighbour list with a good-to-bad sequence. The lower *preference factor* is, the better a neighbour's performance is. So the lower *feedback rate*, the more possible a member becomes a representative. When a receiver sends out a *Neighbour Inquiry* to a neighbour, it will include its *inquired number, feedback rate, preference factor* of this neighbour, *the flag of feedback, feedback no* information.

One follower's item in the Inquiry Set is defined below:

### 4.2 Representative Set

A representative set is builded for each sender which want to get feedback information from the group. A representative set is described below.

Every representative acknowledged by the sender should be allocated one feedback no to uniquely identify it. A special value may be used to indicate a unallocated feedback slot in the representative set.

For building a representative set, at startup, a sender will multicast a FEEDBACK_SELECT message to the group, then wait for FEEDBACK_SELECT_REPLY messages from those representatives which are already selected among the receivers. As soon as the sender receive FEEDBACK_SELECT_REPLY message, it then allot a feedback no to the representative and inform it by a SET_FEEDBACK message. After collect a number of representatives, the sender may begin transmission.

During the transmission, the representative set may change in some cases:

- an ordinary receiver can upgrade to be a new representative by sending a APPLY_FEEDBACK to the sender to apply a feedback no from it. The sender then issues a SET_FEEDBACK to inform it the feedback allocated to it.
- a representative member is suppressed by other representative members which are more suitable than it. The suppressed representative will send a CANCEL_FEEDBACK message to the sender to let the sender delete it from the feedback set and reuse the feedback no allocated to it.
- The sender issues a INVOKE_FEEDBACK to an ordinary member in some cases. The member then is upgraded to a representative. This is the only way that a sender can effect the representatives selection.

### 4.3 Representative Selection

The representatives are selected from all the group members. The selection procedure begins when the first member join the group. It is rather irrelevant with the sender. Actually it is up to the group topography and network traffic. A desirable representative selection should make:

1. The *knots* in the group will be selected as representatives.
2. The representatives are evenly distributed in the group and can reflect the entire group's situation.
3. Every member which is not representative can find at least one representative in its neighbour list.

So we get following rules for these purposes. A member will decide whether it is a representative by these rules:

| feedback no allocated from the sender |
| Is it a representative |
| Is it waiting due to some events |
| inquired number |
| preference factor |
| feedback rate |
| The last acknowledged series number |
| ...... |

**Table 2.** Inquiry set

| |
|---|
| Feedback no allocated by the sender |
| The RTT time to the representative |
| The last acknowledged packet no |
| ...... |

**Table 3.** Representative set

1. Initiate state: All receivers are representatives but may have not got confirmed by the source.
2. If there is no neighbor around me, I am a representative and apply a feedback confirm when a sender appears.
3. If there are some neighbors around me, and there is no representative among them, I am a representative and apply a feedback confirm when a sender appears.
4. If there are some neighbors around me, and some of them are representative, but all of representatives have a lower *inquired number* than me or has a same *inquired number* but a higher *feedrate* than me, I am a representative and apply a feedback confirm when a sender appears.
5. If there are some neighbours around me, and some of them are representative, and some of representatives have a higher *inquired number* than me or have a same *inquired number* but a lower *feedrate* than me, I am suppressed by them and set suppressed flag with these neighbours. I will not apply a feedback confirm when a sender appears and don't response to the FEEDBACK_SELECT message.
6. Those representatives which get a feedback confirm from a sender can function as a representative to send feedback.

## 5  Simulation of Self-Organization Algorithm

We simulate our representative selection algorithm with arbitrary network topology and loss rate. The simulator randomly generates a network topology with given member number. For clearly illustrating the distribution of selected representatives, we assume link delay has a linear relation to the length of link. The simulator randomly changes the loss rate of members every a short interval. The average loss rate is limited in 18% according to [XT98]. We show the simulation parameters and result in table 4.

From distribution diagram of selected representatives, we can look that the representatives are evenly distributed in the group so that they can reflect the situation around them quickly with little overhead. By adjusting the neighbour number in the neighbour list, we can increase or decrease the neighbour searching scope, then adjust the size of representative set. In real network, there may be other factors to effect the neighbour number in the neighbour list such as error recovery, the density of member distribution. The simulation also show

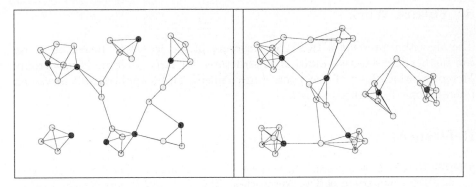

**Fig. 3.** 31 members, 3 neighbours each member

**Fig. 4.** 37 members, 5 neighbours each member

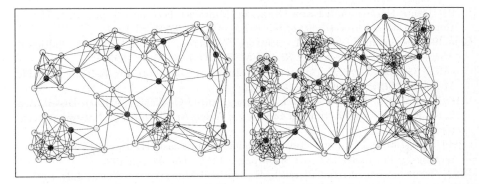

**Fig. 5.** 75 members, 7 neighbours each member

**Fig. 6.** 150 members, 9 neighbours each member

this algorithm can make the group quickly get to a stable state and keep in it with occasional changes. Frequent changes in the group state may seriously weaken the transmission effectivity.

| Member | Neighbour | Representative |
|--------|-----------|----------------|
| 31     | 3         | 8              |
| 37     | 5         | 6              |
| 75     | 7         | 11             |
| 150    | 9         | 20             |

**Table 4.** Representatives selection result

# 6 Future Work

The algorithm presents a two-layer structure in the receivers. Based on it, we can further find effective multi-layer structures for large scale multicast group. Moreover, which kind of flow control mechanism can be applied on it is also an interest topic for future research.

## References

[CGDSL97] Liu, C.-G., Estrin, D., Shenker, S., and Zhang, L., Local Error Recovery in SRM: Comparison of Two Approaches, *USC Technical Report* 97-648, January 1997.

[Dan89] P. B. Danzig. Optimally Selecting the Parameters of Adaptive Backoff Algorithms for Computer Networks and Multiprocessors. Ph.D thesis, University of California, Berkeley, December 1989.

[DO97] Dante DeLucia and Katia Obraczka. Multicast Feedback Suppression Useing Representatives. *Proceedings of the IEEE Infocom'97*, April 1997.

[Gem.J97] Gemmell, J., ECSRM - Erasure Correcting Scalable Reliable Multicast, *Microsoft Research Technical Report*, MSR-TR-97-20, June 1997.

[Gro.M96] Grossglauser, M., Optimal Deterministic Timeouts for Reliable Scalable Multicast, *Infocom 1996*, pp. 1425-1432.

[LDW97] Matthew T. Lucas, Bert J. Dempsey, Alfred C. Weaver, MESH: Distributed Error Recovery for Multimedia Streams in Wide-Area Multicast Networks, Technical Report.

[LP96] J. C. Lin and S. Paul. Rmtp: A Reliable Multicast Transport Protocol. *Proceedings of the IEEE INFOCOM'96* pages 1414-1424, March 1996.

[JE97] Nonnemacher, J., and Biersack, E., Optimal Multicast Feedback, July 1997.

[PDSL98] Sharma, P., Estrin, D., Floyd, S., and Zhang, L., Scalable Session Messages in SRM, *Technical report*, February 1998.

[PSLB94] S.Paul, K. K. Sabnani, J. C. Lin, and S. Bhattacharyya, Reliable Mulitcast Transport Protocol(RMTP): A Detailed report

[SAK92] S. Armstrong, A. Freier, K. Marzullo, Multicast Transport Protocol, *Internet RFC1301* February 1992

[SD89] S.Deering Host Extensions for IP Multicasting. *RFC1112*, August 1989

[SD91] S.Deering Multicast Routing in Datagram Internetwork. *Ph.D. Dissertation*, Stanford University. December, 1991

[SJD96] Kasera, S., Kurose, J., and Towsley, D., Scalable Reliable Multicast Using Multiple Multicast Groups, *CMPSCI Technical Report* TR 96-73, October 1996.

[SVCSL97] Floyd, S., Jacobson, V., Liu, C., McCanne, S., and Zhang, L., A Reliable Multicast Framework for Light-weight Sessions and Application Level Framing, *IEEE/ACM Transactions on Networking*, December 1997, Volume 5, Number 6, pp. 784-803.

[XMZY97] X. R. Xu, A. C. Myers, H. Zhang, and R. Yavatkar. Resilient Multicast Support for Continuous-Media Applications Proceedings of the *NOSSDAV1997*, 1997.

[XT98] Hong Xiang, Sun Teck Tan, End-to-End Network Path Performance Evaluation, Technical Report of Dept of CS, National University of Singapore, 1998

# Deadlock Prevention in Incremental Replay of Message-Passing Programs

Franco Zambonelli

Dipartimento di Scienze dell'Ingegneria – Università di Modena
franco.zambonelli@unimo.it

**Abstract.** To support incremental replay of message-passing applications, processes must periodically checkpoint and must log some of the messages. The paper shows that known adaptive logging algorithms are likely to introduce deadlocks in replay and presents a new algorithm that prevents deadlocks and achieves better performance.

## 1 Introduction

High-performance applications are typically long-running. Then, their debug requires incremental replay techniques, which allow to replay selected intervals of an execution without re-executing the whole application.

To support incremental replay, processes must periodically checkpoint, so that their execution can be restarted from one of their checkpoints and not from the beginning [1, 9]. However, in the case of message-passing programs, further problems arise. On the one hand, the order of message delivery must be traced and preserved during the replay, to ensure deterministic re-execution. On the other hand, all the messages received by a process in the replayed interval need to be reproduced. The paper assumes tools for preserving the delivery order – widely studied in past works [3, 4] – available and focuses on the latter problem. Two main techniques are possible: *(i)* all the messages received by a process are logged and restored during the replay [8]; *(ii)* the intervals of those processes from which messages have been received during the interval of interest are re-executed too, in order to re-compute the messages and send them again [7]. The former technique is ineffective, because logging a message is an expensive operation. The latter technique does not grant any bound on the amount of computation that must be re-executed.

An alternative technique, known as *adaptive message logging*, introduces on-line algorithms to dynamically detect whether a message received by a process needs to be logged or not, on the basis of the dependencies the message introduces on the receiver process [6]. Dependency information can be made on-line available to processes by piggybacking it into the application messages.

In the past, several adaptive logging algorithms have been proposed which significantly reduce the amount of computation needed to replay an interval by logging only a limited percentage of application messages [6, 5]. These algorithms exploit a little amount of on-line information, in order to limit their overhead. However, this can make it impossible to obtain, after the execution, all the information necessary to effectively perform a replay. The paper shows that this is likely to introduce deadlocks in replay and defines a new logging algorithm that not only permits to prevent deadlocks, but also exhibits better performance, as confirmed by tests on a set of message-passing applications.

## 2 The Model

A message-passing program can be modeled by a set of processes $P_1, \ldots, P_n$ that can execute either internal or communication events. Internal events of a process can include local checkpoint events. The paper indicates as $C_{i,x}$ the $x^{th}$ checkpoint taken by process $P_i$ and as $I_{i,x}$ its $x^{th}$ checkpoint interval, i.e., all the events included between $C_{i,x}$ and $C_{i,x+1}$. Communication events include message sending and receiving. Message delivery is not required to be FIFO.

Given a program, several different executions are possible, depending on both the order in which messages are delivered to process and the time at which processes take local checkpoints.

### 2.1 The Replay Dependence Relation

Given an execution of a checkpointed message-passing program, the replay dependence relation ($\xrightarrow{RD}$) shows how events depend on one another during a replay [6]. An event $b$ is said to be *replay dependent* on an event $a$ ($a \xrightarrow{RD} b$) if and only if $a$ must be re-executed before $b$ can be re-executed, either because $a$ precedes $b$ in the same process and no checkpointing occur between $a$ and $b$ or because a sequence of *unlogged* messages was sent from $a$ (or a following event) to $b$ (or a preceding event). The relation is transitive.

By considering the execution of figure 1 – where horizontal arrows represent the execution of each process, black dots represent local checkpoints, solid interprocess arrows unlogged messages and dashed ones logged messages – the receipt of $m2$ makes the successive events of $P_2$ until $C_{2,2}$ replay dependent on the events of $P_3$ preceding the sending of $m2$. The receipt of $m4$ from $P_1$ makes it replay dependent on events of $P_2$ and, transitively, on events of $P_3$.

The $\xrightarrow{RD}$ relation is causal and included in the *happened before* ($\xrightarrow{HB}$) one [1]: all replay dependencies represent causal connections between events. For this reason, replay dependencies are detectable on-line.

### 2.2 The Replay Set

The replay dependence relation can be extended to checkpoint intervals. The notation $I_{i,x} \xrightarrow{RD} I_{j,y}$ indicates that there are events in $I_{j,y}$ that are replay dependent on $I_{i,x}$. With reference to figure 1, $I_{2,1}$ is replay dependent on $I_{3,0}$, because of $m2$. Then, $I_{3,0}$ must be re-executed (at least partially) to replay $I_{2,1}$.

Given one interval $I$ of an execution, there exists a set of checkpoint intervals that introduce replay dependence on it and need to be re-executed when

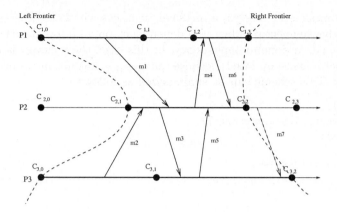

**Fig. 1.** The replay set of interval 1 in process $P_2$

replaying $I$. We call this the *replay set* associated with the interval. Formally, for an interval $I$ of an execution, its *replay set* $RS[I]$ is the set of intervals $RS[I] \equiv \{I_{i,x} : I_{i,x} \xrightarrow{RD} I\}$

The two sets of the earliest and of the latest checkpoints that delimitate the extension of the replay set can be defined as its *left* and *right frontiers*, respectively. For example (figure 1), the replay set of the interval $I_{2,1}$, (emphasized by width lines) includes the intervals $I_{1,0}, I_{1,2}, I_{2,1}, I_{3,0}, I_{3,1}$; its left frontier is composed of $C_{1,0}, C_{2,1}, C_{3,0}$; its right frontier is composed of $C_{1,3}, C_{2,2}, C_{3,2}$.

## 3 The Domino Logging Algorithm

The domino algorithm for adaptive logging, presented in [6], is fully distributed: a process locally decides whether to log a message or not when the message is received. The decision is based on information locally stored by each process and piggybacked with each message.

The basic idea is to log a message if it introduces dependencies on past intervals of the receiver. In the execution of figure 1, the algorithm logs $m3$, that would make $I_{3,1}$ replay dependent on $I_{3,0}$, and $m4$, that would make $I_{1,2}$ replay dependent on $I_{1,0}$. To achieve this goal, the algorithm does not need to exactly detect the replay set, but only its left frontier. A vector of $N$ checkpoint indexes (where $N$ is the number of application processes) is stored by each process and piggybacked with messages, to keep track of the earliest checkpoint interval of each process, if any, onto which the current interval is replay dependent.

At any new local checkpoint in a process $P_i$, all the components of the vector but the $i^{th}$ one are voided. The $i^{th}$ is set to the current local checkpoint index, because any interval is replay dependent on itself. At any received and unlogged message, the local vector is updated on a component wise minimum with the piggybacked one, to compute the left frontier of the new replay set.

When a process $P_i$ receives a message, it checks whether the $i^{th}$ component of the piggybacked vector indicates a dependency on an interval earlier than the current one, i.e., a domino dependency. In this case, the message is logged and the local vector is not updated, because the message has not introduced any new dependency. This scheme can be summarized as follows:

```
m=receive();
if((local_vector[i]<m.vector[i])
      log(m);
else
      for (j=1;j<=N;j++)
            if(j!=i) local_vector[j]=min(local_vector[j],m.vector[j]);
fi
```

where *m.vector* indicates the piggybacked information and *local_vector* the locally stored one.

The domino algorithm is very effective in breaking dependencies and limiting the amount of computation needed to replay, as shown in [6]. However, it is likely to introduce deadlocks in replay, as shown in the following section.

## 4 Preventing Deadlocks in Replay

After an execution, to replay a given interval, its execution must be resumed from the corresponding checkpoint, together with the execution of all the intervals that belong to its replay set. The most trivial solution is to independently resume all the intervals of the replay set from their corresponding checkpoints. With reference to figure 2, to replay the interval $I_{1,1}$ one has to resume independently the execution of $I_{2,0}$ and $I_{2,2}$, together with $I_{1,1}$ itself, from $C_{2,0}, C_{2,2}$ and $C_{1,1}$, respectively.

In the domino algorithm, each process has available on-line only the information about the left frontier of the replay set. It has the availability neither of the exact shape of the replay set nor of its right frontier. Clearly, if this information is not collected on-line and stored at each checkpoint it cannot be available off-line, after the execution. For this reason, having applied the domino algorithm, one cannot execute one replay by resuming independently the execution of all the intervals of the replay set, because one cannot know which intervals are included in it. The same consideration applies to any other logging algorithm which exploits a similar or a minor amount of on-line information [5].

The alternative solution is to resume the execution of the processes from their checkpoints on the left frontier of the replay set, and let them proceed until the replay of the interval of interest completes, without resuming independently each interval. With reference to figure 2 and the replay of $I_{1,1}$, one can resume (in addition to $P_1$ from $C_{1,1}$) $P_2$ from $C_{2,0}$ and let its execution proceed until needed, i.e., until the replay of $I_{1,1}$ completes. This is the replay technique proposed in [6] and in [5].

Though very simple and elegant, the above scheme tends to waste execution resources. In fact, the replay does not skip the execution of those intervals that

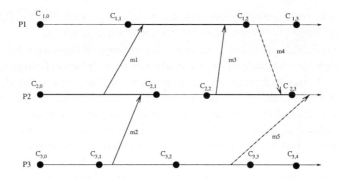

**Fig. 2.** A deadlock in the replay of interval 1 of process P1

are not included in the replay set, though included between its left and right frontiers, and could also proceed over the right frontier of the replay set.

Apart from wasting resources, the above replay scheme is not generally applicable because of possible deadlocks. In figure 2, the replay of $I_{1,1}$ requires the replay of $I_{2,0}$, because of $m1$, and of $I_{2,2}$, because of $m3$. However, after having sent $m1$, the execution of $P_2$ blocks because the message $m2$ from $P_3$ will never arrive. In fact, no interval of $P_3$ is included in the replay set of $I_{1,1}$ and, then, $P_3$ is not replayed at all. $P_2$ will not proceed with its execution through $I_{2,1}$ and $I_{2,2}$, and will never be able to send $m3$, deadlocking the replay.

In general terms, the deadlock problem from which the above replay scheme suffers can be stated as follows:

**Theorem 1.** given the replay set $RS[I]$ of one interval $I$, if this replay set includes several intervals of the same process $P_j$, $I_{j,x}, \ldots, I_{j,x+n}$, and one of these intervals but the latest one, i.e., $I_{j,x+i}, i \neq n$, has a replay set $RS[I_{j,x+i}]$ not included in $RS[I]$, i.e., $RS[I_{j,x+i}] \not\subseteq RS[I]$, the replay deadlocks.

**Proof.** To replay the interval $I$, one has to replay $I_{j,x}, \ldots, I_{j,x+i}, \ldots, I_{j,x+n}$. Suppose that $I_{j,x+i}$ has a replay set that includes the interval $I_{k,y}$. Then, to complete the re-execution of $I_{j,x+i}$, let it proceed over $C_{j,x+i}$ and arrive until $I_{j,x+n}$, one has to re-execute $I_{k,y}$. Otherwise, the execution of process $P_j$ blocks, transitively blocking the replay of $I$, because the intervals of $P_j$ in its replay set included between $I_{j,x+2}$ and $I_{j,x+n}$ will never be re-executed. However, if $I_{k,y}$ is not included also in the replay set of $I$ it will not replayed and the replay deadlocks. □

To solve this problem and avoid deadlock in replay, different solutions can be sketched.

One can think of computing the transitive union of the left frontiers of all the intervals identified by the left frontier of the interval of interest. The union computed in this way identifies a left frontier which, by construction, does not suffer from the problem identified in the above theorem. Then, one can execute the replay from the left frontier obtained in that way. The problem of this solution is that it does not grant any bound on the amount of computation required to replay a given interval: the replay is likely to roll-back in the past and, in most cases, the identified left frontier would represent the beginning of the computation itself. This obviously nullifies any logging effort.

Another possible solution is to evaluate, at replay time and for any receive event, whether the message has to be waited from an interval belonging to the replay set or not. In the former case, the message can be waited and the execution can proceed afterwards. In the latter case, the execution of the process must be stopped and resumed from its next checkpoint, to avoid deadlock. This solution has the drawback of requiring the trace of every message received by a process. Though tracing messages is necessary to grant deterministic re-execution, optimal schemes exist that require to trace only a limited amount of messages [4]. If, as advisable, these tracing schemes are applied during execution, it may be impossible to know, during the replay, from where a message has to be waited.

Finally, the only general solution to prevent deadlocks in replay, is to re-execute all the intervals of the replay set independently. However, the domino algorithm lacks the necessary information. One could think of applying the domino algorithm by making it available the additional on-line information about the replay set that can permit to prevent deadlocks during the replay. In this case, however, there would be no reason not to exploit the additional information available to implement a more informed logging algorithm.

## 5 A Deadlock-free Logging Algorithm

To avoid the above identified deadlock problem, one must store and piggyback with messages the information about the exact shape of the replay set. One could criticize that to store and piggyback exact information about the replay set is likely to intolerably increase the on-line overhead, because the replay set is likely to grow in size as the execution proceeds in a checkpoint interval and new messages are received. This criticism does not apply if a logging algorithm different than the domino one is applied.

The adaptive logging algorithm I propose detects the exact shape of the replay set (for this reason, I call it *full informed*) and has the goal of limiting its size, i.e., the number of intervals it is composed of: a message is logged if it would make the size of the replay set increase over a tolerated bound. Consequently, the maximum amount of information one process will ever be forced to store and piggyback is $N + Bound$ indexes, where $N$ is the number of application processes and *Bound* the bound on the size of the replay set imposed by the algorithm. In fact, for each of the $N$ application processes, one has to indicate the process index and the indexes of its intervals that introduce replay dependencies, at most *Bound*.

At any new checkpoint, replay dependencies are voided, i.e., for an interval $I$, $RS[I] = I$, because an interval is always replay dependent on itself. At any received and not logged message $m$, the new replay set of the receiver is computed as the union of the current replay set and of the sender replay set (piggybacked with $m$ and indicated as $m.RS$), i.e., $RS[I] \leftarrow RS[I] \bigcup m.RS$. The intervals that are included in both replay sets are counted once in the union. This scheme can be summarized as follows:

```
m=receive();
if(size_of_union_of_replay_sets(RS, m.RS) > Bound)
        log(m);
else
        RS=compute_union_of_replay_sets(RS, m.RS));
fi
```

The proposed algorithm not only permits to prevent deadlocks in replay but it is also more flexible than the domino one. The user can choose the preferred trade-off between on-line replay costs (logging) and off-line ones (amount of replay intervals) by selecting the most appropriate bound on the size of the replay set. If a user wants to minimize the on-line logging costs, (s)he can choose a large bound for the replay set, tolerating a slower and more expensive replay activity. If a user is in need of fast and cheap replay, (s)he can impose a very strict bound on the size of the replay set, paying the price of higher on-line logging costs. In the domino algorithm, the logging function cannot be parameterized, thus lacking the possibility of tuning the algorithm behavior to user needs. In addition, the domino algorithm exhibits worse performance, as shown in the following section.

## 6 Performance Evaluation

To evaluate the effectiveness of the presented algorithms, I adopted 5 message-passing programs as testbeds and simulated the execution of the algorithms from message traces of executions of each program, with different checkpoint periods (from the 1% to the 50% of the global execution time). The test programs have been developed for a 16-node $iPSC860$ hypercube and include: programs to compute the determinant of a matrix, the fast fourier transform and finite differences over a grid; a circuit test generator and a $VLSI$ channel router. The data relative to the different programs have been aggregated and averaged, to alleviate the presentation.

Three indicators are significant towards the evaluation of the algorithms: *(i)* the percentage of logged messages measures the on-line replay cost, i.e., the logging overhead; *(ii)* the average and *(iii)* the maximum number of replay intervals per process required to replay the intervals of an execution (i.e., by considering all the intervals, the average size of the associated replay sets and the size of the largest one, divided by the total number of processes) measure the average and the worst case off-line replay costs, respectively.

Both algorithms are effective in reducing, with a limited logging effort, the amount of computation needed to replay over the case in which no logging algorithm is applied.

The behavior of the full informed algorithm depends on the imposed bound on the size of the replay set. A strict bound forces the algorithm to log a high

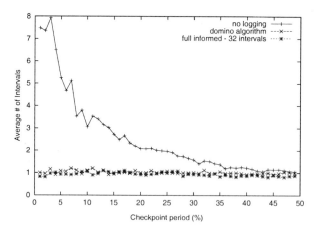

**Fig. 3.** Average number of replay intervals per process when logging about 15% of messages

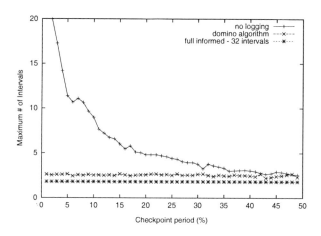

**Fig. 4.** Maximum number of replay intervals per process when logging about 15% of messages

amount of messages, but grants fast and low-cost incremental replay. For example, a bound of 16 interval for the size of the replay set causes the log of about 25-35% of the messages and bounds the average and the maximum number of replay intervals per process to about 0,5 and 1, respectively. Larger bounds reduce the percentage of logged messages and permit anyway to limit the computation required for replay. For example, a bound of 32 intervals for the size of the replay set reduces the percentage of logged messages to 10-15% and limits the average and the maximum number of replay intervals per process to about 1 and 2, respectively.

To compare the performance of the full informed algorithm with the one of the

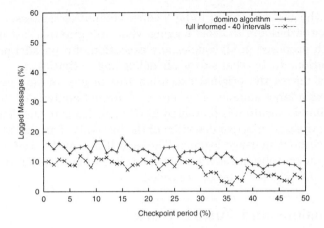

**Fig. 5.** Percentage of logged messages when bounding the maximum number of replay intervals per process to 2.5

domino algorithm, I have firstly set the bound of the fully informed algorithm to 32, to make it log about the same percentage of messages of the domino algorithm. In this case, the two algorithms behave comparably w.r.t. the average number of replay intervals per process (figure 3). Instead, the maximum number of replay intervals per process is lower for the full informed algorithm (figure 4). In other words, with the same on-line costs, the full informed algorithm grants a lower worst case for the off-line replay costs.

Then, I have set the bound in the full informed algorithm to 40, to make it achieve the same worst-case off-line cost of the domino algorithm, i.e., the same maximum number of replay intervals per process. In this case, the full informed algorithm logs a lower percentage of messages than the domino one (figure 5). In other words, the full informed algorithm induces lower on-line replay costs to grant the same worst case for the off-line replay costs. The behavior of the algorithms is almost independent on the checkpoint period.

## 7  Related Work

Simple approaches to incremental replay propose to log the content of all messages [8]: the run-time cost of logging could be extremely high and could even exceed storage capability. More recent papers propose adaptive logging algorithms for incremental program replay, as the already discussed domino algorithm [6]. The algorithm described in [5] logs a message if its delivery would create a sequential execution path exceeding the specified bound. This requires a single scalar information to be piggybacked with messages, but makes the algorithm suffer the identified deadlock problem. In addition, the algorithm does not grant any bound on the global amount of computation required for replay. Then, it can be effective only if parallel execution resources are fully available for debug, which is not the case of most of today's super-computing centers.

Other approaches aim to avoid logging while still granting fast re-execution. The approach proposed in [2] couples any execution of a parallel program with a twin execution, to be charged of all debugging activities, i.e., tracing and logging. This makes the original execution free of any on-line overhead but, again, requires a large amount of resources. A formal analysis of the properties of a checkpointed execution is presented in [7]. This leads to the definition of an algorithm for the post-mortem detection of the intervals of an execution that can be replayed without message logging. The algorithm can be useful to optimize message logging on subsequent executions, but it is of no help if fast incremental re-execution of a selected interval is needed.

## 8 Conclusions and Future Work

The paper focuses on incremental replay of message-passing programs and shows that known algorithms for adaptive logging are likely to introduce deadlocks in replay. A new algorithm is proposed that prevents deadlocks by exploiting additional on-line information and also achieve better performance. I am currently studying the relationship between adaptive logging and consistent checkpointing [9], towards the definition of a logging algorithm that integrates the capability of forcing additional checkpoints in processes to reduce the logging effort.

## References

1. O. Babaoglu, K. Marzullo, Consistent Global States of Distributed Systems, Distributed Systems, ACM press, Editor S. J. Mullender, Chapter 4, 55-96, 1993.
2. O. Gerstel, et al., On-the-Fly Replay: a Practical Paradigm and its Implementation for Distributed Debugging, 6th IEEE Symposium on Parallel and Distributed Processing, Dallas (TX), Oct. 1994.
3. T. J. LeBlanc, J. M. Mellor-Crummey, Debugging Parallel Programs with Instant Replay, IEEE Transactions on Computers, 36 (4), April 1987, 471-482.
4. B. P. Miller, R. H. B. Netzer, Optimal Tracing and Replay for Debugging Message Passing Programs, The Journal of Supercomputing, 8, 1995, 371-388.
5. R. H. B. Netzer, S. Subramanian, J. Xu, Critical-Path-Based Message Logging for Incremental Replay of Message Passing Programs, International Conference on Distributed Computing Systems, Poznan (P), June 1994.
6. R. H. B. Netzer, J. Xu, Adaptive Message Logging for Incremental Program Replay, IEEE Parallel and Distributed Technology, Vol. 1, No. 3, November 1993.
7. R. H. B. Netzer, Y. Xu, Replaying Distributed Programs Without Message Logging, 6th IEEE Symposium on High-Performance Distributed Computing, Portland (OR), Aug. 1997.
8. L. D. Wittie, Debugging Distributed C Programs by Real-Time Replay, ACM Workshop on Parallel and Distributed Debugging, Madison (WI), May 1988.
9. F. Zambonelli, On the Effectiveness of Distributed Checkpoint Algorithms for Domino-free Recovery, 7th IEEE Symposium on High-Performance Distributed Computing, Chicago (IL), July 1998.

# Remote and Concurrent Process Duplication for SPMD Based Parallel Processing on COWs[1]

M. Hobbs and A. Goscinski
{mick,ang}@deakin.edu.au

School of Computing and Mathematics
Deakin University
Geelong, Victoria 3217, Australia.

**Abstract**. The increasing popularity of a Cluster of Workstations (COW) for the execution of parallel applications can be attributed to its impressive price to performance ratio. Unfortunately, currently available software to manage the execution of parallel applications on COWs do not provide satisfactory levels of performance, nor do they provide the application developer with a friendly programming environment. This final problem can be attributed to the lack of transparency provided by these systems when managing distributed resources such as parallel processes, processors and memory. This paper presents a unique process instantiation approach that addresses the shortfalls in currently available SPMD parallelism management systems for COWs by employing a combination of process duplication, process migration and group communication services; all built on and supported by a distributed operating system.
**Keywords**: Parallel Processing on COWs, Operating Systems, Remote and Concurrent Process Duplication, Performance Evaluation.

## 1 Introduction

A key feature which makes Clusters of Workstations an attractive architecture for parallel processing is that they are composed of commodity processors and networks. The use of commodity components greatly reduces the price, while the rapid increase in performance of these components provide COWs with impressive price to performance ratios, especially when compared to traditional supercomputers and MPPs. Harnessing the computational capacity found within a COW is an ongoing problem, with current attempts primarily based on an execution environment built on top of existing network operating systems, e.g., PVM [3]. A number of serious problems exist with such approaches. Firstly, they lack transparency which increases the programming burden on the user. Secondly, unnecessary overheads are introduced since many of the services provided by the environment are also offered by the underlying operating system. Finally, such environments only support process creation as the method of instantiating a set of parallel processes. It is this final problem that is addressed by this paper.

In the course grain Single Program Multiple Data (SPMD) model of parallelism, each parallel child process must be instantiated on a particular workstation. Process creation is commonly employed for this purpose. An alternative and simple method of instantiating parallel processes involves the duplication of parallel processes. Process duplication is commonly used in shared memory parallel architectures to implement the simple 'divide and conquer' programming model (falls within the SPMD model), where each parallel process of the application is formed as a copy of the master (parent) process and execution begins at the point of duplication. In the process duplica-

---
1. This work was partly supported by the Small ARC Grant 0504003157.

tion method, the memory used to form the text and data regions of the child process may be shared (either fully or copy-on-write), thus improving memory usage. However, in a COW environment traditional process duplication on its own is not able to instantiate processes remotely, new processes must be moved to selected remote workstations within the COW. For this purpose, process migration is employed. Traditionally, process duplication results in a single child process, which is not enough to efficiently support parallel processing of a SPMD application. Thus, more than one child process should be duplicated concurrently.

We propose in this paper a unique and advanced process instantiation, remote and concurrent process duplication, that combines traditional process duplication with process migration and group communication. In order to design and implement this advanced process duplication method to support SPMD based parallel processing on COWs, the services of an enhanced distributed operating system should be employed [6]. A distributed operating system enables parallelism to be managed efficiently, allows parallel applications to achieve high performance, and provides users with simple programming environments. Such an approach has been advocated and demonstrated that it is feasible in both [2], where MOSIX based load balancing and process migration services were used to allocate processes to workstations to support PVM applications; and [8], where RHODOS based load balancing and remote process creation services were used to allocate processes to workstations, and load balancing and process migration to dynamically balance load.

## 2 Design of a Remote and Concurrent Process Duplication Service

The course grained SPMD model of parallelism is ideal for execution on COWs due to the high ratio of computation to communication. An SPMD parallel applications can be divided into three distinct phases: initialisation, execution and termination. The initialisation phase involves the instantiation of a set of identical processes and the distribution of the data (work) to each of these child processes. The next phase involves each of the child processes executing in parallel on their allocated data and increasing the number of child processes executing in parallel on the problem data can improve the overall execution time. In the termination phase each child completes the work on their data and then returns the result back to the parent process before exiting. The focus of this work relates to the initialisation phase of the parallel program, more specifically, employing process duplication to instantiate a set of parallel child processes.

### 2.1 Related Systems

The common approach taken by many of the currently available systems supporting parallel execution on COWs is to only provide single process instantiation based on process creation. Only the work performed by the Open Group Research Institute [9] have investigated process duplication as a method of instantiating parallel processes. Two projects were carried out. The first involved implementing a **rfork-multi()** call, based on the Unix **fork()** system call, to concurrently create processes on multiple nodes. The second project involved the implementation of concurrent remote tasking methods and the effect of remote paging/memory sharing techniques have on the instantiation time. However, the architectures involved are multicomputer systems with high speed bus based interconnection networks.

Our research involves the use of COWs. Therefore, it is also relevant to present systems run on such hardware. PVM [3] is composed of specialised library and server processes running on top of an existing network operating system and provides the user with a number of parallelism management services, such as process instantiation. Other well known systems capable of managing parallelism on COWs are NOW [1] and MOSIX [2]. These systems are developed as extensions and enhancements to existing network operating systems.

A common and serious problem exists with PVM, NOW and MOSIX. Each of these systems only supports the creation of single processes. Although the primitives provided by each of these systems enable the user to request multiple processes to be created, internally they are only created one at a time. This problem is the result of the reliance each of these systems have on the underlying network operating systems, which were designed and implemented to only support the creation of single processes. Therefore, if a given SPMD parallel application requires 100 child processes to be created, then the single creation service needs to be called 100 times.

## 2.2 Remote and Concurrent Process Duplication

To design a remote and concurrent process duplication service specifically for SPMD based parallel processing, a number of requirements need to be addressed, including:
- Multiple Duplication of Processes — it must be possible to duplicate concurrently many instances of a process on a workstation or over many workstations;
- Scalability — the proposed service, to take full advantage of available parallelism, must be scalable to many workstations; and
- Complete Transparency — the proposed service must hide from the user all location and management elements, in particular process location.

The SPMD model has two main characteristics which simplify the provision of these requirements. Firstly, a SPMD parallel application is composed of multiple identical copies of the one child process. Each of these child processes begins execution at the same point within the program. This enables memory sharing between the parent and child processes (the text regions of processes can be shared 'read only', while the data regions can be shared 'copy on write'). Secondly, child processes do not interact with each other, simplifying the provision of the transparency requirement.

The duplication service must not only be capable of duplicating many processes on a single workstation, but must also be able to duplicate them on many workstations. The placement of the child processes on the various workstations within the COW can greatly influence the performance of the overall application due to load imbalances [7]. If child processes are mapped to workstations that are heavily loaded whilst other workstations are lightly loaded or idle the overall performance of the program executed will be diminished. Thus, the child processes should be to mapped to a set of workstations such that the overall load of the COW remains balanced, and thus achieve higher performance.

## 2.3 Logical Design of a Remote and Concurrent Process Duplication Service

It is the duplication of many child processes over many remote workstations that forms the critical problem. Each child process is composed of a copy of the parents physical and logical resources. The address space forms the major physical resource component, where locally it can be shared using local memory operations, remotely it must either be shared copy on reference or the address space has to be transferred to each remote workstation. We propose that to improve the flexibility, modularity and implementability of a process duplication service by employing the client/server model, where different services (management of resources) are provided by different server processes [5]. Furthermore, a process duplication server could also be provided to coordinate the reproduction of each of the resources to form a new child process.

The logical design of a remote and concurrent process duplication service would require the interaction of the Memory, Process, Process Mapping and Process Duplication Servers. The role of the Process Mapping Server is to map processes to workstations at their instantiation, thus maintain a COW-wide balanced load. Figure 1 illustrates the interaction between these servers, which are involved in the duplication of a child process on a local workstation. A fourth server is also required if the child process is to be duplicated on a workstation remote to the parent process. This can be achieved either by duplicating the child process locally and then moving it to the selected remote workstation, or by moving the parent process to the remote workstation and then duplicating the child process. Both methods involve a currently executing process being moved from the local workstation to a remote workstation (process migration) and is considered the responsibility of the Process Migration Server.

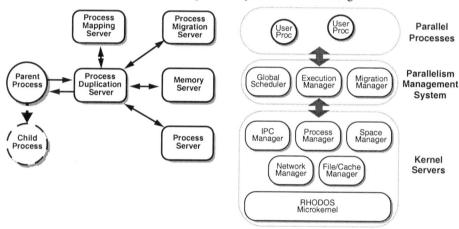

**Figure 1: Server Interaction to Duplicate a Single Process**

**Figure 2: RHODOS Parallelism Management System**

It was stated earlier that to efficiently support SPMD parallel processing on COWs more advanced duplication services are required. Firstly, the remote process duplication service must be able to duplicate multiple child processes on a workstation. When this workstation is the same as the parent's workstation, simple memory sharing primitives can be used to duplicate multiple instances of the child process. When the given workstation is remote to the parent process, either the set of child processes are duplicated locally and then individually migrated to the remote workstation, or the parent process is migrated to the remote workstation where it is duplicated the required number of times and then migrated back to the local workstation. The migration of a process between workstation takes a considerable amount of time to complete, due to the movement of the process' memory [4]. Of the two options, it can be seen that if there are more than two child processes to be duplicated remotely, the migration of the parent process to the remote workstation is simpler and more efficient than migrating each of the child process separately.

When many processes are required to be duplicated on many workstations, the multiple process duplication operation can then be performed on each workstation selected by the Process Mapping Server. The parent process could be migrated sequentially to each remote workstation and the respective number of child process duplicated. Unfortunately, as the number of remote workstations increases, the total instantiation time also increases since each remote workstation must have the parent process migrated to it individually. To enhance the multiple process duplication service when many workstations are involved requires the employment of group migration exploiting group communication. Instead of migrating the parent process sequentially to each remote workstation, the parent can be sent using group communication to all remote workstations where it can be used to concurrently duplicate the respective number of child processes.

## 3 A Remote and Concurrent Process Duplication Service

Unlike the current approaches of building environments on top of existing networked operating systems or the partial extension to an existing network operating system, the approach taken here was to build the process duplication service as inherent part of a purpose built distributed operating system [5].

The RHODOS process duplication service is a component of the RHODOS Parallelism Management System [6], [7], presented in Figure 2. The process mapping, process duplication and process migration servers (Section 2.2) are embodied in the Global Scheduler, RHODOS Execution Manager (REX) and Process Migration Manager, respectively. These servers form the core of the RHODOS Parallelism Management system but rely on the services of the Inter Process Communication (IPC), Process, Space (Memory) and File Managers to perform the required operations.

The role of the Global Scheduler is to ensure the load of the entire COW remains balanced. In particular, it allocates processes to workstations at their instantiation. To achieve this the Global Scheduler, if centralised, uses load information collected about the load conditions of each of the workstations within the COW on which it bases its mapping decisions, or if distributed, exchanges information about computational load among peer Global Schedulers running on each workstation of the COW. The current implementation of the RHODOS Parallelism Management System employs a single, centralised Global Scheduler that uses event notification to record the current loads of the workstations within the COW. The events are a process enters an executable state (duplicated or created) or leaves an executable state (exits).

The Global Scheduler cooperates with the REX Manager to enact the process duplication decisions. The REX Manager exists on each workstation within the COW and provides the interface between the parent process and the RHODOS Parallelism Management System. The parent process sends the REX Manager a request to duplicate a given number of child processes. The REX Manager obtains from the Global Scheduler the location of the workstations on which these child processes should be duplicated and then cooperates with the Migration Manager and its peer REX Managers on the selected remote workstations to perform the process duplication.

The REX Manager is capable of duplicating many copies of a child process on the local workstation with the support of the Process and Space Managers. With the use of the group communication and process migration services, the REX Manager is also able to concurrently duplicate child processes over many workstations with a single migration.

## 3.1 Single Local and Remote Process Duplication

The interaction between the parent process (P), REX Manager, Global Scheduler (GS), Process Manager (PM), Space Manager (SM) involved in the duplication of a new, single, child process on the same workstation as the parent is presented in Figure 3a. This figure also shows the relative order of the messages required to duplicate a process with the RHODOS Parallelism Management System. The Global Scheduler is shown on a separate workstation. In this instance the Global Scheduler has chosen the local workstation for the duplication of the child process, no migration is required.

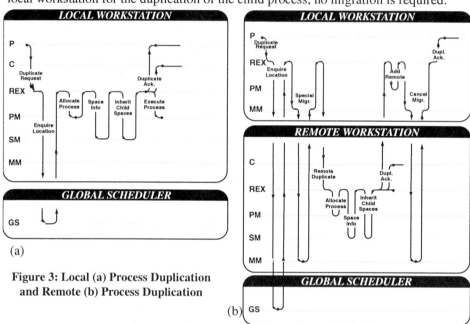

Figure 3: Local (a) Process Duplication and Remote (b) Process Duplication

The interaction (and order of messages) between the parent process and the respective managers involved in the remote duplication of a single process is presented in Figure 3b. In this case the Global Scheduler responds to the location request of the local REX Manager with the address of a remote workstation. The local REX Manager then requests the Migration Manager (MM) to transparently migrate the parent process to the selected remote workstation. Once complete, the local REX Manager then forwards on the request to its peer remote REX Manager. The actions undertaken by the REX Manager on the remote workstation are identical to that of a local process duplication. The only difference is that the acknowledgment of the completed duplication is sent back to the REX Manager on the local workstation, which registers the new remote child process with the local Process Manager, cancels the migration of the parent process and then replies with the duplication result back to the parent process. The RHODOS Parallelism Management System ensures that the location of the child process is hidden from the parent process, all interaction (communication and coordination) between the parent and child process is completely transparent.

## 3.2 Multiple Local and Remote Process Duplication

The basic single process duplication service was extended in the RHODOS Parallelism Management System to support the duplication of multiple child processes on one workstation by employing memory sharing. Figure 4a shows the interaction and order of messages involved in the duplication of $n$ multiple child processes on two workstations. These workstations are selected by the Global Scheduler and in this case are the local workstation which is required to have $m$ child processes and a remote workstation which is required to have $n$-$m$ child processes duplicated (where $n > m$). The order of operations are similar to the single process duplication, although the child processes are formed from duplicates of the parent process' memory. This involves an extra call to the Space Manager to duplicate each of the children's memory.

For each subsequent remote workstation selected by the Global Scheduler, the local REX Manager requests the parent process to be migrated to that workstation and then forwards on the remote duplication request to it's respective peer REX Manager. The remote REX Manager then performs the same operations to duplicate the number of child processes as the local multiple duplication service. Once complete, the remote REX Manager replies to the local REX Manager where the same operation is performed for the next remote workstation. This is repeated until all selected workstations have had the respective number of child processes duplicated on them. When many child process must be duplicated on a number of remote workstations, the overall instantiation time can be considerable due to the sequential nature of this method.

### 3.3 Group Process Duplication

The limitation found with the multiple process duplication service was solved with the employment of group communication in the migration of the parent process to the remote workstations. In the group process duplication method, a single group process migration allows the parent process to be migrated to all participating remote workstations, dramatically reducing the overall instantiation time as the number of workstations involved in the process duplication increases.

Following this method, the Migration Manager on each workstation within the COW, at the initialisation of the workstation, joins a communication group where a single message sent to the group is received by all members of the group. Each of the REX Managers also joins their own group. In the group duplication method, the local REX Manager acts as a coordinator for the overall group duplication by requesting the group migration of the parent process to all participating workstations and forwards on the remote duplication details using group communication to each of its participating peer REX Managers. Once the local REX Manager has received replies from all remote REX Managers, it then proceeds to inform the Process Manager of the remotely duplicated child process, cancels the migration of the parent process and returns the result of the group duplication to the calling parent process.

The interaction and order of the messages involved in the group duplication of a set of child processes is shown in Figure 4b. The distinction between this operation and the multiple duplication operation shown in Figure 4a is that only a single migration of the parent process and only the forwarding of a single remote duplication request to all remote REX Managers is required. Therefore, the performance concern faced with the previous method is avoided as the number of workstations increases.

## 4 Experiment Performance

The implemented system was tested on the RHODOS COW, which is composed of 13 Sun 3/50 workstations connected by a shared 10 Mb/s switched ethernet network. One of the workstations within the COW is dedicated as a File Server and the remaining 12

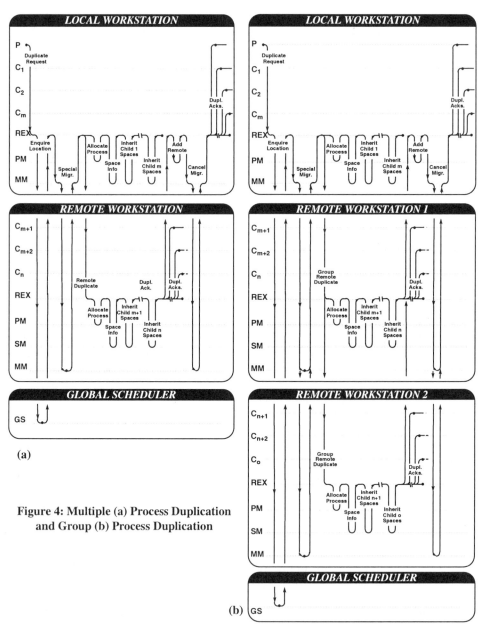

Figure 4: Multiple (a) Process Duplication and Group (b) Process Duplication

workstations are used for normal user computation. To examine the performance of the three duplication methods a sample SPMD based parallel program was developed and the time to instantiate a number of child processes was measured. The experiment was performed with the number of child processes being varied from 1 through to 20 child processes. The number of workstations involved was also varied from 1, 2, 4, 8 and 12. Each experiment was performed 20 times and the average result presented.

The performance results obtained from using the single process duplication service is presented in Figure 5. From these results it is clear that the single process duplication service provides, as expected, the worst performance. This can be attributed to the extremely high overhead of having to repeatedly call the process duplication code for each child duplicated. Therefore, a linear performance result is observed as the number of child processes duplicated increases. A slight overall increase can be observed when the number of workstations increases which can be attributed to the overhead required to forward on the duplication request from the local REX Manager to the peer REX Manager on the remote workstation.

A large performance improvement is observed when memory sharing between multiple child processes on the same workstation involved the multiple process duplication service (Figure 6). A dramatic levelling off is obtained when more than one process is duplicated on a given workstation. As the number of workstations used in the experiment increases, the execution time also increases which can be attributed to the migration of the parent process to each new workstation.

The group process duplication service provides the best performance results (Figure 7). The overall performance is inherently linear with only a slight step found when two or more workstations are used.

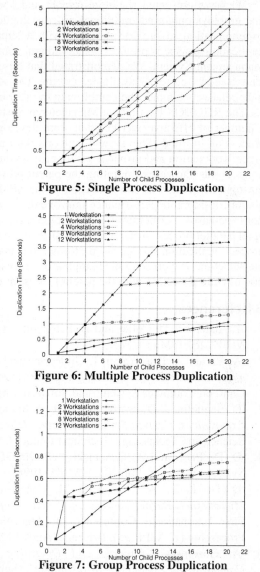

**Figure 5: Single Process Duplication**

**Figure 6: Multiple Process Duplication**

**Figure 7: Group Process Duplication**

This step can be attributed to the change from a local process duplication to a remote process duplication and therefore the need to perform the single group process migration of the parent and the associated forwarding on of the remote duplication request.

## 5 Conclusion

The benefits of employing COWs for the execution of parallel applications revolve around very high performance to price ratios, especially when compared to traditional supercomputer and MPP systems. Unfortunately, current operating software do not provide the user with satisfactory levels of service or performance. One important service that is missing form current systems is that of efficient process duplication.

We have presented a unique remote and concurrent process duplication service that employs enhanced traditional process duplication, process migration and group communication to provide the user with a simple and efficient instantiation. This duplication service is provided as an integral component of the RHODOS Parallelism Management System. We demonstrated that this system addresses many of the problems faced by currently available systems. The solution to these problems was shown to be relatively simple when employing the services of a distributed operating system.

The RHODOS Parallelism Management System, presented in this work, provides a number of effective services that were developed specifically with SPMD parallel processing in mind. This system supports the multiple duplication of processes on a given workstation and provides group process duplication over a number of workstations. It was shown by migrating the parent process to all workstations of the virtual machine involved with the execution of the SPMD parallel program, it is possible to obtain considerable performance improvements, especially as the number of workstations increases. Furthermore, users/programmers benefit enormously from the developed parallel process instantiation because all operations of the remote and concurrent process duplication service are performed transparently. The users are not involved in the selection of workstations, mapping of processes to workstations, or the coordination of child processes with the parent process at the end of process instantiation.

## 6 References

1. T. Anderson, D. Culler and D. Patterson. *"A Case for Networks of Workstations: NOW"*. IEEE Micro, Feb., 1995.
2. A. Barak, A. Braverman, I. Gilderman and O. La'adan. *"Performance of PVM with the MOSIX Preemptive Process Migration"*. Proceedings of the 7th Israeli Conf. on Computer Systems and Software Engineering, Herzliya, June 1996.
3. D. Beguelin, J. Dongarra, A. Geist, R. Manchek, S. Otto and J. Walpole. *"PVM: Experiences, Current Status and Future Directions"*. Oregon Graduate Institute of Science and Technology. Technical Report. CSE-94-015. April 1994.
4. D. De Paoli and A. Goscinski. *"The RHODOS Migration Facility"*. Journal of Systems Software. Vol. 40, pp.51-65, 1998.
5. D. De Paoli, A. Goscinski, M. Hobbs and G. Wickham. *"The RHODOS Microkernel, Kernel Servers and Their Cooperation"*. Proceedings of The IEEE First International Conference on Algorithms and Architectures for Parallel Processing (ICA3PP-95). Brisbane, April 1995.
6. A. Goscinski. *"Towards and Operating System Managing Parallelism of Computing on Clusters of Workstations"*. (Submitted to) Parallel Computing.
7. M. Hobbs. *"The Management of SPMD Based Parallel Processing on Clusters of Workstations"*. School of Computing and Mathematics, Deakin University. PhD Thesis, August 1998.
8. M. Hobbs and A. Goscinski. *"The RHODOS Remote Process Creation Facility Supporting Parallel Execution on Distributed Systems"*. Journal of High Performance Computing, Vol. 3, No. 1, Dec. 1996.
9. D. Milojicic, A. Langerman, D. Black, M. Dominijanni and R. Dean. *"Concurrency: A Case Study in Remote Tasking and Distributed IPC in MACH"*. IEEE Concurrency. Vol. 5, No. 3, April-June 1997.

# Using BSP to Optimize Data Distribution in Skeleton Programs

Andrea Zavanella and Susanna Pelagatti

Dipartimento di Informatica
Università di Pisa Italy
{zavanell,susanna}@di.unipi.it
http://www.di.unipi.it/ ~{zavanell,susanna}

**Abstract.** Parallel programming can be made easier by means of a skeleton based methodology, such as $P^3L$, which helps programmers to compose their applications by using a set of fixed parallel patterns. Such kind of approach is also useful to obtain portability because the "structured" nature of the language can be used to devise a composable support for each parallel pattern so that the complexity of finding an "optimal" implementation on different parallel architectures can be reduced. In this work, we show how we can conjugate the BSP abstract model and its related cost analisys to provide an implementation strategy "abstract enough" for being also machine independent. We hope this can be a first step towards the idea of a portable set of optimization rules. The first results show how an implementation template for the Map constructor, able of to be tuned automatically, can be designed. A validation of the technique is given for a Cray T3E and a cluster of PC-Linux.

## 1 Introduction

The structured approach to parallel programming exploited by the skeleton languages allows a rapid design and development of complex parallel applications [1]. Skeletons free the programmer from low level details of parallel hardware and provide a set of pre-defined parallel "patterns" (also known as "templates" or "constructors") to build his program. The main reasons to use skeletons are explained in many papers already published [5, 7, 15] while practical implementations have been presented in [8, 3, 6]. Despite the benefits of the approach shown by the first academic and industrial experiences, it is now clear that the diffusion of skeleton methodologies as a general purposes programming style is limited by the lack of portability of their supports, which in general highly depend on the structure of a particular target machine. A partial exception to this rule is the $P^3L$ language which has been ported on a MPI layer [4]. Unfortunately, the current implementations of MPI lack a common cost model, thus the support and the optimisations to obtain a reasonable performance must be rewritten for

---
[0] This work has been partially supported by the italian M.U.R.S.T. in the framework of the Project "MOSAICO".

each platform. This paper is a first attempt to overcome these problems using a portable target abstract machine for the $P^3L$ back-end. The idea is to use the BSP abstract machine introduced by Valiant and others [17, 14]. The simple cost model for a BSP computation can be exploited to make reliable prediction of the execution costs on different machines. This simplifies the design of a portable support which is able of to produce automatically optimised implementations for different platforms. Sec. 2 explains how data parallel computations are expressed in $P^3L$, how the $P^3L$ compiler works and the basic idea under the compilation on BSP. Sections 3–5 details our approach to data distribution using BSP and some machine independent optimizations. Eventually, in Section 6, we demonstrate that the automatic optimisations we have found can produce optimised data-distribution respect to the parameters of the target machine and discuss the results of some experiments performed on a Cray T3E and a clustr of PC-Linux. Section 7 concludes the apper and summarizes some related work and further developments.

## 2  $P^3L$ and Anacleto on the BSPC

In this section, we introduce how data parallel computations are expressed in $P^3L$ and describe how they are compiled by anacleto, the current $P^3L$ compiler. There are three kinds of skeletons in $P^3L$: *task parallel* skeletons, exploiting parallelism among independent tasks (such as pipeline or task farming); *data parallel* skeletons, abstracting typical patterns of data parallel computations (such as for-all or reduction); and control skeletons which allow the encapsulation of fragments of the host language in a sequential process (seq skeleton) or the iteration (loop skeleton). In the following, we will focus on the data parallel skeletons. The map skeleton expresses data parallel computations on independent sets of data. The reduce skeleton provides the usual reduce operation on vectors [2, 13] and on multi-dimensional arrays. The comp skeleton expresses the composition of a sequence of data parallel skeletons. A $P^3L$ program consists of a set of skeleton instances (*modules*) combined and nested to form the global structure.

As an example, consider the $P^3L$ data parallel program in Fig. 1 computing the square of a matrix a n by n. Here, we have two skeleton instances: a map and a seq. Each skeleton instance has a *header* describing the module interface. The header follows the skeleton specifier (seq, map) and consists of a module name (inner_product) and of the specification of formal parameters taken in input (the inlist in(...)) and produced in output (the outlist in(...)). The inner_product module encapsulates some plain C code computing the inner product of two vectors sequentially. The input parameters a and b are passed by value and the value returned in c is the value assigned to c before leaving the sequential section. The map instance takes in input a matrix a n by n and builds a result matrix b=a$^2$. This is done computing in parallel all the elements of b. The computation of each b[i][j] depends on a[i][] and a[][j] (the $i$th row and the $j$th column of a) and is computed by calling in parallel n$^2$ instances

```
map Squaremat in(int a[n][n]) out(b[n][n])
    inner_product in(a[*i][], a[][*j]) out(b[*i][*j])
end map

    seq inner_product in(int a[n],int b[n]) out (int c)
        ${
          int k;
          c=0;
          for (k=0;k<n;k++)
             c=c+a[k]*b[k];
        }$
    end seq
```

**Fig. 1.** A simple $P^3L$ program computing the square of a matrix a.

of the sequential code defined in inner_product. All this is stated in the body of the map definition using the two *free variables* *i and *j. The free variables define an array of potentially parallel activities (the computations of all b[i][j] in this example) which are called *virtual processors* in $P^3L$ terminology. For each activity b[*i][*j], the positions of the corresponding free variables in the input list define which are the data needed. Several optimization techniques have been implemented in the $P^3L$ compiler in order to take automatically decisions on the degree of parallelism, program restructuring, data-arrangement strategies etc. Unfortunately, deriving performance models which are valid for different architectures is difficult and this forces to develop many different implementation templates and strategies for different machines.

The main problem is that, at the moment, the code generated uses collective MPI primitives, which have different performance models for different target machine. The BSP model has been selected as an alternative intermediate level on which the compiler can devise a portable implementation. Portability in this case has a stronger sense than simple code-compatibility. In fact, the claim of this paper is that enhancing the abstraction level we can also achieve the goal of the performance portability simply compiling skeletons on a BSPC and exploiting the cost model to decide among a limited set of already optimized implementations. A BSPC is an abstraction of a parallel machine which consists in a set of four parameters: $s$ (the speed of the processors), $g$ (the permeability of the network to continuous traffic for uniformly-random destinations), $l$ (the cost of a barrier synchronisation) and $p$ the number of available processors. A BSP computation is a sequence of *supersteps* which are compounded of three phases: the first one for local computation, then a communication step and eventually a barrier synchronisation. The model provides a simple formula to account for the cost of a superstep:

$$T_{sstep} = W + h \star g + l \qquad (1)$$

Where $W$ is the maximum work done by a processor in the computation phase and $h$ is the maximum number of messages sent (or received) by a processor during the communication phase [1]. Further details can be found in [14, 17].

## 3 A BSP Study on P3L Data-Distribution

In this section, we analyze how a worker in the data parallel support can be implemented and optimized exploiting the BSP collective operations, in the simple case in which the data parallel computation is defined by a single `map` instance.

The generic structure of a data parallel worker is the following

```
DP::    < distribution >
        < computation & re-distribution >
        < collection >
```

If we have a single map (as happens in Fig. 1) the redistribution is not needed and the collection is a simple gather of the results computed, so we focus on the < distribution > phase.

Consider a generic $P^3L$ data parallel `map` skeleton:

```
map map_name in(in-list) out(out-list)
    body_name in(body-in-list) out(body-out-list)
end map
```

the data distribution is expressed by means of the *free variables* specified by the ⋆ prefix in the `body-in-list`. Using the free variables in the `body-in-list` we can define three different kind of arrangements for an array:

- *broadcasting*: when the array appears without free variables (e.g., `a[][]`)
- *scattering*: when all the free variables appear in a given array
- *multicasting*: when a subset of the free variables appears a given array.

Parsing the `body-in-list` we can build the sets of data structures to distribute in broadcast, scatter and multicast respectively, as follows: $\mathcal{B} = \{b_1, \ldots b_{k_1}\}, \mathcal{S} = \{s_1, \ldots s_{k_2}\}, \mathcal{M} = \{m_1, \ldots m_{k_3}\}$.

The simplest BSP implementation for the distribution of all data structures in the three sets consist in a sequence of three communication supersteps. Each superstep distributes all the data structures in a set according to the distribution required (broadcast, multicast, scatter). The cost of the distribution phase is given by the sum of the costs of the three steps:

$$T^{bsp}_{dist} = T^{bsp}_{broad} + T^{bsp}_{scatter} + T^{bsp}_{multi} \quad (2)$$

the three terms can be further expanded in:

$$T^{bsp}_{bro/scat/mul} = T^{bsp}_{bro/scat/mul}(\sum_{i=1}^{k_{1/2/3}} |x_i|) \quad (3)$$

$$(4)$$

---

[1] The parameters $g$ and $l$ are normalized to the speed of the processor $s$.

where $|x_i|$ denotes the size of structure $x_i$. We note that the optimal cost of broadcasting a structure according to the BSP model depends both on the target architecture and on the size of the structure [10]. In practice, when:

$$n > -\frac{gp^2 N_{1/2} - gp N_{1/2} + lp}{3gp - 2g - gp^2} \quad (5)$$

Note that here we use the following formulas to account for the different size of the messages: $g(size) = g_\infty(\frac{N_{1/2}}{size} + 1)$ Where $g_\infty$ is the asymptotic permeability for very large messages while $N_{1/2}$ is the size of messages that produces half of the optimal bandwidth so $g(N_{1/2}) = 2g_\infty$. In the rest of the paper, we use $g = g_\infty$ while $l$ and $p$ are the usual BSP parameters for the cost of a barrier and for the number of processors. In this case, the minimum cost is given by the two-phase algorithm which first scatters all the data on the available processors and the performs a global all-to-all communication. The corresponding cost is:

$$\frac{2n(p-1)g}{p}(\frac{pN_{1/2}}{n} + 1) + 2l \quad (6)$$

In the other case, the best strategy is to send directly a copy of the structure to all the processors, and the cost is simply: $ng(p-1)(\frac{N_{1/2}}{n} + 1) + l$. A scatter distribution can only be implemented by sending to all processors the appropriate partition, which results in the following cost

$$T_{scatter}^{bsp} = \sum_{i=1}^{k_2}(|s_i|)g(\frac{N_{1/2}}{\sum_{i=1}^{k_2}(|s_i|)} + 1) \quad (7)$$

The cost for a multicast distribution is more complex to be computed as several strategies can be used for its implementation in BSP. The easier one is the following. We first compute the union set $\mathcal{IN}(p)$ of all the data in the `body-in-list` which are used by the virtual processor assigned to a given worker $p$. Then we send in parallel $\mathcal{IN}(p)$ to all the processors. The cost of this 'basic' implementation is:

$$T_{multi1}^{bsp} = |\mathcal{IN}| g(\frac{N_{1/2}}{|\mathcal{IN}|} + 1)(p-1) + l \quad (8)$$

where $|\mathcal{IN}|$ denotes the sum of the sizes of all the $\mathcal{IN}$ sets computed. This cost is similar to the cost of a scatter operation, however it can in principle be reduced because the data sent to different processors are in general partially overlapped. In the following sections, we describe a technique to exploit such an overlapping in order to optimize the cost of a multicast operation.

## 4 Computing data partitions

The first step to optimize the distribution strategy is having a calculus to handle the *data sets*. In this section, we define a method to describe the data set to be gathered by each virtual processor and a set of rules to obtain a compact description of the $\mathcal{IN}$ set for each physical processor. The notation for the data distributions admitted by a $P^3L$ map can be described as in the following rules:

- intervals of rows and columns:

$$a[\star i - k_1 \ldots \star i + k_2][] \to \mathcal{IN}(i,j) = a[i - k_1, i + k_2][] \qquad (9)$$
$$a[][\star j - k_1 \ldots \star j + k_2] \to \mathcal{IN}(i,j) = a[][j - k_1, j + k_2] \qquad (10)$$

- rectangles:

$$a[\star i - k_1 \ldots \star i + k_2][\star j - k_3 \ldots j + k_4] \to \qquad (11)$$
$$\mathcal{IN}(i,j) = a[i - k_1, i + k_2][j - k_3, j + k_4]$$

We see that, when computing, the input set for a physical processor $\mathcal{IN}(p)$ we need some rules to simplify the potential overlapping. We introduce a set of equations which help to obtain a minimum convex set including the multiple elements of a data distribution:

- (to unify overlapped rows or columns)

$$a[k_1, k_2][] \cup a[z][] = a[k_1, k_2][] \text{ if } k_1 \leq z \leq k_2 \qquad (12)$$
$$a[][k_1, k_2] \cup a[][z] = a[][k_1, k_2] \text{ if } k_1 \leq z \leq k_2 \qquad (13)$$

- (to unify intervals)

$$a[k_1, k_2][] \cup a[k_3, k_4][] \subset a[k_5, k_6][] \qquad (14)$$
$$\text{with } k_5 = Min(k_1, k_3), k_6 = Max(k_2, k_4)$$
$$a[][k_1, k_2] \cup a[][k_3, k_4] \subset a[][k_5, k_6] \qquad (15)$$
$$\text{with } k_5 = Min(k_1, k_3), k_5 = Max(k_2, k_4)$$

## 5 How to Implement Multicasts with Multibroadcasts

In this section we present a possible technique to exploit the redundancy within a multicast distribution to reduce the communication costs. The technique consists in clustering the set of processors $\mathcal{P}$ in broadcast subgroups $\pi_1 \ldots \pi_k$. For each subgroup we have a substructure $\sigma_i$ to be broadcasted to every processor of $\pi_i$.

**Definition 1.** *A multibroadcast is a communication pattern defined by the pair $(\Pi, \Sigma) = (\pi_1 \ldots \pi_k, \sigma_1 \ldots \sigma_k)$ in which $\sigma_i$ is broadcasted to $\pi_i$.*

**Definition 2.** *A set of multibroadcasts $(\Pi, \Sigma)_j$ is a cover for the multicast: $(\mathcal{IN}(p_1) \ldots \mathcal{IN}(p_p))$ when for each $p_i$ the collection of the structure broadcasted to $p_i$ given by $\bigcup_j \sigma_i$ includes $\mathcal{IN}(p_i)$.*

We claim that finding a set of multibroadcasts covering a given multicast can reduce the overall cost. In this section we first present an algorithm to find a covering set of multibroadcasts and then we derive the conditions under which it improves the communication cost. The point is that reducing a multicast to a set of broadcast we can devise a set of superstep more balanced and then we can obtain a reduced distribution cost. We show that such a transformation can always be done providing a factorising algorithm.

## 5.1 Factorising in multibroadcast

Assuming a multicast distribution $(\mathcal{IN}(p_1), \ldots \mathcal{IN}(p_p))$, the following algorithm can find a set of multibroadcast $(\pi_{u,v}, \mathcal{IN}(\pi_{u,v})$ covering the multicast.

0  $u = 1$

1  $v = 1$ and consider $p_v \in \mathcal{P}$ and $\pi_{u,v} = \{p_v\}$ and $\mathcal{IN}(\pi_{u,v}) = \mathcal{IN}(p_v)$

2  Choose a $p_j$ such that $\forall_i \mid \mathcal{IN}(p_i) \cap \mathcal{IN}(\pi_{u,v}) \mid \leq \mid \mathcal{IN}(p_j) \cap \mathcal{IN}(\pi_{u,v}) \mid$

3  $\mathcal{IN}(\pi_{u,v}) = \mathcal{IN}(\pi_{u,v}) \cap \mathcal{IN}(p_j)$

4  $\pi_{u,v} = \pi_{u,v} \cup \{p_j\}$ and $\forall p \in \pi_{u,v} : \mathcal{IN}(p) = \mathcal{IN}(p) - \mathcal{IN}(\pi_{u,v})$

5  Repeat 2-4 until $\forall p \in \mathcal{P}\ p \notin \pi_{u,v} \Rightarrow \mathcal{IN}(p) \cap \mathcal{IN}(\pi_{u,v}) = \emptyset$

6  Choose a $p$ such that $p \notin \pi_{u,v}$

7  $v = v + 1$

8  $u = u + 1$ Repeat 1-7 until $\forall p \in \mathcal{P}\ \mathcal{IN}(p) = \emptyset$

The algorithm builds a cover for the multicast using a strategy which first selects the larger overlapped substructures (due to the choice in the row 2). Other strategies and other "clustering" algorithms can be used to obtain different kinds of covers.

## 5.2 Costing the Multibroadcast

The cost of a multibroadcast distribution depends on how we implement the single broadcast. In this section we derive the cost for a two-phase multibroadcast based on the two phase broadcast. In the first step each $\sigma_i$ is scattered in the corresponding $\pi_i$. In the second step we have an all-to-all communication within each $\pi_i$. The cost of a two-phase multibroadcast for $(\Pi, \Sigma) = (\pi_1, \ldots \pi_k, \sigma_1, \ldots \sigma_k)$ is given by the following formula:

$$mb_1^{bsp} = \sum_{i=1}^{k} \mid \sigma_i \mid \mid \pi_i \mid g(\tfrac{N_{1/2}}{|\sigma_i|} + 1) + l \qquad (16)$$

Assuming: $\mathcal{H} = Max_i(\mid \sigma_i \mid \tfrac{(\pi_i - 1)}{\pi_i})$:

$$mb_2^{bsp} = \sum_{i=1}^{k} \mid \sigma_i \mid g(\tfrac{N_{1/2}}{|\sigma_i|} + 1) + \mathcal{H}g(\tfrac{N_{1/2}}{\mathcal{H}} + 1) + 2l \qquad (17)$$

## 5.3 Multibroadcasting instead of Multicasting

Consider the example in Fig. 1 and suppose the virtual processors are distributed block, block on the $p$ processors available. Each processor has a partition of virtual processor having shape $[d][d]$ where $d^2 = \frac{n^2}{p}$. The input set for each processor is:

$$\mathcal{IN}(p[i][j]) = a[I][] \cup a[][J] \qquad (18)$$

Where: $I = i \cdot d, [(i+1) \cdot d - 1]$ and $J = j \cdot d, [(j+1) \cdot d - 1]$ Applying the algorithm discussed before, we factorize the multicast in two multibroadcast having $\sqrt{p}$ subsets each. The first distributes the rows:

$$\pi_{1,1} = p[1][1] \ldots p[1][\sqrt{p}] \ldots \quad \pi_{1,\sqrt{p}} = p[\sqrt{p}][1] \ldots p[\sqrt{p}][\sqrt{p}] \qquad (19)$$
$$\sigma_{1,1} = a[1,d][] \qquad \ldots \qquad \sigma_{1,\sqrt{p}} = a[n-d,n][]$$

In order to decide when factorizing multibroadcasts may improve the distribution cost, we compare the costs for the two implementations. Considering that: $\forall p \in \mathcal{P} \mid \mathcal{IN}(p) \mid = 2dn$

$$T^{bsp}_{multi} = 2dn(p-1)g(\tfrac{N_{1/2}}{2dn} + 1) + l \qquad (20)$$

Using the two-phase multibroadcast:

$$t^{bsp}_{mb2} = 2g(\tfrac{N_{1/2}}{\frac{dn}{\sqrt{p}}} + 1)\frac{dn}{\sqrt{p}}(p + \sqrt{p} - 2) + 4l \qquad (21)$$

The two-phase multibroadcast is better than the multicast when:

$$n > \sqrt{p\frac{gN[p + 2\sqrt{p} - 3] + 3l}{2g(p\sqrt{p} - 2\sqrt{p} - p + 2)}} \qquad (22)$$

In the following section, we check this method and the predictions obtained on two completely different parallel architectures: a cluster of PC running Linux and a CRAY T3E.

## 6 Experiments

The optimization technique presented in the previous section has been validated using the BSP-lib on a cluster (Backus) of 10 PCs clocked at 233Mhz connected with a two-directional 100Mbit switch and running Linux, and on a 160 nodes Cray T3E. The idea is to show that on such different targets optimizations can be guided by the same formulas thanks to the cost portability of BSP. The curves in Fig. 3 show that predicted and real values are extremely compatible.

We noticed that in practice, for all of the relevant values, the multibroadcast works better on both the considered architectures. This is due to the high communication bandwidth and the low latency of their communication networks which accounts a low-cost for complex communication patterns. However we expect the multicast solution can be employed on realistic size of data input on architectures with slower communication mean as a normal Ethernet 10Mbit.

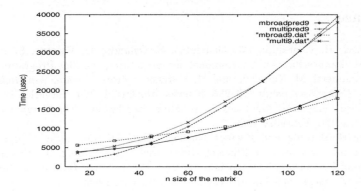

**Fig. 2.** Predicted and measured times on Backus (9 PCs)

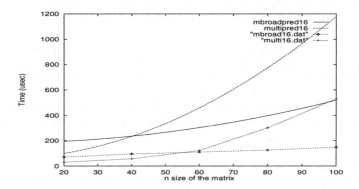

**Fig. 3.** Predicted and measured times on the T3E (16 nodes)

## 7 Conclusions and Related works

Implementing the $P^3L$ language on top of an abstract machine with a realistic cost model, as the BSPC, we can accurately predict the costs for data distribution. A pre-defined set of already-optimized collective operations can be used as well as more complex patterns as the multibroadcast can be built, being sure that the analytical formulas derived using the BSP cost model are still valid using different kinds of parallel machines or varying the number of processors on the same machine. This result is a first encouraging step towards the design of a full portable support for $P^3L$ . We think that such a kind of support requires also a machine-independent mechanism for programs transformations equivalent to the approach proposed in [12], and a global optimiser as the one proposed in [9]. A related approach to provide a machine-independent implementation for skeleton constructs has been presented by Osoba et al. in [11]. In this work a support for an iterative skeleton is optimised using the BSP cost model . The idea of combining a data-distribution algebra with skeleton programming is also present in [16].

# References

1. B. Bacci, B. Cantalupo, M. Danelutto, S. Orlando, D. Pasetto, S. Pelagatti, and M. Vanneschi. An Environment for Structured Parallel Programming. In L. Grandinetti, M. Kowalick, and M. Vaitersic, editors, *Advances in High Performance Computing*, pages 219–234. Kluwier, Dordrecht, The Netherlands, 1997.
2. R.S. Bird. Algebraic identities for program calculation. *The Computer Jurnal*, 32(2):122–126, February 1989.
3. George H. Botorog and Herbert Kuchen. Skil: An Imperative Language with Algorithmic Skeletons for Efficient Distributed Programming. In *Proceedings of the Fifth International Symposium on High Performance Distributed Computing (HPDC-5)*, pages 243–252. IEEE Computer Society, 1996.
4. S. Ciarpaglini, M. Danelutto, L. Folchi, C. Manconi, and S. Pelagatti. ANACLETO: a Template-based p3l Compiler. In *Proceedings of the Seventh Parallel Computing Workshop (PCW '97)*, Australian National University, Canberra, August 1997.
5. M. Cole. *Algorithmic Skeletons: Structured Management of Parallel Computation.* The MIT Press, Cambridge, Massachusetts, 1989.
6. M. Danelutto, R. Di Meglio, S. Orlando, S. Pelagatti, and M. Vanneschi. A Methodology for the Development and the Support of Massively Parallel Programs. In D.B. Skillicorn and D. Talia, editors, *Programming Languages for Parallel Processing*. IEEE Computer Society Press, 1994.
7. J. Darlington, A.J. Field, P.G. Harrison, P.H.J. Kelly, D.W.N. Sharp, and Q. Wu. Parallel Programming using Skeleton functions. In *Parallel Languages and Architectures Europe*. Springer Verlag, 1993.
8. J. Darlington, Y.K Guo, H.W. To, and J. Yang. Functional Skeletons for Parallel Coordination. In *Proceedings of Europar*, 1995.
9. D.B.Skillicorn, M. Danelutto, S. Pelagatti, and A. Zavanella. Optimising Data-Parallel Programs Using the BSP Model. In *Proceeding of EUROPAR98*, LNCS. Springer, 1998.
10. A.V. Gerbessiotis and C.J. Siniolakis. Primitive Operations on the BSP Model. Technical Report PRG-TR-23-96, Oxford Unversity Computing Laboratory, October 1996.
11. F.O. Osoba and F.A. Rabhi. A Parallel Multigrid Skeleton using BSP. In *Proceeding of EuroPar98*, LNCS. Springer, 1998.
12. S.Gorlatch and S. Pelagatti. A Transformational Frame for Skeleton based Parallelism: An Oveview and a Case Study. submitted, October 1998.
13. J.M. Sipelstein and G.E. Blelloch. Collection-oriented languages. *Proceedings of the IEEE*, 79(4):504–523, April 1991.
14. David B. Skillicorn, Jonathan M. D. Hill, and W. F. McColl. Questions and answers about BSP. Technical Report 15-96, Programming Research Group, Oxford University Computing Laboratory, August 1996.
15. S.Pelagatti. *Structured Development of Parallel Programs.* Taylor & Francis, 1997.
16. M. Sudholt, C. Piepenbrock, K. Obermayer, and P. Pepper. Solving Large Systems of Differential Equations using Covers and Skeletons. In *50th IFIP WG 2.1 Working Conference on Algorithmic Languages and Calculi*. Chapman- Hall, Feb 1997.
17. L.G. Valiant. A Bridging Model for Parallel Computation. *Communications of the ACM*, 33(8):103–111, August 1990.

# Swiss-Tx Communication Libraries

Stephan Brauss[1], Martin Frey[2], Anton Gunzinger[1], Martin Lienhard[1], and Josef Nemecek[1]

[1] Swiss Federal Institute of Technology Zurich (ETHZ),
{brauss, gunzinger, lienhard, nemecek}@ife.ee.ethz.ch
[2] Supercomputing Systems AG (SCS), frey@scs.ch

**Abstract.** The goal of the Swiss-Tx project is to develop, build and install a series of new supercomputers which are mostly based on commodity parts. Only the communication devices and the communication libraries are custom because available products (e.g. Ethernet with the standard socket interface) do not offer the necessary functionality, bandwidth and latency. This paper presents the high-performance communication libraries for the new Swiss-Tx supercomputer series and shows some early results.

## 1 Introduction: The Swiss-Tx Project

Today's state-of-the-art general-purpose supercomputers normally use custom hard- and software. Due to the resulting long and expensive development time, they often show a bad price to performance ratio. A solution to overcome this problem is using off-the-shelf workstations and standard operating systems. To show that it is possible to realize a supercomputer with standard hard- and software, a new project called Swiss-Tx [2] [4] [5] was launched end of 1997 at the Swiss Federal Institute of Technology in Lausanne (EPFL) and Zurich (ETHZ), by Supercomputing Systems (SCS) and by Compaq. The goal is to develop, build and install a series of new commodity based supercomputers until beginning of year 2000. The final machine will have 504 processors, 252 Gbytes of main memory and 5 Tbytes of disk memory. Peak performance of the so-called Swiss-T2 supercomputer will be approximately 1 Tflop/s. All machines are mostly based on commodity parts. Only the communication devices and communication libraries are custom because available products (e.g. Ethernet in combination with the socket interface) do not offer the mandatory functionality, bandwidth and latency. For further information see [3] and [7].

## 2 Communication Libraries

The highlights of the Swiss-Tx communication libraries are low-latency, high-bandwidth, compatibility, and portability. Compatibility means that user programs should run without any modifications on different platforms. Therefore, we developed a high-level communication library offering the standardized Message Passing Interface (MPI) [6]. Portability means that all our libraries run on various hardware platforms (currently PCs and Compaq Alpha-based Workstations and Servers) and operating systems (currently Linux, DIGITAL UNIX, and Microsoft Windows NT).

The communication software consists of a set of libraries:

- MPI (Message Passing Interface) [6]
- ADI (Abstract Device Interface)
- FCI (Fast Communication Interface) [1]

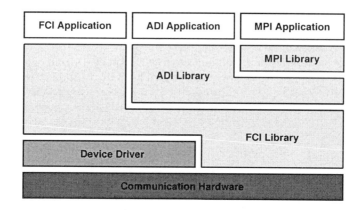

**Fig. 1.** MPI is built on top of ADI, ADI on top of FCI. FCI directly talks to the communication device and uses a device driver for startup and maintenance

ADI is a small library doing some memory management and the MPI data-type management [6]. The core routines are all located in the FCI library. FCI offers the following functionality:

- environmental management
- process group management
- blocking and non-blocking message passing sends/receives/probes (point-to-point operations)
- broadcasts and barriers within process groups (collective operations)
- the *Remote Store Routines* to transfer data with very low overhead
- lock management

### 2.1 The Remote Store Concept

All data transfers in FCI are based on the so-called *Remote Store Concept*. To understand its functionality, take a look at figure 2 which shows the architecture of a state-of-the-art parallel machine. The processing elements are connected to each other by a communication network. At this point, no assumptions are made concerning the physical layout and the topology of the network. Maybe Ethernet, ATM, or Fiber Channel is used. A single processing element (PE) consists of a control and computation unit (CPU), a bus, memory, and a communication device (I/O). Typically the memory is divided into application and system spaces.

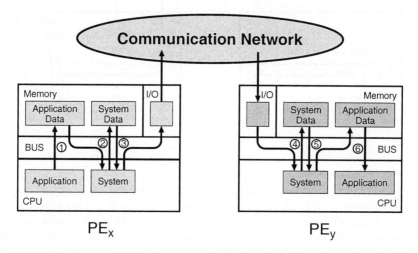

**Fig. 2.** Two processing elements of a state-of-the-art parallel machine. The operating system is involved and data is copied many times

Assume that an application wants to transmit some data from processing element $PE_x$ to processing element $PE_y$. First of all, the part of the application running on $PE_x$ stores the data in its memory (1). Then, the operating system on $PE_x$ must be notified that data has to be transferred. It copies the data into a save memory region (2) and passes it to the communication device (3). The communication device on $PE_y$ will receive the data. The operating system on $PE_y$ transfers it into a save region in system memory (4). As soon as the part of the application running on $PE_y$ wants to receive the data, it informs the operating system, which copies the data from the system memory to the application memory (5). At last, the application can fetch the data (6).

Copying data and calling the operating system are usually expensive operations that decrease bandwidth and increase latency. For example, systems using Ethernet with the TCP/IP stack show latencies of up to 1 ms. The *Remote Store Concept* avoids any system calls and avoids any data copying. This increases the bandwidth and reduces the latency to the hardware limits. The basic idea is that we have some sort of shared memory where a PE can only write to the memory of a distant PE but not read from it. For an example, see figure 3: a process that wants to send some data directly writes it to the communication hardware (1) and the communication hardware in the receiving PE is capable of storing it directly into the application memory (2). Each PE has a set of windows $w$ in the so-called *Global Communication Space (GCS)*. This GCS is represented by a virtual address space whose maximum size only depends on the communication

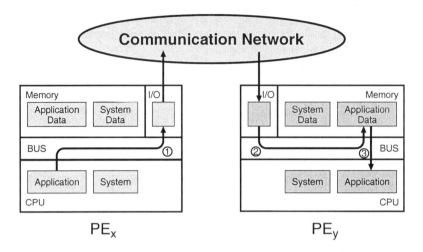

**Fig. 3.** Two processing elements equipped with a *Remote Store* communication hardware. Data is generated by the application and directly written to the communication device (1) (it is also possible that data is first written to the application memory and then fetched by the communication device which inserts one additional copy step). The receiving communication device copies data directly to the application memory (2) from where the application can fetch it (3)

hardware. Therefore, a 32 Bit machine is not limited to a 4 Gbyte GCS which makes it possible to build massively parallel workstation clusters with standard workstations. The windows define address ranges in the GCS where data will be received. This windows can be static or dynamic which is hardware dependent. Basically, they can be arranged in any possible variation. Overlapping in the GCS is allowed which makes it possible to define broadcast and multi-cast windows. Depending on the communication hardware, it is possible to define the set of processing elements that is allowed to receive the data.

In figure 4, two Processing Elements ($PE_x$ and $PE_y$) are involved. Each of them has two open windows in the GCS ($PE_x$: $w_{x,1}$, $w_{x,2}$, $PE_y$: $w_{y,1}$, $w_{y,2}$). $PE_x$ maps windows $w_{x,1}$ and $w_{x,2}$, $PE_y$ $w_{y,1}$ and $w_{y,2}$. Corresponding to these windows, other windows of the same size exist in the main memories. Window $w'_{x,1}$ corresponds to window $w_{x,1}$, window $w'_{x,2}$ corresponds to window $w_{x,2}$ and so on. Each window has a base address named $w\_base_{i,j}$ or $w'\_base_{i,j}$. $w\_base_{i,j}$ is the base address of window $w_{i,j}$ in the GCS, $w'\_base_{i,j}$ the base address of window $w'_{i,j}$ in the main memory of $PE_i$. If data is written to an address in the GCS that matches window $w_{i,j}$ and $PE_i$ is allowed to receive, this data is copied into main memory in window $w'_{i,j}$ at the corresponding relative location. Assume that we want to write to address $addr$ in the GCS. Thus, the destination address $addr'$ in the main memory of the receiver window is calculated as follows:

$$addr' = (addr - w\_base_{i,j}) + w'\_base_{i,j} \qquad (1)$$

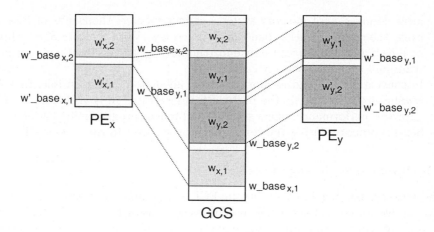

**Fig. 4.** Each PE maps two GCS windows in the local main memory

For example, if $PE_x$ writes to an address that is in window $w_{y,2}$, $PE_y$ will receive the written data if it is allowed to. The communication device copies the data into window $w'_{y,2}$ starting at address $addr'$ in the main memory of $PE_y$ without any CPU interaction.

A machine using the *Remote Store Concept* is like a shared memory system and offers the same advantage in comparison to other architectures: Shared memory systems are easy to program. But there is one basic difference: It supports only write but no read instructions to the global space which is in fact a great advantage: read instructions are always local! If you take a closer look at it, you will recognize that it is easy to implement the communication hardware. Conventional shared memory systems have some problems concerning data consistency and read instructions:

- It is difficult to cache global data: It is necessary to have a global synchronization mechanism that guarantees data consistency.
- Normally, a write instruction is a uni-directional operation which is fast (the write command followed by the data is sent by the writer). A non-cached read instruction is bi-directional and needs more time (the read command is sent by the reader, a controller is returning the requested data).
- Each PE that wants to read global data wastes communication bandwidth.

The *Remote Store Concept* combines many advantages of shared memory and message passing and adds some special features:

- A *Remote Store* write instruction can have several destinations if windows are overlapping which is similar to broad- and multi-cast functionality of a message passing system.
- As a conventional message passing system, it is easier to scale than conventional shared memory systems. In shared memory systems, read instructions are more often than write instructions and read instructions are often not

local. Many shared memory systems cannot manage thousands of PEs because the communication bandwidth is too low. With *Remote Store*, this is no longer a problem because read instructions are always local main memory reads and are automatically cached, too.
- In conventional shared memory systems, all shared data is visible. In a *Remote Store* system, only the part a PE is interested in is visible for it. This improves performance, offers security and makes programming easier.
- Several processes using the communication device can run on one PE.

## 2.2 Intelligent Message Passing

The message passing functionality in FCI is a minimum subset of functions necessary to implement the widely used message passing library MPI [6]. It offers routines to startup, shutdown, and maintain an application, to manage process groups, to send, receive, and probe for messages in blocking or non-blocking mode, and to perform broadcasts and barriers. Normally, sending and receiving

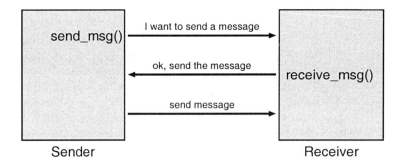

**Fig. 5.** State-of-the-art message-passing sending and receiving. The sender tells the receiver that it wants to send him a message. When the receiver is ready to receive this message, it will tell the sender. Then the message is transferred

a message works as shown in figure 5. A handshake takes place before the real message is sent. This is necessary because normally a process is not allowed to send a message until a matching request is posted at the receiver. Without this matching receive request, probably no space is allocated for the message at the receiver. Figure 6 shows how messages are transferred in FCI. The receiver tells the sender that it is ready. As soon as the receive request is caught by the sender, it will send the message if it has a matching one. This mechanism minimizes the handshake overhead and the necessary network transmissions which reduces the message passing latency.

All FCI message passing routines use the *Remote Store Concept* to transfer data between the processes. This makes hardware design easy (no special message-passing functionality is necessary) but demands sophisticated software: the *Intelligent Sender Concept* with its efficient message handling guarantees

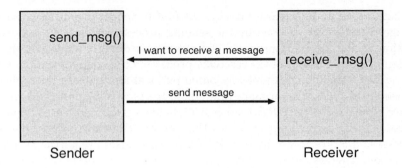

**Fig. 6.** Optimized message-passing sending and receiving. The receiver tells the sender that it is ready to receive the message and the sender will send it if it has a matching one. The two network transmissions of the handshake shown in figure 5 are replaced by one transmission

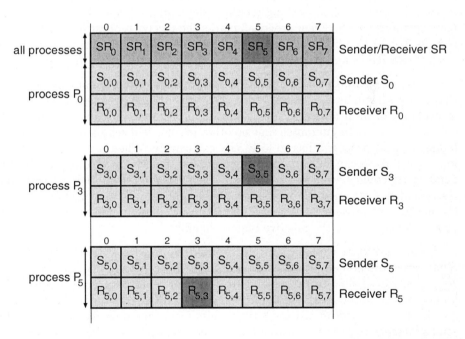

**Fig. 7.** *A part of the Receive Request Tables (RRTs)* in the GCS for an application with 8 processes

high-bandwidth and decreases latency. *Intelligent Sender* means that the whole message transfer can be done by the sending process without any software interaction at the receiving process. The message is directly copied at the right location in main memory of the receiving process by the communication device[1]. The message handling is completely based on local and global tables which are mapped in the GCS. The primary tables hold the receive requests and are called *Receive Request Tables (RRTs)*. Each RRT holds a fixed number of requests. A process puts its receive requests into the remote RRTs of the possible sender processes and stores it also in local RRTs. If $n$ processes are participating in an application, $n$ RRTs are necessary to maintain receive requests that can be satisfied by any process and $2n^2$ RRTs are used to maintain receive requests where the sender is explicitly given. Figure 7 shows a clipping of the GCS where the RRTs are held ($n = 8$). The uppermost line is used to maintain receive requests that can be satisfied by any process, the others are involved when the sender is explicitly given.

Assume that process 5 wants to receive a message from process 3 and that process 3 wants to send a message to process 5. Process 5 inserts a receive request in RRT $R_{5,3}$ (local in process 5) and also in RRT $S_{3,5}$ (remote in process 3). Process 3 waits for the matching receive request in $S_{3,5}$, finds it and sends the message to process 5. Then, process 3 marks the receive request in $S_{3,5}$ (local in process 3) and also in $R_{5,3}$ (remote in process 5) inactive. Later, process 5 can clean up its RRTs and remove the inactive receive request from $R_{5,3}$ and $S_{3,5}$.

Now, assume that process 5 wants to receive a message from any process and that process 3 will be the sender. Process 5 inserts a receive request in $SR_5$ which is available for all processes. Process 3 waits for the matching receive request in $SR_5$ and finds it. To guarantee that no other process will send a message for the found request, process 3 has to lock $SR_5$ exclusively. When it has attained the lock, it has to rescan $SR_5$ in order to guarantee that the receive request is still pending (it could be marked inactive or even removed if another process was faster and has attained the lock first). If the request is still available, it sends the message to process 5 and marks the receive request in $SR_5$ inactive. Later, process 5 can remove the inactive request handle.

Some additional explanations: $R_{x,0..7}$ and $S_{x,0..7}$ are stored in process $x$. $SR_{0..7}$ are stored on all processes. $R_{5,3}$ and $S_{3,5}$ have exactly the same contents. $R_{5,3}$ is only accessible by process 5 (if it inserts or removes a receive request for a message from process 3) and by process 3 (to mark a receive request inactive). $S_{3,5}$ is only accessed by process 3 (if it waits for a matching receive request to send a message to process 5 or to mark a receive request inactive) and by process 5 (if it inserts or removes a receive request for a message from process 3).

---

[1] This is only true for contiguous and aligned messages (the alignment for the current hardware is 4 Bytes). Structured messages or miss-aligned messages must be copied by the sender and/or the receiver.

# 3 Early Results

Basic measurements are available for a prototype installed at the Swiss Federal Institute of Technology in Zurich. The machine consists of 8 standard PCs with Intel Pentium processors running at 233 MHz and LINUX as operating system. The PCs are connected by Fast Ethernet for standard communication and for

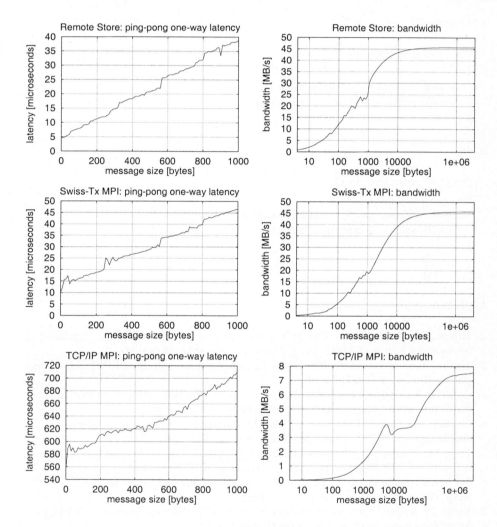

**Fig. 8.** Ping-pong one-way latency and bandwidth as a function of the message size of *Remote Store*, TCP/IP MPI, and Swiss-Tx MPI

a TCP/IP bases MPI implementation. The Swiss-Tx communication hardware transfers 32 Bit in parallel and runs at 12 MHz. Therefore, the maximum possible bandwidth is 48 MB/s. We have measured latency and bandwidth as a function of the message size. Figure 8 shows the results. The one-way latency for the TCP/IP based MPI implementation is about 550 $\mu s$, for Swiss-Tx MPI 10 $\mu s$ and for *Remote Store* 5 $\mu s$. Maximum bandwidth for TCP/IP MPI is less than 8 MB/s, for Swiss-Tx MPI and *Remote Store* 46 MB/s. The TCP/IP MPI bandwidth graph is smoothed. In reality, a lot of large spikes are present.

*ping-pong algorithm:*

```
if(rank==0)
{
  for(i=0;i<count;i++)
  {
    SendMessage(1,message,size);
    ReceiveMessage(1,message,size);
  }
}
else
{
  for(i=0;i<count;i++)
  {
    ReceiveMessage(0,message,size);
    SendMessage(0,message,size);
  }
}
```

# References

1. Brauss, S., Nemecek, J.: The FCI Reference Manual. Swiss Federal Institute of Technology Zurich (ETHZ), http://www.ife.ee.ethz.ch/hpc/fci
2. Gruber, R., Gunzinger, A.: The Swiss-Tx Supercomputer Project. EPFL Supercomputing Review, **9** (1997) 21–23
3. de Vita, A., Gruber, R., Kuonen, P., Volgers, P.: Parallel Computer Architectures for Commodity Computing. HPCN'99 Keynote lecture (1999)
4. Gruber, R., Dubois-Pèlerin, Y., Swiss-Tx Group: Swiss-Tx: First Experiences on the T0 System. EPFL Supercomputing Review, **10** (1998) 19–23
5. Frey, M., Gruber, R., Kuonen, P.: Swiss-T0 Installation and Acceptance Report. Swiss Federal Institute of Technology Lausanne (EPFL) (1998)
6. MPI: A Message-Passing Interface Standard. University of Tennessee, http://www.mcs.anl.gov/mpi/index.html
7. Swiss-Tx Architecture. Swiss Federal Institute of Technology Lausanne (EPFL), http://capawww.epfl.ch/swiss-tx/index.html

This project was funded by the Swiss Commission for Technology and Innovation (CTI).

# Finding the Optimal Unroll-and-Jam*

N.Zingirian and M.Maresca

Dipartimento di Elettronica ed Informatica – University of Padua, Italy

**Abstract.** Reducing the traffic between CPU and main memory is one of the main issues in the optimization of programs for load/store architectures. It is the register allocation module of optimizing compilers that keeps this traffic low by cleverly associating the program variables to the CPU registers. Since register allocation takes place during code generation and works on the intermediate code produced by the compiler front-end, the structure of such a code, which closely depends on the structure of the source code, heavily affects the effectiveness of register allocation. Proper techniques can be used to restructure the source programs in such a way to produce intermediate code able to take advantage of advanced register allocation schemes.
In this paper we analyze one of these techniques called *unroll-and-jam*. In particular we find the fractional optimal unroll-and-jam transformation valid for a large class of computing intensive programs. The paper presents the analytical model of the optimal unroll-and-jam and a method to compute the unrolling parameters numerically.
**Keywords**: Load/Store architectures, Register Utilization, Unroll-and-Jam.

## 1 Introduction

Most of today's high performance architectures are based on the instruction level parallel RISC paradigm. These are load/store architectures in the sense that their instructions can exclusively process operands present in the CPU internal register file[5]. A well known source of inefficiency of load/store architectures is the *register spilling* phenomenon. Register spilling takes place when an operand is swapped out from registers, although it is still to be used by future instructions, because other operands are more urgently required to be in registers and no more registers are available. This fact increases the traffic between memory and CPU.

The register spilling reduction is obtained through the combined action of two kinds of optimizations:

- machine language level techniques corresponding to *register allocation* techniques [2][3] which take place during code generation and consist of associating the program variables on the limited pool of physical registers as much efficiently as possible and

---

* Work carried out under the financial support of the Ministero dell'Universita' e della Ricerca Scientifica e Tecnologica (MURST) in the framework of the Project "*Methodologies and Tools of High Performance Systems for Multimedia Applications*".

– source code level techniques [8][7], which consist of restructuring the computation at a source level, in such a way to group the operations which use as much variables as possible in common.

Register allocation is known to be an instance of a basic graph problem, namely graph coloring, the optimal solution of which is known to be NP-complete. Heuristics are used to find acceptable solutions in optimizing compilers. Program restructuring, on the contrary, is not an instance of any basic problem, as it depends both on the original program structure, which is highly variable, and on the specific transformation applied. In particular in this paper we concentrate on a well-known restructuring technique for the reduction of register spilling in programs consisting of nested loops, i.e. the *unroll-and-jam* transformation[4]. In previous works[6] we have shown, for a rather strict class of programs, that the unroll-and-jam allows for the absolutely best register utilization asymptotically. The present work proposes an analytical investigation and a related numerical resolution method that allows finding the fractional optimal organization of unroll-and-jam transformation for a large class of programs.

The paper is organized as follows. First we introduce the basic concepts of the unroll-and-jam transformation and define the class of programs for which the results of the paper are valid (Sect. 2). Then we present the analysis which leads to the fractional optimality conditions (Sect. 3) and the numerical (Sect. 4) approach which allows computing the parameters which match the optimality conditions for a given program. Some remarks conclude the paper (Sect. 5).

## 2 Concepts and Terminology

This section introduces the basic concepts related to the unroll-and-jam transformation (Sect. 2.1) and defines the class of programs for which the optimal unroll-and-jam has been identified (Sect. 2.2)

### 2.1 The Unroll-and-Jam Transformation

In this section we show that a program consisting of a number of nested loops and running on a load/store architecture is affected by high inefficiencies from the point of view of register utilization.

Let us consider the generic program of Figure 1(A), as an example. In this case the innest body contains three functions f, g, h which respectively depend on the external loop index, on the internal loop index and on both. Of course each function can be thought of as a memory access or as any other expression. All the values of function h are loaded on registers once because each loop iteration uses a different value of h. As a consequence register utilization of h is optimal, i.e. no register spilling takes place. Function f also allows for optimal register utilization, because the compiler can factorize the computation of f out of the innest loop. On the contrary function g causes bad register utilization, because each value of g is loaded into registers and swapped out $N\_i$ times, causing register spilling

```
[tbH]
   for ( i=0 ; i < N_i; i++ )
      for ( j=0 ; j < N_j; j++ )
         body(f(i),g(j),h(i,j));
```

(A)

```
   for ( i=0 ; i < N_i; i+=U ) {
      r_1=f(i);
      r_2=f(i+1);
      ...
      r_U=f(i+U-1);
      for ( j=0 ; j < N_j; j++ ){
         r = g(j);
         body(r_1,r,h(i,j));
         body(r_2,r,h(i+1,j));
         ...
         body(r_U,r,h(i+U-1,j));
      }
   }
```

(B)

**Fig. 1.** Example of unroll-and-jam: (A) flat program; (B) unrolled program

overhead. Of course loop permutation simply exchanges the roles of f and g, without removing the inefficiency.

The unroll-and-jam transformation reduces *register spilling* overhead as shown in Figure 1(B). If the external body is unrolled U times then U values of f can be kept simultaneously in registers. This code organization allows the same value of g, loaded once on a register associated to r, to be used for U iterations without being spilled. As a consequence register spilling of g is reduced by factor $U$ [2]. The reduction of register spilling is counterbalanced by the increased number of registers used simultaneously, as a consequence the value of the *unrolling factor* U is limited by the number of registers available in a given load/store architecture. It is worth noticing that the internal loop unrolling does not reduce register spilling.

While it is easy to find the optimal unrolling factor in simple cases such as the example above, the problem becomes much more complex if the program to be unrolled consists of $N$ nested loops and contains functions which depend on arbitrary subsets of the $N$ loop indices. In this case the set of the $N-1$ unrolling degrees of the $N-1$ external loops which minimize register spilling must be determined.

In the following we present the results of an analytical approach to the computation of the optimal unrolling factors for a large class of programs.

---

[2] Note that in the unrolled code functions f and h are not involved by register spilling again.

## 2.2 Program Model

In this section we define a class of programs to which we apply the approach to the computation of the optimal unroll-and-jam factors. We consider programs consisting of $N$ nested loops and in which the body includes a number of functions each depending on an arbitrary subset of loop indices. The restrictions of the programs considered are that

- the iterations of the program can be performed in any arbitrary order;
- each function takes the same time to be executed and each value occupies the same number of registers, for any arguments.

The first condition can be verified in many application domains such as low level image processing, linear algebraic computations, etc. This hypothesis does not require the absence of loop-carried dependencies but only the commutativity of iterations, since the unroll-and-jam transformation changes the execution order of the iterations.

The second condition makes the innest body functions able to model load/store instructions rather than arbitrary program expressions, neglecting caching effects.

In the following we provide a definition of the model of the programs addressed in the rest of the paper.

**Definition 1.** Let be $P^{(N,m)}(D_1,\ldots,D_m,T,K)$ a program consisting of $N+1$ nested loops without iteration precedence constraints and organized as follows,
```
for (i₁=0; i₁<B_i;i₁++)
   ...
      for (i_N=0; i_N<B_N;i_N++)
         for(i = 0;i<B;i++)
            body(F₁(D₁,i),...,F_m(D_m,i))
```
where $D_k \subseteq I = \{i_1,\ldots,i_N\}^3$ is the domain of the functions $F_k$, for $k = 0,\ldots,m$. $T$ is the execution time of each function $F_k$ and $K$ is the number of registers occupied by the output value of each $F_k$.

## 3 Analysis of Optimality

In this section we present the analysis aimed at determining the necessary conditions for which the program $P^{(N,m)}$ defined in the previous section is optimally unrolled. First we present and prove the conditions of optimality (Sect. 3.1) and then we discuss them (Sect. 3.2).

---

[3] In the following we adopt an abuse of notation considering $I = \{1, 2, \ldots, N\}$ instead of $I = \{i_1, \ldots, i_N\}$ and the $D_f$ as subsets of such an $I$. In this way an element $k \in D_f$ can be used as pedix of other values, e.g. $U_k$, i.e. the unrolling degree of the $k-th$ loop

## 3.1 Conditions of Optimality

Given the program $P^{(N,m)}$ of Definition 1 it is hard to decide which degree of unrolling $U_k$ should be applied to the index $i_k$ (of the $k-th$ loop) using rules of thumb. In fact the unrolling of the index $i_k$ has in general a benefit and a cost. The benefit is that the number of invocations of the functions which do not depend on $i_k$ is reduced by factor $U_k$. The cost is that the registers are occupied by $U_k$ copies of each function which depends on $i_k$.

In addition, the unrolling factors of different loops of the nest combine their costs and benefits by multiplying each others to yield the overall cost in terms of register used and benefits in terms of spilling reduction, as it is illustrated in the following.

The following theorem determines the conditions which must be necessarily fulfilled by the unrolling degrees of each external loop to achieve the optimal unroll-and-jam, i.e. the one which minimizes the register spilling.

**Theorem 1.** *In a program $P^{(N,m)}(T, K, D_1, \ldots, D_N)$ (see Definition 1) is optimally unrolled if the $i-th$ loop is unrolled by degree $U_i$ for each $i = 1, \ldots, N$ where $U_i$ satisfies the following conditions: being*

$$x_f = \prod_{l \in D_f} U_l \quad f = 1, \ldots m \tag{1}$$

$$Z_k = \sum_{f : k \notin D_f} x_f \quad k = 1, \ldots, N \tag{2}$$

*then*

$$\begin{cases} Z_i = Z_j \\ Z_i \geq Z_j \ only \ if \ U_j = 1 \end{cases} \tag{3}$$

*for each $i, j = 1, \ldots, N$, and*

$$\sum_{i=0}^{m} x_i = R/K \tag{4}$$

*where $R$ is the number of registers available.*

*Proof.* The proof of Theorem 1 is based on the necessary conditions of minimum stated by the theory of Lagrange Multipliers[1]. Vector $\mathbf{U} = (U_1, \ldots, U_N)$ is the set of variables which must be determined as argument of minimum.

The overall execution time is the function to be minimized. The execution time of the function $F_i$ depends on the values of those $U_k$ such that $k$ does not belong to the domain $D_i$. More in details, the function $F_i$ is invoked $B_1 \cdot \ldots \cdot B_N \cdot B \cdot (1/\prod_{j \notin D_i} U_j)$ times. As a consequence the overall execution time is

$$t(\mathbf{U}) = B_1 \cdot \ldots \cdot B_N \cdot B \sum_{i=1}^{m} \left( T_i \prod_{j \notin D_i} \frac{1}{U_j} \right) \tag{5}$$

where $T_i$ is the execution time of function $F_i$[4] and $B_1, \ldots, B_N$ are the bounds of the $N$ external loops. We will remove the bounds from the cost function in the following steps of the proof because they are not relevant in the minimization.

The unrolling-and-jam is effective if the values are stored in registers. The number of values returned by $F_i$ that are stored in the registers is equal to the product of all the unrolling degrees related to the index belonging to $D_i$. The overall number of register occupied simultaneously cannot exceed the total number $R$ of registers, as a consequence

$$r(\mathbf{U}) = \sum_{i=1}^{m} K_i \cdot \prod_{j \in D_i} U_j \leq R \qquad (6)$$

where $K_i$ is the number of registers occupied by each value of $F_i$. Of course the unrolling degree $U_k$ is equal to 1 when no unrolling is applied to the index $i_k$ and cannot be lower than 1, as a consequence, the following constraints hold:

$$-U_k \leq -1 \; for \; any \; k = 1, \ldots, N$$

The values of $U_1 \ldots U_N$ that determine the minimum of function $t(\mathbf{U})$ satisfy the following Lagrange equations:

$$\frac{\partial}{\partial U_i} \left( t(\mathbf{U}) + \mu_0 \cdot \left( r(\mathbf{U}) - R \right) + \sum_{j=1}^{N} \mu_j U_j \right) = 0 \; for \; i = 1, \ldots, N \qquad (7)$$

$$\mu_i \geq 0 \; for \; i = 0, \ldots, N \qquad (8)$$

$$\mu_0 \cdot \left( r(\mathbf{U}) - R \right) = 0 \qquad (9)$$

$$\mu_i \cdot (U_i - 1) = 0 \; for \; i = 1, \ldots, N \qquad (10)$$

Replacing Eq (5) and Eq (6) into Eq (7) we obtain

$$-\sum_{f: i \notin D_f} \frac{T_f}{U_i} \cdot \frac{1}{\prod_{j \notin D_f} U_j} + \mu_0 \cdot \sum_{f: i \in D_f} \frac{K_f}{U_i} \cdot \prod_{j \in D_f} U_j - \mu_i = 0$$

which can be combined with Equation (10), to obtain, for any index $i \in I$

$$-\sum_{f: i \notin D_f} T_f \cdot \frac{1}{\prod_{j \notin D_f} U_j} + \mu_0 \cdot \sum_{f: i \in D_f} K_f \cdot \prod_{j \in D_f} U_j - \mu_i = 0 \qquad (11)$$

It is easy to see that if $\mu_0 = 0$ the equation above cannot be solved because it is the sum of strictly negative components. As a consequence, for Eq (9), $r(\mathbf{U}) = R$.

---

[4] In the hypotheses of this theorem the execution time $T$ is equal for each function: we carry out the proof relaxing this hypothesis (i.e., considering different $T_i$ for each function $F_i$) until the last step in order to obtain more general intermediate results. We will do the same for the size of the function $K$.

This fact allows writing the following equation

$$\sum_{f:i\in D_f} K_f \cdot \prod_{j\in D_f} U_j = R - \sum_{f:i\notin D_f} K_f \cdot \prod_{j\in D_f} U_j \qquad (12)$$

In addition it is also immediate to verify that

$$\frac{1}{\prod_{j\notin D_f} U_j} = \frac{\prod_{j\in D_f} U_j}{U_{prod}} \qquad (13)$$

where $U_{prod} = \prod_{j}^{N} U_j$, i.e. the product of all the unrolling degree of all external loops. By combining equations (12)(13) and (11) we obtain

$$-\sum_{f:i\notin D_f} \left((T_f/U_{prod} + \mu_0 K_f)\prod_{j\in D_f} U_j\right) + \mu_0 R - \mu_i = 0 \qquad (14)$$

The hypothesis that $T_i = T$ and $K_i = K$ for any $i = 0,\ldots N$ allows subtracting two instances of equation (14)

$$\sum_{f:i\notin D_f}\left(\prod_{j\in D_f} U_j\right) - \sum_{f:k\notin D_f}\left(\prod_{j\in D_f} U_j\right) + \frac{\mu_i - \mu_k}{(T/U_{prod} + \mu_0 K)} = 0 \qquad (15)$$

that, written as

$$Z_i - Z_k + \frac{\mu_i - \mu_k}{(T/U_{prod} + \mu_0 K)} = 0 \qquad (16)$$

which, combined with equations (8) and (10) yields the equation (3).

## 3.2 Discussion of the Conditions of Optimality

Theorem 1 states that each function $F_f$ is to be associated to a cost $x_f$ which counts the number of instances of $F_f$ that are stored simultaneously on registers, for a given set **U** of unrolling degrees. Then, for each index $i$ of the external loops, a new cost $Z_i$ is built which accumulates the costs $x_f$ of the functions which do not depend on $i$. The optimal values **U** of unrolling degrees are such that costs $Z_i$ are equal, unless a constraint is violated. If a constraint is violated the theorem tells that the indices $i$ that are in the border of the constraint (i.e. for which $U_i = 1$) can be associated to a cost $Z_i$ which is less than the cost associated to the indices which are not in the border of the constraints. On the contrary all the indices $j$ that are not in the border of the constraint must be characterized by the same value of $Z_j$.

The problem of finding the optimal unrolling degree of a program can be often reduced by means of a useful corollary:

**Corollary 1.** *If there is an index $i_k$ which appears as an argument of a superset of the functions in which another index appears then the $k-th$ loop is not to be unrolled. In addition the optimal unrolling degree of the other indices are equivalent to the optimal unrolling degrees of the program obtained removing the $k-th$ loop as well as each occurrence of the index $k-th$.*

*Proof.* The hypothesis can be expressed as follows: there exists $k$ and $h$ such that if $k \notin D_f$ then $h \notin D_f$. This fact implies that $Z_h > Z_k$ (see eq (2)) and, as a consequence, $U_k = 1$ for the equation (3). In addition, it easy to verify that for $U_k = 1$ the cost and constraint function (see eq (6) and (5)) are equal to the cost functions of the program obtained removing the $k-th$ loop as well as each occurrence of the index $i_k$.

## 4 Numerical Method

In this section we show that the conditions stated by Theorem 1 allows identifying a set of unrolling factors that optimize the use of CPU registers which is unique under some hypotheses.

The points that make the solution not immediate are:

- the conditions of Theorem1 are not expressed by linear equation because each $Z_i$ consists of the sum of products of variables, i.e., the $x_j$.
- the $Z_i$ are not always equal to each other, but they can be different whenever their equality could violate the constraint of $U_i \geq 1$ fore some $i \in [1, N]$.

First we suppose that all the $Z_i$ are equal for any $i = 1, \ldots N$, and we show that *i)* there exists one or more equivalent solutions and *ii)* provide an iterative set of equations that rapidly converge to that solutions. Then we discuss the case of constraint violations.

The condition $Z_i = Z_j$ (see eq. (2)) for each $i, j$ can be conveniently rewritten as follows

$$M\mathbf{x} = k(1, \ldots, 1)^T \qquad (17)$$
$$M = \{m_{i,j} | m_{i,j} = 0 \text{ if } i \in D_j, \qquad (18)$$
$$m_{i,j} = 1 \text{ otherwise for } i = 0, \ldots, N, j = 0, \ldots, m\}$$
$$\mathbf{x} = (x_1 \ldots x_m)^T \qquad (19)$$

where $k$ is a real number to be determined in the following.

The idea is now to express the logarithm of $\mathbf{x}$ as a linear combination of the logarithms of $U_i$, as follows

$$\ln \mathbf{x} = (\ln x_1 \ldots \ln x_m)^T = R\mathbf{L} \qquad (20)$$
$$R = (\mathbf{r_1} | \ldots | \mathbf{r_m})^T = \{r_{i,j} | r_{i,j} = 1 - m_{j,i}\} \qquad (21)$$
$$\mathbf{L} = (\ln U_1 \ldots \ln U_n)^T \qquad (22)$$

In this way matrix $R$ is derived from matrix $M^T$ by replacing null elements with 1 and vice-versa.

Combining equations (17) and equations (20) we obtain that

$$f(\mathbf{L}) = M\left(e^{\mathbf{r}_1^T \cdot \mathbf{L}}, \ldots, e^{\mathbf{r}_m^T \cdot \mathbf{L}}\right)^T = k(1, \ldots, 1)^T \qquad (23)$$

The first order approximation of $f(\mathbf{L})$ around $\mathbf{L}^*$, namely $f(\mathbf{L}^*) + \nabla f(\mathbf{L}^*)^T (\mathbf{L} - \mathbf{L}^*)$ yields

$$f(\mathbf{L}) \approx M\mathbf{x}^* + M\,diag(\mathbf{x}^*)R(\mathbf{L} - \mathbf{L}^*) \qquad (24)$$

$$\mathbf{x}^* = (x_1^*, \ldots, x_m^*)^T = \left(e^{\mathbf{r}_1^T \cdot \mathbf{L}^*} \ldots e^{\mathbf{r}_m^T \cdot \mathbf{L}^*}\right)^T \qquad (25)$$

where $diag(\mathbf{x})$ denotes a matrix $m \times m$ which is null except for the diagonal which contains the values $x_1, \ldots, x_m$. The interesting property of this approximation is that the product of matrix $M\,diag(\mathbf{x}^*)R$ yields a quadrate matrix. This fact assures that the problem is not overconstrained, i.e. there exist one ore more solutions. A more accurate analysis of the singularity properties of matrix $M \cdot R$ is still in progress. In addition, since the gradient is always monotonic (being a combination of exponential curves of $\mathbf{L}$) then the function is convex and allows finding a solution rapidly with the tangent method.

The idea is to compute the first order approximation for $\mathbf{L}^* = (0, 0, \ldots, 0)^T$ and to find the value of $\mathbf{L}(k)$ which intersects the vector $k \cdot (1, 1, \ldots, 1)^T$, for a generic value of $k$. The value of $k$ can be determined by means of the constraint (4), that can be expressed as

$$\mathbf{x} \cdot (1, 1, \ldots, 1)^T = R/K \qquad (26)$$

The equation is not linear but is scalar and monotonic; as a consequence the result can be found in a few steps of dichotomy. At this point $\mathbf{L}$ is completely determined, and a new first order approximation can be computed around the new value of $\mathbf{L}$. The computation can be repeated until the values of $\mathbf{L}$ obtained at two succesive steps are enough close to each other. In the following we present formally the steps necessary to compute the values of $U_i$.

1. Let $\mathbf{L}^{(0)} = (0, \ldots, 0)$, and let be the initial step $p = 0$
2. compute $\mathbf{x}^{(p)}$ from $\mathbf{L}^{(p)}$ using (25)
3. compute the new value of $\mathbf{L}^{(p+1)}(k)$ as

$$\mathbf{L}^{(p+1)}(k) = (M\,diag(\mathbf{x}^{(p)})R)^{-1}\left((k \cdot (1, 1, \ldots, 1)^T) - M\mathbf{x}^{(p)}\right) + \mathbf{L}^{(p)}$$

obtained from eq (24)
4. find $k^*$ such that $\mathbf{x}^{(p+1)} \cdot (1, 1, \ldots, 1)^T = R/K$ through dichotomy.
5. $\mathbf{L}^{(p+1)} = \mathbf{L}^{(p+1)}(k^*)$
6. if $(|\mathbf{L}^{(p+1)} - \mathbf{L}^{(p)}| > \epsilon)$ then $p = p + 1$ go to step 2

This method can be preceded by a phase in which matrix $M$ is simplified on the basis of Corollary 1. If the matrix $M \cdot R$ has rank $n$, then the solution is unique for each point of linearization. In addition the unrolling degree yielded are greater than zero because are expressed as exponential functions of $\mathbf{L}$. The only limitation is that in some cases the optimal unrolling degrees can result between 0 and 1. This fact violates the constraints of the problem and the method does not take into account of it. However, while work is still in progress to deal with this issue, it is worth noticing that this limitation is acceptable under the hypothesis of fractional solution. This computation provides a lower bound for register spilling, tailored on a detailed description of the problem, as well as provides a solution reasonably close to the integer optimal solution of the problem.

## 5 Concluding Remarks

In this paper we presented a novel analytical approach to the computation of the optimal loop unrolling factors valid for a large class of computing intensive programs. First necessary conditions of optimality have been extracted by means of Lagrange Multipliers, then a numerical method has been presented to compute the optimal unrolling degrees. The optimality has been obtained for fractional values. Unfortunately, the integerization of such values does not necessarily lead to the optimal integer solution. Nevertheless the results obtained are an excellent hint for highly efficient unroll-and-jam based restructurations and provide accurate lower bounds for register spilling in programs running on architectures equipped with a limited number of registers.

## References

1. R. Bertsekas. "Constrained Optimization and Lagrange Multipliers". Academic Press, 1982.
2. G.J Chaitin. "Register Allocation and Spilling via Graph Coloring". "Proceeding of the ACM SIGPLAN Symp. on Compiler Construction", 17(6):98–105, June 1982.
3. F.C. Chow and Hennessy. J.L. "Priority-Based Coloring Approach to Register Allocation". "ACM Trans. on Programming Language and Systems", 12(4):501–536, 1990.
4. K. Dowd. "High Performance Computing". O'Reily & Associates, Inc, Sebastopol, Ca 95472, 1988.
5. J.L. Hennesy and D.A. Patterson. "Computer Architecture:A Quantitative Approach". Morgan Kaufmann Publishers, Inc., San Mateo, Ca 94403, 1990.
6. N.Zingirian and M.Maresca. "External Loop Unrolling of Image Processing Programs: Optimal Register Allocation for RISC Architectures". " Proc. of IV Int. Workshop on Computer Architecture for Machine Perception, Boston ", October 1997.
7. M. Maresca P. Baglietto, M. Migliardi and N. Zingirian. "Image Processing on High Performance RISC System". "Proceeding of the IEEE", 84(7):917–930, 1996.
8. N. Zingirian and M. Maresca. "Scheduling Image Processing Activities on Instruction Level Parallel RISC Systems Through Program Transformations". "Lecture Notes in Computer Science, HPCN97", 1225, Vienna, April 1997.

# A Linker for Effective Whole-Program Optimizations

Andrea G. M. Cilio and Henk Corporaal

Delft University of Technology
Computer Architecture and Digital Techniques Dept.
Mekelweg 4, 2628CD Delft, The Netherlands
A.Cilio@its.tudelft.nl    H.Corporaal@its.tudelft.nl

**Abstract.** The use of a standard binary format in the later part of code generation promotes efficiency and interchangeability of tools, but leaves little information on the source file in the machine code representation. We propose a new approach to code generation, based on a single, highly structured internal format used during proper compilation, machine code generation and linkage. This format offers new opportunities for *whole-program* optimizations. We have implemented and tested a code generator based on this format. Although the use of traditional a binary format is more efficient, we believe that the increase in code size and compilation times are largely compensated by the opportunities offered by this new trajectory. To support this assertion, we discuss some of its potential applications.

## 1 Introduction

A traditional compilation path is split into three phases. The first phase, the actual compilation, transforms the source program into assembly code. To support the many optimizing code transformations performed during compilation, a rich internal intermediate representation, which usually preserves most of the source-level information, is used. The intermediate representation is then translated into code for the target architecture, and a text assembly file is written out. During this step, a large part of the source-level information is lost. In the second phase, the assembler parses the assembly file and translates the code into the binary format of the target machine. During this process, an offset relative to a program section is assigned to every variable and machine instruction, and relocation information is built up. The resulting output file is called *object file*. The third step is the linkage of one or more object files and libraries into an executable binary file. During linkage, program sections from different modules are merged into a single file and symbolic references to data objects or instructions are *resolved*, i.e., replaced with the actual physical address of the referenced symbol.

This well-established code generation path offers three advantages for general purpose target processors. First, it is *efficient*: the binary format is a compact representation; linking modules and libraries does not require expensive code transformations. Second, it becomes possible to distribute library code without the source files. Third, since assembly syntax and binary format are standardized with respect to target architecture and operating system, it promotes *interchangeability* of the tools that perform assembly and linkage, provided that they comply to these standard formats.

This compilation approach is also common for Application Specific Instruction-set Processors (ASIPs). Code generators are often derived by re-targeting tools originally designed for general purpose processors [9]. This approach, although popular, has some drawbacks. Interchangeability of binary tools is less relevant, because these are usually ad-hoc designs for a single (possibly *templated*) architecture. Moreover, the concern is now shifted towards the "quality" of the code produced rather than the speed of code generation. At the same time, a number of new requirements become critical when generating code for ASIPs and, more generally, for instruction-level parallel processors. Altogether, these requirements force us to completely change the compilation trajectory.

The rest of this paper is organized as follows. Section 2 describes the new requirements of a code generator for ASIPs and the guidelines for its design. In Sect.3 we present our implementation of the proposed compilation path, while Sect.4 describes the linker in detail. Section 5 summarizes some of the opportunities offered by this new trajectory. In Sect.6 we review related work. Finally, Sect.7 concludes this paper.

## 2 Requirements and Design of a New Compilation Trajectory

This section details the requirements of a code generation path geared towards ASIPs. From these requirements we derive the guidelines for the design of a new compilation path. In short, the semantic level of object file representation has to be risen. The intermediate format used in the first compilation phase has to integrated with the format used to represent the program at the machine-code level.

*Requirements of code generation for ASIPs.* To effectively exploit the resources offered by a modern architecture, a compiler must apply all known state-of-the-art optimizations. This need is even more critical when compiling for ASIPs, as every code improvement not only reduces the execution time, it may also contribute to reducing the overall cost of the system being designed.

Inter-procedural optimizations are normally restricted to functions in the same compilation unit. No information is available on functions that have external linkage during compilation. When it is guaranteed that no more code is to be included, the optimizer is not forced to make conservative assumptions about code not yet linked and applying optimizations on *whole programs* has great potential for improving the code. Since whole-program optimizations require that all modules be linked together, they are best performed during or after linkage. However, the binary formats commonly used at linktime are inconvenient for complex code transformations. The new format, destined to supersede the binary format, must thus allow to easily perform analyses and transformations on the machine code after assembly and linkage.

To reduce the design cost, ASIPs are often built according to a template. A target processor is obtained by instantiating the architectural parameters of the template, and that the code generator must address a possibly huge space of target processors. How does linkage of libraries fit into this scheme? Clearly, generating library code for every possible target configuration in advance is impossible. Linkage of source files (referred to as *monolithic compilation*) or at the machine-independent intermediate level has also a severe drawback. Compilation of large files is highly expensive and can unacceptably stretch the design cycle. A short design cycle is essential for automatic exploration of the design space of an ASIP architecture. In this context, finding a cost-effective

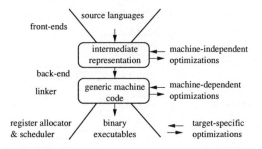

**Fig. 1.** Code generation for a templated target architecture; the application can be coded in one of the supported source languages and compiled to a binary for one of the many possible targets.

solution may require to compile for hundreds of different target configurations [2]. A good solution is to generate object code and libraries at the level of a *generic* machine. Generic machine code should capture the aspects of the templated architecture common to every instantiation, like the base instruction set. Aspects dependent on the processor instantiation, like hardware resources and parallelism available, are dealt with subsequently, in a separate scheduling and allocation phase. The generic machine code is an ideal level of abstraction for object files and linkage, because it takes the best of the two previous strategies: it limits the amount of target-specific passes to perform after linkage while keeping the number of pre-compiled libraries at a minimum. Figure 1 tries to give a conceptual view of this scenario.

*Guidelines for a new code generation path.* The requirements exposed above naturally lead to the organization of a new format representing machine code. To facilitate symbolic analysis and code manipulation, the machine code must be represented in symbolic form. E.g., while the binary representation is bound to the *memory layout* of the program, in our representation the memory layout is just an additional piece of information. References to data and code are pointers to symbol table entries. As a result, a code optimization pass needs not keep track of offset changes to addresses and branch displacements during the code transformation. The scope hierarchy is represented by means of a tree of symbol tables. A crucial piece of information for some optimizations, like memory disambiguation of pointer-based data structures [6], is type information. The intermediate format must be capable of representing the type system of the source language. An effective way to preserve high-level information down to the machine level, is to derive the new format from the machine-independent intermediate format. The semantics of the target instruction set and the architectural registers are added to the new format, while its overall structure is left unmodified.

## 3 The Compilation Trajectory

This section presents the retargetable code generator that implements our new approach. As stated in Sect.2, we generate code for an ample spectrum of target processors sharing the same templated architecture. This architecture offers explicitly programmed

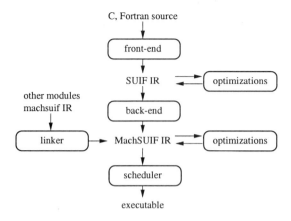

**Fig. 2.** Scheme of the proposed code generation trajectory.

instruction-level parallelism, in a fashion similar to that of VLIW architectures [2]. The code generation is coarsely split in two phases: (1) compilation to the generic-machine instruction set and (2) target-specific, integrated scheduling and register allocation [8]. The first phase will be the focus of this paper.

*The SUIF representation and the front-end.* Our code generator is based on the SUIF compilation system [5] and its intermediate representation, the Stanford University Intermediate Format (SUIF). Every pass of the fist phase is implemented as a SUIF-to-SUIF transformation. The SUIF format can represent the source program semantics at different, machine-independent levels of abstractions. Figure 2 gives an overall view of our code generation trajectory. SUIF instructions capture source-level information like the array indexing, multi-branch statements, and the types used in the program. At this level, several analyses and optimizations that require source-level information, like loop restructuring and memory disambiguation, are performed.

*Code generation and section offset assignment.* The compiler back-end maps SUIF instructions into generic machine code. The machine instructions are represented in *MachSUIF*, a format that extends the SUIF representation for machine-specific compilation and code optimizations [10]. Differently from most compilers, the translation to machine code fully preserves the high-level information present: only the instruction representation is extended, to encapsulate machine-specifics like opcodes, registers and addressing modes.

In our compilation path, the typical assembly step is bypassed, in that the machine instructions are not translated into their actual binary encoding yet. Symbols are tagged with their program section and the offset within this section by means of a *SUIF annotation*, called "section_offset", shown in Fig.3. The section ID number uniquely defines the section. As only three IDs are currently used, it is possible to define additional sections for special purposes. The offsets are expressed in the addressable unit relative to their section; e.g., instruction words for the TEXT section. The symbols that have an

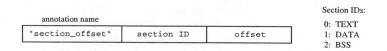

**Fig. 3.** The format of a "section_offset" SUIF annotation.

offset are: global and *private* variables, procedure and label symbols. By private symbol we mean a global symbol that visible only in the source file where it is defined.

This organization allows to conveniently lookup a symbolic reference and its section offset in a given symbol table. Offset assignment is implemented by iterating over all symbols defined in a symbol table and recursing over the tree of symbol tables. The resulting MachSUIF output file corresponds to a binary object file of the traditional trajectory. This object file can then be linked with other library or object modules, by our newly designed MachSUIF linker. An important benefit of using SUIF annotations to record section offsets is that it is *transparent* to other MachSUIF passes. As a result, code optimization passes can be reordered before or after section assignment or linkage.

## 4 Linking MachSUIF Files

This section describes the implementation of the linker. Apart from the alignment restrictions and the word sizes, the approach presented does not depend on the target architecture. Since alignments and word sizes are encapsulated under a uniform interface in MachSUIF, any trajectory based on MachSUIF can use this linker, provided that the section offsets are computed as described in Sect.3.

### 4.1 Basic approach for linking MachSUIF files

The MachSUIF linker performs two functions. First, it merges symbol tables coming from different modules and, for each program section, merges the offsets into a single offset space. Second, it resolves references to external, library symbols.

Section offset merging is easily implemented. Every module occupies a segment of the section offset space starting from zero; the linker simply adds to the offset of every symbol the base offset relative to its section. The base offsets are the section sizes resulting from summing the offset space taken by the files linked before the current one. This means that the offset spaces of each module are shifted by a constant amount.

Unresolved references to an external symbol $s$ are resolved by function Resolve() as follows. The name of $s$ is looked up in a given library symbol table. If the symbol, $s'$, is found, all the references (i.e., pointers) to $s'$ are replaced with references to $s$. This guarantees that any library function or variable that will be linked references the symbol in the output file. If $s$ is a variable, a unique section offset and the initializer of $s'$, are assigned to $s$.[1]

---

[1] For procedures, this process is postponed.

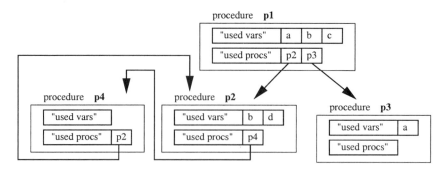

**Fig. 4.** Reference graph of library procedures.

When linking a library file, only the symbols which are referenced in one of the object files have to be linked to the output file. Therefore, the simple offset computation described for object files is not applicable. A MachSUIF library file differs from normal MachSUIF object files. First of all, the offset of global and private symbols is stripped off, as there is no need to maintain a relative order between symbols' offset. Second, the segments of offset space relative to each procedure are 'relocated' to base offset zero. The motivation for this is that we want to maintain the relative offset order between procedure's symbols, and avoid to recompute the alignments.

An external symbol that is resolved in a library may contain references to other library symbols which need to be resolved. This process must iterate until no new unresolved references are added. Note that some of these references may be to private symbols of the library. Detecting symbol references within SUIF procedures is time-consuming, as it requires scanning all symbol table entries and MachSUIF instructions of the procedure. Therefore, we perform it once and for every procedure when we *build* the library MachSUIF file. For each procedure we record all the non-local symbol references found in two *reference lists*; one list contains references to variables, one references to procedures. The lists are attached to the procedure's symbol table by means of two annotations, called "used_vars" and "used_procs", respectively. This organization defines a directed graph of procedure references, as shown in Fig.4. Starting from a procedure, all references to the library symbols that need to be resolved can be captured by traversing the graph. Note that, as opposed to the call graph, this reference graph includes procedures that are not called, but whose addresses are used.

### 4.2 Algorithm for resolving references to library symbols

Figure 5 shows the algorithm used to resolve global symbols defined by a library. First, the symbol table of the output file is scanned. Every external symbol defined in the library symbol table is resolved. A resolved variable $v$ may contain symbolic references in its initializer, VarDef($v$). A resolved procedure $p$ has its symbol references recorded in the reference list. The function FindRef() retrieves all the unresolved references of the variable initializer, while the function References($p$) selects the symbols in $p$'s reference lists as-yet to be resolved. Unresolved references are recorded in the temporary

```
procedure ResolveLibraryReferences(library_symtab, out_symtable)
beginproc
    reflist := ∅
    foreach extern variable v in out_symtab do        /* resolve external variables */
        if v is defined in library_symtab then
            Resolve(library_symtab, v)
            reflist := reflist ∪ FindRef(VarDef(v))
        endif
    endfor
    foreach extern procedure p in out_symtab do       /* resolve external procedures */
        if p is defined in library_symtab then
            Resolve(library_symtab, p)
            reflist := reflist ∪ References(p)
        endif
    endfor
    while reflist ≠ ∅                                  /* resolve indirectly referenced library symbols */
        foreach global library variable v' ∈ reflist do
            Resolve(library_symtab, v')
            reflist := reflist − v'
        endfor
        foreach library procedure p' ∈ reflist do
            if p' has external linkage Resolve(library_symtab, p')
            reflist := (reflist − p') ∪ References(p')
        endfor
    endwhile
endproc
```

Fig. 5. Algorithm for resolving all direct and indirect references to global library symbols.

list *reflist*. After all external symbols defined in the library are resolved, *reflist* contains all the library symbols that still have to be linked to the output file. First, every global variable $v'$ is resolved[2] and removed from *reflist*. This time, we do not need to search for possible references in $\text{VarDef}(v')$, because such references have been recorded in the reference lists of the same procedure that referenced $v'$. Then, every procedure $p'$ is resolved and removed from *reflist*, while its unresolved references are added to *reflist*. This process is iterated until *reflist* is empty.

## 5 Possible Applications

This section motivates our approach by presenting a number of applications taken from the literature that benefit from the opportunities offered by the new trajectory. We also describe a novel application that we intend to implement. This application is made possible by the combination of source-level type information and the possibility to conveniently analyze and modify linked code.

---
[2] Private symbols are resolved in a later pass.

*Inter-module optimizations.* By using the MachSUIF format for object files and linkage, all optimizations requiring inter-procedural analysis can be easily extended across compilation units. One such optimization is *static allocation* of local variables [9]. A variable is statically allocated if its location is known at compile time. Local variables of different procedures can share the same location if it is guaranteed that these procedures belong to disjoint branches of the call graph. This means that every possible call path leading to a function execution must known. When applicable, static allocation is preferable to stack allocation, because it eliminates address computations and instructions to save and restore the stack pointer. Note that in a traditional compiler static allocation is restricted to private functions, because the calls to a function with external linkage in other source files are not known at compile time. Another important optimization is inlining of external functions. When performed at machine level, function inlining requires redoing register allocation and scheduling of the inlined code. Our trajectory avoids this drawback, because inlining is performed on the generic machine code before scheduling and register allocation.[3]

*Optimizations using high-level information.* A number of low-level transformations, particularly those increasing the instruction-level parallelism of the program, are more effective when the information computed by source-level analyses is available [7]. Our proposal is to compute the required information at the point in compilation which is most suitable for the analysis, and to maintain this information throughout compilation until it is needed by some code transformation pass. Dependence information is one of them. When reordering machine instructions that access the memory, for example, it is useful to know about the *certainty* of a memory dependence. If the dependence is certain, the compiler can remove load-store pairs, while if two memory accesses can be proved to be independent, the instructions can be rearranged. Neglecting high level semantic of abstract data structures like lists, binary trees and sparse matrices limits the effectiveness of memory dependence analysis reduces the exploitable parallelism. Hummel et al. [6] classify the structural properties of dynamically allocated, pointer based data structures. These properties may allow the compiler to disambiguate references to pointer-based data structures. This approach requires analyses that need type information about the data structures at level of machine code. In [3] a suite of heuristics for improved static branch prediction is presented that makes use of source-level and variable type information.

*Link-time conversion from floating point to fixed point.* Due to cost/performance constraints, ASIPs often do not have floating-point hardware. Software emulation is also impractical, because it is too slow. A common compromise between speed and precision is the use of fixed-point arithmetic. One of the features that we intend to implement in our code generator is data type conversion from floating- to fixed-point. Currently, our converter is a source-to-source transformation tool working on a SUIF representation and processing one file at a time [1]. Our converter statically determines the fixed-point format of variables by propagating the type information already available along the data

---

[3] The generic machine uses an infinite register set model, therefore assigning new registers to a piece of inlined code is straightforward.

flow graph and by applying a set of fixed-point arithmetic rules. When a static function call is encountered, the fixed-point format information can flow in two directions: from a converted actual argument to a formal argument or vice versa.

The current converter can not propagate fixed-point information across separate compilation units through external function calls. To overcome this limitation, we propose to complete floating- to fixed-point conversion of external functions at link-time. This strategy allows to maintain one floating-point set of libraries instead of several fixed-point versions. Moreover, the conversion of a library function can be tailored to the fixed-point format of the call sites of the program being linked. This allows to take advantage of a priori information, like the fact that the result type of several mathematical functions has a limited range and therefore a certain, optimal fixed-point format.

## 6 Related Work

Wall and Srivastava [12, 11] dedicated much work on optimizations performed at link-time. They found that even simple transformations that do not move or insert code, can greatly improve the code quality, if performed on whole programs. In [12] Wall proposes to delay register allocation until link-time, when all the call sites to a procedure are known and an exact call graph can be computed. This allows to consistently allocate global variables in registers and to remove save and restore code around function calls. This work remarks the difficulty of performing extensive code transformations on a binary format at link time. The author's solution is to encode additional ad-hoc information into special relocation tables of the binary files, to direct the subsequent link-time code transformations. This approach is difficult to generalize.

In a later work [11] the authors propose *OM*, a complete modification system that supports intermodule code optimizations. *OM* translates binary code into a Register Transfer Language and a symbol table. Code transformations are performed on this symbolic representation, which is then translated back to binary. A major shortcoming of this approach is that the source-level information available for optimizations is limited: a substantial part, like type information, has been lost when the binary code was generated. Even the part that can be retrieved requires considerable effort. In practice, *OM* must undo the work done by the compiler system. These work suggest that a structured format that preserves the source-level information could be used to more effectively perform whole-program optimizations.

The idea of exploiting MachSUIF to skip the assembly step has been demonstrated by Greenbaum. In [4] he describes a code generator for experimental Digital Signal Processors that directly translates the MachSUIF representation to binary object files.

## 7 Conclusions and Future Work

A new strategy to code generation and linkage, superseding the traditional path based on binary tools, has been described. We have motivated its application to ASIP code generation, which would benefit from the source-level information that we maintain throughout compilation. We implemented this approach in our code generator and tested it on a SPARC 20 running Solaris 2.4. The results of linkage demonstrate that our approach is feasible; e.g. the size of MachSUIF binaries of *libgcc* (500KB) and *libc* (4.7MB) libraries are 2.4 and 3.3 times larger than the equivalent binary code on the host machine. This is in part due to the wealth of information not present in a normal binary format, like source-level type information. We consider the increase in code size an acceptable trade-off for the increased flexibility offered by the MachSUIF representation. The approach is applicable to code generators based on MachSUIF for any target architectures.

Future research will focus on applications that take advantage of the new code generation path. Code transformations that require analysis techniques crossing the boundary of single compilation units will be experimented. An interesting application is semi-automatic program conversion from floating- to fixed-point format. We intend to integrate conversion and linkage of floating-point library functions to fixed-point programs, a sort of adaptive library linkage.

## References

[1] Andrea G. M. Cilio, Ireneusz Karkowski, and Henk Corporaal. Fixed-point Arithmetic for ASIP Code Generation. In *Proceedings of the 4th conference of the Advanced School for Computing and Imaging*, June 1998.

[2] Henk Corporaal. *Microprocessor Architectures; from VLIW to TTA*. John Wiley, 1997. ISBN 0-471-97157-X.

[3] Brian Deitrich, Ben-Chung Cheng, and Wen-mei W. Hwu. Improving static branch prediction in a compiler. In *Proceedings of the International Conference on Parallel Architectures and Compilation Techniques*, Paris, France, October 1998.

[4] Jack Greenbaum. Generating Object Files Directly from SUIF/MACHSUIF Using GNU libbfd.a. In *Proceedings of the Second SUIF workshop*, August 1997.

[5] Stanford Compiler Group. *The SUIF Library*. Stanford University, 1994.

[6] Joseph Hummel, Laurie J. Hendren, and Alexandru Nicolau. A general data dependence test for dynamic, pointer-based data structures. In *Proceedings of the ACM SIGPLAN Conference on Programming Language Design and Implementation*, June 1994.

[7] Wen-mei W. Hwu and et al. Compiler technology for future microprocessors. *IEEE Transactions on Computers*, 41(12):1625–1640, December 1995.

[8] Johan Janssen and Henk Corporaal. Registers on demand: an integrated region scheduler and register allocator. In *Conference on Compiler Construction*, April 1998.

[9] Clifford Liem. *Retargetable Compilers for Embedded Core Processors*. Kluwer Academic Publishers, 1997. ISBN 0-7923-9959-5.

[10] Michael D. Smith. Extending SUIF for Machine-dependent Optimizations. In *Proceedings of the First SUIF Workshop*, January 1996.

[11] Amitabh Srivastava and David W. Wall. A practical system for intermodule code optimization at link-time. Technical Report 6, Western Research Laboratory, Digital Equipment Corporation, December 1992.

[12] David W. Wall. Global register allocation at link time. Technical Report 6, Western Research Laboratory, Digital Equipment Corporation, October 1986.

# The **Nestor** Library: A Tool for Implementing Fortran Source to Source Transformations

Georges-André Silber and Alain Darte

LIP, ENS-Lyon, 69007 Lyon, France
[gsilber,darte]@ens-lyon.fr

**Abstract.** We describe Nestor, a library to easily manipulate Fortran programs through a high level internal representation based on C++ classes. Nestor is a research tool that can be used to quickly implement source to source transformations. The input of the library is Fortran 77, Fortran 90, and HPF 2.0. Its current output supports the same languages plus some dialects such as Petit, OpenMP, CrayMP. Compared to SUIF 2.0 that is still announced, **Nestor** is less ambitious, but is light, ready to use (http://www.ens-lyon.fr/~gsilber/nestor), fully documented and is better suited for Fortran to Fortran transformations.

## 1 Introduction and motivations

Several theoretical methods that transform programs to gain parallelism or to improve memory locality have been developed (see [23, 3] for surveys). Unfortunately, there is a gap between the bunch of known parallelism detection and code optimization algorithms, and those implemented in real compilers. Indeed, these algorithms are often difficult to implement, because they use graph manipulations, linear algebra, linear programming, and complex code restructuring (see for example [10, 22, 14, 8] for some parallelism detection algorithms). Consequently, their implementation is a research problem by itself and must be ease by a simple but powerful representation of the input program. This representation should provide all the basic blocks to let the researcher concentrate on algorithmic implementation problems, and hide the classical low level representation of the program (low level abstract syntax tree).

The **Nestor** library provides such a framework, focusing on tools for source to source transformation of Fortran and its dialects (Fortran 77, 90, HPF 2.0 [11], OpenMP [16], CrayMP, SunMP, etc.). We chose HPF and its variants because it offers useful means to express parallelized codes, such as directives for parallel loops, privatized arrays, data distributions, task parallelism, etc. Moreover, the obtained codes can be executed on a parallel computer after compilation by an HPF (or equivalent) compiler. Furthermore, the parallel code is still readable for the programmer and by **Nestor** itself. We believe that this high level parallelizing approach (through directives insertion) is important to improve the relationship between the programmer and the compiler, and to enable semi-automatic parallelization.

Nestor is a C++ library that provides classes and methods to parse a source code in Fortran, to analyze or modify this source code easily with a high level representation, and to unparse this representation into Fortran with parallelization directives. The representation that can be manipulated with Nestor is a kind of AST (Abstract Syntax Tree) where each node is a C++ object that represents a syntactical element of the source code, and that can be easily transformed. The front-end of Nestor is the robust front-end of Adaptor [6]

Nestor aims to be used by researchers that want to implement high level transformations of Fortran source codes. It provides all the features for building a source to source Fortran transformer. The library is easy to use and is portable: it is written in GNU C++ and uses the STL (Standard Template Library) that implements classes for manipulating lists and other container objects. There is a full documentation in postscript and an on-line documentation in HTML and Java that describes all the classes of the library.

This paper is organized as follows. In Section 2, we present the differences between Nestor and existing related tools, in particular SUIF. In Section 3, we describe Nestor in details. Then, we give some concluding remarks and some future work.

## 2 Related tools

Several research tools implementing program analysis or parallelization algorithms have been developed (Bouclettes [4], Petit [18], LooPo [12]). Their main objectives were to prove that these algorithms could be implemented in a real compiler. For example, Petit demonstrates the power of the Omega test and of the Omega library. Nevertheless, they were not conceived to handle real codes: they all take as input a toy language, usually a subset of Fortran. But theoretical methods have to be validated also by testing their behaviors on real applications, such as benchmarks or scientific codes. Several more ambitious tools have been developed in the past: PIPS [17] developed at the Ecole des Mines de Paris, Polaris [20] developed at Urbana-Champaign, SUIF [19] developed at Stanford, PAF developed at PRiSM. These tools have been developed over more than ten years for some of them: they are now huge systems (difficult to maintain and extend) and their internal representations begin to be at a too low level for new developers. For example, SUIF 1.0 is currently moving to SUIF 2.0 with deep changes. People from Polaris think of changing their internal representation. We believe that Nestor could be an interesting platform for that.

One could argue that SUIF 2.0 has the same objectives as Nestor concerning the simplicity of use of its internal representation. Compared to SUIF 1.0, the main transformation is indeed to make the representation more object-oriented. Although this feature is already available in Nestor, when SUIF 2.0 is only announced, Nestor should not be seen as a rival tool for SUIF 2.0. First of all, Nestor is devoted to Fortran like languages, whereas SUIF has been designed for manipulating C programs (Fortran is handled through f2c and cannot be unparsed). Fortran is a simpler language than C, easier to optimize at a high level (in particular for dependence analysis), and thus it leads to a simpler internal representation.

In addition to this main difference in the input and output languages, **Nestor** is far less ambitious. It is not a full compiler, but just a kernel for source to source transformations of Fortran programs. **Nestor** does nothing! It only provides means to do something. But this limited goal gives it some advantages.

– **Nestor** is small. The library provides only the basic blocks for building source to source transformation systems. Its size allows an easy and quick installation on every system. It is developed and maintained by a single person.

– **Nestor** is fully documented. We think that this feature is maybe even more important than any other, as far as implementing algorithms is the main issue. A prostscript and an HTML/Java documentation is automatically generated from the source files and is always up to date, thanks to doc++, a public domain software [24].

– **Nestor** offers a Fortran90/HPF input and output. All the other tools support only Fortran 77. It is impossible for example to insert parallelizing directives a la HPF as easily as **Nestor** does.

## 3 Description of Nestor

**Implementation choices.** The first choice was to choose a language to develop the library. This choice is important, because users that are going to use the library will have to write their transformations in this language. For several reasons, we have chosen C++ to develop **Nestor**.

– C++ is widely used, and a lot of existing libraries are written with this language (or in C). This means that these libraries can be used together with the **Nestor** library.

– C++ is object-oriented. The internal representation of a Fortran program fits very well in the object-oriented world. For instance, the inheritance principle gives several views of the internal representation. All objects composing it can be seen as a tree organization of **NstTree** objects, which is the base class of the library (each class inherits from it). But, the programmer can have a higher view of the representation by using the actual type of each object. There are no real **NstTree** objects, but only child objects of this class (**NstStatement**, **NstDeclaration**, etc.). This is explained in Section 3.

– C++ offers a lot of useful templates with the Standard Template Library (STL), included in the ANSI/ISO Standard C++ Library. This library is now widely available, especially in the GNU C++ compiler that is our compiler of development. This compiler is available on every usual platform. The choice of using the STL is then a reasonable choice, instead of investing a lot of time developing and maintaining new templates for lists, containers, etc.

– C++ offers operator overloading. This feature is useful, especially for writing the unparse: the operator << is overloaded for each class of the library, allowing to write custom unparse and to write to a file or to standard output easily.

– C++ offers virtual methods, for example, to define cloning methods.

**Writing a source to source transformation pass.** When developing Nestor, we wanted to create a development platform for writing a source to source parallelizer. We had a lot of algorithms to implement and evaluate, and we wanted to automatically generate HPF programs starting from Fortran sequential programs. With Nestor, each transformation can be written independently, because the output of one pass can be used by another pass. It is one of the advantages of working at the source level. For instance, one pass can take a Fortran source code, insert HPF directives in front of DO loops and unparse the internal representation. One second pass can take the result of this pass and can generate the distributions of the arrays. The main idea is that each programmer can write its own optimization and test it immediately by compiling the result with a Fortran compiler that supports parallelization directives. Moreover, the result is easy to read because it is still written in a high level language instead of a tricky internal representation or in a low level language like C.

*Example 1 (A simple pass).* The following code gives the scheme of a typical transformation pass written with our library. The first statement creates an object from the class NstComputationUnit that gives a starting point to the internal representation of the program. The last statement unparses the internal representation in Fortran on the standard output.

```
#include <nestor.H>
void main(int argc, char** argv) {
  if (argc == 2) {
    NstComputationUnit prog(argv[1]);
    // Transformations on 'prog' here
    cout << prog << endl;
  }
}
```

**A quick look at the internal representation.** Usually, compilers use program internal representations that are too low-level for user's manipulation. They do not fully retain the high-level language semantics of a program written in Fortran. One of the advantage of Nestor is that its internal representation is intuitive because there is a one-to-one correspondence between a syntactical Fortran element and the corresponding Nestor class.

*Example 2 (DO statement).* Consider a DO statement in Fortran: it is composed of an index variable, a range, and a list of statements. The following source code is an excerpt of the C++ definition of the Nestor class NstStatementDo, representing DO statements of Fortran.

```
class NstStatementDo : public NstStatement {
public:
  NstVariableLoop* index() const;
  NstVariableLoop* index(NstVariableLoop& new_var);
  NstExpressionSlice* range() const;
  NstExpressionSlice* range(NstExpressionslice& new_exp);
  NstStatementList body;
  int independent; ... };
```

Each element of a DO can be accessed or modified by the corresponding method in the class. A call to the method index() returns an object representing the index variable of the loop. A call to the method index(j) replaces the old index variable by a new object j. Each element of the DO can be accessed/modified the same way. The statements in the body of the loop are stored in the list of statements body. This is a doubly-linked list that can be modified with usual operations on a list: add, delete, traversal, ... The flag independent tells if the loop is parallel or not and unparses an !HPF$ INDEPENDENT in front of the loop if the flag is set (or the equivalent directive for other dialects). The class NstStatementDo inherits from the class NstStatement and then has all the methods and attributes of this class. The class NstStatement contains the information common to all types of statements in a Fortran code: line number in the source code, label of the statement, ... This class inherits from the class NstTree, like all the other classes of the Nestor library.

**Memory management.** Each class has a number of children. For instance, the DO statement of the Example 2 has three children: the index, the range, and the body. These classes have children too, etc. The Nestor library uses an all-copy scheme: when an object is plugged into another one, it is recursively copied. For instance, using the method statement_do.range(slice_expression) will clone the object slice_expression before replacing the current range of the DO statement. The old value is recursively deleted. Thus, each object can only have one father in the code structure: we maintain a pointer to it and some information about the location of the object. In the case of a slice expression that is into a DO statement, a string RANGE_OF_DO_STATEMENT is stored in the expression. This scheme makes the creation of new objects very easy.

Each class has a method cut() that removes the object from its current location (replacing it with a default object). A cut object can be placed once in another object without creating a copy, as in a cut/paste mechanism. Similarly, the method replace(NstTree& t) cuts the object and replaces it by t.

**Traversals.**

*Example 3 (Traversal).* This example is a traversal of a list of statements that prints the number of specified parallel loops. It illustrates the use of types and the use of the lists defined in the Standard Template Library. Lists can be traversed with iterator objects.

```
void print_number_of_par(NstStatementList& sl)
{ int num = 0;
  NstStatementList::iterator sl_it;
  for (sl_it = sl.begin(); sl_it != sl.end(); sl_it++)
    if (NstStatementDo* do_loop = dynamic_cast<NstStatementDo*>(*sl_it))
      if (do_loop->independent) ++num;
  cout << "There are " << num << " parallel loops." << endl; }
```

The class NstTree gives a mechanism to write recursive traversals of the AST. The class NstTree provides the two virtual methods init() and next()

that gives respectively the first and the next child of a `NstTree` (or a derived class) object.

**The front-end.** The front-end of Nestor is a slightly modified version of the front-end of Adaptor [6] that recognizes HPF 2.0 [11][1]. Adaptor has been written by Thomas Brandes and is an excellent public domain HPF compiler. Its front-end is robust, publicly available, and has been written with the GMD compiler toolbox [6] (a high level language to easily describe grammars). We have added the directives of OpenMP, Cray and our own directives for semi-automatic parallelization. This front-end allows Nestor to handle the real codes that Adaptor handles, instead of only considering a subset of language or an *ad-hoc* language. This part is also useful because it checks the syntax and the semantics of the code. Once the code has been parsed with the class `NstComputationUnit`, the resulting object represents a correct Fortran program.

**The back-end.** The internal representation of Nestor can be unparsed in Fortran 77, Fortran 90, HPF 2.0, OpenMP, CrayMP directives and in the Petit language. Each object has its own unparse methods, one for each of the Fortran dialect. Unparsing recursively an object is a very simple task by the overload of the C++ operator <<. The unparsed language can be chosen by a global flag. By default, the unparsed language is HPF 2.0.

**Dependences and graphs.** Dependence analysis is the first step before any optimization that modifies the order of computations in a program. Without a sophisticated dependence analyzer, code transformations such as loop transformations, scalar expansion, array privatization, dead code removal, etc. are impossible. Therefore, any parallelizing tool must contain a dependence analyzer. Nevertheless, it is well known that the development of a dependence analyzer both powerful and fast is a very hard task. For this reason, we decided to rely on a free software tool, named Petit [18], developed by Bill Pugh's team at the University of Maryland. Petit's input is a short program, written in a restricted language, close to - but different than - Fortran 77. Its output is a file that describes pairs of array references involved in a dependence, and this dependence is represented by a (sometimes complicated) relation based on Presburger arithmetic.

Following Bill Pugh's advice, we chose to use Petit as an independent tool through its input and output files. This strategy is not only simpler to implement, it is also more portable: potential bugs in Petit and potential bugs in Nestor are separated, and furthermore updating Petit to new versions will be easier. Two problems still remained: feeding Petit with a correct input, and plugging Petit's output at the right place into the original Fortran code.

---

[1] Historically, Nestor has been written as a parallelizing pre-phase for Adaptor. Starting from a sequential Fortran program, Nestor+Adaptor can transform it into a parallel program with message passing. The name Nestor comes from the term *loop nest* and the name Adaptor.

The first task was easy to complete thanks to the clean design of **Nestor**. We just had to redefine the output operator (the C++ operator <<) for all C++ classes that have their equivalence in the Petit language. For example, the operator <<, applied to the class that corresponds to a Fortran DO loop, automatically generates a loop in Petit's format, and recursively applies the operator << to the body of the loop. In **Nestor**, there is a global flag that determines the output language chosen by the unparse function, and that switches from one to another.

The second task was twofold. First, we had to make the correspondence between array references in Petit and the original array references of the Fortran program. Line numbers are not sufficient because both languages can be formatted in a different manner, and furthermore, only a part of the original code may be sent to Petit. Therefore, we slightly modified Petit's grammar so as to number array references in the same order as they appear in **Nestor**'s abstract syntax trees. Second, we modify the way dependences are represented in Petit's output. Indeed, in most parallelizing algorithms, what we need is an approximation of distance vectors, and not a too complicated Presburger formula. We wrote a small tool, based on the Polylib [21], a library for manipulating polyhedra, developed at IRISA in Patrice Quinton's team (mainly by Doran Wilde and Hervé Le Verge). This tool extracts, from a Presburger formula, a description of dependences by level, direction vector, and polyhedral approximation, the three representations used respectively by the parallelizing algorithms of Allen and Kennedy [1], Wolf and Lam [22], and Darte and Vivien [8].

We point out that we don't need to send the full program to Petit. Indeed, we use Petit only to analyze small portions of codes that we want to parallelize: the unparse function of **Nestor** builds the corresponding code in Petit's format, and also creates the declaration part of this small program, based on all variables that are used in this portion. For example, if we decide to analyze a single loop, surrounded by an outer loop, then the loop counter of the outer loop becomes a parameter that must be declared in Petit's input. This "local" unparsing technique allows us to manipulate large codes, even if Petit is limited to the analysis of small codes.

Building a dependence graph is a very easy task and is completed by a call to the constructor of the class `NstRDGVar`. This class contains the list of edges and the list of vertices of the dependence graph. The constructor builds the Petit input, calls Petit and retrieves the output of Petit to build the dependence graph. Each vertex is linked to the corresponding variable access in the AST of **Nestor**. Classical graph manipulation algorithms, such as computations of connected components, of strongly connected components, topological ordering, etc. are provided.

**Automatic parallelization.** Nestor already implements two algorithms for parallelism detection. These algorithms are very simple and were implemented to validate the internal representation of **Nestor** and to check its ease of use. The first algorithm only detects if the loop is parallel without any modification, the following example gives its principle.

*Example 4 (Checking parallelism).* This example builds a dependence graph from a statement and checks if there are no dependences carried by the loop (in this case, the loop is parallel). To make it run recursively on any object, call the method: traversal(parallel_loop). Note that this simple example is a parallelizer from Fortran to HPF for a shared memory machine (no distributions are generated). This example illustrates the use of a dependence graph that contains a list of edges labeled with dependences and a list of vertices representing statements. The code is not optimized for space reason, a more efficient version will not recompute the dependence graph for each nested loop.

```
int parallel_loop(NstTree* t, void* ignored)
{ NstEdgeList::iterator el_it;
  int parallel = TRUE;
  if (NstStatementDo* stdo = dynamic_cast<NstStatementDo*>(t)) {
    NstRDGVar dep_graph(*stdo);
    if (dep_graph.built()) { // graph successfully built
      for (el_it = dg.edges.begin(); el_it != dg.edges.end(); el_it++)
        if (NstDependence* dep = dynamic_cast<NstDependence*>(*el_it))
          if (dep->level == 1) parallel = FALSE;
      stdo->independent = parallel;
    }
    else
      stdo->independent = FALSE;
  }
  return 0; }
```

The second algorithm is a modified version of the Allen and Kennedy algorithm [1]. Our goal with this modified version was to have at least one robust algorithm, able to handle complex loops with conditionals and possibly non constant loop bounds, in other words structured codes that may contain control dependences. Many extensions of the Allen-Kennedy algorithm have been proposed in the literature that are able to handle control dependences. All of them rely on the creation of "execution variables" (scalar or array variables) that are used to pre-compute and store the conditionals, and on the conversion of control dependences into data dependences [2, 13, 15].

While implementing such an algorithm, we found out that it was difficult, in general, to determine the size of these new arrays, especially in parameterized codes. Furthermore, in the context of High Performance Fortran and distributed memory systems, the problem of aligning and distributing these new arrays arises. To avoid these two problems, it may be better to re-compute the conditionals (when it is possible) instead of using a stored value. It may also be better to manipulate privatized arrays or scalars than to manipulate distributed arrays. We therefore tried to understand how these two new constraints – the control of the new array dimensions, and the re-computations of conditionals – can be handled, since no previously proposed algorithm can take them into account.

For that, we explored a new strategy for taking control dependences into account. The technique is to pre-process the dependence graph, and once this process is achieved, any version of the Allen-Kennedy algorithm can be used: the dimensions of the new arrays are guaranteed to satisfy the desired constraints. To make things simpler, the automatic version that is currently implemented in Nestor is a version that guarantees that all new variables are at worst privatized scalar variables (thus, with no size to declare). A semi-automatic version offers to the user the choice of the array dimension he tolerates for his program.

## 4  Conclusion and future work

This paper provides a description of the Nestor library. We think that this library is very useful for the researcher who wants to implement and test new source to source transformations. Our library has a front-end and a back-end that totally supports Fortran and its dialects, and an object-oriented internal representation that eases the process of implementing new algorithms. Furthermore, it is fully documented, small, robust, and easy to install on every system.

Several researchers are already interested by Nestor, especially by the fact that it is both light and practical. We hope that Nestor is going to be effectively widely used by researchers for implementing new parallelization strategies. For the time being, Nestor is used at LIP by researchers involved in automatic parallelization and high level transformations. It is used in the project Alasca for automatic insertion of HPF data redistributions, it is used for inserting automatically low overhead communication and computation subroutines in Fortran codes [9], it is used in high level loop transformations before compilation to VHDL, and it is used in the project HPFIT [5] to implement parallelization algorithms. Nestor is now publicly available with its source code and its documentation at the address http://www.ens-lyon.fr/~gsilber/nestor. We are implementing new parallelization algorithms [7] into it. These parallelization algorithms could be included in the base Nestor package and then transform it into a more powerful source to source automatic parallelization kernel.

## References

1. John Randy Allen and Ken Kennedy. Automatic Translation of Fortran Programs to Vector Form. *ACM Transactions on Programming Languages and Systems*, 9(4):491–542, October 1987.
2. John Randy Allen, Ken Kennedy, Carrie Porterfield, and Joe Warren. Conversion of Control Dependence to Data Dependence. In *Conference Record of the Tenth Annual ACM Symposium on the Principles of Programming Language*, Austin, Texas, January 1983.
3. David F. Bacon, Susan L. Graham, and Oliver J. Sharp. Compiler transformations for high-performance computing. *ACM Computing Surveys*, 26(4), 1994.
4. Pierre Boulet. Bouclettes: A fortran loop parallelizer. In *HPCN 96*, pages 784–791, Bruxelles, Belgium, June 1996. Springer Verlag Lecture Notes in Computer Science.
5. T. Brandes, S. Chaumette, M.-C. Counilh, A. Darte, J.C. Mignot, F. Desprez, and J. Roman. HPFIT: A Set of Integrated Tools for the Parallelization of Applications Using High Performance Fortran: Part I: HPFIT and the TransTOOL Environment. *Parallel Computing*, 23(1-2):71–87, 1997.

6. Thomas Brandes. ADAPTOR, High Performance Fortran Compilation System. World Wide Web document, http://www.gmd.de/SCAI/lab/adaptor/adaptor_home.html.
7. Alain Darte, Georges-André Silber, and Frédéric Vivien. Combining Retiming and Scheduling Techniques for Loop Parallelization and Loop Tiling. *Parallel Processing Letters*, 7(4):379–392, 1997.
8. Alain Darte and Frédéric Vivien. Optimal fine and medium grain parallelism detection in polyhedral reduced dependence graphs. *International Journal of Parallel Programming*, 25(6):447–497, 1997.
9. F. Desprez and B. Tourancheau. LOCCS: Low Overhead Communication and Computation Subroutines. *Future Generation Computer Systems*, 10(2&3):279–284, June 1994.
10. Paul Feautrier. Some efficient solutions to the affine scheduling problem, part II: multi-dimensional time. *Int. J. Parallel Programming*, 21(6):389–420, December 1992.
11. High Performance Fortran Forum. High Performance Fortran Language Specification. Technical Report 2.0, Rice University, January 1997.
12. The group of Pr. Lengauer. The loopo project. World Wide Web document, http://brahms.fmi.uni-passau.de/cl/loopo/index.html.
13. Ken Kennedy and Kathryn S. McKinley. Loop Distribution with Arbitrary Control Flow. In *Supercomputing'90*, August 1990.
14. Amy W. Lim and Monica S. Lam. Maximizing parallelism and minimizing synchronization with affine transforms. In *Proceedings of the 24th Annual ACM SIGPLAN-SIGACT Symposium on Principles of Programming Languages*. ACM Press, January 1997.
15. Kathryn S. McKinley. *Automatic and Interactive Parallelization*. PhD thesis, Department of Computer Science, Rice University, 1992.
16. OpenMP Standard for Shared-memory parallel directives. World Wide Web document, http://www.openmp.org.
17. PIPS Team. PIPS (Interprocedural Parallelizer for Scientific Programs). World Wide Web document, http://www.cri.ensmp.fr/~pips/index.html.
18. William Pugh. Release 1.10 of Petit. World Wide Web document, http://www.cs.umd.edu/projects/omega/.
19. Stanford Compiler Group. The SUIF Compiler System. World Wide Web document, http://suif.stanford.edu/suif/suif.html.
20. The Polaris Group. Polaris Project Home Page. World Wide Web document, http://polaris.cs.uiuc.edu/polaris/polaris.html.
21. Doran K. Wilde. A library for doing polyhedral operations. Master's thesis, Oregon State University, Corvallis, Oregon, Dec 1993. Also published in IRISA technical report PI 785, Rennes, France; Dec, 1993.
22. Michael E. Wolf and Monica S. Lam. A loop transformation theory and an algorithm to maximize parallelism. *IEEE Trans. Parallel Distributed Systems*, 2(4):452–471, October 1991.
23. Michael Wolfe. *High Performance Compilers for Parallel Computing*. Addison-Wesley, 1996.
24. Roland Wunderling and Malte Zöckler. A documentation system for C/C++ and Java. World Wide Web document, http://www.zib.de/Visual/software/doc++/index.html.

A long version of this paper under the form of a research report is available at the address:
ftp://ftp.lip.ens-lyon.fr/pub/LIP/Rapports/RR/RR1998/RR1998-42.ps.Z

# Performance Measurements on Sandglass-Type Parallelization of Doacross Loops

Motoyasu Takabatake, Hiroki Honda, Toshitsugu Yuba

Graduate School of Information Systems, The University of Electro-Communications,
1-5-1 Choufugaoka Choufu Tokyo, Japan
{takabatake, honda, yuba}@yuba.is.uec.ac.jp

**Abstract.** In this paper, we propose the sandglass-type parallelization technique for a doacross loop which has the characteristics of iteration-based parallelizing and software pipelining. We prove its effectiveness by comparing the sandglass-type to well-known three parallelization techniques: iteration-based, software pipelining, and a combination of doall-type parallel and sequential techniques. We conclude that the sandglass-type parallelization technique is the most effective among the techniques mentioned above in cases in which there are less than ten processing elements and the size of tasks with loop-carried dependences is smaller than the size of tasks lacking loop-carried dependence.

## 1 Introduction

For effective execution of a parallel program on a parallel computer, it is important to correctly and suitably partition a program into parallel executable program parts (tasks) and schedule them to processing elements (PEs). We propose a new partitioning and scheduling, i.e., parallelizing technique for a program with a doacross loop, which is a time-consuming part of numerical applications.

One common technique involves allocating each iteration of a doacross loop to a PE as a task [1][4]. This technique is defined as *Do-across* and requires a large amount of communication among PEs caused by loop-carried dependence of a doacross loop. Therefore the parallelization technique is not effective to parallel computers with large communication overhead.

Another technique parallelizes a doacross loop by partitioning the loop and using software pipelining [2][6]. This technique is defined as *Do-pipeline* and also requires a large amount of communication among different stages of pipelines. It is generally difficult to make pipeline pitches the same in parallelizing.

In another technique, a doacross loop is divided into loops with loop-carried dependence and those lacking loop-carried dependence. We define this technique as *Do-all & seq*. Each divided loop is distributed to a PE, and then loops with loop-carried dependence are executed sequentially and loops lacking loop-carried dependence are executed in parallel as doall-type loops.

In this paper, we propose the sandglass-type parallelization technique, defined as *Do-sandglass*, which has the characteristics of both iteration-based parallelization and software pipelining. These four parallelization techniques are investigated by executing some benchmark programs on actual parallel machines, measuring and comparing the performance. The advantage of our proposed technique is shown as well as its disadvantage.

## 2 Sandglass-type parallelization

### 2.1 Model of dependence graph

A *dependence graph* [9] has the vertices(tasks) which can correspond to statements or to blocks of code in the program, and the edges which correspond to constraints that prevent reordering the statements. There are two kinds of parts in the dependence graph for a doacross loop; a cyclic part with loop-carried dependence and acyclic parts without it. In both parts, a source task and a target task on a edge of loop-independent dependence are combined into a larger task. As a result, the dependence graph for a typical doacross loops is modeled as shown in Fig. 1. We define the cyclic part (such as $S_2$) with loop-carried dependence as a *loop-carried dependence task* and the acyclic part (such as $S_1$ and $S_3$) as a *non loop-carried dependence task*. Loop-carried dependence relations for a doacross loop is classified as *LFD (Lexically-Forward Dependence)* and *LBD (Lexically-Backward Dependence)* pointing forward or backward in the program source [1], respectively.

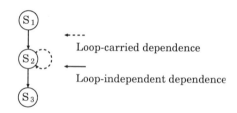

**Fig. 1.** Dependence graph for a doacross loop

### 2.2 Sandglass-type parallelization technique

In parallel computing, loop-carried dependence tasks and non loop-carried dependence tasks are allocated to different PEs. When the size of a non loop-carried dependence tasks is larger than that of a loop-carried dependence task, non loop-carried dependence tasks are allocated to different PEs by each iteration. In other words, loop-carried dependence tasks and non loop-carried dependence tasks form a software pipeline, and each non loop-carried dependence tasks is executed in parallel by different iteration. This is called the sandglass-type parallelization technique (Fig. 2). Later we will show that the sandglass-type parallelization technique is effective when non loop-carried dependence tasks are larger than the loop-carried dependence tasks.

Fig. 3 shows a Gantt chart of four parallelization techniques. The Horizontal axis is time and the vertical axis is PE. Fig. 3(a) shows *Do-across*. All communication is included in the critical path. Fig. 3(b) shows *Do-pipeline*. In this case,

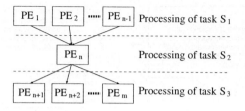

**Fig. 2.** Sandglass-type parallelization

The number of pipeline stages is five. $S_1$ is partitioned into $S_{11}$ and $S_{12}$. $S_3$ is partitioned into $S_{31}$ and $S_{32}$. Fig. 3(c) shows *Do-all & seq*. During the task $S_2$ is executed, all PE except PE0 do nothing. Fig. 3(d) shows *Do-sandglass*. The critical path of this technique includes only two communication phases.

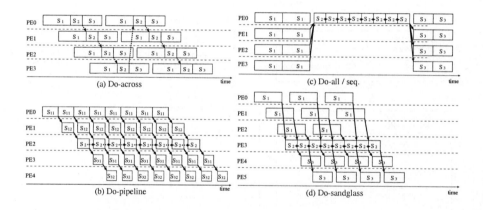

**Fig. 3.** A Gantt chart of four parallelization techniques for a doacross loop

## 3 Doacross loops for benchmarking

The benchmark programs used for the evaluation of each technique are a part of the NAS kernel benchmark program (Fig. 4(a)), a part of the spline interpolation program (Fig. 5(a)), and the Livermore Fortran kernel 23 (Fig. 6(a)).

We classify these programs by the number of loops, the loop-carried dependence relation, and the ratio of size of each task. A part of the NAS kernel benchmark program (NAS) has a single loop, the loop-carried dependence relation is LBD and the dependence distance is 1. This program is partitioned as shown in Fig. 4(b). The ratio of $S_1 : S_2$ is 3 : 1.

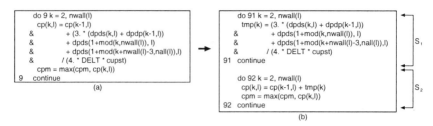

**Fig. 4.** NAS kernel benchmark and its loop partition

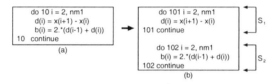

**Fig. 5.** Spline interpolation program and its loop partition

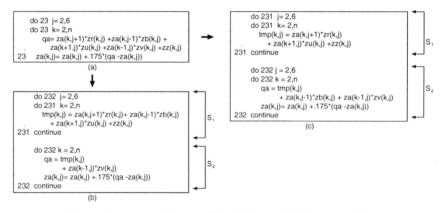

**Fig. 6.** Livermore Fortran Kernel 23 and its loop partition

A part of the spline interpolation program (Spline) has a single loop, the loop-carried dependence relation is LFD, and the dependence distance is 1. This program is partitioned as in Fig. 5(b). The ratio of $S_1 : S_2$ is 2 : 3.

The Livermore Fortran Kernel 23 (Livermore) has a double loop, both loop-carried dependence relations are LFD, and both the dependence distances are 1. As shown in Fig. 6(b), the inner loop is partitioned and the ratio $S_1 : S_2$ is 3 : 2. As shown in Fig. 6(c), Not only the inner loop but also the outer loop is partitioned and the ratio $S_1 : S_2$ is 2 : 3.

The characteristics of doacross loops used for benchmarking are summarized in Table 1, where $S_1 : S_2 : S_3$ means the ratio of the task sizes. Note that no loop contain a task $S_3$ in these cases.

**Table 1.** Doacross loops for benchmarking

|          | relation | $S_1 : S_2 : S_3$ | nesting |
|----------|----------|-------------------|---------|
| NAS      | LBD      | 3 : 1 : 0         | single  |
| Spline   | LFD      | 2 : 3 : 0         | single  |
| Livermore| LBD      | (b) 3 : 2 : 0     | double  |
|          |          | (c) 2 : 3 : 0     |         |

## 4 Performance measurement

### 4.1 Environment of measurement

The number of iterations for each loop is 10,000. Note that Livermore is a double loop, the number of iterations for the outer loop is 10 and for the inner loop is 1,000.

For performance evaluation, we used three parallel machines; EM-X, SPARCserver 1000, and Cenju-3. The EM-X [3], developed at the Electrotechnical Laboratory, is a distributed memory-type parallel machine with a fine-grain architecture. The EM-X has 80 PEs. The SPARCserver1000 is a shared memory-type parallel machine and is able to communicate by *read* and *write* operations to the memory. The Cenju-3, developed by NEC, is a distributed memory-type parallel machine.

For the measurement, all PEs of the distributed memory-type parallel machine have the same data. We dose not consider data distribution in order to observe the difference of four parallelization techniques.

The benchmark programs are executed on EM-X, and the most effective program for sandglass-type parallelization technique is executed on the other machines.

### 4.2 Performance measurement on the EM-X

Message passing between PEs of the EM-X is carried out by executing a direct remote memory write instruction. Synchronization between message communication uses the I-structure operation with hardware matching mechanism. If the sending data is one word, only I-structure operation is used to send and synchronize the data.

Fig. 7 shows the measurement result of NAS executed on the EM-X. The speedup means performance scaling up to sequential execution time. The most effective technique is *Do-sandglass*. The speedup is about 3.4 at four PEs. Because the ratio $S_1 : S_2$ is 3 : 1, *Do-sandglass* is not effective in cases in which the number of PEs is larger than five. The EM-X makes possible fine-grain communication, then the speedup of *Do-across* is approximately three. For *Do-pipeline*, the number of PEs is from one to four as a result of the difficulty of partitioning of the loop into stages. For *Do-all & seq*, the speedup is about 3.2 in cases in which the number of PEs is less than ten. In these cases, *Do-all & seq* is less effective than *Do-sandglass*; when the number of PEs is 80, the speedup is approximately 4.2. When the number of PEs is large, *Do-all & seq* is more effective than *Do-sandglass*. On the other hand, *Do-sandglass* is effective in cases in which there are four to ten PEs.

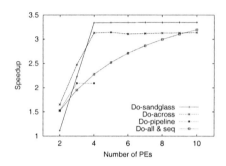

**Fig. 7.** Results of NAS executed on the EM-X

Fig. 8 shows the measurement result of Spline executed on the EM-X. The most effective technique is *Do-all & seq* in this case. Because the loop-carried dependence relation is LFD, both $S_1$ and $S_2$ of Spline are doall-type loops. *Do-across* increases its speed in proportion to the number of PEs, because the loop-carried dependence relation is LFD and it is possible for each iteration to be executed at the same time.

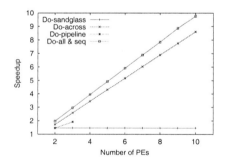

**Fig. 8.** Results of Spline executed on the EM-X

There are two reasons that *Do-sandglass* has no good effect on Spline compared with the other techniques. The first is that the Spline's loop-carried dependence relation is LFD, where the other techniques can increase their speed in proportion to the number of PEs. The second is that $S_1 : S_2$ is 2 : 3, in case in which $S_2$ becomes a bottleneck and no execution time is shortened. Note that it is impossible to examine *Do-pipeline* on more than three PEs, because more partitioning the loop into pipeline stages cannot be done in the case of Spline.

Fig. 9 shows the measurement result of Livermore executed on the EM-X. It is effective that $S_2$ is smaller than $S_1$ for *Do-sandglass*, where the loop partition is as shown in Fig. 6(b). For *Do-all & seq*, the loop partition is as shown in Fig. 6(c).

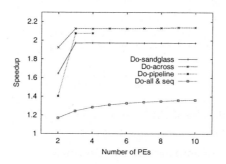

**Fig. 9.** Results Livermore 23 executed on the EM-X

The most effective one among the four techniques is *Do-across*. On *Do-across*, an inner loop is executed in parallel by an iteration. When the inner loop is finished, the data carried by the loop-carried dependence of the outer loop is within the same PE. Therefore, it is not necessary to communicate with the loop-carried dependence of the outer loop. When the loop-carried dependence tasks of the inner loop are executed on a PE by using *Do-sandglass*, data defined on the PE are distributed to other PEs by the loop-carried dependence of the outer loop.

Because the ratio $S_1 : S_2$ is 3 : 2 on *Do-sandglass*, the speedup is 2 at three PEs and it is not better effect on larger numbers. The reason for this small effect on *Do-all & seq* is that $S_1$ of the loop is smaller than $S_2$ of it. Almost all the execution time is that of $S_2$ of the loop in case of Livermore.

### 4.3 Performance measurement on SPARCserver1000

A Pthreads [5] library is used for parallel programming and a spin lock technique is used for synchronization.

Fig. 10 shows the measurement result of NAS executed on SPARCserver1000. The number of threads is the same as the number of PEs. On SPARCserver1000, the most effective technique is *Do-sandglass*. When the number of PEs is five, the speedup is about 1.6. The difference between this and the result on the EM-X is that when the number of PEs is larger than five, the performance is decreased. When the number of threads increases, it is predicted that memory accesses increase at same time as well.

*Do-all & seq* speeds up by increasing the number of threads. When $S_1$ of NAS as the doall-type loop is finished, synchronization among the threads leads to a smaller parallelization effect.

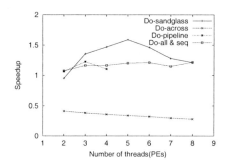

**Fig. 10.** Results of NAS executed on the SPARCserver1000

### 4.4 Performance measurement on Cenju-3

Fig. 11 shows the measurement result of NAS executed on Cenju-3. An MPI library is used for message passing programming. Because communication overhead of Cenju-3 is large, the performance of all parallelization techniques but *Do-all & seq* are far less than that of the sequential execution. Parallelization is not effective in this case.

We can make messages among PEs blocking by communication vectorization [9] in order to decrease the communication overhead. *Do-pipeline* and *Do-sandglass* are able to utilize the communication vectorization. *Do-across* cannot use the communication vectorization, because each iteration corresponds to one task. In Fig. 11, the block size (B) is 100 and 1,000.

The communication vectorization is effective for parallelization, whenever it can be utilized. The most effective technique is *Do-sandglass* as shown in Fig. 11. When the block size (B) is 1,000, the speedup at four PEs is 2.0.

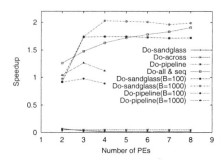

**Fig. 11.** Results of NAS executed on the Cenju-3

## 4.5 Results of performance measurement

Table 2 shows peak speedup ratios of each benchmark program on each machine by each technique. A number of used PEs is different by each result.

**Table 2.** Results of performance evaluation(Peak speedup ratio)

|  | Parallelization techniques | NAS LBD | Spline LFD | Livermore LBD |
|---|---|---|---|---|
| EM-X | *Do-sandglass* | 3.35 | 1.46 | 1.97 |
|  | *Do-across* | 3.14 | 8.61 | 2.14 |
|  | *Do-pipeline* | 2.09 | 1.92 | 2.08 |
|  | *Do-all & seq* | 3.20 | 9.77 | 1.36 |
| SPARCserver1000 | *Do-sandglass* | 1.59 | 0.96 | 0.93 |
|  | *Do-across* | 0.41 | 0.32 | 0.83 |
|  | *Do-pipeline* | 1.23 | 0.94 | 1.03 |
|  | *Do-all & seq* | 1.22 | 0.79 | 0.68 |
| Cenju-3 | *Do-sandglass* | 2.03 | 0.34 | 0.30 |
|  | *Do-across* | 0.05 | 0.03 | 0.01 |
|  | *Do-pipeline* | 1.26 | 0.33 | 0.59 |
|  | *Do-all & seq* | 1.91 | 4.16 | 0.51 |

For the benchmark programs, *Do-sandglass* is effective for NAS on all machines. The reasons are that the loop-carried dependence tasks is larger than the non loop-carried dependence tasks, and the loop-carried dependence relation is LBD. When the loop-carried dependence relation is LFD as Spline, *Do-all & seq* is more effective than *Do-sandglass*. *Do-sandglass* is not effective for double loops as Livermore 23. The sandglass-type parallelization becomes hard by the loop-carried dependence of inner and outer loop.

*Do-sandglass* dose not depend on a difference in the machines. When the communication overhead is large, the communication vectorization is useful.

## 5 Conclusion

In this paper, we described the sandglass-type parallelization technique. And we show the effectiveness used by three benchmark programs on three parallel machines. As a result, the sandglass-type parallelization technique is effective under the following conditions.

- The size of a task with loop-carried dependence is smaller than the size of a task lacking loop-carried dependence. When the ratio of the task size is large, the effect is large.
- The loop-carried dependence relation is LBD.

- A number of PEs of a parallel machine is about ten.

Disadvantage of the sandglass-type parallelization technique is following.

- The size of a task with loop-carried dependence is larger than the size of a task lacking loop-carried dependence.
- The Loop-carried dependence relation is LFD.
- The loop is nested.

Future works will address whether dependence distance can become longer by loop restructuring. We also consider the effectiveness of the sandglass-type parallelization technique in cases in which the dependence distance is larger than two. Furthermore, it is necessary to estimate our technique by using loops with task $S_3$. Another proposed parallelization technique is Loopacross [8]. We will need to compare our technique to this technique.

## Acknowledgments

We would like to thanks members of the Computer Science Division in the Electrotechnical Laboratory for permitting us to use the EM-X parallel computer. We would like to thanks the Real World Computing Partnership for permitting us to use the SPARCserver1000. This research was partly supported by the Ministry of Education through the Grant-in-Aid for Scientific Research No.(B)(2)10480057.

## References

1. Cytron, R.: Doacross: Beyond Vectorization for Multiprocessors, Proc. of the Int. Conf. on Parallel Processing, (1986) 836–844
2. Kanamaru, T., Koseki, A., Komatsu, H. and Fukazawa, Y.: Loop Staging: A Loop Parallelization Technique for Shared Memory Multiprocessors, Proc. of Joint Symposium on Parallel Processing 1997, (1997) 197–204 in Japanese
3. Kodama, Y., Sakane, H., Sato, M., Yamana, H., Sakai, S. and Yamaguchi, Y.: The EM-X Parallel Computer: Architecture and Basic Performance, Proc. 22nd Annual Int. Symp. on Computer Architecture, (1995) 14–23
4. Nakanishi, T., Joe, K., Polychronopoulos, C. D., Araki, K. and Fukuda, A.: Estimating Parallel Execution Time of Loops with Loop-Carried Dependences, Proc. of Int. Conf. on Parallel Processing, Vol. 3, (1996) 61–69
5. Nichols, B., Buttlar, D. and Farrell, J. P.: Pthreads Programming, O'Reilly & Associates, Inc. (1996)
6. Padua, D. A., Kuck, D. J. and Lawrie, D. H.: High-Speed Multiprocessors and Compilation Techniques, IEEE Trans. on Comp., Vol. C-29, No. 9, (1980) 763–776
7. Takabatake, M., Honda, H., Osawa, N. and Yuba, T.: A Parallel Technique for a Single Doacross Loop, Proc. of Joint Symposium on Parallel Processing 1998, (1998) 367–374 in Japanese
8. Yamana, H., Tatebe, O., Koike, H., Kodama, Y., Sakane, H. and Yamaguchi, Y.: Loopacross: Beyond Doacross for Distributed Memory Multiprocessors, Proc. of the Second IASTED International Conference, (1998) 229–235
9. Zima, Z. and Chapman, B.: Supercompilers for Parallel and Vector Computer, Addison Wesley (1990)

# Transforming and Parallelizing ANSI C Programs Using Pattern Recognition

Maarten Boekhold, Ireneusz Karkowski and Henk Corporaal

Delft University of Technology
Mekelweg 4, P.O. Box 5031, 2600 GA Delft, The Netherlands
{maartenb,irek,heco}@cardit.et.tudelft.nl

**Abstract.** Code transformations are a very effective method of parallelizing and improving the efficiency of programs. Unfortunately most compiler systems require implementing separate (sub-)programs for each transformation. This paper describes a different approach. We designed and implemented a fully programmable transformation engine. It can be programmed by means of a transformation language. This language was especially designed to be easy to use and flexible enough to express most of the common and more advanced transformations. Its possible applications range from coarse-grain parallelism exploitation to optimizers for multimedia instruction sets. It has been successfully tested on various applications.

## 1 Introduction

Due to advances in IC technology, multiprocessor systems are becoming ever more affordable these days. While most of these systems are used in either the server market or in scientific research, it can be expected that multiprocessor systems will also show up in embedded systems. Especially, because of the feasibility of single-chip multi-processor implementations.

To be able to use the full computing power that is available in such systems, it is necessary to execute the embedded applications in a parallel mode. Unfortunately most of the existing embedded codes are written in a sequential languages. Also the programmers usually feel more at ease in writing sequential programs. (Semi-)automatically transforming sequential programs to their parallel equivalents represents therefore an attractive alternative. Direct parallelization however often does not lead to an efficient implementation. A series of code transformations [15] are necessary to enable efficient parallelization. Since the number of standard transformations (and combinations of them) is large, writing separate (sub-)programs for each of them represents a tedious task. Even if finished, any extension to the already implemented set of parallelizations requires each time a substantial effort, due to the very low code reuse. Tools, which make writing such programs easier ([8,14]), represent only a partial solution. They make programming faster, but still the transformations have to be coded separately. An alternative is to design a programmable transformation engine, which could be easily configured for most useful code transformations.

This paper presents one such source-to-source code transformation tool, targeted at translation of ANSI C programs. It can be configured with new transformations by means of a dedicated *transformation language*. The language has been carefully designed to enable efficient specification of most common code transformations. Its syntax has been derived from ANSI C. Thanks to that the language is easy to learn and powerful to use. Note that the tool does not decide *if* a possible transformation should be applied. This decision is currently left to the user.

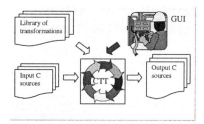

**Figure 1** Schematic overview of the transformation program.

A top view on our transformation system is presented in Figure 1. The block in the middle represents the transformation engine itself (called **ctt**, **C**ode **T**ransformation **T**ool). Transformations, which are specified in the *transformation language,* form a library of transformations. Input files are translated using a subset of transformations from this library. Finally output C source files are generated. The design of the program allows its use in any ANSI C translation context. Potential applications include coarse-grain parallelism exploitation [11], ILP enhancement (for example for data locality/low power [16]), and optimizations for multi-media instruction sets [5].

The remainder of this paper is organized as follows. Section 2 describes related work, and in particular the MT1 system [2,3], which has been an inspiration for us. Section 3 introduces the transformation language that is used to specify transformations. The implementation details of the transformation program are briefly discussed in section 4. Section 5 presents the results of experiments with the system, confirming the power and flexibility of the transformation language. Finally, section 6 concludes the paper.

## 2  Related work

The traditional approach to code transformations has been different than ours. For each transformation, separate program code had to be written. Although there exist a myriad of tools that try to make this task as easy as possible (Suif [14] and Sage++ [8] for example), the amount of coding required remains significant. Such tools represent only a partial solution. They make programming faster, but still every transformation has to be implemented separately.

This work has been inspired by a system called MT1 [2,3], from the University of Utrecht and the University of Leiden, both in the Netherlands. MT1 is a system that performs a translation from Fortran77 to Fortran90. Its name should be pronounced as 'empty', which describes its basic state: it is a compiler, which contains almost no intelligence. In this state, it does an identity transformation of its input. All additional knowledge about transformations should be added to MT1 at a later stage.

Unfortunately MT1 has some drawbacks making it useless in our context. First of all, the research within our laboratory is mainly concerned with embedded applications (which are most often written in C). Secondly, with MT1 it is impossible to apply a number of interesting types of transformations:
- MT1 cannot apply *inter-procedural* transformations.
- It has no explicit control over variables and it is impossible to create *new* variables in a transformation.
- It can only operate in an interactive way, and it is therefore not possible to use MT1 from within other applications.

Our work aims not only at being an ANSI C version of MT1, but also at adding the above capabilities.

## 3 Transformation language

This section will introduce our transformation language. Section 3.1 shortly introduces the reader to the topic of code transformations. The general overview of the syntax of our transformation language is given in section 3.2. The remaining subsections will present features specific to different stages of a transformation. Because of the space limitations we present only the most important elements of the transformation language. The interested reader is referred to [4] for more details.

### 3.1 Preliminaries

As a short introduction to the world of the transformations, let us start with two examples. For the educational purposes relatively simple loop transformations were selected. We will use them in the remainder of the paper to explain both the way our system works, and the transformation language itself (see [4] for much more elaborate examples).

**Example 1** Loop interchange *transformation exchanges the loops within a set of tightly nested loops. Consider a loop nest shown in upper part of Figure 2. When applied to these 2 tightly nested loops, the transformation exchanges the inner and outer loops. As the result code in the lower part of Figure 2 is obtained. Before the transformation however may be applied, the program should check if there exist no ( <, > ) dependencies [15] within the loop body. Only if this condition is met, the inner and outer loop may be exchanged.*

```
for ( i = 0 ; i < 200 ; i++ ) {
    for (j = 0 ; j <= 100 ; j++ ) {
        A[i][j] = A[i-1][j] + B[i][j];
    }
}
```

```
for ( j = 0 ; j <= 100 ; j++ ) {
    for ( i =0 ; i < 200 ; i++ ) {
        A[i][j] = A[i-1][j] + B[i][j];
    }
}
```

**Figure 2** Code before and after loop interchange transformation.

Some transformations may be applied to the code, which is spread among different procedures. Inter-procedural transformations form a very important class, and are fully supported by our transformation engine. The following example presents an example transformation belonging to this category.

```
....
for (i=0;i<100;i++){
    do_something();
}
....
void do_something() {
    procedure body
}
```

```
....
do_something2();
....
void do_something2() {
    int i;
    for (i=0;i<100;i++){
        procedure body
    }
}
```

**Figure 3** Code before and after loop embedding transformation.

**Example 2** *Consider the piece of code on the left in Figure 3. The loop embedding inter-procedural transformation moves the* `for` *loop to the body of the procedure* `do_something`*. As a result the code on the right in Figure 3 is generated. The transformation is always legal, and may be used to obtain more efficient code* *(smaller procedure call overhead), or to enable parallelism exploitation in the loop (data dependencies in the transformed code can be more easily analyzed). Note that we still may need the old version of the procedure* do_something *if it is called from more than one place.*

Both above transformations differ quite substantially, and therefore, using standard approaches, two separate programs would have to be written for each of them. This can be avoided if we properly organize the process of applying a transformation. Our engine uses such an organization. The translation is divided into 3 distinct stages:
- **Code selection stage:** In this stage the engine searches for code that has a strictly specified structure (that matches a specified *pattern*). Each fragment that matches this *pattern* is a candidate for the transformation.
- **Conditions checking stage:** Transformations can pose other (non-structural) restrictions on a matched code fragment. These restrictions include, but are not limited to, conditions on data dependencies and properties of index variables.
- **Transformation stage:** Code fragments that matched the specified structure and additional conditions are replaced by new code, which has the same semantics as the original code.

The structure of the transformation language closely resembles these steps, and is shown in Figure 4. As can be deduced, there is a one to one mapping between blocks in the transformation definition and the translation stages.

```
PATTERN {
    description of the code selection stage
}
CONDITIONS {
    additional constraints
}
RESULT {
    description of the new code
}
```

**Figure 4** Structure of a transformation description.

### 3.2 Language overview

The choice of the transformation language has a very large impact on the applicability of the tool. The following requirements should be considered:
1. The language should allow the specification of most useful transformations.
2. It should be powerful enough to express the functionality of typical embedded applications (It must be possible to add sections of code in the result code).
3. The language should be easy to learn and efficient.
4. It should be closely related to one of the languages used in the embedded systems community.

While a large fraction of the embedded systems are still programmed in assembly language, the ANSI C has become a widely accepted language of choice for this domain. Therefore, to satisfy the above requirements we decided to derive our transformation language from the ANSI C. As result, all C language constructs can be used to describe a transformation. Using only them would however be too limiting (and not satisfy the first requirement above). The patterns specified in the code selection stage would be too specific, and it would be impossible to use one pattern block to match a wide variety of input codes. Therefore we extended our transformation language with a number of *meta-elements*. Among others the following *meta-elements* were added and can be used to specify *generic* patterns, i.e. patterns that represent more than one element in the input C sources:

- *Statements:* keyword STMT represents any statement.
- *Statement lists*: keyword STMTLIST represents a list of statements.
- *Expressions*: keyword EXPR represents any expression.
- *Variables:* keyword VAR represents any variable (of any type).
- *Procedure calls:* keyword PROCCALL represents any procedure, which satisfies specific requirements.

Due to implementation details, a few variants of these meta-elements exist. For example there exist separate meta-elements for upper bound and step expressions in for loops (keywords BOUND, STEP_EXPR). As can be seen, we included also meta-elements for the variables. This was motivated by the desire of avoiding direct relationship between variable names in the transformation definition and the variable names in the input C sources.

Meta-element must be assigned a number (except for variables), which should be included in braces behind their keyword (STMT(1) and STMT(2) represent then different C statements). Some meta-elements take also additional arguments (for example BOUND and STEP_EXPR take also as argument the name of the loop index variable). In the following sections we will describe the use of transformation language constructs in each sub-block of the transformation definition (recall Figure 4).

### 3.3 The Pattern Block

The pattern block contains a description of the code selection stage. It describes the structure of code (a *pattern*) that may be a candidate for a transformation. The transformation engine uses a *pattern-matching* algorithm to find code fragments in the input C source that match it.

The block may be specified using all introduced (meta-)elements. On the first occurrence of a meta-element in the pattern block, the code at the corresponding location in the input is *bound* to this meta-element's number. If a meta-element is used more than once with the same number (for variables the same meta-variable name), the corresponding input code on subsequent occurrences must be identical to the input code that was bound to this meta-element at its first occurrence. The examples below illustrate that.

**Example 3** *A pattern block that describes the code selection stage of* the loop interchange *transformation is shown in Figure 5. This pattern has the following meaning. "Look for 2 tightly nested loops, of which the inner loop can contain any statement list". The expression 1 has been used twice (in both loops) and therefore the lower bounds in both loops must be the same. Clearly this* pattern *matches the loop nest from Figure 2.*

```
PATTERN {
    VAR x, y;

    for (x=EXPR(1);BOUND(1,x);STEP_EXPR(2,x)) {
        for (y=EXPR(1);BOUND(2,y);STEP_EXPR(3,y)) {
            STMTLIST(1);
        }
    }
}
```

**Figure 5** Pattern block for the loop interchange transformation.

**Example 4** *Consider again the loop embedding transformation. The pattern block for it can be specified using the* PROCCALL *meta-element (see Figure 6). The presence of this meta-element means that at that particular location in the input code, a procedure call is expected, where the called procedure must satisfy a specified pattern* p_pat. *The whole pattern means: "Look for a* for *loop which contains a* *single call to a procedure (which takes no real arguments). The called procedure may contain any list of statements as its body". Of course more complex procedure patterns (with arbitrary number of procedure arguments) are possible.*

```
PATTERN {
    VAR x;

    for (x=EXPR(1);BOUND(1,x);STEP_EXPR(2,x)) {
        PROCCALL(1, p_pat);
    }
}

void p_pat() {
    STMTLIST(1);
}
```

**Figure 6** Pattern block for the loop embedding transformation.

As could be seen in Example 4 the pattern for the called procedure must be specified outside of any block in Figure 4 (C does not allow procedure definitions inside any compound statement). Also other definitions, which may be necessary to define the behavior of the transformation stages, must be defined before the transformation definition blocks. Typically there is a single file for each transformation. It starts with inclusion of standard header files, which is followed by local definitions and the definitions of the transformation stages. For all of them the ANSI C scope rules apply.

### 3.4 Conditions block

A conditions block contains additional conditions that are imposed on code that was selected in the previous stage. These conditions can be divided into 3 categories:
- **Type properties:** This category includes requirements such as "is this expression a constant?", "does this statement contain a procedure call?" or 'does this expression have site effects?".
- **Structural properties:** In case of some structural properties it is not desirable to have to specify them in the pattern block. Most importantly, if we had to specify the existence or absence of control transfer statements (break, continue and return statements) in the pattern block, we would severely restrict the generality of the descriptions of transformations. Therefore, extra conditions that test for these structural properties are necessary.
- **Data dependencies**: Data dependencies are vital in testing for the legality of code transformations. Depending on the type of transformation we must be able to test for different following properties of data dependencies.

The following example demonstrates these concepts.

```
CONDITIONS {
    stmtlist_has_no_unsafe_jumps(1);
    not(dep("* direction=(<,>)
            between stmtlist 1 and stmtlist 1"));
}
```

**Figure 7** Conditions block for loop interchange transformation.

**Example 5** *Recall* the loop interchange transformation *example. We can see that for this transformation to be legal the following two requirements must be fulfilled. The first one is that the loop-body should not contain* break *or* continue *statements. Second, that the inner loop-body is not allowed to have a (<,>) data dependencies within itself. These conditions may be specified as shown in Figure 7. The '\*' in the second statement denotes any dependence.*

## 3.5 The Result Block

The result block uses the same part of the transformation language as used in the pattern block. Also a number of the result block specific elements are allowed. These include for example support for creation of new variables in the transformed code. Also strict control over the data type of these newly created variables is possible. The examples below demonstrate the transformation block specification.

**Example 6** *Consider the* loop interchange transformation *example. The result block describing it is shown in example in Figure 8. This description says: "Replace the matched code with 2 tightly nested loops, where the body of the new inner loop is the same as the body of the inner loop of the original code".*

**Example 7** *To specify the transformation stage of an inter-procedural transformation, we use the* PROCCALL *meta-element again. The result block for the* loop embedding *transformation is shown in Figure 9. It means: "Replace the original* for *loop and the call with a single new call. The called procedure should contain a for-loop whose body is equal to the body of the original called procedure".*

```
RESULT {
    VAR x, y;

    for (y=EXPR(1);BOUND(2,y);STEP_EXPR(3,y)) {
        for (x=EXPR(1);BOUND(1,x);STEP_EXPR(2,x)) {
            STMTLIST(1);
        }
    }
}
```

**Figure 8** Result block for loop interchange transformation.

```
RESULT {
    PROCCALL(1, p_transformation);
}

void p_transformation() {
    VAR x;

    for (x=EXPR(1);BOUND(1,x);STEP_EXPR(2,x)) {
        STMTLIST(1);
    }
}
```

**Figure 9** Result block for loop embedding transformation.

## 4 Implementation

The transformation program has been written in C++, using the SUIF compiler toolkit. The decision of using SUIF was dictated mainly by the existence of a good ANSI C front end, a convenient internal representation (IR) with C++ interface, and by the existence of an IR to ANSI C conversion utility (s2c). Thanks to that we could concentrate more on the design of the transformation engine itself, which works entirely on the IR level only. The whole transformation trajectory is presented in Figure 10. Both input source and transformation definitions are compiled to the IR by the front-end. After linking them the translation process takes place. The user decides which possible transformations are applied. Once the transformations have been applied, the s2c program is used to convert the IR back to ANSI C.

All functionality needed to perform a transformation (i.e. code selection, conditions checking and transforming) has been implemented as a collection of C++

classes, which can be accessed through a single C++ class interface. This makes it easy to embed the full functionality in other C++ programs.

Although efficient algorithms (using Finite State Machines) exist for finding patterns in tree-structures (``pattern matching'' [10,9,1]), these proved to be difficult to use with SUIF. The main problem was that another intermediate representation of the program would have to be used. While it was possible to derive it from the SUIF IR, the possible advantages did not, in our eyes, justify the large effort of implementing it. Therefore, a *straightforward* pattern-matching algorithm has been used [4]. The nodes of program are visited in the DFS order, and at each place all transformations are sequentially tried.

A static data dependency analyzer of SUIF combined with our dynamic data dependence analysis package [12] is used to check the data dependence conditions that are specified in the condition blocks.

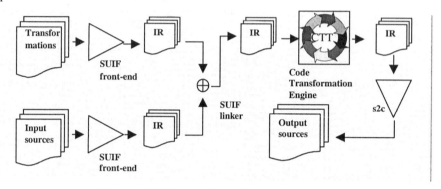

**Figure 10** The transformation trajectory.

The Graphical User Interface (GUI) has been written using the Qt toolkit [13]. The GUI allows users to experiment with different sets of transformations and provides an easy interface to each of the 3 transformation stages. While the translation process may proceed completely automatically, an interactive mode allows the user to *override* the decision made in the conditions checking stage (e.g. it is possible to apply a transformation even though the conditions checking stage says that this would be illegal).

## 5 Experimental results

In order to test the flexibility of the transformation language, we have specified a large number of transformations (and their combinations). All of them have been tested successfully on typical embedded applications, including 10 DSP programs from [6] and the MPEG-2 encoder from [7].

As example, table in Figure 11 presents results which we obtained using our transformation engine for mapping embedded programs to architectures with multimedia instruction sets. As can be seen, about 85% of the SIMD suitable loops have been automatically recognized and transformed. We must mention that our approach is different than the one described in [5]. Their methodology involved first converting the code to the vector code (working on infinite vectors), and then mapping vector instructions back to parallel loops (using vectors of length 4). In contrast to their approach, we performed the conversion directly.

| Benchmark | Description | #loops | #SIMD | SIMD mapped |
|---|---|---|---|---|
| arfreq | Autoregressive freq. estim. | 2 | 0 | 0 |
| g722 | Adaptive differential PCM | 12 | 7 | 6 |
| instf | Frequency tracking | 9 | 3 | 3 |
| interp3 | Sample rate convertion | 3 | 0 | 0 |
| mulaw | Speech compression | 1 | 0 | 0 |
| music | Music synthesis | 4 | 0 | 0 |
| radproc | Doppler radar processing | 7 | 1 | 1 |
| rfast | Fast convolution using FFT | 9 | 5 | 4 |
| rtpse | Spectrum analysis | 10 | 3 | 3 |
| mpeg2enc | Video/MPEG-2 encoder | 171 | 26 | 22 |
| Total | | 228 | 45 | 39 |

**Figure 11** Numbers of loops mapped to SIMD instructions.

Also a large set of well known and general transformations have been tried (see [4] for detailed specifications):

- Loop peeling.
- Index set splitting.
- Loop reversal.
- Loop skewing.
- Loop fusion.
- Wave fronting.
- Inlining.
- Loop fission.
- Strip mining.
- Code sinking.
- Unswitching.
- Loop embedding and extraction.

All the experiments proved the viability of our approach and the strengths of our transformation language. Also the learning time for a C programmer was short. Most specifications of transformations turned out to be simple to write and understand.

Some remarks are in order. There is a class of transformations, which is "by nature" more suitable to be implemented in an algorithmic way; this mainly for efficiency reasons. For example, while specification of inlining transformation is fully supported in our system, it could be implemented much more efficiently using an algorithmic approach (i.e. writing a special program for it). The reason for it is that separate patterns must be written for procedures with different numbers of arguments. Similarly, a single specification of the *wave fronting* transformation does not cover all possible situations in which this transformation could be applied. This is due to the fact that the exact transformation depends on the dependence distances in the loop-body. It is not yet possible to use this information to specify the transformation. Still our system can be used. By "pre-calculating" the transformations for specific dependence distances, patterns for these distances can be separately generated.

## 6 Conclusions

In this paper we presented a programmable engine for code transformations on ANSI C programs. The knowledge about the transformations is added by means of a convenient and efficient *transformation language*. Using this language to specify new transformations is much easier and faster than having to write separate (sub-) programs for each of them. A very large subset of possible transformations is supported, including the inter-procedural ones. All of them (and their combinations) have been successfully specified using the transformation language, thereby proving the viability of its concept.

In the future we plan to further extend the transformation language to obtain an even larger coverage (of a single transformation specification). We believe that, adding the following extra features is feasible:

- Algorithmic calculation of the expressions and statements used in the transformed code.
- Support for conditional construction of the code (by adding conditional statements to the result block).
- Support for iterated construction of the code (by adding loop statements to the result block).

Implementing these elements should bring us even further ahead of any of the existing transformation tools.

# References

1. Marianne Baudinet and David MacQueen. *Tree pattern matching for ML*. Available from David MacQueen, AT&T Bell Laboratories, 600 Mountain Avenue, Murray, Hill, NJ 07974, 1986.
2. Aart J.C. Bik. *A Prototype Restructuring Compiler*. Technical Report INF/SCR-92-11, Utrecht University, Utrecht, the Netherlands, November 1994.
3. Aart J.C. Bik and Harry A.G. Wijshoff. *Mt1: An interactive prototype restructuring compiler*. In Proceedings of the second annual conference of the Advanced School for Computing and Imaging, pages 78--83, June 1996.
4. Maarten Boekhold, Ireneusz Karkowski and Henk Corporaal. *A Programmable ANSI C Code Transformation Engine*. Technical Report no. 1-68340-44(1998)-08, Delft University of Technology, Delft, The Netherlands, August 1998.
5. Gerald Cheong and Monica S. Lam. *An Optimizer for Multimedia Instruction Sets*. In the Proceedings of the Second SUIF Compiler Workshop, Stanford University, USA, August 21-23, 1997.
6. P.M. Embree. *C Language Algorithms for Real-Time DSP*. Prentice-Hall. 1995.
7. MPEG Software Simulation Group. *MPEG-2 Video Codec*. http://www.mpeg.org/.
8. Dennis Gannon et al. *Sage*. http://www.extreme.indiana.edu/sage/, 1995.
9. Josef Grosch. *Transformation of attributed trees using pattern matching*. Lecture Notes in Computer Science, 641:1-15, 1992.
10. Christoph M. Hoffman and Michael J. O'Donnell. *Pattern matching in trees*. Journal of the ACM, 29(1): 68--95, 1982.
11. Ireneusz Karkowski and Henk Corporaal. *Design Space Exploration Algorithm For Heterogeneous Multi-processor Embedded System Design*. In 35th Design Automation Conference Proceedings, June 1998, San Francisco, USA.
12. Ireneusz Karkowski and Henk Corporaal. *Overcoming the limitations of the traditional loop parallelization*. Future Generation Computer Systems, 13(4-5), 1998.
13. Troll Tech. *The Qt tool*kit. http://www.troll.no/products/qt.html, 1998.
14. Robert Wilson, Robert Franch, Christopher Wilson, Saman Amarasinghe, Jennifer Anderson, Steve Tjiang, Shin-Wei Liao, Chau-Wen Tseng, Mary Hall, Monica Lam, and John Hennessy. *An Overview of the SUIF Compiler System* http://suif.stanford.edu/suif/suif.html, 1995.
15. Michael Wolfe. *High Performance Compilers for Parallel Computing*. Addison-Wesley Publishing Company, 1996.
16. Nikos D. Zervas, Kostas Masselos and C.E. Goutis. *Code Transformations for Embedded Multimedia Applications: Impact on Power and Performance*. In the Proceedings of the Power Driven Microarchitecture Workshop (in conj. with ISCA'98). June 1998, Barcelona, Spain.

# Centralized Architecture for Parallel Query Processing on Networks of Workstations

Sijun Zeng[1] and Sivarama P. Dandamudi[2]

[1] CrossKeys System Corporation, Kanata, Canada, `szeng@crosskeys.ca`
[2] Center for Parallel and Distributed Computing, School of Computer Science, Carleton University, Ottawa, Canada, `sivarama@scs.carleton.ca`

**Abstract.** Network of workstations (NOW) is a cost-effective alternative to a multiprocessor system. Here we propose a centralized architecture for parallel query processing on network of workstations. We describe a three-level processing strategy and evaluate its performance. The top two levels use a space-sharing technique to assign a partition to a query. The third-level uses a chunk-based load sharing policy to accommodate dynamic local load present on the workstations. To evaluate the performance of this three-level strategy, we have implemented it using PVM on a Pentium-based NOW.

## 1 Introduction

Networks of workstations (NOWs) are cost-effective alternative to multiprocessors systems [1]. The use of networks of workstations for parallel query processing has been proposed for improving query performance of database systems [2]. Query processing benefits from a NOW-based system not only because it provides additional processor cycles but also adds substantial main memory, which reduces the number of disk accesses. The improved performance is often obtained without any additional cost by using existing client workstations.

For query processing, there are significant differences between a parallel database system and a NOW-based system. The main differences are:

- In a NOW-based system, the database itself is not partitioned across the nodes as in a parallel database system. This causes additional overhead as the data have to be sent to the workstations to work on.
- The workstations in a NOW-based system are typically heterogeneous.
- In parallel database systems, load on each processing module is stable. In NOW-based systems, load on each workstation can vary dynamically even while a query is being processed due to local load variations. This poses several problems to balance query load across the system.
- Parallel database systems have a dedicated, special-purpose fast communication network, whereas the NOW-based systems use general-purpose LANs, which are slow and involve high communication overhead.

These differences clearly indicate that performance issues in NOW-based systems are different from those in a parallel database system. This paper presents a three-level centralized architecture for parallel query processing on NOWs. Note that the centralized architecture was mentioned in [5] as a possible architecture for query processing on NOW-based systems. We make two main contributions to the area of parallel query processing. First, we present the design of a three-level strategy for query processing. The second contribution is the implementation of this three-level architecture on an experimental Pentium-based NOW. Our NOW system is heterogeneous, consisting of four classes of Pentium nodes (from Pentium 133 to Pentium 200). This allows us to test how the proposed design handles node heterogeneity. We have used PVM [6] to facilitate our implementation. Our implementation takes communication overheads, buffer copying overheads, etc. into account. This, we believe, increases the utility of the results presented here.

## 2 Centralized Architecture for Parallel Query Processing

In the centralized architecture, there are three types of nodes.

**Coordinator node:** This is also called the "master" node. All queries generated in the system are first sent to this node for scheduling. To facilitate scheduling, this node maintains state information on the whole system. This node is responsible for receiving and scheduling queries from the other nodes. When there are no processors available, queries are put into a waiting queue. The system state information maintained includes computational capacity, background load, and current state of the nodes.

**Repository node:** This is a special node where the database is physically located. Upon request, it sends the tuples of the required tables in the database to the requesting node. Note that the Repository and Coordinator nodes are logical and could be mapped to the same physical node in a system.

**Ordinary node:** These "slave" nodes perform the join operations of the query. This node gets the tuples of the tables involved from the Repository node and schedules join operations on the nodes of the partition.

When a new query is submitted or a completed query leaves the system, the Coordinator recalculates the partition size based on the current system state. It then assigns processors to queries in the waiting queue until no more processors are available. Here, for the sake of simplicity, we assume that the requests are processed in first-come-first-served order. Among the processors of a partition, one is designated as the Scheduler. The partition information along with the query is sent to the Scheduler. The Scheduler schedules the tasks among the nodes in the partition to complete the query and sends the results to the originator.

Our scheduling procedure is a three-level strategy. The task at the first level determines the partition size based on the current query load on the system. This task is done by the Coordinator. The task at the second level, also performed by the Coordinator, finds a specific processor partition that is to be allocated. Though this task is easy to implement in multiprocessor systems because of the absence of external load, it is more complicated in a NOW due to dynamic local load on the workstations. The third-level strategy determines how the allocated processors work on the query to achieve effective load sharing.

The reminder of this section gives details about these three levels.

## 2.1 Level 1: Partition Size Calculation

In multiprocessor scheduling area, several scheduling strategies have been proposed to determine the partition size [4, 8–10]. An adaptive policy is adopted as our processor allocation policy because of its flexibility and efficiency demonstrated in multiprocessor systems [4, 8]. In this policy, the partition size depends not only on the jobs in the waiting queue, but also on those currently being executed [4]. This means the jobs of the whole system are taken into account. This policy has been shown in [4] to provide better performance than the one that is based only on the number jobs in the waiting queue proposed in [8, 9]. The partition size in our system is calculated as

$$PartitionSize = Max\left(1, \left\lfloor \frac{P}{Q + f * E + 1} + 0.5 \right\rfloor\right) \quad (1)$$

where $P$ is the number of workstations in the system, $Q$ is the number of queries waiting to be executed in the query wait queue, and $E$ is the number of queries currently being executed. The parameter $f$ ($0 \leq f \leq 1$) is used to vary the weight of $E$ in the partition size calculation. This policy also reserves a partition for future query arrivals. This is the reason for adding 1 to the denominator in Eq. 1.

In NOW-based systems, the computational capacity may differ significantly among the nodes due to heterogeneity. To get an accurate system state information, a uniform criterion must be established among the nodes. A new term, *processor unit* (PU), is introduced in place of processor. The node with the lowest computational capacity (or speed) is considered to be one PU. Computational capacity of other nodes is expressed relative to this node (e.g., 2 PUs, 3.45 PUs, etc.). In the remainder of the paper, we use the term processor to refer to processor units.

In the first level of our strategy, we consider the system capacity as *static* with no local load on the workstations. Thus we simplify the computation of partition size by taking only the query load into account. The total number of processors $P$ and their "speed" in the NOW system can be known at the configuration time. Workload issues are left to be handled later within the next two levels.

## 2.2 Level 2: Processor Partitioning

Once the partition size is determined, the next step at the second level is to actually partition the available nodes into subsets and assign them to each query. By available we mean the node's non-query local load should be lower than a pre-assigned threshold value (similar to the load distribution techniques used in distributed systems [11, 3]) and it should not have been assigned to another query in the system. As discussed before, each partition and hence each node is assigned at most one query. In a multiprocessor system, it is easy to partitioning as all nodes are identical. In a NOW-based system, due to heterogeneous computational capacity and local workload variations, finding a partition is more complex. Note that a candidate partition obviously should have a size of at least the required PartitionSize. Among these candidate partitions, our goal is to find the best partition (i.e., the sum of whose processor units has the least difference from the required size). If two or more partitions have exactly the same size, we choose the partition with the fewest nodes due to the following considerations:

- The communication overhead tends to be smaller with smaller number of nodes
- Better performance is obtained with faster nodes (two faster nodes is better than four slower nodes of equivalent processing power)

When no partition that meets the PartitionSize requirement is available, queries will remain in the waiting queue until the system state changes (due to the arrival of a new query or departure of a completed query).

### 2.3 Level 3: Operation Scheduling

Our policy is designed to successfully exploit three types of parallelism in processing queries on NOW-based systems: concurrent query execution, inter-operation parallelism, and intra-operation parallelism. In our three-level strategy, concurrent query execution is implemented by dividing the system nodes into partitions and executing a query in each partition. This is done at the first and second levels as discussed in the previous two subsections. Within a partition, intra-query and intra-operation parallelism are exploited by parallelizing join operations in a query job and executing partial operations of a join. These functions are done at the third level between the Scheduler node and the partition nodes.

A query typically consists of several join operations. The basic idea for operation scheduling is to determine

- how the nodes in the assigned partition work together on a query's join operations and
- how partial operations of a join operation are managed.

These issues are handled by the Scheduler within the partition. The Scheduler determines how many join operations are scheduled and which nodes will work on a particular join. For example, in a 8-relation query, there are four joins at the bottom level. Depending on the number of nodes in the partition, these four joins could be executed concurrently by different sets of nodes within a partition; or alternatively, all nodes may work on one join at a time. When executing a particular join, partial operation is processed in load shared fashion by "chunking" the involved records and assigning chunks of records in turn to each node for execution (as in parallel loop scheduling [7]). Initially, each node is given the same chunk size records. Upon receiving a partial result, the Scheduler assigns another chunk, if there is one, to the node. Thus, the query load is shared by all nodes in the partition to achieve good overall system performance. In our experiments, fragment size parameter determines the minimum chunk size.

## 3 Experimental System and Implementation

Parallel Virtual Machine (PVM) on a network of workstations has been used to study the performance of parallel query processing. The Pentium-based NOW system used in our experiments has four different types of nodes: Pentium 133 with slow cache, Pentium 133 with fast cache, Pentium 166, and Pentium 200.

The database we used in our experiments consists of sixteen base tables. Each table consists of between 5000 to 9000 records of data. There are two kinds of key fields in the tables of the database. In two of the sixteen tables, the keys are unique in the tables. There is no repetition of the key values in these tables and the values of the keys are sequential long integers. For the remaining fourteen tables in the database model, the key values in each table are normally distributed with the same mean value but with different variance for the keys in each table. By having these two kinds of tables, we obtain different selectivities for the join operations between different tables (see [12] for details).

Selectivities are used to determine the query types in the database model. Each query type in our model contains information on how many tables and which specific tables in the database are involved in the query. For each query type, the number of original tables in the database is either 2, 4, or 8. Sixteen query types are created based on the selectivities of the join operations among the data tables in the database model. In order to model queries with growing-sized result tables, in half of the query types, tables of join operations with higher selectivities are selected. In the other half of the query types, tables whose join operations have lower selectivities are selected to produce shrinking-sized result tables. For more details, see [12].

## 4 Performance Evaluation

In this section, performance of the centralized architecture for parallel query processing on a NOW is evaluated. The average response time is used to represent the performance. Each experimental value represents an average of 30 batches with each batch consisting of 50 to 100 queries. 95% confidence intervals were obtained for the results to give an indication of the variability. The value of the confidence interval is less than 10% of the mean value reported when the system is heavily loaded and less than 5% when the system is lightly loaded. In our simulation, the inter-arrival times of queries are exponentially distributed. The default $f$ value is set to 0.75.

We have conducted several experiments to determine the computational capacity of each class of nodes in our experimental system. These experiments were done by executing 16 different types of queries sequentially (without queueing) on each node. These results show that the Pentium 200 node is nearly 2.8 times faster than the Pentium 133 node (slow cache). Faster cache Pentium 133 and Pentium 166 nodes are about 16% and 38% faster than the slow cache Pentium 133 node, respectively.

### 4.1 Performance Sensitivity to Query Load

Figure 1 shows the mean response time as a function of query load. These results were obtained on a 8-node system with fragmentation size fixed at 200,000 bytes.

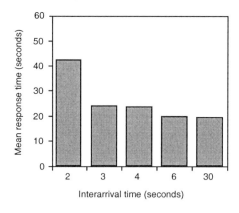

**Fig. 1.** Performance sensitivity to query load

Our experimental data presented in [12] suggest that saturation occurs when the query rate is below 4.4 queries per second when no load sharing is used (i.e, query load is equally distributed among the node of a partition). Since all 8 nodes are given the same workload, fast nodes finish their work earlier than the slow nodes. Further improvement in performance is possible by using load sharing techniques as the Pentium 200 is almost three times faster than the slow Pentium 133.

When load sharing is incorporated as in our policy, results presented in Figure 1 show that the system can handle queries at the rate of one query every two seconds (as opposed to 4.4 seconds when query load is equally distributed). Experimental data (not presented here) suggest that the increase in the mean response time is mainly caused by the increase in query waiting time. The system saturates when the inter-arrival time is less than two seconds.

## 4.2 Sensitivity to Fragmentation Size

The fragmentation size determines the number of partial operations allowed. The larger the fragmentation size, the fewer the number of partial operations. A smaller fragmentation size leads to more partial operations and larger scale intra-operation parallelism. Thus, load can be distributed to achieve good load sharing, which is the key to achieve good performance in NOW-based systems. On the other hand, with smaller partial operations, the overhead of communication and operation scheduling increases. This trade-off implies that the system response time may not always decrease as the fragmentation size decreases.

Figure 2 shows performance sensitivity to fragmentation size when the system is lightly loaded (inter-arrival time for queries is set at 24 seconds) and relatively heavily loaded (inter-arrival time is 6 seconds). It can be observed that a fragmentation size of 10KB gives the best performance under both system loads. As the fragmentation size decreases from 200KB, response time improves as better load sharing can be achieved due to smaller chunk size used by the Scheduler

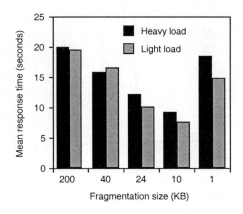

**Fig. 2.** Performance sensitivity to fragmentation size

to distribute work to other nodes. However, as the fragmentation size becomes smaller (in our experiments smaller than 10KB), performance improvement obtained from load sharing is more than offset by the increase in communication and scheduling overheads. This results in performance deterioration as shown in Figure 2.

### 4.3 System Speedup and Scaleup

Ideal parallel systems demonstrate two key properties, namely *linear speedup* and *linear scaleup*. If increasing the size of a parallel system (e.g., number of nodes) results in a proportional improvement in performance (e.g., response time), the parallel system is said to be successful in achieving linear speedup.

Two experiments have been done to evaluate the system speedup obtained by using the centralized architecture for parallel query processing. In the first experiment, systems with 1, 2, 4 and 8 nodes are given the same amount of light load that each system can handle. Figure 3a shows the performance results of this experiment. The fragmentation size used is 200,000 bytes. Queries arrive into the system every 24 seconds on average. With this workload, the mean response time on 1, 2, 4 and 8 nodes systems are 89.5, 26.9, 23.4 and 19.3 seconds, respectively. The speedup obtained is 3.4 instead of 2 when the number of nodes in the system increases from 1 to 2. One reason for this is that the 1-node system is heavily loaded with an inter-arrival time of 24 seconds. In the 2-node system, there is query generating process only on one of the nodes while in the 1-node system the only node handles all the scheduling, query processing, and query generation. The overhead of the query generator also affects the average response time of queries.

In the second experiment (Figure 3a), the inter-arrival time is set to 6 seconds. With this increase in query load, the 1-node and 2-node systems saturate. The 4-node system achieves mean response time of 25.1 seconds while the 8-nodes

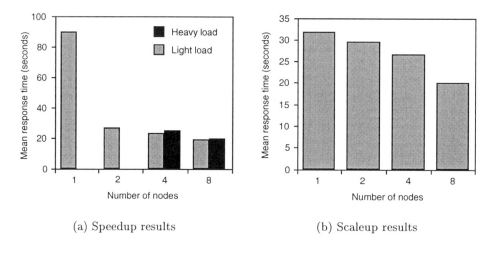

**Fig. 3.** Speedup and scaleup results

system achieves a response time of 19.9 seconds. Thus, the speedup of 8-nodes system to 4-nodes system is 1.3.

We did not get good speedup from 4-node to 8-node system mainly because of the small size of the database tables involved. For example, increased speedups can be obtained even with this data set if we increases the query load (by decreasing the query interarrival time). Furthermore, the 4-node system in the experiments has a total computational capacity greater than half of that of the 8-node system (due to processor heterogeneity). This is another reason for the small speedup we obtained when the system is expanded from 4 nodes to 8 nodes. Overall, we conclude that the centralized system is successful in providing good speedups.

Linear scaleup refers to the characteristic that with increasing system size (e.g., number of nodes), the system can perform proportionally larger amount of work during the same time period. While speedup holds the problem size constant and increases the system size to reduce execution time, scaleup measures the ability to increase both the system and problem sizes.

In our scaleup experiments, we assume that twice of the inter-arrival time leads to half the amount of work. Figure 3b shows the response time for different system sizes. For the 8-, 4-, 2-, and 1-node systems, the inter-arrival times are set at 6 seconds, 12 seconds, 24 seconds, and 48 seconds, respectively. From the results shown in Figure 3b, it is clear that the centralized architecture is successful in achieving linear scaleups. The 2-node system, 4-node system and 8-node system can all complete 2 times, 4 times, and 8 times the amount of work as a 1-node system does in increasingly smaller amount of time. It shows that

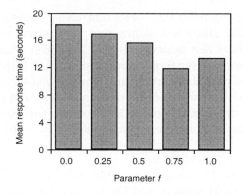

**Fig. 4.** Performance sensitivity to parameter $f$

the centralized architecture not only succeeds in linear scaleup but also succeeds in combining it with good speedups.

### 4.4 Sensitivity to Parameter $f$

Equation 1 determines the partition size in our strategy. The parameter $f$ in the formula affects the partition size. When the other parameters remain the same, a larger $f$ leads to a smaller partition size. A small partition size gives a better chance for the processor allocator to find a free partition of nodes for a query. A large partition size indicates that, though there might be some low capacity nodes that are idle, they could not be assigned to a query. So queries may have to wait in the queue for a longer time. This increases the query waiting time, which increases the response time. On the other hand, assigning smaller partition to a query may reduce its waiting time, but execution time might increase.

Figure 4 shows the performance sensitivity to parameter $f$. The fragmentation size used in the experiment is 24000 bytes, which means partial operations are used. The system inter-arrival time is 6 seconds. Our experimental data (not shown here) indicate that as $f$ increases, the query waiting time decreases. This leads to decreased response time. When $f = 1$, though the waiting time is the smallest, the increase in its execution time increases the response time. Based on this data, we observe that the centralized architecture obtains the best performance when $f = 0.75$. For this reason, we have used $f = 0.75$ in all our previous experiments. It has also been observed that $f = 0.75$ provides the best performance even in distributed-memory multicomputer systems that schedule only CPU-intensive jobs [4].

## 5 Conclusions

This paper studied parallel query processing on network of workstations. Our main objective was to design a parallel query processing strategy for NOW-based systems. We have presented a three-level centralized architecture and evaluated its performance. Our study has shown that centralized architecture achieves good system speedups and scaleups. It has also been found that the system performance is sensitive to the fragmentation size. Small fragmentation size leads to a better chance of partial operation and increased parallelism. In this case, the mean response time decreases because of the decrease in query execution time. However, when the fragmentation size is too small, the overhead in scheduling large number of partial operations dominates the benefits obtained from this parallelism.

**Acknowledgements** The authors gratefully acknowledge the financial support provided by the Natural Sciences and Engineering Research Council of Canada and Carleton University. This work was done while the first author was at the School of Computer Science, Carleton University.

# References

1. T. Anderson, David Culler, David Patterson, and the NOW team, "A Case for NOW (Networks of Workstations)," *IEEE Micro*, February 1995, pp. 54–64.
2. S. Dandamudi, "Using Networks of Workstations for Database Query Operations," *Int. Conf. Computers and Their Applications*, Tempe, March 1997, pp. 100–105.
3. S. P. Dandamudi, "Sensitivity Evaluation of Dynamic Load Sharing in Distributed Systems," *IEEE Concurrency*, July-September 1998, pp. 62–72.
4. S. Dandamudi and H. Yu, "Performance Sensitivity of Space-Sharing Processor Scheduling in Distributed-Memory Multicomputers," *12th Int. Parallel Processing Symp.*, March 1998, Orlando, Florida, pp. 403–409.
5. S. Dandamudi and G. Jain, "Architectures for Parallel Query Processing on Networks of Workstations", *Int. Conf. Parallel and Distributed Computing Systems*, New Orleans, October 1997, pp. 444–451.
6. A. Giest, A. Beguelin, J. Dongarra, W. Jiang, R. Mancheck, V.Sunderam, *PVM: Parallel Virtual Machine—A Users' Guide and Tutorial for Networked Parallel Computing*, The MIT Press, 1994.
7. D. J. Lilja, "Exploiting the Parallelism Available in Loops," *Computer*, February 1994, pp. 13–26.
8. E. Rosti, E. Smirni, L. W. Dowdy, G. Serazzi, B. M. Carlson, "Robust Partitioning Policies of Multiprocessor Systems," *Performance Evaluation*, Vol. 19, 1994, pp. 141–165.
9. E. Rosti, E. Smirni, L. W. Dowdy, G. Serrazi, and K. C. Sevcik, "Processor Saving Scheduling Policies for Multiprocessor Systems," *IEEE Trans. Computers*, Vol. 47, No. 2, February 1998.
10. K. C. Sevcik, "Application Scheduling and Processor Allocation in Multiprogrammed Parallel Processing Systems", *Performance Evaluation 19*, 1994, pp. 107–140.
11. N. G. Shivaratri, P. Krueger and M. Singhal, "Load Distributing for Locally Distributed Systems," *Computer*, December 1992, pp. 33–44.
12. S. Zeng, *Centralized Architecture for Query Processing on Networks of Workstations*, MCS Thesis, School of Computer Science, Carleton University, 1998.

# Object-Oriented Database System for Large-Scale Molecular Dynamics Simulations

Jacek Kitowski[1,2], Dariusz Wajs[1] and Piotr Trzeciak[1]

[1] Institute of Computer Science, AGH, al. Mickiewicza 30, 30-059 Cracow, Poland
[2] ACC CYFRONET, ul. Nawojki 11, 30-950 Cracow, Poland
email: kito@uci.agh.edu.pl

**Abstract.** In the paper a model of the object-oriented database system is presented for archiving results generated with particles simulations and for retrieving simulation results from the database system for further processing.

## 1 Introduction

One of the most exciting tools for large scale computing is metacomputing. Metacomputer traditionally is characterized as a meta-tool offering three kinds of resources: high performance computing power, scientific visualization and graphics resources as well as storage capability. According to the progress in computer technology the first two elements have been rather well developed, while both efficient storage (including hierarchical management) and database management systems for archiving simulation results are still in development stage. For some kinds of large-scale simulations, especially those using particles applied for mezoscopic simulations (like Molecular Dynamics, MD, e.g. [1], Dissipative Particle Dynamics, DPD, e.g. [2], or Smoothed Particle Hydrodynamics, SPH, e.g. [3]), it is highly profitable to make postprocessing on a base of snapshots obtained during the simulation.

Effective management of the simulation results allows one:

- to study time-dependent snapshots,
- to choose the most representative snapshots for further analysis or for a movie,
- to calculate aggregate quantities on the base of one or more simulation runs,
- to perform more detailed insight into the results with scientific visualization packages (e.g. zooming, rotating, [4, 5], etc.).

The use of object-oriented paradigm in database systems is increasing due to the modeling power offered by the paradigm, which is not found in relational modeling [6]. Also progress in information technology makes possible use of complicated, abstract data types. Development of database systems for applications of more complex data in comparison to those applied in traditional relational database systems can be characterized with three threads:

- extended relational database systems to use abstract data with identifiers but without methods, e.g. [7],
- relational-object systems, in which abstract data are defined as classes; this pragmatic approach is applied for example in system Illustra [8],
- object-oriented systems, in which objects and methods are defined; this class of systems is the most popular at present in scientific applications, see, e.g. papers in [9].

The main purpose of the simulations using particles is to obtain trajectory for each particle placed within a simulation box. Since it is useful to store them for further analysis, this results in big quantity of information. For postprocessing and obtaining aggregate results the information has to be effectively archiving and maintained.

The goal of the paper is to present a dedicated Object-Oriented Database Management System (OODBMS) for archiving simulation results obtained by means of computer experiments and for retrieving simulation results from the database system for further processing. Although many OODBMS exist (see for example [10]) and new database architectures emerge (e.g. [11]), implementation of OODBMS for scientific data (potentially of different kinds) is still an issue of interest (e.g. [12]). Alternatively, storing and managing the simulation results can be organized without the object paradigm, using for example HDF for this purpose ([13]). However, the object paradigm is more adequate while complicated data structures and distributed approach are considered.

## 2 System Assumptions

The pilot version of the system for MD simulations has been developed with main elements presented in Fig. 1.

A simulation program stores raw simulation results to a disk. Each simulation run generates one or two disk files in which a simulation header and particles positions or particles momenta are stored in chosen subsequent time steps. The positions or momenta can be graphically presented using a visualization interface. Both, the simulation program and the interface have already been used independently [1, 5]. Within the reported project the object-oriented database and the application have been developed.

The purpose of the object-oriented application is

- to manipulate the data
  - generate a new database,
  - add new items to the system,
  - remove items from the system,
  - extract a given item,
  - browse the data.
- to preview and to transform the data
  - preview raw data,
  - choose the most representative part of the data for further archiving.

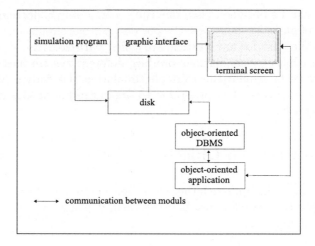

**Fig. 1.** Main elements of the system

Database management system (DBMS) incorporates object-oriented paradigm. Both, the DBMS and the application are written in C++. The presented solution is suitable for results obtained from 2D and 3D simulations; in the following description we will concentrate mostly only on 2D data for clarity.

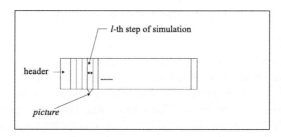

**Fig. 2.** A *system* file

### 2.1 Raw simulation results

The structure of the raw simulation results obtained from one run of the simulation program is presented in Fig. 2.

Such a set of results is called a *system* file. It consists of a header (free text, like a description of the run, simulation parameters, etc., and a table representing the particle kinds) and a collection of output data for chosen steps of simulation (in binary representation). Typically, for one timestep the following data are stored sequentially: the timestep number, $x$-coordinates for all particles and $y$-coordinates for all particles. Optionally, the second set of results (the second *system* file) can be obtained from one run – containing collection $v_x$ and $v_y$ as coordinates of particles momenta. Output data for one simulation step are called a *picture*, while data for any particle are called *particle* data.

Different computer simulations produce different *system* files. Since the experiments could be similar (for example simulation of a chosen phenomenon at different temperatures) it is profitable to organize *system* files into one set of results called a *file*.

## 3 Access Plan and Objects

In the object-oriented approach there is a choice between inheritance and encapsulation properties. The balance between them depends on types of the objects and the access plan.

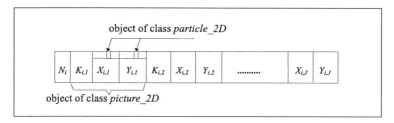

**Fig. 3.** Object of class *system_2D*

For complicated queries a consolidation operator may be required to arrange the results in an appropriate final form. The task of consolidation operator can vary from collecting the results of two operators at a time to collecting the results of all operators at once. The access plan depends on that requirement and consequently it can be classified into four categories (see e.g. [11]): *left-deep tree*, *right-deep tree*, *bushy tree* and *flat tree*.

One way of optimization is to reduce the height of the tree. The flat tree is a tree of height one. The consolidation operator can be heavily loaded. In simple cases with no complicated consolidation required it would be the best choice. Also, this category has been adopted in the presented system.

### 3.1 Classes of the objects

We define the following classes of the objects [14]

1. *file*,
2. *system_2D*, *system_3D*,
3. *picture_2D*, *picture_3D*,
4. *particle_2D*, *particle_3D*.

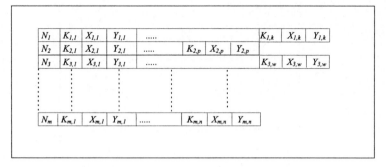

**Fig. 4.** Object of class file

Difference between classes <class_name>_2D and <class_name>_3D consists in dimensionality of the computer simulation – 2D and 3D respectively.

Schema of the class system_2D is presented in Fig. 3. $N_i$ is the header of the class system_2D being the $i$-th element of the object class (cf. Fig. 4). $K_{i,j}$ for $j = 1, \ldots, J$, is $j$-th timestep number, while $J$ is the simulation horizon for the $i$-th simulation run. Elements of the class picture_2D are vectors $X_{i,j}$ and $Y_{i,j}$ representing $x$ and $y$ particles coordinates respectively. Components of the particles momenta, $v_x$ and $v_y$, can be optionally stored into $X_{i,j}$ and $Y_{i,j}$ vectors.

Object of the class file is presented schematically in Fig. 4. It consists of a set of objects system_2D ($i = 1, \ldots, m$) for different simulation time horizons ($J \in \{k, p, w, \ldots, n\}$).

### 3.2  Class and query schema

The graph of the class file is presented in Fig. 5. There is attribute/domain dependence between objects representing the same simulation dimensionality (i.e. 2D or 3D). The inheritance property applies for 2D and 3D classes.

The query graph consists of the classes for which predicates are defined. The set of queries is depicted in Fig. 6. In our case the simple predicate of the form <attribute_name operator data> is applied with operator being the equality operator. Q1 represents more general query which allows to browse the general system parameters (like the header elements, cf. Figs. 2, 3). Q2 is a more detailed query which gives access to the desired picture. The query graph is a subgraph of the graph class.

### 3.3  Instance structure

The instance data format is presented in Fig. 7. Although originally the data access format represents data structure stored on the disk (cf. [6]), in our pilot implementation the original data structure is maintained (see Fig. 2), thus the data format in Fig. 7 defines the logical data format only. Such a strategy protects the simulation results against errors in the development stage of DBMS and saves disk space.

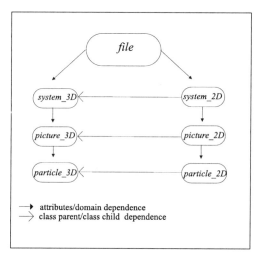

**Fig. 5.** Class schema

Q1: **select**:V **from** file: V
    **where**: V = { 2D, 3D }
    **and**: V system = "no.of system"

Q2: **select**:V **from** file: V
    **where**: V = { 2D, 3D }
    **and**: V system = "no.of system"
    **and**: V picture = "no.of picture"

**Fig. 6.** The queries

The UID field describes an unique identifier of an object. It consists of two parts. These are <class-identifier> and <instance-identifier>. <Class-identifier> describes membership to the class while <instance-identifier> represents position of the object in the class. For example for the objects of class file, UID is {<class_file>, <file_name>}, while for the class system_2D – {<class_system_2D>, <address>}. <address> is an element address of <class_system_2D>.

The field *length of object* is the size of the object. *Counter of attributes* represents number of attributes stored on the disk. Length of *vector of attributes*, $q$, is described by the *counter of attributes*, while its elements, $V_i$, are names of data or methods defined for the class. *Vector of addresses* consists of relative addresses, $P_i$, for $V_i$ values stored in *data* field which have been requested by the query.

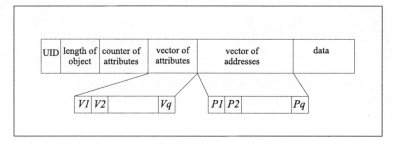

**Fig. 7.** Data access format

## 4 Presentation Level

The purpose of the presentation level is to preview raw simulation results or to choose representative results for further archiving. The view of the header of *system_2D* or *system_3D* (cf. Fig. 3) allows one to see a simulation description (like simulation parameters, temperature, number of particles of different kinds, potential parameters, external simulation field, and a comment). A simple visualisation procedure is applied for displaying particles positions with pixel colors representing the particles kinds. For momenta visualisation two methods are used. In the first method, pixel colors specify momenta values. In the second – momenta vectors are plotted on the screen for each particle separately while colors define the particles kinds, thus this method is suitable for rather small ensembles only.

In visualisation stage one has posibility to determine a part of the screen (a frame) with results of special interest. This frame can be static, i.e., the object of class *system_2D* is generated (for $j = 1, \ldots, J$) from particles whose coordinates belong to the frame area. In the dynamic frame approach the initial frame (for $j = 1$) is used for defining identifiers of the particles to be archived into the database despite their coordinates in next timesteps ($j = 2, \ldots, J$). The similar approach applies to 3D results with frame to cuboid extension.

## 5 System Evaluation

The purpose of the evaluation is to analyze the effectiveness of object access using the pilot implementation of the presented model and test data. The system is running on an Intel Pentium (100MHz) based PC computer with 16 MB RAM and 32-bit file system.

For the test case the number of objects of the class *system* is the same ($J = k = p = w = \ldots = n$, cf. Fig.4). The number of *particle* objects is mentioned by $\nu$.

In Tables 1 and 2 the average access time for objects is presented. For the *system* objects we examined the access to four different objects, each tested four times. No substantial difference in those measurements has been observed. For

| System kind | $\nu$ | |
|---|---|---|
| | 5000 | 15000 |
| system_3D | 0.6 | 1.9 |
| system_2D | 0.4 | 1.3 |

**Table 1.** Average access time (in s) for any *system* object ($m = 10$ and $J = 5$)

| System kind | $\nu$ | |
|---|---|---|
| | 5000 | 15000 |
| system_3D | 0.35 | 1.1 |
| system_2D | 0.26 | 0.8 |

**Table 2.** Average access time (in s) for any *picture* object from $i$-th *system* object ($m = 10$, $i = 5$ and $J = 5$)

the *picture* objects we also tested the access to four different pictures, again with no substantial differences in the access time. This feature makes possible averaging of the results and confirms the object paradigm.

The access time takes into account the access to the *system* header, $N_i$. In the structure of the header the table of the particle kinds is introduced (see section 2.1), with dimensionality equal to $\nu$. This results in (almost linear) dependency of the access time on $\nu$. High cost of removal of a *system* object is observed (equal to 48s for *system_3D* and 33s for *system_2D* for $\nu = 15000$) while addition of a *system* object is more efficient (taking 2.1s and 1.8s respectively).

## 6 Conclusions

We have presented a model and its pilot implementation of the object-oriented database for storing and managing results of computer simulations using particles. The proposed DBMS can be used for different kinds of simulation models, for example MD, DPD and SPH.

The simplest class of the object is the *particle*; the *picture*, the *system* and the *file* formulate other classes with attribute/domain dependence. The inheritance property is applied for classes concerning different simulation dimensionality (2D or 3D). Performance evaluation shows that the access time is independent of physical data organization and displays high cost of data removal. The first feature is typical for OODBMS while the second depends on the physical data organization. We suppose that the low measured performance suffers from the poor computer configuration used for the evaluation.

The plans for future development concern move the system to Unix servers, extension of object classes to GIFF or TIFF pictures of chosen simulation results, MPEG video for some simulation runs and introduction of more complicated queries. Probably these extensions will require distributed approach.

## 7 Acknowledgments

The work has been sponsored by Polish Committee for Scientific Research (KBN) Grant No. 8T11C 006 15.

# References

1. Mościński, J., Alda, W., Bubak, M., Dzwinel, W., Kitowski, J., Pogoda, M., Yuen, D., Molecular Dynamics Simulations of Rayleigh-Taylor Instability, in: Annual Review of Computational Physics, vol. V, 97-136, 1997.
   Alda, W., Dzwinel, W., Kitowski, J., Mościński, J., Pogoda, M. and Yuen, D.A., Complex fluid-dynamical phenomena modeled by large-scale molecular dynamics simulations, Comput.in Physics, **12**, 6 (1998) 595-600.
2. Dzwinel, W., Alda, W., and Yuen, D.A., Cross-scale numerical simulations using discrete particle models, *Molec.Simul.*, submitted.
3. Wingate, C.A., Dilts, G.A., Mandell, D.A., Crotzer, L.A., Knapp, C.E., and Libersky, L.D., Progress in smooth particle hydrodynamics, in: Idelsohn, S., Onate, E. and Dvorin, E. (Eds.), Proc.Int.Conf. Computat.Mechanics, Buenos Aires, June 29-July 2, 1998, CIMNE, Barcelona, 1998.
4. William, S.A., Sawley, M.L. and Cobut, D., Distributed run-time visualization and solution steering of parallel flow simulations, Comput.in Physics, **12**, 5 (1998) 493-502.
5. Kitowski, J. and Boryczko, K., Parallel visualization for molecular dynamics simulations, in: Garrido, P.L. and Marro, J. (eds.), Europhys.Conf. on Computat.Physics, Granada, Sept.2-5, 1998, Europhys.Conf.Abstracts, **22F** (1998) 244.
6. Kim, W., Introduction to Object-Oriented Databases, MIT press, Cambridge, Mass.,1990.
7. Delobel, C., Lecluse, C., Richard, P., Databases: From Relational to Object-Oriented Systems, Int.Thomson Publishing, 1995.
8. Stonebraker, M., Object-Relational Database Systems, Techncal Report, Illustra Information Technologies, Inc. (1990).
9. Sloot, P., Bubak, M. and Hertzberger, B. (eds.), Proc.Int.Conf. on High Performance Computing and Networking, April 21-23, 1998, Lecture Notes in Comput.Sci. **1401**, Springer (1998).
10. Chaudhri, A.B., Workshop report on experiences using object data management in the real world (extended version), http://www.soi.city.ac.uk/~akmal/oopsla97.dir/report.html.
11. Taniar, D., and Jiang, Y., A high performance object-oriented distributed parallel database architecture, in: Sloot, P., Bubak, M. and Hertzberger, B. (eds.), Proc.Int.Conf. on High Performance Computing and Networking, April 21-23, 1998, Lecture Notes in Comput.Sci. **1401**, Springer (1998), pp. 498-507.
12. Kimura, A., Sasaki, T., Shibata, A., and Takaiwa, Y., Perfmormance of an object-oriented database at a HEP experiment, KEK Preprint 97-23 (Japan).
    Le Goff, J.-M., Willers, I., McClatchey, R., Kovacs, Z., Martin, F., Zach, F., and Dobrzynski, L., Getting physics data from the CMS ECAL construction database, CMS Note 1998/087 (CERN).
13. Mościński, J., Nokolow, D., Pogoda, M., and Słota, R., Storing large volumes of structured scientific data on tertiary storage, Proc.Int.Conf. on High Performance Computing and Networking, April 12-14, 1999, Lecture Notes in Comput.Sci. (this volume), Springer (1999).
14. Trzeciak, P., and Wajs, D., Object database model, M.Sc. Thesis, Department of Electrical Engineering, Academy of Mining and Metallurgy, Kraków (1998) (in Polish).

# Virtual Engineering of Multi-disciplinary Applications and the Significance of Seamless Accessibility of Geometry Data

Vaibhav Deshpande [1], Luciano Fornasier [2], Edgar A. Gerteisen [1], Nils Hilbrink [1,3], Andrey Mezentsev [1], Silvio Merazzi [3] and Thomas Woehler [4]

[1] Swiss Center for Scientific Computing (CSCS), 6928 Manno, Switzerland
{vaibhav, egerteis, hilbrink, mezentse}@cscs.ch
[2] Daimler–Chrysler Aerospace AG, P.O. Box 80 11 60, 81663 Munich, Germany
luciano.fornasier@lm.otn.dasa.de
[3] SMR SA, C.P. 4014, 2500 Bienne 4, Switzerland
{hilbrink, merazzi}@smr.ch
[4] FhG–IPK, Pascalstrasse 8–9, 10578 Berlin, Germany
Thomas.Woehler@ipk.fhg.de

**Abstract:** The concept of virtual engineering (VEng) can be understood as a generalization of "multi–disciplinary problem solving", an ever more used term in scientific computing. An abstract space consisting of the physical, the geometrical, and the cost function directions, called CGP, is introduced. The VEng problem can be seen as a complex manifold embedded in this space. Common standard data formats, unified data access, as well as open, non monolithic systems, are discussed. These contribute to smoothing the sharp edges and closing the gaps of the manifold. The significance of seamless accessibility of data is illustrated by means of the tightly coupled fluid–structure interaction in aero–elasticity.

## 1 Introduction

The complexity of today's engineering simulations requires that HPC (High–Performance Computing) applications can be carried out directly by end–users. The engineer needs access to all simulation tools – ideally in a seamless environment – in order to meet the demand for significant reduction in design time as product complexity increases. The increased performance and lower prices of HPC platforms allow for much broader access to numerical simulation, and at the same time as numerial methods keep advancing, computational sciences move from focusing on distinct well–separated and closed problems to more end–product oriented simulation tasks. Virtual engineering (VEng) is an emerging field of scientific computing and many applications are of multi–disciplinary nature. Examples of such applications are the simulation of processes involving different physical models [11], fluid–structure coupling problems in aero–elasticity [3, 9, 21], or macro–level optimization combining micro–level simulations with multiple objective design parameters [10, 28]. However, the multitude of tools needed for a single complex simulation creates a number of bottlenecks and associated limitations, the major question here being how well these tools work together, and how to obtain optimal data transfer between them. Full benefits of HPC systems can only be obtained if greater efficiency is introduced into the interfaces at the end–user and the applications developer level.

The importance of *problem solving environments* (PSEs) for addressing present and future challenges in computational science and engineering is clearly identified [27, 28]. However, while emphasizing on computational steering, the practical significance of an efficient data management and common data formats as demonstrated in [11, 13, 15, 16, 29, 25] remains underestimated. In contrast, the backbone of the European project JULIUS [19] rigorously regards the data flow throughout the entire simulation process chain. Cutting edge industrial end–user HPC applications give the impetus to the developments within the JULIUS project [23], all of which require perfect access to all the involved data on every different level.

This paper presents ongoing activities with respect to the implementation of seamless accessibility of geometry data. The essential role of the data manager is discussed. It provides for modularity by enabling efficient data interaction in multi–disciplinary simulations. Accessibility of geometry data in these applications is crucial, and the corresponding preparation process is described in section 4 (CAD–repair). The related functional CAD interface, as discussed in section 5, allows for opening the mesh generation algorithms to the mathematics of any underlying geometry engine. Finally, the data preparation process for fluid–structure coupling illustrates the significance of data standards in an open environment. The corresponding multi–disciplinary application consists of tightly coupled fluid–structure interaction in aero–elasticity.

## 2 Virtual Engineering

The task of engineering a real–world object can be described as a complicated process where a set of trajectories pass through an abstract space spanned by the physical direction, the geometric direction, and the direction of a performance or cost function (see Fig. 1, [26]), henceforth called CGP space. The complete engineering problem then covers a certain domain within the CGP space and a specific solution is defined by a point in the space. The general goal of an engineering task will consist of searching for performance optima and cost function minima.

The engineering simulation task tries to follow a smooth path in order to find an optimum solution in the complete set of all feasible technical solutions. These are mathematically represented by a homomorphism in the CGP space. Figure 1 shows a typical example of the process chain of an engineering application. Initially, the geometry may be given in a specific machine–readable format or in the form of drawings resulting from 'artistic brain storming'. The domain known as pre–processing encompasses the whole preparation process for enabling a virtual description of the product (CAD), and for building a representative numerical model that can be simulated on the computer, e.g. a finite element or finite volume model. A parametrized computer model with geometric as well as with physical parameters can then be used as the basis on which to perform an optimization cycle with respect to a defined cost function. The optimization process finally results in a new geometry possibly together with physical parameters,

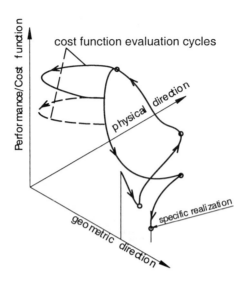

**Fig. 1.** Illustration of the computational virtual engineering (VEng) problem as a complicated constraint manifold (homomorphism) in the abstract CGP space which is formed by the geometric, physical and cost function coordinates

or with a set of physical and geometric parameter ranges according to the constraints of a multiple objective cost function [29]. In the ideal case, the new geometry may be used directly by a computer aided machining device for producing the final product.

By constantly enhancing HPC platforms, the simulation of models with a highly increased number of degrees of freedom, embedded in optimization or transient analysis cycles, becomes possible. However, while technical applications tend to become increasingly interactive with respect to the geometry, i.e. dynamic mesh adaptation, fluid–structure coupling, or geometric optimization, the corresponding tools for interfacing to the geometry do not keep in pace with the changing requirements. Thus, the virtual engineering scenario as described before is a highly idealized view. Each of the CGP space directions does already have its own specific problem areas with certain discontinuities, and the global abstract view of the complete system exhibits additional discontinuities or even disconnected regions. While some discontinuities can be resolved by the introduction of interfaces, the presence of disjoint regions – as happens with regard to geometry – leads to essentially ill–posed situations. Under this background, standard methods for the seamless access of geometry data are highly desirable.

The first attempts towards a definition of unified data formats were made already in the late 80's [4]. Standard data formats for CAD, meshes, and simulation results together with semi–interactive (batch executable) service modules led to a powerful tool for the integration of Problem–Solving Environments (PSEs) [13, 15, 16, 11]. As data carrier for these systems the MemCom database [30] proved highly efficient. A powerful tool in itself, the self–contained approach advocated by the environment of [4] however exhibited some incompatibility with respect to widely accepted geometry standards and formats. Apart from this minor flaw, the aforementioned modular data management approach clearly demonstrated the advantage of using common data formats in multi–disciplinary simulations. Here, cooperative design and development become prerequisites, and in addition to the traditional one–way pipeline (from preprocessing – to computation– to postprocessing) several feedback loops may appear.

## 3 Data Management

The Six–Sense (6S) environment of the JULIUS project [19] consists of a large number of different modules, e.g. CAD repair module, mesh generation and partitioning modules, Computer Supported Collaborative Work (CSCW) modules, and computational service libraries such as the fluid–structure interface. These modules are inter–operable through a common data format and a data management system for storing, retrieving, and manipulating data. While the data base is accessed by means of the MemCom data management system [30], data are distributed among the modules by a CORBA [24] based data–bus which steers and monitors in a heterogeneous network, and by Mem-Com for tightly coupled HPC applications, MemCom being a proven high–performance solution on HPC platforms.

### 3.1 The MemCom System

The MemCom data manager has been developed for numerical simulations like finite element analysis and pre– and post–processing in modular distributed environments. The requirements upon a data manager in numerical analysis are specific: It must be capable of dealing with large objects like vectors, matrices, and hyper–matrices. These typical data objects found in numerical analysis can be manipulated very efficiently within MemCom. MemCom stores data in standard binary form according to the ANSI/IEEE 754 standard either in 'big–endian' or 'little–endian' format, thus ensuring effi-

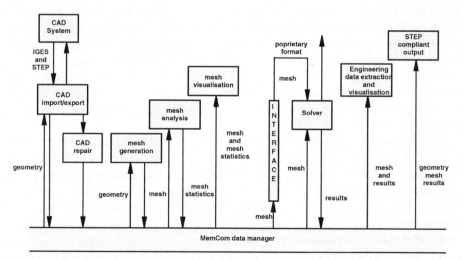

**Fig. 2.** Schematic data flow for the 6S environment. Although showing a typical scenario, the view is largely simplified. More complex feedback loops between the different modules occur in real world multi–disciplinary applications (see also section 6 and [12]).

cient access to binary data as well as portability. A MemCom data base can be accessed in stand–alone mode or by means of a client–server access model. In stand–alone mode an application is linked to the data manager server library, thus forming one single program. Data are then exchanged directly, i.e. without any intermediate control. This is the most efficient way of working with the data base, the only drawback being limited concurrent access. In the client–server model the application program attaches itself to one or several servers which in their turn will access the data base, thus ensuring concurrent data access, however at the expense of efficiency. A CORBA–based server is under preparation as part of the JULIUS project.

In contrast to other data management systems, MemCom does not only handle pure data base access operations, like store and retrieve, but it also allows for data management in the application program. Specific functions automatically load data objects to the application and data objects loaded in applications can be controlled one by one or in groups. Data can also be accessed in paged i/o mode, this feature being particularly important for i/o optimization or in cases where the data sets are larger than the available memory in a specific platform. MemCom does not provide a data design language: Applications interact directly with the data manager to create objects such as vectors, matrices, or associative arrays. Interaction occurs by means of the application program interface (API). Currently, a C/C++ and a FORTRAN API are provided. Likewise, a data browser allows for interactive monitoring of data bases.

## 3.2 6S Extensions

An outline of a typical data flow through 6S is shown in Fig. 2. Simultaneous access to the computational and geometrical data of the model is provided via the MemCom backbone. Inter–operability between the JULIUS CORBA architecture and the traditional client–server is achieved by introducing an additional controller. In practical terms the MemCom database has been hidden behind a hierarchical setup of classes which can be transferred, in parts or whole, to the requesting CORBA client [24]. Hence, the design is suited for both procedural and object oriented programming. It is to be stressed that CORBA in itself does not provide the means for efficiently implementing the required data structures, and that, in its present form, CORBA is not suited for high speed HPC communication. MemCom and CORBA are therefore to be considered complementary.

Additional extensions consist of application program interfaces to end–user customized data formats. These include, besides some specific traditional FE (finite element) data objects and the input of IGES or STEP files [22], an advanced geometry format derived from boundary representation (section 4). It is also worth mentioning the possibility to keep track of the relation between the original CAD data and the subsequent surface meshes (see Fig. 4, section 5). This information is established automatically during the mesh generation process. Thus, the open architecture, together with the CAD repair functionality and its advanced geometry data format accessible via MemCom, constitute an essential step towards virtual engineering.

## 4 CAD–Repair

The CAD model is imported into the 6S CAD–repair module by converting it to the ACIS boundary representation (BREP) geometry format [1], see Fig. 2. Due to their limitations, the exchange formats (such as IGES), presently do not exploit the high level model semantic of BREP and hence different errors such as gaps, incorrect topology of trimming curves, etc. get introduced. These errors lead to certain problems in the VEng downstream applications, the latter requiring continuity of the object boundary (FE, rasterization, CAM). Based on user supplied tolerances the repair software detects and heals defects semi–automatically [5, 2, 20], though even a completely waterproof model may still contain artifacts which are undesirable, here referred to as *badly meshable* geometries (BMG) as shown in Fig. 3. Another issue is model reduction and domain specific preparation. In fact, some of the details of a complete geometry model are of no interest for a domain of application, like fluid dynamics, but of great significance for another one, like structural mechanics. Thus, the initial CAD model is naturally split into a heterogeneous model. Care must then be taken to ensure that the original model hierarchy can be reassembled after having traversed a simulation cycle.

### 4.1 Software Design

As a consequence of the complexity and the multiple objectives of CAD–repair, automatic features are supplemented by direct user interaction. The CAD–repair module provides for three different and complementary user interfaces, a Graphical User Interface (GUI), a Scripting User Interface (SUI) and a 3D Interface (3DI), see (see [7, 17] for more detail). The reason for providing three different User Interfaces (UIs) is that each of these has its own strength: The 3DI provides proximity information for the area which requires user intervention, the GUI contains an object browser to provide complementary topological and non–viewable information of the same area, and the SUI provides access to the CAD–repair kernel functions. In addition, the SUI is a powerful tool for steering the CAD–repair process, allowing for modifying, extending, and tailoring the CAD–repair functionalities, and for interfacing with an external control or assistance module such as the Active Engineering Advisor of the 6S environment.

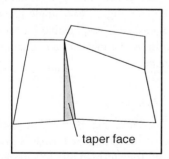
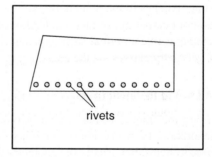

**Fig. 3**. Two examples of *badly meshable geometry* (BMG) parts. The taper face (left) is a typical example leading to undesirable point distribution in the mesh, and it can impact the p–completeness of typical incremental meshers. The rivets (right) may need to be resolved when targeting CSM, whereas a carpeting may be necessary for a CFD calculation

## 4.2 Model Healing

The absence of topological structure information for geometric elements in current transfer formats (such as IGES) leads to model semantic loss. However, the complete assemblage can also consist of several genuinely disconnected parts. In such cases, model reduction and preparation (MRP) is imperative for a reasonable definition of topological healing. In addition, the use of a 'pre–healed' model can be beneficial in the MRP phase. Thus, the different CAD repair functions are concurrent and all of them may depend on the downstream applications.

The geometric validity itself is attained by means of a two stage verification and repair scheme. The essential step during the verification stage consists of detecting face adjacency within the user–defined tolerances [20] discovering geometrical and topological errors, of BMGs, and of amending attributes to the data structure by marking the type of defect. After completion of the subsequent automatic repair process, the remaining model deficiencies need to be eliminated by direct user interaction. A multitude of different algorithmic solutions are available, and modularity is ensured by separating invariant verification from context dependent modules reflecting the requirements of specific applications.

## 4.3 Complementary Model Object Tree (CMOT)

While the practical engineer prefers to think in constructive units such as fuselage, wing, etc. (in case of a aircraft design), the healing algorithms are based on local proximity considerations. For combining both viewpoints, a complementary model object tree (CMOT) is introduced which, in addition to the ordinary geometry based entities (patches, lines, etc.) typifies two alternative elements – the constructive and the super–element. Typically, the constructive element is generated during the MRP phase, whereas the super–element is closely related to BMGs. The new elements also define a hierarchy, from constructive element via super–element to the trimmed surface patch.

Both new elements contribute to significantly improve the CAD repair process. Evidently, the proximity detection can be enhanced by assigning different tolerances to distinct constructive–elements. The role of the super–element consists of combining several distinct patches in the neighborhood of a BMG (see Fig. 3) to a single geometric entity without carpeting. In fact, the CAD–mesh interface is responsible for handling super–elements like a normal surface patch. In addition, the mesher can be tailored to specific geometry classes via the constructive element.

## 5 CAD–Mesh Interface

Maintaining the fidelity to the original geometry model is one of the basic issues of mesh generation. In theory, the accessibility of the native geometry format is given in monolithic systems, but usually the geometry data are transformed into an internal representation by converters (e.g. via IGES), if no access to the native format of the different geometry engines is available [18]. This conversion is commonly carried out with very high order splines to avoid too many dissimilarities from the original design. The other issue is that – although a certain fidelity of the geometry can be assured within the system – the relational information between the surface grid points and the corresponding geometry patches is lost when the mesh is written into the format of a specific downstream application program.

In fact, this information loss creates some immanent problems further down from the initial monolithic mesh generation. For instance, high fidelity of the geometry can not be assured during the application of mesh adaptation or alteration techniques, and the actual discrete representation potentially diverges from the original geometry. Interrelations are maintained by the CAD–Mesh interface in accordance with an open VEng system which offers smooth and seamless data access, along with the basic surface mesh and the original geometry data objects. This complementary information is referred to as reference tuples (Fig. 4). The surface mesh generation is typically carried out in the parametric space of generally trimmed surface patches. Thereby, the mesh density and the shape of elements in the physical space is controlled by the introduction of distribution functions and sometimes by taking into account the metric of the local mapping [6, 14]. Examples of typical geometry operations are the projection from parametric to physical space and vice versa, the computation of the normals in physical space, etc. The current interface design suggests a purely functional approach where the actual underlying geometry engine is hidden, access to the geometry being provided to the mesh generation developer in the form of a library.

Although the present developments are based on the standard 6S data formats – actually NURBS [22] for the geometry (Fig. 4) – the more abstract functional approach of the mesh interface should allow for establishing the connection between any mesh generator and any geometry engine. Having in place the standard interface specification, all tool developers can then independently develop their specific mathematical algorithms.

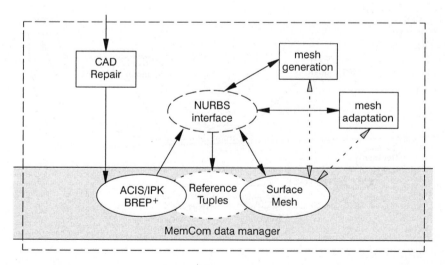

**Fig. 4.** Schematic data flow for the mesh generation CAD interface and its embedding into the JULIUS 6S environment. Note that the relation between the CAD and mesh data objects is maintained by the interface.

## 6 CFD–CSM Data Processor

As an example – others are given in [11, 12, 13, 15, 16] – of the significance of seamless data access throughout the CGP space, the fluid–structure (CFD–CSM) coupling is considered. A tightly coupled CFD–CSM approach is required in aero–elasticity where the interaction between computational fluid dynamics (CFD) and computational structural mechanics (CSM) takes place during the transient simulation. The interaction is such that the aerodynamic loads induce elastic deformations in the structure, the structural deformations induce a change in the external aircraft geometry, and the geometrical shape modifications lead to modified aerodynamic forces (Fig. 5). In addition, the modularity of the integrated application needs to be maintained. Accordingly, a separate processor in the form of an interface data preparation library is envisioned which makes use of the common data carrier MemCom [30].

Many references can be found which deal with the simulation of aero–elastic deformations, see for instance [3, 9, 21]. One common problem in the general case of more complex geometry models, however, is the link between the different domain specific representations of the respective solutions on the shared geometry parts. The points and elements of the CFD and CSM meshes are in general not coincident, usually they differ largely in density, size, and type. Simultaneously, ad–hoc interpolation of solutions leads to undesirable non–conservative effects during the transient simulation. A neutral layer is introduced to incorporate the original surface geometry. The subsets of the CFD grid and the CSM grid on the shared geometry parts are both mapped to the NURBS [22] parametric space. Subsequently, an integration scheme can be implemented which, to a certain extent, ensures the conservation of moments and forces. Vice versa, the structural deformations are represented in the geometry model by refitting the NURBS

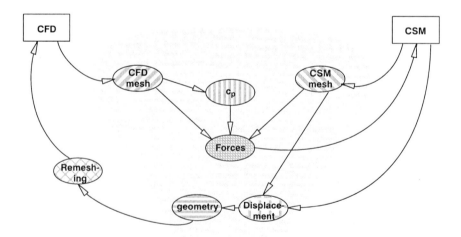

**Fig. 5.** Illustration of fluid–structure coupling (CFD–CSM). The communication process is depicted for one iteration cycle where the data flow is indicated by the arrows. The shaded region denotes the geometry based neutral interface realized in the NURBS parametric space.

according to the given displacements at structural nodes. After re–matching the mesh of the flow domain, one feedback loop is accomplished (Fig. 5). The illustrated coupling cycle, in turn, can be iterative if applied in an implicit fashion.

Prototype implementations have been carried out on the basis of a public domain geometry library [8] especially designed for multi–disciplinary applications. The tests were carried out on a standard wing geometry (Onera–M6) discretized by means of a structured mesh. The representation of the mesh and aerodynamic loads on a common parametric NURBS space is operational. Further work dealing with unstructured meshes is under investigation.

## 7 Conclusion

The present paper shows that multi–disciplinary applications require a higher abstraction level (CGP space) compared to the traditional task of establishing a theoretical model for a distinct physical problem. It has been discussed how common standard data formats, unified data access and open non–monolithic systems can contribute to smoothing the sharp edges and disconnected regions of a complete engineering problem.

The application of the different data management and geometry related tools has been illustrated by an aero–elasticity example. The extension of the MemCom data manager towards widely accepted standards in combination with the tools developed in the JULIUS project could potentially develop into a standard European framework. The BREP$^+$ geometry representation is an extension to presently available standards such as IGES. The JULIUS consortium strives to influence the emergence of future STEP developments.

## Acknowledgements

JULIUS is embedded in the Fourth Framework Programme, RTD in Information Technology Domain 6, HPC Simulation and Design, Primarily Task 6–3. The support of JULIUS by the CEC, as well as the financial contribution of the CEC, of the Swiss Government, and of the industrial partners is acknowledged.

## References

1. ACIS – 3D Toolkit User Guide, Spatial Technology, Inc., Version 4.0, 1998.
2. G. Barequet; Using Geometric Hashing to Repair CAD Objects, IEEE Computational Science & Engineering, 4(4), pp. 22–41, October–December 1997.
3. John T. Batina; Unsteady Euler Algorithm with Unstructured Dynamic Mesh for Complex Aircraft Aeroelastic Analysis; AIAA–89–1189–CP, 1989.
4. E. Bonomi et al., Astrid: A Programming Environment for Scientific Applications on Parallel Vector Computers, Plenum Press, New York , 1990.
5. G. Butlin, C.Stops; CAD data repair, Proceedings 5th International Meshing Roundtable, Sandia National Laboratories, pp. 7–12, October 1996.
6. O. Hassan, N.P. Weatherill, J. Peraire, K. Morgan and E.J. Probert; Generation and adaptation of unstructured meshes, COSMASE Course on Grid Generation, CERFACS, February, 1995.
7. V. Deshpande, E. A. Gerteisen, N. Hilbrink, S. Merazzi, A. Mezentsev; Sixth Sense (6S) – An integrated Problem Solving Environment, To appear in Proceedings of Swiss CAD/CAM Conference, 1999.

8. Spline Geometry Subprogram Library, Theory Document, Version 3.5, Boeing Information & Support Services, with contributions of D. R. Ferguson, Th. A. Grandine, F. W. Klein, R. Ames, Sept. 1997.

9. Computational Fluid Dynamics '98 – Session on *CFD in Fluid/Structure Interaction*; Fourth European Fluid Dynamics Conference held in September 1998, John Wiley & Sons, Ltd., 1998.

10. Computational Fluid Dynamics '98 – Session on *Impact of Evolutionary Computing in Flow Engineering*, Special Technology Session on *Real World CFD Optimization Problem*, Special Technology Sessions on *Industrial Design using Parallel CFD*; Fourth European Fluid Dynamics Conference held in September 1998, John Wiley & Sons, Ltd., 1998.

11. W.Egli, U. Kogelschatz, E.A. Gerteisen and R. Gruber, 3D Computation of Corona, Ion Induced Secondary Flows and Particle Motion in Technical ESP Configurations, J. of Electrostatics 40&41, pp. 425–430, 1997.

12. L. Fornasier, E. A. Gerteisen, S. Merazzi; Multi–Disciplinary Applications using the MemCom Data Management System, To appear in Proceedings of Swiss CAD/CAM Conference, 1999.

13. R. Gruber, W. Egli, E. Gerteisen, G. Jost, S. Merazzi; Problem–Solving Environments: Towards an Environment for Engineering Applications, SPEEDUP Journal 9, 1995.

14. P.L. George; Automatic mesh generation. Application to Finite Element Methods. Wiley, 1991.

15. E. Gerteisen, R. Gruber and S. Merazzi; Domain Decomposition: Theory and Practice, Computers in Physics, 11 (1), Jan/Feb 1997.

16. E.A. Gerteisen and Ralf Gruber; Computational Environment based on Domain Decomposition Approach, Proc. of 9th Int. Conf. on Domain Decomposition, John Wiley & Sons Ltd., 1996.

17. E. A. Gerteisen, N. Hilbrink, A. A. Mezentsev, Th. Woehler; CAD Repair for Discrete Simulation Methods, To appear in Proceedings of Swiss CAD/CAM Conference, 1999.

18. ICEM CFD/CAE, Installation Guide, ICEM CFD Engineering, CA, USA, 1996.

19. JULIUS (Joint Industrial Interface for End–User Simulations) Project, ESPRIT 25050, Technical Work Programme, Parts 1 & 2, October 1997 [ see also www.6S.org].

20. Krause, F.–L., Stiel, Ch., Lüddemann, J.; Processing of Data Conversion, Verification and Repair, Proc. of Fourth Symp. on Solid Modelling and Applications, 248–254, 1997.

21. R. Löhner, C. Yang, J. Cebral, J.D. Baum, H. Luo, D. Pelessone C. Charman; Computational Fluid Dynamics Review 1995, John Wiley & Sons, 1995 (ISBN 0–471–95589–2).

22. L. Piegl and W. Tiller; The NURBS book, 2nd Edition, Springer Verlag, 1997 (ISBN 3–540–61545–8).

23. John Murphy, David Rowse, Luciano Fornasier, Philippe Thomas, Argiris Kamoulakos and Edgar A. Gerteisen; JULIUS Project – ESPRIT 25050 Joint Industrial Interface for End–User Simulations;, CossCuts – the newsletter of CSCS/SCSC, Vol. 7, No. 2, October 1998.

24. R. Orfali, D. Harkey, J. Edwards; Instant CORBA, John Wiley & Sons, 1997 (ISBN 0–471–18333–4)

25. Diane Poirier, Steven R. Allmaras, Douglas R. McCarthy, Matthew F. Smith, Fancis Y. Enomoto; The CGNS System, AIAA–98–3007, 1998.

26. Pierre Perrier, New data–base concepts for CFD validation; ECCOMAS 1998, J. Wiley & Sons, 1998.

27. S. G. Parker, C. R. Johnson, and D. Beazley; Computational Steering Software Systems and Strategies, IEEE Computational Science & Engineering, 4(4), pp. 50–59, October–December 1997.

28. J. R. Rice, R. F. Boisvert; From Scientific Software Libraries to Problem–Solving Environments, IEEE Computational Science & Engineering, 3(3), pp. 44–53, 1996.

29. D. Spicer, J. Cook, C. Poloni, P. Sen; EP20082 Frontier: Industrial multiobjective design optimization, Fourth European Fluid Dynamics Conference, pp. 546–551, John Wiley & Sons, 1998.

30. S. Merazzi; The MemCom User manual, B2000 Data Access and Data Description Manual, SMR Corporation, Bienne, Switzerland, 1995.

# Some Results from a New Technique for Response Time Estimation in Parallel DBMS

Neven Tomov[1], Euan Dempster[1], M. Howard Williams[1], Albert Burger[1], Hamish Taylor[1], Peter J.B. King[1], and Phil Broughton[2]

[1] Department of Computing and El. Engineering,
Heriot-Watt University, Edinburgh EH14 4AS, UK
[neven,euan,howard,ab,hamish,pjbk]@cee.hw.ac.uk
[2] International Computers Limited,
High Performance Technology,
Wenlock Way, West Gorton, Manchester M12 5DR, UK
pb@wg.icl.co.uk

**Abstract.** The need for tools for performance prediction of parallel database systems is generally recognised. One such tool which has been developed (Steady) is based on analytical techniques to obtain a rapid estimate of performance. The approach to predicting response time involves a heuristic approximation coupled with standard queueing solutions. This paper reports on preliminary results for both maximum transaction throughput and response time obtained in comparing this approach against actual measurements.

## 1 Introduction

The growth of commercial interest in the potential and use of parallel computers for running relational database applications has been noted by Norman and Thanisch [1]. Such applications offer rich amounts of exploitable parallelism that database management systems running on parallel platforms can take advantage of. Several well known commercial DBMSs such as Oracle [2], Informix [3], Ingres [4], Sybase [5] and DB2 [6] are now available on popular and dedicated SMP and MPP machines.

The ability to predict the performance of parallel relational databases is important for their application sizing, capacity planning, data placement and performance tuning. To assist in these processes, an analytical technique for estimating the performance of parallel relational database systems has been developed. The technique has been incorporated in a tool [7, 8] which can rapidly estimate how relational database applications will behave without running lengthy simulations or actually trying them out.

Part of the performance estimation technique involves prediction of query response time, which is a non-trivial problem. This paper reports on experiments in which the results of the technique are compared with actual measurements obtained from a parallel platform. The experiments are performed as part of the process of validation of the analytical approach.

The remainder of this paper is organised as follows. Section 2.1 briefly describes the hardware platform (the ICL GoldRush MegaSERVER [9]) and DBMS (Informix XPS [3]) experimental setup. This is followed by a description in section 2.2 of a subset of the queries and data used in the comparisons. Section 2.3 provides a brief introduction to Steady – the parallel DBMS performance prediction tool. The response time prediction technique incorporated in Steady is discussed in section 2.4. Section 2.5 explains the methods used in obtaining performance figures. Some comparison graphs of actual vs. predicted transaction throughput and query response time are given in section 3. Finally, section 4 provides a summary and conclusions.

## 2 Obtaining Performance Measures

The ICL GoldRush MegaSERVER [9] is a parallel platform developed as a backend database server to host several different database systems (Ingres, Oracle, Informix). Steady is a performance prediction tool designed specifically for shared-nothing parallel architectures [10] and calibrated for the ICL GoldRush platform running Informix XPS.

### 2.1 GoldRush and Informix XPS

The basic GoldRush hardware architecture consists of a number of Processing Elements (PEs) and Communication Elements (CEs) linked by a high speed DeltaNet network. Each PE is connected to its own disc storage subsystem and runs a UNIX operating system and DBMS code. A CE provides external links for GoldRush to clients via LANs. The particular GoldRush setup used here has 1 CE, 8 PEs, 6 discs per PE and a cache of 16MBytes on each PE.

This type of architecture is an ideal platform for the Informix Extended Parallel Server(XPS) [3]. It contains a set of internal components called co-servers which are installed on each of the PEs of GoldRush. A co-server provides a user entry point to GoldRush on a given PE. Locks on data items and all data processing are managed locally within each co-server. Where deemed suitable, single queries are broken into subtasks and processed concurrently by threads within a single co-server and across co-servers. Data can be partitioned across discs and PEs so that parallel I/O operations can take place. Additionally, certain partitions, known to be irrelevant for a particular query, can be skipped altogether.

### 2.2 Tables and Queries

The experiments detailed in this paper are carried out on a subset of the AS3AP benchmark [11] tables. In particular, a number of variations of the *uniques* relation are used; the attributes of *uniques* are given in Table 1.

There are four *uniques* relations used. The first is called *uniques30k*. It has 30000 tuples and is placed on one disc of one PE (PE 7). The second relation

Table 1. Attributes of AS3AP *uniques* relation

| key | integer(4) | decim | numeric(18,2) |
|---|---|---|---|
| int | integer(4) | date | datetime(8) |
| signed | integer(4) | code | char(10) |
| float | real(4) | name | char(20) |
| double | double(8) | address | varchar(20) |

is called *uniques80* and has 80 tuples which are also stored on one disc of PE 7. The other two relations are *uniques270k* and *uniques540k*. They contain 270000 and 540000 tuples respectively and are placed across all eight PEs. Each of these two tables is fragmented into 8 fragments by a simple hash function on the *key* primary key attribute with one hash fragment placed on a single disc of each of the PEs. Each tuple has a unique value for attributes *key* and *int*. Attribute *signed*, however, is modified from the benchmark specification so that tuples have *signed* values in the range 1 to 10 with an equal number of tuples for each value.

A number of experiments have been performed and two will be presented here as typical. The first query performs a simple aggregation of a *uniques* relation, looking for the maximum and minimum values of the *int* attribute, for those tuples whose *signed* value is not 1. Thus the aggregation is carried over $9/10^{ths}$ of the *uniques* tuples. Tables *uniques30k* and *uniques270k* are used:

SELECT max(*int*), min(*int*)
FROM *uniques30k* (*uniques270k*)
WHERE *signed* not in (1)

The second query finds the maximum value of the *int* attribute from the result of a join of *uniques80* with *uniques540k* where the *int* values are the same. The tuples of *uniques80* are taken from *uniques540k* so that the size of the resulting join is 80 tuples:

SELECT max(*uniques80.int*)
FROM *uniques80*, *uniques540k*
WHERE *uniques80.int* = *uniques540k.int*

Although this could be handled with a much simpler query, this form was used in order to perform a join of this type and which would facilitate measurements of this operation.

Table *uniques30k* is placed on one PE and queries involving only it are not parallelised. Those involving tables *uniques270k* and *uniques540k*, however, are performed in parallel by all co-servers containing the table data. The second query employs a hash-join algorithm, which builds a hash table across all co-servers using *uniques80*, and subsequently probes this with the tuples from *uniques540k*.

## 2.3 Steady

STEADY [7, 8] (System Throughput Estimator for Advanced Database sYstems) is an analytical tool for performance estimation of parallel relational database systems. It has been designed to handle a range of different platforms although thus far it has only been calibrated against the GoldRush platform. Apart from a graphical user interface, it consists of five major modules:

1. The **Profiler** is a statistical tool primarily responsible for generating base relation profiles and estimating the number of tuples resulting from data operations;
2. **DPTool** is used to generate various data placement schemes for a parallel database using different strategies;
3. The **Modeller** is responsible for producing the profile of the tasks required for a particular benchmark or query with assistance from the Profiler, query paralleliser and the cache model [12];
4. The **Evaluator** takes the task profiles and produces resource usage profiles, maximum system throughput values and system bottlenecks;
5. The **Response Time Estimator** takes the resource usage profiles and from these estimates response times.

Relations are partitioned into fragments by DPTool using declustering methods such as *hash*. These fragments are then allocated to the PEs using placement methods such as *size, bubba, hua, etc.*. These relation fragments can then be assigned to the discs of PEs according to various methods. Based on the generated data placement and the chosen DBMS architecture, a profile of the tasks required for a particular benchmark or query is generated by the Modeller which includes an estimation of disk I/O requirements in terms of the number of pages read and written to each disk. This is derived on the basis of the estimated cache hit ratios for the particular relation fragments in the cache of each PE and is carried out by the cache model. The task profiles are then converted by the Evaluator into resource utilisation profiles, from which the system bottlenecks and the maximum throughput rate are determined. This gives the user an indication of the upper limit of system capacity in terms of throughput. The resource usage blocks are then fed into the Response Time Estimator which estimates the response time of the query, as described briefly in the following section.

## 2.4 Response Time Technique

The input to the Response Time Estimator module of Steady consists of the original query task profiles represented as patterns of resource consumption incurred on the various GoldRush resources such as CPUs, discs, and interconnect. This representation is equivalent to an open multi-class queueing network [13]. In general, however, the networks obtained are not in product form, due to the non-exponential service times required at each resource. This means that exact solutions for quantities such as the mean response time cannot be obtained analytically.

To overcome this problem, a heuristic is proposed in [14]. It specifies a procedure for labeling each resource in the queueing network as either an M/M/1 or an M/G/1 resource. The decision is based on knowledge of the resource's utilisation and relative visit ratio [13]. With each resource labeled, the network can be solved to obtain the mean response time of individual queries.

## 2.5 Taking Measurements

Calibration of the models required running a selection of queries, using the parallelised query execution plans to determine the way in which they were broken down by the system, and reconciling these against measurements obtained using the Informix XPS performance measuring tool *onstat*.

Once this was complete, a transaction generator was created to emulate a parallel database system workload with many users independently querying the data. It was used to fire single-query transactions from the communication element (CE) to a given co-server at a specified rate. Two generators were created. The first fired transactions with constant inter-arrival times and was used to determine the arrival rate for which the maximum throughput of the machine can be achieved.

The second version of the transaction generator fired transactions with exponentially distributed inter-arrival times and was used to obtain transaction response times.

Figure 1 shows the response times for query 1 on *uniques30k*. The query is fired 100 times with exponentially distributed inter-arrival times at a rate of 0.04 transactions per second which corresponds to 11% of the predicted maximum throughput.

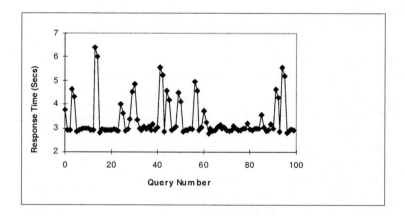

**Fig. 1.** Individual response times for 100 queries of type 1 with a table of 30K tuples. Query arrivals are Poisson with rate 0.04 (11% of predicted max throughput).

## 3 Results

Figures 2 and 3 show the throughput results for query 1 with the 30K tuple table and the 270K tuple table, respectively. Note that since the 270K table is spread across 8 nodes (i.e. each node has approximately 30375 tuples) the performance achieved is similar to that for the 30K table which is placed on a single node. Figure 4 shows the results for query 2. Similarly, Figures 5 and 6 show the response time results for query 1 with the 30K tuple table and the 270K tuple table, respectively. Figure 7 shows the response time results for query 2.

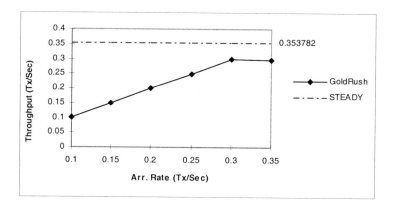

**Fig. 2.** Throughput of query 1 with a table of 30K tuples.

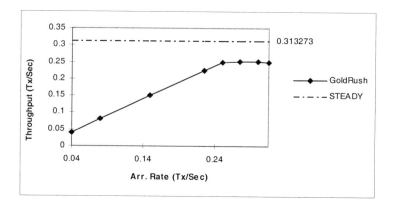

**Fig. 3.** Throughput of query 1 with a table of 270K tuples.

The figures show that the maximum throughput achieved on GoldRush is slightly less than that predicted by Steady by about 13%. We believe that this is due to additional operating system overhead which we were not able to isolate during the calibration process. Similarly response times observed were slightly larger than those predicted by Steady — less than 20% for arrival rates less than 70% of maximum throughput (as predicted by Steady), and less than 30% for arrival rates less than 90% of maximum throughput.

A number of other examples which have been investigated produce results which are similar to those for the queries presented here. Error margins are roughly the same for the cases investigated so far.

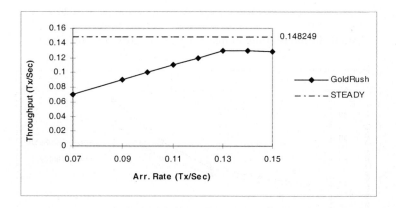

**Fig. 4.** Throughput of query 2 with a table of 540K tuples.

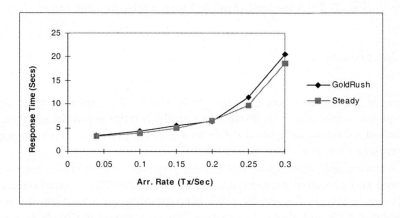

**Fig. 5.** Response time of query 1 with a table of 30K tuples.

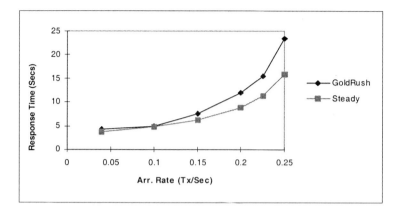

**Fig. 6.** Response time of query 1 with a table of 270K tuples.

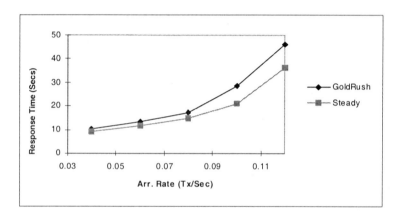

**Fig. 7.** Response time of query 2 with a table of 540K tuples.

## 4 Conclusions

This paper has reported on experiments in which the results of an analytical technique for performance estimation in parallel relational DBMS are compared with actual measurements obtained from a parallel platform. Both throughput and response time results are presented.

The results are encouraging and generally the relative difference between the measured and predicted performance is less than 20%. The relative error in the predictions is largest for arrival rates corresponding to a bottleneck resource utilisation of 80% and above. In practice, however, one is not likely to load the system much beyond 70% utilisation – thus 80% is a reasonable upper limit for performance prediction beyond which queues rapidly become unmanageable. The fact that Steady tends to predict lower response times and higher maximum throughputs can be attributed to additional resource consumption by the OS which is presently not taken into account by the analytical model.

## Acknowledgements

The authors acknowledge the support received from the UK Engineering and Physical Sciences Research Council (EPSRC) under the PSTPA programme (GR/K40345) and from the Commission of the European Union under the Framework IV programme (Mercury project). They also wish to thank Arthur Fitzjohn and Monique Mitchell of ICL(Manchester).

## References

1. Norman, M.G., Thanisch., P., Parallel Database Technology, Bloor Research Group, Netherlands (1995) 1-546
2. Oracle Corporation., Oracle7 Parallel Server Concepts & Administration, Release 7.3, Oracle, Server Products, California, USA (1996)
3. Informix Software Inc., Informix OnLine Dynamic Server administrator's guide (1994)
4. The ASK Group Inc., ASK OpenINGRES database administrator's guide (1994)
5. Sybase Inc., Sybase, http://www.sybase.com (1998)
6. IBM Corp., DB2, http://www.software.ibm.com/data/db2 (1998)
7. Zhou, S., Williams, M.H., Taylor, H., Practical throughput estimation for parallel databases, IEE Software Engineering Journal (July 1996)
8. Williams, M.H., Dempster, E.W., Tomov, N.T., Pua, C.S., Taylor, H., Burger, A., Lü, J., Broughton, P., An Analytical Tool for Predicting the Performance of Parallel Relational Databases, submitted to Concurrency: Practice and Experience (1998)
9. Watson, P., Catlow, G., The architecture of the ICL GoldRush MegaSERVER, Proceedings of the 13th British National Conference on Databases (BNCOD 13) Manchester U.K. (July 1995) 250–262
10. Stonebraker, M., The case for shared nothing, Database Engineering 9(1) (March 1986)
11. Turbyfill, C., Orji, C., Bitton, D., AS3AP: An ANSI SQL Standard Scaleable and Portable Benchmark for Relational Database Systems, The Benchmark Handbook, Second Edition (1993), Gray, J. (editor)
12. Zhou, S., Tomov, N., Williams, M.H., Burger, A., Taylor, H.: Cache Modelling in a Performance Evaluator of Parallel Database Systems Proceedings of the Fifth International Symposium on Modelling, Analysis and Simulation of Computer and Telecommunication Systems (January 1997) 46-50
13. Molloy, M.: Fundamentals of Performance Modeling Macmillan Publishing Company (1989)
14. Tomov, N.T., Dempster, E.W., Williams, M.H., King, P.J.B., Burger, A.: Approximate Estimation of Transaction Response Time, to appear in The Computer Journal

# PastSet – A Distributed Structured Shared Memory System

Brian Vinter, Otto J. Anshus, Tore Larsen

Department of Computer Science, University of Tromsø
{vinter, otto, tore}@cs.uit.no

**Abstract.** The architecture and performance of a structured distributed shared memory system, PastSet, is described. The PastSet abstraction allows programmers to write applications that run efficiently on different architectures from four-way SMP nodes to larger clusters. PastSet is a tuple-based three-dimensional structured distributed shared memory system, which provides the programmer with operations to perform causally ordered reads and writes of tuples to a virtual structured memory space called the PastSet. PastSet is specially implemented to utilize physical shared memory where available and distributed memory otherwise. The contribution of the PastSet model is good performance combined with ease of programming new and porting existing applications. It has been show in [1] that a shared memory version of PastSet is able to outperform System V IPC on the same platform, both multiprocessors and uni-processor systems. We have also previously shown that running on a cluster of multiprocessors, PastSet was able to outperform MPI and PVM when running real applications.

## Introduction

Structured distributed shared memory was first introduced in Linda [2], but the interest into structured shared memory has since been low. The general consensus is that it is not possible to achieve performance comparable to that of simpler systems. We started the work on PastSet with the idea that bad performance is not inherent to the concept of structured shared memory, and that good performance can be achieved by paying close attention to design and implementation. As we believed that the ability to utilize physical shared memory if available was essential, we started out by first developing an efficient version of PastSet on symmetrical multiprocessors. We then used this implementation as the basis for the distributed version described in this work. The non-distributed version is implemented entirely at kernel level in the Linux 2.0.30 kernel, while the distributed version is implemented at user level and utilizes the kernel level version.

A small set of applications has been ported to the PastSet model. Their performance using PastSet as well as the corresponding performance of the same applications using several other approaches (sequential, multithreaded, MPI and PVM) on two different cluster architectures can be found in [3]. We go on to briefly describe PastSet, then we describe the architecture of the distributed version and continue to describe the different data-distribution models that have been implemented. We go on to describe a set of performance experiments and their outcome, and finally we briefly discuss related work, future work and we end with a conclusion.

## PastSet

PastSet was first introduced as an interprocess communication paradigm in [4]. The paradigm resembles that of Linda, but with some significant differences. Tuples are generated dynamically based on tuple templates that may also be generated dynamically. A tuple template is an ordered set of types. Tuples based on a particular template has an ordered set of variables matching the types in the template. Each type in a tuple template has an associated value-space, or dimension, describing the set of all possible values for that type in this template. Taken together, the types in a template spawns a space encompassing all conceivable type-value combinations for tuples based on that template. A tuple with all singular values represents a singular point in this space.

As with Linda, PastSet supports writing (called `move`) tuples into tuple-space and reading (called `observe`) tuples that reside in tuple space. Contrary to Linda's `in` operation, PastSet `observe` does *not* remove tuples from tuple space, the tuple that is observed is marked as observed but remains in the PastSet so that it can be read again if specified so directly. No mechanism is provided to remove individual tuples from PastSet. All PastSet operations return only when the operation has completed, or an error has been detected, no asynchronous calls exist.

In PastSet, each set of tuples based on identical templates is denoted an element of PastSet. An element may be seen as representing a trace of interprocess communications in the multidimensional space spawned by the tuple template. PastSet preserves the causal order among all operations on tuples based on the same or identical templates. There is no ordering between tuples of different elements. Tuples that match the same type, but which the programmer does not wish to place in the same element can be differentiated by an initial flag.

In effect, PastSet keeps a causally ordered log of all tuples of the same or identical templates that has existed in the system. This also allows the processes to re-read previously read tuples.

It is the intention that the added semantics of PastSet will allow programmers to more easily create parallel programs that are not of the traditional 'bag of tasks' type.

Two pointers `First` and `Last` are associated with each element in PastSet. `First` refers to the elements oldest unobserved tuple. `Last` refers to the tuple most recently moved into the element. A parameter, `DeltaValue`, associated with each element in PastSet defines the maximum number of tuples allowed between `First` and `Last` for that element. A process may change `DeltaValue` at any time. The `move` and `observe` operators update `First` and `Last`, and obey the restrictions imposed by `DeltaValue` for each element in PastSet.

Functionality is provided to truncate PastSet on a per element basis, permanently removing all tuples that are older than a given tuple

## Architecture

The distributed version of PastSet is based on the shared memory implementation, and is focused on achieving high performance. To minimize overhead the architecture comprises two parallel paths, one that executes PastSet operations that may be completed locally and one for operations that requires remote operations.

Since the nature of PastSet focuses on the elements of PastSet, the distribution also focuses on the elements. Each element exists on one node only and is not replicated. An operation on elements that are located on the same node as the process that performs an operation is called a local operation. Operations on elements that are located on other nodes are called remote operations. Thus PastSet supports remote observes and remote moves, similar to remote read and remote write on other DSM systems. Remote operations are wrapped and communicated to a PastSet server on the remote node using TCP/IP.

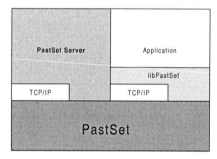

**Fig. 1.** Layout of a node that supports PastSet

The PastSet application library handles access to PastSet, including multiplexing between local and remote execution paths. Two versions of the library are available, one single threaded version and one that supports multi-threaded applications.

## Data Location

To achieve low overhead on operations it is imperative that applications can easily determine, at runtime, whether an operation should execute locally or remotely, and in the latter case, which node should perform the remote operation. To this end each element is named by an integer of which the first 16 bits denotes the server that holds the element. All PastSet operations, except the Enter operator, are called via macros that test whether the element is local or remote. In the case of a local element the macro calls the PastSet extension to the kernel directly. For a remote element, the call is redirected to a PastSet application library procedure, which send the operation to the respective server and then reads and returns the answer. The use of a macro for this multiplexing essentially reduces the overhead for a local operation to a single 'if' statement.

## The PastSet Server

Each node in the cluster that holds parts of PastSet[1] runs a PastSet server that services remotely issued PastSet operations on local elements. As the PastSet operations move and observe are potentially blocking, the server must be able to serve several blocking operations simultaneously. It is easily seen why simply performing one operation at a time and queuing other operations would allow false deadlocks[2] to occur. This problem is solved by making the PastSet server multi-threaded, so that once a thread blocks, another thread is activated. The PastSet server starts out with a small pool of threads. When the number of idle threads drops below a certain threshold, by default 1, a new thread is created. The scenario where creation of a new thread fails is currently just detected and reported. None of the applications ported and tested so far has approached the limit of 256 threads.

As elements are denoted via integers, one could easily imagine the case where a malicious process could guess the number of an element thus ruining the security of the system. This is solved by a translation table in the PastSet server between elements that the calling process has entered, and the actual element ids. Assuming the identity of another process is not feasible as the system uses connected communication. By default PastSet does not provide additional security to the system, thus the security of PastSet is based on the security of the remaining system, so that if a process can eavesdrop on the network security is non-existing.

## PastSet Application Library

A user level library is available to interface between applications and PastSet. As previously mentioned this library provides the execution path multiplexing between operations for local and remote execution. As local operations are simply identified by an 'if' statement and then performed as a direct kernel call, this case is trivial. When an operation can not perform locally, a library call completes the operation. Tuples are also generated via the library, and are allocated in such a way that additional space for a communication header is automatically added in front of the actual tuple data. This way operations need not two send operations, i.e. one for operation header and one for data.

The library exists in two versions, one for sequential workers and one for multithreaded workers. The non-threaded version is straight forward and synchronous, i.e. when an application issues a remote operation, control shifts to the library which sends the request and then performs a blocking read for the result, once it has arrived the function returns. The multi threaded version can not perform like this, as results need not arrive in the order requests were sent. Instead, once a request has been sent, the thread goes to sleep. A service thread is constantly listening to all channels to the servers, once a reply comes in, it reads the id of the thread that issued the corresponding request and wakes the thread, which then reads the reply data if any

---

[1] The design of PastSet allows nodes that do not support PastSet locally to participate in the PastSet virtual machine.

[2] By false deadlocks we mean a deadlock that occurs due to the runtime system, rather than a deadlock in the application.

exists and continues its execution. This approach has been chosen for two reasons, first of all the service thread cannot know the size of reply before it arrives, thus the reply must be read in two operations. First waking the destination thread, which then reads the reply data and then signals the service thread that the channel is available again, does add extra overhead relative to the alternative, which is that the service thread reads the whole reply and then wakes the destination thread. However, on multiprocessors the latter scheme would read the data onto the cache of one CPU, and schedule the receiver on another. Thus the second reason for the choosen approach id that migrating the data easily becomes more expensive that the synchronization overhead.

## Distribution

A central part of the distributed PastSet is the distribution of elements. Elements are placed on a server the first time a process enters the dimensions that specify the element. To this end every application that uses PastSet is connected to a central name-server that keeps track of all existing elements. When an application issues an EnterDimension operation, the PastSet library first contacts the name server to find out where the element is or should be located, and then it sends the EnterDimension operation to the corresponding server. This way different data-distribution models can easily be implemented by changes in the name-server only. Currently we have implemented three distribution models: Central-Server, Round-Robin Server and First-Touch.

The Central-Server model does not distribute data at all, but centralizes PastSet on one machine. This model does not scale well, but serves as a reference for the other models. The Round-Robin Server tries to balance the load of the elements by letting each node hold every $n$'th element. This way the load on the servers is equally distributed, assuming the elements are used in a uniform way. In the First-Touch model an element is created on the node of the process that first enters the element. Under the assumption that not all elements are accessed uniformly by all processes, the programmer can take advantage of the First-Touch policy by having an element placed at the node where the most activity of the element will be initiated, thereby improving both local and global performance by reducing unnecessary network traffic.

## Performance

Good performance is vital to all programming paradigms. To measure the basic performance of PastSet we have designed three micro-benchmarks that measure the latency and bandwidth of operations on the PastSet. We have previously shown that PastSet can outperform MPI and PVM at application level, but in this work we focus on identifying the cost of the raw-operations, so that one may consider them when writing applications.

All experiments are run on a cluster of 8 HP LX-Pro Net-servers, each having 4 166MHz Pentium Pro CPUs and 128MB of main-memory. The nodes in the cluster are connected via a 100VG net card (100Mb/sec) connected to a hub.

**Latency**

In distributed shared memory systems it turns out that the latency of communication, i.e. the time it takes from a package is sent until it is received, is of significance to the task grain size that can be supported efficiently by the system. The higher the communication latency is the coarser the grain size must be for good performance. A lower latency allows finer grained parallelism to perform efficiently. For instance, the Internet project of solving the original RSA cryptography challenge communicated via an e-mail system over a global network. In this case latency was extremely high, but the parallelism was trivial, e.g. no communication required during execution. On a system like an SMP machine where the communication latency is lower than access to main memory, i.e. cache to cache, parallelism can be very fine grained.

To determine the latency of PastSet operations we look at two basic cases, one where two processes passes a token of varying size back and forth, and one where an increasing number of nodes form a ring that passes a token around. The first experiment will be executed in three scenarios. Firstly where the elements used for communicating and the two processes are placed on the same node. Secondly, where one of the processes is on a remote node. Thirdly, where the elements and the processes are all placed at different nodes. For the ring experiment we will let the PastSet place the processes automatically and repeat the experiment with a few different package sizes. Pseudo code for the ring experiment core looks something like:

```
observe();  //Observe the initial token
get_start_time();  //Get timestamp
for(i=0; i<1000; i++){
  move();  //Send token to right neighbor
  observe();  //Get token form left neighbor
}
get_stop_time();
```

**Fig. 2.** Pseudo-code for the ring latency test.

Figure 3 shows the round-trip latency for a package of varying size cycled between two processes. Not surprisingly the local ping-pong is extremely fast, while having one process located on a remote node, which means passing the data twice over the network, is 13 times slower, and having both processes remotely, e.g. passing the data four times over the network, is 24 times slower. The relative relations remain fairly stable over increasing package sizes.

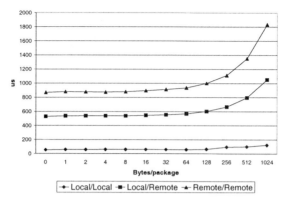

**Fig. 3.** Round trip latency for two process synchronization.

In figure 4 we see how the latency of communication scales fairly linear with the number of clients, however we clearly see a 'staircase' in the latency as the latency goes up steeply every four clients, e.g. every time an extra network operation is necessary. It is clear from the graph that once more than one node is in use, the latency grows dramatically, and the latency grows more per process even if no extra nodes are used. This is the result of interference between PastSet and the socket operations at the kernel level and the server threads at the user level. At 512 bytes the latency of the full ring is approximately 65 us pr process plus 281 us pr node, when more than one node is used.

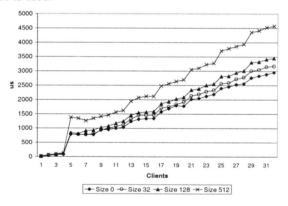

**Fig. 4.** Latency of passing a package in a process ring.

## Bandwidth

The purpose of testing the communication bandwidth of the PastSet system is to define the limits for writing applications that communicate large amounts of data. Applications that require high bandwidth are frequently found in scientific processing and data mining. The bandwidth experiments are similar to the latency setup, although somewhat simpler. A process first writes, and then reads varying block sizes. The experiment is done first with the target element being local and then remote. To eliminate noise, and partly to test the behavior when the element grows, each packet size is written and read 1000 times.

```
get_start_time();
for(i=0; i<1000; i++)
   move(); //Write data-block to PastSet
get_stop_time();
get_start_time();
for(i=0; i<1000; i++)
   observe(); //Read a data-block from PastSet
get_stop_time();
```

**Fig. 5.** Pseudo code for the bandwidth test.

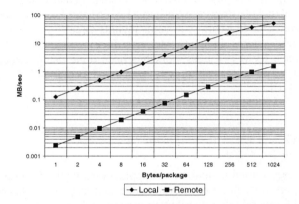

**Fig. 6.** Bandwidth of PastSet local and remote operations

As one would expect we see a large difference between local and remote bandwidth, approximately 1.5 orders of magnitude. The difference between reading and writing is very small as writing requires only slightly more time when new memory need to be allocated for the PastSet. Again the difference between local and remote operations remains fairly constant as the package size grows. At 1024 bytes per package the local bandwidth is just over 51 MB/sec and the remote bandwidth just over 1.5 MB/sec. From a package size of 32KB the bandwidth levels out and local operations yields 82.54 MB/sec (average on read and write) and remote bandwidth is 6.6 MB/sec. As a comparison FTP of a large file which is cached at the server side and goes to /dev/null on the client side gives just under 5.2 MB/sec and tcpblast, which ignores the data at the reciver side, peaks at 8.5 MB/sec on the same platform. Testing local bandwidth with memcpy yields app. 130 MB/sec when copying 32KB blocks, within one process.

**Previously reported performance**

The PastSet DSM system has been tested with three applications [3]. A matrix multiplication was implemented so that it had very small communication frequency and very large data blocks. A successive over relaxation, SOR, kernel that exhibited frequent communication of fairly small data blocks, and a particle based wind tunnel simulator that uses results in frequent communication of large data blocks. The PastSet DSM system managed to outperform identical implementations using MPI and PVM. The matrix multiplication and the SOR kernel were both 44% faster with PastSet than MPI. The wind tunnel application was 25% faster with PastSet than PVM, which on this application was much faster than MPI.

## Related Work

Much effort has been put into cluster communication using either a shared memory model or an explicit communication model. Distributed Shared Memory implementations include Princeton Shrimp SVM [5] and Rice TreadMarks [6]. Object based Distributed Shared Memory systems include Orca [7]. Noteworthy examples of message-passing systems includes Message Passing Interface, MPI [8], Parallel Virtual Machines, PVM [9]. Less work has been done using Structured Distributed Shared Memory. The most well known systems include Linda [2], and more recently, Global Arrays [10].

## Future Work

We have currently ported a small set of applications to the PastSet API [3] and are continuing to port more in order to evaluate PastSet as a programming paradigm. We are investigating several other distribution models and are evaluating the performance of applications under these models. As a part of this we are considering adding extra semantics to the EnterDimension operator to allow the programmer to influence the placement of an element to a greater extent than what the First-Touch model allows. We are also looking into a closer integration with the interconnection network as well as supporting PastSet in heterogeneous and widely distributed environments. A port of PastSet into Linux v. 2.2.0 is also under way, and we expect that the improved SMP performance will reflect on the performance of PastSet as well.

## Conclusion

In this paper we have briefly described the design, implementation, and performance of a high performance structured distributed shared memory system, PastSet. We have showed how it is possible to make a unified programming model for shared and distributed memory that is able to utilize the existence, where available, of physical shared memory for better performance. Through performance measurements we have documented the latency and bandwidth performance of PastSet. The latency grows linearly with the issued number of communication operations, thus the latency of one operation remains constant as the system scales. Compared to native memory copy, within one process, PastSet manages 60% of the bandwidth available. When comparing the network performance PastSet outperforms FTP, and achieves 50% of the theoretical bandwidth and 78% of the achievable bandwidth.

## Acknowledgements

We would like to thank the Research Council of Norway for supporting this work with grants no. 107625/24, 116591/410 and others.

## References

1. Vinter, B., Anshus, O.J., Larsen, T.: PastSet - An Efficient High Level Inter Process Communication Mechanism. In the Proceedings of International Conference on Parallel Processing 1998
2. Gelernter, D.: Generative Communication in Linda. ACM. Transactions On Programming Languages and Systems, vol 7. 1985.
3. Vinter, B., Anshus, O.J., Larsen, T.: An Empirical Evaluation of Symmetrical Multiprocessors in a Cluster Environment. Norsk Informatik Konferense 1998
4. Anshus, O.J., Larsen, T.: MacroScope: The Abstractions of a Distributed Operating System. Norsk Informatikk Konferanse 1992, October 1992.
5. Blumrich, M., Li, K., Alpert, R., Dubnicki, C., Felten, E., Sandberg, J.: A virtual memory mapped network interface for the shrimp multicomputer. In Proceedings of the $21^{st}$ Annual Symposium on Computer Architecture, pages 142–153, Apr. 1994.
6. Keleher, P., Cox, A., Dwarkadas, S., Zwaenepoel, W.: TreadMarks: Distributed shared memory on standard workstations and operating systems. In Proceedings of the Winter USENIX Conference, pages 115–132, Jan.1994.
7. Bal, H.E., Kaashoek, M.F., Tanenbaum, A.S.: Orca: A Language For Parallel ProgrammingOf Distributed Systems. IEEE Computer 25(8), pp. 10-19 (Aug. 1992).
8. Walker, D.W.: The Design of a Standard Message-Passing Interface for Distributed Memory Concurrent Computers. Parallel Computing, Vol. 20, No. 4, pages 657-673, April 1994
9. Sunderam, V.S.: PVM: A Framework for Parallel Distributed Computing. Concurrency: Practice and Experience, Vol. 2, No. 4, Dec. 1990.
10. Nieplocha, J., Harrison, R.J., Littlefield, R.J.: Global Arrays: A Portable Shared-Memory Programming Model for Distributed Memory Computers. Proceedings of the conference on Supercomputing '94, pages 340-ff.

# Optimal Scheduling of Iterative Data-Flow Programs onto Multiprocessors with Non-negligible Interprocessor Communication

D. Antony Louis Piriyakumar[1], Paul Levi[1], and C. Siva Ram Murthy[2]

[1] Institute of Parallel and Distributed High-Performance Systems
University of Stuttgart, 70565 Stuttgart, Germany
{piriyaku,levi}@informatik.uni-stuttgart.de
[2] Department of Computer Science and Engineering
Indian Institute of Technology, Chennai 600 036, India
murthy@iitm.ernet.in

**Abstract.** The problem of optimal compile-time multiprocessor scheduling of iterative data-flow programs with feedback (delay elements) is addressed in this paper, unlike the earlier studies assumed the availability of a large number of processors and complete interconnection among them along with the interprocessor communication (IPC) to be non-negligible to be more realistic. We first explain the effects of including IPC in *non-overlapped, overlapped, fully-static,* and *cyclo-static* multiprocessor schedules with LMS filter as a realistic example. The effect of IPC in the rate-optimal schedules with the transformation techniques viz. *unfolding* and *retiming* in scheduling data-flow programs with optimal unfolding is discussed with an example. We then propose an algorithm, based on the well-known $A^*$ algorithm, for optimal scheduling of data-flow programs onto multiprocessors, which uses only minimum number of processors. To alleviate the impediments of large requirements of memory space and CPU time for the optimal scheduling algorithm, we present an effective technique, *branch join path isomorphism* (BJP) which relies on our previously defined *processor isomorphism, task isomorphism,* and *node isomorphism* apart from the *lower bound theory* and *upper bound* on the completion time. The schedules produced by our algorithm are superior to those obtained by the earlier algorithms despite considering IPC as non-negligible and also not completely connected multiprocessor systems.
Index Terms - Data-flow programs, intra and inter iteration precedences, interprocessor communication, multiprocessor schedules, iteration period, iteration bound, optimal scheduling, unfolding and retiming. ...

## 1 Introduction

Data-flow graphs representing iterative, deterministic, digital signal processing algorithms are used for scheduling and resource allocation during high-level VLSI synthesis, compilers for parallel machines like VLIW or Data-flow machines. The data-flow graphs (programs) describing several signal and image processing problems, possess a high degree of multiprocessing due to their inherent iterative and nonterminating nature. As these problems are applied widely, it is essential to solve them efficiently by exploiting the parallelism available in the problems so that the iteration period (the time to execute all tasks of the data-flow program once) is minimized. As the data-flow program (DFP) is iterative with the feedback, it has a fundamental lower bound on the iteration period [1]. Thus, the tasks of DFP have to be scheduled optimally so as to obtain the iteration period asymptotically close to the lower bound.

Critical path schedulers produce a schedule where the minimum possible iteration is equal to the critical path length of DFG [2]. Loop schedulers [2] generate schedules where the minimum possible iteration is bounded above by the loop critical path length of DFG. The cyclo-static schedulers in [3] utilize exhaustive search to generate cyclo-static schedules, which may reduce the iteration periods most of the times. To exploit the hidden parallelism available in the DFP, transformation techniques such as unfolding and retiming have been applied to the corresponding DFG [4]. The retiming technique minimizes the critical path length of a DFG but does not guarantee a critical path time less than a specified iteration period. In fact, the DFG tasks need to be scheduled optimally to minimize the iteration period, which was not given adequate focus previously. Moreover, the interprocessor communication (IPC) in multiprocessor systems was not considered earlier for DFG excepting for the general scheduling [10], [9] and it leads to unrealistic schedules. Recently, heuristic algorithms for scheduling iterative task computations are discussed in [7]. This algorithm is mainly for distributed memory machines whereas such a constraint is not needed for our algorithm and unlike our algorithm this algorithm is not guaranteed to produce optimal solutions always.

In the remainder of the paper, we explain the data-flow program model and the relevant definitions in section 2. In section 3, the effects of interprocessor communication on periodic schedules are studied with iteration period as the parameter. The effects of IPC on rate-optimal schedules obtained from the transformation techniques including optimal unfolding and retiming are explained with examples in section 4. Section 5 discusses the need of optimal scheduling of DFG, and an optimal compile-time scheduling algorithm for multiprocessor systems is proposed in section 6. The restricting effects of IPC on these transformation techniques and the superiority of our algorithm are summarized in section 7.

## 2 Iterative Data-Flow Program Model

A nonterminating, iterative, data-flow program is represented as a weighted, directed acyclic graph [12], $G_t = \{V_t, E_t\}$, where $V_t = \{\ v_i : \text{i=1,2,...,n}\ \}$ the set of vertices (tasks) with associated service demand $s_i$, and $E_t = \{< v_i, v_j > : \text{i,j} = 1,2,...,\text{n}, \text{i} \neq \text{j}\ \}$ the set of directed edges with associated intertask communication (data) between task $T_i$ and task $T_j$, imposing the partial order that task $T_j$ can be executed only after the execution of task $T_i$ as in Fig. 1.

The main difference between general DFGs and the signal processing DFGs is the associated delay elements (registers) in the directed edges [1]. An edge without a register represents precedence between tasks within iteration. If an edge has n registers, it describes the precedence between tasks of different (i,n+i) iterations which differ by n iterations. A simple example of a nonterminating, iterative, data-flow program with feedback is given in program 1. The DFG in Fig. 1 corresponds to program 1. A complete discussion with appropriate examples for explaining the data-flow model is available in [1].

Program 1: Initial conditions : db(-1),db(0)

for (i=1 to $infinity$)
  ab(i) = $f_{ab}$ [x(i)]
  bc(i) = $f_{bc}$ [ab(i) db(i-2)]
  cd(i) = $f_{cd}$ [bc(i)]
  db(i) = $f_{db}$ [cd(i)]
  y(i) = $f_y$ [cd(i)]

## 3 The Effects of IPC on Periodic Multiprocessor Schedule

The data-flow programs can be scheduled onto multiprocessors in overlapped or non-overlapped manner with two other methods viz. fully-static and cyclo-static [1]. A multiprocessor schedule is said to be non-overlapped if the execution of the $(n+1)^{th}$ iteration begins only after the completion of all the tasks of the $(n)^{th}$ iteration, otherwise it is overlapped. A periodic schedule is said to be fully-static, if all the iterations of some task are scheduled on the same processor. A periodic schedule is said to be cyclo-static, if the iteration n of some task $T_i$ is scheduled in processor $V_p$ at time t in the $n^{th}$ iteration, then in the $(n+1)^{th}$ iteration the task $T_i$ is scheduled in processor $V_{(p+K) modulo m}$ at the time (t+T), where T is the time displacement (iteration period) and m is the total number of processors and K is the processor displacement.

Unlike in [1], the multiprocessor system onto which tasks are scheduled is not assumed to be completely connected which in practice may not always hold. It is assumed to be either homogeneous (all processors have the same service rate, memory capacity, link capacities, etc) or heterogeneous (the processors may differ in service rates). A processor is assumed to perform both computation and interprocessor communication at the same time like an INMOS transputer. The multiprocessor system is represented as a weighted undirected graph, $G_p = \{V_p, E_p\}$, where $V_p = \{ v_q : q = 1,2,...,m\}$ set of processors with associated service rates $\mu_q$ and $E_p = \{ (p,q) : p,q = 1,2,...,m, p \neq q\}$ set of links with associated link capacities $L_{pq}$. We assume that the data communication between a pair of processors follows the shortest path as defined in [12]. The execution of a task on a processor is nonpreemptive.

Here, we consider the iteration period as the parameter as it plays a vital role in the multiprocessor periodic schedules especially when IPC is included. An important point to note is that if iteration period is considered as defined earlier, it will not suffice to account for inter-iteration precedences more specifically when IPC is non-negligible. Hence, when IPC is included in the multiprocessor scheduling, the average iteration period is taken into consideration as it represents the steady state in the nonterminating programs. We have considered here the DFG to be same as in Fig. 1 and the processor graph in Fig. 2. We have given the schedule only for overlapping fully-static without IPC in Fig. 3 and with IPC in Fig. 4. Similarly, the cyclo-static without IPC in Fig. 5 and with IPC in Fig. 6 as the rest are similar. It is evident from the Table 1 that IPC increases the iteration period as it is more realistic than just neglecting IPC. With a realistic example, LMS filter, it is shown that IPC increases the iteration period from Fig. 7, Fig. 8 and Fig. 9.

**Table 1.** Comparison of Iteration Periods with and without IPC

| *Method* | *Without IPC* | *With IPC* |
|---|---|---|
| non-overlapping fully-static | 4.0 | 7.0 |
| non-overlapping cyclo-static | 4.0 | 8.0 |
| overlapping fully-static | 2.0 | 3..75 |
| overlapping cyclo-static | 2.5 | 3.75 |

## 4 The Effects of IPC on Transformation Techniques in Periodic Multiprocessor Schedules Demanding Optimal Scheduling

To exploit the inter-iteration parallelism which other techniques viz. non-overlapping failed, the standard transformation, namely program unfolding is used in [1]. An unfolded DFG with an unfolding factor J describes J consecutive iterations of the original DFG. A detailed study on systematic unfolding, an algorithm for constructing unfolded programs, and some properties of unfolded data-flow programs are presented in [1]. For the sake of simplicity, we take a simple example to explain the impacts of IPC in various cases.

Consider the example graph given in Fig. 10. From the following theorem proved in [1], we can determine the unfolding factor J. If an arbitrary DFG is unfolded by the least common multiple of the number of loop registers (delay elements) in the original DFG, the unfolded DFG is a perfect-rate DFG. A perfect-rate DFG is a DFG having one register in each loop. Moreover, any unfolded DFG with the unfolding factor equal to the least common multiple of the register counts (delay elements) in all the loops can be scheduled rate-optimally. A multiprocessor schedule is rate-optimal if the iteration period equals the iteration bound. Thus to achieve rate-optimal schedule, all that we need is to unfold the DFG with J factor equal to the least common multiple of the number of loop registers, in this case J = 4. One of the major problem in scheduling is fixing the number of processors needed to get the optimal schedules. Fortunately in this case, as proved in [1], the upper bound on number of processors required for fully-static, rate-optimal schedule equals the number of loops in the graph unlike in [5] demanding unbounded number of processors. As the unfolded graph has 4 loops, we have taken 4 processors. Unlike assuming complete connection among the processors, we have considered them as linearly connected as shown in Fig. 2 with m=4. A fully-static schedule and overlapped cyclo-static schedules with IPC are given in Fig. 11 and Fig. 12 with the iteration periods as 20.0 and 17.5 units respectively. However, the iteration bound is 11.25 units. Even though, we have rate-optimally unfolded, we are not able to achieve this due to non-negligible IPC on incompletely connected processors.

This in fact has motivated us to try to asymptotically push the iteration period towards iteration bound, which resulted in demanding optimal schedules.

To reduce the iteration period yet another transformation, retiming was proposed by Leiserson et al in [8]. Retiming involves moving around the registers in the DFG such that the total number of registers in any loop remains unaltered. One example for a retiming transformation is the removal of a fixed number of registers from all the incoming edges of a node, and addition of the same number of registers in all the outgoing edges of the node. This is referred as cutset transformation. We have retimed the DFG in Fig. 10 as shown in Fig. 13. The corresponding overlapped cyclo-static rate-optimal schedule with IPC on linearly connected multiprocessors with m = 4, is shown in Fig. 14 which has its iteration period at the steady state as 18.75 units. The optimal schedule, shown in Fig. 15 as per our algorithm defined in section 5, is 15.00 units only which is closer to the iteration bound. It may also be recalled irrespective of any number of processors and even with complete connection, with IPC one can not get a better iteration period. Here, in all these cases, the IPC is 5 units only. Thus, IPC neccesiates the need for optimal scheduling at compile-time which is NP-Hard [6] to eliminate the overheads associated with dynamic scheduling demanding better iteration periods closer to iteration bounds.

## 5 The New $A^*$ Based Optimal Compile-time Multiprocessor Scheduling Algorithm

### 5.1 New Techniques for Reducing Space and Time

The $A^*$ algorithm described [11], [12], can be used to solve our problem of multiprocessor task scheduling. But the main impediment with the $A^*$ algorithm is the requirement of large memory space and computational time. So, to reduce the space and time requirements of $A^*$ algorithm, we have developed a new technique *Branch Join Path isomorphism*.

**Branch Join Path Isomorphism** This isomorphism is well pronounced in DFGs especially in digital signal processing applications. However, this isomorphism is only for homogeneous multiprocessor systems. The branch join path (BJP) isomorphism is defined as follows,

1. In a DFG, consider a task having 2 children. say $c_1$ and $d_1$. There can be more than 2 children but consider them in pairs.
2. Now, let $c_1$ has only one child $c_2$ and $c_2$ has only one child $c_3$ like this till some $c_x$ for some positive $x > 2$.
3. Similarly, let $d_1$ has only one child $d_2$ and $d_2$ has only one child $d_3$ like this till some $d_y$ for some positive $y > 2$.
4. Let the tasks $c_x$ and $d_y$ be same.
5. Let sum-c be the sum of the execution times of c tasks and sum-d be for d tasks.
6. If sum-c = sum-d, then it is sufficient that either the search is tried with c tasks or with d tasks, preferable with c tasks when $x < y$ or vice versa.

In this case in Fig. 1 and Fig. 2 with m = 2 after 2-unfolding, there are task $T_0(A_1)$ and task $T_4(A_2)$ as the start task with static level of 4 units which can be considered as children of a fictitious node. It is logical to consider only in the case when the tasks are start tasks. Similarly, the tasks $T_3(D_1)$ and $T_7(D_2)$ are the end tasks which can be considered as the parents of another fictitious task. It is also logical to consider only in the case when the tasks are end tasks.

Now, BJP isomorphism exists as sum-c = sum-d = 4 and all other conditions also are fulfilled. Hence it is sufficient to try with one branch itself to get the optimal schedule. In fact, when x = y = 2, this reduces to node isomorphism. Hence, the original DFG itself can be modified so that there will be less number of tasks to schedule at the same time optimality is maintained. All that the modification required is to merge all nodes $c_1$, $c_2$, ..., $c_{x-1}$ into one node and similarly for d tasks. It is interesting to note that even when there is only one branch, one can merge these type of tasks forming a chain into a single task thereby not only reducing the number of tasks to be scheduled but also some of the isomorphism previously mentioned to exhibit voluntarily. This portion of algorithm is just of $O(n^2)$ only.

## 5.2 The New Algorithm for Optimal Task Scheduling

Our new $A^*$ algorithm is explained succinctly as follows.

First, for rate-optimal solution depending upon the J factor, the given DFG is J times unfolded as per the algorithm given in [1]. If the given multiprocessor system is homogeneous, then for handling BJP isomorphism, possible chains of tasks are merged. After this, using a heuristic algorithm, find a schedule and set the schedule length to UB. Find the static level of each task and set LB as the static level of the start task. The basic idea behind the algorithm is that given a node (initially empty), find all the ready tasks. Assign one ready task from each task isomorphic group in every processor excepting for the start task or for the node isomorphism. In case of the start task, just assign it to only one member from all the isomorphic groups of processors. As we have explained above, trying all isomorphic tasks which are ready is futile and one is sufficient to guarantee an optimal solution. In the same vein, isomorphic nodes are also deleted without impeding an optimal solution as the property ensures optimal solution. Compute the value of the heuristic evaluation function f(c) for each of these nodes. If the node does not occur earlier and $f(c) < UB$, add the node in the search tree as child of the recently expanded node. Check whether the node is a goal node for reaching a solution and if f(c)=LB, then also stop by producing the optimal solution. If the node to be expanded is a goal node, just output the schedule as optimal schedule and stop. Otherwise, repeat the process until no more node could be expanded. We now present our new $A^*$ algorithm using the notations as specified earlier.

**Algorithm Optimal-Scheduler;**

1. First, unfold the given DFG, J times so as to get the perfect-rate DFG.
2. Next, tasks which exhibit BJP isomorphism are merged accordingly.
3. Using a heuristic algorithm, find a schedule. Let UB be the schedule length obtained by the heuristic algorithm.
4. Calculate the static levels $\lambda_i, \forall T_i$.
5. Let LB be the lower bound which is the static level of the start task. If there is more than one start task, LB is the maximum of the static levels of such start tasks.
6. If (UB = LB) then stop as the heuristic solution itself is optimal.
7. Otherwise, construct the isomorphic groups of processors.
8. Construct the isomorphic groups of tasks.
9. Initially those tasks which do not have any predecessor will be ready. c = 0 (* node count *).
10. Build the initial node $N_0$ and insert it in the list with $f(N_0) = 0$.
11. Repeat
12. Select the node $N_k$ with smallest f value.
13. If ($N_i$ is not a solution) then
    (a) Generate the successors by finding the ready tasks.
    (b) Do the following for only one ready task from each of the isomorphic task groups
    (c) If ($T_i$ is the first task) i.e., $\Gamma_{N_c} = \phi$ then
        assign task $T_i$ in only one processor from each isomorphic group of processors
        else
        assign the task on all processors.
    (d) For each such assignment do the following
        – Check whether it is already there in the list to eliminate the duplication
        – If (already available) then
          don't add the node

- Check whether this node is isomorphic with one of the already available nodes in the list, i.e.,
If (the node is isomorphic) then
don't add the node
else
compute f($N_i$) = g($N_i$) + h($N_i$) for this node $N_i$.
If ($f(N_i) < UB$) then c = c + 1
insert it in the list
endif
If (all tasks have been allocated and f($N_i$) = LB)
Print $N_i$ as the optimal solution and stop.
endif
endif
endif
else
print the schedule and quit
14. Until ($N_k$ is solution OR list is empty).

## 6 Performance Evaluation

Table 2. Comparison of Previous $A^*$ and New $A^*$ Algorithms

| Arch | Standard $A^*$ | | Isomorphisms | | BJP Isomorphism | |
|---|---|---|---|---|---|---|
| | Nodes | CPU Time | Nodes | CPU Time | Nodes | CPU Time |
| l2 | 747 | 5.5 | 22 | 0.12 | 3 | 0.11 |
| c3 | 2896 | 38.15 | 32 | 0.13 | 4 | 0.11 |
| h2 | 11464 | 346.7 | 42 | 0.14 | 5 | 0.12 |
| l4 | 11464 | 346.6 | 44 | 0.14 | 6 | 0.12 |

In order to evaluate the performance of our algorithms, we implemented them on the network of SUN workstations with SPARK stations 10 and 20 using the programming language C. In Table 2, l2 is linearly connected multiprocessor system with 2 processors, c3 represents completely connected with 3 processors, h2 indicates a hypercube of dimension 2. The values under the column Isomorphism includes processor, task and node isomorphisms introduced in [12]. It is very obvious from this table that there is drastic reduction in the number of nodes say from the order of thousands to just single digits. From the figures 5,6,8 and 9, it is easy to understand that due to the restriction posed on the assignment of tasks onto processors in both the cases viz. fully-static and cyclo-static, it is not possible to attain the results obtained by our algorithm. It may be also recalled that our algorithm used lesser number of processors to get better schedules than the methods so far available as our algorithm is guaranteed to produce always optimal schedules. When the iteration period is very crucial and which will be repeated many times ought to be scheduled optimally relieving the restrictions. Even though the principal performance metric is the number of nodes generated by each algorithm, the CPU times are given to appreciate the average case time and space complexities of the algorithms. Our algorithm makes use of the isomorphism available in the processor architecture, task pattern, node formation and path symmetries to reduce the computations.

# 7 Conclusion

In this paper, we presented an efficient algorithm for the multiprocessor scheduling problem of the iterative programs optimally, based on the well-known state space reduction algorithm, $A^*$. We employed a new technique, *branch join path isomorphism*, based on our previous techniques *processor isomorphism, task isomorphism*, and *node isomorphism* and the effective use of *upper bound* and *lower bound theory* in our algorithm to drastically reduce the space and the computational time requirements. We demonstrated the effectiveness of our algorithm with the examples which clearly portrays the reduction not only in computational time but also in the reduction of iteration periods with minimal processors despite considering the interprocessor communication to be non-negligible. Currently, we are implementing our optimal scheduler on Intel Paragon multiprocessor system.

# References

1. K.K. Parhi and D.G. Messerschmitt, "Static rate-optimal scheduling of iterative data-flow programs via optimum unfolding", IEEE Transactions on Computers, vol. 40, no. 2, pp. 178-194, Feb. 1991.
2. L.E.Lucke and K.K.Parhi, "Data-flow transformations for critical path time reduction in high-level DSP synthesis", IEEE Transactions on Computer-Aided Design, vol. 12, no. 7, July 1993.
3. D.A.Schwartz and T.P.Barnwell, "cyclo-static solutions: Optimal multiprocessor realizations of recursive algorithms", VLSI Signal processing II, IEEE Press, 1986.
4. K.K.Parhi, "Algorithm transformation techniques for concurrent processors", Proceedings of the IEEE, vol. 77, no. 12, Dec. 1989.
5. T. Yang and A. Gerasoulis, "DSC: Scheduling Parallel Tasks on an Unbounded Number of Processors", IEEE. Trans. on Parallel and Distributed Systems, vol. 5, no. 9, pp. 951-967, September. 1994.
6. M.R.Gary and D.S.Johnson, *Computers and Intractability: A Guide to the Theory of NP-Completeness*, W.H.Freeman and Co., 1979.
7. T. Yang and C, Fu, "Heuristic Algorithms for Scheduling Iterative Task Computations on Distributed Memory Machine", IEEE Transactions on Parallel and Distributed Systems, vol. 8, No. 6, pp. 608-622, June 1997.
8. C.E. Leiserson, F. Rose, and J.Saxe, "optimizing synchronous circuitry by retiming", Proc. Third Caltech Conf. VLSI, Pasadena, CA, pp. 87-116, March 1983.
9. J.J. Hwang, Y.C. Cow, F.D. Anger, and C.Y. Lee, " Scheduling Precedence Graphs in Systems with Interprocessor Communication times", SIAM Journal of Computing, pp. 244-257, 1989.

10. S.Selvakumar and C. Siva Ram Murthy, "Scheduling Precedence-constrained Task Graphs with Non-negligible Intertask Communication onto Multiprocessors", IEEE Transactions on Parallel and Distributed Systems, vol.5, no.3, pp.328-336, 1994.
11. N.J.Nilson, *Principles of Artificial Intelligence*, Springer-Verlag, 1980.
12. D. Antony Louis Piriyakumar, C. Siva Ram Murthy, and Paul Levi, "A New $A^*$ Based Optimal Task Scheduling in Heterogeneous Multiprocess or Systems Applied to Computer Vision", HPCN-98, Amsterdam, April, 1998.

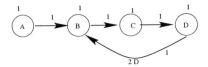

**Fig. 1.** Data-Flow Graph

**Fig. 5.** Overlapping Cyclo-static Schedule without IPC

**Fig. 2.** Linearly Connected Multiprocessors

**Fig. 6.** Overlapping Cyclo-static Schedule with IPC

**Fig. 3.** Overlapping Fully-static Schedule without IPC

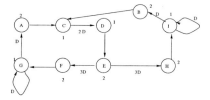

**Fig. 7.** DFG of LMS Filter

**Fig. 4.** Overlapping Fully-static Schedule with IPC

|    | 1  | 2  | 3  | 4  | 5  | 6  | 7 |
|----|----|----|----|----|----|----|---|
| P0 | A1 | A1 | C1 | D2 | E2 | E2 |   |
| P1 | B1 | B1 |    | F2 | F2 | G2 |   |
| P2 | D1 | E1 | E1 | H2 | H2 | I2 |   |
| P3 | F1 | F1 | G1 | A2 | A2 | C2 |   |
| P4 | H1 | H1 | I1 | B2 | B2 |    |   |

**Fig. 8.** Cyclo-static Schedule for LMS without IPC

|    | 1  | 2  | 3  | 4  | 5  | 6  | 7  |
|----|----|----|----|----|----|----|----|
| P0 | A1 | A1 | D2 | C1 | E2 | E2 |    |
| P1 | B1 | B1 | F2 | F2 |    | G2 |    |
| P2 | D1 | E1 | E1 | H2 | H2 | I2 |    |
| P3 | F1 | F1 | G1 | A2 | A2 |    | C2 |
| P4 | H1 | H1 | I1 | B2 | B2 |    |    |

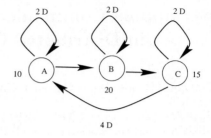

**Fig. 9.** Cyclo-static Schedule for LMS with IPC

**Fig. 13.** Retimed Data-Flow Graph

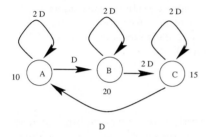

**Fig. 10.** Data-flow Graph

|    | 0  | 10 | 20 | 30 | 40 45 |    | 60 |    | 75 |
|----|----|----|----|----|-------|----|----|----|----|
| P0 | A1 |    | B1 |    | C1    |    |    |    |    |
| P1 | A2 |    | B2 |    | C2    |    |    |    |    |
| P2 |    |    |    | A3 |       |    | B3 | C3 |    |
| P3 |    |    |    | A4 |       |    | B4 | C4 |    |

**Fig. 14.** Cyclo-static Schdule with IPC for Fig. 13

|    | 0  | 10 | 25 | 35 | 40 | 50 55 | 65 | 80 |
|----|----|----|----|----|----|-------|----|----|
| P0 | A1 |    | A2 |    | A3 |       | A4 |    |
| P1 | B1 |    | B2 |    |    | B3    |    | B4 |
| P2 | C1 |    | C2 |    | C3 | C4    |    |    |
| P3 |    |    |    |    |    |       |    |    |

**Fig. 11.** Fully-static Schedule with IPC for Fig. 10

|    | 0  | 10 | 20 | 30 | 45 |    | 60 |
|----|----|----|----|----|----|----|----|
| P0 | A1 |    | B2 |    | C3 | C4 |    |
| P1 | B1 |    | A2 |    | B3 |    | A4 |
| P2 | C1 |    | C2 |    | A3 | B4 |    |

**Fig. 15.** The Optimal Schedule by Our Algorithm for Fig. 10

|    | 0  | 10 | 30 | 45 | 60 | 70 |
|----|----|----|----|----|----|----|
| P0 | A1 | B2 |    | C3 |    |    |
| P1 | B1 | C2 |    |    | A4 |    |
| P2 | C1 |    | A3 | B4C |   |    |
| P3 |    | A2 | B3 |    | C4 |    |

**Fig. 12.** Cyclo-static Schedule with IPC for Fig. 10

# Overlapping Communication with Computation in Distributed Object Systems

Françoise Baude, Denis Caromel, Nathalie Furmento, and David Sagnol

SLOOP - Joint Project CNRS / INRIA / University of Nice Sophia Antipolis
INRIA - 2004 route des Lucioles - B.P. 93 - 06902 Valbonne Cedex, France
FirstName.LastName@sophia.inria.fr

**Abstract.** In the framework of distributed object systems, this paper presents the concepts and an implementation of an overlapping mechanism between communication and computation. This mechanism allows to decrease the execution time of a remote method invocation.

## 1 Introduction

The idea to overlap communication with computation is attractive but not new. As far as we know, this idea has never been investigated in the area of distributed object-oriented languages based on remote service invocations through method calls as RMI [12] in Java or RPC in C/C++ [1]. Optimization of the parameter copying process, as in [15] is a different but complementary approach.

The general idea is that during a remote service dealing with large data requiring transmission, communication and computation are automatically split in steps with a smaller data volume; then, it is only a question of pipelining these steps in order to achieve overlapping between the current step of the remote computation and the data transmission related to the next step of the remote computation. This requires to execute a computation and a transmission step **at the same time**. One way to achieve this is to use asynchronous communications.

Several problems have to be solved, including:

1. design and implement elementary mechanisms, mainly data splitting and computation with partial data;
2. make it as much as possible a transparent mechanism for programmers, but give them the possibility to guide the data splitting, if they wish;
3. try to determine in an automatic way the appropriate size for data packets (i.e. try to estimate the duration of the different steps).

The implementation of this technique is generally restricted to the field of the compilation of data parallelism language for a parallel architecture with distributed memory: HPF [2], FortranD [14], but also LOCCS [6], a library for communication routines and computation. Our contribution is to design, implement, and evaluate it within the context of an object-oriented language extended with mechanisms for parallelism and distribution, C++// [4]. Only points 1 and 2 are resolved in this paper. Dynamically solving point 3 will require more precise information about the computation (a strategy is developed in [6]).

In Sect. 2, various solutions for point 1 are proposed. Then, solutions for splitting requests (point 2) are presented. Sect. 3 introduces one implementation for this technique using the C++// language on top of the SCHOONER library. In Sect. 4, we present some benchmarks. Finally, we conclude in Sect. 5 with the benefits and applications of this technique.

## 2 Communication/Computation Overlap

This section presents our overlapping technique and the necessary requirements for its implementation.

### 2.1 Elementary Mechanisms

They should allow to resolve point 1:

- send a request in pieces (without taking into account the strategy used for splitting);
- be able to rebuild a partial request in such a way that service execution can be started;
- be able to update a missing part when it comes even if service execution has started;
- be able to block the computation if it tries to use a missing piece.

*Step for Request Creation.* In every system that proposes a RPC mechanism, the remote service request has to contain the method ID and the different parameters of the call which are marshalled using a deep copy of the objects graph[1]. Then, the request is sent asynchronously. To obtain overlapping, we dissociate the splitting-flattening operations with the sending ones.

**Requirement 1.** Gain access to the runtime code that sends requests in order to be able to decide when to send a request piece.

*Step for Request Rebuilding.* Once arrived in the remote system, the request is rebuilt: each parameter is reconstructed with the corresponding data and then the service can start. For implementing the overlapping technique, we have to be able to put a mark for the missing data. This mark informs the service that data are, temporarily, unavailable.

**Requirement 2.** Gain access to the runtime code that deals with the unmarshalling of the request in order to manage marks of missing pieces.

When the remote context receives a new part of a request that is already partially rebuilt, the context has to be able to deal with it in an *automatic* and *transparent* way regarding the service that is already executing.

**Requirement 3.** A mechanism that receives and manages messages transparently.

---

[1] If a field of an object is a reference to a remote object, i.e. is a proxy, we just flatten a copy of this proxy.

*Step for Service Execution.* The service can run without any problem as long as it does not attempt to access missing data. An automatic and transparent blocking mechanism is required when it tries to use a missing data. In the same way, resumption has to be transparent and automatic. This requires a wait-by-necessity mechanism [3]. Such mechanism is provided by the classical *future* mechanism, so our solution is the following: each missing data at the instantiation time of the request object is replaced with a *future* data type. In order to do this replacement - preferably in an automatic way - we also require:

**Requirement 4.** A way of transparently adding a *future* semantics to class data type.

### 2.2 Strategies for Splitting a Request

This section deals with the point 2 presented in the introduction. The crucial idea is to break, in the most transparent way for the programmers, the request parameters (by inserting breakpoints). It requires a modification of the marshalling/unmarshalling routines of objects. Whether these routines are generic or not, we have to be able to overload them.

**Requirement 5.** Be able to change the default marshalling/unmarshalling routines.

Strategies can be split in two groups whether they modify or not the class of the object involved in a request.

*Without Class Modification.* We define a new marshalling routine which, for example, inserts a breakpoint: (1) between each data member of the object graph we want to flatten (given the recursive characteristic of the flattening, it means that each data item of a fundamental type is separated from other types); (2) between each level of the object graph (if the marshalling algorithm runs through this graph breadth first).

On the other hand, if the language allows that a member function be used for the flatten operation instead of the standard one, a class can define a suitable routine. This can be used to customize the insertion of breakpoints, for example, by inserting less breakpoints than strategy 1 does (see Code 1).

*Example of code 1 (*`flatten()` *function with breakpoints inserted manually).*

```
Buffer *Matrix::flatten(Buffer *buff) {
  *buff << nb_line << nb_column;
  buff->cut(); // insert a breakpoint
  for(i=0 ; i<nb_line ; i++) {
    *buff << line(i);
    buff->cut(); // insert a breakpoint
    }
  return(buff);
}
```

*With Class Modification.* In this case, we modify explicitly the class of the objects that we want to send latter in another message. The behavior we want is that objects from these classes are move apart during the first inspection of objects belonging to the request. According to the previous requirements, the expected behavior for these objects is the same as that of *future* objects. Such classes can simply inherit from the *future* class to obtain such behavior. With this solution, the problem with the location of breakpoints is implicitly solved: breakpoints are where *future* objects are found.

## 3 Prototype Environment

We present in this section an implementation of the communication/computation overlapping mechanism. We use for this a parallel and distributed extension of C++, called C++//, and its runtime support provided by the SCHOONER library.

### 3.1 C++//

*The C++// Model.* The C++// language [4] was designed and implemented with the aim of importing reusability into parallel and concurrent programming. It does not extend the language syntax, and requires no modification of the C++ compiler, since C++// is implemented as a library and a preprocessor (a Meta-Object Protocol [10]). Two operational versions exist, one of them uses the NEXUS library [7] as the runtime support [5], the other one uses SCHOONER [8].

We first present C++// principles. C++// provides an heterogeneous model with both passive and active objects. Active objects act as sequential processes serving requests (i.e. method invocations) in a centralized and explicit manner by default (such objects are instances of subclasses of the specific C++// class Process). There is a systematic asynchronous communications towards active objects. C++// offers also a wait-by-necessity (transparent *futures*), especially used in dealing with service invocation replies. Finally, there is no shared passive objects (only call-by-value between processes, implying making deep copies of request/reply parameters, like serialization in Java RMI).

*The C++// MOP.* The MOP is centered around reification points concerning RPC: request send, request receive, reply send, reply receive. These points manipulate requests or replies as first-class objects. Generic flatten and rebuild functions are used for these objects. All the necessary object type information is generated during the preprocessing phase. The reply of a service invocation is transparently built as a *future*. Access through method invocation to any object of *future* type is reified and blocks the caller if the result is not back yet.

### 3.2 The SCHOONER Runtime Library

SCHOONERis a library of C++ classes used for writing either coarse grained or more lightweight parallel and distributed applications in C++. This library is mainly an object-oriented distributed runtime support.

Its main purposes are: (1) to be independent on the underlying runtime, in particular the communication library; (2) to provide the user with a structured and abstract view of its application (collection of available computers, set of runtime execution units called *clusters*, set of computation nodes called *communicating objects* and clustered onto *clusters*); (3) to allow a straightforward redefinition of its behaviors.

*The* SCHOONER *Programming Model.* Distributed communicating objects support the effective application, whereas computers and clusters act as the runtime support. Communicating objects interact through asynchronous and "active messages"-like messages, inheriting from a specific C++ class (`Data_Comm_Object`). Flatten and rebuild functions are defined on a per-message class basis. On the sending side, the flatten function is transparently applied when calling the asynchronous `send()` method defined on a proxy (acts as the local representant of the target communicating object). On receipt of an object representing a message, the `rebuild()` and the `process()` methods (also defined on the `Data_Comm_Object` class) are automatically triggered by the hosting cluster with the receiving communicating object reference as parameter. The coarse grained version of SCHOONER is used here to implement C++//, because conducted experiments presented below do not require that C++// active objects be executed by lightweight processes.

## 3.3 C++// on top of SCHOONER

The C++// active objects class inherits from the SCHOONER communicating objects class. The classes for request and reply objects inherit from the `Data_Comm_Object` class, whose `flatten()` and `rebuild()` methods essentially encapsulate the generic ones defined at the C++// level. The aim of the `process()` method of requests is to store these requests in the corresponding queue of the target C++// active object. This C++// object main activity consists in picking - according to the service policy - one request in the list and serving it.

## 3.4 Where the Overlapping Technique is Implemented

We recall our solution according to the requirements presented in Sect. 2. The objects - sent in such a way as to obtain overlapping with the remote service - are instances of a class which inherits from the *future* class. It is the only modification the C++// programmers have to introduce in their applications (see Code 2). These objects are marshalled and sent later on (see 2.1, *Step for Request Creation*). After that, the reception on the remote site has to update the missing data, and to release the service execution potentially blocked on these data.

*Example of code 2 (Definition of a future class).*

```
class Matrix_Future : public Future, public Matrix {
...
};
```

Implementing the overlapping technique requires only minor modifications in the language runtime support. But unlike classical distributed extensions of object-oriented languages (such as Java RMI) the C++// implementation philosophy has always been to act as an open language runtime support, which moreover is object-oriented. As such, it is possible and quite straightforward to modify the behavior of the building and transport request steps.

At the MOP level, the main modification is to write a new generic function to flatten requests: this function builds a first fragment which holds the request header and the non-*future* parameters, and then one fragment for each parameter of *future* type.

The main modification at the SCHOONER level is to write a new `send()` method defined on the communicating objects proxy class. In case of a request presented as a list of fragments, it simply sends them one after the other. Fragments of requests are embedded in instances of a new subclass of the class `Data_Comm_Object` whose `process()` method updates the corresponding awaited request parameter.

## 4 Performances

### 4.1 Benchmark

The test is based on the call of the method `OpMatrix::rang()` (see Code 3) which takes two matrices of *future* type, squares the first one, and adds the two.

*Example of code 3 (Definition and use of a C++// remote service with future parameters).*

```
class OpMatrix : public Process {
  virtual int rang(Matrix *m1,
                   Matrix *m2) {
    m1->square(); // t8 or t'8
    m2->plus(m1);
    int res = m2->result();
    return (res);
  }
};
```

```
OpMatrix *dom = CppLL_new(("host"),OpMatrix,());
Matrix *m1 =
    CppLL_new (Matrix_Future, (COLUMN, LINE));
Matrix *m2 =
    CppLL_new (Matrix_Future, (COLUMN, LINE));
// set the values for m1 and m2
CLOCK_Call_Time_START;
int res = dom->rang(m1, m2);
CLOCK_Call_Time_STOP;
```

The technique presented above allows to overlap the execution of the method m1→square() with the transmission and reception of the matrix m2. Compared with an execution without overlapping, the duration of m1→square() will increase, since, at the same time, the processor also handles the reception and the update of some packets of the matrix m2.

*Test Bed.* In order to evaluate the benefits of this technique, we have to modify both the computation time (by varying the matrices size) and the communication time (we use 2 Linux Pentium Pro connected by a LAN or a WAN). Moreover, we will test the 2 communication routings provided by PVM [9] which is the communication library encapsulated by SCHOONER: the first one uses a direct task-to-task link between clusters, and in the second one, the PVM daemon is in charge of the communication.

## 4.2 Results

We want to maximize the time benefit of the call of the remote service including the reply reception (for example res=dom→rang(), see Code 3). Those call times will be denoted $Call\_Time_{with\_overlapping}$ and $Call\_Time_{without\_overlapping}$.

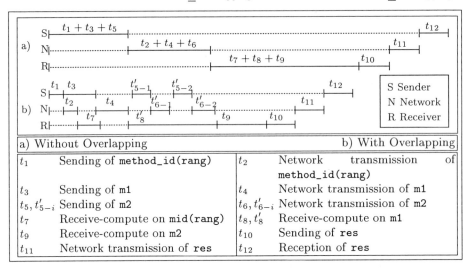

**Fig. 1.** One schematic example of a communication / computation overlap (transmission of m2 requires 2 packets).

Note that the overlapping duration depends on various parameters: (1) load of the network and the hosts used; (2) computation time of m1 ($t_8$ on Fig. 1); the shorter this time is, the less significant the benefit is; (3) transmission time for m2 ($t_6$ on Fig. 1); the shorter this time is, the less significant the benefit due to the overlapping of a part of the computation time on m1 is. We also have to evaluate the time overhead due to the use of the technique (we will say more about it later). Let's define the following value:

$$benefit = \frac{Call\_Time_{without\_overlapping} - Call\_Time_{with\_overlapping}}{Computation\_Time\_on\_M1_{without\_overlapping}(=t_8)}. \quad (1)$$

In theory, the highest *benefit* is 1: this value is obtained when the whole computation time on m1 is overlapped.

We experiment our technique with two kinds of networks: one LAN using Ethernet 10 Mb (see Fig. 2) and one WAN using ATM (see Fig. 3). These figures show that, in practice, this benefit is:

1. sometimes, higher than the theorical value if the network load occurring for executions not using the overlapping mechanism is unusually higher than the one for the executions using it (see the peaks above 1 on Fig. 3);

2. sometimes, symmetrically, lower than 0 if the network load is higher for the executions using the overlapping mechanism (see the peaks below 0 on Fig. 3);
3. generally in the interval $]0, 1[$ since an overlapping is obtained in each case, but the overhead is not insignificant: the computation time for m1→square() is always incremented which proves that although this method has started (m1 is available), there are perturbations due to the reception of the second matrix. Referring to Fig. 1, this means that $t'_8 > t_8$. We will see later that these perturbations depend on the size of m2.

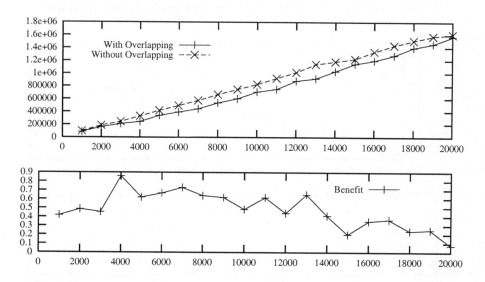

**Fig. 2.** Execution of the remote service (caller side, $\mu s$) and obtained benefit in function of matrices size (number of integers). Direct task-to-task communication between clusters through a LAN.

The overhead of the mechanism is subdivided in:

- a *fixed cost* independent on the data size. It mainly consists in additional function calls (for SCHOONER and for communication) to manage the delayed parameters (2 in our experiment, m1 and m2). Since the number of parameters for a service invocation is generally small, this cost can be seen as a constant.
- a *variable cost* dependent on the data size. This cost is not due to SCHOONER (there is no additional work dependent on the data size), but to an increase of the transmission time of the matrix m2, because the receiver cluster has in the same time to receive parts of m2 and to perform computation on m1. We can explain this overhead by watching carefully what happens in PVM:
  - *if the PVM daemon is in charge of the communication*: the sender daemon does not send a packet while the previous one has not been acknowledged (by default, the acknowledgment window is 1). So, the longer the sending of an acknowledgment is delayed, the latter the sending for the

subsequent packet of m2 occurs ($\sum_i t'_{5-i} > t_5$), thus increasing the whole transmission time for m2. In this case, we see that the time overhead depends on m2 size (i.e. the number of packets used for its transmission).
- *if there is a direct task-to-task link between clusters*: the transmission of m2 has been stopped because the receiving cluster has not acknowledged the bytes written on its socket by the sender cluster. This mechanism can be noticed thanks to tcpdump [13].

To overcome this kind of problem, a solution should be to modify the window size (e.g by using pvm_setopt() or setsockopt() as the case may be), but this is not desirable since in this case, the solution is no more portable.

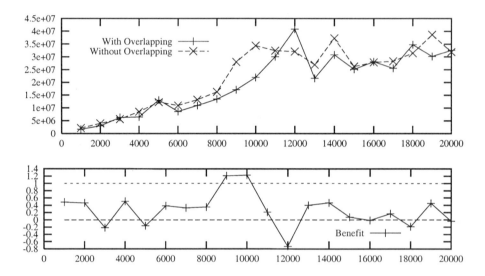

**Fig. 3.** Execution of the remote service (caller side, $\mu s$) and obtained benefit in function of matrices size (number of integers). Direct task-to-task communication between clusters through a WAN.

## 5 Conclusion

In each case (LAN and WAN), we obtain a speedup since the remote computation and the transmission of the *future* parameters are overlapped. But the overhead of the mechanism is not insignificant, in particular due to *variable costs*. The situations for which the overlapping mechanism is relevant have the following characteristics: (1) the computation time using the already received parameters must be similar or superior to the transmission time of the subsequent ones; (2) this transmission time must be high enough to make insignificant a potential increase of it.

As for the second point, the speed and the distance of the network link between the two clusters is also significant. Although in case of a WAN (see

Fig. 3), the gain will be more variable due to high and variable transmission time, in case of a LAN (see Fig. 2), it will be almost constant.

The requirement to implement the overlapping technique in an object-oriented distributed language is mainly to have free access to the transport layer and a MOP for the language. If so, essentially only the flatten and rebuild phases of remote procedure call need to be modified: the object representing the remote call has to be fragmented into several pieces independently managed. Those phases need only to use a *future* mechanism. Automatic message processing is required at the runtime support level. Its aim is to transparently receive and manage late fragments. Such a mechanism is of widespread use, and is in particular available in SCHOONER, in NEXUS, and in PM$^2$ [11], all of them acting as "low-level" runtime supports for parallel and distributed computations.

*Future Work.* We plan to test others fragmentation policies. At the moment, only the simplest one requiring that late fragments be objects of *future* type has been tested. Others policies could maybe help to expose new overlapping opportunities in the progress of the remote service.

# References

1. A.D. Birrell and B.J. Nelson. Implementing Remote Procedure Calls. *ACM Transactions on Computer Systems*, 2(1): 39–59, Feb. 1984.
2. T. Brandes and F. Desprez. Implementing Pipelined Computation and Communication in an HPF Compiler. In *Euro-Par'96*, J:459-462, Aug. 1996.
3. D. Caromel. Towards a Method of Object-Oriented Concurrent Programming. Communications of the ACM, 36(9):90-102, Sep. 1993.
4. D. Caromel, F. Belloncle, and Y. Roudier. *Parallel Programming Using C++*, chapter The C++// System, p 257-296. MIT Press, 1996. ISBN 0-262-73118-5.
5. D. Caromel and D. Sagnol. C++// home page. http://www.inria.fr/sloop/c++ll/
6. F. Desprez, P. Ramet, and J. Roman. Optimal Grain Size Computation for Pipelined Algorithms. In *Euro-Par'96*, T:165-172, Aug. 1996.
7. I. Foster, C. Kesselman, and S. Tuecke. The Nexus Approach to Integrating Multithreading and Communication. *JPDC*, 37:70-82, 1996.
8. N. Furmento and F. Baude. Schooner: An Object-Oriented Runtime Support for Distributed Applications. In *PDCS'96*, 1:31-36, Dijon, France, Sep. 1996. ISBN: 1-880843-17-X.
9. A. Geist *et al.* PVM *Parallel Virtual Machine: a user's guide and tutorial for networked parallel computing*. MIT Press, 1994.
10. G. Kiczales, J. des Rivières, and D.G. Bobrow. *The Art of the Metaobject Protocol*. MIT Press, 1991.
11. R. Namyst and J-F. Méhaut. PM$^2$: Parallel Multithreaded Machine. A computing environment for distributed architectures. In *ParCo'95*, Gent, Belgium, Sep. 1995.
12. Sun Microsystems. Java RMI Tutorial, Nov. 1996. http://java.sun.com.
13. W.R. Stevens. *Advanced Programming in the UNIX Environment*. Addison-Wesley Publishing Company, 1992.
14. C.W. Tseng. *An Optimizing Fortran D Compiler for MIMD Distributed-Memory Machines*. PhD thesis, Rice University, Jan. 1993.
15. C. Videira Lopes. Adaptive Parameter Passing. In *ISOTAS'96*, Kanazawa, Japan, Mar. 1996.

# Exploiting Speculative Thread-Level Parallelism on a SMT Processor

Pedro Marcuello and Antonio González

Universitat Politècnica de Catalunya, Departament d'Arquitectura de Computadors
c/ Jordi Girona 1-3, Mòdul D6; 08034 Barcelona (Spain)
Email: {pmarcue,antonio}@ac.upc.es

**Abstract.** In this paper we present a run-time mechanism to simultaneously execute multiple threads from a sequential program on a simultaneous multithreaded (SMT) processor. The threads are speculative in the sense that they are created by predicting the future control flow of the program. Moreover, threads are not necessarily independent. Data dependences among simultaneously executed threads may exist. To avoid the serialization that such dependences may cause, inter-thread dependences as well as the values that flow through them are predicted. Speculative threads correspond to different iterations of the same loop, which may significantly reduce the fetch bandwidth requirements since many instructions are shared by several threads. The performance evaluation results show a significant performance improvement when compared with a single-threaded execution, which demonstrates the potential of the mechanism to exploit unused hardware contexts. Moreover, the new processor architecture can achieve an IPC (instructions per cycle) even higher than the peak fetch bandwidth for some programs.

## 1. Introduction

Several studies on the limits of the instruction-level parallelism (ILP) that current superscalar organizations can attain show that it is rather limited when a realistic configuration is considered (see for instance [18]), especially for non-numeric applications. Three of the most important bottlenecks that cause this limitation are: the serialization imposed by data dependences, the instruction window size and the fetch bandwidth. On the other hand, some predictions based on current trends indicate that by the year 2010, several hundreds of millions of transistors will be available in a single chip processor, which will enable a huge amount of innovative microarchitectural issues.

Whereas a lot of effort has been devoted to reduce the penalties caused by control and name dependences, techniques to relieve the serialization caused by data dependences have been practically ignored so far. *Data speculation* techniques are emerging as a new family of techniques that can provide a significant boost in ILP (see [4][6] among others). *Data speculation* can be classified into two categories: *data value speculation* and *data dependence speculation*. Data value speculation is based on predicting either the source or destination operands of some instructions in order to execute speculatively the instructions that depend on them. Data dependence speculation refers to the execution of a code according to some predicted dependences.

The amount of ILP that a superscalar processor can exploit is highly dependent on the instruction window size. However, increasing it poses new problems that limit its feasibility or its effectiveness. First, branch prediction accuracy limits the average window size. To go beyond a single basic block, superscalar processors rely on predicting the outcome of unresolved branches. However, this process is sequential in nature because the instruction window is composed of a contiguous region of the dynamic instruction sequence, which is called a *thread of control*[1] (or thread for short) in this paper. In consequence, a single mispredicted branch prevents the instruction window from growing further until the branch is resolved. Second, the complexity and delay of the issue logic grow with the instruction window size. It has been shown in a recent study that the issue logic is likely to be one of the most critical parts in the future [9].Finally, in addition to the branch prediction accuracy, the two main factors that limit the instruction fetch bandwidth are: the branch prediction throughput and the potential to fetch noncontiguous instructions.

On the other hand, simultaneous multithreaded processors (SMT) have been recently proposed as an effective microarchitecture to exploit high degrees of ILP [15]. However, they completely rely on software techniques to extract multiple threads to run on their multiple hardware contexts. When the software fails to provide enough threads, hardware contexts become unused and the performance significantly drops. This may be the case when a non-numeric application that is difficult to parallelize is executed alone.

In this paper, we propose an extension to the SMT microarchitecture in order to use the otherwise idle contexts when a single-threaded application is executed. The resulting microarchitecture can improve the performance of single-threaded programs by relieving the three bottlenecks mentioned above in the following way.

First, the processor implements an effective large instruction window that is made up of several nonadjacent smaller windows. Each one corresponds to a different thread of control of the same program. These threads, which are not necessarily independent, are created from a single sequential program by hardware without compiler intervention.

Second, the execution ordering constraints imposed by inter-thread dependences are avoided through the extensive use of data speculation. The proposed architecture speculates on inter-thread data dependences and the values that flow through them. That is, for each new speculative thread it predicts which register and memory dependences it has with previous threads in the control flow and which values are going to flow through such dependences. The thread is then executed obeying the predicted dependences and using the predicted values to avoid waiting for the actual data.

Third, since simultaneously active threads correspond to different iterations of the same loop, they share the same code (the loop body), so a simple fetch engine can feed all the threads by fetching each instruction just once and dispatching it to all the threads.

The extended SMT microarchitecture, which is referred to as *Data Speculative Multithreaded Architecture (DaSM)*, does not require any modification in the ISA: any program compiled for a superscalar processor can run in this new processor architecture.

The rest of this paper is organized as follows. The DaSM architecture is presented in section 2. Some performance figures of the DaSM architecture are discussed in section 3. The related work is reviewed in section 4. Finally, section 5 summarizes the main contributions of this paper.

## 2. The Data Speculative Multithreaded Architecture

The DaSM hardware architecture (see figure 1) is based on a simultaneous multithreaded processor [15] with extensions to speculate on multiple threads of control obtained from a single sequential program, and to speculate on inter-thread data dependences/values through registers and memory. Thus, the distinguishing features of the novel architecture are the extensive use of speculation techniques to deal with both control and data

---

[1] Regardless of the particular approach used to obtain it (i.e. the partition of a program into threads of control could be done by the hardware, as proposed in this work).

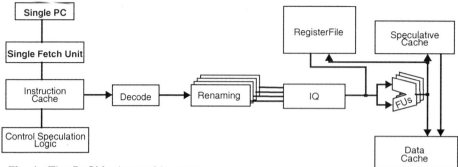

**Fig. 1.**: The DaSM microarchitecture

dependences. Both types of speculation are completely performed by hardware without requiring any compiler support. Thus, the object code can be specified in any conventional instruction set architecture (ISA), without any additional extension.

The DaSM processor fetches instructions from a single program counter (PC). Fetched instructions are decoded and treated by the control speculation logic as it is described in Section 2.1. The instructions are then renamed separately for each thread (every thread has its own register map table although all the threads share the register file) and dispatched to the instruction queue from where they are issued out-of-order. The threads simultaneously executed by the DaSM processor may have dependences among them. The enforcement of inter-thread dependences through registers is described in Section 2.2.2, and dependences through memory are managed by a special first level data cache that is called Speculative Cache, which is described in Section 2.2.3.

### 2.1. Speculative Thread Generation

A DaSM processor executes a program out-of-order by means of a large instruction window that consists of several non-contiguous small windows, each one corresponding to a different thread of control. These multiple threads of control are generated at run-time through control speculation. A key issue to build a large effective instruction window is to speculate only on highly predictable branches. Branches that close loops can be predicted very accurately since they are usually taken and have the same target address. Thus, the multiple threads of control of the DaSM execution model correspond to different iterations of loops. The mechanism used to dynamically extract multiple threads of control from a sequential program is described in detail in [14].

Among all the threads that proceed in parallel at any given time, there is only one of them that is not control speculative on previous threads, which is called the *non speculative thread*. The remaining ones are called the *speculative threads*. When a speculative thread reaches the closing branch of its corresponding iteration, it is suspended until it is either committed or squashed. When the non speculative thread finishes an iteration of a loop that has some speculative threads allocated to following iterations, if the branch is not taken all the speculative threads of this loop are squashed. Otherwise, the thread allocated to the next iteration becomes the new non speculative one.

### 2.2. Data Speculation

Inter-thread dependences both through registers and memory are predicted based on the history of each loop. In particular, dependences through memory are predicted by taking advantage of the highly predictability of memory addresses. The memory address

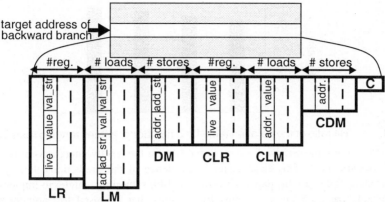

**Fig. 2.:** The iteration table.

prediction mechanism is based on keeping track of the last effective address generated by each memory instructions and its last stride. Using this scheme, about 75% of the memory instructions executed by the SPEC95 can be correctly predicted [4]. Predicting dependences through registers is based on the observation that the architected registers that are live at the beginning of every iteration of a given loop are usually the same. This obviously holds for loops without conditionals in the loop body and it may also hold in the presence of conditionals, for instance, when the *then* and *else* parts read the same live variables, or when conditional branches repetitively have the same outcome. Prediction of data values is performed by a stride-based predictor.

#### 2.2.1. Data structures

Data dependences and data values are predicted by means of a table that is called the *iteration table* (see figure 2). Each entry of this table keeps the information regarding the dependences among iterations of a loop (usually called loop-carried dependences) as well as the data that flow through such dependences. It is indexed by the loop identifier (the target address of a backward branch) and each entry contains the following fields:

- LR (Live registers): For each architected register it indicates whether it was first read (LIVE), or it was first written (DEAD) or it was not accessed at all (UNUSED) in the last iteration. For every live register it also contains the value that it held at the beginning of the last iteration and the stride. The stride is the difference between the values at the beginning of the last two iterations.
- LM (Live memory): List of memory addresses that contained live values in the last iteration of the loop[1] in program order. For each entry it also contains an address stride, which is the difference between the effective address of the two memory references that occupy the same relative location in the LM array for the last two iterations of the loop. For those entries whose stride is equal to zero it contains the live value that was read in the last iteration and the value stride, which is defined as the difference between the values of the last two iterations.
- DM (Dead memory): List of memory addresses that contained dead values in the last iteration of the loop[2] in program order. For each entry it also contains a stride, which is the difference between the effective address of the two memory references that

---

[1] A live memory location in an iteration is a location that is read before being written.
[2] A dead memory location in an iteration is a location that is written before being read.

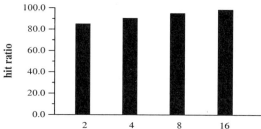

**Fig. 3.:** Hit ratio of the iteration table for a number of entries ranging from 2 to 16.

occupy the same relative location in the LM for the last two iterations of the loop.
- C (Confidence): This field assigns confidence to the predictions done using the previous fields. In this paper we assume a 2-bit saturating counter to implement it.

In addition, each entry contains three fields that are used to compute the data corresponding to the iteration in progress: CLR (Current live registers), CLM (Current live memory), and CDM (Current dead memory). These fields are used to update the LR, LM and C fields when an iteration finishes. A new entry in the iteration table is allocated for every new started loop (this event is detected by means of the mechanism described in [14]). The entry corresponding to the loop with the least recently started iteration is chosen for replacement. At the end of a loop, the LR, LM and DM fields corresponding to its entry are updated by means of the CLR, CLM and CDM fields.

Each entry of the iteration table is quite large. However, very few entries are enough to obtain significant benefits since programs usually spend large periods of time in the same few loops. For instance, figure 3 shows the percentage of iterations that find their history in the iteration table for a number of entries from 2 to 16 averaged for the whole SPEC95 benchmark suite. This percentage is in average 90% for 4 entries.

### 2.2.2. Dependences through registers

Architected registers (also called logical registers) are dynamically renamed to physical registers using a different *register map table* for each thread (Rmap for short). Notice that each map table reflects a different mapping from logical to physical registers, each one corresponding to a different point of the execution. A Rmap entry may contain a special value, NIL, that indicates that logical register is not currently mapped to any physical one.

When speculative threads are created, their Rmap tables are initialized as follows. A new physical register is allocated for each live logical register, and it is initialized with its predicted value. Thus, all instructions that depend on that logical register are allowed to be issued immediately. Dead registers are mapped to NIL and unused registers are initialized with the same mapping that is currently used by the non speculative thread.

For each decoded instruction with destination register, a new free physical register is allocated by each thread. Logical source registers are renamed to physical registers by means of the corresponding Rmap table of each thread. If a source operand was mispredicted as a dead register, it will be mapped to a physical register whose contents is NIL. In this case, if the instruction is not in an intra-thread control speculative path, the current thread and all subsequent ones are squashed. Also, when an instructions whose destination register was classified as UNUSED finishes, all succeeding threads (if any) are squashed since they may have read an incorrect value for that register.

When the non-speculative reaches the beginning of the next thread (if any), all physical registers assigned to the logical ones that are not UNUSED are freed, since for

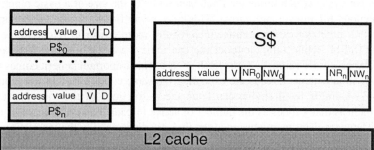

**Fig. 4.:** The Speculative cache organization.

those that are live in following threads, other physical registers have already been allocated.

### 2.2.3. Dependences through memory

In the same way as the processor provides support to store multiple states of the registers, each one corresponding to the view that a different thread has, a similar type of support has to be provided for memory. This is achieved by means of a particular organization of the cache hierarchy (see fig. 4). The first level cache of the DaSM processor consists of several *private caches* ($P\$_i$), each one being associated to one of the different contexts supported by the architecture, plus a *shared cache* ($S\$$), which is shared by all the contexts. Private caches are initialized with those memory values that are live and predictable (according to the LM field of the corresponding iteration table entry). Also, they are used to keep all memory updates performed by speculative threads until they are committed.

Dependences among memory references with a stride different from zero are managed by the shared cache. Each entry of the shared cache has the following fields: address; value; and a valid bit (V). In addition, it has as many NR and NW fields as number of contexts. Each NR/NW field contains the number of remaining reads/writes that the corresponding thread is expected to perform from/to such address before it finishes. When the processor shifts to speculation mode, the shared cache is initialized through the LM and DM of the iteration entry corresponding to the speculated loop.

A thread can perform a read to a data that is not in its private cache when all the writes from previous threads to the same address have been performed. Similarly, it can perform a write when all previous reads and writes to the same address have been carried out.

When a thread performs a write, the corresponding NW entry is decreased and if it becomes negative (the number of predicted writes was wrong), the succeeding threads are squashed since they may have violated a dependence through memory. Similarly, for each read operation, the corresponding NR entry is decreased and if it becomes less than zero, then the current thread and the succeeding ones are squashed[1].

When a thread finishes, if the corresponding NR/NW field of any line of the shared cache is greater than zero, it is reset to zero. This occurs when some predicted read/write did not actually occur. In this case, all dependences have been obeyed but there may be loads of succeeding threads waiting for a non existent memory operation.

### 2.3. Instruction Fetching

The instruction fetch bandwidth is known to be one of the critical factors of superscalar processors. Fetch units of current commercial microprocessors perform a single branch

---

[1] More powerful misspeculation mechanism based on selective squashing are feasible but beyond the scope of this paper.

prediction per cycle, which limits the fetch bandwidth to the size of a basic block (around 4-5 instructions for integer programs). This may be sufficient for a 4-way superscalar processor but more aggressive architectures require more powerful fetch engines. In the case of the DaSM architecture, the processor can achieve a much higher performance than a superscalar processor using the same fetch engine. This important benefit comes from the fact that all simultaneous threads are executing the same code (different iterations of the same loop). A single fetch engine can fetch the instructions of the loop just once and replicate them as many times as the number of active threads. Each instruction is renamed using a different register map table and then they are dispatched to the instruction queue.

## 3. Performance Evaluation

In this section we present a performance evaluation of the DaSM architecture. The objective is to demonstrate the potential of the new architecture to exploit thread-level parallelism, in addition to ILP, without any compiler support. Evaluation of different configurations as well as the tuning of critical parameters of the architecture like the prediction scheme, size of register file, etc., are beyond the scope of this paper.

### 3.1. Experimental Framework

The DaSM architecture has been evaluated through trace-driven simulation of the SPEC95 benchmark suite. The programs have been compiled for a DEC AphaStation 600 5/266 with a 21164 processor with full optimization and instrumented by means of the Atom tool. A cycle-by-cycle simulation is performed in order to obtain accurate timing results. Because of the detail at which simulation is carried out the simulator is slow, so we have simulated 100 million of instructions for each benchmark after skipping 500 million.

We have assumed a DaSM processor that has 4 contexts, an issue bandwidth of 4 instructions per cycle for each context, 4 entries in the iteration table, a shared cache with 128 entries and 4 private caches with 64 entries each. Every context has a local reorder buffer with 64 entries. Also, we assume a fetch bandwidth up to 4 instructions or one basic block, whichever is shorter, 256 physical registers, perfect branch prediction just for intra-thread branches and an ideal L2 cache memory. The existent functional units (latency in brackets) are: 8 simple integer (1), 4 integer multiplication (2), 5 memory (2), 6 simple FP (1), 3 FP multiplication (4) and 2 FP division (17).

### 3.2. Performance Figures

The evaluation of the DaSM architecture is summarized in Table 1. The first column shows the average number of committed instructions per cycle. It can be seen that for the FP programs the IPC is significantly higher than for integer codes. For most of the FP programs, the DaSM architecture achieves an IPC even higher than the fetch bandwidth. This confirms the potential benefits of the fetch mechanism in terms of reduction in fetch bandwidth requirements.

The second column shows the average number of active threads per cycle (TPC) that are correctly speculated. The TPC is the main source of additional parallelism exploited by the novel features of the DaSM architecture. The average TPC for FP programs is 3, out of a maximum of 4. For integer programs, the amount of thread-level parallelism is much lower (1.6 on average) but still significant since these programs are known to be hardly parallelizable. The third column shows the percentage of correctly speculated threads. This percentage is in general quite high, even for integer programs, which confirms that the speculation mechanism is quite accurate in identifying predictable threads.

Fig. 5 shows the speedup achieved when compared with a single-threaded execution.

**Table 1:** Performance statistics

|    |         | IPC | TPC | hit ratio |
|----|---------|-----|-----|-----------|
| FP | tomcatv | 7.9 | 3.6 | 0.98 |
|    | swim    | 5.6 | 2.7 | 0.80 |
|    | hydro2d | 5.8 | 3.4 | 0.96 |
|    | mgrid   | 8.2 | 3.6 | 0.97 |
|    | applu   | 3.8 | 1.7 | 0.86 |
| int | m88ksim | 3.9 | 2.3 | 0.92 |
|    | vortex  | 3.5 | 1.2 | 0.78 |
|    | ijpeg   | 4.1 | 1.2 | 0.72 |

It can be seen that exploiting speculative thread-level parallelism provides a 80% performance increase on average for FP codes and 25% for integer codes, so we can conclude that the DaSM architecture has significant potential to exploit resources already available on a SMT processor that are unused due to the lack of processes.

## 4. Related Work

There are few proposals in the literature dealing with the dynamic management of a large window that consists of several threads of control not necessary independent among them obtained from a sequential program. Pioneer work on this area was the Expandable Split Window paradigm [3] and the follow-up work on Multiscalar processors [12]. Other proposals are the he SPSM architecture [2]; the Superthreaded architecture [13]; Trace processors [10][11]; the Dynamic Multithreaded processors [1]; and the Dependence Speculative Multithreaded [7], and the Speculative Multithreaded Architecture [8]. There are important differences between the DaSM and those previous proposals:

- The Multiscalar, SPSM and Superthreaded architectures require some addition/ extension to the ISA. Moreover, in those architectures data dependences are always enforced by executing the producer instruction before the consumer one.
- Data speculation has been used in previous proposals, mainly in the context of a superscalar processor with a single thread of control (see [4][6] among others). Data speculation is also used by Trace processors. However, the approach to build the instruction window is different: whereas in Trace processors it is based on a trace cache [11], in DaSM processors it is based on a loop prediction technique [14].
- Data dependence speculation is used by the Dependence Speculative Multithreaded Architecture but it does not perform data value speculation. Data dependence speculation is also used by the Multiscalar and SPSM architectures, but they just

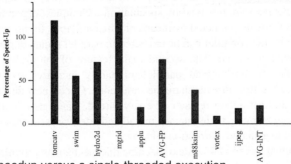

**Fig. 5.:** Speedup versus a single-threaded execution.

implement an "always-independent" prediction scheme, whereas the approach used by the DaSM is more powerful since it is based on memory address prediction.
- The Dynamic Multithreaded Processor spawns a new thread whenever a loop is encountered. The non-speculative thread proceeds with the loop and the speculative one executes the code that is expected to be executed after the loop is terminated. It also spawns a thread at subroutine invocations. Moreover, it does not predict dependences and the value prediction scheme is very simple: the register file of the parent thread is copied into the child register file
- The Speculative Multithreaded Processor uses the same mechanism as the DaSM processor to spawn threads, but the SM has a clustered design and also implements data dependence speculation through registers.

Recently several microarchitectural organizations that support multiple threads of controls and are oriented to reduce the branch misprediction penalty have been proposed (see [19] among others). The common feature of these architectures is that threads are spawn at conditional branches that are difficult to predict.

The DaSM architecture reduces the fetch bandwidth requirements by taking advantage of the fact that simultaneously active threads process the same code with different data. A similar feature is exploited by the CONDEL[16] architecture and the dynamic vectorization approach proposed in [17]. Those approaches are more restrictive than the one used by DaSM processors since the former is limited to loops whose static body does not exceed its instruction window, whereas the latter is feasible only if the dynamic sequence of instructions of the loop are the same for all the iterations and they fit into a single instruction trace cache line.

Finally, several approaches to spawn speculative threads on a multiprocessor environment have been recently proposed (see [5] among others). In addition to the important differences due to the underlying architecture, another important difference with the DaSM architecture is that those works use an "always-independent" dependence prediction scheme and do not include value prediction mechanisms.

## 5. Conclusions

We have presented a run-time mechanism to exploit speculative thread-level parallelism on a simultaneous multithreaded architecture. The main objective is to exploit the otherwise idle resources of the machine when the number of threads is lower than the number of hardware contexts. A novel feature of such mechanism is its ability to dynamically extract and execute multiple speculative threads from a single sequential program written in a conventional ISA and without any compiler intervention. Multiple concurrent threads execute different iterations of the same loop. These threads may be dependent but interthread data dependences are resolved by speculation techniques: dependences and values that flow through them are predicted by means of a history table. In this way, loops that are not parallelizable by the compiler can be executed in parallel if data dependences and data values are correctly predicted. The second main feature of the architecture is that the additional ILP due to inter-thread parallelism hardly increases the fetch bandwidth requirements since multiple threads share the same code. Once a new instruction is fetched, it is copied into the instruction register of every thread. Then, its operands are renamed using a different register map table for each thread and afterwards, the renamed instructions can be dispatched to a shared instruction window.

We have shown that for a 4-context machine this mechanism provides a 80% speedup for FP codes and 25% for the integer ones. It has also been shown that in some programs the IPC can be even higher than the maximum fetch bandwidth, due to the reuse (replication) of fetched instructions.

## Acknowledgments

This work has been supported by the Spanish Ministry of Education under grants CICYT TIC 511/98 and AP96-52274600. The research described in this paper has been developed using the resources of the European Center for Parallelism of Barcelona (CEPBA).

## References

1. H. Akkary and M.A. Driscoll, "A Dynamic Multithreading Processor", in *Proc. 31st. Ann. Int. Symp. on Microarchitecture*, 1998
2. P.K. Dubey, K. O'Brien, K.M. O'Brien and C. Barton, "Single-Program Speculative Multithreading (SPSM) Architecture: Compiler-Assisted Fine-Grained Multithreading", in *Proc. Int. Conf. on Parallel Architectures and Compilation Techniques*, pp. 109-121, 1995.
3. M. Franklin and G.S. Sohi, "The Expandable Split Window Paradigm for Exploiting Fine Grain Parallelism", in *Proc. of Int. Symp. on Computer Architecture*, pp. 58-67, 1992.
4. J. González and A. González, "Speculative Execution via Address Prediction and Data Prefetching", in *Proc of 11th. Int. Conf. on Supercomputing*, 1997.
5. L. Hammond, M. Willey and K. Olukotun, "Data Speculation Support for a Chip Multiprocessor", in *Proc. of Int. Conf. on Architectural Support for Prog. Lang. and O.S.*, 1998
6. M.H. Lipasti and J.P. Shen, "Exceeding the Dataflow Limit via Value Prediction", in *Proc. of Int. Symp. on Microarchitecture*, pp. 226-237, 1996.
7. P. Marcuello and A. González, "Control and Data Dependence Speculation in Multithreaded Processors", in *Proc. of the Workshop on Mulithreaded Execution Architecture and Compilation* held in conjuction with HPCA-4, 1998
8. P. Marcuello and A. González, "Speculative Multithreaded Processors", in *Proc. of the 12th. Int. Conf. on Supercomputing*, pp.77-84, 1998.
9. S. Palacharla, N.P. Jouppi and J.E. Smith, "Complexity-Effective Superscalar Processors", in *Proc. of Int. Symp. on Computer Architecture*, pp. 206-218, 1997.
10. E. Rotenberg, S. Bennet and J.E. Smith," Trace Processors", in Proc. of the Int. Symp. on Microarchitecture, 1997.
11. E. Rotenberg, Q. Jacobson, Y. Sazeides and J E. Smith,"Trace Cache:a Low Latency Approach to High Bandwidth Instruction Fetching", *Proc.Int. Symp. on Microarchitecture*, 1996.
12. G.S. Sohi, S.E. Breach and T.N. Vijaykumar, "Multiscalar Processors", in *Proc. of the Int. Symp. on Computer Architecture*, pp. 414-425, 1995.
13. J-Y. Tsai and P-C. Yew, "The Superthreaded Architecture: Thread Pipelining with Run-Time Data Dependence Checking and Control Speculation", in *Proc. Int. Conf. on Parallel Architectures and Compilation Techniques*, pp. 35-46, 1996.
14. J.Tubella and A.González,"Control Speculation in Multithreaded Processors through Dynamic Loop Detection", *Proc. of the Int. Symp. on High-Performance Computer Architecture*, 1998.
15. D.M. Tullsen, S.J. Eggers and H.M. Levy, "Simultaneous Multithreading: Maximizing On-Chip Parallelism", in *Proc. of the Int. Symp. on Computer Architecture,* 1995.
16. A.K. Uht, "Concurrency Extraction via Hardware Methods Executing the Static Instruction Stream", *IEEE Trans. on Computers*, vol. 41, July 1992.
17. S. Vajapeyam , T. Mitra, "Improving Superscalar Instruction Dispatch and Issue by Exploiting Dynamic Code Sequences", *Proc. the Int. Symp. on Comp. Architecture,* 1997.
18. D.W. Wall, "Limits of Instruction-Level Parallelism", Tech. Report WRL 93/6, Digital Western Research Laboratory, 1993.
19. S. Wallace, B. Calder and D. Tullsen, "Threaded Multiple Path Execution", in *Proc. of Int. Symp. on Computer Architecture,* pp. 238-249, 1998

# Network Interface Active Messages for Low Overhead Communication on SMP PC Clusters

Motohiko Matsuda, Yoshio Tanaka, Kazuto Kubota and Mitsuhisa Sato

Real World Computing Partnership
Tsukuba Mitsui Building 16F, 1-6-1 Takezono
Tsukuba, Ibaraki 305-0032, Japan

**Abstract.** NICAM is a communication layer for SMP PC clusters connected via Myrinet, designed to reduce overhead and latency by directly utilizing a micro-processor equipped on the network interface. It adopts remote memory operations to reduce much of the overhead found in message passing. NICAM employs an Active Messages framework for flexibility in programming on the network interface, and this flexibility will compensate for the large latency resulting from the relatively slow micro-processor. Running message handlers directly on the network interface reduces the overhead by freeing the main processors from the work of polling incoming messages. The handlers also make synchronizations faster by avoiding the costly interactions between the main processors and the network interface. In addition, this implementation can completely hide latency of barriers in data-parallel programs, because handlers running in the background of the main processors allow reposition of barriers to any place where the latency is not critical.

## 1 Introduction

Symmetric multiprocessor PCs (SMP PCs) have recently attracted widespread attention, and clusters of SMP PCs with fast networks have emerged as important platforms for high performance computing. While the bus bandwidth sometimes limits the computation performance, SMP PCs easily reveal a bottleneck when multiple processors are accessing the bus simultaneously. Even worse, the network interfaces for clustering further burden the bus.

Thus, we designed a communication layer NICAM which reduces the communication overhead by utilizing a micro-processor on the network interface. Overhead reduction is important, because the overhead, the involvement of main processors in communication, wastes bus bandwidth as well as the processing power of the processors. In addition, a common technique of overlapping of computation and communication tends to make the communication grain-size finer, which results in larger total cost of overhead. As researchers show [1], while latency reduction in data transfer is not so relevant for performance, overhead reduction directly affects the utilization of the processing power. Thus, overhead reduction by the network interface will be fruitful while it incurs larger latency resulting from the relatively slow micro-processor on the network interface.

**Table 1.** Aggregate bandwidth of the memory bus on an SMP PC.

| threads | read | write | copy |
|---|---|---|---|
| 1 | 360.0 | 91.1 | 85.6 |
| 2 | 258.2 | 106.6 | 88.9 |
| 3 | 251.7 | 106.6 | 87.7 |
| 4 | 250.3 | 99.7 | 88.3 |

(MB/s)

**Table 2.** Barrier synchronization time for threads in a node.

| threads | SMP PC | Sun SMP |
|---|---|---|
| 1 | 0.11 | 0.05 |
| 2 | 1.09 | 1.84 |
| 3 | 1.78 | 2.23 |
| 4 | 1.83 | 2.70 |

($\mu$sec)

NICAM also reduces latency in synchronization primitives. While latency is not the first issue in data transfer, latency is the only issue in synchronization. Direct handling of messages by the network interface not only frees the processors from polling overhead, but also eliminates the costly interaction between the processors and the network interface.

NICAM employs an Active Messages framework [2] for flexibility in programming the network interface. Active Messages provide extensibility by simple additions of new handlers, because they run almost mutually independently. This flexibility allows combination of data transfer primitives with synchronizations, which compensates for large latency.

This paper reports on the design of NICAM and its basic performance. We present our platform PC cluster in Section 2, and the NICAM primitives in Section 3. Then, we present a technique of latency hiding and two sets of experimental results in Section 4. We briefly discuss related work in section 5 and conclude in section 6.

## 2 Background

### 2.1 SMP PC Cluster Platform

Our research platform, COMPaS [3], is a PC-based SMP cluster, consisting of eight server-type PCs. Each node contains four Pentium Pro's (200 MHz, 512 KB L2 cache, 450GX chip-set, 256 MB main memory), and a Myrinet network interface card. They are connected by a single Myrinet switch. The operating system is Solaris 2.5. This section presents the performance characteristics of the node which guided our design of the communication layer.

Table 1 shows the memory bus bandwidth for read, write, and copy when multiple threads execute operations simultaneously. Each figure shows the aggregate bandwidth, that is, the sum of the measured bandwidth of each processor. Notice that the aggregate bandwidth is almost independent of the number of threads. This means that a single processor can consume all the bandwidth available to the memory bus for these simple operations.

Table 2 shows barrier performance among the threads in a node to compare the cache coherency performance. The column *Sun SMP* shows the results of the Sun Enterprise 4000 for comparison. The algorithm of this barrier uses write operations to a location dedicated to each processor, and detects the condition that all locations are written. A single processor checks the condition and notifies the others. This algorithm does not need any atomic operations and is regarded as reflecting the cache coherency performance. The result shows that the SMP PCs have a good cache coherency implementation comparable to a more sophisticated SMP machine.

## 2.2 Myrinet

Myrinet is a Giga-bit LAN system from Myricom Inc. [4]. It consists of communication links, switches, and host interfaces. The communication link is a bi-directional 8-bit data path, whose speed is 160 MB/s in each direction. The switch is an 8-by-8 crossbar and uses cut-through routing. Switches can be cascaded in an arbitrary topology, and the route is statically specified in the header of a packet. The host interface is implemented as an I/O adaptor, and performs data transfers from/to the host memory system using a DMA (Direct Memory Access).

Each Myrinet board contains three DMA engines, each dedicated to the main memory, transmitter and receiver, respectively, and a micro-processor to control these DMA engines. The on-board SRAM is used both to buffer messages and to store the program of the micro-processor. The following features guided the design:

**On-board micro-processor**: The micro-processor is a 32-bit custom RISC CPU core with a general purpose instruction set.
**Reliable link hardware**: Links are reliable to assume they are error-free.
**Unrestricted DMA capability**: The DMA engine is capable of accessing the whole physical memory in the host PC.
**Large SRAM size**: The SRAM size is up to 1 MB depending on the board type.

## 3 NICAM Primitives

### 3.1 Communication Layer Design

NICAM provides primitives for remote memory operations and synchronization primitives. We based communication primitives on remote memory operations, because they are preferable to message passing with respect to overhead. That is, message passing suffers from flow-control and buffer management tasks for handling incoming messages, and sometimes requires copying messages which sacrifices bus bandwidth. In addition, message passing may need mutual exclusions to coordinate processors in an SMP node.

In NICAM, all events to the main processors, such as completion of a data transfer or a barrier, are notified via a flag in the main memory, because it reduces much of the overhead of the main processor. Some primitives take a flag argument which points to a memory location. The flag is set by the network interface when a condition is satisfied.

|  |
| --- |
| *Initialization* |
| `nicam_init()` |
| `nicam_lock_memory(addr, range)` |
| *Simple Data Transfer* |
| `nicam_bcopy(src_node, src_addr, dst_node, dst_addr, size)` |
| `nicam_sync(flag_addr)` |
| `nicam_write1(dst_node, flag_addr, val)` |
| *Data Transfer with Synchronization* |
| `nicam_bcopy_notify(src_node, src_addr, dst_node, dst_addr, size,`<br>`    flag_addr, onoff)` |
| `nicam_set_counter(flag_addr, count)` |
| `nicam_bcopy_countup(src_node, src_addr, dst_node, dst_addr, size)` |
| *Broadcast* |
| `nicam_bcast(src_node, src_addr, dst_addr, size, flag_addr, onoff)` |
| `nicam_bcast_discard(onoff)` |
| *Barrier Synchronization* |
| `nicam_barrier(flag_addr)` |
| *Message Passing Support* |
| `nicam_bcopy_src(key, dst_node, src_addr, size, flag_addr, onoff)` |
| `nicam_bcopy_dst(key, src_node, dst_addr, size, flag_addr, onoff)` |

**Fig. 1.** NICAM primitives.

NICAM is designed for a single job, and makes exclusive use of the resources for communication. This is not a problem practically, since we assume a dedicated environment for a parallel job to investigate the utilization of the resources in a cluster. In addition, it requires that the remotely accessed regions of memory should be pinned-down in advance of remote memory operations. The pin-down operation protects the region of memory from the paging system in a virtual memory environment. Some other systems take alternative approaches and use only limited pinned-down areas [5].

### 3.2 Remote Memory Operations

Figure 1 lists the set of primitive operations supported in NICAM. Remote memory operations have a similar interface to the local copy (`bcopy`) operation, but have additional arguments to specify source and destination nodes. If the source is specified as a node other than the local one, it can act as a remote read. An Active Messages mechanism forwards the request to the specified source node. `nicam_sync` is used by the invoking node to know the completion of currently issued copies.

A variant of a copy operation, `nicam_bcopy_notify`, provides a data transfer primitive combined with point-to-point synchronization. It notifies the destination node about the completion of a copy by setting a flag. This is useful for programs taking a message passing style. While NICAM does not directly support message passing, many message passing programs can be rewritten using

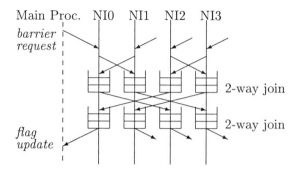

**Fig. 2.** Steps in a multi-stage barrier performed between network interfaces (NI0–NI3).

this primitive. It is also useful in optimization to reduce the number of synchronizations in data-parallel programs, where it is assumed that completions of data transfers are notified to the destination [6].

Another variant, `nicam_bcopy_countup` provides a counted completion. Sometimes synchronization points are known by the number of data items exchanged. For example, in all-to-all communication, the synchronization point is reached when the count reaches the number of nodes involved. This completely avoids explicit synchronizations. `nicam_set_counter` specifies the count and the flag address to signal completion.

### 3.3 Barrier Synchronization

The barrier primitive signals the completion by setting a flag in memory, too. This not only lowers the notification overhead, but also makes it a fuzzy barrier [7]. In addition, its implementation uses a relaxed algorithm which allows concurrent progress of multiple instances of barriers. That is, barriers can be invoked multiple times before the completion of a previously issued one. This feature tolerates latency and is useful when the environment has a large variance of loads, or the scale of a cluster is large.

Figure 2 shows the steps in a barrier execution. The barrier is implemented using a multi-stage $\log(P)$ step algorithm, and each stage of the barrier performs a 2-way join and forwards a message to the next stage. The node $n$ performs a join with the node $n \oplus 2^i$ at the $i$-th stage. Small queues are placed in the join to accommodate multiple instances of barriers. It is beneficial to avoid the involvement of the main processors, where the algorithm needs multiple stages. It would be worthless in relaxing barriers if the execution required polling by the processors.

### 3.4 Implementation

Incoming requests of Active Messages are handled using a simple polling on the network interface. There are three sources of requests: the main processors

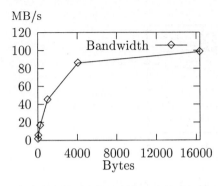 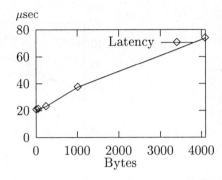

**Fig. 3.** Basic Performance of NICAM. Bandwidth (throughput) and latency (one-way) of remote memory data transfer.

and the two DMA engines of the receiver and the main memory, respectively. Active Messages are also used for local requests from the main processors, which simplify the implementation by making the local and remote requests be handled in the same way.

Hand-shaking is used between the main processors and the network interface to manage requests. The main processor makes a request by writing a flag in the SRAM on the network interface, and the network interface makes an acknowledgment by writing a flag in the main memory. In order to avoid mutual exclusions among the main processors, there are multiple pairs of flags, one for each processor.

NICAM passes only virtual addresses between nodes, and they are translated to physical addresses by the network interface. It maintains its own copy of the address translation table in the SRAM. The SRAM on the network interface is large enough to hold the whole table for the entire physical memory.

### 3.5 Basic Performance

Figure 3 shows the bandwidth (throughput) and the latency of `nicam_bcopy`, up to sizes of 16 KB for bandwidth and 4 KB for latency. The maximum bandwidth is about 105 MB/s observed at 64 KB, and the start-up parameter $N_{\frac{1}{2}}$ is a little below 2 KB. The minimum latency for small messages is about 20 $\mu$sec (one-way). Since NICAM uses pinned-down areas and does not need any buffering or flow-control, these figures show the actual performance observable from applications.

Figure 4 compares the barrier synchronization time between NICAM and a message passing library PM [5]. We implemented a barrier on top of PM, which is performed by the main processor using message passing. It uses essentially the same algorithm as the one used in NICAM. The figure shows that, while NICAM is slower than PM for two nodes, it becomes faster as the number of nodes increases. This is because the cost of interaction between the main

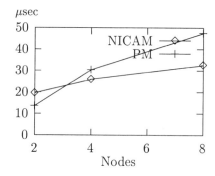

**Fig. 4.** Barrier synchronization time between nodes using NICAM and a message passing library PM.

processor and the network interface is larger in NICAM, but the interaction occurs only once in NICAM.

The call overhead of most NICAM operations is about 5.7 μsec. The break down of the overhead is: (1) argument checks, (2) copying of a handler address and arguments into the SRAM on the network interface, and (3) a word write to a flag in the SRAM to start processing. Serializing instructions (the CPUID instruction) are inserted after steps (2) and (3) to flush the write buffer. Although the serializing instruction has a very large cost, it is necessary because the write buffer of Pentium Pro reorders write requests.

## 4 Hiding Latency and Experimental Results

### 4.1 Hiding Latency of Barriers in an Array Class

The main target application of NICAM is a scientific computing library in C++. The array class library may be considered in a variant of the BSP (Bulk Synchronous Parallel) model. However, while synchronization points are explicitly specified in the programs in the BSP model, they are implicit in the array class and it is necessary to synchronize at each expression or statement.

Some researchers have been successful in avoiding explicit barriers at the end of the *super-steps* in a message passing environment [8]. However, synchronizations may still be required at the beginning when the communication is based on remote writes, because there is a possibility of overwriting the memory contents on which local processing is currently in progress. Since the library has no knowledge of the use of arrays, it would have to strictly lock-step in each super-step, and this would make the library sensitive to latency.

The barriers at the beginning of super-steps can be eliminated as mentioned in [6], where a new area is always allocated to store the results. It is clear that the remote writes are directed to those newly allocated locations and never overwrite the existing one. However, incorporating this technique requires modification to the storage reclaimer because the storage cannot be freed locally. That is, the

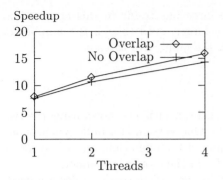 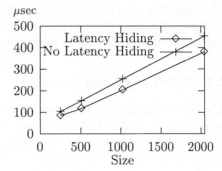

**Fig. 5.** Difference of speed-up between Laplace solvers with/without overlapping.

**Fig. 6.** The effect of latency hiding in the cshift operation. Size $N$ for a two-dimensional array of $N \times N$.

storage could be overwritten if a storage were reclaimed and then reallocated independently while other nodes still had a reference to it.

The storage should be reclaimed when all nodes are ready to release it. This condition can be checked by the use of barriers. The reclaimer first signals a barrier on a location specific to the freed storage, then checks the completion of the barrier later. Completion indicates that all the nodes are ready to release it.

The implementation of the barrier has the preferable properties of background execution and the concurrent progress of multiple instances. The background execution is important, because the barriers are triggered at the storage reclamation and it is hard to find insertion points for polling. It would otherwise degrade performance when the polling by the main processors were scattered in every operation. The concurrent progress is also important, because the variance of issuing of barriers is large and the barriers may overlap.

### 4.2 Overlapping Communication and Computation

Overlapping of computation and communication is a common technique in exploiting hybrid distributed/shared-memory programming on SMP clusters [3]. The communication overhead must be small enough for overlapping to be effective. An overlapping effect is presented for an explicit Laplace equation solver using the Jacobi method. In each iteration, a new array is computed from the old array by averaging the four neighbors in a two-dimensional space.

The array is partitioned into equally-sized strips and these are distributed among the nodes. Although the values of the boundary elements need to be exchanged between nodes, computation can be started without waiting for the completion of the exchanges, because the internal elements have no dependency on the ones exchanged.

Figure 5 shows the result of the experiments, when the number of threads is varied by 1, 2, and 4. The number of nodes is fixed to eight. The result shows the speed-up relative to the single processor, for the array of size $1024 \times 1024$.

The speed-up is not good because of the large bus traffic in this application. However, the gain of overlapping becomes larger, even though the number of communications increases in proportion to the number of threads.

### 4.3 Effect of Hiding Latency

The effect of hiding latency of the barrier synchronization is shown using the two versions of a cshift (cyclic shift) operation. One version is a bulk synchronous version which uses a barrier at the beginning of the operation, and the other is a latency hiding version which uses a barrier in the storage reclaimer.

Figure 6 shows the effect of the latency hiding. All the eight nodes are used, but only a single processor on a node is used. The shift size is one. The figure shows that the whole barrier cost is hidden (the gain is over 30 $\mu$sec). Also, slightly more gain is achieved by the relaxation effect of the synchronization condition in each step, because no explicit barriers appear in the code.

## 5 Related Work

Schauser et al reported on experiments running Active Messages on a network interface [9]. Also, Krishnamurthy et al reported on running the Active Messages handlers for Split-C primitives on various network interfaces [10]. They reported that low latency is achieved on platforms such as the Paragon and Berkeley NOW. However, NICAM further exploited the benefits of utilizing the network interface.

There are a number of fast message passing layers for the Myrinet network, such as BIP [11], FM [12], and PM [5]. These are concerned mainly with the bandwidth and latency. In contrast, NICAM aims at reducing overhead, but it shows some disadvantage in bandwidth and latency. This is due to the fact that NICAM runs much more work on the relatively slow micro-processor.

## 6 Conclusion

NICAM makes use of the micro-processor on the network interface to reduce the overhead in data transfer, and also to reduce the latency in synchronization. In addition, background execution of barriers can completely hide latency. It employs an Active Messages framework for flexibility in programming the network interface, which allows an easy integration of new primitives, such as the ones to help rewrite message passing by remote memory operations.

While NICAM is based on remote memory operations, it is not considered restrictive in data-parallel programming. The one-sided communications are often more suitable to implement data-parallel operations than message passing. Many data-parallel operations are straightforwardly implemented using remote memory operations, including shift, scan, and exchange operations. In contrast, it is sometimes necessary in message passing to swap the order of code to make a sender-receiver pair, or to use asynchronous operations to avoid a pair and to tolerate latency.

While many distributed memory MPP machines provide remote memory operations and fast barrier synchronizations in hardware, NICAM attempts to provide these operations using commodity network hardware. In NICAM, the data transfer primitives are optionally combined with actions for synchronization and the synchronization primitives take latency hiding into account, which will compensate for the relatively large latency in a software implementation.

# References

1. R. P. Martin, A. M.Vahdat, D. E. Culler, T. E. Anderson: Effects of Communication Latency, Overhead, and Bandwidth in a Cluster Architecture. *Int'l Symp. on Computer Architecture* (ISCA'97) (1997).
2. T. von Eicken, D. E. Culler, S. C. Goldstein, K. E. Schauser: Active Messages: a Mechanism for Integrated Communication and Computation. *Int'l Symp. on Computer Architecture* (ISCA'92), pp.256-266 (1992).
3. Y. Tanaka, M. Matsuda, M. Ando, K. Kubota, M. Sato: COMPaS: A Pentium Pro PC-based SMP Cluster and its Experience. *IPPS Workshop on Personal Computer based Networks of Workstations* (PC-NOW'98) (1998).
4. N. J. Boden, D. Cohen, R. E. Felderman, A. E. Kulawik, C. L. Seitz, J. N. Seizovic, S. Wen-King: Myrinet – A Gigabit-per-Second Local-Area Network. *IEEE MICRO*, Vol.15, No.1, pp.29–36 (1996). (http://www.myri.com).
5. H. Tezuka, A. Hori, Y. Ishikawa, M. Sato: PM: An Operating System Coordinated High Performance Communication Library. *High-Performance Computing and Networking*, LNCS 1225, pp.708–717, Springer-Verlag (1997).
6. M. Gupta, E. Schonberg: Static Analysis to Reduce Synchronization Costs in Data-Parallel Programs. *Symp. on Principles of Programming Languages*, pp.322–332 (1996).
7. R. Gupta: The Fuzzy Barrier: A Mechanism for High Speed Synchronization of Processors. *Int'l Conf. on Architectural Support for Programming Languages and Operating Systems* (ASPLOS-III) pp.54–63 (1989).
8. A. Fahmy, A. Heddaya: Communicable Memory and Lazy Barriers for Bulk Synchronous Parallelism in BSPk. Boston University Technical Report BU-CS-96-012 (1996).
9. K. E. Schauser, C. J. Scheiman, J. M. Ferguson, P. Z. Kolano: Exploiting the Capability of Communications Co-processor. *Int'l Parallel Processing Symposium* (IPPS'96) (1996).
10. A. Krishnamurthy, K. E. Schauser, C. J. Scheiman, R. Y. Wang, D. E. Culler, K. Yelick: Evaluation of Architectural Support for Global Address-Based Communication in Large-Scale Parallel Machines. *Int'l Conf. on Architectural Support for Programming Languages and Operating Systems* (ASPLOS-VII) (1996).
11. L. Prylli, B. Tourancheau: BIP: a New Protocol Designed for High Performance Networking on Myrinet. *IPPS Workshop on Personal Computer based Networks of Workstations* (PC-NOW'98) (1998).
12. S. Pakin, M. Lauria, A. Chien: High Performance Messaging on Workstations: Illinois Fast Messages (FM) for Myrinet. *Supercomputing'95* (1995).

# Experimental Results about MPI Collective Communication Operations

Massimo Bernaschi
IAC-CNR
V.le del Policlinico, 137
00161 Roma-ITALY
massimo@iac.rm.cnr.it

Giulio Iannello and Mario Lauria
Dipartimento di Informatica
e Sistemistica
v. Claudio, 21 – 80125 Napoli-ITALY
iannello,lauria@grid.unina.it

**Abstract.** Collective communication performance is critical in a number of MPI applications, yet relatively few results are available to assess the performance of mainstream MPI implementations. In this paper we focus on two widely used primitives, broadcast and reduce, and present experimental results for the Cray T3E and the IBM SP2. We compare the performance of the existing MPI primitives with our implementation based on a new algorithm. Our tests show that existing all-software implementations can be improved and highlight the advantages of the Cray hardware-assisted implementation.

## 1 Introduction

The development of parallel codes based on message passing is usually simplified if common operations like the evaluation of a global minimum can be executed by means of a single collective communication primitive. In the MPI standard [7] great importance is reserved to the definition of the collective communication.

In the same document, the implementors are *adviced* to use tree communication patterns to improve the performance. With such patterns (hereafter called *spanning* trees) messages can be forwarded by intermediate nodes instead of travelling directly from the source to the target node.

Many sophisticated algorithms are available to determine the *optimal* spanning tree, but they are hardly ever considered for inclusion in libraries for message passing like PVM or MPI. This is due to their complexity compared to the binomial tree which is regarded as an acceptable trade-off between efficiency and complexity.

To overcome this difficulty, we have implemented a *simplified* version of optimal algorithms (i.e. *quasi optimal*) for both PVM and MPI. In [2] we reported the details of our approach and a large set of results obtained with the PVM library. In the present work, we report the experimental results obtained with the MPI version of the same algorithms. The paper is organized as follows: section 2 is a short reminder of the theory of *quasi optimal* algorithms; section 3 describes the hardware/software components of the experiments along with our timing methodology; section 4 discusses the results, and finally section 5 concludes the work with a look at the future perspectives.

## 2  α trees

Our work is based on the *LogP* model [6] which is aimed to characterize a distributed memory parallel system by means of four parameters: the number of processor-memory pairs $P$; the lower bound on the time between two consecutive communication operations of the same processor called *gap* ($g$); the maximum delay associated with delivering a message from the source processor to the target processor called *latency* ($L$) and finally the *overhead* ($o$) that is the amount of time for which a processor is busy during a communication operation. The model assumes that the communication network is represented as a complete graph.

Our idea is to make use of a class of spanning trees (hereafter called $\alpha$-trees), that closely approximate the optimal schedule within the *LogP* model.

For instance, with a number of processors equal to $P$, the algorithm **bcast**$(S, P, s, m)$, that broadcasts a message $m$ in an $P$-node set $S$ from a source node $s \in S$ using an optimal spanning tree, works as follows.

If $P = 1$, **bcast** terminates; otherwise it performs four steps: (i) partition S into two subsets $S'$ and $S''$ of sizes $P' = \min(round(\alpha P), P-1)$ and $P'' = P - P'$, respectively, and such that $s \in S'$ (here *round* is the round-off function); (ii) select a leader $s' \in S''$; (iii) send the message $m$ from $s$ to $s'$; (iv) perform **bcast**$(S', P', s, m)$ and **bcast**$(S'', P'', s', m)$ concurrently and recursively.

The coefficient $\alpha$ must be chosen in the range $0.5 \leq \alpha < 1$ as a function of the $L, o, g$ and $P$ parameters. In [2] we show how the proper value of $\alpha$ can be determined in different cases. An analysis of the results reported in the cited paper suggests the idea that $\alpha$ is substantially independent from $P$. In such a way $\alpha$ becomes a constant for a given platform, which makes the implementation of the algorithm straightforward. The simplified algorithm generates a tree whose structure is dependent on just the value of $\alpha$.

An $\alpha$-tree is wider and shallower than a binomial tree with the same number of nodes, and this difference grows as $\alpha$ increases. Besides the broadcast, other collective operations can be implemented by exploiting $\alpha$-trees. In this work we report results about the *broadcast* and the *reduce* operations. Reduce looks like the inverse problem of broadcast. Actually the broadcast is a pure communication operation whereas the reduce entails the evaluation of a function. Further details about the implementation of both algorithms can be found in [2].

## 3  Hardware and software components

The hardware platforms we have used to run the tests are an IBM SP2 parallel system with 48 computing nodes and a Cray/T3E (same number of nodes).

The SP2 nodes (based on Power2 or PowerPC processors) are connected by means of the so called *High Performance Switch* (HPS). The HPS topology is a variant of an *Omega* network. Further details about this platform can be found in our previous works [2],[4]. Here, we recall that the behavior of the SP2 is represented pretty well by the LogP model.

The Cray T3E provides a shared physical address space of up to 2048 processors over a 3D torus interconnect. Torus links provide a raw bandwidth which exceeds 600 Mbytes/sec in each direction whereas the "payload" bandwidth is about 480 Mbytes/sec. The network implements fully adaptive, minimal-path routing[9]. The system is enhanced by a number of novel features aimed to tolerate latency and increase its scalability. The most important of these features is the set of external registers (*E-registers*) which augment the memory interface of the single node (that is an Alpha 21164 processor). The global E-register operations, called *Puts* and *Gets*, are used to transfer data to/from global (meaning remote or local) memory. Since there are a large number of E-registers, Gets and Puts may be highly pipelined [10]. In addition, a series of low level state flags (one per register) protects against the read-after-write hazard thus providing an implicit low-level synchronization mechanism. The Cray T3E network latency is about 1 microsecond if the basic shared memory primitives are used. However, the effective latency for message passing libraries is pretty higher (between 10 and 15 microseconds). The overhead is mostly due to the mechanisms of buffering and deadlock detection [10].

Our code makes use of the basic point-to-point communication primitives of MPI and all the information required to build the $\alpha$-trees (i.e. number of tasks, rank of a task) are retrieved by using MPI primitives. In such a way the code is immediately usable in any implementation of the MPI standard. For the *local* reductions, we have our own general routine. While this choice might somehow limits the absolute performance achievable with respect to an implementation fully integrated in the library, we had no alternatives for lack of access to the source of the proprietary (IBM and Cray) versions of MPI.

Moreover, we had to resort to *local* temporary buffers for the implementation of the reduce operation since a general buffering scheme is not available in MPI (for the broadcast operation no temporary buffer is required). The reduce operation is a simple product of integer numbers.

Among the results we have always included those obtained with $\alpha = 0.5$. Note that 0.5 corresponds to a "special" $\alpha$-tree: the binomial tree, a spanning tree widely used by many algorithms.

The timing of collective communication operations presents difficulties and requires some attention. It is difficult to measure accurately the time taken by a single operation because the timer call overhead may be large compared to the time to be measured. It is common practice to reduce such overhead by timing a long enough sequence of operations. To eliminate the effect of spurious runs, either the minimum or the average of a set of such measurements is taken as the final result.

However in some cases (notably the broadcast operation) such approach can induce a "pipeline" effect due to the overlapping of successive executions, which clearly leads to inaccurate results. To avoid such problems the following scheme has been used here: all the timings have been obtained by taking the minimum of a set of, at least, 20 independent runs. For each run, the collective operation has been performed 10 times and we have considered, among all the nodes, the greatest of the total elapsed times. The potential "pipeline" effect is eliminated by changing the *root* of the operation each time.

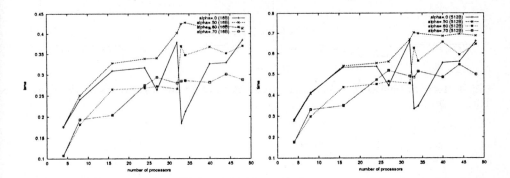

**Fig. 1.** Time in milliseconds for a reduce operation of 16 and 512 bytes on the IBM SP2.

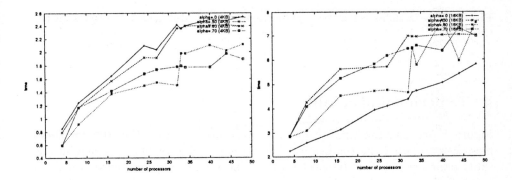

**Fig. 2.** Time in milliseconds for a reduce operation of 4096 and 16384 bytes on the IBM SP2.

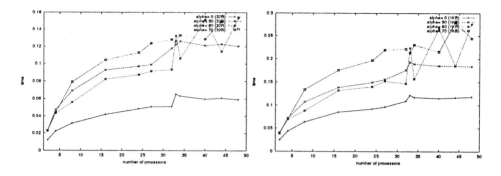

**Fig. 3.** Time in milliseconds for a reduce operation of 32 and 1024 bytes on the Cray T3E.

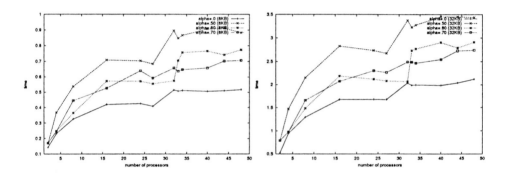

**Fig. 4.** Time in milliseconds for a reduce operation of 8192 and 32768 bytes on the Cray T3E.

## 4 Experimental results

Figure 1 presents the timings of the reduce operation on the SP2 for short messages; Figure 2 the timings for long messages. Figures 3 and 4 present the same data for the T3E (respectively for short and long messages). Figure 5 presents the timings of the broadcast of short messages on the SP2; Figure 6 the timings for long messages. Figures (7 and 8) report the corresponding results obtained on the T3E.

We like to stress that all the results with $\alpha$ spanning trees are obtained with the *same* code. This is a major strength of our approach: a single code implements a number of algorithms by changing a single paramenter. Besides the binomial

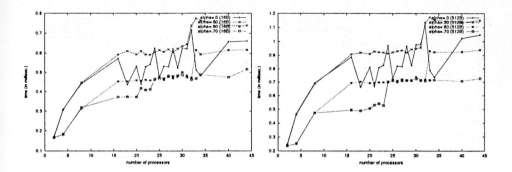

**Fig. 5.** Time in milliseconds for a broadcast operation of 16 and 512 bytes on the IBM SP2.

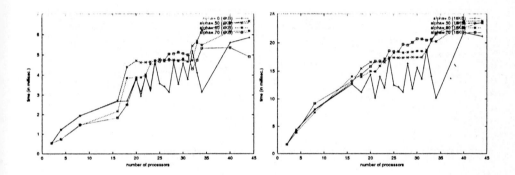

**Fig. 6.** Time in milliseconds for a broadcast operation of 4096 and 16384 bytes on the IBM SP2.

tree, another "well-known" algorithm that can be seen as a special $\alpha$-tree is the *single-loop* algorithm where the root is involved in all communications required by the collective operation. This case corresponds to a tree with $\alpha = 1$. It is no surprise, that the timings for $\alpha = 1$ scale linearly as the number of tasks increases. This behavior makes the *single-loop* algorithm pretty useless for any practical purpose. For such reason, results with $\alpha = 1$ are not presented here.

Data presented as $\alpha = 0$ are actually those of the *MPI_Bcast* and *MPI_Reduce* primitives available in the proprietary versions of MPI.

It is readily apparent that on the SP2, spanning trees with $\alpha > 0.5$ show better performances compared to the binomial tree ($\alpha = 0.5$) and to the IBM

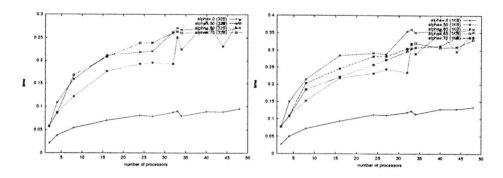

**Fig. 7.** Time in milliseconds for a broadcast operation on 32 and 1024 bytes on the Cray T3E.

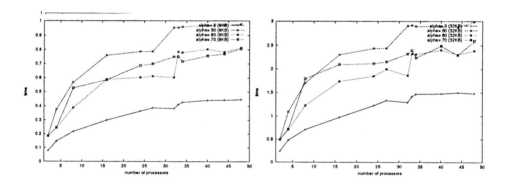

**Fig. 8.** Time in milliseconds for a broadcast operation of 8192 and 32768 bytes on the Cray T3E

implementation ($\alpha = 0$). Actually, the overlapping between the results with $\alpha = 0$ and $\alpha = 0.5$ in figures 1 and 5 it is a clear indication that the IBM library makes use of binomial trees, at least for short messages (less than 16 Kbytes). An unexpected *feature* of the IBM library is the instability of the results obtained by changing the number of nodes involved in the operation. For the *broadcast* operation, it looks like the code is highly tuned for cases with a number of tasks close to a power of 2.

For the Cray T3E, the picture is completely different. The Cray collective primitives perform significantly better than our code based on $\alpha$ trees. We were puzzled by such results and we have tried to understand them more in depth. After a number of tests we found that on two T3E nodes the following fragment of MPI code:

```
MPI_Bcast(message0,n,MPI_BYTE,0,MPI_COMM_WORLD);
MPI_Bcast(message1,n,MPI_BYTE,1,MPI_COMM_WORLD);
```

is twice as fast as an equivalent code based on point-to-point primitives:

```
if(rank==1) {
   MPI_Recv(message0, n, MPI_BYTE, MPI_ANY_SOURCE,
           10, MPI_COMM_WORLD, &status);
   MPI_Send(message1, n, MPI_BYTE, 0, 20, MPI_COMM_WORLD);
} else {
   MPI_Send(message0, n, MPI_BYTE, 1, 10, MPI_COMM_WORLD);
   MPI_Recv(message1, n, MPI_BYTE, MPI_ANY_SOURCE,
           20, MPI_COMM_WORLD, &status);
}
```

(obviously the two codes are "equivalent" only on two nodes). We double checked that on the SP2 these codes take approximately the same time (the point-to-point code is sligthly faster). The difference on the T3E can hardly be explained unless the *MPI_Bcast* is implemented directly on top of the shared memory support of the T3E. This "guess" is confirmed by [8].

However, figures 3,4,7,8 indicate that spanning trees with $\alpha > 0.5$ perform better than the binomial tree also on the T3E. So, within the limits of our assumptions (i.e. the collective operations make use of basic point-to-point primitives) the experimental data confirm, once again, the theory.

The MPI standard defines a number of "communication modes". As a further investigation, we have tried to observe their effect on the broadcast and reduce operations. On the T3E, a scheme which makes use of non-blocking send primitives (*MPI_Isend*) is always slower than an equivalent algorithm (same value of the $\alpha$ parameter) based on blocking send operations. Figure 9 shows that the difference increases as the message size becomes bigger.

Interesting results come from the tests on the reduce operation. Non-blocking communication primitives are supposed to allow a better overlapping between communication and computation. Actually we found that on the Cray T3E a non-blocking scheme is, in general, slower than a blocking one. Just in case of pretty large messages the performance is comparable (see Figure 10) but the expected advantage of the non-blocking primitives does not appear.

## 5 Conclusions

The collective communication primitives offered by vendors' implementations of the MPI standard have been compared with our routines based on *quasi-optimal* spanning trees. The results show that our approach, which does not introduce additional complexity, gives, compared to the classic binomial spanning tree, a clear advantage, unless special hardware primitives are available to support the collective communication.

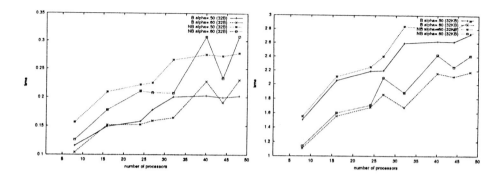

**Fig. 9.** Time in milliseconds for a broadcast operation of 32 and 32768 bytes on the Cray T3E using blocking (B) and non blocking (NB) primitives

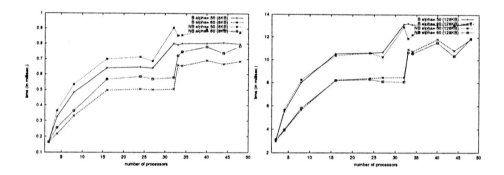

**Fig. 10.** Time in milliseconds for a reduce operation of 8192 and 131072 bytes on the Cray T3E using blocking (B) and non blocking (NB) primitives

In a forthcoming paper [3], we present a similar work for the *MPI_Reduce_scatter* primitive. This primitive is important because it can be used as a base for the development of other primitives like the *MPI_Allreduce*.

Our final goal is to release a library of collective communication primitives that can be *plugged* in any implementation of the MPI standard. Currently we are working on the *MPI_Scan* which is used to perform a prefix reduction. Note that we do not pretend to be able to improve the performance of *any* collective communication primitive. For instance, the all-to-all exchange (*MPI_Alltoall*), that is very interesting being the base of the parallel Fast Fourier Transform (FFT), requires each task to send (and receive) as many distinct items as the number of tasks. This leaves little room for the optimization of existing algorithms [5].

The results we have obtained with the non-blocking primitives for the reduce operation show that overlapping computation and communication does not give an immediate advantage. More in general they show how, in a complex field like parallel computing, any theory must be double checked with experimental data.

## Acknowledgements

This work has been carried out partially under the financial support of the Ministero dell'Università e della Ricerca Scientifica e Tecnologica (MURST) in the framework of the Project MOSAICO (Design Methodologies and Tools of High Performance Systems for Distributed Applications).

## References

1. A. Bar-Noy and S. Knipis,
   "Designing Broadcasting Algorithms in the Postal Model for Message-Passing Systems", *Procs. of the 4th Annual ACM Symp. on Parallel Algorithms and Architectures*, June 1992, pp. 11–22.
2. M. Bernaschi and G. Iannello,
   "Collective Communication Operations: Experimental Results vs. Theory", *Concurrency: Practice and Experience*, vol. 10, No. 5, pp. 359-386, April 1998.
3. M. Bernaschi, G. Iannello, M. Lauria,
   "Efficient Implementation of Reduce-Scatter in MPI", *Quaderno IAC, n.6/1998* submitted to *Parallel Computing*.
4. M. Bernaschi, G. Iannello and F. Papetti,
   "Efficient Collective Communication Operations for Parallel Industrial Codes", Proceedings of HPCN96, H. Liddell, A. Colbrook, B. Hertzberger and P. Sloot editors, Lecture Notes in Computer Science (Springer) n. 1067.
5. J. Bruck, C.T. Ho, S. Kipnis and D. Weathersby,
   "Efficient Algorithms for All-to-All Communications in Multi-Port Message Passing Systems", *Procs. of SPAA 94*, 1994, pp. 298–309.
6. R.M. Karp *et al.*, "Optimal Broadcast and Summation in the LogP Model", *Procs. of the 5th Annual ACM Symp. on Parallel Algorithms and Architectures*, June 1993, pp. 142–153.
7. Message Passing Interface Forum, "Document for standard message-passing interface",
   *The International Journal of Supercomputer Applications and High Performance Computing*, vol. 8, No. 3/4, 1994.
8. G. J. Miller (Cray Research-Global Product Support), *private communication*
9. S. Scott, "Synchronization and Communication in the T3E Multiprocessor", Procs. of ASPLOS-VII, Cambridge, 1996.
10. E. Anderson, J. Brooks, C. Grassl, S. Scott,
    "Performance of the Cray T3E Multiprocessor", Procs. of Supercomputing 97, *http://www.supercomp.org/sc97/proceedings/TECH/ANDERSON/INDEX.HTM*

# MaDCoWS: A Scalable Distributed Shared Memory Environment for Massively Parallel Multiprocessors[1]

Dimitris Dimitrelos and Constantine Halatsis

Athens High Performance Computing Laboratory and
Department of Informatics, University of Athens
ddimitr@hpcl.uoa.gr, halatsis@di.uoa.gr

**Abstract.** In this paper we present MaDCoWS, a software implementation of a Distributed Shared Memory (DSM) runtime system, specifically designed for massively parallel 2-D grid multiprocessors. The system takes advantage of the network topology in order to minimise the paths of the message sequences realising the shared operations. As a result its performance is increased and the system becomes scalable even to very large processor numbers. We present the basic ideas for 2-D optimisations, the implementation structure and results from synthetic and application benchmarks executed on a 1024 processor Parsytec GCel.

## 1. Introduction

Distributed Shared Memory (DSM) tries to eliminate the gap that traditionally existed between the Shared Variable programming paradigm and the Distributed Memory architecture model. DSM offers the Shared Variable programming model, avoiding the intuitively obscure Message Passing paradigm, and at the same time is applicable to low-cost Distributed Memory multiprocessors which can scale to a large number of processors without adding further complexity levels to the hardware.

2D-grid interconnection networks have been very popular with system designers as they scale to a large number of processors with low cost material and relatively low complexity designs. As a result many 2D grid parallel systems have appeared employing thousands of processors.

MaDCoWS (Massively Distributed Configurable Variable Validation System) is a software implemented DSM environment (offering shared variables or data structures of arbitrary size, global semaphores and global barriers) specifically aimed to cover very large 2D-grid multiprocessor systems. MaDCoWS has been implemented under the PARIX parallel operating environment and tested and

---

[1] This work was partly supported by the Greek General Secretariat of Research and Technology under the YPER-94 program.

benchmarked on a Parsytec GCel-1024, a 1024 transputer machine. The system is currently being ported to the Intel Paragon and to the Parsytec GC-PowerPlus.

MaDCoWS has been implemented in a configurable way regarding many of its operational parameters. As a result the programmer can exploit his program's specific behaviour configuring the system in terms of

- Coherency (consistency) model used (sequential consistency or processor consistency)
- Distribution of Directory Data (Centralised Directory distribution, Full Distribution, Reciprocal Distribution, User Defined distribution)

## 2. MaDCoWS design issues

The first decision that a DSM system designer faces is the level of implementation. Leaving hardware (high cost, high complexity) aside, the decision has to be made between preprocessor-compiler, operating system or runtime system. Compilers-preprocessors can not yet devise a complete dynamic communication structure hiding all implementation details from the user and providing a straightforward shared variable environment.

Implementation on operating system level was not feasible since PARIX[1], for which MaDCoWS was originally written, does not provide virtual memory support. Therefore our options were limited to a light-weight implementation on top of the message passing language itself. Apart from the simplicity of the design and its light weight, the big advantage of the runtime system is that it can be used in parallel with the message passing system, a combination which has been demonstrated to work well in many applications[2].

After that, choosing the programming object as the sharing unit comes as a natural choice. In MaDCoWS, the user can share any kind of variable, as well as any spatially continuous data structure. This gives a feeling of 'object-oriented' sharing and avoids one of the most common problems in DSM implementations: false sharing[3]. The only drawback of this approach is the fact that the DSM system can not identify or exploit the spatial locality of the shared variables.

The coherence model is not really chosen by the DSM system designer. It is dictated by various implementation restrictions and optimisations. In fact, hardware and communication pattern optimisations were the driving force for inventing weaker coherence model. This is also the case with MaDCoWS. Specific implementation optimisations, which will be discussed in detail later, forced us to weaken the coherence model from sequential consistency, under which "the result of any execution is the same as if the operations of all processors were executed in some sequential order, and the operations of each individual processor appear in this sequence in the order specified by its program"[4] to processor consistency (formally defined in [5]) under which the order of observed writes between different processors may differ. Nevertheless, in MaDCoWS the programmer can choose between the two coherence models balancing the trade-off between simpler, risk-free design and lower shared operation latency.

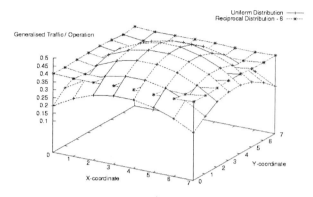

Fig. 1. Simulated communication traffic for Uniform and Reciprocal Directory Distribution. X and Y correspond to the processor's position in the grid and the Z value to its traffic

Finally, the protocol used was a directory-based, invalidate on write coherence protocol, that allows overlapping write-permission and invalidation messages in order to reduce shared write latency. The novelties of the protocol are

- The way the read and write accesses of the shared data are implemented, in order to take advantage of the 2D spatial locality of involved processors
- The distribution of directory data, that is discussed in the next paragraph.

Most commercial 2D grid systems use the X-Y (messages travel first along the X and then along the Y axis) message routing algorithm. X-Y routing has the disadvantage of sending more traffic through the central than through the edges of the grid. This property is inherited to any DSM system that distributes the directory entries uniformly (as the majority of directory-based DSM system do). As a result the communication load on the central processors is considerably (up to 41%) higher and the centre of the grid gets congested easier. MaDCoWS employs a non-uniform static directory distribution algorithm, called *Reciprocal Directory Distribution* in order to eliminate the message centralisation caused. Acoording to the proposed scheme bigger parts of the global object directory are allocated to the processors that are far from the centre of the grid. As a result the communication traffic is equalised throughout the grid as simulation (figure 1) and implementation[6] results show. Reciprocal directory distribution is discussed in detail in [6].

## 3. Implementation Details

MaDCoWS was implemented over PARIX[1] as a runtime C-library. PARIX offers a typical message passing system environment with synchronous and asynchronous communication, thread creation and handling, optioning (selecting) and local timers. The user of MaDCoWS is provided with a set of library calls which he/she can use to access shared variables, global semaphores and barriers. Our system was developed as a set of modules each of which corresponds to a set of threads. All the threads are executed concurrently in all processors that take part in the DSM system. The module structure is depicted in figure 2. In more detail, the modules are:

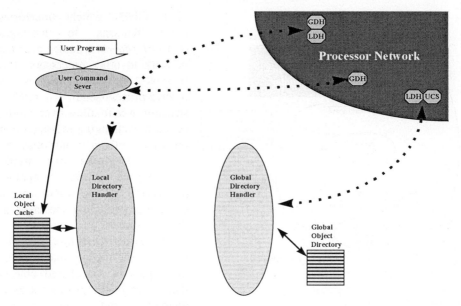

**Fig. 2.** Internal structure of MaDCoWS

The *Global Directory Handler* (GDH). The global directory includes all information that is relevant to the shared objects. Respectively, each shared object has a directory entry associated with it. The directory entry for each object, which is dynamically created, keeps the object's id, its status (exclusive, copied or invalidated), its copy owners and its size. GDH is the module that takes over the task of handling the global directory entries that exist in each processor. It is also responsible for iniating the sequences that serve the user requests for shared read or shared write accesses. Each shared object, semaphore or barrier has a GDH associated with it. The GDH of each shared object is statically assigned and known at all times to all the processors in the system. Internally GDH consists of 3 separate modules (figure 3):

The *incoming request handler*(GDH_IRH), which receives shared access requests from the user program, preprocesses them and places them on an internal FIFO queue for the request server. An important point is that the GDH_IRH performs rotating optioning (selecting) watching for incoming messages from a different starting processor each time. This is done in order to avoid the inherently unfair implementation of the select mechanism in PARIX which favours the first processors of the selected (or waiting-for-their-messages) processors in the option list. The *request server* (GDH_RS) does the main job of processing the requests. It finds out which processor should send a copy to the requesting one, in the case of a read access, or it grants write-permission to the requesting processor, after invalidating or initiating an invalidation sequence for older copies, for the write case. GDH_RS also handles global semaphores and global barriers. When the request processing is finished, GDH_RS generates a number of outgoing request which are placed on the outgoing FIFO queue. This queue is then read by the *outgoing request handler*

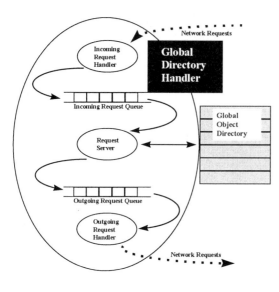

**Fig. 3.** Data flow on the Global Directory Handler

(GDH_ORH) which transforms these requests to messages (*network requests*) passed over the network to other processors. The reasons for implementing a receive-process-send module structure was to allow overlapping between exchanging messages and processing them, and also to prevent deadlocks that would inevitably occur when 2 separate processor's GDHs would try to exchange conflicting messages at the same time.

The *Local Directory Handler (LDH)* is the module that handles the copies of the shared objects that reside in the processors local memory and are owned temporarily by that processor (in the local 'object cache'). The LDH invalidates copies of outdated objects, or sends read copies to requesting processors (on GDH's request ). LDH maintains a structure very similar to the GDH with an incoming request handler, a request sever and an outgoing request handler.

The *User Command Server (UCS)* is the interface between the single command access requests by the user program and the rest of the DSM modules. It verifies the ownership of the requested shared objects and it performs the memory transfers between the user code and the DSM system.

## 4. Optimising Read and Write Accesses

### 4.1 Read Accesses

When a shared read access operation occurs in some processor (denoted as R in figure 4), and the object is not sufficiently owned locally (i.e. no valid copy exists in the local 'object cache'), then the processor holding the directory entry (DH) for the specific object is contacted through a read-access message. Consecutively, DH sends a send-copy request to one of the owners (for example O1, O2 and O3) of the shared object. The selected owner then forwards a copy of the message back to the original processor. In this way, a read sequence path is created by consecutive messages.

The length of the read sequence path is the dominant factor of the time that the messages travel on the network and therefore of the latency of the operation

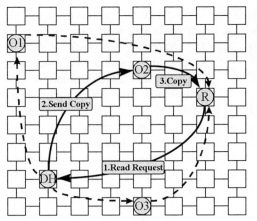

**Fig. 4.** Optimised Read Sequence

(compared to the network transmission times the a few commands processing times in the processor are negligible). Also, the third part of the sequence (i.e. the message carrying the variable copy to the requesting processor) has greater relative weight than the first two, if the variable has bigger size than a normal system message (4 bytes).

MaDCoWS takes advantage of the processor's proximity in the grid to minimise the length of the read sequence path, as well as the distance that the last message of the sequence has to travel. When a read access request is sent from a processor to the object's DH, the latter checks all owners of the object, and sends the send-copy request to the one that is closer to the requesting processor, or the one that creates the shortest read sequence path, depending on the size of the shared object. As figure 4 shows, the owner that has the best relative position is owner O2, therefore the solid line path is chosen from the dashed line paths which would have a bigger total length and would place the owner farther from the requesting processor.

### 4.2 Write Accesses

In a similar fashion, the path of the write sequence must be minimised. Since in this case there is no actual data transfer (only DSM system messages are exchanged) the path length is the only factor determining the latency of the operation. However, the message sequence is not as straightforward as in the read case, as the coherence model assumed also plays an important role. The sequential consistency model requires that all existing copies of the shared object must be verified to have been invalidated before the actual write can take place locally. On the other hand, if processor

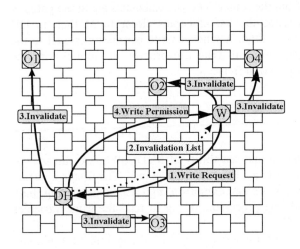

**Fig. 5.** Optimised write sequence (sequential consistency)

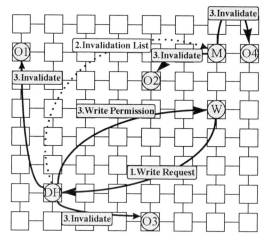

**Fig. 6.** Optimised Write Sequence (Processor Consistency)

consistency is assumed then the permission to write can be given to the requesting processor by the DH before the invalidation messages have reached their destination. The techniques that are employed by MaDCoWS for the two separate cases are the following:

Assume 6 processors: W is the processor requesting a shared write on a variable (whose directory entry is handled by processor DH) that is copied by processors O1, O2, O3 and O4. If the user chooses to use the strong consistency model (sequential consistency), then on arrival of the write-request (that was sent by processor W) and if the number of copies-to-be-invalidated is significant (>2), the DH calculates the paths that the invalidation messages have to travel. If there are processors that reside closer to the requesting processor than to the DH (O2 and O4 in figure 5), then the DH sends to the requesting processor their id's. At the same time, the DH sends invalidation signals to the rest of the owners (O1 and O3). The requesting processor (W), after invalidating the processors on the invalidation list, waits for the write permission signal from the DH. The sequence is depicted in figure 5.

In the case of the relaxed consistency model (processor consistency) the DH is assisted by its *mirror* (M) processor, i.e. the processor residing in the antidiametrical position of the grid, measuring from the center. The DH calculates again the paths of the invalidation messages and sends the list of copies closer to M than to DH (invalidation list) to M. At the same time, it sends a write permission signal to the requesting processor (W), so that the local write can procede in parallel with the invalidations. The sequence is depicted in figure 6.

## 5. Benchmark Results

In order to evaluate the performance and scalability of MaDCoWS, we run several synthetic benchmarks and two real life applications -two games. The first is a virtual odd-even game played by the processors that exhibits a high degree of sharing, both temporally and spatially, with all the processors trying to read the same variables concurrently.

The second application is a Connect-4 game played by the system vs. a human. The system plays using a relatively simple decision-tree recursive algorithm. The decision tree is split into many branches which are allocated to a processor waiting in a processor pool. The algorithm is load balanced by further splitting decision

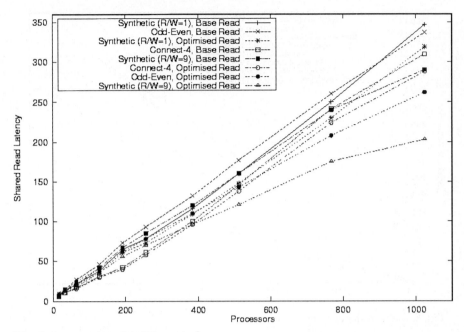

**Fig. 7.** Average shared read latencies[2]

branches and re-allocating them to idle processors. Connect-4 exhibits a low degree of sharing since variables are usually shared between 2 processors in an unsynchronised, data-flow driven pattern. The synthetic benchmarks tried to imitate an average application program behaviour by accessing random variables with different read/write ratios. The base model for measuring our proposed read sequence assumed the use of the first owner from the owner list as the actual sender of the object copy to the requesting processor. For the write case, the base model assumed DH to invalidate all copies (waiting for invalidation acknowledgment or not, according to the consistency model) before granting write permission.

In Figures 7 and 8 the results from the execution of the benchmarks under the processor coherency model are presented[3].

The Odd-Even game is the application that exhibited the most intensive use of the DSM system, as it is based on 'broadcast' type operations for several variables. Therefore, the read/write ratio is very high. As a result, at the time of the shared access there many copies available in the grid. That means that if the access is a read operation most probably there will be a copy near to the requesting processor, and our grid optimisation scheme will perform well. Also, in the case of a write access, there are many copies to be invalidated therefore our invalidation scheme performs better than the traditional one-by-one invalidation scheme. Consequently, in this benchmark the performance gains from using the proposed access sequences were

---

[2] Times are in thousands of the transputer high frequency counter units.
[3] We chose to visualise this coherence model since it performs better. Benchmark results under sequential consistency are similar.

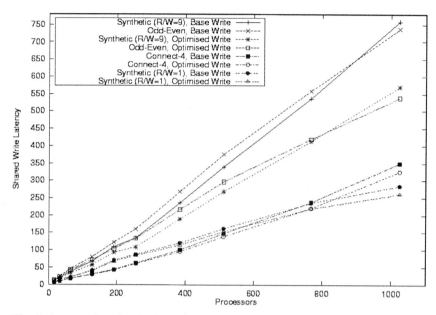

**Fig. 8.** Average shared write latencies

high: 22% for read, and 27% for the write case. The algorithm implementing the odd-even game is rather tightly synchronised, therefore relaxing the consistency model from sequential consistency to processor consistency did not result to any significant performance improvement.

| Consistency Model | Operation | Odd-Even | Connect-4 | Synthetic r/w ratio=9 | Synthetic r/w ratio=1 |
|---|---|---|---|---|---|
| Processor Consistency | Read | 22% | 8% | 30% | 2% |
|  | Write | 27% | 7% | 25% | 8% |
| Sequential Consistency | Read | 25% | 7% | 35% | 2% |
|  | Write | 30% | 7% | 25% | 5% |
| Proc-Seq rel. performance |  | 105% | 115% | 105% | 120% |

**Table 1.** Performance Improvements for the executed benchmarks

Unlike the Odd-Even case, the Connect-4 game exhibits a sparse sharing behaviour, with most shared variables being read only once by another processor before they are rewritten. As a result both the read and the write optimised sequences do not improve the performance dramatically (up to 8% and 7% for read and write respectively). In this case the algorithm is less strictly synchronised, therefore the relaxation of the consistency models allows more communication/computation overlapping and the algorithm performs better.

Finally, the efect of the optimisations on the synthetic benchmark depends, as expected, greatly on their r/w ratio and varies from 2-25%.

The actual improvements[4] in performance for the average shared read and shared write latency are presented in Table 1(in the last row of the table the relative performance of the two consistency models is shown).

The overall impression obtained by the benchmark execution experiments, is that MaDCoWS is a highly scalable DSM system that does not cause congestion to the system, as is obvious by the linear-like graphs, even when operating on very large numbers of processors.

## 6. Conclusions

In this paper we presented MaDCoWS, a multi-parametric and scalable DSM system for massively parallel multiprocessors. MaDCoWS has been optimised for use with 2D grid topologies. The shared accesses are using the information on the processors' position in the grid in order to minimise the total distance that the messages realising these operation have to travel. Benchmark results indicate that the proposed message sequences improve the basic model by up to 35% in large processor grids.

## 7. Acknowledgments

The authors would like to thank Dr.Alexander Reinefeld and the PC2 centre of the University of Paderborn for allowing us to use their parallel computers for benchmark runs.

## References

1. Parsytec Computer GmbH, '*PARIX 1.2: User Manual*', 1993.
2. Cordsen, J. et al., W. '*On the Coexistence of Shared-Memory and Message Passing in the Programming of Parallel Applications*'. In Proc. of the HPCNE'97, pp. 718-727, Apr 1997.
3. Freeh, V. W. and Andrews, G. R. '*Dynamically Controlling False Sharing in Distributed Shared Memory*' . In Proc. of the Fifth IEEE Int'l Symp. on High Performance Distributed Computing (HPDC-5), pages 403-411, August 1996.
4. Lamport, L. '*How to Make a Multiprocessor Computer That Correctly Executes Multiprocess programs*'. IEEE Transactions on Computers, C-28(9):690-691, Sept. 1979.
5. Gharachorloo, K., Gupta, A., and Hennessy, J. L. *Revision to 'Memory Consistency and Event Ordering in Scalable Shared-Memory Multiprocessors*'. Technical Report CSL-TR-93-568, Computer Systems Laboratory, Stanford University, April 1993.
6. Dimitrelos, D and Halatsis, C. '*Improving the Performance of Distributed Shared Memory Environments on Grid Multiprocessors*', Submitted to EUROPAR'99.

---

[4] Improvement percentages for 1024 processors.

# Workshop:

# Virtual Reality

# VisualExpresso : Generating a Virtual Reality Internet

David Cleary[1], Diarmuid O'Donoghue[2]

Department of Computer Science,
National University of Ireland, Maynooth,
Co.Kildare,
Ireland.
eeidcley@eei.ericsson.se[1] dod@cs.may.ie[2]

**Abstract.** Finding knowledge from vast quantities of data is a difficult task, made simpler by visually representing this information. The Internet can be considered a vast (global) database, but whose unstructured format frequently leads to a feeling of being "lost in hyperspace". Visualising this structure can make browsing a more productive endeavour. In this paper we introduce a means of "lossy" translation from HTML to VRML (Virtual Reality Modeling Language). This allows for the generation of a virtual world containing the structure with some content information about the site. This implements a new metaphor for intranet and internet browsing.

## 1 Introduction

When presented with an overwhelming quantity of information, it is difficult to know where to start, especially when trying to gain an understanding of the overall content of the information. This problem is complicated even further when the information is distributed and inter-linked by dynamic references. The WWW (World Wide Web) is an example of such a system, where a collection of electronic documents are connected by hyperlinks across numerous machines around the world. The current means for navigating and retrieving information from the WWW is by using conventional web browsing technology as offered by Microsoft and Netscape. These operate by manipulating graphical interfaces with the aid of search engines. However this method of interacting with the Internet using a page view is frustrating and forces the user into a DFS (Depth First Search) of the electronic documents which often leads to a fruitless search for information. This could be overcome by viewing the web site in its entirety.

One possible solution is to understand its structure, and allowing meaningful random access to it. A structure map for a subset (web site) of the WWW can be accomplished by firstly constructing an abstract representation of the site, presenting its meaning to the user in a graphical form. Many web sites try to accomplish this by providing static site maps in the form of two-dimensional picture. VisualExpresso is a powerful visualisation tool written entirely in Java, that allows site structure maps to browsed in three dimensions, and unlike most other site map tools can be used in a number of different ways:

- A Network Management tool allowing Webmasters overview sites.
- A Site Design tool, enabling the visualisation of web site structures in a breadth first manner from any starting position. This is of great importance when designing an effective and navigable site.
- A navigation tool, the virtual environment allow the user to select any page at any level in the site structure, immediately jumping to the page by simply clicking the mouse.
- Being able to visualise the neighbouring HTML pages from your current position allows greater freedom when searching a site for a particular piece of information. It also avoids revisiting pages when re-searching a site.

Another major problem when dealing with information is the format in which it is presented to the user. One possible solution is to present the information in graphical from. Various techniques exist for visually understanding large information spaces [1]. The most realistic way of understanding abstract representations is to map them to the physical world, by manipulating three-dimensional graphics that allow interaction with constituent graphical objects.

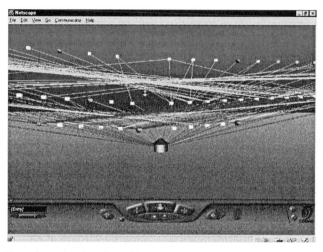

**Fig. 1.** Screen Shot taken of a virtual world representing a web site created by VisualExpresso.

VisualExpresso generates virtual worlds representing the structure and contents of intranet sites. This requires extensive computing power to firstly retrieve the entire contents of a web site, second to parse the entire contents of this site, third create intermediate structure and finally generate the final virtual world. The almost unlimited expense of such operations limits our tool to generating Intranet worlds, where "off site" internet locations are detected via their URLs and are depicted by spheres that are texture-mapped with globes - as in Figure 1. Visualisations of large sites also imposes heavy requirements on the browser's platform. For a true virtual internet world, each site requires an equivalent virtual world representation. Thus, constructing a virtual reality internet would require extensive computation, distributed across many (or all) internet servers. The solution we propose to this seeming impasse is to have each site generate its own intranet world, which is lined to other sites via the existing URL locations. This distribution of computation makes possible the creation of a Virtual Reality interface to the internet. Without this distributed computation, the overwhelming complexity and cost of generating an almost unbounded Internet world is intractable. Furthermore, by a standard representation of nodes between visual worlds, we facilitate navigation between intranet worlds/sites. Thus is born the three dimensional internet.

## 2   The Need for New Browsing Techniques

Currently, most users of the Internet use conventional browsers like Netscape or Internet Explorer to extract information from the WWW, with heavy reliance on search engines. Although these methods are easy to use and have their place, using them to find exact pieces of information is nearly futile unless you know the exact location of or description of the data. One of the main failings of the conventional browser is its interface, which encourages depth first searching. Every time you descend a level the previous levels links disappear and the user is faced immediately with the choice of the next level. By continuing in this fashion down the information tree, the user is forced into a depth first search. This searching technique often leads to a *dead link* (one of no interest to the user), forcing backtracking [2] and resultant confusion.

One possible solution to these problems is to obviate the structural information, acting as a "road map" outlining the location of web pages contained within the web site and how to get to them. This information can act as an invaluable navigation tool to the user. Site maps let the user see the current position in relation to other web pages, supporting the identification of other pages from the current position. This information is crucial to any *productive* general purpose internet browsing.

## 3   Three Dimensional Web Browsing using VRML

VRML (Virtual Reality Modeling Language) is a standardised means of creating three-dimensional virtual environments over the Internet. It is an interpreted language, and is viewed through a VRML browser. The VRML worlds can also be viewed through a conventional web browser with the aid of a *plugin*. These VRML browsers have various navigational tools to let the user interact with the virtual environment. The most exciting features of VRML is that it allows the creation of dynamic worlds and sensory-rich virtual environments. The following features are very desirable when developing a new metaphor for browsing the Internet
- Animate objects in your worlds
- Play sounds and movies within worlds
- Allows users to interact with worlds
- Control and enhance worlds with scripts - small programs that are created to act on VRML worlds.

## 3.1 The Structure of VRML

VRML is hierarchical in structure and is built up of *nodes* describing shapes and their properties. Nodes are the building blocks for virtual worlds, and individual nodes are used to describe the shape, colour and appearance of the objects within the world. They also deal with the positioning, orientation and interaction of objects. A VRML file can contain any number of the following components:
1. The VRML header (compulsory at the beginning of the file)
2. Prototypes, a means of defining nodes.
3. Shapes, interpolators, sensors, and scripts that enhance the Haptic / Kinaesthetic system [3].
4. Routes (wired instructions) to make your world dynamic.

VRML has a pre-defined set of primitive shapes (box, cone, cylinder, line and sphere), and these are combined to create more complex shapes. The language also provides the means for defining custom shapes. VRML also supports the concept of *inline files* which allows the worlds to be defined in numerous files and connected together using *anchors*. Anchors are attached to shapes within the world and specify the location of the inline file redefining that section of the world. The location of the inline file is specified by a URL contained within the anchor node.

## 3.2 Why use VRML?

*"Virtual reality, or virtual environment techniques, will change the way in which man interacts with computer systems"* [3]. By displaying the hyper-documents structure in a virtual three-dimensional environment, we gain a better understanding of its content. Any interacting with the three-dimensional worlds we must address the following:
- **Navigation.** The world is viewed through a browser and provides a set of navigation tools to interact with the environment. These navigation tools allow total freedom within the VRML world, and is accomplished by the combined use of both mouse and keyboard. Although this interaction does not provide a great deal of haptic responses to the user, it offers enough feed back to give the impression of navigating through a virtual world.
- **Users perspective.** The visualisation site is comprised of distinct stages:
1. The set up stage specifies the particular web site to be viewed and the depth to which the site is to be analysed, combined with the required protocols to be displayed in the virtual world.
2. Once the set up stage is complete the user is free to interact with the world through a VRML browser. Unlike other web site visualisation tools, the user can create any number of views of the site by interacting with the virtual environment and manipulating the objects contained within it.
3. Content view. Under certain VRML viewers, we can see a pages URL or title (etc.). This allows the user to view a page, without the usual exhaustive DFS steps.

# 4 VisualExpresso (for 3D web browsing)

Software visualisation is a human computer interfacing technology devised to improve human understanding of complex computer software using graphical design, typography, animation and cinematography techniques [4]. The application that was developed did not try to analyse all aspects of a web site, but concentrate on certain aspects of it. Incorrect or malformed *href* are ignored. For efficiency reasons, only pages that are contained in the same domain are parsed. A subset of the protocols supported by URLs [5]; HTTP, FTP, GOPHER, and MAILTO are represented. The structure of the web site is static reflected in the structure of the site at a given moment in time.

## 4.1 The Problems VisualExpresso addresses

When working with hypertext and other large information repositories, we encounter a variety of related problems. Among these problems are:
- How to conceptually understand the context and the expansiveness of the virtual environment?
- How to display large quantities of information in a confined space?
- How to provide a clear and consistent navigational cue throughout the virtual environment?

The World Wide Web is a vast collection of electronic documents, however it suffers from numerous problems when one tries to visualise it:

1. Organisation. The web is built upon a network of hypertext documents which has no discernible structure. This randomness provides a flexible way of presenting information, however it makes developing a context for a site almost impossible.
2. No physical context. As the user of a web site can only inspect one page at a time, developing an overall perspective of the site context can not be accomplished without extensive navigation through the site - and even so a complete picture is difficult to imagine.
3. Navigation and Orientation. Due to the nature of web sites and the limited navigation tools offered by conventional browsers, one can easily get the feeling of being "lost in hyperspace".

The above outlined problems are mostly dealt with in an *ad hoc* manner by the use of static site maps, frames or block diagrams created by hand. These methods are not data driven and are difficult and costly to maintain. Furthermore, they only capture a small section of the overall structure of the web site.

## 4.2 Abstract Graphical Objects Requirements

Abstract graphical objects provide the foundation for our visualisation. They are the mappings from the internet domain to the three-dimensional world. They are required to satisfy the following criteria [6]:
1. **Completeness.** Represent the entire target domain.

2. **Extensibility.** As the system will grow, it must be able to accommodate the unforeseen introduction of new graphical objects.
3. **Hierarchy.** Each visualisation must be broken down into reusable components. This helps in the logical mapping from abstract objects to graphical objects.
4. **Parameterization.** All properties of the graphical object should be parameterized to support declarative mappings, and ease of adaptation and expansion.
5. **Abstraction.** The graphical mapping should be represented independently in case they need to be changed.

VMRL provides a hierarchical language that allows construction of graphical objects in a manner that satisfies these criteria. In the next section we will examine how these abstract requirements can be mapped onto a VRML representation of site structure.

### 4.3 Three Dimensional Representation Of a web Site

The representation of the web site takes the graphical form of an inverted BFS (Breath First Search) tree, with each level in the tree an equivalent level of indirection from the base page. Before we are able to represent the structure of a web site, we must first establish the following:

1. **Node Representation.** The type of nodes in a web site are identified by their protocol as specified in the URL. Each supported protocol was assigned a VRML primitive shape, with a texture mapping applied to the shape, made key nodes easier to identify. A special node identifies the base page. Nodes located outside the DNS (Domain Name Server) of the base page are represented by a sphere. It should be noted that the above elements are defined in inline VRML files and are independent of the world generation application.
2. **Representation of hyperlinks.** The representation of hyperlinks between nodes was divided into two categories depending on their position within the web site:
   - Links from a node to a deeper node, further down the parse tree
   - A link coming from a node lower down the site tree back to the current node.
   - Recursive links are not graphically represented.
3. **Dead Nodes.** These are elements contained within the web site that could not be connected too. These are depicted by black spheres.
4. **Scalability within world.** Each element within the world with the exception of the base node has the same VRML dimensions. The reason for making the base node larger was to make its identification easier in large web visualisations.

## 5 Overview of VisualExpresso's Architecture

VisualExpresso is a JAVA application that creates a VRML world to represent any web site on the internet. The application structure is designed on a component object oriented framework, that facilitates expansion.

## 5.1 Framework for creating three dimensional Worlds

The VisualExpresso tool can be thought of as a collection of abstraction objects viewed in a graphical manner [6]. This integrates three major functional units.
1. Generate abstraction objects.
2. Store the extracted objects in a coherent manner.
3. Manipulate the extracted objects to produce a visual environment.

The VisualExpresso Use Case Diagram is depicted in Figure 2.

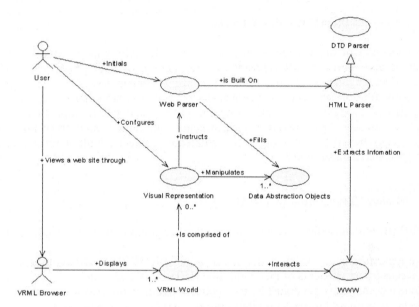

**Fig. 2.** Use Case Diagram of the interaction of the user with VisualExpresso.

## 5.2 Information Extraction Process

It is clear form our initial analysis that the extraction of the information is going to be of major concern. This process will have to provide a flexible interface for the raw extraction of structural data. It will also have to be abstracted to a level that allows easy extension, supporting different types of information from the data domain.

The World Wide Web structure extraction process, can be thought of as a hierarchical problem. Each subset of the WWW to be analysed is a collection of smaller entities referenced by URLs [7]. So the extraction of data will adhere to two main constraints.

1. Parsing a given subset of the WWW in a breath first manner to an arbitrary dept, from an initial start point.

2. Extracting information from the individual entities within the subset of the WWW, this was be limited to HTML pages

HTML pages are parsed to extract structural information contained within them. The parsing was limited to URLs contained within *href* tags. This extraction process was accomplished by building an HTML parser. The HTML parser used was constructed by creating a package containing four objects and integrating these objects with a DTD parser developed by Sun Microsystems.

### 5.3 Storage of the Abstraction Objects

This collection of objects will store and manage the information extracted by the parser. It also stores details relevant for the creation of a graphical visualisation system depicting the structure of the web site. The main issues arising from the implementation of this aspect concerned the type of data structure to be employed. After careful thought to both the extraction process and the creation of a VRML world, a complex object model was constructed. The implementation of this object model relies heavily on holding and manipulating objects, built upon a complex object oriented data structure.

### 5.3 "Loosy" Translation to VRML

The VRML worlds generated by this application are composed of two major sections:

1. Pre-generated code, which is defined in Inline VRML files. This is used to draw the constituent elements of the world, like HTTP, FTP nodes etc.
2. Code automatically generated by the application, this in turn has to two parts:
– Nodes in the world to represent the elements contained within the web site.
– Links connecting the nodes within the world, depicting site structure.

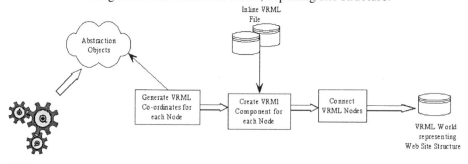

**Fig. 3.** Overview of the process of creating a VRML world to represent a web site.

Organising a cohesive world structure is achieved as follows. Each abstract object is allocated a particular shape depending on its type. Each VRML node then, is integrated and allocated a VRML co-ordinate to position it in the world. The final process involves a two-pass technique to connect the nodes representing the site structure. This connection process involves distinguishing the connection of nodes to a greater parse depth and via versa. These events are summarised in Fig 3

### 5.4 Virtual World Layout, Orientation and Sense of Direction

We position nodes within the world corresponds to the breadth first manner in which the web pages were parsed. This arrangement conveys a similar perspective to that of a user visiting a site using a conventional web browser. However, the user is not forced into searching the site in a depth first manner as with normal browsers. The position equations were experimentally determined by examining web sites of various sizes and depths.

$$X = \frac{N+100}{20} \qquad (1)$$

$$Y = level \frac{N+100}{20} \qquad (2)$$

$$Z = - level \frac{N+100}{20} \qquad (3)$$

Where *N* is the number of nodes at a given level, and *Level* is the parse depth relative to the starting position (or base address).

**Fig. 4.** An alternative screen shot of a world created by VisualExpresso, which shows how navigating the world is assisted by the uses of colour and directional aids.

VisualExpresso denotes direction by representing connections from a node to one at a greater parse depth by a grey line and the reverse by a red line Fig.4. An alternative approach of using multiple colours for the connections between levels was also tried, but this was rejected in favour of the previous representation, as it was deemed to be cleaner and less messy. The application also provided the option of enhancing the world by adding texture mapping to the nodes within the world to make it easier to identify the different node types.

# 6 Conclusion

To conclude we have proved that VRML can be used successfully as a three-dimensional visualisation analyse tool allowing virtual interaction with a three-dimensional representation of a complex networks of information. We have also thrown light into the dark abyss of Internet browsing suggesting one possible solution with the advent of high performance computing for the simplification and navigation of large volumes of distributed data found on the WWW.

## References

1. Jearding, Dean F. and Stasko, John, T. The Information Mural: Atechnique for Displaying and Navigating Large Information Spaces. Proceedings of the IEEE Symposium on Information Visualisation, Atlanta, GA., pp 43-50, Oct. 1995.
2. Lieberman Henry. Letizia: An Agent That Assists Web Browsing. International Joint Conference on Artificial Intelligence, Montreal, August 1995.
3. Kalawsky Roy S. The Science of Virtual Reality and virtual Enviroments. Addison-Wesley, 1994.
4. Price, B. A. Baecker, R. M. Small, I.S. "A Principled Taxonomy of Software Visualisation", Journal of Visual Languages and Computing, 3, 4, 211-266.
5. Network Working Group T. Berners-Lee. Request for Comments: 1738, *Uniform Resource Locators (URL)*. Editors L. Masinter Xerox Corporation,M. McCahill University of Minnesota. December, 1994.
6. Resiss, S.P. A Framework For Abstract 3D Visualisation. Proceedings of the 1993 IEEE Symposium on Visual Languages, pp 108-115, Aug. 24-27, 1993.
7. Network Working Group R. Fielding UC Irvine. Request for Comments: 1808, *Relative Uniform Resource Locators*. June, 1995.

# VIVRE: User-Centred Visualization

D R S Boyd, J R Gallop, K E V Palmen, R T Platon, C D Seelig

CLRC Rutherford Appleton Laboratory

Chilton, Didcot, Oxon OX11 0QX, United Kingdom

**Abstract**. A new approach to visualization is described which places the user at the centre of this interactive, investigative process. This is achieved by integrating the immersive user interaction environment provided by a virtual reality system with the comprehensive visualization capability provided by two established data visualization systems. Users' existing investment in visualization applications is thus protected while offering them a powerful new way of steering their investigations of complex data. This integrated environment has been specified and is being developed in a user-driven EU ESPRIT demonstrator project, VIVRE.

## 1. Introduction

Data visualization today is commonly carried out using dataflow visualization systems such as AVS/Express [1] or IRIS Explorer [2]. These provide the user with a rich set of functions to read, filter, map and render his data and a somewhat more limited range of facilities to view the generated (typically 3D) data representation within a 2D window on the screen. The viewing operations usually supported are intrinsically data-centred and involve rotating the data and zooming in to magnify features of interest within the data. Exploration of complex datasets using these mechanisms can become a tedious task and in consequence the user may never find that unique perspective on his data which affords him the crucial insight he is seeking.

The main idea behind the work described here is to harness the user-centred navigation and interaction capabilities of virtual environment (VE) systems to exploit the user's exploratory skills learnt in the real world and provide a more interactive and responsive environment for data visualization. By empowering users to explore "inside" their data, subtle features may be observed and new relationships previously hidden in the data may be revealed. Users are able to "steer" the course of their visualization in a way not easily possible before. By coupling the complementary strengths of visualization and VE systems, the VIVRE project aims to provide a new way to explore data and gain insights into a wide range of industrial and business processes, thereby delivering measurable benefits for the participating organisations.

First, previous work in this area is discussed briefly followed by a summary of the main characteristics of the VIVRE project. A short outline of the user requirements leads into a more extensive description of the technical approach adopted by the project. This is followed by a preliminary indication of the operational state of the system and the paper ends with an outline of the project exploitation plans and conclusions.

## 2. Previous work

There have been several attempts to bring together visualization and VE techniques. Various approaches have been explored:

- taking an existing visualization system and adding improved navigation and interaction capability [3,4,5],
- taking a VE system and adding visualization capability [6,7,8],
- developing a custom-designed integrated visualization and VE system [9,10].

Each of these approaches has its strengths and weaknesses. Extending a visualization system exploits existing user expertise in that package but does not solve the problem that real time interaction needs a system architecture designed from first principles to support this capability. Efforts to add visualization functions to a VE system have been able to take advantage of the fast response to user interactions but have tended to focus on the needs of particular application areas where the functionality required is well defined and domain-specific. Custom-designed solutions have demonstrated some of the benefits of closer integration but are not widely available so there is not a large base of experience of using them within industry. For a summary of work in this area see [11].

The VIVRE project seeks to take advantage of the existing investment and experience of the user partners in the established general purpose visualization systems AVS/Express and IRIS Explorer and of the availability of a relatively mature VE system dVISE [12]. All three systems permit extensions through the addition of user-developed software. The project is using this to integrate these environments and provide the user partners with a demonstrator system which will enable them to assess the practical benefits of this approach to their present and future visualization requirements.

## 3. The VIVRE project

VIVRE is an EU ESPRIT HPCN Preparatory, Support and Transfer Activity Demonstrator Project No 26008 [13]. VIVRE stands for Visualization through an Interactive Virtual Reality Environment. Its main objectives are:

- to demonstrate the integration of data visualization and VE software systems to provide interactive visualization steering for industrial and commercial applications,
- to identify the resulting business benefits,
- to disseminate the results into European industry.

The first of these objectives has been realised by designing a generic framework for integrating visualization and VE systems and by developing prototype implementations of this framework using commercially available visualization and VE software. To validate the generic nature of the framework, working prototypes have been developed using two separate visualization systems. Details of the technical approach taken are given later in the paper.

The project partners are Tessella (the project co-ordinator), CLRC, NAG, Tethys, Air Liquide, BNFL, BSSI, Labein and Unilever. The latter five organisations are end users whose role in the project is to specify requirements, provide application scenarios and evaluate the prototype implementations using these scenarios in the context of their own business processes, thus addressing the second objective. The project is an activity within the ESCALATE Technology Transfer Node (TTN) at the Parallel Applications Centre, Southampton [14]. Dissemination is taking place through various routes including presentations, the TTN mechanism and the normal contact channels of the commercial partners.

The application scenarios provided by the user partners cover a wide range of domain and scale. They include:

- visualising combustion and gas flow in industrial processes,
- visualising potential leakage paths surrounding underground storage caverns,
- visualising the dynamic operation of a flexible fluid control valve,
- visualising complex natural microstructures and their properties.

## 4. User requirements

The initial project task was to capture the functional requirements of each end user and to amalgamate these into a single coherent list of prioritised user requirements. This then guided subsequent project development work. These requirements will also provide the basis for user evaluation of the integrated environment developed by the project.

The user requirements divided into the following categories:

- visualization functions to be provided,
- user control of visualization to be available within the virtual environment,
- user navigation to be available within the virtual environment,
- data display options to be available within the virtual environment,
- data properties to be accessible from within the virtual environment,
- operational environment and performance.

An important requirement which evolved during the project was the primary delivery platform for the project software. While this was initially specified as SGI/IRIX, the end users and exploiters subsequently argued strongly for this to be extended to include PC/NT in recognition of its rapidly growing market share for this type of work.

## 5. Technical approach

First, an example is given of a typical user scenario, then the technical approach taken by the project is described in some detail. Interesting technical issues encountered are highlighted.

### 5.1 Typical user scenario

- First the user sets up a visualization network. An example of this may include an isosurface calculation for two different thresholds and three planar slices, on each of which an image is to be calculated representing the distribution of a parameter.

- The visualization network is evaluated and the resulting geometrical objects are sent to the VE system. Depending on the underlying data, the result can be a highly complex scene. Working within the VE, the user controls the reception and display of these objects.

- Using the navigation facilities of the VE system, the user steers to a position and orientation offering a good view of a region of interest in the data. To achieve this, the VE system supports a variety of devices for user interaction. On a low cost desktop system, a mouse and keyboard could be used. This involves learning the convention for mapping between these controls and the 6 degrees of freedom (DOF) required for navigation. For moderate cost, an input device directly controlling 6 DOF could be used, making navigation more convenient. Alternatively a head-mounted display system could be used, where the user's head movements control the direction of movement. These alternative devices are already supported by the dVISE system and need only minimal configuration.

- The slices are represented as separate objects in the VE. The user selects one of these and manipulates its position and orientation. Typically the application dataset is large, so recalculation is deferred until the user releases the slice.

- When the user releases the slice, the changed position and orientation and the identity of the slice are transmitted to the visualization system, which uses this information to recalculate the image and send it for display on the slice in the VE.

- Similar interactions are repeated by the user until he is finished.

If the dataset is large and complex, recalculation may be a lengthy process. Operations local to the VE can be carried out while this is happening. Thus navigation can be carried out concurrently with more expensive visualization operations.

Most user interactions will follow the same basic sequence of selecting a visualization object, performing an operation on it, allowing the visualization system to recalculate and displaying the recalculated object in the VE. Our intention is that such iterative operations should take place without the VE user needing to interact directly with the visualization system. An immersed user could therefore stay immersed throughout this process.

In our present design, however, some operations do require direct interaction with the visualization system. For instance, we do not attempt to provide access to the visualization network editor from within the VE system. Introducing a different visualization technique, such as streamlines, would therefore require a direct interaction with the visualization network. Similarly, reading a different data file is regarded as a less frequent operation which would be performed directly using the visualization system.

## 5.2 Extension mechanisms provided by the underlying systems

The two visualization systems and the VE system being used in the project all permit extensions to their functionality through the addition of new software. This is in the form of new network modules in the case of AVS/Express or IRIS Explorer. These project-specific modules can be interconnected with modules provided by the base visualization system. The mechanism for extending dVISE is via a plug-in which contains a collection of project-specific action functions. Action functions, whether system- or project-supplied, are associated with events which occur in the VE. These extension mechanisms have been used to develop a socket-based communication library for handling the exchange of information between the visualization and VE systems, geometrical conversion tools to turn visualization output geometry into discrete, correctly rendered objects within the VE and a facility for the user to control the visualization system from within the VE. This is shown schematically in Figure 1.

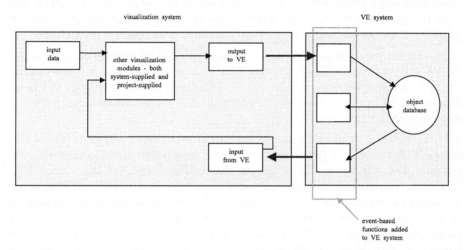

**Fig. 1.** Outline of the VIVRE system architecture

Initial attempts to use shared memory for data transfer were suspended as we did not have a consistent scheme that would work for both UNIX and NT. For this demonstrator project sockets provided an acceptable communication mechanism.

dVISE also supports the development of new user interaction tools which are needed in this project to enable users to control visualization functionality from within the VE. These make use of the dVISE mechanism known as the *Toolbox*. This is a floating hierarchical 3D menu which can be programmed to display the control options available to the user at any point in the interaction. The user can call up the Toolbox anywhere in the VE and select the required command using whatever pick mechanism is being used. The Toolbox is available in both desktop and immersive modes of interaction thus ensuring that the user can transfer easily between these according to the hardware resources available or the demands of the application.

Since the VIVRE software makes use of standard extension mechanisms supported by the software vendors, it can be easily adopted by users who already have the underlying software. This will be an important point in exploiting the project results.

### 5.3 Dynamic changes

Unlike many VE applications, in which the major components of the virtual world are created when it is first set up, with visualization applications some significant components of the scene, such as isosurfaces, can be changed by the user at any time. This requires careful optimisation of the mechanisms involved if these changes are to happen with minimum delay and loss of context by the user. In VIVRE, the initial VE set up in dVISE is simply a skeleton to which objects subsequently received from the visualization system are attached.

For example, *isosurfacelat* in IRIS Explorer has a geometry output port which is normally connected to the *render* module. VIVRE has developed a set of Explorer modules which captures this geometry description, writes a new geometry file and sends a message to the VE system telling it where to find it. The VE system reads from the geometry file, decides whether this is a new object and then adds the geometry to the scene. Similar functionality has been developed for AVS/Express.

At present the user is required to explicitly acknowledge the passage of new data into the VE system from the visualization system. This is so that the user is not surprised by the scene changing unexpectedly as he is navigating around it. Receipt of this new data is carried out by the user from within the VE so he can maintain his current viewpoint and context.

This highlights a significant issue in the design of VIVRE which was the need to reconcile the two quite different time cycles operating within the visualization and VE systems. The VE system must update the rendered view of the virtual world continuously as the user navigates around it while the visualization system waits until a change in a visualization parameter occurs and then re-evaluates the appropriate part of the visualization network. This may take some time as some functions are computationally expensive. It is important that the response of the VE system is not affected by this delay thus leaving the user free to continue navigation. When the visualization output is available, the user can accept it into the virtual world.

### 5.4 Multiple objects

In a visualization application, it is common for several different types of visualization representation, and multiple instances of each, to be present in the scene. The scenario presented earlier uses multiple isosurface thresholds and multiple slices. So that these can be independently manipulated by the user, it is necessary that they be recognised in the VE as separate objects.

The visualization network, which may contain several modules of the same type, must therefore hand over separately identified visualization objects to the VE. Existing visualization modules which generate geometrical objects do not necessarily generate these as separately identified objects. These visualization modules are therefore preceded and followed by VIVRE modules which are responsible for the management of user input from the VE, the generation of identification information that must be provided about the objects and the extraction and conversion of separate geometrical objects to a format recognised by dVISE. This is shown in Figure 2.

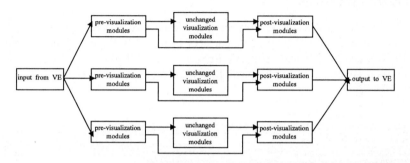

**Fig. 2.** Managing multiple visualization objects

### 5.5 Geometry conversion

Although geometry conversion from the visualization systems to the VE system was available via Inventor for IRIS Explorer and VRML for AVS/Express, these routes were not found to be adequate. The major problem was the representation of colour that varied across a surface. Normally this is achieved by attaching a colour to each vertex. Unfortunately the way dVISE worked did not allow for such colours to be shaded so they did not appear as intended within the VE.

The solution has been to produce new conversion software capable of accepting the geometry produced by the visualization modules and using texture co-ordinates to encapsulate the colour information. Texture co-ordinates are interpreted in dVISE prior to the shading calculation and therefore are rendered as intended.

### 5.6 Support of multivariate data

The user application scenarios rely heavily on multivariate data. One typical CFD example dataset uses 12 parameters over a 3 dimensional base. Users motivated to use VE for visualization are likely to have complex data and the presence of multiple variables is just one aspect. A particular visualization representation (one of the objects in the scene) is likely to use a subset of the parameters, in practice commonly one or two. The ability for the user to specify and vary the parameters to be displayed in association with each object within the VE is regarded as essential.

### 5.7 Interrogation of the visualization scene

It should not be necessary for the user to have to continually switch the focus of his attention from the VE to the conventional visualization GUI and back. To assist with this, it is essential that while a user is interacting within the VE (particularly if immersed), the VE provides him with sufficient information and control to enable him to achieve his immediate objectives. In the user requirements, access to several items of information was identified as necessary to meet the objectives of the project, for example the identity of the parameters being displayed, their current values and the maximum and minimum values represented. This is achieved via the Toolbox.

## 6. Operational results

At the time of writing, working prototypes of the integrated VIVRE system have been demonstrated and delivered to project partners using IRIS Explorer and dVISE on both SGI/IRIX and PC/NT and using AVS/Express and dVISE on SGI/IRIX. Figure 3 shows a screen image of the Explorer+dVISE VIVRE environment running in desktop mode on SGI/UNIX.

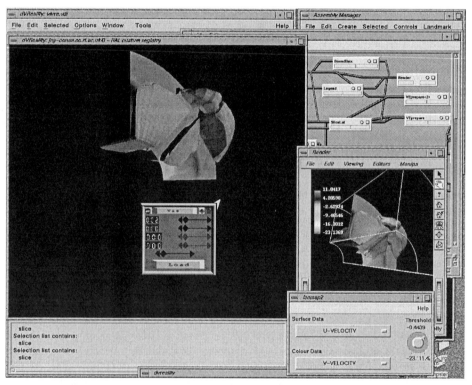

**Fig.3.** VIVRE software with Explorer and dVISE running on SGI/IRIX

The example is from one of the user-supplied application problems involving visualization of the dynamic operation of a flexible control valve. The largest window belongs to dVISE and shows a quadrant of the valve with a cut-plane coloured by a flow velocity component, together with the Toolbox control object. Also visible are the Explorer renderer window and a part of the Explorer network. In immersive use, only the dVISE window is visible to an HMD user and all control is effected using the Toolbox whose content changes depending on the interactive context.

At this stage of the project some preliminary feedback from users has been obtained. This indicates that for applications where the number of data items being displayed is large, the interactive performance of a VE system can be an order of magnitude faster than that of a visualization system. An example of this was the exploration of a 3D dataset generated by confocal microscopy of a biofilm (actually bugs growing on teeth!). This consisted of over a thousand separate isosurfaces. Within the VE, the user was able to explore this dataset interactively looking for patterns in the distribution and interrogating the attributes of individual bugs. It had not been possible to work interactively with datasets of this size before.

For the example shown in Figure 3, the advantages were less dramatic in performance terms but were more through the ability to navigate around the application data and interact with it immersively. This highlighted a clear distinction between desktop use of the VE system, where the interface controlling this interaction was considered not very easy to use, and the more intuitive experience provided by fully immersive use. More detailed evaluations of users' experience of the VIVRE environment will take place over the next two months.

## 7. Exploitation plans

It is intended that the integrated VIVRE software environment will be made available to users outside the project through co-operation between NAG, the supplier of IRIS Explorer, and Division, the supplier of dVISE. NAG will make the VIVRE software available as an extension to Explorer and are willing to act as a dVISE reseller. This will provide potential users with a single point of contact for access to the complete integrated system. Other partners will use the VIVRE software in their on-going commercial contract or internal research and development work.

## 8. Conclusions

The VIVRE project has demonstrated the technical feasibility of linking commercial data visualization and VE software systems to provide users with an integrated environment offering the capabilities and strengths of both systems. This new environment can exploit the computational performance and graphics capability of high-end HPC systems while also being accessible to PC users. While the initial prototype system does not yet provide complete access to all the visualization systems' functionality, it does provide a sufficient range of functions that users are able to develop useful application scenarios. Initial reaction indicates that the user-centred experience of navigating and interacting with their data using an immersive VE offers them commercially useful benefits compared with the techniques previously available to them.

## 9. Acknowledgements

The authors thank the European Commission for supporting the VIVRE project and the other project partners for their co-operation and contribution to the project.

## 10. References

[1] AVS/Express, see http://www.avs.com/products/expovr.htm

[2] IRIS Explorer, see http://www.nag.co.uk/Welcome_IEC.html

[3] Sherman, W.R., Integrating Virtual Environments into the Dataflow Paradigm, Proceedings of $4^{th}$ Eurographics Workshop on Visualization in Scientific Computing, 1993

[4] Haase, H., Goebel, M., Astheimer, P., Karlsson, K., Shroeder, F., Fruehauf, T., Zeigler, R., How Scientific Visualization Can Benefit from Virtual Environments, CWI Quarterly, 7(2), pp159-174, 1994

[5] Fuhrmann, A., Loeffelmann, H., Schmalsteig, D., Gervautz, M., Collaborative Visualization in Augmented Reality, IEEE Computer Graphics and Applications, 18(4), pp54-59, 1998

[6] Fruehauf, T., Dai, F., Scientific Visualization and Virtual Prototyping in the Product Development Process, Proceedings of the $3^{rd}$ Eurographics Virtual Environments Workshop, pp223-233, Springer-Verlag, 1996

[7] Sastry, L., Boyd, D.R.S., Fowler, R.F., Sastry, V.V.S.S., Numerical flow visualization using virtual reality techniques, Proceedings of the $8^{th}$ International Symposium on Flow Visualization, 1998

[8] Benoelkan, P., Niemeier, R., Lang, U., Collaborative Volume Rendering in a Distributed Virtual Reality Environment, Proceedings of the $4^{th}$ Eurographics Virtual Environments Workshop, 1998

[9] Bryson, S., Levit, C., The Virtual Windtunnel: An Environment for the Exploration of Three Dimensional Unsteady Flows, Proceedings of IEEE Visualization '91, 1991

[10] Haase, H., Symbiosis of Virtual Reality and Scientific Visualization System, Proceedings of Eurographics '96, 1996

[11] Gallop, J.R., Virtual Reality - its Application to Scientific Visualization, Tutorial at Eurographics '96, 1996

[12] dVISE, see http://www.division.com/2.sol/a_sw/sol_a.htm

[13] VIVRE, see http://www.tessella.co.uk/projects/vivre/vivre.htm

[14] ESCALATE, see http://www.pac.soton.ac.uk/escalate.html

# GEOPROVE: Geometric Probes for Virtual Environments

R.G. Belleman, J.A. Kaandorp, D. Dijkman, P.M.A. Sloot

Parallel Scientific Computing and Simulation Group
Faculty of Mathematics, Computer Science, Physics and Astronomy
University of Amsterdam
Kruislaan 403, 1098 SJ Amsterdam
the Netherlands

**Abstract.** We present a software architecture that can be used to instrument interactive virtual environments with virtual probes to obtain quantitative information from geometric presentations. This architecture provides tools by which measurements can be obtained from multiple levels of data presentations, ranging from graphically displayed geometry to the underlying raw data sets.

## 1 Introduction

In many scientific computing problems, the level of complexity in the generated data is too vast to analyze numerically. For these situations, interactive scientific visualization is an essential method to present and explore the data in a way that allows a researcher to comprehend the information it contains. Immersive virtual environments such as the CAVE [5] further enhance a researcher's perception and are therefore often used to obtain better insight in multi-dimensional datasets for which desktop visualization environments are too restrictive.

In most scientific visualization environments, a visualization pipeline transforms numerical data into geometric constructs that are rendered into visual presentations. These presentations allow researchers to qualitatively analyze their data. Many visualization environments stop at this point and provide little means to obtain quantitative information on what is being presented. This may be acceptable to some applications, for others however, an instrument for obtaining quantitative information from the visualization is a valuable asset. Examples of this are applications where simulations are verified to the real-life phenomena that are being modeled, applications for diagnostic purposes based on medical data obtained from medical scanners (i.e. CT, MRI, etc.), or computer aided design tasks.

This paper describes GEOPROVE, a geometric probing software architecture for interactive data exploration environments, virtual environments in particular. This architecture allows researchers to probe visual presentations in order to obtain quantitative information. We will show the application of this probing system in a test-case which we describe next.

## 1.1 A test-case: coral growth simulation

In a previous paper we have described an interactive virtual exploration environment that allows us to explore a simulation model for the investigation of diffusion and flow limited biological growth [2]. We have used data sets resulting from simulations of growth processes (aggregation processes) in which an aggregate consumes nutrients from its environment and where nutrients are dispersed by a combined process of flow and diffusion. Details about the simulation model are given elsewhere [10].

Although our exploration environment provides methods by which the data sets can be explored visually, an important aspect in the development of any simulation model is its verification against the system that is modeled. A major problem in the quantitative comparison of the simulation results with actual phenomena is that in many cases there is no single discriminative feature by which they can be differentiated. In our test-case for example, a property such as the fractal dimension [11] gives some insight in the global resemblance of different structures but is inadequate in describing the quality of the simulation model as only a limited aspect of the overall morphology of an object is captured. Therefore, it is often more suitable to obtain measurements on multiple properties in local areas of the data sets that together form a discriminative measure.

In case of the growth model we wish to compare the shape of the resulting structures to those found in nature. In previous work it has been demonstrated that morphological properties such as for example the thickness of branches and the shortest distance between neighbouring branch points ("branch spacing") provides relevant biological information and can be used to compare simulated and actual growth forms [8, 9].

In addition, when comparing simulated coral objects with actual corals, the comparison procedure should be non-destructive to the real coral as most of these are valuable and irreplaceable specimens. However, this makes many measurements difficult if not impossible since the complex shape of these structures prohibits the use of instruments that may damage the coral. One possible solution is to acquire a sufficiently accurate three-dimensional scan of the coral. Since conventional photographic or laser scanning techniques are only suitable for obtaining surface models of objects which have no obstructing components, these devices are unsuitable for scanning complex and irregularly structured objects such as corals. Fortunately, we have obtained digital 3D data sets of a number of corals which we can use for our purposes through the assistance of the Radiology department of Leiden Academic Hospital, who have graciously offered to scan the corals with a computed tomography (CT) scanner.

Although the properties we want to measure could be obtained automatically using data analyses techniques, this often requires designing and implementing specialized algorithms that are dedicated to the specific task. Quite often these techniques rely on heuristic algorithms that are difficult to design, implement and control. An interactive virtual environment equipped with a geometric probing system described in this paper can provide the techniques needed to acquire quantitative properties from data sets which would have been difficult to obtain otherwise.

## 1.2 Related work

While most scientific visualization environments provide some probing functionality, most of these act as subset selectors that extract selected regions from larger data sets for localized visualization, complementing global visualization methods [12, 16]. The Visualization Toolkit (Vtk) for example provides probe filters for the computation of point attributes in local areas. Point attributes are computed at input points specified by a probe consisting of a geometric structure by interpolating into the source data. Vtk also contains methods that calculate properties such as the volume, surface area and normalized shape index of closed triangle surfaces [14].

The work by van Leeuw et al. takes this method one step further by using visual probes that consist of a set of geometric primitives. Multiple characteristics of a small area in a flow field transform the geometric primitives in the probe to visualize velocity and local change of velocity [6]. Although these visual probes provide excellent means for exploring local properties of a dataset, they are often used for localized visualization only, and not for obtaining quantitative measurements.

The work by Brady et al. shows a CAVE application for the visualization of biomedical images obtained from medical scanners which allows features to be manually traced and labeled [3]. This software has been used for obtaining the lengths of biological structures and for segmenting medical images.

## 2 Geometric probing

In our system, geometric probes consist of one or more markers. These markers are used to sample properties of the presentations in the virtual environment. A property can either be the coordinates of a position in the environment, or a value obtained from data at this position. The property obtained from the markers are used in an evaluation function producing the result of a measurement. The evaluation function defines a relation between markers in a probe which in this paper we illustrate by connections between markers.

Determined by the spatial configuration of the markers, probes have dimensions and a certain degree of freedom by which they can be positioned over an area of interest. For example, a probe consisting of exactly one marker has 3 degrees of freedom in a 3D environment (translation in 3 directions) and can thus either be used to obtain the position $(x, y, z)$ of a feature in 3D space or a mapped quantity $f(x, y, z)$ at this position, where $f$ provides a mapping of a position to a quantity (i.e. a scalar, vector, tensor). An evaluation function takes this sample and produces the result of the measurement.

A measurement procedure with probes in a virtual environment requires the following course of actions (see also Fig. 1):

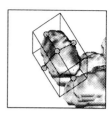

**Fig. 1.** Probing procedure, as an example applied here to determine the bounding volume of part of a structure. From left to right: calibration, interactive placement of the probe, registration of the probe to the data, calculation of bounding box.

1. Calibration - As in any measurement, the properties that are calculated based on sampled quantities need to be referenced to a well defined unit. This unit of reference is especially required when the measurements from two different objects are to be compared. For the same type of measurements, calibration will only have to be performed once.
2. Placement of the probe - Interactive placement of a probe in a virtual environment makes use of devices whose position and orientation are tracked in three-dimensional space. Through these devices, a user is able to place a probe roughly over the region of interest.
3. Registration of the probe - The interactive placement of a probe is not accurate in most cases because of inaccuracies in the tracking hardware or inexperience of the user. Registration of the probe involves refining the position, scale and rotation of the probe, either interactively or aided through registration functions.
4. Calculation of the result - Once the probe is in place, the result can be calculated. Depending on the type of probe, the calculation is either performed purely based on the position and orientation of the probe, or the positions of each marker is first mapped to a quantity.
5. Presentation of the result - When the calculation is finished, the results need to be presented to the user in some meaningful way. In addition, the user should be able to log the measurement on file for later analyses and some method of annotation is required so that the user can relate back to the measurement once they are analyzed elsewhere.

Table 1 shows some examples of probes consisting of a number of markers and examples of properties that can be obtained. Most of the properties in this table can be relatively easily obtained. For the first implementation of this architecture, with the coral growth data exploration system as a test-case, we limit ourselves to measurements that require probes consisting of 2 markers (length, distance) and 3 markers (angles). However, in the design of GEOPROVE we have attempted to keep the architecture generic so that the addition of probes for the acquisition of other properties can be achieved with little effort.

| # markers | 1 | 2 | 3 | 4 | 5 |
|---|---|---|---|---|---|
| probe | • | •—• | ∧ | ◇ | ✦ |
| positional property | position | length, distance | angle, curvature | bounding rectangle | saddle point |
| mapped property | value | derivative | extrema | surface | Gaussian distribution |

**Table 1.** Examples of probes with different number of markers and examples of positional and mapped properties that can be measured with these probes.

### 2.1 Software architecture

Interactive visualization applications benefit from a design in which the computation and visualization processes are implemented by separate communicating threads of control [4]. The library used in CAVE environments supports primitives for this design [17]. If implemented carefully, this configuration allows interactive virtual environments to be built that have a high frame rate and minimal interaction delay. However, it does have implications for the design of the GEOPROVE architecture. The software architecture that we have developed is shown schematically in Fig. 2.

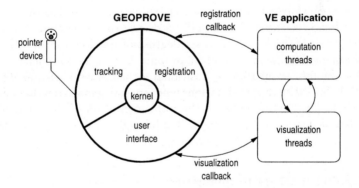

**Fig. 2.** The software architecture for GEOPROVE and its interface to a virtual environment application.

The heart of GEOPROVE consists of the *kernel* that logs the positions and annotations of the probes, performs calculations and maintains a record of the measurements for storage and retrieval. The *tracking* part provides the kernel with position and orientation information of pointer devices with which probes can be positioned in the virtual world. The *registration* part manages the positioning of probes based on information provided by the application. This allows probes to be "snapped" in real time to computed data so that accurate measurements can be obtained. The *user interface* part provides access to GEOPROVE, including the rendering of probes, the presentation of results and annotations, and methods to drive GEOPROVE.

Measurements can be obtained on two levels; on the first, most basic level, measurements are based purely on the position and orientation of pointer devices that are tracked in three-dimensional space. On the second level, measurements take place based on structures defined by the application. This two-level scheme is described in more detail in the following two sections.

## 2.2 Tracker based probing

On a most basic level, probes are positioned based solely on the location of pointer devices that are being tracked in the visualization environment domain. Note that in this situation measurements are taken without a direct relation to the application other than the user's perception of the environment. This allows simple measurements to be made such as distances and angles without any feedback from the application.

As this method of probing is independent of what is being presented, probes can not be registered on data from the application. Therefore, its accuracy and usefulness is inherently dependent on the quality of the tracking systems, the experience of the user and the application for which it is used. Most tracking systems are sensitive to noise, therefore GEOPROVE supports scaling the coordinates obtained from the tracker system such that a more accurate positioning of a probe can be performed. In this research, the physical devices that are tracked consist of a head tracker and a "wand", a device that is similar to the desktop mouse but which is tracked using a six degrees-of-freedom sensor [1].

The main benefit for this kind of probing lies in its simplicity and its allowance for the fast acquisition of positional measurements such as lengths, angles, spatial derivative approximations and special geometries like the fractal box dimension [7]. Furthermore, as the implementation of this kind of probing can be isolated into GEOPROVE itself, it minimally interferes with existing VE software.

## 2.3 Mapping markers to quantities

Quantities that relate to properties that are defined by the application can only be obtained by interrogating the visualization or computation thread. Since GEOPROVE does not have direct access to the data maintained in the application, the quantification of a marker from its position has to be handled by the application via a callback function.

Some quantities can be best obtained from the abstractions that are made from application data when these are visualized. An example of this is the calculation of a surface area which can be approximated using isosurfaces that are extracted from grid based data for visualization through e.g. a surface extraction algorithm. For other measurements, the data contained in the computation may be of higher quality.

## 2.4 Registration

For some types of measurements it may be necessary to perform calculations on specific features of the underlying data. In these cases the probe needs to be aligned to these features before calculations can be performed. Depending on the probe's shape and its degrees of freedom, the registration of a probe to the underlying data sets takes the same form as the techniques that are used in *geometric hashing* [18]. In short, this method first takes two markers of the probe as a handle to match "points of interest" in the geometric dataset. Using geometric transformations (the most common being translation, rotation and scaling) a basis is constructed which determines the exact position of all markers in the probe. Through a voting mechanism a histogram is then constructed of candidate bases from which the best candidate for registration is chosen. Using this method, positions acquired from tracker sensors can be registered to data structures that have been used for visualization or computation. Although this technique is in general computationally expensive, we may exploit "focus locality" by disregarding the geometry which is far from the user's focus.

## 2.5 Presentation, logging and annotation

GEOPROVE does not render its own user-interface. Instead, it relies on the VE application to do this. This allows GEOPROVE to be seamlessly integrated into existing applications. Most of our applications contain standard user interface components that can be used by multiple processes at the same time (see also Fig. 3). Existing software can be instrumented with probing facilities with minimal effort. In the current version, only four function calls need to be added to the source code of an existing application to obtain the most rudimentary features. These functions consist of: (1) the initialization of GEOPROVE, (2) a display handler that renders feedback to the user for both interaction and presentation of results, (3) a user interaction handler, and (4) a registration function that allows markers to be snapped to visualized geometry.

Probe locations and measurement results are stored in a log file. During runtime, the user can browse through this log via the user interface and view entries or delete entries. The user also has the option to add annotations to a measurement via a "snapshot"; a virtual photograph made from the perspective of the user. Both low resolution (for runtime inspection) and high resolution (for off-line inspection) pictures are supported. We are currently adding the option to record a speech-sample to annotate measurements with extra information.

When the application is terminated, the log is written to a file, containing the measurements and references to possible snapshots and speech annotations. The log can then be analyzed on a workstation, saving valuable CAVE time.

## 3 Results

We instrumented the interactive coral growth exploration environment described in [2] with GEOPROVE to obtain measurements of the shortest distances between neighbouring branch ends ("branch spacing") (see also Fig. 3). The importance of this spatial observable is described in [15].

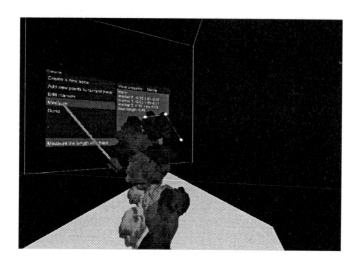

**Fig. 3.** A CAVE simulator snapshot of an application instrumented with GEOPROVE. Three markers (shown in white) of a trace are registered (or "snapped") to the visual geometry. The window in the back shows the user interface to GEOPROVE and presents the length of the trace.

Fig. 4 shows histograms of 106 measurements obtained from a CT scan of a real coral structure (*Pocillopora damicornis* in a sheltered environment) and 155 measurements obtained from 8 simulated coral structures under similar conditions. We will discuss these measurements in the following section.

## 4 Discussion and future work

We have presented a software architecture that allows us to instrument interactive virtual exploration systems with probes to obtain quantitative information based on visual presentations. We have used this system to instrument an existing application with little effort.

It can be observed from the measurements shown in Fig. 4 that there is a remarkable difference between the measurements of the branch spacing done in the CT scan of *Pocillopora damicornis* and the simulated growth forms. Three observations can be made from these measurements:
(1) The measurements from the simulated structures seem to be bimodally distributed while the CT scan measurements are unimodal. This may be the result

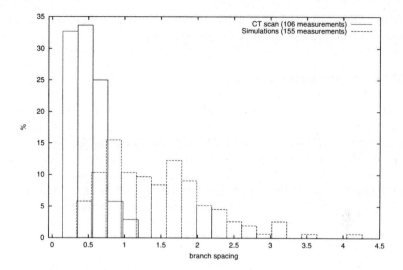

**Fig. 4.** Comparison of branch spacing in a CT scan of *Pocillopora damicornis* in a sheltered environment with 8 simulated structures under similar conditions.

of measurement artifacts or it may be caused by the simulation mechanics. To fully understand the reasons for this we would need to obtain a larger number of more detailed measurements, which we have not yet been able to acquire.

(2) The mean branch spacing of the two distributions differs significantly ($\approx 0.5$ versus $\approx 1.5$). This may be due to the scaling of the measurements: in order to obtain a scale that makes the simulated structures comparable to the real objects, the measurements from the simulated structures have been scaled using a factor obtained from the ratio in dimensions of the simulated structures and the real coral. We will reconsider this scaling method in the near future.

(3) In real *Pocillopora damicornis* the variability is relatively low compared to the simulated forms. These measurements seem to indicate that there is a mechanism which regulates growth of branches in the immediate vicinity of other branches, this mechanism is not present in the current simulation models. In the study by Rinkevich and Loya [13] it is proposed for the branching stony coral *Stylopora pistillata* that there is a chemical signal mechanism which regulates growth of branches, resulting in a relatively low variability and even a remarkable uniformity of the branch spacing. Their experiments indicate the possible appearance of a chemical signal which is being secreted into the water column and works as repellent, growth suppressing, agent. A similar mechanism might be present in *Pocillopora damicornis*. The morphological measurements indicate that in future versions of simulation models of stony corals a regulation mechanism is required in order to obtain a better approximation of the actual growth process.

Until now we have limited our system to simple measurements with probes that consist of one, two or three markers. In the future, we will enhance our system with probes that consist of more complex markers, enabling more advanced measurements such as surfaces, volumes and more.

## 5 Acknowledgements

We gratefully acknowledge the people at SARA for their CAVE support and patience during the course of this work. Thanks to dr R.W.M. van Soest of the Department of Coelenterates and Porifera, Institute for Systematics and Population Biology, University of Amsterdam, for allowing us to use collection material for generating the CT scans. Thanks to Dick Bakker of the Academic Hospital Leiden, the Netherlands, for making the CT scans. Thanks to Arjen Schoneveld, University of Amsterdam, for suggesting the name GEOPROVE.

## References

1. Academic Computing Services Amsterdam (SARA), Amsterdam, the Netherlands. *SARA - CAVE Homepage*, 1998. http://www.sara.nl/hec/vr/cave/.
2. R.G. Belleman, J.A. Kaandorp, and P.M.A. Sloot. A virtual environment for the exploration of diffusion and flow phenomena in complex geometries. *Future Generation Computer Systems*, 14(3-4):209–214, 1998.
3. Rachael Brady, John Pixton, George Baxter, Patrick Moran, Clinton S. Potter, Bridget Carragher, and Andrew Belmont. Crumbs: a virtual environment tracking tool for biological imaging. In Murray Loew and Nahum Gurshon, editors, *Proceedings of the IEEE Symposium on Frontiers in Biomedical Visualization*, pages 18–25, Los Alamitos, CA, October 30 1995. IEEE Computer Society Press. http://mayflower.ncsa.uiuc.edu/crumbs/crumbs.html.
4. Steve Bryson and Sandy Johan. Time management, simultaneity and time-critical computation in interactive unsteady visualization environments. In *Proceedings of Visualization '96*, page 255. IEEE Computer Science Press, Los Alamitos, CA, 1996.
5. C. Cruz-Neira, D.J. Sandin, and T.A. DeFanti. Surround-screen projection-based virtual reality: The design and implementation of the CAVE. In *SIGGRAPH '93 Computer Graphics Conference*, pages 135–142. ACM SIGGRAPH, August 1993.
6. Willem C. de Leeuw and Jarke J. van Wijk. A probe for local flow field visualization. In R.D. Bergeron G.M. Nielson, editor, *IEEE Visualization '93*, pages 39–45, Los Alamitos, CA, 1993. IEEE Computer Society Press.
7. J. Feder. *Fractals*. Plenum Press, New York, London, 1988.
8. J.A. Kaandorp. Analysis and synthesis of radiate accretive growth in three dimensions. *J. Theor. Biol.*, 175:39–55, 1995.
9. J.A. Kaandorp. Morphological analysis of growth forms of branching marine sessile organisms along environmental gradients. *Mar. Biol.*, (in press).
10. J.A. Kaandorp, C. Lowe, D. Frenkel, and P.M.A. Sloot. The effect of nutrient diffusion and flow on coral morphology. *Physical Review Letters*, 77(11):2328–2331, 1996.
11. B.B. Mandelbrot. *The fractal geometry of nature*. Freeman, San Francisco, 1983.
12. The Numerical Algorithms Group Ltd., Oxford, UK. *Iris Explorer User's Guide*, 1998. http://www.nag.co.uk/visual/IE/iecbb/DOC/UG/CONTENTS.html.

13. B. Rinkevich and Y. Loya. Coral isomone: a proposed chemical signal controlling interclonal growth patterns in a branching coral. *Bull. Mar. Sci.*, 36:319–324, 1985.
14. Will Schroeder, Ken Martin, and Bill Lorensen. *The Visualization Toolkit, an object-oriented approach to 3D graphics (2nd edition)*. Prentice Hall, Upper Saddle River, NJ, 1997. ISBN 0-13-954694-4.
15. K.P. Sebens, J. Witting, and B. Helmuth. Effects of water flow and branch spacing on particle capture by the reef coral *madracis mirabilis* (duchassaing and michelotti). *J. Exp. Mar. Biol. Ecol.*, 211:1–28, 1997.
16. C. Upson, T. Faulhaber Jr., and D. Kamins et al. The Application Visualization System: a computational environment for scientific visualization. *IEEE Computer Graphics and Applications*, 9(4):30–42, July 1989.
17. Virtual Reality Consulting (VRCO) Inc., Chicago, IL. *CAVE User's Guide*, 1998. http://www.vrco.com/CAVE_USER/index.html.
18. Haim J. Wolfson and Isidore Rigoutsos. Geometric hashing: An overview. *IEEE Computational Science and Engineering*, pages 10–21, October-December 1997.

# Workshop:

# Distributed Computing and Metacomputing

# A Gang-Scheduling System for ASCI Blue-Pacific

José E. Moreira[1], Hubertus Franke[1], Waiman Chan[1], Liana L. Fong[1],
Morris A. Jette[2], and Andy Yoo[2]

[1] International Business Machines Corporation
Armonk, NY 10504, USA
{jmoreira,frankeh,waimanc,llfong}@us.ibm.com
[2] Lawrence Livermore National Laboratory
Livermore, CA 94550, USA
{jette,ayoo}@llnl.gov

**Abstract.** The ASCI Blue-Pacific machines are large parallel systems comprised of thousands of processors. We are currently developing and testing a gang-scheduling job control system for these machines that exploits space- and time-sharing in the presence of dedicated communication devices. Our initial experience with this system indicates that, though applications pay a small overhead, overall system performance as measured by average job queue and response times improves significantly. This gang-scheduling system is planned for deployment into production mode during 1999 at Lawrence Livermore National Laboratory.

## 1 Introduction

The Accelerated Strategic Computing Initiative (ASCI) program is creating a new state-of-the-art in large scale computations. Through this program, three national laboratories in the United States (Lawrence Livermore, Los Alamos, and Sandia) have each received machines capable of Teraflops performance. Combined with new simulation techniques these machines will help guarantee the safety of the United States nuclear stockpile while preventing the need for actual tests. In addition, machines in the ASCI program will be available to academic research in a variety of areas.

Effective and efficient job scheduling in such large environments, which can easily have hundreds of simultaneous users, represents a new challenge: How to provide service for production codes that execute for tens of hours on hundreds of processors while at the same time catering to the needs of numerous scientists debugging and developing new codes? In this paper we describe a new job scheduling system that we are developing for the ASCI Blue-Pacific machines, in the Lawrence Livermore National Laboratory (LLNL). These machines are based on the RS/6000 SP, but they have their own peculiarities. Our system is called GangLL and it supports both time- and space-sharing of parallel jobs using gang-scheduling. We have already developed and deployed the first generation of GangLL. Although it is currently running only on a small test system, we will show some encouraging results, both from actual runs as well as simulations. It is important to acknowledge that the development of GangLL has been motivated by excellent performance results obtained by LLNL with gang-scheduling in other large systems [2,5,7].

The rest of this paper is organized as follows. Section 2 gives some details of the ASCI Blue-Pacific machines, and describes the gang-scheduling approach to space- and time-sharing. Section 3 describes the architecture and organization of GangLL. Section 4 presents the results from our experiments on an actual machine and simulations. Finally, Section 5 presents our conclusions and discusses some future work.

## 2 The ASCI Blue-Pacific Machines

The main machine in the ASCI Blue-Pacific program is the *Sustained Stewardship TeraOPS* (SST) "Hyper-Cluster". Figure 1 shows the high-level organization of the SST machine. It consists of three 488-node *sectors* that are connected via high performance gateway links. Each node is comprised of a 4-way SMP with 332 MHz PowerPC 604e processors and up to 2.5 GB of main memory. Each node executes its own operating system (AIX) image. The nodes in each sector are interconnected via a TB3MX high-performance switch that delivers a bidirectional bandwidth of 150 MB/s per node.

While the SST machine is for use on classified applications, the ASCI Blue-Pacific program also provides computing resources to academic research through the ASCI Strategic Alliance Program. In this program the *Combined Technology Refresh* (CTR) machine provides 320 nodes of similar characteristic as the SST nodes, organized in a single sector. Features of both machines are summarized in Table 1.

**Table 1.** Characteristics of the ASCI Blue-Pacific machines.

| characteristic | SST | CTR |
|---|---|---|
| Peak speed | 3.9 teraOPS | 892 gigaOPS |
| Memory | 2.6 terabytes | 504 gigabytes |
| Local Disk | 17.3 terabytes | 3.0 terabytes |
| Min/Max memory/node | 1.5-2.5 gigabytes | 1.5 gigabytes |
| Number of compute nodes (4-way SMPs) | 1,464 | 320 |
| Node-to-node bandwidth (TB3MX switch, bidirectional) | 150 MB/s | 150 MB/s |
| Total number of processors | 5,856 | 1,344 |
| Processor-to-memory bandwidth | 2.1 TB/s | 2.1 TB/s |
| Compute node peak performance | 2.656 gigaOPS | 2.656 gigaOPS |
| Delivered RAID I/O bandwidth | 6.4 GB/s | 320 MB/s |
| Delivered I/O bandwidth to local disk | 10.5 GB/s | 4.7 GB/s |
| RAID storage | 62.5 terabytes | 10.0 terabytes |

Despite the size of the ASCI Blue-Pacific machine, we know from LLNL's experience [2,5,7] with large systems that the demand for processors to execute jobs will at times be larger than the total number of processors available. The traditional solution to this problem in a space-sharing environment is simply to queue the excess jobs: They wait to start execution until some running jobs have finished. It is well known in the literature that this can be highly detrimental to system and job performance. Simply queueing jobs until enough resources are available leads to low system utilization and

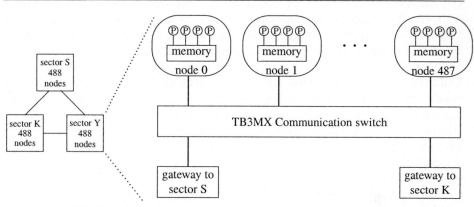

**Fig. 1.** High-level organization of the ASCI Blue-Pacific machine.

high job response time. The problem can be alleviated in part by clever scheduling techniques, such as backfilling [8]. These scheduling technique cannot, however, handle the also important issue of *preemption*: Some times it is desirable to suspend the execution of a lower priority job so that a high-priority job can run.

A more flexible solution to the problem of executing a large number of jobs is to share the machine's resources not only spatially but also temporally [2, 4, 6, 10]. The examples in the literature illustrate the benefit of adding this other axis of partitioning for parallel systems. It can result in better system utilization and reduced response time.

Time-sharing on a large scale distributed parallel machine is complicated by two issues: (i) the native schedulers of the operating systems executing on each node are not coordinated, and (ii) tasks of distributed jobs interact tightly through the communication switch. The tasks of a parallel job must be coscheduled [9] (*i.e.*, must run concurrently on all nodes) or inefficient communication behavior results. Without coscheduling, receivers may not be ready when senders are and vice-versa. As a result, we adopt a coarse grain time-sharing strategy that allows us to guarantee that tasks of the same job execute simultaneously despite the fact that the operating system images are not coordinated. We accomplish this by (i) partitioning the time axis into large time slices (order of seconds to minutes), (ii) populating the time slices with tasks from parallel jobs so that all tasks of a job occupy the same time-slice, and (iii) implementing the schedule at the individual nodes. The schedule is represented by an Ousterhout matrix [9], where the columns correspond to processors and the rows correspond to revolving time slices.

In the next section we describe the organization of GangLL, which is designed to provide the job execution environment described here for the ASCI Blue-Pacific machines. At first, we are designing GangLL to operate on each sector independently. Section 3 discusses some issues related to multi-sector operation.

## 3  Architecture and Organization of GangLL

We build our GangLL research prototype system as an extension to IBM LoadLeveler. The high-level organization of GangLL is shown in Figure 2. We follow a hierarchical scheduling scheme. A *Central Manager* performs resource allocation and overall job scheduling. Using a hierarchical distribution scheme, schedules are distributed to local autonomous node-level schedulers, which are responsible for implementing the columns of global schedules that pertain to their designated node.

**Central Manager:** The *Central Manager* is responsible for receiving all job submissions from users and it enqueues the jobs into a central job queue. The *Central Manager* then selects jobs from the queue for execution and performs spatial and temporal scheduling. The schedule is represented by an Ousterhout matrix that precisely identifies the assignment of tasks to processors both in space and time. The matrix is then distributed to the individual node-level schedulers. Note, that each node-level scheduler only needs the columns of the matrix that correspond to the processors in its node. It is also the responsibility of the *Central Manager* to interact with a site-specific external scheduler. This external scheduler can perform some of the scheduling decisions, thus implementing policies that are specific to a particular site.

Jobs are ordered in the queue according to their priority. A job's priority (also known as *sysprio*) is computed by the central manager based on rules specified by the system administrator. Each job has a set of attributes: submission time, owner, processor requirement, memory requirement, estimated execution time, etc. The system administrator defines a formula that computes the priority of a job as a function of those attributes. For example, the following formula would give higher priority to older jobs that need more processors: $sysprio = NumProcs * 1000000 - QDate$, where *NumProcs* is the number of processors requested and *QDate* is the time stamp (seconds since epoch) of the job submission. Higher values of *sysprio* correspond to higher priorities.

After the Ousterhout matrix is built, it has to be distributed from the *Central Manager* to the corresponding node-level schedulers. Although the matrix itself is relatively small (a few kilobytes), it has to be distributed efficiently to hundreds of nodes. Furthermore, we do want to avoid the undesirable situation in which some nodes start to use a newly distributed version of the matrix while other nodes are still using an old version. We solve this problem through a *two-phase, hierarchical distribution scheme*.

The *Central Manager*, instead of contacting each node-level scheduler individually, uses a binary-tree scheme to distribute the matrix. It sends the whole matrix to one node-level scheduler, which extracts its own columns and forwards the remaining columns to two other node-level schedulers. This operation continues recursively until all node-level schedulers have received their columns. With this approach, we accomplish the distribution of the matrix in $O(\log n)$ time, as opposed to the $O(n)$ in the naive approach.

The distribution scheme is two-phase: in the first phase a new Ousterhout matrix is distributed to all node-level schedulers. The new matrix is committed only after all node-level schedulers have received it, thus replacing the old one. If the distribution fails, the matrix is not committed, and the nodes continue to use the old one. The replacement actually happens at a specified wall-clock time, thus guaranteeing that all node-level schedulers start using the new matrix at the same time.

**Node-Level Scheduler:** Each node in the system has a node-level scheduler, responsible for implementing the schedule as defined by the columns of the matrix that cor-

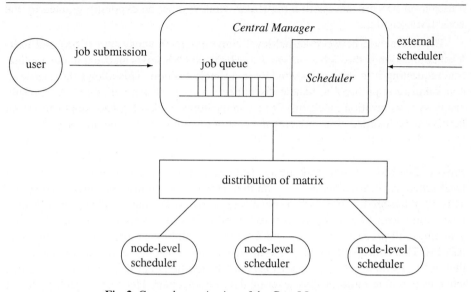

**Fig. 2.** General organization of the GangLL prototype.

respond to the processors in that node. For the node-level scheduler to perform coordinated context-switches at the time-slices boundaries, it suffices that they have some form of synchronized wall-clocks (*e.g.*, by running NTP). No explicit synchronization among node-level schedulers is necessary. Note that the synchronization does not have to be strict. The wall-clocks only need to be synchronized within a very small fraction of the time-slice (*e.g.*, a part in a thousand).

Each node-level scheduler has to perform the context-switch for all the tasks running in its node. There are two components of a context-switch: (i) context switching of the processors and (ii) context switching of the communication device. Context switching of the processors is achieved through suspending and resuming execution of user processes. The context-switch of the communication device is necessary because on the RS/6000 SP applications have user-level direct access to the communication device. These connections are a scarce resource: In the ASCI Blue-Pacific machine, each node can have four connections, one for each processor. The node-level scheduler interacts with two other system services to accomplish proper context switching of tasks: (i) the communication subsystem (CSS) and (ii) the process tracking service (PTS).

*Communication Device Control:* In the RS/6000 SP, communication protocols such as MPI transfer data packets to a non-pageable communication buffer and activate a DMA engine (on the TB3MX adapter) to transmit the packets to the network. Due to the memory requirements of the communication buffer as well as of those in the adapter, there is a limited number of user-level processes that can be granted direct connection to the communication device at any given time. Each connection is called a *window*. Thus, before suspending a user-level process, we force it to release its window, which then can be reassigned in the next time-slice to another process. Similarly, before resuming a user-level process, access to a particular window must be explicitly granted by the node-level scheduler.

The interaction between kernel-level communication subsystem (CSS) and user-level application is through a *call-back* mechanism: CSS sends an event to the application requesting it to release its window in a timely fashion. This allows the user-level communication protocol to handle disconnections at well defined points. As a result, costly synchronization mechanism (e.g., semaphores, locks) between application and kernel can be avoided, thus reducing the cost of communication operations.

*Process Tracking:* A new task always starts as a single process created by the node-level scheduler itself. This process can then create new children processes and so on. This set of descendant processes is not always a tree, since intermediate processes may have terminated. The relationship between processes and tasks is defined by *process genealogy*: all processes that descend from the initial, or root, process are in the same task [3]. This concept is not provided in UNIX. Individual processes can explicitly cut their relationship to *groups* and *sessions* [11]. In addition, a process becomes orphaned when its parent terminates, thus severing the creation chain. Hence, the resulting forest of processes cannot be adequately identified by the initial process alone.

To solve these deficiencies, we have implemented a new kernel extension that utilizes process-change events (in particular, process creation and process termination) to maintain proper genealogy of each task. When the node-level scheduler creates the first process of a new task, that task and its process are registered with the kernel extension. After that, the kernel extension maintains information on all the processes that constitute a task. The kernel extension is also responsible for suspending and resuming task execution atomically. Hence, there is no risk of a process creation/termination event being lost. Task suspension and resumption is accomplished from the kernel extension through standard UNIX signal mechanisms, using SIGSTOP to suspend execution and SIGCONT to resume it.

**External Interfaces:** Large supercomputing centers are likely to have their own job scheduling systems, either to handle site specific conditions or to organize multiple machines into a larger, metacomputing system. The goal is usually to increase system throughput and/or decrease response time for the particular workload in those centers.

To accomplish some level of control by the site scheduler, we introduce the concept of an *execution factor* associated with each job. Let $e_i$ be the execution factor of job $J_i$. Let $J_1, J_2, \ldots, J_n$ be the set of jobs sharing a particular processor $p$. Job $J_i$ will receive a fraction of the non-idle processor time of $p$ equal to $e_i / \sum_{j=1}^{n} e_j$. That is, the fraction of processor time that each job receives is proportional to its execution factor.

By manipulating the execution factors of jobs, a site scheduler can perform its own job control, albeit at a coarser granularity. By setting $e_i = 1$ and $e_j = 0, j = 1, \ldots, n, j \neq i$, the site scheduler enables job $J_i$ for execution while suspending all other jobs. Note that this approach can be used even across sectors with a single *meta-job* represented as three different jobs, one in each sector.

# 4 Experiments

There is a cost associated with time-sharing, due mostly to: (i) the cost of the context-switches themselves, (ii) additional memory pressure created by multiple jobs sharing nodes, and (iii) additional disk space pressure caused by more jobs executing concurrently. For that reason, the degree of time-sharing is usually limited by a parameter that we call, in analogy to uniprocessor systems, the multiprogramming level (MPL). A time-sharing system with multiprogramming level of 1 reverts back to a space-sharing system.

Using a 10-node, 40-processor development system for GangLL, and also a simulator for a larger system, we have performed experiments to assess the efficacy of our gang scheduling approach. First, we measure how GangLL impacts the performance of parallel applications. Then, we analyze the performance characteristics of the entire system under GangLL. Finally, we study the negative impact of paging, when the memory footprint of the active jobs exceeds the physical memory size of the nodes.

**Application performance under GangLL:** To measure application performance under GangLL, we use mid-size (class B) versions of the three pseudo-applications (BT, LU, and SP) in the NAS Parallel Benchmark suite 2.3. Each benchmark is written in Fortran with calls to the MPI message-passing library. The benchmarks are explicitly compiled to run on 36 (BT, SP) or 32 (LU) tasks, thus using most of the processors in the development machine and preventing space sharing. Each task consists of exactly one process. The name, a brief description, number of tasks, the memory footprint (per task), and execution time on a dedicated environment for each benchmark are listed in Table 2.

**Table 2.** Three NAS parallel benchmarks used in our experiments.

| benchmark | description | number of tasks | footprint (MBytes) | execution times (s) |
|---|---|---|---|---|
| BT | CFD block tridiagonal solver | 36 | 38.6 | 565 |
| LU | CFD LU solver | 32 | 8.8 | 371 |
| SP | CFD pentadiagonal solver | 36 | 15.9 | 433 |

We first run each benchmark on a dedicated machine (multiprogramming level of 1) and measure its execution time. We then run two, three, and four instances of the same benchmark, time-sharing the machine through our gang-scheduling system with a fine-grained time-slice of 5 seconds. Results from these experiments are shown in Figure 3.

In an ideal case, running $m$ instances of a benchmark should slow their execution by a factor of $m$. We note that the execution overhead created by the time-sharing is between 0 and 10% of this ideal execution time, for multiprogramming levels between 2 and 4. Overall, our gang-scheduling system does an effective job of providing the illusion of a slower virtual machine for the execution of each job.

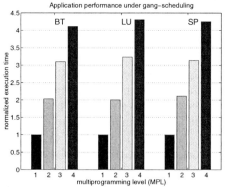

**Fig. 3.** Application performance under gang-scheduling for three NAS benchmarks.

**System performance:** The goal of this section is to quantify the performance benefits delivered by gang-scheduling in a large multiprocessor system. We examine the behavior of the average job *queue* and *response* (turn-around) times in the presence of varying degrees of multiprogramming. The job queue time is defined as the elapsed time between job submission and the time the job starts execution. The job response time is defined as the elapsed time between job submission and job completion. We show these performance parameters for different multiprogramming levels.

We use a week-long trace from a production 256-processor system. The trace logs submission time and actual processor time for over 1000 jobs. The jobs range in size from 2 to 128 processors with an average execution time, on a dedicated system, of 3150 seconds. We simulate the execution of these jobs on a 256-processor system under the control of GangLL. Jobs are queued in order of arrival and selected for execution on a first-come-first-serve (FCFS) basis as dictated by the availability of processors and multiprogramming level. The system is homogeneous on processors: any arbitrary set of processors (of the appropriate size) can be combined in a pool to run a parallel job, and the performance of any such pool is the same. This is typical of RS/6000 SP systems comprised of a homogeneous set of nodes. We consider two different values of an *overhead* parameter: 0% and 5%. An overhead of 5% means that during 5% of the time slice no work is accomplished (due to both context-switch time and paging time).

The results for multiprogramming levels 1 to 5 are shown in Figure 4. Analyzing first the case of zero overhead, the average time a job waits in the queue goes from almost a day to a few minutes when the multiprogramming level goes from 1 to 5. With a 0% overhead time-sharing, the average execution time of jobs increase from 3150 seconds with an MPL of 1 to 7070 seconds with an MPL of 5. Nevertheless, we are able to deliver a better than 10-fold improvement in job turn-around time. When the overhead climbs to 5% we still observe substantial reduction in average job queue and response times, even though the average execution time for MPL of 5 increases to 8330 seconds. Similar positive impact of gang-scheduling on FCFS scheduling has also been reported in [10].

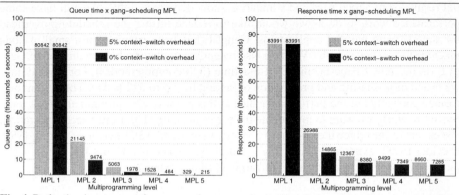

**Fig. 4.** Reduction in average job queue and response (turn-around) time through gang-scheduling.

**Impact of paging:** Paging effects can significantly hurt performance if jobs have large memory footprints. Figure 5 shows the behavior of the sequential (single task) NAS benchmark LU class A under control of GangLL. The results were obtained through actual experimentation. Each instance of this job has a 45 MB memory footprint. In both the 128 MB node (dark bars) and the 256 MB node (light bars), time-sharing two instances of LU class A (multiprogramming level of 2) behaves close to the ideal case (twice the time for a dedicated execution). However, paging effects are severe under a multiprogramming level of three for the 128 MB node. The average execution time for the three jobs is 3.5 times as large as in the 256 MB node under the same conditions. Strategies to reduce the impact of paging on gang-scheduling are discussed in [12].

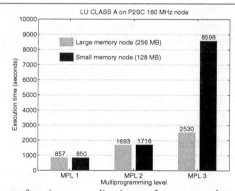

**Fig. 5.** The impact of paging on application performance under gang-scheduling.

## 5 Conclusion and Future Work

We have designed and implemented the first generation of GangLL, a gang scheduling system for the RS/6000 SP intended for use in the ASCI-Blue Pacific machine. Initial results from actual runs and simulation show definite benefits. We still have some performance work to do as we would like to reduce further the scheduling overhead for time-sharing parallel applications. GangLL has been running on a small 4-node machine for several months. We have also recently completed a beta evaluation of GangLL on the ASCI Blue-Pacific CTR machine (320 compute nodes). For several days we tested time-sharing and preemption of large jobs (up to 512 tasks). The system performed correctly but the performance data at this point is very preliminary. As a topic for future work, we want to investigate the use of *flocking* [1] for inter-sector operation. We expect to introduce the system to production on large ASCI machines during 1999. Previous LLNL experience with gang scheduling indicates that GangLL will be a major factor in improving usability and utilization of these systems.

## References

1. D. H. J. Epema, M. Livny, R. van Dantzig, X. Evers, and J. Pruyne. **A worldwide flock of Condors: Load sharing among workstation clusters**. *Future Generation Computer Systems*, 12(1):53–65, May 1996.
2. D. G. Feitelson and M. A. Jette. **Improved Utilization and Responsiveness with Gang Scheduling**. In *IPPS'97 Workshop on Job Scheduling Strategies for Parallel Processing*, volume 1291 of *Lecture Notes in Computer Science*, pages 238–261. Springer-Verlag, April 1997.
3. H. Franke, J. E. Moreira, and P. Pattnaik. **Process Tracking for Parallel Job Control**. Technical Report 21290, IBM Research Division, September 1998.
4. H. Franke, P. Pattnaik, and L. Rudolph. **Gang Scheduling for Highly Efficient Multiprocessors**. In *Sixth Symposium on the Frontiers of Massively Parallel Computation, Annapolis, Maryland*, 1996.
5. B. Gorda and R. Wolski. **Time Sharing Massively Parallel Machines**. In *International Conference on Parallel Processing*, volume II, pages 214–217, August 1995.
6. N. Islam, A. L. Prodromidis, M. S. Squillante, L. L. Fong, and A. S. Gopal. **Extensible Resource Management for Cluster Computing**. In *Proceedings of the 17th International Conference on Distributed Computing Systems*, pages 561–568, 1997.
7. M. Jette, D. Storch, and E. Yim. **Timesharing the Cray T3D**. *Cray User Group*, pages 247–252, March 1996.
8. D. Lifka. **The ANL/IBM SP scheduling system**. In *IPPS'95 Workshop on Job Scheduling Strategies for Parallel Processing*, volume 949 of *Lecture Notes in Computer Science*, pages 295–303. Springer-Verlag, April 1995.
9. J. K. Ousterhout. **Scheduling Techniques for Concurrent Systems**. In *Third International Conference on Distributed Computing Systems*, pages 22–30, 1982.
10. U. Schwiegelshohn and R. Yahyapour. **Improving First-Come-First-Serve Job Scheduling by Gang Scheduling**. In *IPPS'98 Workshop on Job Scheduling Strategies for Parallel Processing*, March 1998.
11. U. Vahalia. **UNIX Internals: The New Frontiers**. Prentice-Hall, Inc, 1996.
12. F. Wang, M. Papaefthymiou, and M. Squillante. **Performance Evaluation of Gang Scheduling for Parallel and Distributed Multiprogramming**. In *IPPS'97 Workshop on Job Scheduling Strategies for Parallel Processing*, volume 1291 of *Lecture Notes in Computer Science*, pages 277–298. Springer-Verlag, April 1997.

# Towards Quality of Service for Parallel Computing: An Overview of the MILAN Project

Holger Karl

Humboldt-University of Berlin
Institut für Informatik
Unter den Linden 6
10099 Berlin, Germany
karl@informatik.hu-berlin.de

**Abstract.** Parallel computing is faced with many practical difficulties, e.g. the gap between simple, high-level programming models and complex, real execution environments like a cluster of workstations, and the unpredictability of program execution. The MILAN project addresses these problems and aims at increased Quality of Service (QoS) for parallel programs. This paper presents an overview of the current research results of MILAN. For clusters of workstations, the Calypso system provides a simple programming environment that leverages theoretical results on fault-tolerant execution of parallel programs. A resource management system implements the necessary resource contracts for QoS, particularly for parallel applications. The concepts of Calypso have been applied to Web-based computing with the Charlotte system.

## 1 Introduction

High-performance computing is a classic problem of computer science. Traditionally, supercomputer architectures of various kinds have been used (e.g. vector computers or massively parallel machines). In recent years, researchers have considered commercial off-the-shelf (COTS)-based machines as an alternative to custom-built supercomputers because of the short innovation cycles of today's hardware and the ability to take advantage of mass production economics. Many research projects have investigated such clusters of standard workstations as a basis for high-performance computing. The Metacomputing in Large Asynchronous Networks (MILAN) project of New York University and Arizona State University is one of these projects.

One of the key issues in cluster computing, apart from mere performance, is the ease of use of and satisfaction with a cluster system for all parties involved: the programmer and the user of a parallel program on the hand, and interactive users of the cluster on the other hand. A programmer finds himself faced with a more complex situation than what is typically found in a supercomputer environment. Unlike in a supercomputer, individual nodes in a cluster can be of arbitrary speed, might be taken away from the parallel application, or additional machines might become available. It is a difficult task for a programmer to allow for such unpredictabilities in the execution environment. Also, complicated programming paradigms, e.g. elaborate consistency models, can make a programmer's life unnecessarily difficult. Hence, it would be ideal for a programmer to write programs for an idealized virtual machine with a very simple semantic. The Calypso system [3, 12] provides such a virtual machine with straightforward distributed shared memory (DSM) semantics; Calypso's run-time system is responsible for executing a program written for this virtual machine on a set of real, imperfect machines.

For a user of a parallel program, a simple setup and installation is one of the first expectations. Ideally, the user should not have to install any additional software. MILAN systems fulfill this requirement as far as possible. Other demands like efficiency or predictability of program execution are also important for a user's satisfaction.

Interactive users of a cluster do not want to be impeded by the parallel computations. Or at least, they want to rely on a certain minimum degree of service still available to them. This gives rise to the notion of a resource contract between a parallel application and the environment it executes in. On the basis of such a contract, a parallel application is granted a certain maximum amount of resources, e.g. CPU time. Interactive users are guaranteed access to the remaining CPU share.

Such a resource contract is also important to newer types of parallel programs, namely those with Quality of Service (QoS) requirements. Applications like image or signal processing must complete their execution before a given time, and to do so, they need guarantees for minimum available resources. MILAN's resource management system gives a framework for various types of applications as well as the necessary guarantees of minimum and maximum resource share, in particular for distributed applications. As a concise QoS metric, we use (among others) the notion of *responsiveness*: the probability of a program meeting its deadline, even in the presence of faults [19, 24]

Clusters of workstations are not the only environment for parallel programs. Since the techniques developed in the MILAN project are concerned with hiding complexities of the execution environment from a programmer, they are naturally suited to a particularly complex environment: the World Wide Web (WWW). The WWW has been described as an enormous source of idle computing capacity. The Charlotte system [6] leverages Calypso's techniques to make this capacity usable to parallel applications written in Java, allowing standard Web browsers to serve as the execution environment. KnittingFactory [5] provides infrastructure support for such applications.

This paper gives an overview of the MILAN project and describes the system (in particular, Calypso and Charlotte) and methodologies (especially eager scheduling and TIES) developed within this project. Further information on the MILAN project can be found at the project home pages http://www.cs.nyu.edu/milan and http://milan.eas.asu.edu.

The rest of this paper is organized as follows. Sect. 2 gives a brief overview of some related work, Sect. 3 introduces the basic concepts of the MILAN project and the Calypso system. Sect. 4 shows the resource management systems of MILAN. Sect. 5 discusses how the concepts of MILAN can be used to make metacomputing on the WWW feasible. Finally, Sect. 6 considers possible future work and gives some conclusions.

## 2 Related Work

Cluster computing is by now so large an area that it is impossible to give a comprehensive list. Some systems that provide DSM are Midway [7], Treadmarks [1], and Munin [10]. Web-based computing also has received a lot of attention, usually centered around Java. Some projects like ParaWeb [8] attempt to modify the Java Virtual Machine to endow Java with a distributed memory semantics. Others, e.g. Javelin [9], choose not to modify Java and tradeoff efficiency against flexibility.

While there is an enormous amount of work for QoS for network connection, multimedia applications, etc., the notion of QoS for parallel programs is fairly new. As one example, MPI/RT [15] is a standardization proposal for a middleware-implemented application programming interface that provides real-time capabilities.

## 3 Calypso: A System for Cluster Computing

### 3.1 Theoretical Background

The theoretical background for the MILAN project is work on efficient asynchronous execution of large-grained parallel programs (cf. e.g. [2]). Consider a program $\mathcal{P}$ that is written for an idealized, synchronous machine, using a bulk synchronous parallelism (BSP)-like style [23]: it consists of a number of parallel steps, where each step has a number of parallel routines. Assume further that in one parallel step, any shared variable is updated by at most one routine. It is a challenging problem to execute such a program on a realistic, asynchronous machine (where a processor can also become infinitely slow, i.e. fail). This problem is solved in [2] by compiling $\mathcal{P}$ into a semantically equivalent program $C(\mathcal{P})$ that can be efficiently executed on an asynchronous machine.

### 3.2 The Calypso System

Calypso is a software system, based upon this theoretical work, that allows writing and executing parallel programs on COTS-based clusters of workstations [3, 12]. One of the objectives of Calypso is to evaluate the theoretical results in a practical prototype for their suitability to issues like programmability, high performance, scalability, load balancing, and fault masking.

**The Calypso programming model.** In the Calypso system, a programmer writes a program for an ideal virtual machine with shared memory and infinitely many processors. This allows the programmer to concentrate on expressing the parallelism inherent in the algorithm, and separates the algorithmic parallelism from the available parallelism in the execution environment (which might be different for various runs of a program).

The Calypso Source Language (CSL) is used to write programs. CSL extends C++ by four keywords to express parallelism: shared, parbegin, parend, and routine. In a CSL program, sequential steps alternate with parallel steps. The beginning and end of a parallel step are marked by parbegin and parend, respectively, and within a parallel step, routine is used to denote one of possibly many parallel routines. An example for a parallel step would look like this:

```
parbegin
   routine[int-expr] (int width, int id) { routine body }
parend
```

The *routine body* of a routine statement is the code executed as a parallel routine. The optional *int-expr* is the number of instantiations of such a routine. The two arguments *width* and *id* are passed to the routine as actual arguments at invocation time: *width* is the number of routines, *id* is the unique number for each routine. Therefore, one routine is able to identify its own identity relative to sibling routines. A parallel step ends once all routines have terminated.

Data that has been marked as shared is accessible in parallel routines. All other data is strictly local, either to the sequential part of the program or to each parallel routine. Data access to shared data follows the CR&EW (concurrent read and exclusive write) policy: a data item can be read by any number of routines, but only written by at most one. The granularity for sharing is the primitive data type of C (e.g. int). During a parallel step, a routine reads the value of a variable as it was at the beginning of the step, and write updates occur atomically at the end of it. Thus all routines execute in isolation.

A CSL program is compiled by a preprocessor to standard C++. This program is then compiled with any off-the-shelf C++ compiler and linked with the Calypso library, producing a stand-alone executable.

**Executing a Calypso program.** How can this semantics be implemented on real, unreliable machines? A Calypso program executes in a master-worker fashion. The master process executes all sequential steps and manages the execution of the parallel steps. The routines are executed by any number of worker processes, usually residing on remote machines. At the beginning of a parallel step, the master waits for workers requesting work. Upon such a request, the master assigns a routine to a worker, which the worker will then execute. During execution, the worker will access data in the shared memory. The DSM is implemented at the page level: at the beginning of a parallel step, all the pages of the shared memory are access protected. If a worker reads a variable located on such a page, the operating system generates a page fault. The Calypso library catches this exception and requests the corresponding page from the master. A write access to a protected page marks it as dirty, and at the end of a routine, all dirty pages are sent back to the master.

The master, however, can not yet integrate such a dirty page in the shared memory, since a request for this page by another worker would possibly result in a value different from the one at the beginning of a parallel step. Hence, the page updates from the workers can only be stored in the master's shared memory when a parallel step has been completed. This memory management is called two-phase integrated execution strategy (TIES).

TIES allows another important technique: *eager scheduling*. Consider what happens if one worker fails: the routine it has been assigned will not be finished, and the master waits for this routine, even though other workers become idle. In Calypso, the master can re-assign routines to idle workers (once all routines have been assigned at least once), since TIES guarantees an exactly-once semantics of routine execution. This implies that worker failures can be masked, and that slow machines do not stall a computation, since eager scheduling entails automatic load balancing. Even intermittently available machines can be used by Calypso.

**Extensions to Calypso.** [13] adds nested parallelism to Calypso, Chime [21, 22], Calypso's successor developed at Arizona State University, adds arbitrary synchronization and supports the entire CC++ language [11]. One of Chime's outstanding features is sharing the program stack across a network.

### 3.3 Improving Calypso's QoS

While Calypso is able to mask worker faults, the master process is a single point of failure. One obvious way to overcome this is checkpointing. Since one of the objectives of the MILAN project is QoS for parallel programs, the QoS attribute (here: responsiveness) has large influence on the checkpointing procedure. An important question for checkpointing is the choice of the checkpointing interval: it should be chosen so as to maximize the responsiveness of a given program. An analysis for this problem can be found in [18], and a corresponding checkpointing mechanism is implemented in Calypso.

An alternative approach to checkpointing is replication of the master process, which also promises better load balancing between these replicated processes. On the other hand, keeping the replicas synchronized with each other imposes additional overhead. We are currently implementing such a replicated master, and the trade-offs between checkpointing (as an example for redundancy in time) and replication (redundancy in space) are a question of active research — in particular with respect not only to performance, but to predictability of program run time.

An additional requirement for a QoS-capable version of Calypso is an analysis of the eager scheduling mechanism, since it is necessary to know how long a parallel step will take in order to argue about the execution time of an entire program. While there are some preliminary results (for general distributions of routine lengths and faults), this is still an area of active research.

## 4 Resource Management in Clusters

For QoS of parallel programs, guaranteeing sufficient resources is an important problem. MILAN provides the following two mechanisms.

### 4.1 Just-in-time Resource Management

The first mechanism is a general resource management scheme that monitors resources and dynamically allocates them to concurrently executing jobs [4]. This system is especially designed to handle *adaptive* parallel programs — programs that can take advantage of added resources or tolerate their removal. It is implemented at the level of Unix's rsh command, enabling it to work for programming systems like PVM, MPI, or Calypso. The basic idea is to add a pseudo-machine name "ANY" telling rsh to start the process on any machine that the resource management system deems appropriate. For adaptive programs, the resource management system can shrink the amount of allocated resources to better accommodate other programs.

The system is plug-in based and extensible to other parallel systems. It can be used to implement sophisticated resource allocation schemes.

### 4.2 Implementing Resource Contracts for Distributed Applications

The resource management system just described is concerned with mapping machines to resource requests. To be able to give guarantees on the execution time of programs, it is additionally necessary to ensure that a program will actually receive a certain CPU share of the machines it has been mapped to. Since we do not consider exclusive use of machines by a parallel program, uncontrolled background load (e.g. by interactive users) can void any resource assignments.

There exists previous work on how to dynamically partition local resources in a COTS environment, without modifying the operating system (e.g. [14,20]). These systems are often called "scheduling server"; they exploit fixed-priority scheduling classes of the underlying operating system. The scheduling server is a daemon that runs with the highest available priority and controls the execution of some client processes. These controlled processes run at second-highest priority. The scheduling server cyclically resumes and suspends the controlled program, leaving enough free capacity for the rest of the system to make progress. These scheduling servers guarantee a minimum (and, if desired, a maximum) CPU share to an application, regardless of the background load.

All existing systems, however, are only concerned with applications on a single machine and do not consider parallel programs. Directly using scheduling servers for parallel programs results in unacceptable performance. This is due to the fact that the execution of the distributed processes is not coordinated in time, and hence, communication or synchronization can take very long. The need for such coordination is well-known for parallel programs and usually called gang- or co-scheduling. In [17], scheduling servers are extended by a synchronization mechanism to combine the guaranteed resource allocation of scheduling servers with the beneficial effects of coscheduling. This is achieved via additional scheduling events, generated by control messages exchanged between the distributed scheduling servers.

## 5 Embracing the WWW

Hiding complex execution environments from a programmer is one of the main objectives of MILAN. The WWW is perhaps the most complex execution environment for distributed programs today, and it also promises access to a vast number of idle workstations. It is therefore interesting to see how the concepts of Calypso carry over to parallel computing in the WWW.

### 5.1 Charlotte

Parallel computing in the WWW is faced with other challenges than in clusters. Among these are heterogenous clusters, security, the undesirability of installing programs on remote machines, and the inherent unreliability of the Internet.

Java has successfully addressed the problems of heterogeneity and — to some extent — of security, and the ubiquitous web browsers form a generally available execution environment for Java. It is therefore a prime candidate for implementing a parallel computing system for the WWW. Charlotte [6] is such a system that uses Calypso's programming model and eager scheduling techniques. In particular, the fault masking properties are indispensable for computing in the WWW.

Charlotte pursues the concept of *volunteer computing*. Similar to Calypso, a Charlotte program is structured in a master-worker style. The master, a Java application, initiates an application; the workers are Java applets. A volunteer, anywhere on the WWW, can download this applet from a server and start executing it in any Java-enabled browser. The worker applet will contact the master, request routines and execute them.

The main difference between Charlotte and Calypso is the memory management concept: since Charlotte is implemented in Pure Java, and Java does not give access to low-level abstractions like memory pages, the shared memory in Charlotte is realized at the object level. For every primitive data type in Java, e.g. `int`, there is a corresponding distributed class `Dint`, that implements the distributed memory semantics. Access to the actual data happens via `get()` and `set()` methods.

These distributed classes allow a simple programming style. However, the overhead associated with data access via method invocations is comparably large, as is the cost of requesting data from a remote master. In [16], an annotation scheme for Charlotte programs is introduced that allows a stepwise incorporation of semantic knowledge into a program. Charlotte's run-time system can use this information to improve execution efficiency. On the simplest level, program correctness is assured even if the annotations are wrong, yet data can be present avoiding costly requests from a remote manager. With complete trust in the annotations, the objects representing distributed memory can be replaced by their primitive counterparts. This results in efficiency comparable to message-passing programs, without abandoning Charlotte's main advantages like fault masking and adaptive load balancing.

### 5.2 KnittingFactory — An Enhanced Infrastructure for Web Applications

Charlotte makes it easy for a volunteer to contribute his idle CPU time to a parallel application, but it does not answer the question how a volunteer can find such an application. This problem is solved by KnittingFactory [5]. Applications looking for work can register with a directory service, and volunteers can point their browsers to well-known entry points to this directory. Unlike other directories, KnittingFactory's search is off-loaded from the directory servers and is realized by Javascript programs executing in a volunteer's Web browser. This allows to implement, e.g., breadth-first search to favor topologically close applications. Once a program is found, the applet is started automatically.

KnittingFactory also removes another limitation of typical Java applications. For an applet to connect to an application, the application process must run on a Web server. KnittingFactory provides a core Web server component that can be trivially used by any Java application, allowing it to run on any host.

## 6 Conclusions

The MILAN project started out investigating questions of programmability, performance, load balancing, and fault masking for parallel applications in COTS environments. Hiding the complexities of the execution environment from the programmer, and using the run-time system to implement an ideal virtual machine on a set of imperfect, real machines is one of the basic tenets of MILAN. The main techniques to achieve this are the TIES memory management scheme for DSM in combination with the eager scheduling principle. This results in automatic load balancing and fault masking for failing workers. The feasibility of this approach has been shown by the Calypso system for cluster computing and by Charlotte for Web-based computing.

The current focus of MILAN is QoS for parallel applications. Among many possible QoS attributes, responsiveness looks promising. To achieve responsive execution, the Calypso system has been enhanced with added fault tolerance mechanisms (like checkpointing). Also, resource management schemes have been developed that provide a framework for arbitrary management decisions, and that support the implementation of resource contracts, especially for parallel applications.

Yet a lot of work remains to be done. In particular, resource management for multiple programs running on the same cluster and competing for resources is still a challenging problem. Also, a simplified model of MILAN's eager scheduling principle is necessary to facilitate an effective analysis, upon which a resource management system can base its decisions.

# Acknowledgements

This paper would have not been possible but for the work of the entire MILAN research group of New York University and Arizona State University, particularly Zvi Kedem and Partha Dasgupta as principal investigators, as well as Arash Baratloo, Ayal Itzkovitz, Mehmet Karaul, Donald McLaughlin, Shantanu Sardesi, Peter Wyckhoff, Yuanyuan Zhao, among others.

This research was sponsored by the Defense Advanced Research Projects Agency and Rome Laboratory, Air Force Materiel Command, USAF, under agreement number F30602-96-1-0320; by the National Science Foundation under grant number CCR-94-11590; by Deutsche Forschunggemeinschaft; and by Microsoft. The U.S. Government is authorized to reproduce and distribute reprints for Governmental purposes notwithstanding any copyright annotation thereon. The views and conclusions contained herein are those of the authors and should not be interpreted as necessarily representing the official policies or endorsements, either expressed or implied, of the Defense Advanced Research Projects Agency, Rome Laboratory, or the U.S. Government.

# References

1. C. Amza, A. L. Cox, S. Dwarkadas, P. Keleher, H. Lu, R. Rajamony, W. Yu, and W. Zwaenopoel. TreadMarks: Shared Memory Computing on Networks of Workstations. *IEEE Computer*, 29(2):18–28, February 1996.
2. Y. Aumann, Z. Kedem, K. Palem, and M. Rabin. Highly Efficient Asynchronous Execution of Large-grained Parallel Programs. In *Proc. 34th IEEE Ann. Symp. on the Foundations of Computer Science*, pages 271–280, 1993.
3. A. Baratloo, P. Dasgupta, and Z. M. Kedem. CALYPSO: A Novel Software System for Fault-Tolerant Parallel Processing on Distributed Platforms. In *Proc. 4th IEEE Intl. Symp. on High-Performance Distributed Computing*, pages 122–129, Washington, D.C., August 1995.
4. A. Baratloo, A. Itzkovitz, Z. Kedem, and Y. Zhao. Just-in-time Transparent Resource Management in Distributed Systems. Technical Report 1998-762, Courante Institute of Mathematical Sciences, New York University, March 1998. http://www.cs.nyu.edu/milan/publications/tr1998-762.ps.gz.
5. A. Baratloo, M. Karaul, H. Karl, and Z. Kedem. KnittingFactory: An Infrastructure for Distributed Web Applications. *Concurrency: Practice and Experience*, 10(11–13):1029–1041, 1998.
6. A. Baratloo, M. Karaul, Z. Kedem, and P. Wyckhoff. Charlotte: Metacomputing on the Web. In *Proc. 9th Intl. Conf. on Parallel and Distributed Computing Systems*, pages 181–188, Dijon, France, September 1996.
7. B. Bershad, M. J. Zekauskas, and W. A. Sawdon. The Midway Distributed Shared Memory System. In *Proc. of COMPCON 93*, pages 528–537, 1993.
8. T. Brecht, H. Sandhu, M. Shan, and J. Talbot. ParaWeb: Towards World-Wide Supercomputing. In *7th ACM SIGOPS European Workshop*, pages 181–188, Connemara, Ireland, September 1996. http://cs.yorku.ca/~brecht/papers/html/paraweb/paraweb.html.

9. P. Cappello, B. Christiansen, M. F. Ionescu, M. O. Neary, K. E. Schauser, and D. Wu. Javelin: Internet-Based Parallel Computing Using Java. *Concurrency: Practice and Experience*, 9(11):1139–1160, November 1997.
10. J. B. Carter, J. K. Bennet, and W Zwaenepoel. Implementation and Performance of Munin. In *Proc. 13th ACM Symp. on Operating System Principles*, pages 152–164, October 1991.
11. K. M. Chandy and C. Kesselman. CC++: A Declarative Concurrent, Object Oriented Programming Notation. Technical Report CS-92-01, California Institute of Technology, 1992.
12. P. Dasgupta, Z. M. Kedem, and M. O. Rabin. Parallel Processing on Networks of Workstations: A Fault-Tolerant, High Performance Approach. In *Proc. 15th Intl. Conference on Distributed Computing Systems*, pages 467–474, 1995.
13. S.-C. Huang and Z. M. Kedem. Supporting a Flexible Parallel Programming Model on a Network of Workstations. In *Proc. 16th IEEE Intl. Conf. on Distributed Computing Systems*, pages 75–82. IEEE, 1996.
14. J. Kamada, M. Yuharo, and E. Ono. User-level Realtime Scheduler Exploiting Kernel-level Fixed Priority Scheduler. In *Intl. Symposium on Multimedia Systems*, Yokohama, Japan, March 1996.
15. A. Kanevsky, A. Skjellum, and A. Rounbehler. MPI/RT — An Emerging Standard for High-Performance Real-Time Systems. In *Proc. HICSS '98*, January 1988. http://www.mpirt.org/documents/hicss31_paper.pdf.
16. H. Karl. Bridging the Gap between Distributed Shared Memory and Message Passing. *Concurency: Practice and Experience*, 10(10–13):887–900, 1998.
17. H. Karl. A Prototype for Controlled Gang-Scheduling. Technical Report Informatik Bericht 112, Institut für Informatik, Humboldt-Universität, Berlin, Germany, August 1998.
18. H. Karl and M. Werner. An Optimal Checkpointing Interval for Real-Time Systems. In H. R. Arabnia, editor, *Proc. of Intl. Conf. Parallel and Distributed Processing Techniques and Applications*, pages 604–612, Las Vegas, NV, July 1997.
19. M. Malek. Responsive Systems (The Challenge for the Nineties). *Microprocessing & Microprogramming*, 30:9–16, 1990.
20. A. Polze. How to Partition a Workstation. In *Proc. Eigth IASTED/ISMM Intl. Conf. on Parallel and Distributed Computing and Systems*, pages 184–187, Chicago, IL, October 1996.
21. S. Sardesi. *CHIME: A Versatile Distributed Parallel Processing System*. PhD thesis, Arizona State University, Tempe, AZ, May 1997.
22. S. Sardesi, D. McLaughlin, and P. Dasgupta. Distributed Cactus Stacks: Runtime Stack-Sharing Support for Distributed Parallel Programs. In H. R. Arabnia, editor, *Proc. Intl. Conf. Parallel and Distributed Processing Techniques and Applications*, pages 57–65, Las Vegas, NV, July 1997.
23. L. G. Valiant. A Bridging Model for Parallel Computation. *Communications of the ACM*, 33(8):103–111, August 1990.
24. M. Werner and H. Karl. Towards a Definition of Responsiveness. Technical Report Informatik-Berichte Nr. 91, Institut für Informatik, Humboldt-Universität, Berlin, 1997.

# Resource Allocation and Scheduling in Metasystems *

Uwe Schwiegelshohn and Ramin Yahyapour

Computer Engineering Institute, University Dortmund
44221 Dortmund, Germany
uwe@ds.e-technik.uni-dortmund.de
yahya@ds.e-technik.uni-dortmund.de

**Abstract.** In this paper we present *NWIRE*, a new management architecture for metacomputing systems. After the general introduction of the architecture we first describe the properties that are relevant for scheduling. Then we derive general requirements for scheduling strategies in metasystems and point out differences to conventional job scheduling for parallel processors. This leads to the metacomputing scheduling concept of *NWIRE* which is based on a brokerage and trading approach.

## 1 Introduction

A metasystem or a metacomputer consist of numerous independent computing resources which are linked by interconnection networks. The term *independent* emphasizes the fact that the individual resources are not controlled by a single entity. While this definition does not include performance details of the individual resources forming a metacomputer, it is typically assumed that those resources are high performance computing components like large parallel processors, networks with high speed and high bandwidth or huge databases. Only the component at the user access point may not necessarily belong to this category. However, the user of such a metasystem may not be aware of the internal system structure but has the illusion of a single virtual machine. She may even use several components of the system concurrently to solve a large problem. With other words, it seems possible to avoid the expense for very big machines by connecting several smaller ones with a fast network. However, it must be noted that this concept may require modified application software in some cases.

The metasystem needs specific software to provide this view. This software has further the goal to simplify administration and to improve efficiency of the system. Presently, there are a number of projects that work on the realization of a metacomputing environment. These projects include Globus, Legion, Codine, Condor, AppLeS and LSF. While many of those systems have been especially developed for a certain class of applications there are a few which use a more general approach. Among them is the Globus project. It addresses many aspects of metacomputing like for instance account management, message passing

---

* Supported by the NRW Metacomputing grant

communication, and security. Information on the configuration of the available resources is accessible via the Light Weight Directory Access Protocol (LDAP). The scheduling is done by a multi-level hierarchy of resource brokers, resource co-allocators, and local managers. Globus also uses a special resource specification language to define resources and generate requests.

In the *NWIRE* project we have also developed an architecture to link the various computing resources to a metacomputing system. While this architecture has similar features as the Globus architecture [2] in general, there are also several important differences e.g. in scheduling, resource allocation, and the information management. In this paper we will mainly focus on these issues. Our solution has been designed to allow the inclusion of additional management functions like the management of network devices. As in a metasystem remote resources are used in parallel, its performance depends on the network condition. Therefore, network management cannot be ignored in a metasystem. This is not limited to forecasting and arbitrating the available bandwidth, but also includes the actual configuration of network devices.

## 2 Architecture of NWIRE

In this section we first describe our metacomputing architecture *NWIRE*. As this paper mainly addresses the scheduling aspects of metacomputing the architectural description concentrates on those components which are important in the scheduling context. Then we briefly discuss the relationship between *NWIRE* concepts and general requirements of metasystems.

### 2.1 Description of the Architecture

On the top logical level *NWIRE* consists of several so called *MetaDomains* which are linked by an interconnection network. Note that no specific structure for the network has been specified. It is only required that all *MetaDomains* can directly or indirectly exchange and share information among each other. New *MetaDomains* can be introduced into the metasystem at any time. Similarly, it is possible to temporarily or permanently remove some *MetaDomains* from the metasystem. A *MetaDomain* is controlled by a *MetaManager*. This *MetaManager* accesses a set of *ResourceManagers*. Such a *ResourceManager* is the interface to a specific resource in the metasystem. Typically, the *ResourceManager* will connect to the local management system of a multiprocessor that participates in the metasystem. For instance in case of an IBM RS6000/SP, the corresponding *ResourceManager* will translate LoadLeveler and machine information into data for the *MetaManager* and vice versa. Therefore, the *ResourceManager* generates a common information layer. A separate *ResourceManager* must be developed for each different resource type in the metasystem. Fig. 1 shows a simple example of an *NWIRE* system.

Note that we did not specify the term resource. In *NWIRE* it is simply required that

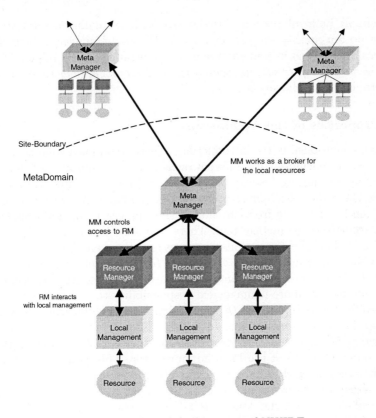

**Fig. 1.** Distributed Architecture of NWIRE

- certain methods can be applied to a resource and
- a resource provides status information to the attached *ResourceManager*.

For instance, in many metasystems the interconnection network between two multiprocessor systems is considered as a resource.

From a functional point of view a *MetaManager* permanently collects information about all resources associated with it. Further, it handles all requests inside its *MetaDomain* and works as a resource broker to the outside world, that is to other *MetaDomains*. Therefore, requests to a *MetaManager* can either be submitted from users in its *MetaDomain* or from other *MetaManagers*. Each *MetaManager* contains a scheduler that maps requests for resources to a specific resource in its *MetaDomain*. The scheduler may split up complex resource requests into more specific requests.

After a suitable allocation for a request has been found, the *MetaManager* can be used to start the execution of the job associated with the request. To this end, the *MetaManager* connects to the participating *ResourceManagers*, sets them up and starts the job. The actual meaning of the expression 'starting a job' depends on the job and the implementation of the *ResourceManager*. For some systems it would simply be a submission of the job to the local management system. Note that we do not specify the properties of a *ResourceManager*, that is arbitrary resources are allowed in a metasystem. If a special feature is requested but not available on a resource as represented by its *ResourceManager*, this resource is not suitable for an allocation.

## 2.2 Properties of the Architecture

As already mentioned in the introduction, a metasystem is a voluntary combination of several machines with different owners. Hence, computers belonging to a metasystem will not exclusively be used for the purpose of metacomputing. For instance, the owner or administrator of such a machine may decide to temporarily exclude her resource from the metasystem or use it for additional requests not received from the metasystem. There may also be different allocation and scheduling policies for the various resources in the system. The administrator of an individual machine can also choose *NWIRE* to simply act as another user. Then *NWIRE* will not enforce any specific use of the resource. Alternatively, he may decide to give more management responsibility to *NWIRE*. Consequently, the scheduling strategy of *NWIRE* must significantly differ from those used for large parallel processor systems. In addition, *NWIRE* is easy to set up once the *ResourceManager* has been developed for this type of resource and does not require much effort from the local administrator. This property will hopefully lower the acceptance threshold for many administrators.

As metasystems may include many independent and geographically separated components and users, the issue of reliability becomes very important. To this end *NWIRE*, is based upon a highly distributed architecture. This architecture also supports scalability of the metasystem and avoids the potential performance bottleneck of central servers. Unfortunately, this leads to complex scheduling problems due to the lack of global system information at each *MetaManager*.

In addition, a user is mainly interested in the issues performance, ease of use and security. In a typical scenario she simply wants to submit a job to the metasystem and to receive the result of her computation as soon as possible. Ideally, this submission process should be similar to the submission of a job to a local workstation. Although this goal cannot always be achieved due to the necessary description of the required compute resources, a well designed user interface is a key element to get a high degree of user acceptance. On the other hand, the user expects that her data are secure from outside interference. This is especially important as the user may not know where her job is actually executed. Similarly, the owner of a compute resource may wish to restrict access to his machine. To address these demands the *NWIRE* architecture is implemented in CORBA. This allows the integration of appropriate security technology like for instance Secure Socket Layers (SSL) for secure communication or DCE or Kerberos for authentication. A future transfer to other technologies like IP-sec or LDAP is also possible. Use of CORBA as the communication platform further has the advantage that a variety of services like trading services are already available and can be used to build a scheduler.

# 3 Scheduling Considerations

In this section we describe general properties of a generic metasystem scheduler. This includes *variable scheduling objectives*, the existence of additional *independent schedulers*, the availability to generate *arbitrary resource requests*, the support of *resource reservation* including the ability to provide *guarantees for job execution*.

**Variable Scheduling Objectives**: In common job scheduling there usually exists a single scheduling objective or performance metric that is fixed for a parallel computer and all its jobs. For example, this can be the minimization of the average response or turnaround time [1]. The objective is typically determined by the local management system or by the administrator. In metacomputing this objective is variable. As we assume a distributed system that is not controlled by a single instance, the objective should further be adaptable for each resource in the metasystem. While the schedule target for some machines may for instance be the maximization of the throughput, others have the objective to minimize the response time. Besides the objectives for the resources, we must also take the needs of the user into account. Some user may favor the availability of specific resource properties like the size of the main memory while other may have additional constraints about the execution of a job. A typical example would be a deadline for a job that must be met while it is of no particular interest if the job is completed as fast as possible. For this user the minimization of the response time would not reflect her demands. In metacomputing it is necessary that user objectives are considered. For instance, the user may only be interested in resources that fit her needs better than any local resources. Therefore, the scheduling must be adaptable to generate the most appropriate result.

**Independent Schedulers**: As mentioned in Section 2, the scheduler in metacomputing cannot demand exclusive control over all resources. For scheduling in metasystems, we have to cope with the situation that jobs may not only be submitted via the metacomputing interface. Hence, any limitations of the local management must also be considered. For instance, one problem is the non-deterministic scheduling of most management systems which do not provide any information about the expected completion time of a job. Unfortunately, availability of this kind of information is important in distributed metasystems to allow future allocation planning. Of course, if the local management provides additional information, the metacomputing scheduler should be able to utilize it. In this case the metacomputing management does not perform any local scheduling but relies on the existing system scheduler. The resulting schedule efficiency highly depends on the features of the lower-level scheduler. If a resource does not provide the requested features like a guaranteed completion time, it cannot be considered suitable for some job requests. This limits the usability of this resource for the metasystem. Nevertheless, the metacomputing scheduling should support all kind of local management systems.

**Arbitrary Resource Requests**: As job requirements and resources in a metasystem may vary according to type and application, there is a need for the description of complex requests. For instance, assume two different users: The first user does not provide a very detailed request as she wants to get as many computing resources as possible. More restrictive requirements would only reduce the possible resource set for her job. The other user is looking for very specific resources. He may have access to an alternative set of local resources for the execution of his job and is therefore only looking for a better resource allocation. Consequently, he formulates special requirements and preferences. The metascheduler must support both approaches. The individual user should be able to influence the resource selection and the scheduling so that she gets the best suited set of resources. The attributes of a resource and therefore the available fields in a request should not be considered invariant. Different resources may have different attributes and features that may not be known to the scheduler at the time of implementation. But the system should still be able to handle them.

**Resource Reservation**: This feature is necessary for some applications as well as for the consideration of resource maintenance. For instance, demonstrations may require the reservation of a resource allocation for a dedicated time span. It is also advantageous for the schedule to consider system downtime or restricted usage that is known in advance. Reservations are further needed for multi-site applications. As there is no global scheduler instance, it must be possible for the local scheduler to reserve resources for a specific time span in order to guarantee the concurrent availability of resources at different locations.

**Job Execution Guarantees**: In metacomputing it would be inefficient to schedule jobs on an ad-hoc strategy as it is difficult to respect several objectives by not assuming a central scheduler. For example, if a job does not need to be executed as soon as possible, this flexibility can be used to improve the schedule. Assume again the mentioned case of a job with an execution deadline. Typically, the user needs immediate feedback whether his requirements can be met. It is therefore necessary for the user to receive in advance guarantees about the schedule of his job so that he can react accordingly. The scheduler need not always provide such guarantees, but it should be capable of giving them if they are required. Those guarantees are additional constraints in the scheduling of a job.

## 4 Scheduling in NWIRE

Section 3 has shown that metasystem scheduling requirements are more difficult than those for normal job scheduling. This must be taken into account when deciding on the scheduler architecture. It is possible to ignore most of the requirements and to implement a queuing scheduler that works similar to standard job schedulers with some extensions to allow cooperation with remote schedulers or maybe in a hierarchy of schedulers. Alternatively, an approach from economics based on a trading and brokerage system can be used. Both concepts are quite different as trading allows more freedom in selecting resources, implementing arbitrary complex strategies and providing guarantees. On the other hand, a hierarchical scheduler is expected to produce more efficient schedules. The schedule quality in the trading approach relies on some market mechanism. There, the description of the objectives is key element for the resource allocation.

We assume that a metacomputing system will only be accepted, if it meets most or all of the requirements defined in the previous section. Especially the ability to supply a scheduling objective for every resource is important as many owners of computing resources are only willing to share their resources with someone else under certain conditions. This includes political scheduling [3] as well as strategies on the overall machine utilization. Currently, this seems only to be feasible with a trading system. In addition, this approach provides the highest degree of flexibility and is usually rather fault-tolerant. If a single trader fails only some jobs may be affected but not the whole metasystem.

### 4.1 Resource Request

According to the architecture of *NWIRE*, as presented in Section 2, a request for resources is directed to the *MetaManager* of a *MetaDomain*. This request is formulated in a resource description language which allows arbitrary combinations of requirements and attributes about the job and the user. An objective function can be part of this request. This function is then used to calculate a value for the utility of a resource allocation, see also [4]. The request is parsed to determine whether a resource is suitable. While it is not necessary that the scheduler recognizes all attributes in a request, some other properties are known by management system. This method allows arbitrary resource definitions. But resources of the same class must have the same set of obligatory and optional attributes. Otherwise if the interfaces differ, users and programs would be unable to specify or even access those resources. However, this definition must not be known by the scheduler and can therefore later be modified and standardized to meet the requirements of a metasystem.

### 4.2 Resource Determination

A request is generated by an application or a user. It is then submitted to the *MetaManager* in the local *MetaDomain*. The *MetaManager* tries to find suitable resources and makes allocations if requested. The search for a resource match always starts in the local *MetaDomain*. Giving priority to local resources typically leads to shorter response time, less network load and may result in shorter startup times. The *MetaManager* may split up requests into more detailed requests which are then used to determine if a specific resource fits the requirements. Moreover, the scheduler connects to known remote *MetaManagers* and forwards requests to them. The reply to a request is a resource offer. As there may be several possible allocations that fit a request, there may also be several offers. Even a single resource can provide several offers which differ in its attributes. Note that every *MetaManager* can ask any other *MetaManager* for offers. To prevent network flooding we therefore limit the scope of requests by special attributes: maximum number of hops to search, maximum number of collected offers etc. The whole search mechanism is similar to the trading

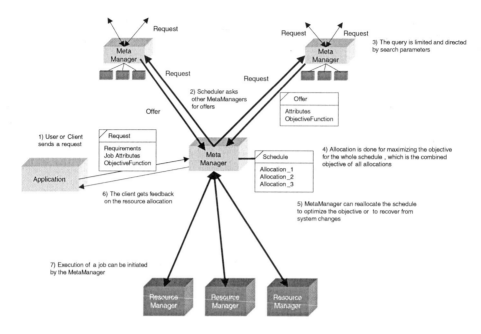

**Fig. 2.** Scheduling in NWIRE

service as defined in the CORBAservices. Therefore, the performance impact for searching is usually limited and does not query the whole network. Optimization with caching and lookup-strategies further minimizes the number of involved *MetaManagers*. Note that a suitable link configuration between *MetaManagers* in this trading scheme is essential for utilizing network locality. For instance, *MetaManagers* should be linked if the corresponding *MetaDomains* are connected by a network with high bandwidth and low latency.

Each trading step results in a set of resource offers. Depending on the request attributes these offers may only be valid for a specific time. Further, obtaining and reserving an offer may already result in costs. The *MetaManager* is responsible to guarantee the validity of its offers. In order to find possible allocations for a request a scheduler may combine several offers. Finally, it is up to the scheduler or the requesting user to accept an allocation.

### 4.3 Scheduling and Allocation

The selection of the best suited allocation is based on a comparison of the provided objective functions. The objective function of a request is applied to an allocation in order to generate a value for the utility from the user's point of view. Similarly, the offers also provide an objective function or a value for its utility to represent the resource's point of view. Now, the responsible *MetaManager* combines the objectives and determines an allocation that maximizes the overall objective with respect to its full schedule. It is also possible to provide the user with a front-end that allows interactive selection of allocations. Such a front-end can also be used to obtain status information about the metasystem with the help of the request mechanism. This information may help a user to generate a request which results in the best suited set of resources depending on the current condition in the metasystem.

Note that our method is not an auction system as we do not provide a market where several jobs compete for a resource. Instead our schedulers select allocations that fit a request best at a particular time instance. However, the selected allocation must not necessarily be executed. The *MetaManager* maintains a schedule with all current allocations in its domain. Its scheduler is free to modify the current schedule at any time. However, changes in the current scheduling are only allowed as long as they do not violate any guarantees that have been given for a job. Requested guarantees are additional constraints that limit future requests for rescheduling. This procedure is used to improve the current schedule and to cope with resource failures or cancellations of jobs. The rescheduling requires new requests for offers if other allocations are not active anymore. Note that a valid schedule exists at every moment. Also, there is a tentative schedule for a job after each request and a following allocation.

Any improvement of the schedule is measured by combining all objective values. To this end, the scheduler attempts to maximize the overall objective value of the schedule. As an objective function is received for every request and for every offer, there is a combined objective function or value for each allocation. The objective functions of all allocations together define the optimization problem. An improvement can be achieved for instance by moving existing allocations while all constraints are observed. Alternatively, the scheduler can look for new allocations. The NWIRE scheduling concept further supports multi-site scheduling and co-allocation. However, this requires the inclusion of network management as just another high performance computing resource to provide guaranteed communication bandwidth between participating resources. In addition the local resource managers must provide offers with schedule guarantees which must be exactly met. While this scheduling strategy does not guarantee an optimal schedule in general, it meets all requirements of Section 3 as separate objectives are allowed for each resource in the metasystem.

Let us consider an example to better illustrate the scheduling architecture. Assume a user submitting a job with the following specific requests to a metasystem managed by NWIRE at noon: 16 nodes of type Power3 with operating system AIX 4.3.1. and at least 8 GByte of total main memory. The execution time of the job will not exceed 1 hour and the user would like to obtain the results as soon as possible. Unfortunately, there are no AIX nodes on the local system. Therefore, the local *MetaManager* forwards the request to other known *MetaManagers*. If the computing system attached to such a *MetaManager* is able to provide the required resources and also permits access of the requesting user, then its *Meta Manager* generates an offer, temporarily reserves the resources and sends the offer to the first *MetaManager*. In addition, *MetaManagers* may further forward the request. The validity of those offers will typically expire after some time. Assume in our example that there are two offers: The first offer provides 8 GByte memory and guarantees completion of the job by 3 pm. The other offer does not include any information about the completion time but allows the use of 16 GByte. Based on the criterion of the user the first *MetaManager* selects the first offer, informs the user, and notifies the *MetaManagers* that generated the offers. If a scheduling criterion is not detailed enough or cannot be completely met, the *MetaManager* may also leave the final decision up to the user. After acceptance of its offer a local *ResourceManager* can rearrange the schedule as long as the conditions of the original offer are observed. For instance, in our case the job must not be started before 2 pm. Note that the process of data and code submission is not addressed in this simple example.

## 5 Conclusion

In this paper we have described a scheduling concept for a metacomputing environment. This method is based on a brokerage and trading approach and provides a high degree of flexibility. For instance, administrators of resources in the metasystem are still able to apply arbitrary scheduling and allocation strategies for their resource. The system also supports guarantees, reservation and multi-site applications. Similarly, a dedicated objective can be defined for each request by a user. An additional feature is the instant feedback to a request. This allows the user to adapt his job requirements and to explore the metasystem with informal requests. He can also monitor the planned allocation and execution time of his job at any time.

A prototype which allows actual job submission and execution with the presented features is under development in the NWIRE project. The management infrastructure has already been implemented and is currently combined with the trading and brokerage facilities.

## References

1. D.G. Feitelson, L. Rudolph, U. Schwiegelshohn, K.C. Sevcik, and P. Wong. Theory and practice in parallel job scheduling. In D.G. Feitelson and L. Rudolph, editors, *IPPS'97 Workshop: Job Scheduling Strategies for Parallel Processing*, pages 1–34. Springer–Verlag, Lecture Notes in Computer Science LNCS 1291, 1997.
2. I. Foster and C. Kesselman. Globus: A metacomputing infrastructure toolkit. *International Journal of Supercomputer Application*, 11(2):115–128, 1997.
3. R.N. Lagerstrom and S.K. Gipp. PscheD: Political Scheduling on the CRAY T3E. In D.G. Feitelson and L. Rudolph, editors, *IPPS'97 Workshop: Job Scheduling Strategies for Parallel Processing*, pages 117–138. Springer–Verlag, Lecture Notes in Computer Science LNCS 1291, 1997.
4. P. Tucker and F. Berman. On market mechanisms as a software technique. Technical Report CS96-513, University of California – San Diego, Department of Computer Science and Engineering, December 1996.

# Workshop:

# Java in HPC

# The Use of Java in High Performance Computing: A Data Mining Example

David Walker and Omer Rana

Department of Computer Science,
Cardiff University, POBox 916,
Cardiff CF2 3XF, UK

**Abstract.** The role of Java in high performance computing is discussed with particular reference to the efforts of the Java Grande Forum to develop Java into the environment of choice for large-scale, resource-hungry applications. The support provided in HPJava for the design of parallel Java programmes is outlined, and its use demonstrated with a data mining application based on a neural network.

**Keywords**: *Java Grande, High Performance Java, Parallel Computing, Neural Networks, Distributed Objects*

## 1 Introduction

High performance computing (HPC) refers to any application that is sufficiently large-scale to require more computing resources than are typically directly available on a desk-top computer. This includes, but is by no means limited to, (1) numerical computing applications in the physical sciences, engineering, and mathematics; (2) commercial applications such as data mining and financial modelling; (3) immersive virtual environments for training, simulation, and the exploration of large data sets. The distinctions between these categories are becoming increasingly blurred.

A number of hardware and software technologies are driving high performance computing away from a model based around a stand-alone computer, such as a high-end workstation or multiprocessor, towards a distributed model in which network-accessible computing resources comprise a meta-computer providing computational, transactional, archiving, and visualisation services. The main hardware advances driving this trend are the improving cost-performance of processors, memory, and networks, set against the backdrop of a society that is becoming increasingly "digital". The overriding trend in software technology is towards the use of a programming model based on distributed objects. The impact of these trends taken together has the potential for significantly changing how computing resources are used in high performance applications. One such change already evident is that the Web is now becoming genuinely object based.

Key technologies for this transition are CORBA for network transparency, Java for implementation, and XML for specifying interfaces to these Java implementations. These interfaces define what parameters need to be passed to the called Java objects, and what is to be sent back to the calling object. These emerging technologies make use of the current emphasis on component based software development, with components implemented as Java objects, connected via CORBA middleware and invoked via Java/CORBA IDL or XML interfaces. XML is particularly useful when component interfaces need to be published on the Web. Enabling customised tags to be defined by a user, for a particular application as a document type definition (DTD) enables XML to be used both as a mark-up language and as a language to define component interfaces. The DTDs can be referenced over the Web, and need not be locally stored at each user site. Furthermore a DTD may be hierarchically constructed by combining DTDs from various vendors. This facilitates the establishment of standardising component interfaces within particular application domains.

The goal of this move towards distributed object-based computing is the provision of plug-and-play software components that may be linked together through high-level intelligent interfaces to build applications that run transparently on distributed, network-accessible resources. The long-term vision is to make computational power on the Internet as easy to access and use as electricity on a power grid.

The aim of this paper is to review the impact that Java is having on high performance computing, and in particular the work of the Java Grande Forum (JGF) in seeking to develop Java as the best environment for high performance computing (or *Grande Applications* in JGF terminology). As a concrete demonstration of how Java might be used in a distributed Grande Application a data mining example that uses HPJava will be presented. The rest of this paper is structured as follows. A summary of the aims and achievements of the JGF is given in Section 2. In Section 3, an overview of HPJava is presented. Section 4 introduces the data mining application and its implementation, and performance results are presented and discussed also. Section 5 summarises the main points made in the paper.

## 2 The Java Grande Forum

The Java Grande Forum was established in February 1998, and the current status of its work is summarised in a report presented at the SC98 conference in November 1998 [8]. The impetus for the formation of the JGF was the realisation that Java has the potential for providing a better development environment for HPC applications than any other existing language. Foremost among Java's attractive features is the promise of platform independent programming – a Java code will run (with a few caveats) on any machine with a Java Virtual Machine (JVM) resident. Other attractive features stem from Java's object oriented programming model, such as modularity, maintainability, and the ability to reuse software components. Furthermore Java's automatic memory management, operating system abstractions, and C-like syntax make it easy to learn and use. Since its inception about four years ago, Java has progressed from being used mainly for graphical enhancements to web pages and user interfaces to being a mainstream programming language. However, it has not been widely adopted yet for HPC applications. The main reasons for this relate to Java's suitability for numerical computing and its performance. A large fraction of HPC applications involve numerical computing and hence Java must provide the appropriate semantics for these types of scientific application. The Java Grande Numerical Working Group (JGNWG) has been formed to address issues relating to the use of Java in numerical computing. A second group, the Java Grande Concurrency and Applications Working Group (JGCAWG), is largely concerned with the performance of Java applications (both sequential and parallel/distributed), Java benchmarks, and the provision of a Java message passing API.

The JGNWG seeks to improve Java's support for numerical computing without degrading performance or requiring any substantial changes to the JVM [1]. The key issues being discussed within the JGNWG are efficient support for complex numbers and multi-dimensional arrays, the ability to exploit unique features of the floating-point hardware on certain machines, operator overloading, and the creation of lightweight classes. The JGNWG proposes to support complex numbers through lightweight classes and operator overloading. This is done in a way that avoids changes to the JVM, or unnecessary overheads, and makes complex numbers look and behave like floats and doubles. Lightweight classes and operator overloading are also important in supporting other arithmetic systems, such as interval arithmetic and multiple-precision arithmetic.

The JGNWG's proposal on the use of floating-point hardware is mainly directed at its efficient and predictable use on the x86 family of processors, and the use of the fused multiply-accumulate (mac) instruction on other architectures, such as the PowerPC. The x86 family of processors operate most naturally on 80-bit floating-point values, and on such processors a strict implementation of Java's floating-point semantics is extremely inefficient. A proposal by Sun addressing the floating-point behaviour of Java on x86 processors [9] was rejected by the JGNWG on the grounds that it compromised Java's predictability. JGNWG's counter proposal provides for three kinds of floating-point semantics: strict, default, and associative. Strict semantics correspond to the current Java floating-point semantics and do not allow fused mac instructions. For x86 processors, default semantics allows anonymous `float` and `double` values created as intermediate results in expression evaluation to use an extended exponent range. In general, default semantics on PowerPC processors allows the the use of fused mac instructions. Associative semantics allows compilers to optimise floating-point operations by re-arranging them as if the were associative.

The Java Grande Concurrency and Applications Working Group (JGCAWG) has so far concentrated on Java performance issues, particularly in the context of scientific and engineering applications. Recently there has been much interest in the broader research community in overcoming some of Java's sequential performance problems [10, 12]. Since Java is an interpreted language the scope for compiler optimisation is limited, and performance degrades because of the high level of abstraction separating the Java application from the hardware. Just-in-time (JIT) compilers seek to enhance performance by converting Java's stack-based intermediate representation into machine code immediately prior to execution. This is particularly effective when the application requires little garbage collection and thread synchronisation. Another recent approach is the HotSpot virtual machine. HotSpot optimises methods which it judges from runtime analysis to be most heavily used. This analysis is based on constructing a dynamic call graph for method invocations within a Java program. The call graph is subsequently pruned based on actual program execution. Hence, the use of an interpreter enables the HotSpot virtual machine to make optimisation that would not be possible in static compilers, such as in C++. HotSpot also utilises program analysis to perform automatic background compilation to native methods of the most executed code. It also uses a sophisticated approach to garbage collection. [3].

The JGCAWG has focused on enhancing the parallel and distributed performance of Java, rather than its sequential performance, through improvements to Java's remote method invocation (RMI) mechanism (see [11] for some recent work in this area). An area identified by JGCAWG where RMI performance can be improved is object serialisation and parameter marshaling. One approach to this is to reduce the amount of type data that must be marshaled and transmitted for each remote method invocation through compile time optimisation [13]. It is also important to reduce the number of copy operations wherever possible. Another enhancement seeks to improve the efficiency with which floats and arrays of floats are serialised since these data types are of particular importance is scientific computing. The current implementation of RMI uses the Java Native Interface (JNI) to convert the floats into their byte representation. This incurs a large overhead, and for arrays of floats the JNI routine must be invoked for each array element. The JGCAWG advocates an approach that minimises the number of JNI calls. A second area identified by the JGCAWG for improving RMI performance is RMI's management of network resources. An instance of this is the re-creation of socket connections used for object serialisation streams, rather than leaving them open throughout the user program. Another problem is that RMI currently creates a new port and a new thread for each remote object. To avoid the overhead associated with this the JGCAWG proposes that RMI should re-use sockets and threads from a pool, and to allow one thread to monitor several socket connections. The difficulty of improving RMI performance is compounded by the fact that much of RMI's source code is not available.

The Edinburgh Parallel Computing Centre (EPCC), under the auspices of the JGF, is co-ordinating the development of a suite of benchmarks for comparing Java execution environments [2]. This suite is made up of low-level operations, kernels, and large-scale applications. Another group within the JGF is discussing the design of an MPI interface for Java to use for message passing in parallel applications instead of the packages Java currently provides for communication

(BSD sockets and RMI). Several research groups are working in this area, and have developed prototype MPI-Java interfaces.

## 3 High Performance Java

As described above, there has been recent interest in linking Java with a message passing framework for distributed memory parallel computing. The motivation being to provide syntax extensions to the Java language, such as distributed arrays, available in traditional high performance languages but not supported in Java. The code containing extensions is subsequently translated by a preprocessor to standard Java code and calls to a runtime kernel written in C++, which makes use of an MPI library for communication. This is equivalent to generating a C/C++ based dynamic link library, which may then be invoked from a Java program, using native methods. The MPI calls are implemented via a runtime system for performing collective communications, such as *broadcasts*, and also supports gather/scatter operations for irregular data access, for instance.

We concentrate on the distributed memory paradigm, and consider additional keywords provided within *HPJava* [5], to aid a programmer in developing parallel Java applications – shared memory parallelism can be achieved using the *JavaSpaces* API. HPJava extends the *Java* programming language to provide support for scientific parallel programming, and combines tools, class libraries and language extensions to support parallel processing paradigms such as *shared memory programming, message passing* and *array-parallel programming*. Once such a framework is in place, bindings to higher level libraries and application specific codes such as *CHAOS* [7], and *ScaLAPACK* [6] may also be developed.

The first step in developing extensions to Java for parallel programming, is to introduce characteristic ideas of other high performance languages, such as the distributed array model and array intrinsic functions and libraries of *HPF*. The resulting programming model would be SPMD (Single Program Multiple Data), allowing direct calls to MPI or other communications packages from the HP-Java program. Providing distributed arrays as language primitives would allow the programmer to simplify error-prone tasks such as converting between local and global array subscripts and determining which processor holds a particular element. The compiler for HPJava would make calls to a run-time library and generate underlying Java code. The translator is being implemented in a compiler construction framework developed by members of the *Parallel Compiler Runtime Consortium (PCRC)* [4] .

Within *HPJava* the PCRC runtime Kernel is referred to as *Adlib*, and implemented as a C++ library, involving a hybrid of the SPMD and data parallel approaches. The following types of functionalities are provided in *Adlib*:

1. *Multidimensional Arrays:* Multidimensional arrays allow regular section subscripting, similar to Fortran 90 arrays. Such arrays are a language extension and coexist with ordinary Java arrays, e.g.

```
float [[,]] a = new int [[5,5]] ;
int [[,,]] b = new int [[10,n,20]] ;
```

An example of section subscripting is:

```
int [[]] e = a [[2,:]] ;
```

where *e* becomes an alias for the 3rd row of *a*. *Adlib*, unlike Fortran, does not however permit vectors of integers as subscripts. An array can be split across a number of processes, each process acting on a subset of the total array. An array range may also be collapsed, indicating that a complete array may be mapped to a single process, or an array may be identically replicated across a number of processes.

2. *Distributed Arrays:* Distributed arrays may be viewed as coherent global entities, but their elements are divided across a set of cooperating processes. Distributed arrays are based on the concept of a *Process Array*, over which elements of a distributed array are scattered.

For instance, a 2 by 2 process array can be defined as:

```
Procs2 p = new Procs2(2,2) ;
```

whereas a 6 element, one dimensional process array is:

```
Procs1 q = new Procs1(6)
```

3. *Ranges:* A Range object defines a range of integer subscripts, and how such subscripts are to map into a process array dimension. Each value in the range is mapped to the process (or slice of processes) with that coordinate. Hence, a distributed range object may appear in place of an integer extent in the constructor of the array, as follows:

```
float [[*,*,]] a =
   new float [[x,y,100]] on p;
```

which defines *a* as an *x* by *y* by 100 array of floating point numbers. As the first two dimensions of the array are distributed ranges (the dimensions of *p*), *a* is realised as four segments of 100 elements, one in each of the processes.

4. Other extensions include *BlockRange* which is a subclass of *Range* and describes a simple block distributed range of subscripts, *Distributed Parallel Loops* similar in concept to the *FORALL* construct of Fortran, *Subranges* and *Subgroups* which can be viewed as a slice of a process array, and formed by restricting the process coordinates in one or more dimensions to single values. Hence, the active process group or the group over which an array is distributed may be just some slice of a complete process array.

## 4 A Parallel Neural Network in HPJava

There are many overlaps in enterprise computing, generally available in the commercial sector, and large scale scientific computing. For instance, applications in both areas can be large – demanding high network throughput and I/O services. In both cases, large data sets may be employed, though the data distributions and access patterns may be different. Similarly, there are application programs which can be employed in problems in both domains, such as the use of particular data analysis tools. In both cases, data may be stored at various places on a network, and data may be generated from multiple runs of the same program, or constitute transactional data over a predefined time interval. One such application is data mining, which involves the detection of patterns in large data sets for forecasting or model construction. Data mining can be applied to various problem domains, such as financial forecasting or predicting river levels to detect the onset of a flood. We implemented a neural network data mining algorithm in HPJava as a proof-of-concept application. In this study, *network*, *node* and *data* parallelism are used for a three layer *Multi-Layer Perceptron (MLP)* neural network architecture with the backpropagation learning rule. *Network* parallelism corresponds to dividing a single neural network into clusters (layers), and running each cluster in parallel. *Node* parallelism occurs when data necessary for each node to compute its output is local to a node – enabling all nodes to perform a local update in parallel. *Data* parallelism corresponds to distributing a large data set across a number of processors, enabling either each processor to run the same neural network on a different portion of the data set, or dividing a single neural network across processors and employing a suitable data distribution. We change the number of data samples to determine the scalability in data set size, obtained by adding more training samples and also increasing the number of processors in the host pool.

### 4.1 Changing data set size

The neural network used has 30 input layer neurons, 16 hidden layer neurons and 2 output neurons. The neural network is trained on *sunspot* activity data obtainable from various web sites[1]. The number of samples in the training set was varied, and the time to learn measured. Some training samples were repeated to reach a reasonable sized training set. The computer configuration used consisted of four *Sun* workstations in a cluster, running the Solaris 2.5 operating system. All Sun workstations used JDK 1.1.6 and MPICH software from Argonne National Labs. The results are plotted in Fig. 1, and are average times obtained after each processor had been used as the master. Training was repeated by swapping the master processor for each of the four Sun machines. The results indicate that as more samples are added to the data set, the training time does not increase significantly. Communication delays in the HPJava implementation we used were significant, and the values provided in the figure have been normalised to the fraction components, to highlight the general trend.

---

[1] http://www-isis.ecs.soton.ac.uk/research/nfinfo/fzdata.shtml

## 4.2 Changing number of processors

Changing the number of processors leads to reasonable improvements in performance. The speedup is not linear however, primarily due to the additional costs of communication between processors. We used a logical array of 12 tasks on two, three, and four Sun workstations (with one workstation acting as the master in each scenario), connected on the local network. We used the Unix ping command to calculate the time to send data between the host machine and other workstations to measure network delays. All Sun machines used the same software as outlined in the previous section. The results are plotted in Fig. 2. The results indicate a speedup of 29% with 4 additional processors. These results are quite encouraging considering each processor is assumed to have background workload in addition to the HPJava tasks. We could improve performance by using a better data distribution and by employing a load balancing approach.

## 5 Conclusions and Future Work

The role of Java in high performance computing has been outlined. The progress made by the Java Grande Forum in its goal of making Java the best environment for Grande Applications has been discussed, and an approach to parallel data mining using HPJava, with support for parallelism provided via MPI libraries, has been described. The use of HPJava in a neural network application distributed across a workstation cluster has been demonstrated as a proof-of-concept application. The current neural network application can be easily scaled, by employing network parallelism – enabling a different neural network to be used for analysing different data sets. The data set may be local to a workstation cluster, and avoids having to transfer large quantities of data over a network. The results can then be combined for visualisation, in decision support applications for instance. Future work will involve developing a visualisation environment for combining results from various neural networks, to enable better control of learning parameters and selection of data sets. The developer may use a variety of data mining algorithms, and is not restricted to the use of neural networks. Hence, various analysis algorithms may be combined in this way, some of which may come from third parties.

The release of various HPJava implementations from JGF participants has now made it possible to develop new parallel applications in Java. These implementations have been ported to various platforms such as the IBM SP2, the SGI Origin-2000, the Fujitsu AP-3000, the Hitachi SR2201 parallel platforms, and clusters of workstations running Solaris or Windows NT operating systems. Most of these implementations at present rely on the use of native methods, written in C/C++, although current efforts are focused on improving the Java language to enable a pure Java implementation. Java also supports communication between objects through sockets or the Remote Method Invocation (RMI) mechanism, but these are generally too low level for most parallel programming applications, which generally employ an abstraction of a large set of interacting

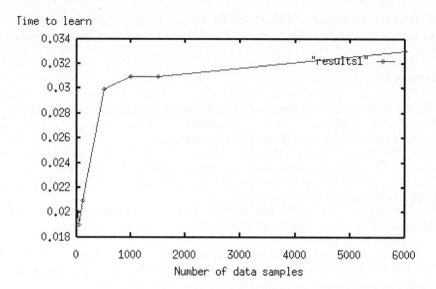

**Fig. 1.** *Effect of changing data set size*

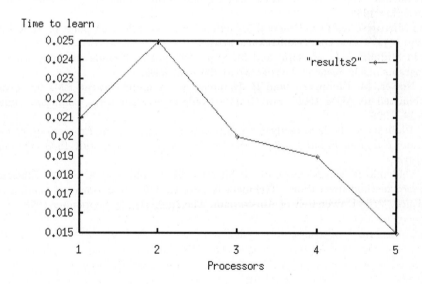

**Fig. 2.** *Effect of changing number of processors*

peers arranged as a group. One track for future work within the forum is to create more applications using some of the libraries developed by participants.

## References

1. See the Java numerics web page at http://math.nist.gov/javanumerics/.
2. See the EPCC Java bencmark suite web page at http://www.epcc.ed.ac.uk/research/javagrange/.
3. The HotSpot Virtual Machine, 1998. See the web page at http://www.developer.com/journal/techfocus/052598_hotspot.html.
4. The Parallel Compiler Runtime Consortium, 1998. See the PCRC web page at http://www.npac.syr.edu/projects/pcrc/.
5. B. Carpenter, G. Fox, D. Leskiw, X. Li, and Y. Wen. Language bindings for a data-parallel runtime. Technical report, NPAC, Syracuse University, Syracuse, New York 13244, 1997.
6. J. Choi, J. J. Dongarra, S. Ostrouchov, A. Petitet, D. W. Walker, and R. C. Whaley. The design and implementation of the scaLAPACK LU, QR, and Cholesky Factorization routines. *Scientific Programming*, 5:173–184, 1996.
7. R. Das, M. Uysal, J. Salz, and Y.-S. Hwang. Communication optimizations for irregular scientific computations on distributed memory architectures. *Journal of Parallel and Distributed Computing*, 22(3):462–479, 1994.
8. Java Grande Forum. Java Grande Forum report: Making Java work for high-end computing. Available at http://www.javagrande.org/ together with other JGF documents. Report presented at Java Grande Forum Panel at SC'98 conference, November 1998.
9. Sun Microsystems Ltd. Proposal for extension of Java floating point in JDK 1.2. http://java.sun.com/feedback/fp.html.
10. S. P. Midkiff, J. E. Moreira, and M. Snir. Optimizing bounds checking in Java programs. *IBM Systems Journal*, 37(3):409–453, 1998.
11. C. Nester, M. Philippsen, and B. Haumacher. A more efficient RMI for Java. Submitted to ACM 1999 Java Grande Conference, Palo Alto, California, June 12–14, 1999.
12. M. Philippsen. Is Java ready for computational science? In *Proceedings of the Second European Parallel and Distributed Systems Conference*, July 1998. Vienna, Austria.
13. R. Veldema, R. van Nieuwpoort, J. Maassen, H. E. Bal, and A. Plaat. Efficient remote method invocation. Technical Report IR-450, Department of Computer Science, Vrjie Universiteit of Amsterdam, The Netherlands, September 1998.

# Interfaces and Implementations of Random Number Generators for Java Grande Applications

P.D. Coddington, J.A. Mathew and K.A. Hawick

Advanced Computational Systems Cooperative Research Centre
Department of Computer Science, University of Adelaide
Adelaide, SA 5005, Australia
{paulc, jm, khawick}@cs.adelaide.edu.au

**Abstract.** The Java Grande Forum aims to drive improvements to the Java language and its standard libraries in order that Java may be efficiently used for large-scale scientific applications, particularly on high-performance computers. Random number generators are one of the most commonly used numerical library functions in applications of this kind. For the current random number generator provided within Java, neither the implementation nor the interfaces are adequate to meet the needs of some Java Grande applications, such as Monte Carlo simulations. We present a preliminary proposal for an API for accessing a random number generator within a Java scientific software library for supporting Java Grande applications. A reference implementation of the proposed API is described, and we discuss some implementation and performance issues. Mechanisms for efficiently handling concurrency are also discussed.

## 1 Introduction

Java has the potential to be an excellent language for developing large science and engineering applications, particularly on distributed and high-performance computers. The Java Grande Forum [9] aims to drive improvements to the Java language and its standard libraries in order that Java may be efficiently used for such "Grande Applications". Random number generators are commonly used in these types of applications, so it is important to provide access to an efficiently implemented, high-quality generator through a standardized Java application programming interface (API).

In this paper, we investigate the requirements for interfaces and implementations of random number generators for Java Grande applications. We explain why the existing random number generators available in Java are inadequate for use in some scientific applications, and propose a new API that might be used as part of a Java numerical library, or scientific software library (SSL).

Random number generators use iterative deterministic algorithms for producing a sequence $X_i$ of pseudo-random numbers that approximate a truly random sequence. The main algorithms used to implement random number generators in numerical libraries are:

**Linear congruential generators** (LCGs), with $X_i = (A * X_{i-1} + B) \bmod M$, which we denote by $\mathrm{L}(A, B, M)$;

**Lagged Fibonacci generators** (LFGs), with $X_i = X_{i-P} \odot X_{i-Q}$, which we denote by $\mathrm{F}(P, Q, \odot)$, where $P$ and $Q$ are the lags, and $\odot$ is any binary arithmetic operation, such as addition or multiplication modulo $M$.

**Shift register generators** can usually be defined in terms of LFGs using the bitwise exclusive OR function XOR, however these are of lower quality than equivalent LFGs using addition or multiplication;

**Combined generators** that combine (usually by addition modulo $M$) the results of two or more generators, usually two LCGs, or an LFG combined with an LCG or some other algorithm.

For more information on these algorithms, see one of the many review articles on random number generators [5, 8, 11, 12, 18].

A generator provided in a numerical library for general use should be both fast and of high quality, i.e. have been rigorously tested and passed a variety of standard statistical and empirical tests of randomness. However, determining the quality of a random number generator is a notoriously difficult problem, and there have been many instances of poor quality random number generators being provided in numerical libraries [5, 8, 11, 12, 18].

For many applications, the quality of the pseudo-random numbers is not that important, and any generator that is reasonably random will do the job. However in many of the Grande Applications for which random number generators are most heavily used, such as Monte Carlo simulations [1], the quality of the random number generator is crucial, since an inadequate random number generator can produce incorrect results. New applications, or more powerful computers that can produce longer sequences, may also show up flaws in generators that were previously thought to be first rate [4, 5]. So providing an adequate random number generator for Java Grande applications is a challenging problem.

## 2 Random Number Generators in Java

Java has a number of advantages for implementing random number generators. Most algorithms use integer arithmetic requiring specific precision, either 32 or 64 bits. Java implementations of these algorithms are portable since `int` and `long` integer types have well-defined precision, unlike the corresponding types in C. Some algorithms can be implemented more efficiently if the handling of overflow in integer arithmetic is well defined, as it is in Java. Random number generators require initialization, which needs to be done before they are called. In procedural languages it is up to the programmer to remember to call the initialization routine, however in an object-oriented language like Java, the initialization can be incorporated into the constructor of the generator class.

Another advantage of Java is that it is designed to use standard class libraries with well-defined APIs, and a huge variety of such libraries are available, including random number generators. Fortran 90 and ANSI C both offer standard interfaces to random number generators in the form of intrinsic functions or standard libraries, however the interfaces are somewhat restrictive, in terms of both the functionality and the algorithms available. Java offers the potential for providing a full-featured, high-quality, portable class library for random number generation.

There are currently three different classes providing access to random number generators in Java [19]:

- `java.security.SecureRandom` provides an interface to cryptographically strong random number generators, which are special-purpose generators that use hardware devices and/or non-linear algorithms to produce sequences that are not predictable (unlike the simple linear generators commonly used in numerical libraries). Cryptographic generators are not used in scientific simulations requiring large quantities of random numbers, since they are much too slow, or in the case of hardware devices, are not reproduceable.
- `java.util.Random` provides quite a well-structured random number generator interface. There is a fundamental method called `next` which returns up to 32 random bits, and the other methods use this to produce random numbers of different types – `nextBoolean`, `nextInt`, `nextLong`, `nextFloat` and `nextDouble`.
- `java.Math` provides a `Random` method, which provides a simpler interface to the `nextDouble` method of `java.util.Random`, for applications that just need to access a stream of pseudo-random floating point numbers.

These APIs all have shortcomings that make them inadequate for supporting Grande scientific applications. The generators used in the `SecureRandom` class are targeted at cryptography applications, which have different requirements than scientific simulations, and consequently this class has an API that is unsuitable for a Java SSL. The other APIs are aimed at general applications, but do not have some special features or provide the quality and choice of algorithms required for scientific applications such as Monte Carlo simulation. The pros and cons of the existing APIs are discussed in more detail in the next section, which outlines some requirements for a random number generator API for a scientific software library to be used by Java Grande applications.

## 3 API Requirements for Java Grande Applications

There has been considerable work recently by the Java Grande Forum on developing proposed APIs and reference implementations for object-oriented numerical libraries within Java [10, 2], to provide what is effectively a Java scientific software library.

The random number generator API used in `java.util.Random` is adequate for most purposes, and provides a reasonable basis for a more comprehensive API that could be used within a Java SSL. We have therefore used it as the basis for our initial proposal, although alternate approaches are possible and may prove to be more practical as more work is done on requirements and implementation details.

## 3.1 A choice of algorithms

The main problem with random number generators is that they do not produce truly random sequences, so no generator can be guaranteed to work adequately for a given application. It is therefore essential, particularly for Grande applications which tend to be more sensitive to the quality of the generator, that a choice of algorithms is available, so users can run their application using at least two different generators, and check that the results are the same within statistical errors.

The documentation for the java.util.Random class points out that the API is structured so that a programmer can implement an alternate random number generator algorithm by creating a sub-class that overrides the basic next method, but inherits all the others. However no other algorithms are provided. A Java SSL should provide a choice of algorithms for the random number generator, and the API should be general enough to support any generator algorithm. Ideally, it should also allow different generators to be called without having to modify the application code.

The SecureRandom class in the java.security package provides this capability, allowing the user to select from a set of standard algorithms [19]. The recommended mechanism for doing this is to provide a static method getInstance which can be used to obtain an instance of a random number generator class. The Java classname for the particular algorithm can be optionally supplied to the method to specify which algorithm should be used. The method will attempt to instantiate a class with the name supplied and throw an exception if it is not found. If no parameter is supplied the default generator is returned. This mechanism has been incorporated into our proposed API.

As with java.util.Random, users can also provide their own random number generator by simply sub-classing the Java SSL random number generator class, and providing an implementation of the methods that are algorithm-dependent, such as next. All the other methods can be inherited. In some cases it may be more efficient to override other methods, such as nextFloat or nextDouble.

## 3.2 An adequate period

Random number generators are different to standard mathematical functions in that they must maintain a state between calls to the generator, since they use iterative functions. The state of the generator is finite, so the sequence of numbers produced by the generator must repeat after a certain period. Generators provided as part of a Java SSL should have periods that are much larger than the amount of pseudo-random numbers that might be produced in a Java Grande application. How large is this likely to be?

The 32-bit linear congruential generators commonly used in the past have a period of $2^{32}$, which corresponds to roughly a Megaflop-hour of operations. This is a trivial amount of computation for today's desktop machines, so these generators are no longer adequate for scientific applications. The period of the default generator used in java.util.Random is $2^{48}$, corresponding to a few Gigaflop-days, which is commonly exceeded by current Monte Carlo simulations in computational physics, and within 20 years will seem as trivial as a Megaflop-hour seems today. 64-bit or combined 32-bit LCGs have a period of around $2^{64}$, corresponding to about a Teraflop-year, which is a huge amount of computation, but within the grasp of grand challenge applications on current supercomputers, and in 20 years will be commonplace for large-scale scientific applications.

However, it is not difficult to implement generators with periods that are so large that they will be adequate for any Java Grande application in the forseeable future. Lagged Fibonacci generators can have arbitrarily long periods, depending on the size of the lag that is used. Combining two 64-bit LCGs, or four 32-bit LCGs, gives a period of around $2^{128}$, which should be adequate since it corresponds to roughly a Petaflop-age-of-the-universe computation!

### 3.3 A better default algorithm

The API for java.util.Random mandates the use of a specific algorithm for the next method, so that it will give the same results across any Java Virtual Machine (JVM). The algorithm chosen as the default is the 48-bit LCG used in the C library function drand48. This is quite a good algorithm, however it has some significant problems that make it inadequate for use as the default generator for a Java SSL. The first is that rather than using a prime modulus, which gives better quality random numbers, it uses a modulus that is a power of 2, which is faster to implement. A more serious problem is that the period of this generator is too small for many Java Grande applications.

A Java SSL should provide a higher quality, longer period default random number generator. But should we mandate a particular algorithm for the sake of portability, as is done in java.util.Random, or opt for flexibility and allow any algorithm to be used, as is done in SecureRandom? If we do not mandate a default algorithm, then we sacrifice portability, and some implementations of the Java SSL might use a substandard generator. However if we do mandate an algorithm, it may prove to be inferior as new algorithms, applications and statistical tests are developed. A better option would be to mandate a specific default algorithm for a particular version of the Java SSL, but allow the possibility that the default will change in future versions.

The Java SSL should provide a suite of different algorithms covering a range of speed and quality trade-offs to suit different applications. The default algorithm should provide a general-purpose happy medium, with good performance but high quality. Good candidate algorithms include a multiplicative lagged Fibonacci generator with a long lag, or a combined generator using either two 64-bit or four 32-bit LCGs to provide a large enough period.

### 3.4 Initializing and checkpointing the generator

In java.util.Random, initializing (or seeding) the generator is done by passing a single long integer (the seed) to the constructor, or to a setSeed method.

If no seed is provided, the generator is initialized using a value taken from the clock time. This is fine for initializing the state of the drand48 generator, which is just a 48-bit integer value, but other algorithms such as combined LCGs have multiple integers for their state variables, while LFGs require the initialization of arrays of hundreds or thousands of integer or floating point values used to keep their state.

It is unwise to leave the initialization of the state variables of these generators up to the user, since in some cases (particularly LFGs) this is a subtle process, and a naive initialization may result in a correlated (non-random) sequence. A better approach is to follow the existing API, with the user providing just a single long integer, and require the generator implementation to provide a sound mechanism for initializing its state based on that single seed. This approach also has the advantage that the interface to setSeed is independent of the generator used.

Large-scale scientific applications, particularly Monte Carlo simulations, often require multiple runs of the program to produce a final result. To do this, the user will checkpoint the state of the simulation so that the run can be restarted later on. This includes checkpointing the state of the random number generator, by writing it to a file. The API should therefore provide a getState method that returns the current state of the generator. This should return a general Object, since different algorithms have different state variables. A corresponding setState method would allow such an object to be used to initialize the generator after checkpointing. These are algorithm-dependent methods that must be implemented for each generator sub-class.

Since in most cases the user will want to checkpoint the generator's state directly to a file, it is helpful to also provide a writeState method that writes the state to a specified file after obtaining the object using getState, and a readState method that reads in the state from a specified file and returns a general Object that can be used by the setState method. These methods can be implemented in the base class in a generator-independent way. Java's object serialization mechanism can be used to provide portable data files.

## 3.5 Generating arrays of random numbers

Since generating a random number only takes a few floating point operations, the overhead of the method call can be relatively high. If the application requires an array of random numbers, it may be more efficient to call a method that fills up the whole array at once, rather than making multiple method calls. Some numerical libraries provide routines to support this.

Advanced compilers should be able to inline the method calls to avoid this overhead, which would appear to obviate the need for providing such a method. However Fortran 90 provides such an interface not just to avoid the overhead of the function call, but primarily to allow for the possibility that the generation of the array of random numbers may be vectorized or parallelized, so a method of this kind may still be useful to support data parallel implementations of Java, such as proposed by the HPJava Project [7].

Within Java, we can offer a simple interface to such a method by overloading the method call for generating a single random value if it is made with an array as an argument. So a call to `nextInt()` will return a single random integer, whereas a call to `nextInt(int rand[])` will fill the array `rand[]` with random integers.

## 3.6 Concurrency and synchronization

The reference implementation for `java.util.Random` suggests that the `next` method be synchronized, so that it is thread-safe. Unfortunately, synchronization in Java can have a substantial performance overhead, which may reduce the speed by up to a factor of 3. It would be possible to run different instances of the random number generator in different threads, thus avoiding the synchronization overhead. However we need to avoid having overlapping or correlated pseudo-random sequences in different threads.

The problem is similar to implementing a random number generator on a parallel computer. To avoid all the processors accessing a single random number stream from one process, which has a costly communications overhead, parallel random number generators are designed to provide different random number streams for each processor (see ref. [5] for a review of parallel random number generators). A similar approach could be adopted for the random number generator in a Java SSL, to handle concurrency from using multiple threads or a parallel Java program (using Java with MPI [3], for example). For each concurrent process, the programmer could instantiate a new random number generator, which would synchronize only in the initialization procedure. After that, all calls to the generator could be unsynchronized.

In some cases the programmer may not wish to explicitly manage the creation of a new instance of the unsynchronized random number generator for each thread in the program. The API could provide synchronized generators as default, but allow the user to call unsynchronized versions if required.

## 3.7 Some other issues

We have not addressed alternate probability distributions in our reference implementation. For random number generators in SSLs, the default is always to provide uniformly distributed random numbers, however some applications require other distributions, such as Gaussian, or Poisson. A `nextGaussian` method is provided by `java.util.Random`, and it would be straightforward to also provide a `nextPoisson` method, as well as other distributions. It may be possible to make this more general, by enabling the user to specify or implement the required probability distribution.

The `java.util.Random` API provides a useful method for returning an integer between 0 and n, by overloading `nextInt()` to allow `nextInt(int n)`. Curiously, the documentation does not list a corresponding `nextLong(long n)` method, which should be added.

## 3.8 The proposed API

The outline of our proposed API is given below. It includes the methods available in java.util.Random, as well as the additions outlined in this section (which are marked with an asterisk). This should be viewed as a first pass at an API for a Java SSL random number generator, which will hopefully be subject to much discussion and improvement. A more detailed version of the API (using javadoc) is available on the Web [6].

```
public class Random extends java.util.Random {

   Random();
   Random(long seed);

*  public static Random getInstance(String type) throws RandomException;
*  public static Random getInstance();

   public void setSeed(long seed);
*  public Object getState();
*  public void setState(Object seeds);
*  public Object readState(String filename);
*  public void writeState(String filename);

   protected int next(int bits);
   public void nextBytes(byte[] bytes);
   public boolean nextBoolean();
   public float nextFloat();
   public double nextDouble();
   public int nextInt();
   public int nextInt(int n);
   public long nextLong();
*  public long nextLong(long n);

*  public void nextInt(int[] random_ints);
*  public void nextLong(long[] random_longs);
*  public void nextFloat(float[] random_floats);
*  public void nextDouble(double[] random_doubles);

   public double nextGaussian();

}
```

## 4 Implementation and Performance Issues

We have developed a reference implementation of the proposed random number generator API outlined above, which is available on the Web [6]. We have implemented five different algorithms, all of which are believed to be excellent generators for scientific applications.

- LCG64 – a 64-bit LCG with a prime modulus, $L(2307085864, 0, 2^{63} - 25)$, recommended by L'Ecuyer [14].
- Ranmar – a variation of Marsaglia's commonly-used RANMAR generator combining a simple Weyl generator with an additive LFG [17,8], however we have increased the lags and used F(4423,1393,+) to improve the quality.
- MultLFG – a multiplicative lagged Fibonacci generator, F(1279,418,*).
- Ranecu – the popular combined 32-bit LCG of L'Ecuyer [13,8].
- CLCG4 – a combination of four (rather than two) 32-bit LCGs, to give a longer period [15].

Table 1 shows timings for the random number generator implementations in both C and Java (JDK 1.2) on two different platforms — a 300 MHz Sun UltraSPARC under Solaris and a 300 MHz Pentium II under NT. The algorithms we have implemented are similar in performance to the default generator for java.util.Random, but are of higher quality.

The synchronization overhead is only about 50% for Solaris on the SPARC, but as much as a factor of 3 on Pentium under NT. However, the performance of the synchronized methods is about the same for each JDK, since NT on the Pentium gives correspondingly better performance than Solaris for SPARC on the basic unsynchronized methods.

**Table 1.** Results of timings of Java (JDK 1.2) and C implementations of different random number generators, given as millions of pseudo-random numbers produced per second (so higher values are better). Java results are given for both synchronized and unsynchronized method calls. The Sun results are for a 300 MHz Sun UltraSPARC under Solaris and the NT results are for a 300 MHz Pentium II under Windows NT. drand48 refers to the implementation in the C library and in java.util.Random. Some algorithms have no C timings since they require 64-bit longs.

|  |  | drand48 | LCG64 | Ranmar | MultLFG | Ranecu | CLCG4 |
|---|---|---|---|---|---|---|---|
| Sun C | float | 0.75 | — | 3.46 | — | 1.32 | 0.62 |
| Sun Java unsynch | int | — | 0.69 | 0.98 | 2.53 | 0.56 | 0.20 |
|  | float | — | 1.11 | 2.44 | 2.09 | 1.26 | 0.41 |
|  | double | — | 1.09 | 0.63 | 1.79 | 0.55 | 0.21 |
| Sun Java synch | int | 2.03 | 0.56 | 0.70 | 1.63 | 0.44 | 0.18 |
|  | float | 1.85 | 0.89 | 1.50 | 1.20 | 0.97 | 0.38 |
|  | double | 0.86 | 0.84 | 0.49 | 1.30 | 0.49 | 0.20 |
| NT Java unsynch | int | — | 1.81 | 1.20 | 6.14 | 1.27 | 0.48 |
|  | float | — | 3.04 | 3.88 | 5.08 | 4.11 | 1.14 |
|  | double | — | 3.04 | 0.78 | 5.24 | 1.51 | 0.48 |
| NT Java synch | int | 7.97 | 1.11 | 0.83 | 1.93 | 0.69 | 0.36 |
|  | float | 6.17 | 1.44 | 1.70 | 1.88 | 1.80 | 0.81 |
|  | double | 3.17 | 1.49 | 0.62 | 1.89 | 0.97 | 0.42 |

The use of improved just-in-time (JIT) compilers in JDK 1.2 has greatly improved the performance over JDK 1.1. Results obtained using older compilers (JDK 1.1.3) showed the Java implementations to be around 10 times slower than C code, however the latest JDK gives results that are only about 50% slower than C. The `drand48` routine has somehow been optimized so that it is actually faster in Java than in C.

Many of the routines in scientific software libraries, such as linear algebra solvers, involve substantial amounts of Fortran code, so in developing scientific software libraries for Java, it is much easier to just provide a Java interface to these existing routines and to call them as native methods [2]. This usually provides better performance than a pure Java implementation, however it sacrifices the portability of the implementation, which is one of the advantages of using Java. Fortunately, random number generator algorithms only require a small amount of coding, and are easy to convert to pure Java. It would also be easy to provide an implementation using Java interfaces to native methods, however since the pure Java version is not much slower than C, the additional overhead of a native method call may outweigh the small performance advantage of the native code.

## 5 Conclusions

Java currently provides a random number generator in `java.util.Random` that is adequate for many applications, although both the interface and the implementation lack many of the qualities required for the type of large-scale scientific applications being addressed by the Java Grande Forum.

Some of the interface issues include mechanisms for checkpointing the state of the random number generator; generating arrays of random numbers; handling concurrency; and providing a convenient way to choose between a variety of different algorithms. We have presented a proposal for an improved API, and provided a reference implementation, for a random number generator package that could be used as part of a Java scientific software library.

Implementation issues include choosing a good general-purpose default algorithm that provides an adequate period and a reasonable trade-off between speed and quality. The default algorithm in `java.util.Random` does not have a large enough period of repetition for some scientific applications such as large-scale Monte Carlo simulations, and it is unfortunate that a 48-bit linear congruential generator was selected as the default, rather than a superior 64-bit generator. For a random number generator within a Java SSL, a better default algorithm would be required. Good candidate algorithms include a multiplicative lagged Fibonacci generator or a combined linear congruential generator.

Both the proposed interface and the reference implementation are still in a preliminary stage, and further user input, implementation tests, experimentation and discussion will be required to improve them. However, we have highlighted a number of the issues involved, and suggested some possible solutions.

# Acknowledgements

This work was carried out under the Distributed High-Performance Computing Infrastructure (DHPC-I) project of the Research Data Networks (RDN) and Advanced Computational Systems (ACSys) Cooperative Research Centers (CRC). RDN and ACSys are established under the Australian Government's CRC Program.

# References

1. K. Binder ed., *Monte Carlo Methods in Statistical Physics*, Springer-Verlag, Berlin, 1986.
2. R.F. Boisert *et al.*, Developing numerical libraries in Java, *Proc. of ACM Workshop in Java for High-Performance Network Computing*, Stanford, February 1998, http://www.cs.ucsb.edu/conferences/java98/program.html.
3. Bryan Carpenter et al., MPI for Java - Position Document and Draft API Specification, Java Grande Forum Technical Report JGF-TR-03, November 1998, http://www.npac.syr.edu/projects/pcrc/reports/MPIposition/position.ps.
4. P.D. Coddington, Analysis of Random Number Generators Using Monte Carlo Simulation, *Int. J. Mod. Phys. C* **5**, 547 (1994).
5. Paul D. Coddington, Random Number Generators for Parallel Computers, *The NHSE Review*, http://nhse.cs.rice.edu/NHSEreview/, 1996 Volume, Second Issue.
6. P.D. Coddington, J.A. Mathew and K.A. Hawick, Random number generators for Java Grande, http://acsys.adelaide.edu.au/projects/javagrande/random/.
7. HPJava Project, http://www.npac.syr.edu/projects/pcrc/HPJava/.
8. F. James, A review of pseudorandom number generators, *Comp. Phys. Comm.* **60**, 329 (1990).
9. The Java Grande Forum, http://www.javagrande.org/.
10. Java Grande Forum, Making Java Work for High-End Computing, Java Grande Forum technical report JGF-TR-1, http://www.javagrande.org/reports.htm.
11. D.E. Knuth, *The Art of Computer Programming Vol. 2: Seminumerical Methods*, Addison-Wesley, Reading, Mass., 1981.
12. P. L'Ecuyer, Random numbers for simulation, *Comm. ACM* **33:10**, 85 (1990).
13. P. L'Ecuyer, Efficient and portable combined random number generators, *Comm. ACM* **31:6**, 742 (1988).
14. P. L'Ecuyer, F. Blouin, and R. Couture, A Search for Good Multiple Recursive Generators, *ACM Trans. on Modeling and Computer Simulation* **3**, 87 (1993).
15. P. L'Ecuyer and T.H. Andres, A Random Number Generator Based on the Combination of Four LCGs, *Mathematics and Computers in Simulation* **44**, 99 (1997).
16. G.A. Marsaglia, A current view of random number generators, in *Computational Science and Statistics: The Interface*, ed. L. Balliard, Elsevier, Amsterdam, 1985.
17. G.A. Marsaglia, A. Zaman and W.-W. Tsang, Toward a universal random number generator, *Stat. Prob. Lett.* **9**, 35 (1990).
18. S.K. Park and K.W. Miller, Random number generators: Good ones are hard to find, *Comm. ACM* **31:10**, 1192 (1988).
19. Sun Microsystems Inc., Java Platform 1.2 API Specification, http://java.sun.com/products/jdk/1.2/docs/api/index.html.

# Java as a Basis for Parallel Data Mining in Workstation Clusters

Matthias Gimbel, Michael Philippsen, Bernhard Haumacher,
Peter C. Lockemann, and Walter F. Tichy

Universität Karlsruhe, Department of Computer Science
Am Fasanengarten 5, D-76128 Karlsruhe, Germany
gimbel@ira.uka.de
http://wwwipd.ira.uka.de/RESH/

**Abstract.** The exploitation of hidden information from large datasets by means of data mining techniques suffers from long response times. We address this problem by using the processing power of workstation clusters and have studied the performance of OLAP queries as a first step towards a portable data mining platform.

The results of our study suggest that with the availability of parallel workstation clusters that are equipped with high performance communication networks, fine-grained and communication-intensive parallelizations of queries are promising – even though they are considered too costly in traditional database systems.

The paper describes our Java framework for parallel OLAP-type query execution, necessary optimizations to the standard Java implementation, and analyzes the performance of non-standard parallel execution schemes on a workstation cluster.

## 1 Introduction

The need for and the benefits of data mining have been commonly accepted by the business and scientific communities. There is no dearth either of preprocessing techniques, mining methods, and postprocessing facilities. Unfortunately though, there are still huge difficulties to exploit the potential of data mining.

One difficulty is the processing time. It can take hours or even days for an algorithm to produce a result. Data mining is computationally expensive due to the massive amounts of data and due to the complexity of knowledge discovery algorithms. Even worse, because of the explorative nature of the knowledge discovery process (the preprocessing steps and the learning algorithms offer many parameters to experiment with), there exists a strong interest to make knowledge discovery an interactive process – with according performance requirements.

One answer to the performance demands is the use of parallelism. In traditional data base systems coarse-grained concurrency has been demonstrated to be appropriate for inter- and intra-transaction parallelism. It is an open question whether this design is well-suited for the different nature of data mining, where the algorithmic nature of of data exploration seems to call for fine-grained parallelism. Indeed, we suspect that parallelism in data mining covers the entire spectrum from coarse to fine granularity, where the former is more suited to the early preprocessing stages which handle vast amounts of data, whereas the latter is better suited when it comes to the complex processing of data.

Since networks of workstations with high-performance communication hardware have made significant progress, not only do they become an alternative to expensive parallel machines but at the same time they promise the desired continuous spectrum in data mining granularity. On the other hand they are a fairly new concept that may require new answers to parallelism in data processing.

As a rough first approach one may equate the building blocks of data mining with OLAP[1] queries. Therefore, in order to explore the impact of parallelism in networks of workstations, we have studied the performance behavior of OLAP queries on this platform. We have developed a Java-framework for query execution with various execution strategies. The results are promising: since latency and bandwidth of current networks of workstations are no longer the main problem, fine-grained parallelism has an important place in data mining.

At first glance, the choice of Java may sound surprising. However, we targeted a portable solution and we expected to benefit from the optimized versions of Java's object serialization and RMI (remote method invocation) that have been developed along our way. Moreover, the results provide further directions in the development of the Java platform as well as parallel query processing and data mining techniques in general.

The paper is organized as follows: First, we give a short survey of related work. In Section 3 we describe the execution of OLAP-Queries. Section 4 introduces JavaParty, the parallel Java infrastructure we have used, and its central performance features, i.e. fast serialization and RMI. In Section 5 we discuss the implementation of our execution engine in JavaParty. The benchmarks of Section 6 suggest that non-standard query parallelization techniques might work.

## 2 Related Work

Parallel database systems are in broad commercial use today and have been the subject of research for several years [6, 5, 15]. The way queries are executed in these systems still mainly depends on the underlying machine architecture. Three types of parallel architectures can be distinguished:

- Shared everything (shared memory)
- Shared disk (each processor has its own memory, but they have shared access to the data on disk)
- Shared nothing (SN, each processor has its own memory and disk. Networks of workstations belong to this class)

Whereas fine-grained, communication-intensive techniques (e.g. pipelining between nodes) are employed mainly in shared-memory systems [1, 9] this work

---

[1] OLAP is a business term and stands for Online Analytical Processing, in contrast to OLTP (Online Transaction Processing)[11].

suggests that it can be advantageous to use them as well on SN-architectures with high performance communication networks. Furthermore, on SN architectures the static allocation of the data on disk typically drives the execution of the query [13, 10]. We suggest that it may not be necessary to tie allocation of data and execution too tightly if the network is sufficiently fast.

The architectures described above have foremost been studied for traditional query processing which tends to be I/O-bound. More balance between processing and I/O needs is observed in data warehousing. Although those applications differ from data mining in that they are more static and rely heavily on preplanning, data mining could benefit from approaches to speed up data warehousing. Here much work is done in the fields of data and index structures [3] as well as in developing parallel system designs [11]. Indeed there has been first work towards extendig parallelism to data mining [7]. However, the approach to simply map the data analysis steps of a data mining algorithm to SQL-statements executed by a database system can lead to insufficient performance [16]. Hence, a closer coupling of parallel mining algorithms and parallel DBMS seems reasonable. In this environment we concentrate on the various granularities of parallelism.

## 3 Execution of Data Mining and OLAP-Queries

OLAP-queries are different from traditional OLTP transactions because they touch large parts of the data, are complex (i.e., they employ many expensive operators), and unpredictable (i.e. the part of the data, which will be used in the further process and this process are determined on short notice). Consequently, classical performance-enhancing techniques, e.g. indexes, play a minor role. The efficiency of query execution is the main factor. In the remainder of this section we describe the execution of OLAP queries both in a sequential and in a parallel environment and identify the degrees of freedom and the critical parameters.

### 3.1 Sequential Execution

The following type of SQL query frequently occurs in OLAP tasks:
SELECT $Att_1, \ldots, Att_k$ FROM $R_1, \ldots, R_l$
WHERE $Cond_1, \ldots, Cond_m$ GROUP BY $Att_1, \ldots, Att_n$
The query produces tuples with attributes $Att_1, \ldots Att_k$ (that can also be aggregates such as SUM(), COUNT(), etc.) from a given set of tables $R_1, \ldots, R_l$. The WHERE-clause is used to mask data items and to combine items of several tables. The GROUP BY-clause specifies attributes on which the items are grouped before being aggregated. Figure 1a shows a simple operator tree, that combines and processes data from two input tables. When the query is executed, the data flows through this tree. The data is accessed from two tables in the physical storage, then the selection operators ($\pi$) filter out the data specified by the WHERE-clause. The join operator ($\bowtie$) combines the resulting data from the two tables. Finally, an aggregation (agg) is performed on the joined data.

Here no use is made of the inherent parallelism (pipelines, independent branches), this parallelism is just used for sequential re-ordering of the execution.

## 3.2 Parallel Execution

In a parallel environment the inherent parallelism can be exploited. There even exists a new potential for parallelism due to the distribution of the data. However, parallel query processing is more complex. It is even more difficult to achieve performance that scales well with growing numbers of processors.

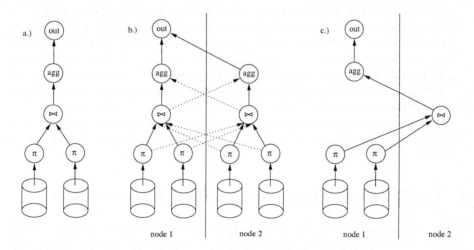

**Fig. 1.** Query trees for sequential and parallel execution

When operators are duplicated to multiple nodes, one has to consider global-scope operators. These operators have to correlate tuples they have seen earlier with those currently received and therefore need a complete history regarding the partition of the data they process. E.g. in case of a join, corresponding tuples from the joined relations have to "meet" within the same operator node in order to produce the correct result. In general, data is redistributed during query execution: Preceding operators have to send their tuples to the appropriate instance of the global-scope-operator according to the attribute values this operator uses to correlate the tuples. The resulting execution scheme (Figure 1b) is typical for parallel databases on SN-architectures and rests on the assumption that communication cost is the dominating factor: Each node keeps a part of the data on its disk, but possesses a full set of operators, and therefore communication between nodes (dotted lines) is employed only when necessary.

## 3.3 Alternative Parallelization Strategies

In clusters of workstations with high performance communication networks, the underlying assumptions no longer hold. Inter-node communication is much better (lower latency and higher bandwidth) so that one could expect that there are new degrees of freedom to be exploited for optimized parallelization strategies:

- The *degree of parallelism*: Under the traditional strategy the degree of parallelism is restricted by the growing cost of data redistribution and, hence, data communication. Moreover, it is heavily influenced by the static distribution of the raw data across the nodes. Once communication cost is low, both are no longer the limiting factors: The number of nodes should be determined by the needs of the algorithms, and initial distribution of the data may be arbitrary. Instead of replicating the operators to all nodes, the number of parallel operators performing an action can be chosen by the optimizer.
- The *parallelism paradigm* can be chosen dynamically, ranging from pure data parallelism to the extreme use of pipelining, where every operator is located on a different node and the data is piped through the nodes. Pipelining may work if less performance is lost by extensive communication than is gained by increased parallelism. An example for the moderate use of pipeline parallelism is shown in Figure 1c: The file-I/O, the selection, and the final aggregation are performed on one node, the expensive join is performed on the other node. Compared to the SN-approach, twice the amount of data is sent over the network.

## 4 Optimized JavaParty as Distributed Environment

While our long-term goal is to place OLAP and data mining activities within a continuum of function and data granularity, the present studies explore the suitability of networks of workstations and Java for parallel OLAP processing. The software platform we used is the JavaParty system for transparent parallel and distributed programming in Java. This section briefly introduces JavaParty and presents the central features that allow JavaParty to perform efficiently.

### 4.1 Standard JavaParty

JavaParty [12, 14] is a programming layer on top of Java that transparently adds remote objects purely by declaration, and avoids exposing the programmer to sockets, RMI, and message passing libraries. JavaParty code is preprocessed into Java code with RMI hooks, both are then compiled by regular Java and RMI compilers into platform independent and secure ByteCode.

JavaParty extends Java with a new class modifier `remote`. By this modifier, the programmer can distinguish between objects that are local and objects that may be instantiated on a remote node. Since Java's threads are implemented by means of objects as well, the programmer can create remote threads that run on remote processors. JavaParty implements Java's object semantics, i.e., the programmer has the impression of writing regular multi-threaded Java programs. The source code size does not change when moving from Java to JavaParty, but JavaParty programs are portable between single-processor workstations, shared memory parallel computers, and distributed memory platforms. Since the topology and the number of processor nodes of the underlying parallel computer is completely transparent to JavaParty programs, they automatically adapt to changing configurations. The JavaParty preprocessor and the rest of the JavaParty environment are 100% pure Java and freely available [12].

## 4.2 Reducing the Cost of Remote Object Access in JavaParty

Being built on top of Java's Remote Method Invocation, JavaParty's performance depends on an efficient RMI. Unfortunately, in current Java implementations RMI is too slow for high performance computing since a remote method invocation between two workstations that are connected through 100 MBit-Ethernet takes about 4.6 milliseconds (one object with 32 `int` values as argument, JDK 1.1.7, two 500 MHz Digital Alphas).

In order to plan for improvements, we have analyzed where this time is spent: About one third of the time is spent in serialization, one third is spent within the current implementation of RMI, and one third is spent in the communication network. Hence, we have optimized all three parts: by an optimized serialization, an optimized RMI implementation, and by using the ParaStation communication network to couple the workstations. We now achieve a remote method invocation that takes about 350 microseconds (with the same object as argument).

*Better Serialization.* We have discussed the individual problems of Java's object-serialization and built an optimized version of it, see [8]. For several benchmarks our serialization outperforms the official serialization by a factor of up to 35.

*Better RMI.* We have re-designed and re-implemented RMI. Similar to the official RMI design, we have three layers (stub/skeleton, reference, and transport). In contrast to the official version however, our design features clear interfaces between the layers. This has two essential advantages. First, a performance advantage: The layers communicate by means of method invocation instead of passing around an object which describes a remote invocation. Hence, in our design a remote method invocation requires just two additional method invocations at the interfaces between the layers. No costly helper object creation is needed. The second main advantage is that alternative transports can be used. Whereas the official RMI is not designed to work with non-TCP/IP-networks, the clean interface between reference layer and transport layer allows for implementations that can work with high performance communication hardware.

For the ParaStation Network (see below) we have implemented a packet based transport that directly uses the hardware communication ports. We exploit the fact that packets are guaranteed to be delivered in order. Moreover, since either the whole cluster of workstations is working or not, there is no need to protect against network errors, e.g., temporary unavailability of some nodes, connection failure, etc. Therefore, the ParaStation transport is very slim.

*ParaStation Network.* ParaStation [17] is a communication technology for connecting off-the-shelf workstations into a supercomputer. The current ParaStation system is based on Myrinet [2], fits into a PCI slot, employs technology used in massively parallel machines, and scales up to 4096 nodes. ParaStation's user-level message passing software preserves the low latency of the communication hardware by taking the operating system out of the communication path, while still providing full protection in a multiprogramming environment. On the Alpha platform, ParaStation achieves end-to-end (process-to-process) latencies as low as 15 $\mu$s and a sustained bandwidth of more than 50 Mbyte/s per channel.

## 5 OLAP Queries in JavaParty

### 5.1 System Design

Figure 2 shows the general design of our parallelization framework.

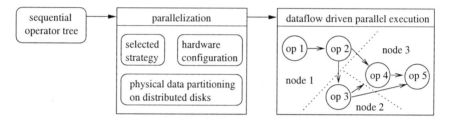

**Fig. 2.** OLAP queries in JavaParty

The query statement provided by the user is parsed and translated into an operator tree. This tree is then parallelized according to three factors: First, the static distribution of the data on the distributed disks must be taken into account. Second, the configuration of the network of workstations must be considered: How many nodes are available. Finally, the parallelization can work according to alternative parallelization strategies, as discussed in section 3.3.

From these input factors, a parallel execution scheme is generated that is driven by dataflow. The operators of the parallel execution tree are distributed to and instantiated on the available nodes. The operators are connected to implement the required dataflow. Figure 2 shows an example where five operators are distributed to three nodes.

### 5.2 System Implementation in JavaParty

For the implementation, we made use of JavaParty's remote objects. Every operator (select, join, etc.) is implemented as a remote object. These objects can be distributed and instantiated on the network of workstations at will. To instantiate a select operator on node n, the corresponding code looks like:

```
DistributedRuntime.SetTarget(n);
node = new jSelectNode();
```

The operator nodes contain an array of references to their successors in the dataflow graph. They are initialized after node creation. All the operators are subclasses of the general operator class jNode shown below. This class is an active class since it has a run() method that is executed by a single thread per object. The run() method executes an infinite loop that processes the data arriving in the input buffer inbuff.

```
remote class jNode implements Runnable{
    jNode[] Succs;      // Successor nodes
    byte[][] inbuff;    // receive buffers

    void run(){
        while(true)
            processTuple();
    }
    void processTuple(){...} // read, process and send to successor
    void sendTuple(data){    // called to send data
        Succs[k].put(data);
    }
    synchronized void put(data){...} // take data and buffer it
}
```

The sendTuple() routine is used to push results to the successor nodes. The index k is calculated according to the redistribution scheme (see section 3). To coordinate multiple senders, the put-method is declared synchronized.[2]

Note, that the data is transfered by means of a simple remote method invocation (Succs[k].put()). This approach has several benefits compared to the use of a dedicated socket for each connection: (1) It is much simpler, (2) flow control can be performed more easily, (3) it consumes fewer resources, since there is no need for a special thread object per incoming connection, and (4) optimizations of remote method invocation result in improved performance of OLAP queries.

## 6  Experimental Results

### 6.1  Benchmark Setup

The measurements were performed on a Cluster of 8 Alpha-workstations running Digital Unix with 500MHz clock rate connected with the ParaStation Network. We used the JDK 1.1.6 for our experiments. As our test dataset we used the 1-GB-TPC-D Dataset [4]. This dataset is part of a standard decision support benchmark and provides large sample datasets from a typical retail environment. The two tables used here hold data from 1.5 Mio. orders and 150000 customers and have the following structure (key attributes are emphasized):
table ORDER (*Orderkey*, Custkey, Orderstatus, Totalprice, Orderdate,
             Orderpriority, Clerk, Shippriority, Comment)

---

[2] We do not show the complete code to keep it simple. Special care must be taken to avoid buffer overflow etc.

```
table CUSTOMER (Custkey, Name, Address, Nationkey, Phone, Acctbal,
                Mktsegment, Comment)
```
Each query was repeated several times; we took the arithmetic mean of the execution times.

### 6.2 Enhanced Communication

In the first experiment we measured the effects of optimized JavaParty implementation. As a basis for this test we used Query 1, see Fig. 3.

This very resource-intensive query that is well-suited for SN-architectures lists all nationkeys, the number of orders placed by customers in that nation, the total volume and the average price per order for all orders placed before 1997.

*Query 1*
```
SELECT NATIONKEY,
COUNT(NATIONKEY),
SUM(TOTALPRICE),
AVG(TOTALPRICE)
FROM ORDER, CUSTOMER
WHERE ORDER.CUSTKEY =
CUSTOMER.CUSTKEY
AND ORDERDATE < 1997-01-01
GROUP BY NATIONKEY
```

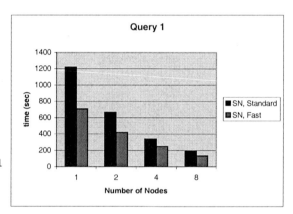

**Fig. 3.** The impact of fast Serialization and RMI

We ran an implementation of this query based on a SN-approach on 1, 2, 4, and 8 nodes and compared the execution times before/after the optimization of JavaParty. As depicted in Figure 3, we see a performance gain of up to 45%, on 8 nodes we have an improvement of 33%. For more nodes, the improvement is less prominent. The reason is that under the classic shared-nothing-strategy, the amount of local communication is reduced, when more nodes are added. (The amount of data that remains on the same node during redistribution is anti-proportional to the number of nodes to which the data is spread.) Since our optimized version of RMI takes special care of local communication, this optimization must have less effect for more nodes.

### 6.3 Alternative Parallelization Strategy

To compare a fine-grained pipeline-based execution strategy with a traditional SN-approach, we've studied the performance of Query 2, see Table 1.

The data is skewed, since it only has five different values for the priority. Due to that fact, only 5 nodes can be used for the aggregation, the others do only file-I/O and selection. That means that the scaling is pretty good until it comes to more than 5 nodes.[3] For two nodes, the pipeline-based implementation used an execution scheme similar to that shown in Figure 1c: The file-I/O is performed on node A, the selection is performed on node B, and the final aggregation is performed on node A again. So there is no local pipelining. The data is sent over the network to the selection. The result is sent back to the aggregation node. For 4 and 8 nodes we took proportionally more operator nodes of each type (node splitting), but the placement scheme was the same: half the nodes perform file-I/O and aggregation, the other half performs the selection. Again there is no local pipelining; all the data is communicated to and from the selection nodes.

*Query 2*
SELECT ORDERPRIORITY,
COUNT(ORDERPRIORITY)
FROM ORDER
WHERE ORDERDATE < 1997-01-01
GROUP BY ORDERPRIORITY

| nodes | SN | Pipeline | **Improvement** |
|---|---|---|---|
| 1 | 269 s | 269 s | 0% |
| 2 | 181 s | 172 s | 5% |
| 4 | 119 s | 103 s | 14% |
| 8 | 95 s | 71 s | **25%** |

**Table 1.** Different execution strategies

Table 1 shows the benchmark results. With a growing number of nodes, the improvement that comes with the pipelining strategy increases. This has two reasons: First, the SN-approach suffers from growing redistribution costs with more nodes (as explained in the previous section). Second, the SN-approach is more affected by data skew. Hence, the pipeline strategy pays off even for just 2 nodes. Our pipeline-based strategy outperforms the traditional SN-approach by 25% on eight nodes, even though more than twice the amount of data is communicated.

## 7 Conclusion

We presented our Java-framework for the parallel execution of OLAP queries on networks of workstations. Our results show Java(Party) as a promising approach for parallel data mining on this platform. Moreover, the results demonstrate, that communication costs are no longer the only crucial factor in parallel query-optimization for this platform. Therefore, alternative parallelization strategies should be studied to achieve efficient execution of data mining tasks.

In the future, we will investigate fine-grained parallelism for object-relational query-execution and the coupling with parallel data mining algorithms.

---

[3] In this example, the problem could be solved by a two-phase GROUPBY, but in other redistribution cases, e.g. joins, that will not help.

## Acknowledgments

We would like to thank the JavaParty team, especially Christian Nester, Daniel Lukic, and Joachim Blum for their work on problems caused by porting the JavaParty environment to ParaStation.

## References

1. B. Bergsten, M. Couprie, and M. Lopez. DBS3: A parallel data base system for shared store (synopsis). In *Proc. Parallel and Distr. Inf. Sys.*, San Diego, CA, January 1993.
2. Nanette J. Boden, Danny Cohen, Robert E. Felderman, Alan E. Kulawik, Charles L. Seitz, Jarov N. Seizovic, and Wen-King Su. Myrinet: A Gigabit-per-Second Local Area Network. *IEEE Micro*, 15(1):29–36, February 1995.
3. Chee-Yong Chan and Yannis E Ioannidis. Bitmap index design and evaluation. In *Proceedings of the SIGMOD International Conference on Management of Data*, SIGMOD Record. ACM Press, 1998.
4. Transaction Processing Council. http://www.tpc.org/dspec.html.
5. David DeWitt and Jim Gray. Parallel database systems: The future of high-performance database systems. *Comm. of the ACM*, 35(6):85–98, June 1992.
6. D.J. DeWitt, R.H. Gerber, G.Graefe, M.L.Heytens, K.B.Kumar, and M. Muralikrishna. Gamma-a high performance dataflow database machine. In *12th Conference on Very Large Data Bases (VLDB)*, pages 228–237, Kyoto, Japan, August 1986.
7. A.A. Freitas and S.H. Lavington. *Mining very large Databases with parallel processing*. Kluwer Academic Publishers, 1998.
8. Bernhard Haumacher and Michael Philippsen. More efficient object serialization. In *International Workshop on Java for Parallel and Distributed Computing*, Puerto Rico, April 12–16 1999.
9. Wei Hong. Exploiting inter-operation parallelism in XPRS. In *Proceedings of the SIGMOD International Conference on Management of Data*, volume 21-2 of *SIGMOD Record*, pages 19–28, New York, NY, USA, June 1992. ACM Press.
10. Informix dynamic server v7.3. White paper, Informix Corp., 1998.
11. W.H. Inmon, Ken Rudin, C.K. Buss, and R. Sousa. *Data Warehouse Performance*. Wiley Computer Publishing, New York, USA, 1998.
12. JavaParty. http://wwwipd.ira.uka.de/JavaParty.
13. Oracle7 server. scalable parallel architecture for open data warehousing. White paper, Oracle Corp., 1995.
14. Michael Philippsen and Matthias Zenger. JavaParty: Transparent remote objects in Java. *Concurrency: Practice and Experience*, 9(11):1225–1242, November 1997.
15. P.Valduriez. Parallel database systems: Open problems and new issues. *Distributed and parallel Databases*, 1(2):137–165, April 1993.
16. Sunita Sarawagi, Shiby Thomas, and Rakesh Agrawal. Integrating association rule mining with relational database systems: Alternatives and implications. *SIGMOD Record (ACM Special Interest Group on Management of Data)*, 27(2), 1998.
17. Thomas M. Warschko, Joachim M. Blum, and Walter F. Tichy. ParaStation: Efficient parallel computing by clustering workstations: Design and evaluation. *Journal of Systems Architecture*, 44:241–260, December 1997. Elsevier Science Inc., New York.

# Garbage Collection for Large Memory Java Applications

Andreas Krall and Philipp Tomsich

Institut für Computersprachen, Technische Universität Wien
Argentinierstraße 8, A–1040 Wien, Austria
{andi,phil}@complang.tuwien.ac.at

**Abstract.** The possible applications of Java range from small applets to large, data-intensive scientific applications allocating memory in the multi-gigabyte range. As a consequence copying garbage collectors can not fulfill the requirements, as large objects can not be copied efficiently. We analyze the allocation patterns and object lifespans for different Java applications and present garbage collection techniques for these. Various heuristics to reduce fragmentation are compared. We propose just–in–time generated customized marker functions as a promising optimization during the mark–phase.

## 1 Introduction

The programming language Java is used for a wide range of applications ranging from small applets running in a browser to large scientific programs taking hours of computation time and gigabytes of memory. It is difficult to design a garbage collector which performs well under different workloads. The garbage collector for CACAO – a 64 bit JavaVM for Alpha and MIPS processors [KG97,Kra98] – has been designed for large objects and large memory spaces but also performs well for small objects. In this study we evaluate the behavior of different garbage collection schemes under different workloads to find common patterns which can be used to design efficient garbage collection heuristics.

There exist hundreds of different garbage collection algorithms [JL96,Wil94] which can be largely divided into copying and non–copying collectors. The non–copying mark–and–sweep collectors are very efficient but it is widely believed that they suffer of memory fragmentation problems. Recent studies [JW98] showed that the fragmentation problem is small. For large object spaces copying collectors are unusable because of the copying overhead. It is impractical to copy objects which reach Gbyte sizes. Feasible solutions are treadmill and mark–and–sweep collectors.

The remainder of this paper is structured as follows: Section 2 lists related work. In section 3 we briefly present an overview of the garbage collector and present the result from the lifespan analysis for objects. We also discuss different heuristics to improve garbage collection. Customized marking using just–in–time generated marker methods is introduced in section 4. Section 5 presents the experimental results from our tests. We draw our conclusions in section 6.

## 2  Related Work

Precise garbage collection for Java virtual machines has been described by Ageson et al. [ADM98]. In contrast to conservative (or partially conservative) collectors precise collectors have exact information about objects on stack and heap. The computation of the stack maps in a JavaVM is complicated by the fact that local variables can contain any type across the call of a local subroutine (jsr) which is used to implement exception handler routines. During compilation the Java byte-code has to be rewritten to rename variables which are live across a call of an exception handler with different types. Using the precise collector the heap size can be reduced by 4% on average in comparison with a partial conservative collector.

A recent study by Johnstone and Wilson [JW98] evaluated the fragmentation of conventional dynamic storage allocators. The study analyzed 8 big C and C++ programs using 16 different implementations of malloc/free. Most of the live objects are very small (less than 64 bytes). Usually only very few large objects exist. Large objects have a long life time. The best and also efficient algorithms showed an average fragmentation below 3%. These fragmentation numbers are not directly comparable to garbage collection fragmentation behavior since life times and allocation/deallocation patterns are different.

Hicks et al. [HHMN98] studied the garbage collection times for large object spaces. Using a separate non–copy collected space for large objects results in significant performance improvements for copying garbage collectors. This study evaluated varied size thresholds as well as whether or not large objects may contain pointers. A treadmill collector was compared with a mark–and–sweep collector. As benchmarks different programs written in Java and SML are used. For Java programs optimal threshold values are smaller than for SML. There is no measurable difference in performance between the treadmill and the mark–and–sweep collector.

Colnet et al. [CCZ98] describe compiler support to customize the mark–and–sweep algorithm in the SmallEiffel compiler. The SmallEiffel garbage collector is a classical partially conservative mark–and–sweep collector. Type inference is used to compute the necessary information to segregate objects by type and statically customize most of the GC code. In this study 22 different implementations of Othello (both leaky (which rely on a garbage collector) and non leaky versions) have been evaluated. The results show both a reduction in run time and memory footprint compared with either no GC at all or the Boehm–Weiser collector [BW88].

## 3  The garbage collector

We implemented a conservative garbage collector using a mark–and–sweep algorithm. During the mark phase the contents of the stack are considered references to root objects. Starting with these root objects, objects on the heap are marked recursively. Every heapblock contained in a marked object is considered a potential reference and has to be examined: if it points to the start of an unmarked object, that object is marked in turn.

In order to store an indication of where objects start and whether objects are marked, we use bitmaps where each bit represents one heapblock. These bitmaps encode whether an object starts at a heapblock, whether it is marked and whether it may contains references to other objects.

The sweep phase is currently implemented using the start and free bitmaps. Continuous free space resulting from neighboring objects is automatically detected and freed as one block of memory. We are currently working on an incremental method of collecting and releasing this garbage, which may reduce the cost of sweeping by up to an order of magnitude as our preliminary tests indicate.

Allocation is performed from a list of free memory blocks. First, we attempt to satisfy the allocation request with an exact block; if that fails, a part of a larger block is split off and used. If no exact or large block can be found, we grow the heap up to a maximum limit. As soon as that limit is reached, a garbage collection is performed.

### 3.1 Object lifespans

To collect information on object lifespans, we performed a full garbage collection during each allocation request. The results confirmed our hypothesis, that objects remain valid for either a very short period of time or for a very long one. Our experiments show that only about 20% of all objects live longer than 512 bytes (which equals 10 allocations in most programs) (see fig. 1). This implies that the best method to reduce overall fragmentation, which is very significant for the total memory required for the execution of a program, is to collect early and often at the cost of a small increase in garbage collection time.

### 3.2 Generational garbage collection

Recently, generational garbage collection [HM92] gained in popularity: it exploits the typical allocation patterns found in modern languages to optimize garbage collection. Assuming, that memory allocated on the heap either becomes garbage very quickly or remains valid for most of the program's runtime, garbage collection should differentiate between *young* and *mature* objects. While *young* objects have a high probability to die with the next collection, *mature* objects will likely survive it. The generational approach separates the allocated objects can into different generations according to the number of collections they survived. Whenever a garbage collection is triggered, they youngest generations are observed first, which reduces the number of memory accesses required. Surviving objects are promoted to older generations.

Using a generational garbage collector may appear a good solution to the small lifespans observed, but it leads to severe problems: the write–barrier necessary to record cross–generational references imposes an overhead on all as-

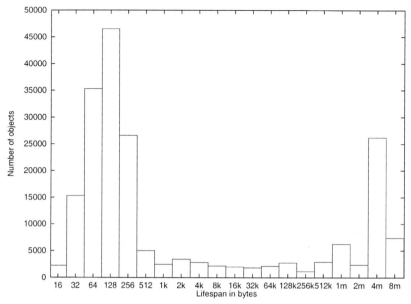

**Fig. 1.** Lifespan graph for a compilation of JavaLex with javac

signments of objects. In addition, generational garbage collection uses a larger amount of heap space, as garbage in older collections is released only infrequently. This memory, which may remain uncollected for some time even after becoming garbage, can amount to a considerable memory area in applications with large memory requirements.

### 3.3 Heuristical threshold values

Fragmentation can be reduced by scheduling collections on the extent of heap (i.e., the highest address used within the heap at any given time). Our garbage collector collects as soon as a heuristically chosen threshold for the extent of the heap is exceeded. For the implementation we considered 3 heuristics:

1. **Adding a constant value.**
   This naive technique ignores the smaller size of the heap after the collection and grows the threshold very quickly.
2. **Adding a multiple of the current extent.**
   Adding a multiple of the heap extent resulting from the garbage collection ignores the fact, that non–copying garbage collectors leave a fragmented heap (i.e., the extent of the heap is considerably larger than the heap size necessary to hold all the objects on the heap). This leads to a quickly expanding heap for non–copying garbage collectors, such as the mark–and–sweep collector used in CACAO.
3. **Adding a fraction of the unused space.**
   This is a very effective heuristic, which works well with the fragmented heaps

resulting from non–copying garbage collection. It reduces the overall fragmentation, such that programs can be run with less memory than needed for the other two strategies.

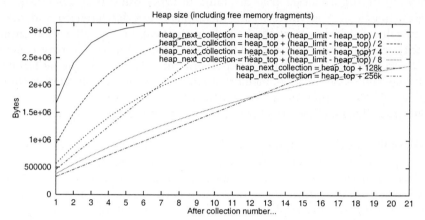

**Fig. 2.** Heap size for different threshold heuristics (javac)

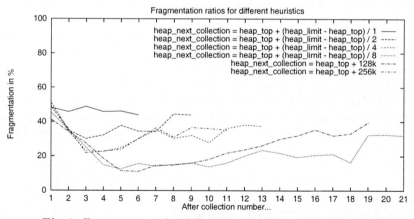

**Fig. 3.** Fragmentation for different threshold heuristics (javac)

Figures 2 and 3 show that the method of adding a fraction of the unused space is preferable, since it maintains a far more compact heap than the other variants. Additionally, it offers superior better performance during garbage collection by collecting "early and often". This fact is due to the observed pattern of small objects with small lifespans; frequent collections keep the heap compact and reduce the cost of garbage collecting further by improving the caching behavior of the mark–and–sweep algorithm.

## 4 Customized marking

The data gathered during our experiments indicates that about 60% of all heapblocks examined (for numerical applications this percentage raises to more than 95%) during a marker pass are built–in types other than references (i.e., integers, doubles). Nonetheless, these heapblocks are dereferenced and the resulting value checked against the bounds of the heap and an allocation–bitmap to verify whether an object actually begins at this address. We may conclude from these findings, that a very large potential improvement in the performance of garbage collection for large memory application would result from excluding those heapblocks within an object, which will never contain pointers, from the marker pass. Two methods to store and evaluate this information exist:

1. **Bitmap based methods.** The type–information for the physical components of classes can be stored as bitmaps within the class-info structures. These bitmaps encode boolean values indicating whether the heapblock at a certain offset within the object heapblock may contain a reference or not. During the mark–phase, these bitmaps are interpreted to customize the marking on a per–class basis.

    The disadvantages of this approach are the overheads introduced by the interpretation of the bitmaps and the additional memory accesses necessary to retrieve the bitmaps. In addition, large objects containing only few pointers can neither be represented efficiently nor be marked without examining the entire bitmap.

2. **Just–in–time generated marker methods.** The bitmap–based methods described above may be modified by translating the information encoded within these bitmaps into executable code. For a portable implementation byte code versions of the mark methods are generated during class loading which are translated on demand into native code during garbage collection. This provides just–in–time generated marker methods customized for every class. During garbage collection these methods are called for every live object.

    This solution offers almost optimal execution times for the marker, because only those heapblocks are examined which may contain references (i.e., all heapblocks, for which the compiler can determine that they will never contain a reference, are excluded). While additional code needs to be generated and stored for every class, the storage overhead involved is negligible in size compared to the other information in class-info. Generally the resulting marker method will require less than 50% of the memory accesses needed for the naive approach.

### 4.1 Conserving stack space

Recursive marking algorithms require large amounts of stack space, particularly for deeply nested structures and long lists. This may cause stack overflows.

Pointer reversal techniques provide an alternative to recursive marking, but impose far more memory accesses.

A method to save both stack space, as well as improve performance by reducing the number of necessary recursive calls is to optimize for tail recursion. This allows the last recursive call to reuse the current stack and return address. This is especially useful in the context of lists, where a `next`–pointer can be detected and placed at the tail of the structure by the compiler. This optimization is particularly beneficial in the context of large memory applications, where large lists are processed.

## 5 Experimental results

To evaluate fragmentation and different heuristics we used following test applications:

| | |
|---|---|
| javac | the Java compiler |
| jess | Java Expert Shell System based on CLIPS |
| db | memory resident data base system |
| raytrace | a raytracer rendering a dinosaur |
| scimark | a mix of numeric applications |
| linpack | the famous linpack benchmark |

Besides the small `linpack` benchmark, all applications allocate between 100 and 500 Mbyte of objects. Our results show that most applications allocate only small objects with average sizes between 30 and 60 bytes (see table 1). Only the scientific/numerical applications `scimark` and `linpack` used big arrays and have an average object size of 810 respectively 1353 bytes. The distribution of the objects shows the peak at the small sizes with most objects smaller than 128 bytes.

Table 2 gives the number of references which have to be checked if they are valid object pointers. For scientific programs nearly 100% of the pointers are either false pointers (i.e., potential pointers that fall outside of the heap) or null pointers. This result shows the potential performance improvement of the JIT-marker. The high percentage of null pointers demonstrates the importance of checking the null pointer at the call site of the mark method.

The data in table 1 and the fragmentation data in the plots and in table 3 show fragmentation between 10 and 30% of the heap. Since the fragments are large enough for allocation of new objects the fragmentation is not a real problem for Java garbage collectors.

## 6 Conclusion

Java applications with large objects require special techniques for garbage collection. Only non–copying collectors can provide acceptable performance both in

| benchmark | javac | jess | db | raytrace | scimark | linpack |
|---|---|---|---|---|---|---|
| heap size | 62667832 | 468663656 | 124944496 | 187717936 | 43611752 | 334328 |
| objects | 1340767 | 7918647 | 3202929 | 6338943 | 53817 | 247 |
| object size | 46.7 | 59.1 | 39 | 29.6 | 810 | 1353 |
| 16 | 0 | 48 | 0 | 0 | 0 | 0 |
| 32 | 458205 | 1425465 | 3061233 | 4693884 | 563 | 24 |
| 64 | 584415 | 4474601 | 98875 | 1419488 | 535 | 10 |
| 128 | 224339 | 997584 | 42442 | 225550 | 268 | 4 |
| 256 | 13493 | 1019885 | 17 | 15 | 16 | 4 |
| 512 | 783 | 395 | 6 | 4 | 515 | 1 |
| 1024 | 2279 | 214 | 2 | 0 | 51914 | 1 |
| 2048 | 225 | 290 | 1 | 0 | 0 | 203 |
| 4096 | 379 | 113 | 1 | 1 | 0 | 0 |
| 8192 | 180 | 47 | 1 | 0 | 0 | 0 |
| 16384 | 33 | 4 | 1 | 0 | 0 | 0 |
| 32768 | 2 | 1 | 1 | 0 | 0 | 0 |
| 65536 | 1 | 0 | 1 | 0 | 2 | 0 |
| 131072 | 0 | 0 | 346 | 0 | 3 | 0 |
| 262144 | 0 | 0 | 1 | 1 | 1 | 0 |

Table 1. Object counts

| benchmark | javac | jess | db | raytrace | scimark | linpack |
|---|---|---|---|---|---|---|
| all references | 12637996 | 6448116 | 41470054 | 41133853 | 478324 | 47842 |
| false pointers | 5630274 | 2330767 | 29954232 | 26416277 | 445394 | 912 |
| null pointers | 2225825 | 1089709 | 3087413 | 2968714 | 31372 | 46620 |
| false percentage | 44.5 | 36.1 | 72.2 | 64.2 | 93.1 | 1.9 |
| null percentage | 17.6 | 16.8 | 7.4 | 7.2 | 6.5 | 97.4 |

Table 2. Reference checking

| benchmark | javac | jess | db | raytrace | scimark | linpack |
|---|---|---|---|---|---|---|
| number of fragments | 46148 | 8449 | 212 | 599 | 5 | 0 |
| average size of fragments | 147.3 | 1692 | 19394 | 5145 | 268041 | 0 |

Table 3. Fragmentation

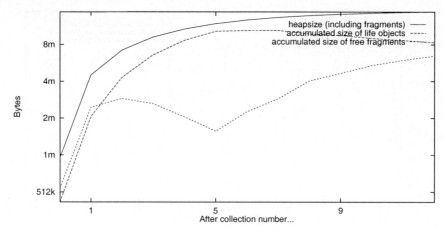

**Fig. 4.** Fragmentation of javac

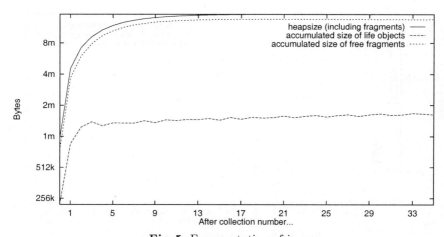

**Fig. 5.** Fragmentation of jess

applications with small objects and those with large objects. Our analysis of object lifespans demonstrated that almost all objects have very short lifespans. The fragmentation analysis showed that fragmentation is not a significant problem for non–copying garbage collectors for Java. Just–in–time generated customized marker methods reduce the runtime overhead of marking by more than 60%.

# References

[ADM98]   Ole Agesen, David Detlefs, and J. Eliot B. Moss. Garbage collection and local variable type-precision and liveness in Java virtual machines. In *Conference on Programming Language Design and Implementation*, volume 33(6) of *SIGPLAN*, pages 269–279, Montreal, 1998. ACM.

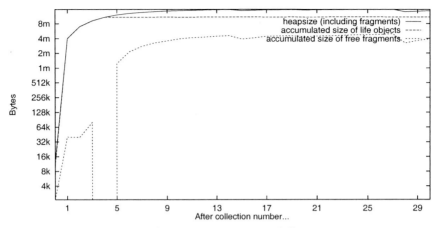

**Fig. 6.** Fragmentation of db

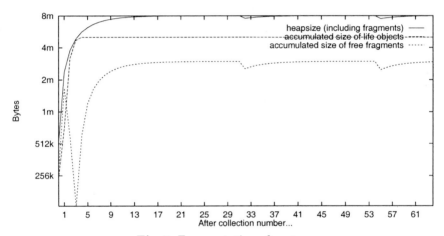

**Fig. 7.** Fragmentation of raytrace

[BW88]  Hans-Juergen Boehm and Mark Weiser. Garbage collection in an uncooperative environment. *Software Practice and Experience*, 18(9):807–820, 1988.

[CCZ98]  Dominique Colnet, Philippe Coucaud, and Olivier Zendra. Compiler support to customize the mark and sweep algorithm. In *1998 International Symposium on Memory Management*, pages 154–165, Vancouver, 1998. ACM.

[HHMN98]  Michael Hicks, Luke Hornof, Jonathan T. Moore, and Scott Nettles. A study of large object spaces. In *1998 International Symposium on Memory Management*, pages 138–145, Vancouver, 1998. ACM.

[HM92]  Richard L. Hudson and J. Eliot B. Moss. Incremental collection of mature objects. In *Proceedings of the International Workshop on Memory Management*, pages 388–403, September 1992.

[JL96]  Richard Jones and Rafael Lins. *Garbage Collection*. John Wiley, 1996.

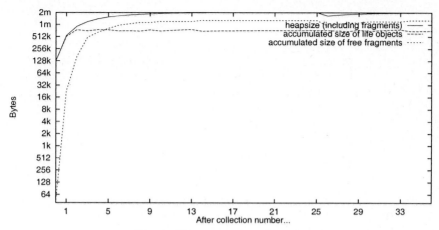

**Fig. 8.** Fragmentation of scimark

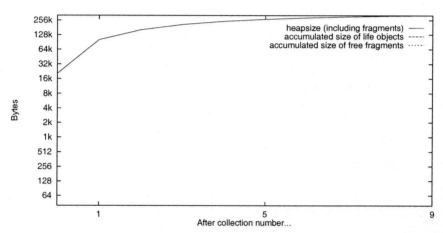

**Fig. 9.** Fragmentation of linpack

[JW98] Mark S. Johnstone and Paul R. Wilson. The memory fragmentation problem: Solved? In *1998 International Symposium on Memory Management*, Vancouver, 1998. ACM.

[KG97] Andreas Krall and Reinhard Grafl. CACAO – a 64 bit JavaVM just-in-time compiler. *Concurrency: Practice and Experience*, 9(11):1017–1030, 1997.

[Kra98] Andreas Krall. Efficient JavaVM just-in-time compilation. In Jean-Luc Gaudiot, editor, *International Conference on Parallel Architectures and Compilation Techniques*, pages 205–212, Paris, October 1998. IFIP,ACM,IEEE, North-Holland.

[Wil94] Paul R. Wilson. Uniprocessor garbage collection techniques. In *ACM Computing Surveys*, page to apear. ACM, 1994.

# Workshop:

# IEEE EMBS ITIS-ITAB '99

# The Emergence of Virtual Medical Worlds

Andy Marsh[1], Tuomo Kauranne[2], Gudrun Zahlmann[3]
Ad Emmen[4], Leslie Versweyveld[4]

[1]Institute of Communications and Computer Systems
National Technical University of Athens
9 Iroon Polytechniou Street, GR-15773 Zografou, Athens, Greece.
e-mail marsa@phgasos.ntua.gr

[2]University of Joensuu,
P.O. Box 111, FIN-80101 Joensuu, Finland.
Email kauranne@csc.fi

[3]GSF, medis
Ingolstaedter Landstr. 1, Neuherberg 85764, Germany.
e-mail zahlmann@gsf.de

[4]Genias Benelux,
James Stewartstraat 248,
NL-1325 JN Almere, Netherlands.
Email vmw@hoise.com

**Abstract** - The impact of Information Technology has proliferated the development and uptake of Telecommunication, Compunetic (Computer and Networking) and Simulation technologies. The accessibility and interoperability of (Tele-)medical systems, performing Biomedical and Biomechanical applications, is one of the grand challenges for future healthcare. A new market is being developed dealing in Telemedical technologies. To support the integrated development of such an International Telemedical Information Society (ITIS), a supportive community of web-services has been identified and being setup. This paper presents the framework and building blocks of the emerging Virtual Medical Worlds.

## 1. Introduction

The objective of the EUROMED initiative aims to integrate national initiatives to setup an International Telemedical Information Society (ITIS) to provide a primary level of healthcare to all European citizens including dispersed and isolated regions of the community. Such a society will involve the cooperation and collaboration of a number of leading edge technologies ranging from data storage to Virtual Reality and from Electronic Commerce to International Law in a range of Biomedical and Biomechanical applications. The EUROMED initiative is planned for a 10-year duration. The feasibility study began in 1995 and investigated the concept of an ITIS based on Internet and WWW technologies. The derived conclusions, presented in October 1995, indicated the fundamental requirement for a modular standardised secure environment. The development stage, began in January 1996, with a duration of 3 years and included an EC ISIS '95 supported project called EUROMED (2 MECU) which proposed a standard called *Virtual Medical Worlds* (VMW) and a modular infrastructure which consists of 20 fundamental building blocks [http://euromed.iccs.ntua.gr]. Additionally, the EC INFOSEC funded project called EUROMED-ETS (0.5 MECU) has proposed a secure infrastructure based on Trusted Third Party (TTP) services.

Simultaneously with the development phase an awareness programme has been established. An on-line monthly magazine called *Virtual Medical Worlds Magazine* [http://www.hoise.com/vmw] was setup in November 1997 to report on ITIS related issues. In 1998, a new conference was established called *ITIS* [http://www.hoise.com/vmw/conference/ITIS98] which promotes forum discussions, inspired by invited speakers and paper presentations, tackling the current technical issues related to an ITIS. The first ITIS event was sponsored by the EC HPCNET project and took place in Amsterdam April $22^{nd}$-$24^{th}$ April 1998 as part of the International HPCN '98 conference. The ITIS '98 forum consisted of three sessions and tackled the issues of *"What facilities should be offered ?"*, *"Is the Web the best platform ?"*, and *"What obstacles need to be overcome ?"*. The second ITIS event took place in Hong Kong on October $29^{th}$ 1998 as part of the IEEE EMBS '98 conference and concentrated on the issue of *"Integration of technologies"*. After the success of the ITIS events the IEEE-EMBS will sponsor ITIS '99 which will be held in Amsterdam as part of the HPCN '99 conference but this time will be joined with the second international conference on Information Technology Applications in Biomedicine (ITAB) [http://www.hoise.com/vmw/conference/ITIS99]. To support the magazine and conference and with an emphasis on research publications, in October 1998, a new concept in publishing research material was setup called the *Journal of ITIS Letters* [http://www.hoise.com/vmw/science/ITIS]. This new journal was sponsored by Elsevier Science and has the objective to promote the dissemination of leading edge research though the use of web technologies. Also, in 1998, a new company called *VMW Solutions*, has been setup to promote industrial collaborations and bridge the gap between research and industry. The results of research projects are then commercialised into a common framework, namely Virtual Medical Worlds, and presented to the market as a collection of co-operating services.

On the $1^{st}$ of January 1999 the Virtual Medical Worlds Community (VMWC) initiative began. This working group's goal is to create a dynamic synergy between users operating in the telemedical sector and technology suppliers to advance the European industry in both areas. VMWC will try to stimulate exchange of information and expertise to generate awareness in the medical sector via the organization of workshops, technical meetings, participation in major telemedical events, and a virtual community environment on the Internet for active brainstorming and discussions among the members. The market validation stage, of the EUROMED initiative, will also begin in 1999 with a planned duration of 2 years and has the objective to define a business plan that will detail the expansion of a pilot ITIS to other EC member states. In conjunction with the market validation phase a research programme will be established. This will consist of the establishment of a research centre which will collaborate and coordinate a number of research initiatives and international groups acting as a catalyst to promote the uptake of telemedical research activities in the newly developing industrial market place. After the Market validation phase a 4-year deployment phase is planned to begin in the year 2001. The *Virtual Medical Worlds*, as depicted in figure. 1, will include collaborating entities namely; international societies & bodies (e.g. IEEE-EMBS), press (e.g. Elsevier), international conferences (e.g. ITIS), publications (e.g. Journal of ITIS Letters), industrial links (e.g. VMW Solutions), funding organisations (e.g. EC), academic links, industrial links and most importantly end users namely hospitals and medical centers. The emerging Virtual Medical Worlds is multi-national and utilises all the currently developing technologies, integrated through web technologies, to support the promotion and proliferation of the uptake of leading edge ITIS research.

**Fig. 1** : The emerging Virtual Medical Worlds

## 2. Conclusions

The aim of the Virtual Medical Worlds is to establish a global community of collaborating research groups and industrial products. From technological aspects a Virtual Medical World can be developed to incorporate the latest IT developments into the healthcare environment. With a modular framework and use of standards such a society, if coordinated on a worldwide scale, will support the Tele-medical usage of 21$^{st}$ healthcare services and provide an advanced infrastructure for future healthcare.

## 3. References

[1]. Virtual Medical Worlds Magazine [http://www.hoise.com/vmw]
[2]. ITIS '98 [http://www.hoise.com/vmw/conference/ITIS98]
[3]. ITIS '99 [http://www.hoise.com/vmw/conference/ITIS99]
[4]. Journal of ITIS Letters [http://www.hoise.com/vmw/science/ITIS].

# Characteristics of Users of Medical Innovations

Mary Moore

Arkansas State University, Dean of Library and Information Resources, P.O. Box 2040, State University, Arkansas, USA
mmoore@choctaw.astate.edu

**Abstract.** Some innovations are adopted quickly and widely used. Others never seem to find widespread acceptance. Among the many factors related to diffusion of innovations are sociodemographic, psychological and communications variables. This examination of three groups of physicians who adopted innovations and changed some aspect of their medical practice identifies common characteristics of younger age or younger professional age, board certification or specialization, increased information seeking behavior and increased contact with opinion leaders.

## 1 Introduction

Those working in medical informatics discipline often find themselves responsible for designing and introducing innovations in technologies and services. Sometimes the innovations are adopted quickly and used extensively by health professionals. Other times implementation is difficult, as time and resources are wasted attempting to provide services that never catch on. If informatics professionals knew who would be more likely to adopt the innovations, they might be able to target their efforts better.

Rogers [5,6] has been one foremost among those researchers studying the diffusion of innovations. His theories have sparked hundreds of research studies. He grouped the characteristics of those who were earlier to adopt innovations into a number of sociodemographic, psychological, and communications variables. Rogers speculated that earlier adopters would have the following characteristics:

- greater intelligence, abstraction, and rationality;
- higher aspirations for education, achievement and social mobility;
- more favorable attitudes toward risk, change and science;
- less dogmatism;
- more information seeking behavior;
- more social and communication channels,
- greater empathy,
- more social participation,
- more opinion leadership.

Whether the generalizations are consistent with the characteristics of those who are early to adopt of medical informatics innovations remains to be proved. Three examples of the characteristics of physicians who changed their practice behavior are presented: a study of pediatrician innovators, one of anesthesiologists, and one of family practitioners using telemedicine.

## 2 Pediatricians

Weiss, Charney, and Baumgardner [7] studied characteristics of pediatricians motivated to change some aspect of their practice. These researchers asked 200 pediatricians whether they used three innovations in pediatric care; 156 agreed to participate in the telephone survey. "The characteristics significantly associated with using the innovations were board certification, group rather than solo practice, teaching, medically related publications, academic appointment, younger age, and caring for a greater number of patients per week." For two of the innovations the research confirmed the hypothesis that ". . . personal communication between colleagues catalyzes innovative practice more often than do other sources of medical information, such as conferences or journals . . ." Contacts with local specialists and subspecialists were most likely to motivate the adoption of changed behavior. This study confirmed the experiences of Coleman, Katz, and Menzell [1], which showed that physician's social networks were more important than personal characteristics in explaining adoption of innovations.

## 3 Anesthesiologists

Fineberg, Gabel, and Sosman [2] studied the dissemination of innovation among anesthesiologists. In 1976 they mailed a four-page questionnaire to the 631 members of the Massachusetts Society of Anesthesiologists. They asked how and when the anesthesiologists had learned about three specific innovations studied and whether they had adopted the innovation. They also asked about methods the physicians used to keep current with changing medical practice. Fifty-six percent (56%) responded, and 51% of the responses were usable. Respondents were more likely to be recent graduates and to be board-certified than were nonrespondents. Younger anesthesiologists were more likely to rely on colleagues for information on innovations, and board-certified anesthesiologists were more likely to read journals for information on innovations. Each of the three innovations displayed different patterns of adopter awareness. A published journal article accelerated awareness for only one of the innovations. The other innovations exhibited more gradual awareness patterns.

# 4 Telemedicine

Telemedicine is the use of interactive video for the purpose of patient consultation across distances. Although this type of telemedicine usage was begun in the 1950's, it is only recently that costs of hardware, software and telecommunications have allowed more general use of telemedicine.

Moore [3] studied the characteristics of rural family practitioners who were early adopters of telemedicine. The theoretical foundation for this study was Rogers' generalizations regarding the sociodemographic, psychological, and communications characteristics of earlier adopters of innovations [5]. This study used the NEO-PI-R, a personality instrument, the Opinion Leadership Scale, and a detailed questionnaire to examine the generalizations with a population of 38 rural, general practitioners who received telemedicine consultations from specialists.[1]

## 4.1 Demographics

The rural physicians were relatively young; mode age was 40-45. Physicians reported they belonged to a mean of 3.3 professional organizations, with a standard deviation of 1.2. Two-thirds reported specialty board certification:

Most respondents (82%) reported their primary work place was a single hospital or clinic. Bed size of the hospital, population of the community, and distance from the closest academic medical center ranged widely, but all physicians were determined to be in rural locations.

## 4.2 Information Seeking Behavior

On the average, these physicians spent a large amount of time in information, education and communication behavior, however the responses showed considerable disparity. Physicians reported their most common source of information to be that derived from receiving or providing patient consultations (mean 9 hours per week, standard deviation of 5), followed by reading professional literature (mean 8 hours per week, standard deviation of 4). They reported spending approximately 9 hours per

---

[1] In the study 300 packets were distributed to individuals, telemedicine sites and conferences. Respondents returned a total of 57 packets. Cluster analysis identified an internally homogenous group for the study. Among the questionnaires completed, family medicine practitioners who had requested consultations from distant locations completed 43. Diverse specialists submitted an additional 14 forms with 8 from specialists who only provided consultations from academic health centers. To allow a more uniform sample these 14 cases were excluded. Among the 43 family practitioners, 41 completed all three instruments but only 38 had no critical missing data. Those 38 cases form the basis for results reported here.}

week communicating about patients. Meetings consumed more than 4 hours per week. Physicians attended formal continuing education courses a mean of 26 hours per year, and professional meetings a mean of 10 hours per year. In addition, they spent at least 96 hours per year in other education-related activities. Based on their reports these physicians spent a mean total of over 23 hours per week in the information and education related activities listed on the questionnaire.

Physicians reported spending a total of 75 hours per week (standard deviation of 34) when all information, education, and mass communication activities reported were summed, or more than 10 hours per working day. This is not unreasonable when we consider that medicine is, by definition, an information-intensive discipline.

### 4.3 Telemedicine Use

Physicians' primary motivations for implementing telemedicine included the lack of local specialty physicians and the desires to serve their patients, remain up-to-date, retain patients locally, and remain competitive.

All the respondents had used telemedicine to seek a consultation for a local patient. These physicians used telemedicine for consultations a mean of 7 hours per month, standard deviation of 4, with a range from less than one hour to 20 hours per month. The respondents reported that they had been using telemedicine for less than two years. Half reported using the services within the first two weeks of installation. All individuals were considered earlier adopters because they had used telemedicine within a month of working installation, and they reported continued use of the services.

### 4.4 Social Diffusion Networks

On the general questionnaire 66% of physicians reported they had helped influence others to use telemedicine services. 58% reported that most often they had influenced a local generalist physician. In only three cases (8%) did physicians report influencing a distant physician. However physicians reported that distant physicians most frequently influenced their involvement with telemedicine. Other influences included local physician colleagues, a personal desire to learn, and contact with sales representatives.

Discussion of Information Seeking Behavior and Social Networks Responses:
Earlier adoption of telemedicine among this group showed a positive correlation with the likelihood of consulting the journal literature for patient care problems and an inverse correlation with consulting local colleagues. Since these physicians have few local colleagues to call on (the study found physicians had contact with a mean of 4.4 local colleagues in a typical day), it follows that they would be more likely to consult

journal literature for patient care problems. The journal literature has often been criticized by physicians as being difficult to navigate and evaluate, as well as being inefficient and time consuming. Logically, these physicians might be quicker to embrace other alternatives, such as telemedicine. Early adoption was also negatively correlated to the total time spent in educational activities. This might also imply that although physicians might want to keep current with information, they have less opportunity to do so, and therefore are especially quick to adopt telemedicine as an alternative. Alternatively, the inverse correlation might reflect a preference for learning through individual contacts, such as those offered by telemedicine, as opposed to formal educational offerings.

### 4.5 Personal Characteristics

The NEO-PI-Revised was used to measure personal characteristics. This instrument measures the five main factors of personality: Neuroticism or emotional maturity, extraversion, openness, conscientiousness and agreeableness. Psychological researchers have achieved consensus within recent years that these five factors constitute the majority of personality variance. The instrument has been widely recognized as one of the most valid and consistent existing measure of the five factors, and of the 30 individual facets of personality that compose the five factors.

With the NEO-PI-R, general population scores have been normed on T scores, with the mean at 50 and standard deviation of 10. The scores for this group of physicians generally fell within the average range of 45 to 55. The only exceptions were Neuroticism and Modesty (lower than average); and Achievement and Openness (higher than average).

This group of physicians differed from the general population, however, in the standard deviation of many scores. Perhaps because of the homogeneity of the group, standard deviations were generally smaller. When calculated against the population norm (one sample t-test) the physicians scored higher than the general population on the factors of Openness and Conscientiousness. They scored lower than the general population in Neuroticism and Agreeableness, and within the norm on Extraversion. These physicians were higher than the norm on the following facets of personality: Assertiveness, Activity, Openness to Action, Openness to Ideas, Openness to Values, Altruism, Competence, Order, and Achievement. They were lower than the norm in the following facets: Anxiety, Angry Hostility, Depression, Self-consciousness, Vulnerability, Openness to Aesthetics, Openness to Feelings, Modesty, and Tender-mindedness.

Discussion of Personality Trait Results : One might have predicted that earlier adopters would score higher than the norm on the personality domain of Openness. They did on three of the Openness facets (Actions, Ideas, and Values). The other two facets of Openness (Aesthetics and Feeling) were lower than the general population, however. One explanation may be found in the rationality that is required by the medical profession. Rural physicians may not have much opportunity to cultivate an appreciation for the arts, explaining the lower score on Openness to Aesthetics. Those low in Openness to Feeling are described as being less emotional. This trait may be a characteristic of the profession. The low scores on Neuroticism and the associated factors are somewhat surprising. Is it possible that earlier adopters might be more emotionally mature? Or is this sample of physicians more likely to recognize questions related to emotional stability and to answer favorably? A third consideration might be that, generally, those living in rural areas score higher in emotional stability.

## 4.6 Telemedicine Study Limitations

This study was an initial attempt to identify the characteristics of earlier adopters of telemedicine. Any discussion of the results must begin with recognition that telemedicine was still relatively new and no large samples were readily available for study. Currently groups of early adopters cannot readily be compared with later adopters, since telemedicine is still unfolding. Most of the physicians in this study reported that they had been using telemedicine for a year or less. This study was also limited by the fact that all measures used self-reporting instruments. Subsequent research on telemedicine might compare these results to those from individuals who had refused the opportunity to adopt telemedicine or who had discontinued usage. A comparative study on the characteristics of later adopters might be especially rewarding.

# 5 Conclusions

What are commonalities among the three studies of pediatricians, anesthesiologists and rural family practitioners? Not all of Rogers generalizations were confirmed in these studies. Only one study specifically examined personality characteristics of earlier adopters. But some generalizations are confirmed implicitly by the nature of the samples studies. Physicians are, almost by definition, more intelligent, better educated, more rational, more scientific, and more motivated by achievement. In addition the following commonalities emerge:

1. Younger age or younger professional age (lower number of years since graduation)
2. Board certification or specialization
3. Increased information seeking behavior, using either print materials or colleagues
4. Contact with colleagues and/or opinion leaders. Telemedicine adopters may be motivated by the desire for larger contact with these individuals.

Does this mean that medical informatics professionals, seeking to introduce new information resources and services should market only to these individuals? While this may seem like a logical assumption, there are many elements influencing the diffusion of an innovation. The intercorrelations among personal traits, features of the particular innovation, organizational culture, and the changing health care environment must also be systematically explored to provide a complete perspective.

# References

1. Coleman, J., Katz, E., Menzell, H:. Medical Innovation: A Diffusion Study. Bobbs-Merrill, New York (1966)
2. Fineberg, H. V., Gabel, R. A., Sosman, M. B. (1978). Acquisition and Application of New Medical Knowledge by Anesthesiologists: Three Recent Examples. Anesthesiology, 48 (1978) 430-36
3. Moore, M. Characteristics of Earlier Adopters of Telemedicine. University of Texas at Austin, Austin, Texas (1995)
4. Morris, W. C. The Information Influential Physician: The Knowledge Flow Process Among Medical Practitioners. The University of Michigan, Ann Arbor, Michigan (1970)
5. Rogers, E. M. Diffusion of Innovations. $4^{th}$ edn. Free Press, New York (1995)
6. Rogers, E. M., Cartano, D. G. Methods of Measuring Opinion Leadership. Public Opinion Quarterly 26 (1962) 435-441
7. Weiss, R., Charney, E., Baumgardner, R.A. Changing Patient Management: What Influences the Practicing Pediatrician? Pediatrics 85 (1990) 791-795

# Security Analysis and Design Based on a General Conceptual Security Model and UML

Bernd Blobel[1], Peter Pharow[1], and Francis Roger-France[2]

[1]University of Magdeburg, Medical Faculty, Institute of Biometrics and Medical Informatics, Leipziger Str. 44, D-39120 Magdeburg, Germany
bernd.blobel@mrz.uni-magdeburg.de
peter.pharow@medizin.uni-magdeburg.de
[2]Catholique Université de Louvain, Cliniques Universitaires St. Luc,
10 Avenue Hippocrate, Box 3716, B-1200 Brussels, Belgium
roger@infm.ucl.ac.be

**Abstract.** To facilitate the different users' view for security analysis and design of health care information systems, a toolset has been developed using the nowadays popular UML approach. Paradigm and concepts used are based on the general security model and the concepts-services-mechanisms-algorithms-data scheme developed within the EC "ISHTAR" project. Analysing and systematising real health care scenarios using appropriate UML diagrams, only 7 use case types could be found in both the medical and the security-related view. Therefore, the analysis and design might be simplified by an important degree. The understanding of the approach is facilitated by (incomplete) examples. Based on our generic scheme and with the results described, the security environment needed can be established by sets of such security services and mechanisms.

## 1 Introduction

Shared care as the answer to the challenge for efficient and high quality health care systems must be supported by appropriate information systems' architectures as health care networks, distributed health record systems etc. Dealing with sensitive, personal medical data and often communicating these data across organisational, regional and even national borders, such information systems have to meet comprehensive security requirements to respond threats and risks in distributed health information systems. Regarding security in general, we have to look for the concepts of security, safety and quality [10]. To keep the approach feasible, the consideration in most of the chapters is restricted on the concept of security only.

Considering nowadays health information and communication systems, most of them do not fulfil the security requirements or provide partial add-on solutions only.

## 2 Methods

For a systematic and open analysis, design and implementation of security services and mechanisms in interoperable health information systems, an agreed or even standardised methodology is inevitable. The popular object-oriented paradigm as well as the further development and harmonisation of the corresponding tools for analysis, design and implementation based on the Unified Modelling Language (UML) provide the open and comprehensive solution to respond to these challenges [11].

## 3 General Security Model and Scheme

For an appropriate granularity of security issues in distributed information systems to provide feasible solutions, on the one hand the domain concept is used defining the domain concerns via the Security Policy. Beside the definition of Security Policy Domains, Security Environment Domains and Technology Domains have been specified. For details see, e.g., [2, 3]. On the other hand, a concepts-services-mechanisms-algorithms-data scheme has been developed to systematise and support aspects and views or different user groups [5, 6].

## 4 Modelling of Users' Security Needs

In general, analysis and design of systems in hardware and software is based on a model describing state and/or behaviour of that system. Also the currently popular OO modelling techniques of Grady Booch, James Rumbaugh and Ivar Jacobson provide such an overall model consisting of the components classes, class categories, objects, subsystems, modules, processors, devices, and the relationships between them. These model components mentioned possess properties which identify and characterise them. They can appear in none, one, or several of a model's diagrams associated with other components. Thus, looking for the different components,
 - the class category contains class diagrams and scenario diagrams associated with its components: classes and their objects, and nested class categories,
 - the subsystem contains module diagrams associated with its components: modules and nested subsystems,
 - the class contains its state diagrams,
 - a model's top level contains the diagrams for its top level components as class categories, classes, subsystems, and modules, and its process diagram.
 - In OMT-2, four partial models allow capturing as well as analysis and design of the considered system or domain: the logical, the physical, the static, and the dynamic model. Contrary to other approaches the UML methodology, which is based upon the Booch methods, the OMT-2 methods of Rumbaugh, and the OOSE and Objectory methods of Jacobson, facilitates different views of the overall model described verbally by specifications and through different diagrams (e.g., logical dia-

grams, class diagrams, class structure diagrams, scenario diagrams, collaborations diagrams, component diagrams, distribution diagrams, activity diagrams, use-case diagrams, sequence diagrams).

### 4.1 The UML-Methodology

The UML views are:
- The use case view showing the functionality of the system as perceived by external actors. The use case view is described in use case diagrams and activity diagrams. Use case diagrams are basic descriptions influencing the other views. While the use case looks from outside the system using natural languages to describe the use case, the collaboration (context and interaction) diagram has an inside the system perspective to describe interactions in time (sequence diagram), in space (collaboration diagram) and concerning the work (activity diagram). Finally, the scenario diagram describes a scenario in time (sequence diagram), in space (collaboration diagram) and concerning the work (activity diagram) via an execution path through the system.
- The logical view showing how the functionality is designed inside the system, in terms of the system's static structure and dynamic behaviour.
- The component view showing the organisation of the code components.
- The concurrency view showing concurrency in the system, addressing the problems with communication and synchronisation present in a concurrent system.
- The deployment view showing the deployment of the system into the physical architecture with computers and devices called nodes.

A scenario is a sequence of important interactions between objects as instances of classes within concrete application environments. Scenarios are used to represent critical requirements, depict the action of key mechanisms, and demonstrate desired series of operational cases. The scenarios can be described, considered and manipulated by two types of isomorphic scenario diagrams: the object message diagram and the message trace diagram. An object message diagram illustrates the existence of objects and the communication as the flow of messages among them.

Responding to the enduser requirements for security enhancement, only a part of the UML methodology is really needed. In that context, the use case diagram (and sometimes the sequence diagram as well as the activity diagram) must be mentioned.

The use case defines a framework for using a (information) system. Starting with an abstract use case type, the use case instances describe concrete application scenarios in the sense of the description of .business processes and their communication/interaction with actors. Actors in the healthcare domain are health professionals (doctors, nurses, administrators, technical staff, management, ...) and patients, but also people from other domains. Often, the domain-specific description of the use case is done verbally. Looking for security in information systems, specially security-related use case instances must be mentioned.

To model the needs of the health professionals (medical users, medical and technical staff, administration, management, legal experts), the use of the UML toolset should be recommended. Depending on the different user groups' need, an appropriate granularity of the model may be depicted. The specific components can be described by abstract types using the OO properties like inheritance etc. Complex scenarios may be created combining the abstract or basic types needed. Investigating legal implications on security solutions within the „TrustHealth-2" project (a TAP project of the 4$^{th}$ Framework Programme), very simple abstract types and complex scenarios are used fulfilling the legal experts' needs.

### 4.2 Medical Use Cases

Analysing and grouping the real-world scenarios, basic scenarios or abstract use cases may be defined, which enable the description on any real scenario by compositions of use cases types specified.

Regarding the last 2 years activities results of the „ISHTAR" project funded by the European Commission, administrative tasks (use cases) and medical tasks (use cases) might be distinguished. Grouping these tasks, the following abstract administrative and medial use cases can be found (the relationship to the Swedish approach described in the next paragraph is mentioned by reference numbers):

**Table 1.** Abstract Administrative and Medical Use Cases

| Administrative Use Cases | Medical Use Cases | Ref. # |
|---|---|---|
| Admission, discharge, transfer | | 1 |
| | Diagnosis, assessment, decisions, conclusions | 2 |
| Scheduling and appointments | | 3 |
| Financial transactions | Activities: Visits, Diagnostic procedures, Treatments, Care procedures | 4 |
| Non-medical communications: Insurance Communications, Supplier communications | Medical communications: Order entry, Result reporting, Access to patient information | 5 |
| | Reports (medical documentation) | 6 |

Currently modelling and developing an Swedish Electronic Health Record, the groups involved have found the following abstract use case types:
1. Establishment of contact between patient and health care professional
2. Assessment/conclusion by the health care professional
3. Creation of a specific health care plan for the patient
4. Activities are initiated, performed and looked after
5. Access to patient information
6. Record of health care information
7. Conclusion

### 4.2.1 Medical Use Case Examples – Request Patient Information

The medical use case types mentioned above might be illustrated by some practical examples of real-world systems implemented in the Magdeburg cancer registry. The security solution for this architectural approach of the first German regional cancer registry has been described in detail in [1, 4,].

Considering the very complex challenges to such regional Electronic Health Record (EHR) systems as the Clinical Cancer Registry Magdeburg/Saxony-Anhalt, the first distributed and secured EHR in Germany, only one typical example will be presented in the following paragraphs. As such example, the doctor's request of patient information from the cancer registry or from other Health (Care) Professionals (H(C)P) have been selected. However, this example covers the complete set of the security-related use case types explored in our studies.

A typical scenario in the shared care environment is the doctor's request of information about his currently cared patient.

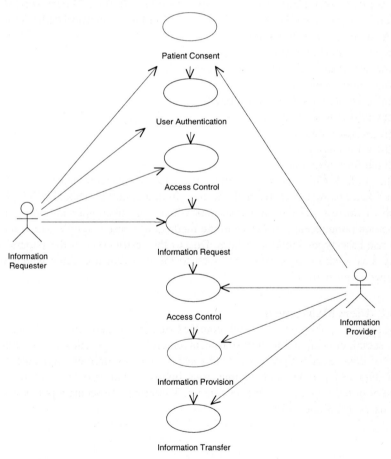

**Fig. 1.** Use Case "RequestPatientData"

This medical use case may occur to receive needed information from former diagnosis or treatment as well as to ask for a second opinion. The requested party could be, e.g., another HCP or a documentation system like an electronic archiving system or as in our case a clinical cancer registry. In the telemedicine/telematics framework, such request is considered as "remote" independent of the real distances bridged over by the communicating and co-operating systems, which could be located even on the same sever. Figure 1 presents that medical use case expressed in the UML notation. To respond to the different possibly communicating parties, the principals in the use case diagrams are neutrally tagged as information requester and information provider.

### 4.3 Security-Related Use Cases

To describe security-related use cases for open systems communication and co-operation, a set of abstract use case types have been defined. Afterwards, the different security-related use cases can be created combining the appropriate basic use cases.

Abstract use case types are:
- the users management
- the user authentication,
- the patient consent
- the communication initialisation,
- the information request,
- the access control,
- the information provision and
- the information transfer.

In the following paragraphs, these use case types mentioned will be described in more detail using the UML methodology. In that context it should be emphasised that each of the use case components may consist of subcomponents and can establish supercomponents etc., as shown in the medical use case example which represents all the use case types mentioned here. Due to the restrictions in the papers' extension, only a very short explanation is given to the different use cases which will be explained in more detail in [7].

#### 4.3.1 Users Management
The security policy describes the complex of legal, organisational, functional, medical, social, ethical, and technical aspects which have to be considered within the context of data security and privacy. The security policy defines the framework, rights, and duties of persons and organisations involved, but also the consequences in the case of non-compliance to the agreement. Therefore, the security policy also specifies the users' roles and rules.

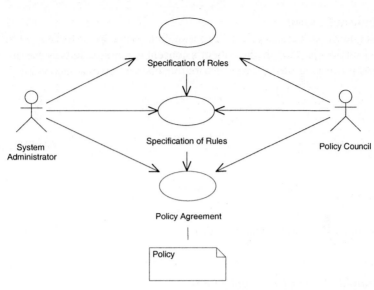

**Fig. 2.** Abstract Use Case "UsersManagement"

### 4.3.2 User Authentication

The authentication of principals communicating and co-operating via information systems is the basic service also needed for other security services and mechanisms as authorisation, access control, accountability etc. Authentication in health information system must be provided mutually and in a strong way using cryptographic algorithms. In our context, we consider human users keeping in mind the generalisation to principals. The TTP provides the user's identity certificate.

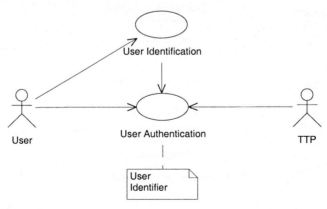

**Fig. 3.** Abstract Use Case "UserAuthentication"

### 4.3.3 Patient Consent

According to the principles of the European data protection directive [8, 9] and German laws on data protection, the patient's consent is required in the case of collecting, recording, processing, storing, and distributing his/her personal medical information.

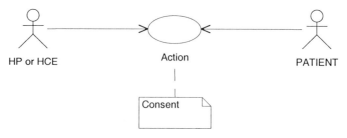

**Fig. 4.** Abstract Use Case "Patient Consent"

### 4.3.4 Initialisation of Communications

For bilateral and multilateral communication and co-operation, the mutual identification and authentication of the partners (principals) involved is needed. The authentication must be verified by the certificates provided by the TTP.

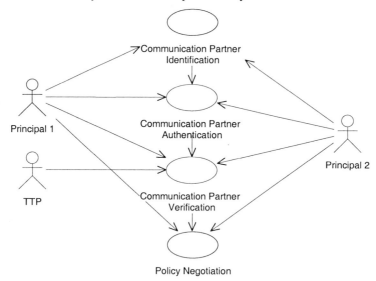

**Fig. 5.** Abstract Use Case "CommunicationInitialisation"

### 4.3.5 Information Request

After the initialisation of any communications and co-operations, the information requested has to be specified.

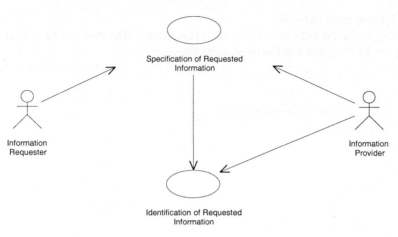

**Fig. 6.** Abstract Use Case "InformationRequest"

### 4.3.6 Access Control

Fulfilling the need to know principle and the privacy rights of the patient, the access to and the use of patient's information must be restricted and controlled according to the underlying mandatory and discretionary access control models. On that way, the functional and data access rights of the different user or user groups respectively in correspondence to their functional and organisational (structural) roles are defined and decided according to the rules agreed.

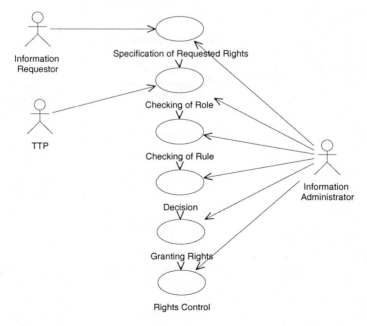

**Fig. 7.** Abstract Use Case "AccessControl"

### 4.3.7 Information Provision

According to the security policy and its access control decision, the permitted information can be provided and finally transferred.

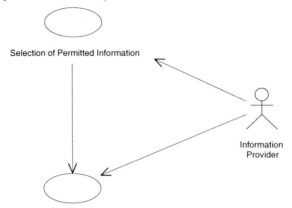

**Fig. 8.** Abstract Use Case "InformationProvision"

### 4.3.8 Information Transfer

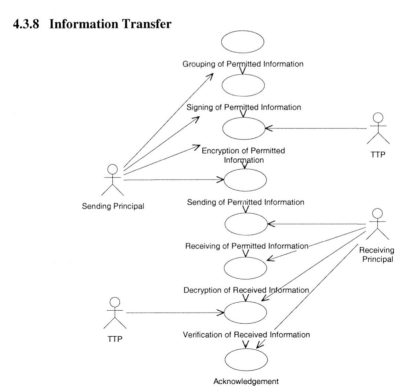

**Fig. 9.** Abstract Use Case "InformationTransfer"

The abstract use case "InformationTransfer" is defined in an very generic way also including the record of information and its transfer between user and system. Therefore, both application and communication security services dealing with integrity, confidentiality, and accountability (in the context of communication security also dealing with non-repudiation of origin and receipt) are reflected in the model presented.

To fulfil the policy agreed, beside the users also the information has to be classified and grouped.

## 5 The Layered Security Model

Based on the definition of a general security model within the „ISHTAR" project funded by the European Commission in the Fourth Framework "Telematics Applications Programme" context, a layered extension of this model has been developed. It allows the selection of appropriate concepts-services-mechanisms-algorithms-data relationships according to the analysis and design results for the health information system under consideration [6, 7]. Interacting with the UML approach results partially presented in this paper, the complex and generic methodology supports the comprehensive analysis, design and implementation of secure health information systems. Within the EC „TrustHealth-2" project, the methodology developed is also successfully used by the demonstration sites of the different participating countries. The complete results will be presented on other places [7].

## 6 Acknowledgement

The authors are in dept to the European Commission and to the Ministry of Education and Science of the German Federal State Saxony-Anhalt for funding as well as to the partners of both the „ISHTAR" and the „TrustHealth-2" project for kind co-operation.

## References

1. Blobel, B.: Clinical Record Systems in Oncology. Experiences and Developments on Cancer Registries in Eastern Germany. In: Preproceedings of the International Workshop "Personal Information - Security, Engineering and Ethics" pp 37-54, Cambridge, 21-22 June, 1996, also published in: Anderson, R. (edr): Personal Medical Information - Security, Engineering, and Ethics. Springer, Berlin (1997) 39-56
2. Blobel, B, Bleumer, G., Müller, A., Flikkenschild, E., Ottes, F.: Current Security Issues Faced by Health Care Establishments. Deliverable of the HC1028 Telematics Project ISHTAR, October 1996
3. Blobel, B. and Pharow, P.: Results of European Projects Improving Security of Distributed Health Information Systems. In: Cesnik, B, McCray, A.T., Scherrer, J.-R. (eds.) MEDINFO '98. IOS Press Amsterdam, Berlin, Oxford, Tokyo, Washington DC (1998) 1119-1123

4. Blobel, B., Pharow, P.: Security Infrastructure of an Oncological Network Using Health Professional Cards. In: Broek, L. van den, Sikkel, A.J. (eds.): Health Cards '97. Series in Health Technology and Informatics, Vol. 49. IOS Press, Amsterdam (1997) 323-334
5. Blobel, B., Pharow, P., Spiegel, V.: Shared Care Information Systems Based on Secure EDI. In: Moorman, P.W., Lei, J. van der, Musen, M.A. (eds.): EPRiMP – The International Working Conference on Electronic Patient Records in Medical Practice. IMIA Working Group 17, Rotterdam (1998) 164-171
6. Blobel, B., Roger-France, F.: Healthcare Security View Based on the Security Services Concept. ISHTAR Project HC 1028, Deliverable, August 1998
7. Blobel, B., Roger-France, F., Pharow, P.: A Systematic Approach for Secure Health Information Systems. (submitted to the International Journal of Medical Informatics)
8. Committee of Ministers: European Recommendation (Draft) No. R(96) of the Committee of Ministers to Member States on the Protection of Medical Data (and Genetic Data). CJ-PD (96). Strasbourg (1997)
9. Council of Europe: Directive 95/46/EC on the Protection of Individuals with Regard to the Processing of Personal Data and on the Free Movement of such Data. Strasbourg (1995)
10. Laske, C.: Legal Issues in Medical Informatics: A Bird's Eye View. In: Barber, B., Treacher, A. and Louwerse, K. (eds.): Towards Security in Medical Telematics – Legal, and Technical Aspects. Studies in Health Technology and Informatics, Vol. 27. IOS Press, Amsterdam (1995) 53-78
11. Eriksson, M., Penker, S.: UML Toolkit. Wiley Computer Publishing, New York (1998)

# 3DHeartView: Introducing 3-Dimensional Angiographical Modelling[1]

The 3DHeartView team[2]

E-mail: 3dheartview@hpcl.uoa.gr

**Abstract.** The 3DHeartView project produced an advanced prototype of a system capable of modelling 3-dimensional cardiac structures based on a single sequence of clinical angiographic X-rays. That can be used as an add-on to any modern digital angiographer. The system's accuracy has been measured and our approach validated with static and dynamic phantom object studies. The prototype has been in used many real-patient cases in a clinical routine environment. In this paper we present the main features of the system, the methodology of 3D angiographical modelling, and certain testing and operation results. We also describe the application of HPC technology in the modelling process.

## 1 Introduction

Heart diseases are the most severe illness in industrial countries. Statistically, about 50% of the population will suffer from cardiovascular diseases (followed by cancer with 20%). For the majority of clinical cases (coronary stenosis, congenital heart defects, etc.) X-ray angiography is the diagnostic method of choice for a variety of reasons (accuracy, high temporal acquisition, relatively low equipment cost).

Contrary to the majority of similar diagnostic techniques (Magnetic Resonance Tomography, Computer Tomography, Cardiac Ultrasound) angiography has not yet advanced into the area of 3-dimensional reconstruction and visualisation. The reasons lie with the empirical nature of the diagnostic process: The acquisition of angiographic images is a manually driven procedure aiming to provide the cardiologist with an ad-hoc view of the cardiac structure of interest. The procedure can not be repeated since X-rays are harmful to the patient, the cardiologists and the operators of the angiographic equipment. As a result, the angiographic sequences produced do not provide well ordered quantities of spatial information that could be combined in order to create 3D models of the cardiac structures of interest. This is the reason most systems that attempt 3D angiographic modelling, either depend on a

---

[1] This work was partly supported by the European Commission under the ESPRIT program (project 3DHeartView - 24484).

[2] Technical project manager: Dimitris Dimitrelos (AHPCL), Lambis Tassakos, Nikos Papazis (Skoutas SA), Georgios Sakas, Jürgen Jäger (ZGDV), Hans Gerd Kehl (WWU-Münster), Zenon Kyriakides, Theofilos Kolettis, Katerina Karaiskou (OCSC), Andy Marsh, Kostas Delibasis, Christian Michael (NTUA-ICCS).

priori information about the model, or require extensive user interaction during pre-processing of the images and/or modelling. Overall, 3D modelling from angiographic images can be classified as a highly complicated and computationally intensive procedure.

The 3DHeartView project produced a system that can produce 3D models of several cardiac structures, based on a single sequence of 2D X-ray angiographic images. The result is produced with minimal user interaction and with no assumptions about the geometric properties of the model or any usage of a priori information. The 3-dimensional model is produced in a few seconds and is then directly visualised by a multi-parametric 3D visualisation tool, which allows the user-cardiologist to manipulate and measure it in a variety of ways. One of the most important aspects of the system is that it allows fully automatic measurement of the volume of specific heart chambers, a measurement that up to now was done using unsafe geometric assumptions. Our system can model up to 10 different phases of the cardiac cycle, thus extending its functionality into the 4-dimensional space (3 spatial dimensions + 1 temporal).

The 3DHeartView prototype has been used in a large number of clinical cases and has proved its medical added value. Some of the clinical cases it was applied on are:

1. Congenital heart defects. To estimate the severity and advance of the patient's disease, the spatial heart shape the blood volume in both main chambers and the fraction of ventricle volume from diastole to systole are important factors. All 3 of these parameters are adequately generated or calculated by the 3DHeartView system. The accuracy of the calculated volume was evaluated by modelling static or dynamic phantom objects of known volume and then measuring the system produced model's volumes. Our studies confirmed an accuracy of 93-97%.

2. Myocardial Infarction. After a myocardial infarction, a 4D model of the heart chambers can be extremely useful in detecting diskinetic or akinetic regions of the interior chamber wall. The 4D model of the chamber produced by the 3DHeartView system allows the cardiologist to view a fully pulsating heart over the entire cardiac cycle from an arbitrary viewing point, or freeze it at any given instance of the cycle.

In order to make the system useful in everyday hospital routine, the modelling time should be decreased to clinically acceptable levels (less than 2 minutes per cardiac phase). To achieve this we employed HPC technology in the form of low-cost off-the-shelf PC components integrated into a housing that could be easily accommodated in an average cath-lab.

The system was developed within the 3DHeartView project by the Computer Graphics Centre (ZGDV e.V.) in Darmstadt, Skoutas SA and the Athens High Performance Computing Laboratory (AHPCL). It was also tested and evaluated under clinical routine conditions by two separate clinics (the department of Paediatric Cardiology of the Westfälische Wilhelms-Universität Münster and the $2^{nd}$ department of cardiology of the Onassis Cardiac Surgery Centre in Athens) and an institute (Institute of Communications and Computer Systems - Athens).

## 2 Methodology

### 2.1 Definition, Differentiation, and References

In general the term 3D-image reconstruction describes a method of extracting implicit object information to an explicit object description. This means the shape information about the object of interest, which is coded in image density values in sequences of 2-D projected X-rays, is translated to a 3D object boundary description structure, e.g. into a spatial image matrix.

In our case the input data are radiographic projections, where the opaqueness of all anatomic structures is accumulated along the X-ray beam – no structures are hidden but superimposed –, there is a need for different X-rays form various angle of vision to separate object depths. Using this kind of input data 3D reconstruction methods are divided in two main groups: density-based approaches like back-projection and edge-based approaches as active contour segmentation. The visualisation receptively its rendering technique (e.g. ray casting or triangle shading) depends on the different types of output data formats, a discrete volume or a polygonal mesh. Most of the edge-based algorithms additionally use a priori information about the object of interest, e.g. in form of a shape template.

Back-projection is used in some types of computer tomography (e.g. CT or SPECT). Johann Radon was the first 1917 who published the mathematical theory of back-projection and introduced his so-called radon-transformation [1]. Feldkamp *et al* extend this to a practical cone-beam variant [2] respecting the cone-shape of the acquiring beam geometry. Beside the analytical there exists also an algebraic approach [3], where the result will be iteratively improved by projecting, comparing with the origin frames, and back-projecting a correction term upon the previous result. And finally, the back-projection can be calculated voxel- or pixel-driven. In the project 3DHeartView we are using a voxel-driven analytical cone-beam back-projection approach for 3D modelling the heart ventricles from rotational angiography.

### 2.2 The Reconstruction Pipeline

The whole process of 3D modelling in cardiology in the 3DHeartView prototype can be explained in a five-step reconstruction pipeline (see figures 1 to 4):

At the start the clinical image acquisition is performed. While a patient is connected to an electrocardiograph, a cardiologist introduces a hollow catheter into the heart chamber of interest. After this preparation the cardiologist simultaneously injects contrast agent through this catheter, radiates the patient with a bi-plane angiographic device, and rotates the gantries around the patient's torso. The X-ray image sequences are stored in a digital image archive medium (e.g. CD-ROM) in DICOM-3 format and act as the input of the next pipeline step.

In the second step the digital X-ray frames are pre-processed on a PC system. For generating 3D models of the cardiac structure the system has to acquire some attributes of each X-ray frame: the beam-geometry, the gantry rotation angle, and the cardiac phase. For example, after labelling a frame sequence to a calibration acquisition, which was acquired under the same operational device settings (i.e. camera distance, magnification factors), the system knows the X-ray beam-geometry (i.e. pixel size, ISO-centre location). After the labelling of the cardiac cycle for instance, the system can group frames showing the heart in the same cardiac phase and passes them to the next pipeline step. With the 3DHeartView prototype this labelling task can be done semi-automatically by previewing the patient's ECG curve and some intensifier information, and offering interpolation features.

The third step is the 3D modelling itself, the execution of the cone-beam back-projection algorithm for each frame group once. This step needs the highest computational effort, so it is executed on a small-scale parallel system. The algorithm projects all frames back into a 3D-image space formed as a stack of pixel slices. The images are smeared back along a large number of virtual rays into the volume grid following the perspective and rotation angles, thus leading to higher accumulations at locations where all X-ray projections show higher object density.

The forth step consists of the visualisation of the back-projected slice stack. A volume rendering like the minimal intensity projection (MinIP) can be immediately performed on the volume data sets. For surface rendering, boundary features like gradients have to be calculated on the volumes before. Both rendering modes can be arbitrary mixed in the 3DHeartView software.

The last step in the reconstruction pipeline is the volume calculation in space. Since a ventricle boundary is already found by applying the gradient filter in the previous step, there is only a need for separating the ventricle edges from other structures by using a 3D region-growing algorithm, which flood fills the ventricle and counts the interior voxels. As the contrast agent in the blood is also visible in the aortic arch or others the user has to draw a cut to adjoining cardiac spaces and click into the object of interest.

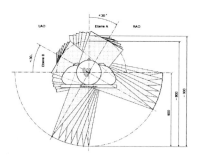

**Fig. 1.** Example of an acquisition geometry of a bi-plane rotational angiography

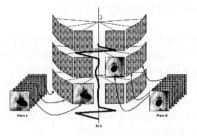

**Fig. 2.** Labelling of the cardiac phase and the gantry rotation angle

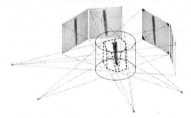

**Fig. 3.** Back-projection of a frame group

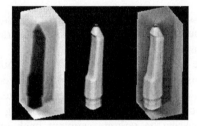

**Fig. 4.** Visualisation of the result as volume rendering (left), surface rendering (middle), and a mixture (right)

## 3  Parallelisation Aspects

The modelling time of the 3DHeartView methodology should not exceed a few tens of seconds in order to be applicable for everyday medical practice. Therefore we decided to employ HPC technology in order to make the system response time acceptable. Furthermore the parallel system should be easy to install and accommodate in a hospital environment, e.g. produce low level noise and have small dimensions, should exhibit a robust and reliable nature of operation, should be fast and at the same time relatively inexpensive.

## 3.1 Hardware and Software Platform

CCi-D by Parsytec GmbH was chosen to be the underlying hardware platform of our system. It comprises of 4 dual Pentium Pro nodes which run the Windows NT operating system, along with other hardware and software modules necessary for this kind of application. All nodes are interconnected via a 100 MBPS Ethernet connection that imposes certain restrictions in the performance of the overall system. The fact that the system is configured with off-the-self industrial PC technology gives us the chance to adopt to future improvements.

The nature of our hardware and the desire to achieve portability of the implementation, compliance with the state-of-the-art standards, and a high degree of efficiency led us to choose the MPI standard for message passing as the underlying software platform. In particular we used the *WMPI* implementation of MPI, a Win32 implementation of the MPI standard, developed at the *Departamento de Endenharia Inmformatica, University of Coimbra Portugal* [5], that proved to be more reliable and of higher performance than other available Windows NT implementations at that time [6].

## 3.2 Architecture

The parallel modeller consists of two components, the *volume-writer* that deals with I/O and initialisation operations and the *partial-modeller* that carries out the parallel computation of the 3D model.

After invocation, the volume-writer reads the 2D input x-ray images and then broadcasts the data to instances of the partial-modeller over the 100 MBPS Ethernet module. Upon reception of the data, each instance computes the appropriate segment of the 3D volume, which is uniformly distributed among them. When an instance completes its task, it passes the volume segment back to the volume-writer. The volume-writer receives each segment and forms the total 3D volume. In figure 5, aspects of the data parallelism technique employed in the described methodology are depicted.

The algorithm makes no assumptions about the existing number of instances of the component partial-modeller. Each instance first determines the number of its peers and then which segment of the total 3D volume to compute. Thus the algorithm is scalable, in that it can deal with different hardware configurations without the need of alterations at source-level. In the case of changes in the hardware configuration the only required change in the 3DHeartView parallel system would be in a few lines in the configuration file that WMPI uses to determine the architecture of the parallel application and consequently to spawn it [4].

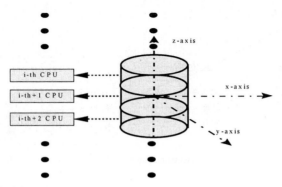

**Fig. 5.** Data partitioning

## 3.3 Results

The performance tests were conducted at the premises of the Athens High Performance Computing Laboratory and of Skoutas SA. The experiments were carried out using real patient data, as well as validation (phantom) object data.

It is important to state that originally the 3D modelling procedure took over 1 hour to complete on an average workstation. Aggressive code optimisation that was carried out during the project resulted in a decrease in the sequential modelling time to 15 -20 minutes. Thus a better basis for the parallelisation task was accomplished.

Times are measured in seconds and volume resolutions in units per axis. To ensure statistical validity for each measurement 10 runs were executed. The maximum and minimum values of each set were discarded and the remainder values were kept. Each measurement was repeated 5 times and the overall mean value was computed. For each experiment a data set of 9 2D x-ray images was used, which is an average number of frames for clinical practice.

In the context of the presented work, we were interested in developing a system that would fulfil the needs of the collaborating medical doctors. The main aspect, that needed the utilisation of HPC technology, was the overall system response time, which the physician is witnessing. The latter is the main feature that would prove the proposed methodology applicable in every day clinical practice or not. *The metric of the efficiency of the parallelisation strategy should be the decrease of this delay*, rather than other features.

The current system implementing the 3DHeartView methodology is supporting volume resolutions from 61 up to 255 units in each axis. Resolutions over 150 units per axis are the most suitable for medical applications. Consequently the measurements were done on representative volumes, i.e. 166, 200 and 255 voxels per axis.

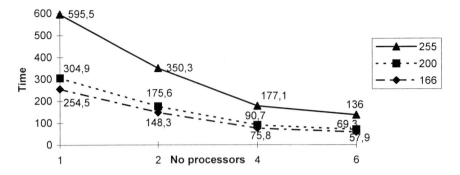

**Fig. 6.** System response time relative to volume resolution and number of CPUs

In figure 6 the system response time in relation to the existing number of CPUs and the volume resolution is depicted. The horizontal axis holds the number of CPUs and the vertical axis holds the system response time in seconds. There are three curves corresponding to volume resolutions of 166, 200 and 255 voxels per axis respectively. It is highly evident the decrease in overall system response time through the increase in the number of CPUs used.

## 4 Medical Impact

### 4.1 Applications

The appreciation in value for medical diagnoses by using the 3DHeartView prototype can be seen fore-fold:

At first, the surface visualisation of a 3D model of a cardiac structure eases the imagination and gives a spatial impression in a more natural way of seeing objects as the radiography does. With the possibility to rotate, zoom, and cut the 3D model in arbitrary manner, interactively and in real-time, it is possible to examine the shape and size of a pathological defect (e.g. an aneurysm) in space.

A second application is reached by the virtual X-ray visualisation mode. With this it is possible to rotate the model of a pathological deformation (e.g. a vessel constriction) to an optimal point of view in order to acquire the scene with these rotation and angulation angles with the angiographical device again.

Further the quantitative measurement of a cardiac enlargement (i.e. dilatation and hypertrophy), that often indicates indirectly the severeness of a disease, supplies another aspect on which the cardiologist can ground their diagnosis (e.g. the interior blood volume in a ventricle).

The forth and advanced use-case is the possibility of visualising a 3D motion study. After a cardiac infarction the heart muscle often shows akinetic contraction. This can be investigated with a qualitative dynamic replay, where the user still has interactive influence on all visualisation parameters (e.g. the viewing angle).

## 4.2 Validation

The system passed some phantom studies to estimate the accuracy of the overall system. The volume calculation on convex static phantoms (few post-mortem casts of ventricles and laboratory tubes filled with contrast agent) has an error of less the 5%. Concave phantoms will be overestimated dependent from the size of the dent. Dynamic phantoms (an artificial heart assist system) leads to an error of 5 to 10%. Patient ventricles approximately 10%.

**Fig. 7.** Few concave static phantoms

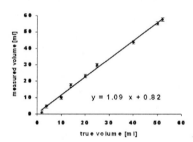

**Fig. 8.** Measurement results of these phantoms

## 5 Conclusion & Future Work

The 3DHeartView system is an advanced prototype for the production of 3D models of cardiac structures based on a single standard angiographic X-ray image sequence. The system has been proved to reconstruct accurately complex volumes and offers several advantages to the diagnostic procedure. Moreover, the system can be added to any modern angiographer, requires minimal user interaction, and, through the use of HPC technology, has extremely limited response times.

Our future work plans include 3D modelling of the coronary artery and the vessel tree. These structures are very small, compared to heart ventricles, and their relative pulse movement is extensive, making them an extremely difficult modelling target. However, an adequately accurate model of the vessel tree could be integrated with the angiographer gantry-movement and zoom mechanism and therefore minimise the number of injections during examination and potentially change the way angiographic examinations are done in the future.

# References

1. Johann Radon, "Über die Bestimmung von Funktionen durch ihre Integralwerte längs gewisser Mannigfaltigkeiten." Ber. Verh. Sächs. Akad. Wiss. Leipzig, Math-Nat. 69 (1917) 262-277.
2. L.A. Feldkamp, L.C. Davis, and J.W. Kress, "Practical Cone-Beam Algorithm," Optical Society of America, Vol. 1 (No. 6), pp. 612–619, 1984.
3. K. Mueller, R. Yagel, and J.F. Cornhill, "Accelerating the anti-aliased Algebraic Reconstruction Technique (ART) by table-based voxel backward projection," Proceedings EMBS'95 (The Annual International Conference of the IEEE Engineering in Medicine and Biology Society), pp. 579-580, 1995.
4. Home page of WMPI: http://dsg.dei.uc.pt/wmpi/intro.html.
5. Message Passing Interface Forum: MPI: A Message-Passing Interface Standard.
6. Home page of MPICH/NT: http://www.erc.msstate.edu/mpi/mpiNT.html.

# WWW Based Service for Automated Interpretation of Diagnostic Images: The AIDI-Heart Project

Mattias Ohlsson, PhD[1], Andreas Järund[2], Lars Edenbrandt, MD, PhD[2]

[1] Department of Theoretical Physics, Lund University, Sweden, mattias@thep.lu.se
[2] Department of Clinical Physiology, Lund University, Sweden
*http://ism.thep.lu.se/*

**Abstract.** This paper presents AIDI-Heart, a computer-based decision support tool for automated interpretation of diagnostic heart images, which is made available via the web. The tool is based on image processing techniques, artificial neural networks, and large and well validated medical databases. The performance of the tool has been evaluated in several recent papers and the results show the high potential for the tool as a clinical decision support system. The tool has now been integrated into a WWW environment for easy access and operation using your favourite web-browser. This first version of AIDI-Heart is evaluated by three hospitals, two in Denmark and one in Sweden.

## 1 Introduction

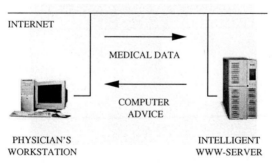

**Fig. 1.** The information flow between the physician and the server.

The practice of medicine is to a great extent an information-management task. A physicians decision making is based upon expert knowledge, information from the individual patient, and information from many previous patients, the latter known as experience. Decision making is often very difficult due to the fact that not only is the required expert knowledge in each, of many different, medical fields enormous, and growing daily, but also, the information available for the individual patient is multi-disciplinary, imprecise and very often incomplete. The interpretation of all data available from a patient is made by a physician, who may have a limited knowledge, and experience, in analysis of some of the data. In this situation physicians can consult a more experienced colleague at the clinic.

With the aid of modern information technology, it is also possible for a physician to contact an experienced colleague at another hospital and transfer data to him or her. For example, a physician at a remote hospital can send a diagnostic image to an experienced physician at a university hospital. Thereafter they can discuss the image over the phone. A problem with this technique is that experienced physicians are not always available when the advice is needed. Therefore computer-based decision support systems available via the web is an interesting alternative. With this technique decision support is available 24 hours per day, 365 days per year.

## 2 Automated Interpretation of Heart Images

### 2.1 Heart Images

The blood flow to the heart can be examined by injecting a radioactive tracer (technetium-99m sestamibi) and thereafter acquiring scintigraphic images with a gamma camera. The patient is examined both at rest and after exercise and the results are presented as two so called bull's-eye images, see fig. 2. Dark areas

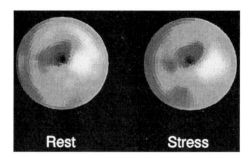

**Fig. 2.** Heart images in the form of so called bull's-eye images, one obtained at rest and the other after exercise.

in these images represent parts of the heart with reduced blood flow, generally caused by coronary artery disease (CAD). The interpretation of these images is a pattern recognition task. The physician must rely on his or her experience rather than on simple rules of how to interpret the image. A less experienced physician can benefit from a computer-based decision support system when a more experienced colleague is not present. Also a very experienced physician can use such a system for a second opinion.

## 2.2 Automated Interpretation

Automated interpretation of heart images using artificial neural networks was introduced a few years ago [1–4]. It was shown that the best neural networks detected CAD as good as or even better than experienced observers [4]. In clinical practice this type of intelligent computers will not replace, but assist the physicians by proposing an interpretation of the studies. It was therefore of interest to find that physicians interpreting heart images benefit from the advice of neural networks measured both as an improved performance and a decreased intra- and inter-observer variability [5]. It has also been shown that these neural networks can maintain a high accuracy in a hospital separate from that in which they were developed [6].

## 2.3 Databases

The performance of decision support tools such as artificial neural networks depends largely on the size and the composition of the training databases. Therefore, four different sets of data were pooled to a training database consisting of heart images obtained from 441 subjects. The true diagnosis (CAD or not) for each of the patients, was obtained by performing an invasive and more expensive examination, coronary angiography. The healthy subjects all had a $< 5\%$ likelihood of CAD. A total of 221 patients had signs of CAD while the remaining 220 patients and healthy subjects were defined as normal.

The following four data sets were used in the first version of AIDI-Heart:

- 135 patients examined at the University Hospital in Lund, Sweden during the period from November 1992 to October 1994 [4–6].
- 110 patients examined at the University Hospital in Lund, Sweden during the period from June 1995 to May 1997.
- 68 patients examined at Rigshospitalet, Copenhagen, Denmark during the period from December 1991 to March 1994 [6].
- 128 healthy subjects in the Copenhagen City Heart Study Denmark [7].

## 2.4 Image Processing

Artificial neural networks are used as the computer-based classification tools for the heart images. The size of the images are $17x64$ pixels and an image reduction method is needed in order to limit the complexity of the neural network. A Fourier transform technique was used as follows: Each of the two $17x64$ images was expanded by mirroring about row 17, and then discarding the last row (i.e. the first row of the succeeding Fourier period), to produce $32x64$ matrices.

The two 32x64 matrices were input as the real and imaginary parts of a complex 32x64 matrix in a fast Fourier transform [8]. A selection of 30 values constituting the real and imaginary part of the coefficients for 15 of the lowest frequencies were used as inputs to the neural networks. For more details see [4].

## 2.5 Artificial Neural Networks

A 3-layer perceptron neural network architecture [9] was used. The input consisted of 30 Fourier frequency components and it was found that 4 hidden neurons was enough for this application. The single output neuron encoded a possible CAD (1) or normal (0) subject. A weight decay term was used during the training to further regularize the network in order to optimize the generalization performance.

The performance of the artificial neural networks was evaluated using a cross-validation procedure. The output values for the test cases were in the range from 0 to 1. A threshold in this interval was used above which all values were regarded as consistent with CAD. By varying this threshold a receiver operating characteristic curve was obtained. Areas under these curves were calculated as measures of performance. Figure 3 shows the ROC-curves for the territories LAD and RCA/LCX (see below for an explanation). Both territories had an area under the ROC-curve of 0.81.

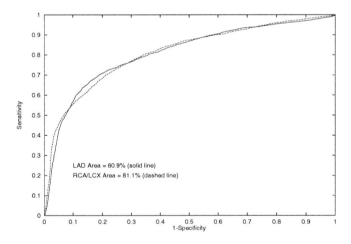

**Fig. 3.** Receiver operating characteristic curves for the LAD and RCA/LCX territories.

## 2.6 Presentation of Computer Advice

The AIDI-Heart interprets all images regarding the presence/absence of CAD in two vascular territories of the heart, the LAD and the RCA/LCX territories.

Therefore two different sets of neural networks were trained to detect CAD, one in the LAD territory and one in the RCA/LCX territory. The output values of the networks were in the range from 0 to 1. Three thresholds were used to transform the output values into four different statements. For example, output values below the lowest threshold were regarded as "definitely not CAD". The advice of AIDI-Heart was presented to the physicians as two statements, one for each of the two territories, together with the corresponding output values. The output value can be regarded as an estimate of the probability for CAD.

## 3 The WWW Service

The functionality of AIDI-Heart is based on a client-server paradigm via Internet and integrated into a WWW environment. The server is able to accept queries from clients and return an advice based on a computer-based decision support tool. The client can, besides sending images, also return feedback concerning the computer advice. Collecting and storing this feedback on the server will enable us to further enhance the decision support tool and thereby decrease the number of misinterpretations.

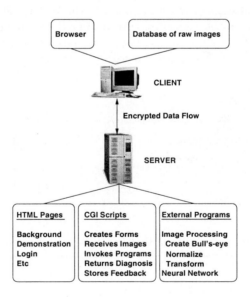

**Fig. 4.** Overview of the AIDI-Heart server for automated interpretation of heart images.

Figure 4 shows an overview of the functionality of the server. A typical session can be described as follows: The client, i.e. a physician with a workstation connected to Internet, starts his favourite browser and opens the home-page of

AIDI-Heart. This page is maintained by the server, physically located in Lund, Sweden. When the client requests the page containing the actual user interface for the service, the server demands a correct user name and password from the client. If the login succeeds a CGI-script is launched at the server. This script presents a form to the client to fill out with the path and filename of the raw image residing on the client computer. The heart image is then sent over the Internet in an encrypted state (40 bits) to avoid patient data to be read by a third party. The script, still running on the server, receives the image and starts manipulate it with the help of several external programs. When all programs are executed, a diagnostic advice will appear in the clients browser. This answer contains the evaluation result from the decision support tool (i.e. the artificial neural network) along with bull's-eye images for a visual feedback. The client can also return feedback which can be used for evaluation of the system.

### 3.1 Demonstration

This demonstration will show how the AIDI-Heart service works. There are three steps to complete in order to obtain an interpretation of the heart images.

**Login** Start your favorite browser and point to the WWW-address for the AIDI-Heart service. You will be presented with a "LOGIN" button. After clicking it you have to enter your personal username and password (assigned by ISM) in order to access the actual send and answer page (figure 5).

**Fig. 5.** Username and password form that has to be completed in order to access the full service.

**Locate Data** You must locate the file(s) containing your data. The supported file format is currently the Interfile Version 3.3 format. There are two possibilities here:

1. One file contains both the header and the data. The filename usually end with .ifl. This is a safe way since the patient name and patient data cannot be mixed up.

2. Two separate files, one usually ending with .ihd (or .hdr) contains the header and one ending with .img (or .dat) which contains the raw image data. It is important here to make sure that the header and data belong to the same patient.

In order to send the data file(s) you will be presented with a filebrowser interface, figure 6.

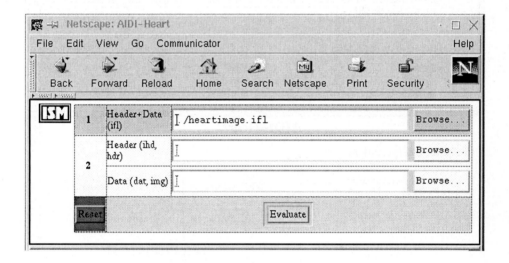

**Fig. 6.** The interface used to send the data file(s).

**Send Data** Simply click the "Evaluate" button and the chosen file(s) will be sent to the server. Note! that the files being sent to the server will be encrypted because of the nature of the information sent.

The interpretation, figure 7, will be presented as one out of four different statements for each of the two vascular territories (LAD and LCX/RCA) together with the corresponding network output values. The bull's-eye images and some patient data are shown at the lower part of the figure. There are also the possibility to send feedback to the server regarding the diagnostic interpretation.

### 3.2 System Evaluation

During the winter 1998/1999 the system is evaluated by three university hospitals, two located in Denmark and one in Sweden. The evaluation process is primarily focused on the user interface and the security issues. The accessibility of a decision support system is very important. Most physicians are non-IT

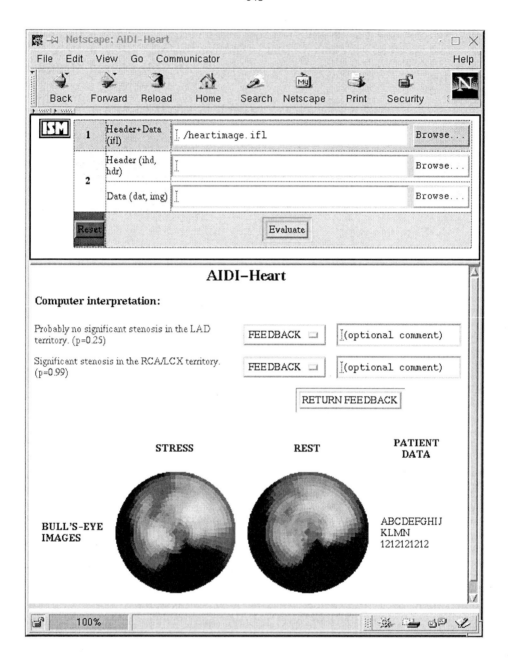

**Fig. 7.** The output from the AIDI-Heart. For these heart images the neural network detected a significant stenosis (CAD) in the RCA/LCX territory, while the upper region (LAD) appeared to be normal. The client can return feedback and some additional comments using the FEEDBACK-buttons. The lower part of the figure shows the bull's-eye images.

specialist who only will adopt a system which is perceived as being easy, and not time consuming, to use. Another important issue that is evaluated is security. The data transferred from the client (physician) to the server includes data which is attributable to an individual patient. Therefore, techniques were developed to ensure data security at all times. These techniques include encryption of the data transferred.

## 4 Discussion

The are four main reasons to believe that the use of computer-based decision support systems in the medical field have the potential for rapid expansion within the immediate future.

1. The development of interfaces can make computer-based decision support systems easily available to non-IT specialists such as the physicians, by employing state of the art information transfer, and assimilation techniques via web-browsers and java-scripts. These techniques make it possible to reach many physicians world-wide, as well as facilitating distribution of the latest version of the software.
2. Emerging techniques such as Data Mining and Artificial Intelligence make it possible to develop more accurate decision support systems than would have been possible some years ago. It is crucial that different methods of data analysis and decision support system designs are studied in order to find optimal solutions for different diagnostic problems in various medical domains.
3. Vast medical databases in digital form can now be developed. These large information resources are important in order to create and control high quality decision support systems.
4. The infrastructure required for a widespread introduction of intelligent information systems, i.e. physicians workstations, networking, common standards for data-exchange, data storage, has been put in place in many hospitals, clinics, and surgeries, even in remote areas. Thereby, patient data, history, findings from physical examination as well as data from laboratory tests, and diagnostic imaging and so on can easily by used as input to an intelligent system.

## References

1. Fujita, H., Katafuchi, T., Uehara, T., Nishimura, T.: Application of artificial neural network to computer-aided diagnosis of coronary artery disease in myocardial SPECT bull's-eye images. J. Nucl. Med. **33** (1992) 272–276
2. Porenta, G., Dorffner, G., Kundrat, S., Petta, P., Duit-Schedlmayer, J., Sochor, H.: Automated interpretation of planar thallium-201-dipyridamole stress-redistribution scintigrams using artificial neural networks. J. Nucl. Med. **35** (1994) 2041–2047

3. Hamilton, D., Riley, P.J., Miola, U.J., Amro, A.A.: A feed forward neural network for classification of bull's-eye myocardial perfusion images. Eur. J. Nucl. Med. **22** (1995) 108–115
4. Lindahl, D., Palmer, J., Ohlsson, M., Peterson, C., Lundin, A., Edenbrandt, L.: Automated interpretation of myocardial SPECT perfusion images using artificial neural networks. J. Nuc.l Med. **38** (1997) 1870–1875
5. Lindahl, D., Lanke, J., Lundin, A., Palmer, J., Edenbrandt, L.: Computer-based decision support system with improved classifications of myocardial bull's-eye scintigrams. J. Nucl. Med. **40** (1999) 1-7
6. Lindahl, D., Toft, J., Hesse, B., Palmer, J., Ali, S., Lundin, A., Edenbrandt, L.: Inter-institutional validation of an artificial neural network for classification of myocardial perfusion images. (submitted)
7. Nyboe, J., Jensen, G., Appleyard, M., Schnohr, P.: Risk factors for acute myocardial infarction in Copenhagen. I. Heriditary, educational and socioeconomic factors. Eur. Heart. J. **10** (1989) 910–916
8. Press, W.H., Teukolsky, S.A., Vetterling, W.T., Flammery, B.P.: Numerical Recipies in C, Second edition. Cambridge University Press. (1995) 521–525.
9. Rumelhart, D.E., McClelland, J.L., eds. Parallell distributed processing. Volumes 1 & 2. Cambridge, MA: MIT Press. (1986)

# Decision Trees - A CIM Tool in Nursing Education

Peter Kokol[1], Milan Zorman[1], Vili Podgorelec[1], Ana Habjanič[2], Tanja Medoš[2], and Majda Brumec[3]

[1] Faculty of Electrical Engineering and Computer Science, University of Maribor
Smetanova 17, 2000 Maribor, Slovenia
Kokol@uni-mb.si
[2] University College of Nursing Studies, University of Maribor
Žitna 15, 2000 Maribor, Slovenia
[3] House of Elderly
2000 Maribor, Slovenia

**Abstract.** Better health care for all and an overall health care restructuring are two of the most important processes that will, in addition to many existent assignments, expose many new tasks and responsibilities for the nursing personnel. Each of these responsibilities entails information as input and also generates information as the output, and in the near future it is to be expected that most of the information processing including data gathering, decision making support, data mining and information generation will be performed by computers. To cope with this new situation, the nurses have to be armed with proper knowledge and thereafter the proper education will play a very important role. In this manner we started an EU project called NICE, which one of the main objectives is to develop new computer managed educational tools for nursing.

## 1 Introduction

Better health care for all and an overall health care restructuring are two among most important processes which should be introduced in the new millenium. These two processes will, in addition to many existent assignments, expose many new tasks and responsibilities for the nursing personnel providing in many different working environments like community settings, clinics, schools, workplace and hospitals. Each of these responsibilities entails information as input and also generates information as the output, and in the near future it is to be expected that most of the information processing including data gathering, decision making support, data mining and information generation will be performed by computers. To cope with this new situation, the nurses have to be armed with proper knowledge and thereafter the proper education will play a very important role. In this manner we started an EU project called NICE, which one of the main objectives is to develop new computer managed educational tools for nursing.

The aim of this paper is to briefly present the NICE project, and a tool developed by NICE, to support the education of nurses based on the concept of decision trees and automatic learning.

## 2 The NICE Project

As a response to the need for the nurses educated in nursing informatics we started an EU Project financed by Phare Tempus grant called NICE (Nursing Informatics and Computer Aided Education). The aim of the NICE project is:

The development and introduction of a new short cycle degree courses in Nursing Informatics at the university colleges in the partner countries. The project will also produce teaching materials, computer aided tools for nursing education, and establishment of a multimedia teaching laboratory enabling teleconferencing, telemedical applications and distant learning.

The NICE consortium consists of following institutions and countries:

- The University College of Nursing, Maribor, Slovenia, contractor
- The Faculty of Electrical Engineering and Computer Science, Maribor, Slovenia,
- coordinator
- Primary Health Center Adolf Drolc, Maribor, Slovenia
- The University of Klagenfurt, Klagenfurt, Austria
- Centre Hospitalier Universitaire de Nantes, Nantes, France
- The National Technical University of Athens, Athens, Greece
- Politecnico di Milano, Milano, Italy

The NICE project is currently in the final year and most of the main outcomes have been already achieved. The curriculum is agreed and will be given into the experimental use shortly. In addition, the drafts of the teaching materials have been produced, as have been plans for the classroom. Educational tools have been developed, and one of them the DecTree will be presented in following sections.

## 3 Computer Managed Tools for Nursing Education

Computer managed instruction (CMI) has been used in nursing education since the late 1960's [14, 15]. It is due to the accessibility and self paced format that CMI is very well suited for both students and practising nurses [14, 15], while learning can occur at the learner's own pace and time. In addition, CMI supports also continuing education and distant learning.

The early applications of CMI employed room sized mainframes at large institutions, but the proliferation of microprocessors in the 1990 expanded the depth and breadth of instructional computing [16, 17]. Recent studies [15, 18 - 20] show that students using CMI have better average examination scores, improved ability for critical thinking [16] and enhanced computer literacy, facilitated decision making skills and positively affected achievements [12]. In spite of these advantages still many students and faculty [23] are still reluctant to utilise CMI, but Haus [22] reports that the perception and attitude toward computer managed instruction positively changes after actual use of CMI software packages.

Athappilly [21] lists three benefits for using CMI and multimedia tools in the educational process:
1. quality multimedia presentation reduces the cost, in spite that initial investments are large the reduction of participants and instructors time are significant;
2. the effectiveness of the teaching and learning is improved because of greater motivation, retention, and mastery of learning;
3. production is improved because of increased satisfaction and enjoyment of learning.

## 4 Decision Trees in Education of Nurses

The majority of the current teaching tools for nursing education are based on the so called concept of Drill and practice [24], motivated by the research of Skinner [25]. The major advantage of such type of learning is the immediate feedback to a student. There is no waiting period for correction and therefore students do not practice their mistakes. But some researchers suggest that after the novelty effect of drill and practice wears off and the motivational power is lost. The wear effect can be overcome if the educational package is adaptive and can be individualized. This can be achieved with the use of artificial intelligence and automated learning, which in addition offers the possibility to analyze the mistakes and explain the problem to a student. Thereafter we decided to employ the concept of decision trees to improve the learning process in nursing education.

### 4.1 Decision Trees – a brief Overview

The use of techniques from artificial intelligence especialy machine learning is a common practice in various medical applications. Machine learning deals with the discovery of hidden knowledge, unexpected patterns and new rules. Therefor decision trees have been and are still being used to extract knowledge from data, in order to make decisions in cases, where explicit human knowledge cannot be used efficiently [26,27].

The algorithm for learning a decision tree is trivial and the representation of accumulated knowledge can be easily understood. Namely, the decision trees don't give us just the decision in a previously unseen case. They also give us the explanation of the decision, and that is essential in medical applications.

A decision tree is induced on a *training set*, which consists of *training objects* (*instances*). Each training object is completely described by a set of *attributes* and a *class* label (*category, outcome*). Classes are mutually exclusive, what means that the training object can belong to only one class. Attributes can be *continuous* (*numeric*) or *discrete*. Continuous attributes are not suitable for learning a tree, so they must be mapped into a discrete space. A decision tree contains *nodes* and *edges* (*links*). There are two types of nodes. Each *internal node* (*non-terminal node*) has a *split*, which *tests* the value of the chosen attribute for the training objects, that have come into this node and according to that splits the training set. Each internal node has at least two child nodes. *External nodes*, also called *leaves* or *terminal nodes*, are labelled with outcomes. Nodes (internal and external) are connected with edges. Edges are labelled with different outcomes of test, performed in the source node. Number of edges that come out of the node depends on the number of possible outcomes of the test.

Decision tree in figure 1 is built hierarchically from left to right. Leaves are marked with outcomes. Internal nodes are marked with attributes and have successors.

The procedure of constructing a decision tree from a set of training objects is called *induction* of the tree. In induction of a decision tree, we start with an empty tree and an entire training set. The procedure of induction is recursive and has four steps:

1. If all the training objects in the training (sub)set have the same outcome, create a leaf labelled with that outcome and go to step 4.
2. With the help of heuristic evaluation function find the best attribute out of all attributes that have not been used on the path from the root to the current node. Create an internal node with split based on the selected attribute. Split the training set to subsets.
3. For each subset of training objects go to step 1.
4. Go one level up in recursion.

Unlike to some other approaches the representation of a decision tree can be easily understood by a human. All tests in internal nodes of a tree can be determined, so the importance of attributes can be obtained from the decision tree. And that is the way to take advantage of the decision trees even without using them for their primary task – decision making.

This decision tree characterisation becomes next a basis for:
1. forecasting in which class an object previously unseen belongs; and
2. the hierarchical representation of the most important attributes of the concept being investigated.

According to above, decision trees can support the nursing education process in four ways:
1. to represent the knowledge and decision making as a simple two-dimensional hierarchical model;
2. to outline important factors needed for successful decision making
3. to enable a nurse to use the decision tree (in the paper form or as a computer program) to support their own decision making in new situations and with new cases
4. to construct the decision tree (using automatic learning) for their own cases, and then use the generated tree as described in topics 1 – 3.

## 5 A Sample Decision Tree

As an example of a theory presented above we generated a decision trees for deciding about the nursing diagnose called incontinence. The tree is shown bellow

```
NeuropathyPreventingTransmissionOfReflex
|
|____[0] LessenCapacityOfBladder
```

```
|   |
|   |____[0] SensoyCognitiveOrMobilityDeficits
|   |   |
|   |   |____[0] WeakPelvicMuscles
|   |   |   |
|   |   |   |____[0] DegenerativeChangesInPelvicMuscles
|   |   |   |   |
|   |   |   |   |____[0] NeurologicalImpairment
|   |   |   |   |   |
|   |   |   |   |   |____[0] Without
|   |   |   |   |   |
|   |   |   |   |   |____[1] Reflex
|   |   |   |   |
|   |   |   |   |____[1] UrinaryUrgency
|   |   |   |   |   |
|   |   |   |   |   |____[0] Total
|   |   |   |   |   |
|   |   |   |   |   |____[1] Urgent
|   |   |   |
|   |   |   |____[1] NeurologicalImpairment
|   |   |   |   |
|   |   |   |   |____[0] InabilityToReachToiletInTime
|   |   |   |   |   |
|   |   |   |   |   |____[0] HighIntraabdominalPressure(Obesity_GravidUterus)
|   |   |   |   |   |   |
|   |   |   |   |   |   |____[0] Reflex
|   |   |   |   |   |   |
|   |   |   |   |   |   |____[1] Stress
|   |   |   |   |   |
|   |   |   |   |   |____[1] Stress
|   |   |   |   |
|   |   |   |   |____[1] Age(years)
|   |   |   |   |   |
|   |   |   |   |   |____[60 .. 80] Reflex
|   |   |   |   |   |
|   |   |   |   |   |____[80 .. 100] Total
|   |   |
|   |   |____[1] HighIntraabdominalPressure(Obesity_GravidUterus)
|   |   |   |
|   |   |   |____[0] SpontaniusVoiding
|   |   |   |   |
|   |   |   |   |____[0] NoAwarenessOfBladderFilling
|   |   |   |   |   |
|   |   |   |   |   |____[0] Without
|   |   |   |   |   |
|   |   |   |   |   |____[1] Functional
|   |   |   |   |
```

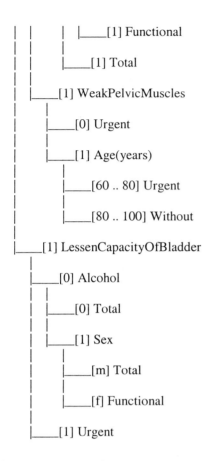

Some interesting facts have been found out. To decide about the various forms of incontinence the tree does not use all 44 different related factors, but just a subset of them (in average 10). This fact was not evident to students (not even to lecturers) by using traditional teaching methods, but has become very clear after using our approach – as a consequence the decision making has been much simplified.

Using the above decision tree we classified 12 test objects (Table 1.). 11 were classified correctly (92%), and just one classification failed.

**Table 1.** The use of a decision tree on the test objectsTable 1.

| Incontinence | Stress | Reflexive | Urgent | Total | Functional |
|---|---|---|---|---|---|
| Stress | 2 | | | | |
| Reflexive | | 2 | | | |
| Urgent | | | 3 | | |
| Total | | | | 2 | 1 |
| Functional | | | | | 2 |

# 6 Conclusion

To improve and better support nursing education with the use of Computer Managed Instruction concept we employed artificial intelligence – more specifically machine learning and decision trees. It is our belief that in such way the use of computers in nursing education can become still more successful.

# References

1. Checkland P. "Systems Thinking, System Practice". John Wiley & Sons, Chichester, (1981).
2. Checkland P and Scholes J. "Soft Systems Methodology in Action". John Wiley &Sons, Chichester (1990).
3. Kokol P, Brumec V, The first report about the NICE Project -Nursing informatics and computer aided education in Nursing informatics / edited by U. Gerdin, M. Tallberg and P. Wainwright, IOS Press : Ohmsha, (1997).
4. Gregory W J, Designing Educational Systems: A Critical System Approach. *System Practice* Vol 6 No 2.
5. Maruyama G, Application and Transformation of Action Research in Educational Research and Practice. *System Practice* Vol 9 No 1.
6. Bathany B. H (Ed.), Transforming Education By Design. *System Practice* Vol 8 No 3, Special issue on education.
7. Jeffcutt P (Ed.): Management Education and Critical Practice. *System Practice* Vol 10 No 6. Special issue on education.
8. Kokol P, Zazula D, Brumec V, Kolenc L, Šlajmer Japelj M: New Nursing Informatics Curriculum – an Outcome from the NICE Project in Proceedings of HTE 98 / edited by J. Mantas, University of Athens, (1998).
9. Hovenga E J S: Global Health Informatics Education, Proceedings of HTE 98 / edited by J. Mantas, University of Athens, (1998).
10. Ball M. J. et al. "Nursing Informatics". Springer, New York, (1995).
11. Hardin L, Patrick T.B: Content Review of Medical Educational Software Assesmets, Medical Teacher, Vol. 20, No. 3 (1998).
12. J. M. Belfry, A Review of the Effect of Computer Assisted Instruction in Nursing Education, Computers in Nursing, March 1988. (http://www.nursing.ab.umd.edu/students/~jkohl/cai.htm).
13. C. M. Curtis, M K Howard, A Comparison Study of the Effectiveness of Computer Assited Instruction and Traditional Lecture Techniques as Methods of Staff Development for Nurses, Research report, 1990. (http://www.widomaker.com/~ccurtis/cai.study.html).
14. J.E. Kohl, M. C. Su, Computer Assisted Instruction: Implications for Achevement and Critical Thinking, Research report, 1995.
15. T. Hebda, A Profile of the Use of CAI within baccalaureate nursing education, Computers in Nursing, November 1988.
16. M. Williamson, High teach training, Byte, December 1994,
17. M. Cambre, L. J. Castner, The status of interactive video in nursing education environments, FITNE 1993, Atlanta, 1993.
18. F. R. Jelovsek, Abebonojo, Learning principles as applied to computer – asssisted instruction, MD Computing, October, 1993.
19. K. K. Athappilly, C. Durben, S. Woods, Multimedia Computing. Multimedia Computing edited by S. Reisman, Harrisburg 1994.

20. K. Haus, Changes in Perception of Computer Aided Instruction, Research report, 1996 (http://parsons.ab.umd.cdu/~khaus/paper.html)
21. Gleydura, J. Michelman, C. Wilson, Multimedia Training in Nursing Education, Computers in Nursing, July, 1995.
22. Khoiny, F. Factors that contribute to computer assisted instruction effectiveness, Computers in Nursing, July 1995.
23. Habjanič A, Kokol P, Zorman M, Japelj M: CArE – A Software Package for Computer Aided Nurses Education, paper in preparation.
24. Conrick M.: Computer Based Education: more than just a package. AEJNE Volume 4 - No.1 October, 1998.
25. Skinner, B. 1953 Science of Human Behaviour. New York, MacMillan.
26. Kokol Peter, Završnik Jernej, Zorman Milan, Malčić Ivan, Kancler Kurt. Participative design, decision trees, automatic learning and medical decision making. In: Brender J... (ed.). Medical Informatics Europe `96, (Studies in health technology and informatics, Vol. 34). Amsterdam [etc.]: IOS Press; Tokyo: Ohmsha, 1996, pp./ A 501-505.
27. Stuart J.Russel, Peter Norvig, et al.: Artificial intelligence: a modern approach, Englewood cliffs, Prentice-Hall, ch.18, pp. 525-562, 1995.

# HealthLine: Integrated Information Provision to Telemedicine Networks

Yiannis Samiotakis[1], Sofia Anagnostopoulou[1], and Antonis Alexakis[1]

[1] ATKOSoft SA, 3 Romanou Melodou St, Marousi 151 25, Athens, Greece,
{yiannis, sana, aale}@atkosoft.com

**Abstract.** The needs of the developing, existing telemedicine networks have encouraged the HealthLine EU project to proceed in creating an Internet Corner where health professionals will be able to find the needed informative material and training on telemedicine issues. Information and expertise circulation among the professionals will be strengthened, results of other EU projects will be disseminated and suitable software links are being developed from existing telemedical software to the HealthLine Internet Corner, allowing on-the-job support of the working health expert. It is thus hoped that HealthLine will promote the success of telemedicine in Europe and facilitate health professionals in their work.

## 1 The Needs Observed

Current telematics facilities and technologies propose new forms of Health service provision, but in practical terms they are far from being widespread used. The *combination of computing and telematics in health* provides an increasingly powerful solution to a variety of problems so far encountered in health care provision, yet its *use is limited*.

Telemedicine networks are starting to spawn, however their success is still restricted because of the difficulties both health carers and general public have in getting acquainted, accepting and learning to use the new tools. By studying the expansion of the existing networks, it becomes apparent the users of such systems need to be instructed new ways of conducting their business, of taking advantage of the offered services and perceiving health care provision.

HealthLine aims to support the telemedicine networks
- by uniting results (content, know-how, technology) of existing projects on relevant issues under a common umbrella,
- by providing the necessary high-quality information and training courses on telemedicine issues for health-professionals,
- by implementing the necessary software to assist in pan-European information dissemination,
- by implementing the necessary links from telemedical software to the information repository created.

Web technology is a suitable mean to reach health professionals working with telematics tools, thus the core of HealthLine is an *Internet Corner*, available to all users with Internet access. It is hoped that the project will contribute towards assisting telemedicine in Europe to be strengthened.

## 2  Description of HealthLine – the Suggested Solution

*HealthLine*[1] *is an Internet Corner which aims to the strengthening of telemedicine networks*, by providing:
- Suitable training on telemedicine related issues for health professionals,
- Information dissemination services,
- The necessary tools for the health-professional to acquire on-the-job support for his/her work,
- and Information provision services (for example Articles, Discussion Groups, Directories).

*Four independent sites* are being built (Ireland, Italy, Sweden, Greece), all of them under the same quality assurance and operational/functioning rules. Each site hosts the same set of services (different, local, content) which includes training courses, articles, discussion lists, directories, organised library of hyper-links, help desk, career-opportunities section and other complementary services. The sites share part of the services (for example some training courses) and are constantly exchanging information so that a *cross-fertilisation* is made possible and an optimal functionality achieved.

### a)  On-line training of health professionals

The *training* provided includes multimedia on-line courses, to be used in conjunction with the articles, discussion lists and the various other services specifically targeted to the needs of the telemedicine network, as well as to telemedicine practices in general. Didactic methodology (approach and strategy), used is based on the results of the Nightingale project[2]. The content for the Health Informatic skills courses will be provided by the IT-EDUCTRA project[3].

---

1 The *HealthLine Internet Corner* is the set of services developed by the EU funded HealthLine project – "Securing the success of health Telematics projects implementation of Telehealth through information dissemination and training" (HC 4007). Under the Telematics Applications Programme (European Commission, DG XIII C/E), the project started in January 1998 and has a two-year duration. For more information, please contact the authors.
[2] *NIGHTINGALE* is a project in the planning and implementation of strategy in Training the Nursing Profession in using and applying healthcare information systems. For more information, please see http://nightingale.dn.uoa.gr/index2.htm .
[3] *IT-EDUCTRA* stands for Information Technologies Education and Training, a project within the Telematics for Health Care work sector of the Telematics Applications Programme (4th R&D Framework). For more information, please see http://www.fundesco.es/inteme/ .

**Table 1:** Training will be provided in three levels, the first two in English and shared between all HealthLine sites.

1. Health Informatics skills, including:
   - Basic computer skills, hardware and software, networks, security,
   - Patient record, virtual/distributed patient record, modelling information in health,
   - Computer-aided diagnostic support,
   - Global Information Society,
   - (and other themes)
2. Telemedicine network specifics (currently NIVEMES and RISE are covered – see later for details on these two networks)
3. Local site modules. These courses cover various local (national) needs, as for example the courses given to the nurses of the Irish Southern Health Board on caring of the elderly using telematic means.

### b) Information and Expertise Dissemination

Robust *information and expertise dissemination routes* have been established in each site (for example from the discussion lists to the articles service) and among the sites, to ensure optimum trans-European circulation of needed information. A central database, covering all sites, is storing summarised information for each posting (a posting can be an article, a directory, a discussion list, or other forms of data) publicised – thus supporting a powerful searching mechanism built specifically for the HealthLine sites.

### c) Links from Telemedicine Application Software to HealthLine

*Smart links have been developed from the telemedicine software* used, over the Internet (WWW) to HealthLine, thus allowing instantaneous access of the person using the software to the desired (depending on the actual screen/field of software currently chosen) to suitable material in HealthLine.

*Two existing telemedicine networks*, NIVEMES and RISE[4], *have already developed* strong links with HealthLine. *NIVEMES* is an international network of Health Service providers which offer telemedicine - teleconsultation services to remote, isolated places and to ship vessels for both routine and emergency situations. So far it has nodes in Ireland, Sweden, Greece, Portugal. *RISE* targets home-confined patients and the elderly by building the necessary network for the health professional visiting at home the beneficiary, to retrieve/store data from the central server. Existing nodes reside in Italy, UK, Greece, Ireland and Sweden. Both NIVEMES and RISE are currently being installed suitable smart links to the HealthLine, so that information/training will be available by "plainly right-clicking your

---

[4] NIVEMES and RISE are EU funded projects (Telematics Applications Program). For more information see http://www.atkosoft.com/ .

mouse"[5], over any Internet connection. The interested NIVEMES/RISE user will be able to find information on the subject of his/her interest easily, *gaining part of the services offered by HealthLine on top of the services offered by NIVEMES or RISE*. The topics covered will vary, including both medical and technical information. The search will be context-sensitive, depending on the current focused field/screen the user of NIVEMES/RISE is operating (for example the specified diagnosis, the use/updating of the available ICD code, and a variety of issues of interest to a health professional and in particular to a health professional using telemedical tools).

**Table 2:** On-the-job support offered: linking from NIVEMES and RISE, *to* HealthLine. *Context-sensitive* information/training will be provided.

|   | Action Conducted | Nature (Technical/ Medical) | From (application) sensitivity | To (entity linked) | Position (environment linked) |
|---|---|---|---|---|---|
| 1 | Electronic contacting | Technical | Current screen | Technical experts on application | NIVEMES or RISE national co-ordination centre |
| 2 | Electronic contacting | Medical | Current screen | Medical experts | NIVEMES or RISE national co-ordination centre |
| 3 | Acquisition of information on application | Medical | Current field (if not available, on current screen) | Related information pages | HealthLine knowledge basis |
| 4 | Acquisition of (tele)medical information | Medical | Current field's value (if not available, on current field) | Related information pages | HealthLine knowledge basis |
| 5 | Acquisition of training | Technical | Current field (if not available, on current screen) | Related training courses | HealthLine knowledge basis |

## 3 Equity of Information and Health Services Provision

HealthLine, by providing
- Training,
- Information dissemination routes (on national and international level),
- Links from NIVEMES and RISE (and later from other telemedical application), and
- Information provision services (for example Articles, Discussion Groups, Hyperlinks Libraries),

---

[5] G. Gianniotis, director of Agia Eleni Hospital in Piraeus-Greece, on his presentation of the Greek HealthLine site (Athens, 1998).

on telemedicine issues, aims to on-the-job support the working telemedicine expert.

Additional telemedicine systems are welcomed to join by providing themselves links to HealthLine and *inter-operate* as NIVEMES and RISE do; the existing linkage system is easily usable (added) to additional applications as well, and it is hoped that it will thus provide the same support as it does for NIVEMES and RISE.

The same holds for *other Internet sites* offering medical information - they can develop links to and from HealthLine, minimising thus the duplication of information and extending the limits of both sites. A *cross-fertilisation* is possible between the pool of information built in HealthLine and all health-related tele-services.

Through HealthLine and other similar activities, it is hoped that *health professionals* in remotely located areas *will achieve higher quality standards of information provision and training*, matching the ones available in an urban centre. By strengthening and supporting the health-professionals interested/using telemedicine, it is hoped that enhanced health care services will be provided. *Remotely located, mobile and home-confined populations* (as well as any other beneficiary of telemedicine) **will acquire improved services** since the tools the health professionals use will be better supported. The standard of living and health care of these persons will be improved, and equity of information provision and health services, irrespective of the location/country of residence, will be strengthened.

# Multi Modal Presentation in Virtual Telemedical Environments

Emil Jovanov[1], Dusan Starcevic[3], Andy Marsh[4],
Zeljko Obrenovic[5], Vlada Radivojevic[6], Aleksandar Samardzic[2]

[1] The University of Alabama in Huntsville, EB 213, 35899 Huntsville
[2] School of Electrical Engineering, POB 816, 11001 Belgrade
[3] Faculty of Organizational Sciences, Belgrade
[4] National Technical University of Athens, Athens
[5] Military Academy, Belgrade
[6] Institute of Mental Health, Belgrade
e-mail: jovanov@ece.uah.edu

**Abstract** - Telemedicine can be used to create virtual environments for the collaboration of patient, physicians, and medical staff. Multi modal presentation is increasingly used to improve human-computer interface. In this paper we present a multimodal interactive environment for EEG/MEG data presentation based on Internet technologies and a virtual reality user interface (VRUI). In addition to visualization, VRUI takes advantage of other input/output modalities such as audio. Animation on the 3D models is used to give insight into spatio-temporal patterns of activity. In addition, sonification is applied to emphasize the temporal dimension of the selected visualized scores. Our proposed approach requires only a standard Web browser to run the multi modal viewer which presents 3D EEG topographic maps using a VRML head model which is animated with Java applets.

## 1. Introduction

Conventionally healthcare services rely on the physical presence and collaboration of the patient, multiple physicians, and skilled medical stuff. Globalization and higher people mobility (business, tourism, etc.), has lead to fragmented care, delivered at scattered locations. However, the healthcare trend is now to provide the latest available information technology to all segments of the community including urban and rural sites. The increased performance of telecommunication infrastructures now facilitates real-time execution of remote applications, establishing a basis for a new medical discipline "telemedicine" [1]. Telemedicine can be used to create virtual environments for collaboration, independently of the participant's physical presence. Therefore, high-quality medical services now become available for distant patients and urgent cases.

The Internet and World Wide Web (WWW) as a global information infrastructure offer a low cost environment for telemedical applications [2]. At the present state of technology, Web based medical applications represent a natural way of creating interactive collaborative environments. Also, Virtual reality (VR) technologies can be used to shift the human-computer interaction paradigm from a graphical user interface (GUI) to a VR-based user interface (VRUI) [3-5]. The main characteristic of a VRUI

is multi-modal presentation. The most frequently used presentation modalities, in addition to visualization, are acoustic and haptic rendering. Immersive environments are particularly appropriate to improve insight into complex biomedical phenomena.
In this paper we present the development of a telemedical environment for multimodal EEG visualization and sonification of brain electrical activity. We propose examination of raw or derived EEG data by using a set of Virtual Medical Devices (VMD). In telemedical applications a VMD has specialized I/O (at patient side), processing (at Web application server), and presentation (at physician side). A VMD allows different views of the same data set. In our example the supported views are either standard waveform electroencephalogram or animated 3D topographic maps of brain electrical activity.

## 2. Multi-modal presentation

The main issue of multi modal presentation is the design of a VRUI, having in mind the characteristics of human perception. Bernsen [6] proposes the model of human-computer interface with physical, input/output and internal computer representation layers [6]. A two step transformation process is then required for human-computer interaction. For the input it is abstraction and interpretation, and for the output representation and rendering.
Technology and tools for multi modal presentation are commercially available due to the progress of multimedia and VR hardware and software. However, multimedia and VR technology applied simply as a human-computer interface does not guarantee successful presentation.
Conventional applications use uni-modal presentation, minimizing the use of resources, to mediate the information. Simultaneous presentation of the same information in different modalities appears to be a loss of resources. However, our natural perception is based on redundancy. Redundancy in the human-computer interface should be realized using multi-modal presentation. Therefore the main issue in the design of multi-modal presentation is the level of redundancy. A low level of redundancy increases cognitive workload, while a high level of redundancy irritates the user. There is always however an appropriate measure of multi modal redundancy for a given application.
The visualization of large number of data streams leads to visual overloading, and therefore an additional sensory modality, such as acoustic presentation is required to introduce new data streams and increase human-computer communication bandwidth.

## 3. Sonification

Multi-modal data presentation is a complex problem, due to nature of cognitive information processing. Efficiency of sonification, as an acoustic presentation modality, depends on other presentation modalities. The most important advantages of acoustic data presentation are [7]:

- Faster processing than visual presentation.
- Easier to focus and localize attention in space (appropriate for sound alarms).
- Good temporal resolution (almost an order of magnitude better than visual)

- Additional information channel, releasing visual sense for other tasks
- Possibility to present multiple data streams

Disadvantages of acoustic rendering are:

- Difficult perception of precise quantities and absolute values.
- Limited spatial resolution.
- Some sound parameters are not independent (pitch depends on loudness)
- Interference with other sound sources (like speech).
- Absence of persistence.
- Dependent on individual user perception.

It could be seen that some characteristics of visual and acoustic perception are complementary. Therefore, sonification naturally extends visualization. The system must provide the ability to extract the relevant diagnostic information features. The most important sound characteristics affected by sonification are:

- *Pitch* is the subjective perception of frequency. For pure tones it is basic frequency, and for sounds it is determined by the mean of all frequencies weighted by intensity.
- *Timbre* is characteristic of instrument generating sounds that distinguishes it from other sounds of the same pitch and volume. The same tone played on different instruments will be perceived differently. It could be used to represent multiple data streams using different instruments.
- *Loudness* or subjective volume is proportional to physical sound intensity.
- *Location* of sound source may represent information spatially. Simple presentation modality may use *Balance* of stereo sound to convey information.

Early sonification applications have been mostly using the so-called "orchestra paradigm", where every data stream has been assigned its instrument (flute, violin, etc.). Data values are then represented by notes of different pitch. The main advantage of this approach is the possibility to apply standard MIDI support, using a system Application Programming Interface (API). Unfortunately, the proposed approach often leads itself to cacophony of dissonant sounds.

## 4. Multi modal viewer - mmViewer

One of the most important challenges of contemporary science is how the brain operates. Insight into brain operation is possible using the electrical (EEG) or the magnetic (MEG) component of the brain electrical activity. We implemented an environment for multi-modal presentation of brain electrical activity using 3-D visualization synchronized with data sonification of EEG data. Visualization is based on animated topographic maps projected onto the scalp of a 3-D head model. Our first environment was developed in Visual C++ to test the most important perceptual features [8].

Virtual reality techniques facilitate multiple data stream presentation and navigation through huge data sets. Virtual environments are particularly appropriate to improve insight into complex biomedical phenomena, which are naturally multidimensional.

The Virtual Reality Modeling Language (VRML) is a file format for describing interactive 3D objects and worlds, applicable on the Internet, intranets, and local client systems [9]. VRML is also intended to be a universal interchange format for integrated 3D graphics and multimedia. VRML is capable of representing static and animated dynamic 3D and multimedia objects with hyperlinks to other media such as text, sounds, movies, and images [10]. VRML browsers, as well as authoring tools for the creation of VRML files, are widely available for many different platforms. Therefore we have chosen VRML as widely accepted platform for Internet based information systems. In our system a VRML world is controlled by Java applets. Sonification is implemented as the modulation of natural sound patterns to reflect certain features of processed data, and to create a pleasant acoustic environment. This feature is particularly important for prolonged system use.

Principally, there are two possible multi-modal data presentations. The simplest one is signaling of state transitions or indication of certain states, which is often implemented as sound alarms. The second one is acoustic rendering of a data stream. Additional modes of presentation may be employed either as a redundant mode of presentation emphasizing certain data features or to introduce new data channels. Redundant presentation is intended to create artificial synesthetic perception of the observed process [11]. Artificial synesthesia generates sensory joining in which the real information of one sense is accompanied by a perception in another sense. Multi sensory perception could improve understanding of complex phenomena, by giving other clues or triggering different associations. In addition, an acoustic channel could facilitate new information channels without information overloading.

In distributed medical systems, and telemedical environments, acquisition, archiving and presentation are performed on different physical locations. After acquisition, the original record is archived in a standard format, such as ASTM for EEG [12].

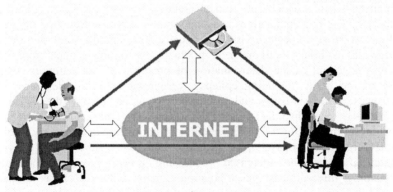

Figure 1. Virtual Medical Devices VMD

Physicians can examine the EEG data either on-line, in real-time during recording, or off-line from the archive, and add their findings to the patient medical record, as depicted in figure 1. The presentation of raw or derived data could be performed using tools that allow different views of the data set, as shown in figure 2. We call this set of tools Virtual Medical Devices, or VMD. An example VMD could present a standard electroencephalogram as a waveform plot. We have integrated *mmViewer* as a VMD in a distributed medical information system called *DIMEDAS* [2]. It provides a view of an animated 3D topographic map of selected EEG scores.

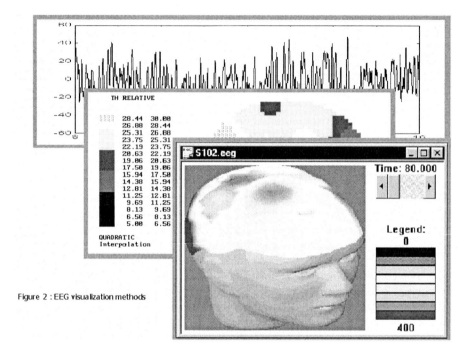

Figure 2 : EEG visualization methods

Topographic maps of different parameters of brain electrical activity have been commonly used in research and clinical practice to represent spatial distribution of activity [13]. The first applications used topographic maps representing the activity on two dimensional scalp projections. They usually represented a static picture from the top view. EEG brain topography is gradually becoming clinical tool. Its main indication is to aid in the determination of presence of brain tumors, other focal disease of the brain (including epilepsy, cerebrovascular disorders and traumas), disturbances of consciousness and vigilance, such as narcolepsy (the abrupt onset of sleep and other sleep disorders grading the stages of anesthesia or evaluation of coma, intraoperative monitoring of brain function in carotid endarterectomy, etc). It is a valuable tool in neuropsychopharmacology and psychiatry. Recent advances in computer graphics and increased processing power provided the means of implementing three-dimensional topographic maps with real time animation, as presented in figure 3. 3-D visualization resolves one of the most important problems in topographic mapping namely the projection of the scalp surface onto the plane. The other problems of topographic mapping are interpolation methods, number and location of electrodes, and score to color mapping.

While in CT and PET images every pixel represents actual data values, brain topographic maps contain observed values only on electrode positions. Consequently, all the other points must be spatially interpolated using known score values calculated on electrode positions. Therefore, a higher number of electrodes produces a more reliable topographic mapping. Electrode setting is usually predefined (like the

International 10-20 standard), although for some experiments custom electrode setting could be used to increase spatial resolution over a certain brain region. Finally, color representation of data values may follow different paradigms like *spectrum* (using colors of the visible spectrum, from blue to red) and *heat* (from black through yellow

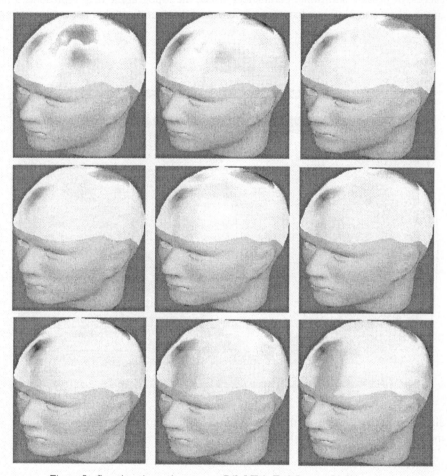

Figure 3 : Sample animated sequence; P(2.5 Hz); T=0.5sec; t=20-24sec

to white) [13]. Isochronous representation of observed processes preserves genuine process dynamics, and facilitates perception of intrinsic spatio-temporal patterns of brain electrical activity.

The multi modal viewer *mmViewer* (Figure. 4) is based on a VRML head model, animated with Java applets. Currently we use a synthetic head model, but we are planning to integrate support for a user-specific head model derived from MRI recording. That model could be archived in the patient medical record, and could be accessed by the web.

Sonification is supported by VRML, using standard VRML nodes *Sound* and *AudioClip*. A Sound pattern is stored in a predefined audio file. A Java applet then modulates the sound according to values in the sonified data stream during animation. We applied sonification to emphasize temporal dimension of selected visualized scores, or present additional parameters. Since the topographic map represents a mass of visual information, we have found very useful to sonify global parameters of brain

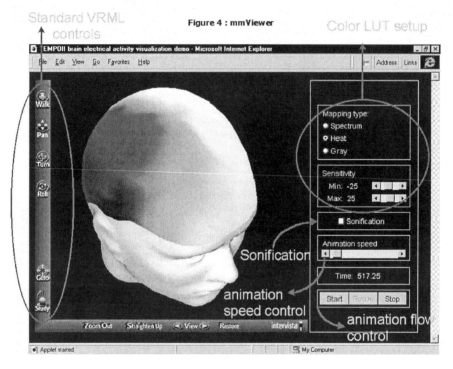

Figure 4 : mmViewer

electrical activity, such as the global index of left/right hemisphere symmetry. The index of symmetry (*IS*) is calculated as:

$$IS = (P_1 - P_2) / (P_1 + P_2)$$

where $P_1$ and $P_2$ represent power of symmetrical EEG channels, like O1 and O2 for example. We have sonified this parameter by changing the position of the sound source in the VRML world. Therefore, activation of a hemisphere could be perceived as the movement of the sound source toward that hemisphere.

## 5. Discussion

The developed system was applied in the analysis of EEG records at the Institute for Mental Health. The performance of the mmViewer on a standard PC/workstation was satisfactory, where the animation speed was often limited by capacity of user perception rather than computer performance.

It was difficult to find the most appropriate multi modal presentation for a given application. We have evaluated different visualization and sonification methods to find out perceptually the most admissible presentation. Moreover, the creation of user-specific templates is highly advisable, as perception of audio-visual patterns is personal.

The selection of scores for multi modal presentation is another delicate issue relying on human perception. The score selected for acoustic rendering may be used either as new information channel (sonification of symmetry in addition to visualization of EEG power) or redundant channel of visualized information. By introducing additional channels one should be careful to avoid information overloading. Redundant multi modal presentation offers the possibility to choose the presentation modality for a given data stream, or to emphasize temporal dimension for a selected stream.

A possible application of sonification could be long-term EEG monitoring in an outpatient clinical practice or intensive care units (ICU). During examination of long EEG records the physician needs to sustain a high level of concentration, sometimes for more than one hour. Monotonous repetition of visual information induces mental fatigue, so that some short or subtle changes in the EEG signal would be probably omitted. Additional acoustic information could help in alerting the observer.

## 6. Conclusion

Multi modal presentation significantly improves the quality of user interface in virutal telemedical environments. Visualization is usually enriched using sonification and tactile feedback. We presented use of visualization and sonification in the telemedical environment for EEG analysis. We have implemented EEG multi-modal viewer using visualization and sonification in telemedical environment. Since visualization efficiently represents spatial activity distribution, sonification was used either to improve temporal dependency or present new data channels. The proposed multi-modal viewer requires only a standard Web browser and therefore it is applicable both to a stand-alone workstation and distributed Internet based telemedical applications. The lack of general insight into the design of multi-modal presentation, restricts wider acceptance of multi-modal applications. Therefore it is necessary, for a given application, to find the combination of parameters and their presentation modalities to accentuate the changes in a perceptual domain. We are currently investigating the most appropriate presentation paradigm for telemedical EEG analysis.

## Acknowledgement

The authors would like to thank Professor Zoran Jovanovic, Branko Marovic, and Nebojsa Djukanovic for their help in integration *mmViewer* into *DIMEDAS* information system.

# References

[1] J.E. Cabral, Y. Kim, "Multimedia Systems for Telemedicine and Their Communication Requirements", *IEEE Communications*, 34(7):20-27, 1996.

[2] A.Marsh, "EUROMED - Combining WWW and HPCN to supoort advanced medical imaging", *Int Conf on High Performance Computing and Networking HPCN 1997*, Vienna, Austria, 1997.

[3] W.J. Greenleaf, "Developing the Tools for Practical VR Applications," *IEEE EMBS*, 15(2), 1996, pp. 23-30.

[4] M.Haubner, A.Krapichler, A.Losch, K.Englmeier, W.van Eimeren, "Virtual Reality in Medicine-Computer Graphics and Interaction Techniques", *IEEE Trans on Information Technology in Biomedicine*, 1(1):61-72, 1997.

[5] G.Burdea : *Force and Touch Feedback for Virtual Reality*. John Wiley & Sons, Inc., New York, 1996.

[6] N.O. Bernsen, "Foundations Of Multimodal Representations: A taxonomy of representational modalities," Interacting with Computers, 6, 1994, pp. 347-371.

[7] G.Kramer (Ed.), *Auditory Display, Sonification, Audification and Auditory Interfaces*, Addison Wesley, 1994

[8] A.Samardzic., E.Jovanov., D.Starcevic., *"3D Visualisation of Brain Electrical Activity"*; Proceedings of 18$^{th}$ Annual Int'l Conf. IEEE/EMBS, Amsterdam, October 1996.

[9] The Virtual Reality Modeling Language, http://www.vrml.org/Specifications/VRML97

[10] D.R. Begault, *3D Sound for Virtual Reality and Multimedia*, Academic Press, Inc., Boston, 1994.

[11] R.E. Cytowic, "Synesthesia: Phenomenology And Neuropsychology A Review of Current Knowledge," *Psyche*, 2(10). 1995.

[12] A Society for Testing and Materials: Standard Specification for E1467-94 Transferring Digital Neurophysiological Data Between Independent Computer Systems.

[13] F.H. Duffy, Ed, *Topographic Mapping of Brain Electrical Activity*, Butterworth Publishers, 1986.

# Using Web Technologies and Meta-Computing to Visualise a Simplified Simulation Model of Tumor Growth in Vitro

Georgios S. Stamatakos, Evangelia I. Zacharaki, Nikolaos A.Mouravliansky,
Konstantinos K.Delibasis, Konstantina S. Nikita, Nikolaos K. Uzunoglu,
and Andy Marsh[*]
Department of Electrical and Computer Engineering, Division of Electroscience,
National Technical University of Athens,157 80  Zografou, Athens, Greece.
[*]Fax: +301772 3557, Tel: +3017722287, Email: euromed@naxos.esd.ece.ntua.gr

**Abstract** - The aim of this paper is to demonstrate the impact that Web technologies and meta computing can have on the simulation of biological processes such as tumor growth. A client-server architecture allowing real time surface and volume rendering using a standard Web browser is proposed. A simplified three-dimensional cytokinetic simulation model of tumor growth in vitro is developed and results are obtained concerning the development of a small cell lung cancer (SCLC) tumor spheroid in cell culture. A Gaussian distribution of the cell cycle phase durations is considered. The behavior of the model is compared with both published data and laboratory experience. The  application of Web technologies and meta computing leads to a spectacular three-dimensional visualisation of  both the external and the internal structure of a growing tumor spheroid.

## 1. Introduction

Tumor growth modeling has proved to be a particularly useful means of both gaining insight into the biological process of tumor development [16,23,26,30] and optimising therapeutic techniques in  oncology [16,19,26]. The fact that cancer is characterised by uncontrolled cell proliferation has stimulated biomathematicians to construct cell proliferation models (continuous, discrete, deterministic, stochastic) usually based on differential equations describing growth and kinetics of abnormal cell multiplication [2,3,16]. Based on these findings, W.Duechting formulated  the hypothesis that *'cancer can be interpreted as structurally unstable, negative feedback control loops'* [4-17]. By applying systems analysis, control and automata theory, heuristics and computer science, Duechting and his collaborators managed to produce elaborate, flexible and efficient simulation models, revolutionising tumor growth modeling [4-17, 20, 29].  Duechting's basic philosophy has been adopted in the development of the biophysical part of this paper. An improved algorithm for the simulation of the cell to cell communication (assumption 13 in Section 2) has also been introduced.

As the main purpose of the present work is to demonstrate the impact that modern visualisation techniques and the use of remote computing resources can have on tumor growth simulations rather than to improve the simulations themselves, a simplified three-dimensional cytokinetic model of tumour growth in vitro is developed. The model takes into account only the very basic characteristics of the process, sometimes even in a rather qualitative manner [28]. External actions, such as therapeutic schemes, or space limitations are not considered in this model. Simulation results concerning the development of small cell lung cancer (SCLC) tumor spheroids in cell culture are obtained. Initially, an equatorial section of the spheroid in different moments is visualized using a standard mathematical software package e.g. MATLAB. The results are compared with both published simulation data and laboratory experience. A satisfactory agreement regarding at least the gross features of the process is observed.

For the three-dimensional (3D) visualization of the results, a standard Web browser is used allowing real time surface and volume rendering on inexpensive computer hardware. Its application to the model leads to a spectacular three-dimensional visualization of both the external surface and the internal structure of a growing tumor spheroid. In addition, a procedure to animate volumes or triagulated surfaces is suggested as an extension to the proposed visualization system. Finally, the issue of utilising remote computational resources for calculating the tumor growth prediction in a client-server architecture is addressed.

## 2. Assumptions of the model

The following simplifying assumptions have been used for the development of the simulation model.
1. A three-dimensional mesh discretizing the volume of the cell culture (including tumor and nutrient medium) is used. Each geometrical cell of the mesh can be occupied by a single biological cell of cubic shape, by nutrient medium or by the products of cell lysis.
2. The total space of the modeled cell culture is limited to $100 \times 100 \times 100$ geometrical cells. This type of limitation depends on the computer memory and power available as well as on the maximum tolerated simulation run time .
3. Time is discretized and measured in appropriate units such as hours (h).
4. Vascularization is not considered. This can refer to either tumor growth in vitro or to the early stages of avascular tumor growth in vivo.
5. Side effects, immunologic reactions, heterogeneity, drug resistance and the formation of metastases are neglected.
6. The development of a tumor spheroid starts immediately after the placement of a single tumour cell in the phase of mitosis at the centre of the mesh.
7. The simplified cytokinetic model of a tumor cell shown in Fig.1 is considered. A probability of 0.01 for each cell to undergo lysis every hour due to apoptosis is adopted.
8. The duration of each cell cycle phase follows a Gaussian distribution.
9. Only horizontal and vertical communication between neighboring cells is possible.
10. The simulation may be considered as a row-to-row computation of the cell algorithm for each individual cell. At each time step, the time remaining for the current phase of each cell is reduced by one unit. The configuration obtained in this way serves as the initial step of the subsequent calculation step.
11. The following heuristic cell production and interaction rule is employed in order to describe the cell-to-cell communication: 'If the distance between the dividing tumor cell and the nutrient medium is more than three cell layers, the tumor cell will transfer to the resting (dormant) phase G0 and later to the phase of necrosis and to lysis'.

12. A tumor cell can divide even if there is no empty space available for a daughter cell; this rule is restricted by the previous distance-dependent statement.
13. The position of a conventionally 'newborn cell' is chosen in such a way that the number of cells that will have to shift (in a straight line) in order to give space to the 'newborn cell' is the least possible. In case that cell shifting in more than one direction is permissible, the selection of the shifting direction is made using a random number generator (Monte Carlo technique).

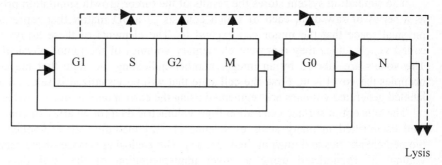

Fig.1 A simplified cytokinetic model of a tumor cell

As an example of tumor growth in vitro, the case of small cell lung cancer (SCLC) tumour growth in cell culture has been considered. SCLC is a rapidly growing neoplasm. The mean durations and deviations of its cell cycle phases are as follows [13, 14]: necrosis duration $T_N=40\pm 2$ h, resting phase duration $T_{G0}=25\pm 5$h, first gap phase duration $T_{G1}=3\pm 0$ h, DNA synthesis duration $T_S=5\pm 1$ h, second gap phase duration $T_{G2}=1\pm 0$ h, mitosis duration $T_M=1\pm 0$ h. The time unit has been taken equal to 1 h. Six mesh snapshots corresponding to the simulated instants 1 h, 40 h, 80 h, 120 h, 160 h, 200 h (after the placement of the initial SCLC cell at the centre of the nutrient medium) were stored for further visualization processing.

## 3. The visualization system

The visualization of 3D anatomical or biological structures is traditionally performed by displaying a series of 2D transverse intersections (slices) or contours from which the observer tries to conceive the 3D shape of the structure. The facts that:
a) there is a number of states that a cell can be in (e.g. G0, M etc.) and
b) what needs to be visualized is not just a 3D volume but a dynamically evolving structure, equivalent to a 4D modality

make the use of new visualization techniques imperative, if the expert is to comprehend the full width of the information produced by the tumor growth simulation. The requirements that the proposed visualization system satisfies can be summarized as follows:
a) The input of the system is raw data with voxels labeled by a code number corresponding to a specific cell state.

b) The output of the system is a set of triangulated surfaces that allow real time surface and volume rendering to be performed on hardware as inexpensive as a high end Pentium based PC compatible computer. The output files are WWW compatible (i.e. in VRML format).

## 4. Detailed description of the visualization system

The simulation system stores the results of the tumor growth simulation procedure as a series of raw data, each of which consists of a 3D matrix that represents the physical space that the tumor can expand to. The elements of these arrays will be called voxels, since they represent elementary volumes of the tumor physical space. The voxels are labeled by an integer number indicating the state that the cell that occupies the voxel is in. First, the cell state that will be visualized is selected and the labeled generated volumes are segmented using the corresponding integer value.

The concept of surface generation from volumetric data (or an array of contours) is well established in many areas of science [1, 22] including many applications in biomedical physics and imaging [e.g. 27, 31]. The enclosing surface of the segmented volume is triangulated using a novel implementation of the well documented Marching Cubes (MC) algorithm [18, 24, 32]. The version of MC employed has been developed and implemented with a substantial difference from traditional implementations [25]. In the specific implementation, a generic rule is used to optimally triangulate the image voxels, instead of employing prespecified voxel configurations. This approach compares favourably to the existing, traditional ones, in terms of both the number of triangles/polygons and the execution time. The novel implementation of the MC algorithm produces an optimized number of polygons, small enough to allow high end PC compatible computers to perform real time volume rendering. The triangulated surfaces produced by the employed MC algorithm can be easily visualized for a number of $10^6$ evolving cells. Therefore, it is evident that the limiting factor is the speed of the server performing the actual tumor growth simulation rather than the volume rendering process, assuming that the later takes place on a high end Pentium based PC.

The visualization system stores the triangulated surface in VRML (Virtual Reality Modelling Language) 2.0 format. The procedure can be repeated if another cell state is to be visualized and a new triangulated surface is produced. An arbitrary number of surfaces can be included in a single VRML file, since the use of color, transparency and volume rendering assists the visualization of complex cell structures.

The proposed visualization procedure is not only efficient in visualizing complex 3D structures, but can also be used as a tool to handle 4D images, such as the dynamically changing tumors that evolve from the growth simulation system. VRML 2.0 allows the creation of animations that demonstrate the volume evolution in a graphical and interactive way [21]. The procedure that has been developed to animate volumes or triangulated surfaces can be summarized as follows.

I) A series of cell volumes is produced in time steps sufficiently small so that the shape change between consecutive volumes can be tolerated by the algorithm.

II) Starting with the volume with the largest number of cells $V_t$,

II.A) Volumes $V_t$ and $V_{t-1}$ are registered considering only the spatial translation of their centres of mass. A new surface $V'_{t-1}$ is constructed.

II.B) For each vertex $i$ of the triangulated surface of $V_t$

II.B.1) The vertex $j$ from the triangulated surface of $V_{t-1}$ for which the Euklidean distance from $i$ is minimal is located.

II.B.2) The Euclidean coordinates of vertex $i$ of $V'_{t-1}$ are set equal to the Euclidean coordinates of vertex $j$ of $V_{t-1}$ i.e. $V'_{t-1}(i) = V_{t-1}(j)$

II.C) After the mapping is finished, the triangulated surface of $V_{t-1}$ is replaced by $V'_{t-1}$.

II.D) If spatial registering took place, the inverse transform is applied to $V'_{t-1}$.

II.E) $t = t - 1$

II.F) If $t > 0$, go to step II.A, else end.

The mapped series of the triangulated surfaces of $\{V_t, V'_{t-1}, ..., V'_0\}$ is now adequate to be introduced into the VRML 2.0 so that the animation may be performed.

## 5. The client - server architecture

The simulation of tumor growth becomes more realistic as the number of tumor cells increases. However, this results in a very fast increase in the computing time demands of the application. Therefore, in the case of in vivo tumor growth it may not be feasible to predict the development of a tumor or its response to specific therapeutic schemes with the computing power installed in a hospital. In such a case the calculation of tumor growth could be provided to hospitals or to research laboratories as an external service. An architecture schematically, shown in Figure 2, implementing this approach can be summarised as follows:

a) The tumor simulation/prediction program executes on a fast, remote server, possibly dedicated to this purpose, for example the Onyx Infinite Reality[2] with 4 processors located at the ICCS/NTUA (Greece). The client submits a request for the prediction program through a Web page, which invokes the execution of a CGI (Common Gateway Interface) script on the server.

b) The generated volumes are triangulated, as it is described in the previous section and the corresponding VRML files are produced. At present the results are in two formats ASCI-Numeric and VRML, the envisaged 3rd format is a stereographic image that can be viewed by a Virtual Reality headset.

c) The client receives the generated VRMLs through an e-mail, or can access them through a password protected Web page. The use of a Web compatible format, like VRML, allows great flexibility in exchanging data between the client and the server.

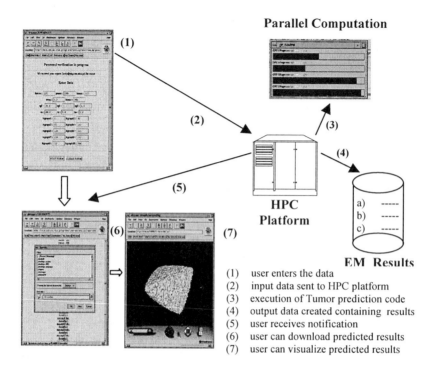

Figure 2 : A client-server architecture coupling meta-computing and web technologies

It becomes evident that while the actual prediction can take place in a remote, powerful server, the visualization may be performed on the client's computer, which may be as inexpensive as a high end PC. The small size of the generated VRMLs permits the use of inexpensive hardware running just a shareware Web Browser. Figures 3 and 4 have been produced by applying the visualization procedure described in sections 3 and 4. The execution time for the presented examples takes about 20 mins on the four processor Onyx platform. Fig. 3 demonstrates the evolution of the external surface of the tumor. Fig. 4 gives a 3D view of the internal structure of the tumor. It consists of a series of 3D sections of the tumor. The simulation model predictions compare favorably with those of Duechting [13,14]. Agreement with laboratory experience, at least as far as the gross features of the simulated process are concerned, has also been observed.

**Fig. 3** Small cell lung carcinoma (external surface)
Sequence of visualization instants: 1h, 40h, 80h, 120h, 160h, 200h
[Color code: **yellow:** proliferating phases (G1, S, G2), **red**: mitosis ,**blue**: products of cell lysis.]

On the external surface of the spheroid most of the cells are in the proliferating phases (G1,S,G2) whereas cells in the process of mitosis or products of cell lysis can also be seen.

**Fig.4** Small cell lung carcinoma (section)
Sequence of visualization instants: 1h, 40h, 80h, 120h, 160h, 200h
[Color code: **yellow:** proliferating phases (G1, S, G2), **red**: mitosis, **blue**: products of cell lysis, **pink**: G0, **green**: necrosis]

In the right snapshot the lysis products mainly lie in the central region of the tumor and are enclosed by the majority of necrotic cells. The next tumor shell basically contains resting cells (in the G0 phase), whereas the proliferating cells are to be found in the external layer of the tumor. Products of lysis can be found anywhere in the tumor as they can be produced either by the process of necrosis or by apoptosis.

## 6. Discussion

The simulation model presented seems to be a useful platform for the theoretical investigation of the biological process of tumor growth. It should be noted that graphs giving the temporal dependence of quantities such as the number of cells in a certain phase can be readily produced from the quantities calculated at each step of the model. Obvious improvements would include, the introduction of differential equations in order to describe the diffusion of oxygen and glucose in a quantitative manner [20], the consideration of non cubic shapes of the cells etc. It should be pointed out that the octahedral rather than spheroidal shape of the growing tumor that is apparent in Fig. 3 and Fig. 4 results from the restriction of the cell communication to horizontal and vertical directions.

The usefulness of the simulation model would be greatly enhanced if the effects of various therapeutic schemes such as chemotherapy and radiation therapy were taken into account. A simulation of tumor growth in vivo where the effects of angiogenesis would be considered in addition to its response to various therapeutic modalities would certainly be of much more practical interest. Such an advanced tool might substantially contribute to approaching the ultimate goal of cancer modeling which is the optimization of treatment prior to any therapeutic or paliative intervention. Our team is currently working on the extention of the present simulation model so that the above mentioned factors will be taken into account. In addition to the modeling itself, the proposed VRML visualization system proved capable of substantially contributing to an efficient 3D perception of the biological growth process. It is expected that its contribution to the visualization of tumor growth in vivo will be of even greater importance to clinicians. The use of VRML represents the first stage of utilising Virtual Reality techniques to visualize tumor growth [33]. Techniques based on Augmented Reality and Fully Emersive Virtual Reality are currently being explored. The ultimate goal of this work is to present the tumor growth as a holographic image. Furthermore, the proposed client-server architecture of the entire system will bring tumor growth modeling closer to the medical practice.

## 7. Conclusions

A simplified cytokinetic simulation model of tumor growth in vitro has been developed. The model has been used to simulate the growth of a SCLC tumor spheroid in cell culture. An equatorial section of the spheroid has been visualized at different instants using the software package MATLAB. For the three-dimensional visualization of the results, a special procedure allowing real time surface and volume rendering on inexpensive computer hardware has been proposed and applied. A spectacular 3D virtual reality visualization of both the external surface and the internal structure of a growing SCLC tumor spheroid has been achieved. Satisfactory agreement of the model predictions with both published simulation results and laboratory experience has been established. A procedure to animate volumes or triangulated surfaces has been suggested as an extension to the proposed visualization system. A client-server architecture of the entire simulation-visualization system has also been proposed and implemented. In conclusion, it has become clear that the application of advanced visualization techniques can significantly enhance the potentiality of tumor growth simulation models.

## 8. Acknowledgments

The authors would like to thank both the European Commission DGIII/B for supporting the EUROMED Project under the ISIS '95 Programme and the State Fellowship Foundation of Greece (IKY) for supporting G.S.Stamatakos' post doctoral research project. The authors would also like to thank Prof. Nikoloaos Zamboglou and Prof. Dimos Baltas of the Staedtische Kliniken Offenbach, Germany, for their contribution to the macroscopic evaluation of the model.

# References

[1] J. D. Boissonat, Shape Reconstruction from Planar Cross Sections, Computer Vision, Graphics and Image Processing 44 (1988) 1-29.

[2] C. J. Clem, J.P. Rigaud, Computer simulation modelling and visualization of 3D architecture of biological tissues. Simulation of the evolution of normal, metaplastic and dysplastic states of the nasal epithelium, Acta Biotheor. 43(4) (1995) 425-442.

[3] R. Demicheli, R. Foroni, A. Ingrosso, G. Pratesi, C. Soranzo, M. Tortoreto, An exponential-Gompertzian description of LoVo cell tumor growth from in vivo and in vitro data, Cancer Res, 49(23) (1989) 6543-6.

[4] W. Duechting, Krebs, ein instabiler Regelkreis. Versuch einer Systemanalyse, Kybernetik 5(2) (1968) 70-77.

[5] W. Duechting, Modelling and simulation of normal and malignant tissue, in: M.H.Hamza and S.G. Tzafestas eds., Advances in Measurement and Control, Vol. 3 (ACTA Press, Anaheim, 1979) 909-918.

[6] W. Duechting and G. Dehl, Spatial growth of tumors. A simulation study, in: L.Fedina, B.Kanyar, B.Kocsis, M.Kollai eds., Adv. Physiol. Sci., Vol. 34 (Akademiai Kiado, Budapest, 1981) 123-131.

[7] W. Duechting and T. Vogelsaenger, Aspects of modelling and simulating tumor growth and treatment, J. Cancer Res. Clin. Oncol 105 (1983) 1-12.

[8] W. Duechting, Computer models applied to cancer research, in: M. Thoma and A. Wyner eds, Lecture Notes in Control and Information Sciences, 121 (Springer-Verlag, Berlin, 1988) 397-411.

[9] W. Duechting, Recent progress in 3-D computer simulation of tumor growth and treatment, Acta Applicandae Mathematicae 14 (1989), 155-166.

[10] W. Duechting, R. Lehring, G. Rademacher, W. Ulmer, Computer simulation of clinical irradiation schemes applied to in vitro tumor spheroids, Strahlenther. Onkol. 165 (1989) 873-878.

[11] W. Duechting, Computer simulation in cancer research, in:D.P.F. Moeller, ed., Advanced Simulation in Biomedicine (Springer-Verlag, NY, 1990) 117-139.

[12] W. Duechting, Tumor growth simulation, Comp. & Graph. 14(1990) 505-508.

[13] W. Duechting, W. Ulmer, R. Lehring, T. Ginsberg, E. Dedeleit, Computer simulation and modelling of tumor spheroid growth and their relevance for optimization of fractionated radiotherapy, Strahlenther. Onkol. 168 (1992), 354-360.

[14] W. Duechting, W. Ulmer and T. Ginsberg, Modelling of tumor growth and irradiation, in: A.R. Hounsel, J.M. Wilkinson and P.C. Williams, eds, Proceedings of the Xith International Conference on the Use of Computers in Radiation Therapy (North Western Medical Physics Department, Christie Hospital, Manchester, UK, 1994) 20-21.

[15] W.Duechting, T. Ginsberg, W. Ulmer, Modelling of radiogenic responses induced by fractionated irradiation in malignant and normal tissue, Srem Cells 13(suppl 1) (1995) 301-306.

[16] W. Duechting, W. Ulmer and T. Ginsberg, Cancer: A challenge for control theory and computer modelling, European J. of Cancer, 32A(1996) 1283-1292.

[ 17 ] W. Duechting, W. Ulmer and T. Ginsberg, Computer models for optimizing radiation therapy, in: H.U.Lemke, K.Inamura, M.W.Vannier, A.G.Farman, eds, CAR' 96, Computer Assisted Radiology (Elsevier, Amsterdam, 1996)

[ 18 ] M.J.Durst, Additional reference to "Marching Cubes", Computer Graphics 22(2) (1988) 72-73.

[ 19 ] J.F. Fowler, Review of radiobiological models for improving cancer treatment,in: K.Baier and D. Baltas, eds, Modelling in Clinical Radiobiology, Freiburg Oncology Series, Monograph No. 2 (Albert-Ludwigs-University Freiburg, Germany, 1997)

[ 20 ] T. Ginsberg, Modellierung und Simulation der Proliferationsregulation und Strahlentherapie normaler und maligner Gewebe (Fortschr.-Ber. VDI, Reihe 17, Nr 140, VDI Verlag, Duesseldorf,1996).

[ 21 ] Lemay L, Couch J and Murdock K, 3Dgraphics and VRML 2.0 (Sams.net Publishing, 1996)

[ 22 ] W.C. Lin, S.Y. Chen and C.T. Chen, A new surface interpolation technique for reconstructing 3D objects from serial cross sections, Computer Vision, Graphics and Image Processing 48 (1989) 124-143.

[ 23 ] H.Lodish, D.Baltimore, A. Berk, S. L.Zipursky, P.Matsudaira, J.Darnell, Molecular Cell Biology (Scientific American Books, NY, 1995) 1247-1294.

[ 24 ] W.E. Lorensen and H.E. Cline, Marching Cubes: High resolution 3D surface construction algorithm, Computer Graphics 21(3) (1987) 163-169.

[ 25 ] G.K.Matsopoulos, N.Mouravliansky, K.K.Delibasis and K.S. Nikita, A novel and efficient implementation of the Marching Cubes Algorithm, Computer & Graphics, accepted, October 1998.

[ 26 ] H.I. Pass, J.B.Mitchell, D.H. Johnson, A.T.Turrisi, Lung Cancer, Principles and Practice (Lippincott-Raven, Philadelphia, New York, 1996).

[ 27 ] B.A. Payne and A.W. Toga, Surface mapping brain function on 3D models, IEEE Computer Graphics and Applications 10(2) (1990) 33-41.

[ 28 ] G.S.Stamatakos, N.K.Uzunoglu, K.Delibasis, M.Makropoulou, N. Mouravliansky, A.Marsh, A simplified simulation model and virtual reality visualization of tumour growth in vitro, Future Generation Computer Systems 14(1998) 79-89.

[ 29 ] T. Vogelsaenger, Modellbildung und Simulation von Regelungsmechanismen wachsender Blutgefaessstrukturen in normalen Geweben und malignen Tumoren, Ph.D. Thesis, Department of Electrical Engineering, University of Siegen, Germany, 1986.

[ 30 ] J.D.Watson, N.H.Hopkins, J.W.Roberts, J.A.Steitz, A.M.Weiner, Molecular Biology of the Gene, 4th Edition (The Benjamin/Cummings Publishing Company, Inc., Menlo Park,California, 1987) 1058-1096.

[ 31 ] S.B. Xu and W.X. Lu, Surface Reconstruction of 3D Objects in computerized Tomography, Computer Vision, Graphics and Image Processing 44 (1988) 270-278.

[ 32 ] C. Zhou, R. Shu and M. S. Kankanhalli, Handling small features in isosurface generation using Marching Cubes, Comp. & Graphics 18(6) (1994) 845-848.

[ 33 ] B. Grant, "Virtual Reality gives medicine a powerful new tool", Biophotonics International, November/December 1997, p. 40-45.

# The Electronic Commerce Component in Telemedicine

Despina. Polemi, Andy Marsh
Department of Electrical and Computer Engineering, Division of Electroscience,
National Technical University of Athens,157 80 Zografou, Athens, Greece.
Fax: +301772 3557, Tel: +3017722287, Email: euromed@naxos.esd.ece.ntua.gr
http://secgroup.iccs.ntua.gr

**Abstract** - The goal of this paper is to provide definitions and key ideas for the interplay of electronic commerce and telemedicine. The natural interaction of these Web applications is exploited in this paper

## 1. Introduction

In [8] 20 building blocks were indicated and 39 practical steps were identified in order to built an Internet based global telemedical information society. Among these blocks were security and privacy required for the communication of telemedical participants and medical data.

EUROMED-ETS an INFOSEC/DGXIII project [5] provided a secure solution based on the establishment of Trusted Third Party Services (TTPs) over the World Wide Web (WWW) using the Secure Session Layer (SSL) protocol. TTPs were also proposed for the security of medical Java applets [19] and the secure billing and payment process of telemedical services.

The arrival of the Next Generation Internet will bring the Web based telemedicine into a wider audience providing them with even more benefits. Numerous advantages were described in [13] including the linkage of health records, data pooling and analysis, high quality virtual reality models. These will translate into the increase of collaborative research, interactive patient consultation and home-education.

But the benefits gained by the high speed and bandwidth will make telemedical participants more demanding. They would require handling all their healthcare-related activities electronically. Such activities include, subscription to a health insurance company, contract signing, payment and billing of telemedical services and insurance fees, regulation of damages, on-line purchase of telemedical information, consultation, purchase of health care products (equipment, supplies) mail-order pharmacy.

But these former activities are typical scenarios of electronic commerce in which security is still a barrier. In [11] TTPs, biometric smart cards and SET were proposed for the secure payment of telemedical services.

In this paper basic electronic commerce definitions translated in telemedicine applications are provided (Section 2). Electronic commerce actions in a telemedical information society are identified (Section 3). Typical electronic commerce scenarios in telemedicine are covered (Section 4). Finally conclusions and further research directions are drawn (Section 5).

## 2. Electronic Commerce in Telemedicine

Electronic Commerce is a general term for the conduct of business with the assistance of telecommunications and of telecommunications-based tools. There are two major parties in an electronic commerce environment, the *seller* who offers its *goods* and the *byer* who purchases them.

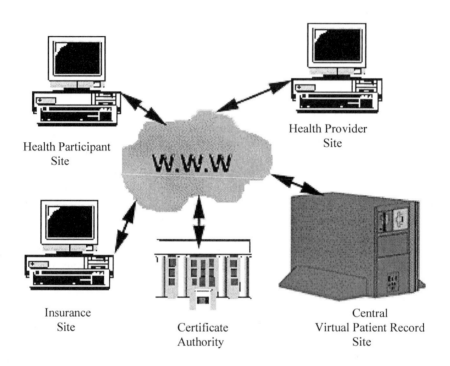

**Fig 1.** Telemedicine Network

In a telemedicine society the (telemedicine) seller is a hospital (or any health care institution), a doctor, a medical laboratory, an insurance company or any entity that offers health care goods. The (telemedicine) byer on the other hand is a patient, an insured entity (hospital, person), or any health care participant that receives these goods. These are the typical components of a telemedicine society (see Fig.1).

### *2.1 Telemedical Goods*

In a telemedical society the goods that are being traded can be categorized as follows:

*Custom-Built Health Goods and Services:* These are standardized custom-built health care products and services. This category includes standard health care products, all telemedical services (e.g. electronic stethoscopes, teleradiology, telepathology, radiology, cardiology, telepsychiatry, dermatology), consultation services (telehealth videoconferencing), homecare services and equipment used for the provision of such services.

*Medical Data*: Medical data such as medical Java applets, medical images (structured medical graphics), healthcare medical records, telemedical statistics, projected sound (medical speeches), medical performances (surgeries), medical video, medical video conferencing, booking and tickets for hospital submission.

*Insurance Data:* Bills, fees and payments among insurance companies or persons and healthcare participants (hospitals, doctors, patients).

*Telemedicine Publishing:* Medical documents in electronic form (medical articles, books), medical statistics, low-volatility reference information (medical dictionaries, encyclopaedias), high-volatility reference information (medical breakthroughs, news), projected sounds (medical speeches, medical performances), projected video and animation (medical films, video-clips), education, consultation via multi-media.

*Healthcare products:* These include standardized health products, equipment and supplies produced by industries.

*Health Care Commodities*: This particular class of products exists in identifiable form, in considerable quantity and is available from a variety of sources. Such commodities are insurance stocks and shares of a hospital or a healthcare organization.

*Customised Health Services, Products and Tools:* This category includes base healthcare products and services in which are modified according to healthcare participant needs as well as clinical decision-support tools (e.g. patient-specific consultation tools, tools for focusing attention).

## 3 Commerce Actions in Telemedicine

In a telemedical environment similar commerce actions occur as in the physical health care environment. Such actions are:

*Offering of Goods:* The telemedicine seller (hospital, doctor, medical laboratory, insurance company) *offers* its goods to the telemedicine byer (patient, doctor) who *orders* them; this is a two-part negotiation sometimes ending with an *agreement* or *contract*.

*Certification of agreement:* Both telemedicine sellers and byers need to *certify* such agreements or contracts. For example in order for a patient to be submitted to a hospital or use its facilities, the hospital requires from the patient or doctor respectively their *accreditation* with well-known payment system providers or they require their insurance information. In the latter case the insurance company declares them trustworthy (since they certify that the patient is their member).

*Fair exchange:* Telemedicine sellers deliver their goods and telemecine byers make payments (either directly using a payment instrument or indirectly via their insurance). This is a *two party* fair exchange.

*Dispute:* Telemedicine byers or sellers might be dissatisfied with the realization of their agreement or contract, and legal actions may be taken place.

*Co-operation:* Some telemedicine services require third party assistance. For example the assistance of *high performance computing centers, computer laboratories, universities* is required for the development of medical images or applets. Since payments are necessary for buying goods, *financial institutions* are part of a telemedical information society. *Electronic brokerage* centers utilizing technologies such as mobile agents may enter the society for assisting healthcare participants to make appropriate telemedical choices according to their needs. For example a patient wishes to find the most qualified, economical medical group that transplant islets (insulin producing cells).

The above actions may take place between:
- *Healthcare entity-to-Healthcare entity:* Any entity that offers medical goods (hospital, clinic, medical laboratory, insurance company, medical office)
- *Healthcare participant- to-Healthcare entity:* A *healthcare participant* is a health professional, doctor, patient, technician, and nurses.
- *Healthcare entity-to-Cooperative entity*: Cooperative entity is an entity with which a healthcare participant or entity co-operates with. Such entities are financial and administration institutions, brokers, suppliers, technology providers and vendors, universities, computing centers.
- *Healthcare participant-to-Cooperative entity*

For the realization of all above actions among all participants, the provision of all four components (confidentiality, integrity, authenticity, availability) of security are required.

## 4. Commerce Scenarios in Telemedicine

Typical electronic commerce scenarios that are covered in a telemedical information society are:

- *Mail-Order Retailing:* A retailer accepts electronic orders and payments, based on digital conventional catalogue and deliver physical goods.
- *On-line Purchase of Medical Information and telemedical goods:* Mail-order retailing of any digital medical data, information or medical goods that are delivered on line.
- *Telemedicine Mall:* An organization that offers services from several service providers such services are: directory services, medical consultation and training, billing services.
- *Medical Subscriptions:* An organization offers services on a time bounded subscription basis, e.g. subscription to insurance services, consultation and training services, medical database services, pharmacy, medical publications.
- *Statements:* Transfer of medical documents, data, applets or records supporting security requirements.
- *Contract Signing:* Two or more healthcare participants exchange signed copies of the same statement (insurance contract, hospital submission contract).

- *Healthcare Insurance:* Submission to an insurance company, payment of fees, regulation of damages.
- *Auctions:* Healthcare participants may enter an anonymous auction of healthcare products or equipment.
- *Ticketing:* healthcare participant byes tickets used to access certain services for a limited period of time.
- *Electronic Mail:* Speedy communication with healthcare participants, remote sales representatives, co-operated entities.
- *Healthcare participants' support:* constant presence for healthcare participants to turn to for information on their medical records, insurance status etc.
- *Corporate marketing:* press releases, medical product data and contact information available to healthcare participants and medical suppliers and vendors.
- *Electronic Publishing:* On-line publishing of medical journals, videos, tapes, and other publications.

## 5. Conclusions and Further Research

In this paper basic electronic commerce definitions, actions and scenarios were adopted to telemedicine revealing the natural interconnection of these two Web based applications. Security has been identified [8], [1] as an important component for the realization of both applications. The establishment of Trusted Third Party services has become a mature security solution through various pilot EC projects [5], [15], [1] for each of these two applications.

For the exploitation of a complete, secure telemedical information society, the security threats, risks and failures, considering all electronic commerce scenarios in telemedicine should be analyzed and security countermeasure should be provided.

## References

[1] ACTS/SEMPER Secure Electronic Marketplace for Europe http://ww.zurich.ibm.com

[2] Adams C., Chain P., Pinkas D., Zuccherato R., «*Internet Public Key Infrastructure, Part V: Timestamp protocols*», PKIX Working Group Draft, (via ftp://ftp.ietf.org/internet-drafts/draft-adams-time-stamp-01.txt), March 1998

[3] Davies and W.L. Price. *Security for Computer Networks*. John Wiley & Sons, 1994

[4] INFOSEC. T H I S: Tru*sted Third Party Services Spri, Sweden*, 2.0, 1995

[5] INFOSEC/ETS-I/E U R O M E D - E T S: *Trusted third party services for health care in Europe*», http://secgroup.iccs.ntua.gr/, 1996-1997

[6] INFOSEC/ETS-II/BESTS: http://www.cordis.;u/infosec/, 1997-1998

[7] ISIS/EUROMED 1995-1998

[8] Marsh, A. «*The creation of a Global Telemedical Information Society*» International Journal of Medical Informatics 49:2 pp. 173—193, 1998

[9] Menezes, P. van Oorschot, and S.A. Vanstone. «*Handbook of Applied Cryptography*» CRC Press, 1997

[10] Polemi, D. and Marsh, A.``*Secure telemedicine applications*", High Performance Computing and Network Europe (HPCN'98), 1998

[11] Polemi, D. "*TTPs and biometric technologies for securing the payment of telemedical services*" Future Generation Computer Systems, Special Issue on the telemedical information society, Elsevier Science B.V.(accepted for publication)

[12] Polemi D. and Varvitsiotis A."*A framework for the security of telemedical services* " Advanced Infrastructures for Future Health Care" IEEE EMBS (to appear)

[13] Shortliffe, E.H «*Health Care: The next generation Internet*» Annals of Internal Medicine 129:2, pp. 138-140, 1998

[14] Schneier, B. «*Applied Cryptography, Protocols, Algorithms and Source Code in C*» John Wiley and Sons, Second Edition, N.Y, 1996

[15] TELEMATICS *COSACC Coordination of Security Activities between Chambers of Commerce*, http:// secgroup.iccs.ntua.gr/, 1998-2000

[16] Waidner, M. «*Initial Model and Architecture*» http://www.infm.ulst.ac.uk/research/wwwstats/proc/mkurki-suonio/derrypt.html, 1998

[17] Varvitsiotis A., Polemi, D.and Marsh, A.``*Securing Web-based medical applications using Trusted Third Party Services*", International Conference on Parallel and Distributed Processing Techniques and Applications (PDPTA'98), 1998

[18] A. Varvitsiotis, D. Polemi, A. Marsh ``*Using Trusted Third Party Services to provide a secure framework for telemedical interaction*", International Conference of the IEEE Engineering in Medicine and Biology Society (EMBS'98), 1998

[19] A. Varvitsiotis, D. Polemi, A. Marsh ``*EUROMED-Java: Trusted Third Party Services for securing medical Java applets*", Proceedings of 5th European Symposium on Research in Computer Security ESORICS 98, Lecture Notes in Computer Science, Springer Verlag, vol. 1485, pp. 209-220, 1998

# Efficient Implementation of the Marching Cubes Algorithm for Rendering Medical Data

Konstantinos K. Delibasis, George K. Matsopoulos*, Nikolaos A. Mouravliansky and
Konstantina S. Nikita

Institute of Communication and Computer Systems
National technical University of Athens, 157 80 Athens, Greece
Fax: +3017723557, Tel: +3017722288, Email: gmatso@naxos.esd.ece.ntua.gr

**Abstract.** In this paper, a novel and efficient implementation of the Marching Cubes (MC) algorithm is presented for the reconstruction of anatomical structures from real three-dimensional medical data. The proposed approach is based on a generic rule, able to triangulate all 15 standard cube configurations used in the classical MC algorithm as well as additional cases presented in the literature. The proposed implementation of the MC algorithm can handle the type A "hole problem" which occurs when at least one cube face has an intersection point in each of its four edges. Theoretical and experimental results demonstrate the ability of the new implementation to reproduce standard MC results, resolving type A "hole problem". Finally, the proposed implementation was applied to real medical data to reconstruct anatomical structures. The output of the proposed technique is in WWW compliant format.

## 1. Introduction

Three-Dimensional (3D) Medical Images are routinely produced in current clinical practice. Anatomical structures can be segmented and identified as a stack of intersections with a number of parallel planes, corresponding to the 3D image slices. Viewing these slices leaves too much to the viewer's imagination, if the shape and morphology of the object is to be comprehended. After image segmentation and reconstruction, the structure should be visualised and further manipulated in such a way that realism is maximised. This can be considered as the first step towards Virtual Reality, which is the current trend in Medical Imaging Systems.

The techniques available for reconstructing a structure from 3D images are based on the following principles:
a) produce contours of the required object and then solve the triangulation problem using De Launey triangulation - Voronoi diagrams [1] or graph techniques [2].
b) produce the appropriate triangles from the binary image, working in cubic neighborhoods of (typically) 8 voxels (e.g. the Marching Cubes Technique, [3]).
c) face the problem as a functional minimisation procedure [4, 5].

The first and third techniques require closed, non intersecting contours from a grey scale image, a problem that can be quite difficult to handle, especially in complicated 3D images where many structures may exist.

## 2. The standard Marching Cubes (MC) implementation and the "TYPE A" hole problem

Given a grey scale 3D image, the MC algorithm produces an *isosurface* of value $t$, according to [3]. The algorithm operates on a standard length (usually one voxel) cubic region of the image, which occupies 8 adjacent voxels. This cubic region will be called *"the cube"* in this paper. The vertices of the cube are set to 1, if the value of the corresponding image voxel is greater than or equal to the threshold $t$ and 0, otherwise. The pattern produced in this way will be called *"cube configuration"* whereas the points that belong to the isosurface will be called *isopoints* in this paper.

There are a number of 256 possible cube configurations in each of which, the isosurface is triangulated. In [6], the Weaving Wall algorithm is proposed, in which all 256 cube configurations are explicitly defined. This method, however, is tedious and error prone. In the standard MC implementation, the use of symmetry reduces the number of cases to 15. In [3], the complementary symmetry is defined as the equivalence between complementary configurations. Two configurations are defined as complementary, if the action of the logical NOT operator on the one of them generates the other. The MC algorithm can be summarised in pseudocode as follows:

*FOR each image voxel*
    *a cube of length 1 is placed on 8 adjacent voxels of the image*
    *for each of the cube's edge{*
        *if (the one of the node voxels is above t and the other below t)*
            *{calculate the position of a point on the cube's edge that*
            *belongs to the isosurface, using linear interpolation}*
    *}*
    *for each of the predefined cube configurations{*
        *for each of the 8 possible rotations{*
            *for the configuration's complement{*
                *{compare the produced cube configuration of the*
                *above calculated isopoints to the set of predefined*
                *cube configurations and produce the*
                *corresponding triangles}*
            *}*
        *}*
    *}*

In the case of a binary image with threshold $t$, the isopoints necessarily lay on the middle of the cube's edges, for which the two voxel-nodes have values of 0 and 1. When the input to the algorithm is a grey scale image, and the two cube vertices (image voxels) have values $v_1$ and $v_2$ satisfying the inequality: $(t-v_1)(t-v_2)<0$, the location of the isopoint is calculated using linear interpolation, along the edge connecting the two voxels. Non linear interpolation models can be also used at the expense of execution time.

The essence of the standard MC algorithm is that for each cube configuration encountered in the image, a match has to be found in the predefined cube configurations, so that the surface triangles can be determined. In order to decide which cube pattern matches the cube in the current image position, the cube is extracted, rotated in both directions by step of $90^0$ and compared to every one of the predefined cube configurations. Fig. 1 shows the 15 cube configurations (Case 0 at the top left and Case 14 at the bottom right) as defined in [3]. The spheres represent image voxels with values above threshold and the lines the produced triangles.

The classical MC algorithm implementation drawbacks include great number of produced triangles and computational overhead imposed by the cube rotations. From the programming point of view, this method of predefining cube configurations and their connected triangles is tedious and error prone. Furthermore, it suffers from occasional "hole problem" [7].

The use of symmetry that reduces the number of cube configurations can produce topologically incoherent surfaces, or "holes" in certain cases of two adjacent cubes. This is indicated as type A "hole problem" in [3] and [7]. A certain topology problem with an ambiguity in the surface connection is presented in [7] causing type A holes to appear in certain pairs of cube configurations. A modified MC algorithm has been developed in [8] in order to tackle the type A "hole problem". This approach introduced the problem of 'inconsistent surface construction' as shown in [9]. Furthermore, a total of 84 cases (Table 1 in [10]) and 6 additional cube configurations [7] have been introduced in order to tackle the "hole problem".

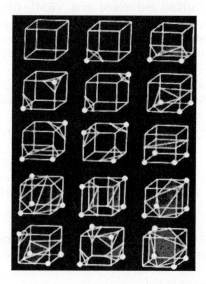

**Fig. 1.** The triangulation of the 15 predefined cube configurations (Case 0 top left up to Case 14 bottom right), as employed be the standard MC algorithm [3]. The spheres represent image voxels with values above threshold and the lines the produced triangles.

## 3. The proposed algorithm

A novel implementation of the classical MC algorithm is proposed based on a single rule capable of generating all predefined cube configurations. The new MC algorithm *generates the resulting triangles in every possible cube configuration, without using any predefined cases.* A lookup table of predefined cube configurations is the common element of every variation of the standard MC algorithm. The proposed approach orders the isosurface points directly in polygons rather than triangles, thus produces less triangles.

The isosurface intersects a cube edge only if the two end voxels of the edge lay above and below the value $t$ of the isosurface. Therefore, an isopoint can only exist on a cube edge, which contains one and only one voxel with grey value above the threshold $t$. We define this voxel as the isopoint's *associated* voxel.

The algorithm starts by finding all the isopoints lying on a cube according to the above criterion and stores them into a list. These isopoints can form up to four different polygons (case 13, Fig. 1). However, each isopoint has to belong to only one polygon. The isopoints are ordered into a polygon according to the following rule:

> given the current isopoint, the next isopoint in the polygon is the one which *lies on the same cube face with the curren AND its associated voxel shares a common cube edge with the current iso-point's associated voxel OR the cube face containing the current and the unmarked iso-point contains 3 voxels with values above t.*

The polygon is traced with the above rule until the initial isopoint is found. The polygon isopoints are marked and stored in a VRML file. The same procedure is repeated until every isopoint of the list is placed into a polygon. Since a triangle is the simplest polygon, the polygon tracing rule always returns more than two isopoints.

The proposed algorithm can be described in pseudocode as follows:

**Step 0** FOR *each image voxel*
  *a cube of length 1 is placed on 8 adjacent voxels of the image*
  FOR *each of the cube's edge*
    IF *(the grey value of one of the edge node voxels is above t and the other below t)* THEN
      *{calculate the position of the isopoint on the cube's edge using linear interpolatio;*
      *place the isopoint into a list*
    *}*
  *p=0*

**Step 1** *scan the list of isopoints until the first unmarked isopoint is found*
  IF *no unmarked isopoints exist in the list* THEN GOTO **Step 0**
  *p=p+1*
  *set this isopoint as initial*
  *mark initial isopoint as belonging to polygon p*
  REPEAT
    *if ((there is another unmarked isopoint in the list that lies on the same cube face with the current isopoint) AND (its associated voxel shares a common cube edge with the current isopoint's associated voxel) OR (the cube face containing the current and the unmarked isopoint contains 3 voxels with grey value above t))* THEN
      *{set this isopoint as the current one*
       *mark the current isopoint as belonging to polygon p}*
  UNTIL *current_isopoint=initial isopoint*
  *store the polygon into the VRML file*
  GOTO **Step 1**

The above algorithm generates all the 15 cases that are predefined by the standard MC algorithm, according to [3] and [7], as it will be shown in the next section. The only difference is that the output is produced in polygons instead of triangles, but with their points equivalently ordered. For instance, the outputs of the standard MC algorithm and the proposed method, for the Case 9, are shown in Fig. 2. The standard MC algorithm produces four triangles whereas the suggested implementation produces a hexagon. Since all the four triangles are coplanar, it is much faster to treat them as a hexagon rather than individual triangles. Furthermore, the algorithm can handle cases where more than one polygon is present in the same cube.

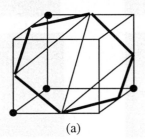 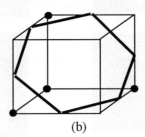

(a)            (b)

**Fig. 2.** Output of the standard MC algorithm. (a) and the proposed implementation. (b) for Case 9 cube configuration of Fig. 1. Note the replacement of the four coplanar triangles by an equivalent hexagon.

The output of the algorithm is in a WWW compliant format, more specifically in VRML 2.0 format [12]. A three-dimensional structure in VRML format can be rotated and scaled in real time even using hardware as inexpensive as a typical PC. VRML viewers are offered with every computer platform, usually with Internet Browsers, allowing interaction with JAVA programs as well as creation of VRML worlds where sensors and multiple users from remote sites can interact, thus enhancing realism. Furthermore, most VRML viewers support co-planar and non-planar polygon rendering.

## 4. Comparison of the proposed implementation with previous methods

An important advantage of proposed technique lie on the fact that definition of different cube configurations is not required. Rotations and comparisons between them and the predefined cube configurations that are met in the actual image are no longer necessary. Execution time is therefore reduced. In Fig. 3, the proposed algorithm generates all the 15 predefined cases with the same order as they appear in Fig. 1. The display of the triangulated isosurface is in VRML format. The orientation may not always coincide with the one of Fig. 1 to achieve better visualisation.

As it has already been mentioned, the proposed method does not depend on the complementary symmetry. Thus, by definition, it does not suffer from a specific type of holes that appears due to this symmetry. In [7], this surface problem is called the type A "hole problem" and three extra cases, additional to the 15 standard ones, have been introduced to tackle it. The three extra cube configurations, effectively defining the complementary configurations of Cases 3, 6 and 7, noted as 3b, 6b and 7b respectively and can be seen in Fig. 4 (a), (c) and (e). The proposed algorithm can generate these additional cases, as it shown in Fig. 3 (b), (d) and (f), respectively.

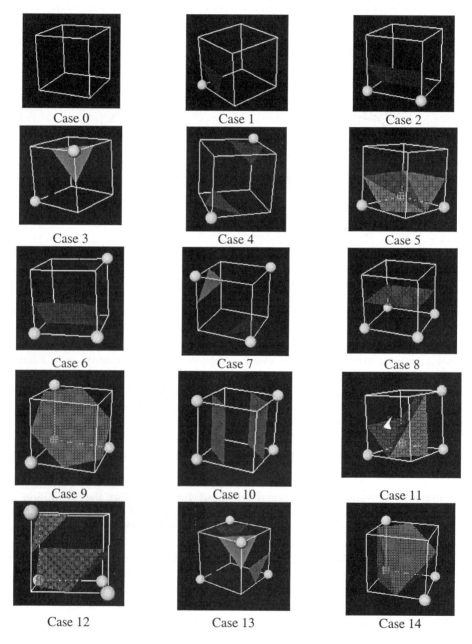

**Fig. 3**. The triangulation of the 15 predefined cube configurations by the proposed implementation. The display of the triangulated isosurface is in VRML format.

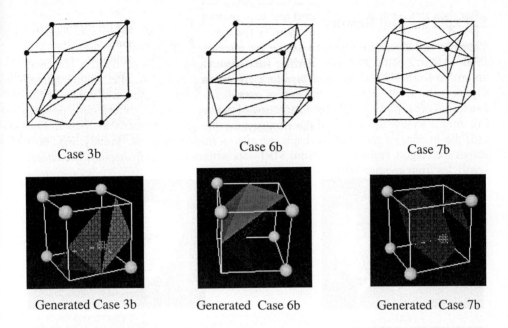

**Fig. 4.** Three extra cube configurations effectively defining the complementary configurations of Cases 3, 6 and 7, noted as Generated Cases 3b, 6b and 7b respectively.

Furthermore, in [13, 14], six additional predefined cases were introduced to tackle the type A "hole problem". The proposed algorithm can generate all these additional cases.

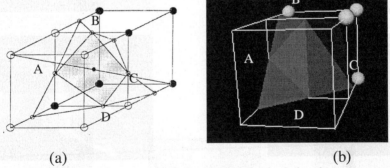

**Fig. 5.** (a) Resolution of a type A "hole problem" (shaded polygon), according to [11]. (b) Resolution of the problem by the proposed implementation (polygon ABCD).

In Fig. 5, one case of a type A "hole problem" is demonstrated, according to [11]. In Figure 5(a), the standard MC algorithm creates a hole (shaded polygon). The proposed algorithm also resolves the type A "hole problem", as it can be seen in Fig. 5(b) where the polygon ABCD corresponds to hole created in Fig. 5(a).

## 5. Experimental Results

The proposed algorithm is able to handle full resolution CT and MRI of human heads, (SGI Indigo 2 with 128 MB RAM). Rendered views produced by the proposed algorithm from 3D real anatomical data are presented in Fig. 6, after segmentation. Fig. 6(a) shows the 3D reconstruction of a skull from full resolution CT data of the head, Fig. 6(b) shows the 3D reconstruction of the bone structure of the same patient, Fig. 6(c) is a close up view of the bone structure showing fine details whereas in Fig. 6(d) the interior of the skull is displayed after a mid-sagittal cut. Finally, Fig. 6(e) shows the largest brain vessels from MRI data whereas Fig. 6(f) displays the lateral ventricles from another MRI brain study.

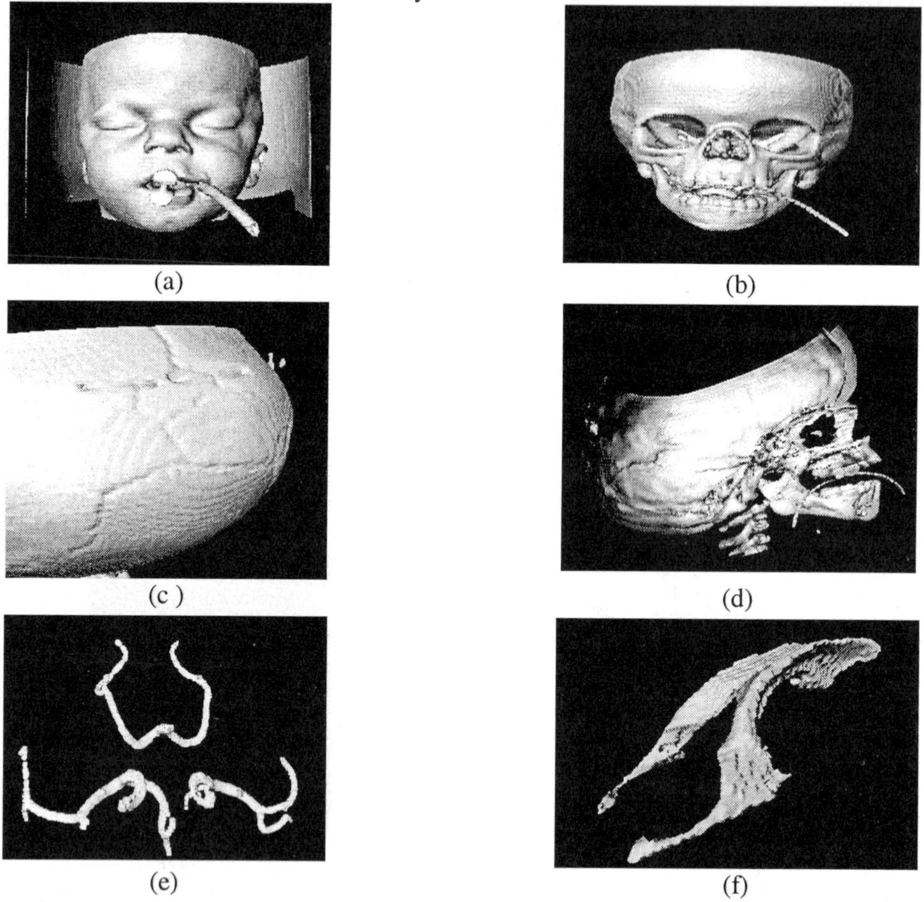

**Fig. 6.** Render views from 3D real medical data (CT and MRI).

It is important to note that the proposed algorithm produced VRML files with reduced number of points and polygons. This is considered a great advantage since it allows real time rotation and rendering to be performed in less expensive hardware than the other algorithms.

## 6. Conclusions

A novel and efficient implementation has been presented, based on the standard MC algorithm in order to produce triangulated surfaces from volumetric data. The proposed algorithm uses a generic rule capable of generating the correct polygons in *any case* of cube configurations. It has been shown that the proposed algorithm can reproduce the 15 predefined cases of cube configurations of the standard MC as well as additional cases presented in literature. Furthermore, it does not suffer from the type A "hole problem", which occurs in the standard MC. The proposed algorithm was implemented in JAVA, generating output files in VRML format from 3D real medical data; it is therefore suitable for platform independent applications (such as the Telemedicine applications). Future research will be focused on the aspect of reducing the number of triangles produced by the proposed implementation.

## References

1. J. D. Boissonat, Shape Reconstruction from Planar Cross Sections, *Computer Vision, Graphics and Image Processing* **44**, 1-29, (1988).
2. H. Fuchs, Z.M. Kedem and S.P. Uselton, Optimal surface reconstruction from planar contours, *Communication of the ACM*, **20**(10), 693-702 (1977).
3. W.E. Lorensen and H.E. Cline, Marching Cubes: High resolution 3D surface construction algorithm. *Computer Graphics* **21**(3), 163-169 (1987).
4. S.B. Xu and W.X. Lu, Surface Reconstruction of 3D Objects in Computerized Tomography, *Computer Vision, Graphics and Image Processing* **44**, 270-278 (1988).
5. W.C. Lin, S.Y. Chen and C.T. Chen, A new surface interpolation technique for reconstructing 3D objects from serial cross sections, *Computer Vision, Graphics and Image Processing* **48**, 124-143 (1989).
6. H.H. Baker, Building surfaces of evolution: the Weaving Wall. *International Journal of Computer Vision*, **3**,51-71 (1989).
7. C. Zhou, R. Shu and M. S. Kankanhalli, Handling small features in isosurface generation using Marching Cubes *Computer & Graphics* **18**(6), 845-848 (1994).
8. B.A. Payne and A.W. Toga, Surface mapping brain function on 3D models, *IEEE Computer Graphics and Applications* **10**(2), 33-41 (1990).
9. A.D. Kalvin, Segmentation and Surface based modeling of objects in 3 Biomedical images, *Ph.D. Thesis*, Courant Institute of Mathematical Sciences, New York University (1991).
10. J. Wilhems and A. van Gelder, Topological considerations in isosurface generation, *Computer Graphics* **24**(5), 79-86 (1990).
11. M.J.Durst, Additional reference to "Marching Cubes", *Computer Graphics* **22**(2), 72-73 (1988).
12. R. Lea, K. Matsuda and K. Miyashita, *JAVA for 3D and VRML Worlds*, New Riders Publ., Indiana, (1996).
13. J. Sharman,, The Marching Cubes Algorithm, www.exaflop.org/docs/marchcubes, (1998).
14. Shoeb, "Improved Marching Cubes", enuxsa.eas.asu.edu/~shoeb/graphics /improved.html, (1998).

# Workshop:

# High Performance Numerical Computation

# and Applications

# Multilevel Algebraic Elliptic Solvers[*]

Tony F. Chan[**] and Petr Vaněk[***]

Department of Mathematics, University of California at Los Angeles, 405 Hilgard Ave, Los Angeles, CA, 90024

**Abstract:** We survey some of the recent research in developing multilevel algebraic solvers for elliptic problems. A key concept is the design of a hierarchy of coarse spaces and related interpolation operators which together satisfy certain approximation and stability properties to ensure the rapid convergence of the resulting multigrid algorithms. We will discuss smoothed agglomeration methods, harmonic extension methods, and global energy minimization methods for the construction of these coarse spaces and interpolation operators.

## 1 Introduction

There has been a recent resurgence of interest in algebraic multilevel elliptic solvers. Part of the motivation for considering the *multilevel* approach is that it is the only general method for deriving scalable solvers for large problems. Another reason is that multilevel methods are often naturally parallelizable. The main motivation for considering *algebraic* solvers is ease of use as often the user need only supply the problem in purely matrix-vector form without referring to (often very complicated) grid and other physical and geometric information.

The interest in algebraic multilevel solvers has been heightened by the recent emergence of *unstructured grids* as a powerful and flexible approach in scientific computing for modelling complex geometries. This popularity has led to increased interest in developing fast, efficiently parallelizable elliptic (and some non-elliptic) solvers on such grids. In particular, there has been a lot of research activities in adapting "optimal" multilevel elliptic solvers for structured grids, such as multigrid and domain decomposition methods, to unstructured grids [4]. The absence of a natural coarse grid hierarchy has led to several approaches to developing multilevel methods, ranging from computational geometric approaches for explicitly constructing appropriate coarse grids and the associated intergrid transfer operators, to more abstract agglomeration approaches of constructing appropriate nested or non-nested coarse spaces, together with the associated interpolation operators, over agglomerated coarse grid elements. The resulting methods are often completely *algebraic*, in the sense of not needing to know about geometric information about the underlying

---

[*] This research has been supported by NSF grant ASC-972057, Sandia National Laboratory grant LG-4440 and NASA Ames grant NAG2-1238.
[**] E-mail address: chan@math.ucla.edu
[***] E-mail address: pvanek@math.ucla.edu

grids. Thus, in summary, the recent interest in algebraic multilevel methods are due to two factors: the emergence of unstructured grids and the need for parallelizable, scalable and easy-to-use elliptic solvers for increasingly large problems in scientific computing.

In this paper, we'll give a brief survey of some of the key ideas and approaches in this field and also discuss in some more details several approahes that we (with collaborators) have been developing based on energy minimization and smoothed aggregation. The plan of the paper is as follows, after a brief review of the basic multigrid algorithm in Sec. 2, highlighting the essential role of the interpolation operator, we summarize in Sec. 3 the essential desirable properties that the interpolation operator should satisfy in order to lead to rapidly convergent and efficient multilevel algorithms. In Sec. 4, we discuss algorithms which computes *locally* the interpolation operator and in Sec. 5, we discuss methods which compute all of the interpolation operators in a global, coupled, manner.

## 2 Variational multigrid algorithm

In this section we briefly recall the standard variational (Ritz-Galerkin) coarsening scheme and the multigrid cycle for solving the linear system

$$A\mathbf{x} = \mathbf{b}$$

with symmetric, positive definite matrix $A$.

Given an $n \times n$ matrix $A$, we create prolongators $I_{l+1}^l$ and coarse level matrices $A_l$ following this general scheme:

**Algorithm 1 (setup)** *Initialize $l = 1$, set $n_1 = n$, $A_1 = A$.*
**Reapeat**
- *create $n_l \times n_{l+1}$ $n_{l+1} < n_l$ prolongator $I_{l+1}^l$,*
- *get the Ritz-Galerkin coarse-level matrix*

$$A_{l+1} = (I_{l+1}^l)^T A_l I_{l+1}^l \quad \text{and set } l \leftarrow l + 1 \tag{1}$$

**until** $A_l$ *is sufficiently small to be treated by a direct solver.*
*Set the number of levels $L = l$.*

The key step of the above variational coarsening scheme is the construction of prolongators $I_{l+1}^l$. There are two main issues involved: a) specifying the nonzero structure of prolongators and, b) finding appropriate values of nonzero entries. In a finite element context, the nonzero structure of prolongators corresponds to the supports of coarse-level "macroelements" (shape function supports.) The values of nonzero entries then determine shape functions on given supports.

The purpose of this paper is to clarify the requirements on prolongators that lead to an efficient multilevel solver and present algorithms that strive to match those requirements.

One iteration $\mathbf{x} \leftarrow MG(\mathbf{x}, \mathbf{b})$ of the multigrid algorithm is as follows:

**Algorithm 2 (iteration)** *Let* $R_l : \mathbb{R}^{n_l} \to \mathbb{R}^{n_l}$, $l = 1, \ldots, L-1$ *be given smoother preconditioners and* $\nu, \gamma > 0$ *be a given smoothing and cycle parameter, respectively and* $A \equiv A_1$ *a stiffness matrix. Set* $MG = MG_1$, *where* $MG_l(\cdot, \cdot)$, $l = 1, \ldots, L-1$ *is defined by:*

**Pre-smoothing:** *Perform* $\nu$ *iterations of* $\mathbf{x}^l \leftarrow (I - R_l A_l)\mathbf{x}^l + R_l \mathbf{b}^l$.
**Coarse grid correction:**
- *Set* $\mathbf{b}^{l+1} = (I_{l+1}^l)^T (\mathbf{b}^l - A_l \mathbf{x}^l)$,
- *if* $l+1 = L$, *solve* $A_{l+1}\mathbf{x}^{l+1} = \mathbf{b}^{l+1}$ *by a direct method, otherwise set* $\mathbf{x}^{l+1} = 0$ *and perform* $\gamma$ *iterations of* $\mathbf{x}^{l+1} \leftarrow MG_{l+1}(\mathbf{x}^{l+1}, \mathbf{b}^{l+1})$,
- *correct the solution on level* $l$ *by* $\mathbf{x}^l \leftarrow \mathbf{x}^l + I_{l+1}^l \mathbf{x}^{l+1}$.

**Post-smoothing:** *Perform* $\nu$ *iterations of* $\mathbf{x}^l \leftarrow (I - R_l A_l)\mathbf{x}^l + R_l \mathbf{b}^l$.

## 3 Construction of prolongators - objectives

Following the considerations in [5], this section specifes requirements on prolongators $I_{l+1}^l$ needed to design an efficient multigrid solver.

Let us start the discussion by considering a simple two-level method. It is well-known that commonly used smoothers eliminate efficiently fine-level errors **e** which satisfy

$$\frac{\mathbf{e}^T A \mathbf{e}}{\mathbf{e}^T \mathbf{e}} \approx \varrho(A_l). \tag{2}$$

This behavior can be demonstrated on a case of a simple Richardson-type smoothing procedure

$$\mathbf{x} \leftarrow (I - \varrho(A)^{-1} A) + \varrho(A)^{-1} \mathbf{b} \tag{3}$$

(that is, $R_1 = \varrho(A)^{-1} I$ in our notation.) Setting $\hat{\mathbf{x}} = A^{-1}\mathbf{b}$, the smoothing procedure (3) replaces the error $\mathbf{e} = \mathbf{e}(\mathbf{x}) \equiv \mathbf{x} - \hat{\mathbf{x}}$ by a smoothed error $(I - \varrho(A)^{-1} A)\mathbf{e}$. Let us demonstrate that the smoother (3) is efficient in suppressing errors satisfying (2).

Denoting $\langle \mathbf{x}, \mathbf{y} \rangle = \mathbf{x}^T \mathbf{y}$ and $\|\mathbf{x}\| = \langle \mathbf{x}, \mathbf{x} \rangle^{1/2}$, elementary manipulations give

$$\|(I - \varrho(A)^{-1} A)\mathbf{e}\|^2 = \|\mathbf{e}\|^2 - 2\varrho(A)^{-1} \langle A\mathbf{e}, \mathbf{e} \rangle + \varrho(A)^{-2} \|A\mathbf{e}\|^2.$$

Using the estimate $\|A\mathbf{e}\|^2 = \langle A\mathbf{e}, A\mathbf{e} \rangle \leq \varrho(A) \langle A\mathbf{e}, \mathbf{e} \rangle$, we get

$$\|(I - \varrho(A)^{-1} A)\mathbf{e}\|^2 \leq \|\mathbf{e}\|^2 - \frac{1}{\varrho(A)} \langle A\mathbf{e}, \mathbf{e} \rangle = \left(1 - \frac{1}{\varrho(A)} \frac{\langle A\mathbf{e}, \mathbf{e} \rangle}{\|\mathbf{e}\|^2}\right) \|\mathbf{e}\|^2.$$

As one can see from the last estimate, the "efficiency" of the smoother is guided by the ratio on the left-hand side of (2); the convergence speeds up as the left-hand side of (2) approaches its maximum $\varrho(A)$.

Two-level methods strive to split the computational work between a fine-level smoothing and a coarse-level correction. To accomplish this goal succesfully, one has to assure that error components not suppressed by the smoother significantly can be approximated well by coarse-level vectors $I_2^1\mathbf{v}$. In other words, we need a coarse space $\mathrm{Rng}\,(I_{l+1}^l)$ that approximates well fine-level $\mathbf{e}$ vectors which satisfy

$$\frac{\langle A\mathbf{e}, \mathbf{e}\rangle}{\|\mathbf{e}\|^2} \ll \varrho(A_l).$$

The smaller the ratio $\langle A\mathbf{e}, \mathbf{e}\rangle/\|\mathbf{e}\|^2$ is, the better approximation $I_2^1\mathbf{v}$ is needed. This requirement can be seen in a key assumption of the classical variational two-level method theory [2] (when we divide both sides of the inequality by $\|\mathbf{e}\|^2$):

*There is a positive constant $C_A$ such that for every fine level vector $\mathbf{e}$ we can find a coarse-level vector $\mathbf{v}$ such that*

$$\|\mathbf{e} - I_2^1\mathbf{v}\|^2 \leq \frac{C_A}{\varrho(A)} \langle A\mathbf{e}, \mathbf{e}\rangle. \qquad (4)$$

Assuming the smoother (3) is used, the convergence theory in [2] gives an $A$−norm rate of convergence estimate ($\mathbf{x}^i$ denotes the result after $i$ iterations)

$$\|\mathbf{e}(\mathbf{x}^{i+1})\|_A \leq (1 - 1/C_A)\|\mathbf{e}(\mathbf{x}^i)\|_A$$

In the general multilevel case, the mechanism is essentially the same. For each level $l < L$ and every vector $\mathbf{u}_l$ on the level $l$, we need coarse approximation $\mathbf{u}_{l+1}$ such that

$$\|\mathbf{u}_l - I_{l+1}^l\mathbf{u}_{l+1}\|^2 \leq \frac{C_A}{\varrho(A_l)} \langle A_l\mathbf{u}_l, \mathbf{u}_l\rangle. \qquad (5)$$

There are two natural ways to satisfy (5), and their combination is needed to accomplish it cheaply:
a) minimizing the left-hand side of (5) (i.e. controlling the approximation properties of the range of $I_{l+1}^l$),
b) maximizing the right-hand side (making (5) easier to satisfy).
To make the right-hand side of (5) large, we will minimzize $\varrho(A_l)$ subject to certain constraints. Note that the easiest (but useless) way of making $\varrho(A_l)$ small is to multiply the prolongator $I_l^{l-1}$ by a small constant $\alpha$. Then $A_l \equiv (I_l^{l-1})^T A_{l-1} I_l^{l-1}$ becomes $\alpha^2 A_l$, $\varrho(A_l)$ becomes $\alpha^2 \varrho(A_l)$ and the right-hand side of (5) remains unchanged for all vectors $\mathbf{u}_l$. Such a degeneration of a minimization process will be prevented by constraints aimed to a).

We sum up the above considerations in a form of list of requirements on $I_{l+1}^l$ that lead to an efficient multigrid solver.

**Zero-energy modes (local kernel) preserving property.** Let $B^1$ be the matrix consisting of columns that generate the local kernel of $A$. That is,

$$(AB^1)_{ij} = 0$$

for all columns $j$ of $B^1$ and all degrees of freedom $i$ that are not located on elements with Dirichlet boundary conditions.

We require columns of $B^1$ (the zero energy modes) to be represented on each level. More precisely, we want prolongators $I_{l+1}^l$ such that there are coarse-level reperesentations $B^2, \ldots, B^L$ of $B^1$ satisfying (aside from Dirichlet boundary conditions, see Rem. 1.)

$$B^1 = I_2^1 B^2 = \ldots = I_2^1 \ldots I_L^{L-1} B^L \text{ or equivalently, } B^l = I_{l+1}^l B^{l+1} \; \forall l < L. \quad (6)$$

Note that the finest level zero-energy modes $B^1$ must be supplied as input data. To create prolongators $I_{l+1}^1$ satisfying (6), we first generate simultaneously $I_2^1$ and $B^2$ such that $B^1 = I_2^1 B^2$. Then $B^2$ is used to get $I_3^2$ and $B^3$, etc.

The motivation arise from the need to capture the low-energy vectors by coarse levels. On the theoretical front, (6) together with bounded overlaps of basis-function supports allows to prove approximation properties of the form: *For every* $\mathbf{u}_l$ *there is a coarse level vector* $\mathbf{u}_{l+1}$ *such that*

$$\|\mathbf{u}_l - I_{l+1}^l \mathbf{u}_{l+1}\|^2 \leq C_A(l) \langle A_l \mathbf{u}_l, \mathbf{u}_l \rangle \quad (7)$$

using Bramble-Hilbert lemma type arguments, see [6].

**Small spectral radii of coarse-level matrices.** Assume $B^l$ is available when constructing $I_{l+1}^l$ and its coarse-level representation $B^{l+1}$ has been chosen. We want

$$\varrho(A_{l+1}) = \varrho\left((I_{l+1}^l)^T A_l I_{l+1}^l\right)$$

to be as small as possible; the requirement (6) prevents the "scaling degenearion" mentioned above.

Note that from (7), the inequality (5) holds with $C_A = C(l)\varrho(A_l)$. Our goal is to create the prolongators $I_{l+1}^l$ so that $C(l)\varrho(A_l) \leq C$ with a small constant $C$ independent of the level $l$. As the level $l$ increases, coarse level approximation properties deteriorate and $C(l)$ in (7) grows. To guarantee a good convergence rate, this "approximation" loss must be compensated by a decrease of the "maximal energy meassure" $\varrho(A_l)$.

*Note 1.* (Relaxing (6) near Dirichlet boundary conditions). For nodes that are close to Dirichlet boundary conditions, the objective (6) contradicts to the smoothness requirement (small $\varrho(A_l)$.)

As zero-energy modes do not respect Dirichlet boundary conditions in general, (6) causes the function analogues of coarse-level vectors to be nearly discontinuous close to Dirichlet constraints.

For the above reason, we relax (6) as follows:

$$(B^l)_{i,j} = (I_{l+1}^l B^{l+1})_{ij} \quad \text{for all} \quad i, j \text{ such that } (A_l B^l)_{ij} = 0. \quad (8)$$

In other words, we require (6) only for the degrees of freedom where the column of $B^l$ is a valid local kernel.

**Sparsity of coarse-level matrices.** The coarse-level matrices $A_l$ should be sparse. More precisely, the number of nonzero entries per column of $A_l$ should be bounded by a small constant independent of the level. The reasons for this requirement are mostly practical; the excessive fill-in of coarse-level matrices increases the amount of computational work needed both in the setup and the iterations.

In a finite element context, nonzero off-diagonal entries of $A_l$ indicate the overlap of basis-function supports. The assumption on bounded overlaps of supports simplify a convergence analysis, but can be avoided ([7].)

**Geometry of coarsening and coupling.** Prolonagtors $I_{l+1}^l$ must have only bounded number of nonzeroes per column and the nonzero structure of the columns should follow large entries of $A_l$.

More specifically, two entries $(I_{l+1}^l)_{ij}$, $(I_{l+1}^l)_{kj}$ of the $j$-th column of the prolongator are allowed to be nonzeroes only if $i,k$ are two degrees of freedom on the level $l$ that are strongly coupled in the matrix $A_l^\alpha$, where $\alpha$ is a small integer (eg. 1,2 or 3.)

The motivation for this objective comes again from considerations about underlying (continuous) basis function supports. The bounded number of nonzeroes per column of $I_{l+1}^l$ together with the "distance" constraint (small $\alpha$) reflects the need to control the meassure and diameter of supports.

The need to follow strong connections is important for solving anisotropic problems, where the coarsening should be performed only in the direction that corresponds to a large coefficient (semicoarsening.)

The need to follow strong connections in the coarsening can be avoided if more sophisticated (colored or line) smoothers are used.

## 4 Local energy minimizatiom methods

The zero-energy modes preservation constraint (14) is global in a sense that one cannot update two columns of the prolongator indepndently without (potentialy) violating (14). To avoid data dependencies that come with such a globality, two coarsening techniques secribed here use a following princile: *If all the columns of the prolongator are updated (independently) via a multiplication by (a same) linear mapping that behaves like the identity for zero energy modes, the updated prolongator satisfies (14) again.* As those methods do not need explicit means to establish a global communication between the columns, we call them local.

### 4.1 Smoothed aggregation method

The smoothed aggregation method (proposed by Vaněk in [9,8] and further developed in [5,10,6]), builds the prolongator in the form

$$I_{l+1}^l = S_l P_{l+1}^l, \tag{9}$$

where $P_{l+1}^l$ is a very simple *tentative prolongator* satisfying

$$B^l = P_{l+1}^l B^{l+1} \tag{10}$$

and $S_l$ is a Richardson-type *prolongator smoother* derived from matrix $A_l$, e.g.,

$$S_l = I - \frac{\omega}{\varrho(A_l)} A_l. \tag{11}$$

Note that on the finest level, the zero-energy modes $B^1$ must be given. To satisfy (10), we build simultaneously $P_2^1$ and $B^2$ so that $B^1 = P_2^1 B^2$. Then using $B^2$, we construct $P_3^2$ and $B^3$, etc. For details, see Alg. 3.

Before specifying $P_{l+1}^l$ and $S_l$, we briefly discuss their purpose and relationship to requirements on $I_{l+1}^l$ listed in Sec. 3.

The requirement (8) is enforced through (10). Assume a prolongator smoother of the form (11) is being used. Then the smoothed prolongator satisfies the relaxed constraint (8). Indeed, from (10) (setting $\alpha = \omega/\varrho(A_l)$ for convenience),

$$I_{l+1}^l B^{l+1} = (I - \alpha A_l) P_{l+1}^l B^{l+1} = (I - \alpha A_l) B^l = B^l - \alpha A_l B^l,$$

and one can see that the matrices $I_{l+1}^l B^{l+1}$ and $B^l$ differ only in entries where $A_l B^l \neq 0$.

The purpose of the prolongator smoother $S_l$ is to minimize $\varrho(A_{l+1})$. As shown at the beginning of Sec. 3, simple pointwise smoothers eliminate efficiently high-energy errors. The prolongator smoother (11) is an error propagation operator of a Richardson-type iteration. By appling it to the range of $P_{l+1}^l$, we suppress high-energy vectors, which in turn reduces the "maximal energy measure" $\varrho(A_{l+1})$. Rigorous explanation is given in Remark 2.

The geometrical requirements are respected through a choice of aggregates that determine a nonzero structure of $P_{l+1}^l$, see Alg. 3.

*Note 2.* (PROLONGATOR SMOOTHING EFFECT) Consider a tentative prolongator $P_{l+1}^l$ such that $\|P_{l+1}^l \mathbf{x}\| = \|\mathbf{x}\|$ for all $\mathbf{x} \in \mathbb{R}^{n_{l+1}}$. Note that Alg. 3 creates $P_{l+1}^l$ satisfying above isometry, see [6]. Consider a prolongator smoother in the form (11). We will show that if $\omega$ is chosen properly, the spectral radius $\varrho(A_{l+1})$ is at least 9 times smaller then $\varrho(A_l)$.

As $S_l$ is a polynomial in $A_l$, $S_l$ is symmetric and commutes with $A_l$. Hence (1) and (9) gives $A_{l+1} = (S_l P_{l+1}^l)^T A_l (S_l P_{l+1}^l) = (P_{l+1}^l)^T S_l^2 A_l P_{l+1}^l$. This identity gives

$$\varrho(A_{l+1}) = \max_{\mathbf{x} \in \mathbb{R}^{n_{l+1}}} \frac{\langle A_{l+1} \mathbf{x}, \mathbf{x} \rangle}{\|\mathbf{x}\|^2} = \max_{\mathbf{x} \in \mathbb{R}^{n_{l+1}}} \frac{\langle S_l^2 A_l P_{l+1}^l \mathbf{x}, P_{l+1}^l \mathbf{x} \rangle}{\|P_{l+1}^l \mathbf{x}\|^2} \leq \varrho(S_l^2 A_l).$$

The spectral mapping theorem applied to the last estimate yields ($\sigma$ denotes the spectrum)

$$\varrho(S_l^2 A_l) = \max_{t \in \sigma(A_l)} \left(1 - \frac{\omega}{\varrho(A_l)} t\right)^2 t \leq \max_{t \in [0, \varrho(A_l)]} \left(1 - \frac{\omega}{\varrho(A_l)} t\right)^2 t = \varrho(A_l) \cdot \max_{\tau \in [0,1]} (1 - \omega \tau)^2 \tau.$$

It is easy to verify that for $\omega = 4/3$, the expression on the right-hand side of the last estimate equals to $\frac{1}{9} \varrho(A_l)$.

Based on Rem. 2, we use a prolongator smoother

$$S_l = I - \frac{4}{3}\bar{\varrho}(A_l)^{-1}A_l, \tag{12}$$

where $\bar{\varrho}(A_l)$ is an available upper bound of $\varrho(A_l)$.

It remains to specify the tentative prolongator $P_{l+1}^l$.

Assume $B^l$ is available and denote the number of its columns by $r$. Our goal is to create the tentative prolongator $P_{l+1}^l$ and the coarse-level representation $B^{l+1}$ of $B^l$ satisfying (10).

Our construction is based on the supernodes aggregation concept. On each level, degrees of freedom are organized in small disjoint clusters called supernodes. On the finest level, these clusters have to be specified, e.g., as the sets of degrees of freedom associated with the finite element vertices, the coarse level supernodes are then created by our aggregation algorithm. The prolongator $P_{l+1}^l$ is constructed from a given system of aggregates $\{\mathcal{A}_i^l\}_{i=1}^{N_l}$ that forms a disjoint covering of level $l$ supernodes. A simple greedy algorithm for generating aggregates based on the structure of the matrix $A_l$ is given in [5]. The property (10) is enforced aggregate by aggregate; columns of $P_{l+1}^l$ associated with the aggregate $\mathcal{A}_i^l$ are formed by orthonormalized restrictions of the columns of $B^l$ onto the aggregate $\mathcal{A}_i^l$. For each aggregate, such a construction gives rise to $r$ degrees of freedom on the coarse level forming a coarse level supernode.

The detailed algorithm follows. For ease of presentation, we assume that the fine level supernodes are numbered by consecutive numbers within each aggregate. This assumption can be easily avoided by renumbering.

**Algorithm 3** *For the given system of aggregates $\{\mathcal{A}_i^l\}_{i=1}^{N_l}$ and the $n_l \times r$ matrix $B^l$ satisfying $P_l^1 B^l = B^1$, we create a prolongator $P_{l+1}^l$, a matrix $B^{l+1}$ satisfying (10) and supernodes on level $l+1$ as follows:*

1. *Let $d_i$ denote the number of degrees of freedom associated with aggregate $\mathcal{A}_i^l$. Partition the $n_l \times r$ matrix $B^l$ into blocks $B_i^l$ of size $d_i \times r$, $i = 1, \ldots, N_l$, each corresponding to the set of degrees of freedom on an aggregate $\mathcal{A}_i^l$ (see Fig. 1).*
2. *Decompose $B_i^l = Q_i^l R_i^l$, where $Q_i^l$ is an $d_i \times r$ orthogonal matrix, and $R_i^l$ is an $r \times r$ upper triangular matrix.*
3. *Using the blocks $Q_i^l$, $i = 1, \ldots, N_l$, create the prolongator $P_{l+1}^l$ as shown by Fig. 1.*
4. *Create $B^{l+1}$ consisting of the blocks $R_i^l$, $i = 1, \ldots, N_l$, (see Fig. 1.)*
5. *For each aggregate $\mathcal{A}_i^l$, the coarsening gives rise to $r$ degrees of freedom on the coarse level (the $i-$th block column of $P_{l+1}^l$). These degrees of freedom define the $i-$th coarse level supernode.*

The abstract convergence bounds for smoothed aggregation methods are given in [6]. Under weak assumptions on the aggregates it is shown that each iteration reduces the $A-$norm of the error at least by a factor $1 - C/L^3$, where $L$ is the number of levels and $C$ is a constant independent of $L$. When applied to a linear system obtained by discretizing a second order elliptic problem, $C$ is independent of the meshsize.

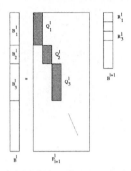

**Fig. 1.** The construction of a tentative prolongator $P^l_{l+1}$ and $B^{l+1}$

### 4.2 Macroelement based coarsening technique

The method outlined here creates – by algebraic means – coarse levels that are very close to finite element spaces, while overcoming their geometrical limitations. For more details, see [3].

We restrict our brief explanation to case a of coarsening the finest level finite element stifness matrix $A$. As the method reproduces essential properties of finite element spaces on the next level, the further coarsening is analogous. The main steps follow:

**1.** Using a nodal adjacency information contained in the stiffness matrix, the elements are clustered into small nonoverlapping groups called coarse-level *macroelements*. Then the coarse points (macroelement verices) are specified, see Fig.2.

**2.** The basis functions (columns of $I_2^1$) satisfying (14) are first defined only on interfaces (boundaries of macroelements). Efficient constructions of the interface basis are given in [3].

**3.** The basis is then extended inside the element *harmonically*, i.e. by setting macroelement interior degrees of freedom so that the energy (evaluated over the macroelement) is minimal. This operation is performed by solving Dirichlet problems corresponding to the macroelemnts.

The harmonic extension enforces the smoothness of the coarse level basis. Further, since the harmonic extension returns the vector with minimal energy, the harmonic extension of a zero-energy mode restricted to the interface is the zero-energy mode. Hence (14) is satisfied on all macroelements where $B^1$ is a valid local kernel.

## 5 Global energy minimization method

This section describes a method of constructing prolongators by minimizing the trace of the coarse-level matrix. In a finite element context, diagonal entries of coarse-level matrices correspond to energies of coarse-space basis functions and the trace minimization can be interpreted a minimization of their sum.

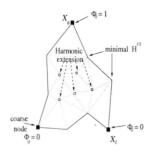

**Fig. 2.** The macroelement

Using the requirements on prolongators formulated in [5], the energy minimization method has been developed independently in [12] and [11]. This section follows the slightly more general variant proposed in [11].

Based on the objectives given in Sec. 3, the most straightforward way of constructing the prolongator $I_{l+1}^{l}$ is the following:

Assume $B^l$ has been cretated during the construction of $I_2^1, \ldots I_l^{l-1}$.

1. Define the nonzero structure of $I_{l+1}^{l}$ using the geometrical objectives in Sec. 3.
2. Choose the coarse-level representation $B^{l+1}$ of $B^l$.
3. Find $I_{l+1}^{l}$ of a given nonzero structure minimizing $\varrho((I_{l+1}^{l})^T A_l I_{l+1}^{l})$ subject to (8).

Direct minimization of a functional by an iterative method usually requires the calculation of its value and steepest descent direction in every iteration. Such calculations are very expensive for a spectral radius objective function.

On the other hand, it is very cheap to evaluate a trace of the matrix (sum of eigenvalues=sum of diagonal entries.) As the contribution of large eigenvalues prevails in their sum, the trace minimization is aimed mainly to the higher part of the spectrum. Further, as demonstrated at the beginning of Sec. 3, simple iterative methods tend to eliminate "energetically reach" error components within a few iterations.

For the above heuristical reasons, one can expect to get a good prolongator by solving approximately the minimization problem:

Let $Z$ be a 0/1 matrix defining a nonzero structure of $I_{l+1}^{l}$ and the symbol $*$ denotes the matrix multiplication entry by entry ($Z * I_{l+1}^{l} = I_{l+1}^{l}$ means that $I_{l+1}^{l}$ has nonzeroes only in positions $i,j$ marked by a flag $Z_{ij} = 1$.)

$$\text{find } I_{l+1}^{l} \text{ minimizing } J(\cdot) : I_{l+1}^{l} \mapsto \text{tr}((I_{l+1}^{l})^T A_l I_{l+1}^{l}) \tag{13}$$

subject to

$$I_{l+1}^{l} B^{l+1} = B_{mod}^{l} \text{ and } Z * I_{l+1}^{l} = I_{l+1}^{l}, \tag{14}$$

Here, $B_{mod}^{l}$ is a (properly) modified matrix $B^l$ such that

$$(B^l - B_{mod}^l)_{ij} \neq 0 \text{ only if } (AB^{l+1})_{ij} \neq 0.$$

Note that if the prolongator satisfies the above constraints, it will also satisfy (8).

Following the considerations in Remark 1, we modify the zero-energy modes $B^{l+1}$ to eliminate "discontinuities" near Dirichlet boundary conditions. The natural way consists in replacing all columns $\mathbf{b}$ of $B^l$ by modified columns $\mathbf{b}'$

minimizing $\langle A_l \mathbf{b}', \mathbf{b}' \rangle$ subject to $\mathbf{b}_i = \mathbf{b}'_i$ for all $i : (A_l \mathbf{b})_i = 0$.

The above procedure typically requires to solve a well-conditioned system of a modest size. Few iterations of a simple pointwise iterative method usually gives a satisfactory approximation.

In what follows we omit the subscript in $B^l_{mod}$; the original matrix $B^l$ is not needed anymore.

To solve (13), we use a projected gradient type iterative method that requires an initial guess $I^l_{l+1}$ and $B^{l+1}$ satisfying (14). Those data can be obtained by the aggregation Algorithm 3 or by smoothed aggregations.

Once the initial guess $I^l_{l+1}$ satisfying (14) is available, we search for correction $G$ such that $I^l_{l+1} - G$ satisfies (14) again and $J(I^l_{l+1} - G)$ is as small as possible. Obviously, *admissible corrections* form a subspace of the space $\mathbf{M}^{n_l \times n_{l+1}}$ of all $n_l \times n_{l+1}$ matrices,

$$\mathbf{U} = \{U \in \mathbf{M}^{n_l \times n_{l+1}} : UB^{l+1} = 0 \text{ and } Z * U = U\} \quad (15)$$

and their columns $U_{i*}$, $U \in \mathbf{U}$, form linear spaces

$$\mathbf{U}_i = \{U \in \mathbf{M}^{1 \times n_{l+1}} : UB^{l+1} = 0 \text{ and } Z_{i*} * U = U\}. \quad (16)$$

Here, $Z_{i*}$ denotes the $i$-th row of $Z$.

First, we calculate the gradient of $J(I^l_{l+1})$ [1] in the space $\mathbf{Z} \equiv \{U \in \mathbf{M}^{n_l \times n_{l+1}} : Z * U = U\}$ by

$$G = Z * (A_l I^l_{l+1}),$$

see [11], Lemma 4.1. In general, $G$ is not an admissible correction as $GB^{l+1}$ is not guaranteed to be zero. We construct its (best) admissible approximation

$$G' \in \mathbf{U} \text{ such that } \operatorname{tr}\left((G - G')(G - G')^T\right) \text{ is minimal.}$$

In other words, we project $G \in \mathbf{Z}$ onto $\mathbf{U}$ using a projection orthogonal in Hilbert space equipped with the Frobenius norm $\|\cdot\|_{fr} : U \mapsto \operatorname{tr}(UU^T)^{1/2}$. Note that the Frobenius norm is related to Euclidean norms of rows $U_{i*}$ by

$$\|U\|^2_{fr} \equiv \operatorname{tr}(UU^T) = \sum_{j=1}^{n_l} \|U_{i*}\|^2. \quad (17)$$

The "orthogonality" of the columns in (17) together with the constraint (16) allows to implement the projection operator row by row. Indeed, one can easily see that to get

$$G' \in \mathbf{U} \text{ minimizing } \operatorname{tr}\left((G - G')(G - G')^T\right) = \sum_{i=1}^{n_l} \|G_{i*} - G'_{i*}\|^2, \quad (18)$$

---

[1] That is $G \in \mathbf{Z}$ maximizing $\frac{d}{dt}\operatorname{tr}((I^l_{l+1} + tG)^T A_l (I^l_{l+1} + tG))$ at $t = 0$.

it is sufficient to find (independently) its rows

$$G'_{i*} \in \mathbf{U}_i \text{ minimizing } \|G_{i*} - G'_{i*}\|, \quad i = 1, \ldots, n_l. \tag{19}$$

To do so, we further reformulate the row constraints (16) using the fact that $G \in \mathbf{Z}$,

$$(R^i)^T G^T_{i*} = 0, \quad \text{where } R^i_{k*} = \begin{cases} B^{l+1}_{k*} \text{(copy the line) if } Z_{ik} = 1, \\ 0 \quad \text{otherwise.} \end{cases} \tag{20}$$

The Euclidean orthogonal projection of a column vector $\mathbf{x} \in \mathbb{R}^n$ onto the nullspace of a matrix $R^T$, $R \in \mathbf{M}^{n \times m}$, is given by the well-known formula

$$\mathbf{x} \mapsto [I - R(R^T R)^+ (R)^T] \mathbf{x}, \tag{21}$$

where $+$ denotes a pseudoinverse.

Using the equivalence of the minimization problems (18) and (19), the row constraints in the form (20) and the formula (21), the projection of a gradient can be performed as follows:

**Algorithm 4 (Projection of a gradient)** *For given matrix $G \in \mathbf{Z}$, construct $G' = PG \in \mathbf{U}$ minimizing (18) as follows:*
**For each row $G_{i*}$ of $G$, $i = 1, \ldots, n_l$, do**
- $R = R_i$, where $R_i$ is given by (20),
- $\mathbf{y} = [I - R((R)^T R)^+ (R)^T](G_{i*})^T$,
- create the $i-$th row $G'_{*i} = \mathbf{y}^T$
**end for.**

Now we are ready to write down the minimization algorithm:

**Algorithm 5 (Energy minimization - iteration)** *Given a matrix $A_l$, initial guess $I^l_{l+1}$, (post-processed) zero energy modes $B^l$, their coarse-level representation $B^{l+1}$ and the nonzero structure $Z$ satisfying (14), find $I^l_{l+1}$ minimizing $J(.)$ subject to (14) as follows:*
**Repeat**
- Compute the gradient of $J(I^l_{l+1})$ in the space $\mathbf{Z}$ by $G = Z * (A_l I^l_{l+1})$,
- Get the projected gradient $G' = PG \in \mathbf{U}$ using Alg. 4.
- update $I^l_{l+1} \leftarrow I^l_{l+1} - \alpha G'$, where $\alpha \in (0, 2/\varrho(A_l))$ is a given parameter.
**until convergence.**

Note that $(R_i)^T R_i$ is an $r \times r$ matrix, where $r$ is the number of columns of $B^1$ ($r = 1$ for scalar elliptic problems, 3 for planar elasticity, 6 for solids, plates and shells.) Hence its pseudoinverse is not expensive.

## 6 Numerical experiments

The experiments we run on 4 and 15 R10000 processors of a 16-processor SGI Origin/2000. The results of the experiments are presented in Tab. 6. The first line in the table corresponds to the smoothed aggregation technique with a smoother of degree 1. The lines 2–6 demonstrate the effect of the additional energy minimization. In those experiments, the smoothed prolongator has been used as an initial guess and its nonzero strauctur has been for defining $Z$ in (14).

In all the experiments below, the method is used as a preconditioner for the conjugate gradient method, and the stopping criterion used was

$$\|\mathbf{r}^i\|_B \leq (10^{-5}/\sqrt{\mathbf{cond}(B,A)}) \cdot \|\mathbf{r}^0\|_B$$

where $B$ is the preconditioner, $\mathbf{r}^i$ denotes the residual after $i$ steps of the iteration, and $\mathrm{cond}(B,A)$ is a condition number estimate computed at run time.

| min st | solid, (see Fig. 3) 75,174 dofs, 4 CPUs. | | | | solid, 407,277 dofs, 15 CPUs. | | | |
|---|---|---|---|---|---|---|---|---|
| | setup/iter/tot time [s] | num it. | cond est | tim/tim sm aggr | setup/iter/tot time [s] | num it. | cond est | tim/tim sm aggr |
| 1 | 8.1/5.3/13.4 | 8 | 2.58 | 1.00 | 12.7/36.3/49.1 | 13 | 6.16 | 1.00 |
| 2 | 8.6/5.3/13.9 | 8 | 2.55 | 1.03 | 15.4/34.0/49.4 | 12 | 5.04 | 1.01 |
| 3 | 8.8/5.4/14.2 | 8 | 2.50 | 1.06 | 18.1/32.0/50.1 | 11 | 4.45 | 1.02 |
| 4 | 8.8/4.7/13.7 | 7 | 2.20 | 1.02 | 20.2/31.1/51.4 | 11 | 4.23 | 1.04 |
| 5 | 9.0/4.7/13.8 | 7 | 2.21 | 1.03 | 22.4/28.4/50.8 | 10 | 4.00 | 1.03 |
| 6 | 9.5/4.7/14.3 | 7 | 2.17 | 1.06 | 24.9/28.5/53.5 | 10 | 3.90 | 1.09 |

**Fig. 3.** The mesh of an automobile steering knuckle. Courtesy of Charbel Farhat, University of Colorado at Boulder.)

# References

1. J. H. BRAMBLE, J. E. PASCIAK, J. WANG, AND J. XU, *Convergence estimates for multigrid algorithms without regularity assumptions*, Math. Comp., 57 (1991), pp. 23–45.

2. A. BRANDT, *Algebraic multigrid theory: The symmetric case*, Appl. Math. Comput., 19 (1986), pp. 23–56.
3. T. F. CHAN, J, XU, AND L. ZIKATANOV, *An agglomeration multigrid for unstructured meshes.*, In: Domain Decomposition Methods 10, (Proceedings of the tenth international conference on domain decomposition methods) Mandel, Farhat, Cai Eds., AMS 1998.
4. T. F. CHAN, S. GO, AND L. ZIKATANOV, *Lecture Notes on Multilevel Methods for Elliptic Problems on Unstructured Grids*, UCLA CAM Report 97-11, March 1997. Lectures notes for the lecture series "Computational Fluid Dynamics", von Karman Inst., Belgium, March 3-7, 1997. An abridged version has been published as CAM Report 97-36, August, 1997 and appeared in "Computational Fluid Dynamics Review 1997", Hafez and Oshima (eds.), Wiley.
5. P. VANĚK, J. MANDEL, AND M. BREZINA, *Algebraic multigrid based on smoothed aggregation for second and fourth order problems*, Computing, 56 (1996), pp. 179–196.
6. P. VANĚK, M. BREZINA, AND J. MANDEL, *Convergence of Algebraic Multigrid Based on Smoothed Aggregation*, Submitted to Num. Math.
7. P. VANĚK, A. JANKA, AND H. GUILLARD, *Convergence of Petrov-Galerkin Smoothed Aggregation Method*, To appear.
8. P. VANĚK, *Fast multigrid solver.* Applications of Mathematics, to appear.
9. *Acceleration of convergence of a two-level algorithm by smoothing trarnsfer operator*, Applications of Mathematics, 37 (1992), pp. 265–274.
10. P. VANĚK, M. BREZINA, AND R. TEZAUR, *Two-Grid Method for Linear Elasticity on Unstructured Meshes*, To appear in SIAM J. Sci. Comp.
11. J. MANDEL, M. BREZINA, AND P. VANĚK, *Energy Optimization of Algebraic Multigrid Bases*, To appear in Computing
12. W. L. WAN, T. F. CHAN, AND B. SMITH, *An energy-minimizing interpolation for robust multigrid methods*, UCLA CAM Report 98-6, Department of Mathematics, UCLA, February 1998.

# Parallel Performance of Chimera Overlapping Mesh Technique

Jacek Rokicki[1,3], Dimitris Drikakis[2], Jerzy Majewski[3], and Jerzy Żółtak[4]

[1] Zentrum für Hochleistungsrechnen, TU Dresden, D-01062 Dresden, Germany,
jacek.rokicki@zhr.tu-dresden.de
[2] UMIST, PO Box 88, Manchester M60 1QD, United Kingdom,
drikakis@umist.ac.uk
[3] Warsaw University of Technology, Nowowiejska 24, 00-665 Warsaw, Poland,
jmajewsk@meil.pw.edu.pl
[4] Aviation Institute, Al. Krakowska 110/114, Warsaw, Poland,
jzoltak@meil.pw.edu.pl

**Abstract.** In the paper we present the Chimera overlapping mesh technique applied to the solution of compressible flow problems. This technique is used to facilitate the grid generation in complex geometries allowing at the same time for natural parallelisation of the problem. The presented algorithm is particularly suitable for cases with large number of meshes overlapping in almost arbitrary manner (including multiple overlaps). The parallel implementation is based on the PVM approach.

## 1 Introduction

Simulation of fluid flow in complex geometries continues to be one of the major research areas in Computational Fluid Dynamics (CFD). Several CFD methods have been developed over the last two decades but improvement of their accuracy and efficiency remains a challenging problem. Nowadays a popular way to reduce the cost of computations is the use of parallel computers in conjunction with implicit solvers and techniques for the acceleration of the numerical convergence e.g multigrid.

The overlapping mesh techniques (e.g. [4, 1, 5, 6]) can be employed as an efficient tool for grid generation in complex geometries and as a natural method for parallelisation of the existing solvers. This technique is based on the subdivision of the physical domain into overlapping subdomains. Subsequently the system of flow equations is solved on each subdomain separately. The global solution is obtained by iteratively adjusting the boundary conditions on each subdomain.

In the present study we investigate the parallelisation of the method via PVM approach. We discuss the efficiency of the algorithm and the issues related to the load balancing. We perform also the grid convergence study for the case of compressible flow past the two-element aerofoils.

## 2 Grid generation and the implicit unfactored solver

The 2D geometry in the physical domain is defined by prescribing the shape of its boundary. This boundary is divided into overlapping pieces in such a way that each part can form one *side* of a local grid with rectangular topology.

After the whole boundary is covered with local grids, the farfield grid is added. Then, the remaining area is covered with simple rectangular grids. During this procedure the size and also the shape of each subdomain must be chosen such that this subdomain remains in the interior of the flowfield. Care must also be taken to assure that each point of the physical domain is sufficiently overlapped by at least one of the local grids.

An example of a complicated grid system around the two-element GA(W)-1 aerofoil is shown in Fig. 1.

The compressible flow of an inviscid fluid is governed by the Euler equations. These equations can be written in dimensionless conservation form and for a general curvilinear coordinate system, as:

$$(JU)_t + E_\xi + G_\eta = 0 \, . \tag{1}$$

where $J$ is the Jacobian of the transformation from curvilinear to Cartesian coordinates; the vector $U$ contains the conservative variables $U = (\rho, \rho u, \rho v, e)^T$ where $\rho$ is the density, $u,v$ are the Cartesian velocity components and $e$ is the total energy per unit volume. The inviscid fluxes are denoted as $E$ and $G$. The system of the governing equations is completed by the perfect gas equation of state.

The simplest implicit discretisation of equation (1) can be written as

$$J \frac{U^{n+1} - U^n}{\Delta t} + E_\xi^{n+1} + G_\eta^{n+1} = 0 \, . \tag{2}$$

The system of equations (2) is solved by a Newton-type method. . The Riemann solver [2, 3] calculates the inviscid fluxes at the cell faces using the characteristic values of the conservative variables. A Godunov-type upwind scheme up to third-order accurate has also been employed for the calculation of the characteristic cell face values.

## 3 The overlapping-mesh algorithm

### 3.1 General structure of overlapping

We consider here the general case in which the flow domain $\Omega$ is fully covered by several overlapping subdomains $\Omega_1, \ldots, \Omega_K$, $\Omega = \Omega_1 \cup \ldots \cup \Omega_K$. If $\Gamma$ and $\Gamma_j$ denote the boundaries of $\Omega$ and $\Omega_j$ subdomains, respectively, then $\gamma_j = \Gamma_j \setminus \Gamma$ will denote the interface between the subdomains in which boundary information takes place.

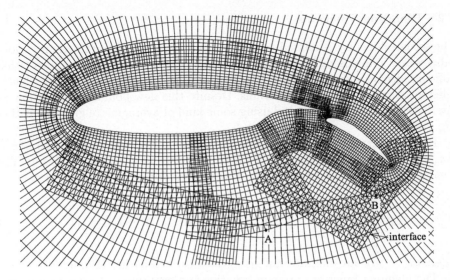

**Fig. 1.** The system of 22 overlapping grids around the two-element GA(W)-1 aerofoil

In order to obtain convergence of the global solution, sufficient overlapping between subdomains must be applied. This can be expressed using the *internal distance* function $d_j(A)$ [5,6] defined as:

$$d_j(A) := \begin{cases} 0 & A \notin \Omega_j \\ \rho(A, \gamma_j) & A \in \Omega_j \end{cases} \quad (3)$$

where $\rho(A, \gamma_j)$ stands for the distance between point $A$ and the interface $\gamma_j$. The characteristic size of the overlap structure is defined as

$$\delta := \inf_{A \in \Omega} d(A) := \inf_{A \in \Omega} \max_{j=1,\ldots,K} d_j(A) \quad (4)$$

The overlapping is called uniform if $\delta$ is strictly positive. In the present study non-uniform overlapping is also considered. In such cases $d(A)$ is allowed to vanish at isolated points on the physical boundary (usually in the corners of $\Gamma$) remaining positive in the whole interior of $\Omega$. The internal distance function $d_j(A)$ is used subsequently to define the blending functions.

Each subdomain $\Omega_j$ is covered by a suitable regular grid. However, the grid points in the overlap region do not necessarily coincide (e.g. see point A in Fig.1). As a result each interfacial grid point $C$ has to be localised in the reference frame of each grid in the system. This procedure is performed once in the beginning of the calculation, the computational cost is not significant in comparison with the computational cost of a single sweep of the Euler solver.

Once the interfacial points are properly identified one needs to apply an interpolation procedure in order to prolongate the local solution onto the whole subdomain. In contrast to [6] where high-order interpolation was employed, in the present study bilinear or at most biquadratic interpolation formulas were found to provide satisfactory results.

## 3.2 Blending functions

The second difficulty is related to the fact that the solution obtained on various subdomains remains different in the overlapping regions. This is due to either different discretisation or the fact that the solution procedure has reached a different stage in various subdomains. Despite this new boundary condition on interfaces needs to be generated using some kind of averaging (e.g. at point B of Fig.1).

For this purpose weight functions $\chi_p$ are defined, (henceforth called *blending functions*), and subsequently are used to construct the global solution. The blending function $\chi_p$ associated with the $p$-th subdomain $\Omega_p$ is defined as

$$\chi_p(C) := \frac{d_p(C)}{S(C)}, \qquad S(C) := \sum_{j=1}^{K} d_j(C), \quad p = 1, 2, \ldots, K. \tag{5}$$

where $d_1, \ldots, d_K$ are the *internal distances* described earlier (Eq. 3).

The blending functions are non-negative and continuous, forming the partition of unity ($\chi_1 + \ldots + \chi_K \equiv 1$ on $\Omega$). Other properties are described in [6]. One should notice that equations (3) and (5) have a simple algebraic form and as a result the blending functions can be easily evaluated numerically. Their value can be computed once the overlapping structure is identified at the beginning of all calculations.

## 3.3 Global solution and full algorithm

Let assume that $U_{[p]}$ is the solution vector defined on the subdomain $\Omega_p$ and extended to the whole $\Omega$ by assuming that $U_{[p]}(C) \equiv 0$ for $C \notin \Omega_p$ (hence $U_{[p]}$ has discontinuity at $\gamma_p$). The global solution $U(C)$ is defined as

$$U(C) = \sum_{p=1}^{K} \chi_p(C) U_{[p]}(C), \qquad C \in \Omega. \tag{6}$$

It should be noted that $U(C)$ remains continuous at all interfaces despite the fact that each $U_{[p]}$ is discontinuous on the corresponding interface $\gamma_p$. On the other hand, the discontinuities such as shock waves will remain intact in the global solution.

The full algorithm encompasses the following steps:

1. Given the grid in each subdomain $\Omega_p$, *identify* interfacial grid points in the cells of remaining grids and evaluate all blending functions on the interfaces.
2. Initialise boundary conditions at the interfaces $\gamma_p$ ($p = 1, \ldots, K$).
3. Perform few iterations using the Euler solver on each local grid to obtain a new local solutions $U_{[p]}$, ($p = 1, \ldots, K$).
4. Interpolate each local solution on the prescribed interfaces.
5. Evaluate the global function $U(\cdot)$ on each interface using Eq. (6).

6. Check the convergence on all interfaces and if there is no convergence return to step 4.

It is important to point out that the step 5 cannot be performed in parallel. It requires interprocessor communication if different subdomains are served by different processors. However the computational cost associated with the steps 4 and 5 is negligible compared to the cost for step 3. The number of floating point numbers transferred between processors is proportional to the number of boundary points, while the cost of step 4 is one order of magnitude higher. As a result one should expect that the benefits of parallelism would increase when increasing grid size.

## 4 Numerical results

In order to asses the numerical quality of the method two test cases were chosen.

The first case is the subsonic flow around the GA(W)-1 aerofoil with a 29% flap deflected by 40°. The free-stream Mach number is equal to 0.2 and the angle of attack is 0°. Two different grid systems B1 and B2 were considered here to prove that solution converges when grids are refined (grid system B2 contains roughly twice as many grid points as B1).

The total number of grids was 22 with 17 grids actually in contact with one of the aerofoil profiles. (see Figs. 1 showing the grid system B1).

The Mach number distribution obtained on the grid system B2 is shown around the flap region in Fig. 2. The solid lines in the flowfield correspond to the interfaces of the overlapping grids. The smooth transition of Mach values between the various subdomains can easily be observed in the above figures.

Quantitatively, the accuracy of the algorithm can be assessed by comparing (Fig. 3) the experimental results of [8] with the computed pressure coefficient distributions. The $c_p$ values are very similarly predicted on both grids and both solutions are in a very good agreement with the experiment. In all cases the iterations converge to almost the machine accuracy.

The second case is the transonic flow around the supercritical SKF-1.1 aerofoil with the deflected maneuver-flap [7]. The free-stream Mach number is $M_\infty = 0.60$ and the angle of attack is $\alpha = 2°$.

Here again three grid systems B1, B2, and B3 with increasing degree of refinement, are chosen to demonstrate the convergence of the numerical solution. The systems consist of 20 grids (25 for B3) as shown in Fig. 4. The number of grid points is 18410 for B1, 42183 for B2 and 55304 for B3.

The Mach-number field corresponding to the grid system B3 is shown in Fig. 5. The shock wave is well visible in the middle part of the main aerofoil, while a second much weaker shock appears over at the flap. Figure 6 presents the comparison of the computed $c_p$ distributions on the main aerofoil and on the flap for all grid systems used. The results are very similar with the sharpest shock wave obtained on the finest grid.

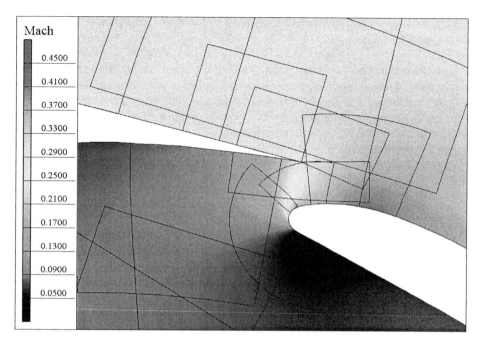

**Fig. 2.** Mach number distribution around the GA(W)-1 aerofoil for $M_\infty = 0.2$ (zoom near the flap); solid lines denote interfaces.

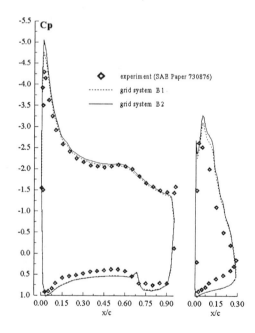

**Fig. 3.** Pressure coefficient distribution $c_p$ on the main aerofoil and on the flap of the GA(W)-1 aerofoil ($M_\infty = 0.2$).

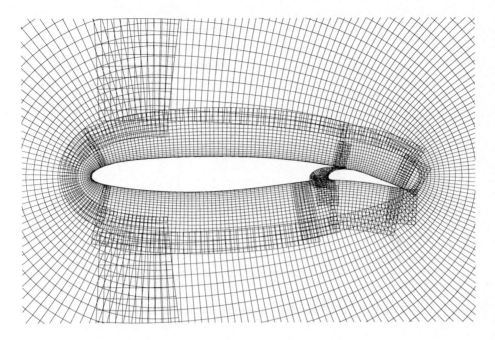

**Fig. 4.** Grid system B1 around the SKF-1.1 aerofoil.

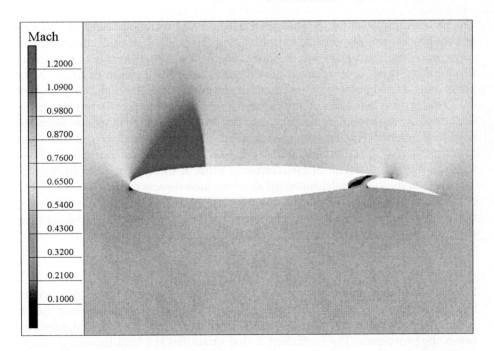

**Fig. 5.** Mach number distribution around the SKF-1.1 aerofoil for $M_\infty = 0.60, \alpha = 2°$ (case B3).

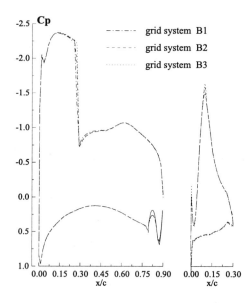

**Fig. 6.** SKF-1.1 aerofoil - pressure coefficient distribution.

## 5 Parallel implementation

The parallel implementation was based on the PVM technique (see Fig. 7). The *master*-process was responsible for initialising the calculations and controlling all *slave*-processes. Each *slave* was an independent flow solver capable of advancing the iterative process on a single grid or on a prescribed group of grids. After performing a fixed number of iterations all *slaves* were sending updated boundary data to the *master*-process. The *master*-process was subsequently re-calculating the boundary data, sending them back to the *slaves*. This procedure was repeated until the numerical convergence was obtained. The total computational effort and total number of iterations did not depend on the number of processors. They were, however, dependent on the number of grids in the system, overlap size etc.

To estimate the parallel speed-up alone we have measured the execution time $\tau_L$ necessary for performing a fixed number of iterations (in our case 100 iterations) on a $L$-processor system. The speed-up factor $\alpha_L$ and the corresponding efficiency $\eta_L$ were defined as

$$\alpha_L = \frac{\tau_1}{\tau_L}, \qquad \eta_L = \frac{\alpha_L}{L} = \frac{\tau_1}{L\tau_L}.$$

In order to obtain maximum speed-up, the workload should be divided evenly between processors. The grids, however, were of different size and did not match the number of processors. As a result, the workload was never fully balanced between the processors. The coefficient $\beta$ that quantifies the load-balancing effects is defined by:

$$\beta_p = \frac{w_p}{\frac{1}{L}\sum_{q=1}^{L} w_q}, \qquad \beta = \max_{p=1,\ldots,L} \beta_p,$$

where $w_p$ stands for the workload assigned to the $p$-th processor ($w_p$ is proportional to the number of grid points). The denominator in the first formula denotes the workload obtained by the ideal balancing. The coefficient $\beta$ is always larger than 1, where $\beta = 1$ corresponds to an ideal load-balancing.

In our analysis we have assumed that all processors have equal speed which is true for the present study, but it may not be the general case e.g. in heterogeneous parallel systems. Results from the parallelisation of the present algorithm are presented for the flow around the GA(W)-1 aerofoil using an overlapping system of 33 grids. In Table 1 the workload, $w_p$, assigned to each processor and the corresponding values of the coefficients $\beta_p$ and $\beta$ are presented. The results for the acceleration (speed-up) factor $\alpha_L$ and total efficiency $\eta_L$, of the computations obtained on a 16-processors CRAY CS6400 machine, are shown in Table 2. These results confirm that the overlapping-mesh technique can be efficiently used on a multi-processor system. The efficiency of such an approach will increase with the complexity of the problem, provided that the workload can be, as evenly as possible, distributed between the processors. This can be achieved by further subdivision of existing grids or by regrouping of grids assigned to each processor.

Similar performance results were obtained for the transonic flow past the NACA 0012 aerofoil.

Table 1

| $L$ | $p$ | $w_p$ | $\beta_p$ | $\beta$ |
|---|---|---|---|---|
| 1 | 1 | 18414 | 1.00 | 1.00 |
| 2 | 1 | 8991 | 0.98 | 1.02 |
|   | 2 | 9423 | 1.02 |      |
| 3 | 1 | 6092 | 0.99 | 1.08 |
|   | 2 | 6611 | 1.08 |      |
|   | 3 | 5711 | 0.93 |      |
| 4 | 1 | 4668 | 1.01 | 1.04 |
|   | 2 | 4369 | 0.95 |      |
|   | 3 | 4572 | 0.99 |      |
|   | 4 | 4805 | 1.04 |      |
| 5 | 1 | 3781 | 1.03 | 1.09 |
|   | 2 | 3402 | 0.92 |      |
|   | 3 | 3556 | 0.97 |      |
|   | 4 | 3675 | 1.00 |      |
|   | 5 | 4000 | 1.09 |      |

Table 2

| $L$ | $T_L$ [sec] | $\alpha_L$ | $\eta_L$ |
|---|---|---|---|
| 1 | 1185 | 1.00 | 1.00 |
| 2 | 640 | 1.85 | 0.93 |
| 3 | 448 | 2.65 | 0.88 |
| 4 | 367 | 3.23 | 0.81 |
| 5 | 311 | 3.81 | 0.76 |

# References

1. Benek, J.A., Buning, P.G., Steger, J.L., A 3-D Chimera Grid Embedding Technique, AIAA Paper 85-1523, 1985.
2. Drikakis, D., Durst, F., Investigation of Flux Formulae in Transonic Shock-Wave/Turbulent Boundary Layer Interaction, *Int. J. Num. Meth. Fluids*, **18**, 385-413, 1994.

3. Drikakis D. and Tsangaris S., On the Accuracy and Efficiency of CFD Methods in Real Gas Hypersonics, *Int. Journal for Numerical Methods in Fluids*, **16**, 759-775 (1993).
4. Henshaw, W.D., A Fourth-Order Accurate Method for the Incompressible Navier-Stokes Equations on Overlapping Grids, *J. Comput. Phys.*, **113**, 1994.
5. Rokicki, J., Floryan, J.M., Domain Decomposition and the Compact Fourth-Order Algorithm for the Navier-Stokes Equations, *J. Comput. Phys.*, **116**, 1995.
6. Rokicki, J., Floryan, J.M., Unstructured Domain Decomposition Method for the Navier-Stokes Equations, *Computers and Fluids*, **28**, 87-120, 1999.
7. Stanewsky, E., Thibert, J.J., Airfoil SKF-1.1 with Maneuver Flap, AGARD Report AR-138, 1979.
8. William, H.W. (Jr.), New airfoil Sections for General Aviation Aircraft, SAE Paper 730876

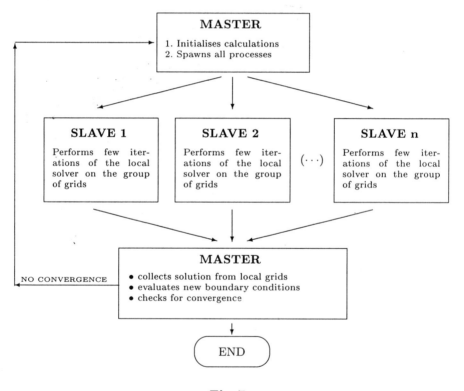

Fig. 7.

# Electromagnetic Scattering with the Boundary Integral Method on MIMD Systems

Thierry Jacques, Laurent Nicolas, Christian Vollaire
CEGELY - UPRESA CNRS 5005 - Ecole Centrale de Lyon
B.P.163 - 69131 Ecully cedex - France
{tjacques, laurent, vollair}@eea.ec-lyon.fr

**Abstract.** This paper deals with parallel computation in electromagnetics. The boundary integral method has been developed to solve scattering by a perfect electric conducting or perfect dielectric bodies. Only parallel computation enables to modelize large devices. We present the implementation of the code on a distributed memory parallel machine. We focus in particular on the parallelization of the BiCGStab solver. Parallel performances are presented with a 2000 degrees of freedom problem.

## 1 Introduction

Numerical computation is more and more used by scientists to modelize physical phenomena or to optimize existing systems. Nowadays, only parallel computers are able to provide the required power to solve today's realistic problems. There are two reasons : large memory is required because of a large amount of data, or speed is required to obtain the solution.

The appearance of parallel computers has caused a revolution in all scientific domains. However electrical engineering seems to be far behind other disciplines. There are several reasons for this delay: first, field problems are generally neither confined to a structure nor set in very regular domains. Second, the geometries involve substantial detail, which cannot be erased. Third, the domain of study have often to be considered to infinity. As proof of this delay between computational electromagnetics and other engineering domains, Table I shows the number of papers dedicated to parallelism in two reference conference on electromagnetic field computation: COMPUMAG'97 (Conference on the Computation of Electromagnetic Fields) and CEFC'98 (IEEE conference on Electromagnetic Field Computation). These numbers are compared to the number of papers presented in a general European conference on numerical methods in engineering ECCOMAS'96 (Paris, 9/9/96 - 9/13/96). In this last one, papers are divided into two classes : Fluid Mechanics and other Engineering sciences.

**Table 1.** Comparison of the number of papers on parallel computation in different conferences on numerical methods since 1996.

| Conference | COMPUMAG'97 | CEFC'98 | ECCOMAS'96 Fluid Mech. | ECCOMAS'96 Eng. Sciences |
|---|---|---|---|---|
| number of papers | 419 | 394 | 168 | 160 |
| papers on parallel computation | 9 | 4 | 25 | 12 |
| ratio | 2,1% | 1,0% | 15,0% | 8,0% |

Several examples of parallelization of the different numerical methods used in computational electromagnetics are described in [1]: the finite difference time domain method, the finite volume time domain method, the integral equation method and the finite element method. In this paper, we are only interested in the boundary integral frequency domain method. This method requires to modelize only the surfaces of the objects and takes implicitly into account the decrease of the field at infinity. On the other hand, it generates a full non-Hermitian matrix that requires a large memory. In a first section, the formulations are described. The parallelization of the code is then presented and some numerical results are shown in section 5.

## 2  Problem Formulation

An incident wave (frequency f, pulsation ω) is perturbed by the object of boundary $\Sigma$. According to the size and the properties of the object, the total magnetic field **H** and electric field **E** are different from the incident fields.

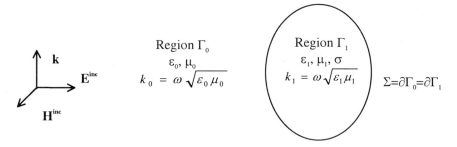

**Fig. 1.** Description of the problem. The electromagnetic properties of the material are the conductivity ($\sigma$), the permittivity ($\varepsilon$) and the permeability ($\mu$).

### 2.1  Perfect electric conducting (PEC) body ($\sigma=\infty$, $\varepsilon_1=\varepsilon_0$)

Since the skin depth is very small compared to the dimensions of the PEC, the fields **H** et **E** are null inside the region $\Gamma_1$. By applying the theorem of Green to Maxwell's equations into the region $\Gamma_0$, we obtain the following equation [2]:

$$\mathbf{J}(\mathbf{x}) = 2\mathbf{n} \times \mathbf{H}^{inc}(\mathbf{x}) + \frac{\mathbf{n}}{2\pi} \times \oiint_\Sigma \mathbf{J} \times \nabla \Phi_0 \, ds \tag{1}$$

where $\Phi_0(x,y) = e^{-jk_0|x-y|}/|x-y|$ is Green's function, **J** is vector current density and **n** is the exterior normal vector.

## 2.2 Perfect dielectric body ($\sigma=0$, $\varepsilon_1 \neq \varepsilon_0$)

By applying the theorem of Green in the regions 0 and 1, we obtain the following equations [3] :

$$\frac{1+\mu}{2}\mathbf{J} = \mathbf{n}\times\mathbf{H}^{\text{inc}} + \frac{\mathbf{n}}{4\pi}\times\oiint_\Sigma k_0 \mathbf{K}(\Phi_0 - \mu\varepsilon\Phi_1) + \mu\mathbf{H}_n(\nabla\Phi_0 - \nabla\Phi_1) + \mathbf{J}\times(\nabla\Phi_0 - \mu\nabla\Phi_1) \quad (2)$$

$$\frac{1+\mu}{2}\mathbf{H}_n = \mathbf{n}.\mathbf{H}^{\text{inc}} + \frac{\mathbf{n}}{4\pi}\oiint_\Sigma k_0 \mathbf{K}(\Phi_0 - \varepsilon\Phi_1) + \mathbf{H}_n(\mu\nabla\Phi_0 - \nabla\Phi_1) + \mathbf{J}\times(\nabla\Phi_0 - \nabla\Phi_1)ds$$

$$\frac{1+\varepsilon}{2}\mathbf{K} = \alpha\mathbf{n}\times\mathbf{E}^{\text{inc}} + \frac{\mathbf{n}}{4\pi}\times\oiint_\Sigma k_0 \mathbf{J}(\Phi_0 - \mu\varepsilon\Phi_1) + \varepsilon\mathbf{E}_n(\nabla\Phi_0 - \nabla\Phi_1) + \mathbf{K}\times(\nabla\Phi_0 - \varepsilon$$

$$\frac{1+\varepsilon}{2}\mathbf{E}_n = \alpha\mathbf{n}.\mathbf{E}^{\text{inc}} + \frac{\mathbf{n}}{4\pi}\oiint_\Sigma k_0 \mathbf{J}(\Phi_0 - \mu\Phi_1) + \mathbf{E}_n(\varepsilon\nabla\Phi_0 - \nabla\Phi_1) + \mathbf{K}\times(\nabla\Phi_0 - \nabla\Phi$$

where $\mu = \mu_1/\mu_0$, $\varepsilon = \varepsilon_1/\varepsilon_0$, $\alpha = j\sqrt{\varepsilon_0/\mu_0}$, $\mathbf{J} = \mathbf{n}\times\mathbf{H}$, $\mathbf{H}_n = \mathbf{H}_1.\mathbf{n}$, $\mathbf{K} = \alpha\mathbf{n}\times\mathbf{E}_1$ and $\mathbf{E}_n = \alpha\mathbf{E}_1.\mathbf{n}$. $\Phi_0$ and $\Phi_1$ are the Green's functions associated in the regions 0 and 1.

These variables and the coefficients of the matrix are of the same order, leading to a better conditioning of the matrix. The numerical discretization is performed using second order surface finite elements. The method proposed in [4] is used to treat the singularities, and half-discontinuous finite elements allow to treat geometrical discontinuities [5].

## 3 Parallel computing

For a n-nodes mesh, the storage of the matrix requires $6n^2$ data for a conducting electrical body et $16n^2$ for a perfect dielectric body. Parallel computation is then essential to carry out a numerical computing about realistic devices.

The global system matrix is computed by degree of freedom : each processor captures data (mesh and physical data) and computes a part of the lines of the matrix relative to nodes. Because of the form of the equations and the use of an iterative method, the contributions of each node are not in the form of matrix. This operation allows to reduce the storage by three for a conducting electrical body and by eight for a perfect dielectric body. Hence no communication between the processors is required during this stage. The algorithms BiCGStab [6] and BiCGStab(l) [7] are then used to solve the resulting linear system.

### 3.1 Preconditioning

We use a diagonal preconditioning. The right preconditioning requires the broadcast and the storage of the diagonal, the computing of a intermediate vector by each processor after each multiplication matrix-vector. Moreover, this intermediate vector must be kept in memory. On the other hand, the left preconditioning can be carried out in the same time as the multiplication matrix-vector. It requires less memory. Moreover we have observed that the left preconditioning is more efficient.

## 3.2  Parallel algorithm of BiCGStab

We use the algorithm BiCGStab of H. Van der Vorst [6,7] to solve the problem $KAx = Kb$, where K is the preconditioning. At each iteration, two multiplications matrix-vector are required. The solution vector is distributed on each processor. The algorithm has been parallelized by minimizing the communications between processors:

**Table 2** Parallelized version of the algorithm BiCGStab - ■: sequential part, O: parallel part.

$x = 0$
$q = p_o = r_o = Kb$
$\omega = 1$
$\rho = <q, r_o>$
Broadcast of vector q and ρ
while($e_j > \epsilon$)
- O     $v_j = K A r_j$
- O     $\Phi_1 = <q, v_j>$

Broadcast of vector $v_j$ and of the local inner product
- ■     $\alpha = \rho \omega / \phi_1$
- ■     $s_j = r_j - \alpha v_j$
- O     $t_j = K A s_j$
- O     $\rho = -<q, t_j>$
- O     $\phi_1 = <s_j, t_j>$
- O     $\phi_2 = <t_j, t_j>$
- O     $\phi_3 = <s_j, s_j>$

Broadcast of vector $t_j$ and of 4 local inner products
- ■     $\omega = \phi_1 / \phi_2$
- O     $x = x + \alpha p_j + \omega r_j$
- ■     $r_{j+1} = s_j - \omega t_j$
- ■     $\beta = \rho / \omega$
- ■     $p_{j+1} = r_{j+1} + \beta (p_j - \omega v_j)$
  - ■     $e_j = \phi_3 - \omega^2 \phi_2$

Before and after each broadcast performed by the message passing library PVM, a synchronization between processors is carried out. For a perfect dielectric body, when the frequency increases, it is necessary to increase the mesh discretization : ten nodes per wavelength are at least required. Because the algorithm BiCGStab shows some difficulty to converge, the algorithm BiCGStab(l) is preferably used. When l increases, the algorithm is more robust. On the other hand, it requires more memory and it is less parallelizable than the BiCGStab.

## 4 Numerical results

The test problem is the scattering of a plane wave at 600 MHz by a cylinder (fig. 2). Length of the cylinder is 2 $\lambda$, and its radius is equal to 0.5 $\lambda$ (where $\lambda$ is the wavelength). In the first example, the cylinder is perfectly conducting. In the second example, it has the following properties : $\mu = 1.1$ and $\varepsilon = 1.5$. It is computed on the Cray T3E of the IDRIS (Institut du Développement et des Ressources en Informatiques Scientifiques, CNRS).

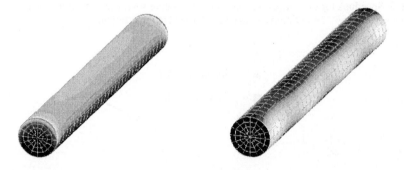

**Fig. 2** 2000 degrees of freedom problem - Left: scattering by a perfect electric conducting cylinder, visualization of the surface current density **J** - Right: scattering by a perfect dielectric cylinder ($\mu = 1.1$, $\varepsilon = 1.5$), visualization of the electric field **E**.

Parallel performances are analyzed on a 2000 lines dense matrix (fig. 3 and fig. 4). Table 3 gives the distribution of the computation between the assembling and the solving stages. Because the assembling time is dominating and no message passing is required during this step, the total speedup is very good. With larger problems, one can expect a smaller time ratio for the assembling, and consequently lower speedups. Further tests are necessary to confirm these results. Note that both of the computed problems are too large to be solved on only one processor of the Cray T3E.

**Table 3**. Distribution of the computation time for both examples.

|  | Total Computation time on 5 processors (Cray T3E) | Ratio of total time for assembling | Ratio of total time for solving |
|---|---|---|---|
| Conducting cylinder | 845 s | 89 % | 11 % |
| Dielectric cylinder | 1680 s | 80 % | 20 % |

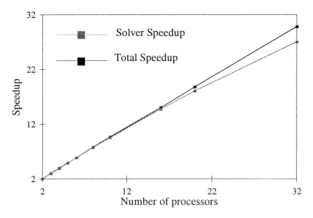

**Fig. 3.** Speedups for the entire code and for the iterative solver, when solving scattering by a perfect electric body - 14 iterations are necessary to solve the problem

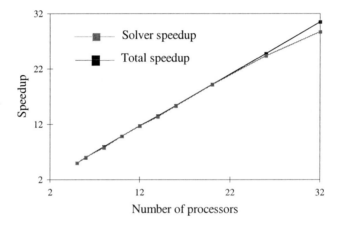

**Fig. 4.** Speedups for the entire code and for the iterative solver, when solving the scattering by a dielectric body. $\mu = 1.1$, $\varepsilon = 1.5$) - 6 iterations to solve the problem.

## 5 Conclusion

The Boundary Integral Method is a computational method whose application demands for parallel computing because this method requires large memory storage. We have implemented a parallel version by distributing the mesh equally over the processors, and the assembling of the full and non-Hermitian matrix is performed by degrees of freedom. This stage represents the main phase of the computation and is carried out without message passing between processors. The use of an iterative solver allows to reduce the storage of the matrix. As shown with the speedups, the iterative solver is efficient, since the parallelized matrix-vector multiplication constitutes its main stage. Furthermore, the computing load seems well distributed between the processors because each processor performs the same number of operations.

## References

1. Nicolas, L., Vollaire, C. :A Survey of Computational Electromagnetics on MIMD Systems. International conference on parallel computing in electrical engineering. Bialystok, 2-5 September 1998. pp. 7-19.
2. Walker, S.P., Leung, C.Y. : Parallel Computation of Time-Domain Integral Equation Analyses of Electromagnetic Scattering and RCS. IEEE Transactions on Antennas and Propagation. Vol. 45.NO.4, April 1997. pp. 614-619
3. Schlemmer, E., Wolgang, M., Rucker, W.M., Richter, K.R. : Calcul des champs électromagnétiques diffractés par un obstacle diélectrique tridimensionnel par des éléments de frontière en régime transitoire. Journal Physique III, november 1992. pp. 2115-2126.
4. Huber, C.J., Rieger, W., Haas, M., Rucker, W.M. :The numerical treatment of singular integrals in boundary element calculations, 7th International IGTE Symposium, 1996. pp 368-373.
5. Rêgo, S., Do, J.J. : Acoustic and Elastic Wave Scattering using Boundary Elements. Computational Mechanics Publications, UK, 1994.
6. Van der Vorst,H.A., : BiCGStab: A fast and smoothly converging variant of BiCG for the solution of nonsymmetric linear systems. SIAM J.Science Statist. Comput., 13, 1992. pp 631-644. Available on http://www.math.ruu.nl/people/vorst/cgnotes/cgnotes.html
7. Sleijpen, G.L.G., Fokkema, D.R.: BiCGStab(l) for linear equations involving unsymmetric matrices with complex spectrum. ETNA, volume1, September 1993, pp. 11-32. Available on http://www.math.ruu.nl/people/sleijpen

# Decomposition of Complex Numerical Software into Cooperating Components

Mark R.T. Roest & Edwin A.H. Vollebregt

Large Scale Modelling Group, dept. of Information Technology and Systems, Delft University of Technology
{roest,edwin}@pa.twi.tudelft.nl

**Abstract.** This paper discusses the approach that is being taken for parallelising the WAQUA/Kalman software of the Dutch Institute for Coastal and Marine Management. This software is used for numerical simulation of flow and transport phenomena in coastal waters and incorporates a Kalman filtering procedure to assimilate observational data into the simulation. In particular the Kalman filtering part of the software is very time-consuming which prohibits its use in e.g. operational storm surge prediction. Parallelisation is being considered to reduce this run-time.
Instead of parallelising the code as it is, it is first decomposed into a number of components which are coupled through an advanced message passing library. After this, each of the components can be parallelised in the way that is most appropriate for the computations in the component. If the original code were parallelised directly, it would not be possible to accommodate the various forms of parallelism that are available in the program. The decomposition into components itself already gives a clear benefit because it enhances the maintainability and reusability of the code.

## 1 Introduction

This paper discusses the approach that is being taken for parallelising the WAQUA/Kalman software of the Dutch Institute for Coastal and Marine Management (RIKZ). This code is effectively an extended version of the WAQUA program that is used at RIKZ for a wide range of applications, including storm surge prediction and the study of flow and transport phenomena in coastal waters. The extra part in the WAQUA/Kalman software is a variant of the Kalman filter [2] that is used to assimilate observations of e.g. water levels into the simulation. The data assimilation is of great value for improving the predictions that are made with the WAQUA model [5].

However, even though the Kalman filtering algorithm that is being used in WAQUA/Kalman (the RRSQRT scheme, [9]) is very efficient, it still takes far too much computing time for operational application. Simulating one week of flow over the North-West European continental shelf takes up to 20 hours which is impractical for practical purposes. Therefore, work is now underway to develop a parallel version of the code which will be used on the multiprocessor system of the Institute. Kalman filtering algorithms have been shown to be well parallelisable [3,4], so there is sufficient confidence that the parallelisation will prove to be a solution to the problem of large computing times.

One of the main difficulties in developing a parallel version of WAQUA/Kalman is to accommodate the various types of computations that occur. For the WAQUA computations the obvious parallelisation strategy is to decompose the computational grid into subdomains that can be handled in parallel. There is already a parallel version of WAQUA, based on this approach. But the Kalman part calls for a very different form of parallelism, in which multiple instances of a full WAQUA computation must be performed in parallel. This issue will be discussed in more detail in Section 2.

To allow these different forms of parallelism within one application, it has been decided to first split the code into several components. Each of the components can then be parallelised in the most appropriate way. The components are coupled through a message-passing library called CouPLE. This library is an extended version of a set of high level communication routines [10] that has been developed for the parallelisation of WAQUA. A discussion of the CouPLE library will be given in Section 3.

The way in which the WAQUA/Kalman software has been split into components will be discussed in Section 4. A component-version of the software is now operational. The work on the actual parallelisation of each of the components is ongoing.

## 2 Parallelising the WAQUA/Kalman Computations

Kalman filtering is a technique to correct the state of a model (for WAQUA e.g. water levels and velocities) with respect to available observations. The filtering procedure takes into account the uncertainties in the observations and in the model itself as well as the model dynamics. The model uncertainties must be specified by the user, usually as noise on the boundary conditions. The adjustments to the model state are made in such a way that the combined uncertainty of the model and the observations is minimal. A short introduction to Kalman filtering is given in Appendix A.1.

Internally, the Kalman filter uses a covariance matrix which gives the covariance of the uncertainty between any two grid points. The full covariance matrix has a size of $N \times N$, where $N$ is the number of state variables. This number can be very large; for WAQUA, it is not uncommon to have 60.000 state variables. Consequently, the full covariance matrix is excessively large for such models. Therefore, a modified Kalman algorithm is applied for WAQUA, which uses a reduced rank approximation of the square root of the covariance matrix. This modified algorithm is known as the RRSQRT scheme and has been developed by Verlaan *et al* [8, 7]. The columns of the square root factor can be seen as vectors that specify noise on each variable in the WAQUA state and are therefore usually referred to as *noise vectors*.

An short discussion of the algorithm in mathematical terms is given in Appendix A.2. The algorithm comes down to the following (referring to the equations in the appendix):

**for** $t = t_{start}$ **to** $t_{end}$ **step** $\delta t$ **do**
    handle noise on WAQUA forcings
    time step for WAQUA state (see Eq. 1)
    time step for noise vectors (see Eq. 2)
    reduction of the covariance matrix (see Eq. 3...4)
    filtering of the state and observations (see Eq. 6 ... 10)
**end for**

The first step handles the noise on the forcings of the model. It performs a time-step for the existing noise (i.e. scaling with a time-correlation factor) and introduces new noise. Then the new WAQUA state is computed by a time step of the WAQUA model. Likewise, the noise vectors in the RRQSRT algorithm are propagated in time, again with the WAQUA model. The time-propagation of the noise vectors gives a covariance matrix that is slightly larger than the original covariance matrix. In order to avoid growth of the covariance matrix, it must be reduced in each time step. After that, the actual filtering of the WAQUA state takes place using the available observations. This filtering uses a so-called gain-matrix that is derived from the covariance matrix.

It can be shown that virtually every step of this algorithm takes a substantial amount of time. Therefore, each step will have to be parallelised in order to get an acceptable and scalable overall speedup. But the steps exhibit different forms of parallelism. The time step for the WAQUA state is best parallelised by dividing the computational grid into subgrids that can be handled in parallel. But parallelising the time step for the noise vectors is best done by computing the different noise vectors in parallel. As propagation of the a noise vector is actually the execution of a WAQUA time step on the noise vector, this means that several instances of a WAQUA time step must be executed concurrently. Finally, the reduction of the covariance matrix involves several linear algebra operations (inner products, eigenvalue decomposition, see also Appendix A.2) which are best performed by decomposing the covariance matrix in a blockwise fashion.

If all these forms of parallelism had to be accommodated within a single SPMD program, then this program would become extremely complex. For instance, it would have to be capable of performing a WAQUA time step both for a part of the grid (for the normal WAQUA time step) and for the full grid (when propagating noise vectors).

Therefore, in order to ease the parallelisation, it has been decided to first decompose the software into several components so that each component can then be parallelised in the way that is most appropriate. The complexities of the data exchange are hidden in the **CouPLE** library. By doing this componentisation, the original WAQUA/Kalman code can be reused almost entirely. Also, it makes it easy to do the parallelisation in a step-wise fashion, parallelising one component after the other.

## 3 The CouPLE Environment

As was mentioned above, the **CouPLE** environment originated from a set of high-level communication operations that has been developed for the parallelisation of RIKZ's TRIWAQ package [10]. It has evolved into a very powerful mechanism for run-time coupling of arbitrary processes. It can be used for parallel computing, where the coupled processes are essentially instances of the same code, operating on different data. But it has been demonstrated to be equally applicable for coupling e.g. a flow simulation program based on finite differences to a program that computes transport phenomena using a particle approach.

An essential characteristic of the **CouPLE** environment is that it considers each process not as part of a set of processes but rather as a fully operational, independent process that has the capability to cooperate with other processes. This view eases the implementation of processes, because the programmer is not dealing with a whole set of processes at the same time but only with a single process that may interact with arbitrary other

```
MODULE covmat
   SETPAR( nstep )
   SETPAR( nmode )
   DO ( nstep )
      DO ( nmode )
         AVAIL av_mode_cv
         OBTAIN ob_mode_cv
      ENDDO
      SETPAR( nmeas )
      DO ( nmeas )
         AVAIL mea_av_gain
      ENDDO
   ENDDO
```

**Fig. 1.** The main part of the coupling algorithm of the COVMAT process. The parameter nstep denotes the number of WAQUA time steps, parameter nmode is the rank of the square root of the covariance matrix and nmeas is the number of measurements that are available for a certain time step. The full coupling algorithm includes some additional details that are not relevant in the context of this paper.

processes. Also, it enhances the reusability of the code, because the programmer will (ideally) not make assumptions about the other processes with which his process will interact.

To support this view of programming, the programmer cannot specify where a certain data item must come from. He can only call one of the obtain routines, which merely state that at the point where the routine is called, some data item has to be received from elsewhere. Likewise, the programmer cannot explicitly send away data, but he can only call one of the avail routines, which only states that the program has some data item available at that point.

Each of the obtain and avail routines has a label. The programmer must specify the sequence of obtains and avails that occur in a program by writing the process's *coupling algorithm*. This coupling algorithm is a list of obtains and avails, possibly parameterised to accommodate conditional and loop-structures in the program. An example of a coupling algorithm is given in Figure 1.

When using a process that has been instrumented with CouPLE, the user can tie together the obtains and avails by specifying a link between them in a coupling file. Note that it is sufficient for a user to know only the coupling algorithm; he does not have to know the exact implementation of the process. Hence, it becomes relatively easy to replace a process by one with a different implementation, as long as it supports the same coupling algorithm.

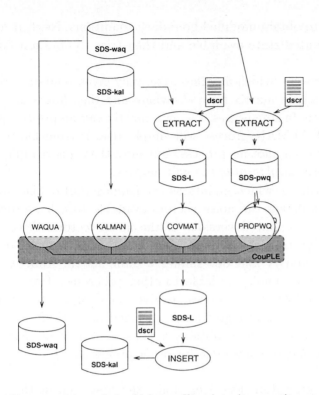

**Fig. 2.** Decomposition of the Kalman filter into its constituent processes. The dark area indicates the coupling through the CouPLE routines.

The CouPLE routines also perform the mapping of data from the 'availing' process onto that of the 'obtaining' process. The programmer specifies which data the program expects through a mechanism that is basically the same as that of the *index sets* and *interfaces* described in [10]. The CouPLE library determines where the data should come from and how it is to be inserted into the the programs data structures.

## 4 Decomposition of the WAQUA/Kalman Program

With the structure of the WAQUA/Kalman program as discussed above, its decomposition is fairly straightforward. The four processes that together make up the entire WAQUA/Kalman computation are (see also Figure 2):

- a WAQUA process, which is essentially a normal WAQUA process, extended with the capability to have its state filtered by an external process. In order to do this, it makes its boundary conditions available

and then tries to obtain modified boundary conditions. Next, it makes its newly computed state available and then tries to obtain a filtered version of it.
- a KALMAN process, which modifies externally given boundary conditions, obtains a gain matrix from elsewhere and then filters an externally given state. In some cases, the gain matrix can be pre-computed and then the KALMAN process will simply read it from file instead of obtaining it from another process. But note that this reading from file is also a form of *obtaining* the gain matrix.
- a COVMAT process, which maintains the (square root of the) covariance matrix. It makes the noise vectors available to other processes and then obtains updated versions of the noise vectors. With these updated noise vectors, it computes the new reduced square root of the covariance matrix and determines a gain matrix from it. This gain matrix is then made available to other processes. The coupling algorithm of the COVMAT process is given in Figure 1 to illustrate the way in which such an algorithm is specified in CouPLE.
- a PROPWQ process, which obtains a noise vector from elsewhere, updates it and makes it available for other processes.

Once this decomposition has been done, the first step in the parallelisation becomes extremely easy. Instead of using just one PROPWQ process, several PROPWQ processes can be used which each take their portion of the noise vectors that must be updated.

The figure also shows an important additional aspect. Each process gets its own input file, which is generated automatically from input files such as those that are now used for the original WAQUA/Kalman software. After the simulation, the results are inserted in the results file. Generic mechanisms for this extracting and collecting automatically have been developed for the parallelisation of TRIWAQ [6] and will be extended for the current purposes.

## 5 Results and Conclusions

Even though the actual parallelisation is still ongoing, the decomposition of the WAQUA/Kalman software into cooperating components has already proven to be a comfortable way to go. Whereas the different forms of parallelism that are needed for the parallelisation cannot possibly be accommodated in the WAQUA/Kalman code as it is, the componentisation has made it possible to still reuse almost the entire code.

The componentisation that is sketched in Figure 2 is now running. There is some overhead from the inter-process communication: the time that is spent in communication is up to 5-10% in representative cases. But this is largely compensated by improved computational performance, which is probably due to the fact that each of the processes needs far less memory than the original, full program, and is therefore able to make better use of the caches. The total execution time for the decomposed software is some 10% shorter than for the original software.

The implementation of the componentisation itself has not taken too much effort. Instead, the major effort has been spent in determining the exact data- and control flow in the program in order to find the most optimal splitting into components. As a side-effect, the componentisation has also revealed the data dependencies between the different parts of the program, which will be of good use when performing the actual parallelisation.

Finally, the decomposition of the software makes it possible to maintain and reuse the various parts of the program independently. For example, it becomes much easier to use the RRSQRT Kalman filter also with other applications. Part of this flexibility is due to the CouPLE approach, which limits the opportunities for a programmer to make assumptions about the set of components that make up the complete program. Consequently, the components will be as much self supporting as possible, which eases their reusability in other settings.

## Acknowledgements

This research is supported by the Dutch National Institute for Coastal and Marine Management/RIKZ.

## References

1. R. E. Kalman. A new approach to linear filter and prediction theory. *J. Basic Engr.*, 82D:35–45, 1960.
2. R.E. Kalman and R.S. Bucy. New results in linear filtering and prediction theory. *J. Basic. Engr*, 83D:95–108, 1961.
3. P.M. Lyster, S.E. Cohn, R. Ménard, L.-P. Chang, S.-J. Lin, and R.G. Olsen. Parallel implementation of a kalman filter for constituent data assimilation. Technical Report DAO Office Note 97-02, Data Assimilation Office, Goddard Laboratory for Atmospheres, NASA, April 1997.
4. M. Morf, J.R. Dobbins, B. Friedlander, and T. Kailath. Square-root algorithms for parallel processing in optimal estimation. *Automatica*, 15:299–306, 1979.
5. M.E. Phillipart et al. Datum2: Data assimilation with altimetry techniques used in a tidal model. Technical Report NRSP-2 98-19, BCRS, 1998.

6. M.R.T. Roest. *Partitioning for Parallel Finite Difference Computations in Coastal Water Simulation.* PhD thesis, Delft University of Technology, 1997.
7. M. Verlaan. *Efficient Kalman Filtering Algorithms for Hydrodynamic Models.* PhD thesis, Delft University of Technology, April 1998.
8. M. Verlaan and A.W. Heemink. Reduced rank square root filters for large scale data assimilation problems. In *Second Int'l Symp on Assimilation of Observations in Meteorology and Oceanography*, pages 247–252. World Meteorological Organisation, march 1995.
9. M. Verlaan and A.W. Heemink. Tidal flow forecasting using reduced rank square root filters. *Stochastic Hydrology and Hydraulics*, 11:349–368, 1997.
10. E.A.H. Vollebregt. Abstract level parallelization of finite difference methods. *Scientific Programming*, 6:331–344, 1997.

## A  The Kalman Filtering Algorithm

### A.1  The Original Algorithm

Kalman filtering is a method to modify the model state $x(t)$ of a linear model, ($x(t+1|t) = A(t)x(t) + B(t)b(t)$, where $A$ and $B$ are matrices and $b(t)$ are the forcings) based on the observations $y(t)$. It is essentially a predictor-corrector method; the prediction and correction steps are usually called *timestep* and *measurement step* respectively. In the time step, an estimate of the new state, $x(t+1|t)$ and the uncertainty therein, $P(t+1|t)$ are obtained from the current state $x(t)$ and current uncertainty $P(t)$ respectively:

$$x(t+1|t) = A(t)x(t) + B(t)b(t)$$
$$P(t+1|t) = A(t)P(t)A(t)^T + F(t)\Sigma_s(t)F(t)^T$$

where $\Sigma_s$ denotes the covariance matrix of the model noise, which is $E[\epsilon_s(t) \cdot \epsilon_s(t)^T]$, and the matrix $F$ specifies how new noise is introduced into the model. Next, the measurement step corrects these predictions. First the *Kalman gain*, $K(t+1)$, is computed that weighs the differences between $x(t+1|t)$ and the observations $y(t+1)$:

$$K(t+1) = P(t+1|t)C(t)^T[C(t+1)P(t+1|t)C(t+1)^T + \Sigma_o(t+1)]^{-1}$$

where $\Sigma_o$ denotes the covariance matrix of the observation uncertainty, and $C$ is the matrix that interpolates the model state to the measurement locations. Then the predictions are corrected:

$$x(t+1) = x(t+1|t) + K(t+1)[y(t+1) - C(t+1)x(t+1|t)]$$
$$P(t+1) = P(t+1|t) - K(t+1)C(t+1)P(t+1|t)$$

## A.2 The RRQSRT Algorithm

In this original form (introduced by Kalman, [1]), Kalman filtering has a very high computational cost for models with a large state space dimension $n$. In that case, approximations must be made to reduce the time-complexity to an acceptable level. One such approximate scheme is the *Reduced Rank Square Root (RRSQRT)* algorithm, developed by Verlaan and Heemink (see e.g. Verlaan and Heemink, [8], or Verlaan [7]). Basically, it uses an approximation $L$ of the square root of the matrix $P$, where the size of $L$ is $n \times q$. The second dimension, $q$, is called the *number of modes* and can be taken much smaller than $n$. In terms of $L$, the Kalman algorithm becomes:

*timestep*:
$$x(t+1|t) = A(t)x(t) + B(t)b(t) \quad (1)$$
$$L(t+1|t) = [A(t)L(t), F(t)\Sigma_s(t)^{1/2}] \quad (2)$$

reduce second dimension of $L$ :

    determine eigendecomposition $V(t)$ of $L^T \cdot L$ (which is relatively small: $q + m \times q + m$):

$$L(t+1|t)^T L(t+1|t) = V(t+1)E(t+1)V(t+1)^T \quad (3)$$

compute the optimal square root of $L \cdot L^T$:

$$L'(t+1|t) = L(t+1|t)V(t+1) \quad (4)$$

reduce $L'$:
$$L'(t+1|t) = [L'(t+1|t)]_{1:n,1:q} \quad (5)$$

*measurement step*
Assuming that observation errors are independent, do for each of the observations in turn:

    determine $K$:

$$H(t+1) = L'(t+1|t)^T C(t+1)^T \quad (6)$$
$$\gamma(t+1) = [H(t+1)^T H(t+1) + \Sigma_o(t+1)]^{-1} \quad (7)$$
$$K(t+1) = L'(t+1|t)H(t+1)\gamma(t+1) \quad (8)$$

    correct predictions:

$$x(t+1) = x(t+1|t) + K(t+1)(y(t+1) - C(t+1)x(t+1|t)) \quad (9)$$
$$L(t+1) = L'(t+1|t)$$
$$\quad - K(t+1)H(t+1)[1 + [\gamma(t+1)\Sigma_o(t+1)]^{1/2}]^{-1} \quad (10)$$

# Multi-block Parallel Simulation of Fluid Flow in a Fuel Cell

S. Baird & J. J. McGuirk

Dept. of Aeronautical and Automotive Engineering and Transport Studies
Loughborough University, Loughborough, LE11 3TU, England

**Abstract.** This paper details the development of a multi-block parallel structured CFD solver to predict the fluid flow in Fuel Cells, thereby facilitating its use as a design tool. A multi-block parallel code is quicker, has greater memory efficiency and has better geometry conformity than a single block code. The multi-block approach allows geometry to be altered easily with only the changed blocks requiring new mesh generation, reducing lead time for predictions of evolving geometry. Efficient use of such a scheme requires parallel implementation using a domain decomposition approach, as described below.

## 1. Introduction

Due to political pressure, the current trend is towards environmentally friendly energy production. The development of fuel cell technology is likely to be a crucial component in the implementation of environmentally friendly electricity supply. To design efficient fuel cells, trade-offs between different design parameters need to be optimized. In order to achieve this, the relative benefits of a particular design need to be ascertained. However, there are many physical phenomena affecting the performance of fuel cells: ion/fuel depletion, rates and reversibility of chemical reactions, shunt currents, fluid distribution, and pressure drops. There are two methods of obtaining the data required in order to estimate the efficiency of a fuel cell, experimental and a computational simulation approach. Experimental tests are expensive and time consuming. A computational simulation can reduce the cost and decrease the time taken for new ideas to be tested and compared. Baird [1] has developed several simulation sub-models to predict the effects of various physical phenomena on a fuel cell. This paper details one of these, which is concerned with the prediction of pressure drops and fluid distribution throughout the entire fuel cell using computational fluid dynamics (CFD) techniques.

### 1.1. Description of a Fuel cell

A bipolar stack, Figure 1, uses two different fluids separated by a membrane flowing between two electrodes. During the discharge the voltage set up across the cell by the electro-chemistry causes ions to pass from solution (A) to solution (B) through the membrane, while electrons pass from solution (A) into the electrode and through the external circuit to into solution (B), thereby maintaining electrical neutrality. During charge an external voltage, opposed to the natural voltage, is imposed on the cell. This forces the electrons and ions to pass in the opposite direction from solution (B) to solution (A). A bipolar stack is made up of many such cells stacked together, the current passing through all the cells before completing the circuit.

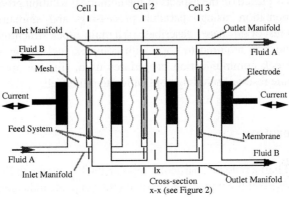

**Fig. 1.** Layout of a three cell bipolar stack

## 1.2. Effects of Pressure drop and Fluid Distribution on Cell Efficiency

The two fluids are pumped around the entire fuel cell and the power required to drive these flows is energy lost. Therefore the lower the pressure drop across the system the greater the efficiency. The fuel cell can only remove electrons from the solution when they are touching the electrode and similarly the ions can only pass through the membrane when they come into direct contact. If the rate at which the electrochemically active species are supplied to the respective surfaces by transport processes is greater than the rate at which they are depleted, then the reaction rate is determined by the current. If less, then the reaction rate is determined by the transport of fluid to the surface. This alters the voltage and causes the efficiency to drop. Hence the quality of the flow distribution and the local processes of convective and diffusive fluid transport processes exert a direct influence on cell behaviour. Often problems arise due to poor fluid distribution, for example in regions of fluid recirculation or slow fluid motion. The electrolyte is only replenished slowly and will eventually run out of reactants. Firstly this reduces the efficiency because the electrode and membrane are not being fully utilised, and secondly other, undesirable, chemical reactions may occur which can damage the electrode and membrane, e.g. by inducing precipitation, when particles can damage the membrane and block off flow. Inadequate fluid distribution also has a detrimental effect on the control systems used in the fuel cell. If each cell in the stack is evenly fed with electrolyte and the distribution within each cell is even then the control equipment can monitor the current/concentrations at just one part of the cell and adjust the control strategy accordingly. If the flow is not even then the control system will have to monitor at several positions in the cell to enable adequate control of the system, increasing production costs.

## 1.3. CFD Approach

In order to simulate the flow process which determines fuel cell pressure drop and the quality of the flow distribution, the governing equations of fluid flow (usually involving a model of turbulent motion) must be solved using CFD techniques. Space does not allow the details of the mathematical model to be provided here (see ref. [1] for details); rather emphasis is placed on the aspects of the numerical solution procedure which will require implementation using, parallel processing and domain decomposition techniques. The rest of this paper describes: (i) a multi-block structured grid generation methodology which is required to cope with complex fuel cell geometry, (ii) some details of the block communication procedures used, (iii) the approach to parallel solution, (iv) some results, including application to a fuel cell.

## 2. Grid Generation

The geometry of half a pole in a particular bipolar stack is shown in Figure 2. Referring to Figure 1 this represents a typical geometry on the cut plane indicated by line x-x. The geometry constraints of a single block structured grid will not be able to model the complex full stack geometry very easily without using many internal blockages. Additionally, memory requirements make it unfeasible to mesh the entire fuel cell with a single grid. Other advantages of the multi-block approach are that only one pole (or half a fuel cell) needs to be meshed at initio because the rest of the stack can be meshed by copying and transforming the grid from this original block. The ability to change just a small portion of the design and only re-grid the changed block will be a huge advantage when using the solver as a design tool, as it reduces the time taken to predict the flow at the next evolutionary stage. A multi-block code also allows different fluids to flow through different blocks, which is essential to modelling fuel cells. The geometry/grid detail shown in Figure 3 is for one half of one bipolar cell and is made up of 32 blocks and clearly demonstrates the requirement for a multi-block code.

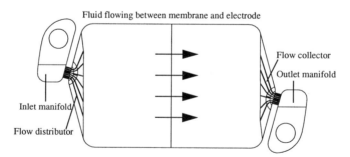

**Fig. 2.** The geometry of a single pole in a bipolar stack.

Capture of the flow distribution features of the cell geometry (see zoomed in grid details) within a single-block mesh could have been accomplished using internal blockages, with associated memory wastage. The multi-block mesh shown in Figure 3 requires 108Mb of memory, a single block grid with the same grid density would use more than 2Gb. A multi-block structured solver can mesh very complex geometry, almost as well as a totally unstructured grid, and is more memory efficient. These two criteria, along with the computational efficiency of the structured solver make it ideal for large, disperse and complex geometry.

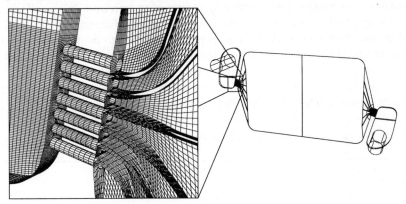

**Fig. 3.** Fuel cell geometry and grid details of flow distributor channels.

## 3. Blocking Scheme

Usually the computational stencil used by structured solvers contains only the finite volume cells adjacent to the cell in question (nearest neighbours). For some higher order discretization schemes the cells next adjacent to the near neighbours are also included in the stencil. Figure 4 shows the stencil used in the present structured 3D solver.

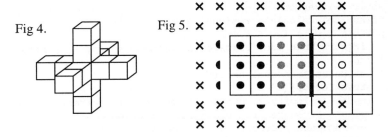

**Fig. 4.** The computational stencil in 3 dimensions

**Fig. 5.** The solution variables *(filled circles)* are stored at the finite volume centres, *(grey circles)* are data transferred to adjacent block, *(empty circles)* transferred from adjacent block, halo cells *(crosses)* and boundary cells *(semi circles)*. The *(filled)* cells are solved implicitly, *(grey)* cells are solved in the implicit sweep but use *(empty circles)* data from the neighbour block from the previous time step

For the structured solver to process the boundary cells of each block in the same manner as "fully internal" cells, without the need for special treatment, then each block must have a copy of the data from adjacent cells in neighbouring blocks, in the solution variable arrays at the correct offsets. This requires each block to have an extra halo of cells around it which a single block code would not require and this can be seen in Figure 5.

Calculations are carried out in each block independently, but a converged global solution requires blocks to communicate with each other, hence the multi-block scheme necessitates a procedure for transferring information into/from halo cells between neighbouring blocks. The transfer algorithm fills a one dimensional array with the boundary data from one block and passes this information to the neighbouring block. This process must be systematically carried out for each boundary on each block; note the data transfer is in both directions because the two blocks on either side of the boundary send data to each other. The details of where and what data is transferred can be seen in Figure 6, which provides the logic of one iterative cycle of the flow solver applied to all dependent variables in turn (i.e. velocity components (u,v,w), pressure correction (dp), and turbulence model variables (k,d)).

```
set number of sweeps for each variable (default=1)
for f = u, v, w, dp, k, d
   set Dirichlet boundary conditions
   set diffusion coefficient
   set wall function
   if(dp)transfer(apu,apv,apw)
   set mass fluxes
      set gradient boundary conditions for velocity components
      set mass fluxes + pressure smoothing
      integrate "fixed" flow across boundary
      integrate "floating" flow across boundary
      scale "floating" flow to match "fixed" flow
   assemble coefficients & sources (Aw,Ae,As,An,Al,Ar,Ap,S)
   if(u,v,w) store Ap as apu,apv,apw respectively
   set implicit boundary conditions
   apply relaxation
   if(not dp) calculate residual errors
   perform 1 or a few S.L.O.R. sweeps
      if(more than one) transfer(f) between sweeps - optional
   if(dp)
      calculate residual errors
      transfer(dp)
      update velocity components from pressure correction
      update pressure from pressure correction
      extrapolate pressure on to boundary
      transfer(u,v,w,p)
   set explicit Neumman boundary conditions
   if(not dp) transfer(f)
```

**Fig. 6.** General code detailing the iterative cycle of the solver. The changes which were required by a multiple-block code are shown in bold

The algebraic equations are solved by a successive line relaxation (S.L.O.R.) technique involving a simple tridiagonal matrix inversion (TDMA) along lines through the structured grid. The tridiagonal matrix solver is swept through just one block. This saves having to store and transfer the line solver coefficients, and means that the matrix inversion can be carried out in parallel. It does however mean that the halo cell information is calculated from values at the previous iterative sweep and this makes the overall iterative process explicit at block boundaries. The cells which are updated in this explicit manner are shown in Figure 5.

## 4. Parallel Approach

The multi-block structured grid approach is here written with the solution variables in each block stored in separate arrays. This enables the blocked scheme to be solved on separate processors using a modified domain decomposition approach. This implementation of the blocked scheme naturally lends itself to a parallel approach and in turn to a reduction in the time required to run a prediction.

The machine used during the development and testing of the present code was a Silicon Graphics Origin 2000 with 2.2Gb of shared memory and twenty four R10000 processors. It was therefore strictly speaking not possible to distribute the memory. However, since the solver is not to be developed for this machine alone, it would still be preferable to adopt a type of distributed memory. The use of separate solution variable arrays for each block leads to the shared memory allocatation being effectively partitioned on a block-by-block basis. Since the code requests solution within each to be carried out on a single processor, then this effectively ties each block partitioned portion of the shared memory to a single processor in a similar way to a "pure" distributed memory architecture although the memory may not be local to the processor running the calculation. This will enable the code to be used on both shared and distributed memory machines without any major change. Only the data transfer routines will require modification as they currently rely on the arrays being accessible by all processors. This will require the use of PVM rather than MPI.

Using MPI the calculations are separated between processors by looping through the blocks and sending each processor a block to compute. When the processor finishes the computation it is given another block to compute, until all blocks have been computed. This implementation does not allow for the prediction to be split up over more processors than there are blocks. However, even for a single pole of a fuel cell the mesh shown in Figure 3 uses 32 blocks and since a full fuel cell may have many hundred bipolar cells this was not deemed a problem. Using this approach, some load balancing occurs naturally, as long as the larger blocks which take longer to solve are computed first. For example, in a 3 block calculation running over two processors, if block 1 is larger than blocks 2 and 3, blocks 1 and 2 start computation, the second processor will finish block 2, then compute block 3. If block 1 takes longer to compute than blocks 2 & 3 combined, then no speed up will be gained by running over more than two processors without splitting block 1 into (possibly) several blocks. This means that for a given number of blocks there is a number of processors above which no speed up will be seen. This number of processors can be calculated from:

$$\text{Max. No. processors for effective speed up} = \frac{\text{Total No. of cells}}{\text{No. cells in largest block}} \quad (1)$$

## 5. Results

In order to validate the code an initial test problem was chosen for which experimental data was available and which was a simple, easy to mesh geometry. The test case chosen was Case 5 from an ERCOFTAC Workshop [2], "Developing Flow in a Curved Rectangular Duct". The grid used 144x24x48 cells and required about 30Mb of memory. The inflow boundary conditions were applied from the experimental data given in [2]. The fluid flows through a duct of uniform rectangular cross section which turns through a 90° bend with a radius three times the width of the duct (see Figure 7). The prediction only models the top half of the duct, and a symmetry condition is applied to the lower face. Figure 8 shows the secondary velocity vectors in the duct at two distances downstream of the bend. The results show that the phenomenon observed in the experiment of the vortex moving down the wall has been captured by the prediction and the agreement with the experiments is reasonable and similar to other predictions presented in [2]. The converged multi-block code gave identical answers, to within machine round-off, to a single block calculation. Running the code across more than one processor produced results exactly the same, iteration for iteration, as when run upon a single processor but the time taken was greatly reduced.

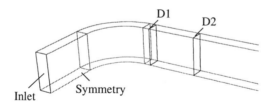

**Fig. 7.** Geometry of curved rectangular duct

Several tests were run, in order to find the most efficient parallel approach. The pressure correction equation (dp) was repeatedly cycled using the TDMA in order to produce a more refined pressure correction before updating the velocities and pressure. This was done with and without transferring the (dp) variable between cycles, the optional transfer indicated in Figure 6. Cycling for the dp variable within the S.L.O.R. matrix solver significantly increased the convergence rate, the more cycles the faster the convergence, although Table 1 shows that this follows a law of diminishing returns. The effect of transferring the (dp) variable between each cycle does increase the convergence per iteration, but not enough to make the extra computational time worthwhile. The time taken to perform the extra data transfer means that the time to achieve convergence is worse.

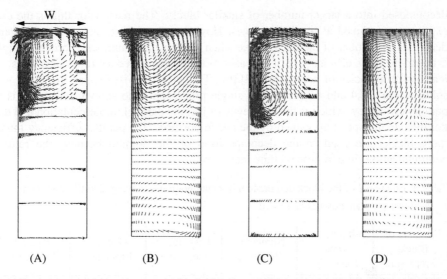

**Fig. 8.** Comparison of experimental and predicted secondary velocity vectors. *A* Experiment at location D1, *B* prediction at D1, *C* Experimental at D2, *D* Prediction at D2, D1 = 0.5xW, D2 = 4.5xW from end of 90° bend

Table 1: Effect of cycling through S.L.O.R. algorithm and transferring between sweeps, 4 blocks were used solved across two processors, converged until maximum residual was permanently under $1.0 \times 10^{-5}$

| Number of cycles | Transfer | Number of iterations | Time Taken (seconds) | Iteration time (seconds) |
|---|---|---|---|---|
| 1 | - | 29957 | 250141 | 8.35 |
| 2 | No | 22883 | 192217 | 8.40 |
| 2 | Yes | 22845 | 192355 | 8.42 |
| 3 | No | 18826 | 159080 | 8.45 |
| 3 | Yes | 18761 | 159093 | 8.48 |
| 4 | No | 16874 | 143429 | 8.50 |
| 4 | Yes | 16807 | 143532 | 8.54 |
| 5 | No | 15697 | 134209 | 8.55 |
| 5 | Yes | 15607 | 134220 | 8.60 |

Based on the above evidence, all further calculations used 5 sweeps with no transfer, Table 2 summarises results obtained for the initial test problem where the number of blocks is processors was varied for fixed problem size and fixed algebraic equation solution procedures. It was possible to run this test problem with a single block and a single processor and this is the first results shown in Table 2. The iterations to achieve convergence of the multi-block parallel solver increases as the solution domain is decomposed into a larger number of smaller blocks. The reason for this is the extra explicitness induced at block interfaces. However, the time to achieve convergence decreases as number of blocks increase when using a fixed number of processors; this is due to the more efficient use of the cashe as the problem size per block decreases, and leads to efficiencies of greater than 100%. This effect is visible also in Figure 9 for solutions on 1,2,4 and 8 blocks on a single processor. Also shown in Table 2 is the parallel efficiency, which is the efficiency of just the parallel implementation and does not include the speed up gained by the decomposition. The parallel efficiency falls as more blocks are used in the decomposition of the domain because the ratio of communication time to process time increases.

Table 2: Efficiency of the block and parallel implementations, converged until maximum residual was permanently under $1.0 \times 10^{-5}$

| Number of Blocks, Processors | Iteration time (seconds) | Number of iterations | Time Taken (seconds) | Over all Efficiency | Parallel Efficiency |
|---|---|---|---|---|---|
| 1,1 | 18.2 | 13961 | 254090 | BASE | |
| 2,1 | 16.5 | 14423 | 237980 | 106.8 | BASE |
| 2,2 | 9.1 | " | 131250 | 96.8 | 90.8 |
| 4,1 | 15.6 | 15697 | 245030 | 103.7 | BASE |
| 4,2 | 8.6 | " | 134209 | 94.6 | 91.3 |
| 4,4 | 5.7 | " | 88960 | 71.4 | 69.3 |
| 8,1 | 15.1 | 16280 | 245830 | 103.4 | BASE |
| 8,2 | 8.1 | " | 131870 | 96.3 | 92.6 |
| 8,4 | 5.4 | " | 87910 | 72.3 | 70.1 |
| 8,8 | 4.0 | " | 65120 | 48.7 | 47.5 |

The results in Figure 9 illustrate the advantages of parallel implementation since the time to achieve convergence using 8 processors is reduced to a quarter of the single processor time.

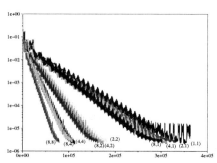

**Fig. 9.** Summation of Maximum Residuals against time. Y-axis is $\sum (res)_{max}$. X-axis is time (secs). *(No. of blocks, No. of processors)*

Since this test case has demonstrated the successful implementation of the multi-block parallel strategy, the CFD code then was applied to the problem of flow through the fuel cell geometry of Figure 3. The results for the flow field are illustrated using particle paths shown in Figure 10.

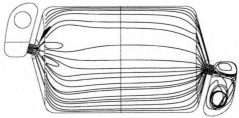

**Fig. 10.** Particle paths through the main flow region showing the effect of the flow distribution

This calculation utilises 32 blocks, occupied 108Mb of memory, and was run across 2 processors. The results show regions of recirculation in the vicinity of some distributor channel exits and not completely even flow distribution across the bulk of the cell. Both of these flow features could be improved by a re-design of the fuel cell geometry.

## 6. Conclusions

The implementation of a multi-block structured solver was successfully achieved through the use of explicit boundary coupling with data being transferred across block boundaries. No special cell treatment was required and the results were, to within machine round off, the same as a single block prediction. The convergence per iteration was hindered by the explicit block coupling, but the iteration time was quicker resulting in an increase in convergence against time.

The extension of the multi-block scheme to allow usage in a parallel-processing environment was achieved on a shared memory machine. The speed up of the calculation when run across multiple processors was very good. However, as expected, the speed up reduced as more processors were used, because the overhead of transferring data block to block increased. Work is continuing on implementing the parallel scheme on a distributed memory system.

The solver has been applied successfully both in a simple geometry test case (curved rectangular duct) and a realistic fuel cell geometry. This highly complex geometry and flow field demonstrate the use of the multi-block parallel solver.

## References

1. S.Baird, "Computational Modelling of Fuel Cells", PhD Thesis, Loughborough University, to be submitted (1999)
2. 4th ERCOFTAC/IAHR Workshop on Refined Flow Modelling, April 3-7, 1995, University of Karlsruhe, Germany

# A Parallel Implementation of the Block Preconditioned GCR Method

C. Vuik[1] and J. Frank[1,2]

[1] Delft University of Technology, Faculty of Information Technology and Systems, Department of Technical Mathematics and Informatics, P.O. Box 5031 2600 GA Delft, The Netherlands, c.vuik@math.tudelft.nl
[2] Center for Mathematics and Computer Science (CWI), P.O. Box 94079, 1090 GB Amsterdam, The Netherlands, frank@math.tudelft.nl

**Abstract.** The parallel implementation of GCR is addressed, with particular focus on communication costs associated with orthogonalization processes. This consideration brings up questions concerning the use of Householder reflections with GCR. To precondition the GCR method a block Gauss-Jacobi method is used. Approximate solvers are used to obtain a solution of the diagonal blocks. Experiments on a cluster of HP workstations and on a Cray T3E are given.

**Keywords:** approximate subdomain solution; parallel Krylov subspace methods; orthogonalization methods

## 1 Introduction

This paper addresses the parallel implementation of a Krylov accelerated block Gauss-Jacobi method for the DeFT Navier-Stokes solver described in [15], and is the continuation of work summarized in [5]. Results from a parallel implementation of a Krylov-accelerated Schur complement domain decomposition method are presented in [3]. We report results for a Poisson problem on a square domain, which is representative of the system which must be solved for the pressure correction method used in DeFT.

Aside from the preconditioning, the main parallel operations required in these methods are distributed matrix-vector multiplications and inner products. For many problems, the matrix-vector multiplications require only nearest neighbor communications, and are very efficient. Inner products, on the other hand, require global communications; therefore, the focus has been on reducing the number of inner products [8, 16], overlapping inner product communications with computation [6], or increasing the number of inner products that can be computed with a single communication [2, 13].

The block Gauss-Jacobi preconditioner is described in Section 1.1. Parallel implementations of orthogonalization procedures for the GCR (Generalized Conjugate

Residual) method are investigated in Section 2. Experiments done on a cluster of workstations and a Cray T3E are given in Section 3.

## 1.1 The block Gauss-Jacobi preconditioner

We consider an elliptic partial differential equation discretized using a finite volume or finite difference method on a computational domain $\Omega$. Let the domain be the union of $M$ nonoverlapping subdomains $\Omega_m$, $m = 1, \ldots, M$. Discretization of the PDE results in a sparse linear system $Ax = b$, with $x, b \in \mathrm{R}^N$. When the unknowns which share a common subdomain are grouped together into blocks one gets the block system:

$$\begin{bmatrix} A_{11} & \ldots & A_{1M} \\ \vdots & \ddots & \vdots \\ A_{M1} & \ldots & A_{MM} \end{bmatrix} \begin{pmatrix} x_1 \\ \vdots \\ x_M \end{pmatrix} = \begin{pmatrix} b_1 \\ \vdots \\ b_M \end{pmatrix}. \tag{1}$$

In this system, one observes that the diagonal blocks $A_{mm}$ express coupling among the unknowns defined on a common subdomain ($\Omega_m$), whereas the off-diagonal blocks $A_{mn}$, $m \neq n$ represent coupling across subdomain boundaries. The only nonzero off-diagonal blocks are those corresponding to neighboring subdomains.

In order to solve system (1) we use the block Gauss-Jacobi preconditioner:

$$K = \begin{bmatrix} A_{11} & & \\ & \ddots & \\ & & A_{MM} \end{bmatrix}.$$

When this preconditioner is used, systems of the form $Kv = r$ have to be solved. Since there is no overlap the diagonal blocks $A_{mm} v_m = r_m$, $m = 1, \ldots, M$ can be solved in parallel. In our method these systems are solved by an iterative method. An important point is the required tolerance of these inner iterations (see [9]). Since the number of inner iterations may vary from one subdomain to another, and in each outer iteration, the effective preconditioner is nonlinear and varies in each outer iteration.

Our choice of approximate solution methods is motivated by the results obtained in [4]. In that paper, GMRES was used as to approximately solve subdomain problems to within fixed tolerances of $10^{-4}$, $10^{-3}$, $10^{-2}$ and $10^{-1}$. Additionally, a blockwise application of the RILUD preconditioner was used. RILUD, a diagonal-restricted variant of the preconditioner introduced in [1], is a weighted average of an ILUD preconditioner [14] and an MILUD preconditioner [10]. The weighting parameter $\omega$, was assigned a value of 0.95 in our experiments. For small problems GCR without a preconditioner converges in a reasonable amount of time. However, when the gridsize increases the CPU time for GCR without preconditioning is much higher than for the preconditioned GCR method.

## 2 Orthogonalization methods for the GCR method

### 2.1 The GCR method

One of the Krylov subspace methods which allows a variable preconditioner is the GCR method [7], [17]. In this paper the Euclidean inner product $\langle x, y \rangle = x^T y$ and associated norm $\|x\| = (x^T x)^{1/2}$ are used.

**Algorithm: GCR**
Given: initial guess $x_0$
$r_0 = b - Ax_0$
for $k = 1, \ldots$, convergence
  Solve $K\tilde{v} = r_{k-1}$ (approximately)
  $\tilde{q} = A\tilde{v}$
  $[q_k, v_k] = $ **orthonorm** $(\tilde{q}, \tilde{v}, q_i, v_i, i < k)$
  $\gamma = q_k^T r_{k-1}$
  Update: $x_k = x_{k-1} + \gamma v_k$
  Update: $r_k = r_{k-1} - \gamma q_k$
**end**

The function **orthonorm()** takes input vectors $\tilde{q}$ and $\tilde{v}$, orthonormalizes $\tilde{q}$ with respect to the $q_i$, $i < k$, updating $\tilde{v}$ as necessary to preserve the relation $\tilde{q} = A\tilde{v}$, and returns the modified vectors $q_k$ and $v_k$.

The primary challenges to parallelization of GCR are parallelization of the preconditioning and parallel computation of the inner products. Inner products require global communication and therefore do not scale. Much of the literature on parallel Krylov subspace methods and parallel orthogonalization methods is focused on orthogonalizing a number of vectors simultaneously. However, this is not possible using a preconditioner which varies in each iteration. For this reason, we need a method for orthogonalizing one new vector against an orthonormal basis of vectors.

### 2.2 Orthogonalization methods

The modified Gram-Schmidt method suffers from the fact that the inner products must be computed using successive communications, and the number of these inner products increases proportional to the iteration number. This is not the case if one uses the classical Gram-Schmidt method. In this algorithm all necessary inner products can be computed with a single global communication. Unfortunately, the classical Gram-Schmidt method is unstable with respect to rounding errors, so this method is rarely used. On the other hand, Hoffmann [11] gives experimental evidence indicating that a two-fold application of the classical Gram-Schmidt method is stable. A third method which has been suggested is the parallel implementation of Householder transformations, introduced by Walker [18]. We shall reformulate that method for GCR in the following section. Additionally, we will present a simple parallel performance analysis for comparison of these three orthogonalization procedures.

## 2.3 Householder orthogonalization

In the following discussion we use the notion $a_k$ to represent the $k$th column of a matrix $A$ and $a^{(i)}$ to represent the $i$th component of a vector $a$. Let a matrix $A \in \mathbb{R}^{n \times m}$, $m \leq n$ with linearly independent columns be factored as $QZ$, where $Q$ is orthogonal and $Z$ is upper triangular. Then the $k$th column of $A$ is given by $a_k = Qz_k$ and the columns of $Q$ form an orthonormal basis for the span of the columns of $A$.

We construct $Q$ as the product of a series of Householder reflections, $Q = P_1 \cdots P_m$, used to transform $A$ into $Z$. The matrices $P_i = I - 2\frac{w_i w_i^T}{w_i^T w_i}$, with $w_i^{(j)} = 0$ for $j < i$ have the property: $P_i(P_{i-1} \cdots P_1)a_i = z_i$.

Suppose one has already produced $k$ orthogonal basis vectors. To compute $w_{k+1}$ one must first apply the previous reflections to $a_{k+1}$ as described in [18]: $\tilde{a} = P_k \cdots P_1 a_{k+1} = (I - 2W_k L_k^{-1} W_k^T) a_{k+1}$, where $W_k$ is the matrix whose columns are $w_1, \ldots, w_k$, and where

$$L_k = \begin{bmatrix} 1 & & & \\ 2w_2^T w_1 & 1 & & \\ \vdots & & \ddots & \\ 2w_k^T w_1 & \ldots & 2w_k^T w_{k-1} & 1 \end{bmatrix}.$$

Note especially that in the $(k+1)$th iteration one must compute the last row of $L_k$, which is the vector $(2w_k^T W_{k-1}, 1)$, as well as the vector $W_k^T a_{k+1}$. This requires $2k - 1$ inner products, but they may all be computed using only a single global communication.

Let $\hat{a}$ be the vector obtained by setting the first $k$ elements of $\tilde{a}$ to zero. The vector $w_{k+1}$ is chosen as: $w_{k+1} = \hat{a} + \text{sign}\,(\hat{a}^{(k+1)}) \|\hat{a}\| e_{k+1}$. In practice, the vectors $w_k$ are normalized to length one. The length of $w_{k+1}$ can be expressed as $\|w_{k+1}\| = \sqrt{2\alpha^2 - 2\alpha \hat{a}^{(k+1)}}$ where $\alpha = \text{sign}\,(\hat{a}^{(k+1)}) \|\hat{a}\|$. The $(k+1)$th column of $Q$ is the new orthonormal basis vector:

$$q_{k+1} = \frac{1}{\alpha}\left[ a_{k+1} - \sum_{i=1}^{k} \tilde{a}^{(i)} q_i \right].$$

Within the GCR algorithm, the same linear combination must be applied to the $v_i$ to obtain $v_{k+1}$.

## 2.4 Performance analysis

In this section the costs of the orthogonalization methods are considered (for the details we refer to [9]).

**Re-orthogonalized Classical Gram-Schmidt (CGS2)**

The $k$th iteration of this method costs $3k$ vector updates and $2k$ inner products. To compute the inner products 2 global messages of length $k$ are sent.

**Modified Gram-Schmidt (MGS)**
The $k$th iteration of MGS costs $2k$ vector updates and $k$ inner products. For the inner products $k$ global messages of length 1 are required.

**Householder (HH)**
In the $k$th iteration of the Householder method, $3k$ vector updates and $2k$ inner products are done. In every iteration 3 communications are necessary: two global messages one with length $k$ and the other with length 1, and a broadcast of $k$ elements.

Comparing the costs we expect that the wall-clock time for CGS2 and HH are comparable. When communication is slow (large latency) with respect to computation one expects that these methods are faster than MGS. Otherwise MGS may be the fastest method because MGS needs less floating point operations.

## 3 Numerical experiments

Firstly we present some time measurements to compare the various orthogonalization methods. Secondly we investigate the scalability of the parallel block preconditioned GCR method. Each processor is responsible for an $n \times n$ subdomain with $n^2$ unknowns.

### 3.1 Measurements of the orthogonalization methods

Tests were performed on a cluster of HP workstations and on a a Cray T3E using MPI communication subroutines. The wall clock times in the orthogonalization part are measured when 60 GCR iterations are performed.

In Figure 1 the parameters

$$\mathcal{F}_{HH} = \frac{\text{orthog. time MGS}}{\text{orthog. time HH}}$$

and

$$\mathcal{F}_{CGS2} = \frac{\text{orthog. time MGS}}{\text{orthog. time CGS2}}$$

are plotted as functions of $n$. In each subdomain an $n \times n$ grid is used. The number of subdomains is equal to the number of processors. On the workstation cluster (HH) and (CGS2) are only advantageous when the number of unknowns is less than 3600 on 4 processors and less than 6400 on 9 processors. On the Cray T3E, the number of unknowns per processor should be fewer than 1000 for 9 or even 25 processors. For larger problems the smaller amount of work involved in modified Gram-Schmidt orthogonalization outweighs the increased communication cost.

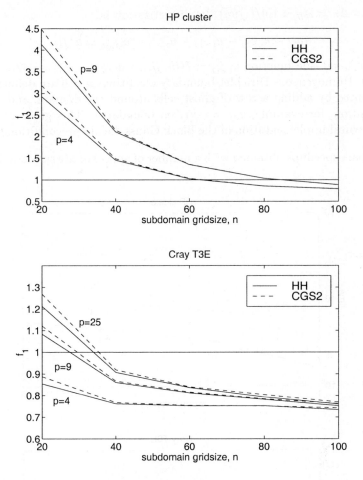

**Fig. 1.** Measured speedup with Householder (HH) orthogonalization and Reorthogonalized Classical Gram-Schmidt (CGS2)

## 3.2 Evaluation of approximate subdomain solvers

As a test example, we consider a Poisson problem, discretized with the finite volume method on a square domain. The pressure correction matrix, which we solve in each time step of an incompressible Navier-Stokes simulation to enforce the divergence-free constraint [12], is similar to a Poisson problem, but with asymmetry arising from the use of curvilinear coordinates. Solution of this system requires about 75% of the computing effort. So that we can obtain a useful indication of the performance of our method on the pressure correction matrix, we do not exploit the symmetry of the Poisson matrix in these experiments. The domain is composed of a $\sqrt{p} \times \sqrt{p}$ array of subdomains, each with an $n \times n$ grid.

With $h = \Delta x = \Delta y = 1.0/(\sqrt{p}n)$ the discretization is

$$4u_{ij} - u_{i+1j} - u_{i-1j} - u_{ij-1} - u_{ij+1} = h^2 f_{ij}.$$

The right hand side function is $f_{ij} = f(ih, jh)$, where $f(x,y) = -32(x(1-x) + y(1-y))$. Homogeneous Dirichlet boundary conditions $u = 0$ are defined on $\partial\Omega$, implemented by adding a row of ghost cells around the domain, and enforcing the condition, for example, $u_{0j} = -u_{1j}$ on boundaries. This ghost cell scheme allows natural implementation of the block Gauss-Jacobi preconditioner as well.

We compare speedups obtained with a number of approximate subdomain solvers

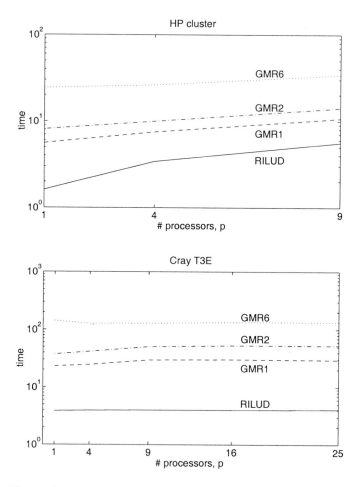

**Fig. 2.** Computation time for fixed subdomain size of $120 \times 120$

to get an impression of which solvers might be effectively used with the Navier-Stokes equations. For these tests a fixed number of GCR iterations (30) are done, and modified Gram-Schmidt was used as the orthogonalization method. The performance measure is wall clock time, after initialization, taken as the minimum achieved over three runs.

The subdomain approximations will be denoted as follows:

- GMR6 = restarted GMRES with a tolerance of $10^{-6}$,
- GMR2 = restarted GMRES with a tolerance of $10^{-2}$,
- GMR1 = restarted GMRES with a tolerance of $10^{-1}$,
- RILUD = one application of an RILUD preconditioner.

Figure 2 shows a comparison of the parallel scalability of the domain decomposition method with approximate subdomain solution. The figure shows computation times on 1, 4 and 9 processors (1, 4, 9, 16 and 25 processors for the Cray T3E) with a fixed subdomain size of 120 × 120. Note that the method scales almost perfectly on the Cray for this range of processors. On the workstation cluster, the scaling is somewhat poorer. The scaling for the GMR variants is reasonable, whereas the scaling of RILUD is bad.

## 4 Conclusions

Wall clock measurements for the modified Gram-Schmidt, re-orthogonalized classical Gram-Schmidt, and Householder orthogonalization methods indicates that classical Gram-Schmidt and Householder require approximately the same amount of work and communication, making the re-orthogonalized classical Gram-Schmidt more attractive, since it is easier to implement and more stable. The Householder and re-orthogonalized classical Gram-Schmidt methods are most effective for relatively small problems: using nine processors, up to about 900 unknowns per processor for a Cray T3E, or 8000 unknowns per processor for a cluster of workstations.

For this type of problem, the best subdomain approximation method in parallel is a simple incomplete factorization restricted to the diagonal: the RILUD factorization. With this preconditioner used as a subdomain approximation, the approximate solves become so cheap (and yet sufficiently accurate) that they offset the increased number of GCR iterations resulting from inaccurate subdomain solution.

## References

1. O. Axelsson and G. Linskog. On the eigenvalue distribution of a class of preconditioning methods. *Numerische Mathematik*, 48:479–498, 1986.
2. Z. Bai, D. Hu, and L. Reichel. A Newton-basis GMRES implementation. *IMA Journal of Numerical Analysis*, 14:563–581, 1994.
3. E. Brakkee, A. Segal, and C. G. M. Kassels. A parallel domain decomposition algorithm for the incompressible Navier-Stokes equations. *Simulation Practice and Theory*, 3:185–205, 1995.

4. E. Brakkee, C. Vuik, and P. Wesseling. Domain decomposition for the incompressible Navier-Stokes equations: Solving subdomain problems accurately and inaccurately. *International Journal for Numerical Methods in Fluids*, 26:1217–1237, 1998.
5. Erik Brakkee. *Domain Decomposition for the Incompressible Navier-Stokes Equations*. PhD thesis, Delft University of Technology, P.O. Box 5031, 2600 GA Delft, The Netherlands, April 1996.
6. E. de Sturler and H. A. van der Vorst. Reducing the effect of global communication in GMRES($m$) and CG on parallel distributed memory computers. *Applied Numerical Mathematics*, 18:441–459, 1995.
7. Stanley C. Eisenstat, Howard C. Elman, and Martin H. Schultz. Variational iterative methods for nonsymmetric systems of linear equations. *SIAM Journal on Numerical Analysis*, 20(2):345–357, April 1983.
8. J. Erhel. A parallel GMRES version for general sparse matrices. *Electronic Transactions on Numerical Analysis (http://etna.mcs.kent.edu)*, 3:160–176, 1995.
9. J. Frank and C. Vuik. Parallel implementation of a multiblock method with approximate subdomain solution. *App. Num. Math.*, 1998. to appear.
10. Ivar Gustafsson. A class of first order factorization methods. *BIT*, 18:142–156, 1978.
11. Walter Hoffman. Iterative algorithms for Gram-Schmidt orthogonalization. *Computing*, 41:335–348, 1989.
12. J. van Kan. A second-order accurate pressure-correction scheme for viscous incompressible flow. *SIAM Journal on Scientific and Statistical Computing*, 7(3):870–891, 1986.
13. G. Li. A block variant of the GMRES method on massively parallel processors. *Parallel Computing 23*, 23:1005–1019, 1997.
14. J. A. Meijerink and H. A. van der Vorst. An iterative solution method for linear systems of which the coefficient matrix is a symmetric M-matrix. *Mathematics of Computation*, 31:148–162, 1977.
15. A. Segal, P. Wesseling, J. van Kan, C.W. Oosterlee, and K. Kassels. Invariant discretization of the incompressible Navier-Stokes equations in boundary-fitted coordinates. *International Journal for Numerical Methods in Fluids*, 15:411–426, 1992.
16. R. B. Sidje. Alternatives for parallel Krylov subspace basis computation. *Numerical Linear Algebra with Applications*, 4(4):305–331, 1997.
17. Henk A. van der Vorst and C. Vuik. GMRESR: a family of nested GMRES methods. *Numerical Linear Algebra with Applications*, 1(4):369–386, 1994.
18. Homer F. Walker. Implementation of the GMRES method using Householder transformations. *SIAM Journal on Scientific and Statistical Computing*, 9(1):152–163, 1988.

# Comparison of Two Parallel Analytic Simulation Models of Inhomogeneous Distributed Parameter Systems

S.W. Brok[1], L. Dekker[2]

[1] Faculty of Applied science, Lorentzweg 1, 2628CJ Delft, The Netherlands
s.w.brok@tn.tudelft.nl
[2] Noordeindseweg 61, 2651LE Berkel en Rodenrijs, The Netherlands
L.Dekker@pa.twi.tudelft.nl

**Abstract.** In this paper parallel modelling of linear inhomogeneous 1D distributed parameter systems is considered. Two modelling methods in relation to parallel simulation are discussed: a two-point series expansion method and an eigenfunction expansion method. Both methods are validated with two different source terms for which the analytical solutions are known. A comparison of both approaches with respect to convergence aspects, accuracy, work load, and explicit parallelism is presented.

## 1 Introduction

Parallel modelling aims at constructing a parallel system model such that the computing time of the parallel model on a parallel computer is considerably smaller than the computing time of an equivalent sequential model on a sequential computer. In parallel simulation of complex and large-scale applications it is a strict demand to have at one's disposal a parallel computer model containing much exploitable parallelism. This means that parallel system modelling must be directed to construct a parallel system model with much explicit parallelism. Then parallel simulation is simple because then the need for data communication is negligible. In case of exclusively explicit parallelism, the parallelism can be exploited with hundred percent efficiency. This occurs when a parallel system model is merely composed of independent subtasks.
Modelling of complex and large-scale systems results in a natural way of partitioning of the model in many interconnected subsystems. This means that system decomposition is often a first mean to enlarge the amount of exploitable parallelism. Parallel modelling must be directed to minimization of the total information processing as well as to maximization of parallelism of data processing, data communication and synchronization. These goals are still difficult to transpose into generally valid guide lines for the construction of parallel system models. This means that often in parallel modelling an acceptable compromise has to be made.

There exists much digital software for simulation of sequential system models. So, in practice parallel modelling mostly implies the transformation of the digital program of a sequential system model in an as good as possible parallel system model. Such a transformation is not easy to realize and moreover, the resulting parallel system model contains as a rule little exploitable parallelism. A basic reason for this is that the functional properties of the architecture as well as the technical properties of the hardware are far from adequate.

An important goal of parallel modelling is to create explicit parallelism as much as possible in the conceptual stage as well as to preserve this parallelism as much as possible in the utilization stage. However, it is also important to increase the robustness of a parallel model with regard to other properties, because increase of robustness also contributes to the applicabillity of a parallel model. How this can be achieved still needs further research.

Parallel modelling will be elucidated briefly for some numerical examples. For linear problems a potential way is to force parallelism by applying the superposition principle. That's to say the solution is formulated as the sum of the solutions of a number of subproblems, such that each subproblem contains much exploitable parallelism. This parallel modelling approach will be illustrated for initial and boundary value problems of linear inhomogeneous 1D distributed parameter systems.

## 2 Two-point Expansion Method

The aim in developing the two-point expansion method was the creation of efficient massively parallel models. The method and its application are illustrated for the 1D diffusion system (2.1) with state $z_d(x,t)$:

$$\frac{\partial z_d}{\partial t} - \frac{\partial^2 z_d}{\partial x^2} = f(x,t); \quad 0 \leq x \leq 1, t \geq 0; \tag{2.1}$$

$$z_d(0,t) = \varphi_0(t); z_d(1,t) = \varphi_1(t); z_d(x,0) = \gamma(x)$$

and the 1D hyperbolic system (2.2) with state $z_h(x,t)$:

$$\frac{\partial^2 z_h}{\partial t^2} - \frac{\partial^2 z_h}{\partial x^2} = f(x,t); \quad 0 \leq x \leq 1, t \geq 0; \tag{2.2}$$

$$z_h(0,t) = \varphi_0(t); z_h(1,t) = \varphi_1(t); z_h(x,0) = \gamma(x); \frac{\partial z_h(x,0)}{\partial t} = \eta(x)$$

The distributed source term $f(x,t)$ and the boundary conditions $\varphi_0(t)$, $\varphi_1(t)$ are assumed to be analytic functions. The solutions of these systems depend on the boundary conditions $\varphi_0(t)$, $\varphi_1(t)$, the inhomogeneous term $f(x,t)$ and the initial conditions of the state and the derivative state: $\gamma(x)$ and $\eta(x)$. Both the inhomogeneous term and the boundary conditions can be conceived to represent input signals of these distributed systems. For a stable system it holds that an input signal, satisfying certain conditions, forces a so-called generalized steady-state solution. For both the diffusion system (2.1) and the hyperbolic system (2.2) the generalized steady-states $z_{de}(x,t)$, $z_{he}(x,t)$ can be conceived to be the superposition of the generalized steady-states $z_d s_e(x,t)$, $z_h s_e(x,t)$, forced by the source term $f(x,t)$ in the case $\varphi_0(t) = \varphi_1(t) = 0$ and the generalized steady-states $z_d b_e(x,t)$, $z_h b_e(x,t)$, forced by the boundary conditions $\varphi_0(t)$, $\varphi_1(t)$ in the case $f(x,t) = 0$. In many preceeding publications, a.o. [1], [4], [5], the way how to determine the generalized steady-state $z_{be}(x,t)$ is discussed for different types of distributed parameter systems.

## 2.1 Source-input generalized steady-state

It is known from literature ([1], [2]) that a function $f(x)$ of one variable, satisfying certain condi-tions, can be expressed within an interval as a two-point expansion of the even derivatives of $f(x)$ in the boundary points of the interval. Applying this to $f(x,t)$, considered as a function of $x$ in the interval $[0,1]$, yields that $f(x,t)$ can be written at each time $t$ as a two-point expansion:

$$f(x,t) = \sum_{k=0}^{\infty} \{g_k(x) f^{(2k,0)}(0,t) + g_k(1-x) f^{(2k,0)}(1,t)\} \quad (2.3a)$$

where $f(0,t)$ and $f(1,t)$ represent the derivatives of $f(x,t)$ with respect to $x$ of order $2k$ in $x = 0$ and in $x = 1$ respectively at each time $t$. The coefficients $g_k(x)$ and $g_k(1-x)$ are polynomials of $x$ of degree $2k + 1$. These polynomials satisfy the recurrent relation:

$$g_k^{(2)}(x) = g_{k-1}(x), \; g_k(0) = g_k(1) = 0, \; k \geq 1; \; g_0(x) = 1 - x \quad (2.3b)$$

Further properties of these polynomials are described in a.o. [1].

*1D diffusion system*
The source-input generalized steady-state $z_d s_e(x,t)$ is the infinite sum of the contributions $z_d s_{ek}(x,t)$ and $z_d s_{ek}(1-x,t)$, corresponding respectively to the terms $g_k(x) f^{(2k,0)}(0,t)$ and $g_k(1-x) f^{(2k,0)}(1,t)$, $k = 0,1,\otimes$ in the two-point expansion (2.3a). Consider now the contributions $z_d s_{ek}(x,t)$ which satisfies for $k = 0,1,\otimes$:

$$\frac{\partial z_d s_{ek}}{\partial t} - \frac{\partial^2 z_d s_{ek}}{\partial x^2} = g_k(x) f^{(2k,0)}(0,t); \quad z_d s_{ek}(0,t) = z_d s_{ek}(1,t) = 0; \quad (2.4a)$$

It can be shown that this contribution $z_d s_{ek}(x,t)$ can be written as:

$$z_d s_{ek}(x,t) = -\sum_{n=0}^{\infty} g_{k+n+1}(x) f^{(2k,n)}(0,t) \quad (2.4b)$$

A similar expression can be constructed for the generalized steady-state $z_d s_{ek}(1-x,t)$ resulting in the source-input generalized steady-state $z_d s_e(x,t)$:

$$z_d s_e(x,t) = \sum_{k=0}^{\infty} \{z_d s_{ek}(x,t) + z_d s_{ek}(1-x,t)\} \qquad (2.5)$$

$$= -\sum_{k=0}^{\infty}\sum_{n=0}^{\infty} \{g_{k+n+1}(x) f^{(2k,n)}(0,t) + g_{k+n+1}(1-x) f^{(2k,n)}(1,t)\}$$

where $f(0,t)$ and $f(1,t)$ represent the derivatives of $f(x,t)$ with respect to $x$ of order $2k$ and with respect to $t$ of order $n$ in respectively $x=0$ and in $x=1$ [1], [8]. This outcome for $z_d s_e(x,t)$ satisfies the homogeneous boundary conditions as well as the inhomogeneous diffusion equation as can easily be verified by substitution.

*1D hyperbolic system*
Following the same procedure as in the case of the 1D diffusion system it appears that the source-input generalized steady-state $z_h s_e(x,t)$ can be formulated as:

$$z_h s_e(x,t) = -\sum_{k=0}^{\infty}\sum_{n=0}^{\infty} \{g_{k+n+1}(x) f^{(2k,2n)}(0,t) + g_{k+n+1}(1-x) f^{(2k,2n)}(1,t)\} \qquad (2.6)$$

This expression for the generalized steady-state has much resemblance with the expression for the generalized steady-state for the 1D diffusion system. It only differs in the derivatives with respect to time of the source term $f(x,t)$ in the points $x=0$ and $x=1$. The time-derivatives of $f(x,t)$ of order $n$ in the expression (2.5) are now replaced by the time-derivatives of $f(x,t)$ of order $2n$.

## 2.2 Boundary-input generalized steady-state

*1D diffusion system*
For the diffusion system (2.1) the boundary-input generalized steady-state $z_d b_e(x,t)$ is the generalized steady-state forced by the boundary conditions $\varphi_0(t)$ and $\varphi_1(t)$ in the case $f(x,t) = 0$. This generalized steady-state $z_d b_e(x,t)$ can be formulated as an infinite linear combination of the time-derivatives of $\varphi_0(t)$ and $\varphi_1(t)$ with space-dependent coefficients $g_k(x)$ and $g_k(1-x)$ which have been derived in several preceeding publications (a.o. [1], [4], and [5]):

$$z_d b_e(x,t) = \sum_{k=0}^{\infty} \{g_k(x)\varphi_0^{(k)}(t) + g_k(1-x)\varphi_1^{(k)}(t)\} \qquad (2.7)$$

*1D hyperbolic system*
Also for the hyperbolic system (2.2) the boundary-input generalized steady-state $z_h b_e(x,t)$ is the generalized steady-state forced by the boundary conditions $\varphi_0(t)$ and

$\varphi_1(t)$ in the case $f(x,t) = 0$. In a similar way as for the 1D diffusion system it can be found:

$$z_h b_e(x,t) = \sum_{k=0}^{\infty} \{g_k(x)\varphi_0^{(2k)}(t) + g_k(1-x)\varphi_1^{(2k)}(t)\} \tag{2.8}$$

Notice again that the only difference with the boundary-input generalized steady-state solution $z_d b_e(x,t)$ (2.7) is that the time-derivatives of the boundary conditions $\varphi_0(t)$ and $\varphi_1(t)$ of order $n$ are replaced by the time-derivatives of order $2n$.

## 2.3 Initial condition(s) and total solution

*1D diffusion system*
Generally, the sum of the source-input generalized steady-state solution $z_d s_e(x,t)$ and the boundary-input generalized steady-state solution $z_d b_e(x,t)$ will in general for $t = 0$ not be equal to the initial condition $\gamma(x)$. Consequently there still exists a transient state $z_d t_r(x,t)$ being an infinite series of eigensolutions satisfying the system equation with zero source term and zero boundary conditions and with a so-called cleared initial condition:

$$z_d t_r(x,t) = \sum_{k=0}^{\infty} a_k \exp(-k^2 \pi^2 t) \sin k\pi x \tag{2.9a}$$

in which the coefficients $a_k$ are determined by:

$$z_d t_r(x,0) = \gamma(x) - z_d s_e(x,0) - z_d b_e(x,0) = \sum_{k=0}^{\infty} a_k \sin k\pi x \tag{2.9b}$$

So, the total solution $z_d(x,t)$ of the 1D diffusion system (2.1) can be formulated as the sum of the generalized steady-state solutions and the transient solution:

$$z_d(x,t) = z_d s_e(x,t) + z_d b_e(x,t) + z_d t_r(x,t) \tag{2.10}$$

*1D hyperbolic system*
For the hyperbolic system (2.2) there also exists besides the two generalized steady-state solutions $z_h s_e(x,t)$ and $z_h b_e(x,t)$ a transient solution $z_h t_r(x,t)$ which again can be formulated as an infinite series of eigensolutions:

$$z_h t_r(x,t) = \sum_{k=0}^{\infty} (a_k \sin k\pi t + b_k \cos k\pi t) \sin k\pi x \tag{2.11a}$$

The coefficients $a_k$ and $b_k$ can be calculated from the cleared initial conditions:

$$z_h t_r(x,0) = \gamma(x) - z_h s_e(x,0) - z_h b_e(x,0) = \sum_{k=0}^{\infty} b_k \sin k\pi x \tag{2.11b}$$

$$\frac{\partial z_h t_r(x,0)}{\partial t} = \eta(x) - \frac{\partial z_h s_e(x,0)}{\partial t} - \frac{\partial z_h b_e(x,0)}{\partial t} = \sum_{k=0}^{\infty} a_k k\pi \sin k\pi x$$

Again the total solution $z_h(x,t)$ of the 1D hyperbolic system (2.2) can be formulated as the sum of the generalized steady-state solutions and the transient solution:

$$z_h(x,t) = z_h s_e(x,t) + z_h b_e(x,t) + z_h t_r(x,t) \tag{2.12}$$

## 3 Eigenfunction Expansion Method

Besides the two-point expansion method for constructing the generalized steady-state solution of distributed parameter systems other approaches are possible. One approach is the so-called eigenfunction method. In the paper this method will be derived for the 1D diffusion system (2.1) and the 1D hyperbolic system (2.2) and homogeneous boundary conditions and zero initial condition(s). Applying the method of separation of variables it is well known that

$$\phi_k(x,t) = \exp(-k^2\pi^2 t)\sin k\pi x \quad k = 0,1,\otimes \tag{3.1}$$

and

$$\begin{aligned}\phi_{1,k}(x,t) &= \sin k\pi t \sin k\pi x \\ \phi_{2,k}(x,t) &= \cos k\pi t \sin k\pi x\end{aligned} \quad k = 0,1,\otimes \tag{3.2}$$

are respectively solutions (eigenfunctions with eigenvalues $k\pi$) that satisfies the system equations (2.1) and (2.2) and homogeneous boundary conditions. Because $\{\sin k\pi x\}_{k=0}$ forms a complete set of orthogonal functions on the interval $[0,1]$ any other function can be expanded into an infinite series of these orthogonal functions.
The source term $f(x,t)$ in the 1D diffusion system (2.1) and the 1D hyperbolic system (2.2) can be expanded into an infinite series of eigenfunctions:

$$f(x,t) = \sum_{k=0}^{\infty} f_k(t)\sin k\pi x \quad \text{with} \quad f_k(t) = 2\int_0^1 f(x,t)\sin k\pi x\, dx \tag{3.3}$$

We now consider the 1D diffusion system (2.1) and try to find a solution $z_d(x,t)$ of the form:

$$z_d(x,t) = \sum_{k=0}^{\infty} c_k(t)\sin k\pi x \tag{3.4}$$

Substitution in the 1D diffusion equation (2.1) taking into account that $\gamma(x) = 0$ results into:

$$z_d(x,t) = \sum_{k=0}^{\infty} \int_0^t \exp\{-k^2\pi^2(t-\tau)\} f_k(\tau) d\tau \sin k\pi x \qquad (3.5)$$

One approach to extract the generalized steady-state solution $z_d s_e(x,t)$ from this expression is by means of the Laplace transformation of $z_d(x,t)$ with respect to the time $t$. For simplicity only the result is given. For a more detailed description see [7]. This generalized steady-state solution $z_d s_e(x,t)$ is given by an infinite series of infinite time integrals of $f_k(t)$:

$$z_d s_e(x,t) = -\sum_{k=0}^{\infty} \left[ \int_t^{\infty} \exp\{(\zeta - k^2\pi^2)(t-\tau)\} f_k(\tau) d\tau \right]_{\zeta=0} \sin k\pi x \qquad (3.6)$$

in which the parameter $\zeta$ must be chosen such that all the time integrals in this expression are bounded.

The generalized steady-state solution can also be obtained in a quite different way. Substituting (3.4) into the 1D diffusion system (2.1) results into an inhomogeneous first order ordinary differen-tial equation for the coefficients $c_k(t)$:

$$\frac{dc_k(t)}{dt} + k^2\pi^2 c_k(t) = f_k(t) \qquad k = 0, 1, \otimes \qquad (3.7)$$

The generalized steady-state solution of this differential equation can be formulated as an infinite series of the derivatives of $f_k(t)$ because $f_k(t)$ can be considered to be the input signal of (3.7):

$$c_k(t) = \sum_{n=0}^{\infty} a_{n,k} f_k^{(n)}(t) \qquad (3.8)$$

in which $f_k^{(n)}(t)$ is the derivative of $f_k(t)$ with respect to time of order $n$. The coefficients $a_{n,k}$ can be determined by substitution of these series expansion in the differential equation (3.7). This leads to the infinite double series:

$$z_d s_e(x,t) = -\sum_{k=0}^{\infty} \sum_{n=0}^{\infty} (-k^2\pi^2)^{-n-1} f_k^{(n)}(t) \sin k\pi x \qquad (3.9)$$

Compared to the expression (3.6) the infinite series of integrals has been replaced by a double infinite series of derivatives of $f_k(t)$ with respect to time.
In the special case that the source term $f(x,t)$ is a function of $x$ only, i.c. $f(x,t) = f(x)$ and $f_k(t) = f_k$, the integral in (3.6) can be evaluated analytically. The infinite double series (3.9) is reduced to a single infinite series (because $f_k^{(n)}(t) = 0$ for $n \geq 1$) which then leads for both represen-tations to the expression:

$$z_d s_e(x,t) = \sum_{k=0}^{\infty} f_k \frac{\sin k\pi x}{k^2 \pi^2} \qquad (3.10)$$

Notice that the double series (3.9) can easily be used for parallel computation of $z_d s_e(x,t)$ by truncating the infinite series.

In the same way we can construct a solution of the 1D hyperbolic system (2.2) satisfying homogeneous boundary conditions and zero initial conditions:

$$z_h(x,t) = \sum_{k=0}^{\infty} \int_0^t \frac{\sin k\pi(t-\tau)}{k\pi} f_k(\tau) d\tau \sin k\pi x \qquad (3.11)$$

While the accompanying generalized steady-state solution $z_h s_e(x,t)$ is given by:

$$z_h s_e(x,t) = -\sum_{k=0}^{\infty} \left[ \int_t^{\infty} \exp\{\zeta(t-\tau)\} \frac{\sin k\pi(t-\tau)}{k\pi} f_k(\tau) d\tau \right]_{\zeta=0} \sin k\pi x \qquad (3.12)$$

in which again the parameter $\zeta$ is needed to bound the integral.

For the 1D hyperbolic system also another formulation can be constructed in a similar way as for the 1D diffusion system resulting in:

$$z_h s_e(x,t) = -\sum_{k=0}^{\infty} \sum_{n=0}^{\infty} (-k^2 \pi^2)^{-2n-1} f_k^{(2n)}(t) \sin k\pi x \qquad (3.13)$$

In case that the source term $f(x,t)$ is a function of $x$ only, i.e. $f(x,t) = f(x)$ and $f_k(t) = f_k$, the integral in (3.12) can be evaluated analytically. Again the double sum in (3.13) is reduced to a single sum. In this special case both approaches lead to the expression:

$$z_h s_e(x,t) = \sum_{k=0}^{\infty} f_k \frac{\sin k\pi x}{k^2 \pi^2} \qquad (3.14)$$

Notice that in this special case $z_h s_e(x,t)$ is equal to the generalized steady-state solution $z_d s_e(x,t)$ of the 1D diffusion system. This can be understood because then both system equations are reduced to the same time-independent system equation: $\partial^2 z / \partial x^2 = f(x)$.

## 4 Implementation Aspects and Results

Considering the source-input and the boundary-input generalized steady-states for the diffusion system and the hyperbolic system it can be seen that in all these expressions the polynomial coefficients $g_k(x)$ and $g_k(1-x)$ appear. So, these coefficients can be calculated in advance for any given grid, which has not to be necessary equidistant.

Although there are several methods to calculate these polynomial coefficients the easiest, the fastest, and the most efficient way is by means of an infinite series, containing only sine terms [1]:

$$g_k(x) = (-1)^k \sum_{n=0}^{\infty} \frac{2\sin n\pi x}{n\pi} \frac{1}{(n^2\pi^2)^k} \tag{4.1}$$

These series is absolutely and uniformly convergent for $0 \le x \le 1$, if $k \ge 1$.
An important property of these polynomial coefficients is that for k sufficiently large $|g_{k+1}(x)| \approx |g_k(x)|/\pi^2$. Numerical experiments have demonstrated that this is already valid for $k \ge 2$.

## 4.1 1D diffusion system

We first focus our attention on a space-independent exponential source term: $f(x,t) = \exp(-\alpha t)$. Using the two-point expansion method for the construction of the source-input generalized steady-state solution $z_d s_e(x,t)$ results then in the formula (2.5). Because in this case $f(x,t)$ is a function of time only, the derivatives in (2.5) with respect to $x$ vanish for $k \ge 1$. So, the solution becomes:

$$z_d s_e(x,t) = -\sum_{n=0}^{\infty} \{g_{n+1}(x) + g_{n+1}(1-x)\}(-\alpha)^n \exp(-\alpha t) \tag{4.2}$$

These infinite series is absolutely and uniformly convergent in [0,1] for $|\alpha| < \pi^2$.
In the eigenfunction expansion method the coefficients $f_k(t)$ in formula (3.3) yield for the function $f(x,t) = \exp(-\alpha t)$:

$$f_k(t) = \frac{4\exp(-\alpha t)}{(2k+1)\pi} \quad k = 0, 1, \otimes \tag{4.3}$$

This implies that the integral in expression (3.6) or the derivatives in (3.9) can also be calculated analytically resulting into the following expression for $z_d s_e(x,t)$:

$$z_d s_e(x,t) = \sum_{k=0}^{\infty} \frac{4\sin(2k+1)\pi x}{(2k+1)\pi\{(2k+1)^2\pi^2 - \alpha\}} \exp(-\alpha t) \tag{4.4}$$

which is absolutely and uniformly convergent in [0,1] for all values of $\alpha$.
Substituting the expression of the generating function for the polynomials $g_k(x)$

$$\frac{\sinh\sqrt{\alpha}(1-x)}{\sinh\sqrt{\alpha}} = \sum_{k=0}^{\infty} g_k(x)\alpha^k \tag{4.5}$$

in the solution 4.2 results into [1], [8]:

$$z_d s_e(x,t) = \left\{ \frac{\sin\sqrt{\alpha}(1-x)}{\sin\sqrt{\alpha}} + \frac{\sin\sqrt{\alpha}x}{\sin\sqrt{\alpha}} - 1 \right\} \frac{\exp(-\alpha t)}{\alpha} \qquad (4.6)$$

The polynomial coefficients $g_k(x)$, the solutions obtained by the two-point expansion method (tem) and the eigenfunction expansion method (eem), and the analytical solution (4.6) are calculated on an equidistant grid in [0,1] with $\Delta x = 0.05$. In table 4.1 the number of terms in the infinite series (4.2) and (4.4) is listed to achieve a prescribed accuracy (i.e. $10^{-3}$, $10^{-4}$, $10^{-6}$, $10^{-8}$, and $10^{-10}$).
The calculations have been carried out for $t = 0$. That means that in case $\alpha > 0$ less terms for $t > 0$ than listed in table 4.1 are needed and when $\alpha < 0$ more terms. As can be seen from this table the parameter $\alpha$ is not very sensitive for the eigenfunction expansion method at a fixed accuracy. This can easily be understood from the form of the terms in the infinite series (4.4).
Concerning the two-point expansion method the behaviour of the needed terms is somewhat more complicated. For values of $\alpha$ in the neighbourhood at the edge of convergence ($|\alpha| < \pi^2$) rather many terms are needed to achieve the prescribed accuracy. This is due to the fact that the decrease of the polynomial coefficient $g_n(x)$ ($|g_{n+1}(x)| \approx |g_n(x)|/\pi^2$) in the infinite series (4.2) is then approximately equal to the increase of the term $(-\alpha)^n$.
Furthermore, it appeared that the solutions of the series (4.2) and (4.4) obtained by truncation of these series after the prescribed accuracy is achieved are also equal to the analytical solution (4.6) within the same accuracy.

**Table 1.** Number of terms for the two methods as a function of the desired accuracy and the parameter $\alpha$ for the diffusion system.

| | | exponential source diffusion system | | | | |
|---|---|---|---|---|---|---|
| | | $10^{-3}$ | $10^{-4}$ | $10^{-6}$ | $10^{-8}$ | $10^{-10}$ |
| $\alpha = 0.5$ | tem | 3 | 4 | 5 | 7 | 8 |
| | eem | 7 | 13 | 60 | 273 | 1253 |
| $\alpha = 2.0$ | tem | 5 | 7 | 9 | 12 | 15 |
| | eem | 7 | 13 | 60 | 273 | 1253 |
| $\alpha = 8.0$ | tem | 36 | 47 | 69 | 91 | 112 |
| | eem | 7 | 13 | 60 | 273 | 1253 |
| $\alpha = 9.5$ | tem | 194 | 255 | 375 | 496 | 617 |
| | eem | 7 | 13 | 60 | 273 | 1253 |

## 4.2 1D hyperbolic system

The source-input generalized steady-state solution $z_h s_e(x,t)$ using the two-point expansion method for the distributed source term $f(x,t) = \exp(-\alpha t)$ is given by:

$$z_h s_e(x,t) = -\sum_{k=0}^{\infty} \{g_{k+1}(x) + g_{k+1}(1-x)\}\alpha^{2k} \exp(-\alpha t) \qquad (4.7)$$

These series is absolutely and uniformly convergent for $|\alpha| < \pi$.
For the eigenfunction expansion method this solution is formulated by:

$$z_h s_e(x,t) = \sum_{k=0}^{\infty} \frac{4\sin(2k+1)\pi x}{(2k+1)\pi\{(2k+1)^2\pi^2 + \alpha^2\}} \exp(-\alpha t) \qquad (4.8)$$

which is absolutely and uniformly convergent in [0,1] for all values of $\alpha$
And the analytical solution can again be derived from the generating function of $g_k(x)$:

$$z_h s_e(x,t) = -\left\{\frac{\sinh\alpha(1-x)}{\sinh\alpha} + \frac{\sinh\alpha x}{\sinh\alpha} - 1\right\}\frac{\exp(-\alpha t)}{\alpha^2} \qquad (4.10)$$

In table 4.2 the number of terms (again for $t = 0$) in the infinte series (4.7) and (4.8) is given for some values of the parameter α to achieve a prescribed accuracy.

**Table 2.** Number of terms for the two methods as a function of the desired accuracy and the parameter $\alpha$ for the hyperbolic system

| | | exponential source hyperbolic system | | | | |
|---|---|---|---|---|---|---|
| | | $10^{-3}$ | $10^{-4}$ | $10^{-6}$ | $10^{-8}$ | $10^{-10}$ |
| α = 0.5 | tem | 3 | 3 | 4 | 6 | 7 |
| | eem | 7 | 13 | 60 | 273 | 1253 |
| α = 2.0 | tem | 9 | 11 | 16 | 21 | 27 |
| | eem | 7 | 13 | 60 | 273 | 1253 |
| α = 8.0 | tem | 81 | 106 | 156 | 206 | 256 |
| | eem | 7 | 13 | 60 | 273 | 1253 |
| α = 9.5 | tem | 278 | 365 | 537 | 710 | 883 |
| | eem | 7 | 13 | 60 | 273 | 1253 |

The number of terms for the two-point expansion method (tem) in table 4.2 are roughly the same as for the diffusion system in table 4.1. And exactly the same for the eigenfunction expansion method (eem). This is due to the fact that in the two series (4.4) and (4.8) the parameter $\alpha$ has not much influence on the number of terms needed to achieve a prescribed accuracy. Again the solution of the series (4.7) and (4.8) are equal to the analytical solution (4.10) within the same accuracy.

## 4.3 1D diffusion/hyperbolic system

When the distributed source term is a sine function in space ( $f(x,t) = \sin \omega \pi x$ with $\omega \neq$ integer ) the source-input generalized steady-state solution of the diffusion system and the hyperbolic system are identical.
The two-point expansion method results then in:

$$z_d s_e(x,t) = z_h s_e(x,t) = -\sin \omega \pi \sum_{k=0}^{\infty} (-\omega^2 \pi^2)^k g_{k+1}(1-x) \tag{4.11}$$

which is absolutely and uniformly convergen for $|\omega|<1$.
While the eigenfunction expansion method leads to the expression:

$$z_d s_e(x,t) = z_h s_e(x,t) = \frac{2 \sin \omega \pi}{\pi^3} \sum_{k=0}^{\infty} \frac{(-1)^k \sin k\pi x}{(\omega^2 - k^2)k} \tag{4.12}$$

which is absolutely and uniformly convergent in [0,1] for all values of $\omega$.
The analytical solution is in this case:

$$z_d s_e(x,t) = z_h s_e(x,t) = \frac{\sin \omega \pi x - x \sin \omega \pi}{\omega^2 \pi^2} \tag{4.13}$$

For some values of $\omega$ table 4.3 shows the number of terms needed in the infinite series (4.11) and (4.12) to achieve a prescribed accuracy (for $t = 0$ ).

**Table 3.** number of terms for the two methods as a function of the desired accuracy and the parameter $\omega$ for the diffusion/hyperbolic system.

|  |  | sine source diffusion/hyperbolic system | | | | |
|---|---|---|---|---|---|---|
|  |  | $10^{-3}$ | $10^{-4}$ | $10^{-6}$ | $10^{-8}$ | $10^{-10}$ |
| $\omega = 0.30$ | tem | 3 | 4 | 6 | 8 | 10 |
|  | eem | 24 | 50 | 230 | 1010 | 4650 |
| $\omega = 0.60$ | tem | 7 | 9 | 14 | 18 | 23 |

|  |  |  |  |  |  |  |
|---|---|---|---|---|---|---|
|  | eem | 24 | 50 | 230 | 1010 | 4650 |
| $\omega = 0.90$ | tem | 32 | 43 | 65 | 87 | 109 |
|  | eem | 24 | 50 | 230 | 1010 | 4650 |
| $\omega = 0.95$ | tem | 66 | 88 | 113 | 178 | 223 |
|  | eem | 24 | 50 | 230 | 1010 | 4650 |

Again in the eigenfunction expansion method (eem) the value of $\omega$ has no influence on the number of terms needed in the infinite series (4.12) at a fixed accuracy. While for the two-point expansion method (tem) the number of terms near the edge of convergence increases.

## 4.4 General remarks

From the tables 4.1, 4.2, and 4.3 it can be seen that for a low to medium accuracy the eigenfunction expansion method becomes favourable with respect to the number of terms needed in the infinite series to achieve a prescribed accuracy. In all other cases the two-point expansion method is preferred in spite of the fact that near the edge of convergence the number of terms increases.

In case of the space-independent source term $f(x,t) = \exp(-\alpha t)$ the number of floating point operations needed to calculate a single term in the infinite series is for the eigenfunction expansion method larger (8) than for the two-point expansion method (4). This holds both for the 1D diffusion system and for the 1D hyperbolic system. For the time-independent source term $f(x,t) = \sin \omega \pi x$ these numbers are respectively 6 and 4.

This implies that the computational workload for the eigenfunction expansion method is twice as much as for the two-point expansion method in case of the space-independent source term and a factor 1.5 in case of the time-independent source term.

In the two-point expansion method the polynomial coefficients $g_k(x)$ has to be calculated for any given grid on $[0,1]$ by means of the (truncated) infinite series (4.1). However, this has to be carried out only once for any arbitrary sourc-input $f(x,t)$.

So, it may be concluded that the two-point expansion method concerning convergence properties and computational workload is preferable.

## 5  Parallel Modelling Aspects

An important issue in developing parallel computer models is the creation of explicit parallelism as much as possible in the conceptual stage of modelling as well as to preserve this explicit parallelism as much as possible in all modelling steps, the parallel computer implementation included. It will be shown that the two-point expansion method meets these requirements. This will be demonstrated in more detail for the 1D diffusion system but the same holds for the 1D hyperbolic system.

As a first step the diffusion problem P is broken up in two subproblems $P_1$ and $P_2$ (see figure 5.1a) with solutions $z_d^*(x,t)$ and $z_d^{**}(x,t)$ such that

$$z_d(x,t) = z_d^*(x,t) + z_d^{**}(x,t) \tag{5.1}$$

For subproblem $P_1$ the solution $z_d^*(x,t)$ is defined as the solution forced upon the diffusion system (2.1) by the boundary conditions $\varphi_0(t)$, $\varphi_1(t)$ and the source input $f(x,t)$. Mostly it holds: $z_d^*(x,0) \neq 0$.

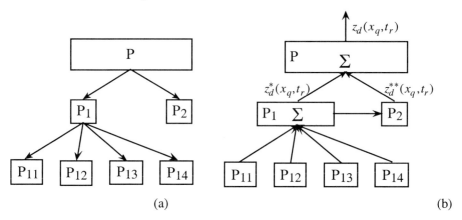

**Fig. 1.** (a) Creation of parallelism by problem splitting. (b) Parallel calculation of the state from the solution of the subproblems

Figure 5.1. (a) Creation of parallelism by problem splitting. (b) Parallel calculation of the
state from the solution of the subproblems

The solution $z_d^{**}(x,t)$ of subproblem $P_2$ equals the remaining transient solution $z_d t_r(x,t)$, i.e. the solution of (2.1) satisfying the homogeneous boundary conditions, the source conditions, and the so-called cleared initial condition:

$$z_d^{**}(x,0) = z_d t_r(x,0) = \gamma(x) - z_d^*(x,0) \tag{5.2}$$

The problem $P_1$ can be broken up in four independent subproblems $P_{11}$, $P_{12}$, $P_{13}$, and $P_{14}$ (see figure 5.1a) concerning the determination of respectively two boundary-input generalized steady-states $z_d b_{1e}(x,t)$, $z_d b_{2e}(x,t)$ and two source-input generalized steady-states $z_d s_{1e}(x,t)$, $z_d s_{2e}(x,t)$:

$$\begin{aligned} z_d^*(x,t) &= z_d b_e(x,t) + z_d s_e(x,t) \\ &= z_d b_{1e}(x,t) + z_d b_{2e}(x,t) + z_d s_{1e}(x,t) + z_d s_{2e}(x,t) \end{aligned} \tag{5.3}$$

The decomposition of $z_d b_e(x,t)$ into the subproblems $z_d b_{1e}(x,t)$ and $z_d b_{2e}(x,t)$ can be understood by inspecting the infinite series (2.7). This infinite series can be interpreted as the sum of two infinite series: one due to the boundary condition $\varphi_0(t)$ and one due to the boundary condition $\varphi_1(t)$. In a similar way the source-input generalized steady-state $z_d s_e(x,t)$ can be split up in the subproblems $z_d s_{1e}(x,t)$ and $z_d s_{2e}(x,t)$.

The evaluation of these four generalized states can be done in parallel for any array of points $(x_q, t_r)$ in the $x,t$-domain ($0 \leq x \leq 1, t \geq 0$) with hundred percent efficiency. The series expansions of the generalized steady-states will be truncated after a finite number of terms when a prescribed accuracy is obtained. The evaluation of a truncated series in a point $(x_q, t_r)$ also contains much parallelism but only partly explicit parallelism. Thus these point evaluations can be performed with large, but not hundred percent efficiency.

As soon as $z_d^{**}(x,0)$ becomes available (see figure 5.1b), solving of problem P2 can be started. Notice that for this reason it is obvious to start the solving of problem P1 at time $t_r = 0$. Then there is only a small time shift between the start of solving problem P1 and the start of solving problem P2. Also for problem P2 the evaluation of $z_d^{**}(x,t)$ is possible in parallel for any array of points $(x_q, t_r)$ in the $x,t$-domain ($0 \leq x \leq 1, t \geq 0$) with hundred percent efficiency.

The above way of modelling yields efficient parallel models for both the diffusion system (2.1) and the hyperbolic system (3.2). Important is that these models are scalable ones. That's to say, enlarging the problem size does not have much influence on the processing time as long as the parallelism of a scalable parallel computer remains sufficiently large.

In general the same holds for the eigenfunction expansion method. But in this case the calculation is a somewhat more complicated because the time-dependent coefficients in the infinite series (3.6) and the time-dependent coefficients $f_k(t)$ are now integral expressions which has to be solved by conventional numerical methods. This problem vanishes using the alternative representation (3.9). However, when the source term $f(x,t)$ and its derivatives vanish at the boundaries of the interval [0,1] (for instance $f(x,t) = \exp(-\alpha t) \sin 2\pi x$) the two-point expansion method cannot be applied. In that case the eigenfunction expansion method may be a reasonable alternative.

Notice that the usual numerical way of modelling by discretization the distributed parameter system results in the necessity to solve many large sets of linear equations. Due to numerical stability conditions the time-step in most of these discretized models must be small. The two-point expansion method and the eigenfunction expansion method has not such restrictions: the solution can be calculated for any arbitrary gridpoint $(x_q, t_r)$ in the $x,t$-domain. The numerical error only depends on how many terms in the infinite series are taken into account.

# 6  Conclusions

In the paper it is elucidated for linear, time-independent inhomogeneous distributed parameter systems that the generalized steady-state is a useful system concept in modelling. It is also exposed that an important consequence of parallel simulation is a quite different approach in system analysis and system modelling. It is based on the creation of explicit parallelism as much as possible in the conceptual phase and in the analysis phase and the preservation of this explicit parallelism as much as possible in both the modelling and in the parallel computer implementation phase.

For 1D distributed parameter systems two approaches are presented: the two-point expansion method and the eigenfunction expansion method. It can be concluded that the two-point expansion method is preferred with respect to convergence properties and computational workload. The results until now look promising that future parallel modelling of problems from the very beginning may often lead in practice to parallel models with a high degree of explicit parallelism.

The concept of the generalized steady-state is also applicable to homogeneous 2D and 3D distributed parameter systems [3], [5]. Future research will be focussed on the applicability to such higher order dimensional inhomogeneous problem formulations.

# 7 References

1. L. Dekker; Numerical aspects of the one-dimensional diffusion equation; PhD thesis, Delft University of Technology, 1964
2. J.M. Whittaker; On Lidstone series and two-point expansions of analytical functions; Proc. London Math. Soc. (2) vol 36, 1934, pp. 451-469.
3. L. Dekker; 4-Point series expansion, a massively parallel formulation of a function of two variables; Report 93-122, ISSN 0922-5641, Fac. of Technical Mathematics & Informatics, Delft University of Technology, 1993.
4. L. Dekker, S.W. Brok, F.J. Lingen; Semi-analytical method for MPP simulation of distributed parameter systems; Proc. EUROSIM Conference Massively Parallel Processing Applications and Developments, Delft, 1994, pp.579-590.
5. L. Dekker, S.W. Brok; A massively parallel simulation method for parabolic and hyperbolic systems; Proc. 1995 EUROSIM Conference, Vienna, 1995, pp. 249-254.
6. L. Dekker, S.W. Brok; Parallel distributed parameter system-adapted simulation and interpolation; Proc. EUROSIM Conference HPCN Challenges in Telecomp and Telecom, Delft, 1996, pp. 93-101.
7. S.W. Brok; Systems, Modelling and Simulation (in Dutch), Graduate Course (Chapter 8 and Chapter 9), Delft University of Technology, August 1994 (soon available in English).
8. L. Dekker, S.W. Brok; An MPP simulation method for inhomogeneous distributed parameter systems; Proc. EUROSIM 1998 Simulation Congress, Helsinki, 1998, pp. 20-26.
9. L. Dekker; Parallelism as a new dimension in science and prospects of parallel simulation; invited paper in the Proc. of the ISCS '97 Conference, Naples, 1997, pp. 1-13.
10. L. Dekker; Modelling and estimation of the generalized steady-state; System Analysis, Modelling, and Simulation, 3 (1986) 1, 3-11.

# Case Studies of Four Industrial Meta-Applications

Tim Cooper

Parallel Applications Centre, 2, Venture Road, Chilworth, Southampton, UK
Tpc@pac.soton.ac.uk

**Abstract.** From improved crash simulation to acoustic optimisation to innovative company-saving designs, distributed computing and meta-applications are enabling European industry to compete more effectively in many areas. Three such meta-applications are PROMENVIR, Optimus and TOOLSHED, all developed in recent ESPRIT projects. The PROMENVIR product provides users with all the functionality needed to perform Monte-Carlo analyses of complex problems, from satellite deployment to crash simulation. Optimus, from LMS also enables the user to perform sensitivity analyses, but with a broader focus encompassing classical design of experiment techniques. TOOLSHED is a problem solving environment (PSE) in the truest sense of the word. The package will automate any analysis process, providing the user with CAD importation, mesh generation, computational steering and visualisation in addition to the standard PSE requirements of seamless data transfer and transparent execution of tasks. The parallel implementation of LUSAS, an FE solver from FEA Ltd. developed in the ESPRIT project PARACOMP has been designed to run on a dual use cluster of NT machines, and has many requirements in common with the other meta-applications to be discussed. All of these products have interfaces to Intrepid, the Intelligent Resource Manager from PAC. Intrepid provides each of these packages with full meta-computing management, from a single point of control for the definition of a meta-computer to a clean, intuitive API through which they can control the execution of jobs. It makes use of performance models of applications to determine the CPU load, disk and memory requirements. These parameters are used to ensure both that tasks have the resources they need to execute and that execution of the entire problem is carried out in the most efficient manner. This paper will present an overview of the design of Intrepid, its interaction with these three packages and industrial examples of its application.

## Intrepid overview

One of the most effective designs for transparent access to resources is the three-tier architecture. In terms of meta-computing, the top tier is the meta-application and the bottom tier is of course the apllication layer where the work is actually carried out. The purpose of the middle tier is to present to the meta-application information about the resources available to it and a set of well-defined methods for controlling those resources.

Intrepid has been designed to act as the bottom two layers in such an architecture, as shown in figure 1 below. The central scheduling module accepts requests for tasks to be executed from the meta-application. Using information about the state of the available resources it makes a best estimate of where and when tasks will execute. This information is then passed back up to the meta-application for inspection by the user. If the user is satisfied with this allocation they may then request that tasks are executed, and Intrepid will control this execution. Execution is carried out to ensure that

1. There is sufficient disk and memory space for tasks to execute.
2. All data dependencies of a task are satisfied.
3. The execution of the entire meta-application is completed in the most efficient manner

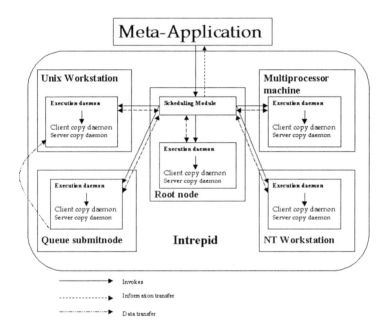

**Fig. 1.** Intrepid Architecture

## Launching Intrepid

Intrepid is launched from the command line. When it is not processing requests, the load it places on a resource is minimal, typically less than 2% cpu load. The first action carried out by Intrepid is to read in a file which describes the resources available to Intrepid which do not change over time.

Intrepid will try and perform an r-shell onto every host defined in the network in order to launch an execution daemon. These daemons not only control the execution of tasks they also return information to the central module about the state of the resource. For every platform where the r-shell was successful, Intrepid will then wait for a connection to be made back from the execution daemon. Once these connections are made, Intrepid is ready to receive requests to launch tasks.

**Launching Tasks using Intrepid.**

Meta-applications typically provide the user with a GUI through which they can define the tasks to be run. Most meta-applications allow the user to define a template from which actual runs will be generated in some way. We shall refer to the full list of tasks that the meta-application generates as a task graph. Intrepid at this stage has no interation with the meta-application, and aims to place minimal constraints on the type of job that may be defined. (In fact the only constraint is that the user provide a command line which may be issued from within a c-shell which will launch the application).

The requirement for interfacing to Intrepid is that the meta-application be able to write out a file containing;

- A full list of the applications to be run.
- For each application,
- The command line (which may be issued within a c-shell)
- The input and output data sets

This file is refered to as a task graph file. When a task graph is submitted to Intrepid, the cpu load, memory requirement as disk space requirements are all used to assign tasks to resources. In calculating these parameters, Intrepid uses one of two methods. For Nastran, SysNoise and some of the codes used in TOOLSHED internal models are used. For other applications, these parameters must be explicitly passed to Intrepid.

Feedback to the user is of vital importance in a meta-application, and Intrepid will allow the user to inspect the initial allocation of tasks before executing the task graph. If the user is satisfied with the allocation, they request that the job be run. Intrepid will then set about launching tasks, following as far as possible the initial allocation. In the event that the state of the resource changes there is a dynamic allocation module which will reassign tasks if their initial resource is not available.

**Data Traffic**

Data traffic is an issue in any meta-applications, and the meta-application which have been integrated with Intrepid utilise a number of techniques to cut down on data transfer. The most obvious way of doing this is to identify when data does not need to be transferred, and this is done in one of three ways. Firstly if the data already resides on the local disk of a machine then Intrepid will catch this and not copy the data. Secondly, if the data resides in a STEP repository then it is assumed that the repository will be able to serve the data to the application and no attempt is made to copy the data. Finally, the meta-application driving Intrepid can request that data transfer be supressed for whatever reason. This can happen because there is an NFS disk capable of serving data to applications, or because the mechanism of launching a task already incorporates the transfer of data back to the user's desktop machine.

One other mechanism that Intrepid can use to reduce data traffic is to compress files before and after copying. Again this can be done in one of two ways. The meta-application can request that instead of performing an r-shell onto a remote machine to launch the daemons, Intrepid launch daemons using the secure shell 'ssh'. Under this configuration the copying of data is carried out using 'scp' which performs automatic data compression. The other method is for the meta-application to request that Intrepid perform compression before copying data.

The final way of reducing data traffic is to copy back only the data that is needed. For instance, in a design of experiment problem or a Monte-Carlo simulation the only information that is needed is the parameter set under investigation. The rest of the data created by the simulation is not needed, and can be deleted. Intrepid enables this by running each task in a separate directory. Once the task is completed, the data sets identified as output from the run are copied to where ever it is they are needed next and the directory and all files in it are deleted. Thus in the instance where an application creates an output file of, say, 1Mb but the meta-application is only interested in one field in that file, then the launch method for the application must contain a method for extracting that parameter to a file. If, then, the output from the application is identified as the parameter file, rather than the full output file, only the parameter file will be copied.

## NT implementation

In terms of implementing Intrepid on NT, there are two important features to be dealt with. Firstly there is no remote shell, and secondly data transfer is non-trivial.

### Remote execution
In order to overcome the problem of remote execution, the Intrepid daemons are implemented as an NT service. Thus rather than Intrepid performing a remote shell onto the platform, launching the daemon and then waiting for connection back again, the NT implementation assumes that the daemon will always be executing on the remote machine. The daemon will therfore listen on a specific port number and wait for a connection from Intrepid.

### Data transfer
The windows API does not allow path names of the form '\\machine_name\shared_dir\file', and this can cause a problem if data transfer is to go over any mechanism other than the Intrepid copy daemons. In order to deal with this, when a data file is to be copied using the Windows API, the Intrepid daemons

will try to make a drive mapping of the relevant directories. This is done by scanning the list of drive letters which have already been mapped, mapping the first unused letter, copying the data and then unmapping the drive letter.

# Meta-applications demonstrated with Intrepid

### Promenvir

### Overview

The Promenvir product is designed to enable engineers to assess the impact of uncertainties in the design process. Numerical simulation is a deterministic process. The same input data will generate the same results every time the simulation is run. In designing a simulation the engineer assumes input parameters that are representative of the physical processes being modelled. Typically the error in these parameters is small in which case it is assumed to have no effect on the result. However in certain situations there may be many parameters that can vary. Although the effect of each individually is small, the engineer cannot be certain that the combined effect is bounded. In such situations, Monte-Carlo simulation provides a way of assessing stochastically the probability of a response given a set of varying parameters.

The Monte-Carlo algorithm is simple to express.
1. Create a stable, reference model of the system
2. Identify the parameters whose scatter is known within certain tolerances
3. For each variable select, randomly, a valuse within the allowed range
4. Run the analysis with the set of variables
5. Extract the response parameter of interest
6. Repeat step 3-5 until the statistics of the response parameters stabilise.

With efficient sampling algorithms, convergence is normally reached in 100 iterations. However for many simulation techniques a single solution may take up to an hour, and it is doubtful whether an engineer is prepared to wait that long for the results of an analysis. The solution used is to build into Promenvir a module that allows it to build and exploit a meta-computer. The commercial release of Promenvir utilises a simple meta-computer that launches tasks onto the next available resource. However Promenvir has been demonstrated with Intrepid, and the results are discussed in the next section.

### The Promenvir WAN experiment

A meta-computer consisting of 102 CPUs was defined using Intrepid. The machines were distributed over 8 sites, 5 of which were protected by firewalls, and ranged from a 64 CPU Silicon Graphics Origin machine to a number of Indigo R3000 platforms. Some of the machines were set aside by their owners for use in the experiment, but a number were used for other tasks. A Promenvir run was created of 1000 shots of the SIMAID solver to examine the sensitivity of a satellite attena deployment to the starting configuration of its joints. The total run would have taken 250hrs on an 'average' CPU in the meta-computer. Statistics for the run shown in table 1 below.

| Partner | CPUs | Availability | | | Shot statistics | | |
|---|---|---|---|---|---|---|---|
| | | In PVC | Accessed | Used | Failed | Successful | Total |
| PAC | 15 | 15 | 15 | 14 | 1 | 150 | 151 |
| Soton University | 10 | 10 | 9 | 6 | 0 | 40 | 40 |
| UPC | 16 | 16 | 16 | 16 | 1 | 275 | 276 |
| RUS | 12 | 12 | 11 | 9 | 0 | 104 | 104 |
| CASA | 15 | 15 | 15 | 14 | 15 | 184 | 199 |
| CEIT | 12 | 12 | 12 | 8 | 0 | 98 | 98 |
| Ital-Design | 11 | 11 | 6 | 5 | 2 | 63 | 65 |
| Blue | 11 | 11 | 11 | 7 | 6 | 61 | 67 |
| **Grand Totals** | 102 | 102 | 95 | 79 | 25 | 975 | 1000 |
| Total cpus installed | 102 | Elapsed Execution Time: | | | | | 4:39:16 |
| Total cpus in PVC file | 102 | | | | | | |
| Total cpus available | 95 | Approx single CPU time: | | | | | 250 hrs |
| Total cpus used | 79 | | | | | | |

Table 1. Results of Promenvir WAN experiment

**TOOLSHED**

TOOLSHED is a Problem Solving Environment that allows the user to plug together all the tools needed to perform a simulation process, from CAD import to automeshing to user-defined simulation processes. It comes equipped with a Visualisation tool, GLView from ViewTech and a Computation Steering tool, CSTool from Sintef. Data transfer between applications is managed using STEP, and data is served to applications by the DEVA repository from CLRC.

The core of the TOOLSHED environment, however, is the TOOLSHED Manager. Within this tool the user defines the sequence of steps to be carried out to solve the problem (called an Activity Template), the location of data sets specific to the problem and the resources to be used to solve the problem. The TOOLSHED Manager is closely coupled to Intrepid and is able to find out information about how to execute parallel processes.

A sample activity templates for a problem defined by Alstom Engines (formerly Ruston Engines) is shown in figures 2. A CAD file has been imported and edited to merge surfaces etc before the data is converted to STEP format. In the activity template the geometry is meshed, and the mesh is written into the STEP repository. A combustion CFD code is executed. The results may then be visualised.

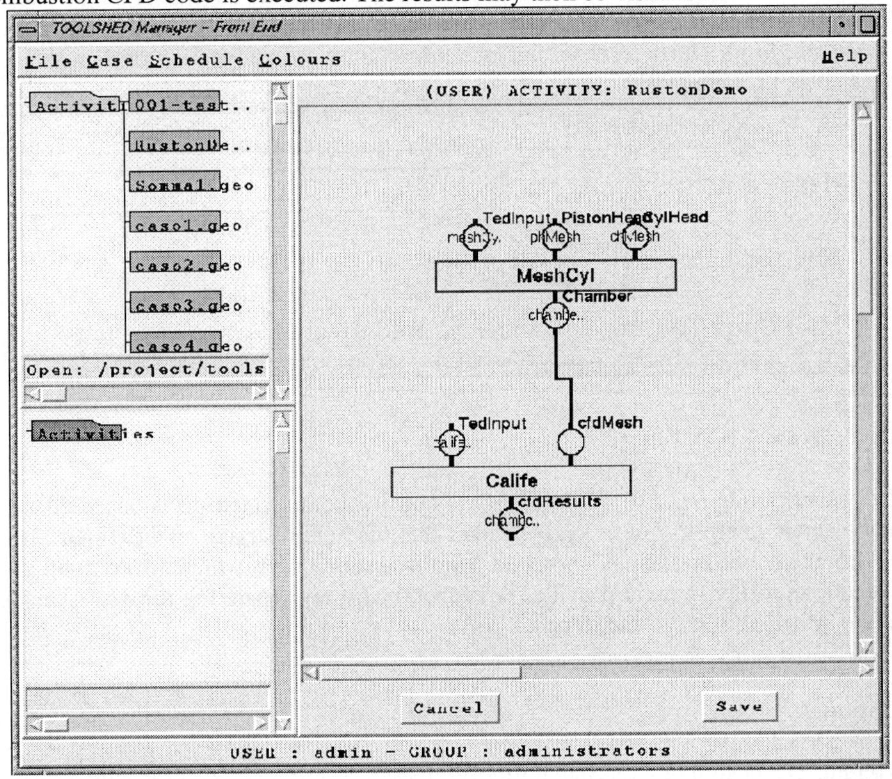

**Fig. 2.** Activity Template for CFD analysis

At the start of the TOOLSHED project, Alsthom Engines had no in-house CFD expertise. However using the process defined above they were able to analyse various design options for a piston head as part of a design effort to reduce emissions below 2001 requirements. The knowledge gained from using CFD has contributed to the patented design of a new piston head that will enable their engines to be competitive into the next millenium.

Intrepid plays two roles in TOOLSHED. In the first instance it allows the expert user to configure the system so that tasks always execute on appropriate machines. The user does not have to make any decision about where tasks will execute. In the classical industrial case where pre- and post- processing is carried out on desktop machines and simulation is performed on a production multiprocessor machine, the user is able to choose how constrained the execution will be. In the limit, they can constrain execution to their own machine or they can allow Intrepid to choose the least loaded machine in their area. Intrepid is also able to provide a significant amount of feedback to the user about the task to be run. Figures 3 shows an example of the type of feedback Intrepid is able to give, in this case advice on the number of CPUs to use for a parallel code. Note that Intrepid here advises not to run the code in parallel. Toolshed assumes that the user is accessing a shared resource, and total throughput is the issue rather than individual throughput. Intrepid will only advise running a code in parallel if this causes each process to run in core (and therefore possibly show super-linear speed-up). Other feedback include information on the location of data sets and the status of running tasks.

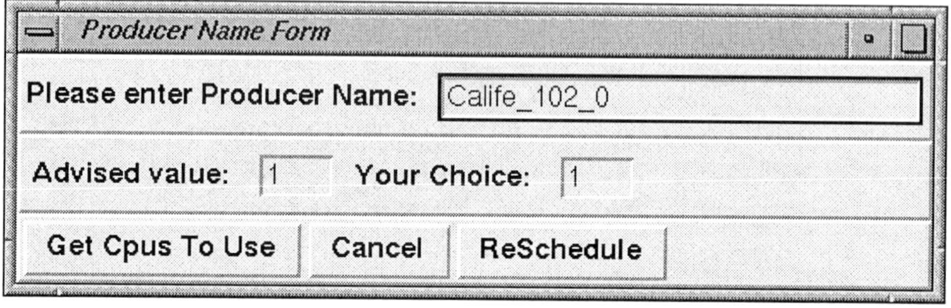

**Fig. 3.** Feedback from Intrepid

The benefit of Intrepid to TOOLSHED is not that complicated decisions need to be made about where to run tasks, but rather that the administrator of the TOOLSHED environment needs a single point of administration from which to configure the execution profile. Intrepid provides this functionality by separating the representation of the system into static and dynamic parts.

**Optimus**

Optimus is a design of experiment tool from LMS. It provides the user with a GUI in which they can
- Define a simulation process.
- Define control parameters of interest within input files.
- Define response variables within output files
- Define constraints within input, output or intermediate files.
- Specify a methodology for varying parameters within a design space.
- View the resulting response surfaces, and associated parameters.

Once the experiment is defined, Optimus then writes out a sequence file specifying the sequence of execution. This file would normally be read back in to Optimus, which then executes task in sequence on the local machine. However, when integrated with Intrepid the file is converted into format needed by Intrepid, is submitted and run.

**In car noise minimisation**

The Refinement Advanced Techniques Group (RATG) at Rover have used the Optimus tool to investigate the effect that engine and gearbox mountings have on in-car noise. Whilst Optimus is able to manage the execution of tasks on the desktop machine, and the existing LSF queue was able to manage the centralised computing resource, Rover needed utilise both desktop and centralised resources coherently. Therefore a series of experiments was carried out to assess how Intrepid could add value to the existing set up at Rover.

Three problems were used to test the system. First a simple box, which had all the characteristics of the full problem, was analysed. The modes of the full structure were calculated using the structural FE code Nastran, before being used as input to the acoustic BE code SysNoise. The Nastran jobs are run on a central HP Convex V class machine which serves the entire organisation. The SysNoise jobs are run on the desktop. Next a larger box model was run to test data transfer robustness before running the full model (the chassis, transmission and pasenger compartment of a generic Land Rover).

Each experiment consisted of running the analysis procedure 9 times with parameters determined by Optimus.

The Convex machine is controlled by an LSF queue where 2 nodes are available to the RATG during the day, and up to 16 during the night. The timings of the runs are shown below in table 2.

| Model | Serial execution time | Execution with Intrepid |
|---|---|---|
| Small Box | 24:18 minutes | 12:21 minutes |
| Large Box | 1:58:08 hours | 40:10 minutes |
| Full vehicle model | 27:40:10 hours | 6:23:38 hours |

**Table 2.** Execution times for DOE runs performed with and without Intrepid

For the small box model and the large box model, execution was carried out during the day when only 2 nodes on the Convex were available. The super-linear speed-up seen for the large box is due to the variable network traffic when recovering data from the Convex.

For the full vehicle model the task ran overnight when all 16 nodes on the Convex were available. In this case, the limitation was the number of SysNoise licences to carry out the final acoustic analysis.

The difference between 27:40 hours and 6:23 hours is hugely important to Rover. An experiment that locks up a workstation for a whole day can really only be carried out over the weekend. By reducing the execution time to 6:23 hours, Intrepid enables the group at Rover to run such an experiment over night.

## LUSAS

The integration with LUSAS has been carried out in the ESPRIT project PARACOMP. The code has been tested by Messier-Dowty on a component from the Airbus landing gear.

The code is executed on a network of dual use NT machines. As with Toolshed, the issue for Intrepid is not to manage complex interdependencies but rather to ensure the consistency of data, that availability periods of machines are respected along with access rights. One of the issues to Messier Dowty is that whilst a serial code might use up one workstation for 7 days, a parallel code which uses 4 workstations for 2 days is no benefit because, in a clusteed environment, it impacts on more people. Therefore a requirement from Messier-Dowty is to specify availability periods for machines so that a job can be executed overnight, suspended when work starts and resumed the next night.

By defining availability periods for machines on a per user basis, Intrepid enables LUSAS to meet this requirement.

## Conclusions

The problems of performing simulation over a distributed heterogeneous computing resource are many, and different organisations implement different systems to enable them to manage their systems. The problem of designing a meta-application that can be deployed on the wide variety of environments to be found in industry is therefore to design an interface which is flexible. Intrepid has shown how a relatively simple resource manager with a well-defined API can interface with a number of different meta-applications and a number of different systems configurations.

## References

1. Marcyzk, J; Meta-computing and Computation Stochastic Mechanics. Computational Stochastic Mechanics in a Meta-computing Perspective (1997) 1-18 Ed. J. Marczyk
2. Meecham, K; Floros, N; Surridge, M. Industrial Stochastic Simulations on a European Meta-Computer, Proceedings 4$^{th}$ International EuroPar Conference, EuroPar 98 Parallel Processing.
3. Hey, A.J.G; Parallel Performance Analysis and the future of PARKbench; 1$^{st}$ NASA workshop on Performance Engineered Information Systems, Spetember 28-29 1998
4. Hey, A.J.G.; Scott C.J; Surrige, M; Upstill, C; Integrating Computation and Information Resources – An MPP Perspective. 3$^{rd}$ Working Conference on Massively Parallel Programming Models (MPPM –97), London, November 12-14 1997

# A Parallel Approach for Solving a Lubrication Problem in Industrial Devices

M. Arenaz[1], R. Doallo[1], J. Touriño[1], and C. Vázquez[2]

[1] Department of Electronics and Systems, University of A Coruña, Spain
arenaz@des.fi.udc.es, doallo@udc.es, juan@udc.es
[2] Department of Mathematics, University of A Coruña, Spain
carlosv@udc.es

**Abstract.** This work presents a parallel version of a complex numerical algorithm to solve a lubrication problem studied in Tribology. The execution of the sequential algorithm on a workstation requires a huge amount of CPU time and memory resources. So, in order to reduce the computational cost, we have applied parallelization techniques to the most costly parts of the original source code. Some blocks of the sequential code were also redesigned for the execution on a multiprocessor. In this paper we describe our parallel version and we also present some results that illustrate its efficiency in terms of execution time.

## 1 Introduction: Description of the Problem

Our work is concerned with the analysis of the lubricant behaviour of an industrial device that arises in Mechanical Engineering. For a wide range of these devices, such as the journal bearing device, the main task is to calculate, for a given imposed load, the pressure distribution of the lubricant fluid and the gap between two surfaces in contact [2].

Most of these devices can be represented by a ball-bearing geometry (Fig. 1(a)) when they are studied in a small region around the contact point. In this geo-

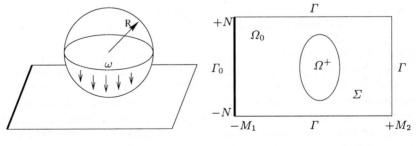

(a) Ball–bearing geometry.    (b) Two-dimensional domain $\Omega$.

**Fig. 1.** Ball–bearing device.

metry, the industrial device surfaces are represented by a sphere and a plane. In order to perform a realistic numerical simulation of the device, the problem is described by a mathematical model of the displacement of the fluid between a rigid plane and an elastic and loaded sphere. The model is posed over a two-dimensional domain which represents the mean plane of contact between both surfaces (Fig. 1(b)). In our mathematical formulation, the goal is to determine the pressure $p$ and the saturation $\theta$ of the lubricant, the gap $h$ between the ball and the plane, and the minimum reference gap $h_0$. The model consists of the following set of partial differential equations, where the unknown vector is $(p, \theta, h, h_0)$:

$$\frac{\partial}{\partial x}\left(e^{-\alpha p} h^3 \frac{\partial p}{\partial x}\right) + \frac{\partial}{\partial y}\left(e^{-\alpha p} h^3 \frac{\partial p}{\partial y}\right) = 12 s \nu \frac{\partial}{\partial x} h, \quad p > 0, \quad \theta = 1 \quad \text{in } \Omega^+ \quad (1)$$

$$\frac{\partial}{\partial x}(\theta h) = 0, \quad p = 0, \quad 0 \leq \theta \leq 1 \quad \text{in } \Omega_0 \quad (2)$$

$$e^{-\alpha p} h^3 \frac{\partial p}{\partial n} = 12 s \nu (1 - \theta) h \cos(\boldsymbol{n}, \boldsymbol{i}), \quad p = 0 \quad \text{in } \Sigma \quad (3)$$

$$h = h(x, y, p) = h_0 + \frac{x^2 + y^2}{2R} + \frac{2}{\pi E} \int_\Omega \frac{p(t, u)}{\sqrt{(x-t)^2 + (y-u)^2}} \, dt du \quad (4)$$

$$\theta = \theta_0 \text{ on } \Gamma_0 \quad (5)$$

$$p = 0 \text{ on } \Gamma \quad (6)$$

$$\omega = \int_\Omega p(x, y) \, dx dy \quad (7)$$

where the two-dimensional problem domain, the lubricated region, the cavitated region, the free boundary, the supply boundary and the boundary at atmospheric pressure are, respectively (see Fig. 1(b)):

$$\Omega = (-M_1, M_2) \times (-N, N)$$
$$\Omega^+ = \{(x, y) \in \Omega \,/\, p(x, y) > 0\}$$
$$\Omega_0 = \{(x, y) \in \Omega \,/\, p(x, y) = 0\}$$
$$\Sigma = \partial \Omega^+ \cap \Omega$$
$$\Gamma_0 = \{(x, y) \in \partial \Omega \,/\, x = -M_1\}$$
$$\Gamma = \partial \Omega \setminus \Gamma_0$$

being $M_1$, $M_2$ and $N$ positive constants. The input data are the velocity field $(s, 0)$, the piezoviscosity coefficients $\nu_0$ and $\alpha$, the unit vector $\boldsymbol{n}$ normal to $\Sigma$ pointing to $\Omega_0$, the unit vector in the $x$-direction $\boldsymbol{i}$, the Young equivalent modulus $E$, the sphere radius $R$, and the load $\omega$ imposed on the device in a normal to the plane direction. Equations (1)-(3) model the lubricant pressure behaviour, assuming the Barus pressure-viscosity relation. Equation (4) establishes the relation between the gap and the lubricant pressure. Equation (7) balances the hydrodynamic and the external loads. Consult [3] for a wider explanation of the physical motivation and the mathematical modelling.

## 2 The Numerical Algorithm

The numerical algorithm corresponding to the mathematical model mainly combines fixed point techniques, finite elements and duality methods [4]. The structure of the algorithm is depicted in Fig. 2.

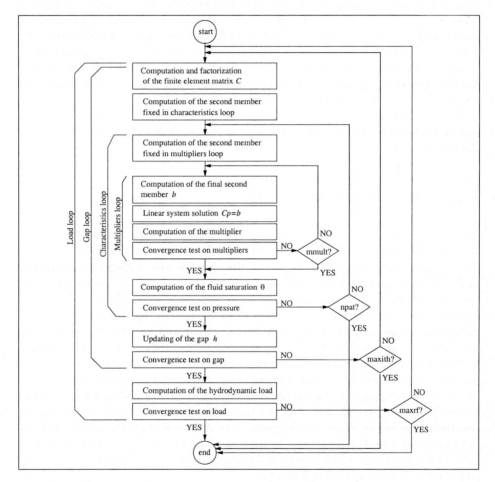

**Fig. 2.** Flowchart of the algorithm. The parameters $mmult$, $npat$, $maxith$ and $maxrf$ represent the maximum number of iterations for multipliers, characteristics, gap and load loops, respectively.

In order to obtain a more accurate approach of different real magnitudes, it is interesting to handle finer meshes. This mesh refinement involves a great increase in storage cost and execution time. So, the use of high performance computing techniques is required in order to reduce the impact of this computational cost problem.

As we can see in Fig. 2, the algorithm consists of four nested loops called load, gap, characteristics and multipliers loops, respectively. The execution begins using a predefined value of the gap parameter $h_0$. In the two innermost loops, namely characteristics and multipliers, the pressure $p$ and the saturation $\theta$ of the lubricant fluid are computed. The computations in these two loops mainly involve the recursive finite element solution of linearized partial differential equations, obtained from the application of a transport-diffusion technique combined with a multiplier method for the saturation variable. In gap loop, the gap $h$ between the sphere and the plane is updated in terms of the pressure following Equation (4). This costly process requires the integration over the whole domain for each finite element node. Finally, the hydrodynamic load generated by the pressure of the fluid is computed by means of numerical integration in (7). If the computed integral is close to the imposed load, the algorithm finishes. Otherwise, a new value $h_0$ is determined by a bisection technique and the algorithm performs a new iteration in load loop. For a wider explanation of this algorithm we address the reader to [4].

## 3 The Parallelization Process

In [7] we developed an efficient version of the algorithm described in the previous section. The target machine was the Fujitsu VP2400 vector processor.

In this section we present a version of the algorithm for a distributed-memory multiprocessor. We have focussed on the most costly parts of the algorithm in terms of execution time. So, firstly, we have performed an analysis of the distribution of the execution time, which led us to the conclusion that approximately 96% of this time concentrates on multipliers loop and on the *updating of the gap* functional block (see Fig. 2). In spite of that, as we will see later in this section, it is necessary to parallelize more parts of the algorithm. Hereafter, we will describe the parallelization process of the most interesting blocks.

Finite element discretization allows to compute the right-hand side $b$ of the linear system in two different ways: by traversing the set of finite elements or by traversing the set of nodes of the mesh. The difference between both methods lies in the order in which the contribution of all finite elements to the final value of $b$ are summed up. In the first method, implemented in the original sequential code, vector $b$ is computed as follows:

$$\forall \text{ triangular finite element } \phi_k$$
$$b[\gamma(\phi_k, i)] = \sum Contrib(\phi_k, i) \quad , \quad i = 1, 2, 3$$

where $\gamma(\phi_k, i)$ represents the number of the node of the mesh that is associated with the $i^{th}$ vertex of the element $\phi_k$, and $Contrib(\phi_k, i)$ is the contribution of the element $\phi_k$ to the node numbered as $\gamma(\phi_k, i)$. Note that vector $b$ is referenced through an indirect index. The parallelization of this algorithm introduces communications overhead that reduces the efficiency of the parallel program.

The second method overrides the problem described in the previous paragraph by using the following algorithm:

$\forall$ *node j of the mesh*
$$b[j] = \sum Contrib(\phi_k, i) \ \forall \ triangular \ finite \ element \ \phi_k \ , \ i = 1, 2, 3$$

The advantage of this method lies in the fact that it is cross-iteration independent, that is, the calculation of the value associated to any node does not depend on the value corresponding to any other node. Therefore, computations can be distributed in the most appropriate way for the subsequent stages of our program.

In order to solve the symmetric positive definite system $Cp = b$ at each step of the innermost loop, we have experimented with a parallel sparse conjugate gradient method [1, Chap.2], and with a parallel sparse Cholesky method [6], combined with reordering strategies to reduce the fill-in in matrix $C$ [5, Chap.8]. Although we have checked that both methods are highly scalable for the characteristics of our algorithm, we have chosen the second one because it was proved to be less costly in terms of execution time. Note that $b$ varies at each step of the multipliers loop, and that the finite element matrix $C$ is the same in each iteration of the gap loop (see Fig. 2).

The *computation of the multiplier* consists of a loop where there are true dependences [8] among the values of the multiplier on the last nodes (those belonging to the right boundary of the domain $\Omega$). This block can be parallelized efficiently by performing the calculations corresponding to the last nodes on the same processor. So, we have used a standard block distribution satisfying this constraint.

At this point, multipliers loop is parallelized but characteristics loop is executed in a sequential manner, which makes necessary to perform two *gather* operations for each iteration in the latter loop. This communications overhead was reduced by parallelizing the blocks included in characteristics loop as follows. First, the *computation of the second member fixed in multipliers loop* has been implemented, on the one hand, by using a cross-iteration independent algorithm that traverses the set of nodes of the mesh and, on the other hand, by imposing a constraint on the data distribution so that the computations corresponding to the nodes located at the supply boundary $\Gamma_0$ are assigned to the same processor. Second, the *computation of the fluid saturation* and *convergence test on pressure* are also cross-iteration independent algorithms, so their computations have been distributed according to the standard block distribution subject to the two previously described constraints.

Regarding the *updating of the gap*, it is one of the most costly stages of our numerical algorithm. For each node of the mesh, the gap is calculated according to (4). This computation is cross-iteration independent, which makes the efficiency of the parallelization process independent of the data distribution scheme. So, in order to avoid data redistributions, we have used the same distribution as in the already described functional blocks.

## 4 Experimental Results

The target machine was the Fujitsu AP3000 distributed-memory multiprocessor. It consists of UltraSparc-II processors at 300MHz interconnected in a two-dimensional torus topology. The parallel program was written in Fortran77 using MPI as message-passing library. Table 1 shows the execution times corresponding to a uniform mesh of the domain $\Omega$ composed of 38400 triangular elements and 19521 nodes. The nodes are ordered up and down, left and right so that the first nodes correspond to the boundary $\Gamma_0$. We disposed of coarser meshes, but it is not worth executing those tests on a multiprocessor because their computational cost is not high enough. We present measures for the most costly parts of the algorithm and for a maximum of eight processors (this limit was imposed by the configuration of our parallel machine). The last two rows make reference to the running time of the whole program and to the corresponding speedups, respectively. The results for one processor are estimations based on the execution times of one iteration of characteristics, multipliers and gap loops. The dimensions of the sparse matrix $C$ are $19521 \times 19521$, and the number of non-zero entries is 96961. The total number of iterations in multipliers loop was 1640221.

**Table 1.** Execution times and speedups

| #PEs | 1 | 2 | 4 | 8 |
|---|---|---|---|---|
| *Characteristics loop* | 10m:02s | 6m:45s | 4m:10s | 2m:58s |
| *Multipliers loop* | 124h:38m:18s | 71h:25m:16s | 38h:01m:34s | 23h:34m:51s |
| *Updating of the gap* block | 8h:33m:59s | 4h:36m:08s | 2h:21m:15s | 1h:12m:47s |
| Whole algorithm | 133h:23m:53s | 76h:09m:04s | 40h:27m:48s | 24h:51m:18s |
| Speedup | 1.00 | 1.75 | 3.30 | 5.37 |

The reduction in the execution time of the algorithm is significant. As we can see, good speedups were achieved by restructuring the sequential code and applying an adequate data distribution to the described dependences in order to reduce the communications overhead of the algorithm.

## 5 Conclusions and Future Work

In this work, we have described the parallelization of a numerical code to solve an elastohydrodynamic piezoviscous lubrication problem. The parallel algorithm reduces the execution time significantly, which will allow us to obtain new numerical results for some tests whose CPU cost is extremely high to be run on a workstation.

As future work, we want to check the scalability of our parallel algorithm by running it on a higher number of processors. To do that, we intend to adapt the parallel code to the architecture of the Cray T3E multiprocessor. Therefore, numerical results for even finer finite element meshes could be computed.

# Acknowledgements

We gratefully thank CESGA (Centro de Supercomputación de Galicia, Santiago de Compostela, Spain) for providing access to the Fujitsu AP3000. This work was partially supported by Research Projects of Galician Government (XUGA 32201B97), Spanish Government (CICYT TIC96-1125-C03), D.G.E.S. (PB96-0341-C02) and European Union (1FD97-0118-C02).

# References

1. Barret R., Berry M., Chan T., Demmel J.W., Donato J., Dongarra J., Eijkhout V., Pozo R., Romine C., van der Vorst H.: *Templates for the Solution of Linear Systems: Building Blocks for Iterative Methods*, SIAM Pub. (1994)
2. Cameron A.: *Basic Lubrication Theory*, John Wiley and Sons, Chichester (1981)
3. Durany J., García G., Vázquez C.: *Numerical Computation of Free Boundary Problems in Elastohydrodynamic Lubrication*, Appl. Math. Modelling, **20** (1996) 104-113
4. Durany J., García G., Vázquez C.: *Numerical Solution of a Reynolds-Hertz Coupled Problem with Nonlocal Constraint*, Scientific Computation Modelling and Applied Mathematics (Eds Sydow) Wissenschaft und Technik Verlag (1997) 615-620
5. Duff I.S., Erisman A.M., Reid J.K.: *Direct Methods for Sparse Matrices*, Clarendon Press (1986)
6. Gupta A., Gustavson F., Joshi M., Karypis G., Kumar V.: *Design and Implementation of a Scalable Parallel Direct Solver for Sparse Symmetric Positive Definite Systems*, Proceedings of 5th SIAM Conference on Parallel Processing, SIAM (March 1997)
7. Arenaz M., Doallo R., García G., Vázquez C.: *High Performance Computing of a New Numerical Algorithm for an Industrial Problem in Tribology*, Proceedings of 3rd Int'l Meeting on Vector and Parallel Processing, VECPAR'98, Porto, Portugal (June 1998) 1021-1034. Also to appear in Lecture Notes in Computer Science
8. Wolfe M.: *High Performance Compilers for Parallel Computing*, Addison–Wesley Publishing Company (1996)

# Workshop:

# High Performance Computing

# on Very Large Datasets

# Restructuring I/O-Intensive Computations for Locality

M. Kandemir[1]   A. Choudhary[1]   J. Ramanujam[2]

[1] ECE Dept., Northwestern University, Evanston, IL 60208 (mtk,choudhar@ece.nwu.edu)
[2] ECE Dept., Louisiana State University, Baton Rouge, LA 70803 (jxr@ee.lsu.edu)

**Abstract.** This paper describes restructuring techniques for out-of-core programs (i.e., those that deal with very large quantities of data) based on exploiting locality using a combination of loop and data transformations. Writing efficient out-of-core program is an arduous task. As a result, compiler optimizations directed at improving I/O performance are becoming increasingly important. We describe how a compiler can improve the performance of the code by determining appropriate file layouts for out-of-core arrays and finding suitable loop transformations. In addition to optimizing a single loop nest, our solution can handle a sequence of loop nests. We also show how to generate code when the file layouts are optimized. Preliminary experimental results obtained on an Intel Paragon distributed-memory message-passing multiprocessor demonstrate marked improvements in performance due to the optimizations described in this paper.

## 1 Introduction

As the speed of the disk subsystem is increasing at a much slower rate than the processor, interconnection network and memory subsystem speeds, any scalable parallel computer system running I/O-intensive applications must rely on some sort of software technology to optimize disk accesses. This is especially true for out-of-core parallel applications where a significant amount of time is spent in waiting for disk access. We believe that the time spent on disk subsystem can be reduced through at least two complementary techniques: (1) reducing the number of data transfers between the disk subsystem and main memory; and (2) reducing the volume of data transferred between the disk subsystem and main memory. A user may accomplish these objectives by investing substantial effort trying to understand the peculiarities of the I/O subsystem, studying carefully the file access pattern, and modifying the programs to make them more I/O-conscious. This poses a severe problem in that user intervention based on low-level I/O decisions makes the program less portable. It is possible and in some cases necessary to leave the task of managing I/O to an optimizing compiler for the following reasons. First, current optimizing compilers are quite successful in restructuring in-core computations (i.e., computations that do not use disk subsystem frequently) to achieve better cache and memory locality; thus, it is reasonable to expect that the same compiler technology can be used (at least partly) in optimizing the performance of the main memory-disk subsystem hierarchy as well. Second, although almost each currennt scalable parallel machine comes with its own parallel file system that comes with a suite of commands to handle I/O, the standardization of I/O interface is only now underway. In fact, MPI-I/O [7] is a result of such an effort. We believe that an optimizing compiler can easily generate I/O code using such a standard interface much like most compilers for message-passing parallel architectures use MPI to generate communication code.

Several techniques exist for optimizing data locality in loop nests; in general, these techniques can be divided into three categories. The first category uses loop transformations, i.e., changing the execution order of iterations as long as it is legal to do so [21, 14, 17, 20]. This affects both temporal and spatial locality characteristics of all arrays accessed in a loop nest, some perhaps adversely. The second category includes techniques that uses only data transformations [1, 10, 15, 13]; i.e., changing the memory layout of multi-dimensional arrays. These are not constrained by data dependencies, but these do not affect temporal locality at all. The third category consists of techniques that use both loop and data transformations to improve locality [6, 9, 11]. None of these are aimed at out-of-core computations. Our approach is one of the first to evaluate the effectiveness of an integrated locality approach—that includes both loop and data transformations—on a large set of out-of-core applications. Compilation of out-of-core codes using explicit I/O has been the main focus of several studies [5, 4, 3, 16]. All these previous techniques were all based on reordering the computation rather than on the re-organization of the data in files. In contrast, we show that locality can be significantly improved using a combined approach which includes both loop (iteration space) and data (file layout) transformationsi, and tiling; this renders our framework potentially more powerful than any existing set of transformations.

This paper presents a compiler approach for optimizing I/O accesses in regular scientific codes. Our approach is oriented towards minimizing the number as well as the volume of the data transfers between disk and main memory. It achieves this indirectly by reducing the number of I/O calls made from within the applications. Our experiments show that such an approach can lead to huge savings in disk access times.

The rest of this paper is organized as follows. Section 2 introduces our technique which is based on modifying both the loop access pattern and the file layouts and shows through an example how an input program can be transformed to out-of-core code. We present the algorithm as well as the additional I/O related issues. Section 3 presents performance results on the Intel Paragon at Caltech. Section 4 concludes the paper with a summary and brief outline of the work in progress.

## 2 Our Approach

In this section we present a compiler-directed approach that employs both loop and data transformations. Our loop and data transformation matrices are non-singular square matrices. To the best of our knowledge, this is the most general framework that uses *linear* transformations for optimizing locality in I/O-intensive codes. Another important characteristic of our approach is that we optimize locality globally, i.e., for several loop nests simultaneously.

Our objective is to optimize the I/O accesses of out-of-core applications through compiler analysis. Since the data sizes in these applications may exceed the size of main memory, data should be divided into chunks called *data tiles*, and the program should operate on one chunk (from each array) at a time that is brought from disk into memory. When the operation on this chunk is complete, the chunk should be stored back on disk (if it is modified). In such applications, the primary issue is to exploit the main memory–disk subsystem hierarchy rather than cache–main memory hierarchy; that is, a data tile brought into memory should be reused as much as possible. Also the number of I/O calls required to bring the said data tile into memory should be minimized. Note that the latter problem is peculiar to I/O-intensive codes.

Notice that the necessity of working with data tiles implies that tiling should be used. Thus, tiling, which is an optional optimization for in-core computations, is a

```
do i = 1, N                    do u = 1, N
  do j = 1, N                    do v = 1, N
    U(i,j) = V(j,i) + 1.0          U(u,v) = V(v,u) + 1.0
  end do                         end do
end do                         end do
do i = 1, N                    do u = 1, N
  do j = 1, N                    do v = 1, N
    V(i,j) = W(j,i) + 2.0          V(v,u) = W(u,v) + 2.0
  end do                         end do
end do                         end do
```

**Fig. 1.** (a) Example code; (b) Optimized version of (a)

"must" for out-of-core programs. Given a series of loop nests that access (possibly different) subsets of out-of-core arrays declared in the program, our optimization strategy proceeds as follows:

**(1)** Transform the program into a sequence of independent loop nests using loop fusion, distribution, and code sinking.
**(2)** Build an *interference graph* and identify the *connected components*. The interference graph is a bipartite graph $(V_n, V_a, E)$ where $V_n$ is the set of loop nests, $V_a$ is the set of arrays, and $E$ is the set of edges between loop and array nodes. There is an edge $e \in E$ between $v_a \in V_a$ and $v_n \in V_n$ if and only if $v_n$ references $v_a$.
**(3)** For each connected component:
　**(3.a)** Order the loop nests according to a *cost* criterion using profile information.
　**(3.b)** Optimize the most costly nest using only data transformations followed by tiling.
　**(3.c)** For each of the remaining nests in the connected component in order:
　　**(3.c.a)** Optimize the nest using loop and data transformations taking the file layouts found so far into account and then tile the nest.
　　**(3.c.b)** Propagate the file layouts found so far to the remaining nests.

A few points need to be noted. First, our method of tiling a nest for out-of-core computations is different from the traditional tiling used to exploit cache locality. We will discuss this issue in detail later on in this paper. Second, the loop transformations found should preserve the data dependences in the program. We ensure this by using Bik and Wjishoff's completion technique [2]. Third, a data transformation in its most general form can increase the space requirements of the original array. For lack of space, we do not present our solution to this problem.

### 2.1 Motivation

In this subsection we illustrate through an example why a combined approach to locality is required. Consider the program fragment shown in Fig. 1(a) assuming that the arrays are out-of-core and the default file layout is *column-major* for all arrays. An approach based on loop transformations alone (e.g., [14]) cannot optimize spatial locality for both arrays in the first nest as there are spatial reuses in orthogonal directions. The same is true for the second nest also. Therefore, two out of four references will go unoptimized. An approach based on data transformations alone, on the other hand, (e.g., [10], [15]) can select row-major layout for $U$ and column-major layout for $W$. Since

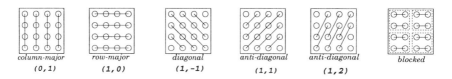

**Fig. 2.** Example file layouts and their hyperplane vectors.

there are conflicting layout requirements for array $V$, one of the references will be unoptimized. The approach discussed in this paper proceeds as follows. Assuming that the first nest is costlier than the second, it first focuses on this nest and (using data transformations) selects row-major layout for $U$ and column-major layout for $V$. Then it moves to the second nest. Since the layout for $V$ has already been determined, it takes this into account and interchanges the loops so that the locality for array $V$ will be good assuming column-major layout. This new loop order imposes array accesses along the rows of array $W$; consequently, our approach selects row-major layout for this array. To sum up, using a combination of loop and data transformations we are able to optimize locality for all the references in the nest.

### 2.2 Technical Details

Our approach uses a simple linear algebra concept called a *hyperplane*. We focus on programs where array subscript expressions and loop bounds are affine functions of the enclosing loop indices and symbolic (loop-invariant) constants. In such a program, a reference to an $m$–dimensional array appearing in a $k$–dimensional loop nest can be represented by an access (or reference) matrix $\mathcal{L}$ of size $m \times k$ and an offset vector $\bar{o}$ of size $m$ [20]. For example, the reference $V(j,i)$ in the first nest of the example fragment shown earlier can be represented by $\mathcal{L}\bar{I} + \bar{o}$, where $\mathcal{L} = \begin{pmatrix} 0 & 1 \\ 1 & 0 \end{pmatrix}$, $\bar{I} = (i,j)^T$, and $\bar{o} = (0,0)^T$. In an $m$-dimensional data space, a *hyperplane* can be defined as a set of tuples $\{(a_1, a_2, ..., a_m) \mid g_1 a_1 + g_2 a_2 + ... + g_m a_m = c\}$, where $g_1, g_2,...,g_m$ (at least one which is nonzero) are rational numbers called *hyperplane coefficients* and $c$ is a rational number called hyperplane constant [18]. We use a row vector $g^T = (g_1, g_2, ..., g_n)$ to denote a hyperplane family (for different values of $c$).

To keep the discussion simple, we focus on two dimensional arrays; the results presented in this paper extend to higher dimensional arrays as well. In a two-dimensional data space, the hyperplanes are denoted by row vectors of the form $(g_1, g_2)$. In that case, we can think of a hyperplane family as parallel lines for a fixed coefficient set (that is, the $(g_1, g_2)$ vector) and different values of $c$. An important property of the hyperplanes is that two data points (array elements) $(a, b)$ and $(c, d)$ lie along the same hyperplane if $(g_1, g_2)(a, b)^T = (g_1, g_2)(c, d)^T$. For example, a hyperplane such as $(0, 1)$ indicates that two elements belong to the same hyperplane if they have the same value for the column index (i.e., the second dimension); the value for the row index does not matter.

It is important to note that a hyperplane family can be used to partially define the file layout of an out-of-core array. In the case of a two-dimensional array, the vector $(0, 1)$ is sufficient to indicate that the elements in a column of the array (i.e., the elements in a hyperplane with a specific $c$ value) will be stored consecutively in file and will have *spatial locality*. The relative order of these columns in file is not as important provided that array size is large enough compared to the memory size which almost always holds true in out-of-core computations. In other words, the vector $(0,1)$ can be used for representing column-major file layout. A few possible file layouts and their associated hyperplane vectors for two-dimensional arrays are shown in Fig. 2. The last layout given is an example of blocked layouts where each dashed square constitutes a block. Our method currently does not handle blocked layouts, although as it is it can be used for determining optimal storage of blocks in file with respect to each other. It should be emphasized that these file layouts are only a handful of the set of all possible layouts, and there are many other hyperplanes which can define file layouts in two-dimensional space. For example, $(7,4)$ also defines a hyperplane family and a file layout such that two array elements $(a,b)$ and $(c,d)$ lie along a same hyperplane (i.e., have spatial locality) if $7a + 4b = 7c + 4d$.

The following claim gives us an important relation between a loop transformation, the file layout, and the access matrix in order for a given reference to have spatial locality in the innermost loop (see [11] for the proof).

**Claim 1** Consider a reference $\mathcal{L}\bar{I} + \bar{o}$ to a 2-dimensional array in a loop nest of depth $k$ where $\mathcal{L} = \begin{pmatrix} a_{11} & a_{12} & \cdots & a_{1k} \\ a_{21} & a_{22} & \cdots & a_{2k} \end{pmatrix}$ and let $Q = \{q_{ij}\}$ $(1 \leq i,j \leq k)$ be the inverse of the loop transformation matrix. In order to have spatial locality in the innermost loop, this array should have a file layout represented by hyperplane $(g_1, g_2)$ such that $(g_1, g_2)\mathcal{L}(q_{1k}, q_{2k}, \cdots, q_{kk})^T = 0$. □

Since both $(g_1, g_2)$ and $(q_{1k}, q_{2k}, \cdots, q_{kk})^T$ are unknown, this formulation is non-linear. However, if either of them is known, the other can easily be found.

If we know the last column of $Q$, $\quad (g_1, g_2) \in Ker\left\{\mathcal{L}(q_{1k}, q_{2k}, \cdots, q_{kk})^T\right\}$ (1)

If we know $(g_1, g_2)$, $\quad (q_{1k}, q_{2k}, \cdots, q_{kk})^T \in Ker\left\{(g_1, g_2)\mathcal{L}\right\}$ (2)

Usually *Ker* sets may contain multiple vectors in which case we choose the one such that the gcd of its elements is minimum. Returning to the example given at the beginning of Section 2.1, we find the access matrices for nest 1 are $\mathcal{L}_U = \begin{pmatrix} 1 & 0 \\ 0 & 1 \end{pmatrix}$, and $\mathcal{L}_{V_1} = \begin{pmatrix} 0 & 1 \\ 1 & 0 \end{pmatrix}$; and for nest 2 are $\mathcal{L}_{V_2} = \begin{pmatrix} 1 & 0 \\ 0 & 1 \end{pmatrix}$, and $\mathcal{L}_W = \begin{pmatrix} 0 & 1 \\ 1 & 0 \end{pmatrix}$. As mentioned earlier, for the first nest we apply only data transformations; that is, $(q_{12}, q_{22})^T = (0, 1)$ ($Q$ is identity matrix). Using Relation (1) given above, for array $U$, $(g_1, g_2) \in Ker\{\mathcal{L}_U (0, 1)^T\}$, or $(g_1, g_2) \in Ker\{(0, 1)^T\}$. A particular solution is $(g_1, g_2) = (1, 0)$; i.e., array $U$ should be stored row-major. For $V$, $(g_1, g_2) \in Ker\{\mathcal{L}_{V_1} (0, 1)^T\}$ means $(g_1, g_2) \in Ker\{(1, 0)^T\}$. Selecting $(g_1, g_2) = (0, 1)$ results in column-major layout for array $V$.

Having fixed the layouts for these two arrays, we proceed with the second nest, assuming again that $Q$ is the inverse of the loop transformation matrix for this nest. First, using Relation (2), we find the loop transformation which satisfies the reference to array $V$ in this nest: $(q_{12}, q_{22})^T \in Ker\{(0, 1)\mathcal{L}_{V_2}\}$, or $(q_{12}, q_{22})^T \in Ker\{(0, 1)\}$. A particular solution is $(q_{12}, q_{22})^T = (1, 0)^T$, which in turn can be completed (using

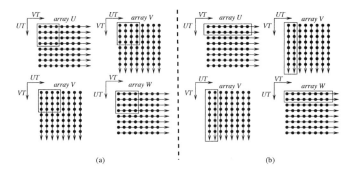

**Fig. 3.** Different tile access patterns.

the approach in [2]) as $Q = \begin{pmatrix} 0 & 1 \\ 1 & 0 \end{pmatrix}$. Notice that this matrix corresponds to loop interchange [21]. The only remaining task is to determine the optimal file layout for array $W$. By taking into account the last column of $Q$ and using Relation (1) once more, we have $(g_1, g_2) \in Ker\{\mathcal{L}_W (1, 0)^T\}$, or $(g_1, g_2) \in Ker\{(0, 1)^T\}$; this means that array $W$ should have a row-major file layout. The resulting program is shown in Fig. 1(b).

### 2.3 Tiling of Out-of-Core Arrays

As mentioned earlier in the paper, tiling is mandatory for out-of-core computations. Several transformations [21, 20] may need to be performed prior to tiling to ensure its legality. In our running example no such a transformation is necessary. A traditional tiling approach can derive the code shown in Fig. 4(a). There, $B$ is the tile size and all the loops are tiled with this tile size. The loops $UT$ and $VT$ are the tile loops whereas the loops $u'$ and $v'$ are the element loops. Notice that tile loops are placed in the outer positions in each nest.

Although such a tiling strategy allows data reuse for the data tiles in memory, its I/O performance might be unexpectedly poor. The reason for this can be seen when we consider the tile access pattern shown in Fig. 3(a). In this figure each circle corresponds to an array element and the arrows connecting the circles indicates file layouts (horizontal arrows for row-major and vertical arrows for column-major). The top two arrays are $U$ and $V$ in the first nest and the bottom two arrays are $V$ and $W$ in the second nest. Assuming that we have a main memory size of 32 elements (for illustration purposes), we can allocate this memory evenly across the arrays in a nest. Traditional tiling causes the tile access pattern shown in Fig. 3(a). Let us focus on array $U$ in the first nest. In order to read a $4 \times 4$ data tile from the file we need to issue 4 I/O calls. Notice that the alternative of reading the entire array and sieving out the unwanted array elements may not be applicable in out-of-core computations. For array $U$, we are able to read 4 elements per I/O call. The same situation also occurs with other array reads (as well as array writes).

Now consider the tile access pattern shown in Fig. 3(b) (the corresponding tiled code is shown in Fig. 4(b)). Focusing on array $U$ in the first nest, we see that in order

```
do UT = 1, N, B                             do UT = 1, N, B
  do VT = 1, N, B                             < read data tiles for U and V from files >
    < read data tiles for U and V from files >  do u' = UT, min(UT+B,N)
    do u' = UT, min(UT+B,N)                     do v' = 1, N
      do v' = VT, min(VT+B,N)                     U(u',v') = V(v',u') + 1.0
        U(u',v') = V(v',u') + 1.0               end do
      end do                                    end do
    end do                                    < write data tile for U to its file >
    < write data tile for U to its file >   end do
  end do
end do                                      do UT = 1, N, B
                                              < read data tiles for V and W from files >
do UT = 1, N, B                               do u' = UT, min(UT+B,N)
  do VT = 1, N, B                               do v' = 1, N
    < read data tiles for V and W from files >    V(v',u') = W(u',v') + 2.0
    do u' = UT, min(UT+B,N)                     end do
      do v' = VT, min(VT+B,N)                 end do
        V(v',u') = W(u',v') + 2.0           < write data tile for W to its file >
      end do                                end do
    end do
    < write data tile for W to its file >
  end do
end do
```

**Fig. 4.** (a) Traditional tiling;        (b) Our approach

to read 16 elements from the file we need to issue only 2 I/O calls (assuming that in a single I/O call at most 8 elements can be read or written). Notice that in both cases (Fig. 3(a) and Fig. 3(b)) we are using the same amount of in-core memory. This small example shows that by being a bit more careful about how to read the array elements from file into memory (i.e., how to tile the loop nests) we might be able to save a number of I/O calls.

The important point is to see that we can achieve this optimized tile access pattern by *not* tiling the innermost loop. This is in contrast with the traditional tiling strategy in which the innermost loop in the nest is almost always tiled (as long as it is legal to do so). Unfortunately, tiling the innermost loop in out-of-core computations (where disk accesses are very costly) can lead to excessive number of I/O calls as this loop exhibits spatial locality. Therefore, as a rule after applying the loop and data transformations to improve locality, tiling should be applied to all but the innermost loop in the nest.

## 3 Experimental Results

In this section we present experimental results obtained on the Intel Paragon at Caltech. Paragon uses a parallel file system called PFS which stripes files across 64 I/O nodes with 64KB stripe units. In the experiments we applied the following methodology. We took ten loop nests from several benchmarks and math libraries. Then we parallelized these nests for execution on the Paragon such that the inter-processor communication is eliminated. This allowed us to focus solely on the I/O performance of the nests and the scalability of the I/O subsystem. After this parallelization and data allocation, we coded *five* different out-of-core versions of each nest using the PASSION runtime library [19]:

**Table 1.** Experimental results on 16 nodes of an Intel Paragon.

| Program  | col    | row   | l-opt | d-opt | c-opt | h-opt |
|----------|--------|-------|-------|-------|-------|-------|
| mat.2    | 257.20 | 93.3  | 65.1  | 56.8  | 60.8  | 54.3  |
| mxm.2    | 220.01 | 181.5 | 100.0 | 112.6 | 79.8  | 67.0  |
| adi.2    | 144.12 | 134.9 | 22.8  | 46.5  | 22.8  | 22.8  |
| vpenta.6 | 135.00 | 47.1  | 100.0 | 47.1  | 47.1  | 29.9  |
| btrix.4  | 91.45  | 66.6  | 100.0 | 61.3  | 61.3  | 42.3  |
| emit.3   | 88.64  | 176.5 | 100.0 | 100.0 | 100.0 | 100.0 |
| syr2k.2  | 215.34 | 86.3  | 52.0  | 77.4  | 52.0  | 47.6  |
| htribk.2 | 248.61 | 110.8 | 127.2 | 81.1  | 81.1  | 72.6  |
| gfunp.4  | 86.05  | 128.4 | 73.3  | 68.0  | 46.9  | 34.0  |
| trans.2  | 181.90 | 100.0 | 100.0 | 48.2  | 48.2  | 48.2  |
| average: |        | 112.5 | 84.0  | 69.9  | 60.0  | 51.9  |

- col: fixed column-major file layout for every out-of-core array
- row: fixed row-major file layout for every out-of-core array
- l-opt: loop-optimized version: no file layout transformations
- d-opt: file layout-optimized version: no loop transformations
- c-opt: integrated loop and file layout transformations
- h-opt: hand optimized version using blocking and interleaving

The col and row are the original (unoptimized) programs. For the l-opt version, we used the best of the resulting nests generated by [14] and [20]. For the d-opt version, we used the best of the resulting nests generated by [15], [13], and [10]. c-opt (compiler optimized) version is the one obtained using the approach discussed in this paper. In obtaining h-opt we used chunking and interleaving in order to further reduce the number of I/O calls. For all the versions except c-opt all the loops carrying some form of reuse are tiled. For the c-opt version we used the tiling strategy explained in Section 2.3. For each nest we set the memory size allocated for the computation to 1/128th of the sum of the sizes of the out-of-core arrays accessed in the nest. Each dimension of each array used in the computation is set to 4,096 double precision elements. However, some array dimensions with very small hard-coded dimension sizes were not modified as modifying them correctly would necessitate full understanding of the program in question.

Table 1 shows the results on 16 processors. For each data set, the col column gives the total execution time of the out-of-core nest in *seconds*. The other columns, on the other hand, give the respective execution times as a fraction of that of col. As an example, the execution time of c-opt version of gfunp.4 is 46.9 percent of that of col. From these results we infer the following. First, the classical locality optimization schemes based on loop transformations alone may not work well for out-of-core computations. On average l-opt brings only a 16% improvement over col. The approaches based on data transformations perform much better. Our integrated approach explained in this paper, however, results in a 40% reduction in the execution times with respect to col. Using a hand optimized version (h-opt) brings an additional 8% reduction over c-opt, which encourages us to incorporate array chunking and interleaving into our technique.

**Table 2.** Results on scalability of optimized versions on the large data set.

| Program | Version | No. of processors | | | |
|---|---|---|---|---|---|
| | | 16 | 32 | 64 | 128 |
| mat.2 | col | 10.9 | 20.6 | 34.8 | 64.3 |
| | row | 11.0 | 20.9 | 35.6 | 66.0 |
| | l-opt | 13.9 | 27.6 | 53.8 | 100.4 |
| | d-opt | 14.5 | 28.1 | 55.0 | 104.2 |
| | c-opt | 14.0 | 27.7 | 54.8 | 102.7 |
| | h-opt | 15.2 | 30.9 | 60.9 | 115.6 |
| mxm.2 | col | 11.1 | 21.2 | 37.6 | 70.0 |
| | row | 8.2 | 15.4 | 30.0 | 52.6 |
| | l-opt | 11.1 | 21.2 | 37.6 | 70.0 |
| | d-opt | 9.7 | 17.0 | 32.1 | 56.4 |
| | c-opt | 13.7 | 24.8 | 56.4 | 106.6 |
| | h-opt | 13.7 | 24.8 | 56.1 | 107.2 |
| adi.2 | col | 12.0 | 22.2 | 51.2 | 70.9 |
| | row | 6.89 | 10.9 | 18.6 | 31.4 |
| | l-opt | 15.3 | 28.2 | 61.4 | 107.5 |
| | d-opt | 13.8 | 24.0 | 55.5 | 74.9 |
| | c-opt | 15.3 | 28.2 | 61.4 | 107.5 |
| | h-opt | 15.3 | 28.2 | 61.4 | 107.5 |
| vpenta.6 | col | 10.0 | 24.2 | 51.3 | 78.9 |
| | row | 14.5 | 28.0 | 60.9 | 109.8 |
| | l-opt | 10.0 | 24.2 | 51.3 | 78.9 |
| | d-opt | 14.5 | 28.0 | 60.9 | 109.8 |
| | c-opt | 14.5 | 28.0 | 60.9 | 109.8 |
| | h-opt | 14.7 | 29.0 | 62.4 | 108.2 |
| btrix.4 | col | 10.0 | 18.1 | 27.0 | 42.7 |
| | row | 12.9 | 23.9 | 45.8 | 87.1 |
| | l-opt | 10.0 | 18.1 | 27.0 | 42.7 |
| | d-opt | 13.9 | 25.1 | 46.2 | 98.1 |
| | c-opt | 13.9 | 25.1 | 46.2 | 98.1 |
| | h-opt | 13.1 | 24.6 | 44.3 | 93.1 |

| Program | Version | No. of processors | | | |
|---|---|---|---|---|---|
| | | 16 | 32 | 64 | 128 |
| emit.3 | col | 12.7 | 23.1 | 45.0 | 89.9 |
| | row | 6.8 | 11.0 | 18.5 | 33.9 |
| | l-opt | 12.7 | 23.1 | 45.0 | 89.9 |
| | d-opt | 12.7 | 23.1 | 45.0 | 89.9 |
| | c-opt | 12.7 | 23.1 | 45.0 | 89.9 |
| | h-opt | 12.7 | 32.1 | 45.0 | 89.9 |
| syr2k.2 | col | 10.3 | 20.0 | 36.5 | 71.5 |
| | row | 11.7 | 22.0 | 38.9 | 78.0 |
| | l-opt | 13.8 | 26.8 | 51.0 | 95.1 |
| | d-opt | 12.5 | 24.1 | 45.6 | 87.4 |
| | c-opt | 13.8 | 26.8 | 51.0 | 95.1 |
| | h-opt | 14.1 | 26.0 | 51.0 | 95.3 |
| htribk.2 | col | 11.7 | 20.3 | 37.7 | 76.6 |
| | row | 9.5 | 16.9 | 30.0 | 55.4 |
| | l-opt | 8.8 | 15.0 | 24.3 | 44.0 |
| | d-opt | 11.9 | 21.5 | 37.9 | 76.9 |
| | c-opt | 11.9 | 21.5 | 37.9 | 76.9 |
| | h-opt | 12.1 | 21.6 | 40.1 | 76.9 |
| gfunp.4 | col | 10.9 | 20.4 | 38.4 | 70.8 |
| | row | 9.5 | 17.0 | 32.6 | 60.6 |
| | l-opt | 8.1 | 15.7 | 28.2 | 52.2 |
| | d-opt | 14.0 | 25.0 | 56.0 | 102.3 |
| | c-opt | 14.0 | 25.0 | 56.0 | 102.3 |
| | h-opt | 14.5 | 24.7 | 57.0 | 105.7 |
| trans.2 | col | 13.0 | 22.7 | 31.6 | 67.7 |
| | row | 13.0 | 22.7 | 31.6 | 67.7 |
| | l-opt | 13.0 | 22.7 | 31.6 | 67.7 |
| | d-opt | 15.4 | 30.9 | 60.2 | 113.0 |
| | c-opt | 15.4 | 30.9 | 60.2 | 113.0 |
| | h-opt | 15.4 | 30.9 | 60.2 | 113.0 |

Table 2, on the other hand, shows the speedups obtained by different versions for processor sizes of 16, 32, 64, and 128 using all 64 I/O nodes. It should be stressed that in obtaining these speedups we used the single node result of the respective versions. For example, the speedup for the c-opt version of emit.3 was computed for $p \in 16, 32, 64, 128$ as $\frac{\text{Execution Time of the c-opt version of emit.3 on 1 node}}{\text{Execution Time of the c-opt version of emit.3 on } p \text{ nodes}}$. Since the execution times of the parallelized codes on single nodes may not be as good as the best sequential version, these results are higher than we expected. Also, since the codes were parallelized such that there is no interprocessor communication, the scalability was limited only by the number of I/O nodes and the I/O subsystem bandwidth.

## 4 Conclusions and Work in Progress

The increasing disparity between the speeds of disk subsystems and the speeds of other components (such as processors, memories, and interconnection networks) has rendered the problem of improving the performance of out-of-core programs (i.e., programs that access very large amounts of disk-resident data) very important and difficult. Programmers usually have to embed code for staging in and out of data between memory and I/O devices explicitly in the program. Often this results in non-portable and error-prone code. This paper presents a technique that an optimizing compiler can use to transform the in-core programs to derive I/O-efficient out-of-core versions. In doing this, the approach uses loop (iteration space) and file layout (data space) transformations. Specifically, this paper uses linear algebra techniques to derive good file layouts along with the accompanying loop transformations. In addition, we address the partitioning of data so that the available memory is best used among the various references in a program. Preliminary results show that our technique substantially reduces the time spent in performing I/O. Currently we are working on extending our approach across procedure boundaries as well as exploring the use of integer linear programming.

## References

1. J. M. Anderson, S. P. Amarasinghe, and M. S. Lam. Data and computation transformations for multiprocessors. In *Proc. 5th ACM SIGPLAN Symp. Prin. & Prac. Par. Prog.*, July 1995.
2. A. J. C. Bik, and H. A. G. Wijshoff. On a completion method for unimodular matrices. Technical Report 94–14, Dept. of Computer Science, Leiden University, 1994.
3. R. Bordawekar, A. Choudhary, K. Kennedy, C. Koelbel, and M. Paleczny. A model and compilation strategy for out-of-core data-parallel programs. In *Proc. SIGPLAN Symp. Prin. & Prac. Par. Pro.*, July 1995.
4. R. Bordawekar, A. Choudhary, and J. Ramanujam. Automatic optimization of communication in out-of-core stencil codes, In Proc. *10th Int. Conf. Supercomp.*, pp. 366–373, 1996.
5. P. Brezany, T. A. Muck, and E. Schikuta. Language, compiler and parallel database support for I/O intensive applications, In *Proc. High Performance Computing and Networking*, 1995.
6. M. Cierniak, and W. Li. Unifying data and control transformations for distributed shared memory machines. Technical Report 542, CS Dept., University of Rochester, 1994.
7. P. Corbett, D. Feitelson, S. Fineberg, Y. Hsu, B. Nitzberg, J. Prost, M. Snir, B. Traversat, and P. Wong. Overview of the MPI-IO parallel I/O interface, *Proc. 3rd Workshop I/O in Par. & Dist. Sys.*, Apr. 1995.
8. M. Kandemir, R. Bordawekar, and A. Choudhary. Data access reorganizations in compiling out-of-core data parallel programs on distributed memory machines. In *Proc. IPPS 97*, pp. 559–564, April 1997.
9. M. Kandemir, J. Ramanujam, and A. Choudhary. A compiler algorithm for optimizing locality in loop nests. In Proc. *11th ACM Int. Conf. Supercomp.*, pp. 269-278, July 1997.
10. M. Kandemir, A. Choudhary, N. Shenoy, P. Banerjee, and J. Ramanujam. A hyperplane based approach for optimizing spatial locality in loop nests. In *Proc. 1998 ACM Int. Conf. Supercomp.*, July 1998.
11. M. Kandemir, A. Choudhary, J. Ramanujam, and P. Banerjee. A matrix-based approach to the global locality optimization problem. In *Proc. Intl. Conf. Par. Arch. & Comp. Tech. (PACT'98)*, Oct. 1998.
12. M. Kandemir, M. Kandaswamy, and A.Choudhary. Global I/O optimizations for out-of-core computations. In *Proc. High-Performance Computing Conference (HiPC)*, Dec. 1997.
13. S. Leung, and J. Zahorjan. Optimizing data locality by array restructuring. Technical Report, CSE Dept., University of Washington, TR 95-09-01, Sep. 1995.
14. W. Li. Compiling for NUMA parallel machines. Ph.D. dissertation, Cornell University, 1993.
15. M. O'Boyle, and P. Knijnenburg. Non-singular data transformations: Definition, validity, applications. In *Proc. 6th Workshop on Compilers for Par. Comp.*, pp. 287–297, 1996.
16. M. Paleczny, K. Kennedy, and C. Koelbel. Compiler support for out-of-core arrays on parallel machines. CRPC Technical Report 94509-S, Rice University, Dec. 1994.
17. J. Ramanujam. Non-unimodular transformations of nested loops. In *Proc. Supercomputing 92*, pages 214–223, Nov 1992.
18. J. Ramanujam, and P. Sadayappan. Compile-time techniques for data distribution in distributed memory machines. In *IEEE Trans. Par. & Dist. Sys.*, 2(4):472–482, Oct. 1991.
19. R. Thakur, A. Choudhary, R. Bordawekar, S. More, and S. Kuditipudi. Passion: Optimized I/O for parallel applications, *IEEE Computer*, (29)6:70–78, June 1996.
20. M. Wolf, and M. Lam. A data locality optimizing algorithm. In *Proc. ACM SIGPLAN 91 Conf. Prog. Lang. Des. & Impl.*, pages 30–44, June 1991.
21. M. Wolfe. *High Performance Compilers for Parallel Computing*, Addison-Wesley, 1996.

# Virtual Memory Management in Data Parallel Applications[*]

Eddy Caron, Olivier Cozette, Dominique Lazure, Gil Utard

LaRIA
Université de Picardie Jules Verne
80000 Amiens, France
paladin@laria.u-picardie.fr

**Abstract.** The PaLaDiN (PArallel LArge Data set In Network of workstations) project is concerned with parallel out-of-core application running on cluster of workstations or PCs. In such architectures, each node has a *virtual memory manager* and a first idea is to use this feature to run "parallel *out-of-core*" application as a parallel *in-core* one. The *out-of-core* part of the problem, i.e. the schedule of data fetch and data write-back, is relegated to the operating system.

In this paper we show that usual *virtual memory manager* is not well suited for parallel out-of-core application. Then, we propose an extension to modern operating system which allow to define application specific virtual memory manager. This extension is made up of one kernel module (MMUSSEL) and one library (MMUM) and run on Linux. We present a new pagination strategy for the LU decomposition program.

## 1 Introduction

Many of important computational applications involve solving problems with very large data sets [14]. For example astronomical simulation [15], crash test simulation [3], global climate modeling [6], and many other scientific and engineering problems can involve data sets that are too large to fit in main memory. Using parallelism can reduce the computation time and increase the available memory size, but for challenging applications the memory is ever insufficient in size: for instance in a mesh decomposition of a mechanical problem, scientist would like to increase accuracy by an increase of the mesh size. This kind of application are referred to as "parallel *out-of-core*" applications.

To solve parallel *out-of-core* problems, the programmer must generally rewrite its code with explicit I/O calls. He has to schedule the data *fetch* in memory and data *write back* to disk with respect to the computation. To achieve good performances, the programmer must generally *restructure* its code, which is a formidable task.

The PaLaDiN[1] project at LaRIA is concerned with parallel out-of-core applications running on cluster of workstations or PCs. In such architectures, each

---

[*] This work is partially supported by the "Pôle de modélisation de la région Picardie".
[1] http://www.laria.u-picardie.fr/PALADIN

node has a *virtual memory manager* and a first idea is to use this feature to run "parallel *out-of-core*" application as a parallel *in-core* one. The *out-of-core* part of the problem, i.e. the schedule of data fetch and data write-back, is relegated to the operating system.

In this paper, we present preliminary results of the PaLaDiN project. We show that usual *virtual memory manager* is not well suited for parallel out-of-core application. Then, we propose an extension to modern operating system which allows to define application specific virtual memory managers. Finally, we present future and related work.

## 2 Pitfalls of LRU Policy

Usual memory manager use a LRU or FIFO like policy to select page to write back when physical memory becomes insufficient. The main paradigm behind this strategy is programs have in general good *temporal locality*, i.e. the data or pages of memory that would be accessed in the near future are the data just accessed. The virtual memory manager keeps in memory pages recently used and write back other one.

We're going to illustrate the impact of such policies on typical data-parallel computation on huge matrices. The first example exhibits the trashing effect which appears on sequential computation. The second example shows the interference of communication on the paging activity in usual SPMD code generated from data-parallel programs.

### 2.1 Computation

The LRU replacement policy is not well suited for linear access to memory which is the usual memory access scheme found in data parallel programs. When the size of data is greater than the physical memory, the pagination system collapses: it is the *hole effect*. In this section we first depict the hole effect on a toy example, and we study its repercussion on the LU decomposition of huge matrix.

**The hole effect** Let's consider the program shown on Figure 1. There is $P$ repetitive linear accesses to a vector $V$ of $N$ elements. Let $M$ be the number of physical memory pages, and let $B$ be the size of a page (in number of reals). Let us consider the situation where $\lceil N/B \rceil > M$, for instance $N = M \times B + 1$, and $V[1..N-1]$ are initially in the physical memory. The access to $V[N]$ raises a page fault, and the LRU policy removes from physical memory the page which contains $V[1]$. Unfortunately, it is the next page to be accessed. The next iteration will generate another page fault. This new page fault removes the page which contains $V[B+1]$

```
real V[N];
for j = 1 to P
    for i = 1 to N
        V[i] = f(i,j);
```

**Fig. 1.** Several linear accesses to the memory.

from memory, i.e. the next page to be accessed... This phenomenon occurs each time a new page is accessed: It is the *hole effect*. The number of disk accesses is equal to $2 * \lceil N/B \rceil * (P-1) + 1$, and is independent of the physical memory size $M$ (whenever $N > M \times B$).

**LU decomposition and LRU policy** Now, let's consider a part of the classical LU decomposition of a matrix of order N. The program is shown on Figure 2. In

```
real A(N,N);
for j = 1 to N
  for i = j+1 to N
    for k=1 to j
      A(i,j)=A(i,j)-A(i,k)*A(k,j)
```

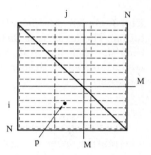

**Fig. 2.** LU decomposition of matrix. The matrix is in row order in memory. Each page is represented by a doted rectangle. The page $p$ holds the element $A(i,j)$. The matrix order $N$ is greater than the number $M$ of physical pages in memory.

this program each element $A(i,j)_{i>j}$ is the dot product of the vector $A(i,1:j)$ by the vector $A(1:j,j)$. Let us estimate the number of page faults when $N$ is greater than the number $M$ of physical pages in memory. Consider the iteration $(i,j)_{i>j}$ and let $p$ the page which holds $A(i,j)$. From the data-dependences, we show that the next reuse of the page $p$ is at the iteration $(i,j+1)$. The number of pages different to $p$ accessed until this iteration is greater or equal to $N-1 \geq M$. Due to the LRU algorithm, the page $p$ is no longer in memory. This is true for all iterations $(i,j)_{i>j}$. Now, let us consider iterations $(i,j)_{i>j}$ when $j > M$. There is is $(N-j)$ linear accesses to the $j$ pages associated to the vector $A(1:j,j)$ in the inner loop. Due to the hole effect, there is $j$ page faults. The number of page faults $p(N,M)$ is greater than $(\frac{N}{2}-1)N + \sum_{j=M+1}^{j=N} j(N-j)$. Since $M$ is constant, $p(N,M)$ is $O(N^3)$. In the next section, we will present a new paging strategy which avoid such crumbling.

## 2.2 Communication

We illustrate the impact of communications on the pagination with the SOR algorithm. The kernel of this algorithm is to compute in each element of a matrix an average of its neighbors. The HPF program and a typical generated code by a compiler is shown on Figure 3. The array is distributed by block of columns on processors. The data needed by/from neighbor processors are sent/received into the first and last column (generally called *ghost area*).

```
        REAL A(N,N)                    Send(A(1:n,1),left)
!HPF DISTRIBUTE A(BLOCK,*)              Send(A(1:n,n),right)
                ...                     Recv(A(1:n,0),left)
        FORALL (I=2:N-1, J=2:N-1)       Recv(A(1:n,n+1),right)
            A(I,J) = (A(I,J+1)
                    + A(I+1,J)          FORALL (i=1:n, j=1:n)
                    + A(I,J+1)              A(i,j)=(A(i,j+1)
                    + A(I-1,J))/4                  +A(i,j-1)
                ...                                +A(i+1,j)
                                                   +A(i-1,j))/4
```

**Fig. 3.** On left, a typical HPF Program Fragment for a Two-dimensional SOR computation. The array A is distributed by block of columns. On right, the generated code from the HPF like SOR algorithm on $p$ processors. Communications of array section are introduced by the compiler. The constant $n = N/p$ is the matrix order of A in each processor.

We ran the generated code on a network of PC with $8192 \times 4KB$ pages each. The matrix order in each processor is equal to 4096. Table 1 gives the number of page faults raised in the different steps of the code. All the physical memory is involved in the communication steps. During the computation some pages sent or received are wrote back to the disk. These pages are fetched one more time when the computation loop reaches them.

**Strip-mining** One way to avoid these unnecessary page-in and page-out is to tightly couple the computations and the communications. The idea is to communicate only data which are fetched in memory by the current computation. This can be obtain by a classical *strip-mining* of the computation part.

The iteration space of the computation part is split into *tiles*. Tiles are such that the number of pages accessed by communication and computation is equal to the number of physical pages available.

The table 1 gives the number of page faults which occur on each step of the algorithm. There is no redundant page faults, and the execution time profit is increased by 30%.

|  | Without tiling | | | With Tiling | | |
|---|---|---|---|---|---|---|
|  | Proc. 0 | **Proc. 1** | Proc. 2 | Proc. 0 | **Proc. 1** | Proc. 2 |
| Send(A[...],left) |  | **9024** | 9106 |  | **10237** | 10220 |
| Send(A[...],right) | 9218 | **9547** |  | 10085 | **10155** |  |
| Recv(A[...],left) |  | **3** | 9291 | 27 | **53** |  |
| Recv(A[...],right) | 9576 | **5715** |  |  | **0** | 31 |
| FORALL | 79713 | **78513** | 79272 | 72288 | **64925** | 72111 |
| Total | 98595 | **102858** | 97734 | 82400 | **85370** | 82362 |
| *Elapsed time* | *720,80 s* | ***771,84 s*** | *777,54 s* | *571,55 s* | ***536,44 s*** | *576,98 s* |

**Table 1.** Number of page fault on the SOR program.

## 3 User Defined Memory Management

We shown that the usual operating system policy for memory management is inadequate for the execution scheme of a typical data-parallel application. Unfortunately, there is no easy way to modify it. The only one is generally to rewrite the kernel part of the OS which is involved in the virtual memory management mechanism.

In this section, we present a new tool which allows the implementation of specific virtual memory manager at the application level. Then, we present a new virtual memory management strategy dedicated to the LU decomposition, and first results.

### 3.1 The MMUM library and MMUSSEL module

The tool is made up of one library, called MMUM[2], and a kernel module, called MMUSSEL[3]. The kernel module is written for the Linux OS, but can be written for every Unix kernel using a MACH like virtual memory organization [1] (BSD, OSF/1, Hurd, ...).

MMUM (Memory Management in User Mode) is a library devoted to write virtual memory management mechanisms at the application level. This library must be use in conjunction with the MMUSSEL module. This module interacts with the OS for the management of physical memory pages. A more complete description of the MMUSSEL module can be found in [5].

This library allows the programmer to define new memory regions with specific virtual memory management policy attached to them. A new memory region is created by a call to the function mmum_create(size,init,nopage). The parameter size is the size in bytes of the new region. The function init is called when the new region is created. The function nopage is called each time a page fault occurs in the region.

Defining a new virtual memory strategy is mainly writing the two procedures init and nopage. Several MMUM functions are provided to manage the physical memory associated to the new memory regions:

- mmum_newpage(void *a): associate a new physical page to the (virtual) address a. The bit Read/Write of the new page is set to 0.
- mmum_releasepage(void *a): free the physical page associated to the (virtual) address a.
- mmum_xchg(void *a,void *b): swap the two physical page associated to the (virtual) addresses a and b. It is not necessary to associate physical pages to the virtual addresses. For instance, if the address a is associated with a physical page and no page are associated to the address b. Then after a call to mmum_xchg(a,b), the address b is associated with the page previously associated with the address a, and no page are associated to the address a.
- mmum_use_rw(void *a): return the Read/Write usage of the page associated to the (virtual) address a and reset it.

---

[2] Memory Management in User Mode.
[3] Memory Management at USer SpacE Level.

## Mapping a file to virtual memory

To illustrate our tool, we present a virtual memory manager which maps a file to a new memory region This virtual manager is used by an application which make some computations on a huge vector of size N, like the program shown on Figure 1. The content of the vector is in fact the content of the file mapped to the new region. The variable V is a pointer to the beginning of the new region. The iteration modifies the vector value with a computation function f. The accesses to the file is transparent and is done by the virtual memory manager, i.e. by the function fm_init and fm_nopage.

In Figure 4 there is the source code of functions fm_init and fm_nopage which implement the file mapping mechanism. In this implementation, there is only one page present at a time. Initially the present page is the first page of the file (function fm_init). When a page fault occurs, i.e. another page of the memory region is accessed, the current page is written back to the file if modified, and the new page is fetched into memory (function fm_nopage).

### 3.2 Back to the LU decomposition

Let us define a new pagination strategy to the LU decomposition program introduced in the section 2.1. We propose to dynamically change the replacement policy at the runtime. We define different phases in the computation.

Since the matrix is row order, columns are grouped together in block of size $B$ as shown on Figure 2. During the computation of all element of one block of columns, $N$ pages of memory are accessed several times. Instead of using LRU algorithm as replacement policy, we use a LIFO (Last In First Out) one, with the first $M-1$ pages fixed in physical memory. Although there is one page fault for each access to elements $A(i,j)$ when $i >= M$, this policy avoids the hole effect and reduce the number of page faults.

```
real A(N,N);
LRU(A)
for j = 1 to N
    if (j>M) and (j mod B)=0 then LRU(A)
    for i = j+1 to N
        Spatial(A(i,1:j))
        for k=1 to j
            if (j>M) and (i=M or k=M) then
                                    LIFO(A)
            A(i,j)=A(i,j)-A(i,k)*A(k,j)
        NoSpatial(A(i,1:j))
```

**Fig. 5.** LU decomposition with pagination directives.

At each iteration $i$, due to spatial locality, there is an extensive use of pages associated to the vector $A(i, 1 : j)$. So these pages are fixed in physical memory, i.e. when a page fault occurs, the virtual memory manager don't replace them.

```
void *fm_start;             /* Address of the new memory region.    */
void *fm_current;           /* Virtual address of current page      */
                            /* loaded in the region.                */
int fm_size;                /* Size of the new memory region.       */
int fm_fd;                  /* File descriptor.                     */

void fm_init(void *v,int size) /* The new memory region is initialized */
{
  fm_start=v;               /* Record address of the new region.    */
  fm_size=size;             /* Record size of the new region.       */
  fm_fd=fopen(FILENAME,"rw"); /* Open the file mapped to the region. */

  new_page(fm_start);       /* Associate a new physical page at     */
                            /* the beginning of the region.         */

                            /* Read the first page of the region.   */
  fread(fm_start,PAGE_SIZE,1,fm_fd);
  fm_current=fm_start;      /* Record the current virtual address   */
                            /* of the physical page.                */
}

void fm_nopage(void *a)     /* The page owning address a is not     */
                            /* present in physical memory.          */
{
                            /* Test if the current loaded page was  */
                            /* modified. If yes write back its      */
                            /* content to the file.                 */
  if (WRITTEN(mmum_use_rw(fm_current)))
    {
      fseek(fm_fd,(fm_current-fm_start)/PAGE_SIZE);
      fwrite(fm_start,PAGE_SIZE,1,fm_fd);
    }
                            /* The physical page associated to      */
  mmum_xchg(a,fm_current);  /* the virtual address fm_current is now*/
                            /* associated to the virtual address a. */

                            /* Read the page corresponding to the   */
                            /* virtual address a.                   */
  fseek(fm_fd,(a-fm_start)/PAGE_SIZE);
  fread(a,PAGE_SIZE,1,fm_fd);
  fm_current=a;             /* Record the current virtual address   */
                            /* of the physical page.                */
}
```

**Fig. 4.** Functions associated with the new memory region for the file mapping implementation.

When a new block of columns is reached, we have to replace the preceding block of column (i.e. physical page) by a new one. To do this, the replacement mechanism switches to the LRU policy for the accesses to the first $M$ elements of the first column of the new block, and switches back to the LIFO policy.

To implement such a dynamically change of replacement policy, we instrumented the LU decomposition program with several directives dedicated to the specific virtual memory manager as shown on Figure 5.

- LRU(A) : when a page fault occurs in the matrix $A$, a LRU like replacement policy is used.
- LIFO(A) : when a page fault occurs in the matrix $A$, a LIFO replacement policy is used.
- Spatial(A(i,1:j)) : the pages which contain the vector $A(i, 1 : j)$ are not replaced by the virtual memory manager.
- NoSpatial(A(i,1:j)) : the pages which contain the vector $A(i, 1 : j)$ can be replaced by the virtual memory manager.

To sum up, all data in the sub-array A(M:N,1:N) are managed by a LIFO replacement policy, unless the data is in a spatial zone. All other data are managed by a LRU like policy. The main benefit of such a strategy is that there is no page faults (except for the compulsory misses) in the sub-array A(1:M,1:N).

We experimented this strategy on a PC under Linux for a matrix order $N = 1024$. The result with different values for M is shown on the following table. It is worthy of note that greater is the ratio $M/N$ greater is the benefit of the optimized strategy.

| M | LRU | Optimized |
|---|---|---|
| 512 | 41 194 616 | 22 761 454 |
| 768 | 7 477 734 | 3 024 878 |
| 921 | 1 651 317 | 281 996 |

## 4 Conclusion and Related Work

In this paper, we observed the behavior of the standard virtual memory manager of operating systems when running a typical data parallel program, like a LU decomposition, on huge data. The first conclusion is that the LRU policy of the virtual memory manager is not well suited for the computation part: the data-parallel program doesn't expose temporal locality. The second is that communication in data-parallel program introduce new threads of memory access. This new threads can be incompatible with the memory access threads of the computation part. A solution is to strip-mine the computation and communication.

Bordawekar and Choudhary proposed different communication strategies for parallel out-of-core programs [2]. The first is called "out-of-core" communication method, where communication and computation are separated: all the data needed are communicated before the computation. The second, from which we extract our solution, is called "in-core" communication method, the communication is coupled with the strip-mined computation.

A more extensive analysis of paging activity of out-of-core parallel program is the work of Marinescu and Wang [8,9]. In this work, there is a statistical analysis of the paging activity of several applications on parallel machines. They observe a non-correlation of the paging activity of SPMD programs between the different nodes. There is no explanation of this phenomenon. Moreover, this non-correlation introduce new difficulty in the derivation of *gang scheduling* algorithms [10] (Gang scheduling is a technique to schedule different independent groups of parallel computations with a better efficiency [12]).

The second part of this paper presents a new library which allows the definition of new virtual memory management algorithms at the application level. This tool was used to define a specific virtual management policy for the LU decomposition program. We obtain significant improvements, in number of page faults.

There are two related works which redefine virtual memory management strategies for large memory-intensive scientific applications. The first one is the work of Park, Scott and Sechrest on the NEC Cenju-3 architecture [13]. In this architecture, the OS provides interface to define user virtual memory manager. They wrote *ad-hoc* virtual memory managers for two applications and get significant improvements. The second is the work of Krueger, Loftesness, Vahdat and Anderson [7] which presents a object oriented interface to a virtual memory manager and a tool to profile virtual memory access. The main idea is to get enough informations from the profiling tool to define new general replacement policy for specific applications.

In this two approaches, the user has to write the virtual memory manager dedicated to its application, like we do for the LU decomposition. Its a non trivial task which clearly must not be at the end-user level.

In the case of parallel out-of-core programs, it will be interesting to reuse informations extracted at compile time. In the work of Mowry, Demke and Krieger, from a static analysis of sequential code, the compiler extract informations on the *spatial and temporal locality* of data. These informations are used to insert prefetch and release directives during the computation [11]. In the Collard and Utard framework [4] the compiler determines *computation tiling* according to the physical memory and the data layout. The *working set* of each tile is less or equal to the physical available memory. Moreover, the compiler tries to reuse data between successive tiles. One of our future work is to define and implement new virtual memory managers (written with the MMUM library), which will be able to capitalize such compiler informations.

# References

1. Vadim Abrossimov, Marc Rozier, and March Shapiro. Generic virtual memory management for operating system kernels. In *Proc. of th 12th ACM Symposium on Operating System Principles*, December 1989.

2. Rajesh Bordawekar and Alok Choudhary. Communication strategies for out-of-core programs on distributed memory machines. In *Proceedings of the 9th ACM International Conference on Supercomputing*, pages 395–403, Barcelona, July 1995. ACM Press.
3. J. Clinckemaillie, B. Elsner, G. Lonsdale, S. Meliciani, S. Vlachoutsis, F. de Bruyne, and M. Holzner. Performance issues of the parallel PAM-CRASH code. *The International Journal of Supercomputer Applications and High Performance Computing*, 11(1):3–11, Spring 1997.
4. Jean-François Collard and Gil Utard. Automatic data layout and code restructuring for out-of-core programs. In *Proc. of the Workshop on Out-of-Core Computation and Adaptative Compilation (COCA'98)*, Cap-Hornu, Baie de Somme, France, September 1998. IEEE Yuforic and GDR ARP, LaRIA and PRiSM.
5. Olivier Cozette. Virtual memory managment in intensive computation. Master's thesis, LaRIA, Université de Picardie Jules Verne, Jully 1998.
6. G. Davis, L. Lau, R. Young, F. Duncalfe, and L. Brebber. Parallel run-length encoding (RLE) compression—reducing I/O in dynamic environmental simulations. *The International Journal of High Performance Computing Applications*, 12(4), Winter 1998. To appear in a Special Issue on I/O in Parallel Applications.
7. Keith Krueger, David Loftesness, Amin Vadhat, and Thomas Anderson. Tools for the development of applications-specific virtual memory management. Technical report, University of California, Berkeley, April 1993.
8. Dan C. Marinescu and Kuei Yu Wang. An analysis of the paging activity of parallel programs, part I: Correlation of the paging activity of individual node programs in the SPMD execution mode. Technical Report CSD-TR-94-042, Purdue University, June 1994.
9. Dan C. Marinescu and Kuei Yu Wang. Characterization of the Paging Activity of NAS Benchmark Programs on the Intel Paragon. Technical Report CSD-TR-95-015, Purdue University, March 1995.
10. Dan C. Marinescu and Kuei Yu Wang. Gang scheduling and demand paging. In *Proc. of the Int. Conf. on High Performance Computing*, pages 180 – 188, New Delhi, India, December 1995.
11. Todd C. Mowry, Angela K. Demke, and Orran Krieger. Automatic compiler-inserted I/O prefetching for out-of-core applications. In *Proceedings of the 1996 Symposium on Operating Systems Design and Implementation*, pages 3–17. USENIX Association, October 1996.
12. J.K. Ousterhout. Scheduling techniques for concurrent systems. In *Proc. of the 3rd Int. Conf on Distributed Computing System*, pages 22–30, October 1982.
13. Yoonho Park, Ridgway Scott, and Stuart Sechrest. Virtual memory versus file interfaces for large, memory-intensive scientific applications. In *Proceedings of Supercomputing '96*. ACM Press and IEEE Computer Society Press, November 1996. Also available as UH Department of Computer Science Research Report UH-CH-96-7.
14. J.M. Del Rosario and A. Choudhary. High performance I/O for massively parallel computers: Problems and Prospects. *IEEE Computer*, 27(3):59–68, 1994.
15. Rajeev Thakur, Ewing Lusk, and William Gropp. I/O characterization of a portable astrophysics application on the IBM SP and Intel Paragon. Technical Report MCS-P534-0895, Argonne National Laboratory, August 1995. Revised October 1995.

# High Performance Parallel I/O Schemes for Irregular Applications on Clusters of Workstations *

Jaechun No[1], Jesús Carretero **, and Alok Choudhary ***

jno@ece.nwu.edu
Dept. of Electrical Engineering and Computer Science,
Syracuse University, USA

**Abstract.** Due to the convergence of the fast microprocessors with low latency and high bandwidth communication networks, clusters of workstations are being used for high-performance computing. In this paper we present the design and implementation of a runtime system to support irregular applications on clusters of workstations, called "Collective I/O Clustering". The system provides a friendly programming model for performing I/O in irregular applications on clusters of workstations, and is completely integrated with the underlying communication and I/O system. All the performance results were obtained on the IBM-SP machine, located at Argonne National Labs.

## 1 Introduction

Due to the convergence of the fast microprocessors with low latency and high bandwidth communication networks, such as ATM, Myrinet, or the Gigabit Ethernet, clusters of workstations are being increasingly used for solving large-scale parallel scientific applications in cost-effective way. Most of those applications have tremendous I/O requirements [10, 7], including checkpointing of large-scale data sets, and writing of periodical snapshots for further visualization. Furthermore, a large subset of those applications are *irregular* applications, where accesses to data are performed through one or more levels of indirection [12]. Sparse matrix computations, particle codes, and many CFD applications where geometries and meshes are described via indirections, exhibit this feature.

In this paper we present the design and implementation of our runtime system for clusters of workstations, *collective I/O clustering*. The I/O architecture of clusters of workstations usually relies on a set of I/O servers, having local disks, and a set of diskless nodes. The design of our runtime system fits this feature, as

---

* This work was supported in part by Sandia National Labs award AV-6193 under the ASCI program, and in part by NSF Young Investigator Award CCR-9357840 and NSF CCR-9509143.
** Arquitectura y Tecnología de Sistemas Informáticos, Universidad Politécnica de Madrid, Spain.
*** Electrical and Computer Engineering, Northwestern University, USA

we distinguish between two kind of processors: I/O servers and compute nodes. All I/O details, such as data exchange, data distribution, and collective I/O, are transparent to the application programmer.

The main objectives for the collective I/O clustering are as follows:

- *Provide flexibility needed to the various I/O configurations for a cluster of workstations.* The collective I/O clustering is designed to support two kinds of I/O configurations: in the first I/O configuration, all processors are clients and I/O servers, and in the second I/O configuration, a subset of processors will only be I/O servers.
- *Provide user-controllable stripe unit.* Appropriate declustering of I/O requests over I/O servers should be addressed to produce high performance I/O bandwidth [5] and has been successfully implemented in the several file systems [8, 1, 4]. In the collective I/O clustering, we use a user-controllable stripe unit which is specified by GF(Group Factor) in the file-creation time.
- *Provide compression facility.* Compression has been traditionally used to reduce disk space requirement [14], but recently it has been applied to parallel applications managing large arrays with the aim of reducing the total execution time [11, 9]. The collective I/O clustering combines compression facility to achieve two major goals: reducing disk space requirement, and reducing total execution time.

The rest of the paper is organized as follows: Section 2 presents an brief overview of the collective I/O clustering on an irregular application. Section 3 presents the implementation details of the collective I/O clustering operation. Section 4 presents the performance results on the IBM/SP machine located at Argonne National Labs. Finally, some conclusions are presented in section 5.

## 2 Motivation

Figure 1(a) describes a typical irregular application, where it sweeps all the edges of an unstructured mesh. In the application, an input mesh file is read, and then the edges and nodes are distributed over processors. We used block distribution to spread them to the processors. *no_of_edges_partitioned_per_proc* represents the number of edges partitioned to a single processor. In the nested loops, *edge[j].V1* and *edge[j].V2* are two nodes connected by an edge *edge[j]*. The reference pattern is specified by *edge[j].V1* and *edge[j].V2*, called *indirection array*, and also these values are used to access to a global array. *X* is a data array which contains the physical values associated with each node. In this application, a node has an array consisting of 3 doubles, and other 2 floats.

Figure 1(b) shows an example of the edge and node partitions by using block distribution. In the processor 0, *4,5,8* are the remote indirection elements whose physical values mush be fetched from processor 1 and 2. All the remote values are fetched before the computation. After the computations are finished, the data (physical values associated with a node) are written to a global array whose

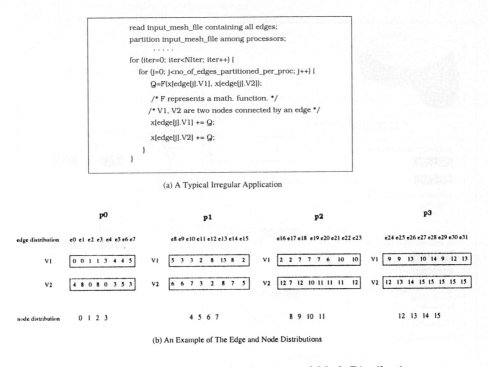

**Fig. 1.** A Typical Irregular Application and Mesh Distribution

access pattern is determined by indirection elements, *edge[j].V1* and *edge[j].V2*, using the collective I/O clustering method.

Figure 2 describes an overview of the collective I/O clustering for an irregular application. In the collective I/O clustering, a global file is organized as a sequence of subfiles, each of which is stored in an I/O server's local disk. When a file is read or written, the appropriate schedule information is constructed. In the schedule information, each processor's data domain for I/O is determined, based on the two-phase method described in [2, 3, 13]. These data domains are logically divided into non-overlapping partitions, which may or may not span several local disks. The logical data domain partition is determined by GF ($1 \leq GF \leq num.ofI/Oservers$), which is given by users at the file-creation time. If GF is 1, the whole data domain of a processor is transferred to the appropriate I/O server. If GF is greater than 1, the data domain of a processor is divided into the contiguous GF number of partitions, and then these partitions are distributed across all I/O servers in round-robin fashion. Hereafter, we call the partition as *block*. Each I/O server stores a subfile, which consists of the blocks received from the processors.

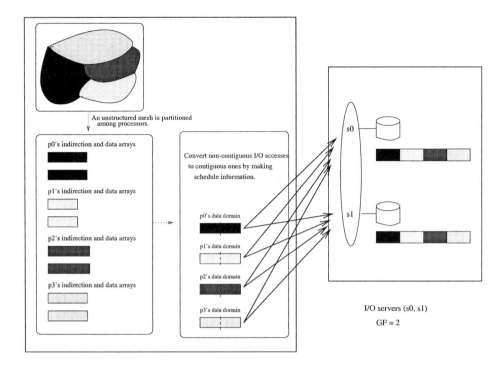

**Fig. 2.** An Overview of the Collective I/O Clustering on an Irregular Application

## 3 Implementation Details of Collective I/O Clustering Operation

In this section, we describe an overview of the collective I/O clustering. The method is organized into three main stages: schedule construction, communications for exchanging the remote indirection and data elements between processors, and I/O operations to write the data to files.

### 3.1 Schedule Construction

Schedule describes the communication and I/O patterns required for each processor. The two main steps are involved in computing the schedule information. First, each processor is assigned a data domain into which it is responsible for reading and writing. To distribute I/O workload evenly, the data domain of each processor is defined to be an almost equal size. Next, the indirection elements in the local memory are individually scanned to decide which processor is responsible for performing I/O for the corresponding data. Based on the scanning, the indirection elements are coalesced into a single message per destination processor. Once the schedule information is constructed, it can be repeatedly used in the irregular application whose access pattern does not change during computations. If the access pattern is changed, the appropriate schedule must be reconstructed, and then be used for the collective I/O clustering.

## 3.2 Communications for The Remote Indirection and Data Elements

In this stage, the remote indirection and data elements in each processor's local memory must be exchanged, according to the processor's data domain determined in the schedule.

Let $\{0,1,3,4,5,8\}$, $\{x(0),x(1),x(3),x(4),x(5),x(8)\}$ be the indirection and data arrays, respectively, stored in processor 0's local memory, and $\{5,3,2,8,13,6,7\}$, $\{x(5),x(3),x(2),x(8),x(13),x(6),x(7)\}$ be the indirection and data arrays, respectively, stored in processor 1's local memory. Also, let 16 be the size of a global array. In the schedule construction, processor 0's data domain is determined as $0 - 3$, processor 1's data domain is determined as $4 - 7$, and so on. In case of writing, processor 0 sends $4,5$ and $x(4),x(5)$ to the processor 1, and $8$ and $x(8)$ to the processor 2. Similarly, processor 1 sends $2,3$ and $x(2),x(3)$ to the processor 0, $8$ and $x(8)$ to the processor 2, and $13$ and $x(13)$ to the processor 3. As a result, each processor collects the data to its data domain. The reverse procedure is employed for read operations.

## 3.3 I/O Operations in the Collective I/O Clustering

Based on the data domain arranged in the previous stage, the I/O portion of the collective I/O clustering is executed. This stage takes buffer offset at which to start I/O, buffer length which is the number of data elements for I/O and pointer to the buffer. The additional information is also passed to this stage, such as the file name which is used for the actual data and meta-data files, compression flag to check if the compression is applied for, number of I/O servers *(NumOfIO)*, and GF size. Among the processors allocated, as many as *NumOfIO* processors are configured as I/O servers, starting from the lowest-rank processor.

The write operation in the collective I/O clustering begins with slicing each processor's data domain into GF size. Each block resulting from the slicing is then converted to be the same size over all processors, while making the block size to fit to a multiple of file system block size. All the blocks are then distributed to the appropriate I/O servers in round-robin fashion. If GF is 1, each processor's entire data domain goes to an appropriate I/O server, exploiting maximum data locality. If GF is the same size as *NumOfIO*, each processor's data domain is distributed over all I/O servers, exploiting maximum data parallelism. Each I/O server gathers all the blocks coming from the processors and writes them to a single subfile to be stored into its local disk. The meta-data describing the characterization of the subfile is sent to the master processor so that it collects all the meta-data and writes them to a single meta-data file.

In case of reading data from the subfiles, the master processor first reads the meta-data file and broadcasts it to all the processors. Each processor then maps its data domain to the blocks stored in the subfiles by using the field of each block's original global offset in the meta-data. After determining from which subfiles and blocks it should read the data, each processor sends the control information to the appropriate I/O servers who are responsible for the subfiles. The control information contains the local block number, starting position in the block and data count to receive. Each I/O server reads a subfile from its own local disk and distributes it to the processors, based on the control information received.

When the compression is applied for the I/O operations, the size of GF is restricted to 1 to maintain data integrity, since one bit of data loss can cause the entire I/O failure. In case of writing, the compressed data domain of each processor goes to the appropriate I/O server. After each I/O server gathers all the compressed blocks coming from the processors, the whole length of the subfile is aligned to be a multiple of file system block size, and the subfile is written to the local disk. In case of reading, the same procedures as those in the read operation without compression are performed, except that each I/O server reads the compressed subfile from its own local disk, and uncompresses it, before distributing the data based on the control information received.

## 4 Performance Evaluation

The results for the collective I/O clustering were obtained on the IBM-SP at Argonne National Labs. The IBM-SP, called Quad, has 80 nodes, each running AIX 4.2.1 with patches. Each node has a 256MB of main memory and two of 4.5 GB disks. The nodes are interconnected by a IBM's high-performance switch.

The collective I/O clustering was combined with the irregular application shown in Figure 1. In the application, a single node has a 32B of physical values consisting of 2 floats and an array of 3 doubles. We obtained all the performance results by using 64 processors, and, after the computation, 256MB of data was written to the files.

Figure 3 shows the time components for the collective I/O clustering, in which the number of I/O servers is 4, 8, 16, 64, respectively. In the Figure 3, **irregular_comp** and **irregular_comm** represent the times spent in the communications for the remote indirection and data elements, before beginning the I/O operation in the collective I/O clustering. In the application shown in Figure 1, after communicating the remote elements for a write operation, each processor is responsible for writing 4MB of data. Since we spread the nodes and edges over processors in block distribution, the times for the schedule and communication to access remote elements do not change significantly, while varying the I/O servers. Usually, irregular applications use a variety of heuristic data mapping to optimize the communication requirements [6]. With such schemes, the times for communicating remote elements, **irregular_comp** and **irregular_comm** in Figure 3, can be greatly reduced by allowing to access the local elements. **I/O operation** shows the time for performing the write operation in the collective I/O clustering. We observed that, as the number of I/O servers increases, the

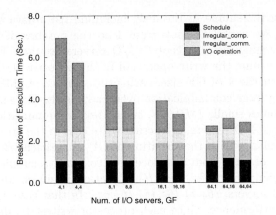

**Fig. 3.** Breakdown of Execution Time for the Collective I/O Clustering as a function of (number of I/O servers, GF.) The execution time includes the times for the schedule, communications for the remote elements and I/O operation.

time for the write operation becomes small, due to the reduced subfile size to be stored in the I/O servers.

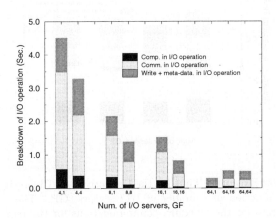

**Fig. 4.** Breakdown of the I/O Operation in the Collective I/O Clustering as a function of (number of I/O servers, GF.)

To illustrate the effect of the GF size on the I/O operation in the collective I/O clustering, in Figure 4, we show the time components of the I/O operation, such as the computation overhead to arrange the partitioned data into the communication buffer, communication overhead to organize subfiles, and actual

I/O cost to write the subfile to the disk, including meta-data organization. In each I/O server, we varied GF size from 1 to the number of I/O servers. In each I/O server configuration except 64 I/O servers, as the GF size increases, the time for performing the write operation in the collective I/O clustering becomes small. With the 1 of GF size, each processor sends 4MB of data to an I/O server, taking large communication overhead. As the GF size increases, the message size becomes small, *4MB/GF*, taking less communication overhead. In the same number of I/O servers, the actual write cost in the I/O operation does not significantly change with the variation of GF, since each I/O server collects the same amount of data from all processors to make a subfile. Figure 4 also shows the time components for the I/O operation in which all processors are configured as I/O servers, *(64,1) (64,16) (64,64)*. In that case, the 1 of GF size shows the best performance. Since each processor writes the data collected in its data domain to the local disk, no communication requires in the I/O clustering. This case exploits data locality as much as possible. When the GF size increases, the I/O operation requires the communication overhead, showing less performance results.

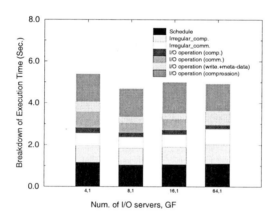

**Fig. 5.** Breakdown of Execution Time for the Collective I/O Clustering combined with Compression as a function of (number of I/O servers, GF.) The execution time is composed of the times for the schedule, communications for the remote elements and I/O operation including the compression time.

Figure 5 shows the performance results with compression, while varying the number of I/O servers from 4 to 64. To maintain the data integrity, only GF of 1 is supported for the compression. We used *lzrw3* compression algorithm. Since the compression is incorporated in the I/O operation of the collective I/O clustering, no effect occurs in the schedule and communication for accessing the remote elements. In the I/O operation of the collective I/O clustering, **I/O operation**, after each processor's data domain is compressed, the communication and actual write operations to organize a subfile to the local disk are performed on the compressed data. Therefore, those times are significantly reduced. However, the compression time takes about 1 sec in every case. This offsets the performance gain obtained by communicating the compressed data and writing it to the I/O servers. In the 4 I/O server configuration, the reduction in the communication and write operations exceeds the compression overhead, resulting in the better performance with the compression. However, as the number of I/O servers increases, the cost for the communication and write operations becomes small, and using the compression rather degrades the performance.

## 5 Conclusions

We developed the runtime system, called Collective I/O Clustering, that can support the irregular applications on clusters of workstations. In the system, two I/O configurations are possible; in the first configuration, all the processors act as I/O servers as well as compute nodes, and in the second configuration, only a subset of processors are dedicated as I/O servers. Further, the system supports a user-controllable stripe unit which is specified by GF in the file-creation time. We evaluated the collective I/O clustering on the IBM-SP at Argonne National Labs. We found that, in the first configuration, exploiting data locality produces the best performance by performing low communication cost. On the other hand, in the second configuration, as the data in each processor's domain is more distributed over I/O servers, the performance results become better due to the less communication overhead. As an optimization, we combined the collective I/O clustering with the compression. As far as the reduction in the communication and I/O costs outperforms the compression overhead, the results with compression are improved, compared with those without compression. However, with the small communication and I/O costs, compression overhead becomes a burden to produce better performance.

## References

1. Fern E. Bassow. Installing, managing, and using the IBM AIX Parallel I/O File System. IBM Document Number SH34-6065-00, February 1995. IBM Kingston, NY.
2. Rajesh Bordawekar, Juan Miguel del Rosario, and Alok Choudhary. Design and evaluation of primitives for parallel I/O. In *Proceedings of Supercomputing '93*, pages 452–461, 1993.
3. Alok Choudhary, Rajesh Bordawekar, Michael Harry, Rakesh Krishnaiyer, Ravi Ponnusamy, Tarvinder Singh, and Rajeev Thakur. PASSION: parallel and scalable software for input-output. Technical Report SCCS-636, ECE Dept., NPAC and CASE Center, Syracuse University, September 1994.
4. Peter F. Corbett, Sandra Johnson Baylor, and Dror G. Feitelson. Overview of the Vesta parallel file system. In *IPPS '93 Workshop on Input/Output in Parallel Computer Systems*, pages 1–16, 1993. Also published in Computer Architecture News 21(5), December 1993, pages 7–14.

5. Thomas H. Cormen and David Kotz. Integrating theory and practice in parallel file systems. Technical Report PCS-TR93-188, Dept. of Math and Computer Science, Dartmouth College, March 1993. Revised 9/20/94.
6. Raja Das, Mustafa Uysal, Joel Saltz, and Yuan-Shin Hwang. Communication optimizations for irregular scientific computations on distributed memory architectures. *Journal of Parallel and Distributed Computing*, 22(3):462–479, September 1994.
7. Juan Miguel del Rosario and Alok Choudhary. High performance I/O for parallel computers: Problems and prospects. *IEEE Computer*, 27(3):59–68, March 1994.
8. Intel Corporation. *Paragon User's Guide*, June 1994.
9. National Center for Supercomputing Applications. HDF reference manual version 4.1. Technical report, University of Illinois, 1997.
10. James T. Poole. Preliminary survey of I/O intensive applications. Technical Report CCSF-38, Scalable I/O Initiative, Caltech Concurrent Supercomputing Facilities, Caltech, 1994.
11. K. E. Seamons and M. Winslett. A data management approach for handling large compressed arrays in high performance computing. In *Proceedings of the Fifth Symposium on the Frontiers of Massively Parallel Computation*, pages 119–128, February 1995.
12. Shamik D. Sharma, Ravi Ponnusamy, Bongki Moon, Yuan shin Hwang, Raja Das, and Joel Saltz. Run-time and compile-time support for adaptive irregular problems. In *Supercomputing 1994*. IEEE Press, November 1994.
13. R. Thakur, A. Choudhary, R. Bordawekar, S. More, and S. Kudatipidi. Passion: Optimized I/O for Parallel Systems. *IEEE Computer*, June 1996.
14. T.A. Welch. A technique for high performance data compression. *IEEE Computer*, 17(6):8–19, June 1984.

# Advanced Data Repository Support for Java Scientific Programming*

Peter Brezany[1] and Marianne Winslett[2]

[1] Institute for Software Technology and Parallel Systems
University of Vienna, Liechtensteinstrasse 22, A-1090 Vienna, Austria
brezany@par.univie.ac.at

[2] Database Research Laboratory, Department of Computer Science
University of Illinois, 1304 W. Springfield, Urbana, IL 61801 USA
winslett@uiuc.edu

**Abstract.** Research in the parallel and scientific computing area has begun to focus on the development of Java-based programming environments. This paper describes the design of an original object-oriented database-style repository interface for high performance storage and retrieval of scientific data. The design is based on the standard interface to object databases that has been defined by the Object Database Management Group. In the paper, we present the mapping of the repository interface into Java constructs.

## 1 Introduction

Since its advent, Java [1] has very quickly become an important language for development of Web based applications. However, at present Java is envisioned as a broad technology that could establish itself as the universal language for creating programs that run on any kind of platform, from a smart card to a PC to a supercomputer.

The possible role of Java as a language for high performance computing has been intensively investigated by G. Fox's research group [2, 3] at the University of Syracuse. Other proposals for Java dialects that address the special needs of high performance computing include Titanium [11] and Spar [9]. Getov, Flynn-Hummel, and Mintchev [5, 6] have developed Java program wrappers for the standard message-passing communication interface MPI and other standard high-performance libraries (e.g., ScaLAPACK). These efforts are helping to make Java a realistic alternative for computationally intensive applications.

Many scientific applications are both compute and I/O intensive. Managing a large dataset on a sequential machine is not an easy task, and is more complicated in a parallel environment, often exceeding the expertise of scientists. Thus there is a need for specialized I/O software, which is efficient and easy to use, to

---

* This research is being carried out as part of the research project "Aurora" supported by the Austrian Research Foundation, and is also supported by NASA under grant NAGW 4244 and the Department of Energy under grant B341494.

facilitate storage management on parallel machines. Moreover, large-scale parallel scientific applications and supporting tools (e.g., visualization, performance analysis, and debugging tools) can benefit greatly from I/O software providing some features which can be found in traditional database management systems (e.g., meta-data management, concurrency control, high application portability, etc.).

Standard Java implementations support I/O by the *java.io* package which allows *sequential-access* and *random-access* file I/O and serialization and deserialization of objects during I/O operations. If the file stores an array, it is possible to access an array element corresponding to a specific index value using the random-access mode. However, a direct access to multi-dimensional array sections is awkward and may be inefficient. The Java object serialization and deserialization support also provides no way to selectively read or write a small part of a large object.

This article describes the design of an object-oriented database-style interface to a *scientific data repository* ("repository" for short) which is a high performance store for scientific data. We intend this interface as a wrapper or native interface for preexisting or newly created repositories (HDF, NetCDF, Panda, PASSION, traditional databases, etc.). In brief, the interface must provide methods for creating, opening, closing, and deleting a repository, creating and finding objects (e.g. arrays) in a repository, operations to read and write repository objects and elements of aggregate repository objects, and for advance disclosure of repository access patterns. (If the underlying repository does not support an interface feature that the user has invoked, exception handling and recovery codes in the interface implementation will address the situation.) The interface design objectives can be summarized as follows.

- **Extensible data types.** The repository interface design should accommodate all types of persistent data accessed by scientific applications, and allow the user to easily add a new data type and methods specific to that type. The repository interface should allow access to the data as well as metadata describing the features and organization of the data on secondary or tertiary storage; in this paper, we focus on the use of secondary storage and on the most important type of data for high-end scientific computation today: multidimensional arrays.
- **Portability.** The repository interface should allow a scientific data repository to be accessed from applications developed in different programming languages. Further, the usage of the interface should be the same regardless of the vendor of the repository accessed through the interface. The same objectives were pursued by the *Object Database Management Group* (ODMG). Therefore our design of the repository interface is based on the current ODMG standard [4]. ODMG provides a language-independent object model; bindings between this model and concrete programming languages are also part of the ODMG standard. This standard only directly supports one-dimensional arrays, which is insufficient for scientific applications. Therefore, we have extended the notation provided by the ODMG by additional fea-

tures to support multidimensional arrays directly. Further, in this paper we only discuss the binding of the interface to Java programs.
- Support for compiler developers and application programmers. The repository interface should be suitable for use by the developer of a compiler of a High Performance Java language dialect, the user writing explicitly parallel Java programs (e.g., in the message-passing style), or the developer of sequential Java programs.
- Elegant interface for the basic interactions with repositories. Creating, opening, and deleting repositories, constructing and deleting objects in repositories, and inquiring about the contents of repositories should be supported by an elegant interface at a high abstraction level.
- Disclosing access patterns. The repository interface should provide means for specification of application access patterns to disclose knowledge of future repository accesses in the form of hints to the repository management system, which can use them to guide aggressive runtime optimizations.
- Support for out-of-core applications. The repository interface should allow operations that require transfer of portions of large objects (e.g., multidimensional array sections) between main memory and repositories.
- Support for parallelism. It should be possible for all repository data and metadata to be stored in a distributed and parallel form across multiple I/O devices. Further, the user and the compiler should have the opportunity to influence the distribution of the repository data, using hints.
- A clear semantics for data visibility and stability. Appropriate facilities for concurrency control and transactional semantics should be provided. For example, the user should be able to ensure that newly written data reaches non-volatile storage and to ensure that that data is visible to applications that subsequently open the repository.

This list represents the goals of our project, rather than an exhaustive list of desirable repository interface features. For example, security features and powerful query languages may be useful for many applications, but are not targeted by our work. A repository interface that meets our design goals will provide the user with a generic interface allowing easy access to data stored in a variety of kinds of repositories. Through its support for hints, the interface will allow the user to control high-performance parallel access to data.

The rest of this paper is organized as follows. Sections 2 to 4 convey the essential concepts of the repository interface, and focus on the special support needed for multidimensional arrays. Section 5 concludes with a brief summary. The Appendix illustrates our ODMG-style object model for a portion of the *Repository* class interface definition.

## 2 Repository Operations

To create a new repository, the user calls the **create** method of the *Repository* class, which returns an instance of that class. Before an application can access data in a preexisting repository, the member function **open** must be called with the location of the repository to be opened. An open repository can be deleted by a call to the **delete** member function, and its physical location can be determined using the **get_location** member function.

For parallel and even sequential repositories, the user may possess information which would be very helpful in choosing physical schemas and other internal parameters of the repository. The user can provide this information to the repository via objects of type *RepInfo*, which can direct repository layout and operation optimizations. For example, the user may specify whether the repository should support parallel accesses and suggest how many input/output processes should participate in repository operations.

The following code[1] creates and opens a repository $r$ associated with file *simulation.dat*. In the hint field *ioproc*, the user suggests to place repository servers on four I/O processors. The user also provides a textual description of the repository contents.

```
RepInfo info = new RepInfo();
info.parallel = true; info.ioproc = 4;
Repository r = Repository.create ("simulation.dat",
                                   Repository.readWrite, info);
r.set_description ("Simulation of Vehicle XW3256 on 09/12/98");
```

Repositories may support sophisticated buffering, caching, and prefetching mechanisms to improve performance for applications that access their data. For that reason, an application does not ordinarily know whether the data it has written to a repository have actually reached disk. If sequential consistency is not being used, the application will not know when its writes became visible to other applications. To give applications a degree of control over these factors, without going so far as to require traditional transaction semantics, the repository class includes a **flush** method. When a call to **flush** returns, the newly written repository data and metadata are stored on non-volatile storage.

Once the application has finished its accesses to a repository, the repository must be closed by a call to the **close** member function.

**The Data in a Repository.** Repositories store application data, and they also store information about the data itself. Subclassing of the *Object* standard class can be used to support new types of data. For example, an array can be stored in the repository as an instance of the class *ArrayInRep* which encapsulates the array data and information about the array, including rank, shape, element type, and (for parallel repositories) distribution type. These metadata are called *properties*. The user may also insert a comment describing the contents of the array. In general, when we want to store an array in the repository, we must

---

[1] **Remark:** In this paper, code examples are marked by $\overline{\nabla}$ and $\underline{\triangle}$.

create an appropriate instance of the *ArrayInRep* class in this repository using the method create_object of the *Repository* class.

The following example illustrates creation of an object[2] storing a two-dimensional float array and inserting a comment describing the contents of the array into the repository object. By the operation set_io_distribution_hint the user expresses a preference for I/O data distribution. The interface supports HPF-style data distributions across the multiple compute nodes and I/O nodes (I/O processors). The example shows the I/O distribution specification for a 2D array which has a (BLOCK,*) distribution across 4 I/O nodes arranged in a 1D array.

---

**ArrayInRep** reparray = ( **ArrayInRep** )r. create_object ( "pressure",
      new **Properties** ("array[1000,1000]<float>"), "ArrayInRep");

**IODistDescriptor** iodist = new **IODistDescriptor** (new **IOProcessors** (4),
      new **IODistType** (BLOCK,NODIST));

reparray. **set_io_distribution_hint** (iodist);
reparray. **set_description** ("Pressure on the Surface WY032");

---

Once an object has been stored in a repository, it can be found again by using one of two different mechanisms supported by the *Repository* class. An object can be found using the lookup member function, by passing in the name that was supplied at the time the object was created (e.g., the name *"pressure"* in the context of the code fragment above). Alternatively, a repository iterator (described below) can be used.

**Repository Iterators.** An *iterator* is used to iterate over all the data objects in a repository. A repository iterator is constructed by invoking the function make_object_iterator of the class *Repository*, which returns an instance of the class *Iterator*. That class provides a set of methods that can be used to navigate through the objects in the repository. The *Iterator* object maintains an internal cursor that keeps track of the current position (object) in the repository. The retrieval methods of the *Iterator* class can be used to get access to the data and metadata of the object that the cursor is currently pointing to.

---

[2] The *"ArrayInRep"* string specifies the creation of an object of the *ArrayInRep* class. The creation of an instance of another class, e.g. *ImageInRep*, would be specified by the *"ImageInRep"* string; if the *ImageInRep* class were not declared, *ClassNotFoundException* would be thrown. This approach relies on the Java **Reflection** API.

## 3 Operations on Arrays in Repositories

Data is moved between repositories and memory by calling the **read** and **write** operations of the class *ArrayInRep*. These operations can transfer whole arrays, array sections, or single data values.

**Transfer of whole arrays.** Transfer of a whole array is illustrated in the following example.

▽

    float [ ][ ] A = new float[1000,1000];
    reparray.**write** ( A );

△

If a data transfer specified in a parallel program involves an array that is or will be distributed across the memories of multiple processors, then a data distribution descriptor for this array must be passed to the operation, as shown in the following example[3]:

▽

    //if B of shape (500,500) is distributed onto a 2 × 2 compute processor mesh:
    float [ ][ ] B = new float[500,500];
    **DistributionDescriptor** dist_B;
    ... initialization of dist_B ...
    reparray.**write** ( B, dist_B );

△

By means of the distribution descriptor the repository is able to figure out which part of the array is stored on each participating processor.

**Transfer of array sections.** So far, Java does not include *array section* notation. For the purpose of specification of data transfers involving subsets of array elements, we use the notation introduced by B. Carpenter et al. in their HP Java proposal [2].

Specification of a read or write operation that transfers an array section has the following form:

    array_in_repository.**read** (*array_identifier, memory_filter, repository_filter*);
    array_in_repository.**write** (*array_identifier, memory_filter, repository_filter*);

where memory_filter specifies the memory section of the array denoted by array_identifier, and repository_filter specifies the section of the array in the repository which is to be involved in the data transfer.

---

[3] The compiler of a High Performance Java dialect can analyze the distribution specifications introduced in the program and automatically generate the code that initializes the data distribution descriptor of each distributed array.

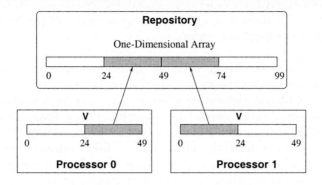

```
float [ ] V = new float[50];
int loclow, lochigh, locstride, globlow, globhigh, globstride;
DataFilter memfilter, repfilter;
if (my_proc() = 0) { loclow = 25; lochigh = 49; locstride = 1;
                     globlow = 25; globhigh = 49; globstride = 1; }
else { loclow = 0; lochigh = 24; locstride = 1;
       globlow = 50; globhigh = 74; globstride = 1; }
memfilter = new DataFilter ( new Section (loclow, lochigh, locstride) );
repfilter = new DataFilter ( new Section (globlow, globhigh, globstride) );
... updating a repository array repvector ...
repvector. write ( V, memfilter, repfilter );
```

**Fig. 1.** Writing Sections of a Distributed Array

In the example on the previous page, the repository object *reparray* encapsulating a (500,500) float array was instantiated. Reading a section of this array onto a (50,50) float array $W$ is illustrated below. The *DataFilter* class provides an interface for constructing memory and repository filters.

```
float [ ][ ] W = new float[50,50];
DataFilter memfilter = new DataFilter ( "[[0:49,0:49]]" );
DataFilter repfilter = new DataFilter ( "[[100:149,200:249]]" );
reparray. read ( W, memfilter, repfilter );
```

Writing a section of a one-dimensional memory array $V$ distributed across two processors onto a section of the one-dimensional repository array is illustrated in Fig. 1. Each processor constructs its memory filter in terms of the parameters of the array segment assigned to it by the array distribution. The corresponding repository filters are constructed in terms of global indices. The *Section* class provides an interface for constructing array section descriptors. It is also allowed to express the memory filter in terms of global indices and pass it, together with the data distribution descriptor of $V$, to a method of the *FilterConversion* class performing global to local conversion.

**Checkpointed arrays.** Many scientific applications require checkpoint and restart capabilities. A checkpoint operation saves the state of a set of program objects to disk to be restored during a restart operation. In case of a system failure, the implementation of a checkpoint operation must guarantee that a correct and consistent set of program objects is stored in the repository. The restart operation reads the data from the most recently fully completed checkpoint to restore the state of the program objects before resuming the computation after a hardware or software failure. Under our approach, checkpointed arrays are instances of the class *CheckpArrayInRep* which is a subclass of *ArrayInRep*.

**Appendable arrays.** Another high-level array I/O operation performed by scientific applications is to append data to an array along an unlimited dimension. For example, this is widely used by applications that simulate the evolution of a physical phenomenon over time. Periodically, the application writes out a snapshot of the current state of interesting arrays. If the arrays have $D$ dimensions, then the write operation can be thought of as adding an array slice to a $D+1$-dimensional array in the repository. This kind of write operation can be thought of as an *"append"* operation, as it adds new data to the end of a sequence.

In the repository, we model appendable arrays with the concept called *linearly versioned array*. All linearly versioned arrays are instances of the class *LVArrayInRep* which is a subclass of the class *ArrayInRep*. All versions of the same object have the same *"object_name"* string. However, they differ in having different version numbers and different data. The first version written out is number 1, and so on.

## 4 Disclosing Access Patterns

The repository management system can effectively apply some optimizations, like prefetching and caching, if it is informed by the user or compiler about repository accesses in advance. No matter how random and unpredictable accesses may appear to the repository management system, they are often quite predictable within the applications.

Hints can be specified using the operation **advance_info** which includes the type of the future operation on the repository object, data items affected, and optionally an estimate of the execution time remaining before the next repository object access.

In the example below, we outline the mechanism which the repository interface provides for specification of hints for a read operation; these hints can be utilized for optimization of prefetching and caching. Our approach is based on work of Patterson et al. [8] on *transparent informed prefetching* of files.

```
float [ ][ ] W = new float[50,50];
DataFilter memfilter = new DataFilter ( "[[0:49,0:49]]" );
DataFilter repfilter = new DataFilter ( "[[100:149,200:249]]" );
...
reparray. advance_info (infoRead, W, memfilter, repfilter);
...
reparray. read  ( W, memfilter, repfilter );
...
```

## 5  Concluding Remarks

I/O support provided by standard Java implementations is not sufficient for large-scale scientific computing. In this paper, we discussed design issues in developing a Java scientific data repository, a new approach to storage and retrieval of scientific data. This approach is based on the current ODMG standard. In the next step, we will complete the Java repository implementation that we have constructed on top of JavaMPI. This version does not support transactional semantics, which is a future research topic.

## References

1. K. Arnold and J. Gosling. The Java Programming Language. Addison-Wesley, Reading, Massachusetts, 1996.
2. B. Carpenter, G. Zhang, G. Fox, X. Li, and Y. Wen. HPJava: Data Parallel Extensions to Java, http://www.cs.ucsb.edu/conferences/java98/program.html.
3. G. Zhang, B. Carpenter, G. Fox, X. Li, and Y. Wen. Considerations in HPJava language design and implementation. In 11th International Workshop on Languages and Compilers for Parallel Computing, August 1998.
4. R.G.G. Cattell, D. Barry, D. Bartels, M. Berler, J. Eastman, S. Gamerman, D. Jordan, A. Springer, H. Strickland, and D. Wade. Object Database Standard: ODMG 2.0. Morgan Kaufmann Publishers, San Francisco, 1997.
5. V. Getov, S. Flynn-Hummel, and S. Mintchev. High-Performance Parallel Programming in Java: Exploiting Native Libraries. http://www.cs.ucsb.edu/conferences/java98/program.html.
6. S. Mintchev and V. Getov. Towards portable message passing in Java: Binding MPI. In M. Bubak, J. Dongarra, and J. Wasniewski (eds.), Recent Advances in PVN and MPI, LNCS vol. 1332, pp. 135-142, Springer-Verlag, 1997.
7. NCSA HDF Reference Manual. Version 4.1. National Center for Supercomputing Applications, University of Illinois, May 1998.
8. R.H. Patterson. Informed Prefetching and Caching. PhD Thesis, Department of Computer Science, Carnegie Mellon University, December 1997.
9. K. van Reeuwijk, A.J.C. van Gemund, H.J. Sips, Spar: A Programming Language for Semi-Automatic Compilation of Parallel Programs. Concurrency: Practice and Experience, 9(11):1193–1205, 1997.

10. K. E. Seamons. Panda: Fast Access to Persistent Arrays Using High Level Interfaces and Server Directed Input/Output. PhD Thesis, Department of Computer Science, University of Illinois at Urbana-Champaign, 1996.
11. K. Yelick, L. Semenzato, G. Pike, C. Miyamoto, B. Liblit, A. Krishnamurthy, P. Hilfinger, S. Graham, D. Gay, P. Colella, and A. Aiken. Titanium: A High-Performance Java Dialect. ACM 1998 Workshop on Java for High-Performance Network Computing, Palo Alto, February 1998.

# Appendix: Illustration of the ODMG Notation

```
class Repository {
    typedef enum AccessStatus{readOnly, readWrite, exclusive,
                    sharedReadOnly, sharedReadWrite};
    typedef enum ObjectType{arrayObjectType, scalarObjectType};
    void create ( in string location, in AccessStatus access,
                in RepInfo info ) raises (RepositoryAlreadyExists,
                    PermissionDenied, MalformedLocation);
    void open ( in string location, in AccessStatus access,
                in RepInfo info ) raises (NoSuchRepository,
                    RepositoryAlreadyOpen, RepositoryOpenFailed,
                    MalformedLocation);
    void set_info ( in RepInfo info )
                raises (RepositoryNotOpen, MalformedInfo);
    RepInfo get_info ( out RepInfo info_used )
                raises (RepositoryNotOpen, MalformedInfo);
    int get_access_status ( ) raises (RepositoryNotOpen);
    void set_description ( in string description )
                raises (RepositoryNotOpen);
    string get_description ( )
                raises (RepositoryNotOpen, NoDescriptionExists);
    void close ( ) raises (RepositoryNotOpen);
    void delete ( ) raises (RepositoryNotOpen, PermissionDenied,
                    RepositoryAccessedByOtherApplication);
    Object create_object ( in string object_name,
                in Properties property, in ObjectType object_type )
                raises (ObjectAlreadyExists, UnknownObjectType,
                    PermissionDenied);
    Object lookup ( in string object_name )
                raises (NoSuchObject, PermissionDenied);
    Iterator make_object_iterator ( ) raises (RepositoryNotOpen);
    void delete_object ( in string object_name )
                raises (NoSuchObject, PermissionDenied);
    void flush ( ) raises (RepositoryNotOpen, PermissionDenied);
    . . .
};
```

# Posters

# Advanced Communication Optimizations for Data-Parallel Programs*
## (Poster Paper)

Gagan Agrawal

Department of Computer and Information Sciences
University of Delaware Newark DE 19716
agrawal@cis.udel.edu

## 1 Introduction

One of the important steps in compiling data-parallel programs for distributed memory machines is communication analysis and optimizations. The existing research in the area of communication analysis and communication optimization can be classified into three levels:

- *Single Loop Nest Level:* The earliest compilers performed analysis within a single parallel loop, or even a single Fortran 90 array statement [3]. Such communication analysis aggregated the off-processor elements required in the different iterations of the same loop, but did not aggregate messages or avoid redundant communication across loops.
- *Single Procedure Analysis:* Several researchers improved upon the single loop level analysis by performing data flow analysis within a single procedure. The Rice Fortran D compiler performed a number of communication optimizations between loops which were related through simple flow of control [9], like message vectorization and message aggregation. Gupta *et al.* used partial redundancy elimination [6] and greedy analysis on the SSA form [4] to improve the performance of regular applications. Hanxleden proposed a data flow framework called *Give-N-Take* for performing communication placement within a single procedure [8].
- *Interprocedural Analysis:* The only work in the area of flow-sensitive interprocedural analysis for communication has been restricted to performing placement of collective communication calls and communication preprocessing calls, after communication aggregation has been done within single procedures [2]. These techniques consider the communication calls as a pure function, and use the mod-ref information on the parameters of the pure function for performing the placement. One particular limitation of this work has been to consider each array as a single entity, and modification or reference to any element of the array as a modification to the entire array. As part of the Fortran D system in Rice university, flow-insensitive analysis

---
* This research was supported by NSF CAREER award ACI-9733520 and NSF grant CCR-9808522

has been implemented for communication optimizations [7]. This work only applies to programs for which the call graph is a directed acyclic graph (i.e. non-recursive programs).

Almost all real applications comprise of several procedures, and it is important to look for opportunities for communication optimizations across procedure boundaries. In this short paper, we describe our work in the area of aggregating communication across procedure boundaries. One of the components of our research is to summarize the result of local analysis for communication in terms of a novel abstraction, called a *communication loop*. We propagate information about communication through flow-sensitive interprocedural analysis, allowing redundant communication elimination and communication aggregation.

## 2 Communication Loop

All the current compilers for distributed memory machines perform analysis within a single parallel loop for determining the communication requirements [3, 9]. We not propose any specific technique for performing such local analysis, but propose a novel abstraction for summarizing the result of this analysis called *communication loop*.

Previous global and interprocedural frameworks have chosen a range of different abstractions for summarizing local communication requirements. We briefly mention these abstractions and then motivate the requirement for a new abstraction.

In our previous work on interprocedural partial redundancy elimination, we described the result of the local communication analysis by a pure function [2]. Typically, this pure function is a collective communication routine, provided by the underlying runtime communication library. This approach has the advantage of simplicity for later analysis. The main disadvantage of this approach is to treat communication requirement of a single loop as an atomic unit. Therefore, only if the communication requirement of two loops are identical, the communication for the later loop can be eliminated.

In the Give-N-Take framework for communication placement, Hanxleden and Kennedy [8] first generate the SPMD code for each processor, and then explicitly evaluate the off-processor elements required, e.g. a processor owning 25 elements of an array requires elements 26 to 30. In incorporating the intraprocedural partial redundancy framework for communication placement, Gupta et al. have used a new detailed abstraction, called *data availability lattice* [5]. In both these approaches, a lattice can only be calculated after SPMD code for each processor has been generated and the number of processors for parallel execution is known.

In contrast to the previous approaches, we believe that communication analysis should be done before knowing the number of processors on which the code is to be executed. There are two main reasons for this:

- There is an increasing trend to use a network of non-dedicated workstations for executing parallel code. For such a configuration, the compiler needs to generate code which is independent of the number of processors.

- We believe that to the extent possible, phases of data parallel compiler should be independent of the number of processors, to allow easy recompilation to a different number of processors.

With the above considerations, we have defined a new abstraction for communication requirements, which we call a *communication loop*. To motivate this abstraction, consider the following parallel loop, which we denote by L1.

```
L1:   forall( i = 1:N-2:1)
        A(i)  =   B(i) + C(i) + D(i+2)
```

If the arrays A, B and C are identically distributed and if owners compute rule is used for loop partitioning, the local communication analysis can determine that the above loop will not involve any off-processor references to the elements of the arrays B and C. However, off-processor references will be required to the elements of array D. The precise set of off-processor references depends upon the number of processors on which this code is to be executed. We can represent the communication requirement of this loop as follows:

```
communicate( i = 1:N-2:1)
  A(i)  =    D(i+2)
```

The above abstraction will allow us to eliminate communication of the elements of array D, if they are being assigned to the elements of array A in another loop. However, if the identical elements of the array D are being assigned to a different array in another loop, the above abstraction not be adequate for removing such redundancy. Consider, for example, the following loop (denote by L2):

```
L2:   forall( i = 1:N-2:1)
        B(i) = C(i) + D(i+2)
```

If the distribution of the arrays A and B are identical, the communication requirements of the two forall loops will be identical.

To allow better reuse of the off-processor data, we improve our abstraction by using the notion of *templates* from languages like HPF and Fortran-D [9]. Typically, arrays with identical distributions are aligned with the same template. For example,

```
Template T(N)
Distribute T(block)
Align  A, B, C, D with T
forall(i = 1:N-2:1)
  A(i)  =   B(i) + C(i) + D(i+2)
```

We represent the communication associated with this forall loop by the following communication loop:

```
communicate( i = 1:N-2:1)
  T(i)  =    D(i+2)
```

## 3  Interprocedural Propagation

Our interprocedural propagation phase is based upon computing *anticipability* of the sets of communication loops. In our previous work, we have shown how flow-sensitive computation of anticipability can be performed, while preserving the calling context of the procedures [1]. In propagating the communication loops at the earliest possible points, redundant communication can be avoided and aggregation can be performed. After determining the placement of each communication loop at the earliest point it is anticipable, a number of heuristics can be used to avoid unnecessary early placement.

## References

1. Gagan Agrawal, Anurag Acharya, and Joel Saltz. An interprocedural framework for placement of asynchronous I/O operations. In *Proceedings of the 1996 International Conference on Supercomputing*, pages 358–365. ACM Press, May 1996.
2. Gagan Agrawal and Joel Saltz. Interprocedural compilation of irregular applications for distributed memory machines. In *Proceedings Supercomputing '95*. IEEE Computer Society Press, December 1995.
3. Z. Bozkus, A. Choudhary, G. Fox, T. Haupt, S. Ranka, and M.-Y. Wu. Compiling Fortran 90D/HPF for distributed memory MIMD computers. *Journal of Parallel and Distributed Computing*, 21(1):15–26, April 1994.
4. Soumen Chakrabarti, Manish Gupta, and Jong-Deok Choi. Global communication analysis and optimization. In *Proceedings of the SIGPLAN '96 Conference on Programming Language Design and Implementation*, pages 68–78. ACM Press, May 1996. ACM SIGPLAN Notices, Vol. 31, No. 5.
5. Manish Gupta, Edith Schonberg, and Harini Srinivasan. A unified data flow framework for optimizing communication. In *Proceedings of Languages and Compilers for Parallel Computing*, August 1994.
6. S.K.S. Gupta, S.D. Kaushik, C.H. Huang, and P. Sadayappan. On compiling array expressions for efficient execution on distributed memory machines. Technical Report CISRC-4/94-TR19, Ohio State University, April 1994. Don't have this one.
7. M.W. Hall, S. Hiranandani, K. Kennedy, and C.-W. Tseng. Interprocedural compilation of Fortran D for MIMD distributed-memory machines. In *Proceedings Supercomputing '92*, pages 522–534. IEEE Computer Society Press, November 1992.
8. Reinhard von Hanxleden and Ken Kennedy. Give-n-take – a balanced code placement framework. In *Proceedings of the SIGPLAN '94 Conference on Programming Language Design and Implementation*, pages 107–120. ACM Press, June 1994. ACM SIGPLAN Notices, Vol. 29, No. 6.
9. Seema Hiranandani, Ken Kennedy, and Chau-Wen Tseng. Compiling Fortran D for MIMD distributed-memory machines. *Communications of the ACM*, 35(8):66–80, August 1992.

# A Cellular Automata Simulation Environment for Modelling Soil Bioremediation

Scott D. Telford

EPCC,
University of Edinburgh, The King's Buildings, Edinburgh EH9 3JZ, UK.
st@epcc.ed.ac.uk

**Abstract.** This paper describes CAMELot, a parallel cellular automata (CA) simulation environment developed as part of the Esprit project COLOMBO. The objective of this project is to apply parallel computing to the simulation of the in situ bioremediation of contaminated soils through the use of CA models. CAMELot provides a programming environment for the development, interactive control and visualisation of CA simulations.

## 1 Background

The technique of in situ soil bioremediation employs bacteria to degrade chemical contaminants in areas of polluted soil. This is regarded as an environmentally-friendly and inexpensive method compared with other techniques. To predict and evaluate the results of field-scale operations, mathematical models describing the geological, chemical and biological phenomena occurring in the remediation process can be used. However, such models are very complex and computationally expensive. The approach taken in the COLOMBO project is to use a 3D cellular automata (CA) [5] representation of the soil volume which is spatially distributed among the cells of the CA. Use of a CA-based model enables established parallel computing techniques to be applied in order to meet the computational requirements.

## 2 The COLOMBO Project

The Esprit HPCN project COLOMBO (parallel COmputers improve cLean up of sOils by Modelling BiOremediation) is a collaboration between Istituto per la Sistemistica e l'Informatica Consiglio Nazionale delle Ricerche (ISI-CNR), Ente Nazionale per l'Energia e l'Ambiente (ENEA), CRA Montecatini SpA, Università della Calabria, University of Edinburgh, Ironside Farrar Ltd, Umweltschutz Nord GmbH, and Quadrics Supercomputers World Ltd.

COLOMBO follows on from an earlier Esprit project, CABOTO (Cellular Automata for the BiOremoval of TOxic contaminants) [2, 4]. This project, which ran from 1995 to 1996, was a collaboration between several Italian partners, some of which were later to collaborate in COLOMBO. CABOTO used a parallel

CA simulation environment called CAMEL (Cellular Automata environMent for systEms modeLing) [1] developed by CRAI (Consorzio per la Ricerca e le Applicazioni di Informatica) and the Università della Calabria.

CAMEL was implemented on a PC-hosted INMOS transputer system in C and occam 2 using a Windows 3 graphical user interface for model development and simulation control and a dedicated framebuffer for visualisation. CARPET (CellulAR Programming EnvironmenT), an extension of the ANSI C programming language, was used for the implementation of the CA models used in the CAMEL environment.

The COLOMBO project comprises two main activities:

- To extend and enhance the existing bioremediation CA models developed in CABOTO to take account of more bioremediation phenomena and validate the models against pilot plant experiments and full-scale field tests.
- To develop an enhanced successor to the CAMEL software (CAMELot) to run on MPP systems. This is initially targeted at the Quadrics (formerly Meiko) CS-2, using the MPI-1 message-passing standard and an X11-based GUI.

## 3 CAMELot

CAMELot (CAMEL with Open Technology) has been implemented at EPCC following a specification from ENEA and ISI-CNR. Most of the code in CAMELot has been written from scratch, and is intended to be portable to various platforms supporting the MPI-1 message-passing standard and running UNIX-compatible operating systems.

CAMELot comprises three main parts:

- The CARPET compiler, implemented as a C preprocessor
- The GUI, providing a simple integrated development environment for CARPET programs, interactive control of CA simulations and realtime visualisation facilities
- The parallel CA simulation environment "engine" which runs the CA model.

Compared to CAMEL, the following functionality enhancements have been added in CAMELot:

- Extensions to the CARPET language: more data types, optimisations, syntactic improvements.
- Improved visualisation capabilities, including multiple visualisation windows and 3D displays, compared to CAMEL's single dedicated 2D display.

## 4 The CARPET Language

CARPET [3] is an extension of ANSI C intended for the implementation of CA algorithms. CARPET code may be compiled into serial or parallel executables without modification and without explicit reference to parallelism in the source.

A CARPET program consists of two parts:

- Global CA declarations, defined in a `cadef{...}` (CA definition) block. This contains declarations defining the state of cells as a set of substate variable declarations, the dimensionality of the CA, symbolic cell neighbourhood mappings and numeric parameters global to the whole CA. A CA model can also be declared deterministic and a threshold condition (see Sect. 6.1) can be specified in order to optimise the execution of the CA.
- The CA transition function, expressed as free-form code.
  The transition function has a number of intrinsic functions available to update the state of the cell in lock-step with all other cells and to return values such as the cell's $x$, $y$ or $z$ coordinates, the iteration number, etc.
  In addition a special syntactic convention is used in the transition function code to reference substates in the local and neighbourhood cells.

The size of the CA is defined externally to the CARPET source.

## 5 The CAMELot Graphical User Interface

The design of the CAMELot GUI follows that of CAMEL, but has been re-implemented as a Motif/X11 application.

The GUI comprises of three types of windows:

- The *CARPET Development* window. This window provides facilities to edit, preprocess and compile a CARPET program into an MPI application. Various external compile-time parameters (such as CA size and number of processors to run on) can be set from a menu.
- The *Simulation* window. This window provides realtime control of the running CA engine.
- The *Visualisation* window(s). These windows show a colour representation of a specified substate in orthographic or isometric projections. These visualisations are updated in realtime at a user-specified iteration interval.

## 6 The CA Engine

The CA Engine comprises one or more SPMD MPI processes, each of which is responsible for a number of CA cells and are hence termed *macrocells*. A "master" macrocell is responsible for communication with the GUI (via TCP connections using the BSD "socket" API) and synchronisation of the other macrocell processes.

### 6.1 Load-Balancing and Optimisation

Many CA models can have large contiguous regions of insignificant or no activity. To avoid potential load-balancing problems with a simple parallel block decomposition, a one-dimensional block-cyclic decomposition is used to distribute CA cells amongst macrocells. This was also employed in CAMEL.

In addition to this decomposition, at compile-time the user can divide the CA in the $x$ dimension into a number of partitions or *folds*. In this case, each fold is then decomposed in a block-cyclic fashion across the macrocells, hence each macrocell is responsible for a sub-block or *strip* of each fold. A feature inherited from CAMEL is a runtime facility to manually select a range of folds as an active region; cells in folds outside this range are then assumed inactive and do not have their state updated.

An alternative facility provided in CAMELot is automatic inactivity detection. If the CARPET program is declared deterministic, and in some generation all cells in a strip satisfy the threshold condition or all cells (and their neighbourhood cells) have not changed state since the previous generation, then the strip is automatically assumed to be inactive. This provides a safer and finer-grained method of optimising the simulation than the previous manual facility.

## 7 Conclusion

This paper has presented the main features of a CA simulation software tool re-engineered for use on a range of open systems platforms, including MPP supercomputers. Although this package in many respects differs and improves upon its predecessor, CA models from the earlier system may be reused and much of the appearance and behaviour of the previous user interface has been preserved.

## References

[1] S. di Gregorio, R. Rongo, W. Spataro, G. Spezzano, and D. Talia. A parallel cellular tool for interactive modeling and simulation. *IEEE Computational Science and Engineering*, 3(3):33–43, 1996.
[2] G. Spezzano and D. Talia. Programming high performance models of soil contamination by a cellular automata language. In *Proc. HPCN Europe '97*, pages 531–540. Springer-Verlag, April 1997.
[3] G. Spezzano and D. Talia. Designing parallel models of soil contamination by the CARPET language. *Future Generation Computer Systems*, 13:291–302, 1998.
[4] D. Talia. Cellular automata thrive on parallel systems. *Scientific Computing World*, October 1998.
[5] J. von Neumann. *Theory of Self Reproducing Automata*. University of Illinois Press, 1966.

# High Efficient Parallel Computation of Resonant Frequencies of Waveguide Loaded Cavities on JIAJIA Software DSMs [1]

Weisong Shi*    Jifu Ma+ and Zhimin Tang*
*Inst. of Comp. Tech.    +EM Comm. Res. Lab.
Chinese Academy of Sciences Pennsylvania State University

**Abstract.** A new approach of parallel implementation of FDTD method on a novel software DSM (Distributed Shared Memory) system is proposed and implemented in this paper. Two unique advantages of our approach are:(1) it is easy to parallelize sequential FDTD codes, (2) it is possible to run programs which requires a memory space larger than the main memory of a single node.

## 1  Introduction

It has agreed that exploiting parallelism of FDTD (Finite-Difference Time-Domain) is the key to overcome two intrinsic challenges of the FDTD:(1) large memory requirement, and (2) long execution time [1, 3, 4, 5]. Currently, the major parallel programming model is message passing, such as PVM and MPI, which is error-prone. Fortunately, software distributed shared memory (DSM) provides a shared address abstraction on distributed memory systems and ease the programming greatly.

In this paper, we present the parallel computation of the resonant frequency of waveguide loaded cavities with FDTD method on JIAJIA[2] software DSM system. Compared to message passing programming platform, it is very convenient to parallelize sequential FDTD codes with JIAJIA system, and high efficiency is achieved. Also of importance is that the problem fully utilize the large memory space potential provided by JIAJIA system.

## 2  FDTD for Waveguide Loaded Cavities

Figure 1(a) shows the schematic of the addressed problem presented by the cavity with its output waveguide, which is considered to be infinite long, and terminated by A.B.C (Absorbing Boundary Condition) in FDTD grids. The transient wave form of any one of the field components at the observation point in the cavity will be:

$$y_k = x_k + n_k = \sum_{t=1}^{M} b_t Z_t^k + n_k \qquad (1)$$

where $k = 0, 1, \cdots (N-1)$ is the time index, $M$ is the number of modes, $Z_t = e^{(\alpha t + jwt)\Delta t}$ are the poles, $\Delta t$ is the time interval, $w$ is the resonate frequency and $\alpha$ is the damping factor, $b_t$ is the residuals of the noiseless time signal $x$, and $n_k$ indicates additional noise.

When the time domain response has been obtained, the resonant frequencies of the cavity can be easily known by applying Matrix Pencil method, which is a spectrum estimation technique [5] used for estimating poles in the transient response in Equation 1.

---

[1] The work of this paper is supported by the National Climbing Program of China, and the National Natural Science Foundation of China, and the President Young Creation Foundation of the Chinese Academy of Sciences.

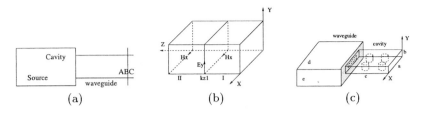

**Fig. 1.** (a) FDTD setup for the waveguide loaded cavity, (b) Space divided with respect to z-axis, and (c) The analyzed structure

## 3 Parallel Implementation

The principle idea of parallel implementation of FDTD algorithm is space decomposition. That is, the computation space is divided into several parts or blocks with field components in each part are stored in local memory and computed by that processor only. In message passing environment, users are required to distribute the data manually and send/receive messages explicitly between adjacent parts when computing the field component located on the block boundary. So the sequential codes should be deeply modified, as demonstrated in following example.

Suppose the problem space is divided into two parts with respect to $z$ direction, as shown in Figure 1(b): Part I (less than $kz1$) and Part II (larger than $kz1$). In message passing model, the field components and coefficients must be represented by two set of variables, such as $Ex1, Ex2$, etc.., and stored in different processors manually. The corresponding computation formula of each field component need to be modified too. For example, the previous sequential iteration formula of $Ey$ for part II is modified from

$$Ey(i,j,kz1) = cey(i,j,kz1) * (-dey(i,j,kz1) * Ey(i,j,kz1) + (Hx(i,j,kz1) - Hx(i,j,kz1-1))/dlz - (Hz(i,j,kz1) - Hz(i-1,j,kz1))/dlx) \qquad (2)$$

to:

$$Ey2(i,j,kz1) = cey2(i,j,kz1)*(-dey2(i,j,kz1)*Ey2(i,j,kz1)+(Hx2(i,j,kz1)-Hx1(i,j,kz1-1))/dlz-(Hz2(i,j,kz1)-Hz2(i-1,j,kz1))/dlx) \qquad (3)$$

In Equation (3), the item $Hx1(i,j,kz1-1)$ should be communicated from the adjacent processor.

It is of importance that the above requirement of storing the data of each block in a separate processor and accordingly communicating strategies are not needed in our parallelization of FDTD codes, which is the substantial advantage of our approach to those previous ones. In JIAJIA, what the programmer need to do is just the same as the case in sequential computer, regardless of the physically distributed memories presented on each of workstations. What we do, for the example of two parts problem in Figure 1(b), is just to replace the origin loop $do \quad k = 0, nz$ with $do \quad k = begin, end$, where $begin$ and $end$ have different values for different processors. The source codes do not change at all.

## 4 Numerical Results

The experimental work of this paper is implemented on Dawning 1000A Cluster system designed by National Center of Intelligent Computing Systems of China.

Currently it includes 8 PowerPC-based nodes connected by 100 Mbps switched Ethernet, each node has 256 MB main memory. JIAJIA system is implemented on Dawning 1000A. The analyzed structure, which consists of a re-entry cavity and its output waveguide, is shown in Figure 1(c).

Firstly the fixed speedup $S_f$ is used to evaluate our implementation, which is a measure of the speedup of a fixed size problem as the number of processors is increased. Then the scaled speedup $S_s$, which is the measure of the speedup of a fixed load per processor, is used for further evaluation of our implementation. The definition of these two speedups are $S_f = T_1/T_n, S_S = n \times T_1/T_n$ respectively, where $T_1$ is the time of single processor running, $T_n$ is the time of $n$ processors.

For simplicity, we compute the same problem in three different scales in terms of gridding space: small ($60 \times 30 \times 208$), middle ($120 \times 60 \times 416$) and large ($240 \times 60 \times 832$). Table 1 shows the running time (for 6000 iteration steps), fixed speedup and memory requirement for three scales respectively.

Keep the computation load in each processor be fixed ($120 \times 60 \times 208$), we measure the scaled speedup. The results are shown in Table 2 together with the running time (also for 6000 iteration steps).

**Table 1.** Execution time, fixed speedup($S_f$) and memory requirement for different scales

| Scale | Seq. time(s) | 4 way /speedup | 8 way /speedup | Memory(MB) |
|---|---|---|---|---|
| $60 \times 30 \times 208$ | 6963.08 | 3410.09/2.09 | 3600.00/1.93 | 20 |
| $120 \times 60 \times 416$ | 44710.66 | 19200.00/2.33 | 10920.01/4.09 | 160 |
| $240 \times 60 \times 832$ | —— | 614760.00/1.00 | 45744.02/13.38 | 660 |

From Table 1 and Table 2, it can be concluded that:

1. For small-scale problem, 2.09 of fixed speedup is achieved in 4-way parallelism. The degradation of fixed speedup with 8-way parallelism is due to the low computation to communication ratio, which effects the parallel efficiency greatly. When the scale of the problem is increased, the speedup is increased too. For the middle scale problem, 2.33 speedup on 4-processors is achieved, which is greater than reported in [6], and slightly less than reported in [17]. Since the speedup is closely related to the problem space, we conclude that our speedup is comparable to that obtained in PVM. We ascribe the main reason to the flexible memory allocation scheme supplied by JIAJIA system, which exploits the potential of memory locality greatly.

2. As we stated in Section 3, JIAJIA system can supply large memory space, which is helpful for solving large scale problems. In Table 1, for the largest scale $240 \times 60 \times 832$ problem, 660 MB memory space is required which prevents the successful running of sequential code on single processor. Therefore, we measure the results on 4-way and 8-way parallelism only. The figure in the table shows that the speedup from 4 to 8 processors is superlinear (13.4 compared to ideal value 2). We ascribe the main reason to the large memory space required by this scale problem. When 4 processors are used, nearly 220 MB (86% of total main memory) memory space are allocated on each node, causing data "ping-pong" frequently between main memory and hard

disk, which will affect it's performance greatly. However, only 43% of main memory space is needed when 8 processors are used.
3. Comparison between Table 1 and Table 2 shows that scaled speedup is more desirable for evaluate the parallel performance. Though the parallelizing effort is reduced greatly compared with others work, our results are comparable to those obtained in [7], which is implemented in PVM environment.

**Table 2.** Execution time, scaled speedup($S_s$) for problem scale $120 \times 60 \times 208$

| Processors | 1 | 2 | 4 | 8 |
|---|---|---|---|---|
| Time(Sec.) | 21661.80 | 23754.60 | 24384.00 | 28650.03 |
| Scaled speedup | 1.00 | 1.82 | 3.55 | 6.05 |

## 5 Conclusions

Aimed at overcoming the two challenges of the large memory requirement and long running time in FDTD simulation of practical EM structures, a parallel implementation of FDTD codes on JIAJIA software DSM system is proposed and implemented. The method is applied for the parallel computation of the resonant frequency of waveguide loaded cavities, and good results are obtained. Performance analysis shows that the speedup we obtained is comparable to those obtained in message passing systems.

**Acknowledgement:** The allocation of computer time from the National Center of Intelligent Computing Systems (NCIC) of China is gratefully acknowledged.

## References

1. Z. Bi, Y. Shen, K. Wu, and J. Litva. Fast finite-difference time domain analysis of resonators using digital filtering and spectrum estimation techniques. *IEEE Trans. Microwave Theory Techniques*, 39(8):1611–1619, August 1992.
2. W. Hu, W. Shi, and Z. Tang. Jiajia: An svm system based on a new cache coherence protocol. In *Proc. of the High Performance Computing and Networking Europe 1999 (HPCN'99)*, April 1999.
3. E. K. Miller. Solving bigger problems by decreasing the operation count and increasing the computation bandwidth. *Proceedings of IEEE Special Issue of Electromagnetics*, 79(10):1493–1504, October 1991.
4. J. A. Pereda, L. A. Vielva, A. Vegas, and A. Prieto. Computation of resonant frequencies and quality factors of open dielectric resonators by a combination of the fdtd and prony's method. *IEEE Microwave Guided Wave Letters*, 2(11):431–433, November 1992.
5. J. Ritter and F. Arndt. Efficient fdtd/matrix-pencil method for the full-wave scattering parameter analysis of waveguiding structures. *IEEE Trans. Microwave Theory Techniques*, 44(10):343–357, October 1997.
6. D. P. Rodohan and S. R. Scuinders. Parallel implementations of the finite-difference time- domain (fdtd) method. In *Proceedings of the 2nd Int. Conference Computation in Electromagnetics*, pages 367–370, April 1994.
7. V. Varadarajan and R. Mittra. Finite-difference time-domain analysis using distributed computing. *IEEE Microwave Guided Wave Letters*, 4(5):144–145, May 1994.

# A Study of Parallel Image Processing in a Distributed Processing Environment

Michio Iikura, Kenichi Kobayashi, Tohru Yoshioka, and Seijiro Ihara

Nippon Institute of Technology, 4-1 Gakuendai, Miyashiro, Saitama 345, Japan
{iikura, s969008, yoshioka, sihara}@nit.ac.jp,
http://www.nit.ac.jp

**Abstract.** Recent performance enhancements, upgrade applications, and rapid advances in computers and networking technologies have all enabled information processing devices to share computer resources. We are interested in image processing that requires a large amount of calculations. If individual pixels forming an image can be calculated independently, the processing speed can be increased by parallel processing, thus reducing the overall processing time. However, it has been previously confirmed that increasing the number of servers beyond a certain point does not raise the parallel processing efficiency in a parallel distributed processing environment using a client-server model. Therefore, we constructed a new processing environment using an agent model in order to investigate parallel image processing. This paper discusses the results of experiments in this environment in comparison with those of parallel processing in a client-server model.

## 1 Introduction

Recently, information processing devices are connected through high-speed networks to form large information processing systems. This kind of networking has resulted in the creation of a new information processing environment. In this information processing environment, computer resources are interoperable, and thus numerous resources distributed on a network can be used simultaneously. Furthermore, if parallel processing is permitted, the simultaneous use of computer resources distributed on a network reduces the processing time.

We are faced with the problem of handling the enormous amount of calculations needed for visualizing Mandelbrot sets and generating images by ray-tracing methods. To reduce the long calculation time required for image generation, we have attempted to carry out processing in the parallel distributed processing environment by Parallel Virtual Machine (PVM) [1]. However, it has been confirmed that there is a limitation to the reduction of processing time in the environment using the client-server model. To overcome this limit, we constructed an agent-based parallel processing environment in order to investigate parallel image processing. The present paper discusses various problems in parallel distributed processing with respect to image generation.

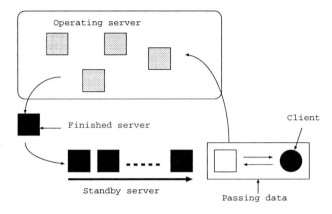

**Fig. 1.** Queue in the Client-Server model

## 2 Parallel Distributed Image Generation Processing in the Client-Server Model

For parallel distributed processing in the client-server model, a client sends calculation area information sequentially to servers and the servers calculate color intensities in the specified areas. After finishing the assigned calculations, the servers transfer the calculation results to the client and receive new calculation area information to restart calculation. This series of processing steps (dynamic load distribution method) can be approximated to the Machine Interference model in queuing theory [2]. In Fig. 1, the servers are machines and the client is an operative. The end of calculation at a server means a machine failure. To repair the machine by operative, the client a receives a calculation result from the server and returns the next calculation instruction to the server. If calculation end reports from servers converge, a server queue will be formed. Since servers waiting in the queue are not actively carrying out calculation, the parallel processing efficiency of the system decreases. If the ratio of the average server processing time to the average client processing time is assumed to be constant, then the greater the number of servers, the longer will be the average queuing length. Therefore, a server increase cannot be expected theoretically to raise the parallel processing efficiency of parallel distributed processing in the client-server model.

## 3 Parallel Distributed Image Generation Processing in the Agent Model

We confirmed that the volume of processing at the client increases with the number of servers used for parallel distributed processing in the client-server model. This means that there is a limitation to the reduction of processing time that can be achieved by parallel processing. To overcome this limit, we constructed an agent-based image processing environment in the form of a parallel distributed processing system and evaluated its parallel processing performance. Here, an agent is defined as an autonomous computing entity for processing computer resources on a LAN and has the following characteristics.

1. The agent status is either idle or busy.
2. When a job is received, the agent becomes busy and looks for an idle agent before starting the job. If an idle agent is found, the busy agent passes half of its job to the idle agent. The busy agent starts the job only after it has confirmed that there are no more idle agents.
3. Once the job has been finished, the agent broadcasts a job-end message and becomes idle.
4. On receiving this job-end message, one of the busy agents passes half of its non-processed job to the idle agent.

An agent in this agent-based processing environment is understood to mean a computing entity having both client and server functions in the client-server model and that this entity then becomes a client or server depending on the circumstances. Unlike the conventional client-server model, the client-server relationship is not fixed statically but changes dynamically.

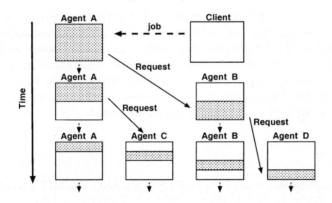

**Fig. 2.** Concept of agent model

Fig. 2 shows the concept of parallel processing using image generation agents. The job given to the first agent is assigned sequentially to other agents. Fig. 3 compares generation of the same image carried out using this method and the client-server model. As the volume of parallel processing increases, this method showed greater processing performance than parallel processing by the client-server model. From the viewpoint of queuing theory, the average number of operating machines under heavy traffic load is greater for parallel distributed processing in the agent model than that in the client-server model.

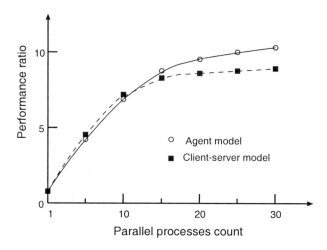

**Fig. 3.** Comparison of performance ratios between the agent model and Client-server model

## 4 Conclusion

We clarified problems related to parallel distributed processing in the client-server model and constructed a new parallel distributed processing environment using an agent model for experimental evaluation. A comparison and discussion of the experimental results confirmed that parallel distributed processing in the agent model was superior to that in the client-server model. In further experimental and theoretical studies, we will construct and analyze a numeric model for parallel distributed processing in the agent model.

## References

1. Al Geist, A.Beguelin, et al.: Parallel Virtual Machine. The MIT Press.(1994)
2. D.R. Cox, Walter L. Smith.: Queues. Butler & Tanner Ltd.(1961)

# Data Prefetching for Digital Alpha

S. Manoharan

Department of Computer Science, University of Auckland, New Zealand.
mano@cs.auckland.ac.nz

**Abstract.** Some of the current microprocessors provide a prefetch instruction, but either the instruction is treated as a NOP, (e.g. Digital Alpha EV4/5), or only a small number of outstanding prefetches is permitted (e.g. MIPS R10K). This paper discusses the design and implementation of the hardware support required to fully support the prefetch instruction for the Digital Alpha architecture. The prefetch support is implemented in a cycle-level functional simulator of the Alpha architecture.

## 1 Introduction

There are two major prefetching schemes. In a hardware scheme the hardware predicts the memory access pattern and brings data into the cache before required by the processor [1, 2]. In a software scheme, the compiler predicts the memory access pattern and places prefetch instructions into the code [6, 4, 5, 3]. Software prefetching, however, requires some hardware support, such as the provision and implementation of a prefetch instruction.

This paper investigates the hardware support required by software prefetching. It bases the investigation on the Digital Alpha EV4 microprocessor [7]. We implement the hardware support for software prefetching in a cycle-level simulator for the Alpha architecture.

## 2 Prefetching under the Alpha Architecture

The FETCH instruction in the Alpha architecture optionally prefetches an aligned 512-byte block surrounding the address specified by FETCH. The instruction is only a hint to the implementation that may allow faster execution [7]. Indeed, the EV4 and EV5 implementations of the Alpha architecture simply ignore this hint, treating the FETCH as a NOP. The programming model of the Alpha architecture specifies that at any instant, there can be at most two outstanding prefetches.

The organization of the Alpha EV4 processor, pertaining to the FETCH instruction, is illustrated in Figure 1. The data cache is a write-through, direct mapped, read allocate cache. It contains 8 KB, organized into 32-byte lines. The write buffer has four entries, each of which has storage for up to 32 bytes. In addition to store instructions, memory barrier (MB) and FETCH instructions go through the write buffer. The buffer has a head and tail pointer. Commands normally enter at the tail end, and exit at the head end.

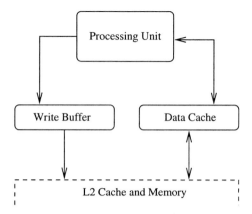

**Fig. 1.** Organization of the Alpha EV4, pertaining to prefetch

## 3 Hardware Support for Prefetching

The main component of the prefetching support is the prefetch issue buffer. It maintains the state of the outstanding prefetches. The prefetched data is placed in the L2 cache. Figure 2 illustrates the placement of the prefetch issue buffer with respect to the processing unit, write buffer and caches.

Prefetch and store instructions follow the same path through to the write buffer. A prefetch is always added to the tail end of the write buffer. A store is also added to the tail end, unless there is already in the write buffer another store to the same block; in this case, the two stores are merged; however, if there is a memory barrier between the two stores, the second store is placed at the tail end.

The prefetch issue buffer keeps a prefetch and the information associated with it until the data corresponding to the prefetch is obtained from the memory and placed into the L2 cache. Rather than prefetching 512 bytes as defined by the Alpha architecture, our implementation prefetches data in terms of L2 cache lines. The number of cache lines to be prefetched, called the prefetch block size, is a configuration parameter, and defaults to one line. The reason for not fixing the prefetch block size to 512 bytes is that prefetching 512 bytes may overly stress the memory system, thereby offsetting the advantages of prefetching. A prefetch thus brings from the memory an L2 cache line (or a number of L2 cache lines, as per the configuration) surrounding the address specified by the prefetch.

A prefetch is placed in the prefetch issue buffer only if the line specified by the prefetch address is not in the L2 cache; otherwise the prefetch request is simply dropped. A prefetch request can also be dropped if the prefetch issue buffer is full (if we do not drop the prefetch, the processor needs to stall until there is room in the prefetch issue buffer; this is not likely to be a performance gain [3]).

Each entry in the prefetch issue buffer holds the prefetch address, a state, the number of words that are requested from memory, and the number of words

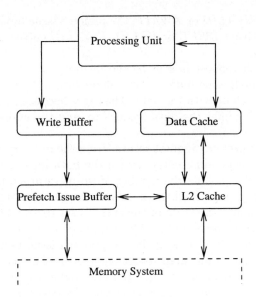

**Fig. 2.** Placement of the prefetch issue buffer

that are returned from memory (and written to the L2 cache). The state of an entry is one of the following:

1. *VALID*: the entry is valid, but no request to memory for this entry has yet been sent,
2. *ISSUE*: some requests to memory for this entry have been sent,
3. *AWAIT*: all memory requests for this entry have been sent and not all the corresponding data from memory have returned,
4. *INVALID*: the entry is not valid.

The prefetch issue buffer, like the write buffer, has a head and tail pointer. Prefetches enter at the tail end, and exit at the head end. At placement, a prefetch entry's state is set to *VALID*.

Prefetches, being not critical, have less priority than any other memory transaction. The prefetch issue buffer sends a memory request only if the L2 cache is not using the memory port. When the prefetch issue buffer sees that the memory port is not busy, it checks up the state of the head entry.

If the entry is *VALID*, it changes this state to *ISSUE*, and a request for the first word of the prefetch block specified by the prefetch address is sent to the memory. The counter that counts the number of words requested from memory is incremented.

If the entry is *ISSUE*, a request for the next word in the prefetch block is sent to the memory. The counter that counts the number of words requested from memory is incremented. If the counter value has reached the number of words in a prefetch block, then the counter is reset, and the state of the entry is set to *AWAIT*.

If the entry is *INVALID* or *AWAIT*, the prefetch issue buffer will look at the next entry that is either *ISSUE* or *VALID*. *INVALID* entries at the head of the buffer are removed.

When a prefetch request is sent to the memory, a tag which identifies its index into the prefetch issue buffer is sent along with it. This tag is returned by the memory system along with the data. The tag is necessary to find which entry in the prefetch issue buffer the data belongs to, and the address corresponding to this data.

When data for a prefetch request returns from the memory, the prefetch issue buffer finds the corresponding entry, writes the data into L2 cache, and increments the counter that counts the number of words returned from the memory. If the counter value has reached the number of words in a prefetch block, then the counter is reset, and the state of the entry is set to *INVALID*.

The read or write misses in the L2 cache, before being sent down to the memory system, result in a look up in the prefetch issue buffer to see if there exists an address that would map to the same cache line as the miss address. If such an address does not exist, then the miss request is processed as normal by the L2 cache. If the address does exist and the entry is *VALID*, this entry in the prefetch issue buffer is invalidated, and a miss request is initiated; but if the entry is either *ISSUE* or *AWAIT*, then the L2 cache stalls until the entry is cleared from the prefetch issue buffer.

## 4 Summary

This paper outlined the design and implementation of a prefetch issue buffer to support software prefetching on the Digital Alpha architecture. Experiments using this prefetching mechanism is the subject of another paper.

## References

1. T.-F. Chen and J.-L. Baer. Reducing memory latency via non-blocking and prefetching caches. In *Fifth International Conference on Architectural Support for Programming Languages and Operating Systems*, pages 51–61. October 1992.
2. Tien-Fu Chen. *Data Prefetching for High-Performance Processors*. PhD thesis, University of Washington, July 1993.
3. Todd Mowry. *Tolerating Latency through Software-Controlled Data Prefetching*. PhD thesis, Stanford University, March 1994.
4. Todd Mowry and Anoop Gupta. Tolerating latency through software-controlled prefetching in shared-memory multiprocessors. *Journal of Parallel and Distributed Computing*, 12(2):87–106, June 1991.
5. Todd Mowry, Monica Lam, and Anoop Gupta. Design and evaluation of a compiler algorithm for prefetching. In *Proceedings of the Fifth International Conference on Architectural Support for Programming Languages and Operating Systems*, pages 62–73. October 1992.
6. Allan Porterfield. *Software Methods for Improvement of Cache Performance on Supercomputer Applications*. PhD thesis, Rice University, May 1989.
7. Richard Sites, editor. *Alpha Architecture Reference Manual*. Digital Press, 1992.

# Circuit-Switched Broadcast in Multi-port 2D Tori*

San-Yuan Wang, Yu-Chee Tseng, Sze-Yao Ni, and Jang-Ping Sheu

Department of Computer Science and Information Engineering
National Central University, Chung-Li, 32054, Taiwan

**Abstract.** This paper studies the *one-to-all broadcast* in a *circuit-switched* 2D torus of any size with $\alpha$-*port* capability. This is a generalization of the one-port and all-port models. Existing results, as compared to ours, can only solve very restricted sizes of tori, and use more numbers of steps.

## 1 Introduction

One primary communication in an interconnection network is the *one-to-all broadcast*, where a source node needs to send a message to all other nodes in the network. This paper studies the scheduling of one-to-all broadcast in a 2D circuit-switched torus. The network is assumed to use the $\alpha$-*port* communication model, in which a node can send up to $\alpha$ messages and simultaneously receive up to $\alpha$ messages at a time, where $1 \leq \alpha \leq 4$. This is a generalization of the *one-port* model ($\alpha = 1$) and the *all-port* model ($\alpha = 2k$). Following the formulation in many works [2, 3, 4, 5], this is achieved by a sequence of *steps*, where a step consists of a set of congestion-free communication paths each indicating a message delivery. The goal is to minimize the total number of steps used.

One-to-all broadcast has been studied for meshes and torus based on different port models and switching models [1, 2, 3, 4, 5, 6]. Some of these results are compared and summarized in Table 1. Our result improves over existing results [1, 2, 4] in both the number of broadcast steps used and its applicability in network sizes.

A 2D torus of size $n_1 \times n_2$ is denoted as $T_{n_1 \times n_2}$. Each node is denoted as $p_{x_1,x_2}$, $0 \leq x_i < n_i$. The torus is assumed to have $\alpha$ ports, in that a node can send up to $\alpha$ messages along any $\alpha$ of its 4 ports.

We map $T_{n \times n}$ into a modulo Euclidean integer space $\mathbb{Z}^2$, where $\mathbb{Z} = \{0, \ldots, n-1\}$. Let elementary vectors $e_1 = (1,0), e_2 = (0,1), e_{-i} = (-1,0), e_{-i} = (0,-1)$. For simplicity, we may write $e_{1,2} = e_1 + e_2$ and $e_{1,-2} = e_1 - e_2$.

**Lemma 1.** *In a 2D $\alpha$-port torus $T_{n_1 \times n_2}$, a lower bound on the number of steps to perform one-to-all broadcast is $\lceil \log_{\alpha+1} n_1 n_2 \rceil$.*

**Definition 1.** *In $\mathbb{Z}^2$, given a node $x$, an $m$-tuple of vectors $B = (b_1, b_2, \ldots, b_m)$, and an $m$-tuple of integers $N = (n_1, n_2, \ldots, n_m)$, we define the* span *of $x$ by vectors $B$ and distances $N$ as a set of nodes* $\mathrm{SPAN}(x, B, N) = \{x + \sum_{i=1}^{m} a_i b_i | 0 \leq a_i < n_i\}$.

---
* This work is supported by the National Science Council of the Republic of China under Grant #NSC88-2213-E-008-014 and #NSC88-2213-E-008-027.

**Table 1.** Comparison on the solvable network sizes and required numbers of steps, assuming a $T_{n_1 \times n_2}$. ($\text{LB}(2)_\alpha$ = the lower bound for $\alpha$-port 2D tori in Lemma 1)

| Algorithm | Ours | Lee-Lee[1] | Park-Choi[2] | Peters[4] |
|---|---|---|---|---|
| Port Model | $\alpha$-port | $\alpha$-port | all-port | all-port |
| Size | $n_1 \times n_2$ | $n_1 = n_2$ | $n_1 = n_2 = 5^p$ | $n_1 = n_2 = 5^p$ or $2 \times 5^p$ |
| Steps | $\begin{cases} \text{LB}(2)_\alpha + 1 & \text{if } n_1 = n_2, \\ \text{LB}(2)_\alpha + 4 & \text{otherwise} \end{cases}$ | $\text{LB}(2)_\alpha + 2$ | $\text{LB}(2)_4$ | $\text{LB}(2)_4$ |

## 2 Broadcasting in a Square 2-D Torus

**When $\alpha = 4$.** The approach is similar to that in [6] except that there is more flexibility as circuit switching is used. There are two stages to achieve broadcast. In the first stage, we recursively spread the broadcast message to the main diagonal of the torus, as shown in Fig. 1(a). In the second stage, we recursively distribute the broadcast message to other diagonals parallel to the main diagonal. This is illustrated in Fig. 1(b).

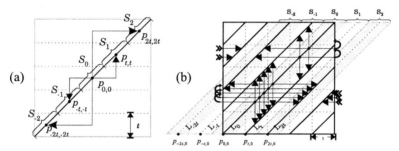

**Fig.1.** Broadcasting in a square 2-D torus: (a) stage 1 and (b) stage 2.

**When $\alpha \leq 3$.** The recursion goes is a slower manner. For instance, in stage 2, the number of diagonals getting the broadcast message will be multiplied by $\alpha + 1$ after each step. The modification is straight-forward.

**Theorem 1.** *In a circuit-switched $\alpha$-port $T_{n \times n}$ torus, broadcast can be done in $2\lceil \log_{\alpha+1} n \rceil$ steps, which number of steps is at most 1 steps more than the lower-bound Lemma 1.*

## 3 Broadcasting in a Non-square 2D Torus

**When $\alpha = 4$.** This case can be solved by modifying the result in [6] for circuit switching. In an all-port $T_{n_1 \times n_2}$ such that $n_1 < n_2$, broadcast can be done within $\lceil \log_5 n_1 \rceil + \lceil \log_5 \frac{n_1}{2} \rceil + \lceil \log_5 \frac{n_2}{n_1} \rceil + c$ steps, where $c = 1$ (resp., 2) when $n_1$ is even (resp. odd), which number of steps is at most 3 (resp. 4) more than the lower bound in Lemma 1.

**When $\alpha = 3$.** We assume that $n_1$ is even, with the understanding that one more step is sufficient if $n_1$ is odd. To avoid the tedium of using floor and ceiling functions, we assume that $n_2$ is a multiple of $n_1$.

**Definition 2.** *Given a non-square torus $T_{n_1 \times n_2}$ such that $n_1 < n_2$, the dilated torus induced by $T_{n_1 \times n_2}$, denoted as $\hat{T}_{n_1 \times n_2}$, is an $n_1 \times n_1$ torus consisting of nodes from the following four $\frac{n_1}{2} \times \frac{n_1}{2}$ tori:*

$$T_{0,0} = \text{SPAN}(p_{0,0}, B_2, N_2), \quad T_{1,0} = \text{SPAN}(p_{1,0}, B_2, N_2),$$
$$T_{0,1} = \text{SPAN}(p_{0,\frac{4}{3}\frac{n_2}{n_1}}, B_2, N_2), T_{1,1} = \text{SPAN}(p_{1,\frac{4}{3}\frac{n_2}{n_1}}, B_2, N_2),$$

*where $B_2 = (2e_1, \frac{2n_2}{n_1}e)$ and $N_2 = (\frac{n_1}{2}, \frac{n_1}{2})$. $\hat{T}_{n_1 \times n_2}$ has $n_1^2$ nodes which are denoted by $\hat{p}_{i,j}$, for $i, j = 0..n_1 - 1$.*

Intuitively, the dilated torus in Definition 2 is partitioned into four sub-tori. For instance, Fig. 2(a) shows the four dilated tori in an $n_1 \times n_2$ torus ($n_1 = 6$). Note that we now do not have "straight" diagonals as in the earlier cases.

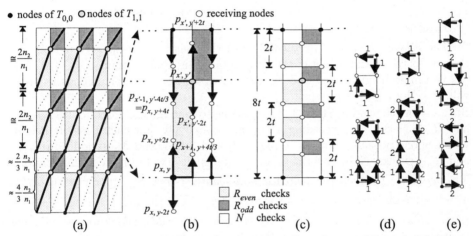

**Fig. 2.** (a) A 3-port $T_{n_1 \times n_2}$ regarded as four dilated $\frac{n_1}{2} \times \frac{n_1}{2}$ tori ($n_1 = 6$). The lines in bold are the alternative diagonals used in Stage 2. (b) The communication pattern in stage 3. (c) The new checkerboard with four smaller rectangles after (b). (d) Stage 4 for $R_{even}$ of heights 1,2,3,and 4. (e) $R_{odd}$ of heights 1, 2, and 3.

**Stage 1:** Spread $M$ to $\text{SPAN}(\hat{p}_{0,0}, (e_{1,2}), (n_1))$, by applying the stage 1 in Section 2. This takes $\lceil \log_4 n_1 \rceil$ steps.

**Stage 2:** Spread $M$ to $\frac{n_1}{2}$ diagonals, $\text{SPAN}(\hat{p}_{2i,0}, (e_{1,2}), (n_1)), 0..\frac{n_1}{2} - 1$, by applying the stage 2 of Section 2. This takes $\lceil \log_4 \frac{n_1}{2} \rceil$ steps. Now, nodes of $T_{0,0}$ and $T_{1,1}$ already have $M$.

**Stage 3:** We first regard the torus $T_{n_1 \times n_2}$ as a checkerboard and classify checks therein as follows (see Fig. 2(a) for an example).

**Definition 3.** *[6] In $T_{n_1 \times n_2}$, each smallest submesh in which the lower-left and upper-right corner nodes are the only two nodes that have received the broadcast message is regarded as a check marked by R (received). Excluding the R-marked checks, the rest of the checkerboards are considered as a number of checks marked by N (non-received).*

**Definition 4.** *A check marked by R is classified as $R_{even}$ if its lower-left node's index along the x-axis is even, and classified as $R_{odd}$ otherwise.*

The recursion should proceed as long as the sum of the heights of two consecutive $R_{even}$ and $R_{odd}$ is $\geq 8$. Let's consider the four consecutive checks in Fig. 2(b). For ease of presentation, let the height $h$ of the rectangle be a multiple of eight, $h = 8t$. We perform the following communications:

- for each $p_{x,y}$ located at the lower-left corner of a $R_{even}$-check, $p_{x,y}$ sends three messages to $p_{x,y+2t}$, $p_{x,y-2t}$, and $p_{x+1,y+\frac{4}{3}t}$, and
- for each $p_{x,y}$ located a the lower-left corner of a $R_{odd}$-check, $p_{x,y}$ sends three messages to nodes $p_{x,y+2t}$, $p_{x,y-2t}$, and $p_{x-1,y+\frac{4}{3}t}$.

After this step, the rectangle will be partitioned into 4 smaller rectangles as shown in Fig. 2(c). The recursion maintains an important invariant:

**I1:** The ratio of the height of $R_{even}$-checks and the height of $R_{odd}$-checks is (or close to) 2 : 1.

The above recursion is repeated until the height of every rectangle is less than 8. As the initial height of the first rectangle is upper-bounded by $\lceil \frac{2n_2}{n_1} \rceil$ and the rectangle height is reduced by a factor of 4 after each recursion, this stage will take $\lceil \log_4 \frac{2n_2}{n_1} \rceil - 1$ steps.

**Stage 4:** At the end of Stage 3, it is possible to manage the height of each $R_{even}$ and $R_{odd}$ checkd not exceeding 4 and 3, respectively. For each possible height, we show one possible solution in Fig. 2(d) and (e) to send $M$ to nodes in $R_{even}$ and $R_{odd}$ checks. Note how the 3-port model is observed in the communication.

**Theorem 2.** *In a circuit-switched non-square 3-port $T_{n_1 \times n_2}$ torus such that $n_1 < n_2$, broadcast can be done within $\lceil \log_4 n_1 \rceil + \lceil \log_4 \frac{n_1}{2} \rceil + \lceil \log_4 \frac{2n_2}{n_1} \rceil + c$ steps, where $c = 1$ (resp. 2) when $n_1$ is even (resp. odd), which number of steps is at most 4 (resp. 5) steps more than the lower bound in Lemma 1.*

**When $\alpha = 1$ and $= 2$.** As our scheme follows a dimension-by-dimension approach, a simple recursive doubling/tripling on rows and columns will do the job.

## References

[1] S.-K. Lee and J.-Y. Lee. Optimal broadcast in $\alpha$-port wormhole-routed mesh networks. In *Int'l Conf. on Paral. and Distrib. Sys.*, pages 109–114, 1997.
[2] J. L. Park and H.-A. Choi. Circuit-switched broadcasting in tori and meshes networks. *IEEE Trans. on Paral. and Distrib. Sys.*, 7(2):184–190, Feb. 1996.
[3] J. L. Park et al. Circuit-switched broadcasting in $d$-dimensional torus and mesh networks. In *Int'l Parallel Processing Symp.*, pages 26–29, 1994.
[4] J. G. Peters and M. Syska. Circuit-switched broadcasting in torus networks. *IEEE Trans. on Paral. and Distrib. Sys.*, 7(3):246–255, March 1996.
[5] Y.-J. Tsai and P. K. McKinley. A broadcasting algorithm for all-port wormhole-routed torus networks. *IEEE Trans. on PDS*, 7(8):876–885, Aug. 1996.
[6] Y.-C. Tseng. A dilated-diagonal-based scheme for broadcast in a wormhole-routed 2d torus. *IEEE Trans. on Comput.*, 46:947–952, Aug. 1997.

# A Distributed Algorithm for the Estimation of Average Switching Activity in Combinational Circuits

Sandeep Koranne

VLSI Design Tools and Technology,
Indian Institute of Technology,
New Delhi, India.
skoranne@giasdla.vsnl.net.in

**Abstract.** We address the problem of estimating the average switching activity in combinational digital CMOS circuits under random input sequences. We have used a probabilistic approach and have given a distributed implementation using PVM. We make use of the Parker-McCluskey heuristic and use bit-vector representation of Signal Probability Expressions to arrive at an efficient implementation. Experimental results on ISCAS '89 circuits show that reliable results can be obtained in considerable less time than required by exhaustive simulation.

## 1 Introduction

Estimation of switching activity has generated a lot of interest not only in the context of power analysis, but also more recently, due to the emergence of a similar problem, that of *crosstalk analysis* [4]. The methods to calculate the switching activity can be divided into two broad classes, *dynamic* and *static* [5]. Dynamic techniques explicitly simulate the circuit under a "typical" input stream. Since their results depend on the simulated sequence, the required number of simulated vectors is usually high. A few yeas ago, the static techniques came into the picture and demonstrated their usefulness by providing sufficient accuracy with low computational overhead. Based on some realistic assumptions about the transistor-level behavior, the problem of estimating switching activity in combinational circuits can be reduced to one of computing signal probabilities of a multilevel circuit derived from the original circuit. For the purpose of this study we have used the approximation suggested by [7], in that, we assume the *switching activity* of a net to be defined as :

$$sw(x) = 2\, p\, (1-p) \qquad (1)$$

Where $p$ is the *static signal probability* of net $x$. In this paper we describe the implementation of SAE, a Switching Activity Estimation Tool, which makes use of distributed computing in the form of PVM [3], to estimate the average switching activity in large combinational circuit blocks.

The rest of the paper is organized as follows. In the next section we describe an efficient method to estimate the static signal probabilities of combinational circuits, keeping in mind the structural correlations arising due to reconvergent fanout. In Sec. 2.3 we give the result of our algorithm on ISCAS '89 benchmark circuits. In Sec. 3 we give our conclusions.

## 2 Implementation of the Parker-McCluskey Heuristic

We want to propagate the probabilities of the primary inputs in a breadth first fashion taking into account the error due to structural correlations due to reconvergent fanout using the Parker-McCluskey (PM) heuristic [6]. Using the PM approximation, we replace any literal of the form $p_x^k$ by $p_x$ in the *signal probability expression*. We have written a class in C++, which does these operations using bit-vector representation.

### 2.1 SAE: Switching Activity Estimator

In this section we describe the implementation of our scheme. The main routine is called SAE, this is responsible for propagating static signal probabilities of the primary inputs.

We give the pseudo-code for SAE below :

```
1. procedure SAE;
2. Input:  Connectivity information of the circuit
3.         Primary Input Static Signal Probabilities
4.         Cell Data Base
5. Output: TRUE  // If all probabilities are computable
6. begin
7.        // Phase I compute the direct probabilities
8.        while not end of connectivity file do in parallel on expr.
9.            Read Connectivity File;    // A=B+C
10.           if RHS known then
11.               Do Computation from Cell Data Base
12.               Update LHS end if;  // Now A is also known
13.           else
14.             Add Expression to Update List; counter++;
15.       end parallel while;
16.       // Phase II Operations on Update List.
17.       ptr=Update List;
18.       while counter > 0 do in parallel on ptr
19.           Do Computation ptr;
20.           if successful counter--; end if;
21.           ptr=ptr->next;
22.       end parallel while;
23. return TRUE;
24. end
```

## 2.2 Analysis of SAE

In line 8 and 18 of SAE we have made use of the while in parallel on x construct. In the NESL [1] language, this construct is available on *sequences*, as the *sieve* (|) construct. We can convert the linked list representation to a sequence one, and then directly use the while in parallel on x construct. If this construct is not supported, we can use the *blackboard* paradigm, where a group of co-operating processes communicate between themselves using a broadcasting mechanism. Whenever a processor or a workstation calculates an expression it broadcasts its value to other processors so that they can use it in their computation. On PVM this is implemented using *message passing*. Since communication costs are high, we must make sure that the input netlist is split in such a manner so as to reduce any mutual dependency. This is done in a preprocessor phase, before the actual computation on the input circuit is done. The cell instances which are not directly computable, as their dependencies are not satisfied, are placed on a linked list on a block by block basis. Since these blocks generally do not have any dependencies internally, the cost of communication is reduced, and the values of *all* the nets computed in a block is sent in one packed message. Using this paradigm we arrive at a recurrence :

$$T(n) = T(\frac{n}{2}) + O(n)$$
$$= \Theta(n) \qquad (2)$$

Eq. 2 follows from the Theorem given in [2].[1] Here, $n$ is the number of module instances in the input circuit.

## 2.3 Experimental Results

We have implemented the above system in the C++ language, compiled with gcc 2.7.2 and with PVM 3.1 running on a mixed system of 8 HP-UX and 8 SunOS systems. The network used was a standard Ethernet, which was also being used by other users at the time of the experiment. We have tested our method on MCNC ISCAS '89 benchmarks. These benchmarks consist of gate-level netlists of primitive gates like AND2, AND3, NAND2 etc. The primary inputs are marked and we have to propagate the signal probabilities of these inputs in the circuit. We compared our results against those provided by a tool called *POSE: Power Optimization and Synthesis Environment* [7]. Our results are shown in Table 1.

# 3 Conclusions

In this paper we describe a distributed implementation of the Parker–McCluskey heuristic to estimate the average switching activity in combinational circuits. Using Parallel Virtual Machine (PVM) we have developed a tool towards this purpose. We have tested our tool with MCNC benchmarks from ISCAS 89' with encouraging results.

---

[1] This theorem states that given the recurrence of type $T(n) = aT(n/b) + f(n)$, if $f(n) = \Omega(n^{\log_b a + \epsilon})$ and if $af(n/b) \le cf(n)$ then $T(n) = \Theta(f(n))$.

Table 1. Results of SAE on MCNC Benchmarks.

| Name | No. Modules | No. PI's | Avg. Sw | Max. Sw | Min. Sw | Time(s) |
|---|---|---|---|---|---|---|
| C17 | 17 | 5 | 0.334 | 0.493 | 0.140 | 0.0 |
| C432 | 432 | 36 | 0.312 | 0.491 | 0.087 | 0.0 |
| C499 | 499 | 41 | 0.314 | 0.494 | 0.023 | 0.0 |
| C880 | 880 | 60 | 0.313 | 0.499 | 0.001 | 0.1 |
| C1355 | 1355 | 41 | 0.249 | 0.499 | 0.001 | 0.3 |
| C1908 | 1908 | 33 | 0.298 | 0.493 | 0.012 | 0.8 |
| C2670 | 2670 | 234 | 0.213 | 0.499 | 0.000 | 1.5 |
| C3540 | 3540 | 50 | 0.341 | 0.494 | 0.010 | 1.8 |
| C5315 | 5315 | 178 | 0.294 | 0.492 | 0.012 | 3.4 |
| C6288 | 6288 | 32 | 0.351 | 0.493 | 0.132 | 4.2 |
| C7552 | 7552 | 207 | 0.287 | 0.499 | 0.001 | 6.3 |

## Acknowledgments

I thank Dr. C. P. Ravikumar of the Indian Institute of Technology, Delhi, for having introduced me to this problem. I would also like to thank the reviewers for their valuable comments. Finally, I would like to express my gratitude to TEXAS INSTRUMENTS INDIA LTD., Bangalore, where this work was carried out as a part of my Master's thesis.

## References

1. G. E. Belloch. "NESL : A Nested Data-Parallel Language",Technical Report, CMU-CS-95-170, Carnegie Mellon University, 1995.
2. T. H. Cormen, C. E. Leiserson, R. L. Rivest. "Introduction to Algorithms", MIT Press, 1990.
3. G. A. Geist, A. Beguelin, J. Dongarra, W. Jiang, R. Manchek and V. Sunderam "PVM: Parallel Virtual Machine, A User's Guide and Tutorial for Networked Parallel Computing", *MIT Press*, 1994.
4. S. Koranne, "XPlan: A methodology to estimate crosstalk in VLSI circuits", M. Tech Thesis Report, Indian Institute of Technology, Delhi, Dec. 1998.
5. R. Marculesu, D. Marculesu and M. Pedram, "Probabilistic modeling of dependencies during switching activity analysis", *IEEE Trans. CAD*, 17(2), pp. 73–83, Feb. 1998.
6. K. Parker and E. McCluskey. "Probabilistic treatment of general combinational networks", *IEEE Trans. on Electronic Computers*, C-24(6), pp. 668-670, 1975.
7. M. Pedram and S. Iman. "POSE: Power Optimization and Synthesis Environment", Technical Report, Univ. South. Cal., 1996.

# BVIEW: A Tool for Monitoring Distributed Systems

Udaya A. Ranawake[1] and John E. Dorband[2]

[1] UMBC/USRA CESDIS/NASA GSFC, Greenbelt, MD 20771, USA
[2] NASA GSFC, Greenbelt, MD 20771, USA

**Abstract.** The advent of distributed computer systems with hundreds of processors has created the need for software tools for monitoring their resource usage and system status. In this paper, we describe the design, implementation, and performance of the BVIEW software tool for monitoring distributed systems. Our tool has been tested on a 64 node PC cluster running the Linux operating system. However, it could be easily ported to any Unix based workstation cluster.

## 1 Introduction

Networks Of Workstations, or NOW [1, 2] technology, is emerging as a powerful computational resource for certain classes of scientific applications. These computers also cost substantially less than the conventional supercomputers. In this paper, we present the BVIEW software tool for monitoring the resource usage and system status of such distributed systems. Some useful attributes of our tool are low overhead, scalability, fault tolerance, and a graphical user interface. It has been implemented on a 64 node PC cluster called *theHive*[2] at NASA Goddard Space Flight Center (GSFC). However, it could be easily ported to any Unix based workstation cluster.

## 2 The BVIEW Monitoring Tool

Fig 1 shows the architecture of the BVIEW monitoring tool. The heart of the monitoring tool is the *bstat* daemon that runs on the server and each client. The daemon on the server is started by the root at boot time. The daemons on the clients are started by the server daemon by calling the *init* function that has a complexity of $O(\log_2 K)$, where $K$ is the number of nodes.

The *init* function is best illustrated by an example. Assume there are 8 nodes (numbered 0 to 7) with node 0 as the server and that node 4 has crashed. The daemon on node 0 calls the *init* function which divides the list of nodes, $\{1, 2, \ldots, 7\}$ into two sublists $\{1, 2, 3\}$ and $\{4, 5, 6, 7\}$ and attempts to spawn a daemon on the first node of the second sublist (node 4). As node 4 has crashed, it is added to the list of crashed nodes of node 0. Node 0 then spawns a daemon on node 5 and adds it to its neighbor list. Then node 0 will recursively call the *init* function with nodes $\{1, 2, 3\}$ as the input list. Similarly, node 5 will call the *init* function with nodes $\{6, 7\}$ as the input list.

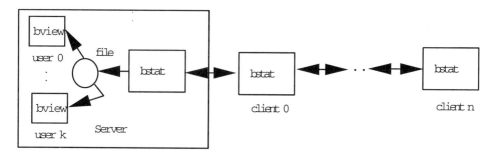

**Fig. 1.** The BVIEW Monitoring Tool

Inter-processor communication for collecting statistics and node synchronization is also handled using $O(\log_2 K)$ based binomial algorithms. This ensures that the monitoring tool would run efficiently on diverse architectures and scale upto hundreds of processors. When collecting statistics, each client reads the statistics from its /proc file system into an array, reads statistics from the children and appends to the array, and sends the array to its parent. Finally, the server writes the statistics to a file. The users run the bview tool that reads the statistics from this file and displays on the screen. Access to the shared file is controlled using semaphores. Synchronization is done by sending a 1 byte message down the tree starting at the root.

The BVIEW monitoring tool is capable of detecting and recovering from node failures in the distributed system. A node failure is assumed if an iteration does not complete in a specified time or if an error condition occurs such as a read or write error on a socket. At the beginning of each iteration, each process turns on an alarm which will cause an interrupt if the iteration does not complete within a specified time. On receiving an interrupt or an error, any client process will branch to a routine which will cause it to shutdown. However, the routine associated with the server process will respawn new *bstat* daemons on each client according to the algorithm described earlier. Any crashed node will be checked once every iteration to determine whether it has recovered. When a node recovers, a bstat daemon will be spawned to collect its statistics.

## 3 Implementation on a PC cluster

The distributed system monitor has been implemented and tested on a PC cluster called theHive at NASA GSFC. The PC cluster consists of 64 nodes where each node is a dual 200 MHz Pentium Pro PC with 448M memory with the nodes interconnected using a fast Ethernet switch. The machines run the Linux operating system - a full-featured clone of the UNIX operating system originally designed for x86 processors and extended recently to support all common desktop architectures.

Figure 3 is a snapshot of theHive system as reported by the BVIEW software tool. The statistics reported are the percentage CPU used, the percentage memory used, the percentage swap space used and the crashed nodes. The information is displayed in the form of a bar chart with one entry for each node in the system. The application is implemented using C and TCL/TK programming languages. The inter-processor communication and the detection of node failures/recoveries are handled using the TCP/IP protocols[3].

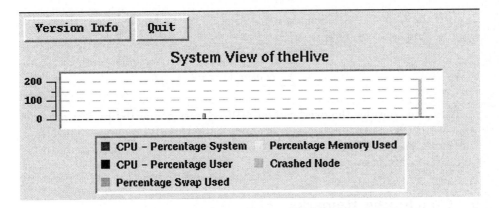

**Fig. 2.** Snapshot of theHive using BVIEW monitoring tool

## 4 Performance Analysis

In this section, we derive analytical expressions for the monitor latency per iteration, $lat_M$ under normal operating conditions where there are no node failures. The *monitor latency* per iteration, $lat_M$ is given by:

$$lat_M = T_{stat} + T_{sync}$$

where $T_{stat}$ is the time to collect statistics, $T_{sync}$ is the synchronization delay.

For a ethernet connected by a switch, communication cost between two nodes is assumed to be a function of the message size and is denoted by $comm(S)$ where $S$ is the message size. It is given by:

$$comm(S) = \alpha + \frac{S}{\beta}$$

where $\alpha$ is the latency and $\beta$ is the bandwidth. $T_{stat}$ includes the time for each client to read the /proc file system for gathering statistics, packing data into an array, sending the array to the server using a binomial tree algorithm, and for the server to write the output to a file. Assume that the number of nodes $K$ is a power of two and $L = \log_2 K$. Let $s$ be the size of the data array per node and let $rw$ be the time to read and write the statistics. During the $i_{th}$ step, pairs of processors will be communicating with a message of length $2^i s$. Then, $T_{stat}$ is given by:

$$T_{stat} = \sum_{i=0}^{L-1} comm(2^i s) + rw$$
$$= \alpha L + \frac{s}{\beta}(2^L - 1) + rw$$

$T_{sync}$ is the time to send a one byte message down the binomial tree starting at the root and is given by:

$$T_{sync} = \sum_{i=0}^{L-1} comm(1) = \alpha L + \frac{1}{\beta}L$$

Next, we consider the overhead on a PC cluster connected by fast ethernet via a switch. For such a network, $\alpha$ is about $0.15mS$ and $\beta$ is about $100Mbits/s$. Therefore, on the 64 processor theHive the total communication overhead per iteration is only between 2 to 3 mS.

## 5 Concluding Remarks

In this paper, we have presented the bview software tool for monitoring distrbuted systems. We have analyzed its performance and shown that it incurs very low overhead on a PC cluster containing as many as 64 nodes. The software including the source code is available freely at *http://newton/gsfc.nasa.gov/theHive*. Future work will extend this program to conform to X/Open Systems Management specification[5] to allow the management of heterogeneous clusters.

## References

1. T. Sterling, et al.: BEOWULF: A Parallel Workstation for Scientific Computation. Proc. of the 1995 International Conference on Parallel Processing, Milwaukee, Wisconsin, August 14-18, 1995.
2. J. Dorband and U. Ranawake: theHIVE: Highly-parallel Integrated Virtual Environment. Accessible on the Internet at World Wide Web URL http://newton.gsfc.nasa.gov/thehive/.
3. W. R. Stevens: UNIX Network Programming. Prentice Hall, 1990.
4. H. Sullivan and T. R. Bashkow: A Large Scale, Homogeneous, Fully Distributed Parallel Machine. Proc. of the 4th Annual Symp. Computer Arch., vol. 5, pp. 105-124, Mar. 1977.
5. Accessible on the Internet at World Wide Web URL http://www.opengroup.org/onlinepubs/.

# The Queue System within PHASE

Sergi Girona[1], Santi Bello[1], Jesús Labarta[1], Pablo Ribes[2], Roman Martin[2], José Soto[3], and Gloria Laffitte[3]

[1]CEPBA-UPC, c/ Jordi Girona 1-3, Mòdul D6, Campus Nord
Universitat Politècnica de Catalunya (UPC)
08034 Barcelona, Spain
{sergi, sbello, jesus}@cepba.upc.es
[2]SAGE, Ronda de la Luna 4
28760 Tres Cantos-Madrid, Spain
{pribes, rmartin}@sag.es
[3]IBERDROLA, Gardoqui 8, 5D-18
48008 Bilbao, Spain
{jose.soto, gloria.laffitte}@iberdrola.es

**Abstract.** This paper describes a system that supports the specification of a large set of simulations and their organization and distributes execution over a PC's network under Windows 95 and NT. Our objective is two faced: maximize resources utilization of IBERDROLA and the possibility of starting experiments that require a high compute power. As an example, we will describe the work done in the ESPRIT project 27238, PHASE, where the total number of analysis increases exponentially, and the maximum resources are required to obtain reasonable order of response time. PHASE is an integrated tool, user oriented, to automatically define, execute and analysis of hundreds of required simulations for a valid study and future electric network planning.

## 1 BACKGROUND

This paper describes the application of distributed computing technology to the extensive analysis of electricity supply networks. The work has been carried out within the PHASE [1] ESPRIT project. In it, technology for the stochastic analysis of structures developed within PROMENVIR [2] project has been ported to networks of PCs running windows 95/NT. Partners in the PHASE project are CEPBA-UPC, a research center with experience in parallel computing, SAGE, a large software house developing solutions for user needs, and IBERDROLA, a large electric utility in Spain.

The task of planning new installations on electrical networks is a time-consuming and costly process. In the last years, this activity has become more difficult since the importance and complexity of various associated factors has increased (e.g. authorization processes, expropriations, environmental studies, etc.). This situation enlarges the period of construction of new installations and demands an exhaustive study for every new constructed one. The requirements for the new installation are to support the foreseen evolution of the network, to extend the working period of the equipment and to optimize the investment.

Other kinds of electrical network studies have to be carried out for maintenance reasons. For instance, when a given plant has to be disconnected for in-depth revision or correction, it is required to simulate and study in detail the network status after the shutdown, and study the behavior when possible contingencies occur. These activities are very important for the electric companies in order to keep the network as secure as possible, and provide a high service quality.

Phase uses PSSE as simulation tool. PSSE is commercial software widely used for these studies. It reads an input file specifying the network components, topology, and computes load distribution.

A planning study requires the simulation of many hypothetical scenarios, considering a large number of input variables and a wide range of variation. Preparing such simulations, as well as managing all the involved data, is a cumbersome task, not to mention the problem of long execution time.

The objective of the PHASE project is to speed up the whole study process by providing a tool that

- Enables the easy specification of scenarios analysis
- Handles the data management and execution of all the simulations on the available network of PCs in a transparent way
- Eases the post processing and presentation of the data obtained

This paper focuses on the second point. More information on whole environment is available at http://dpa.sag.es/~phase. Main differences between other batch systems, LSF [3], CODINE[4], NQE [5] and PHASE queue system is support for data management, global study description in a single script file and directory management for each single study.

## 2 System functionality

The study definition editor generates a list of tasks (scripts) that are sent to the scheduling module. This list of tasks matches the simulations that would be necessary to carry out in sequence if the study was done by hand. The scheduling is in charge of providing transparent access to the underlying network. It is in charge of selecting the PCs where the simulations will be run, handle all needed data copying, remote execution and collection of results.

Finally, the post processing manages the output data providing the proper facilities for data analysis and presentation. To avoid disk space problems, only the requested data for each execution is presented to the user.

## 3 System structure

### 3.1 List of tasks

The list of tasks generated by the editor is included in a single script file. Comments in the script file specify input data (files) or output data (files). From this, the scheduler can infer dependencies between tasks if any. The comments can also specify resource requirements (free memory, free disk space,...).

## 3.2 Data handling

Once a PC is selected for executing a given task, it is necessary to ensure that it has access to the input data and simulator. As well as to ensure that two concurrent executions will not clash in intermediate or results files. For this purpose, a new directory is created in that machine, where the original (sequential) environment is replicated. This directory plus the physical processor constitutes a virtual processor where the task will be executed.

In order to reduce network traffic and considering that all input files can be reused in each different simulation of a single analysis, those directories can be reused, minimizing the file transfer. Thus, when using 8 PCs and running 128 simulations, it will be only necessary to transfer 8 instances of the input dependent files, instead of the expected 128.

## 3.3 Scheduling

PHASE uses a dynamic First Come First Served scheduling as basic algorithm. It is necessary to have a dynamical scheduler because there are hundreds of simulations for each analysis but each analysis can have different processor time consumption, and a dynamic scheduler adapts to the different workload situations. This scheduler also provides good performance when using heterogeneous PCs, because faster machines will get more jobs than slower ones [2].

The scheduler is controllable trough several and simple parameters: time windows, free disk and memory space, current PC workload,... These parameters allow the utilization of desktop PC for the analysis execution, with a negligible interference to the interactive work.

## 3.4 Fault recovery

All queue system does provide a fault recovery mechanism, to solve failures of master node or any of the slave nodes. A typical PHASE analysis contains hundreds of simulations, each of 30 to 60 seconds long. This characteristic determines the selected fault recovery mechanism.

If connection is lost to the execution host, all jobs are rescheduled to a different host, in order to obtain the results back as soon as possible.

If connection is lost to the central scheduler, a new central scheduler is selected using a static algorithm, based on Bullu[6]. There is a list of possible central schedulers and global agreement is used to fix it. Once having a new scheduler, all pending simulations will be rescheduled.

## 4 Results

**Table 1** contains some results of a several PHASE executions using synthetic workload, with different processor time consumption (15, 30, 45 and 60 seconds per job). **Table 1** provides the total execution time when running 128 jobs per analysis.

|         | 15s  | 30s  | 45s  | 60s  |
|---------|------|------|------|------|
| 1 proc  | 1920 | 3840 | 6144 | 7680 |
| 2 procs | 1073 | 2056 | 3011 | 3960 |
| 4 procs | 567  | 1046 | 1527 | 2008 |
| 8 procs | 286  | 526  | 731  | 1015 |

**Table 1.** Time in seconds per analysis

Values corresponding to one processor are ideal ones, only taking into account the execution time of the 128 tasks of the analysis. All other values have been obtained using two, four and eighth PCs, running the central scheduler on a separate, and another one as submission host. As conclusion, the cost of using more processors do not incur in much extra work, and total execution time is nearly reduced to half, when the number of used processors is doubled.

## 5 Conclusions

The system is operational at IBERDROLA facilities, where up to seven PCs distributed between Madrid and Bilbao (300 Km away) cooperate in their current analyses.

## 6 References

[1] "Parallel High-performance Analysis and decision Support for Electrical networks, PHASE", ESPRIT project 27238.
[2] "Monte Carlo workload scheduling". Sergi Girona, Jesús Labarta and Maite Ortega. ISBN 84-89925-04-6. 1997.
[3] "LSF JobScheduler Administrator's Guide", Second Edition, August 1998.
[4] "Codine, Installation and Administrative Guide", Version 4.1, Genias Software GmbH.
[5] "NQE Administration, publication SG-2150"
[6] "Operating System Concepts" A. Silberschatz and P. Galvin. Addison-Wesley Publishing Company, 1994.

# Support Tools for Supercomputing and Networking

V.N. Kasyanov, V.A. Evstigneev, J.V. Malinina,
J.V. Birjukova, V.A. Markin, E.V. Haritonov, S.G. Tsikoza

A.P.Ershov Institute of Informatics Systems
630090, Novosibirsk-90, Russia
E-mail:kvn@iis.nsk.su

**Abstract.** The project PROGRESS being under development at the Institute of Informatics Systems in Novosibirsk is outlined. The information subsystem TRANSFORM aimed at accumulating and processing of knowledge about optimizing and restructuring program transformations is considered. A three-phase networked query interface for information retrieval is discussed. The subsystem SFP intended to support supercomputing and networking on the base of the applicative language SISAL is presented.

## 1 The PROGRESS project

The PROGRESS project [1-5] being under development at the Institute of Informatics Systems in Novosibirsk is aimed at creation of a system that supports:
• Research of systems of optimizing and restructuring transformations; development of new systems of transformations and implementation algorithms; search for the intermediate representations of programs as well as sets, context conditions and strategies of transformations being efficient for a given class of programs and computers.
• Fast prototyping of optimizing compilers for various target architectures, such as VLIW, superscalar and multiprocessors with distributed memory.
• Training students to programming and optimizing compilation methods for parallel architectures.

To achieve these goals, the system is being developed as a tool for manipulation with programs; it includes means for step-by-step program transformation - starting with input text and up to either text representation of the transformed program or executable code to be then run on target architecture model.

Both imperative high-level languages (such as Fortran-77, Modula-2 or C++) and functional languages (such as SISAL) are considered as input languages. The languages are supposed to be extended by program annotation means (formalized comments). Using them an user can control program transformation process (for example, by specifying additional information about properties of the program

---
[0] Partially supported by the Russian Foundation for Basic Research under Grant RFBR 98-01-748.

and the context of its execution), and the system can comment a resultant program with information about transformation process.

## 2 The TRANSFORM information subsystem

The TRANSFORM subsystem is an important component of the PROGRESS system. The subsystem stores information about optimizing and restructuring transformations for parallel computer programs and it is interesting in itself.

Automatic program parallelizing has some disadvantages related to the "conservatism" of compilers that are oriented on an "average" program by both analysis and transformations performed on a program. In other words, when parallelizing a program, it is impossible to completely extract parallelism contained in it. Most of existing parallelizing compilers require prompts from a programmer either before or during translation, or they report segments that need analysis and reprogramming to be successfully parallelized. Thus, information obtained by a programmer from the TRANSFORM system can be used him to improve his program being parallelized.

The main decisions in the information system design include the following:

• implementation of the TRANSFORM system on base of free distributed software (GNU software [6]),

• development the information system for Internet (network) access.

The experimental version of the TRANSFORM subsystem is now accessible on WWW at http://pco.iis.nsk.su/transform/. It uses the Netscape Navigator and MS Internet Explorer as user-agent applications, Apache 1.3 as WWW-server, CGI-gateway (Lisp, C), database system (PostgreSQL 1.6).

## 3 A three-phase networked query interface for information retrieval

The optimizing transformations area is evolving and not strictly classified. There are quite a few parameters which can be considered as classifying parameters, and hence there are many classification schemes based on one or more parameters. Some of these classification subsets of retrieval parameters are mutually exclusive. Therefore, there are alternatives:

• to provide a user with a set of all retrieval attributes; however, thus it is necessary to know the area thoroughly to form a reliable query (user fills in a set of the parameters and gets a fast answer in one query to the server);

• to provide a user with only a constant subset of retrieval attributes reliable in any cases (user chooses the parameters and maybe gets irrelevant information);

• to provide a user with a hierarchical structure of retrieval attributes.

## 3.1 System components

The TRANSFORM system is subdivided into three interacting functional components:
- data visualization tool,
- database manager,
- query builder.

The first of them is intended for data set and query results visualization; it helps to browse the data space, to make conclusions, to analyse and to verify hypotheses.

The second component is intended to control database access, privileges and storage parameters .

The third component consists of the query processor and the user interface for formulating queries. This interface consists of three phases:
- dynamic forms configuration,
- query preview,
- query refinement.

## 3.2 Dynamic forms configuration

Dynamic forms configuration is the first phase of query construction and at this phase attributes subset is selected and passed to query preview phase. In this phase the whole set of retrieval attributes is proposed to a user, but first, only a subset of classification parameters of the highest level is visible to user. As soon as level parameters are selected, the next set of related parameters is proposed, and so on. These actions are performed on a client computer without another queries to the server and to the database. This method is based on the web technology known as Dynamic HTML. At this phase the subset of retrieval attributes for query preview phase is formed.

## 3.3 Query preview

Query preview is an approach for browsing networked data storage. The phase supplies with information about the data distribution and supports a dynamic query user interface where the visual display of summary is updated in one server query in response to users' selections. It consumes modest resource of system because only a small number of attributes from the complete database are used at this phase.

## 3.4 Query refinement phase

Query refinement phase is implemented as a form fill-in interface and the query formed in two previous phases is executed. At query refinement phase, users can select precise values for the attributes. For example, a dynamic query or a form fill-in. Usually, this phase is executed after the query preview phase and can inherits the constraints and the data set from it.

## 4 The SFP functional programming subsystem

Another component being developed within the PROGRESS project is a functional programming system (SFP) based on SISAL-language and aimed at supporting of high performance computing and networking.

The system is supposed to provide a programmer with a convenient environment to develop functional programs targeted for various parallel architectures available via network. This environment must provide means first to write and to debug programs regardless a target architecture and then to adapt the correct program to certain target parallel architecture in order to achieve high efficiency of program execution on supercomputers.

A program adaptation process consists of optimizing cross-translation of functional program being developed to the supercomputer's language program (for example, C++). A programmer controls the process of program optimizations and transformations by supplying additional information or by direct controlling these transformations. In particular, a programmer must be able to visually process a SISAL-program in terms of its internal representation.

Functional language SISAL (Steams and Iterations in a Single Assignment Language) is considered as an alternative to Fortran language for supercomputers [7,8]. Compared with imperative languages (like Fortran), functional languages, such as SISAL, simplifies programmer's work. He has only to specify a result of calculations and it is a compiler that is responsible for mapping an algorithm to certain calculator architecture (including instructions scheduling, data transfer, execution synchronization, memory management and etc.) In contrast with other functional languages, SISAL supports data types and operators typical for scientific calculations.

## References

1. V.N. Kasyanov, V.A. Evstigneev. *A program manipulation system for finegrained architectures,* In: Proc. EURO-PAR'95:Parallel Processing, (1995), 719–722. – (Lecture Notes in Computer Science, Vol. 966).
2. V.N. Kasyanov, V.A. Evstigneev. *Optimizing transformations in optimizing compilers,* Programming, 6 (1996), 12–26. – (in Russian).
3. V.N. Kasyanov, V.A. Evstigneev, L.V. Gorodniaja, J.V. Birjukova, J.V. Malinina, S.G. Tsikoza, T.A. Klimova, E.V. Haritonov, *Parallel processing: problems of training.* In: Proc. III Intern. Conf. on New Informatics Technology in the University Education, Novosibirsk, (1997), 186–187. – (in Russian).
4. *Problems of constructing efficient and reliable programs* / Edit. V.N.Kasyanov, Novosibirsk, (1995). – (in Russian).
5. *Optimizing transformations and program design* / Edit. V.N.Kasyanov, Novosibirsk, (1997). – (in Russian).
6. *GNU's Bulletin.* Free Software Foundation, Inc., Boston, (1997).
7. J.T. Feo, P.J. Miller, S. Skedzielewski, S.M. Denton, C.J. Solomon. *Sisal 90.* In: Proc. High Performance Functional Computing, Denver, Colorado, (1995), 35–47.
8. D. Cann. *Retire Fortran? A debate rekindler,* Comm. ACM 35, 8 (1992), 81–89.

# Ultra High-Speed Superconductor System Design: Phase 2

Mikhail Dorojevets[1], and Konstantin Likharev[2]

[1] Dept. of Electrical and Computer Engineering
State University of New York, Stony Brook, NY 11794-2350, U.S.A.
midor@ee.sunysb.edu
[2] Physics and Astronomy Dept.
State University of New York, Stony Brook, NY 11794-3800, U.S.A.
likharev@rsfq1.physics.sunysb.edu

**Abstract.** A COOL system based on superconductor Rapid Single-Flux-Quantum (RSFQ) technology is being developed at SUNY (Stony Brook, USA) within the framework of the Hybrid Technology MultiThreaded architecture (HTMT) project. The objective of the second phase of the project is the design study of a petaflops scale computer. We have found that a 0.8-μm RSFQ technology would make possible the implementation of the COOL system with multithreaded SPELL processors with an average clock frequency approaching 100 GHz, a pipelined cryo-memory with a 30-ps cycle and 2 TB/s bandwidth per processor, and an inter-processor network with a peak bandwidth of more than 1,000 Petabytes/s. Preliminary results show that an RSFQ system with 4096 processors can achieve a petaflops-level performance with the overall power load of 1 kW at 4 Kelvin, while occupying a physical space as small as $0.6 \text{ m}^3$.

## 1 HTMT Project and Superconductor RSFQ Technology

The goal of the Hybrid-Technology MultiThreaded architecture (HTMT)[1] project [1] is to carry out the development of a multithreaded computer architecture that would utilize novel electronic and optoelectronic technologies to achieve petaflops-level performance (~$10^{15}$ floating-point operations per second). The concept system has a hierarchical organization with multiple levels of distributed memory: holographic data storage, semiconductor SRAM and DRAM, and cryo-memory (CRAM), as well as three types of processors: SRAM- and DRAM-based Processor-In-Memory (PIM) elements operating at room temperature, and RSFQ Superconductor Processing ELements (SPELLs) operating at the temperature of liquid helium (4 to 5 Kelvin).

The Josephson junction RSFQ technology is the first digital technology capable of approaching the 100-GHz frontier. The scaling rules confirmed in experiments with 1.5 μm (370 GHz) and 0.5 μm (770 GHz) devices show that VLSI RSFQ circuits

---

[1] The HTMT project and the work described in the paper are supported by DARPA, NSA, NASA via Jet Propulsion Lab., and in part by NSF under grant No. ECS-9700313.

built with 0.8-μm technology may have the following characteristics: a logic clock rate of 60-160 GHz; a pipelined superconductor memory cycle of 30 ps; an integration scale up to 3M junctions per 2×2 cm$^2$ logic chip, and up to 10M junctions per memory chip of the same size; power dissipation of 0.1 μW per logic gate, and 0.002 μW per memory bit (when operated at the rates specified above).

## 2  COOL Multithreaded Design and Related Work

Our COOL system design [2] is based on a new parallel 64-bit COOL-I instruction set architecture with support for two-level simultaneous multithreading in SPELL processors [3]. The technique of simultaneous multithreading is used to tolerate huge memory latency visible to any SPELL processor. The latency varies from ~30 processor cycles when accessing its local CRAM to several hundred cycles when accessing SRAM.

The technique of two-level simultaneous multithreading has been first implemented in the VLIW processors of the Soviet MARS-M multithreaded computer built in the 1980s [4]. Several commercial high-performance computers using interleaved multithreading, such as HEP [5], and Tera [6], have been built in the USA.

The distinguishing features of our design approach are as follows:

1) Two-level multithreading has to be implemented in multi-gigahertz ultrapipelined SPELL processors where each (coarse-grain) thread has its own register file shared by all (medium-grain) parallel instruction streams (e.g., representing parallel loop iterations) created inside the thread. Total, up to 128 parallel instruction streams can simultaneously issue instructions and run in each SPELL processor.
2) The hardware supports flexible partitioning of each register file into variable-size register frames assigned to instruction streams and provides relative and absolute register addressing at run time. As a result, different instruction streams running simultaneously within the same processor can use a single copy of program code.
3) Streams executing within the same thread can use shared registers for communication and synchronization.
4) There is no single clock generator with global clock distribution, and there is no static pipeline datapath whose latency could be visible to a compiler. Instruction-level parallelism inside each stream is exploited with the assistance of fully distributed scoreboard logic and invisible at the architecture level.

Figure 1 shows a block diagram of the proposed SPELL processor. SPELL has 16 multistream units (MSUs), 5 pipelined floating-point units (FPUs) operating with an average cycle time of 15 ps, and an 8-ported pipelined CRAM with a 30-ps cycle time. MSUs can communicate with the shared floating-point functional units and a processor-memory interface (PMI) via an intra-processor interconnect (PNET).

Each MSU executes control, integer, and floating-point compare operations of up to eight parallel instruction streams. All 8 parallel streams inside one MSU share: a 2KB multi-port instruction cache, 64 general-purpose registers (each of which is able to hold either 64-bit long integer or double-precision floating-point data), a 64-bit integer arithmetic unit, a 64-bit shift/logic unit, a 32-bit address adder, and a PNET interface unit.

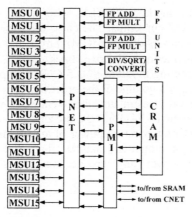

**Fig. 1.** SPELL processor block diagram

Non-shared hardware reserved for each stream include: an instruction fetch unit, an instruction decode/issue unit, 4 condition registers, and reservation stations where operations wait for their operands to become ready before issuing them to integer/floating-point units and memory.

All the units within MSU as well as FPUs are pipelined and have a 15-ps cycle time (for 0.8-μm technology). Although the performance of any individual stream is limited by this 15-ps cycle (equivalent to a 66-GHz "clock" rate), other logic providing access to shared resources (registers and units) inside MSU can work at a much higher rate of up to 160 GHz.

Each SPELL has an eight-ported 1-MB pipelined CRAM module to be implemented with 4 dual-port 256KB chips. CRAM is based on Josephson junction technology, using both RSFQ and dc-biased latching circuits. Our estimates show that the memory chip pipeline may be built of four 30-ps stages. The peak CRAM bandwidth is eight 64-bit words per 30-ps cycle, i.e., more than 2 TB/s.

Communication between each SPELL/CRAM module and local SRAM is provided through a room-temperature interface with two ports (in and out) consisting of almost 1K signal and 1K ground wires and 8-10 Gb/s bandwidth per each signal wire. This gives a bandwidth of one 64-bit data (with additional bits for address and control) packet every 30 ps, i.e., ~512 GB/s in each direction.

SPELL communicates with remote CRAM/SRAM modules through RSFQ CNET to be implemented either as a Banyan network or a multi-dimensional pruned mesh with the peak data bandwidth of each input and output CNET port of ~ 256 GB/s.

The peak performance of a SPELL processor with five floating-point units operating with 15-ps cycle time is more than 300 Gflops. Theoretically, up to eight instructions can be issued by 8 stream instruction decode/issue units and four instructions (control, integer arithmetic, shift/logical, and memory/floating-point) can be completed in one cycle in each of 16 MSUs inside SPELL. Our current design and simulations on simple kernels have shown that an instruction issue rate for an individual single-issue stream will not exceed 1 instruction per twelve 15-ps cycles, thus limiting the peak integer performance of a SPELL processor with 128 streams to 700 Gops.

Physically, each processing module (SPELL with 1MB CRAM) could be implemented as a set of seven 2x2 cm$^2$ chips. The chips are expected to be flip-chip mounted on a 20×20-cm$^2$ multi-chip module (CMCM), physically a silicon wafer with 2 layers of 3-µm wide superconducting microstrip lines. Each CMCM houses more than 60 chips, including 8 processing modules plus 5 CNET chips. A COOL system having 512 CMCMs would occupy a volume of the order 0.6 m$^3$.

Even with the current conservative estimates, the overall power load of the RSFQ COOL system at 4 Kelvin is about 1 kW. With the present-day efficiency of helium cooling, this leads to a total room-temperature power of approximately 300 kW. This number is almost two orders of magnitude smaller than that for a hypothetical petaflops system implemented with any prospective semiconductor transistor technology.

## 3 Conclusions

Our preliminary study has revealed no roadblocks on the way toward a compact superconductor COOL system with petaflops-scale performance and very low power consumption. We have been carried out simulations of the processor and system components on several levels. In particular, we are prototyping and measuring critical parameters of RSFQ circuits built with the commercially available 3.5-µm fabrication technology, as well as evaluating processor performance with a clock accurate simulator. During Phase 3 of the project that is expected to begin in October of 1999, we plan to design and implement a prototype COOL system consisting of two 25-GHz superconductor processors with 1.5 µm RSFQ technology.

## References

1. Gao, G., Likharev, K. K., Messina, P. C., Sterling, T. L.: Hybrid Technology Multithreaded Architecture. Frontiers'96 Annapolis MD (1996) 98–105
2. Dorojevets, M., Bunyk, P., Zinoviev, D., Likharev, K.: Petaflops RSFQ System Design. ETE-04 Report at ASC'98 Palm Desert, CA (1998), to be published in IEEE Trans. on Appl. Supercond. (June 1999)
3. Dorojevets, M.: The COOL-I ISA Handbook: Version 1.00. TR-11, RSFQ System Group, SUNY at Stony Brook (1999)
4. Dorojevets, M., Wolcott, P.: The El'brus-3 and MARS-M: Recent Advances in Russian High Performance Computing. J. Supercomputing, Vol. 6. (1992) 5–48
5. Smith, B.: Architecture and Applications of the HEP Multiprocessor Computer System. SPIE Real Time Signal Processing IV, SPIE, New York (1981) 241–248
6. Tera: Principles of Operation. Tera Computer Company (1997)

# Lilith Lights: A Network Traffic Visualization Tool for High Performance Clusters

David A. Evensky, Ann C. Gentile, and Pete Wyckoff

Sandia National Laboratories, PO Box 969 MS 9011, Livermore CA 94551, USA
{evensky, gentile, pw}@ca.sandia.gov
WWW home page: http://dancer.ca.sandia.gov/Lilith

**Abstract.** High performance computing increasingly utilizes loosely coupled clusters of commodity computers. Performance monitoring data on such platforms is traditionally performed by invasive message passing taps, and examined only after the run is complete. We present an alternative, designed to capitalize on the scalable nature of the platform. *Lilith Lights* is a non-invasive network and processor monitoring tool which collects and processes data from the individual cluster machines. The data is gathered and displayed *concurrently* with the running applications. The display format can correspond to the application's virtual topology or to the physical connectivity of the message passing hardware.

## 1 Introduction

The Lilith Lights visualization tool provides graphical information about CPU usage and communication among nodes of a cluster. Application programmers can use the tool to visually discover problems in their algorithms, such as unexpected interactions between various parallel message passing library components, and the emergence of "hot spots" in the calculation. System resource managers can also use the tool to discover areas in the cluster which are under-utilized.

Typical software implementations for visualizing and debugging parallel codes, e.g., XPVM[1], XMPI[2], Vampir[3], and AIMS[4], insert information gathering calls into the underlying message-passing software of the cluster, require linking the application into special libraries, or adding routines, and hence alter the conditions under which the code is run. The information displayed lags behind the application's state, making the debugging process less direct, especially when attempting to monitor other external devices which are not controlled by the message-passing layer *e.g.,* remote network connections or console messages. This level of tool is also often unable to monitor multiple independent applications on the same machine, as is frequently the case with clusters of multi-user symmetric multi-processors. Also, some tools are restrictive, *e.g.,* Vampir is for MPI applications, TotalView[5] is not for continuous data motion viewing.

Hardware implementations for status gathering exist on some massively parallel processors (MPPs), such as the Intel Paragon. Though non-invasive and real-time, they are inherently platform-dependent and not practical in distributed clusters where the system is not self-contained and the configuration is not static.

Lilith Lights combines the best of the hardware and software implementations, providing a low impact, *low delay* software solution, adaptable to both MPPs and distributed clusters. It is relatively non-invasive with respect to the application under study: it requires *no* hooks into the application and runs independently. Lilith Lights is written in Java providing platform independence; although we capture network and CPU usage data using small hooks placed in the lowest level device driver to record packet transmission and arrival, straightforward modification to message passing libraries would yield a fully portable, though less optimal tool. Lilith Lights consumes a negligible fraction of total CPU cycles. Monitoring traffic can be sent across a secondary network to avoid interference with the application.

Lilith Lights is based on the Lilith[6] tool building framework to capitalize on scalable cluster architectures, *e.g.,* Sandia's Computational Plant (CPlant)[7]. Scalability of both hardware and applications ensures efficiency and reliability.

## 2 Examples

We present two examples of debugging that were made much easier by being able to "see" the traffic. The first deals with an MPI program, and similar information could have been gathered using an existing tool such as XMPI, but the second example uses the network at the lower IP level, which is inaccessible to typical message-passing monitoring tools, and would have been harder to spot from output from tcpdump or router statistics.

The first example involves a code used to verify the status of the software and hardware components of our machines. If everything is working properly, the Lilith Lights display shows a "snake" of lighted communication paths, wandering around a virtual 2D mesh (Figure 1).

At startup, each mesh node discovers its location and the MPI node numbers of its neighbors. No single processor knows the entire path of the snake; each processor only keeps track of the segments involving itself. This avoids global communication that would light up other segments. The code is therefore tricky to debug as there is no single repository of the state information.

The first implementation of the code seemed to work, but the snake confined its travels to the left half of the mesh, even though the parallel code was indeed running on every processor. This information could not be detected from process status information, as they were all consuming CPU cycles. Also it is not an error for a processor to have no lighted segments connected to it, which ruled out normal interactive debugging methods. Without Lilith Lights, we may never have detected that half of the communication links were going unused.

The problem turned out to be a failure to declare the return value of drand48(), which the compiler then mistakenly assumed to be an integer. This biased the "random" decision of where next to pass the head of the snake such that processor numbers 16 and higher were never chosen.

A second example occurred during a demonstration of a parallel application code written to use either MPI or PVM. Three geographically separated clus-

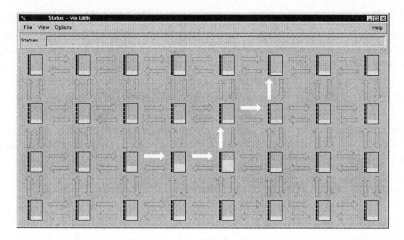

**Fig. 1.** Display of the "snake" program. System load on each node is displayed as a moving vertical bar. Traffic on the network links connecting nodes is represented by colored unidirectional arrows; colors indicate the level of consumed bandwidth.

ter computing installations were connected over the wide area using multiple ATM networks. Within each isolated cluster, the message passing library was configured to use the high-performance network, Myrinet, and messages between nodes in different clusters were sent across the ATM interfaces.

Although proper output was generated, the code was running too slowly given the number of processors and the bandwidths of the network links. XMPI and XPVM showed that the processors were sending and receiving messages as they should, and that there was full connectivity between all the nodes in all the clusters. The Lilith Lights *virtual mesh* display showed the same. The *physical layout* display mode, however, revealed that of the four ATM links connecting each site, only one had significant traffic. Routing decisions were made at each individual cluster node at the IP layer, and the initial installation of the routes forwarded packets destined for remote clusters to only a single node that had an ATM interface. The solution was to rewrite the routing tables to spread the across all four links instead of concentrating it at a single network interface card.

## 3 Design

The major elements in Lilith Lights are: the Lilith framework, which provides scalable data handling on a group of machines; the kernel code that obtains CPU and network activity on each node; and the tool code that interfaces the two.

The Lilith framework[6] spans a tree of machines executing user-defined code. It maintainins the tree and handles communication among nodes. Lilith propagates user code, called "Lilim", down the tree. The Lilim perform user-designated functions on each machine, and Lilith propagates the results back up the tree. Lilim can further process the results from lower nodes during the return. Lilith-based tools are created by writing suitable Lilim. In Lilith Lights, the Lilim code gathers information on the CPU usage and communications of each node.

The tree structure provides logarithmic scaling. Although no application that sends information of limited compressibility from each node can be truly scalable as the number of nodes goes to infinity, for Lilith Lights on a 100Mbit network at typical display rates the theoretical maximum cluster size is of order 100K nodes. A fixed data size is used for communications rather than simple concatenation of each node's results. Note that Lilith's communications structure is independent of that of the application. In fact, on CPlant we route the Lilith traffic across a physical network distinct from that used for application message-passing traffic.

At the lowest level of this data-gathering framework is code in the kernel that records the transfer of every packet. It is called from the physical device driver, and is very fast (less than 10 instructions) so as not to alter the performance of the system as a whole. This technique is efficient and not tied to any particular message passing library. Data gathered at this level is more complete than that from hooks into a library, as it catches messages which may issue from all potential communication sources.

When a Lilim receives a message from its parent requesting data, it passes this message to its children, then reads information from the kernel. System load comes from /proc/stat; traffic delivered from the node is provided by the kernel module *via* /dev/lights0. Each Lilim returns to its parent an accumulated state, consisting of its own information merged in with that of its children.

## 4 Performance

We have measured the CPU utilization for a variety sampling rates in the device driver by averaging over each measurement realization and over all the nodes. For typical to long sampling time periods (3-5 sec.), we see a CPU utilization of 4-7%. We consider this to be a relatively minimal impact. Measurements were made on 12 unloaded 200 Mhz dual Pentium Pros and averaged over 100 samples per node. The load is independent of the position in the tree.

The effect on network utilization depends on the sample rate and the number of hosts. Our worst case is 300 bytes/message sent by each host every 1-5 sec. This is much less than the bandwidth of fast Ethernet, and greatly less than that of Myrinet. On CPlant, message passing traffic is on the Myrinet network while Lilith traffic is on the Ethernet, so we see no contention for network bandwidth.

## References

1. http://www.epm.ornl.gov/pvm.
2. http://ww.osc.edu/Lam/lam/xmpi.html.
3. http://tamon10.softek.co.jp/SPG/pallas/vampir-vis-e.html.
4. http://science.nas.nasa.gov/Groups/Tools/Projects/AIMS.
5. http://www.dolphinics.com/tw/tvover.htm.
6. http://dancer.ca.sandia.gov/Lilith and references therein.
7. http://z.ca.sandia.gov/cplant.

# DSMC of the Inner Atmosphere of a Comet on Shared Memory Multiprocessors

Gregory O. Khanlarov and German A. Lukianov

Institute for High-Performance Computing and Data Bases,
P.O. Box 71, St.Petersburg 194291, Russia
{greg, monte}@fn.csa.ru
http://www.csa.ru/Inst

**Abstract.** The high-performance algorithm of two-level parallelization for direct simulation Monte Carlo of unsteady flows has been described. The algorithm of parallel statistically independent runs is used on the first level of parallelization. The data parallelization corresponds to the second one. The problem of simulation of the inner atmosphere of a comet was used to study the speedup of the algorithm. It has been shown that speedup and efficiency of the algorithm get higher as the problem becomes more complex.

## 1 Introduction

In early 80's there was great interest in study of comets, specifically in problem of their mathematical modeling. It was connected with the comet flyby missions, which had to go through the comet head. Nowadays, scientists still keep taking an interest in this problem in connection with the future space missions dedicated to the study of several comets, e.g. long-duration mission "Rosetta". This mission is focused on the observation of the cometary nucleus and its immediate environment. It plans deeper plunge into the cometary atmosphere and even apparatus landing. Equipment engineers need better comprehension of atmosphere dynamics in order to design probes and measuring tools.

The problem of simulation of the inner atmosphere of a comet has high level of complexity. In general, the flow in the inner atmosphere is spatial, non-equilibrium and unsteady.

The most suitable method for the simulation of such flows is the direct simulation Monte Carlo (DSMC) [1]. On a level with remarkable advantages of this method there is a principal drawback: high requirement for computer resources. This requirement gets higher with the complexity of a given problem. The design of efficient algorithms, providing greater performance of calculations, is very essential. One of the way is the elaboration of parallel algorithms for DSMC, designed for massive-parallel supercomputers [2].

## 2 Parallel Algorithm and Results of Simulation

To simulate unsteady flows by DSMC we have to carry out enough number of statistically independent calculations (runs) $n$ to get the required sample size.

The statistical independence of runs make it possible to execute them parallel (algorithm of parallel statistically independent runs (PSIR) [3]). The efficient

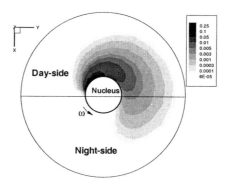

**Fig. 1.** Rotating spherical nucleus. Density contours

usage of computer resources of $p$-processor system can be provided by the implementation of two-level parallelization (TLP algorithm) in case of $n$ being less than $p$. The first level of parallelization corresponds to the PSIR-algorithm [3], the data parallelization [4] is employed for the second level inside each independent run.

This algorithm requires the memory size to be proportional to the number of the first level processors which compute single runs. It can be implemented only on shared memory computers, because it requires the arrays for each run to be stored in the shared memory in order to exclude great data exchange.

This paper presents results of the simulation of unsteady flow in the inner atmosphere of a comet using TLP. The typical values of $n$ for this problem are from 1 to 10. The simulation of 3-D unsteady flow was carried out for a comet with spherical nucleus with one active spot. The evaporation from this spot starts at the moment when it comes out to the day-side. After 180° turn of cometary nucleus the spot gets to the night-side and the evaporation is turned off. The fig. 1 shows gas density contours in the plane of nucleus rotation for the spot with radius equal to $0.35 R_w$ (Knudsen number $\text{Kn} = \lambda/R_w = 0.1$, $\lambda$ — mean free path near the nucleus, $R_w$ — the nucleus radius; Strouchal number $\text{Sh} = (R_w/v)/(\pi/\omega) = 1$, $v$ — the characteristic gas velocity, $\omega$ — the angular velocity of nucleus rotation).

The number of the first level processors $p_1$ was fixed and equal to 6. The number of the second level processors $p_2$ was varied from 1 to 6. The figure 2 depicts the experimental results for speedup as function of the total number of processors $p = p_1 \cdot p_2$ for two cases: with rotational relaxation and without it. We obtained higher speedup and efficiency for simulation of the flow with the additional physical process.

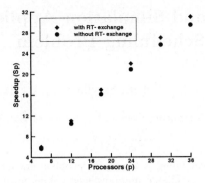

**Fig. 2.** Speedup of TLP algorithm vs total number of processors $p$

## 3 Conclusion

In this paper we presented the solution of problem of moderate complexity. In fact, the problem of simulation of cometary atmospheres is much more complex. The level of problem complexity can get higher in different ways, namely: 1) increase of the number of particles and cells; 2) more complex problem geometry; 3) consideration of additional physical processes;

In the first case we have to do more computational work but the amount of that work is different in parallel and sequential parts of the PSIR algorithm. In the second and third case the computational work is greater only in the parallel part of the data parallelization algorithm. Thus, when the problem becomes more complex a computer spends greater time to execute parallelized subroutines, hence the efficiency of the TLP algorithm will increase anyway.

The study was carried out in the Center of Supercomputing Applications [1] of the Institute for High-Performance Computing and Data Bases.

## References

1. G.A.Bird. Molecular Gasdynamics and Direct Simulation of Gas Flows. Clarendon Press. Oxford. 1994
2. A.V.Bogdanov, N.Y.Bykov, G.A.Lukianov. Distributed and Parallel Direct Simulation Monte Carlo of Rarefied Gas Flows. Lecture Notes in Computer Science, Vol. 1401. Springer-Verlag, Berlin Heidelberg New York (1998)
3. N.Y.Bykov, G.A.Lukianov. Parallel Direct Simulation Monte Carlo of Nonstationary Rarefied Gas Flows at the Supercomputers with Parallel Architecture. St.Petersburg. Institute for High-Performance Computing and Databases. Preprint N5-97. 1997.
4. I.A.Grishin, V.V.Zakharov, G.A.Lukianov. Data Parallelization of Direct Simulation Monte Carlo in Gasdynamics. St.Petersburg. Institute for High-Performance Computing and Databases. Preprint N3-98. 1998.

[1] http://www.csa.ru/CSA

# Data Mining and Simulation Applied to a Staff Scheduling Problem

Kira Smyllie

Edinburgh Parallel Computing Centre, University of Edinburgh, Edinburgh

**Abstract.** Staff scheduling is a complex problem encountered in almost all large businesses. Available manpower and skills must be deployed as effectively as possible, while being flexible enough to cope with the unexpected. This paper shows how high end PC technology in data-mining and simulation, as opposed to super-computing techniques, were applied to staff-scheduling in a large commercial organisation, where randomly fluctuating demand and a requirement for a simple, fixed schedule meant that optimisation techniques were not the most appropriate approach to the problem.

## 1 Introduction

This paper describes a staff scheduling project between Edinburgh Parallel Computing Centre and Kwik-Fit GB Ltd, to develop a system which could be used to test out alternative staff rotas at all levels of the Kwik-Fit operation. It was not within the scope of the project to produce an automatic scheduling tool. Rather, the generated computer model allows Kwik-Fit to test out business hypotheses by inputting Kwik-Fit G.B. new staffing scenarios and analysing the results.

Kwik-Fit operate around 600 centres in the UK, which perform vehicle repairs and supply replacement parts. Customers can drive into a centre and leave their car without a reservation, or make reservations. Customers expect fast, efficient service; generally a delay will mean the customer goes elsewhere. Clearly to satisfy this kind of customer expectation, centres must be adequately staffed, otherwise business will be lost.

The problem is one of resource allocation. Staff need to be allocated to a centre in order to satisfy customer demands which are periodic (on a weekly or longer basis) but with a significant random component, relating to both the number and the nature of jobs to be undertaken Staff allocation itself has a random component due to staff skill-sets, illness, days off, holidays and training. This varying demand coupled with locally varying staffing levels and the scale of the Kwik-Fit operation creates a complex scheduling and planning problem.

This paper concerns itself specifically with staff scheduling in the context of the Kwik-Fit operation, however the techniques and approach used could be applied more widely. Section 2 of the paper explains the scheduling problem, while Section 3 details the data mining. Section 4 shows how the simulation model applies results from the data-mining. Section 5 explains the HPC aspects of the project and Section 6 concludes the paper.

## 2 The Problem

Kwik-Fit centres are open 7 days a week for 10 or 12 hours each day. Centre sizes vary between 4 and 20 staff; typically with a manager, and fitters of 5 varying skill levels. Depending on size there may also be a supervisor and/or assistant manager. Centres, even with the same number of staff, have different skill mixes. Likewise every centre has its own level and pattern of demand and mix of jobs.

Prior to the project a simple rota cycling over 6 weeks was used. The exact details of this rota are not important here, however there were a number of issues to be addressed with the current approach to the rostering. There was no knowledge of the extent to which the current rota was being followed and no understanding of how the current rostering system affected the business. Also the current rota completely ignored demand patterns.

The project aimed to address these issues, firstly, by using data-mining and statistical techniques to assess the performance of centres in relation to parameters such as the size of the centre, the types of job done in the centre, and the skills available. Secondly, key results were used to develop a simulation model of the Kwik-Fit operation. Alternative rotas and scenarios can rapidly be tested and assessed using statistics produced by the model. The resulting software runs on a PC running Windows NT.

## 3 Data-mining

Data-mining concentrated on using statistical techniques both to answer questions Kwik-Fit had about their business, and to generate parameters for the simulation. The data in question came from Kwik-Fit's operational database and related to jobs and staff details in all centres across the period March 1997 - March 1998. The resulting data set is moderately large, over 1 000 000 records.

### 3.1 Data Pre-processing

As with most data-mining projects, a large portion of the effort was spent on data pre-processing. Several problems typical of this area were encountered [1].

**corrupt data:** e.g. staff are clocked in manually from their centres with a paycode indicating if they are present, training etc. Non-existent paycodes present in the database clearly indicated possible incorrect data entry.

**mixture of categorical and numerical data:** e.g. fitter skills and paycodes (indicating if a fitter is present, absent, on holiday, training, or transferred) are categorical, while the numbers of jobs done in centres are numerical.

**missing information:** Fitters are often transferred between centres to compensate for staff shortages. This transfer may not be recorded; or only recorded by the sending centre, in which case the receiving centre is unknown.

**fractured data:** Staff data came from two sources - one set containing past employees and one present employees. However, this data overlapped where any employees had been promoted, resulting in duplicate records.

Some of these pre-processing problems could be solved prior to the data analysis. Corrupt data could either have been discarded or edited. Since inaccurate data entry tended to occur in the same centres, discarding data could have significantly reduced the data on those particular centres. "Correct" entries could be inferred from certain errors e.g. transfers between centres could be matched up where the receiving centre had noted the transfer. This work required careful coding and checking. The size of the data set was reduced by extracting important features. For instance paycodes were reduced to a set of centre based figures e.g. number of days absent, and event based figures e.g. length of absence.

### 3.2 Assessing the Current Situation

The first step in the data-mining was to provide knowledge of the existing situation and understanding of its effects. In essence this was exploratory data analysis and statistical analysis guided by questions provided by Kwik-Fit.

Typical issues considered were the distribution of skills across centres, the number and types of jobs done in centres, and measures of training, absences, holidays and transfers between centres. Hypothesis testing was used to answer specific questions posed by Kwik-Fit while ANOVA was used to test for differences such as absences varying by day of the week.

Of particular use to Kwik-Fit were centre performance indicators produced at this stage. For example, analysis of absence rates (days lost to absence per person) revealed a class of centres where fitters are often absent for short periods. This scenario is of interest as a management issue. By using a statistic calculated for each centre of average absence length, combined with the absence rate, it is possible to pick out these centres very rapidly. Such information based on simple summary statistics is extremely useful to management.

### 3.3 Parameters for the Simulation Model

The second aim of the data analysis was to generate parameters for a model of the Kwik-Fit operation. The model would then be used to predict the daily number of work units done in centres under different rotas, taking into account random fluctuations in staff levels.

A first naive attempt tried to relate productivity of a centre to the number of staff present. The regression analysis failed to produce good results on a daily basis. This was because on a day to day basis, demand in centres was extremely random. Only when the centre was working at full capacity could the number of staff present in the centre have a bearing on productivity.

Instead the model was designed to estimate the capacity of a centre and then compare this to the expected demand. Separate demand curves for each weekday in each centre were generated, based on work units figures. For absences, training and holidays, global frequency distributions were used across all the centres however with different parameters for each centre. This required the assumptions that absences, training and holidays followed the same patterns across the whole of the Kwik-Fit operation, and that work done was representative of demand.

## 4 Simulation

The generated parameters were then used in the simulation of the Kwik-Fit operation. The model was a discrete event simulation following the paradigm in [2]. Events were defined to be the beginning and end of: scheduled days off; absences; holidays; training; and observations of staff numbers and demand.

Scheduled days off are derived from a rota input to the model. Absences, training and holidays are calculated using the distributions produced by the data mining; frequency distributions determining when each starts, and length distributions setting the time for when each ends. At the end of each day modelled, the model counts the number of staff in each centre and simulates that day's demand from the demand distributions. The capacity of a centre can then be calculated from the number of staff present, the maximum number of work units a fitter can do in a day, and the number of hours each centre is open.

Using this approach, rotas can be compared by considering $|capacity-demand|$ Clearly better rotas will bring the result of this closer to 0, that is, when the supply of staff to the centres is matching as closely as possible to demand.

## 5 HPC Aspects of the Project

The work for the project was carried out on a dual processor PC. However, parts of the project were computationally intensive and could equally and more swiftly have been carried out on higher performance computers.

The data cleaning was particularly intensive, especially for the paycodes data. Several of the steps involved required significant computational effort e.g. determining the correct skill level for duplicate entries. Resolving this problem alone took approximately four hours to run on the PC.

There is also scope for the simulation model to use HPC. The structure allows centres to be simulated in parallel, so, for instance, the model could be linked directly to the database and use some data-mining layer to generate up-to-date parameters or distributions for the simulation.

## 6 Conclusions

The approach to staff scheduling described in this paper uses data mining and simulation techniques to assess the effectiveness of different staff rotas. These HPC techniques were used to develop the software, however no investment in HPC technology was required as the resulting application ran on standard PC's. The approach could be used on or extended to effectively use HPC technology.

## References

1. Famili,A., Shen, W.M., Weber, R., Simoudis, E.: Data Processing and Intelligent Data Analysis, Intelligent Data Analysis, Jan. 1997, Vol.1 No.1
2. Pidd, M.: Computer Simulation in Management Science 4th edn., Wiley (1998)

# Neural Network Software for Unfolding Positron Lifetime Spectra

**P. Lindén, R. Chakarova\*, T. Faxén[1] and I. Pázsit**

Department of Reactor Physics, Chalmers University of Technology,
S-412 96 Göteborg, Sweden
[1]National Supercomputer Center, Linköping University,
S-581 83 Linköping, Sweden
rum@nephy.chalmers.se

Abstract

A neural network based program has been developed to unfold lifetimes and amplitudes as well as lifetime distributions from positron spectra. The program is written in Fortran 90 and ported to two different shared memory systems, a Cray Origin 2000 using OpenMP, and a Cray 90 using autotasking. The performance has been examined by profiling tools and the program has been optimized. The achieved speedup will allow a more detailed study of the capability and accuracy of the neural network based method to analyse positron spectra.

## I. Introduction

Positron annihilation lifetime spectroscopy is a technique used to characterize the free volume structure of polymers. The experimental spectra are composed of number of exponential decaying functions attributed to annihilation of positrons at different polymer states. The traditional data analysis is based on fitting the spectra to a sum of exponential terms convoluted with a time resolution function with a constant background added. More sophisticated routines involve extraction of lifetime distributions from the spectra.

Recently, a new method for unfolding lifetimes and amplitudes as well as lifetime distributions from positron spectra has been suggested [1]. It is based on the use of artificial neural networks (ANNs). Its applicability has been tested by selecting a simple case of determining three discrete lifetimes and amplitudes. The method has the potential to unfold an unknown number of lifetimes and their distributions. It has very short identification times and is good alternative to the existing methods. The preliminary investigations show, however, that the task of constructing of a proper and efficient network is not trivial and the training procedure may require weeks long CPU times.

Here we present our work to analyse the program performance and decrease the computation time by implementing the code on a shared memory parallel machine and a vector computer. This will allow a more detailed study of the capability of the ANN based method and its precision limits.

## II. Network model

The network is constructed as a multilayered feed-forward ANN, (Fig. 1. a), with a backward error propagation and the Generalized Delta Rule as a learning algorithm [2]. The principles of such network are briefly described below. The output $O_j^{k+1}$

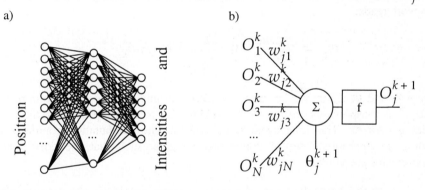

Fig. 1. (a) the structure of a three layered feed-forward neural network
The network used consists of an input and an output layer. In between these layers there is one hidden layer. (b) a graphical representation of a node.

from a node $j$ in the layer $k+1$ can be written as

$$O_j^{k+1} = f\left(\sum_{i=1}^{N} w_{ji}^k \cdot O_i^k + \theta_j^{k+1}\right) \quad (1)$$

where $w_{ji}^k$ is the connecting weight between node $i$ in layer $k$ and node $j$ in layer

$k+1$, and a bias term $\theta_j^{k+1}$ is added for every node. The function $f(x)$ used in this work is the sigmoid, or the logistic function. The layer with $k = 1$ is called the input layer and an input pattern is applied to that layer, i.e. $O_i^1 = X_i = (x_1, x_2, ..., x_N)$. The input signal is propagated through each subsequent layer according to Eq. (1) until a network output $O_l^K$ is generated at the last layer, $k = K$, of the network, the output layer. The signal from the output layer is then compared to the desired output, $Y_l = (y_1, y_2, ...y_L)$, and an error, or cost function $E(w)$, dependent on the connecting weights is defined as:

$$E(w) = \frac{1}{2} \cdot \sum_{l=1}^{L} (y_l - O_l^K)^2 \qquad (2)$$

The essence of the training procedure is to minimize the error function with respect to $w$ and, the usual gradient descent method suggests changing each connecting weight by an amount proportional to the local gradient of $E$. By using the chain rule, the local gradients can be calculated and in that derivation it can be seen that the error is propagating backwards in the network, thus the name error backward propagation algorithm. In the literature, one frequently add a momentum term in the updating procedure. Finally, the updating rule for the connecting weights $w_{ij}^k$ used in this work reads:

$$w_{ij}^k = w_{ij}^{k,old} - \eta \cdot \frac{\partial E}{\partial w_{ij}^k} + \alpha \cdot \Delta w_{kj}^{k,old} \qquad (3)$$

where $\eta$ and $\alpha$ are training parameters called learning rate and momentum rate, and $\Delta w_{kj}^{old}$ is the previous change of the weight. The process repeats encoding all the training patterns. Thus one cycle is completed. The computation continues until finding an appropriate set of connecting weights, such that the error function requirements are satisfied.

In the particular case of positron spectra unfolding, the network size is large. The number of the input nodes, up to 1500, corresponds to the number of the channels in the positron spectrum. The number of the output nodes is determined by the desired output parameters. In the simplest case of three components, the lifetime mean values and intensities, (i.e. 6 parameters), are only needed. However, when the number of components is not predefined, and the lifetime distributions are to be extracted, the network output layer can significantly expand. In that case, the whole region of possible lifetime values should be divided into bins and attached to a large number of nodes, indicating the probability for a lifetime to be in a particular bin. The number of the hidden layers and the nodes in each of them are free parameters.

Positron spectra with known intensities and lifetimes have been generated by a simulation program, originally developed elsewhere, and these have been used as training input/output pairs. The number of the patterns in the simplest case of three component and discrete lifetimes was set to 575.

### III. Parallel and vector computer implementation

A Fortran 90 program has been written and ported to Cray Origin 2000 in Chalmers University, Göteborg. The profiling tool Perfex and the integrated package of performance tools SpeedShop have been used to analyse the program behaviour on one processor and when running in parallel. The most time consuming sections have been identified to include calculations of the output of the first hidden layer as well as the update of the weights between the input and the first hidden layer (Eq. (1) and Eq. (3)).

One processor optimization has been achieved by ordering of the array indexes in a such way, that the data in a cache line are accessed before the line can be displaced from the cache (stride-1 accesses for maximum cache line reuse). The code has been parallelized by inserting OpenMP directives for concurrent computation of the loops of the input and error signals as well as weight update. A static schedule has been specified, where the loop iterations are divided among the threads in contiguous pieces, and one piece is assigned to each thread.

In order to unify the performance measurement on Cray 90 and Cray Origin 2000, the following procedure has been accepted. The number of operations needed per one iteration, (one update of the weights), has been estimated by the hpm-monitor of Cray 90. This number multiplied by the corresponding number of iterations, and divided by the particular running time, gives the performance in units MFlops on the particular system. The performance results are shown in Fig. 2 from start-up and synchronization of the frequent and short parallel regions. The scalability is limited by

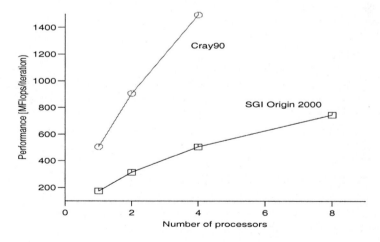

Fig. 2. Performance results for Cray 90 and Cray Origin 2000

the additional overhead incurred from start-up and synchronization of the frequent and short parallel regions.

The dominance of vector operations in the ANN algorithm makes the code suitable to run on a vector computer. The implementation on Cray 90 vector computer, in the National Supercomputing Center at Linköping University, shows one processor performance of 506 MFlops per iteration. Parallel version has been produced by using autotasking, which is very similar to OpenMP (e.g. *!mic$ doall* instead of *!$OMP parallel do* directive). It was found, that the scalability is similar to that on Cray Origin 2000. Thus, 1.47 GFlops per iteration have been measured for 4 processors of Cray 90 (see fig. 2).

One goal of this work is to summarize the results into a Fortran 90 module that should be available for general implementation to problems involving ANNs using parallel computers.

# References

[1]   I. Pázsit, R. Chakarova, P. Lindén and F. H. J. Maurer, *"Unfolding Positron Lifetime Spectra with Neural Networks"*, SLOPOS-8, Cape Town, South Africa, 6-12 September 1998, Submitted to Applied Surface Science (1998)

[2]   I. Pázsit and M. Kitamura, *"The Role of Neural Networks in Reactor Diagnostics and Control"*, Adv. Nucl. Sci. Techn. 24 (1996) 95

# MPVisualizer: A General Tool to Debug Message Passing Parallel Applications

Ana Paula Cláudio[1], João Duarte Cunha[2], and Maria Beatriz Carmo[1]

[1] Faculdade de Ciências da Universidade de Lisboa
Departamento de Informática- Campo Grande - Edifício C5 - Piso 1 - 1700 LISBOA - Portugal
{apc, bc}@di.fc.ul.pt
[2] Laboratório Nacional de Engenharia Civil
Av. do Brasil, nº 101, 1799LISBOA CODEX - Portugal
jdc@lnec.pt

**Abstract.** The paper describes MPVisualizer (Message Passing Visualizer) a general purpose tool for the debugging of message passing parallel applications; its three components are the trace/replay mechanism, the graphical user interface and the central component, called visualization engine. The engine, which plays the main role during the replay phase, can be used with different message passing environments and different graphical environments. This is a major step to make MPVisualizer a general tool. Additionally, the engine is able to recognize potential race conditions and can be easily re-programmed to detect specific predicates.

## 1 Introduction

Performance improvement is the main goal being pursued when a parallel program replaces a sequential one. Before determining if performance improvement was really achieved, the parallel program has to be tested and debugged. However, debugging parallel applications is considerably more difficult than debugging sequential programs because the target is a set of communicating processes. It is usual to classify communication events as external events in contrast with internal events which involve one process only. Therefore, while debugging this kind of applications, two sorts of bugs may be found: bugs related to internal events and bugs related to external events.

The proposed tool, MPVisualizer -Message Passing Visualizer- uses a graphical interface to help programmers in finding and understanding the second sort of bugs. That is, graphically observable events are communication or external events. However, bugs concerning internal events can be detected if a sequential debugger is integrated with our tool. This integration is perfectly compatible with MPVisualizer but is not implemented in the current version.

The tool includes a trace/replay mechanism and a graphical interface. Between these two components, a central component, the visualization engine, makes the tool easily adaptable to different message passing mechanisms and different graphical environments [3]. Besides, the engine detects and notifies the occurrence of race conditions, is capable of detecting predicates and permits observation both during re-execution and post-mortem.

## 2 MPVisualizer

The internal work of a parallel application may be non deterministic, that is, two successive executions of the application with the same input may exhibit different behaviours, even though they may produce the same final output. A trace/replay mechanism makes a particular execution of the parallel application repeatable, allowing cyclic debugging, a frequently used technique in sequential programs. The replay mechanism adopted is similar to the one described in [9]. The mechanism includes two distinct phases: the trace phase and the replay phase. Although minimal, the stored information during trace phase is enough to ensure that, during the replay phase, each process will consume exactly the same messages, in the same order.

It should be emphasized that it is not necessary to modify the code of a parallel application to use our debugging tool, since the monitoring code is inserted in the standard libraries of the message passing software.

During the replay phase, the visualization engine builds an object-oriented model of the application. This model provides the necessary semantic feedback to answer the questions about observable events that the user may ask (using the graphical interface). In MPVisualizer, observable events are not only communication events but also the beginning and ending of a process. Each time an observable event occurs in a process, during replay, a block of information containing the necessary data is sent to a process named spy. During replay phase, there is one spy process running in each machine that is executing processes. Each of these spies receives data blocks from its local processes and sends them to the main process[1]. The main process is a sort of one-way bridge between spy processes and the object oriented model of the application.

The classes in the object oriented model, figure 1, can be categorized in two groups. The first group, the kernel, includes the basic classes that do not depend on the graphical environment and they encapsulate the data and behaviour that are generic for any message-passing application. The second group of classes, comprises the classes that deal with the graphical representation and that derive from the former classes.

---

[1] In the present implementation communication channels between spies and the main process are sockets.

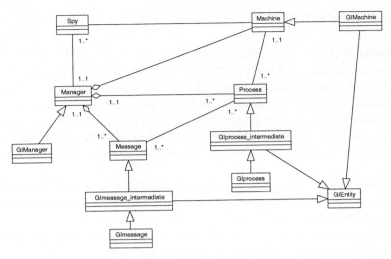

**Fig. 1** Class Diagram

Predicate detection is implemented using inheritance. The kind of predicates that can be detected depends on the granularity of the observation performed by the tool. Since in MPVisualizer detectable predicates are those that concern communication events, as well as beginning and ending of processes, the classes implementing the predicate detection algorithms are derived from one of two kernel classes: "process" or "message".

MPVisualizer automatically signals all potential race conditions. These conditions are responsible for the non determinism in the internal behaviour of parallel applications. Therefore, if race conditions are detected, the programmer is notified of all the points in the application that can be potentially responsible for variations in the behaviour of successive executions. This is a major step in debugging. The detection code for race conditions is encapsulated in only one kernel class named "process".

For the graphical interface, time space diagrams [8] have been adopted. In addition to the graphical representation of the execution, the user can obtain more detailed information about the displayed entities: processes, messages and communication events. This information is contained in pop-up windows that open when the corresponding entity symbol is selected in the graphical interface.

It should be emphasized that dependencies from the message-passing software are strictly restricted to spy processes while dependencies from the graphical software are restricted to graphical classes of the model.

## 3 Present and Future Work

The trace/replay mechanism was implemented and tested on top of PVM [5]. The visualization engine and the graphical interface were implemented and tested using C++ and Motif. So far, MPVisualizer works only in post-mortem mode. It will be developed in order to support also observation during re-execution. To be used in this mode, a breakpointing mechanism should be available. The current graphical interface is not adequate to large volumes of information. There are planes to improve its scalability, namely using zoom in context techniques, e. g. as proposed in [2].

PVM has gained wide acceptance and Geist et al. consider it a *de facto* standard [6]. This, however, does not invalidate our interest in testing MPVisualizer with other message passing systems, such as MPI[4], P4[1] and Parmacs[7]. Description of similar tools and comparison with the one described here would be interesting but is outside the scope of this paper and has been made elsewhere [3].

## References

1. Butler, R., Lusk, E.: User's Guide to the p4 programming system. Tech. Rep. ANL-92/17. Argonne National Laboratory, Mathematics and Computer Science Division (1992)
2. Carmo, M., Cunha, J., Cláudio, A.: Visualization of Geometrical and Non-geometrical Data. In: Proceedings of the WSCG'99, Plzen (1999)
3. Cláudio, A., Cunha, J., Carmo, M.: Debugging of Message Passing Parallel Applications: a General Tool. In: Proceedings of VECPAR '98 - 3$^{rd}$ International Meeting on Vector Parallel Processing, Porto (1998)
4. Dongarra, J., Otto, S., Snir, M., Walker, D.: An Introduction to the MPI Standard. Tech. Rep. CS-95-274. University of Tenesse (1995)
5. Geist, A., Beguelin, A., Dongarra, J., Jiang, W., Manchek, R., Sunderam, V.: PVM: Parallel Virtual Machine. MIT Press (1994)
6. Geist, A., Kohl, J., Papodopoulos, P.: PVM and MPI: a comparison of features. In: Calculateurs Paralleles, Volume 8(2) (1996)
7. Hempel, R., Hoppe, H.-C., Supalov, A.: PARMACS-6.0 library interface specification. Tech. Rep., GMD, PostFach 1316, D-5205 Sankt Augustin 1, Germany (1992)
8. Lamport, L.: Time, Clocks, and the Ordering of Events in a Distributed System. In: Communications of the ACM, Vol. 21( 7) (1978) 558-565
9. Leblanc, T., Mellor-Crummey, J.: Debugging Parallel Programs with Instant Replay. In: IEEE Transactions on Computers, Vol C-36(4) (1987)

# Effect of Multicycle Instructions on the Integer Performance of the Dynamically Trace Scheduled VLIW Architecture

Alberto Ferreira de Souza and Peter Rounce

Department of Computer Science
University College London
Gower Street, London WC1E 6BT - UK
a.souza@cs.ucl.ac.uk, p.rounce@cs.ucl.ac.uk

**Abstract.** *Dynamically trace scheduled VLIW* (DTSVLIW) architectures can be used to implement machines that execute code of current RISC or CISC instruction set architectures in a VLIW fashion, delivering instruction level parallelism (ILP) with backward code compatibility. This paper presents the effect of multicycle instructions on the performance of a DTSVLIW architecture running the SPECint95 benchmarks.

## 1 Introduction

The classic approaches to providing ILP are VLIW and superscalar architectures. With VLIW, the compiler is required to extract the parallelism from the program and to build *Very Long Instruction Words* for execution. This leads to fast and (relatively) simple hardware, but puts a heavy responsibility on the compiler, and object code compatibility[4] is a problem. In Superscalar, the extraction of parallelism is done by the hardware which dynamically schedules the sequential instruction stream on to the functional units. The hardware is much more complex, and therefore slower than a corresponding VLIW design. The peak instruction feed into the functional units is lower for Superscalar. A number of approaches[1][2][3] have been examined that marry the advantages of the contrasting designs: the Superscalar dynamic extraction of ILP, the simplicity of the VLIW architectures. The approach presented here follows that first presented by Nair and Hopkins[3]. Our architecture, the *dynamically trace scheduled VLIW architecture* (DTSVLIW) [4], demonstrates similar results to theirs in providing significant parallelism, but with a simpler architecture that should be much easier to implement. In our earlier work[5], we had zero latency load/store instructions. Here we present results with more realistic latencies.

## 2 The DTSVLIW Architecture

The DTSVLIW has two execution engines: the Primary Processor and the VLIW Engine, and two caches for instructions: the Instruction Cache and the VLIW Cache. The Primary Processor, a simple pipelined processor, fetches instructions and does the first execution of this code. The instruction trace it produces is *dynamically scheduled* by the Scheduler Unit into VLIW instructions, saved in blocks to the VLIW Cache for re-execution by the VLIW Engine. The Primary Engine, executing the Sparc-7 ISA, provides object-code compatibility; the VLIW Engine VLIW performance and simplicity. The Scheduler unit works in parallel with the Primary Engine. Scheduling does not impact on VLIW performance as it does in Superscalar.

## 2.1 The Scheduler Algorithm

The major design problem is the scheduling, which has to maximise the parallelism extracted from the trace, while not extending the machine cycle time. The Scheduler Unit uses a pipelined version of the First Come First Served (FCFS) scheduling algorithm[6]. FCFS has advantages for hardware implementation: it operates with one instruction at a time in execution order; it produces optimum or near-optimum scheduling[6]. We showed it to be suitable for pipelined implementation in [5]. The Scheduler Unit uses a circular *scheduling list,* to build a block of VLIW instructions, using out-of-order execution, register renaming, and speculative execution. An instruction arriving from the Primary Engine is placed at the end of the block, moving up the block on subsequent clock cycles, dependencies allowing. The block starts with one element, increasing to a fixed block maximum. Only one instruction in each block element has to be checked for moving up on a cycle: moving up produces out-of-order execution. Speculative execution moves an instruction up past conditional branches, but delays its write-back stage until the branch outcomes are determined.

For a multicycle instruction, two copies, A and B, are inserted, separated by the instruction latency, into the scheduling list to identify the list region where the instruction is active. B's role is for dependency checking against instructions moving up. A and B are treated partly as other instructions, partly as one instruction, e.g. their separation is kept constant. B is not saved in the VLIW Cache. Scheduling a multicycle instruction lengthens a block by the latency of the instruction, impacting efficiency since the longer block is more difficult to fill, reducing parallelism.

## 3 Methodology and Experiments

A simulator of the DTSVLIW has been implemented in C. All results were produced with the simulator running in *test mode*: a *test machine* is run in tandem with the DTSVLIW. Comparison of the state of the 2 machines after an instruction or a VLIW block completes validates the DTSVLIW machine. The test machine measures the instructions executed to determine the ILP achieved. Model parameters, benchmark programs (SPECint95), together with the input sets used can be found in [5]. Each program was run for 50 million or more instructions, as counted by the test machine.

### 3.1 Effect of the Block Size and Geometry

Fig 1 shows the effect of the block size in terms of the number of instructions and block geometry (instructions per VLIW instruction (width) versus VLIW instructions per block (length)) on performance. The experiments were performed with perfect instruction and data caches, large VLIW Cache (3072-Kbyte), and no next VLIW instruction miss penalty. The performance of machines with the same block sizes and different geometry is significantly different. Thus, the performance with 4x8 geometry is lower than with 8x4 geometry for all benchmarks. The block width and length affect the machine cost in different ways. Block width impacts on the number of functional units, data cache ports, and register file ports; the block length on the number of renaming registers, the length of load/store and checkpoint recovery lists [5], and the required size of the VLIW Cache for the same performance. To increase just the width or just the length of the block does not appear to be the best approach. A DTSVLIW with 8x8-block geometry generally performs better than with 4x16 and 16x4 geometries. Benefits from large block size do not grow linearly. The performance of the 16x16 geometry on the ijpeg benchmark is extraordinary. This benchmark spends most of its execution in one loop. With a large enough block size, more than one iteration of the loop can be scheduled into a single block, allowing instructions from these iterations to be overlapped, extracting much greater parallelism.

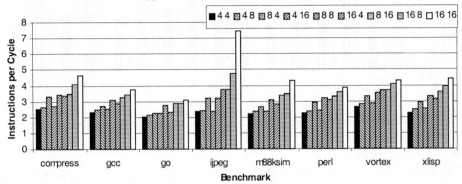

**Fig 1.** Variation of parallelism with the block size and geometry

### 3.2 Effect of the Load/Store Instructions Latency

Fig 2 shows the effect of the load/store instructions latency on an 8x8 geometry: LxSy stands for loads with latency $x$ and stores with latency $y$. Latency is the number of cycles before an instruction's results can be used. Load latency has a severe impact – 25.4% average performance loss with 1-cycle and 50.2% with 2-cycle latency, because loads are frequent and their data is usually required imminently. Store latency does not have such a strong impact, as stored data is usually not imminently required.

With longer blocks the impact of Load/Store latency is smaller. For 8x16 (Fig 3) there is 20.5% average performance loss with 1-cycle latency and 42.6% with 2-cycle latency. With a longer block, there is more opportunity to accommodate instructions in the empty VLIW instructions created by the multicycle scheduling. This results in better scheduling and performance, but the latency impact is still high.

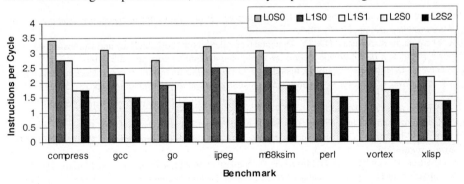

**Fig 2.** Variation of the parallelism with the load/store instructions latency – 8x8-block

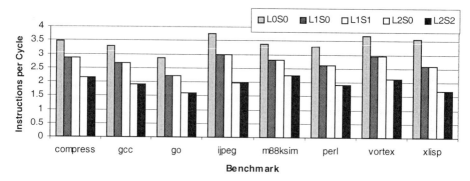

**Fig 3.** Variation of the parallelism with the load/store latency – 8x16-block

## 4 Conclusion

The results show that the DTSVLIW can achieve ILP as high as 7 and average ILP superior to 4 with a large machine geometry. Multicycle load instructions impose a severe performance penalty on the DTSVLIW architecture and it is clear that it is important to get their latency as close to zero as possible: single cycle load operation. Single cycle stores are not so important. Low load/store latency (2 cycles) is achievable with a high frequency clock as demonstrated in the DEC-Alpha[7]. We calculated across all our benchmark results the average number of VLIW cycles per program with 8x16-block geometry of 98.57%, strongly suggesting that the DTSVLIW architecture is effective in taking advantage of its VLIW Engine. The Primary Processor and the VLIW Engine in the DTSVLIW can have high clock rates. The simplicity of the scheduling algorithm means that a similar high clock rate should be achieved for the Scheduler Unit, leading to an overall clocking rate similar to, if not higher than, high clock rate superscalar architectures, but achieving higher ILP.

## References

1. B. R. Rau, "Dynamically Scheduled VLIW Processors", *Proc. of the 26th International Symposium on Microarchitecture*, pp. 80-92, 1993.
2. K. Ebcioglu, E. R. Altman, "DAISY: Dynamic Compilation for 100% Architectural Compatibility", *Proc. of the 24th International Symposium on Computer Architecture*, pp. 26-37,1997.
3. R. Nair, M. E. Hopkins, "Exploiting Instructions Level Parallelism in Processors by Caching Scheduled Groups", *Proc. of the 24th International Symposium on Computer Architecture*, pp. 13-25,1997.
4. A. F. de Souza and P. Rounce, "Dynamically Trace Scheduled VLIW Architectures", *Proceedings of HPCN'98, in Lecture Notes on Computer Science*, Vol. 1401, pp. 993-995, April 1998.
5. A. F. de Souza and P. Rounce, "Dynamically Scheduling the Trace Produced during Program Execution into VLIW Instructions", *To be published in the Proceedings of 13th International Parallel Processing Symposium*, 1999.
6. S. Davidson, D. Landskov, B. D. Shriver, P. W. Mallett, "Some Experiments in Local Microcode Compaction for Horizontal Machines", *IEEE Transactions on Computers*, Vol. C30, No. 7, pp. 460-477, July 1981.
7. J. Keller, "The 21264: A Superscalar Alpha Processor with Out-of-Order Execution", *9th Microprocessor Forum*, 1996.

# MAD - A Top Down Approach to Parallel Program Debugging

Dieter Kranzlmüller, Roland Hügl**, and Jens Volkert

GUP Linz, Johannes Kepler University Linz,
Altenbergerstr. 69, A-4040 Linz, Austria/Europe,
kranzlmueller@gup.uni-linz.ac.at,
http://www.gup.uni-linz.ac.at:8001/

**Abstract.** The MAD environment is a toolset for debugging parallel message passing programs. In contrast to other existing tools and prototypes, where error detection facilities are implemented by combining sequential debuggers, MAD is designed as a top-down approach. The main interface for program analysis is the event graph display, a state-time diagram that shows process interaction. Based on this program execution overview, activities on lower levels of abstraction like event inspection and breakpoint setting can be initiated. The advantage of using a global view as starting point is improved program understanding and simplified coordination of debugging activities.

## 1 Introduction

The testing and debugging phase of the software lifecycle is crucial for the quality of a program. While many debugging tools for sequential programs are available and widely accepted, only few commercial tools and little or proprietary support in terms of tool prototypes exist for parallel software engineers. A well-known commercial representative in that area is the TotalView debugger [2], and several tools exist in academic or research organizations, e.g. p2d2 [3], and pdbg [1]. All these tools have in common, that they assemble several instances of traditional sequential debuggers in order to allow parallel error detection. Additional levels of abstraction are only provided as extensions above these low-level debuggers.

The problem with these bottom-up approaches is the focus of investigation, which is a sequential, textual representation of one or more concurrently executing processes. The missing point is a connection to the parallelism that occurs in the target programs, because text is inadequate to express the complex, multidimensional relationships of executing parallel programs [6]. Therefore it is difficult to steer and manage multiple, concurrently executing tasks and their interaction in a useful way. As a consequence many programmers still rely on `printf`-functions in order to obtain program state information.

The solution of the Monitoring And Debugging environment *MAD* for message passing programs [5] is different from most other parallel debuggers, because

---
** presenting author

it is based on a top-down approach. The goal of this paper is to emphasize the differences and briefly discuss the advantages of MAD's approach.

The starting point for investigations is a global view of the program's execution, which is displayed as event graph representation. This is a state-time diagram, consisting of vertices for occurring events and edges describing either communication or computation of processes. Based on this representation, traditional techniques like breakpointing and program inspection can be performed. Besides that, MAD also contains support for analyzing nondeterministic programs by integrating record&replay and event manipulation.

## 2 Functionality of MAD

In the software lifecycle the starting point for debugging is the testing phase, where programs are executed with different sets of inputs until incorrect behavior is observed. This is either a program failure or computation of wrong results. In that case a debugging tool is used to investigate the program's states during a faulty execution, and to track the errors back to the responsible lines of code.

Thus, the basic features of any debugger are steering possibilities for program execution (like breakpointing and single stepping), as well as state inspection and modification functions. Since parallel programs may be nondeterministic, which means that successive executions with the same input may not yield the same results, a technique called record&replay is required. Only record&replay allows to generate equivalent executions of nondeterministic parallel programs [5].

The debugging strategy of MAD is applied with the following three steps, (1) instrumentation, (2) testing, and (3) debugging. During instrumentation (1) the program is manipulated to include a monitor in order to generate the required data for equivalent re-execution. The testing phase (2) is applied and different sets of inputs are feed into the program. Only if incorrect program behavior is observed during testing, debugging is initiated.

In debugging mode (3), the programs execution is visualized as an event graph (see figure 1, left window). The vertical axis shows the participating processes, while the horizontal axis displays the time. Events that occurred during the execution are displayed as vertices, with different symbols indicating different types of events. The arcs between the events are either sequential computation on one particular process, or communication between two connected events (e.g. corresponding send and receive). Such an event graph provides a global view of the program and may already contain valuable information about the reasons for incorrect behavior [4].

However, the graphical representation also serves as the starting point for more advanced error detection activities. Firstly, the user can inspect the visualization by clicking on objects, which opens event inspection windows and shows all the collected details about the selected event (see figure 1, right window). Of major importance is the connection to the source code, which relates the graphical object to the lines of code responsible for it's occurrence. Secondly, the user can also arrange the display by grouping and hiding of processes,

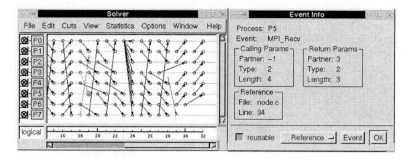

**Fig. 1.** MAD debugging session

as well as collapsing and compressing event patterns, which is especially useful when analyzing many processes [4].

Besides this static post-mortem investigations of the program's execution, the event graph display is also used for re-execution and steering the application. The execution as displayed in the event graph is used as constraint for the equivalent replay, which is required for nondeterministic codes. Additionally, the user can set breakpoints in the graphical view, instead of selecting source code lines for breakpoints in the textual representation. In this case, a breakpoint is called a cut, because more than one process is halted in order to provide a consistent state of the execution.

A cut is generated by selecting one particular event where the execution should stop. All required breakpoints on the other processes are then computed accordingly in order to establish a consistent cut. The example event graph of figure 2 shows a cut, that has been set by selecting a breakpoint on process P7.

During execution (or replay respectively) each process will stop whenever it reaches the breakpoint associated with the next cut. Then the user can attach a sequential debugger to any of the available processes, again by selecting the process from the graphical display, and perform traditional debugging thereafter. In figure 2 the event graph display is connected to an instance of the xxgdb debugger. The source display of xxgdb shows the breakpoint, that has been set at the statement corresponding to the selected event of the event graph.

## 3 Conclusion and Future Work

The current implementation of MAD has been used for message passing systems (MPI and nCUBE2) at our university. Initial evaluation from students working on small projects was encouraging and provided useful feedback for minor improvements. Additionally, the environment was also successfully tested on larger examples, especially from the parallel computer graphics domain.

The goal of improved understanding and usability was achieved and the chosen representation of the event graph was highly accepted. Due to its abstraction, MAD allows to perform activities that are difficult if not impossible with other debugging tools. It seems, that the top-down approach as proposed by MAD is better suited for many activities of the debugging cycle.

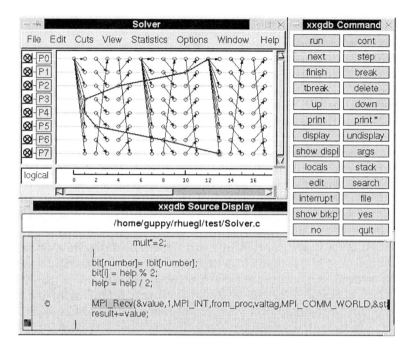

**Fig. 2.** MAD debugging session

For the near future, we are looking to complete a public domain distribution of the toolset, that will be available for other users willing to try our toolset for their applications. Although we believe, that MAD will be successful in most cases, this testing will provide valuable information for future extensions.

# References

1. J. C. Cunha, J. Loureno, J. Vieira, B. Mosco, and D. Pereira, *A framework to support parallel and distributed debugging*, Proc. of HPCN'98, Amsterdam, Netherlands, (1998).
2. Dolphin Interconnect Solutions Inc., *TotalView 3.8*, Document available via WWW at: http://www.dolphinics.com/tw/download/tv3.8.0-webdocs/User_Guide.ps (1998).
3. R. Hood, *The p2d2 Project: Building a Portable Distributed Debugger*, Proc. SPDT'96, ACM SIGMETRICS Symp. on Parallel and Distr. Tools, Philadelphia, USA, pp. 127–136, (May 1996).
4. D. Kranzlmüller, S. Grabner, J. Volkert, *Event Graph Visualization for Debugging Large Applications* Proc. SPDT'96, ACM SIGMETRICS Symp. on Parallel and Distr. Tools, Philadelphia, USA, pp. 108–117 (May 1996).
5. D. Kranzlmüller, S. Grabner, J. Volkert, *Debugging with the MAD Environment*, Parallel Computing, Vol. 23, No. 1–2, pp. 199–217 (Apr. 1997).
6. C. M. Pancake, *Visualization techniques for parallel debugging and performance analysis tools* in: A. Y. Zomaya (Ed.) *Parallel Computing: Paradigms and Applications* Intl. Thomson Computer Press, pp. 376–393 (1996).

# High-Performance Programming Support for Multimedia Document DataBase Management

Pascal Adam[1], Hassane Essafi[1], Michel-Pierre Gayrard[1], and Marc Pic[1]

LETI (CEA – Technologies Avancées)
Marc.Pic@cea.fr

**Abstract.** The european ESPRIT project STRETCH is dedicated to the management of documents by their content. We present in this paper the high-performance developments and the programming support required by STRETCH indexing and retrieval processes.

## 1 Context of the STRETCH Project

One of the aim of the STRETCH project is to develop a pre-production prototype capable of storing and retrieving imaged documents by their pictorial content. The system will allow users to search document bases with series of techniques oriented on texts, images and structures and on some minor categories dedicated to specific applications. Each input document is analysed to produce information blocks in the three main categories and in the minor ones. Each information blocks are themselves processed and archieved in separate databases with appropriate links between the information blocks. Each query provided by the final user of the system is decomposed in relations between basic queries (fragments) corresponding to the previous categories. Each frgment is processed by the specific database to match the corresponding texts, images or structural elements. Answers are composed to produce a multirank-ordered list.

## 2 Computational Characteristics of Algorithms involved in STRETCH

We present in this section the computational phases involved in STRETCH context with the various methods involved in today's implementation.

**Automatic Analysis:** This part contains low-level image processing to achieve document enhancement (using Adaptive threshold, Morphological processing, Smoothing operators, Image sharpening and texture enhancement, Skew detection and correction, Interpolation, and Form removal) and image segmentation (using Connected Component Analysis, Hough Transform, Run-Length Smoothing Algorithm (RLSA) and Texture Analysis) and various other tools including Optical Character Recognition (OCR). Data-parallel programming based on images as primitive data structures is adequate and allows to extract a wide level of parallelism. but is not sufficient. For instance, Wavelet Transform

(WT) uses several filters to perform on the same image. Each filtering can be consider as a very SIMD data-parallel program based on the pixels of the image, but the various filters to perform can be managed concurrently, introducing a SPMD task-parallelism between data-parallel tasks: **a nested parallelism** (NP).

**Image Archiving**: The goal of this part is to extract one or several vectors representing a summary of the image characteristics (geometrical, optical, numerical, statistical, ...) and to store it in the image server database. Computation of the index is a highly computation time consuming process due to the use of Iterated Function Systems (IFS), Zernike moments, irregular triangulation of interest points (Stephen-Harris) and Principal Component Analysis (Karhunen Love transform). Irregularity requires specific data-parallel structures. The introduction of the data-warehousing and clusters tend to produce NP.

**Text Archiving**: The text indexing engine retained for STRETCH is called SPIRIT and is based on a Natural Language Processing facility which allows the indexing of multilingual documents. A text in multiple free languages is processed through a morpho-syntactic analyser to extract basic keywords and relations between those keywords. Extraction of parallelism in SPIRIT is based on the decomposition of the relations between words in analysed statements following a tree structure.

**Structure Archiving**: Text components of structured documents are divided into formal sections (titles, footnotes, ...) and content-bearing sections. The ability to locate and identify structures within a document allows to classify pre-printed or structural information from real content. Object-oriented database to store structure introduces object parallelism

**Multimedia Queries**: This component of the system split the queries (expressed in Document Query Language (DQL)) in basic queries (fragments) for the various primitive components of the database (texts, images, structures,...). Primitive component queries are matched separetely, introducing a client-server parallelism , and the results are fusionned to provide the final answers. Fragment query is based on an indexing phase of the querying fragment matched with the indexes of the corresponding database and requires equivalent high-performance solutions than indexing.

To sum up, the support for STRETCH includes data parallel programming, task-parallel programming for concurrent tasks, nesting of various levels of data and task parallelism, irregular programming for irregular objects, and distribution between servers of the various independant component databases. The speed-up of task parallelism is often more important than the data-parallel speed-up. Sadly, the capacity to extract independant tasks is less important than the data-parallel one.

## 3 Hierarchical Distribution and Parallelism with CORBA and $C_T$++

In the STRETCH prototype, we have chosen CORBA to implement the client-server relations between the various servers. CORBA efficiency is sufficient as long as communications time between the servers can be easily overwhelmed by the time consumption of the algorithms running on each server. However, future developments could be take advantage of remote procedure call systems more dedicated to high-performance computing.

Others optimisations can be managed with a parallel programming support dedicated to nested parallelism like the one we developed, called $C_T$++ [3], based on A++[4] and HPC++[1]. $C_T$++ is composed of two levels, one dedicated to data-parallelism (DP) and one dedicated to task-parallelism (TP). Those two levels are hierarchically embedded in a common syntactic form in which DP levels are nested in TP levels. No language extensions are used (unlike DPCE[5]), which means that compilation could be achieved on a standard sequential workstation only with a standard compiler with sequential libraries and allows to use standard tools. Standard scalar operators (+,*,...) are promoted to handle distributed arrays and parallel conditionals are introduced. In our compilation mechanism, we use a preprocessor called Sage++[2] to identify the statements involving parallelism to improve and reconstruct portion of codes dedicated to a specific target architecture.

- SECTION: a set of integer values characterized by an affine function (begin, end, stride). They represent iterators over parallelized dimensions.
- SHAPE: the geometrical shape of parallel variables.
- TENSORs: the distributed containers of data. Variant arrays (mesh, trees,...) extend containers to irregular data-structure.
- FUNCTOR: scalar or parallel function or method able to be task-parallelised.
- elemental function: promotes a scalar function with scalar arguments to a DP one with conformal parallel arrays as arguments.
- nodal function: concurrently launches several FUNCTORs.

## 4 Conclusion

In this paper we have analysed the various parts of a typical Document Processing and Management tool (STRETCH) in order to identify the needs of efficiency. We recognise five points as necessary for a programming support to be adequate for allowing high-performance computing in the STRETCH context:

1. data-parallel expressivity for low-level regular processing (image filtering,...),
2. irregular structure support in data-parallel expressivity (triangular grid,...),
3. task-parallel expressivity for concurrent management of highly tied tasks,
4. hierarchical imbrication of data-parallelism in task parallelism,
5. ability to interface easily standard distributed programming tools (CORBA,...).

$C_T$++ have been developed in considering those features and uses a powerful restructuring tool to achieve good implementation performances on various target architectures. Its full C++ compatibility helps us to interface it easily with service tools like the CORBA interface used between the servers or standard products like the Application Programming Interface (API) of the OCR.

```
/* First Example :
computation of an histogram */
TENSOR<INT8> ImageBW(512,512);// input
TENSOR<INT16> Histo(256);  // histogram

Histo = 0; //histogram initialization

Histo(ImageBW)++;//histogram computation
```

```
// Third example: Concurrent launch of two
//   sobelfilters of different threshold
//data-parallel imbricated in task-parallel

TENSOR<INT8> ImageBW(512,512); // input
TENSOR<INT8> sob1(512,512); // 1st output
TENSOR<INT8> sob2(512,512); // 2nd output
int thre1, thre2; // threshold values

// initialization of the Image, threshold,

// generation of functors
FUNCTOR SOB1(&sob1,sobel,&ImageBW,&thre1);
FUNCTOR SOB2(&sob2,sobel,&ImageBW,&thre2);

// concurrent processing of sobel filters
nodal(SOB1,SOB2);
// with automatic synchronisation
```

```
// Second example: Sobel Filtering
TENSOR<int> *sobel(const TENSOR<float>& Im,
  int threshold) {
float min, max ;
int S0 = Im.getIterSection(0)->getSize();
int S1 = Im.getIterSection(1)->getSize();
SHAPE SI(S0,S1);        // conform shape
TENSOR<float> Ih(SI),   // after horizontal Sobel
              Iv(SI);   // after vertical Sobel
TENSOR<int> *result=new TENSOR<int>(SI);//result
SECTION I(1,S0-1);      // parallel section in X
SECTION J(1,S1-1);      // parallel section in Y
// convolution with horizontal filter
Ih(I,J) = Im(I-1,J-1) - Im(I+1,J-1)
        + (Im(I-1,J)*2) - (Im(I+1,J)*2)
        + Im(I-1,J+1) - Im(I+1,J+1);
// convolution with vertical filter
Iv(I,J) = Im(I-1,J-1)+(Im(I,J-1)*2)+Im(I+1,J-1)
        - Im(I-1,J+1)-(Im(I,J+1)*2)-Im(I+1,J+1);
Ih = ABS(Ih);       //parallel abs. value of Ih
min = MINVAL(Ih);   // minimum of IH in a scalar
max = MAXVAL(Ih);   // maximum of IH in a scalar
Ih(I,J) =((Ih(I,J)-min)*255)/(max-min);//norm.
Iv = ABS(Iv);       // parallel absolute value Iv
min = MINVAL(Iv);   // minimum of Iv in a scalar
max = MAXVAL(Iv);   // maximum of Iv in a scalar
Iv(I,J) =((Iv(I,J)-min)*255)/(max-min);//norm.
// binarization
WHERE (Ih > threshold) { *result = 255;  }
ELSEWHERE {         *result = 0;         }
ENDWHERE
WHERE (Iv > threshold) { *result = 255;  }
ENDWHERE
return result; }
```

## References

1. HPC++, extreme computing. Technical report, California Institute of Technology and CICA, University of Indiana, 1994.
2. F. Bodin, P. Beckman, D. Gannon, J. Gotwals, and B. Winnicka S. Naranaya, S. Srinivas. Sage++: An object-oriented toolkit and class library for building fortran and c++ restructuring tools. Technical report, Department of Computer Science, University of Indiana, 1992.
3. F. Bodin, H. Essafi, and M. Pic. A specific compilation scheme for image processing architecture. In *CAMP'97*, Boston, USA.
4. Daniel Quinlan. A++/P++ manual. Technical report, 1995.
5. DPCE Subcommittee. Data Parallel C Extensions. Technical Report X3J11/94-068, ANSI/ISO, 1994.

# Behavioral Objects and Layered Services: The Application Programming Style in the Harness Metacomputing System

Mauro Migliardi, Vaidy Sunderam

Emory University[1], Dept. Of Math & Computer Science
Atlanta, GA, 30322, USA
om@mathcs.emory.edu

**Abstract.** Recent advances in hardware and networking have fueled the interest in distributed computing in general and in metacomputing frameworks in particular. Harness is an experimental metacomputing system based upon the principle of dynamic reconfigurability both in terms of the computers and networks that comprise the virtual machine, and in the services offered by the virtual machine itself. In this paper we describe how the capability to reconfigure the virtual machine plugging services on demand can be exploited to design dynamically reconfigurable applications based on behavioral objects and cooperating layered services.

## 1 Introduction

Harness [1]is an experimental metacomputing system based upon the principle of dynamically reconfigurable object oriented networked computing frameworks. Harness supports reconfiguration not only in terms of the computers and networks that comprise the virtual machine, but also in the capabilities of the VM itself. These characteristics may be modified under user control via an object oriented "plug-in" mechanism that is the central feature of the system. At system level, the capability to reconfigure the set of services delivered by the virtual machine allows overcoming obsolescence related problems and the incorporation of new technologies. For example, the availability of Myrinet [2] interfaces has recently led to new models for closely coupled Network Of Workstations computing systems. In traditional metacomputing frameworks the underlying middleware either needs to be changed or re-constructed, thereby increasing the effort level involved and hampering interoperability. On the contrary, a virtual machine model intrinsically incorporating reconfiguration capabilities addresses these issues in an effective manner. At application level the reconfiguration capability of the system allows the incorporation of new capabilities into applications directly at run-time. As a second example we can cite long-lived simulations evolving through several phases. In traditional, statically configured metacomputers, if during the execution the application discovers the need

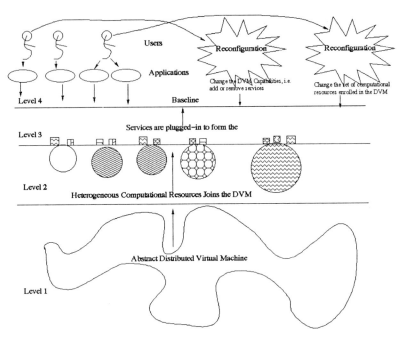

**Fig. 1.** Abstract model of a Harness virtual machine

for a service or a capability that was not accounted for from the very beginning there is no simple way to add this new capability to the system. On the contrary, the capability to dynamically plug-in new services in the virtual machine allows programmers both to adapt the environment to the needs of the application and to update the application itself using behavioral objects.

In this paper we focus on the advantages of run-time reconfigurability at application level and we characterize the programming style that is required in order to maximize these advantages.

The paper is structured as follows: in section 2 we give an abstract overview of the system architecture; in section 3 we describe the application programming style required by the reconfigurability of the system; finally, in section 4, we provide some concluding remarks.

## 2 Fundamental Abstractions and System Architecture

The fundamental abstraction in the Harness metacomputing framework is the **Distributed Virtual Machine** (DVM) (see figure 1, level 1). Any DVM is associated with a symbolic name that is unique in the Harness name space, but has no physical entities connected to it. **Heterogeneous Computational Resources** may enroll into a

DVM (see figure 1, level 2) at any time, however at this level the DVM is not ready yet to accept requests from users. To get ready to interact with users and applications the heterogeneous computational resources enrolled in a DVM need to **plug-in services** (see figure 1, level 3) in order to present a consistent **service baseline** (see figure 1, level 4). Users may **reconfigure** the DVM at any time (see figure 1, level 4) both in terms of computational resources enrolled by having them **join** or **leave** the DVM and in terms of services available by **loading** and **unloading** plug-ins.

The main goal of the Harness metacomputing framework is to achieve the capability to enroll heterogeneous computational resources into a DVM and make them capable of delivering a consistent service baseline to users. This goal require the programs building up the framework to be as portable as possible over an as large as possible selection of systems. The availability of services to heterogeneous computational resources derives from two different properties of the framework: the portability of plug-ins and the presence of multiple searchable plug-in repositories. Harness implements these properties leveraging different features of Java technology. These features are the capability to layer a homogeneous architecture such as the Java Virtual Machine (JVM) [3] over a large set of heterogeneous computational resources, and the capability to re-define the mechanism adopted to load and link new libraries.

## 3 Application Programming Style

The single inheritance constraint imposed by the Java language makes the use of abstract classes to be extended by users extremely restrictive. For this reason in the Harness metacomputing framework we have adopted Interfaces as the mechanism of choice to define the interaction between system and applications components.

The preferred application programming style in Harness requires programmers to minimize the part of the application that is not built as a network of cooperating layered services. In fact this style allows application programmer to take full advantage of the capabilities of Harness in terms of resource availability, fault tolerance, live application updating and code reuse. We describe how it is possible to take advantage of these capabilities by means of an example application, namely a reconfigurable distributed simulation of crystal growth.

We modeled the process of crystal growth as a two-step diffusion-deposition, discrete time process. At each clock tick a particle already present on the growing surface can diffuse if it fulfills a set of constraints, besides there is a small probability that new particles are deposited on the growing surface. The constraints ruling particle diffusion are the core of the physical model, in our application we model a low energy surface on which the attraction between neighbor particles prevents any diffusion.

Our application is almost completely built as a network of cooperating layered services. The only component of the application residing outside the DVM is the GUI that allows the user to control the simulation parameters and visualizes the pattern of crystals growing on the surface in real time. However, the simulation does not need the GUI component in order to run. As a matter of fact a user can launch the simulation, set up the parameters by means of the GUI component, exit the GUI component, let the simulation run autonomously and check it later in order to steer it.

We adopt the farming programming paradigm to exploit the dynamically changing number of resources available to the simulation. The Harness metacomputing framework supports this programming paradigm by means of two interfaces: the Farmer interface and the Worker interface. These interfaces allow the users to reconfigure at run time the computation that is performed by means of behavioral objects and to enroll new computational resources in the computation. Besides, our implementation also allows a Farmer to recover the status of the computation from any surviving worker, thus the application can survive the removal any number of components as long as a single component survives

The capability to plug-in new services at run-time as well as to set behavioral objects into existing ones allows us to perform live upgrade and reconfiguration in our simulation. It is important to notice that none of these new components needs to exist in the system at the time the simulation is started.

## 4 Conclusions and Future Work

Recent advances in hardware and networking have fueled the interest in distributed computing in general and in metacomputing frameworks in particular. However, traditional, statically configured metacomputing framework suffer from rapid obsolescence due to those same advances and tend to force applications to adapt to a fixed environment rather than adapt to them. To tackle these problems we have designed Harness, a dynamically reconfigurable system based on an object oriented, distributed plug-in mechanism. The run-time reconfiguration capabilities of Harness allow incorporating new technologies as well as to adapt to the changing needs of applications. In this paper we have described the programming style that allows the exploitation of these characteristics and we have shown its feasibility by building a crystal growth simulation as a network of layered, cooperating services. Our application shows a high level of fault tolerance as well as the capability of reconfiguring and upgrading itself at run-time.

## References

1 M. Migliardi, V. Sunderam, A. Geist and J. Dongarra, Dynamic Reconfiguaration and Virtual Machine Management in the Harness Metacomputing Framework, proc. Of ISCOPE98, Santa Fe (NM), December 8-11, 1998.
2 N. Boden et al., MYRINET: a Gigabit per Second Local Area Network, IEEE-Micro, Vol, 15, No. 1, February 1995.
3 T. Lindholm and F. Yellin, The Java Virtual Machine Specification, Addison Wesley, 1997.

# Coordination Models and Facilities Could Be Parallel Software Accelerators

## A.E. DOROSHENKO[1], L.-E. THORELLI[2], V. VLASSOV[2]

[1] Institute of Software Systems, National Academy of Sciences of Ukraine
Glushkov prosp., 40, Kiev 252187, Ukraine
[2] Royal Institute of Technology, Electrum 204, S-164 40 Kista, Sweden

**Abstract.** A new coordination model is constructed for distributed shared memory parallel programs. It exploits typing of shared resources and formal specification of a priori known synchronization constraints.

## 1 Introduction

A promising approach to meet demands of performance, flexibility and intelligibility of parallel and distributed computer systems is the development of coordination models and facilities, that allow managing dependencies of parallel systems within a single framework [1, 2]. Parallel programs developed in coordination based style have two components, computation model and coordination model, that are responsible respectively for algorithmic and behavioral aspects of computation. The purpose of this paper is to show that coordination facilities can serve not only as software integrators [3] but also as *software accelerators* in the sense of improving performance of parallel programs. A new coordination model called Co-mEDA is constructed for distributed shared memory parallel programs. It exploits knowledge in two forms of expected behavior of object synchronization: typing of shared resources as in the EDA/mEDA models of multiprocessing [4, 5] and formal specification of a priori known synchronization constraints as in forcing expressions [6].

## 2 Co-mEDA Coordination Model

The model is a triple $(O, S, R)$ where $O$ is a set of *objects* of a program, $S$ is a *coordination space* and $R$ is a set of *rules* that objects follow while interacting in coordination space.

**Objects.** An object is a unit of computation that unifies such concepts as thread, frame and object. An object is a single threaded active entity consisting of *local* variables $L$ (known only to the object), *shared* variables (known also to other objects), and the *script* that defines its thread of control.

The abstract syntax for a command c of an object script is

```
c ::=  skip | c1 ; c2 | if α then c1 else c2 | while α do c1 | b
```

where we do not give further details for the language of basic commands b ∈ B. We presume however that $B$ includes mEDA fetch and store operations on I-, X-, S- and U-typed shared variables, assignments, procedure calls, and other useful commands.

**Coordination space** in the model is a set of *sites* $S = ((N, l, e), t)$, where $N$ is set of nodes, finite or infinite, which are places where objects and shared variables are located by partial many-valued mappings $l : N \to O$, $e : N \to V$. The mapping $t : V \to T$ provides typing synchronization for shared variables.

The synchronization typing of shared variable in Co-mEDA is complemented by another form of coordination — formal specification of communication through U-typed shared variables by means of forcing expressions defined as follows. Let $K_v$ be the set of objects accessing the shared variable $v$. Introduce symbols $f_k$ and $s_k$ where $k \in K_v$ designating fetching and storing operations. A regular expression built upon symbols of fetching and storing and the empty symbol $\varepsilon$ with regular operations of sequencing, branching, and iteration is called *forcing expression* $\phi_v$ *for variable* $v$. We will say that $\phi_v$ is attached to variable $v$.

A forcing expression $\phi_v$ can be conceived as a behavior of accesses by objects to shared variable $v$ that, similar to objects themselves, has its script $\phi(v)$ and activation pointer $ap_v$.

**Rules** in Co-mEDA describe operational semantics of an object in terms of transition systems and object states. States are triples of the form $< sc, ap, env >$ where $sc$ is a script, $ap$ is an activation pointer in the script and $env$ is object's memory valuation. For coordination facilities like synchronization types $env$ is defined by a pair of mappings $env = < \lambda, \sigma >$ where $\lambda : L \to D$ is a valuation of local variables and $\sigma : L \to D^*$ is a valuation of shared variables known to the object. A shared variable is either *empty* or *full*. (We can assume shared variables to take values from the domain $D^*$, where the null sequence corresponds to *empty*.) As an example we show transition rules for coordination of operations fetch $f(X, x, v)$ and store $s(X, x, v)$ where $X$ indicates X-type synchronization, $x$ is a local variable and $v$ specifies a shared variable, in the style of structured operational semantics:

$$\frac{\sigma(v) = d, \quad v = [full]}{< f(X, x, v); c, \lambda, \sigma > \to < c, \lambda[d/x], \sigma[empty/v] >, \quad v = [empty]}$$

$$\frac{\lambda(x) = d, \quad v = [empty]}{< s(X, x, v); c, \lambda, \sigma > \to < c, \lambda, \sigma[d/v] >, \quad v = [full]}$$

If we have a regular program for the script of object $o \in O$ and a set of forcing expressions $\{\phi(v) : v \in V\}$ for shared variables used in the program code of this script, then environment $env$ of the object is defined as $env = < \lambda, \sigma, \phi >$ where $\phi = \{\phi_v : v \in V\}$, $\phi_v = (\phi(v), ap_v)$, is state of control for forcing expressions. The semantics of joint execution of object script and forcing expressions can be described by transition rules below.

$$\frac{\sigma(v) = d, \quad \phi_v = f_v; e_v}{< f(U, x, v); c, \lambda, \sigma, \phi_v > \to < c, \lambda[d/x], \sigma, e_v >}$$

$$\frac{\lambda(x) = d, \quad \phi_v = s_v; e_v}{< s(U, x, v); c, \lambda, \sigma, \phi_v > \rightarrow < c, \lambda, \sigma[d/v], e_v >}$$

Informally the first rule declares that if the next command for running thread is a fetch $f(U, x, v)$ with the current value of $d \in D$ for shared variable $v$ and at this time the observed symbol of corresponding forcing expression is symbol of fetching then the operation is allowed and leads to changing state of control, memory and coordination environment (state of forcing expression) of the object. Thus the value $d$ substitutes the value of $x$ in the current state of local variables and activation pointers in object script and forcing expression are incremented. The second rule similarly describes the behavior while executing a store operation.

## 3  Enhancing Data Parallel Paradigm: A Case Study

As an example of application of coordination facilities for efficient synchronization in parallel programs we will consider the well known linear algebra problem of Cholesky factorization. Let $A = (a_{ij})$ be the $N * N$ input matrix, assumed symmetrical positive definite and $L = (l_{ij})$ the Cholesky factor to be computed. We consider coarse-grained parallelization of the algorithm so the matrices are supposed to be broken into square blocks $B * B$ so that $n = N/B$ is an integer. The coordination space for a parallel program for this problem is defined as a tuple $S = ((\{K(i,j) : 1 \leq j < i \leq n\}, l, e), t)$. The same structure we assume for the set of objects $O = \{O(i,j) : 1 \leq j < i \leq n\}$. So the placement $l : K \to O$ is a one-to-one mapping given that factorization of block $(i, j)$ is performed by object $O(i, j)$ having as input data the block $(i, j)$ of matrix $A$ and the $(j, k)$ and $(i, l)$ blocks of the result matrix $L, 1 \leq j < i \leq n$ that form the set of shared variables of the object.

A comparative study of general purpose PARDO...PAREND barrier-like synchronization facilities, semaphores and U-type synchronization with attached forcing expressions has led to following results.

1. If wavefronts of a parallel algorithm are defined as $Q(k) = \{(i, j) : i + j = const, 1 < i+j \leq 2n\}, 1 \leq k \leq 2n-1$, then PARDO...PAREND synchronization is applicable (Fig. 1a). Assuming the time needed to factorization of a block of the first column as a time unit we have $T_1 = n^3/6 + O(n^2)$ as run time for the sequential algorithm and $T_{n/2} = n^2 + O(n)$ for the parallel algorithm running on $n/2$ processors. In this case, the efficiency is $e(p) = T_1/pT_p = 1/3 + o(1/p)$.

2. In this case using semaphores the wavefronts are organized as shown in Fig. 1b. In spite of increasing the front width from $n/2$ to $n$ this does not improve the order of magnitude for computation time and even decreases the order of efficiency to $1/6 + o(1/p)$ due to extreme workload dispersion for different points of the same wavefront.

3. Forcing expressions $\phi(A_{ij}) = f_{ij}$; and $\phi(L_{ij}) = s_{ij}; (while(u, f_{ij+k});$ $while(v, f_{li}))$; where $0 < j < i \leq n, u = (1 \leq k \leq i-j), v = (i \leq l \leq n)$ provides partial overlapping the computation of wavefronts (Fig. 1c). While

leaving efficiency at the level of order $1/3 + o(1/p)$ they enable a version of the program two times faster than previous ones.

4. The same result that forcing expressions provide we can also obtain by using I-type of synchronization for shared variables of output matrix $L$ of this program and U- type for input matrix $A$ since Cholesky factorization follows a single assignment policy for shared variables of
the output matrix.

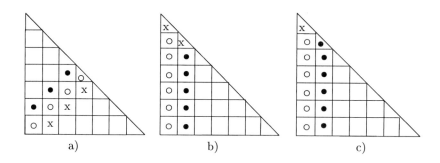

**Fig. 1.** Cholesky factorization: outlook of wavefronts

**Acknowledgments** This work was supported by research project no. 1515 "Models and Environments for Network Based Computing" of Royal Swedish Academy of Sciences within a Program of Cooperation between Sweden and the former Soviet Union.

## References

1. D. Gelernter, N. Carriero, Coordination Languages and Their Significance, *Commun. ACM*, vol. 35, No. 2 (1992), pp. 97-107.
2. T. Malone, K. Crowston, The interdisciplinary study of coordination, *ACM Computing Surveys*, 1994, 26(1), pp. 87-119.
3. P. Ciancarini, Coordination Models and Languages as Software Integrators, *ACM Computing Surveys*, June, 1996, 28(2), pp. 300-302.
4. L.-E. Thorelli: "The EDA multiprocessing model", TRITA-IT-R 94:28, Royal Institute of Technology, 1994.
5. H. Ahmed, L.-E. Thorelli, V. Vlassov, mEDA: a parallel programming environment, in *Proc. 21-st EUROMICRO Conference: Design of Hardware/Software Systems, Como, Italy, Sept 1995*, IEEE Computer Society Press, 1995, pp. 253-260.
6. A.E. Doroshenko, Modeling synchronization and communication abstractions for dynamical parallelization, *High-Performance Computing and Networking, Vienna, Austria, Apr.1997*, Springer Verlag, LNCS ,vol. 1225, 1997, pp.752-761.

# Delays in Asynchronous Communication Domain Decomposition

Marco Dimas Gubitoso and Carlos Humes Jr.

Universidade de São Paulo
Instituto de Matemática e Estatística
R. Matão, 1010, CEP 05508-900 São Paulo, SP, Brazil
{gubi,humes}@ime.usp.br

**Abstract.** This paper addresses the problem of estimating the total execution time of a parallel program based on a domain decomposition strategy.
We consider the situation where a single iteration on each processor can take two different execution times. We show that the total time depends on the topology of the interconnection network and provide a lower bound for the ring and the grid.
The analysis is supported further by a set of simulations and comparisons of specific cases.

## 1 Introduction

Domain decomposition is a common iterative method to solve a large class of partial differential equations numerically. In this method, the computation domain is partitioned in several smaller subdomains and the equation is solved separately on each subdomain, iteratively. At the end of each iteration, the boundary conditions of the each subdomain are updated according to its neighbors.

If the computation always takes the same time to complete the total execution time has a simple expression. If $T_{comp}$ and $T_{exch}$ are the computation and communication (data exchange) times, respectively, the total parallel time, $T_{par}$, is given by: $T_{par} = N \cdot (T_{comp} + T_{exch})$. However, if a processor can have a random delay the type of communication has a great impact on the final time, as will be shown.

In this paper, we will suppose the following hypothesis are valid:

1. The communication time is constant.
2. Any processor can have a delay $\delta$ in computation time with probability $\alpha$.
3. $\alpha$ is the same for all iterations.
4. A processor can communicate with all its neighbors at the same time.

If $T_c$ is the execution time of a single iteration without a delay, the expected execution time for one iteration is $T_c + \alpha\delta$.

For the parallel case with synchronous communication the time of a single iteration is the time taken by the slowest processor. The probability of a global delay is the probability of a delay in at least one processor, that is (1 − probability of no delay)

$$\mathcal{P}(\text{delay in one iteration}) = \rho = 1 - (1 - \alpha)^P \tag{1}$$

and the expect parallel execution is $<T_{par}> = N \cdot (T_{exch} + (T_c + \rho\delta))$

## 1.1 Asynchronous Communication

If the communication is asynchronous, different processors may, at a given instant, be executing different instances of the iteration and the number of delays can be different for each processor.

During data exchanging each processor must wait for its neighbors and the delays *propagate*, forming "bubbles" which expand until they cover all processors. This behavior can be modeled as follows:

1. Each processor $p$ has an positive integer associated, $n_a[p]$, which indicates the total number of delays.
2. Initially $n_a[p] = 0, \forall p$.
3. At each iteration, for each $p$:
   - $n_a[p] \leftarrow \max\{n_a[i] | i \in \{\text{neighbors of } p\}\}$.
   - $n_a[p]$ in incremented with probability $\alpha$.

The total execution time is given by $N \cdot (T_c + T_{exch}) + A \cdot \delta$, where $A = \max\{n_a[p]\}$, after $N$ iterations.

In the remaining of this paper, we will present a generic lower bound for $A$ and validate the model by a set of simulations.

## 2 Lower Bound for the Number of Total Delays

The probability of a delay in any given processor is a combination of the induced and spontaneous possibilities. Let $\sigma$ be this joint probability. It is clear that $\sigma \leq \alpha$, for $\alpha$ corresponds only to the spontaneous delay.

To find a lower bound, we choose on processor among the most delayed and ignore the others. We then compute the expected delay in $l$ successive iterations, to find an approximation for $\gamma$. We call this procedure a '$l$-look-ahead' estimative.

### 2.1 Transition Probability

In order to derive the expression for $\gamma$, we state some definitions and establish a notation:

- A bubble is characterized by an array of levels, called *state*, indicated by $S$, and a propagation function which depends on the topology of the interconnection network.
- The state represents the delay of each processor inside the bubble: $S = s_1 s_2 \cdots s_k \cdots s_C$, $s_k$ is the delay (level) of the $k$th processor in the bubble.

- $s_{max} = \max\{s_1, \ldots, s_C\}$
- $S' = s'_1 \cdots s'_{C'}$ is an expansion of $S$, obtained by an propagation.
- The *weight* of a level in a bubble $S$ is defined as follows:

$$W_S(k) = \begin{cases} \text{number of occurrences of } k \text{ in } S, \text{ if } k = \max\{s_1, \ldots, s_C\} \\ 0, \qquad \text{otherwise} \end{cases}$$

- The number of differences between two bubbles $S$ and $T$ is indicated by $\Delta(S,T)$:
- A *chain* is a sequence of states representing a possible history of a state $S$: $\mathcal{C}(S) = S^0 \to S^1 \to \cdots \to S^a$, $S^a \in G(S)$.

Consider $S^i$ and $S^{i+1}$ two consecutive states belonging to the same chain. For $S^i$ to reach $S^{i+1}$, processors with different levels in $S'^i$ and $S^{i+1}$ must suffer spontaneous delays. The transition probability is then: $P(S^i \to S^{i+1}) = \alpha^{\Delta(S'^i,S^{i+1})} \cdot (1-\alpha)^{|S'^i|-\Delta(S'^i,S^{i+1})}$ and the probability if a specific chain $\mathcal{C}(S)$ to occur is given by: $\mathcal{P}(\mathcal{C}(S)) = \prod_{i=0}^{a-1} \alpha^{\Delta(S'^i,S^{i+1})} \cdot (1-\alpha)^{|S'^i|-\Delta(S'^i,S^{i+1})}$

### 2.2 Effective Delay

The total delay associated with a state $S$ is $s_{max}$. If $S$ is the final state, for a given $S^a \in G(S)$, the final delay can be:

1. $s^a_{max}$ with probability $(1-\alpha)^{W_{s^a}(s^a_{max})}$,
2. $s^a_{max} + 1$, if at least one of these processors has a new delay. The probability for this to happen is $1 - (1-\alpha)^{W_{s^a}(s^a_{max})}$

The expected delay $E$ is computed over all possible histories:

$$E = \sum_{\mathcal{C}(S)} \mathcal{P}(\mathcal{C}(S)) \cdot \left( (s^a_{max} + 1) - (1-\alpha)^{W_{s^a}(s^a_{max})} \right) \qquad (2)$$

and the effective delay, $\gamma$, for the $l$-look-ahead is $E/l$.

For larger values of $l$, the expression for $\gamma$ becomes more complicated as the number of terms grows exponentially. These results are presented in the next section, together with the simulations.

## 3 Simulations

In this section, we present and discuss the simulation results for several sizes of grids and rings. Each simulation had 500 iterations and the number of delays averaged over 20 executions, for $\alpha$ running from 0.0 to 1.0 by steps of 0.05.

Figure 1 presents the comparision for 2 topologies with the same number of processors: $30 \times 30$ grid and a ring with 900 processors. It indicates that, for certain values of $\alpha$ and $\delta$, the ring can be much better.

A test program following the assumptions of the model was developed and figure 2 compares the times of two tipical experimental runs on a 8 nodes *PowerXplorer* with the predicted 2 look-ahead lower-bound. Details about the tests and the program are available upon request.

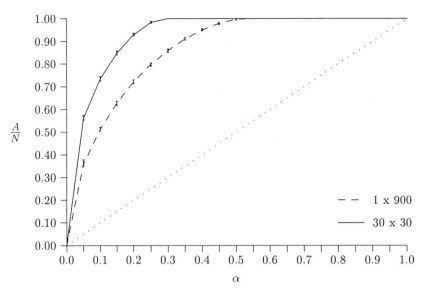

**Fig. 1.** Comparision between a 1x900 array and a 30x30 grid

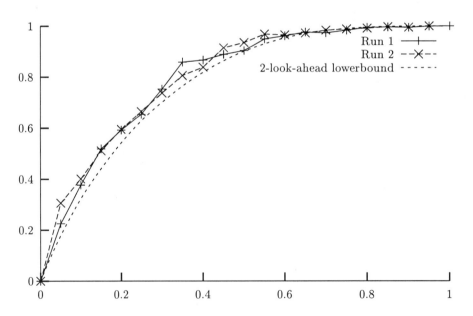

**Fig. 2.** Experimental and lower bound times for the ring

# Experimenting Reflection for Programming Concurrent Objects Scheduling Strategies

L. Bray, J.-P. Arcangeli, P. Sallé

Institut de Recherche en Informatique de Toulouse
118, route de Narbonne, 31062 Toulouse Cedex 04 FRANCE
Tel : +33 (0) 5 61 55 67 65
Fax : +33 (0) 5 61 55 62 58
{bray, arcangel, salle}@irit.fr

**Abstract.** This paper reports a study on the use of reflection as a methodology for expressing concurrent objects, i.e. actors, scheduling strategies. Reflection allows to separate at design time and compose at runtime the applicative program and the placement program, thus providing modularity, portability of applicative programs and reusability of scheduling ones.

## 1 Introduction

Object-based concurrent programming offers a high degree of expressiveness for parallel applications. Efficiency requires scheduling strategies [1] well-adapted to applications. These strategies are all the more dynamic and difficult to implement since applications or hardware supports are irregular. A challenge is to deal with efficiency without complicating applications development, while preserving simplicity of high-level programming, readability of codes, and portability of applications on different architectures. Our purpose is to provide a convenient software methodology to design, implement and maintain efficient and portable distributed actor-based applications. The actor programming system PlasmaII [2] is the framework of this methodological experience. Our objective is not to design or evaluate efficient strategies, but to provide a way for a transparent management of resources.

The actor model [3] is a high-level general-purpose concurrent programming model in which an application is described as a community of cooperative concurrent entities called actors. Actors are objects, communicating by asynchronous messages, and may also be seen as processes : each of them represents a logical autonomous thread of computation. Actors applications use to be strongly dynamic and/or communicative, fine-grained and massively concurrent. An actor language allows to express naturally concurrency (the logical - virtual - parallelism) in the applicative program ("naïve program") without considering management constraints.

Reflection [4] gives to any system (an operating system or an application) the possibility to reason and act upon itself at runtime. In practical use, reflection enables the system to perform dynamic self-observations or changes of, for example, communication protocols, computation modes, resources management, and so on.

We study the use of reflection which allows both a separation of the application description (the "naïve" base-level program) from its execution features at the meta-level, and their composition.

A component-based model of reflection for concurrent objects and actors has been proposed and an experimental reflective actor programming system is available [5]. It is based on an individual operational fine-grained decomposition of actors : each actor is described by a set of reified components. Physically, each component is implemented by a meta-level actor and has a default modifiable behaviour. Among the operational components, the Creation component manages, within each actor, the dynamic creation operation (while others manage message emission, message reception, life cycle, etc.). At runtime, the Creation component of an actor can be modified, and therefore the local placement policy is changed.

## 2 Methodological features

Transparency consists in making the use of a distributed (or parallel) system the simpliest and lightest possible. Transparency means setting aside the underlying runtime system as it is natural to do with high-level programming languages. Our objective is to preserve transparency while dealing with efficiency problems: during the application programming phase, no difference could be done by the programmer between a local execution and a distributed one (if he doesn't want to) and applications could be conceived without management considerations. The programmer could only have to write the "naïve" program.

Nevertheless, high transparency may lead to weak performances, because of a lack of informations or if the runtime system has to get them dynamically. A programmer often knows some of his application characteristics, which are useful for scheduling. To run efficiently his application, a solution consists in inserting placement directives into the source code. But mixing applicative and management codes makes the program more difficult first to write, then to read and maintain, and maybe no longer portable. Low transparency leads therefore to programming difficulties and few portable programs. Moreover, the programmer may neither have all the necessary informations, like dynamic ones, nor all the tools for management. Thus, a compromise between transparency and efficiency must be done.

Two programming levels clearly appear : the applicative one, described using the high-level programming language, and a lower one, the management level. This leads us to a modular approach. On the one hand, the application programmer writes the concurrent "naïve" program without any allocation directive. On the other hand, another programmer - maybe the same - writes or chooses a specialized strategy, adapted both to the application and to the hardware support, which provides efficiency.

## 3  Experience

Let us take as an example a simple application belonging to the class "Divide and Conquer" : the quicksort. In order to simplify our presentation, we consider only binary division of the problem. In the quicksort algorithm, the set of values to sort is recursively divided into two subsets, composed of the elements respectively smaller and greater than a chosen pivot, until the trivial case of a singleton is reached. The computation of the sorted set is then a simple concatenation of the sorted subsets. As the decomposition is not equitable (the sub-problems have different and unpredictable sizes), the application is irregular.

In the actor model, this algorithm may be implemented as follows : any problem or sub-problem corresponds to an actor, and its division corresponds to the dynamic creation of two actors. Each of them is actived by a message containing the subset to sort. If not the trivial case, it processes in two steps: first, it creates two new actors (its children) and sends to each of them a subset; then, it changes its behaviour to wait for its children'results, assemble them, send this result to its own parent, and suicide. If the set is a singleton, the actor returns it directly to its parent, and suicides. Thus, applications are tree-structured : nodes are actors, and the edges are the communication links. Due to the irregularity of the quicksort, the tree is unbalanced and its development depends on the initial set. The "naïve" program describes this functional diagram, no more.

A basic strategy for actors distribution consists in dividing the tree structure in as many subtrees as sites, and assigning one subtree to each site. It is based on a link between a problem (the root actor of a (sub)tree and its descendants) and a dedicated execution domain. This domain can be represented by a list of sites references. The association between a node actor and a domain is realized by message passing. A node receives from its parent its execution domain, and as long as the domain contains several sites, it is divided into two sub-domains whereon the two children are created. When the domain is reduced to a single site, the children and their descendants are created on it.

This strategy is not optimal for irregular applications such as quicksort : due to the unbalance of the tree, some sites may be idle while others are busy. It is possible to refine the strategy for a better load sharing by observing the load on each site and demanding work when it is too low. We propose to associate to every site an actor for computing load and demanding work (CDW), and another one for receiving and processing demands (RPD). The former must be incremented (respectively decremented) when an actor is created (respectively suicides). The latter controls location of all actors created from its site. On a site, when CDW reaches down to a threshold, it sends a demand message to another site RPD. Then, RPD enforces remote creation. While it is not requested, RPD enforces local creation.

In order to implement this strategy, we have to specialize the creation process by setting the target sites of creations. Reflection permits to redefine the creation process within each actor by redefining the Creation component of each actor.

In the reflective implementation, every node of the tree is a reflective actor. Their respective Creation component is got from a single model parametrized by a specific list of sites. A specialized Creation component processes according to its private list of sites. If the list contains several sites, it is divided into two sublists ; so, when the Creation component is invoked to perform an actor creation for the first (respectively second) time, it creates on a site (for example, the first one) which belongs to the first (respectively second) sublist, and sends the first (respectively second) sublist to the (child) Creation component. If the list contains a single element, the Creation component consults the RPD of its site, creates its child on the indicated site, and sends an increment message to the concerned site CDW. The Suicide component (the one which manages self-destruction) has to be customized too, in order to send a decrement message when the reflective actor dies.

The strategy programmer's task consists only in writing the model of Creation component behaviour, and an interface which permits to associate the strategy with the "naïve" program. Interfacing consists in creating a reflective actor root (with the Creation component generic behaviour), and then, sending both, at the meta-level, the list of sites to the Creation component, and, at the applicative level, the set to sort to the root. Then, the list of sites is manipulated at the meta-level only and not at the applicative level.

Thus, the strategy can be written quite easily, by the means of appropriate primitives, apart from the applicative program. Therefore, software complexity is not increased. Modularity could permits to associate different strategies to one application according to the target architecture (portability). Conversely, one strategy specialized for one application can be used for another one with the same characteristics (reusability).

# References

[1] T.L. Casavant, J.G. Kulh, A taxonomy of scheduling in general-purpose distributed computing systems, IEEE Trans. on Software Engineering 14(2), 1988, pp. 141-154.
[2] J.-P. Arcangeli, A. Marcoux, C. Maurel, P. Sallé, La programmation concurrente par acteurs: le système PlasmaII, Calculateurs Parallèles (22), La Boucle Informatique, 1994, pp. 123-150.
[3] G. Agha, Actors : a model of concurrent computation in distributed systems, M. I. T. Press, Cambridge, Ma., 1986.
[4] G. Kiczales, ed., Proceedings of the Reflection'96 Conference, San Francisco, CA, 1996
[5] A. Marcoux, C. Maurel, F. Migeon, P. Sallé, Generic operational decomposition for concurrent systems : semantics and reflection, Parallel and Distributed Computing Practices, Nova Science Publishers, Inc., to appear, 1999.

# Virtual User Account System for Distributed Batch Processing

W. Dymaczewski, N. Meyer, M. Stroiński and P. Wolniewicz

Poznań Supercomputing and Networking Center (PSNC)
ul. Noskowskiego 10, PL-61-704 Poznań, Poland

e-mail: {dymaczew, meyer, stroins, pawelw}@man.poznan.pl

**Abstract.** In this paper we present a simplified version of heterogeneous metacomputing environments for specific applications, allowing ability to run jobs on remote systems without administration overhead connected with maintaining user accounts, and transparently for the real user. We present problems with developing distributed computing environments, the idea of the Virtual User Account on specialised application servers, advantages and disadvantages of this solution, and possible further works.

## 1. Introduction

According to the results of observations conducted at Poznań Supercomputing and Networking Center, we found that most programs ran by users are licensed applications from software packages installed on computing systems by administrators. Using those applications, users specify only input data in a text file and do not modify the source code of the application. Based on this information we developed the idea of a specialised application server system, dedicated to particular types of computation, running jobs only in batch mode. Because the load introduced by different applications changes in time, to provide an adequate number of users, and a satisfactory load for each application server, the system will run jobs requested by users of High Performance Computing sites co-operating in the Polish scientific broadband network POL–34 [4].

## 2. Problems with developing servers system for applications

One of the basic problems of using high performance computing systems is load imbalance. The simplest solution is the usage of a job processing system, which flattens the peak load, disallowing too many jobs to run at one time, and queuing extended ones for future running. It is a very good solution for local site conditions: for systems installed in one place, where there is a constant or slowly changing number of users and computing systems.

However, this solution has a few limits when used to run applications in a distributed environment. Some of them are listed below:
1. **User account problems.** If the queuing system is used to balance the load of a few systems installed in two or more HPC sites, all of the users using this system have to have an account shared or mapped between all computing machines, which is a difficult situation to maintain without compromising security.
2. **File transfer problems.** If computation is made on a remote system, and there is no shared file systems like NFS between machines, a user's file has to be transferred from a local system to a remote one, and then the files, with the results, have to be transferred back.
3. **Security and authorised access problems.** A queuing system which runs jobs on several machines should authorise the access of particular users to particular machines. In the most recent queuing systems, such an authorisation is reduced to checking, whether the account given by the user to run the program on exists or does not on a specified remote machine. If the user has different identifiers on different systems, he has to specify all mapping information when submitting a job to a queuing system.
4. **User accounting problems.** When users are running programs on different machines in geographically distributed sites, with different account identifiers, accounting becomes a real problem. Standard mechanisms included in an operating system are insufficient because they do not provide mapping information, and cannot collect and consolidate data from different machines. Some better features are included in most of the present queuing systems [3], but the data they collect are also insufficient to prepare a bill for a real user.

To handle all specified above problems we suggest a specialised wrapper for any queuing system which helps running jobs in a distributed environment, exploiting the constant number of accounts – a "Virtual User Account System"

## 3. Idea of the Virtual User Account System

The Virtual User Account System is a system allowing a load balance of machines installed in different computing sites without overhead related to creating and maintaining additional user accounts. This system is based on any queuing system. It allows either a better usage of computing systems providing the ability to use free CPU cycles on partially loaded machines, or separate different types of loads, e.g. memory-bound tasks can be sent to bigger machines where they won't be disturbed by smaller, CPU-bound jobs moved to another, smaller system.

There are metacomputing systems (e.g. Legion [1], Globus [2], UNICORE) which allow creating and running jobs in a distributed environment. They are complete distributed environments based on existing user accounts on remote hosts. Those products have a lot of functionality but they are enormously big and hard to configure compared to our needs. The most similar one is UNICORE, based on the Codine queuing system but to use it a user has to have accounts on all machines.

## 4. System structure

The Virtual User Account System consists of few elements:
- any queuing system
- wrapped user interface commands to handle this queuing system
- Virtual User Manager Account, containing mechanisms to communicate with VUS and to manage jobs running on a local machine
- pool of Virtual User Accounts on each machine
- Virtual User Server (VUS) responsible for global translation of user identifiers on all machines, user accounting and authorising a user's access to the machines.

### 4.1. User interface
There is a set of specific commands used by users to submit jobs for execution, or to check the status of the queuing system and submitted jobs. Because forcing users to change the interface they know will be inconvenient to them, this is the only part of the Virtual User Account System which has to be tailored to a particular queuing system's needs. Those overlapped commands are vital to provide data about the user originating the job execution, and this will be necessary in the future when authorising access to particular machines and collecting accounting data. Those commands are communicating with the Virtual User Account Server to save information about users during job submissions or to get data required to properly present queues and jobs status to the user.

### 4.3. Virtual User Account Manager
The main element of a virtual user system is the account of the Virtual User Account Manager, which should exists on every machine in a cluster. This account is used as an interface between the queuing system and the operating system on executive machines. This user owns every task submitted by real users to a queuing system and this account is used to start a task on a remote system. When the queuing system starts a task on a remote machine, a special program is invoked and it manages the Virtual User Accounts on this executive machine and send accounting information to Virtual User Account Server. The manager finds if there are any free Virtual User Accounts and sets an UID on this chosen Virtual User Accounts and runs the task. After the execution is complete, accounting information is collected and sent to a Virtual User Account Server, which keeps information about the total CPU time used by a real user. Number of jobs ran in the same time on a particular machine is limited by the number of created Virtual User Accounts and by the settings of the queuing system.

### 4.4. Virtual User Account
Because users have to be able to run jobs on remote systems without having accounts on those systems, there is a need for a certain number of accounts, which will be used to run jobs. The number of these accounts is stable and independent of the real number of users using Virtual User Account System. An administrator can add or delete accounts and in this way regulate the maximum load introduced by the queuing system.

### 4.5. Virtual User Account Server

An essential element of our system is the Virtual User Account Server daemon. It keeps information about mapping from real users to Virtual User Account identifiers on all systems, provides an access control list to individual systems and queues, collects and processes information about CPU time used by users on all systems. To allow new users to use distributed queues, only a modification to the access control list on the Virtual User Account Server is needed. There is no need to add accounts on all physical systems, which a user is allowed to use.

## Current state and Conclusions

In PSNC we tested basic version of Virtual User Account System on the strength of the queuing system LSF (Load Sharing Facility). We constructed a cluster of Silicon Graphics computers joined through LSF queues. We defined the test queue which can distribute tasks on all computers in cluster. On each machine there are a certain number of Virtual User Accounts. Tasks submitted to LSF are executed on remote machines on virtual user's account independently of the existence or not the account for the user who submitted this task. After the task was completed Virtual User Account Server got completed information about the CPU time used by a user on a particular system.

The system introduced in this article allows the simplification of administration of distributed clusters. Users should have accounts on one of the systems and they should be added to the list of users, who are allowed to use distributed queues, which can send tasks to remote machines. It avoids the trouble of maintaining accounts on all machines joined in a cluster. Particularly it is possible to add easy new machines to a cluster and it is sufficient to add only a certain number of Virtual User Accounts to allow this machine to take part in distributed computing. Because of the joining of HPC centers it is possible to dedicate some machines to run only one kind of the most commonly used application. Then the administrator can deny interactive logins to this machine. The disk space on such a dedicated machine can be reduced, because there are no user accounts and only specific applications are installed.

## References

[1] I. Foster, C Kesselman , *The Globus Project: A Status Report*, Proc. IPPS/SPDP '98 Heterogeneous Computing Workshop, pg. 4-18, 1998

[2] A.S. Grimshaw, W.A. Wulf, J.C. French, A.C. Weaver, P.F.Reynolds Jr, *A Synopsis of the Legion Project*, UVa CS Technical Report CS-94-20, June 8, 1994, http://legion.virginia.edu/papers/CS-94-20.ps

[3] J.A. Kaplan, M.L. Nelson *A Comparison of Queuing, Cluster and Distributed Computing Systems*, NASA technical Memorandum 109025, NASA LaRC, October, 1993

[4] M. Nakonieczny, S. Starzak, M. Stroiński, J. Weglarz, *Polish Scientific Broadband Network POL-34*, Computer Networks & ISDN Systems 30, 1998, s 1669-1676

# Content-Based Multimedia Data Retrieval on Cluster System Environment

Sanan Srakaew, Nikitas Alexandridis, Punpiti Piamsa-nga, George Blankenship

Department of Electrical Engineering and Computer Science
The George Washington University
801 22$^{nd}$ St., N.W., Washington, DC 20052
{srakaew, alexan, punpiti, blankeng}@seas.gwu.edu

**Abstract.** In this paper, we propose a data partitioning scheme for content-based retrieval from a multimedia database using a heterogeneous cluster system. The proposed parallel unified model [6] was used to represent multimedia data. All types of multimedia data in the unified model are represented by k-dimensional(k-d) signals. Each dimension of k-d data is separated into small blocks and then formed into a hierarchical multidimensional tree structure, called a k-tree. The parallel version of k-tree model was introduced in [7]. The previous experimental results show the huge reduction of retrieval time on a cluster of homogeneous workstations. In this paper, we extend our parallel model to a heterogeneous cluster system environment. We demonstrate the experimental results of using a parallel retrieval algorithm for the k-tree unified model on a cluster of heterogeneous system connected via a network. We use system characteristics to help in partitioning the data and balancing the loads of the processors. The experiments of the model with load balancing shows a significant reduction of retrieval time while maintaining the quality of perceptual results.

## 1 Introduction

Multimedia databases have become more important since conventional databases cannot provide the necessary efficiency and performance. Multimedia databases encounter three major difficulties. First, the content is subjective information; that is, intelligence is required to characterize the data. The data recognition requires prior knowledge and special techniques in Computer Vision and Pattern Recognition; the problems are NP-hard. Second, if a method or processing technique is designed and developed for one type of data or feature, it usually not appropriate for others. For instance, a technique designed for indexing audio data may not be usable for image data; or, a technique developed for a color feature may not be useful for a texture feature in image and video data. Third, the usual huge size of multimedia data and the requirement for a similarity search affect the computation. A similarity search is desirable for a multimedia database system. If a picture of a house is used as a query to an image database, we expect to retrieve pictures that contain similar houses in them. The comparison is not pixel by pixel between a query and the records in a database; but rather, closeness to the query. Similarity matching requires the computation of the distance between a query and each record in the database; the best match is chosen from the data set with the smallest distances. To solve these three problems, we use a mathematical model to represent the features; a k-tree model to represent the data structures of the multimedia data; and exploit parallelism to reduce the retrieval time [7].

In this paper, color and texture are the features of interest; they represent the subjective information of the multimedia data. We use a normalization technique to generate the indices. The domain of a feature is reduced to a set of selected values from a universe of potential values for the feature. We use an identification number for each element in the reduced set [7]. When data is inserted into the system, it is converted to the selected domain. The feature is represented by a histogram. For color feature, a few colors are picked from the whole infinite universe of colors. A finite number indexes each color. The color feature of an image or a video is represented by a histogram using the indexed color. For texture feature, we selected a set of textures and assigned an identification number to each texture. The feature of a texture is represented by the histogram of texture identification, which is the same method that was used for the color feature. The comparison of two features is based upon the distance between the histograms that define the features.

To reduce the response time, one may use a parallel model of a homogeneous system to perform a content-based multimedia retrieval. The experimental results were very positive in both qualitative and quantitative metrics. However, in practical, we do not have dedicated machines that always have the same configurations. The homogeneous model may be not used efficiently enough in the real-life heterogeneous environment. In this paper, we investigate a data partitioning scheme for multimedia database retrieval on a heterogeneous cluster system. We use system characteristics, such as processor speed and available storage, to partition data among the processors in the systems. Our computer system environment is composed of Sun Sparc and Pentium-Linux machines, which are connected via a 10Mbit local area network. We evaluate the model by comparing the retrieval results with the previous homogeneous parallel retrieval. The experimental results show the heterogeneous model produces a significant reduction of the retrieval time of an image from a 30,000-record image database.

This paper is organized as follow. Section **2** describes our k-tree parallel model. Section **3** has the details of our cluster system environment and its heterogeneity. The experiment and its results are described in Section **4** The last section concludes our works and proposes future directions.

## 2 The K-Tree Parallel Model

A *k*-tree is a directed graph; each node has $2^k$ incoming edges and one outgoing edge with a balanced structure [6]. A *k*-tree is a *binary tree* for 1-dimensional data and a *quadtree* for 2-dimensional data. Exploiting a k-tree brings three main benefits. First, the k-tree holds the information of spatio-temporal data on the tree structure itself. It reduces distance computation time to a comparison between two tree nodes. Second, a k-tree can accelerate multiresolution processing by calculating small, global information first and then large, local information when precise resolution is needed. Third, the data on a k-tree is unified since only the degree of the tree changes, while the processing algorithm and data structure remain invariant. Therefore, an algorithm for a particular type of feature can be reused for a feature of another media type.

Content-based retrieval of multidimensional signals is done by comparing features extracted from the input query with features extracted from every record in the database. The features of a multidimensional signal are subjective information. They are characteristics that are used to distinguish one signal from others. A 2-dimensional signal, such as an image, is characterized by features such as color, texture, and intensity. The basic algorithms for the searching of data in each of the different domains are quite similar. A matching search requires that the index key (defining feature) be unique and matched to the query. Exactly matched searching requires exhaustive comparisons that are inefficient and unsuitable for multidimensional signals; similarity searching is more appropriate. A similarity-search re-orders the database by distance between each record and the query; the result is selected from the ranking.

Multimedia data retrieval requires similarity searching; exact matching, which is used in conventional database, is not appropriate for this type of application. Similarity searching generally can be done in two steps; 1) finding distances between a query and all records in the database and then 2) sorting the distances and returning the results – the set of data items that have shortest distances. We also call this process "ranking." The details of regular weighted ranking are discussed in [7]. In Figure 1, we show a parallel searching using multi-feature scheme. Prior to the search the database is distributed among processors. Each processor performs the comparison between the query and its portion of the database. The search results based on those features are sorted in parallel to create the final ranking.

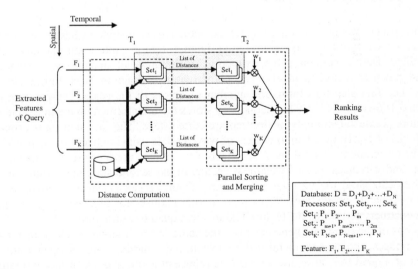

**Figure 1.** Parallel Model Multi-feature Searching

## 3 Cluster System Environment

In this section, we give the details of our experimental platforms, evaluated multimedia database systems, and our proposed load-balancing algorithm for the k-tree model.

### 3.1 Computational platform

A cluster system environment plays an important role in a distributed multimedia database. A simple model of a distributed environment is a homogeneous system, where each workstation is identical and communication links between the workstations are also comparable. In such a system, data partitioning can be evenly subdivided. A static load-balancing scheme can be applied. A system environment is usually heterogeneous; the workstations are different in terms of computational power, storage space, hardware architecture, and so on. In this paper, we establish a uniform framework for content-based multimedia data retrieval on a cluster system environment. In our experiment, one of the workstation serves as a host while others are assigned to be slave nodes. The system is composed of seven machines, which include two models of Sun workstations and Pentium/Linux PCs connected via local area network and use the Message Passing Interface (MPI) library as a message-passing interconnection mechanism [8].

### 3.2 Multimedia database system

In this paper, we use an image database which uses histogram-based features as indices. Two types of histogram-based feature (colors and textures) have been examined. Before beginning the extraction of features, all images are normalized, scaled down to 128x128 pixels. The color feature extraction is performed in two steps. The first step transforms the number of colors of the scaled images to a pre-selected 166-color set [5]. The second step stores the transformed image in a quad tree structure. Texture feature extraction requires three steps. The first step transforms the 64 blocks of 16x16 pixels in to 64 sets of wavelet data using a Quadratic Mirror Filter (QMF) (2 iterations, 7 sub-bands) [7]. Each wavelet data produces seven subbands of means and variances; i.e. a 14-element vector. In the second step, the texture vectors are then compared to 162-reference textures from VisTex [7] in the known texture table to generate 64 texture indices representing textures for blocked data. The third step constructs and stores the texture features in a quad tree structure.

The steps of the quad tree generation are the same for both features. The transformed color images and texture-identification (texture-id) matrices are mapped onto the leaves of a quad tree structure. The leaves represent a single pixel of the normalized image. Histograms of the leaves, which share the same parent nodes, are summed and the results are stored at their parent nodes. The process continues iteratively for each level until the root has been reached.

## 3.3 Data Partitioning and Load Balancing

To exploit heterogeneous parallelism, we use task and machine characteristics to decide which processors the tasks should be allocated to. The heterogeneity in the task level is the differences of the searching into the feature indices and the heterogeneity of the machines includes processor power and communication speed. The general idea is to give more data to a machine whose estimated execution time is smallest. Data given to a particular machine can be locally stored or remotely accessed via a faster network link. For example, suppose a system consists of two machines; one is twice faster than another. Thus, the faster workstation should have two-third of database and the slower one should have the one-third portion. However, if the faster machine doesn't have enough disk space to store all two-third portion, parts of database needed on this workstation can be remotely kept on another one. The size of remotely kept data is dependent of average link latency between these two machines. We can generalize the load distribution in a mathematical model as follows.

Let $S$ be a set of heterogeneous machines; $S = \{M_1, M_2, ..., M_m\}$; m be the number of machines; $M_i(C_i, S_i, L_i)$ be a machine described by computational power $C_i$, an available storage $S_i$, and estimated network bandwidths $L_i$; $L_i = \{\lambda_{ij}, \forall\ i \neq j\}$. $\lambda_{ij}$ is an average network latency between $M_i$ and $M_j$ for all $i \neq j$.

Let B a data base size and $E_I$ be an estimated execution time of task T on $M_i$

$$E_i = C_i + R_i$$

where $C_i$ is an *estimated computational time* and
$R_i$ is an *estimated IO time*

Let $S_{ij}$ be a portion of B on $M_j$ that $E_i$ needs to access, therefore

$$E_i = C_i * \Sigma[p_{ij}] + (R_i/\lambda_{ii}) * \Sigma\ [p_{ij} * \lambda_{ij}]$$

Our goal is to minimize $E_i$ subject to the following constraints:

$$\Sigma[p_{ij}] = 1$$
$$\Sigma[p_{ij}] \leq S_i/B, \quad 1 \leq j \leq m$$
$$E_i = E_j, \quad \forall\ i \neq j$$

## 4 Experiments and Results

The cluster environment consists of seven Unix-based machines; two Pentium/Linux-based PCs, two Sun Sparc-20 workstations, and three Sun Ultrasparc I workstations. One of Sun Ultrasparc is designated as the host. The MPI library is used as the interconnection mechanism. We use comparative computational power to classify workstation types. The computational power ratios are based on the results similarity searching a database of 500 images using single- and multi- feature storage algorithms. The computational power ratio of a Sun Ultrasparc I to a Pentium II PC to a Sun Sparc-20 is 3:2:1.8.

In this experiment, image retrieval is performed using two features; color and texture. The extracted features are derived from database images of 128x128 pixels; each is evenly divided into 64 blocks. The *quadtree* of the histograms for each image is made up of 3 levels; 64 leaf nodes. We perform two data partitioning schemes based on the database of 30,000 images. In the first scheme, the database of color and texture

histograms is evenly divided among workstations, without considering the heterogeneity of cluster environment. In the second approach, data partitioning is based on the ratio of computational power of the workstations and available storage.
The results shown in Figure 2(a) depicts the response time of the system as a function of the number of processors used to perform the ranking; the selection features are both color and texture. Figure 2(b) shows speedup as a function of the number of processors. As the number of processors used to perform the computation increases, the computation time decreases significantly. Moreover, the data partitioning based on heterogeneity information achieves a higher speedup than an even distribution approach. Figure 3 depicts the top-twenty output images on the sorted list when both color and texture are used as selection features.

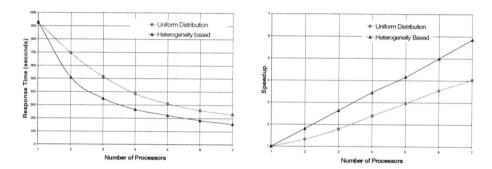

**Figure 2.** (a). Response time, (b) Speedup

**Figure 3.** Perception output; the selection features are color and texture

## 5. Conclusions

We introduced a parallel model for multimedia database content-based retrievals on a cluster of heterogeneous workstations. The model allows the extension of the system for the new types of data, new techniques, and new types of interest contents with less effort. The experimental results show that heterogeneous processing with load balancing can reduce retrieval time over a homogeneous approach. Data partitioning based on system heterogeneity achieves a better response time in comparison with uniform distribution of database over a cluster of workstations. Our future work will focus mainly on classification technique based on multi-resolution structure of the k-tree at different levels. Some load balancing and process migration techniques are also in our future work.

## References

1. P. Chalermwat, N. Alexandridis, P. Piamsa-nga, and M. O'Connell, Parallel image processing on heterogeneous computing network systems, *International Conference on Image Processing*, 1996.
2. T. El-Ghazawi, P. Chalermwat, P. Piamsa-nga, A. Ozkaya, N. Speciale, and D. Wilson, PACET: PC-parallel architecture for cost-efficient telemetry processing, *IEEE Aerospace Conference*, 1998.
3. V. Gudivada and V. Raghavan, "Special issue on content-based image retrieval systems," in *IEEE Computers*, Vol. 28, No. 9, September 1995.
4. Z. Kemp, "Multimedia and spatial information systems," *IEEE Multimedia*, 2(4), 1995.
5. J. R. Smith and S.-F. Chang, SaFe: "A General Framework for Integrated Spatial and Feature Image Search," *IEEE Workshop on Multimedia Signal Processing*, 1997.
6. P. Piamsa-nga, N. Alexandridis, G. Blankenship, G. Papakonstantinou, P. Tsanakas, and S. Tzafestas, "A Unified Model for Multimedia Retrieval by Content," *International Conference on Computer and Their Application (CATA98)*, 1998.
7. P. Piamsa-nga, N. Alexandridis, S. Srakaew, and G. Blankenship, "A parallel algorithm for multi-feature content-based multimedia retrieval," *Seventh International Conference on Intelligent Systems (ICIS98)*, Paris, France, July 1-3, 1998.
8. S. Srakaew, N. A. Alexandridis, P. Piamsa-nga, and G. Blankenship, "A parallel model for multimedia retrieval based on multidimensional signal structure," in *International workshop on systems, signal and image processing (IWSSIP98)*, Zagreb, Croatia, June 3-5, 1998.

# Industrial Supercomputing Center in Hungary - Pre-feasibility Study

Sándor Forrai and Péter Kacsuk
MTA SZTAKI Computer and Automation Research Institute
Hungarian Academy of Sciences
Laboratory of Parallel and Distributed Systems
Victor Hugo u. 18-22, XIII. ker.
Budapest, HUNGARY
E-mail: {forrai, kacsuk}@sztaki.hu

**Abstract.** The main objective of the paper is to highlight the specific characteristics of an industrial supercomputing center, which would be established in Hungary (Budapest). The presented results and conclusions are based on the performed research - in the framework of a pre-feasibility study - required and supported by the National Committee for Technological Development Hungary (OMFB)[1]. The paper will describe in detail each phase of the pre-feasibility study and will focus on the proposed solutions and conclusions.

## 1 Introduction

The primary goal of the pre-feasibility study is to evaluate the risks and the potential that appear at the establishing of an industrial supercomputing center in Hungary. The secondary goal is to produce a feasible development plan for the first years, that can be tracked and evaluated by management, by its Steering Committee, and by funding agencies.

The main objective of the High-Performance Computing and Networking (HPCN) domain is to help all sectors of industry to exploit the opportunities offered by advanced computing and networking systems to add higher levels of intelligence, reach larger throughputs or ensure shorter response times in their products, processes or services.

Advanced networking services have become an integral part of these systems and infrastructures, enabling not only applications which allow sharing and provide interactive use of remote resources, but also applications which support interaction between concurrent activities in geographically dispersed locations.

A wide definition of High Performance Computing (HPC) might be: the technology that is used to provide solutions to problems that require significant computational power either need to access, or process, very large amounts of data quickly, need to operate interactively across a geographically distributed network [1,2].

[1]**Acknowledgement:** The authors gratefully acknowledge the financial support of OMFB (the National Committee for Technological Development Hungary).

## 2 Application fields and benefits

HPC can offer solutions to problems in a wide range of business areas, from traditional large scale engineering to the emerging entertainment markets of the Internet. For example [1,2]:
1. *Optimisation of industrial processes.* Any industrial or manufacturing process probably runs at less than 100% efficiency. In fact, a typical plant may run at only around 80% efficiency, and in today's global marketplace the pressures to improve this is essential.
2. *Computational modelling.* Many firms, particularly in the engineering sector, are making more and more use of computational models as part of their design processes. Computer models provide fast, inexpensive and often safer ways to answer "what if" questions.
3. *On-line transaction processing.* Today's businesses, particularly the commerce and retail sectors, generate staggering amounts of on-line data. Electronic processing of customer orders, credit and debit card transactions, trades in the global capital markets: all these have to be logged, sorted, cross-referenced and retrieved in real time.
4. *Data mining and decision support.* The real advances made in HPC and communications technology means that businesses can now put these databases to work, applying powerful "data-mining" techniques to analyse hidden patterns and provide previously unavailable business information.
5. *Complex visualisation and virtual reality.* Leaps in graphics technology coupled with the power and capacity of HPC now allows engineers and architects to design and render, in real time, virtual "fly-through" of new oil installations, large building projects or even whole streets.

## 3 The pre-feasibility study

As was mentioned above, the pre-feasibility study has been required and supported by the National Committee for Technological Development Hungary (OMFB), in order to evaluate the real market needs concerning with a supercomputing center and to develop a feasible development plan.

The main phases of the performed pre-feasibility study are:
- potential supercomputing users identification;
- users' needs and requirements evaluation based on the developed form (questionnaire);
- information gathering concerning with similar HPCN centers in Europe and United States (hardware and software support, organisation, founding, costs, services, etc.);
- elaboration of different alternative solutions for establishing a HPCN center in Hungary, which contain: detailed technical plan, management plan, financial plan, Gantt chart of the project (all of them based on the potential users needs);

- presentation of a feasible development plan, which can be tracked and evaluated by the management (risks and potentials evaluation of the future HPCN centre).

In order to evaluate the potential users' needs an evaluation form (questionnaire) has been elaborated and is based on interviews technique. The evaluation form contains 29 questions (five of them concern to user identification). Most of questions, 17 concern with technical aspects, which try to evaluate the investigated application field, the available hardware and software tools. Furthermore, future needs both from computing capacity and software support as well as the required courses and seminars in the field of parallel programming are evaluated. The desired future organisation structure is interviewed in three question. Finally, two questions deals with financial aspects and also two with future benefits of a HPCN center.

In the survey process, 17 potential super-computing users have been interviewed. They are grouped in 7 categories and belong to different economic sectors as follows: energy production and distribution (1), automobile industry (2), pharmaceutical industry (2), air and local transportation (2), research and development (2), communications (1), others (7).

The investigation process highlighted that for most of the potential users the most important requirements are: the availability of the software applications as well as the data security. Taking into account the most significant requirements they can be grouped in 5 categories: finite element applications (4), molecular modelling (4), datamining (2), scheduling and optimization (4), others (3).

In figure 1 the evaluated computing needs of the users are presented: in case of 52.9% of users the existing computing capacity is critical, they rent computer resources (11.8%), they charge others (11.8%), they use results developed by others (35.3%) and in case of 41.2% new computing intensive problems might arise.

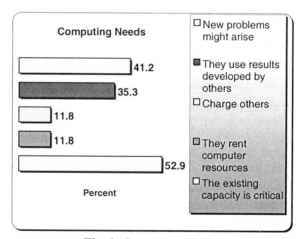

**Fig. 1.** Computing Needs

The developed **technical plan** contains information about: possible supercomputing architectures, hardware and software support of the HPCN center, data security, placement, required facilities (UPS, access control, air-conditioning, lightning and fire protection, etc.). The proposed supercomputing architectures are: classical supercomputing center, workstation cluster and metacomputing technology.

During the survey become obvious that the greatest computing demand come from the Hungarian Meteorological Service and from the oil and gas exploration field.

From software point of view we realised that only in few cases are available own developed software, almost all interviewed partners use well-known (commercially available) software applications. Therefore, the HPCN center should offer general software support (C, FORTRAN compilers, database applications, etc.) as well as special software applications (finite element method, molecular modelling with the related databases, etc.).

The **management and financial plans** are strongly co-related, where important attention is given to the human resources department as well as to the marketing department in the aim to attract new users.

The proposed **organisation structure** is a consortium, where each institution will be represented by one member on the Steering Committee, the project leader will be the president of the Committee. It is expected that the Committee will meet quarterly, to approve the progress reports, and to decide on any matter that can affect the smooth running of the project.

## 4 Results and conclusions

As a conclusion, most of the interviewed potential users support the idea for the establishing of a supercomputer center in Hungary, but only few of them would like to participate actively (financially). Most of them should become active users who should pay for the services offered by the center.

The pre-feasibility study highlighted that the most feasible solution - at that moment, taking into account existing computing resources, the existing financial possibilities and evaluated computing needs - is based on metacomputing technology. This solution seems to be the most flexible and most effective from price/performance point of view. Moreover the Hungarian Supercomputing Center should offer:
- custom software design, development and optimisation for distributed and high-performance systems;
- service and support for industrial and commercial users in the field of high-performance computer systems;
- training courses and materials covering all aspects of high-performance computing;
- general consultation on the application and use of novel and leading-edge computing techniques and technologies.

**References**
1. \*\*\* High Performance Computing for Academia and Industry, Web Pages: www.epcc.ed.ac.uk.
2. \*\*\* European Centre for Parallel Computing at Vienna, 1997, Web Pages: www.vcpc.univie.ac.at.

# Reducing Memory Traffic Via Redundant Store Instructions

Carlos Molina, Antonio González, Jordi Tubella
Departament d'Arquitectura de Computadors,
Universitat Politècnica de Catalunya,
Jordi Girona 1-3, Edifici D6, 08034 Barcelona, Spain
e-mail:{cmolina,antonio,jordit}@ac.upc.es

**Abstract.** Some memory writes have the particular behaviour of not modifying memory since the value they write is equal to the value before the write.These kind of stores are what we call *Redundant Stores*. In this paper we study the behaviour of these particular stores and show that a significant saving on memory traffic between the first and second level caches can be avoided by exploiting this feature. We show that with *no* additional hardware (just a simple comparator) and without increasing the cache latency, we can achieve on average a 10% of memory traffic reduction.

## 1. Introduction

During the last decade, innovation and technological improvements in processor design have outpaced advances on memory design. That is the reason why current high performance processors focus on cache memory organizations, to ease the gap between processor and memory speed [4]. The problem is that many of these techniques used to tolerate growing memory latencies do so at the expense of increased memory bandwidth, and this has been shown to be progressively a greater limit to high performance [1]. The different techniques proposed such as software and hardware prefetching [2], stream buffers [5], speculative load execution [3], and multithreading [6] reduce latency related stalls, but also increase the total traffic between main memory and the processor. In this paper we propose a technique for reducing memory traffic without any special hardware requirement and without affecting the hit/miss ratio of the cache. The only hardware requirement is just a simple comparator.

## 2. Redundant Store Mechanism

This section describes the changes required on different cache configurations in order to take advantage of the redundant stores for memory traffic reduction. Note that the additional hardware is minimal and it is highlighted in each figure.

**Copy Back**: In the regular version, a store accesses cache to write its value and sets the dirty bit. We propose to first check if the current value in cache matches the value that this store is going to write. If so, there is a *Redundant Store* and the dirty bit is not set. Fig. 1.a shows how this mechanism works. Here we have a reduction in the frequency the dirty bit is set, and this consequently reduces cache miss bandwidth because fewer blocks of the cache will have to go to the next level of the memory hierarchy when there is a cache miss. Note that to exploit the benefit of redundant stores we only have to add a simple comparator that tries to detect matching values, and if so, the dirty bit is not set.

**Write Through**: When a store accesses the cache and there is a hit, if the value to be written is the same as the present, we have a *Redundant Store*. Once this store is identified we propose, to do not send to the next level of the memory hierarchy. Fig. 1.b shows how this mechanism works. Note that the output of the comparator decides whether the buffer can send the value to the next level of the hierarchy. Here we have a reduction in memory traffic because memory write values present in cache do not have to go to the next level of the memory hierarchy if they are identified as redundant stores.

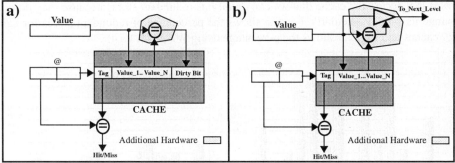

**Figure 1:** Redundant Store Mechanism with **a)** Copy Back **b)** Write Through

A store in a conventional cache first reads the tag and afterwards writes the new value. Our mechanism, reads simultaneously the old value and the tag and then writes the new value if necessary, so we do not increase the cache latency.

## 3. Experimental Framework and Performance Evaluation

We have developed a functional parameterized simulator for cache memories. Mainly, this simulator provides information about required memory bandwidth and hit/miss ratios. Different cache configurations and cache policies have been simulated. For the evaluation we have been considered a subset of the Spec95 benchmark suite. The programs have been compiled with the DEC Fortran and C compilers for a DEC AlphaStation 600 5/266 with full optimizations, and instrumented by means of the Atom tool. Each program was run with the reference input sets and statistics were collected for *1 billion* of instructions after skipping the initial part corresponding to initializations.

|  | COPY BACK- WRITE ALLOCATE | | | | WRITE THROUGH- NO WRITE ALLOCATE | | | |
|---|---|---|---|---|---|---|---|---|
|  | 32 KB | 16 KB | 8 KB | 4 KB | 32 KB | 16 KB | 8 KB | 4 KB |
| Applu | 10.81 | 11.29 | 12.15 | 14.01 | 16.40 | 16.84 | 17.65 | 19.48 |
| Apsi | 12.61 | 13.12 | 15.03 | 29.59 | 14.99 | 15.71 | 18.95 | 35.26 |
| Gcc | 2.40 | 5.13 | 7.81 | 11.61 | 2.85 | 6.57 | 11.58 | 18.90 |
| Go | 3.52 | 5.73 | 10.26 | 17.13 | 7.71 | 9.67 | 12.90 | 17.87 |
| Ijpeg | 1.58 | 2.69 | 7.32 | 15.34 | 1.63 | 3.16 | 7.87 | 15.98 |
| M88ksim | 0.67 | 1.30 | 5.52 | 7.68 | 1.22 | 1.82 | 7.11 | 11.49 |
| Mgrid | 4.97 | 5.21 | 5.35 | 8.46 | 6.89 | 7.12 | 7.25 | 10.35 |
| Perl | 0.27 | 0.55 | 1.64 | 7.35 | 1.80 | 3.37 | 4.91 | 12.86 |
| Tomcatv | 12.06 | 15.98 | 29.13 | 36.26 | 16.07 | 18.76 | 30.87 | 38.87 |
| Turb3d | 5.82 | 7.59 | 8.92 | 10.60 | 5.79 | 7.93 | 10.62 | 13.52 |
| Vortex | 1.88 | 3.45 | 5.40 | 8.04 | 2.96 | 5.83 | 9.09 | 14.22 |
| Wave | 15.33 | 25.85 | 36.02 | 41.97 | 21.36 | 31.93 | 40.90 | 47.93 |
| **A_MEAN** | **5.99** | **8.16** | **12.05** | **17.34** | **8.31** | **10.73** | **14.98** | **21.39** |

**Table 1:** Miss Ratios of *Copy Back* with *Write Allocate* and *Write Through* with *No Write Allocate*

This section evaluates the impact on memory traffic reduction as well as the amount of redundant stores for different cache configurations. We have considered four cache sizes: 32 KB, 16 KB, 8 KB and 4 KB, and 32 byte line size. We have been simulated the following cache policies: *Copy Back with Write Allocate* (CB-WA for short*)* and *Write Through with No Write Allocate* (WT-NWA for short*)*. *Direct mapped* caches have been assumed for all the simulations. Table 1 shows the miss ratio of every combination of the above described sizes and policies.

## 3.1 Amount of Redundant Stores

This section analyses the behaviour of redundant stores in cache. Different cache sizes and policies have been considered. Fig. 2.a and Fig. 2.b show the percentage over all the stores that access the cache that we can qualify as redundant stores because they do not change its state. Particularly, Fig. 2.a shows the percentage of redundant stores for *CB-WA* caches and Fig. 2.b shows the percentage using *WT-NWA*.

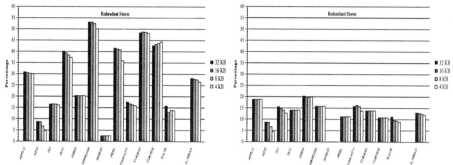

**Figure 2:** Percentage of Redundant Stores of **a)** *Copy Back-Write Allocate* and **b)** *Write Through-No Write Allocate*

Note that the amount of redundant stores in cache is significant in some benchmarks. On average, around 30% of the stores are redundant for a *CB-WA* cache and 15% for a *WT-NWA* cache. This difference is due to the fact that *CB-WA* has better *store miss ratio* than *WT-NWA*. The former has half store miss ratio than the latter.

## 3.2 Amount of Memory Traffic

This section analyses the traffic ratio that exists between the memory cache and its next level of the memory hierarchy. The traffic is computed as the number of bytes transmitted between memories. Fig. 3.a and Fig. 3.b show the memory traffic for each policy and different sizes. Obviously, the memory traffic decreases when we use bigger

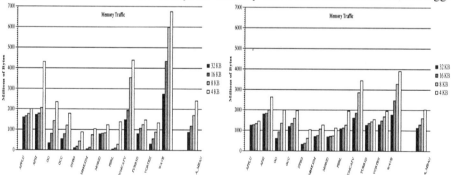

**Figure 3:** Memory Traffic (Millions of bytes) of **a)** *Copy Back-Write Allocate* and **b)** *Write Through-No Write Allocate*

cache sizes due to the fact that the miss ratio decreases. The memory traffic has two main sources: the hit/miss ratio and the size of the cache block. Note that on average *CB-WA* and *WT-NWA* have similar rate of memory traffic. Although *CB-WA* has better hit ratio than *WT-NWA* and intuitively it should have lower memory traffic, it has to transmit the whole cache block on a miss if the replaced block is dirty, while *WT-NWA* only has to transmit a single value per each write.

## 3.3 Memory Traffic Reduction

The important benefit of the proposed technique is shown in Fig. 4.a and Fig. 4.b. They present the percentage reduction in memory traffic after applying the redundant store mechanism. Note that on average we have a 7% of memory traffic reduction for *CB-WA* meanwhile we have a 19% of memory traffic reduction for *WT-NWA* for a 32 KB cache.

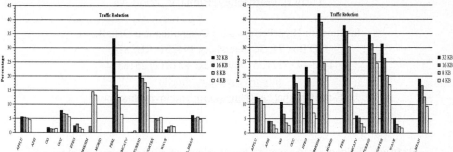

**Figure 4:** Reduction of Memory Traffic of **a)** *Copy Back-Write* Allocate and **b)** *Write Through-No Write Allocate*

Mainly, this difference is due to the fact that all the stores have to go to the next level of the memory hierarchy for *Write Through*, so every redundant store that can be identified results in a saving on memory traffic. On the other hand, using *Copy Back,* every redundant store reduces the frequency of the dirty bit setting, which not always results in a traffic reduction, because for instance two consecutive redundant stores mapped to the same cache line, just reduce the traffic in one block. Specially significant is the traffic reduction that we achieve using *WT-NWA*. Note that some benchmarks achieve more than 30% in traffic reduction.

## 4. Conclusions

This work presents an easy mechanism for reducing the memory traffic between cache and the next level of the memory hierarchy. The novel idea requires *minimal hardware support*. It is based on the particular behaviour of some writes to memory that do not change its contents and the idea can be applied to any of the current cache organizations. These particular stores are what we call redundant stores. We have shown that we can achieve a significant memory traffic reduction. On average we can achieve close to 7% for a cache with *CB-WA* and 19% with *WT-NWA*.

## Acknowledgements

This work have been supported by projects CYCIT 511/98 and ESPRIT 24942, and by the grant AP96-52460503. The research described in this paper has been developed using the resources of the Center of Parallelism of Barcelona and the ECE Department of the Carnegie-Mellon University.

## References

[1] D. Burger, J. R. Goodman, and A. Kägi, "Quantifying Memory Bandwidth Limitations of Current and Future Microprocessors". In *Proc. of the 23rd Int. Symp. on Computer Architecture*, 1996.

[2] T.-F. Chen and J.-L. Baer. "A performance Study of Software and Hardware Data Prefetching Schemes". In *Proc. of the 21st Int. Symp. on Computer Architecture*, 1994

[3] J. González and A. González. "Speculative Execution via Address Prediction and Data Prefetching". In *Proc. of the 11th ACM Int. Conf. on Supercomputing*, 1997.

[4] J. R. Goodman, "Using Cache Memory to Reduce Processor Memory Traffic" In *Proc. of the 10th Int. Symp. on Computer Architecture,* 1983.

[5] N.P. Jouppi. "Improving Direct-Mapped Cache Performance by the Addition of Small Fully Associative Cache and Prefetch Buffers". In *Proc. of the 17th Int. Symp. on Microarchitecture*, 1990.

[6] D.M. Tullsen, S.J. Eggers and H.M. Levy, "Simultaneous Multithreading: Maximizing On-Chip Parallelism". In *Proc. of the 22nd Int. Symp. on Computer Architecture*, 1995

# 3D-Visualization for Presenting Results of Numerical Simulation

Yuriy E. Gorbachev, Elena V. Zudilova

Institute for High-Performance Computing and Data Bases
P/O Box 71, St. Petersburg 194291, RUSSIA
Tel.: +7 (812) 2519092, Fax: +7 (812) 2518314
http://www.csa.ru/ihpcdb
E-mail: gorbachev@hm.csa.ru; zudilova@fn.csa.ru

**Abstract.** The paper is devoted to the problems of presenting results obtained during the numerical simulation. Two different approaches of data visualization: through 2D and 3D objects are described. 3D-visualization requires more computational resources than 2D, especially when the data volumes are large. So the best way for quick getting necessary data during the numerical simulation is to use supercomputers. We are going to present the two animation films demonstrating the results of 3D-visualization designed by the research fellows of the Institute for High-Performance Computing and Data Bases for simulating multidimension stationary/non-stationary molecular gasdynamic problems and for simulating mature convective clouds.

There are two main purposes of using visualization in the sphere of numerical simulation: illustrating results and presenting results. The first one means clarifying obtained data for specialists working with it. The second one means making the data more attractive for demonstration, using sometimes the additional computational and graphic resources. As a rule, for illustrating purpose specialists use ordinary painting programs that are simple for every day usage and can be loaded on common PC. As for the design of professional presentations, the more complicated environment should be used, including not only special design and programming packages, but sometimes also the additional hardware and projection systems.

The methods of visualization are very important for both analyze of numerical simulation results and popularization of supercomputer methods. The dynamics of different physical and chemical processes are better observed in graphic form. Moreover, there are many special information systems based on numerical simulation methods, such as: weather broadcasting, monitoring, etc. where only the visualized data can be used as the output result.

Today the ordinary variant of data visualization is 2D-presentation of investigated objects and their interaction. This approach does not need much computer resources for drawing the process only in two coordinates (x,y). When the object under discovery is not very complicated and the duration of interaction is not so long, digital figures can be obtained by means of PC.

3D-visualization means simulating research conditions in three coordinates (x,y,z) through establishing choreography or movement for the 3D-objects step by step, setting up lights and applying textures. The following approach needs additional computer memory. To obtain the data necessary for 3D-visualization by means of common computer resources is not a simple task. Sometimes this process may cover the period of several months. So usage of supercomputer resources is becoming the only right solution, of course, if these resources are available.

The Institute for High-Performance Computing and Data Bases, St.Petersburg, RUSSIA has a vast potential of supercomputer resources, including HP Convex series, Sun Enterprise cluster, Parsytec massive-parallel computers, special SGI graphic stations Octane, etc.

We are going to present the new 3D-visualized results obtained for such research spheres as simulation of multidimension stationary and non-stationary molecular gasdynamic problems and simulation of mature convective clouds.

One problem that in the sphere of interest is 3D non-stationary side stream interaction with pulse jet. The fig. 1. illustrates the estimation of the power and heat influence at the surface of a spacecraft in the area of this interaction. This process is very important for spacecraft projection. The problem was solved on the base of DSMC (Direct Simulation of Monte Carlo) method using the computational resources of supercomputer Parsytec CC/16. Only elastic elementary processes are simulated [1].

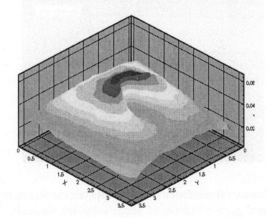

**Fig. 1.** Pressure at the surface of spacecraft for the fourth moment of time

Special parallel algorithm for high-performance computing of non-stationary 3D-flows on a supercomputer systems with distributed memory was developed. The following algorithm makes it possible to solve the same task on several processing nodes at the same time. The application for massive-parallel system Parsytec CC/16 provides simulation efficiency closely to 100%. The speedup of this algorithm has practically a linear dependence on the number of processors. Visualization of 3D fields of total and jet density gives the understanding of flow evolution in time and space.

Another example we are going to present is the 3D-visualization of gas dynamic and physical-chemical processes in modeling of mature convective clouds (fig.2.).

The process of numerical simulation is based on the solution of equations describing the axisymmetric turbulent flow of compressible two-phase two-component medium taking into account condensation of water vapor and precipitation formation. The model includes a module describing the processes of aerosol transfer and scavenging. It provides the ability for calculating the parameters for smoke clouds formed by volcanic eruptions and for estimating the amount of the pollutions deposited on the surface and transported into the upper layers of the troposphere and into the stratosphere [2].

All the calculations were carried out using supercomputers of CONVEX C-series. The algorithms were adapted for vector-parallel architecture of computers. Special program module called CONVEC was elaborated for speedy analysis of obtained large arrays of results. It creates compact databases and allows analyzing their records in convenient graphic form.

We also going to present several 2D-animation films that cover the simulation of hydrodynamic processes and modeling nonlinear processes for comparison with 3D data presentation.

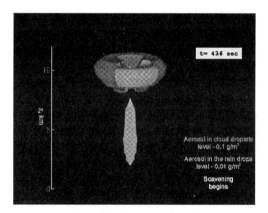

**Fig. 2.** The results of numerical simulation of the atmospheric convection in extreme conditions

We hope that our presentation will help to make the methods of 3D-visualization more popular in different research spheres. Moreover, we have prepared solutions for obtaining the best results using supercomputers of different architectures. We look forward for collaboration with research teams having the experience in numerical simulation and data visualization and with those who only start using supercomputers.

# References

1. N.Y.Bykov, Yu.E.Gorbachev, G.A.Lukianov. Parallel direct simulation Monte Carlo of the expansion of laser-induced plume in vacuum. Report on EUROMECH Colloquium 363 "Mechanics of laser ablation" Novosibirsk, Russia, 23-26 June, Abstract of papers, (1997) p.10.
2. Ramaroson, I.Karol, N.Ozhigina, Y.Ozolin, E.Rozanov, E.Stankova, M.Zatevakhin. Numerical modeling of the chemical, dynamical and radiative effects of the evolution of atmospheric species in the cloudy environment. Proceedings 3-d Conference on Atmospheric Chemistry, Long Beach, California, 2-7 February (1997).

# Optimal Distribution of Loops Containing No Dependence Cycles

Zdzislaw Szczerbinski

Polish Academy of Sciences, Institute for Theoretical and Applied Computer Science,
Baltycka 5, 44–100 Gliwice, Poland
zdzich@iitis.gliwice.pl
http://www.iitis.gliwice.pl/~zdzich.zor

**Abstract.** In the paper, a method is proposed for optimizing parallelization of loops whose dependence graphs are acyclic. The idea of loop distribution as a method for data dependence synchronization is presented. Next, the loop distribution optimization problem is posed and solved with the technique of optimal dependence folding. Theoretical considerations are accompanied by experimental results from applying the proposed optimization method to an example program executed on a distributed-memory parallel supercomputer.

## 1 Introduction

In a distributed-memory environment, a loop is parallelized so that individual iterations or groups of successive iterations are assigned to separate processors. If a *loop-carried dependence* occurs in a parallelized loop, the execution of the loop's iterations must be synchronized so that the source of the dependence is executed before its sink. The synchronization of a loop-carried dependence is ensured through interprocessor communication in the form of a *message transfer*. Since transmitting a message from one processor to another adds a substantial overhead, interprocessor communication is undesirable in loop parallelization.

## 2 Communication optimization by loop distribution

The idea of *loop distribution* consists in decomposing a single loop into a number of more elementary ones. In a distributed-memory environment, loop distribution eliminates the need for explicit synchronization, involving a message transfer, of a *forward* loop-carried dependence, by placing its source and sink in separate loops. In the paper, we limit our attention to loops which contain no dependence cycles i.e. can be restructured into a form where all loop-carried dependences are forward. A simple way to distribute a loop containing $r$ statements and no dependence cycles is to generate $r$ loops, each containing a single statement. However, this method of distribution is inefficient due to the introduced overhead associated with loop organization, repeated for each loop. At this point, the following optimization problem may be posed.

**Optimal distribution problem.** Given a loop containing no dependence cycles, find its distribution which
- ensures communication-free synchronization of all dependences, and
- generates the minimum number of resulting loops.

Loop distribution has been extensively studied in the literature [1, 6, 5]. However, the authors have not addressed the optimal distribution problem posed above.

## 3 Dependence folding and its application to loop distribution

Quite often, synchronizing a dependence simultaneously synchronizes another dependence. An explicit synchronization of the latter dependence is then unnecessary and is said to be *eliminated* by the synchronization of the former one. Consider two forward dependences in a loop: $D_j$ and $D_k$. Let us number the loop body's statements according to their lexical order with consecutive positive integers as subindices, and denote the source and sink of dependence $D_j$ by $S_v$ and $S_w$, respectively, and those of $D_k$ by $S_x$ and $S_y$. Assuming that $S_v$ precedes the remaining statements we distinguish four cases:

- $D_j$ and $D_k$ are *disjoint* if $v < w < x < y$,
- $D_j$ and $D_k$ are *adjacent* if $v < w = x < y$,
- $D_j$ and $D_k$ *overlap* if $v < x < w < y$,
- $D_k$ is *nested in* $D_j$ if $v \leq x < y \leq w$.

According to the general rules for synchronization elimination [3], synchronizing a forward dependence $D_k$ eliminates the synchronization of another forward dependence $D_j$ iff $D_k$ and $D_j$ are of the same distance and $D_k$ is nested in $D_j$. Consider the case of two loop-carried, overlapping dependences of the same distance $d$. No synchronization elimination is directly possible. However, by synchronizing $S_x$ (the source of $D_k$) with $S_w$ (the sink of $D_j$) so that $S_x$ in iteration $i$ of the loop is executed before $S_w$ in iteration $i+d$, which is equivalent to synchronizing an imaginary dependence $D_l$ of distance $d$, we can eliminate the synchronizations of $D_j$ and $D_k$ since $D_l$ is simultaneously nested in $D_j$ and $D_k$. Therefore, instead of two, there is now only one synchronization required. We call the above process *dependence folding*; $D_j$ and $D_k$ are said to be folded into $D_l$ which we call their *condensation* [3]. By applying *statement reordering* [6] one may sometimes turn disjoint dependences of the same distance into overlapping ones and thus achieve additional elimination of synchronizations. These techniques have been utilized in optimizing parallelization of loops in shared-memory programs [3, 4]. In the distributed-memory model of computation, the same method may be applied to optimal synchronization elimination with loop distribution. If we fold the overlapping dependences $D_j$ and $D_k$ into $D_l$ and then "break" $D_l$ by distributing the loop so that $S_v$ and $S_x$ are in one loop and $S_w$ and $S_y$ in another, then $D_l$ has been synchronized with loop distribution i.e. in a communication-free manner. Likewise, transforming a pair of disjoint dependences into overlapping ones (if statement reordering is possible) and then performing dependence folding and loop distribution as described above, eliminates the need for their explicit synchronizations. If a dependence is nested in another one, the two dependences are folded into the "internal" one and loop distribution gives a similar result.

For a given lexical order of statements in a loop, analyzing dependences for the possibility of statement reordering and dependence folding becomes progressively more time-consuming as the numbers of statements and dependences grow. In fact, instead of rearranging the original sequence of statements, it is more efficient to achieve synchronization elimination directly from an analysis of the loop's dependence graph. A method for doing this in an optimal way has been presented in [3, 4]. The method is based on the topological sorting of the loop's acyclic dependence graph in a specific manner. The proposed algorithm ELMAX [4] has been proved to yield the optimal order of statements in a loop for dependence folding and to generate the minimum possible number of dependences' condensations. In this paper, we extend the application of ELMAX to the issue of loop distribution for communication-free synchronization of dependences. The resulting algorithm LOOPDIST performs loop distribution by topologically sorting the loop's dependence graph with ELMAX and, next, building loops around groups of statements so that the dependences' condensations are "broken". The computational complexity of LOOPDIST is $O(n^2)$. It has been proved that the algorithm provides the solution to the optimal distribution problem.

## 4 Experimental results

The optimization method described in the paper has been verified in practice. Execution of an example loop in a parallel computing environment was analyzed. The hardware platform was the Quadrics Computing Surface-2 High Availability (CS-2HA), a 136-processor distributed-memory, massively parallel supercomputer. Several versions of the analyzed loop were written in the form of simple High Performance Fortran (HPF) programs, with cyclic distribution of arrays accessed in the loop. The programs were compiled with pghpf [2], the HPF compiler from Portland Group, Inc., and run on a 16-processor partition. The analyzed loop had 12 statements, with 5 loop-carried and 9 loop-independent dependences between them. The first parallelization variant analyzed was to directly compile the loop with pghpf. Since the loop contained loop-carried dependences, the compiler failed to fully parallelize it and introduced interprocessor message transfers for dependence synchronization. Another variant analyzed was to take advantage of pghpf's option of automatic parallelization. With this option switched on, each do loop in a compiled HPF program is parallelized; in case of loop-carried dependences, loop distribution is performed, based on the compiler's manufacturers' proprietary techniques. The last variant was to apply the algorithm LOOPDIST.

Table 1 shows the timing results for the three variants of parallelization.

| Parallelization variant | n=1000 | n=10000 | n=100000 |
|---|---|---|---|
| Synchronization with message transfers | 2.05 | 18.10 | 177 |
| Automatic parallelization by pghpf | 0.53 | 4.28 | 41.7 |
| Optimization with LOOPDIST | 0.32 | 0.36 | 0.94 |

**Table 1.** Execution times, in seconds, of an example loop on 16 processors of Quadrics CS-2HA

## 5 Conclusion

We have proposed a method for optimizing loop parallelization in distributed-memory programs. The optimal distribution problem has been formulated for the case of a loop containing no dependence cycles. This problem has been solved with the aid of the dependence folding technique. Theoretical considerations have been verified in practice by applying the developed optimization method to an example program, executed in a distributed memory, parallel computing environment. The timing results confirm the method's effectiveness and applicability to practical parallel programming.

## References

1. K. Kennedy and K. S. McKinley, Loop distribution with arbitrary control flow, in: *Proceedings Supercomputing'90*, New York, 1990, 407–416.
2. pghpf *User's Guide*, The Portland Group, Inc., Wilsonville, Oregon, USA, 1997.
3. Z. Szczerbinski, Optimization of parallel loops by elimination of redundant data dependences, Ph.D. thesis (in Polish) (Silesian Technical University, Faculty of Automatics, Electronics and Computer Science, Gliwice, Poland, 1995).
4. Z. Szczerbinski, An algorithm for elimination of forward dependences in parallel loops, in: *Proceedings 2nd International Conference on Parallel Processing and Applied Mathematics PPAM'97*, Zakopane, Poland, 1997, 398–407.
5. M. Wolfe, *High Performance Compilers for Parallel Computing* (Addison-Wesley, Redwood City, California, USA, 1996).
6. H. Zima and B. Chapman, *Supercompilers for Parallel and Vector Computers* (Addison-Wesley, Wokingham, England, 1991).

# Solving Maximum Clique and Independent Set of Graphs Based on Hopfield Network

Y. Zhang, C.H. Chi
School of Computing
National University of Singapore
Lower Kent Ridge Road
Singapore 119260

**Abstract**   Maximum clique and independent set problems are classical NP-full optimization problems, the solutions of which are difficult to obtain from conventional methods. Hopfield network in neural network, which simulates the partial functions of a human brain through the ultra-large scale parallel computation, has been proven to have potentials in solving these problems in a reasonable period of time. The main problem of this approach is the difficulty in defining an efficient energy function and the dynamic equation of motion for the Hopfield model. In this paper, we propose solutions to this problem by solving two typical problems in the coloring of graphs, the maximum clique and independent set, through our refined Hopfield network model. Both the mathematical model and the simulation algorithm are given here. It is found that the time complexity to obtain an optimal solution can approach one order of magnitude lower than the current available solutions.

## 1. INTRODUCTION

Coloring of graphs is a classical and important class of problems in traditional combinatorial optimization research. Just like other traditional optimization problems, this class of problems, which includes the color number $x(G)$, normal $k$-peak, $k$-edge, and $k$-full coloring algorithms of a given graph $G$. is NP-full problem[1]. To obtain optimal solutions to the NP-full problems in a reasonable amount of time period, people turn to neural network and neural computation[2,4,8-14]. One direct consequence of this research direction is the potentials of studying combinatorial optimization and NP-full problems using Hopfield neural network[5].

The basic procedure of using the Hopfield network[5,7] for a combinatorial optimization problem is to map the problem into the Hopfield network. The energy function is defined in accordance to its object function and constrained conditions. Then, it is minimized to give the solution of the problem. Afterwards, the state equation of the Hopfield network motion is constructed from the energy function, and is solved by iterative methods. The key problem of this procedure is the definition and construction of the energy function. For example, Dahl studied the peak coloring problem of graphs using Hopfield network model[6] in 1987. He derived the equation of network motion and obtained the network weight values through the construction of the energy function. However, the energy function was not refined and the required computation was very complex. As a result, optimal solution was not easily found. Currently, this observation is still valid when the Hopfield network is used in solving combinatorial optimization problems.

In this paper, we address the coloring of graphs problem by refining the energy function and the equation of network motion in the Hopfield network model. Two classical problems were investigated. They were the maximum clique and the independent set problems of graphs. The independent set problem was included in the study because it is closely related to the coloring problem. Both the mathematical formulation of the model and its simulation study are presented. It is found that significant speedup in solving the coloring of graphs can be obtained and the result can generally be applied to the normal $k$-Peak coloring, $k$-edge coloring and $k$-full coloring algorithms.

## 2. MAXIMUM CLIQUE ON HOPFIELD NETWORK

To solve combinatorial optimization problems using the Hopfield network model $H = (W, I)$ in which the connection weight coefficient matrix is $W$ and the bias matrix is $I$, the general steps are as follows:
[1] map the problem under study to the output of the Hopfield network,
[2] define the energy function expression in accordance to its goal function and constrained conditions,
[3] compute the minimum value of the energy function that corresponds to the optimal solution of the problem, and obtain the connection weight value and the bias value of the network, and finally
[4] obtain the stable state (optimization) solution to the combinatorial optimization problem under study by running the dynamic equation of the network.

To solve the maximum clique (or maximum independent set) problem of a graph $G$ with $p$ peaks, a Hopfield network with $p$ neurons is established. The stable state output $\upsilon = (\upsilon_1, \upsilon_2, \cdots, \upsilon_p)$ will correspond to a clique $s$ (or an independent set $T$) of $G$. This will be the constrained conditions of the network. The object function here is to make the peak number in the clique (or independent set) maximum. Thus, for the maximum clique problem of a graph, the stable state output of the network $\upsilon = (\upsilon_1, \upsilon_2, \cdots, \upsilon_p)$ is defined as follows:

$$\upsilon_i = \begin{cases} 1 & x_i \in s \\ 0 & x_i \notin s \end{cases} \quad i = 1, 2, \cdots, p$$

## 3. MAXIMUM INDEPENDENT SET ON HOPFIELD NETWORK

Similar to the maximum clique problem of graphs described above, to solve the maximum independent set of $G$ with $p$ peaks, we establish a Hopfield network with $p$ neurons and define the stable state output $\upsilon = (\upsilon_1, \upsilon_2, ..., \upsilon_p)$ of the network as follows:

$$\upsilon_i = \begin{cases} 1 & x_i \in T \\ 0 & x_i \notin T \end{cases} \quad i = 1, 2, ..., p$$

The constrained condition is an independent set of $G$ which the stable state output $\upsilon = (\upsilon_1, \upsilon_2, ..., \upsilon_p)$ is mapped to. Based on the same argument as the maximum clique problem of $G$, the stable state output $\upsilon$ of the network needs to satisfy this condition:

$$\upsilon_i \upsilon_j a_{ij} = 0 \qquad i, j = 1, 2, ..., p, i \neq j$$

The energy function of the Hopfield network is defined as follows:

$$E = -A \sum_{i=1}^{p} \upsilon_i + B \sum_{i=1}^{p} \sum_{j=1, j \neq i}^{p} \upsilon_i \upsilon_j a_{ij}$$

And the connection weight coefficient $W_{ij}$ and the applied bias current $I_i$ of the network are obtained using this relationship:

$$\begin{cases} W_{ij} = B a_{ij} (1 - \delta_{ij}) \\ I_i = -A \end{cases} \qquad i, j = 1, 2, ..., p$$

## 4. Simulation Result

To understand the efficiency of the Hopfield model with our proposed energy function and the dynamic equation of network motion, the algorithms for the maximum clique and maximum independent set problem of graphs using Hopfield network were simulated. The result shows that our proposed Hopfield network model, together with the new energy function, can solve the maximum clique and maximum independent set of graphs problems effectively. The optimization solution can be obtained with an average of less than 1 second in the 20 times runs. Compared to the experimental result of about 60 times runs to obtain optimal solution in reference[3], a speedup factor of 45 is achieved, which is very significant.

## 5. CONCLUSION

In this paper, we investigated the application of Hopfield network on two classical combinatorial optimization problems: maximum clique and maximum independent set of graphs. By proposing a new set of energy function and a dynamic equation of network motion, we showed, by mathematical model and simulation, that the solutions to these two problems can be obtained efficiently. Furthermore, these results can be applied not only to the maximum clique and independent set problems, but also to other problems in the coloring of graphs with little modification.

## References

[1] Bondy, J.A., Murty, U.S.R., *Graph Theory with Applications*, Macmillan Press Ltd., 1976, Chapter 5.

[2]   Takoda, M., Goodman, J.W., "Neural Networks for Computation Number Representations and Programming Completely," *Applied Optics*, Volume 25, Number 18, 1986, pp. 3033-3046.

[3]   Fukushima, K., "Neocoqnitron: a Hierarchical Neural Network Capable of Visual Pattern Recognition," *Neural Networks*, Volume 1, Number 2, 1988.

[4]   Takefugi, Y., Lee, K.C., "Artificial Neural Networks for Four-Coloring Map Problems and K-Colorability Problems," *IEEE Transactions on CAS*, Volume 38, Number 2, 1991, pp. 326-333.

[5]   Tank, D.W., Hopfield, J.J., "Neural Computation by Concentrating In-Formation in Time," *Proc. Natl. Acad. Sci.*, USA, Volume 84, 1987, pp. 1896-1900.

[6]   Dahl, E.D., "Neural Network Algorithm for an NP-Complete Problem: Map and Graph Coloring," *Proceedings of the IEEE First International Conference on Neural Networks,* Volume III, June 1987, pp. 113-120.

[7]   Hopfield, J.J., "The Effectiveness of Analogue Neural Network Hardware," *Neural Networks*, Volume 1, 1990, pp. 27-46.

[8]   Funaliashi, K., "On the Approximate Realization of Continuous Mappings by Neural Networks," *Neural Networks*, Number 2, 1989, pp. 182-192.

[9]   Cybenko, G., "Approximation by Super-Positions of a Sigmoidal Function, Mathematics of Control, Signals and Systems," Number 2, 1989, pp. 303-314.

[10]  Bose, N.K., Garga, A.K., "Neural Network Design Using Variously Diagrams*," IEEE Transactions on Neural Networks*, Volume 4, Number 5, September 1993, pp. 778-787.

[11]  Sethi, I.K., "Entropy Nets From Decision Trees to Neural Networks*," Proceedings of IEEE*, Number 78, 1990, pp. 1605-1613.

[12]  Garga, A.K., Bose, N.K., "Structure Training of Neural Networks*," Proceedings of the IEEE Intl. Conf. Neural Networks*, IEEE World Congress on Computation Intelligence, Orlando, Volume 1, 1994, pp. 239-244.

[13]  Xu, J., Zhang, J.Y., Bao, Z., "Graphic Coloring Algorithms Based on Hopfield Network," *Acta Electronica Sinica*, 1996, Volume 24, Number 10, 1996, pp. 8.

[14]  Ramanujarn, J., Sadayappan, P., "Optimization by Neural Networks," *Proceedings of the IEEE International Conference on Neural Networks,* Volume 2, 1988, pp. 325-332.

# Storing Large Volumes of Structured Scientific Data on Tertiary Storage

J. Mościński[1,2], D. Nikolow[1], M. Pogoda[1,2] and R. Słota[1]

[1] Institute of Computer Science, AGH, al. Mickiewicza 30, 30-059 Cracow, Poland
[2] ACC CYFRONET, ul. Nawojki 11, 30-950 Cracow, Poland
*email:* {darin,jmosc,pogoda,rena}@uci.agh.edu.pl
*phone:* (+48 12) 6173 964, *fax:* (+48 12) 6338 054

**Abstract.** As the scientific data become larger and larger, a method is needed to store and retrieve it in time- and cost-efficient way. NCSA HDF is a data/programming model widely used for storing and sharing scientific data. We present an approach to store large volumes of structured scientific data in HDF format on tertiary storage.

## 1 Introduction

Scientific applications use and produce growing amounts of data in many different types and representations. The process of analysing such data requires merging together many different kinds of data with accompanying meta-data into logical entities. As an example we can give MD particle simulation [1] where the input/output data consist of textual labels, arrays describing positions and momenta of particles, arrays of statistical averages, raster images, etc. It is desirable to have all the data from a simulation experiment in one self-describing file. One of the most popular data formats used for this purpose is *NCSA Hierarchical Data Format* (HDF) [2,3]. HDF supports the most common types of data and meta-data used by scientists. A problem arises when the amount of data to be stored exceeds:

– file size limit imposed by HDF format itself or by underlying file system,
– available secondary storage (disk space) limit.

Later versions of HDF (from 3.2) provide a method of overcoming the first difficulty by allowing placement of data elements (DEs) in external files. A method to cope with the second limit is using tertiary storage [4–7] managed by HSM (*Hierarchical System Management*) software like UniTree, AMASS, SAM-FS, etc. The above methods alone are not sufficient to provide a good efficiency and ease of data access in complex situations. Typical HDF applications would not care about placing large DEs in external files or about data prefetching (staging) from tertiary storage. It would be very useful if the above mentioned operations were performed by an extension to the HDF library rather than by the application itself. This paper presents our design of such an extension.

## 2  System Model and Requirements

HDF library consists of Application Programming Interface (API) and Low Level Interface (LLI). The HDF APIs include several independent sets of routines with each set specifically designed to simplify the process of storing and accessing one type of data. LLI is used to control the physical storage of data (HDF files).

We assume that all available tertiary storage devices are controlled by some HSM software package providing access to the underlying resources through a virtual file system. The relations between applications, HDF interfaces and HSM file system (HSM-FS) are shown in Fig. 1.

HDF files reside on HSM-FS which in theory can fit files of an arbitrary size. In practice, however, only files sized up to a fraction of available disk cache size can be managed efficiently by commonly used HSM packages in multi-user environments. This is why very large data sets should be split into many files.

Another characteristic feature of HSM file systems is that files which have not been recently accessed may be released from disk cache. If an HDF application tries to access one or more DEs residing in such files it may experience a very long latency characteristic of tertiary storage devices. The only way to avoid degradation of the application's performance is to try and stage files back to disk cache some time before they are needed.

We modified some of the API routines in order to be able to control the way HDF files are stored on and retrieved from HSM-FS. Our goal was to meet the following requirements:

- no change to the source code of existing HDF applications is required,
- storing virtually unlimited volumes of aggregate data is allowed,
- large DEs are automatically promoted to external files - we call this functionality Data Splitting (DS),
- external DEs are staged in advance from tertiary storage - we call this functionality Data Prefetching (DP),
- prefetching policies are manageable at user level.

## 3  Concept of Realization, Implementation

In order to add DS and DP functionalities, the following routines have been extended in each of HDF APIs:

- a routine for writing a DE: with automatic placement into an external file if the DE's volume exceeds a certain limit,
- a routine for reading a DE: with prefetching of all external files correlated with the DE to be read.

Below we describe the implementation of DS and DP algorithms for *VS API*. VS interface provides access to *Vdata* ("*vertex data*"), which is a collection of records whose values are stored in fixed-length fields.

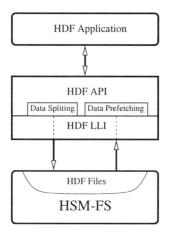

**Fig. 1.** The relations between applications, HDF interfaces and HSM file system.

```
if (current Vdata is external) {
   get name of file for current Vdata;
   stage file;
}
get list of Vdata names
    associated with current Vdata;
foreach name in list {
    find related information;
    if  (Vdata is external)
        stage file;
}
```

**Fig. 2.** DP algorithm

The DS algorithm is easy to implement by altering the *VSwrite* routine (writing a single Vdata to HDF file). When the size of Vdata to be written exceeds a certain limit, the modified VSwrite redirects the data to an external file, by calling VS API routine *VSsetexternalfile()*.

The goal of DP algorithm is to retrieve the files containing external Vdatas, which are likely to be used soon, from tape to disk cache (staging). This is necessary because the files might be purged out of the disk cache. Each Vdata may have a user-defined name and class. In our implementation we require a unique name for each Vdata. This is crucial for the DP algorithm, embedded in VS reading function called *VSread* (see Fig. 2). First we get a list of the names of the Vdatas to be staged based on the current Vdata name by calling a user-provided shell script. For the script we define only the input and output: it takes the current Vdata name as its command line argument and writes the list to its standard output. For each name in this list we check if the corresponding Vdata is external, and if so we obtain the file name. This information we can obtain using LLI calls. Then we start staging of the files. For most HSM systems reading the first byte from a file will cause that the whole file is staged.

## 4 Example of Use

In MD simulation it is often desirable to periodically store "snapshots" (arrays describing the state of the simulated system at given time-steps). Each snapshot contains two Vdatas containing the positions and momenta of particles. In cases where the number of particles is very large, the size of the Vdatas can reach tens or hundreds of megabytes. Due to their large volumes they should be stored in external files placed on HSM file system.

In our example the simulation application gives unique names to all Vdatas (following a regular convention like Pos.000, Pos.001, ... etc. for the subsequent dumps of positions and Mom.000, Mom.001, ... etc. for the subsequent dumps of momenta). The snapshots from MD simulations are usually read and analysed in a sequential manner. Therefore, advance staging of the next snapshot during the time consuming processing of the current one may decrease the overall processing time. In order to achieve such effect (read-ahead) the shell script should take the current Vdata's name (eg. Pos.000), increment the snapshot number by 1 and echo it back (Pos.001).

A user may provide a more sophisticated shell script implementing any other required rules of prefetching.

## 5 Conclusions

In the paper we have presented a simple and efficient way of adapting an existing data/programming model designed for secondary storage to the requirements of storing the growing volumes of structured scientific data on tertiary storage systems. During the preparation of this paper a new version of HDF was released. In our future work we plan to revise our concept with regard to the new features of NCSA HDF v.5.0 [8].

**Acknowledgements** We would like to thank Dr. W. Alda for his valuable help. This research was supported by the KBN grant 8 T11C 006 15.

## References

1. Mościński, J., Alda, W., Bubak, M., Dzwinel, W., Kitowski, J., Pogoda, M., Yuen, D.: Molecular Dynamics Simulations of Rayleigh-Taylor Instability, in: Annual Review of Computational Physics, vol. V, 97-136, 1997.
2. HDF User's Guide, Version 4.0, April 1996, NCSA University of Illinois at Urbana-Champaign.
3. NCSA HDF Specification and Developer's Guide Version 3.2, September 1993, University of Illinois at Urbana-Champaign.
4. Proceedings of the Fourteenth IEEE Symposium on Mass Storage Systems, Monterey, CA, Sept.11-14 1995, IEEE Computer Society Press.
5. Papers From the Sixth NASA Goddard Space Flight Center Conference on Mass Storage Systems and Technologies and The Fifteenth IEEE Symposium on Mass Storage Systems, University of Maryland, College Park, Maryland, March 23-26 1998, http://esdis-it.gsfc.nasa.gov/msst/conf1998.html
6. Hillyer, B.K., Silberschatz, A.: Storage Technology: Status, Issues, and Opportunities, AT&T Bell Laboratories, June 25, 1996, http://www.bell-labs.com/user/hillyer/papers/tserv.ps
7. Nikolow, D., Pogoda, M.: Experience with UniTree Based Mass Storage Systems on HP Platforms in Poland, in: Bubak, M., Mościński, J., (Eds.), Proc. Int. Conf. High Performance Computing on Hewlett-Packard Systems, Cracow, Poland, November 5-8, 1997, CYFRONET KRAKOW, 1997; pp.189-194.
8. HDF5 - A New Generation of HDF, http://hdf.ncsa.uiuc.edu/HDF5/

# Modelling Http Traffic Generated by Community of Users

George Bilchev, Ian Marshall, Sverrir Olafsson, Chris Roadknight

BT Laboratories, Martlesham Heath, Ipswich IP5 3RE, UK
http://www.labs.bt.com/

**Abstract.** A model of the http traffic generated by a community of users connected to the Internet via a proxy cache is described. The model reproduces Internet traffic realistically and is used as input to the Internet cache simulation models developed by British Telecom research laboratories.

## 1. Introduction

Single users generate file requests in a certain manner consistent with their Internet browsing behaviour. When all the requests from the individual users are aggregated at the proxy, they form the incoming file request pattern that is experienced by the caching algorithm. In previous research [1, 2] this pattern has been used in order to extract a relevant single user browsing behaviour model. Results have shown that users typically follow an avalanche of hyper-link clicks until they find the information they have been looking for. Taking each avalanche of hyper-link clicks as a single *browsing session,* and plotting the distribution of session lengths (a browsing session ends if the client is inactive for more than a predefined amount of time, say 5 minutes) we find that the resulting distribution has a long tail (fig. 1). This suggests that there will be a significant degree of auto-correlation in the aggregated requests of all the users in a community. Fig. 2a shows measured http traffic (number of requests) aggregated over a 5 minute time window. We suggest that two main characteristics, illustrated in fig. 2b can be identified from the graph. The first relates to the underlying trend, which reflects the (daily) request pattern of the clients. It can be extracted by calculating a moving average. The second is a stochastic component, which is "superimposed" on the trend. This stochastic component is clearly not random noise and exhibits a certain degree of auto-correlation. Figures 2c and 2d show the prediction of the model for a similar community. It is clear that the main characteristics are reproduced and the predicted traffic is thus realistic.

## 2. The Model

To model the underlying trend we suggest using a superposition of periodic functions:

$$y_i^{trend}(t) = \max\left\{a_i + b_i \sin(2\pi c_i \frac{t}{T} + d_i), 0\right\}$$
$$y^{trend}(t) = \max_i\left\{y_i^{trend}(t)\right\}$$

(1)

where $a_i$ is an amplitude shift, $b_i$ is the amplitude, $c_i$ is the frequency, $d_i$ is the phase and $T$ is the period during which cyclic patterns are observed. The values of the parameters can be tuned by curve fitting the trend model (1) to the approximated trend.

Once the trend has been approximated the stochastic component can be modelled as a Brownian motion:

$$y^{BM}(t) = y^{BM}(t-1) + \eta \tag{2}$$

Two points are worth mentioning. First, since the number of requested files is always non-negative we have to truncate a negative value of $y^{BM}(t)$ to zero. Second, bursts in positive direction are higher than bursts in negative direction. To accommodate for this we define $\eta$ as:

$$\eta = \begin{cases} \eta' & \text{if } \eta' > 0 \\ \dfrac{\eta'}{\lambda} & \text{otherwise} \end{cases} \tag{3}$$

where $\eta' \in \text{Norm}(0, \sigma)$ and $\lambda$ is a parameter determining the ratio between the heights of the positive and negative bursts. The second modification also has the effect of reducing the number of times the series has to be truncated due to negative values.

Since the auto-correlated stochastic component (2) must be superimposed on the trend (1), a way of "guiding" the random walk of the Brownian motion towards the trend without destroying the desired properties is needed. We suggest using a sequence of non-overlapping random walks each starting from around the trend:

$$y(k\Delta t) = y^{trend}(k\Delta t) + Norm(0, \sigma^{trend}) \tag{4}$$

i.e., at each time step $k\Delta t, k = 0,1,2,...$, a Brownian motion process begins for $\Delta t$ steps:

$$y^{BM}(k\Delta t + m) = y^{BM}(k\Delta t + m - 1) + \eta \tag{5}$$

where, $m = 1,2,...,\Delta t - 1$. Then it stops and a new process begins. This completes our model of the intensity of the http requests. But before we can use it in our simulations of Internet caches we also need to define the popularity distribution of the requests. There is significant evidence in the literature [3, 4, 5] suggesting that the popularity distribution follows a Zipf's-like law [6], where the relative popularity of the $i^{th}$ most popular file is given by:

$$p_i^{relative} = \frac{1}{i^\alpha} \tag{6}$$

Therefore, we also need a random number generator that produces Zipf's distributed numbers. We define it in the following way. First the total domain size $N$ and the exponent $\alpha$ must be specified. Then the probability of selecting file $i$ is given by:

$$p_i = \frac{i^{-\alpha}}{\sum_{j=1}^{N} j^{-\alpha}} \quad (7)$$

A uniform random number $n$ is generated in the range between 0 and 1 (most programming languages have already defined uniform random number generators) and an index $k$ is found such that the following inequalities hold:

$$n \leq \sum_{j=1}^{k} p_j$$
$$n > \sum_{j=1}^{k+1} p_j \quad (8)$$

The index $k$ is the desired random number coming from the specified Zipf's-like distribution.

## 3. Discussion and Future Work

To further understand Internet traffic, BT research laboratories are developing models of the caching mechanisms. Current work involves building dynamical models of Internet proxy caches. These models include both knowledge of the inner working of the cache management algorithms and popularity statistics models as observed from real data.

## References

1. R. M. Lukose and B. A. Huberman, Surfing as a Real Option., paper presented at the Computational Economics Symposium, Cambridge, England, June 1998
2. S. Olafsson, A Stochastic Model for Internet Browsing, submitted to IEEE/ACM Transaction on Networks
3. Lee Breslau, Pei Cao, Li Fan, Graham Phillips and Scott Shenker, Web Caching and Zipf-like Distributions: Evidence and Implications, To appear in Proceedings of Infocom'99
4. Margo Seltzer, The World Wide Web: Issues and Challenges", presented at IBM Almaden, July 1996
5. C. A. Cunha, A. Bestravos, and M. E. Crovella, Characteristics of WWW Client-based Traces, Technical Report TR-95-010, Boston University Department of Computer Science, April 1995
6. G. K. Zipf, Human Behavior and the Principle of Least Effort, Addison-Wesley, Cambridge, MA, 1949

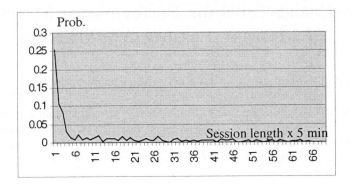

**Fig. 1.** Distribution of browsing session lengths measured over a period of twenty four hours at the university of Pisa, Italy (acknowledgements to Luigi Rizzo for providing the data). The graph comprises data from 114 clients.

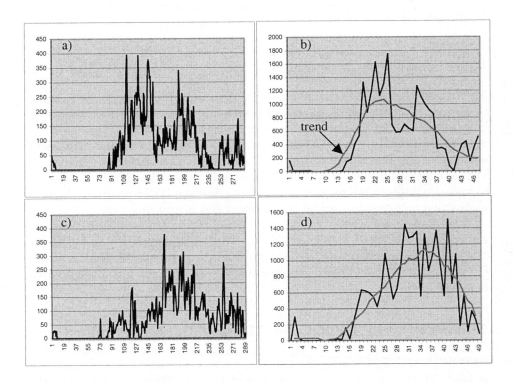

**Fig. 2.** Intensity of file requests aggregated over two non-overlapping time windows of 5(a) and 30(b) minutes. 2b shows an approximated trend using a moving average. The graphs use the same data as in fig. 1. 2c and 2d show the prediction of the model for a similar community.

# OCM-Based Tools for Performance Monitoring of Message Passing Applications

Marian Bubak[1,2], Włodzimierz Funika[1], Kamil Iskra[1], Radosław Maruszewski[1]

[1] Institute of Computer Science, AGH, al. Mickiewicza 30, 30-059 Kraków, Poland
[2] Academic Computer Centre – CYFRONET, Nawojki 11, 30-950 Kraków, Poland
   *email:* {bubak,funika}@uci.agh.edu.pl, {iskra,maruszew}@icsr.agh.edu.pl
   *phone:* (+48 12) 617 39 64, *fax:* (+48 12) 633 80 54

**Abstract.** This paper presents the motivation and insight into modifications and extensions of two performance monitoring tools – PATOP and TATOO – to make them work with message passing applications based on PVM and MPI. Also discussed are the concepts of porting the tools to the OCM monitoring environment, the structure of the modified tools and the extensions made.

## 1 Introduction

Throughout the years, a considerable number of tools that provide performance monitoring of parallel applications have been made available. However, the portability of the tools to popular parallel environments, such as PVM [1] and MPI [2] is generally poor, as is the ability of making two or more tools monitor the same application concurrently. With the recent release of the OMIS specification [3] and its implementation – OCM, the situation is expected to improve. In order to make the tools inter-operable and easier to develop, OMIS provides a layer of abstraction between the parallel environment and monitoring tools.

This paper presents the concepts of porting two existing performance monitoring tools – PATOP and TATOO – to OCM. This makes it possible to use the tools with common programming libraries, in particular with MPI, which OCM is currently being ported to, and with PVM, which is already supported by OCM.

## 2 PATOP and TATOO for Message Passing Applications

PATOP [4] and TATOO [5] are two performance monitoring tools developed at LRR–TUM[1] – the same organisation where the OMIS specification and the OCM have been created. PATOP is an on-line tool, while TATOO is an off-line tool, capable of reading data from a previously created trace file. PATOP currently supports the PARIX parallel operating environment on the Parsytec

---

[1] Lehrstuhl für Rechnertechnik und Rechnerorganisation, Technische Universität München, Munich

systems. It consists of two parts: the target part that monitors the execution of the application and the host part that provides an X11/Motif user interface.

Both tools provide reasonably rich sets of available measurements and display diagrams. It is possible to measure *busy* and *sojourn* time, *delays* in sending and receiving messages, as well as sent and received data volume, at various levels of detail. Available displays include *bargraphs*, *multicurves*, *colorbars*, *distrigrams* and *matrix diagrams*, all with user-selectable update interval. Owing to a carefully thought out, powerful and easy to use interface and the fact that the tools are written in object-oriented programming language (C++), PATOP and TATOO make a good basis for adopting and further extending their capabilities.

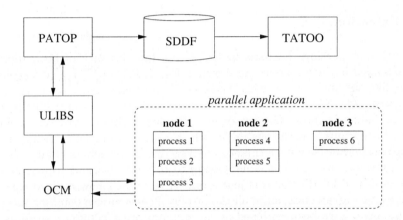

**Fig. 1.** Structure of the modified environment

Fig. 1 shows the structure of the modified monitoring environment. OCM monitors the parallel application, handles requests and sends replies to the on-line tool PATOP.

### 2.1 Porting

As can be seen in Fig. 1, PATOP communicates with OCM through a layer called ULIBS. ULIBS is a library which provides an application programming interface between the tool and the monitoring system. It supports asynchronous communication, callbacks, information on configuration and state of the monitored environment and further debugger-oriented features like breakpoints, single stepping or querying values of variables. The tool issues high-level requests through an abstract interface of functions and classes. ULIBS analyses these requests and, if necessary, translates them to lower level OMIS requests that are sent down to OCM. The reverse actions take place when handling replies from OCM.

The ULIBS layer proved to be very useful when porting PATOP to OCM. Most of the effort spent on porting takes place in this part, since the rest of the code is reasonably environment-independent. Moreover, a large part of ULIBS has already been modified to work with OCM at LRR–TUM. Still, some parts of ULIBS that were not necessary for the DETOP debugger have not been ported to OCM (in particular, the infrastructure which deals with performance analysis has been left out). Also, modifications have been made to existing interfaces, which required changes to the PATOP code itself. For example, the new OMIS interface led to changes in the handling of asynchronous requests. Due to a more flexible and high-level interface, the code could be simplified and also made more readable, since the somewhat cryptic syntax of PARIX requests was replaced by a much more neat and human readable syntax of OMIS requests.

## 2.2 Extensions

In addition to porting the tools to OCM, the functionality of both tools has been extended in certain areas. As shown in Fig. 1, PATOP is meant to generate a trace file, suitable for use with TATOO. Performance data are to be stored and processed in SDDF format [6]. Another extension is an enrichment of the set of available displays. The list of new displays includes statistical displays, such as histograms and various tables which provide min, max, mean, median and variance data. New performance analysis displays, such as Gantt diagrams and scatterplots are also available. Owing to the object-oriented programming model used in PATOP, this was just a matter of adding new derived classes of an abstract `Diagram` class, since all the display classes have a common interface. Measurements have been enriched as far as monitoring communication is concerned. Alternatively to event tracing, monitoring can be performed in *timing* and *counting* mode with user-defined objects specification.

Inspired by the flexibility of the OMIS request language, we are implementing a fixed graphical request language. This will allow users to define their own measurements, better suited to their needs, and display the results on the screen.

# 3 Concluding Remarks

The OMIS specification allows to monitor not only the parallel application itself, but also the environment it runs in. It is possible to monitor system resources usage across processes and threads, as well as various characteristics of node usage, such as the CPU architecture and speed, amount of available memory and disk space. This can be used to show a correlation between the application's behavior and the environment's conditions.

**Acknowledgments.** We would like to thank Dr. Roland Wismüller (LRR–TUM) for his help while making our acquaintance with the features of OMIS and PATOP/TATOO.

This research was partially supported by the KBN grant 8 T11C 006 15.

# References

1. Geist, A., et al.: PVM: Parallel Virtual Machine. A Users' Guide and Tutorial for Networked Parallel Computing. MIT Press, Cambridge, Massachusetts (1994)
2. MPI: A Message Passing Interface Standard. In: Int. Journal of Supercomputer Applications, **8** (1994); Message Passing Interface Forum: MPI-2: Extensions to the Message Passing Interface, July 12, (1997)
   http://www.mpi-forum.org/docs/
3. Ludwig, T., Wismüller, R., Sunderam, V., and Bode A.: OMIS – On-line Monitoring Interface Specification (Version 2.0). Shaker Verlag, Aachen, vol. 9, LRR-TUM Research Report Series, (1997)
   http://wwwbode.in.tum.de/~omis/OMIS/Version-2.0/version-2.0.ps.gz
4. Wismüller, R. and Oberhuber, M. and Krammer, J. and Hansen, O.: Interactive debugging and performance analysis of massively parallel applications, **3**, vol. 22, 415-442, Parallel Computing, Elsevier, North Holland, March (1996)
   http://wwwbode.in.tum.de/~wismuell/pub/pc95.ps.gz
5. Borgeest, R. and Dimke, B. and Hansen, O.: A trace based performance evaluation tool for parallel real time systems, Parallel Computing, vol. 21, **4**, 551 – 564, April (1995)
6. Aydt, R. A.: The Pablo Self-Defining Data Format. Technical Report, University of Illinois, Urbana, Illinois 61801, Department of Computer Science, April (1994).
   http://www-pablo.cs.uiuc.edu/Projects/Pablo/sddf.html

# Enhancing OCM to Support MPI Applications

Marian Bubak[1,2], Włodzimierz Funika[1], Radosław Gembarowski[1],
Paweł Hodurek[1], Roland Wismüller[3]

[1] Institute of Computer Science, AGH, al. Mickiewicza 30, 30-059, Kraków, Poland
[2] Academic Computer Centre – CYFRONET, Nawojki 11, 30-950 Kraków, Poland
[3] LRR-TUM, Technische Universität München, D-80290 München, Germany
{bubak,funika,gebaro,horson}@uci.agh.edu.pl, wismuell@in.tum.de
*phone:* (+48 12) 617 39 64, *fax:* (+48 12) 633 80 54

**Abstract.** The OCM is a universally usable, distributed on-line monitoring system currently implemented for the PVM programming library. Due to the growing use of MPI in parallel programming, there is a need to enhance the OCM in order to support MPI applications development. This paper presents approaches to solve the problems of the mpich-oriented start-up mechanism and profiling MPI library calls in the OCM. While the existing transport layer for the OCM internal communication is preserved, a number of new MPI-related services are added.

## 1 Introduction

The ever growing interest in parallel programming with message passing led to the establishment of MPI as its standard. While there are a number of tools which support MPI applications development, most of them suffer from poor portability and interoperability. The emergence of the *On-line Monitoring Interface Specification* (OMIS) [1] is intended to relieve the problem with providing a well-defined mechanism of interaction between a monitoring system and tools, e.g. for debugging, performance monitoring, load balancing, etc.

For supporting a typical tool functionality, OMIS defines three categories of services: information, manipulation, and event services. To prove the concept, an *OMIS-Compliant Monitoring system* (OCM) has been built using the basic set of services and a PVM-related extension [2]. OCM comprises a set of local monitors, which provide the complete monitoring functionality on *each single node*, and Node Distribution Unit (NDU) which is responsible for managing the local monitors' activities within the *entire configuration*. Due to the substantial differences between PVM and MPI w.r.t. the handling of application processes, the following problems have to be solved when extending the OCM functionality to support MPI applications development: *start-up* mechanism of OCM, realization of *information services*, and *profiling* of MPI library calls. mpich was used as a relatively mature, popular and publicly available implementation of MPI.

## 2 Start-up Mechanism of OCM and MPI-Related Information Services

As defined in OMIS, the information services should gain some information about the monitored MPI parallel application: a *list of nodes* where the program runs, *local UNIX PID*, *rank* in MPI_COMM_WORLD, and *path* to the program's executable. This information can be accessed by the tool using the mpi_get_nodelist() and mpi_get_proclist() services realized within the MPI-related extension of OCM. In contrast to PVM which provides pvm_config() and pvm_tasks() to obtain the above information, in mpich the parallel program to be monitored must be started in the *ptrace mode* to get this information via the system call ptrace(), which allows to read the values of variables from the address space of the application's master process. The data needed for the information services are contained in the table MPIR_proctable (a variable provided by mpich) and the *PID* specification file which is created by mpirun for the p4 device [3].

mpich provides the mpirun start-up script, which being called with a parameter like gdb, ddd, tv or xgdb debuggers, will run the corresponding debugger. Thus, by having modified the start-up script (e.g., with adding OCM to the list of debuggers used for mpich), it has been possible to start the OCM. When mpirun starts an application in this way, it forces the slave processes into an endless loop inside the MPI_Init() function [3]. This provides the time necessary for the OCM to obtain the information about the process IDs and to capture control over all of the application processes. The application's execution can then be forced to leave the loop inside MPI_Init() and to resume by changing the value of a loop control variable using ptrace().

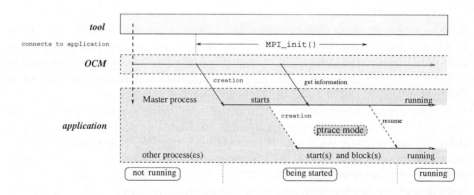

**Fig. 1.** Scheme of the start-up mechanism of OCM

The above mechanism is used by a new function - mpi_config() (used within OCM only), whose purpose is similar to that of pvm_config() in the context of the tool. Thus, we managed, on one hand, to provide a basis for the MPI-related information services, on the other hand, to minimize changes in the existing OCM services. Fig. 1 illustrates the start-up mechanism. First, the master process is created and started, then the other processes follow and get blocked, afterwards the program is enabled to be ptraced. OCM obtains the indispensable data from the master process. Fig. 2 shows how OCM may be used to control

| | |
|---|---|
| mpirun -np 3 -ocm mpi_program | call mpirun with OCM |
| | mpi_program consists of three processes |
| tool | call tool, which uses OMIS functions |
| :mpi_get_nodelist() | return the list of nodes |
| :node_attach([n_1,n_2,n_3]) | attach to nodes |
| :mpi_proc_create() | create all processes of the application |
| :mpi_get_proclist() | return the list of processes |
| :proc_attach([p_1,p_2,p_3]) | attach to processes |
| :thread_continue() | MPI program runs |

**Fig. 2.** Example of the OCM functions calls for MPI program

an MPI program. First, the MPI program is started with OCM and mpirun generates a file with a description of processes of the MPI program. Next, the user can call the tool which uses OMIS functions, and the tool connects to OCM. When the tool calls the services mpi_proc_create and thread_continue, OCM creates the processes and runs the parallel program.

## 3 MPI Library Calls Detection

In the context of the OMIS requirement that the monitoring system should recognize the application events related to the programming library, it is important to be able to detect the entry into and the exit from library functions and to read the values of their input and output parameters. A common method to do this is inserting so called *hooks* at the beginning and the end of all programming library functions. The hooks can inform the monitoring system whenever these events occur. While in PVM programmers are forced to do binary wrapping to detect library calls, in MPI this is realized with the *Profiling Interface* mechanism that implies that all the MPI functions, whose names normally start with a MPI_ prefix, are accessible with a PMPI_ prefix, too. When implementing the detection of the library functions with this mechanism, the profiling versions of MPI functions are written in the style as in Fig. 3. The routines for generating hooks are created only once, and then wrapping them around the PMPI_ calls is done automatically using a wrapper generator with adding a specific code to each function. To use the profiling library, the application has to be linked in a *special way* - the wrapper object code must be linked before the profiling MPI library which, in turn, should be placed before the standard MPI library.

```
int MPI_Function()
{ ...
    if ( OCM_HOOK_IS_ACTIVE(...) )  {   /* is hook active ? */
        ocm_evmgmt_hook_exec(...);       /* inform OCM about entry
        ...                                 into library function */
    }
    retcode = PMPI_Function();           /* call original function
                                            using PMPI_ entry */
    if ( OCM_HOOK_IS_ACTIVE(...) )  {   /* is hook active ? */
        ocm_evmgmt_hook_exec(...);       /* inform OCM about exit
        ...                                 from library function */
    }
    return retcode;
}
```

**Fig. 3.** Profiling version of MPI function

Now the OCM can activate hooks for specific functions and receive notifications about the detected library function calls from the profiling library.

## 4 Conclusion

While enhancing OCM, which up to now worked only with PVM, to support MPI applications as well, two substantial problems were to be coped with: (1) due to differences in starting up applications in PVM and MPI, the *start-up* mechanism was to be modified, (2) to preserve consistency with the PVM *information service*, a new MPI information service was developed. The enhancement presented was carried out for the mpich implementation of MPI. To extend the solutions above to support other MPI implementations, it will be necessary to perform their case studies.

**Acknowledgments.** This research was done in the framework of the German–Polish collaboration.

## References

1. Ludwig, T., Wismüller, R., Sunderam, V., and Bode. A. OMIS – On-line Monitoring Interface Specification (Version 2.0), vol.**9** of LRR-TUM Research Report Series. Shaker Verlag, Aachen, Germany, (1997), ISBN 3-8265-3035-7.
2. Wismüller, R., Trinitis, J., Ludwig, T. A Universal Infrastructure for the Runtime Monitoring of Parallel and Distributed Applications. In EuroPAR'98, Parallel Processing, Lecture Notes in Computer Science, vol. **1470**, 173-180, Southampton, UK, September 1998. Springer Verlag.
3. Wismüller, R. On-Line Monitoring Support in PVM and MPI. In: Recent Advances in Parallel Virtual Machine and Message Passing Interface (Proc. EuroPVM/MPI'98), Lecture Notes in Computer Science vol. **1497**, 312-319, Liverpool, UK, September 1998. Springer Verlag.

# Symbol Table Management in an HPF Debugger

Marian Bubak[1,2], Włodzimierz Funika[1], Grzegorz Młynarczyk[1],
Krzysztof Sowa[3], Roland Wismüller[4]

[1] Institute of Computer Science, AGH, al. Mickiewicza 30, 30-059, Kraków, Poland
[2] Academic Computer Centre – CYFRONET, Nawojki 11, 30-950 Kraków, Poland
[3] Institute for Software Technology and Parallel Systems, University of Vienna, Liechtensteinstrasse 22, A-1090 Vienna, Austria
[4] LRR-TUM, Technische Universität München, D-80290 München, Germany
{bubak,funika,mlynar}@uci.agh.edu.pl, sowa@par.univie.ac.at,
wismuell@in.tum.de
*phone:* (+48 12) 617 39 64, *fax:* (+48 12) 633 80 54

**Abstract.** In this paper we present a functionality and a structure of a symbol table manager for debugging HPF programs. The manager retrieves information from executable files generated by different Fortran 77/90 and HPF compilers and converts it into an internal representation independent of file format, which is accessed by the debugging system.

## 1 Introduction

Debuggers are a critical part of the software engineering process. They allow to control execution of an application and to test its correctness. Most of present debuggers are capable to control only one single process at a time. Though there are also some parallel debuggers, they only support explicit message passing programs and are not sufficient for HPF applications. Programs written in HPF are translated by compilers to run as a set of intercommunicating processes. For rapid, effective development it is important to have a possibility to observe behavior and to track bugs in application interactively at the HPF source code level while leaving the handling of all parallelisation issues to the debugger.

Fig. 1 presents an overview of the *DeHiFo* debugger that is being developed to address source level debugging of HPF programs [2]. The debugging system is dedicated to support programs compiled with the Vienna Fortran Compiler (VFC)[1] which is a source-to-source compilation system translating HPF applications to Fortran 90 (F90) programs. To generate an executable file, a vendor back-end F90 compiler is used. The two-stage compilation process is reflected in the debugger architecture. The main parts of the system are the *Base Debugging Subsystem (BDS)* and the *HPF Dependent Debugging Subsystem (HPF-DS)*. BDS operates as a low level debugger closely related to the machine which it is running on. It should resolve all platform specific issues and hide them from the HPF-DS level. It must also define a clear, simple but unequivocal interface that provides functionality allowing to inspect the state of processes and values of data in the parallel program. When executing, this layer does not consider the consistency of the running application with the HPF source code but provides information on each process of the program. The design relies on the *DETOP* parallel debugger [3] and the *OCM* monitoring system developed at LRR-TUM. HPF-DS works on top of the BDS and provides a higher level functionality which includes the source level (HPF) view of the parallel program execution, interactive control and altering the application data.

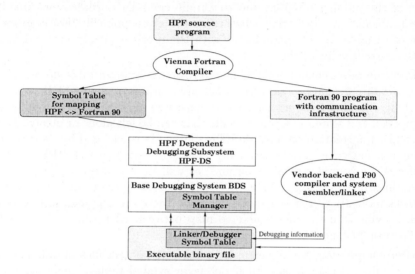

**Fig. 1.** Overview of the HPF debugger architecture

## 2 Extracting Information from Symbol Tables

In a source level debugger, the user operates on the application's objects using only symbol names (e.g variables and procedures names). The debugging system requires two mappings: *target-to-source* (F90 → HPF at HPF-DS level, machine addresses → F90 at BDS level) and *source-to-target* (HPF → F90 at HPF-DS level, F90 → machine addresses at BDS level). Whenever a break occurs in the target code, the original statements related to the current position can be located in the source program using the first mapping type. The latter one, in turn, is used by debuggers for setting breakpoints and obtaining information about the values of variables.

Before the program starts executing, the debugger has to recognize the structure of the program being debugged based on the HPF symbol table file and information stored in the executable binary file. When compilation is performed with debugging options, the HPF compiler produces a symbol table file which contains the relations between HPF and F90 programs and the F90 back-end compiler generates a special symbol table that describes the structure of the source program. Since HPF is a set of directive-based extensions for Fortan 90, designed to allow specification of data parallel algorithms, the scoping structure of the source and target programs is the same. Therefore to avoid redundancy the symbol table file at the HPF level contains information only about HPF specific symbols, e.g., distributed data, processor arrays, and program transformations, i.e., line mapping information, effects of program parallelization and compiler optimizations. This information is used to provide single data and control flow views of the parallel program and to handle compiler optimizations like data value problems, i.e., detecting whether at the given point in the program the value of variable can be inconsistent with the user's expectations, and finding out the current value of the variable.

Since the requirements for the source-to-source symbol table do not closely match existing formats we decided to specify a new format in ASCII to provide portability. Symbol table information is generated for each HPF source file that is processed and stored in a separate file that can be left alone or attached to an object file. Information on a single HPF source file is enclosed in a section containing a sequence of records with a description of variables with HPF attributes or affected by compiler transformations, as well as descriptions of parallel constructs in a program, e.g., INDEPENDENT loops. These records are enclosed in blocks indicating their scope. Additionally there are line translation records which provide a line number table to map between HPF source line numbers and Fortran 90 source line numbers.

After compilation, the linker puts together the object files of different compilation units, combining also their debugger symbol tables. There are several file formats, e.g., ELF, COFF, XCOFF, a.out. Moreover, in the same format there is a variety of extensions in the information structure generated by different compilers. On some platforms, the linker fails to update the address-related information in the symbol tables. Thus, debuggers have to obtain this data from other parts of the executable (related to linker symbols) which are specific to the particular file format.

In order to keep the debugging system portable, a symbol table management library *ST_LIB* has been developed [3]. In this work, we have extended the functionality of this library to support F90/HPF compilers. It is able to read files in various formats (executable, ASCII) and convert the debugging information to the format- and source language-independent, internal structure. The library can manage several symbol tables at the same time, which allows the debugger to handle programs that are executed on heterogeneous clusters as well as programs that consist of different binary files dynamically loaded during execution. symbol table

At the very beginning of the debugging process the executable file is opened by a common reader, which determines the target machine and the file format used to build the file. Next, the format specific reader is called to transform all relevant data to the internal structure. The debugging information is structured according to the program's organization. This structure also influences the internal debugger representation which includes the description of the main program objects: source files associated with the program, procedures (subroutines), functions, variables, data types, mapping between source and target statements.

The internal representation is a tree-like structure. At every level the objects contain not only information specific to them but also the symbol scope and references to all subobjects defined in their range. This is crucial for controlling the program execution. Each statement in the program is associated with its particular scope, and indirectly with the other scope containing that scope. When a user wants to choose a symbol, this is always done using the symbol's name. In this case the debugger has to discover which symbol the name refers to. This process is known as a symbol lookup and is always performed with respect to the current object scope. The interface to access the symbol table's internal representation behaves like a pattern matcher but its matching features are more sophisticated than only mapping names to symbols.

Since for each object the symbol table manager stores a complete specification describing this object, it is capable to return a set of objects that fulfil the given criteria. Apart from peculiarities, all symbol specifications have a common part which consists of symbol name and address, symbol class (e.g. variable, procedure, type definition, constant) and subclass which determines the primary scope of the object (e.g. local/global symbol, function parameter).

It is possible to make requests with various conditions. The system supports wildcards to substitute for each component in the symbol specification. Different search types allow to find all descendant symbols that are, for instance, only local to the current scope or global to the whole program. In this way the symbol manager provides an interface which supports different executable formats, but also allows the debugging system to retrieve information about a specific symbol and to map it backwards and forwards between source and target levels. The manager architecture supports search for a set of symbols with a given class (and/or subclass) which enables to obtain information in the debugging process, for example, about all variables defined in the specified scope.

**Acknowledgement.** We would like to thank Prof. Peter Brezany for valuable discussions. This research was supported by the Austrian-Polish grant.

# References

1. Benkner, S., Sanjari, K., Sipkova, V. and Velkov, B.: Parallelizing Irregular Applications with the Vienna HPF+ Compiler VFC. In: Proc. Int. Conf. High Performance Computing and Networking, vol. 1401 of Lecture Notes in Computer Science, 816-827, Amsterdam, The Netherlands, April 1998. Springer.
2. Brezany, P., Grabner, S., Sowa, K., Wismüller, R.: DeHiFo - An Advance HPF Debugging System. In: Proc. of the 7th Euromicro Workshop on Parallel and Distributed Processing, PDP'99, 226-232, Madeira, Portugal, 3-5 February 1999.
3. Oberhüber, M., Wismüller, R.: DETOP - An Interactive Debugger for PowerPC Based Multicomputers. In: Parallel Programming and Applications, 170–183, IOS Press, Amsterdam, May 1995. LRR-TUM, Germany.
4. Rosenberg, J.B.: How Debuggers Work: Algorithms, Data Structures, and Architecture. John Wiley & Sons, Inc., 1996.

# Tuple Counting Data Flow Analysis and Its Use in Communication Optimization

James B. Fenwick, Jr.[1] and Lori L. Pollock[2]

[1] Computer Science, Appalachian State University, Boone, NC 28608
jbf@cs.appstate.edu
[2] Computer and Information Sciences, University of Delaware, Newark, DE 19716
pollock@cis.udel.edu

**Abstract.** This paper presents a data flow analysis framework which plays a key role in identifying opportunities for communication optimization in tuplespace parallel programs.

## 1 Introduction

Tuplespace, most notably embodied in Linda, is a *structured* distributed shared memory paradigm, which offers programmers a shared space of structures as opposed to a linear array of bytes, and each structure is an individual shared unit[4]. Explicitly created processes share a data space rather than sharing variables. Tuplespace offers the simplicity of shared memory programming and the benefits of distributed memory architectures, without the false sharing and memory consistency concerns of unstructured distributed shared memory systems.

This paper presents a data flow analysis framework which plays a key role in identifying opportunities for communication optimization in tuplespace parallel programs. We have implemented the analysis and several code-improving transformations that use the analysis within our Linda optimizing compiler[1] and integrated the runtime modifications into Deli, our runtime tuplespace system[1]. Another paper describes our work in enabling classical reaching definitions data flow analysis over tuplespace parallel programs[3]. In contrast, this paper presents a new data flow analysis framework to gather information for answering a very important question for tuplespace optimization.

## 2 Tuple Counting Data Flow Analysis

It is impossible at compile-time to know exactly how many tuples, the *tuple count*, may be present in a tuplespace partition at an arbitrary time during execution. A data flow analysis framework is presented that answers the question, *For each tuplespace partition, **may** there ever be more than one tuple in that partition?* This data flow analysis computes a conservative estimate that is at least as large as the true tuple count at runtime.

In the data flow framework TUPLE_CNT = $\langle \mathcal{L}, \wedge, \mathcal{F} \rangle$, let $\mathcal{L} = \{\top, -1, 0, 1, \infty\}$ be the bounded set of potential tuple count values that a partition can have at any given time. The $-1$ element of $\mathcal{L}$ represents the fact that a tuplespace process can have at most one unsatisfied IN or INP operation affecting tuplespace, because these are blocking operations. The $\infty$ element represents the fact that the data flow analysis is not concerned with how much larger a tuple count becomes after it reaches greater than one. The top element, $\top$, is used for initialization. The binary meet operator, $\wedge$, is arithmetic *max*, where $\top$ has the smallest arithmetic value in $\mathcal{L}$ and $\infty$ the largest. This choice for the meet operator reflects the fact that at a join point, there can be no more tuples in a given partition of tuplespace than there are along any one incoming path. Thus, the semilattice is ordered by arithmetic $\geq$; moreover, it is a bounded semilattice.

The TUPLE_CNT framework also consists of a monotone function space, $\mathcal{F}$, which reflects the transfer of information from the beginning to the end of any basic block. We first give a "basis" of three functions that describes the transfer of information for single statements. The entire set $\mathcal{F}$ can then be constructed by composing the functions of this basis set.

1. The identity function $I(x) = x$ is in $\mathcal{F}$, and reflects a statement that does not affect the tuple count of a partition.

2. A generative tuplespace operation (i.e., OUT or EVAL) causes the tuple count estimate for a partition to *increase* by the following function.

3. An extraction tuplespace operation (i.e., IN or INP) causes the tuple count estimate for a partition to *decrease* by the following function.

$$G(x) = \begin{cases} 0 & \text{if } x = \top \\ x+1 & \text{if } x = -1 \text{ or } x = 0 \\ \infty & \text{if } x = 1 \text{ or } x = \infty \end{cases}$$

$$E(x) = \begin{cases} -1 & \text{if } x = \top \text{ or } x = -1 \\ x-1 & \text{if } x = 0 \text{ or } x = 1 \\ \infty & \text{if } x = \infty \end{cases}$$

Computing the tuple count simultaneously for all tuplespace partitions is achieved by maintaining an array of functions for each basic block rather than a single function, and maintaining a corresponding array of tuple count values at each program point.

Computing the data flow information for a particular node involves the application of the appropriate function for each partition on the block's corresponding input data flow values.

Figure 1 shows a tuple counting data flow example. The program utilizes three tuplespace partitions, $P_1$ for the "head" tuples and templates, $P_2$ for the "data" tuples and templates, and $P_3$ for the "worker" tuples. The flow graphs are annotated to indicate two items of information. First, to the left of each node is a function array for that node. For example, the function array $[G, I, I]$ for node $M_4$ indicates that when $M_4$ is visited during the data flow analysis, the $G$ function will be applied to the current tuple count for $P_1$, and the $I$ function will be applied to the current tuple count for the other two partitions. The second annotation is on the edges (or points) between nodes, and indicates the tuple count values computed on each pass of the TUPLE_CNT data flow analysis using the traditional iterative data flow algorithm. For example, the program point located between nodes $M_4$ and $M_5$ shows tuple counts for the initialization pass and four computation passes for each of the three partitions $P_1, P_2$, and $P_3$.

Because more than one process may access a single tuplespace partition, the TUPLE_CNT tuple counting framework is extended to accommodate the explicit parallelism of the tuplespace paradigm. The data flow information for each tuplespace partition in each process is summarized, and this summary information is used at the EVAL operations, which spawn processes. The summary information for a partition is computed as the *max* tuple count for that partition over all the points in the process. If the summary information for a partition is computed to be either $\top$ or $-1$, it is lowered to 0.

Tuple counting is performed by repeating the TUPLE_CNT analysis for each process followed by a process summarization step until there is no change in data flow information at any program point in any process. Because the analysis of each process using TUPLE_CNT is guaranteed to terminate, and the summary information values are bounded by the semilattice, the analysis-summarization repetition also terminates. In the worst case, the summary information for each process assumes each tuple count value on a separate pass. Thus, this analysis is $\mathcal{O}(N^2 \cdot P^2)$, where $N$ is the number of intermediate statements in the program and $P$ is the number of tuplespace process definitions. However, since $P \ll N$ and typically $P$ is a small number, the analysis is effectively $\mathcal{O}(N^2)$.

## 3 Enabled Transformations and Evaluation

We have used the tuple counting data flow analysis in our optimizer to identify tuples being used as shared variables, synchronization, and distributed queues. The use of shared variables can benefit from collapsing an IN and a subsequent OUT into a single operation. When synchronization tuples can be identified, they can be replaced by a more efficient method native to the host architecture. In [2], Fenwick and Pollock present a compiler analysis to detect and transform the set of tuplespace operations acting on a distributed queue, so that triangular messaging[5], an improved handling of distributed queues, is performed during program execution.

Each of these enabled transformations were evaluated in terms of how often the opportunity for the targeted optimization occurs in real programs and how much performance gain can be achieved when the optimization is applied. The results are based on analysis of a set of sixteen application benchmarks including both synthetic and real codes gathered from a variety of sources[1]. Nearly one third of all partitions were used for shared variable tuples that qualified for the IN/OUT collapse transformation. The Deli implementation of this transformation yielded an average 25% decrease in the latency of using a shared variable tuple. About 37% of the programs used distributed queues. Experimental studies of applying the triangular messaging optimization for distributed queue operations showed an average decrease of 18% in the latency associated with using a distributed queue.

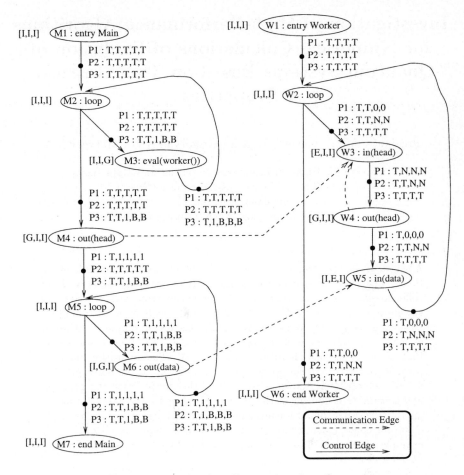

**Fig. 1.** Example of tuple counting data flow

# References

1. James B. Fenwick, Jr. *Compiler Analysis and Optimization of Tuplespace Programs for Distributed-memory Systems.* PhD thesis, University of Delaware, August 1998.
2. James B. Fenwick, Jr. and Lori L. Pollock. Optimizing the use of distributed queues in tuplespace. In *Proceedings of the International Conference on Parallel and Distributed Processing Techniques and Applications (PDPTA)*, volume I, pages 212–217, June 1997.
3. James B. Fenwick, Jr. and Lori L. Pollock. Data flow analysis across tuplespace process boundaries. In *Proceedings of the International Conference on Computer Languages*, Chicago, IL, May 1998.
4. David Gelernter. Generative communication in Linda. *ACM Transactions on Programming Languages and Systems*, 7(1):80–112, January 1985.
5. Gregory V. Wilson. *Practical Parallel Programming.* The MIT Press, 1995.

# Investigation of High-Performance Algorithms for Numerical Calculations of Evolution of Quantum Systems Based on Their Intrinsic Properties

Bogdanov A.V.[1], Gevorkyan A.S., Grigoryan A.G., Matveev S.A.

Institute for High-Performance Computing and Data Bases
P/O Box 71, 194291, St-Petersburg, Russia,
bogdanov@hm.csa.ru,
WWW home page: http://www.csa.ru

**Abstract.** A new method, based on the functional formulation of the theory of quantum mechanical multichannel scattering for three-body collinear systems is proposed. Based on intrinsic properties of scattering system the numerical task was divided into independent subtasks and the parallel algorithm for numerical computations was developed and tested on massive-parallel systems *Parsytec CC/16* and *SPP-1600*. It was shown that efficiency of such algorithm scales linearly. This algorithm makes it possible to carry out converging computations for three-body problem for any energy. It is shown, that even in this simple case the principle of quantum determinism in the general case breaks down and we have a micro-irreversible quantum mechanics. The ab initio calculations of the quantum chaos (wave chaos) for the first time were carried out on the example of an elementary chemical reaction $Li + (FH) \to (LiFH)^* \to (LiF) + H$.

## 1 Formulation of problem

The authors has formulated the new microirreversible quantum representation for a problem of multichannel scattering in collinear three-body system [1]-[2]. The main idea of a new theory consists in investigation of the Schrödinger equation in a local coordinate system that makes a classical movement on a Lagrange surface of bodies system. Such approach allows to consider the process of multichannel scattering with strong interference between channels of products of reaction that results in formation of a transitional complex

$$A + (BC)_n \to \begin{cases} A + (BC)_m \\ (AB)_m + C \\ A + B + C \\ (ABC)^* \to \begin{cases} A + (BC)_m \\ (AB)_m + C \\ A + B + C \end{cases} \end{cases} \quad (1)$$

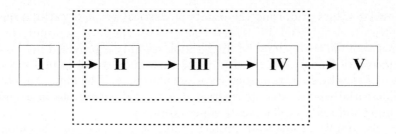

**Fig. 1.** The schematic diagram of the algorithm of numerical calculations without parallelization

It was shown than this task can be transformed into a problem of anharmonic oscillator with non-trivial time (internal time), which in the general case can have a chaotic behaviour. The given approach allows to reject using weakly converging Feynman integrals for calculation of transition probabilities [3]. In our case the transition probabilities are obtained by calculating of a set of independent trajectories, on which then the model equation is solved. The independence of trajectories enables to apply parallel methods of calculations and to use for computations a high-performance parallel computers.

In the present work the study of the problem of quantum chaos is continued using numerical calculations.

## 2 Numerical calculations

The schematic diagram of the algorithm for numerical calculations of our problem is depicted at Fig.1. Each block of the diagram corresponds to a particular stage of calculations:

- **I** - Lagrange surface construction for the system. The curvilinear coordinate system, within which all further calculations are performed, is derived on it;
- **II** - classical trajectory problem solution. At this stage the system of four ordinary non-linear differential equations of the first order is being solved numerically. The problem's parameters are collision energy $E$ and oscillation quantum number of initial stage $n$. This system is being solved by one-step method of $4^{th}$-$5^{th}$ order. This method is conditionally stable (by deviation of initial data and rhs) [4], that's why the standard automatic step decreasing method is implied to provide its stability. It's worth mentioning that initial system is degenerate in certain points. To eliminate this the standard $\sigma$-procedure with differentiation parameter replacement is performed.
- **III** - the results of classical trajectory problem calculation are used for quantum calculations for complete wave function in its final state. At this stage, the numerical problem consists of solution of an ordinary non-linear differential equation of the second order. The computation of this equation is a difficult task due to non-trivial behavior of differentiation parameter. Differentiation algorithm consists of two stages: 1) construction of differentiation

parameter values grid using the results of classical problem calculation and 2) integration of initial differential equation on obtained non-uniform grid by means of multi-step method. Integration stability is provided by choosing the integration step in a classical problem, while control is performed by means of step-by-step truncation error calculation [4]. The obtained solution of differential equation is approximated in a final asymptote in a form of incoming and reflected flat waves superposition [5];

- **IV** - the results of quantum problem solution are used for obtaining the values for matrix elements of transitional probabilities of a reaction and their corresponding cross-sections. Calculation of matrix elements for initial oscillator quantum number $n$ and final oscillator quantum number $m$ is performed with the use of expressions presented in [5]. Let's note that transition probability matrix obtained corresponds to one value of collision energy, stipulated at stage **II**;

- **V** - chemical reaction rate constants calculation. At this stage, the values for reaction cross-section matrix calculated for different collision energies are integrated by Maxwell's distribution.

Let us remind that calculations for steps **II** and **III** are made for specific values of collision energy $E$ and oscillation quantum number of initial state $n$. Results of these calculations allow to obtain one vector of a reaction cross-section matrix, which corresponds to $n$. In order to obtain the entire cross-section matrix, calculations at stages **II** and **III** need to be repeated as many times as dictated by the size of reaction cross-section matrix. As a result the entire probability matrix is obtained. The procedure described needs to be repeated for many values of collision energy $E$ in order to enable further integration and velocity constants finding.

The algorithm described as well as schematic diagram presented at Fig.1 correspond to numerical calculations performing without parallelization. But the algorithm of numerical calculations allows to perform the parallelization and use the multiprocessor supercomputers with massive parallel architecture for calculations.

Further we will show how the algorithm presented can be parallelized for massive parallel supercomputers with distributed and shared memory.

Schematic diagram of parallel calculation algorithm for massive parallel systems with distributed memory is depicted at Fig.2. Calculation parallelization procedure is performed by the values of collision energy. Calculation of classical trajectory problem, quantum calculation and transition probability matrix calculation is performed in each of the parallel branches. Let's note that just as in the case on non-parallelized algorithm all calculations from stages **II** and **III** are performed as many times as it is dictated by the size of transition probability matrix. Due to the fact that calculation in each of thee parallel branches represents a separate problem and does not interact with other branches of calculation, the effectiveness of this parallelization algorithm using vs. relatively unparallelized algorithm is nearly proportional to a number of calculation branches, i.e. to the number of computation nodes.

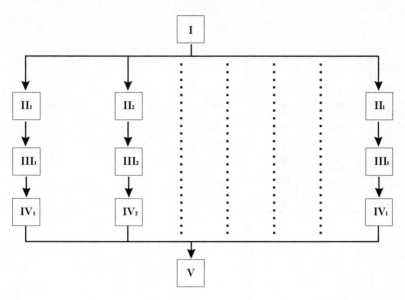

**Fig. 2.** The schematic diagram of the algorithm of numerical calculations on massive parallel supercomputers with shared memory

This algorithm realization was performed on *Parsytec CC/16* supercomputer with massive parallel architecture with distributed memory. As a reaction on which the algorithm was tested, a well studied bimolecular reaction $Li + (FH) \rightarrow (LiFH)^* \rightarrow (LiF) + H$ was taken. The results of testing have shown the calculation efficiency to be nearly proportional to the number of computation nodes.

Now let us review the calculation algorithm for massive parallel systems with shared memory. The schematic diagram for such an algorithm is presented at Fig.3. Just as in the previous algorithm, the first level of parallelization represents the distribution of calculations among the computation nodes in accordance with the values of collision energy. But, as can be seen from a scheme, in each of the parallel branches there is one more parallelization by the values of oscillation quantum number of the initial state as well. The second parallelization is based upon a fact that for classical trajectory problem calculation the same coefficients, that calculated "on-line", are used for different quantum numbers, thus allowing to make such a parallelization.

This algorithm was realized on *SPP-1600* supercomputer with massive parallel architecture with shared memory. The results of testing have shown that just as it was expected, the efficiency of calculations is higher than in the previous example.

Finally we would like to stress one of the important features of parallelization algorithms demonstrated - their scalability. Due to the fact that integration of transition probability matrix and rate constants calculation during stage **V**

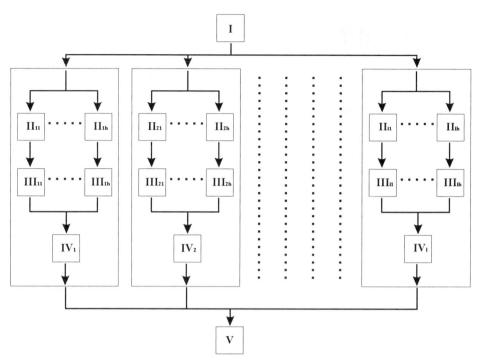

**Fig. 3.** The schematic diagram of the algorithm of numerical calculations on massive parallel supercomputers with distributed memory

requires the values of matrix elements for large number of energy values, one can hardly find a supercomputer with an excessive number of computation nodes.

## 3 Quantum multi-channel scattering problem investigation results

The reduction of multi-channel scattering problem solution to trajectory problem allows to use up-to-date multiprocessor computers with massive-parallel architecture and to calculate more trajectories within short period of time.

In particular, about $10^5$ trajectories were calculated for energy range, within which the resonant state arising is possible. One of the results of this work is a **B-W** map construction (Fig.4) (initial data area is subdivided into rectangles by a grid. The bottom left corner of the rectangle corresponds to initial data of a particular trajectory. The black fields represent reflection, while the white ones show the reaction passing through the barrier). It can be seen that the given representation $(h_{x^2}, h_E) \to \left(X^2 \times \tilde{E}\right)$, where $X^2$ and $\tilde{E}$ represent continuous definition areas of initial data $x^2$ and $E$, while $h_{x^2}$ and $h_E$ represent the uniform grids on $X^2$ and $\tilde{E}$ respectively, shows scale invariancy. Together with

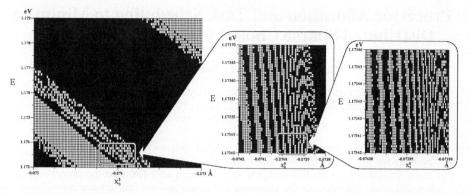

**Fig. 4.** The irregular map of collision energy $E$ and $x_0^2$ coordinate initial values for passing (white fields) and reflecting (black fields) geodesic trajectories. One can see that there is the self-similarity relative to scale transformation in chaotic field.

such commonly recognized methods of stochastic systems research as larger Lyapunov's exponent calculation and Poincarè maps construction, this fact allows to postulate the existence of chaos in trajectory problem and in a motion of a local coordinate system on Lagrange surface within which the problem is being considered as well.

Thus, the possibility of chaos arising in a wave function in a multi-channel scattering problem for three-body system was proved for the first time by means of numerical analyses, which once again confirms the principal importance on numerical analyses methods for physical systems investigations.

## Acknowledgment

The authors would like to thank the International Association for the promotion of cooperation with scientists from the New Independent States of the former Soviet Union, INTAS, for supporting this research work under the grant INTAS-96-0235.

## References

[1] A. V. Bogdanov, A. S. Gevorkyan, *Three-body multichannel scattering as a model of irreversible quantum mechanics*, in Proceedings of the International Symposium on Nonlinear Theory and its Applications, Hawaii, V.2, pp.693-696, (1997).
[2] A. V. Bogdanov, A. S. Gevorkyan, *Multichannel Scattering Closed Tree-Body System as a Example of Irreversible Quantum Mechanics*, Preprint IHPCDB-2, pp. 1-20, (1997); Los Alamos National Laboratory e-Print archive, quant-ph/9712022.
[3] E. E. nikitin, *Itogi Nauki i Techniki* (in russia), VINITI, Moscow, (1985).
[4] A. A. Samarskiy, *Introduction to the Numerical Methods* (in russia), "Nauka", Moscow, (1997).
[5] A. N. Baz', Ya. B. Zel'dovich and A. M. Perelomov, *Scattering reactions and Decays in Nonrelativistic Quantum Mechanics* (in russia), "Nauka", Moscow, (1971).

# Processor Allocation and Task Scheduling to Minimize Distributed Sparse Cholesky Factorization Time

Tsung-Tso Kan, and Chuen-Liang Chen

Department of Computer Science and Information Engineering
National Taiwan University, Taipei 10617, TAIWAN
{taitokan, clchen}@csie.ntu.edu.tw

**Abstract.** In this paper, we discuss the parallel sparse Cholesky factorization problem for distributed memory multiprocessor systems. Although there are already several articles for this problem, most of them only focus on the processor allocation. We discuss the processor allocation and the task scheduling issues simultaneously to minimize the overall parallel execution time. Several existing methods are compared with ours, and experiments conducted show that our method provides the minimum parallel execution time.

## 1 Introduction

Sparse Cholesky factorization is a fundamental process for many numerical problems. Various efforts have been directed at solving it on parallel architectures. Generally, it is difficult to partition and map the sparse matrix to distributed memory multi-processor systems, in a way that minimizes communications and the total execution time of the parallel computation. In this paper, we address the processor allocation and the task scheduling issues simultaneously. The effects of our strategy not only minimize the parallel execution time but also the number of computers used.

## 2 Existing Processor Allocation Methods

There are several processor allocation methods for parallel sparse Cholesky factorization: Sparse-Wrap (SW) [1], Subtree-to-Subcube (SS) [2], Bin-Pack (BP) [3], and Proportional (PR) [4] mappings. Fig. 1 is the illustration for different methods. The used matrix is a 3×4 5-points grid problem, the used fill-reducing algorithm is the minimum degree ordering [5], the used numerical factorization is the multifrontal Cholesky factorization [6], and we assume that there are three processors.

On the elimination trees, the tasks assigned to the same processor have the same color. The character inside a node is the column ID of the original matrix. The value beside a node is its computation cost, i.e., the number of multiplicative operations. A bold edge represents a message between two different processors, and the value within a parenthesis is the size of the contribution block that is transfered from the child to its father. The method to find each node cost can be found in [7].

All of these methods contain wraparound process. The wraparound approach can distribute roughly equal amounts of workload to each processor. However, it also frequently distributes two neighbor nodes to two different processors, hence to introduce heavy communication overhead (the bold edges in Fig. 1). Another common drawback to the previous allocation methods is that they only focus upon processor allocation issue and do not emphasize the execution order of the tasks that are assigned to each processor. The lack of task scheduling may miss the possible reduction of the parallel execution time.

## 3. The New Strategy

Our strategies are to use the quantified computation cost and communication cost of each node, to examine the structure of an elimination tree, and to keep track of the

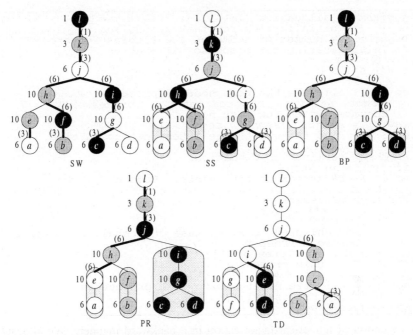

**Fig. 1.** Results for different processor allocation methods

evolution of the parallel numerical Cholesky factorization. We have also scheduled the task sequence by the use of renumber (or reorder) of all the nodes in our algorithm. For example, in Fig. 1, with our Top-Down (TD) method, the character inside a node is the new ordering. Furthermore, all nodes of a subtree that have been allocated into a single processor have been renumbered contiguously. It is reasonable to execute nodes of a subtree contiguously as they are assigned to the same processor. This kind of renumbering has many benefits. For example, it can simplify the data partition in the numerical factorization phase.

There are two directions to examine the elimination tree: from the root to leaf nodes and from leaf nodes to the root. The most difficult problem by the way of from leaf nodes to the root is the bin-pack problem, since there are many available leaf nodes. In order to eliminate the bin-pack problem and to do the numbering described above easily, we decided to examine the elimination trees from the root to leaf nodes. Therefore, we call this algorithm Top-Down (TD) allocation.

Finally, to prevent unnecessary communication, a node is assigned to the same processor that its parent node is assigned to, if such a processor is free. From Fig. 1, it is shown that our method introduces fewer bold edges, i.e., the least amount of communication overhead.

The input of the algorithm is an elimination tree and the number of processors $p$. Assume the computation time for a multiplicative operation is 1, the communication startup time is $\alpha$ and the data transfer time is $\beta$. After the computation of TD algorithm, we have *processor(i)*: the processor to which node $v_i$ is assigned and *new(i)*: the new order of node $v_i$.

Initially, assume that there are $p$ processors, $P_0, P_1, ..., P_{p-1}$. For each processor $P_i$, there are two associated variables $J_i$ and $t_i$. $J_i$ is the task currently allocated to processor $P_i$. $t_i$ is the remaining execution time of $J_i$. "$t_i = \infty$" indicates that processor $P_i$ is free. $\tau$ is a countdown clock with initial value 0. This indicates that all tasks shall be completed before $\tau = 0$. The first allocating task is the root of the elimination tree, because it is the last task in the numerical factorization phase and we keep track

```
Algorithm TD_allocation ():
Step 1: // Initialization for task scheduling.
    seqno = the number of nodes of the elimination tree;
    // Initialization for processor allocation.
    R = ∅;     τ = 0;
    for 1 ≤ i ≤ p - 1
        J_i = ∅;    t_i = ∞;
Step 2: // Allocate the root node to processor P_0.
    J_0 = root, say v_k;    t_0 = cost of v_k;    processor(k) = 0;
    // Renumber it as the last node.
    new(k) = seqno;    seqno = seqno - 1;
Step 3: // Find tasks with the latest beginning time.
    t* = min(t_0, t_1, ..., t_{p-1});
    // Check termination condition.
    if (t* == ∞) stop;
Step 4: // Backtrack t* time slots. Update J_i and t_i.
    τ = τ - t*;
    for each processor P_i
        if (t_i != t*)  t_i = t_i - t*;
        if (t_i == t*)
            // Add children to R, because they shall be completed
            // before current τ.
            R = R ∪ {children of J_i};
            // Free this processor.
            J_i = ∅;    t_i = ∞;
Step 5: // Allocate new tasks to free processors.
    allocate_processor( );
Step 6: goto Step 3
```

of the evolution of numerical factorization in a backward manner. We allocate the root node to processor $P_0$ and renumber it in Step 2. *seqno* is the countdown counter for renumbering.

The loop consisting of Steps 3 to 6 is the kernel of our algorithm. In each iteration, we back the clock $\tau$ to the beginning point of a current task that is recently executed, and update $J_i$ and $t_i$, in Steps 3 and 4. We use a set $R$ to store nodes that are ready to be allocated. Because our evolution is backwards, a node shall be added into $R$ if the beginning point of its parent node is after current $\tau$. This means that it shall be completed before current $\tau$.

Then, in Step 5, we allocate new tasks to a free processor. Here, a heuristic is used. We allocate a node $v_k$ with the largest sequential execution time of the subtree rooted at $v_k$ first. It is reasonable to handle the biggest subtree first. Furthermore, the processor which is doing the father($v_k$) is the first priority candidate processor. If it is busy, we then pick a free processor.

The algorithm repeats the above process, until no more processors have unfinished tasks. This is checked in Step 3.

## 4. Comparisons

Ten test matrices from the Harwell-Boeing collection [8] are used. They are BCSPWR09, BCSPWR10, BCSSTK08, BCSSTK13, BCSSTM13, BLCKHOLE, CAN1072, DWT2680, GR3030 and LSHP3466. The matrices are first reordered by the minimum degree ordering, and then different processor allocation methods are applied. After collecting the data for these ten test matrices, we compute the average results. At the same time, eleven cases of different numbers of processors are tested.

Fig. 2 is the comparison of the execution time. The execution time is normalized. We assume the minimum parallel execution time (MPET) is 1; it is the summation of the computation time of nodes that lie on the critical path of the elimination tree.

Fig. 2(a) is for a communication-free multiprocessor system, i.e., $\alpha = \beta = 0$. The TD algorithm uses the least number of processors to reach MPET, approximately 16. The others should use 512 or an additional number of processors to reach the MPET.

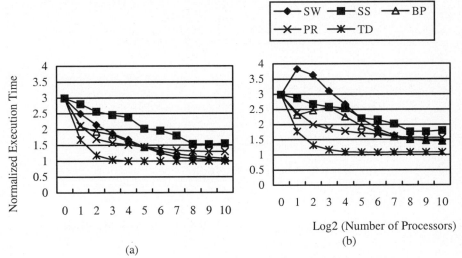

**Fig. 2.** Comparisons of normalized execution time when (a) communication-free and (b) $\alpha = 32$ and $\beta = 0.3$.

The Fig. 2(b) is the case for a multiprocessor system with $\alpha = 32$ and $\beta = 0.3$. The performance variation of TD is very slight, because the communication overhead is being controlled. It is clear that our TD method allows us to use fewer processors to achieve maximum parallelism.

## 5. Conclusion

We conclude that the TD approach has the following advantages from the other previous allocation methods: It is a one step approach, which does processor allocating and task scheduling simultaneously. It also saves the stages of using node amalgamation, supernode, clique tree generation, or bin-pack. It takes better care of parallel completion time, and saves the number of resources used.

## References

1. George, J. A., Heath, M. T., Liu, J. W. H., Ng, E. G.: Sparse Cholesky factorization on a local-memory multiprocessor. SIAM J. Sci. Stat. Comput., vol. 9 (1988) 327-340.
2. George, J. A., Liu, J. W., Ng, E. G.: Communication results for parallel sparse Cholesky factorization. Parallel Computing, vol. 10 (1989) 289-297.
3. Geist, G. A., Ng, E. G.: Task scheduling for parallel sparse Cholesky factorization. International Journal of Parallel Programming, vol. 18, no. 4 (1989) 291-314.
4. Pothen, A., Sun, C.: A mapping algorithm for parallel sparse Cholesky factorization. SIAM J. Sci. Stat. Comput., vol. 14, no. 5, Sep. (1993) 1253-1257.
5. Liu, J. W. H.: Modification of the minimum degree algorithm by multiple elimination. ACM Trans. Math. Software, vol. 11 (1985) 141-153.
6. Duff, I. S., Reid, J. K.: The multifrontal solution of indefinite sparse linear equations. ACM Trans. Math. Software, vol. 9, Sep. (1983) 302-325.
7. Lin, W. Y.: Reorderings of sparse matrices for parallel Cholesky factorization. Ph.D. Thesis, Department of Computer Science and Information Engineering, National Taiwan University, June (1994).
8. Duff, I. S., Grimes, R., Lewis, J.: Sparse matrix test problems. ACM Trans. on Math. Software, vol. 15 (1989) 1-14.

# Late Papers

# Implementation of MPI over HTTP

S.Lakshminarayanan
lakme@meena.iitm.ernet.in
Dept. of Computer Science
Indian Institute of Technology
Madras, India

S.S.Ghosh
ghosh@serc.iisc.ernet.in
Supercomputer Education and Research Center
Indian Institute of Science
Bangalore, India.

N.Balakrishnan
balki@serc.iisc.ernet.in

### Abstract

Message Passing Interface[2] is the de facto standard for multi-computer and cluster message passing. In this paper we explore a new paradigm of high performance distributed computing by implementing a message passing interface over HTTP[1]. This provides a platform independent implementation of MPI and also develops a base for web based computation to achieve global parallel processing. The conventional approach to message passing is through the use of Parallel Virtual Machine(PVM) or MPI software packages that optimize the communication and the synchronization functions for the message passing paradigm needed in high performance distributed computing environments. However the recent trend in Internet communication has been to use the HTTP for inter-linking information and files across the World Wide Web. The HTTP [1], over the years has matured to be a very efficient tool for communication and is fit enough to warrant consideration as a vehicle for inter-processor communication. The performance of our implementation was compared with the standard MPI implementation over a cluster of workstations connected by a Shared Ethernet network and the results are very encouraging.

## 1 Introduction

Nowadays, massively parallel architectures are built by interconnecting a cluster of workstations and accomplishing massive parallel processing using the workstations as the nodes of processing. There is also a growing number of web-enabled machines and the development of web-windows giving the productivity tools for a true distributed high performance computing environment. Message passing is a paradigm used widely on certain classes of parallel machines, especially those with distributed memory. The interface that provides for message passing should establish a practical, portable, efficient and flexible standard for message passing. MPI is the de facto standard for multi-computer and cluster message passing[2].

The Hypertext Transfer Protocol is an application level request-response protocol for distributed, collaborative, hypermedia information systems [1]. HTTP as a protocol has matured over the years and many communication optimization ideas and synchronization routines have been entwined with it. The primitives of HTTP can be used to implement a client-server model over the Web. The recent spurt of enhancements of HTTP like persistent connections[1] and pipelining make HTTP a wonderful protocol for message passing over the World Wide Web.

Present day implementations of MPI are platform specific and optimizations of MPI are performed over a specific platform [3]. By implementing MPI over HTTP one gets a platform independent implementation of MPI. This allows for the development of portable parallel software on different parallel architectures and also provides a wonderful paradigm for web-based computation.

We have implemented the basic MPI calls and other commonly used MPI calls using HTTP primitives and have tested the performance of our implementation with the standard MPI implementation. We have also tested the performance of our MPI_Send and MPI_Recv calls separately. Results indicate that the saturating bandwidth obtained in both implementations are comparable.

In section 2 we present our design and implementation details. In section 3 a detailed performance analysis of our implementation is given. In section 4 we summarize our results and also give a sketch of our future work.

## 2 Design and Implementation

The processors that execute a parallel program are treated as HTTP clients which submit their requests to HTTP servers to accomplish message passing. The Message passing interface developed over HTTP is present as a library at every client executing the MPI programs. Any MPI function that needs to send or receive messages, generates a particular request which is sent to the server for servicing. The server in turn sends a response to the client which is subsequently parsed by the client to get the status of the request. The status indicates a success or a failure in the completion of the request. In order to reduce the connection setup overhead, we follow a connection management policy and maintain long live connections to the servers.

All the basic operations of MPI and other commonly used MPI operations are implemented as part of the client's library. MPI operations like MPI_Send and MPI_Recv which actually perform message passing are implemented with the help of HTTP primitives GET,HEAD and POST. The MPI-operations implemented include the six basic MPI operations namely MPI_Init, MPI_Finalize, MPI_Comm_rank, MPI_Comm_size, MPI_Send and MPI_Recv. All MPI programs can be written using these six basic MPI functions. Other commonly used MPI operations MPI_Wait, MPI_Bcast, MPI_Gather and MPI_Reduce are also implemented.

A list of all the processor nodes is created in the MPI_Init call and the MPI_Comm_rank and MPI_Comm_size calls make use of this list to report the rank and the size for an MPI program. The important communicating MPI calls, MPI_Send and MPI_Recv, implemented efficiently using HTTP primitives. In our framework both the communicating clients agree on a HTTP server with which communication is carried out in the least amount of time, in order that the above objective be satisfied.The message is posted to this server as part of the entity body of the request by a POST request to a cgi-bin executable at the server which stores the message at the server end. As part of the MPI_Recv call, the receiver keeps polling the server for the message, and it through a GET request once it has arrived at the server.

All the collective MPI-operations like MPI_Bcast, MPI_Gather and MPI_Reduce involve one to many or many to one transmissions. In-order to implement them efficiently all the nodes are divided in the beginning into various groups and every group is associated with a group leader. The collective operation is performed in two stages; one between the central node and the group leader and the other among all nodes in a group.

## 3 Performance Results

The above implementation was tested over a 10 Mbps Ethernet link connecting a cluster of machines through a hub. The HTTP daemon is made to run on one of the machines. Message passing is accomplished between two other nodes in the cluster acting as HTTP clients. Various kinds of message passing experiments on this kind of set-up and the results obtained are more than encouraging. From the results of the experiments done below, we can note that the saturating bandwidth obtained using our implementation is comparable to that of conventional MPI implementation. The effective bandwidth obtained for packets of small size increases with the maintenance of long-live connections. The performance of MPI_Recv is poor because of high disk access time at the server end for large packets.

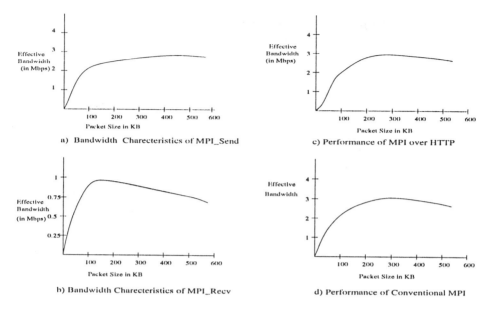

Figure 1: Results of the Experiments

## 4 Conclusion

In this paper we have opened up a new paradigm for Web based computation. MPI over HTTP also provides a pervasive base for global parallel processing and many problems that require huge computation power can be solved using our approach. We have also obtained a platform independent implementation of MPI and HTTP enhancements like pipelining are bound to improve the performance of our implementation tremendously.

## References

[1] R.Fielding,J.Gettys,J.Mogul, H.Frystyk and T.Berners-Lee, Hypertext transfer protocol- HTTP/1.1, RFC 2068, *Internet Request for comments*, Jan 1997.

[2] J.J.Dongarra,S.W.Otto,M.Snir and D.W.Walker, An Introduction to the MPI Standard,*Communications of the ACM* ,Jan 1995.

[3] H.Franke, P.Hochschild,P.Pattnaik and M.Snir, An Efficient Implementation of MPI on IBM-SP1, *Proc. Of the 1994 Int. Conf. on Parallel Processing*, Aug. 1994

# A Distributed Vision Network for Industrial Packaging Inspection

A. Meliones[1], D. Baltas[2], P. Kammenos[3], K. Spinnler[4], A. Kuleschow[4], G. Vardangalos[1], P. Lambadaris[5]

[1]National Technical University of Athens, Dept. of Electrical and Computer Engineering, Division of Computer Science, Zografou Campus, GR-15773, meliones@telecom.ntua.gr
[2]TESEIK Industrial Vision Systems, 27 Irodotou Str., GR-10673 Athens, dbaltas@teseik.gr
[3]KUKAM Industrial Vision Systems, Bois-Genoud 1b, CH-1023 Crissier, Switzerland
[4]Fraunhofer-Institute for Integrated Circuits, Erlangen, D-91058, spk@iis.fhg.de
[5]FAMAR S.A., 7 P. Marinopoulou Str., GR-17465 Athens, Greece

**Abstract:** A distributed vision network is proposed to tackle industrial packaging inspection. The system consists of independent networked inspection stations able to address efficiently parallel inspection tasks such as product identification, character verification, tag inspection and content & packaging quality control at a high production speed. Existing and innovative inspection algorithms such as synergetic classification have been adapted on the smart camera technology of the inspection stations. We present the benefits of the deployment of the system in the production lines of a pharmaceutical packaging facility.

## 1. Introduction

Artificial vision applies digital image processing and analysis to tackle real problems in the industrial production, mainly of standardised products, in real-time conditions [1]. Manual inspection or monitoring of any continuous process is commonly agreed to be inefficient, especially because of its repetitive and tedious nature. For a wide range of inspection tasks, detection proficiency is assumed to start off at 80-90% detection rate, and to deteriorate rapidly after half an hour [2]; additionally, performance of a single inspector degrades rapidly with the number of possible defect types.

Advanced artificial vision systems for the automatic or semi-automatic inspection of the production lines have been developed in several industrial areas from the beginning of the 80's but had been restricted to large organizations. This happened because the high-cost equipment had been a barrier for SMEs. These systems were characterized by high complexity, high cost, high impact on the organization, low flexibility and low reliability. They had a significant role in defining a new paradigm in the control and inspection of the production lines, but their characteristics allowed addressing them only minimally and to a specific group of end-users. From the beginning of the 90's, new technologies are paving the way to a significant change in industrial visual inspection systems, that may now use more sophisticated pre-processing and analysis techniques and more standardised and generic equipment, thus enlarging their application scope. These systems can now be used by a much wider group of industrial users, whose business profitability and efficiency can be improved significantly through the application of new flexible and cost-effective technology.

Nowadays, artificial vision is a mature technology with numerous successful examples in electronics, automotive and pharmaceutical industry, while it penetrates slowly but steadily sectors such as packaging, textiles, ceramics, plastics, cosmetics etc. [3].

Two major obstacles to machine inspection are the difficulty of characterizing defects and the high data rate. Many successful automated visual inspection systems employ signal processing no more sophisticated than intensity thresholding [4]. Other systems involve computationally complex texture algorithms [5], which demand extensive serial computation, and thus are poorly matched to (real-time) high-performance implementation, or involve pattern classification techniques [6].

In this paper we present a distributed vision network for industrial packaging inspection and demonstrate the benefits of the deployment of the system in the production lines of a pharmaceutical packaging facility. This work is supported by the European Union's ESPRIT Program on HPCN (PST activity DIVINE, contract No.26415).

## 2. General Architecture

The proposed system is a set of independent networked inspection stations assigned to production lines. The network uses special hardware, called COMServers. A COMServer is connected via coaxial cable to the managing PC forming an Ethernet backbone. Each COMServer has a unique IP address and can be connected to one, two or four cameras. Each camera connected to a COMServer is assigned a port address under this IP node. The connection from each camera to the COMServer is RS232 at 57600 baud. An inspection station is a VC65 smart camera with embedded ADSP 2181 microprocessor, 8MB DRAM, 2MB FLASH EEPROM, installed in the production line with the appropriate lens and lighting equipment, loaded with the software which implements automatic visual inspection. The PC is a Pentium MMX/233MHz Windows NT workstation or higher. The DIVINE architecture considers the roles of *system administrator*, *product trainer* and *line operator*.

## 3. Software Components

The system software is organised in programs running on the VC camera and programs running on the PC. Those that execute on the camera have been prepared using the GNU C cross compiler. The software is designed with the following components.

The *Inspection Controller* (DVN-C) provides the system console functionality. It helps to maintain the network and its integrity. It provides the user interface, which allows the user to select a particular inspection station in order to train a product inspection operation and set it up to run. DVN-C also allows and controls communication with a higher level of factory process and quality monitoring.

The *Inspection Runtime Station Monitor* (DVN-R) runs on the PC and is activated through the DVN-C when an inspection station is set to run in production. Its purpose is to monitor the station runtime operation and report production/inspection statistics.

The *Inspection Station* (DVN-S) runs on the VC camera and provides the basic functionality that is necessary to each inspection station. It handles image acquisition, controls I/O for the interaction with the related product transport and ejection mechanisms and provides the communication interface between the DVN-C and the Embedded Application Modules (EAM).

The *Embedded Application Modules* provide the functionality for inspection tasks. The EAM software is separated to the PC Client and the VC Server. The client provides the user interface to control training of the product inspection parameters. The server executes the functions requested through the client in the VC camera. These functions concern training a model, setting thresholds, testing tolerances etc.

The *Product Identification* EAM is used in stations where labels, box or instruction insert is inspected. Each packaging element associated with a particular inspection station has a product signature, i.e.a set of patterns arranged on the packaging element reference image in such a way that they uniquely identify the relevant product. The system analyses the camera image and identifies the product by matching the patterns appearing in the runtime image with the signature elements created during training.

The *Character Verification* EAM is used in inspection stations where a string is printed on a label or a box during packaging, giving the expiration or manufacturing date, the lot number of a product etc. During training the system defines synergetic templates for each character that may appear in the respective lot or date code. The collection of these templates is the *station character set* or *font*. The font is independent of a product and is used for all packaging elements that pass through the respective station. Prior to the execution of the inspection, the EAM is informed on the strings that comprise the codes and selects the respective characters from the font. During execution the system analyses the image to locate and then verify the strings.

The *Tag Inspection* EAM is used in stations where the box enclosing the product is inspected to detect presence or absence of a tag and presence or absence of a stamp on the tag. Training concerns the definition of a pattern which allows the detection of the position of a tag in the image, directly or relative to another object fixture, and the measurement of histogram statistics in an area specified to contain the stamp to determine the runtime parameters and the tolerance ranges for runtime inspection.

The *Content & Packaging Quality Control* EAM concerns the actual products that are being packaged. It refers to packs filled with pieces of a product (e.g. blister cells) and allows to verify that all positions are filled with it and that there are no outliers in the pack by mistake. It also involves the detection of broken or faulty pieces of the product when they are already in the pack. During training the inspection system derives from the reference image a feature list with information concerning each part that has to be inspected. In runtime operation the EAM analyses the image checking each position for presence or absence of the respective product and verifying that it is within the acceptable range for color and dimensions. Decision trees relevant to each product are used to provide the answer for each piece position. Local processing makes use of segmentation and histogram analysis. Color classification uses appropriate filters and gray level image analysis. Moments and morphological filtering are used for size and shape calculation. Morphological filtering and local image subtraction are used to detect defects on pieces (breaks, powder etc.). Neural-network based synergetic processing is used for product classification.

## 4. Evaluation of Synergetic Algorithms for Printed Label Inspection

During the production process the labels are stamped with a character string coding important information, such as the lot number, the expiration or manufacturing date etc. There are different types of faults to be detected, such as one or more code characters overlapping the label text (F1), wrong characters in the code (F2), missing characters in the code string (F3) or completely missing code string (F4).

Two different approaches for the inspection of the quality of printed characters based on methods of synergetic computing were evaluated. A first approach was set up to evaluate one line of characters as a whole. In some cases the computed classification coefficient produces overlapping classes and a small, local fault in print quality cannot be distinguished from the overall tolerances of good printing quality. Better results can be achieved with the second approach, classifying the string character by character and then compute a final result. Good and corrupted characters of strings can be classified with the developed algorithm with a recognition rate of 100%. Reference patterns are generated by taking an arbitrary pattern of good quality and training the system with several, slightly rotated copies of the template. Then a sufficient robustness against rotational tolerances with about ±5% is achieved. Extending preprocessing with respect to invariance of scale can further improve performance.

## 5. Pattern Matching Evaluation

The normalized greyvalue pattern matching is designed for doing a position recognition of a previously learned greyvalue pattern. The results of this process are a x-/y-position and a quality measure in the range of 0-100% representing the correlation between the learned and the recognized pattern. Correlation qualities lower than 60..80% normally indicate the absence of the searched pattern. The normalized greyvalue pattern matching can be applied to: (1) position recognition, (2) presence and completeness checks and (3) check for a limited number of discrete rotations.

The pattern size must be in the range of 32 to 256 pixels. Different dimensions in x- and y-direction are possible. For a high repeatability and robustness a pattern size larger or equal to 64*64 pixels is recommended. The pattern should be of a good quality, especially a high image contrast and a homogeneous greyvalue distribution. The number of bright and dark pixels in the pattern should be approximately equal. In COARSE search it is absolutely necessary to have a minimum of one significant coarse structure with dimensions of more than 15 pixels in both directions located inside the pattern. For MEDIUM and FINE search the limits are about 10 and 5 pixels. In COARSE search the number of fine structure elements with less than 5..10 pixels in one direction in the pattern should be minimized. The limitations shown in Table 1 result from the algorithm and the requirement of a strictly restricted run time.

| Search mode | Max. search area [pixel] | Max. pattern area [pixel] | Recognition time | Robustness for critical images |
|---|---|---|---|---|
| Coarse | (1024) | 256 | Smallest | Medium |
| Medium | 480 | 128 | Medium | High |
| Fine | 192 | 64 | Highest | Best |

Table 1. Pattern Matching Evaluation Synopsis.

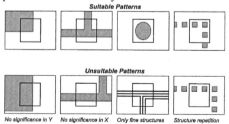

Figure 1. Suitability of Pattern Structures.

## 6. Conclusions and Business Benefit

The system has been deployed and is operational in the production lines of a pharmaceutical packaging facility, namely FAMAR, to address inspection tasks such as verification of lot and of manufacturing and expiration dates, blister, vial, ampoule and syringe inspection etc. in an efficient and reliable - yet user friendly and cost-effective way. The performance of the prototype system meets the requirements for production line throughput of 150 products per minute, error rate and idle time close to zero and false rejection rate less than 1%. At its pilot application the system fulfils the pharmaceutical industry regulations for 100% quality inspection of packaging procedures.

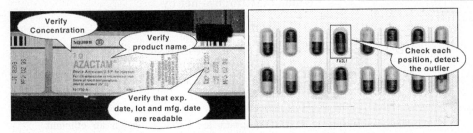

**Figure 2.** Example of DIVINE System Capabilities.

Manufacturing of pharmaceutical products in Europe is spread over many factories of small or medium size. They undertake packaging for the major multinational companies. These factories need an increasing level of automation and automatic process monitoring in order to maintain and improve confidence in product quality as well as comply with strict international regulations. Increased quality control requirements are also affecting industries related to other sensitive products such as cosmetics or food packaging, so that they will become interested in equipment such as that proposed in the DIVINE project. DIVINE followed the Good Automated Manufacturing Practice Guidelines on the Validation of Automated Systems which is promoted by the International Society for Pharmaceutical Engineering. This guide is in accordance with the European Directive 91/356/EEC that governs the use of automated systems in pharmaceutical manufacture and provides a framework of convergence between the developers of the automated system, the pharmaceutical manufacturers and regulators.

## References

1. T. Newman, A. Jain: A Survey of Automated Visual Inspection. Computer Vision and Image Understanding, 61(2), March 1995.
2. M. Sanders, E. McCormick: Human Factors in Engineering and Design. New York: McGraw-Hill, 1987.
3. European HPCN TTN Network, Industrial Groups, Quality Control and Inspection Sector Group: http://www.hpcn-ttn.org
4. A. Thomas, M. Rodd, J. Holt and C. Neill: Real-Time Industrial Visual Inspection: A Review. Real-Time Imaging, 1:139-158, 1995.
5. R. Conners, C. Harlow: A Theoretical Comparison of Texture Algorithms. IEEE Trans. Pattern Analysis and Machine Intelligence, PAMI-2(3), May 1980.
6. C. Therrien: Decision, Estimation and Classification. Chichester, UK: J.Wiley and Sons, 1989.

# Implementation of Montgomery Exponentiation on Fine Grained FPGAs: A Note on Partitioning

Alexander Tiountchik

Institute of Mathematics, National Academy of Sciences of Belarus,
11 Surganova str, Minsk 220072, Belarus
e-mail: aat@im.bas-net.by

Elena Trichina

Department of Computer Science, University of Joensuu,
P.O.Box 111. Joensuu 80101, Finland
e-mail: elena.trichina@joensuu.fi

**Abstract.** Taking as a starting point for FPGAs design an efficient bit-level systolic algorithm facilitates the design process but does not automatically guarantee the most efficient hardware solution. We demonstrate on an example of Montgomery exponentiation a role of partitioning in mapping of linear systolic arrays onto Xilinx XC6000 FPGAs.

## 1 Introduction

Modular exponentiation of long integers is a very slow operation when performed in software on a general purpose computer. On the other hand, a number of efficient bit-level parallel algorithms for modular exponentiation is known which can be implemented directly in Programmable Logic Arrays or in FPGAs. The advantage of using FPGAs is cost and flexibility: they are not expensive, they can provide the speed-up of dedicated hardware, the turn-around time for design of a particular application is comparable with the one for software, and unlike the special-purpose hardware chips, they can be reprogrammed for different applications. This paper describes how cheap and flexible modular exponentiation hardware accelerator for RSA can be achieved using FPGAs. We use complexity of the problem as a benchmark for evaluating computing power of fine grained FPGAs, and for developing a more systematic methodology for their programming.

Bit-level systolic arrays share many limitations and constraints with FPGAs; both favor regular repetitive designs with local interconnections, simple synchronisation mechanisms and minimimal global memory access. It may seem aparent that starting from a parallel algorithm in a form of bit-level systolic array by mapping this array into FPGAs one can obtain an efficient hardware implementation. Our experiment demonstrates that a straightforward mapping does not guarantee an optimal result, although reduces considerably the cost of design. A simple observation emerged that to achieve a high density design, one has to employ some concise *partitioning* strategy, and to support this strategy by designing a few building blocks with the same fuctionality but different layouts. With this simple method we doubled the efficiency of chip area utilisation.

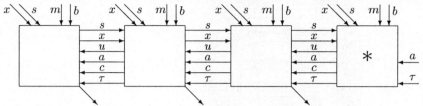

**Fig. 1.** Linear Systolic Array for Montgomery Exponentiation

## 2  Modular Exponentiation of Long Integers

The RSA Laboratories recommended key sizes are now 768 bits for personal use, and 1024 bits for corporate use. Hence, the intermediate results of a modular exponentiation $B^n \bmod m$ are to be reduced modulo $n$ at each step. In 1985, P. L. Montgomery [4] proposed an algorithm for modular multiplication (MM) $AB \bmod m$ without trial division. This algorithm is proved to be extremely well suited for hardware implementation [5, 9, 2].Presented in this paper design is based on systolic array for modular exponentiation described in [8], which in its turn relies on the algorithm described and analysed in [9]. Let numbers $A$, $B$ and $m$ be written with radix 2: $A = \sum_{i=0}^{N-1} a_i \cdot 2^i, B = \sum_{i=0}^{M} b_i \cdot 2^i, m = \sum_{i=0}^{M-1} m_i \cdot 2^i$, where $a_i$, $b_i$, $m_i \in \mathbf{GF}(2)$, $N$ and $M$ are the numbers of digits in $A$ and $m$, respectively. $B$ satisfies condition $B < 2m$, and has at most $M + 1$ digits. $m$ is odd (to be coprime to the radix 2). Extend a definition of $A$ with an extra zero digit $a_N = 0$. The algorithm for MM is:

$$
\begin{aligned}
&s := 0; \\
&\textbf{For } i := 0 \text{ to } N \textbf{ do} \\
&\textbf{Begin} \\
&\quad u_i := ((s_0 + a_i * b_0) * w) \bmod 2 \\
&\quad s := (s + a_i * B + u_i * m) \mathrm{div} 2 \\
&\textbf{End}
\end{aligned}
\tag{1}
$$

Instead of digits $A$ in (1) we can input digits $B$, and calculate an M-multiplication of number $B$ by itself, or $M$-squaring. In general, to raise $B$ to $n$-th power, we can perform $n - 1$ multiplications of $B$ by itself. However, a way of doing this faster is by reducing the computation of $B^n \bmod m$ to a sequence of modular squaring and multiplications [6]. A linear systolic array with $M + 1$ processing elements depicted in Fig. 1 implements a high-to-low method of modular exponentiation, i.e., given initially as input(s) digits of $B$, and a binary representation of $n$, it computes a sequence of operations of M-multiplications and M-squaring, which is defined by the binary representation of $n$. The algorithm runs on inputs which are either both the result of the previous operation (for M-squaring), or one of the inputs ($A$) gets the result of the previous operation, while another one

**Fig. 2.** Automatic allocation and logic connections for 67 PEs before optimisation.

($B$) is the original input integer $B$. Basically, the linear array of PE's in Fig. 1 represents one iteration of the main loop of Algoritm (1); at run time iterations overlap in a pipelined fashion. The starred PE computes the value of $u_i$ besides an ordinary calculations. This systolic array does not use global memory to store intermediate results and initial number $B$; instead they are stored in local registers of each PE, one bit in a register. It also does not use broadcast or global interconnection mechanisms ; instead the digits $s_i$ of the result of the previous operation are propagated to the rightmost vertex so as to be used as values for input $a$, which are streamlined through the array. Links $x$ are used for this purpose. Links $c$ propagate carries. To control the type of operation, a sequence of one-bit control signals $\tau$ is fed into the rightmost PE and propagated through the array; their order is determined by the binary representation of $n$.

## 3 Logic Design and Its Optimisation via Partitioning

Consider the logic design for implementation of an individual PEs. A computational part of the non-starred PE includes control over input data, two logic multiplications, $a_{in} \cdot b_{in}$ and $u_{in} \cdot m_{in}$, and addition of these products with an intermediate sum $s_{in}$ and input carries. Evidently, four-element addition can generate two carries meaning that all main PEs will have two input carries, and produce two output carries. It is not uncommon to implement addition of 5 entries using two full adders and one half adder. An implementation of a full adder can be found in a standard library xc6000 provided by EXACTStep6000. The starred PE selects correct values for its $b$– and $a$–inputs, depending on control signals, propagates data and signal to the left neighbor, computes value

$u$ and the sum $a_{in} \cdot b_{in} + u \cdot m_{in} + s_{in}$. For consistency, two zero carries should also be generated.

The next step is an implementation of a composition of PEs. We found that automatic allocation can provide successful routing for designs with a maximum of 67 PEs (1 rightmost, and 66 main). This design is presented in Fig. 2. A straighforward mapping resulted in a lot of non-local and crisscross logic connections between individual PEs. connections arising in every second row and where the band turns.

When a long and narrow line of PEs has to be mapped onto a square array of logic cells a natural solution is to *partition* this array into blocks of PEs with respect to the width of the board, so that every block can be allocated on a chip in a form of a border-to-border straight line. Pack these blocks in a zig-zag to fill in the whole $64 \times 64$ sea of cells on a chip. However, simlpe partitioning does not eliminate the problem, as Fig. 3 (a) illustrates. To ensure regularity and locality

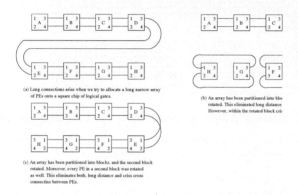

Fig. 3. Illustration of the partitioning method.

of interconnections between blocks, and between PEs in different blocks, one must create a few types of logic designs for PE with identical functionality, but with input/output gates representing some suitable permutation of the gates in the original design. Use these designs to build mirrow images of the original block under rotation and reflection as well as mirrow images of PEs for rotated blocks. New blocks (and PEs) must be used in every second row of the zig-zag, as shown in Fig. 3 (b) and (c). This method allows us to allocate 132 PEs successfully on $64 \times 64$ logic cells. In other words, we can exponentiate a 132-bit long integer on one Xilinx XC6000 chip. An allocation of 132 PEs on a board is presented in Fig. 4. To our knowledge, this is one of the best fine-grained FPGA designs for a modular exponetiation reported so far. 2,615 out of 4,096 gates are used for computations, and about 400 as registers, providing 76% density.

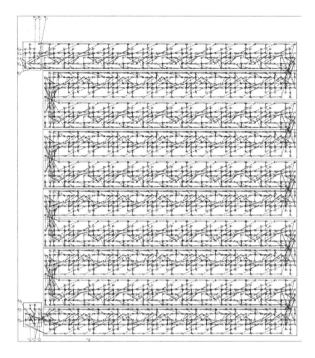

**Fig. 4.** Automatic allocation and logic connections for 132 PEs after optimisation.

## References

1. S. E. Eldridge and C. D. Walter, Hardware implementation of Montgomery's modular multiplication algorithm. *IEEE Trans. on Comput.*, 1993 (42) 693–699.
2. P. L. Montgomery, Modular multiplication without trial division. *Mathematics of Computations*, 1985 (44) 519–521.
3. J. Sauerbrey, A Modular Exponentiation Unit Based on Systolic Arrays, in *Advances in Cryptology – AUSCRYPT'93*, Springer-Verlag, LNCS, vol. 718 (1993) 505–516.
4. M. Shand, J. Vuillemin, Fast Implementation of of RSA Cryptography. In *Proc. of the 11th IEEE Symposium on Computer Arthmetics*, 1993. pp.252–259.
5. A. A. Tiountchik, Systolic modular exponentiation via Montgomery algorithm. J. *Electronics Letters*, 1998 (34).
6. C. D. Walter, Systolic Modular Multiplication. *IEEE Trans. on Comput.*, 1993 (42) 376–378.

# List of Authors

| | |
|---|---|
| 3DHeartView Team | 931 |
| Adam, P. | 1211 |
| Agrawal, G. | 1139 |
| Akarsu, E. | 291 |
| Albada, G.D. van | 300 |
| Alexakis, A. | 959 |
| Alexandridis, N. | 1235 |
| Aloisio, G. | 563 |
| Anagnostopoulou, S. | 959 |
| Ananda, A.L. | 583 |
| Anshus, O.J. | 722 |
| Arcangeli, J.-P. | 1227 |
| Arenaz, M. | 1087 |
| | |
| Baird, S. | 1042 |
| Balakrishnan, N. | 1299 |
| Baltas, D. | 1303 |
| Bartenstein, P. | 430 |
| Baude, F. | 744 |
| Belleman, R.G. | 817 |
| Bello, S. | 1171 |
| Bernaschi, M. | 774 |
| Bilchev, G. | 1266 |
| Birjukova, J.V. | 1175 |
| Blackburn, K. | 513 |
| Blankenship, G. | 1235 |
| Blobel, B. | 919 |
| Bode, A. | 430 |
| Boekhold, M. | 673 |
| Bogdanov, A.V. | 231, 1286 |
| Boyd, D.R.S. | 807 |
| Braun, T.A. | 391 |
| Brauss, S. | 623 |
| Bray, L. | 1227 |
| Brezany, P. | 1127 |
| Briguglio, S. | 241 |
| Brok, S.W. | 1061 |
| Broughton, P. | 713 |
| Brueck, S.R.J. | 109 |
| Brumec, M. | 951 |
| Brune, M. | 270 |
| Bubak, M. | 43, 543, 1270, 1274, 1278 |
| Burger, A. | 713 |

| | |
|---|---|
| Cafaro, M. | 563 |
| Cardinale, Y. | 168 |
| Carmo, M.B. | 1199 |
| Caromel, D. | 744 |
| Caron, E. | 1107 |
| Carretero, J. | 1117 |
| Casavant, T.L. | 391 |
| Celino, M. | 535 |
| Chakarova, R. | 1194 |
| Chan, T.F. | 1001 |
| Chan, W. | 831 |
| Chang, C.-H. | 80 |
| Chapin, S. | 370 |
| Chattratichat, J. | 573 |
| Chen, C.-L | 1292 |
| Chen, S. | 130 |
| Chi, C.H. | 1258 |
| Chopard, B. | 319 |
| Choudhary, A. | 1097, 1117 |
| Cilio, A.G.M. | 643 |
| Cláudio, A.P. | 1199 |
| Cleary, D. | 797 |
| Clement, M. | 440 |
| Clinckemaillie, J. | 300 |
| Coddington, P.D. | 873 |
| Cooper, T. | 1077 |
| Corporaal, H. | 643, 673 |
| Cozette, O. | 1107 |
| Cunha, J.D | 1199 |
| | |
| Dam, A.A. ten | 50 |
| Dandamudi, S.P. | 483, 683 |
| Darlington, J. | 573 |
| Darte, A. | 653 |
| Dawes, W.N. | 90 |
| Dekker, L. | 1061 |
| Delibasis, K.K. | 973, 989 |
| Dempster, E. | 713 |
| Deshpande, V. | 702 |
| Di Martino, B. | 241, 535 |
| Dijkman, D. | 817 |
| Dimitrelos, D. | 159, 784 |
| Divakar, S. | 109 |
| Doallo, R. | 1087 |

| | | | |
|---|---|---|---|
| Domínguez, J.C.C. | 201 | Girona, S. | 1171 |
| Doncker, E. de | 360 | González, A. | 754, 1246 |
| Dorband, J.E. | 1167 | Gorbachev, Y.E. | 1250 |
| Dorojevets, M. | 1179 | Goscinski, A. | 603 |
| Doroshenko, A.E. | 1219 | Grigoryan, A.G. | 1286 |
| Drikakis, D. | 1015 | Grimshaw, A. | 370 |
| Dubbeldam, D. | 339 | Grishin, I.A. | 231 |
| Dupuis, A. | 319 | Gubitoso, M.D. | 1223 |
| Dymaczewski, W. | 1231 | Gunzinger, A. | 623 |
| | | Guo, Y. | 573 |
| Edenbrandt, L. | 941 | | |
| Emmen, A.H.L. | 300, 909 | Habjanič, A. | 951 |
| Essafi, H. | 1211 | Halatsis, C. | 784 |
| Evensky, D.A. | 1183 | Haritonov, E.V. | 1175 |
| Evstigneev, V.A. | 1175 | Harper, J.S. | 473 |
| | | Haumacher, B. | 884 |
| Faxén, T. | 1194 | Haupt, T. | 291 |
| Fenwick Jr., J.B. | 1282 | Hawick, K.A. | 350, 873 |
| Ferrari, A. | 370 | Hedvall, S. | 573 |
| Fichtner, W. | 221 | Heinz, O. | 300 |
| Fogaccia, G. | 241 | Héran, F. | 70 |
| Fong, L.L. | 831 | Heras, D.B. | 201 |
| Ford, R.W. | 420 | Hercus, J.F. | 350 |
| Fornasier, L. | 702 | Hilbrink, N. | 702 |
| Forrai, S. | 1242 | Hobbs, M. | 603 |
| Fox, G. | 291 | Hodurek, P. | 1274 |
| Frank, J. | 1052 | Hoekstra, A.G. | 311, 339 |
| Franke, H. | 831 | Honda, H. | 663 |
| Frey, M. | 623 | Hu, W. | 463 |
| Fritzson, D. | 99 | Hügl, R. | 1207 |
| Fujioka, F. | 130 | Humes Jr., C. | 1223 |
| Funika, W. | 1270, 1274, 1278 | Humphrey, M. | 370 |
| Furmento, N. | 744 | | |
| | | Iannello, G. | 774 |
| Gallop, J.R. | 807 | Ibáñez, M.B. | 168 |
| Gärtner, K. | 221 | Ihara, S. | 1151 |
| Garatani, K. | 133 | Iikura, M. | 1151 |
| Gayrard, M.-P. | 1211 | Iskra, K. | 1270 |
| Gehring, J. | 300 | | |
| Gembarowski, R. | 1274 | Jacques, T. | 1025 |
| Gemund, A.J.C. van | 23 | Järund, A. | 941 |
| Gentile, A.C. | 1183 | Jette, M.A. | 831 |
| Gerteisen, E.A. | 702 | Jézéquel, J.-M. | 260 |
| Gevorkyan, A.S. | 1286 | Jin, M.-H. | 80 |
| Ghosh, S.S. | 1299 | Johnston, W. | 150 |
| Gimbel, M. | 884 | Jovanov, E. | 964 |

| | | | |
|---|---|---|---|
| Juang, H. | 130 | Lauria, M. | 774 |
| Jung, B. | 13 | Lazure, D. | 1107 |
| | | Lazzarini, A. | 513 |
| Kaandorp, J.A. | 817 | Lebas, F.-X. | 60 |
| Kacsuk, P. | 1242 | Lee, J. | 150 |
| Kaliannan, S. | 391 | Lees, M. | 409 |
| Kalinov, A. | 191 | Lenhof, H.-P. | 13 |
| Kammenos, P. | 1303 | Levi, P. | 732 |
| Kan, T.-T. | 1292 | Li, J. | 360 |
| Kandemir, M. | 1097 | Li, S. | 583 |
| Kandhai, D. | 311 | Liang, Z. | 251 |
| Karkowski, I. | 673 | Lienhard, M. | 623 |
| Karl, H. | 841 | Likharev, K. | 1179 |
| Karyadi, E. | 409 | Linden, F. van der | 300 |
| Kasyanov, V.N. | 1175 | Lindén, P. | 1194 |
| Kauranne, T. | 909 | Lockemann, P.C. | 884 |
| Kechadi, M.-T. | 450 | Ludwig, T. | 70, 430 |
| Kellar, W.P. | 90 | Lukianov, G.A. | 231, 1187 |
| Keller, A. | 270 | Lustig, G. | 141 |
| Kerbyson, D.J. | 473 | | |
| Keßler, C.W. | 525 | Ma, J. | 1147 |
| Khanlarov, G.O. | 231, 1187 | Magdoń, M. | 43 |
| King, P.J.B. | 713 | Magotra, N. | 109 |
| Kitowski, J. | 693 | Mahmoud, Q.H. | 281 |
| Kitsuregawa, M. | 553 | Majewski, J. | 1015 |
| Knabe, F. | 370 | Maksymowicz, A.Z. | 43 |
| Kobayashi, K. | 1151 | Malinina, J.V. | 1175 |
| Köhler, M. | 573 | Malony, A.D. | 381 |
| Kokol, P. | 951 | Malyshkin, V.E. | 329 |
| Koponen, A. | 311 | Maniatis, T.A. | 178 |
| Koranne, S. | 1163 | Manoharan, S. | 1155 |
| Kotapati, K. | 483 | Marcuello, P. | 754 |
| Kraeva, M.A. | 329 | Maresca, M. | 633 |
| Krall, A. | 895 | Markin, V.A. | 1175 |
| Kranzlmüller, D. | 1207 | Marsh, A. | 909, 964, 973, 983 |
| Kubota, K. | 764 | Marshall, I. | 1266 |
| Kuleschow, A. | 1303 | Martin, R. | 1171 |
| Kusano, K. | 211 | Maruszewski, R. | 1270 |
| | | Mathew, J.A. | 873 |
| Labarta, J. | 1171 | Matsopoulos, G.K. | 989 |
| Laffitte, G. | 1171 | Matsuda, M. | 211, 764 |
| Lai, F. | 493 | Matveev, S.A. | 1286 |
| Lakshminarayanan, S. | 1299 | McCord, C. | 130 |
| Lambadaris, P. | 1303 | McGuirk, J.J. | 1042 |
| Larsen, T. | 722 | Medoš, T. | 951 |
| Lastovetsky, A. | 191 | Meliones, A. | 1303 |

| | | | |
|---|---|---|---|
| Merazzi, S. | 702 | Pérez, V.B. | 201 |
| Meyer, N. | 1231 | Pharow, P. | 919 |
| Mezentsev, A. | 702 | Philippsen, M. | 884 |
| Migliardi, M. | 1215 | Pic, M. | 1211 |
| Młynarczyk, G. | 1278 | Piriyakumar, D.A.L. | 732 |
| Modersitzki, J. | 141 | Platon, R.T. | 807 |
| Mohr, B. | 503 | Podgorelec, V. | 951 |
| Molina, C. | 1246 | Pogoda, M. | 1262 |
| Moore, M. | 912 | Polemi, D. | 983 |
| Moreira, J.E. | 831 | Pollock, L.L. | 1282 |
| Morse, B. | 440 | Prince, T. | 513 |
| Mościcki, J.T. | 543 | Punpiti, P.-N. | 1235 |
| Mościński, J. | 1262 | | |
| Mouravliansky, N.A. | 973, 989 | Rackl, G. | 70 |
| Müller, P. | 13 | Radhakrishna, H. | 109 |
| Munn, K.J. | 391 | Radivojevic, V. | 964 |
| Munz, F. | 430 | Ramanujam, J. | 1097 |
| Murthy, C.S.R. | 732 | Rana, O. | 863 |
| | | Ranawake, U.A. | 1167 |
| Nakamura, H. | 133 | Reijns, G.L. | 23 |
| Nemecek, J. | 623 | Reinefeld, A. | 270, 300 |
| Ni, S.-Y. | 1159 | Ribes, P. | 1171 |
| Nicolas, L. | 1025 | Riedl, R. | 33 |
| Nikita, K.S. | 178, 973, 989 | Rivas, R. | 168 |
| Nikolow, D. | 1262 | Rivera, F.F. | 201 |
| No, J. | 1117 | Roadknight, C. | 1266 |
| Nordling, P. | 99 | Roads, J. | 130 |
| Nudd, G.R. | 473 | Roest, M.R.T. | 1032 |
| | | Roger-France, F. | 919 |
| Obelöer, W. | 141 | Rokicki, J. | 1015 |
| Obrenovic, Z. | 964 | Rosato, V. | 535 |
| O'Brien, M. | 420 | Rounce, P. | 1203 |
| O'Donoghue, D. | 797 | Rüb, C. | 13 |
| Oguchi, M. | 553 | | |
| Ohlsson, M. | 941 | Sagnol, D. | 744 |
| Okuda, H. | 133 | Sallé, P. | 1227 |
| Olafsson, S. | 1266 | Samardzic, A. | 964 |
| Overeinder, B.J. | 300 | Samiotakis, Y. | 959 |
| | | Sanders, P. | 3 |
| Palmen, K.E.V. | 807 | Sato, M. | 211, 764 |
| Pan, K.-H. | 493 | Sato, N. | 260 |
| Papazis, N. | 159 | Satoh, S. | 211 |
| Pasquarelli, A. | 70 | Savill, A.M. | 90 |
| Patten, C.J. | 350 | Scheetz, T.E. | 391 |
| Pázsit, I. | 1194 | Schenk, O. | 221 |
| Pelagatti, S. | 613 | Schier, J. | 23 |

| | | | |
|---|---|---|---|
| Schmitt, O. | 141 | Tiountchik, A. | 1308 |
| Schoneveld, A. | 409 | Tomov, N. | 713 |
| Schüle, J. | 120 | Tomsich, P. | 895 |
| Schwaiger, M. | 430 | Touriño, J. | 1087 |
| Schwiegelshohn, U. | 851 | Trichina, E. | 1308 |
| Seelig, C.D. | 807 | Trzeciak, P. | 693 |
| Seidl, H. | 525 | Tseng, Y.-C. | 1159 |
| Shang, R.-J. | 493 | Tseng, Y.-O | 80 |
| Sheu, J.-P. | 1159 | Tsikoza, S.G. | 1175 |
| Shi, W. | 463, 1147 | Tubella, J. | 1246 |
| Shiers, J. | 543 | | |
| Silber, G.-A. | 653 | Usländer, T. | 60 |
| Skidmore, J.L. | 381 | Utard, G. | 1107 |
| Sloot, P.M.A. | 300, 311, 339, 409, 817 | Uzunoglu, N.K. | 973 |
| Słota, R. | 1262 | | |
| Smith, D.A. | 402 | Vaněk, P. | 1001 |
| Smith, W. | 130 | Vardangalos, G. | 1303 |
| Smyllie, K. | 1190 | Vázquez, C. | 1087 |
| Snell, Q. | 440 | Versweyveld, L. | 909 |
| Soto, J. | 1171 | Vinter, B. | 722 |
| Sottile, M.J. | 381 | Vlad, G. | 241 |
| Souza, A.F. de | 1203 | Vlassov, V. | 1219 |
| Sowa, K. | 1278 | Voliotis, K. | 178 |
| Spinnler, K. | 1303 | Volkert, J. | 1207 |
| Srakaew, S. | 1235 | Vollaire, C. | 1025 |
| Stamatakos, G.S. | 973 | Vollebregt, E.A.H. | 1032 |
| Starcevic, D. | 964 | Vuik, C. | 1052 |
| Stefani, F. de | 70 | | |
| Sterkenburg, R.P. van | 50 | Wajs, D. | 693 |
| Stevens, D. | 130 | Walker, D. | 863 |
| Stroinski, M. | 1231 | Wang, C.-L. | 251 |
| Sun, Yudong | 251 | Wang, S.-Y. | 1159 |
| Sunderam, V. | 1215 | Waters, A. | 109 |
| Swart, P.J.F. | 23 | Wedelin, D. | 3 |
| Syed, J. | 573 | Wilhelmsson, T. | 120 |
| Szczerbinski, Z. | 1254 | Williams, M.H. | 713 |
| | | Williams, R. | 513, 563 |
| Takabatake, M. | 663 | Windyga, P. | 168 |
| Takkula, T. | 3 | Winslett, M. | 1127 |
| Tanaka, Y. | 211, 764 | Wismüller, R. | 1274, 1278 |
| Tang, Z. | 463, 1147 | Woehler, T. | 702 |
| Taylor, H. | 713 | Wolf, F. | 503 |
| Telford, S.D. | 1143 | Wolniewicz, P. | 1231 |
| Thorelli, L.-E. | 1219 | Wu, H.-K. | 80, 493 |
| Tichy, W.F. | 884 | Wyckoff, P. | 1183 |
| Tierney, B. | 150 | | |

| | | | |
|---|---|---|---|
| Yagawa, G. | 133 | Zacharaki, E.I. | 973 |
| Yahyapour, R. | 851 | Zahlmann, G. | 909 |
| Yoo, A. | 831 | Zając, K. | 43 |
| Yoshioka, T. | 1151 | Zakharov, V.V. | 231 |
| Yu, S. | 440 | Zambonelli, F. | 593 |
| Yuba, T. | 663 | Zavenella, A. | 613 |
| | | Zeng, S. | 683 |
| | | Zhang, Y. | 1258 |
| | | Ziegler, S. | 430 |
| | | Zingirian, N. | 633 |
| | | Żołtak, J. | 1015 |
| | | Zorman, M. | 951 |
| | | Zudilova, E.V. | 1250 |